MATHEMATICAL SYMBOLS

$=$	is equal to
$\neq$	is not equal to
$\approx$	is approximately equal to
$\propto$	is proportional to
$>$	is greater than
$\geq$	is greater than or equal to
$\gg$	is much greater than
$<$	is less than
$\leq$	is less than or equal to
$\ll$	is much less than
$\pm$	plus or minus
$\mp$	minus or plus
x_{av} or $\bar{x}$	average value of x
Δx	change in x ($x_f - x_i$)
$\lvert x \rvert$	absolute value of x
Σ	sum of
$\to 0$	approaches 0
∞	infinity

USEFUL MATHEMATICAL FORMULAS

Area of a triangle	$\frac{1}{2}bh$
Area of a circle	πr^2
Circumference of a circle	$2\pi r$
Surface area of a sphere	$4\pi r^2$
Volume of a sphere	$\frac{4}{3}\pi r^3$
Pythagorean theorem	$r^2 = x^2 + y^2$
Quadratic formula	if $ax^2 + bx + c = 0$, then $x = \dfrac{-b \pm \sqrt{b^2 - 4ac}}{2a}$

EXPONENTS AND LOGARITHMS

$$x^n x^m = x^{n+m}$$

$$x^{-n} = \frac{1}{x^n}$$

$$\frac{x^n}{x^m} = x^{n-m}$$

$$(xy)^n = x^n y^n$$

$$(x^n)^m = x^{nm}$$

$$\ln(xy) = \ln x + \ln y$$

$$\ln\left(\frac{x}{y}\right) = \ln x - \ln y$$

$$\ln x^n = n \ln x$$

VALUES OF SOME USEFUL NUMBERS

$\pi = 3.14159\ldots$	$\log 2 = 0.30103$	$\sqrt{2} = 1.41421$
$e = 2.71828\ldots$	$\ln 2 = 0.69315$	$\sqrt{3} = 1.73205$

TRIGONOMETRIC RELATIONSHIPS

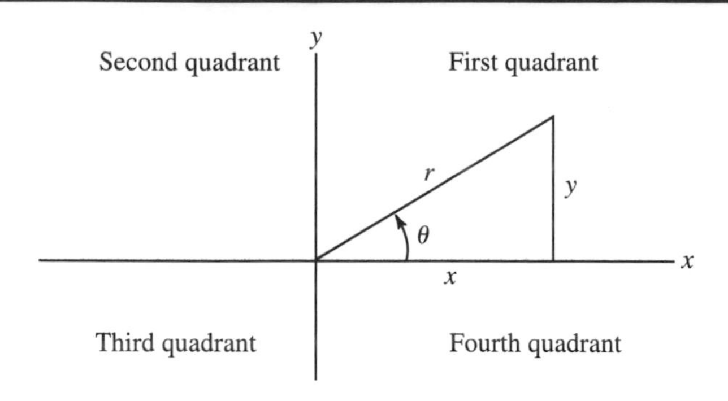

Definitions of Trigonometric Functions

$$\sin\theta = \frac{y}{r} \qquad \cos\theta = \frac{x}{r} \qquad \tan\theta = \frac{\sin\theta}{\cos\theta} = \frac{y}{x}$$

Trigonometric Functions of Important Angles

$\theta°$ (rad)	$\sin\theta$	$\cos\theta$	$\tan\theta$
$0°$ (0)	0	1	0
$30°$ ($\pi/6$)	0.500	$\sqrt{3}/2 \approx 0.866$	$\sqrt{3}/3 \approx 0.577$
$45°$ ($\pi/4$)	$\sqrt{2}/2 \approx 0.707$	$\sqrt{2}/2 \approx 0.707$	1.00
$60°$ ($\pi/3$)	$\sqrt{3}/2 \approx 0.866$	0.500	$\sqrt{3} \approx 1.73$
$90°$ ($\pi/2$)	1	0	∞

Trigonometric Identities

$$\sin(\theta + \phi) = \sin\theta\cos\phi + \sin\phi\cos\theta$$

$$\cos(\theta + \phi) = \cos\theta\cos\phi - \sin\theta\sin\phi$$

$$\sin 2\theta = 2\sin\theta\cos\theta$$

VECTOR MANIPULATIONS

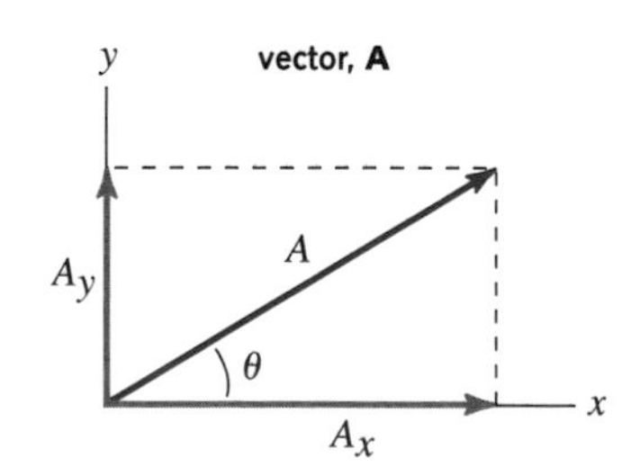

Magnitude and direction	$\longleftrightarrow$	x and y components
A, θ	$\longrightarrow$	$A_x = A\cos\theta$ $A_y = A\sin\theta$
$A = \sqrt{A_x^2 + A_y^2}$ $\theta = \tan^{-1}\left(\frac{A_y}{A_x}\right)$	$\longleftarrow$	A_x, A_y

CONVERSION FACTORS

Mass

$1 \text{ kg} = 10^3 \text{ g}$
$1 \text{ g} = 10^{-3} \text{ kg}$
$1 \text{ u} = 1.66 \times 10^{-24} \text{ g} = 1.66 \times 10^{-27} \text{ kg}$
$1 \text{ slug} = 14.6 \text{ kg}$
$1 \text{ metric ton} = 1000 \text{ kg}$

Length

$1 \text{ Å} = 10^{-10} \text{ m}$
$1 \text{ nm} = 10^{-9} \text{ m}$
$1 \text{ cm} = 10^{-2} \text{ m} = 0.394 \text{ in.}$
$1 \text{ m} = 10^{-3} \text{ km} = 3.28 \text{ ft} = 39.4 \text{ in.}$
$1 \text{ km} = 10^3 \text{ m} = 0.621 \text{ mi}$
$1 \text{ in.} = 2.54 \text{ cm} = 2.54 \times 10^{-2} \text{ m}$
$1 \text{ ft} = 0.305 \text{ m} = 30.5 \text{ cm}$
$1 \text{ mi} = 5280 \text{ ft} = 1609 \text{ m} = 1.609 \text{ km}$
$1 \text{ ly (light year)} = 9.46 \times 10^{12} \text{ km}$
$1 \text{ pc (parsec)} = 3.09 \times 10^{13} \text{ km}$

Area

$1 \text{ cm}^2 = 10^{-4} \text{ m}^2 = 0.1550 \text{ in.}^2$
$= 1.08 \times 10^{-3} \text{ ft}^2$
$1 \text{ m}^2 = 10^4 \text{ cm}^2 = 10.76 \text{ ft}^2 = 1550 \text{ in.}^2$
$1 \text{ in.}^2 = 6.94 \times 10^{-3} \text{ ft}^2 = 6.45 \text{ cm}^2$
$= 6.45 \times 10^{-4} \text{ m}^2$
$1 \text{ ft}^2 = 144 \text{ in.}^2 = 9.29 \times 10^{-2} \text{ m}^2 = 929 \text{ cm}^2$

Volume

$1 \text{ cm}^3 = 10^{-6} \text{ m}^3 = 3.35 \times 10^{-5} \text{ ft}^3$
$= 6.10 \times 10^{-2} \text{ in.}^3$
$1 \text{ m}^3 = 10^6 \text{ cm}^3 = 10^3 \text{ L} = 35.3 \text{ ft}^3$
$= 6.10 \times 10^4 \text{ in.}^3 = 264 \text{ gal}$
$1 \text{ liter} = 10^3 \text{ cm}^3 = 10^{-3} \text{ m}^3 = 1.056 \text{ qt}$
$= 0.264 \text{ gal}$
$1 \text{ in.}^3 = 5.79 \times 10^{-4} \text{ ft}^3 = 16.4 \text{ cm}^3$
$= 1.64 \times 10^{-5} \text{ m}^3$
$1 \text{ ft}^3 = 1728 \text{ in.}^3 = 7.48 \text{ gal} = 0.0283 \text{ m}^3$
$= 28.3 \text{ L}$
$1 \text{ qt} = 2 \text{ pt} = 946 \text{ cm}^3 = 0.946 \text{ L}$
$1 \text{ gal} = 4 \text{ qt} = 231 \text{ in.}^3 = 0.134 \text{ ft}^3 = 3.785 \text{ L}$

Time

$1 \text{ h} = 60 \text{ min} = 3600 \text{ s}$
$1 \text{ day} = 24 \text{ h} = 1440 \text{ min} = 8.64 \times 10^4 \text{ s}$
$1 \text{ y} = 365 \text{ days} = 8.76 \times 10^3 \text{ h}$
$= 5.26 \times 10^5 \text{ min} = 3.16 \times 10^7 \text{ s}$

Speed

$1 \text{ m/s} = 3.60 \text{ km/h} = 3.28 \text{ ft/s}$
$= 2.24 \text{ mi/h}$
$1 \text{ km/h} = 0.278 \text{ m/s} = 0.621 \text{ mi/h}$
$= 0.911 \text{ ft/s}$
$1 \text{ ft/s} = 0.682 \text{ mi/h} = 0.305 \text{ m/s}$
$= 1.10 \text{ km/h}$
$1 \text{ mi/h} = 1.467 \text{ ft/s} = 1.609 \text{ km/h}$
$= 0.447 \text{ m/s}$
$60 \text{ mi/h} = 88 \text{ ft/s}$

Force

$1 \text{ N} = 0.225 \text{ lb}$
$1 \text{ lb} = 4.45 \text{ N}$
Equivalent weight of a mass of 1 kg on Earth's surface = 2.2 lb = 9.8 N
$1 \text{ dyne} = 10^{-5} \text{ N} = 2.25 \times 10^{-6} \text{ lb}$

Pressure

$1 \text{ Pa} = 1 \text{N/m}^2 = 1.45 \times 10^{-4} \text{ lb/in.}^2$
$= 7.5 \times 10^{-3} \text{ mm Hg}$
$1 \text{ mm Hg} = 133 \text{ Pa}$
$= 0.02 \text{ lb/in.}^2 = 1 \text{ torr}$
$1 \text{ atm} = 14.7 \text{ lb/in.}^2 = 101.3 \text{ kPa}$
$= 30 \text{ in. Hg} = 760 \text{ mm Hg}$
$1 \text{ lb/in.}^2 = 6.89 \text{ kPa}$
$1 \text{ bar} = 10^5 \text{ Pa} = 100 \text{ kPa}$
$1 \text{ millibar} = 10^2 \text{ Pa}$

Energy

$1 \text{ J} = 0.738 \text{ ft} \cdot \text{lb} = 0.239 \text{ cal}$
$= 9.48 \times 10^{-4} \text{ Btu} = 6.24 \times 10^{18} \text{ eV}$
$1 \text{ kcal} = 4186 \text{ J} = 3.968 \text{ Btu}$
$1 \text{Btu} = 1055 \text{ J} = 778 \text{ ft} \cdot \text{lb} = 0.252 \text{ kcal}$
$1 \text{ cal} = 4.186 \text{ J} = 3.97 \times 10^{-3} \text{ Btu}$
$= 3.09 \text{ ft} \cdot \text{lb}$
$1 \text{ ft} \cdot \text{lb} = 1.36 \text{ J} = 1.29 \times 10^{-3} \text{ Btu}$
$1 \text{ eV} = 1.60 \times 10^{-19} \text{ J}$
$1 \text{ kWh} = 3.6 \times 10^6 \text{ J}$
$1 \text{ erg} = 10^{-7} \text{ J} = 7.38 \times 10^{-6} \text{ ft} \cdot \text{lb}$

Power

$1 \text{ W} = 1 \text{ J/s} = 0.738 \text{ ft} \cdot \text{lb/s}$
$= 1.34 \times 10^{-3} \text{ hp} = 3.41 \text{ Btu/h}$
$1 \text{ ft} \cdot \text{lb/s} = 1.36 \text{ W} = 1.82 \times 10^{-3} \text{ hp}$
$1 \text{ hp} = 550 \text{ ft} \cdot \text{lb/s} = 745.7 \text{ W}$
$= 2545 \text{ Btu/h}$

Mass–Energy Equivalents

$1 \text{ u} = 1.66 \times 10^{-27} \text{ kg} \leftrightarrow 931.5 \text{ MeV}$
$1 \text{ electron mass} = 9.11 \times 10^{-31} \text{ kg}$
$= 5.49 \times 10^{-4} \text{ u} \leftrightarrow 0.511 \text{ MeV}$
$1 \text{ proton mass} = 1.673 \times 10^{-27} \text{ kg}$
$= 1.007\,267 \text{ u} \leftrightarrow 938.28 \text{ MeV}$
$1 \text{ neutron mass} = 1.675 \times 10^{-27} \text{ kg}$
$= 1.008\,665 \text{ u} \leftrightarrow 939.57 \text{ MeV}$

Temperature

$T_F = \frac{9}{5} T_C + 32$
$T_C = \frac{5}{9}(T_F - 32)$
$T_K = T_C + 273.15$

Angle

$1 \text{ rad} = 57.3°$

$1° = 0.0175 \text{ rad}$	$60° = \pi/3 \text{ rad}$
$15° = \pi/12 \text{ rad}$	$90° = \pi/2 \text{ rad}$
$30° = \pi/6 \text{ rad}$	$180° = \pi \text{ rad}$
$45° = \pi/4 \text{ rad}$	$360° = 2\pi \text{ rad}$

$1 \text{ rev/min} = (\pi/30) \text{ rad/s} = 0.1047 \text{ rad/s}$

APPENDICES IN THE TEXT

Physics

James S. Walker

Washington State University

PRENTICE HALL
Upper Saddle River, New Jersey 07458

Brief Contents

Contents

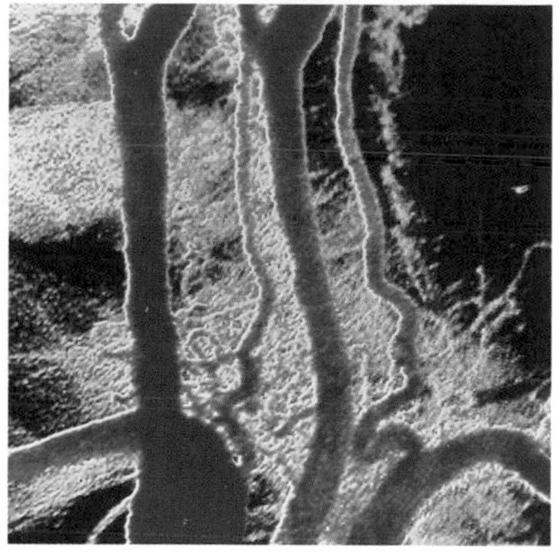

How fast does your blood flow in miles per hour? (Page 10)

Moving upward but accelerating downward. (Page 38)

Changing velocity without changing speed. (Page 68)

Kepler's laws keep this visitor out in the cold for long periods of time. (Page 368)

The timing of the pushes is important. (Page 407)

The sound of music is the sound of standing waves. (Page 449)

Narrowing the nozzle increases the range. (Page 479)

An unusual feature of their fur helps polar bears to stay warm in a cold environment. (Page 517)

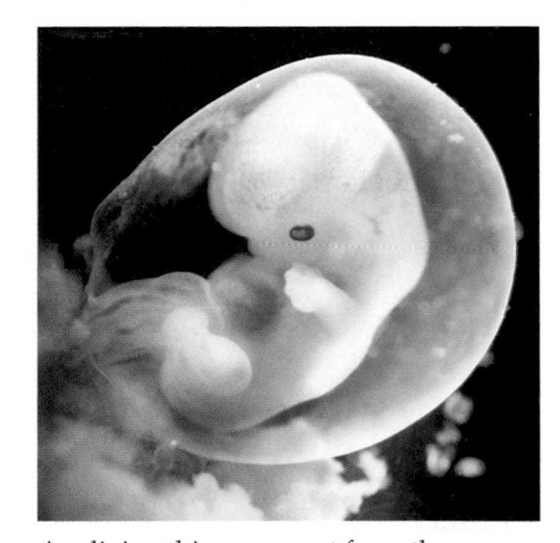

Are living things exempt from the second law of thermodynamics? (Page 599)

Chapter 14

Chapter 15

Chapter 16

Chapter 17

Chapter 18

Preface: To the Instructor

Teaching any subject can be a most challenging—and rewarding—experience. This is particularly true of the introductory algebra-based physics course, where students with a wide range of backgrounds and interests participate in a unique learning experience. With only a limited time at our disposal, we, the instructors, strive not only to convey the basic concepts and fundamental laws of physics, but also to give students an appreciation of its relevance and appeal. This is a tall order, but one that is well worth the effort.

To help with the task, this text incorporates a number of unique and innovative pedagogical features. These features, which evolved from years of teaching experience, have been tested extensively in the classroom and refined on the basis of interviews and discussions with students. The enthusiastic response I receive from students using this material has encouraged my belief that your students, like mine, will find the presentation of physics given in this text to be clear, engaging, and empowering.

Learning Tools in the Text

The goal of this text is to help students improve their conceptual understanding of physics hand in hand with the development of their problem-solving skills. One of the chief means to that end is the replacement of the traditional Examples in the text by an integrated suite of learning tools: fully worked *Examples in Two-Column Format*, *Active Examples*, *Conceptual Checkpoints*, and *Exercises*. Each of these tools performs some of the functions of an Example, but each is specialized to meet the needs of students at a particular point in the development of the chapter's content.

These needs are not always the same. Sometimes students require a detailed explanation of how to tackle a particular problem; at other times, they must be allowed to take an active role and work out the details for themselves. Sometimes it is important for them to perform calculations and concentrate on numerical precision; at other times it may be more fruitful for them to explore a key idea more fully in a non-quantitative context. Sometimes the analysis of a detailed physical context is essential; at other times, practice in using a new equation or relationship is all that is called for.

A good teacher can sense when students need a very patient exposition and when they need only minimal reinforcement; when they need to focus on concepts and when they need an opportunity to practice their quantitative skills. This text attempts to mimic the teaching style of successful instructors by providing the right tool at the right time and place.

Worked Examples in Two-Column Format

Examples provide the most complete and detailed illustration of how to solve a particular type of problem. The Examples in this text are presented in a unique two-column format that focuses on the basic strategies and thought processes involved in problem solving. The aim of this approach is to help students devise a *strategy* to be followed and then implement a clear *step-by-step solution* to the problem. The emphasis is thus on the relationship between the physical concepts and their mathematical expression. This focus on the intimate relationship between conceptual insights and problem-solving techniques encourages students to view the ability to solve problems as a logical outgrowth of conceptual understanding rather than a kind of parlor trick.

Each Example has the same basic structure:

Scope and Organization

Table of Contents

As you will notice from the Table of Contents (pages v–x), the presentation of physics in this text follows the standard practice for introductory courses, with only a few well-motivated refinements.

First, note that Chapter 3 is devoted to vectors and their application to physics. This material could be presented in an Appendix, but my experience has been that students benefit greatly from a full discussion of vectors early in the course. Most students have seen vectors and trigonometric functions before, but rarely from the point of view of physics. Thus, including vectors in the text sends a message that this is important material, and it gives students an opportunity to brush up on their math skills.

Note also that additional time is given to some of the more fundamental aspects of physics, such as Newton's laws and energy. Presenting such material in two chapters gives the student a better opportunity to assimilate and master these crucial topics, which form the foundation for so much of what follows. Given the time constraints we all face in the classroom, the distribution of material presented in this text is designed to give time and emphasis where most appropriate.

Real World Physics

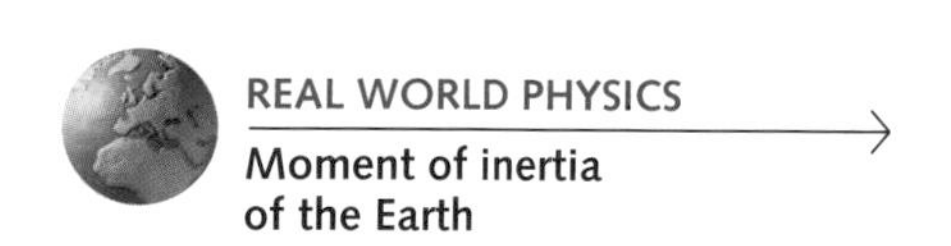

Since physics applies to everything in nature, it is only reasonable to point out applications of physics that students may encounter in the real world. Each chapter presents a number of discussions focusing on "Real World Physics." Those of general interest are designated by a globe icon in the margin. Applications that pertain more specifically to biology and medicine are indicated by a CAT scan icon in the margin. The inclusion of a generous selection of such topics should help to make the course material more interesting and relevant to all students, including the many whose career orientation is toward the life sciences. A full list of the real-world applications in the text is given on pages xi–xii.

The Illustration Program

Drawings

Physics is a highly visual subject, and many physics concepts are best conveyed by graphic means. Figures do far more than illustrate a physics text—often, they bear the main burden of the exposition. Accordingly, great attention has been paid to the figures in this book so as to achieve the optimal balance of realism and stylization, with the primary emphasis always on the clarity of the analysis.

As mentioned previously, *every* Example in the text (as well as most of the Active Examples and many of the Conceptual Checkpoints) is accompanied by a figure. In addition, every Example includes a section, "Picture the Problem," that concentrates on visual representation of the physical situation described in the problem, as well as the use of that representation in devising a problem-solving strategy. Great emphasis has been placed on how to draw a free-body diagram (see pp. 110–111), and all such diagrams in Examples and end-of-chapter problems have been drawn to scale using the data given.

In addition, color has been used consistently throughout the text to reinforce concepts and make the diagrams easier for students to understand. Thus, force vectors are always red, velocity vectors always green, and so on. Similarly, wave fronts, adiabats, heat flow, work output, and many other elements have each been assigned a characteristic color. While the student has not been asked to learn the colors scheme consciously, there can be little question that such consistency is pedagogically helpful on a subliminal level.

Companion Photographs

One of the most fundamental ways in which we learn is by comparing and contrasting. This principle is exploited in the way photographs are used throughout the text. Many photos are presented in groups of two or three that contrast opposing physical principles or illustrate a single concept in a variety of contexts. Grouping carefully chosen photographs in this way emphasizes the universality of physics. To see the similarities and differences between the sand in the bottom of an hourglass and a talus slope at the base of a cliff, for example, is to deepen ones understanding of the effects of static friction.

In conclusion, it is my hope that this text will engage the imagination and interest of the students who use it, and open to them a world governed by the fundamental laws and principles of physics. Those of us who devote ourselves to studying and teaching physics know that it is a subject of great beauty and power. This book strives to share some of that appreciation with others.

Supplements

For the Instructor

Instructor's Solutions Manual (Vol. I: 0-13-027065-2; Vol. II: 0-13-027066-3)
Prepared by Laurel Technical Services. Contains detailed, worked solutions to every problem in the text. An electronic version (0-13-027056-3) is available in CD-ROM (dual platform for both Windows and Macintosh systems) for instructors with Microsoft Word or Word-compatible software.

Instructor's Resource Manual (0-13-040801-8)
This instructor's manual has two parts. The first is a traditional resource manual with lecture outlines, notes, ideas, and resources. The second, by *Just In Time Teaching* developers Gregor Novak and Andrew Gavrin, contains an overview of the JiTT teaching method, strategies for using it in your course, and specific strategies, tips, and feedback for the JiTT material in the Study Guide and Companion Website.

Test Item File (0-13-027059-8)
Contains over 2000 multiple choice questions, many conceptual in nature. All are referenced to the corresponding text section and ranked by level of difficulty.

Prentice Hall Custom Test (Windows: (0-13-027058-X; Mac: 0-13-027057-1)
Based on the powerful testing technology developed by Engineering Software Associates, Inc. (ESA), Prentice Hall Custom Test includes all questions from the Test Item File and allows instructors to create and tailor exams to their own needs. With the Online Testing Program, exams can also be administered on line and data can then be automatically transferred for evaluation. A comprehensive desk reference guide is included along with online assistance.

Transparency Pack (0-13-027062-8)
Includes approximately 400 full color transparencies of images from the text.

Image Viewer CD-ROM (0-13-027055-5)
The CD-ROM contains all text illustrations, digitized segments from the Prentice Hall *Physics You Can See* videotape as well as additional lab and demonstration videos, and animations from the Prentice Hall *Interactive Journey Through Physics* CD-ROM. Instructors can preview, sequence, and playback images, as well as perform keyword searches, add lecture notes, and incorporate their own digital resources. Free to qualified adopters.

***Physics You Can See* Video** (0-205-12393-7)
Contains 11 two- to five- minute demonstrations of classical physics experiments. It includes segments such as "Coin and Feather" (acceleration due to gravity); "Monkey and Gun" (projectile motion); "Swivel Hips" (force pairs); and "Collapse a Can" (atmospheric pressure). One copy is free to qualified adopters.

For the Student

Student Study Guide with Selected Solutions (0-13 027064-4)
David Reid (Eastern Michigan University)
The print study guide provides the following for each chapter:

- Objectives
- Warm-Up Questions from the *Just in Time Teaching* method by Gregor Novak and Andrew Gavrin (Indiana University–Purdue University, Indianapolis)
- Chapter Review with two-column Examples and integrated quizzes
- Reference Tools & Resources (equation summaries, important tips, and tools)
- Practice Problems by Carl Adler (East Carolina University)
- Puzzle Questions (also from Novak & Gavrin's *JiTT* method)
- Selected Solutions for several end-of-chapter problems

An electronic, interactive, media-enhanced version of the print-based study guide, with additional modules, is available on the www at http://www.prenhall.com/Walkerphysics, and on CD-ROM (0-13-027076-8). Please see the description of our companion website below for further details.

Student Pocket Guide (0-13-027063-6)
Biman Das (SUNY Potsdam)
This easy-to-carry 5″ x 7″ paperback contains a summary of the entire text, including all key concepts and equations, as well as tips and hints. Perfect for carrying to lecture and taking notes.

Companion Website (*http://www.prenhall.com/Walkerphysics*)
The Companion Website is an electronic, media-enriched, and interactive version of the print-based Study Guide with additional modules. For each text chapter it provides the following:

- Objectives by David Reid (Eastern Michigan University)
- Warm Up Questions from the *Just in Time Teaching* method by Gregor Novak and Andrew Gavrin (Indiana University–Purdue University, Indianapolis)
- Chapter Review by David Reid, with automated integrated practice quizzes and integrated "Physlet Illustration" Java applets written by Steve Mellema and Chuck Niederriter (Gustavus Adolphus College)
- Chapter Quiz by Carl Adler (East Carolina University)
- Physlet Problems by Wolfgang Christian (Davidson College)
- Algorithmically generated Practice Problems by Carl Adler
- "What's Physics Good For?" applications, with links to related internet sites, by Gregor Novak & Andrew Gavrin
- Puzzle Questions (also from Novak & Gavrin's *JiTT* method)
- An automatically scored MCAT Study Guide by Glenn Terrell (University of Texas at Arlington)
- Selected Solutions for several end-of-chapter problems
- PDF files for *Ranking Task Exercises* by Thomas O'Kuma (Lee College), David Maloney (Indiana University–Purdue University, Fort Wayne), and Curtis Hieggelke (Joliet Junior College)
- On-line Destinations (links to related sites) by Carl Adler

All objective modules are scored by the computer; results can be automatically e-mailed to the student's professor or teaching assistant.

For students without web access, a version of the website is available on CD-ROM (0-13-027076-8). The CD contains the same modules as the website (except for the Applications and Destinations modules, which are internet dependent) and has the same interactive capabilities (except for the ability to e-mail results to the instructor).

Physics on the Internet: A Student's Guide (0-13-890153-8)
Andrew Stull and Carl Adler
The perfect tool to help students take advantage of the Walker Companion Website. This useful resource gives clear steps to access Prentice Hall's regularly updated physics resources, along with an overview of general World Wide Web navigation strategies. Available free for students when packaged with the text.

Interactive Physics Player II Workbook (Windows: (0-13-667312-0; Mac: 0-13-477670-4)
Knowledge Revolution and Cindy Schwarz (Vassar College)
This highly interactive workbook/disk package contains 40 mechanics simulation projects of varying degrees of difficulty. Can be used in conjunction with any text.

Interactive Journey Through Physics CD-ROM (Dual Platform: 0-13-254103-3)
Logal and Cindy Schwarz (Vassar College) with Bob Beichner (North Carolina State University)
This highly interactive resource covers Mechanics, Thermodynamics, Electricity & Magnetism, and Light & Optics. It contains 62 videos, 120 simulations, 39 animations, and 231 problems. Can be used in conjunction with any text.

Prentice Hall/New York Times Themes of the Times—Physics
This unique newspaper supplement brings together a collection of the latest physics-related articles from the pages of the *New York Times*. Updated twice per year and available free to students when packaged with the text.

Acknowledgments

I would like to express my gratitude to my colleagues at Washington State University, as well as others in the physics community for their contributions to this project. In addition to the reviewers and accuracy checkers listed below, I am especially grateful to Jeff Braun (University of Evansville), James Cook (Middle Tennessee State University), Biman Das (State University of New York-Potsdam), Anthony Pitucco (Pima Community College), David Raffaelle (Glendale Community College), Rex Ramsier (University of Akron), Fred Thomas (Sinclair Community College), Jack Tuszynski (University of Alberta), and Karl Vogler (Northern Kentucky University) for their help with applications and problems.

My thanks are due also to many people at Prentice Hall who encouraged and helped me through the long and sometimes arduous process of making this book a reality: Alison Reeves, who sponsored the project and nurtured it at every stage; Tim Bozik and Paul Corey, for their patient support and encouragement; Liz Kell and Christian Botting, for their work on the supplements as well as their coordination of the elaborate reviewing program; Joe Sengotta, for his countless contributions to many aspects of the book, far above and beyond the call of duty; and Dan Schiller, for his help in shaping the final product. In addition, I am grateful for the tireless work of Yvonne Gerin, our indefatigable photo researcher; Jennifer Maughan, for choreographing the production process with diligence and unfailing good cheer; J. C. Morgan, for managing a very complex and demanding art program; and Bill Fellers, for his invaluable assistance with the problem solutions. The patience, dedication, and skill of these people are much appreciated.

I owe a great debt to my students over the years. My experience with them in the classroom provided the motivation and the inspiration for this book.

Finally, I would like to thank my wife, Betsy, for her patience and support during the long years this book was in the making, as well as her expert help in checking the pages.

I would very much appreciate hearing from users of this book. I welcome comments, criticisms, and above all suggestions that might make subsequent editions of the text even more useful to teachers and students of physics.

Reviewers

We are grateful to the following instructors for their thoughtful comments on the manuscript of this text and careful checking of proofs.

Eva Andrei
Rutgers University

Rama Bansil
Boston University

Paul Beale
University of Colorado-Boulder

Mike Berger
Indiana University

David Berman
University of Iowa

Jeff Braun
University of Evansville

Neal Cason
University of Notre Dame

Lattie Collins
Eastern Tennessee State University

James Cook
Middle Tennessee State University

David Curott
University of North Alabama

William Dabby
Edison Community Colege

Robert Davie
St. Petersburg Junior College

Anthony DiStefano
University of Scranton

John Dykla
Loyola University-Chicago

Eldon Eckard
Bainbridge College

Robert Endorf
University of Cincinnati

Raymond Enzweiler
Northern Kentucky University

John Erdei
University of Dayton

Frank Ferrone
Drexel University

John Flaherty
Yuba College

Lewis Ford
Texas A&M University

Asim Gangopadhyaya
Loyola University-Chicago

Barry Gilbert
Rhode Island College

Michael Graf
Boston College

Rainer Grobe
Illinois State University

Mitchell Haeri
Saddleback College

Xiaochun He
Georgia State University

J. Erik Hendrickson
University of Wisconsin-Eau Claire

Zafar Ismail
Daemen College

Dana Klinck
Hillsborough Community College

R. Gary Layton
Northern Arizona University

Michael Lieber
University of Arkansas

Mark Lindsay
University of Louisville

Hilliard Macomber
University of Northern Iowa

Trecia Markes
University of Nebraska-Kearny

Paul Morris
Abilene Christian University

David Moyle
Clemson University

K.W. Nicholson
Central Alabama Community College

Robert Piserchio
San Diego State University

Anthony Pitucco
Pima Community College

William Pollard
Valdosta State University

Robert Pompi
Binghamton University

Earl Prohofsky
Purdue University

David Raffaelle
Glendale Community College

Michael Ram
State University of New York-Buffalo

Rex Ramsier
University of Akron

Bob Rogers
San Francisco State University

Gerald Royce
Mary Washington College

Mats Selen
University of Illinois

Bartlett Sheinberg
Houston Community College

Peter Shull
Oklahoma State University

Christopher Sirola
Tri-County Technical College

Daniel Smith
South Carolina State University

Soren Sorensen
University of Tennessee-Knoxville

Leo Takahashi
Penn State University-Beaver

Harold Taylor
Richard Stockton College

Frederick Thomas
Sinclair Community College

Jack Tuszynski
University of Alberta

Lorin Vant Hull
University of Houston

Karl Vogler
Northern Kentucky University

Toby Ward
College of Lake Country

Lawrence Weinstein
Old Dominion University

Linda Winkler
Moorhead State University

Lowell Wood
University of Houston

Robert Wood
University of Georgia

Student Reviewers

We wish to thank the following students at North Carolina State University for providing helpful feedback during the development of this text. Their comments offered us valuable insight into the student experience.

Margaret Baker
Parker Havron
Jamie Helms
Michelle Lim
Stephanie Starnes
Monique Thomas
Timothy Nathan Witwer
Melissa Wright

Preface: To the Student

As a student preparing to take an introductory, algebra-based physics course, you are probably aware that physics applies to absolutely everything in the natural world, from raindrops and people to galaxies and atoms. Because physics is so wide-ranging and comprehensive it can sometimes seem a bit overwhelming. This text, which is based on extensive classroom experience and testing, is designed to help you deal with a large body of information and develop a working understanding of the basic concepts in physics. As you develop this understanding, you will find that you have enriched your experience of the world in which you live.

Now, I must admit that I like physics, and so I may be a bit biased in this respect. Still, the reason I teach and continue to study physics is that I enjoy the insight it gives into the physical world. I can't help but notice—and enjoy—aspects of physics all around me each and every day. I would like to share some of this enjoyment with you, and that is why I undertook the task of writing this book.

To assist you in the process of studying physics, this text incorporates a number of learning aids, including *Two-Column Examples, Active Examples*, and *Conceptual Checkpoints*. These and other elements work together in a unified way to enhance your understanding of physics on both a conceptual and a quantitative level. The pages that follow will introduce these elements to you, describe the purpose of each, and explain how they can help you.

As you progress through the text, you will encounter many interesting and intriguing applications of physics drawn from the world around you. Some of these, such as magnetically levitated trains or the satellite-based Global Positioning System that enables you to determine your position anywhere on Earth to within a few feet, are primarily technological in nature. Others focus on explaining familiar or not-so-familiar phenomena, such as why the Moon has no atmosphere, how sweating cools the body, or why flying saucer-shaped clouds often hover over mountain peaks even when the sky is clear. Still others, such as countercurrent heat exchange in animals and humans or the use of sound waves to destroy kidney stones, are of particular relevance to students of biology and the other life sciences.

In many cases, you may find the applications to be a bit surprising. Did you know, for example, that you are shorter at the end of the day than when you first get up in the morning? (This is discussed in Chapter 5.) That an instrument called the ballistocardiograph can detect the presence of a person hiding in a truck, just by registering the minute recoil from the beating of the stowaway's heart? (This is discussed in Chapter 9.) That if you hum next to a spider's web at just the right pitch you can cause a resonance effect that sends the spider into a tizzy? (This is discussed in Chapter 13.) That your hearing is sensitive enough to detect sounds that displace the eardrum by as little as the diameter of an atom? (This is discussed in Chapter 14.) That powerful magnets can exploit the phenomenon of diamagnetism to levitate living creatures? (This is discussed in Chapter 22.)

Writing this textbook was a rewarding experience for me. I hope using it will prove equally rewarding to you, and that it will inspire an interest in and appreciation of physics that will last a lifetime.

James S. Walker

In-Chapter Learning Tools

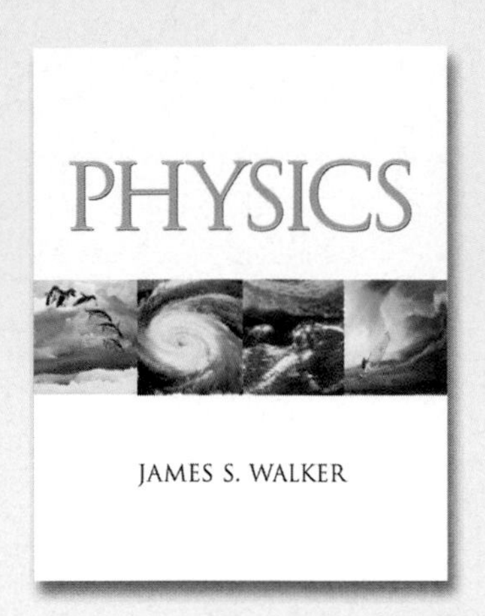

The unique suite of tools in this book fosters understanding of physical principles and active involvement in the learning process. Because one-size-fits-all examples do not sufficiently address the needs of students, the text employs a variety of pedagogical elements, each used where it can contribute most to developing conceptual insight and problem-solving skills.

Worked Examples in Two-Column Format

Worked Examples show students how to apply important concepts to commonly encountered, realistic problems. The structure of the Example not only helps students to master the problem at hand but also gives them greater insight into the physical principles and concepts behind the problem.

Problem Statement

Describes the physical situation to be analyzed, what is known, and what is to be found.

Picture the Problem

Helps students sketch and visualize the processes taking place, identify and label important quantities, and set up coordinate axes.

Strategy

Here, students learn how to analyze the problem, identify the key physical principles at work, and devise a plan for obtaining the solution.

Solution

A unique, two-column format presents the logical thought processes parallel to the mathematical steps, helping students see the relationship between the physical concepts and their mathematical expression.

Insight

Often, solving a problem leads to new insights. This feature points out interesting or significant features of the problem, the solution process, or the result.

Practice Problem

Students are given an opportunity to test their skills on a problem similar to the one just worked. For immediate reinforcement, the answer is provided after the Practice Problem. Related homework problems also direct students to end-of-chapter problems that are similar to the Example.

EXAMPLE 6–4 A Bad Break: Setting a Broken Leg with Traction

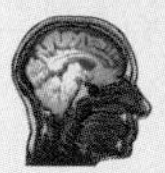

Real World Physics: Bio

A traction device employing three pulleys is applied to a broken leg, as shown in the sketch. The middle pulley is attached to the sole of the foot, and a mass m supplies the tension in the ropes. Find the value of the mass m if the net force exerted on the sole of the foot is to be 165 N.

Picture the Problem

Our sketch shows the physical picture as well as the free-body diagram for the middle pulley. Notice that the rope makes an angle of 40.0° with the horizontal on either side of the middle pulley.

Strategy

We begin by noting that the rope supports the hanging mass m. As a result, the tension in the rope, T, must be equal in magnitude to the weight of the mass; $T = mg$.

Next, the pulleys simply change the direction of the tension without changing its magnitude. Therefore, the tension in each segment of the rope is the same. The net force exerted on the sole of the foot is the sum of the tension T at 40.0° above the horizontal plus the tension T at 40.0° below the horizontal. We will calculate the net force component by component.

Once we calculate the net force acting on the foot, we set it equal to 165 N and solve for the tension T. Finally, we find the mass using the relation $T = mg$.

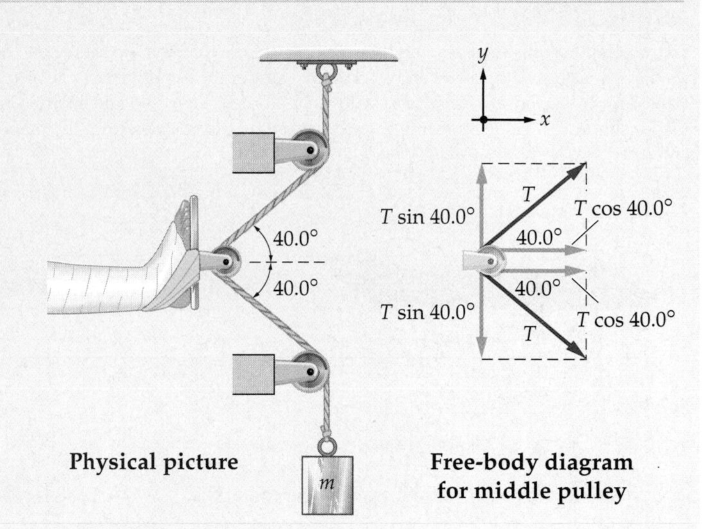

Solution

1. First, consider the tension that acts upward on the middle pulley. Resolve this tension into x and y components:

$$T_x = T\cos 40.0^\circ \qquad T_y = T\sin 40.0^\circ$$

2. Next, consider the tension that acts downward on the middle pulley. Resolve this tension into x and y components. Note the minus sign in the y component:

$$T_x = T\cos 40.0^\circ \qquad T_y = -T\sin 40.0^\circ$$

3. Sum the x and y components of force acting on the foot. We see that the net force acts only in the x direction, as one might expect from symmetry:

$$\sum F_x = T\cos 40.0^\circ + T\cos 40.0^\circ = 2T\cos 40.0^\circ$$
$$\sum F_y = T\sin 40.0^\circ - T\sin 40.0^\circ = 0$$

4. Step 3 shows that the net force acting on the foot is $2T\cos 40.0^\circ$. Set this force equal to 165 N and solve for T:

$$2T\cos 40.0^\circ = 165\text{ N}$$
$$T = \frac{165\text{ N}}{2\cos 40.0^\circ} = 108\text{ N}$$

5. Solve for the mass, m, using $T = mg$:

$$T = mg$$
$$m = \frac{T}{g} = \frac{108\text{ N}}{9.81\text{ m/s}^2} = 11.0\text{kg}$$

Insight

Notice that this pulley arrangement "magnifies the force" in the sense that a 108 N weight attached to the rope produces a 165 N force exerted on the foot. In addition, the net force on the foot produces an opposing force in the leg that acts in the direction of the head (a cephalad force), as desired to set a broken leg.

Practice Problem

(a) Would the required mass m increase or decrease if the angles in this device were changed from 40.0° to 30.0°? **(b)** Find the mass m for an angle of 30.0°. [**Answer: (a)** The required mass m will decrease. **(b)** 9.71 kg]

Some related homework problems: Problem 16, Problem 19, Problem 29

In-Chapter Learning Tools

Active Examples

Active Examples break down a problem into a series of manageable, logical steps, but require the student to play an active role in the solution. These examples act as a bridge between the worked Examples in the chapter, in which the solution process is fully spelled out, and the end-of-chapter problems, where students receive no guidance at all.

ACTIVE EXAMPLE 5–1 Foamcrete

Foamcrete is a substance designed to stop an airplane that has run off the end of a runway, without causing injury to passengers. It is solid enough to support a car, but crumbles under the weight of a large airplane. By crumbling, it slows the plane to a safe stop. For example, suppose a 747 jetliner with a mass of 1.75×10^5 kg and an initial speed of 26.8 m/s is slowed to a stop in 122 m. What is the magnitude of the retarding force **F** exerted by the Foamcrete on the plane?

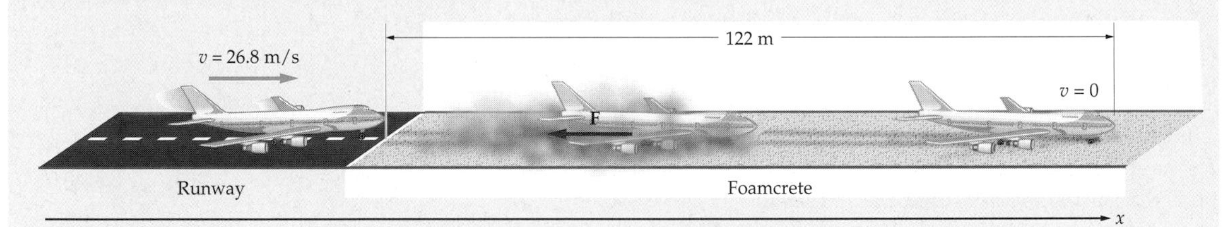

Solution

1. Use $v^2 = v_0^2 + 2a_x\Delta x$ to find the plane's acceleration:	$a_x = -2.94 \text{ m/s}^2$
2. Sum the forces in the x direction. Let F represent the magnitude of the force **F**:	$\Sigma F_x = -F$
3. Set the sum of forces equal to mass times acceleration:	$-F = ma_x$
4. Solve for the magnitude of the force, F:	$F = -ma_x = 5.15 \times 10^5$ N

Conceptual Checkpoints

Conceptual Checkpoints test students' ability to make inferences based on chapter concepts. These qualitative questions are presented in multiple-choice format to guide student response. "Reasoning and Discussion" clearly explains the physical principles behind the answer, while recognizing and addressing students' possible misconceptions.

CONCEPTUAL CHECKPOINT 6–1

A car drives with its tires rolling freely. Is the friction between the tires and the road **(a)** kinetic or **(b)** static?

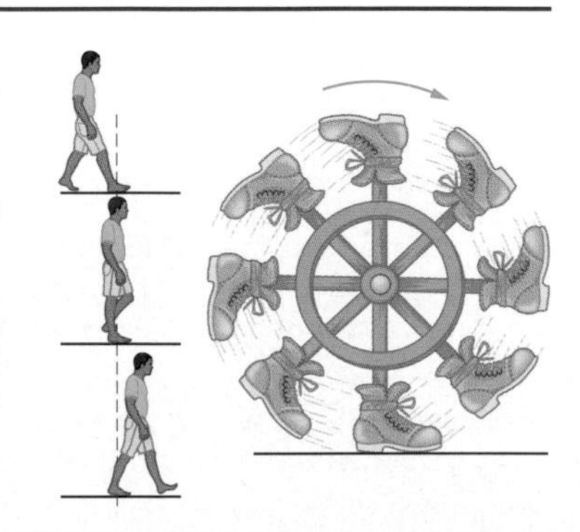

Reasoning and Discussion

A reasonable-sounding answer is that because the car is moving, the friction between its tires and the road must be kinetic friction—but this is not the case.

Actually, the friction is static because the bottom of the tire is in static contact with the road. To understand this, watch your feet as you walk. Even though you are moving, each foot is in static contact with the ground once you step down on it. Your foot doesn't move again until you lift it up and move it forward for the next step. A tire can be thought of as a succession of feet arranged in a circle, each of which is momentarily in static contact with the ground.

Answer:

(b) The friction between the tires and the road is static friction.

REAL WORLD PHYSICS: BIO

Centrifuges and Ultracentrifuges

Finally, we determine the acceleration produced in a **centrifuge**, a common device in biological and medical laboratories that uses large centripetal accelerations to perform such tasks as separating red and white blood cells from serum. A simplified top view of a centrifuge is shown in **Figure 6–14**.

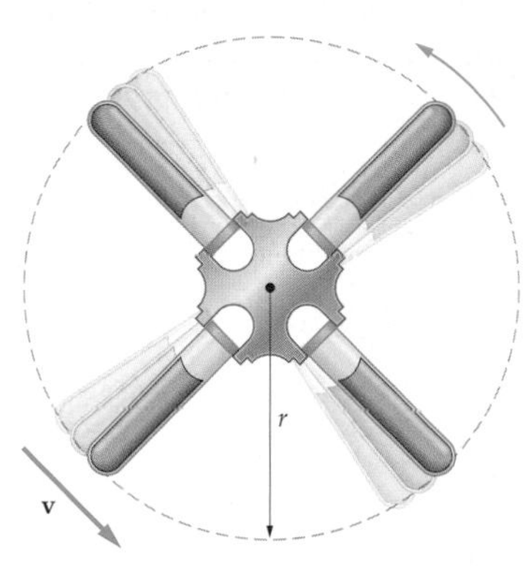

▲ **FIGURE 6–14 Simplified top view of a centrifuge in operation**

▲ A laboratory centrifuge of the kind commonly used to separate blood components.

EXERCISE 6–2

The centrifuge in Figure 6–14 rotates at a rate that gives the bottom of the test tube a linear speed of 89.3 m/s. If the bottom of the test tube is 8.50 cm from the axis of rotation, what is the centripetal acceleration experienced there?

Solution

Applying the relation $a_{cp} = v^2/r$ yields

$$a_{cp} = \frac{v^2}{r} = \frac{(89.3\ \text{m/s})^2}{0.0850\ \text{m}} = 93{,}800\ \text{m/s}^2 = 9560\,g$$

In this expression, g is the acceleration of gravity, 9.81 m/s^2.

Exercises

When a more elaborate device is not needed, simple Exercises give students practice in using a new formula or relationships introduced in the text.

End-of-Chapter Learning Tools

Chapter Summary

The Chapter Summary is organized by section and topic. It provides an easy reference for students to review key terms, concepts, and equations from the chapter.

Chapter Summary

	Topic	Remarks and Relevant Equations	
6–1	**Frictional Forces**	Frictional forces are due to the microscopic roughness of surfaces in contact. As a rule of thumb, friction is independent of the area of contact and independent of the relative speed of the surfaces.	
	kinetic friction	Friction experienced by surfaces that are in contact and moving relative to one another. The force of kinetic friction is given by	
		$f_k = \mu_k N$	6–1
		In this expression, μ_k is the coefficient of kinetic friction and N is the magnitude of the normal force.	
	static friction	Friction experienced by surfaces that are in static contact. The maximum force of static friction is given by	
		$f_{s,\text{max}} = \mu_s N$	6–3
		In this expression, μ_s is the coefficient of static friction and N is the magnitude of the normal force. The force of static friction can have any magnitude between zero and its maximum value.	
6–2	**Strings and Springs**	Strings and springs provide a common way of exerting forces on objects. Ideal strings and springs are massless.	
	tension	The force transmitted through a string. The tension is the same throughout the length of an ideal string.	
	Hooke's law	The force exerted by an ideal spring stretched by the amount x is	
			6–4

Problem-Solving Summary

Type of Calculation	Relevant Physical Concepts	Related Examples
Find the acceleration when kinetic friction is present.	First, find the magnitude of the normal force, N. The corresponding kinetic friction has a magnitude of $f_k = \mu_k N$ and points opposite to the direction of motion. Include this force with the others when applying Newton's second law.	Examples 6–1, 6–2
Solve problems involving static friction.	Start by finding the magnitude of the normal force, N. The corresponding static friction has a magnitude between zero and $\mu_s N$. Its direction opposes motion.	Example 6–3 Active Example 6–1
Find the acceleration and the tension for masses connected by a string.	Apply Newton's second law to each mass separately. This generates two equations, which can be solved for the two unknowns, a and T.	Examples 6–6, 6–7
Solve problems involving circular motion.	Set up the coordinate system so that one axis points to the center of the circle. When applying Newton's second law to that direction, set the acceleration equal to $a_{cp} = v^2/r$.	Examples 6–8, 6–9 Active Example 6–3

Problem-Solving Summary

The Problem-Solving Summary presents an inventory of the types of problems encountered in the chapter. It also provides the relevant physical concepts and related chapter Examples for each.

Conceptual Questions

A separate section of qualitative, conceptual questions allows students to test their understanding of chapter principles. Answers to odd-numbered Conceptual Questions are located in the back of the book.

Conceptual Questions

1. A clothesline always sags a little, even if nothing hangs from it. Explain.
2. In the *Jurassic Park* sequel, *The Lost World*, a man tries to keep a large vehicle from going over a cliff by connecting a cable from his Jeep to the vehicle. The man then puts the Jeep in gear and spins the rear wheels. Do you expect that spinning the tires will increase the force exerted by the Jeep on the vehicle? Why or why not?
3. An object moves stant magnitude. the direction of r Does the object's
4. In a car with rear- less than the max
5. A train typically r for a given initial
6. Give some every beneficial.
7. At the local farm the backseat of th you approach a st you brake a bit ha
8. It is possible to sp have none of the How would you
9. When a traffic accident is investigated, it is common for the length of the skid marks to be measured. How could this information be used to estimate the initial speed of the vehicle that left the skid marks?
10. If you weigh yourself at the equator you get a smaller value than if you weigh yourself at one of the poles. Why?
11. Water sprays off a rapidly turning bicycle wheel. Why?
12. Can an object be in equilibrium if it is moving? Explain.

Problems

Note: **IP** *denotes an integrated conceptual/quantitative problem.* **BIO** *identifies problems of biological or medical interest. Blue bullets (•, ••, •••) are used to indicate the level of difficulty of each problem.*

Section 6–1 Frictional Forces

1. • A baseball player dives into third base with an initial speed of 7.90 m/s. If the coefficient of kinetic friction between the player and the ground is 0.41, how far does the player slide before coming to rest?
2. • A child goes down a playground slide with an acceleration of 1.05 m/s^2. Find the coefficient of kinetic friction between the child and the slide if the slide is inclined at an angle of 35.0° below the horizontal.
3. • Hopping into your Porsche, you floor it and accelerate at 12 m/s^2 without spinning the tires. Determine the minimum coefficient of static friction between the tires and the road needed to make this possible.
4. • When you push a 1.80-kg book resting on a tabletop it takes 2.25 N to start the book sliding. Once it is sliding, however, it takes only 1.50 N to keep the book moving with constant speed. What are the coefficients of static and kinetic friction between the book and the tabletop?
5. • In the previous problem, what is the frictional force exerted on the book when you push on it with a force of 0.75 N?
6. •• **IP** A tie is laid out on a table, with a fraction of its length hanging over the edge. Initially, the tie is at rest. **(a)** If the fraction hanging from the table is increased, the tie eventually slides to the ground. Explain. **(b)** What is the coefficient of static friction between the tie and the table if the tie begins to slide when one-fourth of its length hangs over the edge?
7. •• To move a large crate across a rough floor, you push down on it at an angle of 20°, as shown in **Figure 6–16**. Find

▲ **FIGURE 6–16** Problem 7

8. •• In the previous problem, find the acceleration of the crate if the applied force is 330 N and the coefficient of kinetic friction is 0.45.
9. •• A 45-kg crate is placed on an inclined ramp. When the angle the ramp makes with the horizontal is increased to 23° the crate begins to slide downward. **(a)** What is the coefficient of static friction between the crate and the ramp? **(b)** At what angle does the crate begin to slide if its mass is doubled?
10. •• **IP** A 95-kg sprinter wishes to accelerate from rest to a speed of 12 m/s in a distance of 20 m. **(a)** What coefficient of static friction is required between the sprinter's shoes and the track? **(b)** Explain the strategy used to find the answer to part **(a)**.
11. •• A person places a cup of coffee on the roof of her car while

Integrated Problems

Each set of end-of-chapter Problems contains Integrated Problems, marked **IP**, that combine qualitative and quantitative parts. These problems, which stress the importance of reasoning from principles, show how conceptual insight and numerical calculation go hand in hand in physics.

Companion Website

The Walker Companion Website (at **http://www.prenhall.com/walkerphysics**), with contributions from leaders in physics education research, provides students with a variety of interactive explorations of each chapter's topics, easily accomodating differences in learning styles. These tools help students further their conceptual understanding of physics and provide extra practice in problem solving. Various modules focusing on conceptual understanding and visualization, problem solving, and real-world applications come together to offer a unique additional resource that will help your students succeed.

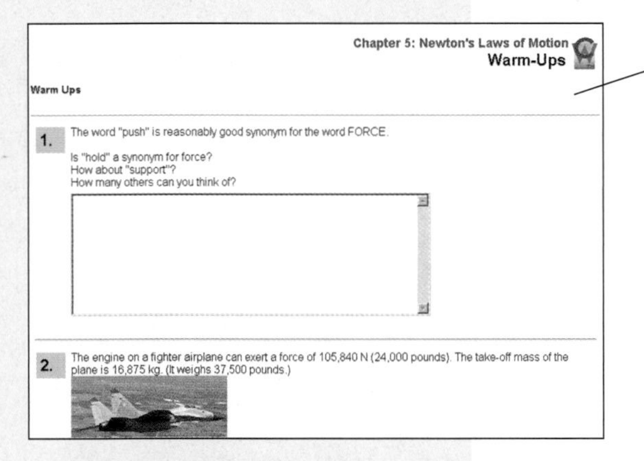

- **Warm-Ups**
  Gregor Novak and Andrew Gavrin (Indiana University–Purdue University, Indianapolis)
  These short-answer questions, also known as "Pre-flights," from the Just-in-Time Teaching (JiTT) method, address concepts about to be presented within an everyday context to pique students' curiosity and provoke discussion for your opening lecture on each topic.

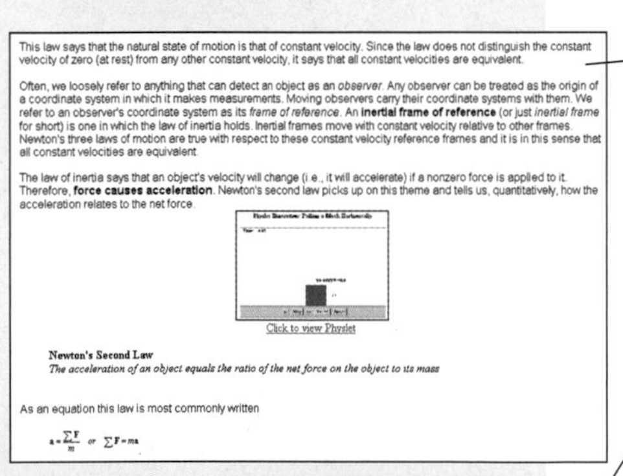

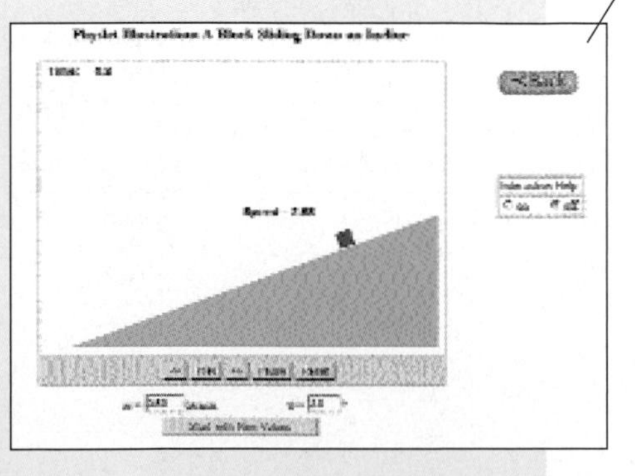

- **Chapter Review**
  David Reid (Eastern Michigan University)
  The Chapter Review provides for each chapter section a brief review followed by several worked Examples. Each section also contains a Practice Quiz with instant feedback. Physlet illustrations by Chuck Niederriter and Steve Melema *(Gustavus Adolphus College)* are integrated throughout the Chapter Review to help students visualize the concepts.

www.prenhall.com/walkerphysics

- **Chapter Quiz**
  Carl Adler (East Carolina University)
  This multiple-choice quiz contains approximately 20-25 questions per chapter; over half are conceptual in nature.

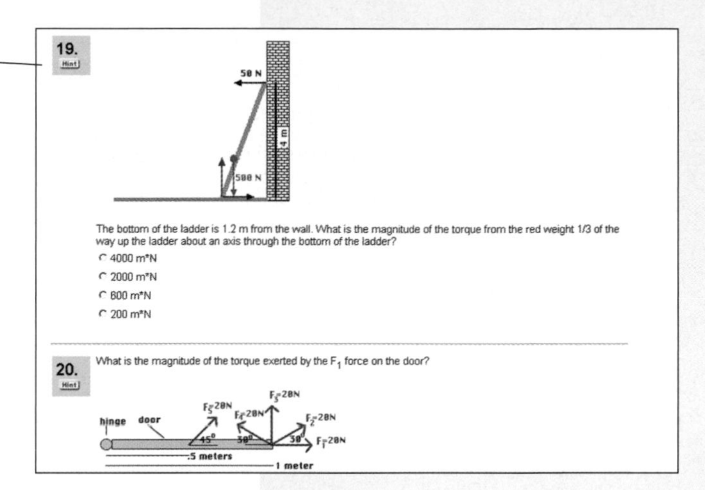

- **Physlet Problems**
  Wolfgang Christian (Davidson College)
  Physlets® are award-winning scriptable Java applets that animate specific physics problems. Students must read the problem, run the Physlet, observe the motion, and make the appropriate measurements, drawing on their knowledge of physics concepts to solve the problem. The scripting for each problem is pre-set and conditions are not alterable by the student, so the student's focus remains on the situation given—and consequently on the physics—not on the media. The Physlet Problems have hints and wrong-answer feedback keyed to Walker's text.

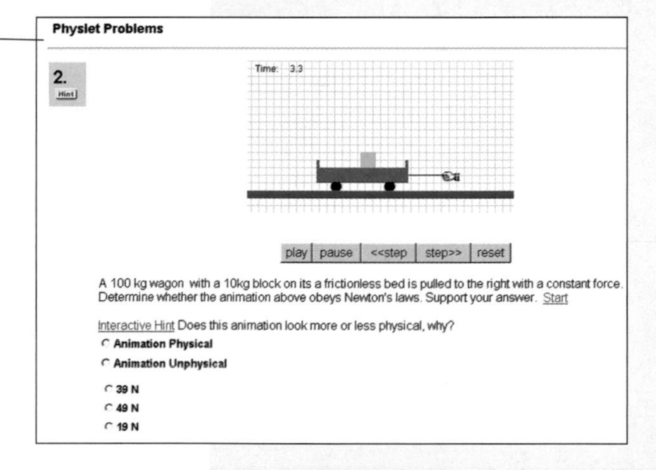

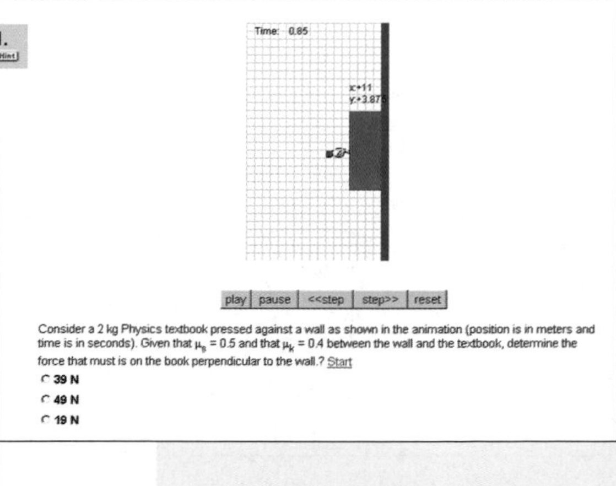

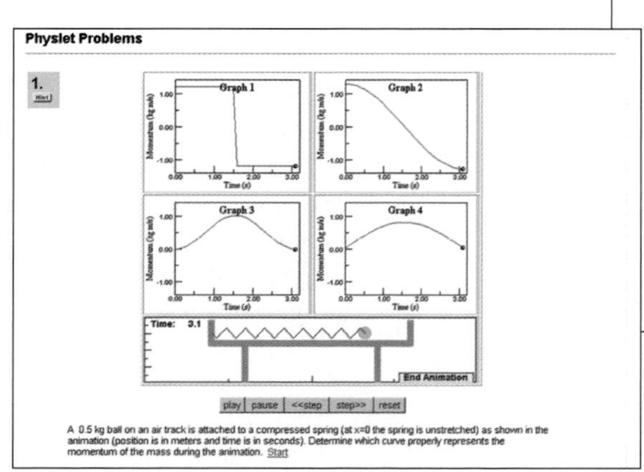

Companion Website

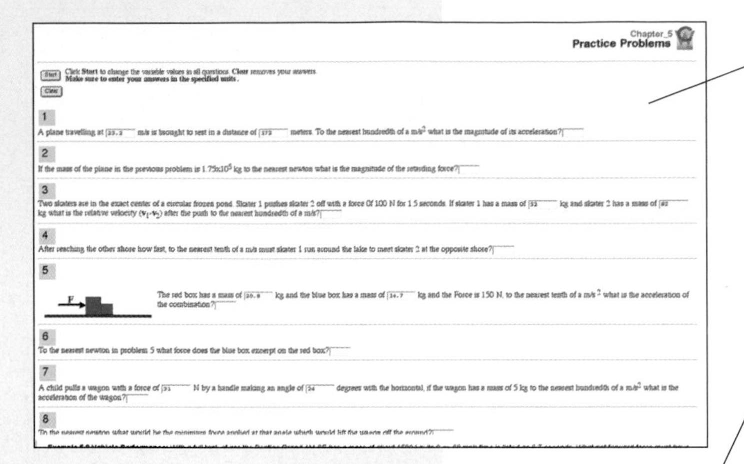

Chapter_5
Practice Problems

Click **Start** to change the variable values in all questions. **Clear** removes your answers.
Make sure to enter your answers in the specified units.

1
A plane travelling at m/s is brought to rest in a distance of meters. To the nearest hundredth of a m/s² what is the magnitude of its acceleration?

2
If the mass of the plane in the previous problem is 1.75×10^5 kg to the nearest newton what is the magnitude of the retarding force?

3
Two skaters are in the exact center of a circular frozen pond. Skater 1 pushes skater 2 off with a force Of 100 N for 1.5 seconds. If skater 1 has a mass of kg and skater 2 has a mass of kg what is the relative velocity (v_1-v_2) after the push to the nearest hundredth of a m/s?

4
After reaching the other shore how fast, to the nearest tenth of a m/s must skater 1 run around the lake to meet skater 2 at the opposite shore?

5
The red box has a mass of kg and the blue box has a mass of kg and the Force is 150 N, to the nearest tenth of a m/s² what is the acceleration of the combination?

6
To the nearest newton in problem 5 what force does the blue box exerpt on the red box?

7
A child pulls a wagon with a force of N by a handle making an angle of degrees with the horizontal, if the wagon has a mass of 5 kg to the nearest hundredth of a m/s² what is the acceleration of the wagon?

8
To the nearest newton what would be the minimum force applied at that angle which would lift the wagon off the ground?

Chapter 6: Further Applications of Newton's Laws
Applications

There is a Chinese story of a legendary government official Wan-Hoo who equipped a chair with 47 large rockets, hoping it would take him to the moon. His 47 assistants ignited the rockets with torches. An enormous explosion followed. When the smoke cleared Wan-Hoo and the chair were gone never to be seen again.

The ancient Chinese, and the ancient Greeks, were familiar with the action-reaction principle, now called Newton's Third Law. An object, such as a rocket can push itself forward, against the inertia of the material that it ejects.
We have all had fun letting go air filled balloons, which propel themselves against the escaping air. A more elaborate version of this idea is a soda bottle rocket filled with compressed air. Check out the bottle rocket instructions presented by no less an authority on rocket flight than NASA.
Unlike airplanes that need air for lift and for oxygen, rockets (or firecrackers) carry their own oxygen and their fuel. They are completely self-sufficient and can propel themselves through vacuum.
Chinese rocketry was supported by another of their clever inventions. Several hundred years B.C. the Chinese invented gunpowder, a mixture of charcoal, sulfur and saltpeter (Potassium Nitrate KNO3.) Many practical applications quickly followed, including fireworks and "fire arrows", bamboo tubes filled with gunpowder. The tubes were made sturdy enough to propel themselves against the exhausted gas, rather than exploding like a grenade.

Rocketry ideas took more than a thousand years to reach Europe. Reports of rocket use by the Mongols date from the 13th century. Seven centuries later, Wan-Hoo's dream became a reality when on July 20, 1969, the first astronauts rode powerful rockets to the Moon.

Copper and Water in Styrofoam Cups—Maximum Temperature 109

You have six styrofoam cups containing the same amount of water at 20°C. You also have six copper blocks whose masses and initial temperatures vary as shown below. One block goes into each cup. (Assume the mass of the water is between 500 g and 1000g.)

Rank these cups according to the maximum temperature of the water after the block is added.

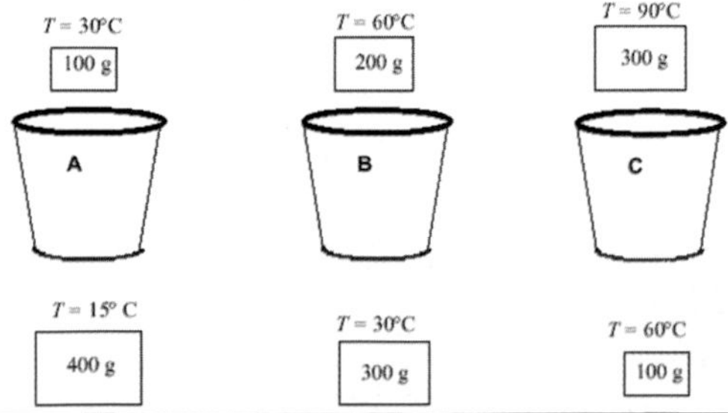

- **Practice Problems**
  Carl Adler (East Carolina University)
  These algorithmically generated numerical problems feature values that change each time a student goes back to the problem set.

- **Applications**
  Gregor Novak and Andrew Gavrin, (Indiana University–Purdue University, Indianapolis)
  These interesting essays address the history of physics or an application of a physics topic in everyday life, giving relevance to the subject by answering the question: "What is physics good for?" (Applications are also part of the JiTT method and are often referred to as "Good-Fors.") They include hyperlinks to other Websites and follow-up essay questions that require research.

- **Puzzles**
  Gregor Novak and Andrew Gavrin (Indiana University–Purdue University, Indianapolis)
  These questions, also from the JiTT method, are posed in a real-world context and are designed to test students' understanding of core concepts after completing a particular topic.

- **MCAT Study Guide**
  Glen Terrell (University of Texas at Arlington)
  This guide offers a multiple-choice practice quiz containing approximately 30-40 questions per chapter that address concepts and problems typically found on the MCAT examination.

- **Ranking Task Exercises**
  edited by Thomas O'Kuma (Lee College), David Maloney (Indiana University–Purdue University, Fort Wayne), and Curtis Hieggelke (Joliet Junior College)
  These downloadable PDF files of acclaimed diagnostic conceptual exercises present students with several contextually similar physical situations, and ask them to rank the situations by a specific quantity. The results reveal common misconceptions and help you tailor your lectures to your students' needs.

- **Destinations**
  Carl Adler (East Carolina University)
  Destinations are links to relevant Websites for each chapter, either about the physics topic in the chapter or about related applications.

- **Syllabus Manager**
  Syllabus Manager provides instructors with an easy, step-by-step process for creating and revising a class syllabus with direct links to the text's Companion Website and other on-line content. Through this on-line syllabus, instructors can add assignments and send announcements to the class with the click of a button. The completed syllabus is hosted on Prentice Hall's servers, allowing the syllabus to be updated from any computer with Internet access.

- **On-line Grading**
  Scoring for all objective questions and problems, as well as responses to essay questions, can be e-mailed back to the instructor. All quiz questions can be submitted for instant grading, and students may arrange for their results to be automatically e-mailed to you or a teaching assistant.

Chapter 16: Temperature and Heat 

The World Wide Web was created for physicists at CERN (European Laboratory for Particle Physics). It remains a major resource for learning about physics, both as a subject and as a community of scholars. Best of all, it's fun!

Chapter 16 Resources

Monthly/Seasonal Weather Predictions
LA NINA
EL NINO
Temperature Scales and All That
World Weather
Make a Thermometer
Clothing and Comfort
More About Temperature
National Mole Day

Major Physics Resources

A Great Place to Learn About the History of Physics
The SuperList of Physics Sites from EINET
The Yahoo Physics Site
How Things Work

Useful and Fun Places

- **Companion Website on CD-ROM**
  For those students without access to the Internet, a version of the Companion Website is available on CD-ROM. It contains all of the modules on the Website (except the Destinations and Applications modules, which are Internet dependent) and has the same interactive capabilities (except the ability to e-mail results to the instructor). As with the Companion Website, questions are automatically scored by the computer, but results are provided immediately rather than tabulated and submitted via e-mail.

1 Introduction

Physics is a quantitative science, based on careful measurements of quantities such as mass, length, and time. In the measurement shown here, an albatross chick is found to have a mass of approximately 1.3 kilograms, corresponding to a weight of just under 3.0 pounds.

The goal of physics is to gain a deeper understanding of the world in which we live. For example, the laws of physics allow us to predict the behavior of everything from rockets sent to the moon, to integrated chips in computers, to lasers used to perform eye surgery. In short, everything in nature—from atoms and subatomic particles to solar systems and galaxies—obeys the laws of physics.

As we begin our study of physics, it is useful to consider a range of issues that underlie everything to follow. One of the most fundamental of these is the system of units we use when we measure such things as the mass of an object, its length, and the time between two events. Other equally important issues include methods for handling numerical calculations and basic conventions of mathematical notation. By the end of the chapter we will have developed a common "language" of physics that will be used throughout this book and probably in any science that you study.

1–1 Physics and the Laws of Nature

Physics is the study of the fundamental laws of nature, which, simply put, are the laws that underlie all physical phenomena in the universe. Remarkably, we have found that these laws can be expressed in terms of mathematical equations. As a result, it is possible to make precise, quantitative comparisons between the predictions of theory—derived from the mathematical form of the laws—and the observations of experiments. Physics, then, is a science rooted equally firmly in theory and experiment.

What makes physics particularly fascinating is the fact that it relates to everything in the universe. There is a great beauty in the vision that physics brings to our view of the universe; namely, that all the complexity and variety that we see in the world around us, and in the universe as a whole, are manifestations of a few fundamental laws and principles. That we can discover and apply these basic laws of nature is at once astounding and exhilarating.

For those not familiar with the subject, physics may seem to be little more than a confusing mass of formulas. Sometimes, in fact, these formulas can be the trees that block the view of the forest. For a physicist, however, the many formulas of physics are simply different ways of expressing a few fundamental ideas. It is the forest—the basic laws and principles of physical phenomena in nature—that is the focus of this text.

1–2 Units of Length, Mass, and Time

To make quantitative comparisons between the laws of physics and our experience of the natural world, certain basic physical quantities must be measured. The most common of these quantities are **length** (L), **mass** (M), and **time** (T). In fact, in the next several chapters, these are the only quantities that arise. Later in the text, however, additional quantities, such as temperature and electric current, will be introduced as needed.

We begin by defining the units in which each of these quantities is measured. Once the units are defined, the values obtained in specific measurements can be expressed as multiples of them. For example, our unit of length is the **meter** (m). It follows, then, that a person who is 1.94 m tall has a height 1.94 times this unit of length. Similar comments apply to the unit of mass, the **kilogram**, and the unit of time, the **second**.

The detailed system of units used in this book was established in 1960 at the Eleventh General Conference of Weights and Measures in Paris, France, and goes by the name Système International, or SI for short. Thus, when we refer to **SI units**, we mean units of meters (m), kilograms (kg), and seconds (s). Taking the first letter from each of these units leads to an alternate name that is often used—the **mks system**.

In the remainder of this section we define each of the SI units.

Length

Early units of length were often associated with the human body. For example, the Egyptians defined the cubit to be the distance from the elbow to the tip of the middle finger. Similarly, the foot was originally defined to be the length of the royal foot of King Louis XIV. As colorful as these units may be, they are not particularly reproducible—at least not to great precision.

In 1793, the French Academy of Sciences, seeking a more objective and reproducible standard, decided to define a unit of length equal to one ten-millionth the distance from the North Pole to the equator. This new unit was named the metre (from the Greek *metron* for "measure"). The preferred spelling in the United States is *meter*. This definition was widely accepted, and in 1799 a "standard" meter was produced. It consisted of a platinum–iridium alloy rod with two marks on it one meter apart.

Since 1983, however, we have used an even more precise definition of the meter, this time based on the speed of light in a vacuum. In particular:

> One meter is defined to be the distance traveled by light in a vacuum in 1/299,792,458 of a second.

No matter how its definition is refined, however, a meter is still about 3.28 feet, which is roughly 10 percent longer than a yard. A list of typical lengths is given in Table 1–1.

TABLE 1–1 Typical Distances

Distance from Earth to the nearest large galaxy (the Andromeda galaxy, M31)	2×10^{22} m
Diameter of our galaxy (the Milky Way)	8×10^{20} m
Distance from Earth to the nearest star	4×10^{16} m
One light year	9.46×10^{15} m
Radius of Pluto's orbit	6×10^{12} m
Distance from Earth to the Sun	1.5×10^{11} m
Radius of Earth	6.37×10^{6} m
Length of football field	10^{2} m
Height of a person	2 m
Diameter of a CD	0.12 m
Diameter of the aorta	0.018 m
Diameter of a period in a sentence	5×10^{-4} m
Diameter of a red blood cell	8×10^{-6} m
Diameter of the hydrogen atom	10^{-10} m
Diameter of a proton	2×10^{-15} m

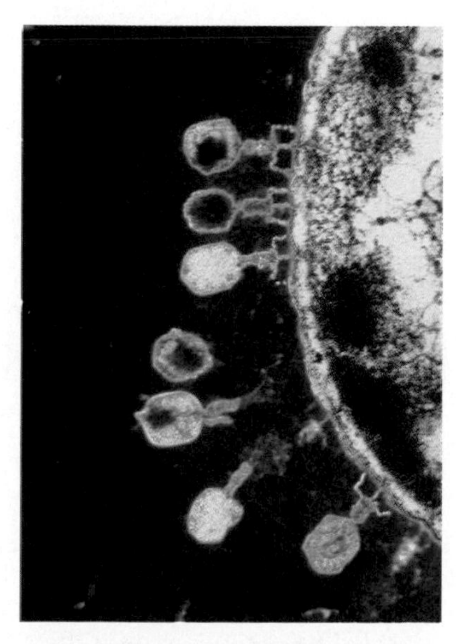

▲ The size of these viruses, seen here attacking a bacterial cell, is about 10^{-7} m.

▲ The diameter of this typical galaxy is about 10^{21} m. (How many viruses would it take to span the galaxy?)

TABLE 1–2 Typical Masses

Galaxy (Milky Way)	4×10^{41} kg
Sun	2×10^{30} kg
Earth	5.97×10^{24} kg
Space Shuttle	2×10^{6} kg
Elephant	5400 kg
Automobile	1200 kg
Human	70 kg
Baseball	0.15 kg
Honeybee	1.5×10^{-4} kg
Red blood cell	10^{-13} kg
Bacterium	10^{-15} kg
Hydrogen atom	1.67×10^{-27} kg
Electron	9.11×10^{-31} kg

▲ The standard kilogram, a cylinder of platinum and iridium 0.039 m in height and diameter, is kept under carefully controlled conditions in Sèvres, France. Exact replicas are maintained in other laboratories around the world.

Mass

In SI units, mass is measured in kilograms. Unlike the meter, the kilogram is not based on any natural physical quantity. By convention, the kilogram has been defined as follows:

> The kilogram, by definition, is the mass of a particular platinum–iridium alloy cylinder at the International Bureau of Weights and Standards in Sèvres, France.

To put the kilogram in everyday terms, a quart of milk has a mass slightly less than 1 kilogram. Additional masses, in kilograms, are given in Table 1–2.

Note that we do not define the kilogram to be the *weight* of the platinum-iridium cylinder. In fact, weight and mass are quite different quantities, even though they are often confused in everyday language. Mass is an intrinsic, unchanging property of an object. Weight, in contrast, is a measure of the gravitational force acting on an object, which can vary depending on the object's location. For example, if you are fortunate enough to travel to Mars someday, you will find that your weight is less than on the Earth, though your mass is unchanged. The force of gravity will be discussed in detail in Chapter 12.

Time

Nature has provided us with a fairly accurate timepiece in the revolving Earth. In fact, prior to 1956, the mean solar day was defined to consist of 24 hours, with 60 minutes per hour, and 60 seconds per minute, for a total of $(24)(60)(60) = 86{,}400$ seconds. Even the rotation of the Earth is not completely regular, however.

Today, the most accurate timekeepers known are "atomic clocks," which are based on characteristic frequencies of radiation emitted by certain atoms. These clocks have typical accuracies of about 1 second in 300,000 years. The atomic clock used for defining the second operates with cesium–133 atoms. In particular, the second is defined as follows:

> One second is defined to be the time it takes for radiation from a cesium–133 atom to complete 9,192,631,770 cycles of oscillation.

A range of characteristic time intervals is given in Table 1–3.

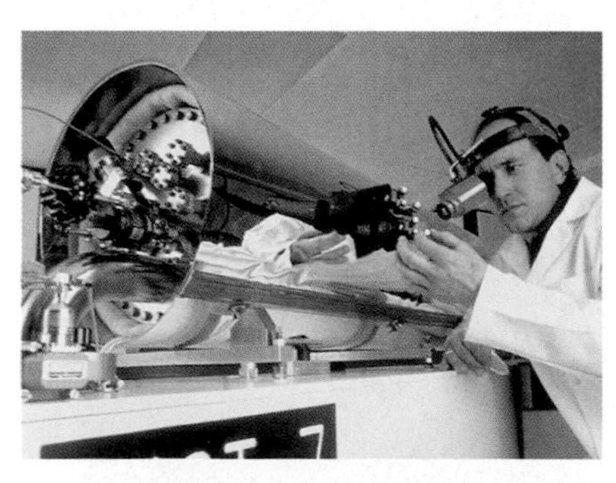

▲ This atomic clock, which keeps time on the basis of radiation from cesium atoms, is accurate to about three millionths of a second per year. (How long would it take for it to gain or lose an hour?)

TABLE 1–3 Typical Times

Age of the universe	5×10^{17} s
Age of the Earth	1.3×10^{17} s
Existence of human species	6×10^{13} s
Human lifetime	2×10^{9} s
One year	3×10^{7} s
One day	8.6×10^{4} s
Time between heartbeats	0.8 s
Human reaction time	0.1 s
One cycle of a high-pitched sound wave	5×10^{-5} s
One cycle of an AM radio wave	10^{-6} s
One cycle of a visible light wave	2×10^{-15} s

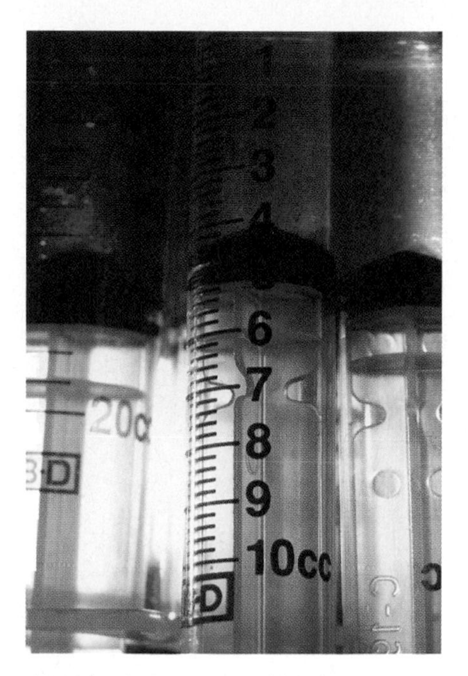

▲ Medical instruments like these syringes are typically graduated in cubic centimeters (cc).

Other Systems of Units and Standard Prefixes

Although SI units are used throughout most of this book and are used almost exclusively in scientific research and in industry, we will occasionally refer to other systems that you may encounter from time to time.

For example, a system of units similar to the mks system, though comprised of smaller units, is the **cgs system**, which stands for centimeter (cm), gram (g), and second (s). In addition, the British engineering system is often encountered in everyday usage in the United States. Its basic units are the slug for mass, the foot (ft) for length, and the second for time.

Finally, multiples of the basic units are common no matter which system is used. Standard prefixes are used to designate common multiples in powers of ten. For example, the prefix *kilo* means one thousand, or, equivalently, 10^3. Thus, 1 kilogram is 10^3 grams, and 1 kilometer is 10^3 meters. Similarly, *milli* is the prefix for one thousandth, or 10^{-3}. Thus, a millimeter is 10^{-3} meter, and so on. The most common prefixes are listed in Table 1–4.

TABLE 1–4 Common Prefixes

Power	Prefix	Abbreviation
10^{15}	peta	P
10^{12}	tera	T
10^{9}	giga	G
10^{6}	mega	M
10^{3}	kilo	k
10^{2}	hecto	h
10^{1}	deka	da
10^{-1}	deci	d
10^{-2}	centi	c
10^{-3}	milli	m
10^{-6}	micro	μ
10^{-9}	nano	n
10^{-12}	pico	p
10^{-15}	femto	f

EXERCISE 1–1

(a) A minivan sells for 33,200 dollars. Express the price of the minivan in kilodollars and megadollars.
(b) A typical *E. coli* bacterium is about 5 micrometers (or microns) in length. Give this length in millimeters and kilometers.

Solution

(a) 33.2 kilodollars, 0.0332 megadollars.
(b) 0.005 mm, 0.000000005 km.

1–3 Dimensional Analysis

In physics, when we speak of the **dimension** of a physical quantity, we refer to the *type* of quantity in question, regardless of the units used in the measurement. For example, a distance measured in cubits and another distance measured in light-years both have the same dimension—length. The same is true of compound units such as velocity, which has the dimensions of length per unit time (length/time). A velocity measured in miles per hour has the same dimensions—length/time—as one measured in inches per century.

Now, any valid formula in physics must be **dimensionally consistent**; that is, each term in the equation must have the same dimensions. It simply doesn't make

TABLE 1–5
Dimensions of Some Common Physical Quantities

Quantity	Dimension
Distance	L
Area	L^2
Volume	L^3
Velocity	L/T
Acceleration	L/T^2
Energy	ML^2/T^2

sense to add a distance to a time, for example, any more than it makes sense to add apples and oranges. They are different things.

To check the dimensional consistency of an equation, it is convenient to introduce a special notation for the dimension of a quantity. We will use square brackets, [], for this purpose. Thus, if x represents a distance, which has dimensions of length [L], we write this as $x = [L]$. Similarly, a velocity, v, has dimensions of length per time [T]; thus we write $v = [L]/[T]$ to indicate its dimensions. Acceleration, a, which is the change in velocity per time, has the dimensions $a = ([L]/[T])/[T] = [L]/[T^2]$. The dimensions of some common physical quantities are summarized in Table 1–5.

Let's use this notation to check the dimensional consistency of a simple equation. Consider the following formula:

$$x = x_0 + vt$$

In this equation, x and x_0 represent distances, v is a velocity, and t is time. Writing out the dimensions of each term, we have

$$[L] = [L] + \frac{[L]}{[T]}[T]$$

It might seem at first that the last term has different dimensions than the other two. However, dimensions obey the same rules of algebra as other quantities. Thus the dimensions of time cancel in the last term:

$$[L] = [L] + \frac{[L]}{[\cancel{T}]}[\cancel{T}] = [L] + [L]$$

As a result, we see that each term in this formula has the same dimensions. This type of calculation with dimensions is referred to as **dimensional analysis**.

EXERCISE 1–2

Show that $x = x_0 + v_0t + \frac{1}{2}at^2$ is dimensionally consistent. The quantities x and x_0 are distances, v_0 is a velocity, and a is an acceleration.

Solution

Using the dimensions given in Table 1–5 we have

$$[L] = [L] + \frac{[L]}{[\cancel{T}]}[\cancel{T}] + \frac{[L]}{[\cancel{T^2}]}[\cancel{T^2}] = [L] + [L] + [L]$$

Note that $\frac{1}{2}$ is ignored in this analysis because it has no dimensions.

Later in this text you will derive your own formulas from time to time. As you do so, it is helpful to check dimensional consistency at each step of the derivation. If at any time the dimensions don't agree, you will know that a mistake has been made, and you can go back and look for it. If the dimensions check, however, it's not a guarantee the formula is correct—after all, dimensionless factors, like 1/2 or 2, don't show up in a dimensional check.

1–4 Significant Figures

When a mass, a length, or a time is measured in a scientific experiment, the result is known only to within a certain accuracy. The inaccuracy or uncertainty can be caused by a number of factors, ranging from limitations of the measuring device itself to limitations associated with the senses and the skill of the person performing the experiment. In any case, the fact that observed values of experimental quantities have inherent uncertainties should always be kept in mind when performing calculations with those values.

Suppose, for example, that you want to determine the walking speed of your pet tortoise. To do so, you measure the time, t, it takes for the tortoise to walk a distance, d, and then you calculate the quotient, d/t. When you measure the distance with a ruler, which has one tick mark per millimeter, you find that $d = 21.2$ cm, with the precise value of the digit in the second decimal place uncertain. Defining the number of **significant figures** in a physical quantity to be equal to the number of digits in it that are known with certainty, we say that d is known to *three* significant figures.

Similarly, you measure the time with an old pocket watch, and as best you can determine it, $t = 8.5$ s, with the second decimal place uncertain. Note that t is known to only *two* significant figures. If we were to make this measurement with a digital watch, with a readout giving the time to 1/100 of a second, the accuracy of the result would still be limited by the finite reaction time of the experimenter. The reaction time would have to be predetermined in a separate experiment (See Problem 67 in Chapter 2 for a simple way to determine your reaction time.)

Returning to the problem at hand, we would now like to calculate the speed of the tortoise. Using the above values for d and t and a calculator with eight digits in its display, we find (21.2 cm)/(8.5 s) = 2.4941176 cm/s. Clearly, such an accurate value for the speed is unjustified, considering the limitations of our measurements. After all, we can't expect to measure quantities to two and three significant figures and from them obtain results with eight significant figures. In general, the number of significant figures that result when we multiply or divide physical quantities is given by the following rule of thumb:

> The number of significant figures after multiplication or division is equal to the number of significant figures in the *least* accurately known quantity.

In our speed calculation, for example, we know the distance to three significant figures, but the time to only two significant figures. As a result, the speed should be given with just two significant figures, $d/t = (21.2\text{ cm})/(8.5\text{ s}) = 2.5\text{ cm/s}$. Note that we didn't just keep the first two digits in 2.4941176 cm/s and drop the rest. Instead, we "rounded up"; that is, because the first digit to be dropped (9 in this case) is greater than or equal to 5, we increase the previous digit (4 in this case) by 1. Thus, 2.5 cm/s is our best estimate for the tortoise's speed.

▲ Every measurement has some degree of uncertainty associated with it. How precise would you expect this measurement to be?

EXAMPLE 1–1 It's the Tortoise by a Hare

A tortoise races a rabbit by walking with a constant speed of 2.51 cm/s for 12.23 s. How much distance does the tortoise cover?

Picture the Problem

The race between the rabbit and the tortoise is shown in our sketch. The rabbit pauses to eat a carrot while the tortoise walks with a constant speed.

Strategy

The distance covered by the tortoise is the speed of the tortoise multiplied by the time during which it walks.

Solution

1. Multiply the speed by the time to find the distance d:

$$d = (\text{speed})(\text{time}) = (2.51\text{ cm/s})(12.23\text{ s}) = 30.7\text{ cm}$$

continued on the following page

continued from the previous page

Insight

Notice that if we simply multiply 2.51 cm/s by 12.23 s, we obtain 30.6973 cm. We don't give all of these digits in our answer, however. In particular, because the quantity that is known with the least accuracy (the speed) has only three significant figures, we give a result with three significant figures. Note, in addition, that the third digit in our answer has been rounded up from 6 to 7.

Practice Problem

How long does it take for the tortoise to walk 17 cm? [**Answer:** $t = (17\text{ cm})/(2.51\text{ cm/s}) = 6.8\text{ s}$]

Some related homework problems: Problem 11, Problem 15

▲ The finish of the 100-meter race at the 1996 Atlanta Olympics. This official timing photo shows Donovan Bailey setting a new world record of 9.84 s. (If the timing had been accurate to only tenths of a second—as would probably have been the case before electronic devices came into use—how many runners would have shared the winning time? How many would have shared the second-place and third-place times?)

Note that the distance of 17 cm in the Practice Problem has only two significant figures because we don't know the digits to the right of the decimal place. If the distance were given as 17.0 cm, on the other hand, it would have three significant figures.

When physical quantities are added or subtracted, we use a slightly different rule of thumb. In this case, the rule involves the number of decimal places in each of the terms:

> The number of decimal places after addition or subtraction is equal to the smallest number of decimal places in any of the individual terms.

Thus, if you make a time measurement of 16.74 s, and then a subsequent time measurement of 5.1 s, the total time of the two measurements should be given as 21.8 s, rather than 21.84 s.

EXERCISE 1–3

You and a friend pick some raspberries. Your flat weighs 12.7 lb, and your friend's weighs 7.25 lb. What is the combined weight of the raspberries?

Solution

Just adding the two numbers gives 19.95 lb. According to our rule of thumb, however, the final result must have only a single decimal place (corresponding to the term with the smallest number of decimal places). Rounding off to one place, then, gives 20.0 lb as the acceptable result.

Scientific Notation

The number of significant figures in a given quantity may be ambiguous due to the presence of zeros at the beginning or end of the number. For example, if a distance is stated to be 2500 m, the two zeros could be significant figures, or they could be zeros that simply show where the decimal point is located. If the two zeros are significant figures, the uncertainty in the distance is roughly a meter; if they are not significant figures, however, the uncertainty is about 100 m.

To remove this type of ambiguity, we can write the distance in **scientific notation**—that is, as a number of order unity times an appropriate power of ten. Thus, in this example, we would express the distance as 2.5×10^3 m if there are only two significant figures, or as 2.500×10^3 m to indicate four significant figures. Likewise, a time given as 0.000 036 s has only two significant figures—the preceding zeros only serve to fix the decimal point. If this quantity were known to three significant figures, we would write it as 3.60×10^{-5} s to remove any ambiguity. See Appendix A for a more detailed discussion of scientific notation.

EXERCISE 1–4

How many significant figures are there in **(a)** 21.00, **(b)** 21, **(c)** 2.1×10^{-2}, and **(d)** 2.10×10^{-3}?

Solution

(a) 4, **(b)** 2, **(c)** 2, **(d)** 3

Round-Off Error

Finally, even if you perform all your calculations to the same number of significant figures as in the text, you may occasionally obtain an answer that differs in its last digit from that given in the book. This is not something to be concerned about—in most cases it is simply due to **round-off error**.

Round-off error occurs when numerical results are rounded off at different times during a calculation. To see how this works, let's consider a simple example. Suppose you are shopping for knickknacks, and you buy one item for $2.21, plus 8 percent sales tax. The total price is $2.3868, or, rounded off to the nearest penny, $2.39. Later, you buy another item for $1.35. With tax this becomes $1.458 or, again to the nearest penny, $1.46. The total expenditure for these two items is $2.39 + $1.46 = $3.85.

Now, let's do the rounding off in a different way. Suppose you buy both items at the same time for a total before-tax price of $2.21 + $1.35 = $3.56. Adding in the 8 percent tax gives $3.8448, which rounds off to $3.84, one penny different from the previous amount. This same type of discrepancy can occur in physics problems. In general, it's a good idea to keep one extra digit throughout your calculations whenever possible, rounding off only the final result. But while this practice can help to reduce the likelihood of round-off error, there is no way to avoid it in every situation.

1–5 Converting Units

It is often convenient to convert from one set of units to another. For example, suppose you would like to convert 316 ft to its equivalent in meters. Looking at the conversion factors on the inside front cover of the text, we see that

$$1 \text{ m} = 3.281 \text{ ft} \qquad \textbf{1–1}$$

Equivalently,

$$\frac{1 \text{ m}}{3.281 \text{ ft}} = 1 \qquad \textbf{1–2}$$

Now, to make the conversion, we simply multiply 316 ft by this expression, which is equivalent to multiplying by 1:

$$(316 \not{\text{ft}})\left(\frac{1 \text{ m}}{3.281 \not{\text{ft}}}\right) = 96.3 \text{ m}$$

Note that the conversion factor is written in this particular way, as 1 m divided by 3.281 ft, so that the units of feet cancel out, leaving the final result in the desired units of meters.

Of course, we can just as easily convert from meters to feet if we use the reciprocal of this conversion factor—which is also equal to 1:

$$1 = \frac{3.281 \text{ ft}}{1 \text{ m}}$$

For example, a distance of 26.4 m is converted to feet by canceling out the units of meters, as follows:

$$(26.4 \not{\text{m}})\left(\frac{3.281 \text{ ft}}{1 \not{\text{m}}}\right) = 86.6 \text{ ft}$$

Thus, we see that converting units is as easy as multiplying by 1—because that's really what you're doing.

▲ From this sign, you can calculate factors for converting miles to kilometers and vice versa. (Why do you think the conversion factors seem to vary for different destinations?)

EXAMPLE 1–2 A High-Volume Warehouse

A warehouse is 20.0 yards long, 10.0 yards wide, and 15.0 feet high. What is its volume in SI units?

Picture the Problem

In our sketch we picture the warehouse and indicate the relevant length for each of its dimensions.

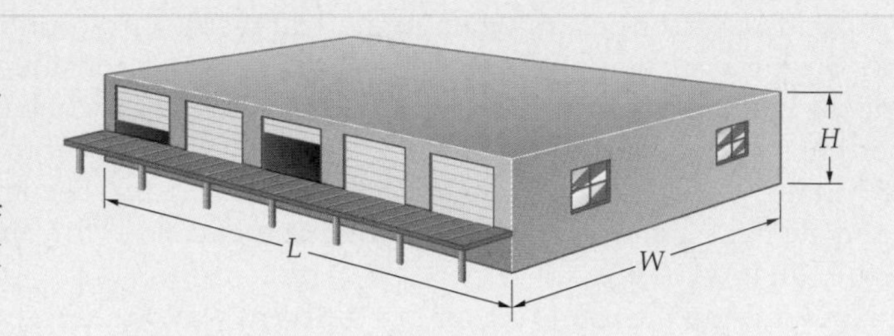

Strategy

We begin by converting the length, the width, and the height of the warehouse to meters. Once this is done, the volume in SI units is simply the product of the three dimensions.

Solution

1. Convert the length of the warehouse to meters: $L = (20.0\text{ yard})\left(\frac{3\text{ ft}}{1\text{ yard}}\right)\left(\frac{1\text{ m}}{3.281\text{ ft}}\right) = 18.3\text{ m}$
2. Convert the width to meters: $W = (10.0\text{ yard})\left(\frac{3\text{ ft}}{1\text{ yard}}\right)\left(\frac{1\text{ m}}{3.281\text{ ft}}\right) = 9.14\text{ m}$
3. Convert the height to meters: $H = (15.0\text{ ft})\left(\frac{1\text{ m}}{3.281\text{ ft}}\right) = 4.57\text{ m}$
4. Calculate the volume of the warehouse: $V = L \times W \times H = (18.3\text{ m})(9.14\text{ m})(4.57\text{ m}) = 764\text{ m}^3$

Insight

We would say, then, that the warehouse has a volume of 764 cubic meters.

Practice Problem

What is the volume of the warehouse if its length is one-hundredth of a mile? [**Answer:** $V = 672\text{ m}^3$]

Some related homework problems: Problem 17, Problem 18

Finally, the same procedure can be applied to conversions involving any number of units. For instance, if you walk at 3.00 mi/h, how fast is that in m/s? In this case we need the following additional conversion factors:

$$1\text{ mi} = 5{,}280\text{ ft} \qquad 1\text{ h} = 3{,}600\text{ s}$$

With these factors at hand, we carry out the conversion as follows:

$$(3.00\text{ mi/h})\left(\frac{5{,}280\text{ ft}}{1\text{ mi}}\right)\left(\frac{1\text{ m}}{3.281\text{ ft}}\right)\left(\frac{1\text{ h}}{3{,}600\text{ s}}\right) = 1.34\text{ m/s}$$

Note that in each conversion factor the numerator is equal to the denominator. In addition, each conversion factor is written in such a way that the unwanted units cancel, leaving just meters per second.

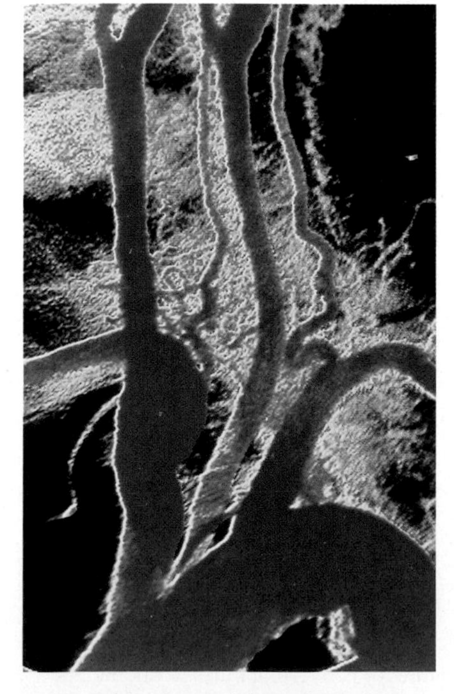

▲ Major blood vessels branch from the aorta (bottom), the artery that receives blood directly from the heart.

ACTIVE EXAMPLE 1–1 The Speed of Life

Blood in the human aorta can attain speeds of 35.0 cm/s. How fast is this in **(a)** ft/s and **(b)** mi/h?

Solution

Part (a)

1. Convert centimeters to meters and then to feet: 1.15 ft/s

Part (b)

2. First, convert centimeters to miles: 2.17×10^{-4} mi/s
3. Next, convert seconds to hours: 0.783 mi/h

Insight

Of course, the conversions in part (b) can be carried out in a single calculation if desired.

1–6 Order-of-Magnitude Calculations

An **order-of-magnitude** calculation is a rough, "ballpark" estimate designed to be accurate to within a factor of about 10. One purpose of such a calculation is to give a quick idea of what order of magnitude should be expected from a complete, detailed calculation. If an order-of-magnitude calculation indicates that a distance should be on the order of 10^4 m, for example, and your calculator gives an answer on the order of 10^7 m, then there is an error somewhere that needs to be resolved.

For example, suppose you would like to estimate the speed of a cliff diver on entering the water. First, the cliff may be 20 or 30 feet high; thus in SI units we would say that the order of magnitude of the cliff's height is 10 m—certainly not 1 m or 10^2 m. Next, the diver hits the water something like a second later—certainly not 0.1 s later nor 10 s later. Thus, a reasonable order-of-magnitude estimate of the diver's speed is $10\text{ m}/1\text{ s} = 10\text{ m/s}$, or roughly 20 mi/h. If you do a detailed calculation and your answer is on the order of 10^4 m/s, you probably entered one of your numbers incorrectly.

Another reason for doing an order-of-magnitude calculation is to get a feeling for what size numbers we are talking about in situations where a precise count is not possible. This is illustrated in the following Example.

▲ Enrico Fermi (1901-1954) was renowned for his ability to pose and solve interesting order-of-magnitude problems. A winner of the 1938 Nobel Prize in physics, Fermi (left) would ask his classes to obtain order-of-magnitude estimates for questions such as "How many piano tuners are there in Chicago?" or "How much is a tire worn down during one revolution?" Estimation questions like these are known to physicists today as "Fermi Problems."

EXAMPLE 1–3 Estimation: How Many Raindrops in a Storm

Real World Physics

A thunderstorm drops a half-inch (~ 0.01 m) of rain on Washington, D.C., which covers an area of about 70 square miles ($\sim 10^8\text{ m}^2$). Estimate the number of raindrops that fell during the storm.

Picture the Problem

Our sketch shows an area $A = 10^8\text{ m}^2$ covered to a depth $d = 0.01$ m by rainwater. Each drop of rain is considered to be a small sphere.

Strategy

To find the number of raindrops, we first calculate the volume of water required to cover 10^8 m^2 to a depth of 0.01 m. Next, we calculate the volume of an individual drop of rain, recalling that the volume of a sphere of radius r is $4\pi r^3/3$. We estimate the diameter of a raindrop to be about 4 mm. Finally, dividing the volume of a drop into the volume of water that fell during the storm gives the number of drops.

Solution

1. Calculate the order of magnitude of the volume of water, V_{water}, that fell during the storm:

$$V_{\text{water}} = Ad = (10^8\text{ m}^2)(0.01\text{ m}) \approx 10^6\text{ m}^3$$

2. Calculate the order of magnitude of the volume of a drop of rain, V_{drop}. Note that if the diameter of a drop is 4 mm, its radius is $r = 2\text{ mm} = 0.002$ m:

$$V_{\text{drop}} = \frac{4}{3}\pi r^3 \approx \frac{4}{3}\pi(0.002\text{ m})^3 \approx 10^{-8}\text{ m}^3$$

3. Divide V_{drop} into V_{water} to find the order of magnitude of the number of drops that fell during the storm:

$$\textit{number of drops} \approx \frac{V_{\text{water}}}{V_{\text{drop}}} \approx \frac{10^6\text{ m}^3}{10^{-8}\text{ m}^3} = 10^{14}$$

Insight

The number of raindrops in this one small storm is roughly 100,000 times greater than the current population of Earth.

Practice Problem

If a storm pelts Washington, D.C. with 10^{15} raindrops, how many inches of rain fall on the city? [**Answer:** About 5 inches.]

Some related homework problems: Problem 31, Problem 33

1–7 Problem Solving in Physics

Physics is a lot like swimming—you have to learn by doing. You could read a book on swimming and memorize every word in it, but when you jump into a pool the first time you are going to have problems. Similarly, you could read this book carefully, memorizing every formula in it, but when you finish, you still haven't learned physics. To learn physics you have to go beyond passive reading; you have to interact with physics and experience it by doing problems.

In this section we present a general overview of problem solving in physics. The suggestions given below, which apply to problems in all areas of physics, should help to develop a systematic approach.

We should emphasize at the outset that there is no recipe for solving problems in physics—it is a creative activity. In fact, the opportunity to be creative is one of the attractions of physics. The following suggestions, then, are not intended as a rigid set of steps that must be followed like the steps in a computer program. Rather, they provide a general guideline that experienced problem solvers find to be effective.

- **Read the problem carefully** Before you can solve a problem you need to know exactly what information it gives and what it asks you to determine. Some information is given explicitly, as when a problem states that a person has a mass of 70 kg. Other information is implicit; for example, saying that a ball is dropped from rest means that its initial speed is zero. Clearly, a *careful* reading is the essential first step in problem solving.
- **Sketch the system** This may seem like a step you can skip—but don't. A sketch helps you to acquire a physical feeling for the system. It also provides an opportunity to label those quantities that are known and those that are to be determined. All Examples in this text begin with a sketch of the system, accompanied by a brief description in a section labeled "Picture the Problem."
- **Visualize the physical process** Try to visualize what is happening in the system as if you were watching it in a movie. Your sketch should help. This step ties in closely with the next step.
- **Strategize** This may be the most difficult, but at the same time the most creative, part of the problem-solving process. From your sketch and visualization, try to identify the physical processes at work in the system. Then, develop a strategy—a game plan—for solving the problem. All Examples in this book have a "Strategy" spelled out before the solution begins.
- **Identify appropriate equations** Once a strategy has been developed, find the specific equations that are needed to carry it out.
- **Solve the equations** Use basic algebra to solve the equations identified in the previous step. Work with symbols such as x or y for the most part, substituting numerical values near the end of the calculations.
- **Check your answer** Once you have an answer, check to see if it makes sense: (i) Does it have the correct dimensions? (ii) Is the numerical value reasonable?
- **Explore limits/special cases** Getting the correct answer is nice, but it's not all there is to physics. You can learn a great deal about physics and about the connection between physics and mathematics by checking various limits of your answer. For example, if you have two masses in your system, m_1 and m_2, what happens in the special case that $m_1 = 0$ or $m_1 = m_2$? Check to see whether your answer and your physical intuition agree.

The Examples in this text are designed to deepen your understanding of physics and at the same time develop your problem-solving skills. They all have the same basic structure: Problem Statement; Picture the Problem; Strategy; Solution, presenting the flow of ideas and the mathematics side-by-side in a two-column format; Insight; and a Practice Problem related to the one just solved. As you work through the Examples in the chapters to come, notice how the basic problem-solving guidelines outlined above are implemented in a consistent way.

Finally, it is tempting to look for shortcuts when doing a problem—to look for a formula that seems to fit and some numbers to plug into it. It may seem harder to think ahead, to be systematic as you solve the problem, and then to think back over what you have done at the end of the problem. The extra effort is worth it, however, because by doing these things you will develop powerful problem-solving skills that can be applied to unexpected problems you may encounter on exams—and in life in general.

Chapter Summary

	Topic	Remarks and Relevant Equations
1–1	**Physics and the Laws of Nature**	Physics is based on a small number of fundamental laws and principles.
1–2	**Units of Length, Mass, and Time**	
	length	One meter is defined as distance traveled by light in a vacuum in 1/299,792,458 second.
	mass	One kilogram is the mass of a metal cylinder kept at the International Bureau of Weights and Standards.
	time	One second is the time required for a particular type of radiation from cesium–133 to undergo 9,192,631,770 oscillations.
1–3	**Dimensional Analysis**	
	dimension	The dimension of a quantity is the type of quantity it is; length (L), mass (M), or time (T).
	dimensional consistency	An equation is dimensionally consistent if each term in it has the same dimensions. All valid physical equations are dimensionally consistent.
	dimensional analysis	A calculation based on the dimensional consistency of an equation.
1–4	**Significant Figures**	
	significant figure	The number of digits reliably known, excluding digits that simply indicate the decimal place. For example, 3.45 and 0.000 034 5 both have three significant figures.
	round-off error	Discrepancies caused by rounding off numbers in intermediate results.
1–5	**Converting Units**	Multiply by the ratio of two units to convert from one to another. As an example, to convert 3.5 m to feet you multiply by the factor (1 ft/0.3048 m).
1–6	**Order-of-Magnitude Calculations**	A ballpark estimate designed to be accurate to within the nearest power of ten.
1–7	**Problem Solving in Physics**	A good general approach to problem solving is as follows: read; sketch; visualize; strategize; identify equations; solve; check; explore limits.

Conceptual Questions

1. Can dimensional analysis determine whether the area of a circle is πr^2 or $2\pi r^2$?
2. If a distance d has units of meters, and a time T has units of seconds, does the quantity $T + d$ make sense physically? What about the quantity d/T?
3. Which of the following equations is dimensionally consistent? **(a)** $x = vt$, **(b)** $x = \frac{1}{2}at^2$, **(c)** $t = (2x/a)^{1/2}$.
4. Which of the following equations is dimensionally consistent? **(a)** $v = at$, **(b)** $v = \frac{1}{2}at^2$, **(c)** $t = a/v$, **(d)** $v^2 = 2ax$.
5. Is it possible for two quantities to **(a)** have the same units but different dimensions or **(b)** have the same dimensions but different units? Explain.
6. Give an order-of-magnitude estimate for the time in seconds of the following: **(a)** a year; **(b)** a baseball game; **(c)** a heartbeat; **(d)** the age of Earth; **(e)** the age of a person.
7. Give an order-of-magnitude estimate for the length in meters of the following: **(a)** a person; **(b)** a fly; **(c)** a car; **(d)** a 747; **(e)** an interstate freeway stretching coast-to-coast.

Problems

Note: **IP** *denotes an integrated conceptual/quantitative problem.* **BIO** *identifies problems of biological or medical interest. Blue bullets (•, ••, •••) are used to indicate the level of difficulty of each problem.*

Section 1–2 Units of Length, Mass, and Time

1. • The movie *Titanic* broke all box-office records by bringing in over \$1,270,000,000 in worldwide distribution. Express this amount in **(a)** gigadollars and **(b)** teradollars.
2. • A human hair has a thickness of about 70 μm. What is this in **(a)** meters and **(b)** kilometers?
3. • The speed of light in a vacuum is approximately 0.3 Gm/s. Express the speed of light in meters per second.
4. • A computer can do 2 gigacalculations per second. How many calculations can it do in a microsecond?

Section 1–3 Dimensional Analysis

5. • Velocity is related to acceleration and distance by the following expression, $v^2 = 2ax^p$. Find the power p that makes this equation dimensionally consistent.
6. • Acceleration is related to distance and time by the following expression, $a = 2xt^p$. Find the power p that makes this equation dimensionally consistent.
7. • Show that the equation $v = v_0 + at$ is dimensionally consistent. Note that v and v_0 are velocities and that a is an acceleration.
8. •• Newton's second law (to be discussed in Chapter 5) states that acceleration is proportional to the force acting on an object and is inversely proportional to the object's mass. What are the dimensions of force?
9. •• The time T required for one complete oscillation of a mass m on a spring of force constant k is

$$T = 2\pi\sqrt{\frac{m}{k}}$$

Find the dimensions k must have for this equation to be dimensionally correct.

Section 1–4 Significant Figures

10. • The first several digits of π are known to be $\pi = 3.141\,592\,653\,589\,79\ldots$ What is π to **(a)** three significant figures, **(b)** five significant figures, and **(c)** seven significant figures?
11. • The speed of light to five significant figures is 2.9979×10^8 m/s. What is the speed of light to three significant figures?
12. • A parking lot is 117.2 m long and 40.14 m wide. What is the perimeter of the lot?
13. • On a fishing trip you catch a 2.65-lb bass, a 10.1-lb rock cod, and a 17.23-lb salmon. What is the total weight of your catch?
14. •• How many significant figures are there in **(a)** 0.000 054, **(b)** 3.001×10^5?
15. •• What is the area of circles of radius **(a)** 5.342 m and **(b)** 2.7 m?

Section 1–5 Converting Units

16. • The Eiffel Tower is 301 m high. What is its height in feet?
17. • **(a)** Calculate the volume of the warehouse in Example 1–2 in cubic feet. **(b)** Convert your result from part (a) to cubic meters.
18. • The Ark of the Covenant is described as a chest of acacia wood 2.5 cubits in length and 1.5 cubits in width and height. Given that a cubit is equivalent to 17.7 in., find the volume of the ark in cubic feet.
19. • How long does it take for radiation from a cesium—133 atom to complete 1 million cycles?
20. • Water going over Angel Falls in Venezuela, the world's highest waterfall, drops through a distance of 3212 ft. What is this distance in km?

▲ Angel Falls, in Venezuela, is over 3000 feet high. (Problem 20)

21. • An electronic advertising sign repeats a message every 8 seconds, day and night, for a week. How many times did the message appear on the sign?

22. • What is the conversion factor needed to convert seconds to years?

23. • The Star of Africa, a diamond in the royal scepter of the British crown jewels, has a mass of 530.2 carats, where 1 carat = 0.20 g. Given that 1 kg has an approximate weight of 2.21 lb, what is the weight of this diamond in pounds?

24. • Many highways have a speed limit of 55 mi/h. What is this in km/h?

25. • What is the speed in miles per hour of a beam of light traveling at 3.00×10^8 m/s?

26. • Kangaroos have been clocked at speeds of 65 km/h. What is their speed in mi/h?

27. •• Suppose 1.0 cubic meter of oil is spilled into the ocean. Find the area of the resulting slick, assuming that it is one molecule thick, and that each molecule occupies a cube 0.50 μm on a side.

28. •• **IP** **(a)** A standard sheet of paper measures 8 1/2 by 11 inches. Find the area of one such sheet of paper in m^2. **(b)** A second sheet of paper is half as long and half as wide as the one described in part (a). By what factor is its area less than the area found in part (a)?

29. •• **BIO** Nerve impulses in giant axons of the squid can travel with a speed of 20.0 m/s. How fast is this in **(a)** ft/s and **(b)** mi/h?

30. •• The acceleration of gravity is approximately 9.81 m/s^2 (depending on your location). What is the acceleration of gravity in feet per second squared?

Section 1–6 Order-of-Magnitude Calculations

31. • Give a ballpark estimate of the number of seats in a typical major league ballpark.

▲ Shea Stadium, in New York. How many fans can it hold? (Problem 31)

32. • Milk is often sold by the gallon in plastic containers. Estimate the number of gallons of milk that are purchased in the United States each year. What approximate weight of plastic does this represent?

33. •• New York is roughly 3000 miles from Seattle. When it is 10:00 A.M. in Seattle, it is 1:00 P.M. in New York. Using this information, estimate **(a)** the rotational speed of the surface of Earth, **(b)** the circumference of Earth, and **(c)** the radius of Earth.

34. •• You've just won the $1 million cash lottery, and you go to pick up the prize. What is the approximate weight of the cash if you request payment in **(a)** quarters or **(b)** dollar bills?

General Problems

35. •• A Porsche can accelerate at 12 m/s^2. What is this in **(a)** ft/s^2 and **(b)** km/h^2?

36. •• **BIO** Type A nerve fibers in humans can conduct nerve impulses at speeds up to 140 m/s. **(a)** How fast are the nerve impulses in miles per hour? **(b)** How far (in meters) can the impulses travel in 5.0 ms?

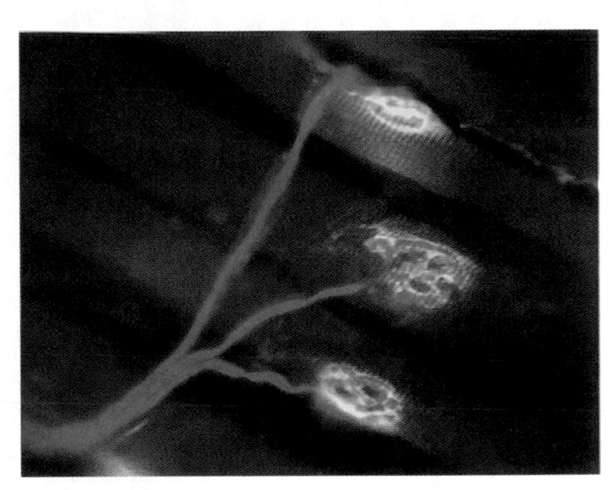

▲ The impulses in these nerve axons, which carry commands to the skeletal muscle fibers in the background, travel at speeds of up to 140 m/s. (Problem 36)

37. ••• Acceleration is related to velocity and time by the following expression, $a = v^p t^q$. Find the powers p and q that make this equation dimensionally consistent.

38. ••• The period T of a simple pendulum is the amount of time required for it to undergo one complete oscillation. If the length of the pendulum is L and the acceleration of gravity is g, then T is given by

$$T = 2\pi L^p g^q$$

Find the powers p and q required for dimensional consistency.

39. ••• Driving along a crowded freeway, you notice that it takes a time t to go from one mile marker to the next. When you increase your speed by 5.0 mi/h, the time to go one mile decreases by 11 s. What was your original speed?

2 One-Dimensional Kinematics

These sprinters, crossing the finish line of the 100-m dash, illustrate one-dimensional motion. At the moment shown in the photograph the runners are moving with constant velocity at a speed of approximately 10 m/s.

We begin our study of physics with **mechanics**, the area of physics perhaps most apparent to us in our everyday lives. Every time you raise an arm, stand up or sit down, throw a ball or open a door, your actions are governed by the laws of mechanics. Basically, mechanics is the study of how objects move, how they respond to external forces, and how other factors, such as size, mass, and mass distribution, affect their motion. This is a lot to cover, and we certainly won't try to tackle it all in one chapter.

Instead, in this chapter we limit ourselves to the **kinematics** of one-dimensional motion. What does that mean, exactly? Well, first, kinematics (from the Greek *kinema*, meaning "motion," as in cinema) is the study of motion and how to describe it, without regard for how the motion is caused. Second, by one-dimensional motion we mean motion that is in a straight line; to the left or to the right, up or down, east or west, and so on.

Furthermore, in this chapter we treat all physical objects as *point particles*. This is a common practice in physics. For example, if one is interested in calculating the time it takes the Earth to complete a revolution about the Sun, it is reasonable to consider the Earth and the Sun as simple particles. In later chapters, we extend our studies to increasingly realistic situations, involving motion in more than one dimension and physical objects with shape and size.

2–1 Position, Distance, and Displacement

The first step in describing the motion of a particle is to set up a **coordinate system** that defines its position. An example of a coordinate system, in one dimension, is shown in **Figure 2–1**.

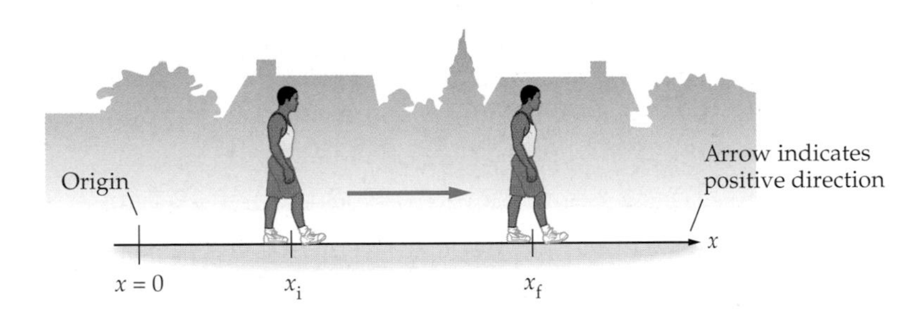

◀ **FIGURE 2–1 A one-dimensional coordinate system**

You are free to choose the origin and positive direction as you like, but once your choice is made, stick with it.

This is simply an x axis, with an origin (where $x = 0$) and an arrow indicating the positive direction—the direction in which x increases. In setting up a coordinate system, we are free to choose the origin and the positive direction as we like, but once we make a choice we must be consistent with it throughout any calculations that follow.

The particle in Figure 2–1 is a person who has moved to the right from an initial position, x_i, to a final position, x_f. Because the positive direction is to the right, it follows that x_f is greater than x_i; that is, $x_f > x_i$.

Now that we've seen how to set up a coordinate system, let's use one to investigate the situation sketched in **Figure 2–2**.

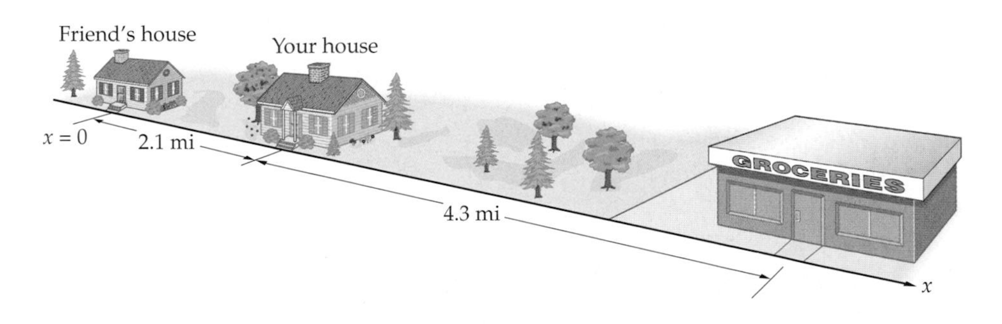

◀ **FIGURE 2–2 One-dimensional coordinates**

The locations of your house, your friend's house, and the grocery store, in terms of a one-dimensional coordinate system.

Suppose that you leave your house, drive to the grocery store, and then return home. The **distance** you've covered in your trip is 8.6 mi. In general, distance is defined as follows:

Definition: Distance

distance = total length of travel

SI unit: meter, m

Using SI units, the distance in this case is

$$8.6 \text{ mi} = (8.6 \text{ mi})\left(\frac{1609 \text{ m}}{1 \text{ mi}}\right) = 1.4 \times 10^4 \text{ m}$$

In a car, the distance traveled is indicated by the odometer. Note that distance is always positive.

Another useful way to characterize a particle's motion is in terms of the **displacement**, Δx, which is simply the change in position.

Definition: Displacement, Δx

$$\text{displacement} = \text{change in position} = \text{final position} - \text{initial position}$$

$$\text{displacement} = \Delta x = x_f - x_i \quad \textbf{2–1}$$

SI unit: meter, m

Notice that we use the delta notation, Δx, as a convenient shorthand for the quantity $x_f - x_i$. (See Appendix A for a complete discussion of delta notation.) Also, note that Δx can be positive (if $x_f > x_i$), negative (if $x_f < x_i$), or zero (if $x_f = x_i$).

The SI units of displacement are meters—the same as for distance—but displacement and distance are really quite different. For example, in the round trip from your house to the grocery store and back the distance traveled is 8.6 mi, whereas the displacement is zero because $x_f = 2.1 \text{ mi} = x_i$. Suppose, instead, that you go from your house to the grocery store and then to your friend's house. On this trip the distance is 10.7 mi, but the displacement is

$$\Delta x = x_f - x_i = (0) - (2.1 \text{ mi}) = -2.1 \text{ mi}$$

where the minus sign means your displacement is in the negative direction, that is, to the left.

ACTIVE EXAMPLE 2–1 Have a Nice Trip

Calculate **(a)** the distance and **(b)** the displacement for a trip from your friend's house to the grocery store and then to your house.

Solution

Part (a)

1. Add the distances for the various parts of the total trip: $2.1 \text{ mi} + 4.3 \text{ mi} + 4.3 \text{ mi} = 10.7 \text{ mi}$

Part (b)

2. Determine the initial position for the trip, using Figure 2–2: $x_i = 0$
3. Determine the final position for the trip, using Figure 2–2: $x_f = 2.1 \text{ mi}$
4. Subtract x_i from x_f to find the displacement $\Delta x = 2.1 \text{ mi}$

2–2 Average Speed and Velocity

The next step in describing motion is to consider how rapidly an object moves. For example, how long does it take for a Randy Johnson fastball to reach home plate? How far does the orbiting space shuttle travel in one hour? How fast do your eyelids move when you blink? These are examples of some of the most basic questions regarding motion, and in this section we learn how to answer them.

The simplest way to characterize the rate of motion is with the **average speed**:

$$\text{average speed} = \frac{\text{distance}}{\text{elapsed time}} \quad \textbf{2–2}$$

The dimensions of average speed are distance per time, or in SI units—meters per second, m/s. Both distance and elapsed time are positive; thus average speed is always positive.

EXAMPLE 2–1 The Kingfisher Takes a Plunge

A kingfisher is a bird that catches fish by plunging into water from a height. If a kingfisher dives from a height of 7.0 m with an average speed of 4.00 m/s, how long does it take for it to reach the water?

Picture the Problem

As shown in the figure, the kingfisher moves in a straight line through a vertical distance of 7.0 m. The average speed of the bird is 4.00 m/s.

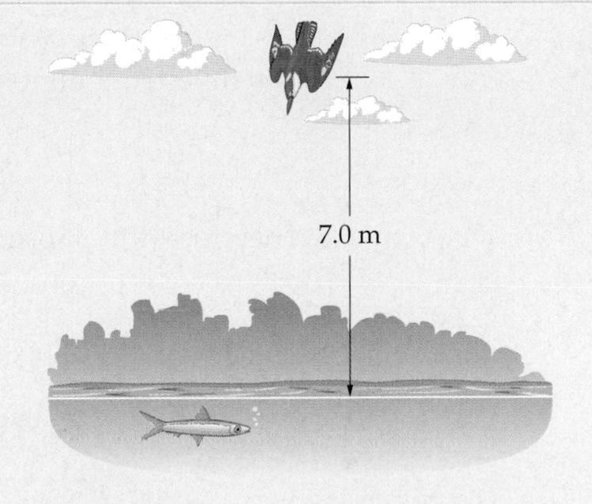

Strategy

By rearranging Equation 2–2 we can solve for the elapsed time.

Solution

1. Rearrange Equation 2–2 to solve for elapsed time: $\text{elapsed time} = \dfrac{\text{distance}}{\text{average speed}}$
2. Substitute numerical values to find the time: $\text{elapsed time} = \dfrac{7.0\ \cancel{\text{m}}}{4.00\ \cancel{\text{m}}/\text{s}} = \dfrac{7.0}{4.00}\ \text{s} = 1.8\ \text{s}$

Insight

Note that Equation 2–2 is not just a formula for calculating the average speed. It relates speed, time, and distance. Given any two of these quantities, Equation 2–2 can be used to find the third.

Practice Problem

A kingfisher dives with an average speed of 4.6 m/s for 1.4 s. What was the height of the dive? [**Answer:** distance = (average speed)(elapsed time) = (4.6 m/s)(1.4 s) = 6.4 m]

Some related homework problems: Problem 11, Problem 13

Next, we calculate the average speed for a trip consisting of two parts of equal length, each traveled with a different speed.

CONCEPTUAL CHECKPOINT 2–1

You drive 4.00 mi at 30.0 mi/h and then another 4.00 mi at 50.0 mi/h. Is your average speed for the 8.00-mi trip **(a)** greater than 40.0 mi/h, **(b)** equal to 40.0 mi/h, or **(c)** less than 40.0 mi/h?

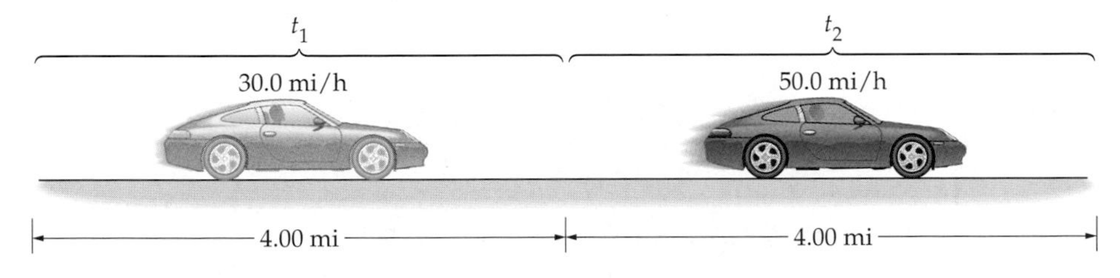

Reasoning and Discussion

At first glance it might seem that the average speed is definitely 40.0 mi/h. On further reflection, however, it is clear that it takes more time to travel 4.00 mi at 30.0 mi/h than it does to travel 4.00 mi at 50.0 mi/h. Therefore, you will be traveling at the lower speed for a greater period of time, and hence your average speed will be *less* than 40.0 mi/h—that is, closer to 30.0 mi/h than to 50.0 mi/h.

Answer

(c) The average speed is less than 40.0 mi/h.

To confirm the conclusion of the Conceptual Checkpoint, we simply apply the definition of average speed to find its value for this trip. We already know that the distance traveled is 8.00 mi; what we need now is the elapsed time. On the first 4.00 mi the time is

$$t_1 = \frac{4.00\ \text{mi}}{30.0\ \text{mi/h}} = (4.00/30.0)\ \text{h}$$

The time required to cover the second 4.00 mi is

$$t_2 = \frac{4.00\ \text{mi}}{50.0\ \text{mi/h}} = (4.00/50.0)\ \text{h}$$

Therefore, the elapsed time for the entire trip is

$$t_1 + t_2 = (4.00/30.0)\ \text{h} + (4.00/50.0)\ \text{h} = 0.213\ \text{h}$$

This gives the following average speed:

$$\text{average speed} = \frac{8.00\ \text{mi}}{0.213\ \text{h}} = 37.6\ \text{mi/h} < 40.0\ \text{mi/h}$$

In many situations, there is a quantity that is even more useful than the average speed. It is the **average velocity**, v_{av}, and it is defined as displacement per time:

Definition: Average Velocity, v_{av}

$$\text{average velocity} = \frac{\text{displacement}}{\text{elapsed time}}$$

$$v_{av} = \frac{\Delta x}{\Delta t} = \frac{x_f - x_i}{t_f - t_i} \qquad \textbf{2–3}$$

SI unit: meter per second, m/s

Not only does the average velocity tell us, on average, how fast something is moving, it also tells us the direction the object is moving. For example, if an object moves in the positive direction, then $x_f > x_i$, and $v_{av} > 0$. On the other hand, if an object moves in the negative direction it follows that $x_f < x_i$, and $v_{av} < 0$. Average velocity gives more information than average speed; hence it is used more frequently in physics.

In the next Example, pay close attention to the positive and negative signs.

EXAMPLE 2–2 Sprint Training

An athlete sprints 50.0 m in 8.00 s, then stops, and walks slowly back to the starting line in 40.0 s. If the "sprint direction" is taken to be positive, what is **(a)** the average sprint velocity, **(b)** the average walking velocity, and **(c)** the average velocity for the complete round trip?

Picture the Problem

In our sketch we set up a coordinate system with the sprint going in the positive direction, as described in the problem. For convenience, we choose the origin to be at the starting line. The finish line, then, is at $x = 50.0$ m.

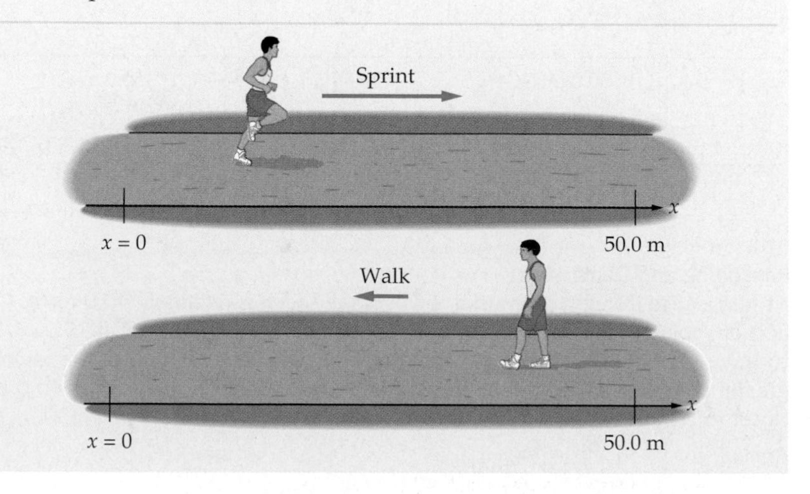

Strategy

Each part of the problem can be solved with a direct application of Equation 2–3. All that is needed is to determine Δx and Δt in each case.

Solution

Part (a)

1. Apply Equation 2–3 to the sprint, with $x_f = 50.0$ m, $x_i = 0$, $t_f = 8.00$ s, and $t_i = 0$:

$$v_{av} = \frac{\Delta x}{\Delta t} = \frac{x_f - x_i}{t_f - t_i} = \frac{50.0 \text{ m} - 0}{8.00 \text{ s} - 0} = \frac{50.0}{8.00} \text{ m/s} = 6.25 \text{ m/s}$$

Part (b)

2. Apply Equation 2–3 to the walk. In this case, $x_f = 0$, $x_i = 50.0$ m, $t_f = 48.0$ s and $t_i = 8.00$:

$$v_{av} = \frac{x_f - x_i}{t_f - t_i} = \frac{0 - 50.0 \text{ m}}{48.0 \text{ s} - 8.00 \text{ s}} = -\frac{50.0}{40.0} \text{ m/s} = -1.25 \text{ m/s}$$

Part (c)

3. For the round trip, $x_f = x_i = 0$; thus $\Delta x = 0$:

$$v_{av} = \frac{\Delta x}{\Delta t} = \frac{0}{48.0 \text{ s}} = 0$$

Insight

Note that the sign of the velocities in parts **(a)** and **(b)** indicates the direction of motion; positive for motion to the right, negative for motion to the left. Also, notice that the average speed (100.0 m/48.0 s = 2.08 m/s) is nonzero, even though the average velocity vanishes.

Practice Problem

If the average velocity during the walk is −1.50 m/s, how long does it take the athelete to walk back to the starting line? [**Answer:** $\Delta t = \Delta x/v_{av} = (-50.0 \text{ m})/(-1.50 \text{ m/s}) = 33.3$ s]

Some related homework problems: Problem 7, Problem 15, Problem 16

Graphical Interpretation of Average Velocity

It is often useful to "visualize" a particle's motion by sketching its position as a function of time. For example, consider a particle moving back and forth along the x axis, as shown in **Figure 2–3**. In this plot, we have indicated the position of a particle at a variety of times.

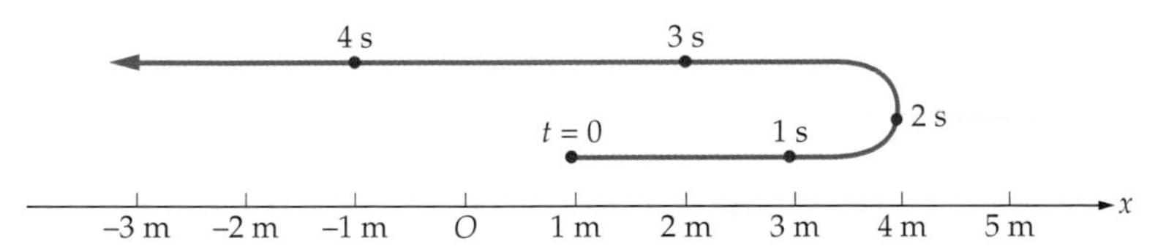

▲ **FIGURE 2–3 One-dimensional motion along the x axis**
The particle moves to the right for $0 \le t \le 2$ s, and to the left for $t > 2$ s. When the particle turns around at $t = 2$ s, we draw its path slightly above the path drawn for $t = 0$ to $t = 2$ s. This is simply for clarity—the particle is actually on the x axis at all times.

This way of keeping track of a particle's position and the corresponding time is a bit messy, though, so let's replot the same information with a different type of graph. In **Figure 2–4** we again plot the motion shown in Figure 2–3, but this time with the vertical axis representing the position, x, and the horizontal axis representing time, t. An **x-versus-t graph** like this makes it considerably easier to visualize a particle's motion.

An x-versus-t plot also leads to a particularly useful interpretation of average velocity. To see how, suppose you would like to know the average velocity of the particle in Figures 2–3 and 2–4 from $t = 0$ to $t = 3$ s. From our definition of average velocity in Equation 2–3, we know that $v_{av} = \Delta x/\Delta t = (2 \text{ m} - 1 \text{m})/(3 \text{ s} - 0) = +0.3$ m/s. To relate this to the x-versus-t plot, draw a straight line connecting the position at $t = 0$ (call this point A) and the position at $t = 3$ s (point B). The result is shown in **Figure 2–5 (a)**.

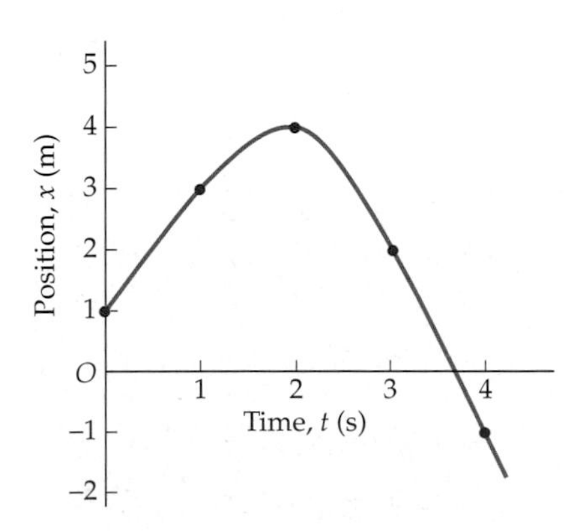

▲ **FIGURE 2–4 Motion along the x axis represented with an x-versus-t graph**
The motion shown here is the same as that indicated in Figure 2–3.

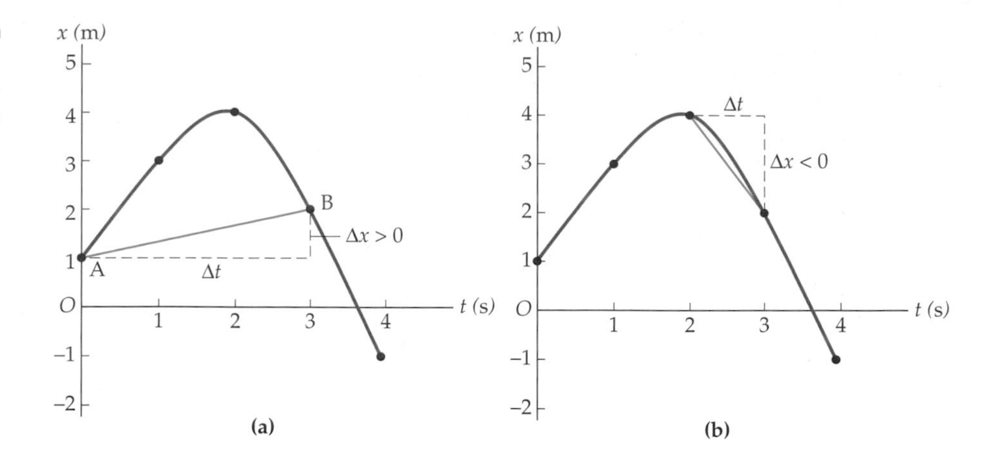

▶ **FIGURE 2–5 Average velocity on an *x*-versus-*t* graph**
(a) The average velocity between $t = 0$ (A) and $t = 3$ s (B) is equal to the slope of the straight line connecting A and B. **(b)** The slope of this straight line is the average velocity between $t = 2$ s and $t = 3$ s. Note that the average velocity is negative, indicating motion to the left.

The slope of the straight line from A to B is equal to the rise over the run, which in this case is $\Delta x/\Delta t$. But $\Delta x/\Delta t$ is the average velocity. Thus we see that:

- The slope of a line connecting two points on an x-versus-t plot is equal to the average velocity during that time interval.

As an additional example, let's calculate the average velocity between times $t = 2$ s and $t = 3$ s in Figure 2–4. A line connecting the corresponding points is shown in **Figure 2–5 (b)**.

The first thing we notice about this line is that it has a negative slope; thus $v_{av} < 0$ and the particle is moving to the left. We also note that it is inclined more steeply than the line in Figure 2–5 (a), hence the magnitude of its slope is greater. In fact, if we calculate the slope of this line we find that $v_{av} = -2$ m/s for this time interval.

Thus, connecting points on an x-versus-t plot gives an immediate "feeling" for the average velocity over a given time interval. This type of graphical analysis will be particularly useful in the next section.

2–3 Instantaneous Velocity

Though average velocity is a useful way to characterize motion, it can miss a lot. For example, suppose you travel by car on a long, straight highway, covering 92 mi in 2.0 hours. Your average velocity is 46 mi/h. Even so, there may have been only a few times during the trip when you were actually driving at 46 mi/h. You may have sped along at 65 mi/h during most of the time, except when you stopped to have a bite to eat at a roadside diner, during which time your average velocity was zero.

To have a more accurate representation of your trip, you should average your velocity over shorter periods of time. If you calculate your average velocity every 15 minutes, you have a better picture of what the trip was like. An even better, more realistic picture of the trip is obtained if you calculate the average velocity every minute or every second. Ideally, when dealing with the motion of any particle, it is desirable to know the velocity of the particle at each instant of time.

This idea of a velocity corresponding to an instant of time is just what is meant by the **instantaneous velocity**. Mathematically, we define the instantaneous velocity as follows:

▲ A speedometer indicates the instantaneous speed of a car. Note that the speedometer gives no information about the *direction* of motion. Thus, the speedometer is truly a "speed meter," not a velocity meter.

Definition: Instantaneous Velocity, v

$$v = \lim_{\Delta t \to 0} \frac{\Delta x}{\Delta t} \qquad \text{2–4}$$

SI unit: meter per second, m/s

In this expression the notation $\lim_{\Delta t \to 0}$ means "evaluate the average velocity, $\Delta x/\Delta t$, over shorter and shorter time intervals, approaching zero in the limit." Note that the instantaneous velocity can be positive, negative, or zero, just like the average velocity. The magnitude of the instantaneous velocity is called the **instantaneous speed**. In a car, the speedometer gives a reading of the vehicle's instantaneous speed.

As Δt becomes smaller, Δx becomes smaller as well, but the ratio $\Delta x/\Delta t$ approaches a constant value. To see how this works, consider first the simple case of a particle moving with a constant velocity of $+1$ m/s. If the particle starts at $x = 0$ at $t = 0$, then its position at $t = 1$ s is $x = 1$ m, its position at $t = 2$ s is $x = 2$ m, and so on. Plotting this motion in an x-versus-t plot gives a straight line, as shown in Figure 2–6.

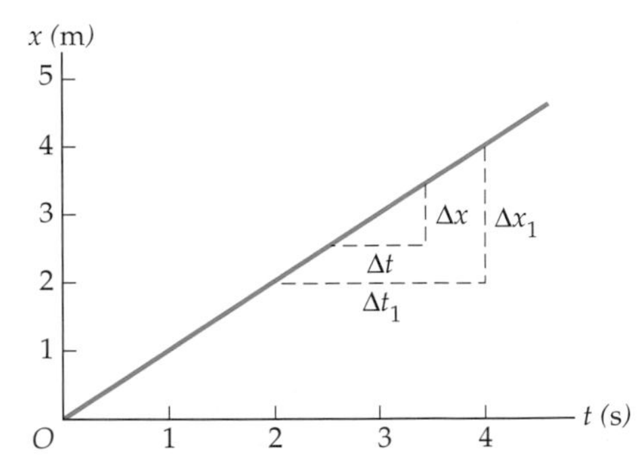

▲ **FIGURE 2–6 Constant velocity on an *x*-versus-*t* graph**
Motion with a constant velocity of $+1$ m/s. The motion begins at $x = 0$ at time $t = 0$. The slope $\Delta x_1/\Delta t_1$ is equal to $(2\text{ m})/(2\text{ s}) = 1$ m/s. Because x-versus-t is a straight line, the slope $\Delta x/\Delta t$ is also equal to 1 m/s for any value of Δt.

Now, suppose we want to find the instantaneous velocity at $t = 3$ s. To do so, we calculate the average velocity over small intervals of time centered at 3 s, and let the time intervals become arbitrarily small, as shown in the Figure. Since x-versus-t is a straight line, it is clear that $\Delta x/\Delta t = \Delta x_1/\Delta t_1$, no matter how small the time interval Δt. As Δt becomes smaller, so does Δx, but the ratio $\Delta x/\Delta t$ is simply the slope of the line, 1 m/s. Thus, the instantaneous velocity at $t = 3$ s is 1 m/s.

Of course, in this example the instantaneous velocity is 1 m/s for any instant of time, not just $t = 3$ s. Therefore:

- When velocity is constant, the average velocity over any time interval is equal to the instantaneous velocity at any time.

In general, a particle's velocity varies with time, and the x-versus-t plot is not a straight line. An example is shown in Figure 2–7, with the corresponding numerical values of x and t given in Table 2–1.

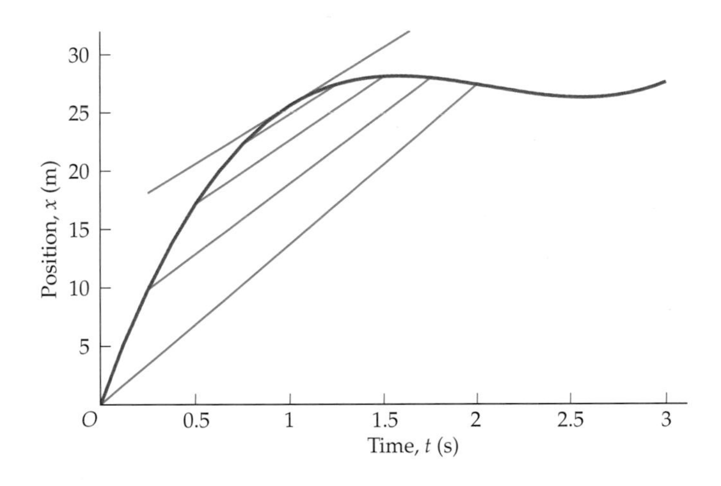

◀ **FIGURE 2–7 Instantaneous velocity**
An x-versus-t plot for motion with variable velocity. The instantaneous velocity at $t = 1$ s is equal to the slope of the tangent line at that time. The average velocity for a small time interval centered on $t = 1$ s approaches the instantaneous velocity at $t = 1$ s as the time interval goes to zero.

TABLE 2–1
***x*-versus-*t* Values for Figure 2–7**

t, (s)	*x*, (m)
0	0
0.25	9.85
0.50	17.2
0.75	22.3
1.00	25.6
1.25	27.4
1.50	28.1
1.75	28.0
2.00	27.4

Now, in this case, what is the instantaneous velocity at, say, $t = 1.00$ s? As a first approximation, let's calculate the average velocity for the time interval from $t_i = 0$ to $t_f = 2.00$ s. Note that this time interval is centered at $t = 1.00$ s. From Table 2–1 we see that $x_i = 0$, and $x_f = 27.4$ m, thus $v_{av} = 13.7$ m/s. The corresponding straight line connecting these two points is the lowest straight line in the figure.

The next three lines, in upward progression, refer to time intervals from 0.250 s to 1.75 s, 0.500 s to 1.50 s, and 0.750 s to 1.25 s, respectively. The corresponding average velocities, given in Table 2–2, are 12.1 m/s, 10.9 m/s, and 10.2 m/s. Table 2–2 also gives results for even smaller time intervals. In particular, for the interval from 0.900 s to 1.10 s the average velocity is 10.0 m/s. Smaller intervals also give 10.0 m/s. Thus, we can conclude that the instantaneous velocity at $t = 1.00$ s is $v = 10.0$ m/s.

TABLE 2–2 Calculating the Instantaneous Velocity at t = 1 s.

t_i(s)	t_f(s)	Δt(s)	x_i(m)	x_f(m)	Δx(m)	$v_{av} = \Delta x/\Delta t$(m/s)
0	2.00	2.00	0	27.4	27.4	13.7
0.25	1.75	1.50	9.85	28.0	18.2	12.1
0.50	1.50	1.00	17.2	28.1	10.9	10.9
0.75	1.25	0.50	22.3	27.4	5.10	10.2
0.90	1.10	0.20	24.5	26.5	2.00	10.0
0.95	1.05	0.10	25.1	26.1	1.00	10.0

The uppermost straight line in Figure 2–7 is the tangent line to the x-versus-t curve at the time $t = 1.00$ s; that is, it is the line that touches the curve at just a single point. Its slope is 10.0 m/s. Clearly, the average-velocity lines have slopes that approach the slope of the tangent line as the time intervals become smaller. This is an example of the following general result:

- The instantaneous velocity at a given time is equal to the slope of the tangent line at that point on an x-versus-t graph.

In the remainder of the book, when we say velocity it is to be understood that we mean *instantaneous* velocity. If we want to refer to the average velocity, we will specifically say average velocity.

CONCEPTUAL CHECKPOINT 2–2

Referring to Figure 2–7, is the instantaneous velocity at $t = 0.500$ s **(a)** greater than, **(b)** less than, or **(c)** the same as the instantaneous velocity at $t = 1.00$ s?

Reasoning and Discussion

From the x-versus-t graph in Figure 2–7 it is clear that the slope of a tangent line drawn at $t = 0.500$ s is greater than the slope of the tangent line at $t = 1.00$ s. It follows that the particle's velocity at 0.500 s is greater than its velocity at 1.00 s.

Answer

(a) The instantaneous velocity is greater at $t = 0.500$ s.

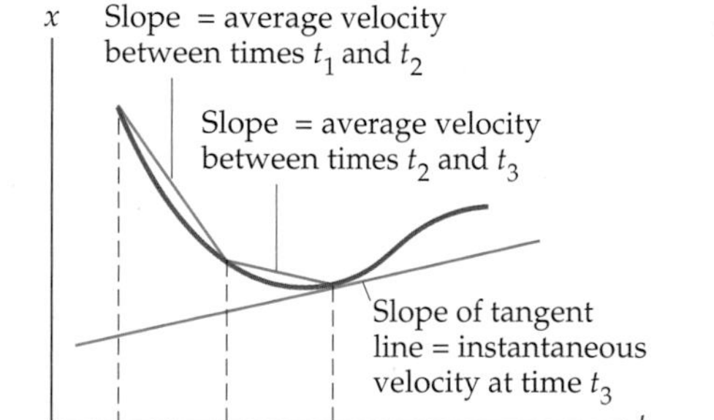

▲ **FIGURE 2–8 Graphical interpretation of average and instantaneous velocity**

Graphical Interpretation of Average and Instantaneous Velocity

Let's summarize the graphical interpretations of average and instantaneous velocity on an x-versus-t graph:

- Average velocity is the slope of the straight line connecting two points corresponding to a given time interval.
- Instantaneous velocity is the slope of the tangent line at a given instant of time.

These relations are illustrated in **Figure 2–8**.

2–4 Acceleration

Just as velocity is the rate of change of *displacement* with time, **acceleration** is the rate of change of *velocity* with time. Of all the concepts discussed in this chapter, perhaps none is more central to physics than acceleration. Galileo, for example, showed that falling bodies move with constant acceleration. Newton showed that acceleration and force are directly related, as we shall see in Chapter 5. Thus, it is particularly important to have a clear, complete understanding of acceleration before leaving this chapter.

We begin, then, with the definition of **average acceleration**:

Definition: Average Acceleration, a_{av}

$$a_{av} = \frac{\Delta v}{\Delta t} = \frac{v_f - v_i}{t_f - t_i} \qquad \textbf{2–5}$$

SI unit: meter per second per second, m/s^2

Note that the dimensions of average acceleration are the dimensions of velocity per time, or (meters per second) per second:

$$\frac{\text{meters per second}}{\text{second}} = \frac{m/s}{s} = \frac{m}{s^2}$$

This is generally expressed as meters per second squared. Typical values of acceleration are given in Table 2–3.

TABLE 2–3 Typical Accelerations (m/s^2)

Ultracentrifuge	3×10^6
Batted baseball	3×10^4
Bungee jump	30
Acceleration of gravity on Earth	9.81
Emergency stop in a car	8
Acceleration of gravity on the Moon	1.62

▲ The Space Shuttle Columbia accelerates upward on the initial phase of its journey into orbit. During this time the astronauts on board the shuttle experience an approximately linear acceleration that may be as great as 20 m/s^2.

EXERCISE 2–1

(a) Saab advertises a car that goes from 0 to 60.0 mi/h in 6.2 s. What is the average acceleration of this car?

(b) An airplane has an average acceleration of 5.6 m/s^2 during takeoff. How long does it take for the plane to reach a speed of 150 mi/h?

Solution

(a) average acceleration $= a_{av} = (60.0 \text{ mi/h})/(6.2 \text{ s}) = (26.8 \text{ m/s})/(6.2 \text{ s}) = 4.3 \text{ m/s}^2$

(b) $\Delta t = \Delta v/a_{av} = (150 \text{ mi/h})/(5.6 \text{ m/s}^2) = (67.0 \text{ m/s})/(5.6 \text{ m/s}^2) = 12 \text{ s}$

Next, just as we considered the limit of smaller and smaller time intervals to find an instantaneous velocity, we can do the same to define an **instantaneous acceleration**:

Definition: Instantaneous Acceleration, a

$$a = \lim_{\Delta t \to 0} \frac{\Delta v}{\Delta t} \qquad \textbf{2–6}$$

SI unit: meter per second per second, m/s^2

Note that average acceleration, like average velocity, always refers to a specific time interval. Instantaneous acceleration and velocity, on the other hand, refer to a specific instant of time. For simplicity, when we say acceleration in the future we are referring to the instantaneous acceleration.

One final note before we go on to some examples. If the acceleration is constant, it has the same value at all times. Therefore:

- When acceleration is constant, the instantaneous and average accelerations are the same.

We shall make use of this fact when we return to the special case of constant acceleration in the next section.

Graphical Interpretation of Acceleration

To see how acceleration can be interpreted graphically, suppose that a particle has a constant acceleration of $-0.50\ \text{m/s}^2$. This means that the velocity of the particle *decreases* by 0.50 m/s each second. Thus, if its velocity is 1 m/s at $t = 0$, then at $t = 1$ s its velocity is 0.50 m/s, at $t = 2$ s its velocity is 0, at $t = 3$ s its velocity is -0.50 m/s, and so on. This is illustrated by curve I in **Figure 2–9**, where we see that a plot of v-versus-t results in a straight line with a negative slope. Curve II in Figure 2–9 has a positive slope, corresponding to a constant acceleration of $+0.25\ \text{m/s}^2$. Thus, in terms of a v-versus-t plot, a constant acceleration results in a straight line with a slope equal to the acceleration.

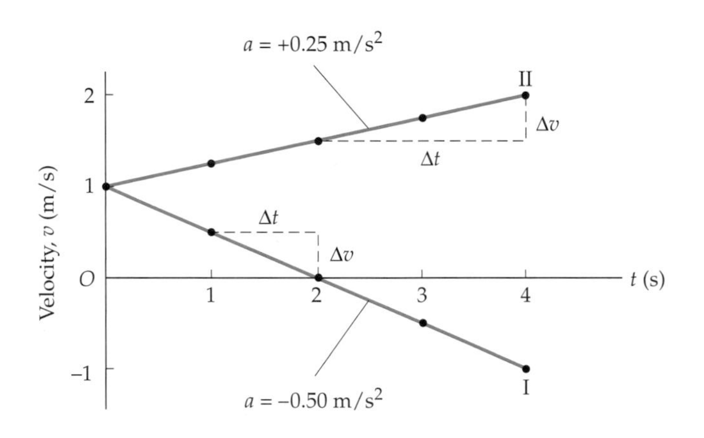

▶ FIGURE 2–9 *v*-versus-*t* plots for motion with constant acceleration

Curve I represents the movement of a particle with constant acceleration $a = -0.50\ \text{m/s}^2$. Curve II represents the motion of a particle with constant acceleration $a = +0.25\ \text{m/s}^2$.

CONCEPTUAL CHECKPOINT 2–3

A particle with the velocity shown by curve II in Figure 2–9 has a speed that increases with time. How does speed change with time for a particle with the velocity shown by curve I in Figure 2–9: **(a)** increases, **(b)** decreases, or **(c)** decreases and then increases?

Reasoning and Discussion

Recall that speed is the *magnitude* of velocity. In curve I of Figure 2–9 the speed starts out at 1 m/s, then *decreases* to 0 at $t = 2$ s. After $t = 2$ s the speed *increases* again. For example, at $t = 3$ s the speed is 0.50 m/s, and at $t = 4$ s the speed is 1 m/s.

Did you realize that the particle represented by curve I in Figure 2–9 changes direction at $t = 2$ s? It certainly does. Before $t = 2$ s the particle moves in the positive direction; after $t = 2$ s it moves in the negative direction. At precisely $t = 2$ s the particle is momentarily at rest. However, regardless of whether the particle is moving in the positive direction, moving in the negative direction, or instantaneously at rest, it still has the same constant acceleration. Acceleration has to do only with the way the velocity is *changing* at a given moment.

Answer:

(c) The speed decreases and then increases.

The graphical interpretations for velocity presented in Figure 2–8 apply equally well to acceleration, with just one small change: Instead of an x-versus-t graph, we use a v-versus-t graph, as in **Figure 2–10**. Thus, the average acceleration in a v-versus-t plot is the slope of a straight line connecting points corresponding to two different times. Similarly, the instantaneous acceleration is the slope of the tangent line at a particular time.

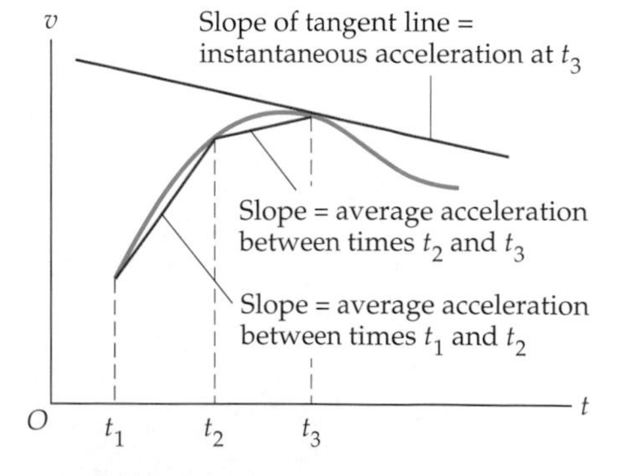

▲ FIGURE 2–10 Graphical interpretation of average and instantaneous acceleration

EXAMPLE 2–3 An Accelerating Train

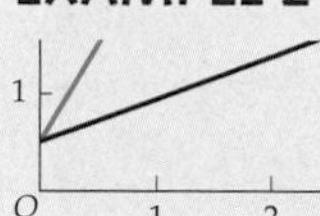

A train with an initial velocity of 0.50 m/s accelerates at 2.00 m/s^2 for 2.0 seconds, coasts with zero acceleration for 3.0 seconds, and then accelerates at -1.5 m/s^2 for 1.0 second. **(a)** What is the final velocity of the train? **(b)** What is the average acceleration of the train?

Picture the Problem

We begin by sketching a v-versus-t plot for the train. The basic idea is that each interval of constant acceleration is represented by a straight line of the appropriate slope.

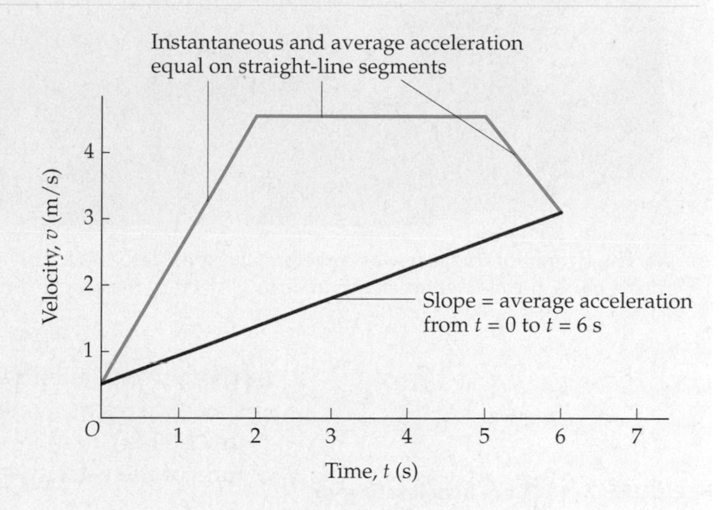

Strategy

During each period of constant acceleration the change in velocity is $\Delta v = a_{av}\,\Delta t = a\Delta t$.

(a) Adding the individual changes in velocity gives the total change, $\Delta v = v_f - v_i$. Since v_i is known, this expression can be solved for the final velocity, v_f.

(b) The average acceleration can be calculated using Equation 2–5, $a_{av} = \Delta v/\Delta t$. Note that Δv has been obtained in part **(a)**, and that the total time interval is $\Delta t = 6.0$ s, as is clear from the graph.

Solution

Part (a)

1. Find the change in velocity during each of the three periods of constant acceleration:

$$\Delta v_1 = a_1\Delta t_1 = (2.00\ \text{m/s}^2)(2.0\ \text{s}) = 4.0\ \text{m/s}$$
$$\Delta v_2 = a_2\Delta t_2 = (0)(3.0\ \text{s}) = 0$$
$$\Delta v_3 = a_3\Delta t_3 = (-1.5\ \text{m/s}^2)(1.0\ \text{s}) = -1.5\ \text{m/s}$$

2. Sum the change in velocity for each period to obtain the total Δv:

$$\Delta v = \Delta v_1 + \Delta v_2 + \Delta v_3$$
$$= 4.0\ \text{m/s} + 0 - 1.5\ \text{m/s} = 2.5\ \text{m/s}$$

3. Use Δv to find v_f, recalling that $v_i = 0.50$ m/s:

$$\Delta v = v_f - v_i$$
$$v_f = \Delta v + v_i = 2.5\ \text{m/s} + 0.50\ \text{m/s} = 3.0\ \text{m/s}$$

Part (b)

4. The average acceleration is $\Delta v/\Delta t$:

$$a_{av} = \frac{\Delta v}{\Delta t} = \frac{2.5\ \text{m/s}}{6.0\ \text{s}} = 0.42\ \text{m/s}^2$$

Insight

Note that the average acceleration for these six seconds is not simply the average of the individual accelerations; 2.00 m/s^2, 0 m/s^2, and -1.5 m/s^2. The reason is that different amounts of time are spent with each acceleration. In addition, the average acceleration can be found graphically, as indicated in the v-versus-t sketch above. Specifically, the graph shows that Δv is 2.5 m/s for the time interval from $t = 0$ to $t = 6.0$ s.

Practice Problem

What is the average acceleration of the train between $t = 2.0$ s and $t = 6.0$ s? [**Answer:** $a_{av} = \Delta v/\Delta t = (3.0\ \text{m/s} - 4.5\ \text{m/s})/(6.0\ \text{s} - 2.0\ \text{s}) = -0.38\ \text{m/s}^2$]

Some related homework problems: Problem 32, Problem 38

In one dimension, nonzero velocities and accelerations are either positive or negative, depending on whether they point in the positive or negative direction of the coordinate system chosen. Thus, the velocity and acceleration of an object may have the same or opposite signs. (Of course, in two or three dimensions the relationship between velocity and acceleration can be much more varied, as we shall see in the next several chapters.) This leads to the following two possibilities:

- When the velocity and acceleration of an object have the same sign, the speed of the object increases.
- When the velocity and acceleration of an object have opposite signs, the speed of the object decreases.

▲ The winner of this race was traveling at a speed of 313.91 mph at the end of the quarter-mile course. Since the winning time was just 4.607 s, the *average* acceleration during this race was approximately three times the acceleration of gravity (Section 2-7).

These two possibilities are illustrated in **Figure 2–11**. Notice that when a particle's speed increases, it means either that its velocity becomes more positive, as in Figure 2–11 (a), or more negative, as in Figure 2–11 (d). In either case, it is the magnitude of the velocity—the speed—that increases.

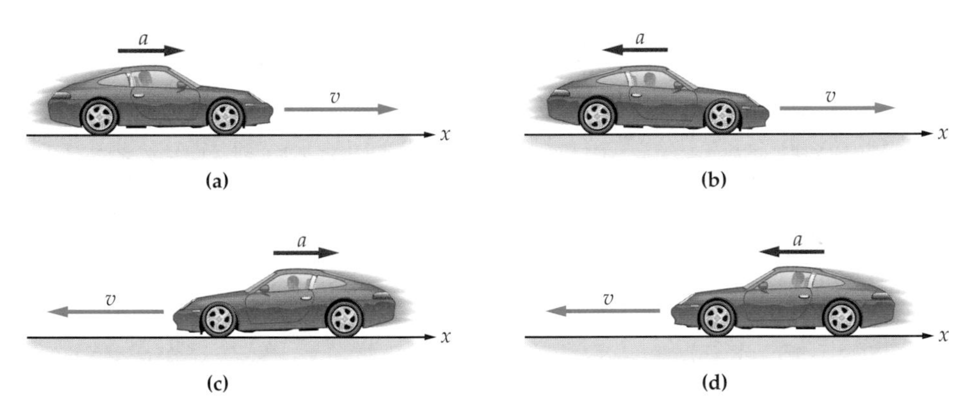

▶ **FIGURE 2–11 Cars accelerating or decelerating**

(a) Initial velocity is positive; velocity becomes more positive with time. **(b)** Initial velocity is positive; velocity becomes less positive with time. **(c)** Initial velocity is negative; velocity becomes less negative with time. **(d)** Initial velocity is negative; velocity becomes more negative with time.

When a particle's speed decreases, it is often said to be *decelerating*. A common misconception is that deceleration implies a negative acceleration. This is not true. Deceleration can be caused by a positive or a negative acceleration, depending on the direction of the initial velocity. For example, the car in Figure 2–11 (b) has a positive velocity and a negative acceleration, while the car in Figure 2–11 (c) has a negative velocity and a positive acceleration. In both cases, the speed of the car decreases. Again, all that is required for deceleration in one dimension is that the velocity and acceleration have *opposite* signs.

Velocity versus time plots for the four situations shown in Figure 2–11 are presented in **Figure 2–12**. In each of the four plots in Figure 2–12 we assume constant acceleration. Be sure to understand clearly the connection between the v-versus-t plots in Figure 2–12, and the corresponding physical motions indicated in Figure 2–11.

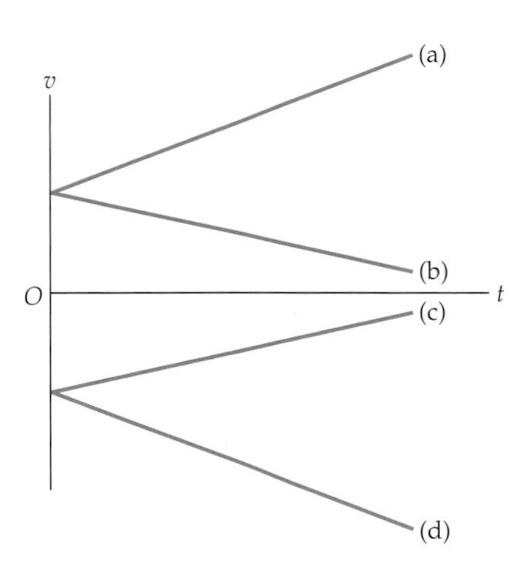

▲ **FIGURE 2–12 v-versus-t plots with constant acceleration**

The four plots correspond to the situations shown in Figure 2–11.

EXAMPLE 2–4 The Ferry Docks

A ferry makes a short run between two docks; one in Anacortes, the other on Guemes Island. As the ferry approaches Guemes Island (traveling in the positive direction), its speed is 7.4 m/s. **(a)** If the ferry slows to a stop in 12.3 s, what is its average acceleration? **(b)** As the ferry returns to the Anacortes dock its speed is 7.3 m/s. If it comes to rest in 13.1 s, what is its average acceleration?

Picture the Problem
Our sketch shows the locations of the two docks and the positive direction indicated in the problem. Note that the distance between docks is not given, nor is it needed.

Strategy
Find the average acceleration using $a_{av} = \Delta v/\Delta t$, being careful to get the signs right.

Solution

Part (a)

1. Calculate the average acceleration, noting that $v_i = 7.4$ m/s and $v_f = 0$:

$$a_{av} = \frac{\Delta v}{\Delta t} = \frac{v_f - v_i}{\Delta t} = \frac{0 - 7.4 \text{ m/s}}{12.3 \text{ s}} = -0.60 \text{ m/s}^2$$

Part (b)

2. In this case, $v_i = -7.3$ m/s and $v_f = 0$:

$$a_{av} = \frac{\Delta v}{\Delta t} = \frac{v_f - v_i}{\Delta t} = \frac{0 - (-7.3 \text{ m/s})}{13.1 \text{ s}} = 0.56 \text{ m/s}^2$$

Insight
In each case, the acceleration of the ferry is opposite in sign to its velocity; therefore the ferry decelerates.

Practice Problem
When the ferry leaves Guemes Island its speed increases from 0 to 5.8 m/s in 9.25 s. What is its average acceleration? [**Answer:** $a_{av} = -0.63 \text{ m/s}^2$]

Some related homework problems: Problem 30, Problem 31

2–5 Motion with Constant Acceleration

In this section, we derive equations describing the motion of particles moving with **constant acceleration**. These "equations of motion" can be used to describe a wide range of everyday phenomena. For example, in an idealized world with no air resistance, falling bodies have constant acceleration.

As mentioned in the previous section, if a particle has constant acceleration—that is, the same acceleration at every instant of time—then its instantaneous acceleration, a, is equal to its average acceleration, a_{av}. Recalling the definition of average acceleration, Equation 2–5, we have

$$a_{av} = \frac{v_f - v_i}{t_f - t_i} = a$$

where the initial and final times may be chosen arbitrarily. For example, let $t_i = 0$ for the initial time, and let $v_i = v_0$ denote the velocity at time zero. For the final time and velocity we drop the subscripts to simplify notation; thus we let $t_f = t$ and $v_f = v$. With these identifications we have

$$a_{av} = \frac{v - v_0}{t - 0} = a$$

Therefore,

$$v - v_0 = a(t - 0) = at$$

or

Constant-Acceleration Equation of Motion: Velocity as a Function of Time

$$v = v_0 + at \qquad \text{2–7}$$

Note that Equation 2–7 describes a straight line on a v-versus-t plot. The line crosses the velocity axis at the value v_0 and has a slope a, in agreement with the graphical interpretations discussed in the previous section. For example, in curve I of Figure 2–9, the equation of motion is $v = v_0 + at = (1 \text{ m/s}) + (-0.5 \text{ m/s}^2)t$. Also, note that $(-0.5 \text{ m/s}^2)t$ has the units $(\text{m/s}^2)(\text{s}) = \text{m/s}$; thus each term in Equation 2–7 has the same dimensions (as it must to be a valid physical equation).

EXERCISE 2–2

A ball is thrown upward with an initial velocity of $+8.2\text{ m/s}$. If the acceleration of the ball is -9.81 m/s^2, what is its velocity after

(a) 0.50 s, and

(b) 1.0 s?

Solution

(a) Substituting $t = 0.50\text{ s}$ in Equation 2–7 yields

$$v = 8.2\text{ m/s} + (-9.81\text{ m/s}^2)(0.50\text{ s}) = 3.3\text{ m/s}$$

(b) Similarly, using $t = 1.0\text{ s}$ in Equation 2–7 gives

$$v = 8.2\text{ m/s} + (-9.81\text{ m/s}^2)(1.0\text{ s}) = -1.6\text{ m/s}$$

Next, how far does a particle move in a given time if its acceleration is constant? To answer this question, recall the definition of average velocity:

$$v_{av} = \frac{\Delta x}{\Delta t} = \frac{x_f - x_i}{t_f - t_i}$$

Using the same identifications given previously for initial and final times, and letting $x_i = x_0$ and $x_f = x$, we have

$$v_{av} = \frac{x - x_0}{t - 0}$$

Thus,

$$x - x_0 = v_{av}(t - 0) = v_{av}t$$

or

$$x = x_0 + v_{av}t \qquad \textbf{2–8}$$

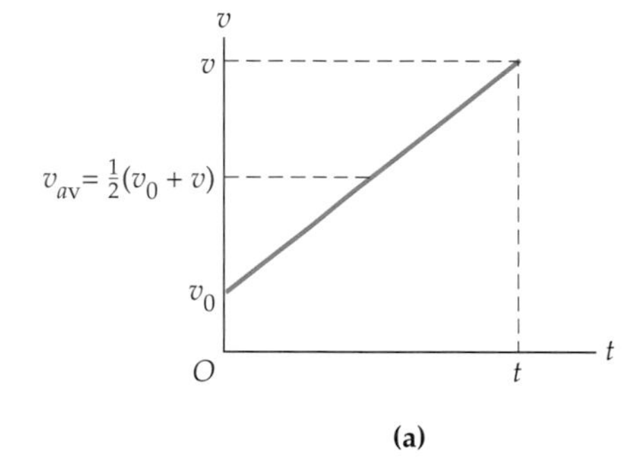

(a)

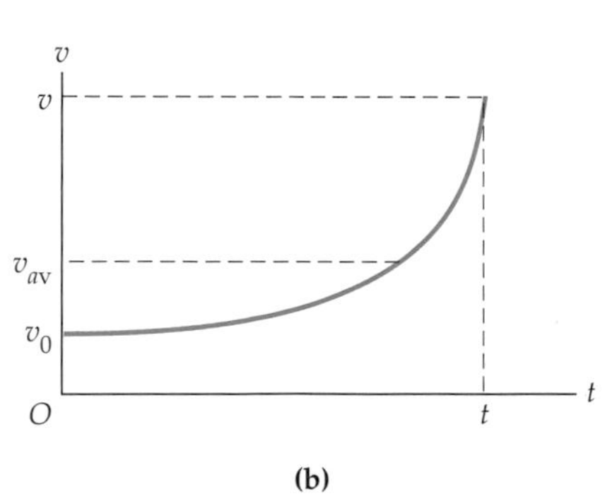

(b)

▲ FIGURE 2–13 ***v*-versus-*t* for (a) constant and (b) non-constant acceleration**

Now, Equation 2–8 is fine as it is. In fact, it applies whether the acceleration is constant or not. A more useful expression, for the case of constant acceleration, is obtained by writing v_{av} in terms of the initial and final velocities. This can be done by referring to **Figure 2–13 (a)**. Here the velocity changes linearly (since a is constant) from v_0 at $t = 0$ to v at some later time t. The average velocity during this period of time is simply the average of the initial and final velocities; that is, the sum of the two velocities divided by two:

$$v_{av} = \tfrac{1}{2}(v_0 + v) \qquad \textit{(constant acceleration)} \qquad \textbf{2–9}$$

The average velocity is indicated in the figure. Note that if the acceleration is not constant, as in **Figure 2–13 (b)**, this simple averaging of initial and final velocities is no longer valid.

Substituting the expression for v_{av} from Equation 2–9 into Equation 2–8 yields

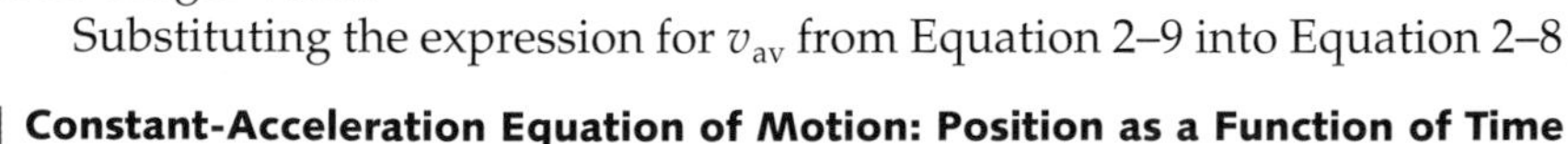

Constant-Acceleration Equation of Motion: Position as a Function of Time

$$x = x_0 + \tfrac{1}{2}(v_0 + v)t \qquad \textbf{2–10}$$

This equation, like Equation 2–7, is valid *only* for constant acceleration. The utility of Equations 2–7 and 2–10 is illustrated in the next Example.

EXAMPLE 2–5 Full Speed Ahead

A boat moves slowly inside a marina (so as not to leave a wake) with a constant speed of 1.50 m/s. As soon as it passes the breakwater, leaving the marina, it throttles up and accelerates at 2.40 m/s^2. **(a)** How fast is the boat moving after accelerating for 5.00 s? **(b)** How far has the boat traveled in this time?

Picture the Problem

In our sketch we choose the origin to be at the breakwater, and the positive direction to be the direction of motion. With this choice the initial position is $x_0 = 0$, and the initial velocity is $v_0 = 1.50\ \text{m/s}$.

Breakwater

$a = 2.40\ \text{m/s}^2$

x

Strategy

In part **(a)** we want to relate velocity to time, so we use Equation 2–7, $v = v_0 + at$. For part **(b)** we relate position to time using Equation 2–10, $x = x_0 + \frac{1}{2}(v_0 + v)t$.

Solution

Part (a)

1. Use Equation 2–7 with $v_0 = 1.50\ \text{m/s}$, and $a = 2.40\ \text{m/s}^2$:

$$v = v_0 + at = 1.50\ \text{m/s} + (2.40\ \text{m/s}^2)(5.00\ \text{s}) = 1.50\ \text{m/s} + 12.0\ \text{m/s} = 13.5\ \text{m/s}$$

Part (b)

2. Apply Equation 2–10, using the result for v obtained in part **(a)**:

$$x = x_0 + \tfrac{1}{2}(v_0 + v)t = 0 + \tfrac{1}{2}(1.50\ \text{m/s} + 13.5\ \text{m/s})(5.00\ \text{s}) = (7.50\ \text{m/s})(5.00\ \text{s}) = 37.5\ \text{m}$$

Insight

Since the boat has a constant acceleration between $t = 0$ and $t = 5.00$ s, its velocity-versus-time curve is linear during this time interval. The connection between the velocity curve and the distance traveled by the boat is analyzed next.

Practice Problem

At what time is the boat's speed equal to 10.0 m/s? [**Answer:** $t = 3.54$ s]

Some related homework problems: Problem 42, Problem 43

The velocity of the boat in Example 2–5 is plotted as a function of time in **Figure 2–14**, with the acceleration starting at time $t = 0$ and ending at $t = 5.00$ s. We will now show that the *distance* traveled by the boat from $t = 0$ to $t = 5.00$ s *is equal to the corresponding area under the velocity curve*. This is a general result, valid for any velocity curve and any time interval:

- The distance traveled by an object from a time t_1 to a time t_2 is equal to the area under the velocity curve between those two times.

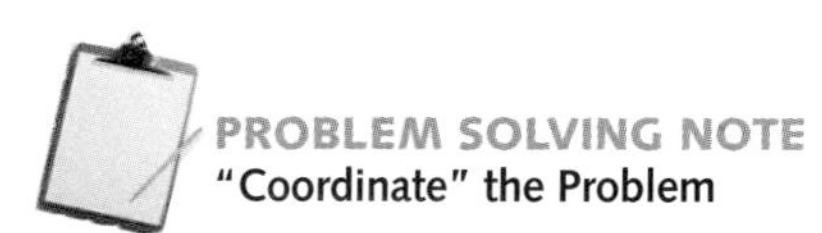

PROBLEM SOLVING NOTE

"Coordinate" the Problem

The first step in solving a physics problem is to produce a simple sketch of the system. Your sketch should include a coordinate system, along with an origin and a positive direction. Next, you should identify quantities that are given in the problem, such as initial position, initial velocity, acceleration, and so on. These preliminaries will help in producing a mathematical representation of the problem.

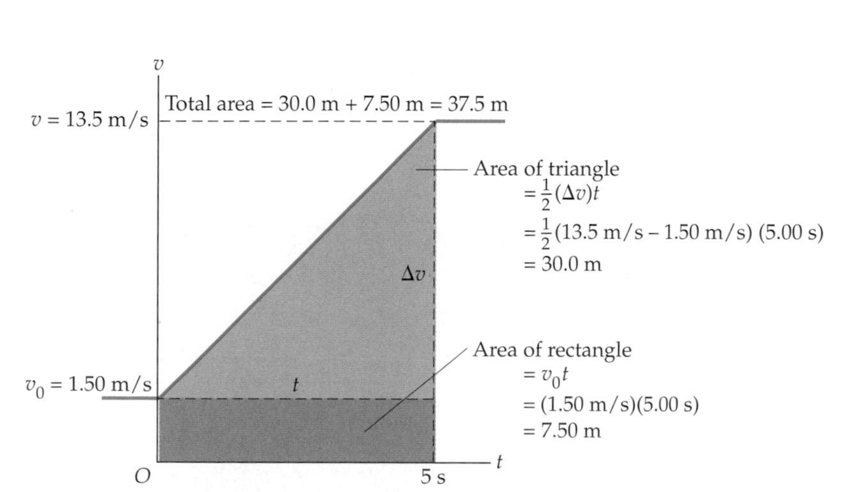

◀ **FIGURE 2–14 Velocity versus time for the boat in Example 2–5**

The distance traveled by the boat between $t = 0$ and $t = 5$ s is equal to the corresponding area under the velocity curve.

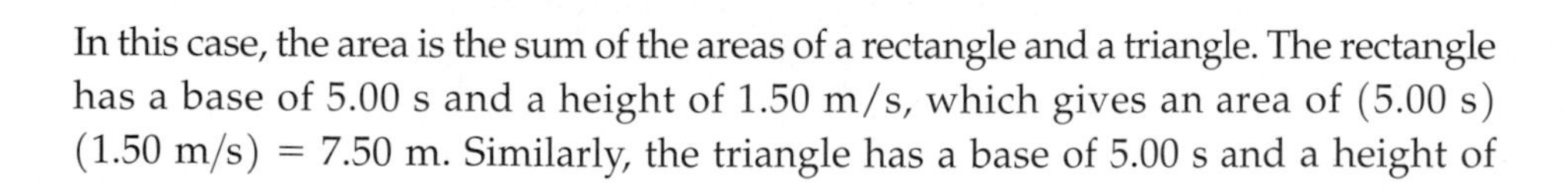

In this case, the area is the sum of the areas of a rectangle and a triangle. The rectangle has a base of 5.00 s and a height of 1.50 m/s, which gives an area of (5.00 s)(1.50 m/s) = 7.50 m. Similarly, the triangle has a base of 5.00 s and a height of

$(13.5 \text{ m/s} - 1.50 \text{ m/s}) = 12.0 \text{ m/s}$, for an area of $\frac{1}{2}(5.00 \text{ s})(12.0 \text{ m/s}) = 30.0 \text{ m}$. Clearly, the total area is 37.5 m, just as found in Example 2–5.

Staying with Example 2–5 for a moment, let's repeat the calculation of part **(b)**, only this time for the general case. First, we used the final velocity from part **(a)**, calculated with $v = v_0 + at$, in the expression for the average velocity, $v_{av} = \frac{1}{2}(v_0 + v)$. Symbolically, this gives the following:

$$\tfrac{1}{2}(v_0 + v) = \tfrac{1}{2}[v_0 + (v_0 + at)] = v_0 + \tfrac{1}{2}at \qquad \textit{(constant acceleration)}$$

Next, we substituted this result into Equation 2–10, which yields

$$x = x_0 + \tfrac{1}{2}(v_0 + v)t = x_0 + (v_0 + \tfrac{1}{2}at)t$$

Multiplying through by t gives the following result:

Constant-Acceleration Equation of Motion: Position as a Function of Time

$$x = x_0 + v_0 t + \tfrac{1}{2}at^2 \qquad \textbf{2–11}$$

Here we have an expression for position versus time that is explicitly in terms of the acceleration, a.

Note that each term in Equation 2–11 has the same dimensions, as they must. For example, the velocity term, $v_0 t$, has the units $(\text{m/s})(\text{s}) = \text{m}$. Similarly, the acceleration term, $\frac{1}{2}at^2$, has the units $(\text{m/s}^2)(\text{s}^2) = \text{m}$.

EXERCISE 2–3

Repeat part (b) of Example 2-5 using Equation 2–11.

Solution

$x = x_0 + v_0 t + \frac{1}{2}at^2 = 0 + (1.50 \text{ m/s})(5.00 \text{ s}) + \frac{1}{2}(2.40 \text{ m/s}^2)(5.00 \text{ s})^2 = 37.5 \text{ m}$

The next example gives further insight into the physical meaning of Equation 2–11.

EXAMPLE 2–6 Put the Pedal to the Metal

A drag racer starts from rest, and accelerates at 7.40 m/s^2. How far has it traveled in **(a)** 1.00 s, **(b)** 2.00 s, **(c)** 3.00 s?

Picture the Problem

We set up a coordinate system where the drag racer starts at the origin and moves in the positive direction. With this coordinate system, $x_0 = 0$. Since the racer starts at rest, the initial velocity is zero, $v_0 = 0$. Incidentally, the positions of the racer in the sketch are drawn to scale.

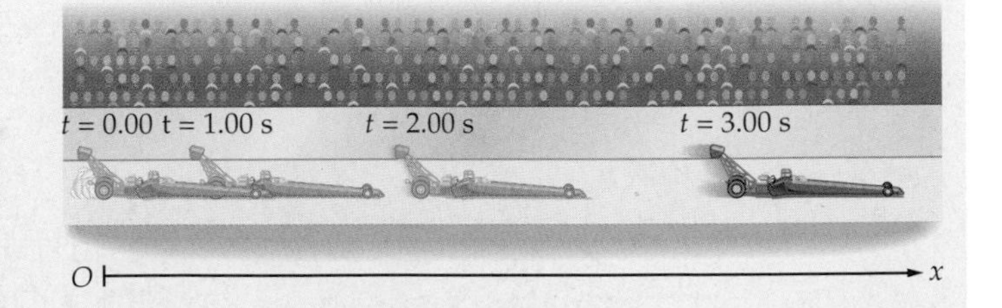

Strategy

Since this problem gives the acceleration and asks for a relationship between position and time, we use Equation 2–11.

Solution

Part (a)

1. Evaluate Equation 2–11 with $a = 7.40 \text{ m/s}^2$ and $t = 1.00$ s:

$$x = x_0 + v_0 t + \tfrac{1}{2}at^2 = 0 + 0 + \tfrac{1}{2}at^2 = \tfrac{1}{2}at^2$$
$$x = \tfrac{1}{2}at^2 = \tfrac{1}{2}(7.40 \text{ m/s}^2)(1.00 \text{ s})^2 = 3.70 \text{ m}$$

Part (b)

2. Note that Equation 2–11 reduces to $x = \frac{1}{2}at^2$ in this case. Evaluate $x = \frac{1}{2}at^2$ at $t = 2.00$ s:

$$x = \tfrac{1}{2}at^2 = \tfrac{1}{2}(7.40 \text{ m/s}^2)(2.00 \text{ s})^2 = 4(3.70 \text{ m}) = 14.8 \text{ m}$$

Part (c)

3. Repeat with $t = 3.00$ s:

$$x = \tfrac{1}{2}at^2 = \tfrac{1}{2}(7.40 \text{ m/s}^2)(3.00 \text{ s})^2 = 9(3.70 \text{ m}) = 33.3 \text{ m}$$

Insight

This example illustrates one of the key features of accelerated motion—position does not change uniformly with time when an object accelerates. In this case, the distance traveled in the first two seconds is 4 times the distance traveled in the first second, and the distance traveled in the first three seconds is 9 times the distance traveled in the first second. This kind of behavior is a direct result of the fact that x depends on t^2 when the acceleration is nonzero.

Practice Problem

In one second the racer travels 3.70 m. How long does it take for the racer to travel 2(3.70 m) = 7.40 m? [**Answer:** $t = \sqrt{2}$ s = 1.41 s]

Some related homework problems: Problem 44, Problem 57

Figure 2–15 shows x-versus-t for Example 2-6. Notice the parabolic shape of the x-versus-t curve, which is due to the $\frac{1}{2}at^2$ term, and is characteristic of constant acceleration. In particular, if acceleration is positive ($a > 0$), then x-versus-t curves upward; if acceleration is negative ($a < 0$), x-versus-t curves downward. The greater the magnitude of a, the greater the curvature. In contrast, if a particle moves with constant velocity ($a = 0$) the t^2 dependence vanishes, and x-versus-t is a straight line.

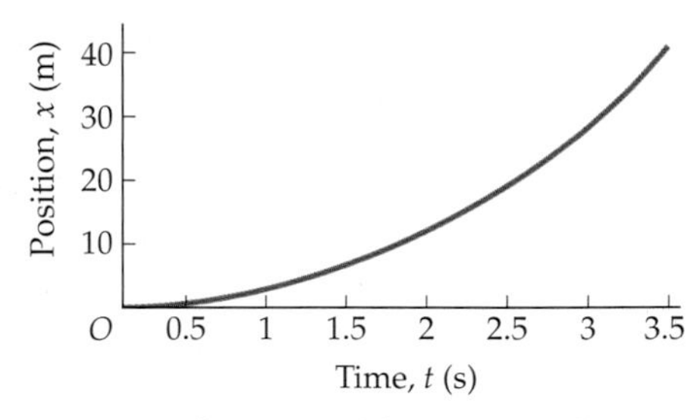

▲ **FIGURE 2–15 Position versus time for Example 2-6**

The parabolic shape, with upward curvature, indicates a constant, positive acceleration.

Our final equation of motion relates velocity to position. We start by solving for the time, t, in Equation 2–7:

$$v = v_0 + at \qquad \text{or} \qquad t = \frac{v - v_0}{a}$$

Next, we substitute this result into Equation 2–10, thus eliminating t:

$$x = x_0 + \tfrac{1}{2}(v_0 + v)t = x_0 + \tfrac{1}{2}(v_0 + v)\left(\frac{v - v_0}{a}\right)$$

Noting that $(v_0 + v)(v - v_0) = v_0v - v_0^2 + v^2 - vv_0 = v^2 - v_0^2$, we have

$$x = x_0 + \frac{v^2 - v_0^2}{2a}$$

Finally, a straightforward rearrangement of terms yields

Constant-Acceleration Equation of Motion: Velocity as a Function of Position

$$v^2 = v_0^2 + 2a(x - x_0) \qquad \textbf{2–12}$$

This equation allows us to relate the velocity at one position to the velocity at another position, without knowing how much time is involved. The next Example shows how Equation 2–12 can be used.

EXAMPLE 2–7 Takeoff Distance for an Airliner

Real World Physics

Jets at JFK International Airport accelerate from rest at one end of a runway, and must attain takeoff speed before reaching the other end of the runway. **(a)** Plane A has acceleration a and takeoff speed v_{to}. What is the minimum length of runway, Δx_A, required for this plane? Give a symbolic answer. **(b)** Plane B has the same acceleration as plane A, but requires twice the takeoff speed. Find Δx_B and compare with Δx_A. **(c)** Find the minimum runway length for plane A if $a = 11.0 \text{ m/s}^2$ and $v_{\text{to}} = 90.0$ m/s. (These are typical values for a 747 jetliner.)

continued on the following page

continued from the previous page

Picture the Problem

In our sketch we choose the origin to be where the plane starts from rest, and the positive direction to be the direction of motion.

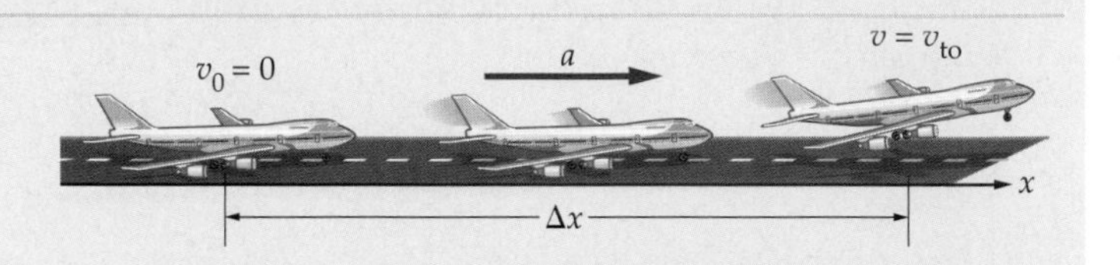

Strategy

From the sketch it is clear that we want to express Δx, the distance the plane travels in attaining takeoff speed, in terms of a and v. Equation 2–12 allows us to do this.

Solution

Part (a)

1. Solve Equation 2–12 for Δx. To find Δx_A, set $v_0 = 0$ and $v = v_{to}$:

$$\Delta x = \frac{v^2 - v_0^2}{2a}$$

$$\Delta x_A = \frac{v_{to}^2}{2a}$$

Part (b)

2. To find Δx_B, simply change v_{to} to $2v_{to}$ in part (a):

$$\Delta x_B = \frac{(2v_{to})^2}{2a} = \frac{4v_{to}^2}{2a} = 4\Delta x_A$$

Part (c)

3. Substitute numerical values into the result found in part (a):

$$\Delta x_A = \frac{v_{to}^2}{2a} = \frac{(90.0 \text{ m/s})^2}{2(11.0 \text{ m/s}^2)} = 368 \text{ m}$$

Insight

There are many advantages to obtaining symbolic results before substituting numerical values. In this case, we find that the takeoff distance is proportional to v^2; hence, we conclude immediately that doubling v results in a fourfold increase of Δx.

Practice Problem

Find the minimum acceleration needed for a takeoff speed of $v_{to} = (90.0 \text{ m/s})/2 = 45.0 \text{ m/s}$ on a runway of length $\Delta x = (368 \text{ m})/4 = 92.0 \text{ m}$. [**Answer:** $a = v_{to}^2/2\Delta x = 11.0 \text{ m/s}^2$]

Some related homework problems: Problem 48, Problem 50

Finally, all of our constant-acceleration equations of motion are collected for easy reference in Table 2–4.

TABLE 2–4
Constant-Acceleration Equations of Motion

Variables related	Equation	Number
velocity, time, acceleration	$v = v_0 + at$	2–7
position, time, velocity	$x = x_0 + \frac{1}{2}(v_0 + v)t$	2–10
position, time, acceleration	$x = x_0 + v_0 t + \frac{1}{2}at^2$	2–11
velocity, position, acceleration	$v^2 = v_0^2 + 2a(x - x_0) = v_0^2 + 2a\Delta x$	2–12

2–6 Applications of the Equations of Motion

We devote this section to a variety of examples further illustrating the use of the constant-acceleration equations of motion. In our first example, we consider the distance and time needed to brake a vehicle to a complete stop.

EXAMPLE 2–8 Hit the Brakes!

A park ranger driving on a back country road suddenly sees a deer "frozen" in his headlights. The ranger, who is driving at 11.4 m/s, immediately applies the brakes and slows with an acceleration of 3.80 m/s^2. **(a)** If the deer is 20.0 m from the ranger's vehicle when the brakes are applied, how close does the ranger come to hitting the deer? **(b)** How much time is needed for the ranger's vehicle to stop?

Picture the Problem

We choose the positive direction to be the direction of motion; thus $v_0 = +11.4$ m/s and $a = -3.80$ m/s^2.

Strategy

In part **(a)** we want to relate distance to velocity, so we use Equation 2–12. In part **(b)** we use Equation 2–7 to relate velocity to time.

Solution

Part (a)

1. Solve Equation 2–12 for Δx:

$$\Delta x = \frac{v^2 - v_0^2}{2a}$$

2. Set $v = 0$, and substitute numerical values:

$$\Delta x = -\frac{v_0^2}{2a} = -\frac{(11.4 \text{ m/s})^2}{2(-3.80 \text{ m/s}^2)} = 17.1 \text{ m}$$

3. Subtract Δx from 20.0 m to find the distance between the stopped vehicle and the deer:

$$20.0 \text{ m} - 17.1 \text{ m} = 2.9 \text{ m}$$

Part (b)

4. Set $v = 0$ in Equation 2–7 and solve for t:

$$v = v_0 + at = 0$$

$$t = -\frac{v_0}{a} = -\frac{11.4 \text{ m/s}}{(-3.80 \text{ m/s}^2)} = 3.00 \text{ s}$$

Insight

Note the difference in the way t and Δx depend on the initial speed. If the initial speed is doubled, for example, the time needed to stop also doubles, but the distance needed to stop increases by a factor of four. This is one reason why speed on the highway has such a great influence on safety.

Practice Problem

Show that using $t = 3.00$ s in Equation 2–11 results in the same distance needed to stop. [**Answer:** $x = x_0 + v_0t + \frac{1}{2}at^2 = 0 + (11.4 \text{ m/s})(3.00 \text{ s}) + \frac{1}{2}(-3.80 \text{ m/s}^2)(3.00 \text{ s})^2 = 17.1 \text{ m}$, as expected.]

Some related homework problems: Problem 50, Problem 51

In the previous example, we saw how to determine the distance necessary for a vehicle to come to a complete stop. But how does v vary with distance as the vehicle slows down? The next Conceptual Checkpoint deals with this topic.

CONCEPTUAL CHECKPOINT 2–4

The ranger in the previous example brakes for 17.1 m to come to rest. After braking for only half that distance, $\frac{1}{2}(17.1 \text{ m}) = 8.55$ m, is the ranger's speed **(a)** equal to $\frac{1}{2}v_0$, **(b)** greater than $\frac{1}{2}v_0$, or **(c)** less than $\frac{1}{2}v_0$?

Reasoning and Discussion

As pointed out in the Insight for Example 2-8, the fact that the stopping distance, Δx, depends on v_0^2 means that this distance increases by a factor of four when the speed is doubled. For example, the stopping distance with an initial speed of v_0 is four times the stopping distance when the initial speed is $v_0/2$.

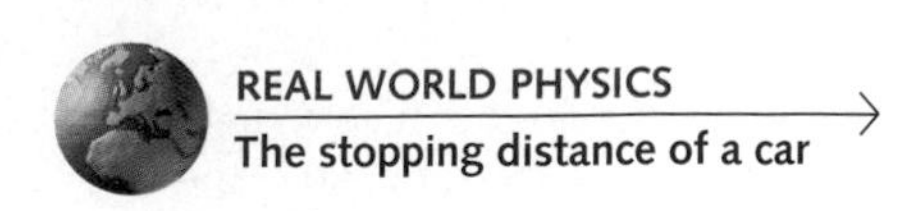
REAL WORLD PHYSICS
The stopping distance of a car

To apply this observation to the ranger, suppose that the stopping distance with an initial speed of v_0 is Δx. It follows that the stopping distance for an initial speed of $v_0/2$ is $\Delta x/4$. This means that as the ranger slows from v_0 to 0, it takes a distance $\Delta x/4$ to slow from $v_0/2$ to 0, and the remaining distance, $3\Delta x/4$, to slow from v_0 to $v_0/2$. Thus, at the halfway point the ranger has not yet slowed to half of the initial velocity—the speed at this point is greater than $v_0/2$.

Answer:

(b) The ranger's speed is greater than $\frac{1}{2}v_0$.

Clearly, v does not decrease uniformly with distance. A plot showing v as a function of x for Example 2-8 is shown in **Figure 2–16.**

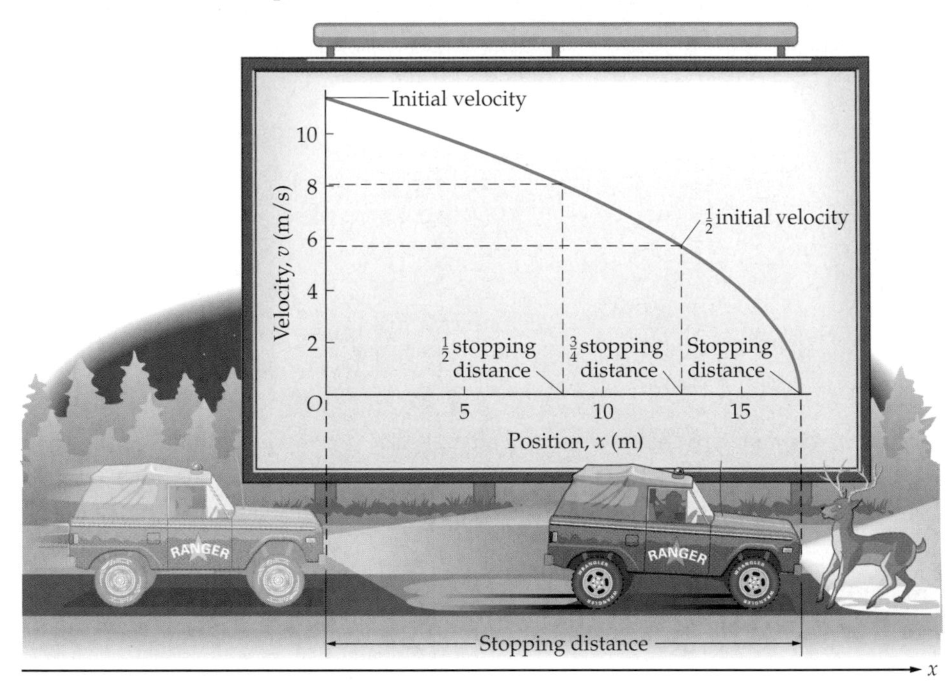

▶ FIGURE 2–16 Velocity as a function of position for the ranger in Example 2–8

As we can see from the graph, v changes more in the second half of the braking distance than in the first half.

We close this section with a familiar, everyday example: a police car accelerating to overtake a speeder. This is the first time that we use two equations of motion for two different objects to solve a problem—but it won't be the last. Problems of this type are often more interesting than problems involving only a single object, and they relate to many types of situations in everyday life.

PROBLEM SOLVING NOTE
Strategize

Before attempting to solve a problem, it is a good idea to have some sort of plan, or "strategy," for how to proceed. It may be as simple as saying, " The problem asks me to relate velocity and time, therefore I will use Equation 2–7." In other cases the strategy is a bit more involved. Producing effective strategies is one of the most challenging—and creative—aspects of problem solving.

EXAMPLE 2–9 Catching a Speeder

A speeder doing 40.0 mi/h (about 17.9 m/s) in a 25 mi/h zone approaches a parked police car. The instant the speeder passes the police car, the police begin their pursuit. If the speeder maintains a constant velocity, and the police car accelerates with a constant acceleration of 4.50 m/s^2, **(a)** how long does it take for the police car to catch the speeder, **(b)** how far have the two cars traveled in this time, and **(c)** what is the velocity of the police car when it catches the speeder?

Picture the Problem

We start with a sketch showing both vehicles at $x = 0$ at time $t = 0$. The speeder has an initial velocity of 17.9 m/s; the police car's initial velocity is zero. In addition, we choose the positive direction to be the direction of motion.

Strategy

To solve this problem, first write down a position-versus-time equation for the police car, x_p, and a separate equation for the speeder, x_s. Next, find the time it takes the police car to catch the speeder by setting $x_p = x_s$ and solving the resulting equation for t. Once the catch time is determined, it is straightforward to calculate the distance traveled and the velocity of the police car.

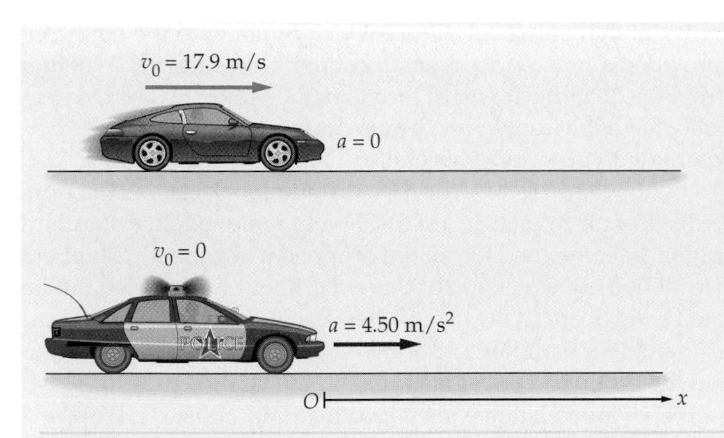

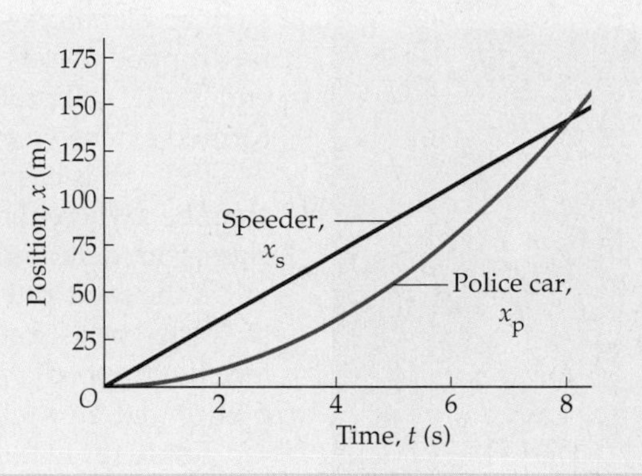

Solution

Part (a)

1. Write equations of motion for the two vehicles. For the police car, $v_0 = 0$, and $a = 4.50\ \text{m/s}^2$. For the speeder, $v_0 = 17.9\ \text{m/s} = v_s$, and $a = 0$:

$$x_p = \tfrac{1}{2}at^2$$
$$x_s = v_s t$$

2. Set $x_p = x_s$ and solve for the time:

$$\tfrac{1}{2}at^2 = v_s t \quad \text{or} \quad (\tfrac{1}{2}at - v_s)t = 0$$

$$\text{two solutions:} \quad t = 0 \quad \text{or} \quad t = \frac{2v_s}{a}$$

3. Clearly, $t = 0$ corresponds to the initial conditions, because both vehicles started at $x = 0$ at that time. The time of interest is obtained by substituting numerical values into the other solution:

$$t = \frac{2v_s}{a} = \frac{2(17.9\ \text{m/s})}{4.50\ \text{m/s}^2} = 7.96\ \text{s}$$

Part (b)

4. Substitute $t = 7.96$ s into the equations of motion for x_p and x_s. Note that $x_p = x_s$, as expected.

$$x_p = \tfrac{1}{2}at^2 = \tfrac{1}{2}(4.50\ \text{m/s}^2)(7.96\ \text{s})^2 = 142\ \text{m}$$
$$x_s = v_s t = (17.9\ \text{m/s})(7.96\ \text{s}) = 142\ \text{m}$$

Part (c)

5. To find the velocity of the police car use Equation 2–7, which relates velocity to time:

$$v_p = v_0 + at = 0 + (4.50\ \text{m/s}^2)(7.96\ \text{s}) = 35.8\ \text{m/s}$$

Insight

When the police car catches up with the speeder its velocity is 35.8 m/s, which is exactly twice the velocity of the speeder. A coincidence? Not at all. When the police car catches the speeder, both have traveled the same distance (142 m) in the same time (7.96 s), therefore they have the same average velocity. Of course, the average velocity of the speeder is simply v_s. The average velocity of the police car is $\frac{1}{2}(v_0 + v)$, since its acceleration is constant, and thus $\frac{1}{2}(v_0 + v) = v_s$. Since $v_0 = 0$ for the police car, it follows that $v = 2v_s$. Notice that this result is independent of the acceleration of the police car, as we show in the following Practice Problem.

Practice Problem

Repeat this example for the case where the acceleration of the police car is $a = 3.20\ \text{m/s}^2$. [**Answer: (a)** $t = 11.2$ s, **(b)** $x_p = x_s = 200$ m, **(c)** $v_p = 35.8$ m/s]

Some related homework problems: Problem 47, Problem 58

2–7 Freely Falling Objects

The most famous example of motion with constant acceleration is free fall—the motion of an object falling freely under the influence of gravity. It was Galileo (1564–1642) who first showed, with his own experiments, that falling bodies move with constant acceleration. His conclusions were based on experiments done by rolling balls down inclines of various steepness. By using an incline, Galileo was able to reduce the acceleration of the balls, thus producing motion slow enough to be timed with the rather crude instruments available.

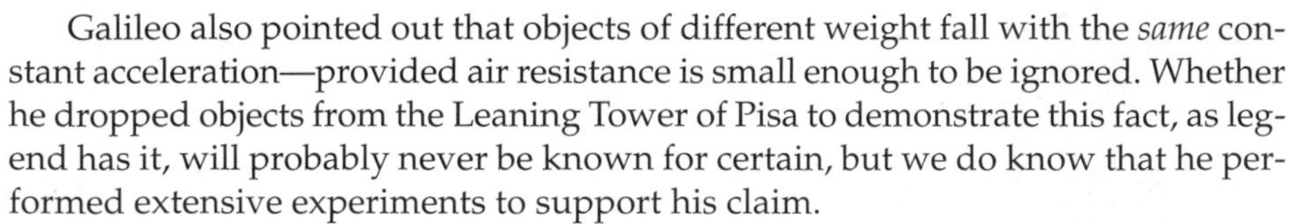

Galileo also pointed out that objects of different weight fall with the *same* constant acceleration—provided air resistance is small enough to be ignored. Whether he dropped objects from the Leaning Tower of Pisa to demonstrate this fact, as legend has it, will probably never be known for certain, but we do know that he performed extensive experiments to support his claim.

Today it is easy to verify Galileo's assertion by dropping objects in a vacuum chamber, where the effects of air resistance are essentially removed. In a standard classroom demonstration, a feather and a coin are dropped in a vacuum, and both fall at the same rate. In 1971, a novel version of this experiment was carried out on the Moon by astronaut David Scott. In the near perfect vacuum on the Moon's surface he dropped a feather and a hammer and showed a worldwide television audience that they fell to the ground in the same time.

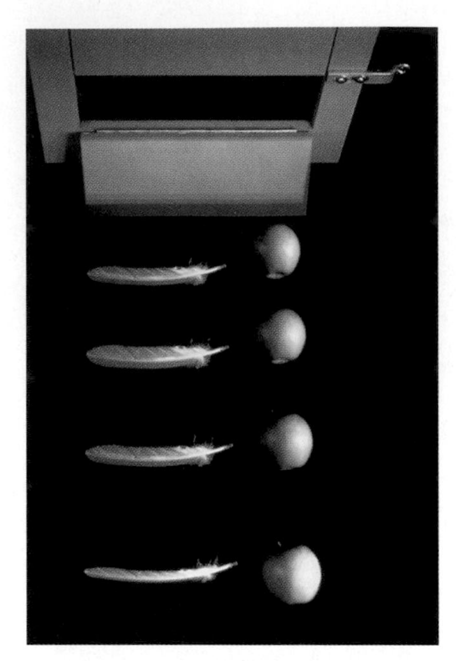

▲ In the absence of air resistance, all bodies fall with the same acceleration, regardless of their mass.

To illustrate the effect of air resistance in everyday terms, consider dropping a sheet of paper and a rubber ball (**Figure 2–17**). The paper drifts slowly to the ground, taking much longer to fall than the ball. Now, wad the sheet of paper into a tight ball and repeat the experiment. This time the ball of paper and the rubber ball reach the ground in nearly the same time. What was different in the two experiments? Clearly, when the sheet of paper was wadded into a ball, the effect of air resistance on it was greatly reduced, so that both objects fell almost as they would in a vacuum.

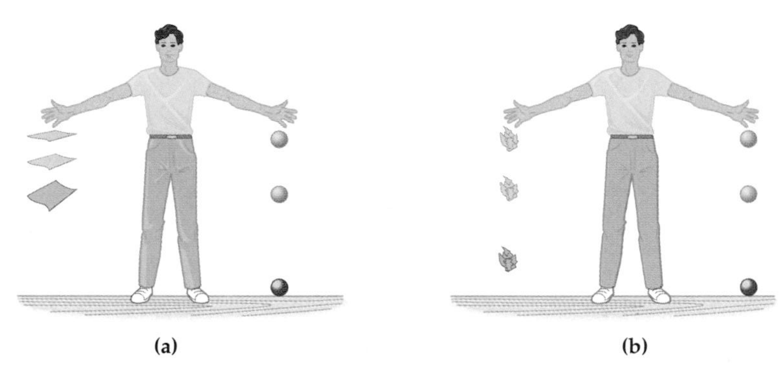

▲ **FIGURE 2–17 Free fall and air resistance**

(a) Dropping a sheet of paper and a rubber ball compared with **(b)** dropping a wadded up sheet of paper and a ball.

Before considering a few examples, let's first discuss exactly what is meant by "free fall." To begin, the word *free* in free fall means free from any effects other than gravity. For example, in free fall we assume that an object's motion is not influenced by any form of friction or air resistance.

- Free fall is the motion of an object subject *only* to the influence of gravity.

Though free fall is an idealization—which does not apply to many real-world situations—it is still a useful approximation in many other cases. In the following examples we assume that the motion may be considered as free fall.

Next, it should be realized that the word *fall* in free fall does not mean the object is necessarily moving downward. By free fall, we mean *any* motion under the influence of gravity alone. If you drop a ball, it is in free fall. If you throw a ball upward or downward, it is in free fall as soon as it leaves your hand.

▲ Whether she is on the way up, at the peak of her flight, or on the way down, this girl is in free fall, accelerating downward with the acceleration of gravity. Only when she is in contact with the blanket does her acceleration change.

- An object is in free fall as soon as it is released, whether it is dropped from rest, thrown downward, or thrown upward.

Finally, the acceleration produced by gravity on the Earth's surface (sometimes called the gravitational strength) is denoted with the symbol g. As a shorthand name, we will frequently refer to g simply as "the acceleration of gravity." In fact, as we shall see in Chapter 12, the value of g varies according to one's location on the surface of the earth, as well as one's altitude above it. Table 2–5 shows how g varies with latitude.

In all the calculations that follow in this book, we shall use $g = 9.81\ \text{m/s}^2$ for the acceleration of gravity. Note, in particular, that g always stands for $+9.81\ \text{m/s}^2$, never $-9.81\ \text{m/s}^2$. For example, if we choose a coordinate system with the positive direction upward, the acceleration in free fall is $a = -g$. If the positive direction is downward, then free-fall acceleration is $a = g$.

With these comments, we are ready to explore a variety of free-fall examples.

TABLE 2–5
Values of g at Different Locations on Earth (m/s^2)

Location	Latitude	g
Quito, Ecuador	0°	9.780
Hong Kong	30° N	9.793
Oslo, Norway	60° N	9.819
North Pole	90° N	9.832

EXAMPLE 2–10 Do the Cannonball!

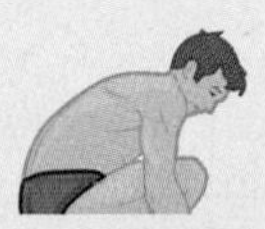

A person steps off the end of a 3.00-m-high diving board and drops to the water below. **(a)** How long does it take for the person to reach the water? **(b)** What is the person's speed on entering the water?

Picture the Problem

In our sketch we choose the origin to be at the height of the diving board, and we let the positive direction be downward. With these choices, $x_0 = 0$, $a = g$, and the water is at $x = 3.00$ m. Of course, $v_0 = 0$ since the person simply steps off the board.

Strategy

For part (a) we want to relate position to time, so we use Equation 2–11. For part (b) we can relate velocity to time by using Equation 2–7, or we can relate velocity to position by using Equation 2–12.

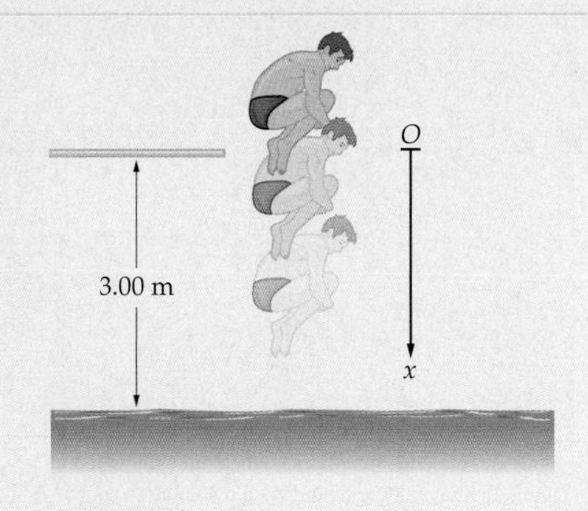

Solution

Part (a)

1. Write Equation 2–11 with $x_0 = 0$, $v_0 = 0$, and $a = g$:

$$x = x_0 + v_0t + \tfrac{1}{2}at^2 = 0 + 0 + \tfrac{1}{2}gt^2 = \tfrac{1}{2}gt^2$$

2. Solve for the time, t, and set $x = 3.00$ m:

$$t = \sqrt{\frac{2x}{g}} = \sqrt{\frac{2(3.00\ \text{m})}{9.81\ \text{m/s}^2}} = 0.782\ \text{s}$$

Part (b)

3. Use the time found in part (a) in Equation 2–7:

$$v = v_0 + gt = 0 + (9.81\ \text{m/s}^2)(0.782\ \text{s}) = 7.67\ \text{m/s}$$

4. We can also find the velocity without knowing the time by using Equation 2–12 with $\Delta x = 3.00$ m:

$$v^2 = v_0^2 + 2a\Delta x = 0 + 2g\Delta x$$
$$v = \sqrt{2g\Delta x} = \sqrt{2(9.81\ \text{m/s}^2)(3.00\ \text{m})} = 7.67\ \text{m/s}$$

Insight

Let's put these results in more common, everyday units. If you step off a diving board 9.84 ft (3.00 m) above the water, you enter the water with a speed of 17.2 mi/h (7.67 m/s).

Practice Problem

What is your speed on entering the water if you step off a 10.0 m diving tower? [**Answer:** $v = \sqrt{2(9.81\ \text{m/s}^2)(10.0\ \text{m})} = 14.0\ \text{m/s} = 31\ \text{mi/h}$]

Some related homework problems: Problem 61, Problem 70

The special case of free fall from rest occurs so frequently, and in so many different contexts, that it deserves special attention. If we take x_0 to be zero, and positive to be downward, as in **Figure 2–18**, then position as a function of time is $x = x_0 + v_0t + \tfrac{1}{2}gt^2 = 0 + 0 + \tfrac{1}{2}gt^2$, or

PROBLEM SOLVING NOTE
Check Your Solution

Once you have a solution to a problem, check to see whether it makes sense. First, make sure the units are correct; m/s for speed, m/s^2 for acceleration, and so on. Second, check the numerical value of your answer. If you are solving for the speed of a diver dropping from a 3.0-m diving board and you get an unreasonable value like 200 m/s ($\approx$450 mi/h), chances are good that you've made a mistake.

$t = 0$ $v = 0$ $x = 0$
$t = 1$ s $v = 9.81$ m/s $x = 4.91$ m
$t = 2$ s $v = 19.6$ m/s $x = 19.6$ m
$t = 3$ s $v = 29.4$ m/s $x = 44.1$ m
$t = 4$ s $v = 39.2$ m/s $x = 78.5$ m
x

▲ **FIGURE 2–18 Free fall from rest**
Position and velocity are shown as functions of time. It is apparent that velocity depends linearly on t, whereas position depends on t^2.

$$x = \tfrac{1}{2}gt^2 \qquad 2\text{–}13$$

Similarly, velocity as a function of time is

$$v = gt \qquad 2\text{–}14$$

and velocity as a function of position is

$$v = \sqrt{2gx} \qquad 2\text{–}15$$

The behavior of these functions is illustrated in Figure 2–18. Note that position increases with time squared, whereas velocity increases linearly with time.

Next we consider two objects dropped from rest, one after the other, and discuss how their separation varies with time.

CONCEPTUAL CHECKPOINT 2–5

You drop a rock from a bridge to the river below. When the rock has fallen 4 m, you drop a second rock. As the rocks continue their free fall, does their separation **(a)** increase, **(b)** decrease, or **(c)** stay the same?

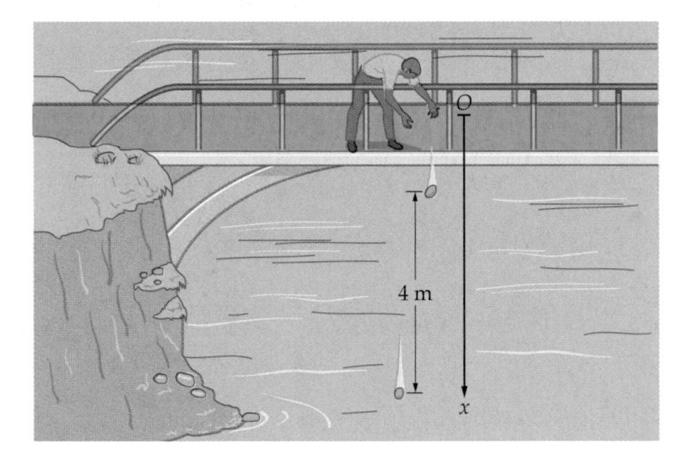

Reasoning and Discussion
It might seem that since both rocks are in free fall, their separation remains the same. This is not so. The rock that has a head start always has a greater velocity than the later one; thus it covers a greater distance in any interval of time. As a result, the separation between the rocks increases.

Answer
(a) The separation between the rocks increases.

An erupting volcano shooting out fountains of lava is an impressive sight. In the next Example we show how a simple timing experiment can determine the initial velocity of the erupting lava.

EXAMPLE 2–11 Bombs Away: Calculating the Speed of a Lava Bomb

A volcano shoots out blobs of molten lava, called lava bombs, from ground level. A geologist observing the eruption uses a stopwatch to time the flight of a particular lava bomb that is projected straight upward. If the time for it to rise and fall back to the ground is 4.75 s, and its acceleration is 9.81 m/s^2 downward, what is its initial speed?

Picture the Problem

Our sketch shows a coordinate system with upward as the positive x direction. For convenience, we offset the upward and downward trajectories slightly and choose $t = 0$ to be the time at which the lava bomb is launched. With these choices it follows that $x_0 = 0$ and the acceleration is $a = -g = -9.81\ \text{m/s}^2$. The initial speed to be determined is v_0.

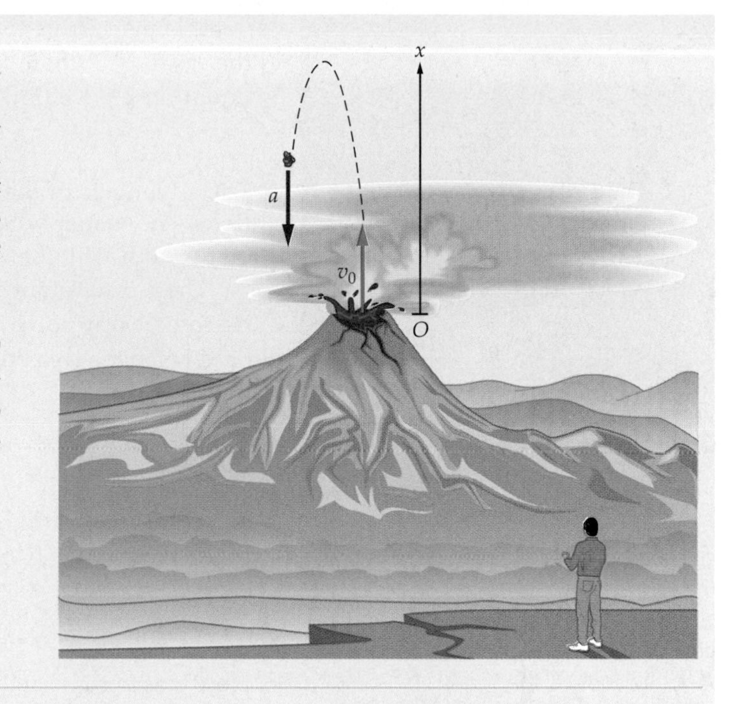

Strategy

We know that the lava bomb starts at $x = 0$ at the time $t = 0$ and returns to $x = 0$ at the time $t = 4.75$ s. It is reasonable, therefore, to consider the kinematic equation that relates position to time: $x = x_0 + v_0 t + \frac{1}{2}at^2$. This equation can be solved for the one unknown it contains, v_0.

Solution

1. Write out $x = x_0 + v_0 t + \frac{1}{2}at^2$ with $x_0 = 0$ and $a = -g$. Factor out a time, t, from the two remaining terms:
   $x = x_0 + v_0 t + \frac{1}{2}at^2 = v_0 t - \frac{1}{2}gt^2 = (v_0 - \frac{1}{2}gt)t$

2. Set x equal to zero, since this is the position of the lava bomb at $t = 0$ and $t = 4.75$ s:
   $x = (v_0 - \frac{1}{2}gt)t = 0$ two solutions: (i) $t = 0$ (ii) $v_0 - \frac{1}{2}gt = 0$

3. The first solution is simply the initial condition; that is, $x = 0$ at $t = 0$. Solve the second solution for the initial speed:
   $v_0 - \frac{1}{2}gt = 0$ or $v_0 = \frac{1}{2}gt$

4. Substitute numerical values for g and the time the lava bomb lands:
   $v_0 = \frac{1}{2}gt = \frac{1}{2}(9.81\ \text{m/s}^2)(4.75\ \text{s}) = 23.3\ \text{m/s}$

Insight

A geologist can determine a lava bomb's initial speed by simply observing its flight time.

Practice Problem

A second lava bomb is projected straight upward with an initial speed of 25 m/s. How long is it in the air? **[Answer:** $t = 5.1$ s**]**

Some related homework problems: Problem 63, Problem 72, Problem 75

What is the speed of a lava bomb when it returns to earth; that is, when it returns to the same level from which it was launched? Physical intuition might suggest that, in the absence of air resistance, it should be the same as the initial speed. To show that this hypothesis is indeed correct, write out Equation 2–7 for this case:

$$v = v_0 - gt$$

Substituting numerical values, we find

$$v = v_0 - gt = 23.3\ \text{m/s} - (9.81\ \text{m/s}^2)(4.75\ \text{s}) = -23.3\ \text{m/s}$$

Thus, the velocity of the lava when it lands is just the negative of the velocity it had when launched upward. Or put another way, when the lava lands it has the same speed as when it was launched; it's just traveling in the opposite direction.

▲ In the absence of air resistance, these lava bombs from the Kilauea caldera on Hawaii's Mauna Loa would strike the ground with the same speed they had when they were blasted into the air.

It is instructive to verify this result symbolically. Recall from Example 2-11 that $v_0 = \frac{1}{2}gt$, where t is the time the bomb lands. Substituting this result into Equation 2–7 we find

$$v = \tfrac{1}{2}gt - gt = -\tfrac{1}{2}gt = -v_0$$

The advantage of the symbolic solution lies in showing that the result is not a fluke—no matter what the initial velocity, no matter what the acceleration, the bomb lands with the velocity $-v_0$.

These results hint at a symmetry relating the motion on the way up to the motion on the way down. To make this symmetry more apparent, we first solve for the time when the lava bomb lands. Using the result $v_0 = \frac{1}{2}gt$ from Example 2–11, we find

$$t = \frac{2v_0}{g} \qquad \textit{(time of landing)}$$

Next, we find the time when the velocity of the lava is zero, which is at its highest point. Setting $v = 0$ in Equation 2–7, we have $v = v_0 - gt = 0$, or

$$t = \frac{v_0}{g} \qquad \textit{(time when } v = 0\textit{)}$$

Note that this is exactly half the time required for the lava to make the round trip. Thus, the velocity of the lava is zero and the height of the lava is greatest exactly half way between launch and landing.

This symmetry is illustrated in **Figure 2–19**. In this case we consider a lava bomb that is in the air for 6.00 s, moving without air resistance. Note that at $t = 3.00$ s the lava is at its highest point and its velocity is zero. At times equally spaced before and after $t = 3.00$ s, the lava is at the same height, has the same speed, but is moving in opposite directions. As a result of this symmetry, a movie of the lava bomb's flight would look the same whether run forward or in reverse.

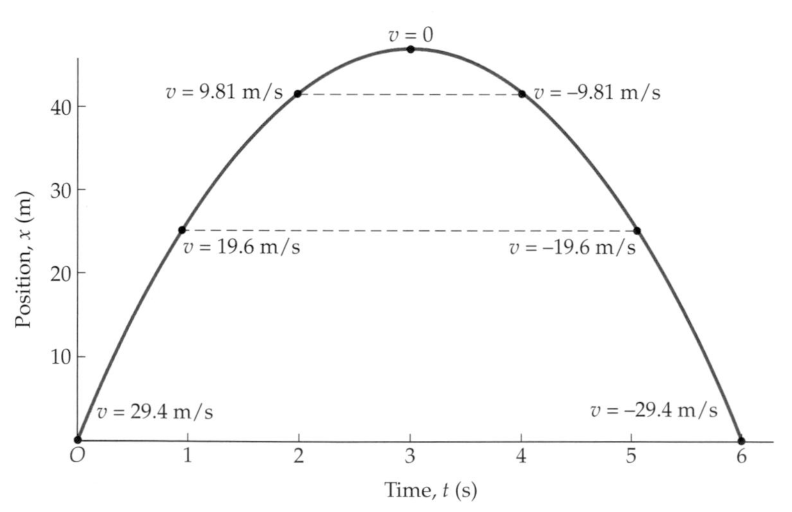

▲ **FIGURE 2–19 Position and velocity of a lava bomb**
This lava bomb is in the air for 6 seconds. Note the symmetry about the midpoint of the bomb's flight.

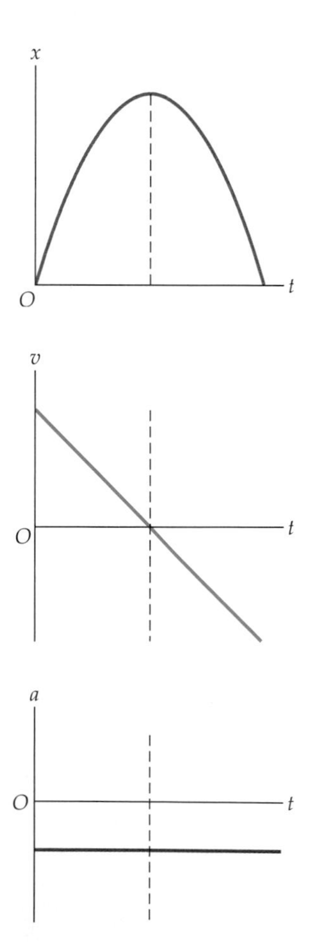

▲ **FIGURE 2–20 Position, velocity, and acceleration of a lava bomb as functions of time**

Figure 2–20 shows the time dependence of position, velocity, and acceleration for an object in free fall without air resistance after being thrown upward. As soon as the object is released, it begins to accelerate downward—as indicated by the negative slope of the velocity-versus-time plot—though it isn't necessarily moving downward. For example, if you throw a ball *upward* it begins to accelerate *downward* the moment it leaves your hand. It continues moving upward, however, until its speed diminishes to zero. Since gravity is causing the downward acceleration, and gravity doesn't turn off just because the ball's velocity goes through zero, the ball continues to accelerate downward even when it is momentarily at rest.

Similarly, in the next Example we consider a sand bag that falls from an ascending hot air balloon. This means that before the bag is in free fall it was moving upward—just like a ball thrown upward. And just like the ball, the sand bag continues moving upward for a brief time before momentarily stopping and then moving downward.

EXAMPLE 2–12 Look Out Below! A Sand Bag in Free Fall

A hot air balloon is rising straight upward with a constant speed of 6.5 m/s. When the basket of the balloon is 20.0 m above the ground a bag of sand tied to the basket comes loose. **(a)** How long is the bag of sand in the air before it hits the ground? **(b)** What is the greatest height of the bag of sand during its fall to the ground?

Picture the Problem

We choose the origin to be at ground level, and positive to be upward. This means that, for the bag, we have $x_0 = 20.0$ m, $v_0 = 6.5$ m/s, and $a = -g$. Our sketch also shows snapshots of the balloon and bag of sand at three different times, starting at $t = 0$ when the bag comes loose. Note that the bag is moving upward with the balloon at the time it comes loose. It therefore continues to move upward for a short time after it separates from the basket, exactly as if it had been thrown upward.

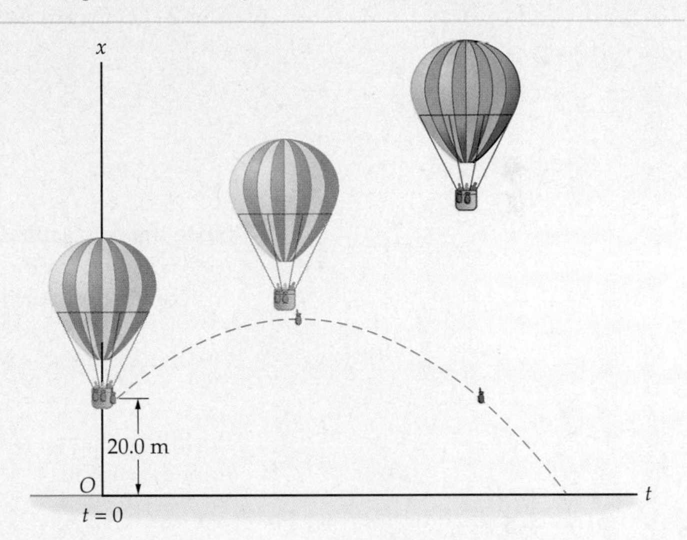

Strategy

In part (a) we want to relate position and time, so we use Equation 2–11. To find the time the bag hits the ground we set $x = 0$ and solve for t. For part (b) we have no expression that gives the maximum height of a particle—so we will have to come up with something on our own. We can start with the fact that $v = 0$ at the greatest height, since it is there the bag momentarily stops as it changes direction. Therefore, we can find the time t when $v = 0$ by using Equation 2–7, and then substitute t into Equation 2–11 to find x_{max}.

Solution

Part (a)

1. Apply Equation 2–11 to the bag of sand, where x_0 and v_0 have the values given. Set $x = 0$:

$$x = x_0 + v_0 t - \tfrac{1}{2}gt^2 = 0$$

2. Note that we have a quadratic equation for t in the form $At^2 + Bt + C = 0$, where $A = -\frac{1}{2}g$, $B = v_0$, and $C = x_0$. Solve this equation for t. The positive solution, 2.8 s, applies to this problem:
(Quadratic equations and their solutions are discussed in Appendix A. In general, one can expect two solutions to a quadratic equation.)

$$t = \frac{-v_0 \pm \sqrt{v_0^2 - 4(-\frac{1}{2}g)(x_0)}}{2(-\frac{1}{2}g)}$$

$$= \frac{-(6.5\ \text{m/s}) \pm \sqrt{(6.5\ \text{m/s})^2 + 2(9.81\ \text{m/s}^2)(20.0\ \text{m})}}{(-9.81\ \text{m/s}^2)}$$

$$= \frac{-(6.5\ \text{m/s}) \pm 20.8\ \text{m/s}}{(-9.81\ \text{m/s}^2)} = 2.8\ \text{s}, -1.5\ \text{s}$$

Part (b)

3. Apply Equation 2–7 to the bag of sand, then find the time when the velocity equals zero:

$$v = v_0 + at = v_0 - gt$$

$$v_0 - gt = 0 \quad \text{or} \quad t = \frac{v_0}{g} = \frac{6.5\ \text{m/s}}{9.81\ \text{m/s}^2} = 0.66\ \text{s}$$

4. Use $t = 0.66$ s in Equation 2–11 to find the maximum height:

$$x_{max} = 20.0\ \text{m} + (6.5\ \text{m/s})(0.66\ \text{s}) - \tfrac{1}{2}(9.81\ \text{m/s}^2)(0.66\ \text{s})^2 = 22\ \text{m}$$

continued on the following page

continued from the previous page

Insight

The positive solution to the quadratic equation is certainly the one that applies here, but the negative solution is not completely without meaning. What physical meaning might it have? Well, if the balloon had been *descending* with a speed of 6.5 m/s, instead of rising, then the time for the bag to reach the ground would have been 1.5 s. Try it! Let $v_0 = -6.5$ m/s and repeat the calculation given in part (a).

Practice Problem

What is the velocity of the bag of sand just before it hits the ground? [**Answer:** $v = v_0 - gt = (6.5\text{ m/s}) - (9.81\text{ m/s}^2)(2.8\text{ s}) = -21\text{ m/s}$; the minus sign indicates the bag is moving downward.]

Some related homework problems: Problem 79, Problem 91

An alternative method to find the time when the bag hits the ground in Example 2–12 is to first calculate the velocity of the bag when it lands, using Equation 2–12. This gives

$$v^2 = v_0{}^2 - 2g(x - x_0) = (6.5\text{ m/s})^2 - 2(9.81\text{ m/s}^2)(0 - 20.0\text{ m}) = 435\text{ m}^2/\text{s}^2$$

or

$$v = \pm 21\text{ m/s}$$

Since the bag is moving downward when it lands, we choose the minus sign; hence $v = -21$ m/s.

Next, solve for the time using Equation 2–7:

$$t = \frac{v - v_0}{a}$$

Recalling that $v_0 = 6.5$ m/s and $a = -g$, and using $v = -21$ m/s as just determined, we find

$$t = \frac{-21\text{ m/s} - 6.5\text{ m/s}}{(-9.81\text{ m/s}^2)} = 2.8\text{ s}$$

as expected.

Chapter Summary

	Topic	Remarks and Relevant Equations
2–1	**Distance and Displacement**	
	distance	Total length of travel, from beginning to end. The distance is always positive.
	displacement	Displacement, Δx, is the change in position: $\Delta x = x_f - x_i$ **2–1** When calculating displacement, it is important to remember that it is always *final* position minus *initial* position—never the other way. Displacement can be positive, negative, or zero.
	positive and negative displacement	The *sign* of the displacement indicates the *direction* of motion. For example, suppose we choose the positive direction to be to the right. Then $\Delta x > 0$ means motion to the right, and $\Delta x < 0$ means motion to the left.
	units	The SI unit of distance and displacement is the meter, m.

2–2 Average Speed and Velocity

average speed — Average speed is *distance* divided by elapsed time:

average speed = distance/time (2–2)

Average speed is never negative.

average velocity — Average velocity, v_{av}, is *displacement* divided by time:

$$v_{av} = \frac{\Delta x}{\Delta t} = \frac{x_f - x_i}{t_f - t_i} \quad \text{2–3}$$

Average velocity is positive if motion is in the positive direction, and negative if motion is in the negative direction.

graphical interpretation of velocity — In an x-versus-t plot, the average velocity is the slope of a line connecting two points.

units — The SI unit of speed and velocity is meters per second, m/s.

2–3 Instantaneous Velocity

The velocity at an instant of time is the limit of the average velocity over shorter and shorter time intervals:

$$v = \lim_{\Delta t \to 0} \frac{\Delta x}{\Delta t} \quad \text{2–4}$$

Instantaneous velocity can be positive, negative, or zero, with the sign indicating the direction of motion.

constant velocity — When velocity is constant, the instantaneous velocity is equal to the average velocity.

graphical interpretation — In an x-versus-t plot, the instantaneous velocity at a given time is equal to the slope of the tangent line at that time.

2–4 Acceleration

average acceleration — Average acceleration is the change in velocity divided by the change in time:

$$a_{av} = \frac{\Delta v}{\Delta t} = \frac{v_f - v_i}{t_f - t_i} \quad \text{2–5}$$

Average acceleration is positive if $v_f > v_i$, is negative if $v_f < v_i$, and is zero if $v_f = v_i$.

instantaneous acceleration — Instantaneous acceleration is the limit of the average acceleration as the time interval goes to zero:

$$a = \lim_{\Delta t \to 0} \frac{\Delta v}{\Delta t} \quad \text{2–6}$$

Instantaneous acceleration can be positive, negative, or zero, depending on whether the velocity is becoming more positive, more negative, or is staying the same. Knowing the sign of the acceleration *does not* tell you whether an object is speeding up or slowing down, and it *does not* give the direction of motion.

constant acceleration — When acceleration is constant, the instantaneous acceleration is equal to the average acceleration.

deceleration — An object whose speed is decreasing is said to be decelerating. Deceleration occurs whenever the velocity and acceleration have opposite signs.

graphical interpretation — In a v-versus-t plot, the instantaneous acceleration is equal to the slope of the tangent line at a given time.

units — The SI unit of acceleration is (meters per second) per second, or m/s^2.

2–5 Motion with Constant Acceleration

Several different "equations of motion" describe particles moving with constant acceleration. Each equation relates a different set of variables:

velocity as a function of time

$$v = v_0 + at \quad \text{2–7}$$

position as a function of time and velocity

$$x = x_0 + \tfrac{1}{2}(v_0 + v)t \quad \text{2–10}$$

position as a function of time and acceleration

$$x = x_0 + v_0 t + \tfrac{1}{2}at^2 \quad \text{2–11}$$

velocity as a function of position $$v^2 = v_0^2 + 2a(x - x_0) = v_0^2 + 2a\Delta x \quad \text{2–12}$$

2–6 Free Fall

Objects in free fall move under the influence of gravity alone. An object is in free fall as soon as it is released, whether it is thrown upward, thrown downward, or released from rest.

acceleration due to gravity

The acceleration due to gravity on the earth's surface varies slightly from place to place. In this book we shall define the acceleration of gravity to have the following magnitude:

$$g = 9.81 \text{ m/s}^2$$

Note that g is always a positive quantity. If we choose the positive direction of our coordinate system to be downward (in the direction of the acceleration of gravity), it follows that the acceleration of an object in free fall is $a = +g$. On the other hand, if we choose our positive direction to be upward, the acceleration of a freely falling object is in the negative direction; hence $a = -g$.

Problem-Solving Summary

Type of Calculation	Relevant Physical Concepts	Related Examples
Relate velocity to time.	In motion with uniform acceleration a, the velocity changes with time as $v = v_0 + at$ (Equation 2–7).	Examples 2–5, 2–8, 2–9, 2–10, 2–11, 2–12
Relate velocity to position.	If an object with an initial velocity v_0 accelerates with a uniform acceleration a for a distance Δx, the final velocity, v, is given by $v^2 = v_0^2 + 2a\Delta x$ (Equation 2–12).	Examples 2–7, 2–8, 2–10
Relate position to time.	The position of an object moving with constant acceleration a varies with time as follows: $x = x_0 + \frac{1}{2}(v_0 + v)t$ (Equation 2–10) or equivalently $x = x_0 + v_0 t + \frac{1}{2}at^2$ (Equation 2–11).	Examples 2–5, 2–6, 2–9, 2–10, 2–11, 2–12

Conceptual Questions

1. You and your dog go for a walk to a nearby park. On the way, your dog takes many short side trips to chase squirrels, examine fire hydrants, and so on. When you arrive at the park, do you and your dog have the same displacement? Have you traveled the same distance?
2. Does an odometer in a car measure distance or displacement?
3. You check your car's odometer before and after a trip. Is it possible that the difference in readings has the same magnitude as your displacement? Explain.
4. An astronaut orbits Earth in the space shuttle. In one complete orbit, is the displacement the same as the distance traveled?
5. After a tennis match the players dash to the net to congratulate one another. If they both run with a speed of 3 m/s, are their velocities equal?
6. Does a speedometer measure speed or velocity?
7. Is it possible for a car to circle a race track with constant velocity? Can it do so with constant speed?
8. Friends tell you that on a recent trip their average velocity was +20 m/s. Is it possible that their instantaneous velocity was negative at any time during the trip?
9. You drive in a straight line at 15 m/s for 10 minutes, then at 25 m/s for another 10 minutes. Is your average velocity (i) 20 m/s, (ii) more than 20 m/s, or (iii) less than 20 m/s? Explain.
10. If the position of an object is zero, does its speed have to be zero?
11. For what kind of motion are the instantaneous and average velocities equal?
12. Two bows shoot arrows with the same initial speed. The string in bow A must be pulled back farther when shooting an arrow than the string in bow B. Which bow gives its arrow a greater acceleration?
13. Assume that the brakes in your car create a constant deceleration, regardless of how fast you are going. If you double your driving speed, how does this affect **(a)** the time required to come to a stop, and **(b)** the distance needed to stop?
14. If the velocity of an object is zero, does its acceleration have to be zero?
15. If the velocity of an object is nonzero, can its acceleration be zero?
16. Is it possible for an object to have zero average velocity over a given interval of time, yet still be accelerating during the interval?
17. A batter hits a pop fly straight up. **(a)** Is the acceleration of the ball on the way up different from its acceleration on the way down? **(b)** Is the acceleration of the ball at the top of its flight different from its acceleration just before it lands?
18. A pop fly goes straight up with an initial speed of 7 m/s. What is its speed when it returns to its initial height above the ground?
19. After winning a baseball game one player drops a glove, while another tosses a glove into the air. How do the accelerations of the two gloves compare?
20. A volcano shoots a lava bomb straight upward. Does the displacement of the lava bomb depend on **(a)** your choice of origin for your coordinate system, or **(b)** your choice of a positive direction?
21. At the edge of a roof you drop ball A from rest, and then throw ball B downward with an initial velocity of v_0. Is the increase in speed when the balls land (i) more for ball A, (ii) more for ball B, or (iii) the same for each ball?

22. At the edge of a roof you throw ball A upward with an initial speed of v_0, and then throw ball B downward with the same initial speed. When the balls hit the ground, which of the following is true, ignoring air resistance: (i) the speed of ball A is greater than the speed of ball B, (ii) the speed of ball A is equal to the speed of ball B, or (iii) the speed of ball A is less than the speed of ball B?

23. A ball is thrown straight upward with an initial speed v_0. When it reaches the top of its flight at height h, a second ball is thrown straight upward with the same initial velocity. Do the balls cross paths (i) at height $\frac{1}{2}h$, (ii) above $\frac{1}{2}h$, or (iii) below $\frac{1}{2}h$. Explain.

Problems

Note: **IP** *denotes an integrated conceptual/quantitative problem.* **BIO** *identifies problems of biological or medical interest. Blue bullets (•, ••, •••) are used to indicate the level of difficulty of each problem. Air resistance should be ignored in the problems for this chapter.*

Section 2–1 Position, Distance, and Displacement

1. • Referring to **Figure 2–21**, you walk from your home to the library, then to the park. **(a)** What is the distance traveled? **(b)** What is your displacement?

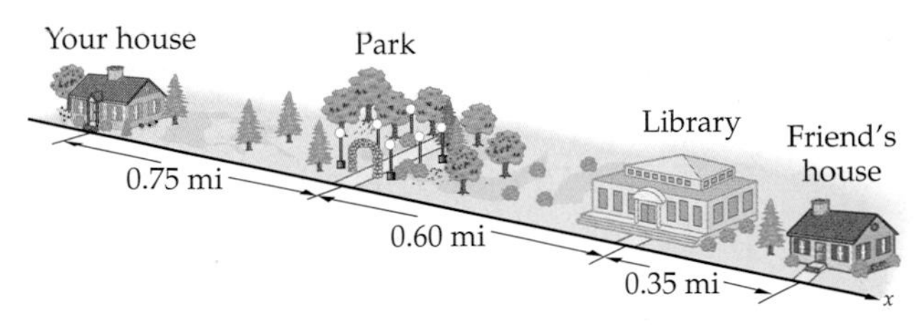

▲ **FIGURE 2–21** Problems 1 and 2

2. • In Figure 2–21, you walk from the park to your friend's house, then back to your house. What is your **(a)** distance traveled, and **(b)** displacement?

3. • The golfer in **Figure 2–22** sinks the ball in two putts, as shown. What is **(a)** the distance traveled by the ball, and **(b)** the displacement of the ball?

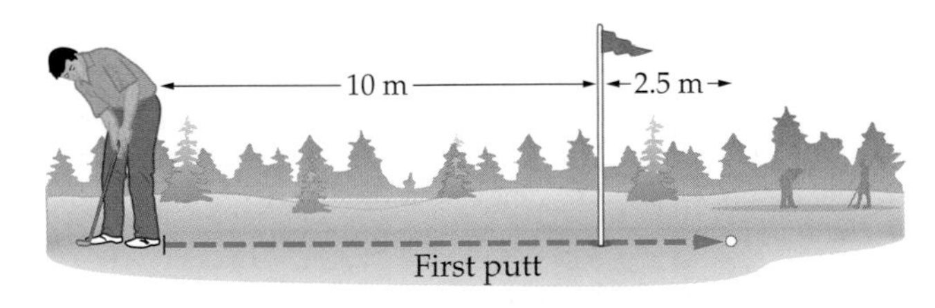

▲ **FIGURE 2–22** Problem 3

4. • The two tennis players shown in **Figure 2–23** walk to the net to congratulate one another. **(a)** Find the distance traveled and the displacement of player A. **(b)** Repeat for player B.

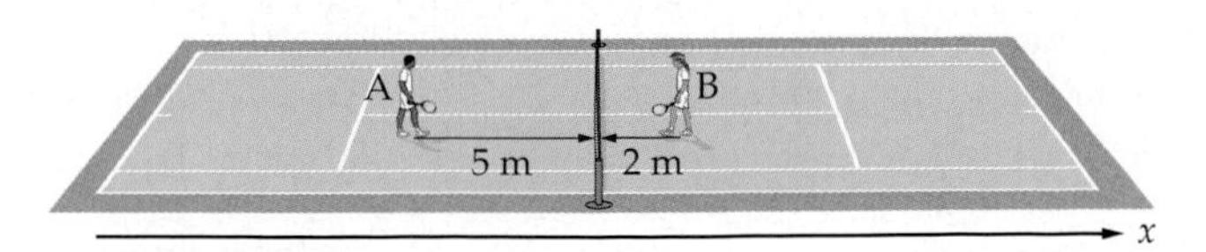

▲ **FIGURE 2–23** Problem 4

5. • A jogger runs on the track shown in **Figure 2–24**. **(a)** What is the distance traveled and the displacement in running from point A to point B? **(b)** Find the distance and displacement for a complete circuit of the track.

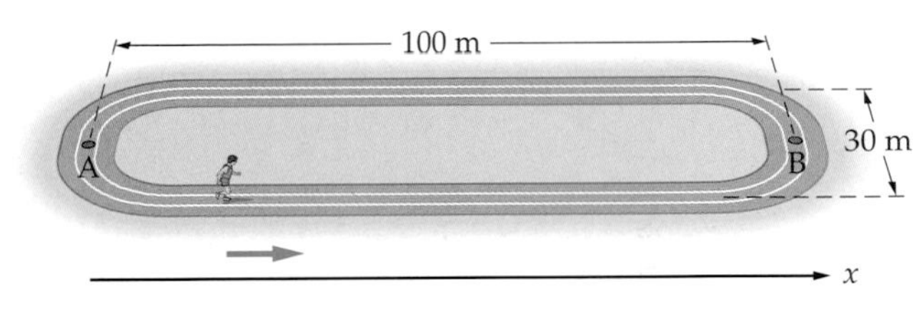

▲ **FIGURE 2–24** Problem 5

6. •• **IP** A child rides a pony on a circular track with a radius of 5.0 m. **(a)** Find the distance traveled and the displacement after the child has gone halfway around the track. **(b)** Does the distance traveled increase, decrease, or stay the same when the child completes one circuit of the track? Does the displacement increase, decrease, or stay the same? Explain. **(c)** Find the distance and displacement after a complete circuit of the track.

Section 2–2 Average Speed and Velocity

7. • Joseph DeLoach of the United States set an Olympic record in 1988 for the 200-meter dash with a time of 19.75 seconds. What was his average speed? Give your answer in meters per second and miles per hour.

8. • In 1992, Zhuang Yong of China set a women's Olympic record in the 100-meter freestyle swim with a time of 54.64 seconds. What was her average speed in m/s and mi/h?

9. • Kangaroos have been clocked at speeds of 65 km/h. How far can a kangaroo hop in 2 minutes at this speed?

10. • A severe storm on January 10, 1992, caused a cargo ship near the Aleutian Islands to spill 29,000 rubber ducks and other bath toys into the ocean. Ten months later, hundreds of rubber ducks began to appear along the shoreline near Sitka, Alaska, roughly 1600 miles away. What was the approximate average speed (in mi/h and m/s) of the ocean current that carried the ducks to shore?

11. • Radio waves travel at the speed of light, approximately 186,000 miles per second. How long does it take for a radio message to travel from the Earth to the Moon? (See Appendix C).

12. • It was a dark and stormy night, when suddenly you saw a flash of lightning. Three-and-a-half seconds later you heard the thunder. Given that the speed of sound in air is about 340 m/s, how far away was the lightning bolt?

13. • **BIO** The human nervous system can propagate nerve impulses at about 10^2 m/s. Estimate the time it takes for a nerve impulse generated when your finger touches a hot object to travel the length of your arm.

14. • Estimate how fast your hair grows in miles per hour.

15. •• A finch rides on the back of a Galapagos tortoise, which walks at the stately pace of 0.060 m/s. After 2.0 minutes the finch tires of the tortoise's slow pace, and takes flight in the same direction for another 2.0 minutes at 12 m/s. What was the average speed of the finch for this 4-minute interval?

16. •• You jog at 6.0 mi/h for 5.0 mi, then you jump into a car and drive for another 5.0 mi. With what average speed must you drive if your average speed for the entire 10.0 miles is to be 11 mi/h?

17. •• A dog runs back and forth between its two owners, who are walking toward one another (**Figure 2–25**). The dog starts running when the owners are 10.0 m apart. If the dog runs with a speed of 3.0 m/s, and the owners each walk with a speed of 1.3 m/s, how far has the dog traveled when the owners meet?

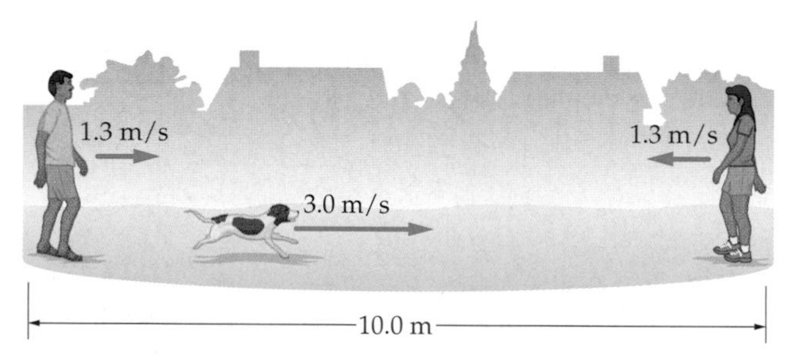

▲ **FIGURE 2–25** Problem 17

18. •• **IP** You drive in a straight line at 20.0 m/s for 10.0 minutes, then at 30.0 m/s for another 10.0 minutes. **(a)** Is your average velocity 25.0 m/s, more than 25.0 m/s, less than 25.0 m/s? Explain. **(b)** Verify your response to part (a) by calculating the average velocity.

19. •• **(a)** Plot a position-versus-time graph for the previous problem. Your plot should extend from $t = 0$ to $t = 20$ minutes. **(b)** Use your plot to calculate the average velocity between $t = 0$ and $t = 15$ minutes.

20. •• **IP** You drive in a straight line at 20.0 m/s for 10.0 miles, then at 30.0 m/s for another 10.0 miles. **(a)** Is your average velocity 25.0 m/s, more than 25.0 m/s, less than 25.0 m/s? Explain. **(b)** Verify your response to part (a) by calculating the average velocity.

21. •• **IP** An expectant father paces back and forth, producing the position-versus-time graph shown in **Figure 2–26**. **(a)** Without performing a calculation, indicate whether the father's velocity is positive, negative, or zero on the segments of the graph labeled A, B, C, and D. **(b)** Calculate the average velocity for each segment, and show that your results verify your answers to part (a).

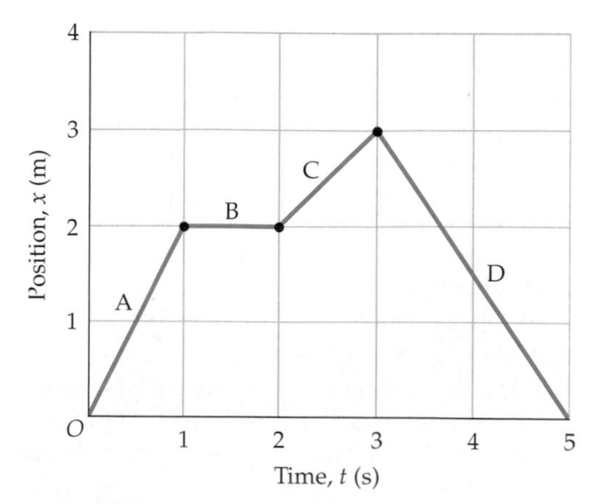

▲ **FIGURE 2–26** Problem 21

22. •• The position of a particle as a function of time is given by $x = (-5\text{ m/s})t + (3\text{ m/s}^2)t^2$. **(a)** Plot x-versus-t for $t = 0$ to $t = 2$ s. **(b)** Find the average velocity of the particle from $t = 0$ to $t = 1$ s. **(c)** Find the average speed from $t = 0$ to $t = 1$ s.

23. •• The position of a particle as a function of time is given by $x = (6\text{ m/s})t + (-2\text{ m/s}^2)t^2$. **(a)** Plot x-versus-t for $t = 0$ to $t = 2$ s. **(b)** Find the average velocity of the particle from $t = 0$ to $t = 1$ s. **(c)** Find the average speed from $t = 0$ to $t = 1$ s.

24. •• **IP** A child rides her tricycle back and forth along the sidewalk, producing the position-versus-time graph shown in **Figure 2–27**. **(a)** Without performing a calculation, indicate on which of the segments of the graph, A, B, or C, the child has the greatest average speed. **(b)** Calculate the average speed for each segment, and show that your results verify your answers to part (a).

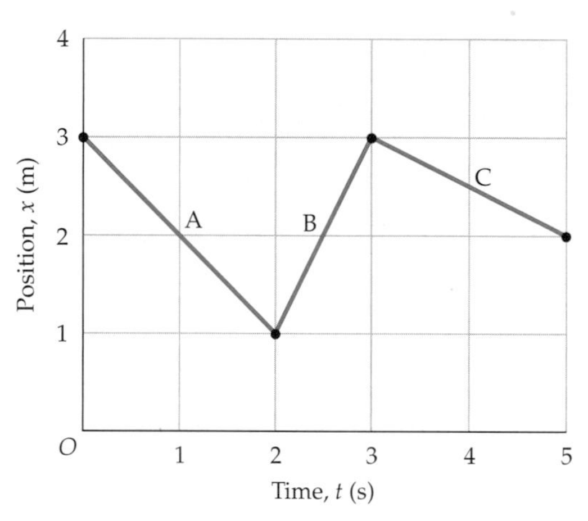

▲ **FIGURE 2–27** Problem 24

25. ••• On your wedding day you leave for the church 30.0 minutes before the ceremony is to begin, which should be plenty of time since the church is only 10.0 miles away. On the way, however, you have to make an unanticipated stop for construction work on the road. As a result, your average speed for the first 15 minutes is only 5.0 mi/h. What average speed do you need for the rest of the trip to get you to the church on time?

Section 2–3 Instantaneous Velocity

26. •• The position of a particle as a function of time is given by $x = (2.0\text{ m/s})t + (-3.0\text{ m/s}^2)t^2$. **(a)** Plot x-versus-t for time from $t = 0$ to $t = 1.0$ s. **(b)** Find the average velocity of the particle from $t = 0.45$ s to $t = 0.55$ s. **(c)** Find the average velocity from $t = 0.49$ s to $t = 0.51$ s.

27. •• The position of a particle as a function of time is given by $x = (-2.0\text{ m/s})t + (3.0\text{ m/s}^2)t^2$. **(a)** Plot x-versus-t for time from $t = 0$ to $t = 1.0$ s. **(b)** Find the average velocity of the particle from $t = 0.15$ s to $t = 0.25$ s. **(c)** Find the average velocity from $t = 0.19$ s to $t = 0.21$ s.

Section 2–4 Acceleration

28. • A 747 airliner reaches its takeoff speed of 180 mi/h in 30.0 s. What is its average acceleration?

29. • At the starting gun, a runner accelerates at 1.9 m/s^2 for 2.2 s. The runner's acceleration is zero for the rest of the race. What is the speed of the runner **(a)** at $t = 2.0$ s, and **(b)** at the end of the race?

30. • A jet makes a landing traveling due east with a speed of 115 m/s. If the jet comes to rest in 13.0 s, what is the magnitude and direction of its average acceleration?

31. • A car is traveling due north at 20.4 m/s. Find the velocity of the car after 5.00 s if its acceleration is **(a)** 1.80 m/s^2 due north, or **(b)** 2.30 m/s^2 due south.

32. •• A motorcycle moves according to the velocity-versus-time graph shown in **Figure 2–28**. Find the average acceleration of the motorcycle during each of the segments, A, B, and C.

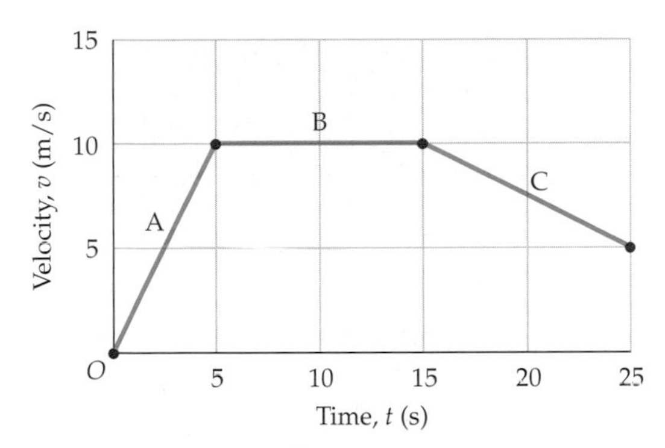

▲ **FIGURE 2–28** Problem 32

33. •• A person on horseback moves according to the velocity-versus-time graph shown in **Figure 2–29**. Find the displacement of the person for each of the segments A, B, and C.

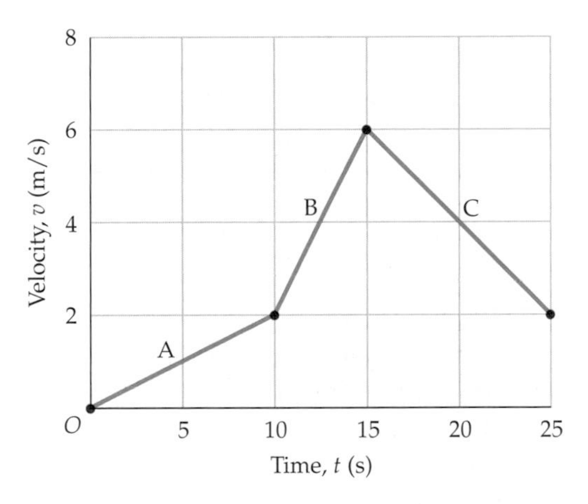

▲ **FIGURE 2–29** Problem 33

34. •• Running with an initial velocity of +11 m/s, a horse has an average acceleration of -1.81 m/s^2. How long does it take for the horse to decrease its velocity to +6.5 m/s?

35. •• **IP** Assume that the brakes in your car create a constant deceleration of 4.2 m/s^2 regardless of how fast you are driving. If you double your driving speed from 16 m/s to 32 m/s, **(a)** does the time required to come to a stop increase by a factor of two or a factor of four? Explain. **(b)** Verify your answer to part (a) by calculating the stopping times.

36. •• **IP** In the previous problem, **(a)** does the distance needed to stop increase by a factor of two or a factor of four? Explain. **(b)** Verify your answer to part (a) by calculating the stopping distances.

37. •• As a train accelerates away from a station, it reaches a speed of 5.2 m/s in 5.0 s. If the train's acceleration remains constant, what is its speed after an additional 6.0 s has elapsed?

38. •• A particle has an acceleration of +6.24 m/s^2 for 0.300 s. At the end of this time the particle's velocity is +9.31 m/s. What was the particle's initial velocity?

Section 2–5 Motion With Constant Acceleration

39. • Landing with a speed of 115 m/s, and traveling due south, a jet comes to rest in 7.00×10^2 m. Assuming the jet slows with constant acceleration, find the magnitude and direction of its acceleration.

40. • When you see a traffic light turn red you apply the brakes until you come to a stop. If your initial speed was 12 m/s, and you were heading due west, what was your average velocity during braking? Assume constant deceleration.

41. •• Suppose the car in the previous problem comes to rest in 35 m. How much time does this take?

42. •• Starting from rest, a boat increases its speed to 4.30 m/s with constant acceleration. **(a)** What is the boat's average speed? **(b)** If it takes the boat 5.00 s to reach this speed, how far has it traveled?

43. •• A cheetah accelerates from rest to 25 m/s in 6.2 s. Assuming constant acceleration, **(a)** how far has the cheetah run in this time? **(b)** How far has the cheetah run in 3.1 s?

Section 2–6 Applications of the Equations of Motion

44. • A child slides down a hill on a toboggan with an acceleration of 1.5 m/s^2. If she starts at rest, how far has she traveled in **(a)** 1.0 s, **(b)** 2.0 s, and **(c)** 3.0 s?

45. • On a ride called the Detonator at Worlds of Fun in Kansas City, passengers accelerate straight downward from zero to 45 mi/h in 1.0 seconds. What is the average acceleration of the passengers on this ride?

▲ The Detonator (Problem 45)

46. • Air bags are designed to deploy in 10 ms. Estimate the acceleration of the front surface of the bag as it expands. Express your answer in terms of the acceleration of gravity g.

47. •• Two cars drive on a straight highway. At time $t = 0$, car 1 passes mile marker 0 traveling due east with a speed of 20.0 m/s. At the same time, car 2 is 1.0 km east of mile marker 0 traveling at 30.0 m/s due west. Car 1 is speeding up with an acceleration of magnitude 2.5 m/s^2, and car 2 is slowing down with an acceleration of magnitude 3.2 m/s^2. Write x-versus-t equations of motion for both cars.

48. •• On October 9, 1992, a 27-pound meteorite struck a car in Peekskill, NY, leaving a dent 22 cm deep in the trunk. If the meteorite struck the car with a speed of 550 m/s, what was the magnitude of its deceleration, assuming it to be constant?

49. •• A rocket blasts off and moves straight upward from the launch pad with constant acceleration. After 3.0 s the rocket

is at a height of 80.0 m. **(a)** What is the acceleration of the rocket? **(b)** What is its speed at this time?

50. •• **IP** You are driving through town at 12.0 m/s when suddenly a ball rolls out in front of you. You apply the brakes and begin decelerating at 3.5 m/s^2. **(a)** How far do you travel before stopping? **(b)** When you have traveled only half the distance in part (a), is your speed 6.0 m/s, greater than 6.0 m/s, or less than 6.0 m/s? Support your answer with a calculation.

51. •• **IP** Referring to the previous problem, **(a)** how much time does it take to stop? **(b)** After braking half the time found in part (a), is your speed 6.0 m/s, greater than 6.0 m/s, or less than 6.0 m/s? Support your answer with a calculation.

52. •• **IP** When a chameleon captures an insect, its tongue can extend 16 cm in 0.10 s. **(a)** Find the acceleration of the chameleon's tongue, assuming it to be constant. **(b)** In the first 0.050 s, does the tongue extend 8.0 cm, more than 8.0 cm, or less than 8.0 cm? Support your conclusion with a calculation.

▲ It's not polite to reach! (Problem 52)

53. •• Coasting due west on your bicycle at 8.0 m/s, you encounter a sandy patch of road 7.2 m across. When you leave the sandy patch your speed has been reduced to 6.5 m/s. Assuming the bicycle slows with constant acceleration, what was its acceleration in the sandy patch? Give both magnitude and direction.

54. •• **IP** Referring to the previous problem, if you had entered the same sandy patch with a lower speed, say 7.0 m/s, would the patch have the same effect on your speed? That is, assuming the sandy patch causes the same acceleration, does your speed decrease by 1.5 m/s, more than 1.5 m/s, or less than 1.5 m/s? Justify your answer with a calculation.

55. •• A boat is cruising at a constant speed of 2.2 m/s when it is shifted into neutral. After coasting 10.0 m the engine is engaged again, and the boat resumes cruising at the reduced speed of 1.6 m/s. How long did it take for the boat to coast the 10.0 m?

56. •• A model rocket rises with constant acceleration to a height of 3.2 m, at which point its speed is 26.0 m/s. **(a)** How much time does it take for the rocket to reach this height? **(b)** What was the rocket's acceleration? **(c)** Find the height and speed of the rocket 0.10 s after launch.

57. •• The infamous chicken is dashing toward home plate with a speed of 6.0 m/s when he decides to hit the dirt. The chicken slides for 1.2 s, just reaching the plate as he stops (safe, of course). **(a)** What is the magnitude and direction of the chicken's acceleration? **(b)** How far did the chicken slide?

58. •• A bicyclist is finishing his repair of a flat tire when a friend rides by at 3.5 m/s. Two seconds later, the bicyclist hops on his bike and accelerates at 2.4 m/s^2 until he catches his friend. **(a)** How much time does it take until he catches his friend? **(b)** How far has he travled in this time? **(c)** What is his speed when he catches up?

59. ••• In a physics lab, students measure the time it takes a small cart to slide a distance of 1.00 m on a smooth track inclined at an angle θ above the horizontal. Their results are given in the following table.

θ,°	10.0°	20.0°	30.0°
time, s	1.08	0.770	0.640

(a) Find the acceleration of the cart for each angle. **(b)** Show that your results for part (a) are in close agreement with the formula, $a = g \sin \theta$. (We will derive this formula in Chapter 5.)

Section 2–7 Freely Falling Objects

60. • Legend has it that Isaac Newton was hit on the head by a falling apple, thus triggering his thoughts on gravity. Assuming the story to be true, estimate the speed of the apple when it struck Newton.

61. • The cartoon below shows a car in free fall. Is the statement made in the cartoon accurate? Justify your answer.

"IT GOES FROM ZERO TO SIXTY IN ABOUT THREE SECONDS."

62. • Referring to the cartoon in Problem 61, how long would it take for the car to go from 0 to 30 mi/h?

63. • Michael Jordan's vertical leap is reported to be 48 inches. What is his takeoff speed?

64. • Seagulls are often observed dropping clams and other shellfish from a height to the rocks below, as a means of opening the shells. If a seagull drops a shell from rest at a height of 14 m, how fast is the shell moving when it hits the rocks?

65. • A volcano ejects a lava bomb straight upward with an initial speed of 23 m/s. Taking upward to be the positive direction, find the speed and velocity of the lava bomb **(a)** 2.0 seconds and **(b)** 3.0 seconds later.

66. • The first active volcano observed outside the Earth was discovered in 1979 on Io, one of the moons of Jupiter. The volcano was observed to be ejecting material to a height of about 2.00×10^5 m. Given that the acceleration of gravity on Io is 1.80 m/s^2, find the initial velocity of the ejected material.

67. • **BIO** Here's something you *can* try at home—an experiment to measure your reaction time. Have a friend hold a meter stick (or a ruler) by one end, letting the other end hang down vertically. At the lower end, hold your thumb and index finger on either side of the stick, ready to grip it. Have your friend release the meter stick without warning. Catch it as quickly as you can. If you catch the meter stick 5.2 cm (~2 inches) from the lower end, what is your reaction time?

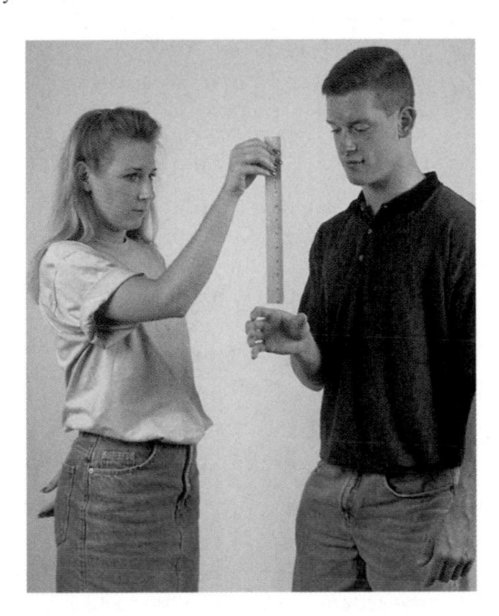

▲ How fast are your reactions? (Problem 67)

68. •• Bill steps off a 3.0-m-high diving board and drops to the water below. At the same time, Ted jumps upward with a speed of 4.2 m/s from a 1.0-m-high diving board. Choosing the origin to be at the water's surface, and upward to be the positive direction, write x-versus-t equations of motion for both Bill and Ted.

69. •• Repeat the previous problem, this time with the origin 3.0 m above the water, and with downward as the positive direction.

70. •• On a hot summer day several swimmers decide to dive from a railroad bridge into the river below. The swimmers step off the bridge and hit the water approximately 1.5 s later. **(a)** How high is the bridge? **(b)** How fast are the swimmers moving when they hit the water?

71. •• **IP** Referring to Problem 64, suppose a crow drops a shell from half the seagull's drop height. When the crow's shell hits the rocks, is it moving half as fast, less than half as fast, or more than half as fast, as the seagull's shell? Support your answer with a calculation.

72. •• A batter pops a ball straight up. If the ball returns to the height from which it was hit 4.0 s later, what was its initial speed?

73. •• The world's highest fountain of water is located, appropriately enough, in Fountain Hills, Arizona. The fountain rises to a height of 560 ft (5 feet higher than the Washington Monument). **(a)** What is the initial speed of the water? **(b)** How long does it take for water to reach the top of the fountain?

74. •• Wrongly called for a foul, an angry basketball player throws the ball straight down to the floor. If the ball bounces straight up and returns to the floor 2.5 s after first striking it, what was the ball's greatest height above the floor?

75. •• To celebrate a victory, a pitcher throws her glove straight upward with an initial speed of 6.0 m/s. **(a)** How long does it take for the glove to return to the pitcher? **(b)** How long does it take for the glove to reach its maximum height?

76. •• **IP** Standing at the edge of a cliff 30.0 m high, you drop a ball. Later, you throw a second ball downward with an initial speed of 10.0 m/s. Which ball has the greater *increase* in speed when it reaches the base of the cliff, or do both balls speed up by the same amount? Justify your answer with a calculation.

77. •• You shoot an arrow into the air. Two seconds later the arrow has gone straight upward to a height of 30.0 m. What was the arrow's initial speed?

78. •• While riding on an elevator descending with a constant speed of 3.0 m/s, you accidentally drop a book from under your arm. **(a)** How long does it take for the book to reach the elevator floor, 1.2 m below your arm? **(b)** What is the book's speed when it hits the elevator floor?

79. •• A hot air balloon is descending at a rate of 2.0 m/s when a passenger drops a camera. If the camera is 45 m above the ground when it is dropped, **(a)** how long does it take to reach the ground, and **(b)** what is its velocity just before it lands? Let upward be the positive direction for this problem.

80. •• **IP** You and a friend, standing side by side, step off a bridge at different times and fall for 1.6 s to the water below. Your friend goes first, and you follow when he has dropped a distance of 2.0 m. When your friend hits the water, is the separation between the two of you 2.0 m, less than 2.0 m, or more than 2.0 m? Verify your answer with a calculation.

81. ••• While sitting on a tree branch 10.0 m above the ground, you drop a chestnut. When the chestnut has fallen 2.5 m, you throw a second chestnut straight down. What initial speed must you give the second chestnut if they are both to reach the ground at the same time?

General Problems

82. • In a well-known Jules Verne novel, Phileas Fogg travels around the world in 80 days. What was Mr. Fogg's approximate average speed during his adventure?

83. • An astronaut on the Moon drops a rock straight downward from a height of 0.95 m. If the acceleration of gravity on the Moon is 1.62 m/s^2, what is the speed of the rock when it lands?

84. • You jump from the top of a boulder to the ground 2.0 m below. Estimate your deceleration on landing.

85. •• **IP** A youngster bounces straight up and down on a trampoline. Suppose she doubles her initial speed from 2.0 m/s to 4.0 m/s. **(a)** By what factor does her time in the air increase? **(b)** By what factor does her maximum height increase? **(c)** Verify your answers to parts (a) and (b) with an explicit calculation.

86. •• At the 18th green of the U. S. Open you need to make a 20.0 ft putt to win the tournament. When you hit the ball, giving it an initial speed of 1.57 m/s, it stops 6.00 ft short of the hole. **(a)** Assuming the deceleration caused by the grass is constant, what should the initial speed have been to just make the putt? **(b)** What initial speed do you need to make the remaining 6.00-ft putt?

87. •• The leader in a bicycle race passes your viewing position at $t = 0$, traveling with a constant speed of 15 m/s. Ten seconds later the next racer goes by with a constant speed of 20.0 m/s. **(a)** Sketch x-versus-t plots for the two bicyclists. **(b)** At what time does the second bicycle catch up with the first? **(c)** How far are the bikes from your position at this time?

88. •• Referring to the previous problem, suppose your viewing position is 565 m from the finish line. What is the minimum speed required of the second bicyclist if he is to catch the leader by the end of the race?

89. •• A popular entertainment at some carnivals is the blanket toss (see photo, p. 38). If a person is thrown to a maximum height of 30.0 ft, how long does she spend in the air during the toss?

90. •• Referring to Conceptual Checkpoint 2–5, find the separation between the rocks at $t = 1.0$ s, $t = 2.0$ s, and $t = 3.0$ s. Verify that the separation increases linearly with time.

91. •• **IP** A seagull, ascending straight upward at 5.4 m/s, drops a shell when it is 14 m above the ground. **(a)** What is the magnitude and direction of the shell's acceleration just after it is released? **(b)** Find the maximum height above the ground reached by the shell. **(c)** How long does it take for the shell to return to a height of 14 m above the ground? **(d)** What is the speed of the shell at this time?

92. •• A doctor, preparing to give a patient an injection, squirts a small amount of liquid straight upward from a syringe. If the liquid emerges with a speed of 1.5 m/s, **(a)** how long does it take for it to return to the level of the syringe? **(b)** What is the maximum height of the liquid above the syringe?

93. •• Watching Old Faithful erupt, you notice that it takes 1.65 s for water to emerge from the geyser and reach its maximum height. **(a)** What is the height of the geyser, and **(b)** what is the initial speed of the water?

▲ Old Faithful (Problem 93)

94. •• **IP** A ball is thrown upward with an initial speed v_0. When it reaches the top of its flight, at a height h, a second ball is thrown upward with the same initial velocity. **(a)** Sketch an x-versus-t plot for each ball. **(b)** From your graph, decide whether the balls cross paths at $h/2$, above $h/2$, or below $h/2$. **(c)** Find the height where the paths cross.

95. •• A hot air balloon has just lifted off and is rising at the constant rate of 2.0 m/s. Suddenly, one of the passengers realizes she has left her camera on the ground. A friend picks it up and tosses it straight upward with an initial speed of 10.0 m/s. If the passenger is 2.5 m above her friend when the camera is tossed, how high is she when the camera reaches her?

96. •• In the previous problem, what is the minimum initial speed of the camera if it is to just reach the passenger?

97. ••• Weights are tied to each end of a 20.0-cm string. You hold one weight in your hand, and let the other hang vertically a height h above the floor. When you release the weight in your hand, the two weights strike the ground one after the other with audible thuds. Find the value of h for which the time between release and the first thud is equal to the time between the first thud and the second thud.

98. ••• A ball, dropped from rest, covers three-quarters of the distance to the ground in the last second of its fall. **(a)** From what height was the ball dropped? **(b)** What was the total time of fall?

99. ••• A stalactite on the roof of a cave drips water at a steady rate to a pool 4.0 m below. As one drop of water hits the pool, a second drop is in the air, and a third is just detaching from the stalactite. **(a)** What is the position and velocity of the second drop when the first drop hits the pool? **(b)** How many drops per minute fall into the pool?

100. ••• You drop a ski glove from a height h onto fresh snow, and it sinks to a depth d before coming to rest. **(a)** In terms of g and h, what is the speed of the glove when it reaches the snow? **(b)** What is the acceleration of the glove as it moves through the snow, assuming it to be constant? Give your answer in terms of g, h, and d.

101. ••• To find the height of an overhead power line you throw a ball straight upward. The ball passes the line on the way up after 0.75 s, and passes it again on the way down after a total of 1.5 s. What is the height of the power line, and the initial speed of the ball?

102. ••• Suppose the first rock in Conceptual Checkpoint 2–5 drops through a height h before the second rock is released from rest. Show that the separation between the rocks is given by the following expression:

$$h + \left(\sqrt{2gh}\right)t$$

In this result, the time t is measured from the time the second rock is dropped.

103. ••• An arrow is fired with a speed of 20.0 m/s at a block of Styrofoam resting on a smooth surface. The arrow penetrates a certain distance into the block before coming to rest relative to it. During this process the arrow's deceleration has a magnitude of 1550 m/s^2 and the block's acceleration has a magnitude of 450 m/s^2. **(a)** How long does it take for the arrow to stop moving with respect to the block? **(b)** What is the common speed of the arrow and block when this happens? **(c)** How far into the block does the arrow penetrate?

104. ••• Sitting in a second-story apartment, a physicist notices a ball moving straight upward just outside her window. The ball is visible 0.25 s as it moves a distance of 1.05 m from the bottom to the top of the window. **(a)** How long does it take before the ball reappears? **(b)** What is the greatest height of the ball above the top of the window?

3 Vectors in Physics

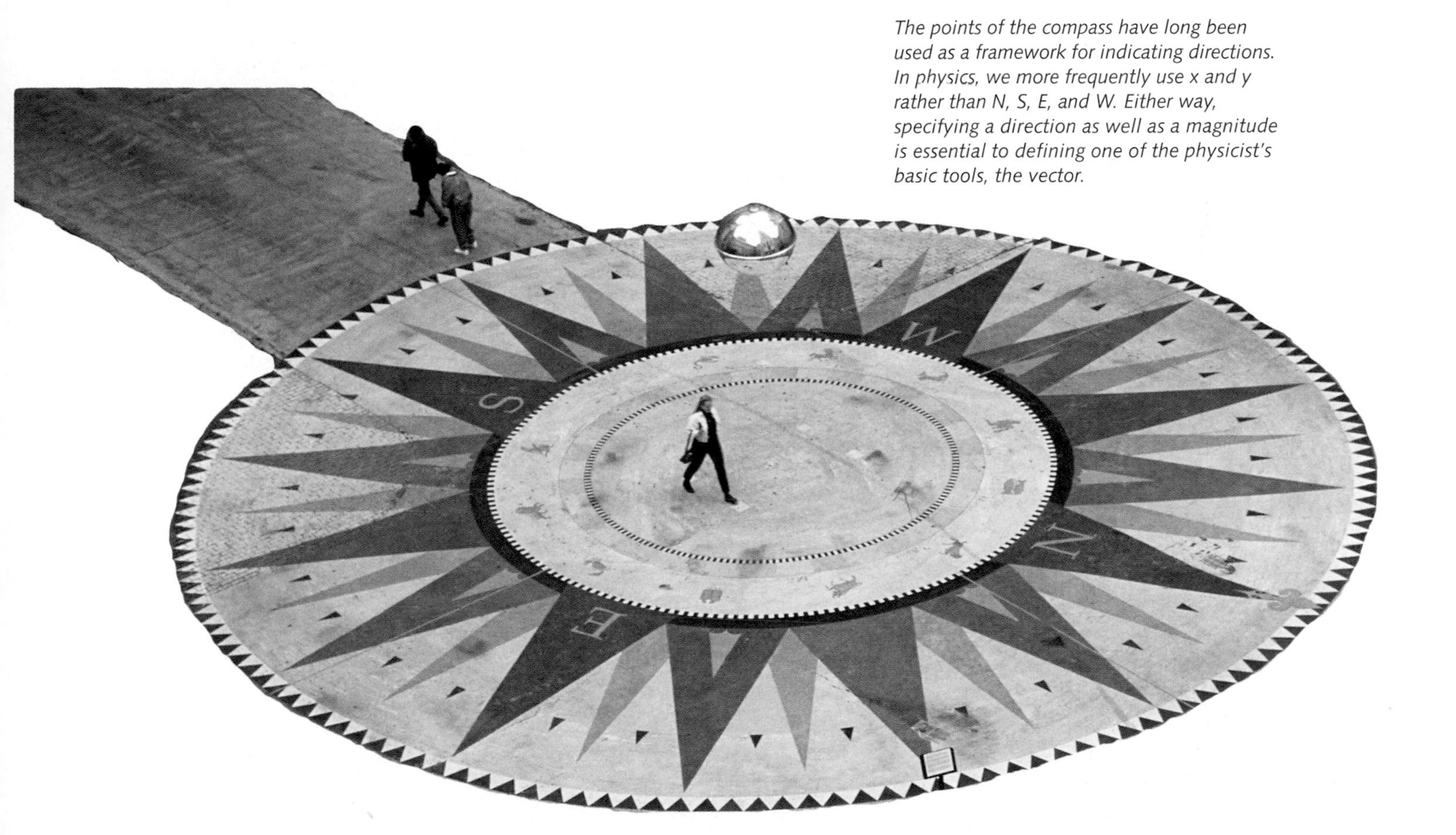

The points of the compass have long been used as a framework for indicating directions. In physics, we more frequently use x and y rather than N, S, E, and W. Either way, specifying a direction as well as a magnitude is essential to defining one of the physicist's basic tools, the vector.

One of the most important mathematical tools used in this book is the vector. In the next chapter, for example, we use vectors to extend our study of motion from one dimension to two dimensions. More generally, vectors are *indispensable* when a physical quantity has a direction associated with it. Suppose, for example, that a pilot wants to fly from Denver to Dallas. If the air is still, the pilot can simply head the plane toward the destination. If there is a wind blowing from west to east, however, the pilot must use vectors to determine the correct heading so that the plane and its passengers will arrive in Dallas and not Little Rock.

In this chapter we discuss what a vector is, how it differs from a scalar, and how it can represent a physical quantity. We also show how to find the components of a vector and how to add and subtract vectors. All of these

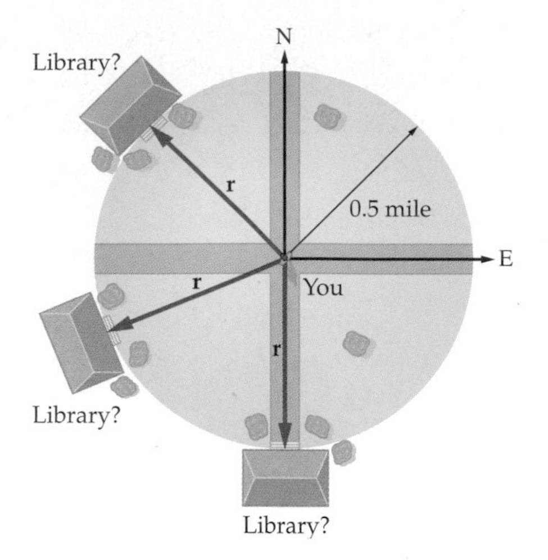

▲ **FIGURE 3–1 Distance and direction**
If you know only that the library is 0.5 mi from you, it could lie anywhere on a circle of radius 0.5 mi. If, instead, you are told the library is 0.5 mi northwest, you know its precise location.

techniques are used time and again throughout the book. Other useful aspects of vectors, such as how to multiply them, will be presented in later chapters when the need arises.

3–1 Scalars versus Vectors

Numbers can represent many quantities in physics. For example, a numerical value, together with the appropriate units, can specify the volume of a container, the temperature of the air, or the time of an event. In physics, a number and its units is referred to as a **scalar**:

- A scalar is a number with units. It can be positive, negative, or zero.

Sometimes, however, a scalar isn't enough to adequately describe a physical quantity—in many cases, a direction is needed as well. For example, suppose you're walking in an unfamiliar city and you want directions to the library. You ask a passerby, "Do you know where the library is?" If the person replies "Yes," and walks on, he hasn't been too helpful. If he says, "Yes, it is half a mile from here," that is more helpful, but you still don't know where it is. The library could be anywhere on a circle of radius one-half mile, as shown in **Figure 3–1**. To pin down the location, you need a reply such as, "Yes, the library is half a mile northwest of here." With both a distance *and* a direction, you know the location of the library.

Thus, if you walk northwest for half a mile you arrive at the library, as indicated by the arrow in Figure 3–1. The arrow points in the direction traveled, and its **magnitude**, 0.5 mi in this case, represents the distance covered. In general, a quantity that is specified by both a *magnitude* and a *direction* is represented by a **vector**:

- A vector is a mathematical quantity with both a magnitude and a direction.

Other vector quantities include the velocity and acceleration of an object. For example, the magnitude of a velocity vector is its speed, and its direction is the direction of motion, as we shall see later in this chapter.

When we indicate a vector on a graph, we draw an arrow, as in Figure 3–1. To indicate a vector with a written symbol, we use **boldface** for the vector itself, and *italic* for its magnitude. For example, the vector in Figure 3–1 is designated by the symbol **r**, and its magnitude is $r = 0.5$ mi. (Sometimes we represent a vector by an arrow labeled with the appropriate magnitude.) It is common to indicate a vector in handwritten material by drawing a small arrow over the vector's symbol, as follows: $\vec{r}$.

▲ The information given by this sign includes both a distance and a direction for each city. In effect, the sign defines a displacement vector for each of these destinations.

3–2 The Components of a Vector

When we discussed directions for finding a library in the previous section we pointed out that knowing the magnitude and direction angle—0.5 mi miles northwest—gives its precise location. We left out one key element in actually *getting* to the library, however. In most cities it would not be possible to simply walk in a straight line for 0.5 mi directly to the library, since to do so would take you through buildings where there are no doors, through people's backyards, and through all kinds of other obstructions. In fact, if the city streets are laid out along north–south and east–west directions, you might instead walk west for a certain distance, then turn and proceed north an equal distance until you reach the library, as illustrated in **Figure 3–2**. What you have just done is "resolved" the displacement vector **r** between you and the library into east–west and north–south "components."

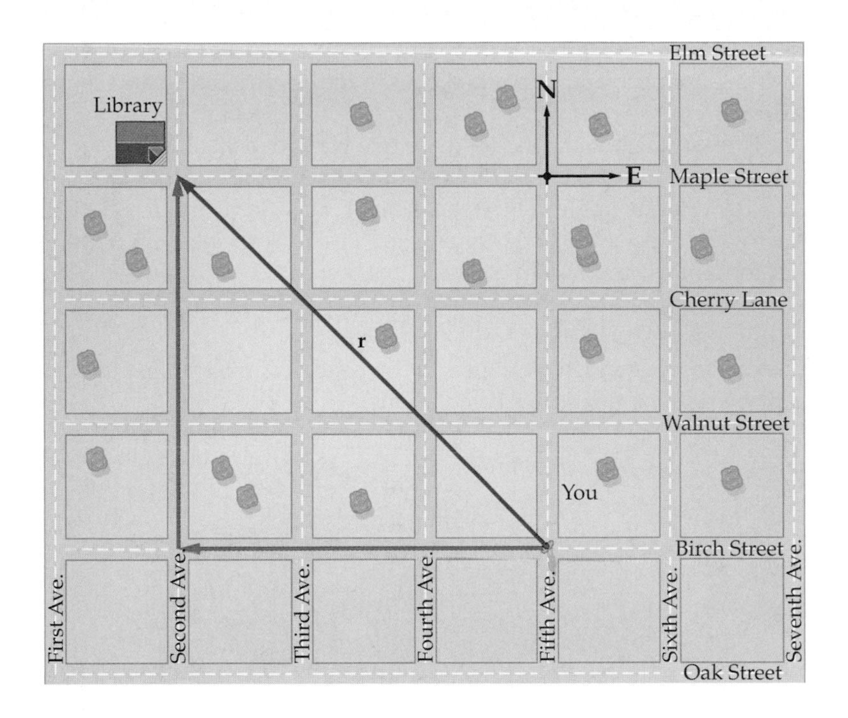

◀ **FIGURE 3–2 A walk along city streets to the library**
By taking the indicated path, we have "resolved" the vector **r** into east–west and north–south components.

In general, to find the components of a vector we need to set up a coordinate system. In two dimensions we choose an origin, O, and a positive direction for both the x and the y axes, as in **Figure 3–3**. If the system were three dimensional, we would also indicate a z axis.

Now, a vector is defined by its magnitude (indicated by the length of the arrow representing the vector) and its direction. For example, suppose an ant leaves its nest at the origin and, after foraging for some time, is at the location given by the vector **r** in **Figure 3–4 (a)**. This vector has a magnitude $r = 1.50$ m, and points in a direction $\theta = 25.0°$ above the x axis. Equivalently, **r** can be defined by saying that it extends a distance r_x in the x direction and a distance r_y in the y direction, as shown in **Figure 3–4 (b)**. The quantities r_x and r_y are referred to as the x and y **scalar components** of the vector **r**.

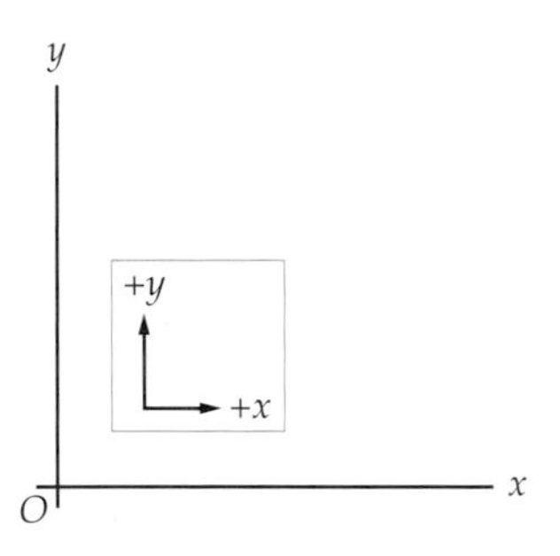

▲ **FIGURE 3–3 A two-dimensional coordinate system**
The positive x and y directions are indicated.

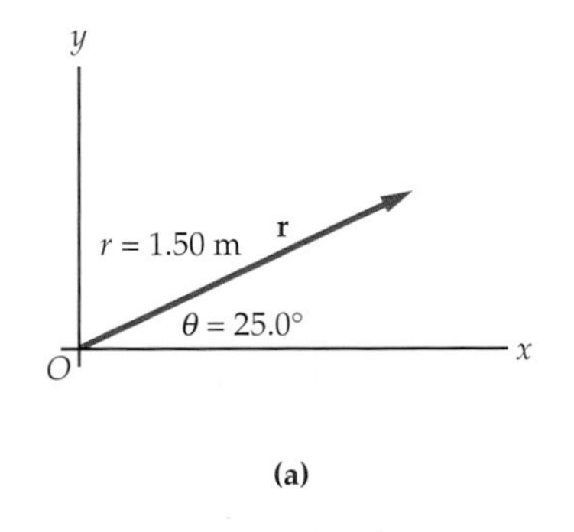

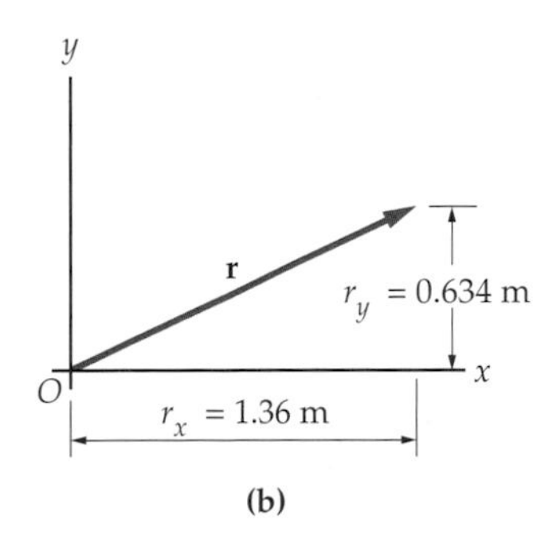

◀ **FIGURE 3–4 A vector and its scalar components**
(a) The vector **r** is defined by its length (1.50 m) and its direction angle (25.0°) measured counterclockwise from the positive x axis. **(b)** Alternatively, the vector **r** can be defined by its x component (1.36 m) and its y component (0.634 m).

We can find r_x and r_y by using standard trigonometric relations, as shown in the Problem Solving Note on p. 56. Referring to Figure 3–4 (b), we see that

$$r_x = r \cos 25.0° = (1.50 \text{ m})(0.906) = 1.36 \text{ m}$$

and

$$r_y = r \sin 25.0° = (1.50 \text{ m})(0.423) = 0.634 \text{ m}$$

PROBLEM SOLVING NOTE
A Vector and its Components

Given the magnitude and direction of a vector, find its components.

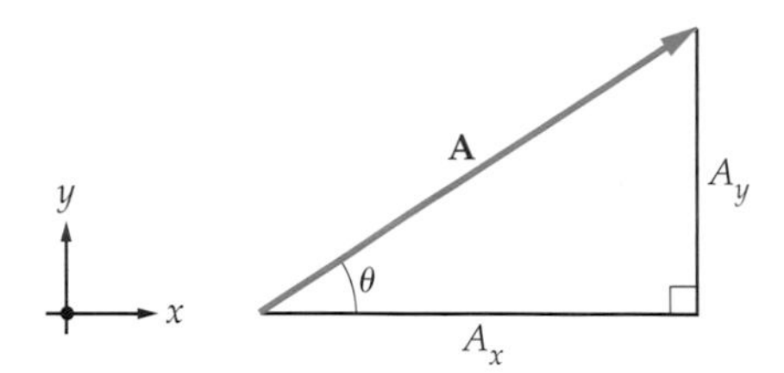

$A_x = A\cos\theta$
$A_y = A\sin\theta$

Given the components of a vector, find its magnitude and direction.

$A = \sqrt{A_x^{\,2} + A_y^{\,2}}$

$\theta = \tan^{-1}\dfrac{A_y}{A_x}$

Thus, we can say that the ant's final displacement is equivalent to what it would be if the ant had simply walked 1.36 m in the x direction and then 0.634 m in the y direction.

To show the equivalence of these two ways of describing a vector, let's start with the components of **r**, as determined previously, and use them to calculate the magnitude r and the angle θ. First, note that r_x, r_y, and r form a right triangle with r as the hypotenuse. Thus, we can use the Pythagorean theorem (Appendix A) to find r in terms of r_x and r_y. This gives

$$r = \sqrt{r_x^2 + r_y^2} = \sqrt{(1.36\text{ m})^2 + (0.634\text{ m})^2} = \sqrt{2.25\text{ m}^2} = 1.50\text{ m}$$

as expected. Second, we can use any two sides of the triangle to obtain the angle θ, as shown in the next three calculations:

$$\theta = \sin^{-1}\frac{0.634\text{ m}}{1.50\text{ m}} = \sin^{-1}0.423 = 25.0°$$

$$\theta = \cos^{-1}\frac{1.36\text{ m}}{1.50\text{ m}} = \cos^{-1}0.907 = 25.0°$$

$$\theta = \tan^{-1}\frac{0.634\text{ m}}{1.36\text{ m}} = \tan^{-1}0.466 = 25.0°$$

In some situations we know a vector's magnitude and direction; in other cases we are given the vector's components. You will find it useful to be able to convert quickly and easily from one description of a vector to the other using trigonometric functions and the Pythagorean theorem.

EXAMPLE 3–1 Determining the Height of a Cliff

Real World Physics

In the Jules Verne novel *Mysterious Island*, Captain Cyrus Harding wants to find the height of a cliff. He stands with his back to the base of the cliff, then marches straight away from it for 5.00×10^2 ft. At this point he lies on the ground and measures the angle from the horizontal to the top of the cliff. If the angle is 34.0°, **(a)** how high is the cliff? **(b)** What is the straight-line distance from Captain Harding to the top of the cliff?

Picture the Problem

Our sketch shows Cyrus Harding making his measurement of the angle, $\theta = 34.0°$, to the top of the cliff. The relevant triangle for this problem is also indicated. Note that the opposite side of the triangle is the height of the cliff, h; the adjacent side is the distance from the base of the cliff to Harding, $b = 5.00 \times 10^2$ ft; and finally, the hypotenuse is the distance, d, from Harding to the top of the cliff.

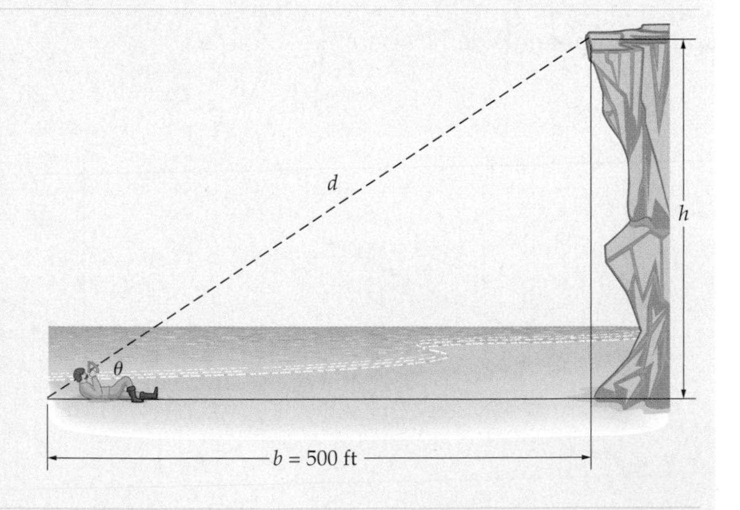

Strategy

The tangent of θ is the height of the triangle divided by the base: $\tan\theta = h/b$. Since we know both θ and the base, we can find the height using this relation. Similarly, the distance from Harding to the top of the cliff can be obtained by solving $\cos\theta = b/d$ for d.

Solution

Part (a)

1. Use $\tan\theta = h/b$ to solve for the height of the cliff, h: $\quad h = b\tan\theta = (500\text{ ft})\tan 34.0° = 337\text{ ft}$

Part (b)

2. Similarly, use $\cos\theta = b/d$ to solve for the distance d from Captain Harding to the top of the cliff: $\quad d = \dfrac{b}{\cos\theta} = \dfrac{500\text{ ft}}{\cos 34.0°} = 603\text{ ft}$

Insight

An alternative way to solve part **(b)** is to use the Pythagorean theorem; $d = \sqrt{h^2 + b^2} = \sqrt{(337 \text{ ft})^2 + (500 \text{ ft})^2} = 603$ ft. Thus, if we let **r** denote the vector from Cyrus Harding to the top of the cliff, as shown here, its magnitude is 603 ft and its direction is 34.0° above the x axis. Alternatively, the x component of **r** is 500 ft and its y component is 337 ft.

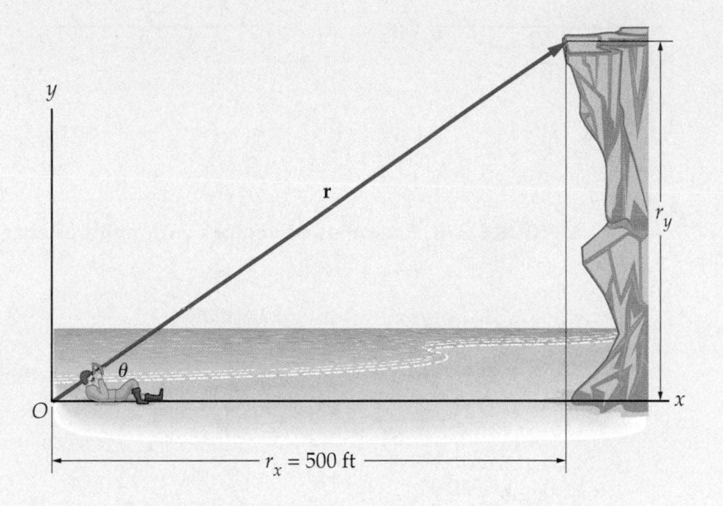

Practice Problem

What angle would Cyrus Harding have found if he had walked 6.00×10^2 ft from the cliff to make his measurement? [**Answer:** $\theta = 29.3°$]

Some related homework problems: Problem 1, Problem 13

EXERCISE 3–1

(a) Find A_x and A_y for the vector **A** with magnitude and direction given by $A = 3.5$ m and $\theta = 66°$, respectively.
(b) Find B and θ for the vector **B** with components $B_x = 75.5$ m and $B_y = 6.20$ m.

Solution

(a) $A_x = 1.4$ m, $A_y = 3.2$ m
(b) B 5 75.8 m, u 5 4.69°

Next, how do you determine the correct sign for the x and y components of a vector? This can be done by considering the right triangle formed by A_x, A_y, and **A**, as shown in **Figure 3–5**. To determine the sign of A_x, start at the tail of the vector and move along the x axis toward the right angle. If you are moving in the positive x direction, then A_x is positive ($A_x > 0$); if you are moving in the negative x direction, then A_x is negative ($A_x < 0$). For the y component, start at the right angle and move toward the tip of the arrow. A_y is positive or negative depending on whether you are moving in the positive or negative y direction.

For example, consider the vector shown in **Figure 3–6 (a)**. In this case, $A_x > 0$ and $A_y < 0$, as indicated in the figure. Similarly, the signs of A_x and A_y are given in **Figures 3–6 (b, c, d)** for the vectors shown there. Be sure to verify each of these cases by applying the rules just given. As we continue our study of physics, it is important to be able to find the components of a vector *and* to assign to them the correct signs.

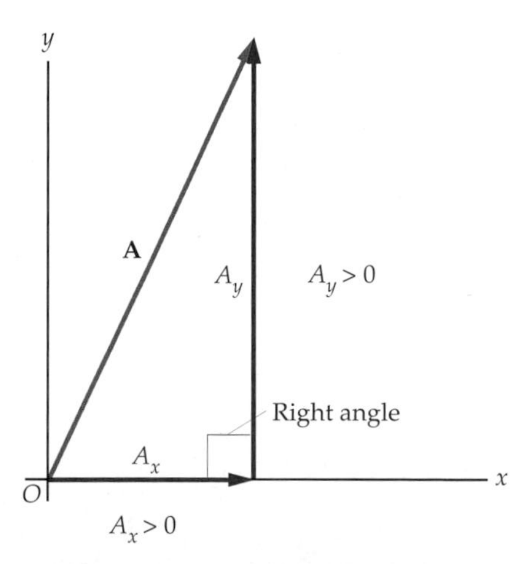

▲ FIGURE 3–5 A vector whose *x* and *y* components are positive

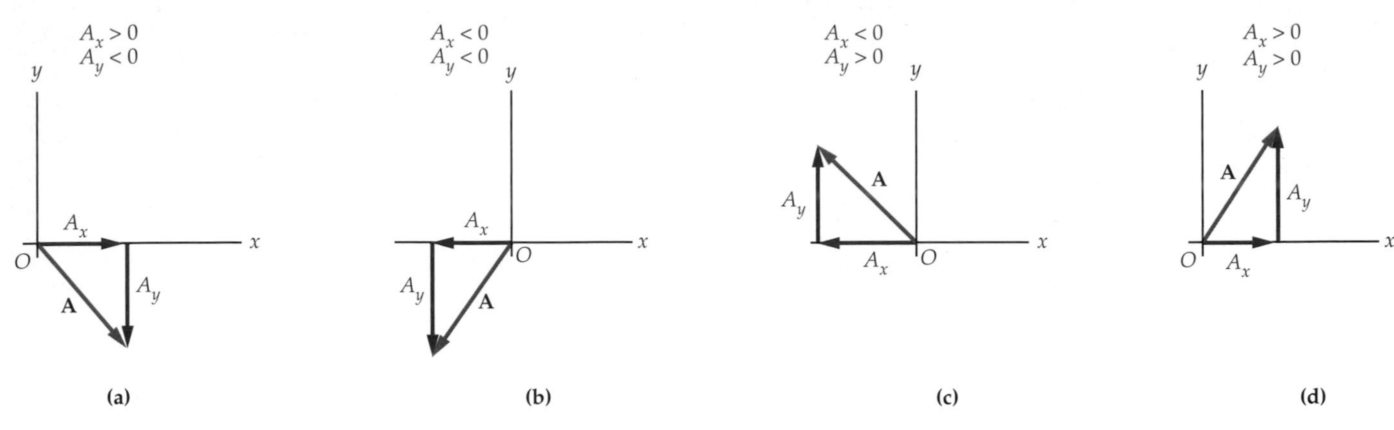

▲ **FIGURE 3–6 Examples of vectors with components of different signs**

EXERCISE 3–2

The vector **A** has a magnitude of 7.25 m. Find its components for direction angles of
(a) $\theta = 5.00°$,
(b) $\theta = 125°$,
(c) $\theta = 245°$, and
(d) $\theta = 340.0°$.

Solution

(a) $A_x = 7.22$ m, $A_y = 0.632$ m
(b) $A_x = -4.16$ m, $A_y = 5.94$ m
(c) $A_x = -3.06$ m, $A_y = -6.57$ m
(d) $A_x = 6.81$ m, $A_y = -2.48$ m

Be careful when you use your calculator to determine the direction angle, θ, because you may need to add 180° to get the correct answer. For example, if $A_x = -0.50$ m and $A_y = 1.0$ m, your calculator will give the following result:

$$\theta = \tan^{-1}\left(\frac{1.0\text{ m}}{-0.50\text{ m}}\right) = \tan^{-1}(-2.0) = -63°$$

Is this angle correct? The way to check is to sketch **A**. When you do, your drawing is similar to Figure 3–6 (c), and thus the direction angle of **A** should be between 90° and 180°. To obtain the correct angle, add 180° to the calculator's result,

$$\theta = -63° + 180° = 117°$$

This, in fact, is the correct direction angle for **A**.

EXERCISE 3–3

The vector **B** has components $B_x = -2.10$ m and $B_y = -1.70$ m. Find the direction angle, θ, for this vector.

Solution

$\tan^{-1}[(-1.70\text{ m})/(-2.10\text{ m})] = \tan^{-1}(1.70/2.10) = 39.0°, \theta = 39.0° + 180° = 219°$

Finally, in many situations the direction of a vector **A** is given by the angle θ, measured relative to the x axis, as in **Figure 3–7 (a)**. In these cases we know that

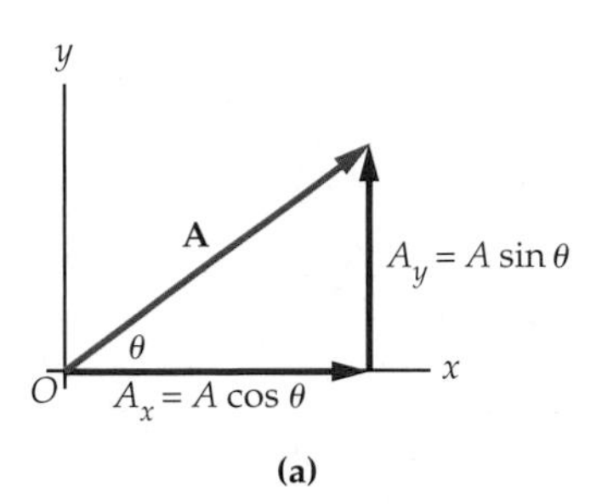

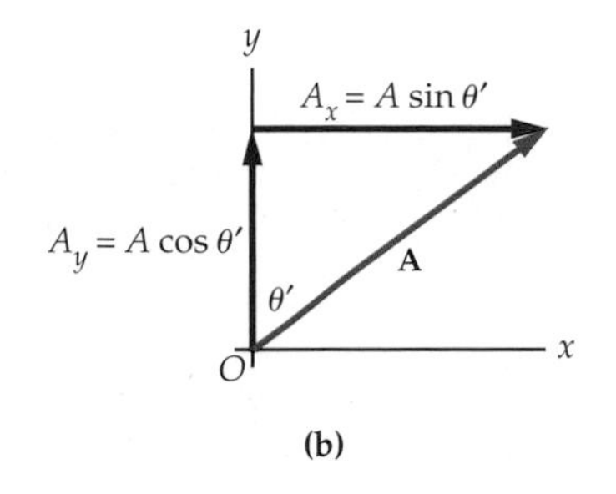

◀ **FIGURE 3–7 Vector angle**
Vector **A** and its components in terms of **(a)** the angle relative to the x axis and **(b)** the angle relative to the y axis.

$$A_x = A \cos \theta$$

and

$$A_y = A \sin \theta$$

On the other hand, we are sometimes given the angle between the vector and the y axis, as in **Figure 3–7 (b)**. If we call this angle θ', then it follows that

$$A_x = A \sin \theta'$$

and

$$A_y = A \cos \theta'$$

These two seemingly different results are actually in complete agreement. Note that $\theta + \theta' = 90°$, or $\theta' = 90° - \theta$. If we use the trigonometric identities given in Appendix A, we get

$$A_x = A \sin \theta' = A \sin(90° - \theta) = A \cos \theta$$

and

$$A_y = A \cos \theta' = A \cos(90° - \theta) = A \sin \theta$$

EXERCISE 3–4

If a vector's direction angle relative to the x axis is 35°, then its direction angle relative to the y axis is 55°. Find the components of a vector **A** of magnitude 5.2 m in terms of
(a) its direction relative to the x axis, and
(b) its direction relative to the y axis.

Solution

(a) $A_x = (5.2 \text{ m}) \cos 35° = 4.3 \text{ m}$, $A_y = (5.2 \text{ m}) \sin 35° = 3.0 \text{ m}$
(b) $A_x = (5.2 \text{ m}) \sin 55° = 4.3 \text{ m}$, $A_y = (5.2 \text{ m}) \cos 55° = 3.0 \text{ m}$

3–3 Adding and Subtracting Vectors

One important reason for determining the components of a vector is that they are useful in adding and subtracting vectors. In this section we begin by defining vector addition graphically, and then show how the same addition can be performed more concisely and accurately with components.

Adding Vectors Graphically

One day you open an old chest in the attic and find a treasure map inside. To locate the treasure, the map says that you must "Go to the sycamore tree in the backyard, march 5 paces north, then 3 paces east." If these two displacements are represented

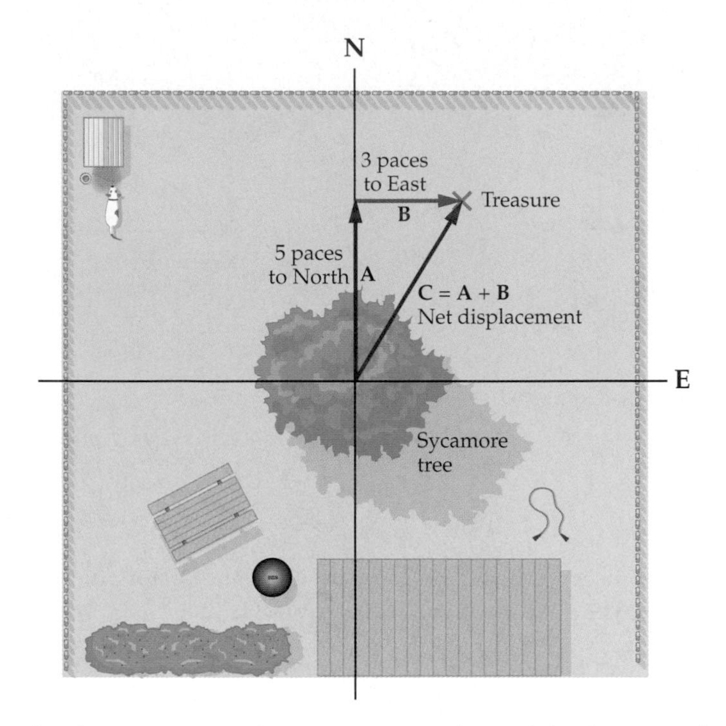

▶ **FIGURE 3–8 The sum of two vectors**
To go from the sycamore tree to the treasure, one must first go 5 paces north (**A**), and then 3 paces east (**B**). The net displacement from the tree to the treasure is **C** = **A** + **B**.

by the vectors **A** and **B** in **Figure 3–8**, the total displacement from the tree to the treasure is given by the vector **C**. We say that **C** is the *vector sum* of **A** and **B**; that is, **C** = **A** + **B**. In general, vectors are added graphically according to the following rule:

- To add the vectors **A** and **B**, place the tail of **B** at the head of **A**. The sum, **C** = **A** + **B**, is the vector extending from the tail of **A** to the head of **B**.

If the instructions to find the treasure were a bit more complicated—5 paces north, 3 paces east, then 4 paces southeast, for example—the path from the sycamore tree to the treasure would be like that shown in **Figure 3–9**. In this case, the total displacement, **D**, is the sum of the three vectors **A**, **B**, and **C**; that is, **D** = **A** + **B** + **C**. It follows that to add more than two vectors, we just keep placing the vectors head-to-tail, head-to-tail, and then draw a vector from the tail of the first vector to the head of the last vector, as in Figure 3–9.

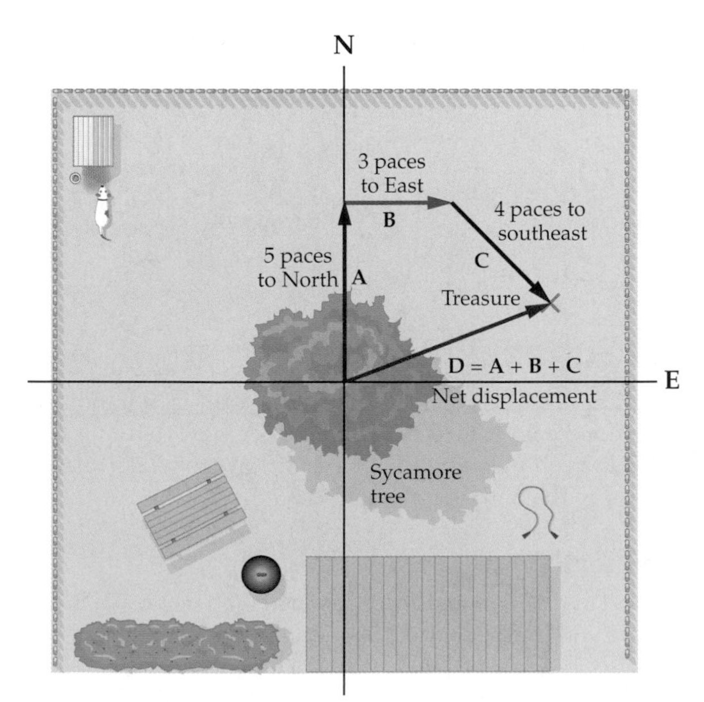

▶ **FIGURE 3–9 Adding several vectors**
Searching for a treasure that is 5 paces north (**A**), 3 paces east (**B**), and 4 paces southeast (**C**) of the sycamore tree. The net displacement from the tree to the treasure is **D** = **A** + **B** + **C**.

In order to place a given pair of vectors head-to-tail, it may be necessary to move the corresponding arrows. This is fine, as long as you don't change their length or their direction. After all, a vector is defined by its length and direction; if these are unchanged, so is the vector.

• A vector is defined by its magnitude and direction, regardless of its location.

For example, in **Figure 3–10** all of the vectors are the same, even though they are at different locations on the graph.

As an example of moving vectors, consider two vectors, **A** and **B**, and their vector sum **C**,

$$\mathbf{C} = \mathbf{A} + \mathbf{B}$$

as illustrated in **Figure 3–11 (a)**. By moving the arrow representing **B** so that its tail is at the origin, and moving the arrow for **A** so that its tail is at the head of **B**, we obtain the construction shown in **Figure 3–11 (b)**. From this graph we see that **C**, which is **A** + **B**, is also equal to **B** + **A**:

$$\mathbf{C} = \mathbf{A} + \mathbf{B} = \mathbf{B} + \mathbf{A}$$

That is, the sum of vectors is independent of the order in which the vectors are added.

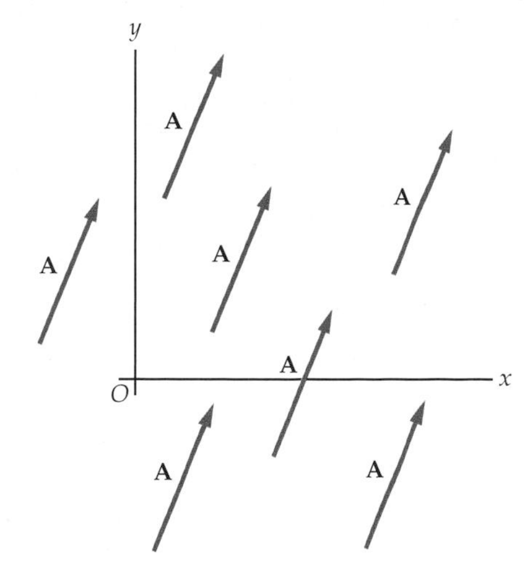

▲ **FIGURE 3–10 Identical vectors *A* at different locations**

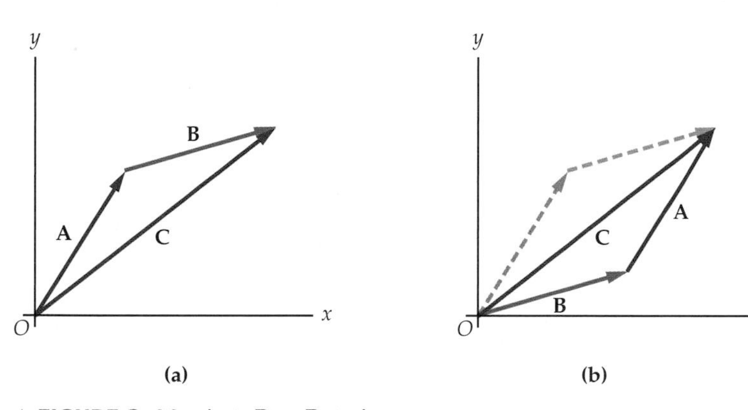

▲ **FIGURE 3–11** **A** + **B** = **B** + **A**
The vector **C** is equal to **(a)** **A** + **B** and **(b)** **B** + **A**.

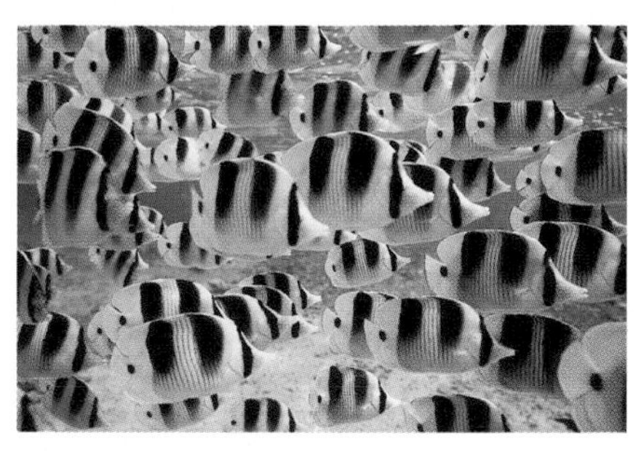

▲ To a good approximation, these butterfly fish are all moving in the same direction with the same speed. As a result, their velocity vectors are equal, even though their positions are different.

Now, suppose that **A** has a magnitude of 5.00 m and a direction of 60.0° above the x axis, and that **B** has a magnitude of 4.00 m and a direction of 20.0° above the x axis. These two vectors and their sum, **C**, are shown in **Figure 3–12**. The question is: What is the length and direction of **C**?

A graphical way to answer this question is to simply measure the length and direction of **C** in Figure 3–12. With a ruler, we find the length of **C** to be approximately 1.75 times the length of **A**, which means that **C** is roughly 1.75(5.00 m) = 8.75 m. Similarly, with a protractor we measure the angle θ to be about 45.0° above the x axis.

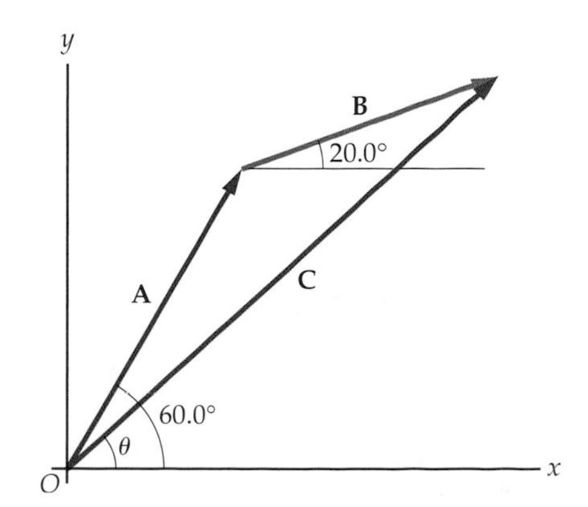

▲ **FIGURE 3–12 Graphical addition of vectors**
The vector **A** has a magnitude of 5.00 m and a direction angle of 60.0°; the vector **B** has a magnitude of 4.00 m and a direction angle of 20.0°. The magnitude and direction of **C** = **A** + **B** can be measured on the graph with a ruler and a protractor.

Adding Vectors Using Components

We can improve on the approximate results just given by adding **A** and **B** in terms of components. To see how this is done, consider **Figure 3–13 (a)**, which shows the components of **A** and **B**, and **Figure 3–13 (b)**, which shows the components of **C**. Clearly,

$$C_x = A_x + B_x$$

and

$$C_y = A_y + B_y$$

Thus, to add vectors, you simply add the components.

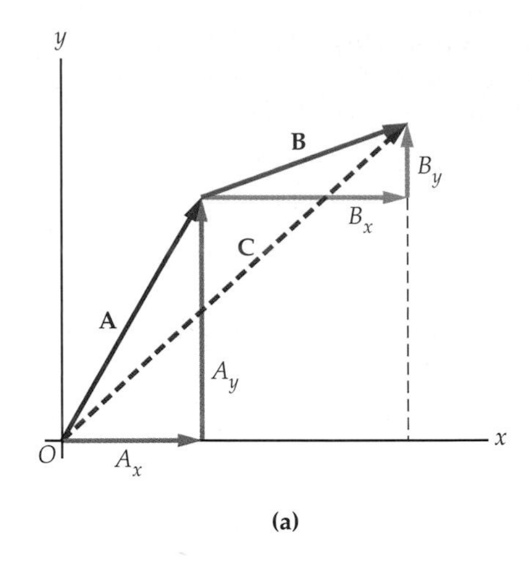

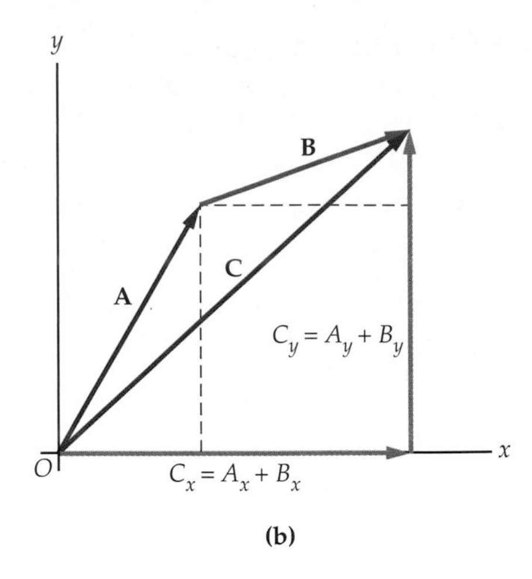

▶ **FIGURE 3–13 Component addition of vectors**
(a) The x and y components of **A** and **B**. **(b)** The x and y components of **C**. Notice that $C_x = A_x + B_x$ and $C_y = A_y + B_y$.

Returning to our example in Figure 3–12, the components of **A** and **B** are

$$A_x = (5.00 \text{ m}) \cos 60.0° = 2.50 \text{ m} \quad A_y = (5.00 \text{ m}) \sin 60.0° = 4.33 \text{ m}$$

and

$$B_x = (4.00 \text{ m}) \cos 20.0° = 3.76 \text{ m} \quad B_y = (4.00 \text{ m}) \sin 20.0° = 1.37 \text{ m}$$

Adding component by component yields the components of $\mathbf{C} = \mathbf{A} + \mathbf{B}$:

$$C_x = A_x + B_x = 2.50 \text{ m} + 3.76 \text{ m} = 6.26 \text{ m}$$

and

$$C_y = A_y + B_y = 4.33 \text{ m} + 1.37 \text{ m} = 5.70 \text{ m}$$

With these results, we can now find *precise* values for C, the magnitude of vector **C**, and its direction angle θ. In particular,

$$C = \sqrt{C_x^2 + C_y^2} = \sqrt{(6.26 \text{ m})^2 + (5.70 \text{ m})^2} = \sqrt{71.7 \text{ m}^2} = 8.47 \text{ m}$$

and

$$\theta = \tan^{-1}\frac{C_y}{C_x} = \tan^{-1}\frac{5.70 \text{ m}}{6.26 \text{ m}} = \tan^{-1} 0.911 = 42.3°$$

Note that these values are in agreement with the approximate results found by graphical addition.

In the future, we will always add vectors using components—graphical addition is useful primarily as a rough check on the results obtained with components.

ACTIVE EXAMPLE 3–1 Treasure Hunt

What are the magnitude and direction of the total displacement for the treasure hunt illustrated in Figure 3–9? Assume each pace is 0.750 m in length.

Solution

To define a convenient notation, let the first five paces be represented by **A**, the next three paces by **B**, and the final four paces by **C**. The total displacement, then, is $\mathbf{D} = \mathbf{A} + \mathbf{B} + \mathbf{C}$.

1. Find the components of **A**:	$A_x = 0$, $A_y = 3.75$ m
2. Find the components of **B**:	$B_x = 2.25$ m, $B_y = 0$
3. Find the components of **C**:	$C_x = 2.12$ m, $C_y = -2.12$ m
4. Sum the components of **A**, **B**, and **C** to find the components of **D**:	$D_x = 4.37$ m, $D_y = 1.63$ m
5. Determine D and θ:	$D = 4.66$ m, $\theta = 20.5°$

Subtracting Vectors

Next, how do we subtract vectors? Suppose, for example, that we would like to determine the vector **D**, where

$$\mathbf{D} = \mathbf{A} - \mathbf{B}$$

and **A** and **B** are the vectors shown in Figure 3–13. To find **D**, we start by rewriting it as follows:

$$\mathbf{D} = \mathbf{A} + (-\mathbf{B})$$

That is, **D** is the sum of **A** and $-\mathbf{B}$. Now the negative of a vector has a very simple graphical interpretation:

- The negative of a vector is represented by an arrow of the same length as the original vector, but pointing in the opposite direction.

The vectors **B** and $-\mathbf{B}$ are indicated in **Figure 3–14 (a)**. Thus, to subtract **B** from **A**, simply reverse the direction of **B** and add it to **A**, as indicated in **Figure 3–14 (b)**.

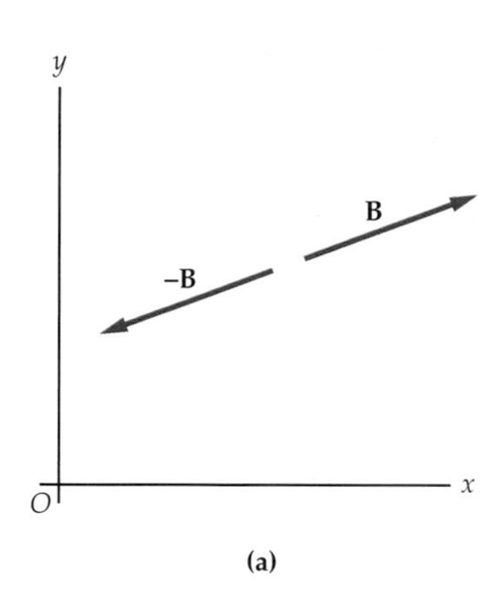

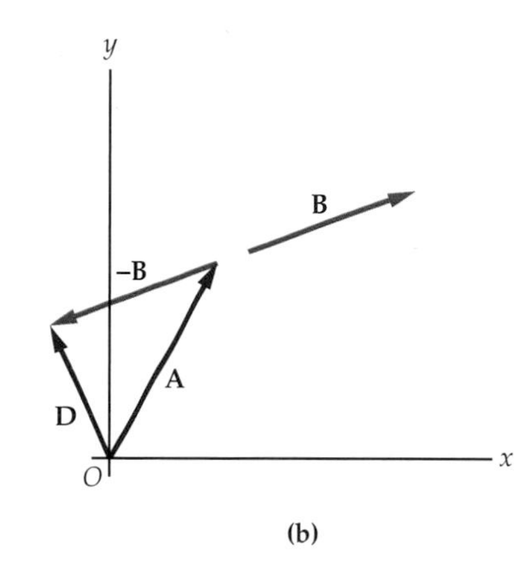

◀ **FIGURE 3–14 Vector subtraction**
(a) The vector **B** and its negative $-\mathbf{B}$.
(b) A vector construction for $\mathbf{D} = \mathbf{A} - \mathbf{B}$

In terms of components, you subtract vectors by simply subtracting the components. For example, if

$$\mathbf{D} = \mathbf{A} - \mathbf{B}$$

then

$$D_x = A_x - B_x$$

and

$$D_y = A_y - B_y$$

Once the components of **D** are found, its magnitude and direction angle can be calculated as usual.

EXERCISE 3–5

(a) For the vectors given in Figure 3–12, find the components of $\mathbf{D} = \mathbf{A} - \mathbf{B}$.
(b) Find D and θ, and compare with the vector **D** shown in Figure 3–14 (b).

Solution

(a) $D_x = -1.26$ m, $D_y = 2.96$ m **(b)** $D = 3.22$ m, $\theta = -66.9° + 180° = 113°$. In Figure 3–14 (b) we see that **D** is shorter than **B**, which has a magnitude of 4.00 m, and its direction angle is somewhat greater than 90°, in agreement with our numerical results.

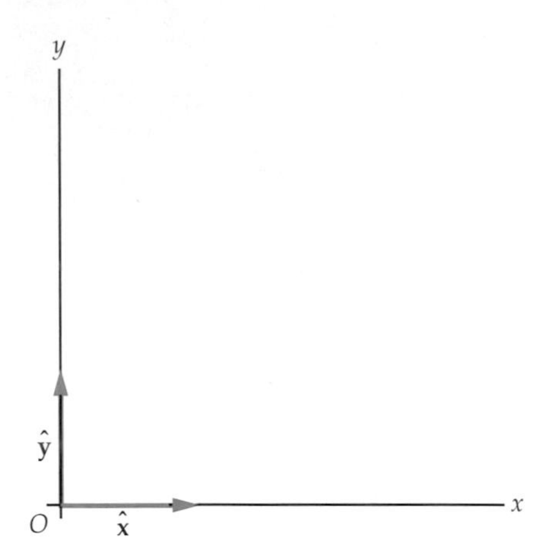

▲ **FIGURE 3–15 Unit vectors**
The unit vectors $\hat{\mathbf{x}}$ and $\hat{\mathbf{y}}$ point in the positive x and y directions, respectively.

3–4 Unit Vectors

Unit vectors provide a convenient way of expressing an arbitrary vector in terms of its components, as we shall see. But first, let's define what we mean by a unit vector. In particular, the unit vectors $\hat{\mathbf{x}}$ and $\hat{\mathbf{y}}$ are defined to be dimensionless vectors of unit magnitude pointing in the positive x and y directions, respectively:

- The x unit vector, $\hat{\mathbf{x}}$, is a dimensionless vector of unit length pointing in the positive x direction.
- The y unit vector, $\hat{\mathbf{y}}$, is a dimensionless vector of unit length pointing in the positive y direction.

Figure 3–15 shows $\hat{\mathbf{x}}$ and $\hat{\mathbf{y}}$ on a two-dimensional coordinate system. Since unit vectors have no physical dimensions—like mass, length, or time—they are used to specify direction only.

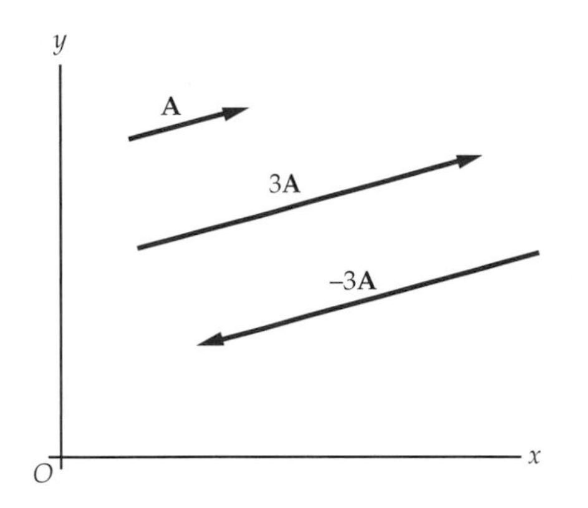

▲ **FIGURE 3–16 Multiplying a vector by a scalar**

Multiplying Unit Vectors by Scalars

To see the utility of unit vectors, consider the effect of multiplying a vector by a scalar. For example, multiplying a vector by 3 increases its magnitude by a factor of 3, but does not change its direction, as shown in **Figure 3–16**. Multiplying by -3 increases the magnitude by a factor of 3 *and* reverses the direction of the vector. This is also shown in Figure 3–16. In the case of unit vectors—which have a magnitude of 1 and are dimensionless—multiplication by a scalar results in a vector with the same magnitude and dimensions as the scalar.

For example, if a vector **A** has the scalar components $A_x = 5$ m and $A_y = 3$ m, we can write it as

$$\mathbf{A} = (5\text{ m})\hat{\mathbf{x}} + (3\text{ m})\hat{\mathbf{y}}$$

We refer to the quantities $(5\text{ m})\,\hat{\mathbf{x}}$ and $(3\text{ m})\,\hat{\mathbf{y}}$ as the x and y **vector components** of the vector **A**. In general, an arbitrary vector **A** can always be written as the sum of a vector component in the x direction and a vector component in the y direction,

$$\mathbf{A} = A_x\hat{\mathbf{x}} + A_y\hat{\mathbf{y}}$$

This is illustrated in **Figure 3–17 (a)**. An equivalent way of representing the vector components of a vector is illustrated in **Figure 3–17 (b)**. In this case we see that the vector components are the *projection* of a vector onto the x and y axes. The sign of the vector components is positive if they point in the positive x or y direction, and negative if they point in the negative x or y direction. This is how vector components will generally be shown in later chapters.

Finally, note that vector addition and subtraction is straightforward with unit vector notation:

$$\mathbf{C} = \mathbf{A} + \mathbf{B} = (A_x + B_x)\hat{\mathbf{x}} + (A_y + B_y)\hat{\mathbf{y}}$$

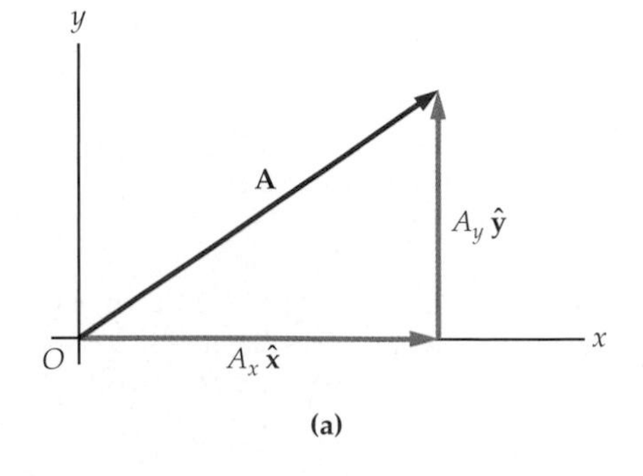

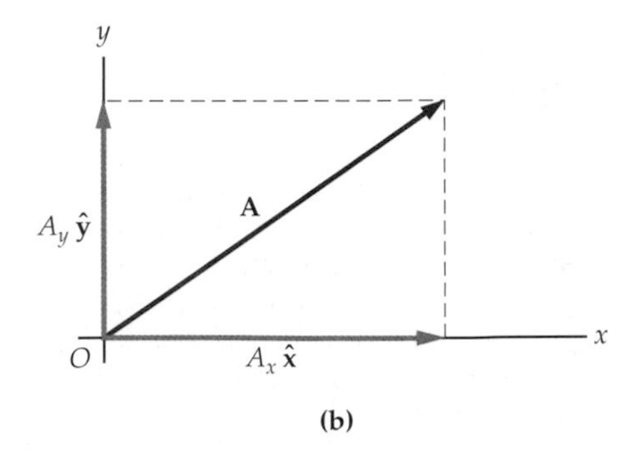

▶ **FIGURE 3–17 Vector components**
(a) A vector **A** can be written in terms of unit vectors as $\mathbf{A} = A_x\,\hat{\mathbf{x}} + A_y\,\hat{\mathbf{y}}$. **(b)** Vector components can be thought of as the projection of the vector onto the x and y axes. This method of representing vector components will be used frequently in subsequent chapters.

and

$$\mathbf{D} = \mathbf{A} - \mathbf{B} = (A_x - B_x)\hat{\mathbf{x}} + (A_y - B_y)\hat{\mathbf{y}}$$

Clearly, unit vectors provide a useful way to keep track of the x and y components of a vector.

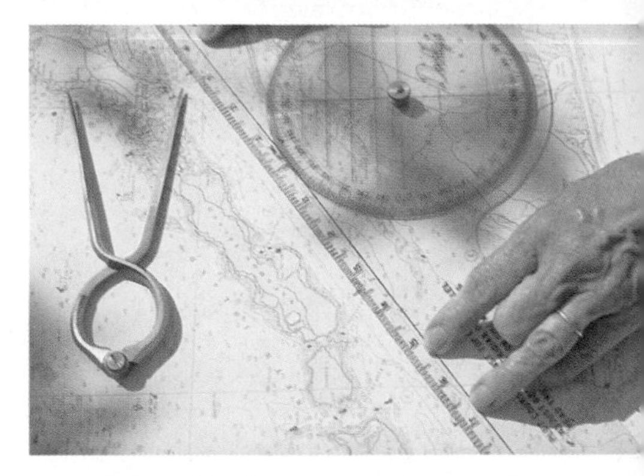

▲ A map can be used to determine the direction and magnitude of the displacement vector from your initial position to your destination.

3–5 Position, Displacement, Velocity, and Acceleration Vectors

In Chapter 2 we discussed four different vectors: position, displacement, velocity, and acceleration. We didn't refer to them as vectors at the time because we were dealing with only one dimension. Still, each of these quantities had a direction associated with it, indicated by its sign; positive meant in the positive direction, negative meant in the negative direction. Now we consider these vectors again, this time in two dimensions, where the possibilities for direction are not so limited.

Position Vectors

To begin, imagine a two-dimensional coordinate system, as in Figure 3–18. Position is indicated by a vector from the origin to the location in question. We refer to the position vector as $\mathbf{r}$; its units are meters, m.

Definition: Position Vector, r

$$\text{position vector} = \mathbf{r} \qquad 3\text{–}1$$

SI unit: meter, m

▲ **FIGURE 3–18 Position vector**
The position vector $\mathbf{r}$ points from the origin to the current location of an object.

In terms of unit vectors, the position vector is simply $\mathbf{r} = x\hat{\mathbf{x}} + y\,\hat{\mathbf{y}}$.

Now, suppose that initially you are at the location indicated by the position vector $\mathbf{r}_i$, and that later you are at the final position represented by the position vector $\mathbf{r}_f$. Your displacement vector, $\Delta\mathbf{r}$, is the change in position:

Definition: Displacement Vector, $\Delta\mathbf{r}$

$$\Delta\mathbf{r} = \mathbf{r}_f - \mathbf{r}_i \qquad 3\text{–}2$$

SI unit: meter, m

Rearranging this definition slightly, we see that

$$\mathbf{r}_f = \mathbf{r}_i + \Delta\mathbf{r}$$

That is, the final position is equal to the initial position plus the change in position. This is illustrated in Figure 3–19, where we see that $\Delta\mathbf{r}$ extends from the head of $\mathbf{r}_i$ to the head of $\mathbf{r}_f$.

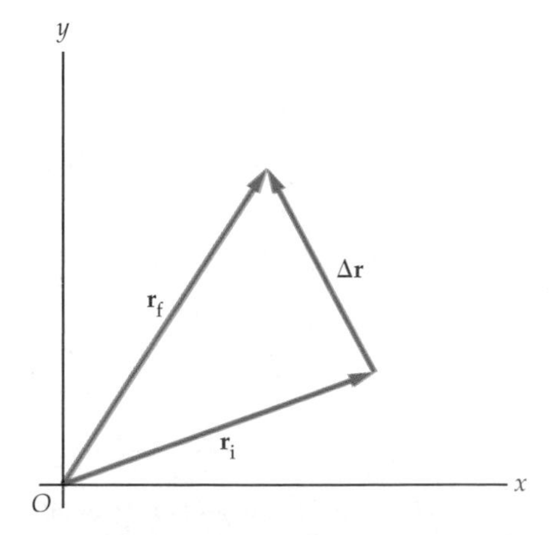

▲ **FIGURE 3–19 Displacement vector**
The displacement vector $\Delta\mathbf{r}$ is the change in position. It points from the head of the initial position vector $\mathbf{r}_i$ to the head of the final position vector $\mathbf{r}_f$. Thus, $\mathbf{r}_f = \mathbf{r}_i + \Delta\mathbf{r}$, or $\Delta\mathbf{r} = \mathbf{r}_f - \mathbf{r}_i$.

Velocity Vectors

Next, the average velocity vector is defined as the displacement vector $\Delta\mathbf{r}$ divided by the elapsed time Δt:

Definition: Average Velocity Vector, $\mathbf{v}_{av}$

$$\mathbf{v}_{av} = \frac{\Delta\mathbf{r}}{\Delta t} \qquad 3\text{–}3$$

SI unit: meter per second, m/s

Since $\Delta\mathbf{r}$ is a vector, it follows that $\mathbf{v}_{av}$ is also a vector; it is the vector $\Delta\mathbf{r}$ times the scalar $(1/\Delta t)$. Thus $\mathbf{v}_{av}$ is parallel to $\Delta\mathbf{r}$ and has the units m/s.

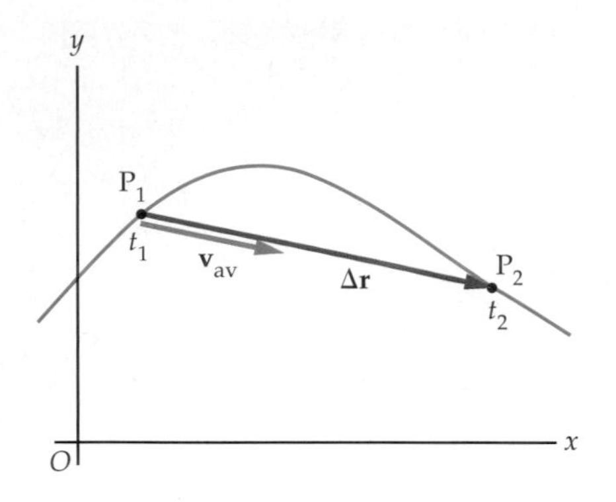

▲ **FIGURE 3–20 Average velocity vector**

The average velocity $\mathbf{v}_{av}$ points in the same direction as $\Delta\mathbf{r}$.

EXERCISE 3–6

A dragonfly is observed initially at the position $\mathbf{r}_i = (2.00\text{ m})\hat{\mathbf{x}} + (3.50\text{ m})\hat{\mathbf{y}}$. Three seconds later it is at the position $\mathbf{r}_f = (-3.00\text{ m})\hat{\mathbf{x}} + (5.50\text{ m})\hat{\mathbf{y}}$. What was the dragonfly's average velocity during this time?

Solution

$\mathbf{v}_{av} = (\mathbf{r}_f - \mathbf{r}_i)/\Delta t = [(-5.00\text{ m})\hat{\mathbf{x}} + (2.00\text{ m})\hat{\mathbf{y}}]/(3.00\text{ s}) = (-1.67\text{ m/s})\hat{\mathbf{x}} + (0.667\text{ m/s})\hat{\mathbf{y}}$

To visualize $\mathbf{v}_{av}$, imagine a particle moving in two dimensions along the path shown in **Figure 3–20**. If the particle is at point P_1 at time t_1, and at P_2 at time t_2, its displacement is indicated by the vector $\Delta\mathbf{r}$. The average velocity is parallel to $\Delta\mathbf{r}$, as indicated in Figure 3–20.

By considering smaller and smaller time intervals, as in **Figure 3–21**, it is possible to calculate the instantaneous velocity vector:

Definition: Instantaneous Velocity Vector, v

$$\mathbf{v} = \lim_{\Delta t \to 0} \frac{\Delta \mathbf{r}}{\Delta t} \qquad \text{3–4}$$

SI unit: meter per second, m/s

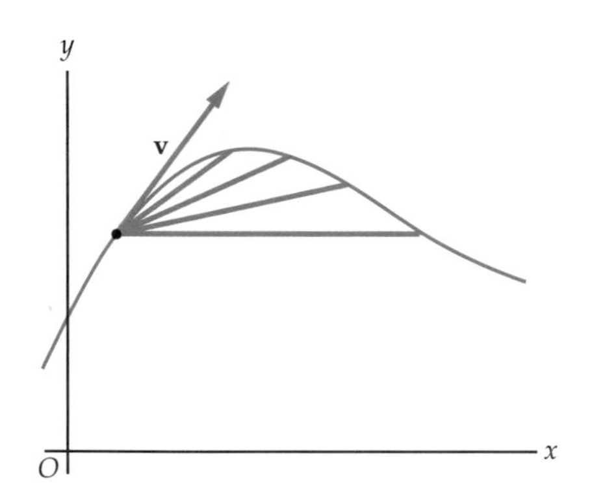

▲ **FIGURE 3–21 Instantaneous velocity vector**

The instantaneous velocity **v** is in the direction of motion at any given time.

As can be seen in Figure 3–21, the instantaneous velocity at a given time is tangential to the path of the particle at that time. In addition, the magnitude of the velocity vector is the speed of the particle. Thus, the instantaneous velocity vector tells you both how fast a particle is moving and in what direction.

EXERCISE 3–7

Find the speed and direction of motion for a rainbow trout whose velocity is $\mathbf{v} = (3.7\text{ m/s})\hat{\mathbf{x}} + (-1.3\text{ m/s})\hat{\mathbf{y}}$.

Solution

speed $= v = \sqrt{(3.7\text{ m/s})^2 + (-1.3\text{ m/s})^2} = 3.9\text{ m/s}$, $\theta = \tan^{-1}\left(\frac{-1.3}{3.7}\right) = -19°$; that is, 19° below the x axis.

Acceleration Vectors

Finally, the average acceleration vector over an interval of time, Δt, is defined as the change in the velocity vector, $\Delta\mathbf{v}$, divided by the scalar Δt:

Definition: Average Acceleration Vector, $\mathbf{a}_{av}$

$$\mathbf{a}_{av} = \frac{\Delta \mathbf{v}}{\Delta t} \qquad \text{3–5}$$

SI unit: meter per second per second, m/s^2

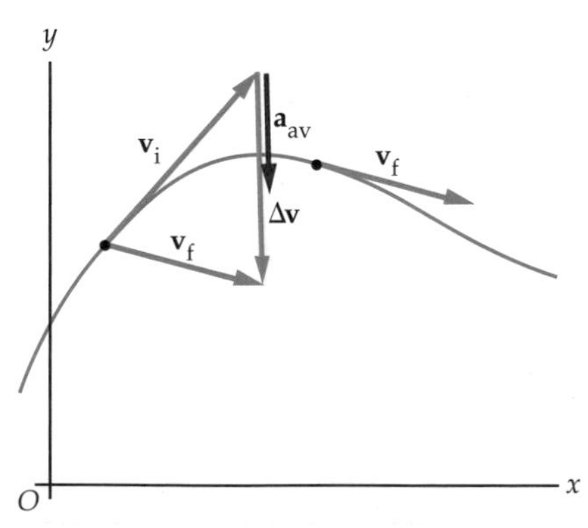

▲ **FIGURE 3–22 Average acceleration vector**

The average acceleration $\mathbf{a}_{av}$ points in the direction of the change in velocity $\Delta\mathbf{v}$. Note that $\mathbf{a}_{av}$ need not point in the direction of motion.

An example is given in **Figure 3–22**, where we show the initial and final velocity vectors corresponding to two different times. Since the change in velocity is defined as

$$\Delta\mathbf{v} = \mathbf{v}_f - \mathbf{v}_i$$

it follows that

$$\mathbf{v}_f = \mathbf{v}_i + \Delta\mathbf{v}$$

as indicated in Figure 3–22. Thus, $\Delta\mathbf{v}$ is the vector extending from the head of $\mathbf{v}_i$ to the head of $\mathbf{v}_f$, just as $\Delta\mathbf{r}$ extends from the head of $\mathbf{r}_i$ to the head of $\mathbf{r}_f$ in Figure 3–19. The direction of $\mathbf{a}_{av}$ is the direction of $\Delta\mathbf{v}$, as shown in Figure 3–22.

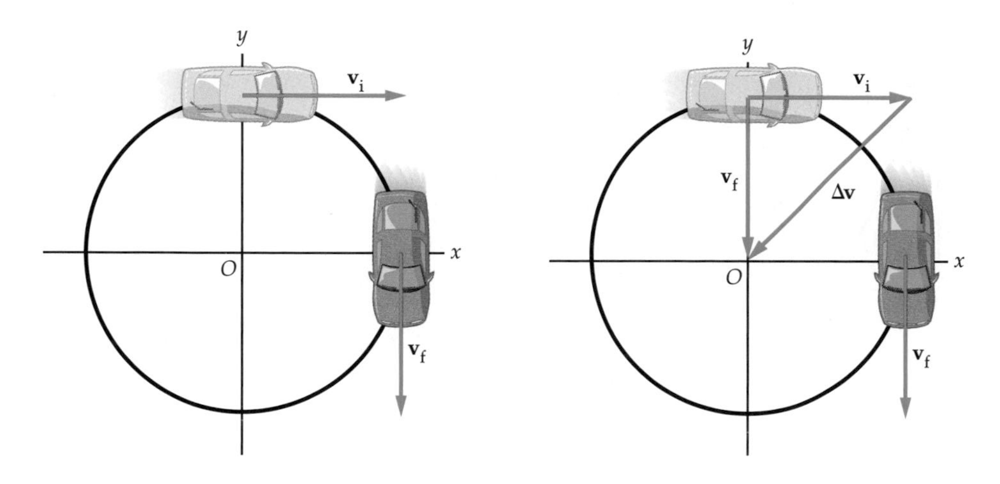

◀ **FIGURE 3–23 Average acceleration for a car traveling in a circle with constant speed**

Can an object accelerate if its speed is constant? Absolutely—if its direction changes. Consider a car driving with a constant speed on a circular track, as shown in **Figure 3–23**. Suppose that the initial velocity of the car is $\mathbf{v}_i = (12\ \text{m/s})\hat{\mathbf{x}}$, and that 10.0 s later its final velocity is $\mathbf{v}_f = (-12\ \text{m/s})\hat{\mathbf{y}}$. Note that the speed is 12 m/s in each case, but the velocity is different because the *direction* has changed. Calculating the average acceleration we find a nonzero acceleration:

$$\mathbf{a}_{av} = \frac{\Delta\mathbf{v}}{\Delta t} = \frac{\mathbf{v}_f - \mathbf{v}_i}{10.0\ \text{s}}$$

$$= \frac{(-12\ \text{m/s})\hat{\mathbf{y}} - (12\ \text{m/s})\hat{\mathbf{x}}}{10.0\ \text{s}} = (-1.2\ \text{m/s}^2)\hat{\mathbf{x}} + (-1.2\ \text{m/s}^2)\hat{\mathbf{y}}$$

Thus, a change in direction is just as important as a change in speed in producing an acceleration.

Finally, by going to infinitesimally small time intervals, $\Delta t \to 0$, we can define the instantaneous acceleration:

Definition: Instantaneous Acceleration Vector, a

$$\mathbf{a} = \lim_{\Delta t \to 0} \frac{\Delta\mathbf{v}}{\Delta t} \qquad \textbf{3–6}$$

SI unit: meter per second per second, m/s^2

ACTIVE EXAMPLE 3–2 **Round a Corner**

A car is traveling northwest at 9.00 m/s. Eight seconds later it has rounded a corner, and is heading north at 15.0 m/s. What is the magnitude and direction of its average acceleration during these 8.00 seconds?

Let the positive x direction be east, and the positive y direction be north.

Solution

1. Write out $\mathbf{v}_i$: $\quad \mathbf{v}_i = (-6.36\ \text{m/s})\hat{\mathbf{x}} + (6.36\ \text{m/s})\hat{\mathbf{y}}$
2. Write out $\mathbf{v}_f$: $\quad \mathbf{v}_f = (15.0\ \text{m/s})\hat{\mathbf{y}}$
3. Calculate $\Delta\mathbf{v}$: $\quad \Delta\mathbf{v} = (6.36\ \text{m/s})\hat{\mathbf{x}} + (8.64\ \text{m/s})\hat{\mathbf{y}}$
4. Find $\mathbf{a}_{av}$: $\quad \mathbf{a}_{av} = (0.795\ \text{m/s}^2)\hat{\mathbf{x}} + (1.08\ \text{m/s}^2)\hat{\mathbf{y}}$
5. Determine a_{av} and θ: $\quad a_{av} = 1.34\ \text{m/s}^2, \theta = 53.6°$ north of east

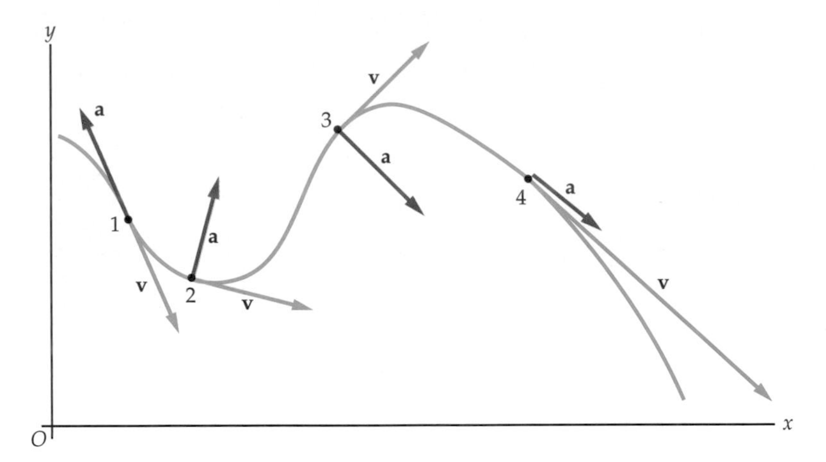

▶ **FIGURE 3–24 Velocity and acceleration vectors for a particle moving along a winding path**
The acceleration of a particle need not point in the direction of motion. At point (1) the particle is slowing down, at (2) it is turning to the left, at (3) turning to the right, and finally, at point (4) it is speeding up.

Note carefully the following critical distinctions between the velocity vector and the acceleration vector:

- The velocity vector, **v**, is always in the direction of a particle's motion.
- The acceleration vector, **a**, can point in directions other than the direction of motion, and in general it does.

An example of a particle's motion, showing the velocity and acceleration vectors at various times, is presented in Figure 3–24.

Note that in all cases the velocity is tangential to the motion, though the acceleration points in various directions. When the acceleration is perpendicular to the velocity of an object, as at points 2 and 3 in Figure 3–24, its speed remains constant while its direction of motion changes. If the acceleration is parallel or antiparallel to the velocity of an object, as at points 1 and 4, its direction of motion remains the same while its speed changes. Throughout the next chapter we shall see further examples of motion in which the velocity and acceleration are in different directions.

▶ The velocities of these cyclists change in both magnitude and direction as they slow to negotiate a series of sharp curves and then speed up agian. Both kinds of velocity change involve an acceleration.

3–6 Relative Motion

A good example of the use of vectors is in the description of relative motion. Suppose, for example, that you are standing on the ground as a train goes by at 15.0 m/s, as shown in Figure 3–25. Inside the train, one of the passengers is walking in the forward direction at 1.2 m/s relative to the train. How fast is the passenger moving relative to you? Clearly, the answer is 1.2 m/s + 15.0 m/s = 16.2 m/s. What if the passenger had been walking with the same speed, but toward the back of the train? In this case, you would see the passenger going by with a speed of −1.2 m/s + 15.0 m/s = 13.8 m/s.

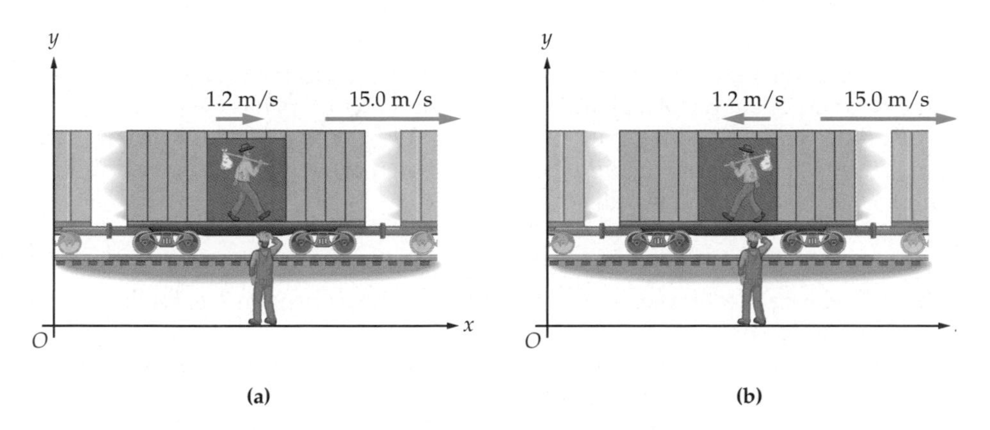

◀ **FIGURE 3–25 Relative velocity of a passenger on a train with respect to a person on the ground**

(a) The passenger walks toward the front of the train. **(b)** The passenger walks toward the rear of the train.

Let's generalize these results. Call the velocity of the *t*rain relative to the *g*round $\mathbf{v}_{tg}$, the velocity of the *p*assenger relative to the *t*rain $\mathbf{v}_{pt}$, and the velocity of the *p*assenger relative to the *g*round $\mathbf{v}_{pg}$. As we saw in the previous paragraph, the velocity of the *p*assenger relative to the *g*round is

$$\mathbf{v}_{pg} = \mathbf{v}_{pt} + \mathbf{v}_{tg} \qquad \textbf{3–7}$$

This vector addition is illustrated in **Figure 3–26** for the two cases we discussed.

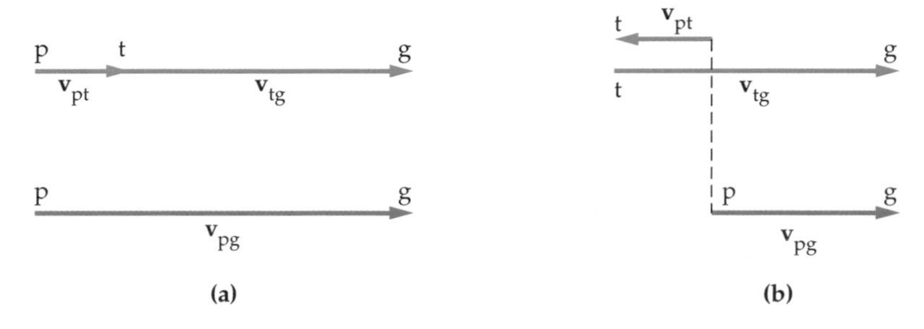

◀ **FIGURE 3–26 Adding velocity vectors**

Vector addition to find the velocity of the passenger with respect to the ground for **(a)** Figure 3–25 (a) and **(b)** Figure 3–25 (b).

Though this example dealt with one-dimensional motion, Equation 3–7 is valid for velocity vectors pointing in arbitrary directions. For example, instead of walking along the aisle, the passenger might be climbing a ladder to the roof of the car. In this case $\mathbf{v}_{pt}$ is vertical, $\mathbf{v}_{tg}$ is horizontal, and $\mathbf{v}_{pg}$ is simply the vector sum $\mathbf{v}_{pt} + \mathbf{v}_{tg}$.

EXERCISE 3–8

Suppose the passenger in **Figure 3–27** is climbing a vertical ladder with a speed of 0.20 m/s, and the train is slowly coasting forward at 0.70 m/s. Find the speed and direction of the passenger relative to the ground.

Solution

$\mathbf{v}_{pg} = (0.70\text{ m/s})\hat{\mathbf{x}} + (0.20\text{ m/s})\hat{\mathbf{y}}$; thus $v_{pg} = 0.73\text{ m/s}, \theta = 16°$

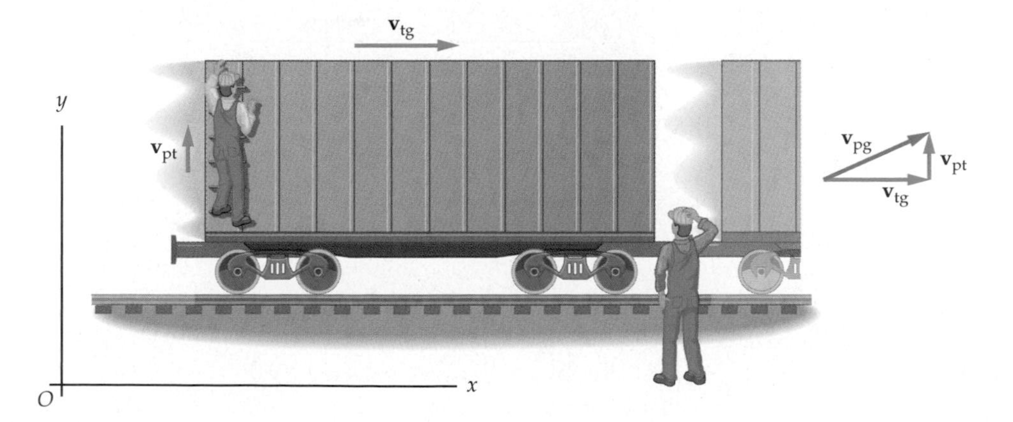

◀ **FIGURE 3–27 Relative velocity in two dimensions**

A person climbs up a ladder on a moving train with velocity $\mathbf{v}_{pt}$ relative to the train. If the train moves relative to the ground with a velocity $\mathbf{v}_{tg}$, the velocity of the person on the train relative to the ground is $\mathbf{v}_{pg} = \mathbf{v}_{pt} + \mathbf{v}_{tg}$.

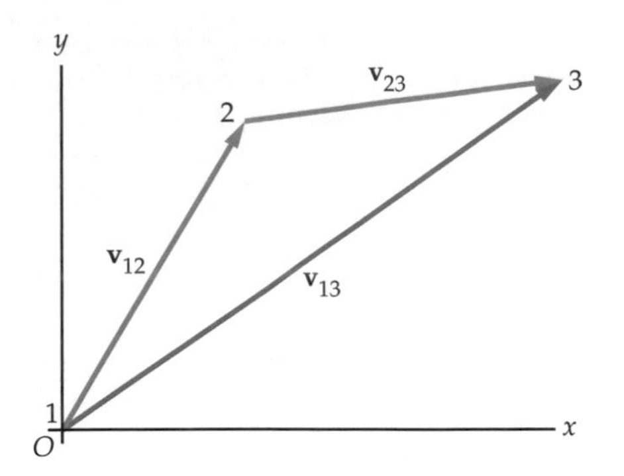

▲ **FIGURE 3–28 Vector addition used to determine relative velocity**

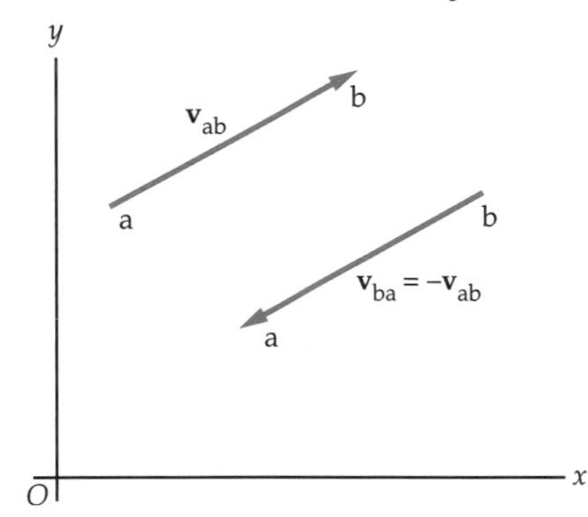

▲ **FIGURE 3–29 Reversing the subscripts of a velocity reverses the corresponding velocity vector**

Note that the subscripts in Equation 3–7 follow a definite pattern. On the left-hand side of the equation we have the subscripts pg. On the right-hand side we have two sets of subscripts, pt and tg; note that a pair of t's has been inserted between the p and the g. This pattern always holds, for any relative motion problem, though the subscripts will be different when referring to different objects. Thus, we can say quite generally that

$$\mathbf{v}_{13} = \mathbf{v}_{12} + \mathbf{v}_{23} \qquad \textbf{3–8}$$

where, in the train example, we can identify 1 as the *p*assenger, 2 as the *t*rain, and 3 as the *g*round.

The vector addition in Equation 3–8 is shown in **Figure 3–28**. For convenience in seeing how the subscripts are ordered in the equation, we have labeled the tail of each vector with its first subscript and the head of each vector with its second subscript.

One final note about velocities and their subscripts: Reversing the subscripts reverses the velocity. This is indicated in **Figure 3–29**, where we see that

$$\mathbf{v}_{ab} = -\mathbf{v}_{ba}$$

for any subscripts *a* and *b*. Physically, what we are saying is that if you are riding in a car due *north* at 20 m/s relative to the ground, then the ground, relative to you, is moving due *south* at 20 m/s.

Let's apply these results to a two-dimensional example.

EXAMPLE 3–2 **Crossing a River**

Real World Physics

You are riding in a boat going 6.1 m/s at an angle of 25° upstream on a river flowing at 1.4 m/s. What is your velocity relative to the ground?

Picture the Problem

We choose the x axis to be perpendicular to the river, and the y axis to point upstream. With these choices the velocity of the boat relative to the water is 25° above the x axis.

Strategy

If the water were still, the boat would move in the direction in which it is pointed. With the water flowing downstream, as shown, the boat will move in a direction closer to the x axis. To find the velocity of the boat we use $\mathbf{v}_{13} = \mathbf{v}_{12} + \mathbf{v}_{23}$ with 1 referring to the boat (b), 2 referring to the water (w), and 3 referring to the ground (g).

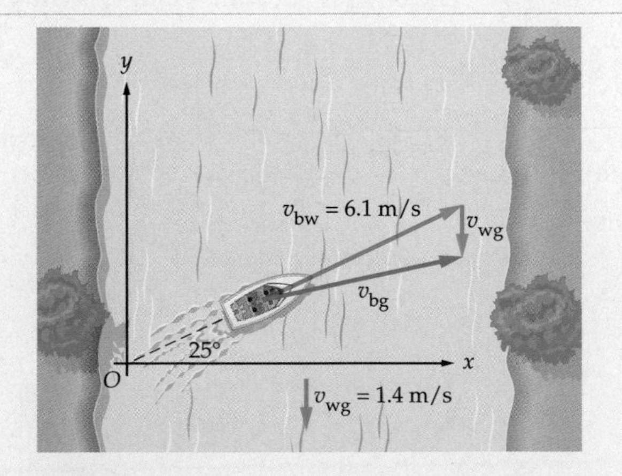

Solution

1. Rewrite $\mathbf{v}_{13} = \mathbf{v}_{12} + \mathbf{v}_{23}$ with $1 \to \text{b}$, $2 \to \text{w}$, and $3 \to \text{g}$: $\quad \mathbf{v}_{bg} = \mathbf{v}_{bw} + \mathbf{v}_{wg}$
2. From our sketch we see that the water flows at 1.4 m/s in the negative y direction relative to the ground: $\quad \mathbf{v}_{wg} = (-1.4\ \text{m/s})\hat{\mathbf{y}}$
3. The velocity of the boat relative to the water is given in the problem statement: $\quad \mathbf{v}_{bw} = (6.1\ \text{m/s})\cos 25°\ \hat{\mathbf{x}} + (6.1\ \text{m/s})\sin 25°\ \hat{\mathbf{y}} = (5.5\ \text{m/s})\ \hat{\mathbf{x}} + (2.6\ \text{m/s})\ \hat{\mathbf{y}}$
4. Carry out the vector sum in step 1 to find $\mathbf{v}_{bg}$: $\quad \mathbf{v}_{bg} = (5.5\ \text{m/s})\ \hat{\mathbf{x}} + (2.6\ \text{m/s} - 1.4\ \text{m/s})\ \hat{\mathbf{y}} = (5.5\ \text{m/s})\ \hat{\mathbf{x}} + (1.2\ \text{m/s})\ \hat{\mathbf{y}}$

Insight

Note that the speed of the boat relative to the ground is $\sqrt{(5.5\ \mathrm{m/s})^2 + (1.2\ \mathrm{m/s})^2} = 5.6$ m/s, and the direction angle is $\theta = \tan^{-1}(1.2/5.5) = 12°$ upstream.

Practice Problem

Find the speed and direction of the boat relative to the ground if the river flows at 3.5 m/s. [**Answer:** $v_{bg} = 5.6$ m/s, $\theta = -9.5°$. In this case, a person on the ground would see the boat going slowly downstream.]

Some related homework problems: Problem 37, Problem 41, Problem 43

Suppose the problem had been to find the velocity of the boat relative to the water so that it goes straight across the river at 5.0 m/s. That is, we want to find $\mathbf{v}_{bw}$ such that $\mathbf{v}_{bg} = (5.0\ \mathrm{m/s})\hat{\mathbf{x}}$. One approach is to simply solve $\mathbf{v}_{bg} = \mathbf{v}_{bw} + \mathbf{v}_{wg}$ for $\mathbf{v}_{bw}$, which gives

$$\mathbf{v}_{bw} = \mathbf{v}_{bg} - \mathbf{v}_{wg} \qquad 3\text{–}9$$

Another approach is to go back to our general relation, $\mathbf{v}_{13} = \mathbf{v}_{12} + \mathbf{v}_{23}$, and choose 1 to be the boat, 2 to be the *g*round, and 3 to be the *w*ater. With these substitutions we find

$$\mathbf{v}_{bw} = \mathbf{v}_{bg} + \mathbf{v}_{gw}$$

which is the same as Equation 3–7, since $\mathbf{v}_{gw} = -\mathbf{v}_{wg}$. In either case, the desired velocity of the boat relative to the water is

$$\mathbf{v}_{bw} = (5.0\ \mathrm{m/s})\hat{\mathbf{x}} + (1.4\ \mathrm{m/s})\hat{\mathbf{y}}$$

which corresponds to a speed of 5.2 m/s and a direction angle of 16° upstream.

▲ Kayakers on a rapidly flowing river. The velocity of each kayak with respect to the shore is the vector sum of its velocity with respect to the water and the velocity of the water with respect to the shore.

Chapter Summary

	Topic	Remarks and Relevant Equations
3–1	**Scalars versus Vectors**	
	scalar	A number with appropriate units. Examples of scalar quantities include time and length.
	vector	A quantity with both a magnitude and a direction. Examples include displacement, velocity, and acceleration.
3–2	**Components of a Vector**	
	x component of vector **A**	$A_x = A\cos\theta$, where θ is measured relative to the x axis.
	y component of vector **A**	$A_y = A\sin\theta$, where θ is measured relative to the x axis.
	sign of the components	A_x is positive if **A** points in the positive x direction, and negative if it points in the negative x direction. The same remarks apply to A_y.
	magnitude of vector **A**	The magnitude of **A** is $A = \sqrt{A_x^2 + A_y^2}$.
	direction angle of vector **A**	The direction angle of **A** is $\theta = \tan^{-1}(A_y/A_x)$.
3–3	**Adding and Subtracting Vectors**	
	graphical method	To add **A** and **B**, place them so that the tail of **B** is at the head of **A**. The sum **C** = **A** + **B** is the arrow from the tail of **A** to the head of **B**. See Figure 3–8. To find **A** − **B**, place **A** and −**B** head-to-tail and draw an arrow from the tail of **A** to the head of −**B**. See Figure 3–14.
	component method	If **C** = **A** + **B**, then $C_x = A_x + B_x$ and $C_y = A_y + B_y$. If $C = A - B$, then $C_x = A_x - B_x$ and $C_y = A_y - B_y$.

3–4 Unit vectors

x unit vector	Written $\hat{\mathbf{x}}$, the x unit vector is a dimensionless vector of length unity in the positive x direction.
y unit vector	Written $\hat{\mathbf{y}}$, the y unit vector is a dimensionless vector of length unity in the positive y direction.
vector addition	$\mathbf{A} + \mathbf{B} = (A_x + B_x)\hat{\mathbf{x}} + (A_y + B_y)\hat{\mathbf{y}}$

3–5 Position, Displacement, Velocity, and Acceleration

position vector	The position vector $\mathbf{r}$ points from the origin to a particle's location.
displacement vector	The displacement vector, $\Delta\mathbf{r}$, is the change in position; $\Delta\mathbf{r} = \mathbf{r}_f - \mathbf{r}_i$.
velocity vector	The velocity vector $\mathbf{v}$ points in the direction of motion and has a magnitude equal to the speed.
acceleration vector	The acceleration vector $\mathbf{a}$ indicates how quickly and in what direction the velocity is changing. It need not point in the direction of motion.

3–6 Relative Velocity

velocity of object 3 relative to object 1	$\mathbf{v}_{13} = \mathbf{v}_{12} + \mathbf{v}_{23}$, where object 2 can be anything.
reversing the subscripts on a velocity	$\mathbf{v}_{12} = -\mathbf{v}_{21}$.

Problem-Solving Summary

Type of Problem	Relevant Physical Concepts	Related Examples
Add or subtract vectors.	Resolve the vectors into x and y components, then add or subtract the components.	Active Example 3–1 Exercise 3–5
Calculate the average velocity.	Divide the displacement, $\Delta\mathbf{r}$, by the elapsed time Δt.	Exercise 3–6
Calculate the average acceleration.	Divide the change in velocity, $\Delta\mathbf{v}$, by the elapsed time Δt.	Active Example 3–2
Find the relative velocity of object 1 with respect to object 3.	Use $\mathbf{v}_{13} = \mathbf{v}_{12} + \mathbf{v}_{23}$ with the appropriate choices for 1, 2, and 3.	Example 3–2 Exercise 3–8

Conceptual Questions

1. For the following quantities, indicate which is a scalar and which is a vector: **(a)** the time it takes for you to run the 100-yard dash; **(b)** your displacement after running the 100-yard dash; **(c)** your average velocity while running; **(d)** your average speed while running.
2. Which, if any, of the vectors shown in Figure 3–30 are equal?
3. Given that $\mathbf{A} + \mathbf{B} = 0$, **(a)** how does the magnitude of **B** compare with the magnitude of **A**? **(b)** How does the direction of **B** compare with the direction of **A**?
4. Can a component of a vector be greater than the vector's magnitude?
5. If each component of a vector is doubled, **(a)** how does the magnitude of the vector change? **(b)** What happens to the direction of the vector?

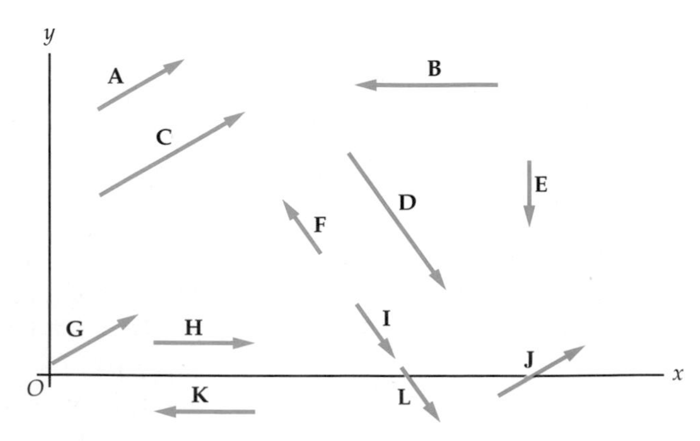

▲ **FIGURE 3–30** Question 2

6. Can a vector with nonzero magnitude still have a component that is zero?
7. Suppose that **A** and **B** have nonzero magnitude. Is it possible for **A** + **B** to be zero?
8. Vector **A** has x and y components of equal magnitude. What can you say about the possible directions of **A**?
9. Given that **A** + **B** = **C**, and that $A^2 + B^2 = C^2$, how are **A** and **B** oriented relative to one another?
10. Given that **A** + **B** = **C**, and that $A + B = C$, how are **A** and **B** oriented relative to one another?
11. Given that **A** − **B** = **C**, and that $A - B = C$, how are **A** and **B** oriented relative to one another?
12. The components of a vector **A** satisfy $A_x < 0$ and $A_y > 0$. What is the range of possible direction angles for **A**?
13. The components of a vector **A** satisfy $A_x > 0$ and $A_y < 0$. What is the range of possible direction angles for **A**?
14. The components of a vector **A** satisfy $A_x = -A_y \neq 0$. What are the possible directions of **A**?
15. Use a sketch to show that two vectors of unequal magnitude cannot add to zero, but that three vectors of unequal magnitude can.
16. When sailing, the wind feels stronger when you sail upwind ("beating") than when you are sailing downwind ("running"). Explain.
17. Rain is falling vertically downward and you are running for shelter. To keep driest, should you hold your umbrella vertically, tilted forward, or tilted backward? Explain.
18. The accompanying photo shows a KC-10A Extender using a boom to refuel an aircraft in flight. If the velocity of the KC-10A is 125 m/s due east relative to the ground, what is the velocity of the aircraft being refueled relative to **(a)** the ground, and **(b)** the KC-10A?

▲ Air-to-air refueling. (Question 18)

Problems

Note: **IP** *denotes an integrated conceptual/quantitative problem.* **BIO** *identifies problems of biological or medical interest. Blue bullets (•, ••, •••) are used to indicate the level of difficulty of each problem.*

Section 3–2 The Components of a Vector

1. • The press box at a baseball park is 38.0 ft above the ground. A reporter in the press box looks at an angle of 15.0° below the horizontal to see second base. What is the horizontal distance from the press box to second base?
2. • You are driving up a long inclined road. After 1.5 miles you notice that signs along the roadside indicate that your elevation has increased by 520 ft. **(a)** What is the angle of the road above the horizontal? **(b)** How far do you have to drive to gain an additional 150 ft of elevation?
3. • A road that rises 1 ft for every 100 ft traveled horizontally is said to have a 1 percent grade. Portions of the Lewiston grade, near Lewiston, Idaho, have a 6 percent grade. At what angle is this road inclined above the horizontal?
4. • Find the x and y components of a position vector, **r**, of magnitude $r = 75$ m, if its angle relative to the x axis is **(a)** 25° and **(b)** 95°.
5. • A baseball "diamond" (Figure 3–31) is a square with sides 90 ft in length. If the positive x axis points from home plate to first base, and the positive y axis points from home plate to third base, find the displacement vector of a base runner who has just hit **(a)** a double, **(b)** a triple, or **(c)** a home run.

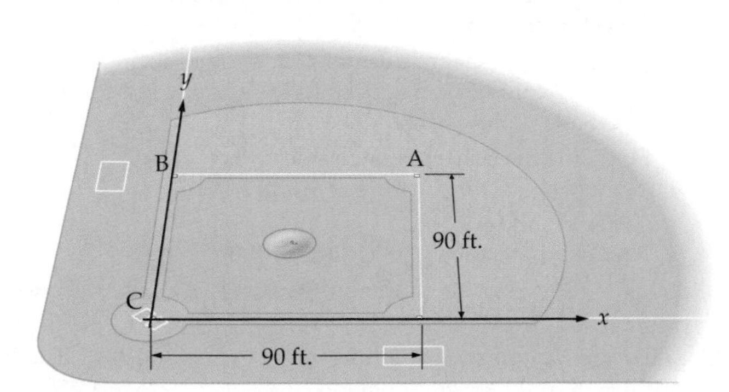

▲ **FIGURE 3–31** Problem 5

6. •• A lighthouse that rises 49 ft above the surface of the water sits on a rocky cliff that extends 19 ft from its base, as shown in Figure 3–32. A sailor on the deck of a ship sights the top of the lighthouse at an angle of 30.0° above the horizontal. If the sailor's eye level is 14 ft above the water, how far is the ship from the rocks?

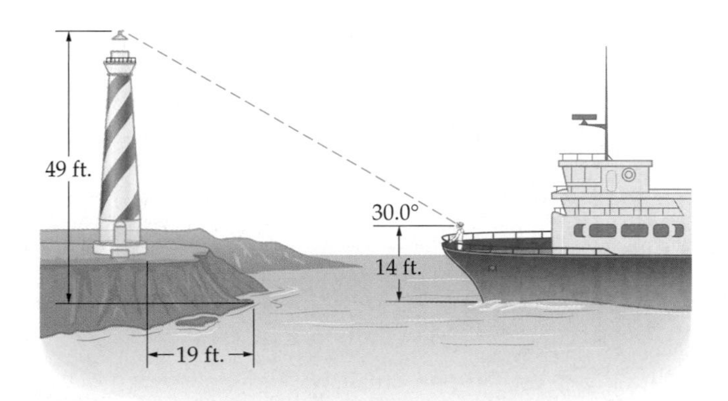

▲ **FIGURE 3–32** Problem 6

7. •• A water molecule is shown schematically in Figure 3–33. The distance from the center of the oxygen atom to the center of a hydrogen atom is 0.96 Å, and the angle between the hydrogen

atoms is 104.5°. Find the center-to-center distance between the hydrogen atoms. (1 Å = 10^{-10} m.)

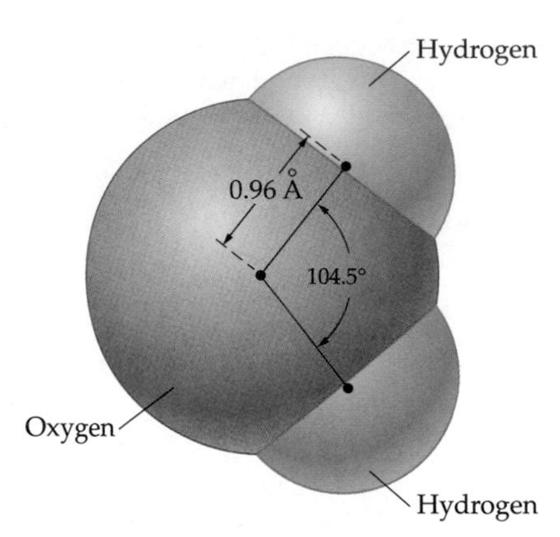

▲ **FIGURE 3–33** Problem 7

8. •• **IP** The x and y components of a vector **r** are $r_x = 24$ m and $r_y = -8.5$ m, respectively. Find **(a)** the direction and **(b)** the magnitude of the vector r. **(c)** If both r_x and r_y are doubled, how do your answers to parts (a) and (b) change?

9. •• The corners of a square with sides 1.5 m long lie on a circle. Find the radius of the circle.

10. •• You drive a car 510 ft to the east, then 250 ft to the north. **(a)** What is the distance you have covered? **(b)** Using a sketch, estimate the direction of your displacement. **(c)** Verify your estimate in part (b) with a numerical calculation of the direction.

11. •• Vector **A** has a magnitude of 50 units and points in the positive x direction. A second vector, **B**, has a magnitude of 120 units and points at an angle of 70° below the x axis. Which vector has **(a)** the greater x component, and **(b)** the greater y component?

12. •• A treasure map directs you to start at a palm tree and walk due north for 10.0 m. You are then to turn 90° and walk 15.0 m; then turn 90° again and walk 5.00 m. Give the distance from the palm tree, and the direction relative to north, for each of the four possible locations of the treasure.

13. •• A whale comes to the surface to breathe, and then dives at an angle of 20.0° below the horizontal (**Figure 3–34**). If the whale continues in a straight line for 150 m, **(a)** how deep is it, and **(b)** how far has it traveled horizontally?

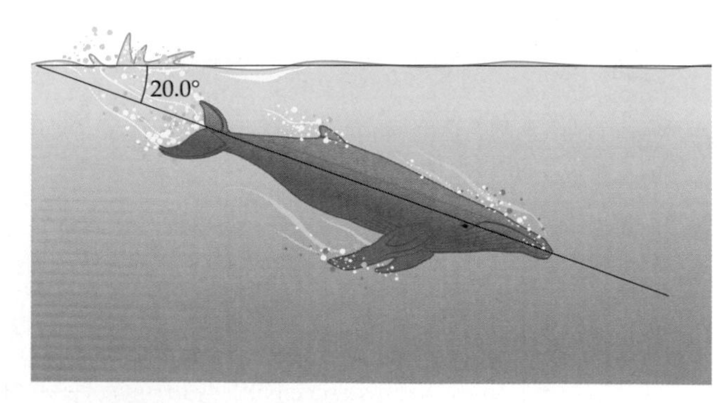

▲ **FIGURE 3–34** Problem 13

Section 3–3 Adding and Subtracting Vectors

14. • A vector **A** has a magnitude of 50.0 m and points in a direction 20.0° below the x axis. A second vector, **B**, has a magnitude of 70.0 m and points in a direction 50.0° above the x axis. **(a)** Sketch the vectors **A**, **B**, and **C** = **A** + **B**. **(b)** Using the component method of vector addition, find the magnitude and direction of the vector **C**.

15. • Referring to Problem 14, **(a)** sketch the vectors **A**, −**B**, and **D** = **A** − **B**. **(b)** Find the magnitude and direction of the vector **D**.

16. • Referring to Problem 14, **(a)** sketch the vectors −**A**, **B**, and **E** = **B** − **A**. **(b)** Find the magnitude and direction of the vector **E**.

17. •• Vector **A** points in the positive x direction and has a magnitude of 75 m. The vector **C** = **A** + **B** points in the positive y direction and has a magnitude of 95 m. **(a)** Sketch **A**, **B**, and **C**. **(b)** Estimate the magnitude and direction of the vector **B**. **(c)** Verify your estimate in part (b) with a numerical calculation.

18. •• Vector **A** points in the negative x direction and has a magnitude of 25 units. The vector **B** points in the positive y direction. **(a)** Find the magnitude of **B** if **A** + **B** has a magnitude of 35 units. **(b)** Sketch **A** and **B**.

19. •• Vector **A** points in the negative y direction and has a magnitude of 5 units. Vector **B** has twice the magnitude and points in the positive x direction. Find the direction and magnitude of **(a)** **A** + **B**, **(b)** **A** − **B**, and **(c)** **B** − **A**.

20. •• A basketball player runs down the court, following the path indicated by the vectors **A**, **B**, and **C** in **Figure 3–35**. The magnitudes of these three vectors are: A = 10.0 m, B = 20.0 m, and C = 7.0 m. Find the magnitude and direction of the net displacement of the player using **(a)** the graphical method and **(b)** the component method of vector addition. Compare your results.

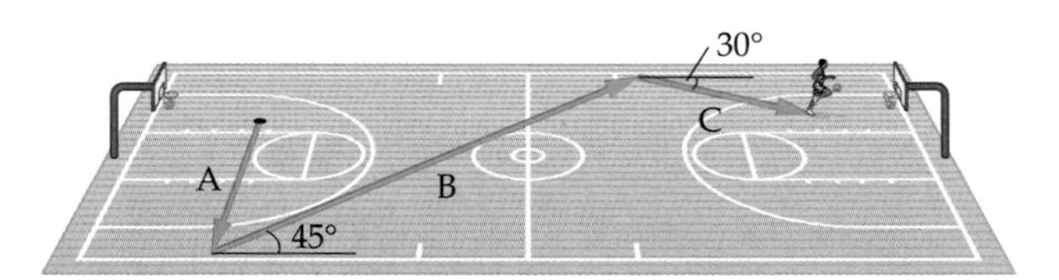

▲ **FIGURE 3–35** Problem 20

Section 3–4 Unit Vectors

21. • A particle undergoes a displacement $\Delta\mathbf{r}$ of magnitude 55 m in a direction 30.0° below the x axis. Express $\Delta\mathbf{r}$ in terms of the unit vectors $\hat{\mathbf{x}}$ and $\hat{\mathbf{y}}$.

22. • A vector has a magnitude of 2.3 m and points in a direction that is 130° counterclockwise from the x axis. Find the x and y components of this vector.

23. • Find the direction and magnitude of the vectors **(a)** $\mathbf{A} = (5.0\text{ m})\,\hat{\mathbf{x}} + (-2.0\text{ m})\,\hat{\mathbf{y}}$, **(b)** $\mathbf{B} = (-2.0\text{ m})\,\hat{\mathbf{x}} + (5.0\text{ m})\,\hat{\mathbf{y}}$, and **(c)** **A** + **B**.

24. • Find the direction and magnitude of the vectors **(a)** $\mathbf{A} = (25\text{ m})\,\hat{\mathbf{x}} + (-12\text{ m})\,\hat{\mathbf{y}}$, **(b)** $\mathbf{B} = (2.0\text{ m})\,\hat{\mathbf{x}} + (15\text{ m})\,\hat{\mathbf{y}}$, and **(c)** **A** + **B**.

25. • For the vectors given in Problem 24, express **(a)** **A** − **B**, and **(b)** **B** − **A** in unit vector notation.

26. • Express each of the vectors in Figure 3–36 in unit vector notation.

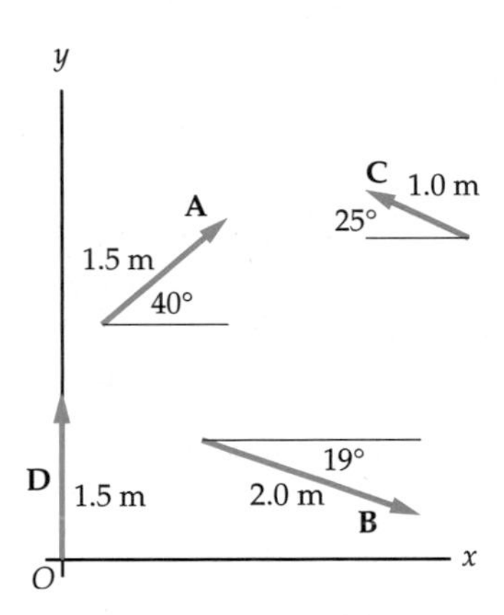

▲ FIGURE 3–36 Problems 26 and 27

27. •• Referring to the vectors in Figure 3–36, express the sum, **A** + **B** + **C**, in unit vector notation.

Section 3–5 Position, Velocity, and Acceleration vectors

28. • Two of the allowed chess moves for a knight are shown in Figure 3–37. If the checkerboard squares are 3.5 cm on each side, find the magnitude and direction of the knight's displacement for each of the two moves.

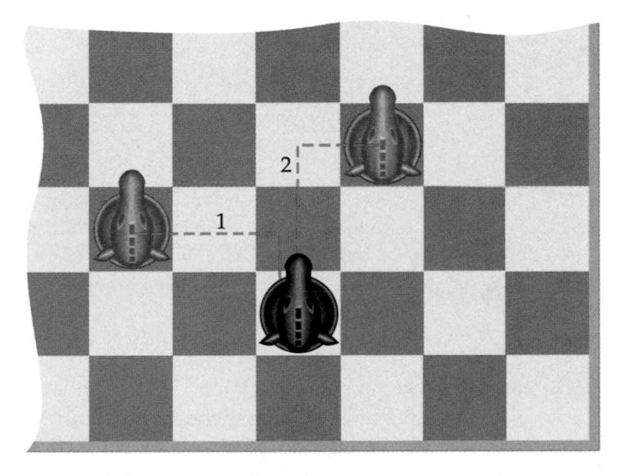

▲ FIGURE 3–37 Problem 28

29. • In its daily prowl of the neighborhood, a cat makes a displacement of 120 m due north, followed by a 72-m displacement due west. Find the magnitude and direction of the displacement required for the cat to return home.

30. • If the cat in Problem 29 takes 45 minutes to complete the first displacement and 17 minutes to complete the second displacement, what is the magnitude and direction of its average velocity during this 62-minute period of time?

31. • What is the direction and magnitude of your total displacement if you have traveled due west with a speed of 20.0 m/s for 120 s, then due south at 15 m/s for 60.0 s?

32. •• You drive a car 1500 ft to the east, then 2500 ft to the north. If the trip took 3.0 minutes, what was the direction and magnitude of your average velocity?

33. •• **IP** A jogger runs with a speed of 3.5 m/s in a direction 20.0° above the x axis. **(a)** Find the x and y components of the jogger's velocity. **(b)** How will the velocity components found in part (a) change if the jogger's speed is halved?

34. •• You throw a ball upward with an initial speed of 4.5 m/s. When it returns to your hand 0.92 s later it has the same speed in the downward direction (assuming air resistance can be ignored). What was the average acceleration vector of the ball?

35. •• A skateboarder rolls from rest down an inclined ramp that is 15 m long and inclined above the horizontal at an angle of $\theta = 20.0°$. When she reaches the bottom of the ramp 3.00 s later her speed is 10.0 m/s. Show that the average acceleration of the skateboarder is $g \sin \theta$, where $g = 9.81 \text{ m/s}^2$.

36. •• Referring to Problem 35, consider a skateboarder who starts from rest at the top of a different ramp that is inclined at an angle of 15° to the horizontal. Assuming that the skateboarder's acceleration is $g \sin 15°$, find his speed when he reaches the bottom of the ramp in 3.0 s.

Section 3–6 Relative Motion

37. • As an airplane taxies on the runway with a speed of 15.5 m/s, a flight attendant walks toward the tail of the plane with a speed of 1.2 m/s. What is the flight attendant's speed relative to the ground?

38. • Referring to Example 3–2, find the time it takes for the boat to reach the opposite shore if the river is 25 m wide.

39. •• As you hurry to catch your flight at the local airport, you encounter a moving walkway that is 85 m long and has a speed of 2.2 m/s relative to the ground. If it takes you 68 s to cover 85 m when walking on the ground, how long will it take you to cover the same distance on the walkway? Assume that you walk with the same speed on the walkway as you do on the ground.

40. •• In Problem 39, how long would it take you to cover the 85 m length of the walkway if, once you get on the walkway, you immediately turn around and start walking in the opposite direction with a speed of 1.3 m/s relative to the walkway?

41. •• **IP** The pilot of an airplane wishes to fly due north, but there is a 75 km/h wind blowing toward the east. **(a)** In what direction should the pilot head her plane if its speed relative to the air is 310 km/h? **(b)** Draw a vector diagram that illustrates your result in part (a). **(c)** If the pilot decreases the air speed of the plane, should the angle found in part (a) be increased or decreased?

42. •• A passenger walks from one side of a ferry to the other as it approaches a dock. If the passenger's velocity is 1.5 m/s due north relative to the ferry, and 4.5 m/s at an angle of 30.0° west of north relative to the water, what are the direction and magnitude of the ferry's velocity relative to the water?

43. •• You are riding on a jet ski at an angle of 35° upstream on a river flowing with a speed of 2.8 m/s. If your velocity relative to the ground is 9.5 m/s at an angle of 20.0° upstream, what is the speed of the jet ski relative to the water?

44. •• **IP** In Problem 43, suppose the jet ski is moving at a speed of 12 m/s relative to the water. **(a)** At what angle must you point the jet ski if your velocity relative to the ground is to be perpendicular to the shore of the river? **(b)** If you increase the speed of the jet ski relative to the water, does the angle in part (a) increase, decrease, or stay the same? Explain.

45. ••• **IP** Two people take identical jet skis across a river, traveling at the same speed relative to the water. Jet ski A heads directly across the river, and is carried downstream by the current before reaching the opposite shore. Jet ski B travels in a direction that is 35° upstream, and arrives at the opposite shore directly across from

the starting point. **(a)** Which jet ski reaches the opposite shore in the least amount of time? **(b)** Confirm your answer to part **(a)** by finding the ratio of the time it takes for the two jet skis to cross the river.

General Problems

46. • You slide a box up a loading ramp that is 10.0 ft long. At the top of the ramp the box has risen a height of 3.00 ft. What is the angle of the ramp above the horizontal?

47. •• Find the x, y, and z components of the vector **A** shown in **Figure 3–38**, given that $A = 75$ m.

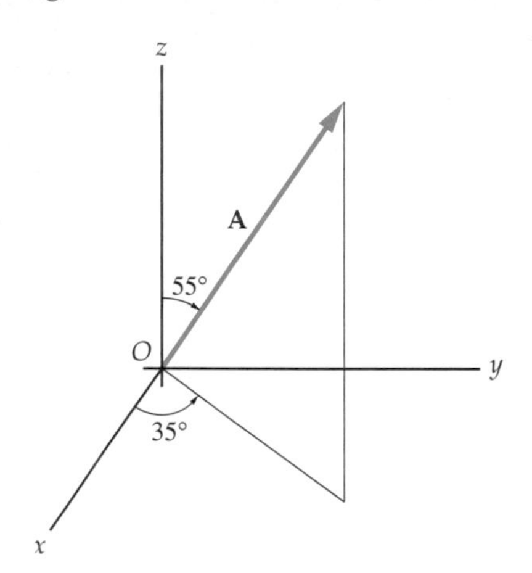

▲ **FIGURE 3–38** Problem 47

48. •• Initially, a particle is moving at 5.0 m/s at an angle of 35.0° above the horizontal. Two seconds later, its velocity is 6.0 m/s at an angle of 50.0° below the horizontal. What was the particle's average acceleration during this time interval?

49. •• A shopper at the supermarket follows the path indicated by vectors **A**, **B**, **C**, and **D** in **Figure 3–39**. Given that the vectors have the magnitudes $A = 51$ ft, $B = 45$ ft, $C = 35$ ft, and $D = 13$ ft, find the total displacement of the shopper using **(a)** the graphical method and **(b)** the component method of vector addition. Compare your results.

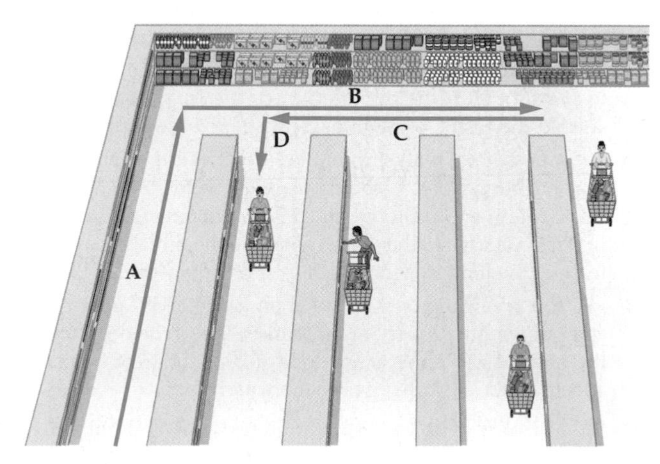

▲ **FIGURE 3–39** Problem 49

50. •• Two airplanes taxi as they approach the terminal. Plane 1 taxies with a speed of 12 m/s due north. Plane 2 taxies with a speed of 7.5 m/s in a direction 20.0° north of west. **(a)** What are the direction and magnitude of the velocity of plane 1 relative to plane 2? **(b)** What are the direction and magnitude of the velocity of plane 2 relative to plane 1?

51. •• A passenger on a bus notices that rain is falling vertically just outside the window. When the bus moves with constant velocity, the passenger observes that the falling raindrops are now making an angle of 15° with respect to the vertical. **(a)** What is the ratio of the speed of the raindrops to the speed of the bus? **(b)** Find the speed of the raindrops, given that the bus is moving with a speed of 18 m/s.

52. •• The corners of an equilateral triangle lie on a circle of radius 2.0 m. Find the length of a side of the triangle.

53. •• **Figure 3–40** shows the velocity of a wave relative to the shore, $\mathbf{v}_{ws}$, and the velocity of a boogie boarder relative to the wave, $\mathbf{v}_{bw}$. The magnitudes of these velocities are $v_{ws} = 1.3$ m/s and $v_{bw} = 2.7$ m/s. Find the velocity of the boogie boarder relative to the shore, using both the graphical and component methods of vector addition. Compare your results.

▲ **FIGURE 3–40** Problem 53

54. ••• Referring to Example 3–2, what heading must the boat have if it is to land directly across the river from its starting point? How much time is required for this trip if the river is 20.0 m wide?

55. ••• Vector **A** points in the negative x direction. Vector **B** points at an angle of 30.0° above the positive x axis. Vector **C** has a magnitude of 15 m and points in a direction 40.0° below the positive x axis. Given that $\mathbf{A} + \mathbf{B} + \mathbf{C} = 0$, find the magnitudes of **A** and **B**.

56. ••• As two boats approach the marina, the velocity of boat 1 relative to boat 2 is 2.3 m/s in a direction 40.0° east of north. If boat 1 has a velocity that is 0.75 m/s due north, what is the velocity (magnitude and direction) of boat 2?

4 Two-Dimensional Kinematics

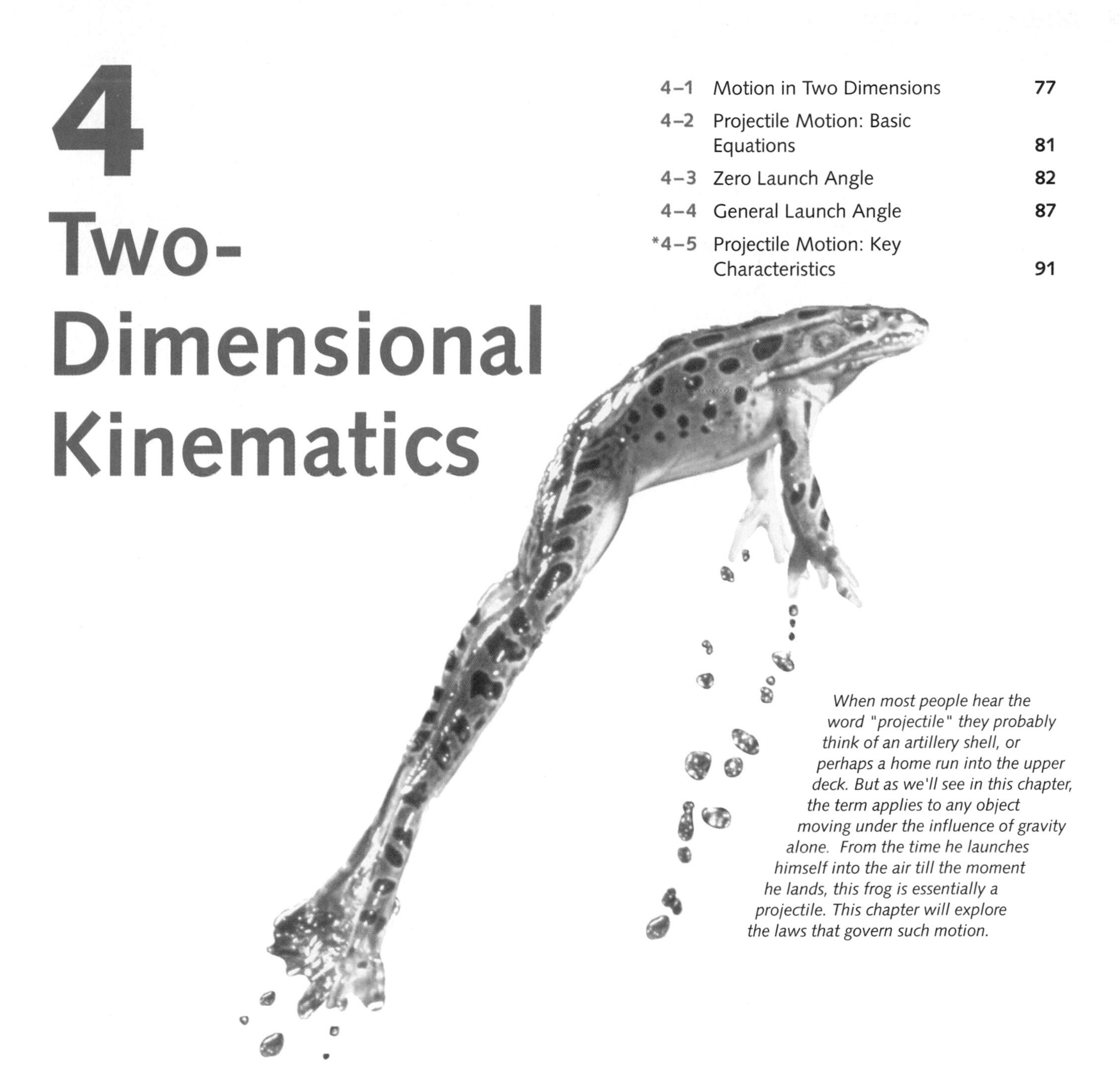

When most people hear the word "projectile" they probably think of an artillery shell, or perhaps a home run into the upper deck. But as we'll see in this chapter, the term applies to any object moving under the influence of gravity alone. From the time he launches himself into the air till the moment he lands, this frog is essentially a projectile. This chapter will explore the laws that govern such motion.

We now extend our study of kinematics to motion in two dimensions. This allows us to consider a much wider range of physical phenomena observed in everyday life. Of particular interest is **projectile motion**, the motion of objects that are initially launched—or "projected"—and which then continue moving under the influence of gravity alone. Examples of projectile motion include balls thrown from one person to another, water spraying from a hose, salmon leaping over rapids, and divers jumping from the cliffs of Acapulco.

The main idea of this chapter is quite simple: Horizontal and vertical motions are independent. That's it. For example, a ball thrown horizontally with a speed v continues to move with the same speed v in the horizontal direction, even as it falls with an increasing speed in the vertical direction. Similarly, the time of fall is the same whether a ball is dropped from rest straight down, or thrown horizontally. Simply put, each motion continues as if the other motion were not present.

This chapter develops and applies the idea of independence of motion to many common physical systems.

4–1 Motion in Two Dimensions

In this section we develop equations of motion to describe objects moving in two dimensions. First, we consider motion with constant velocity, determining x and y as functions of time. Next, we investigate motion with constant acceleration. We show that the one-dimensional kinematic equations of Chapter 2 can be extended in a straightforward way to apply to two dimensions.

Constant Velocity

To begin, consider the simple situation shown in Figure 4–1. A turtle starts at the origin at $t = 0$, and moves with a constant speed $v_0 = 0.26$ m/s in a direction 25° above the x axis. How far has the turtle moved in the x and y directions after 5.0 seconds?

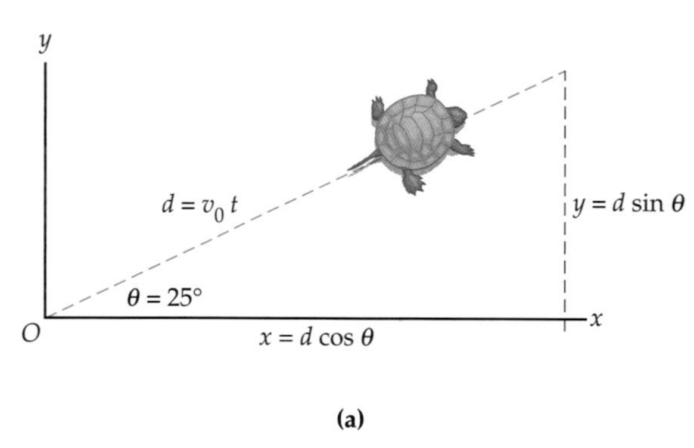

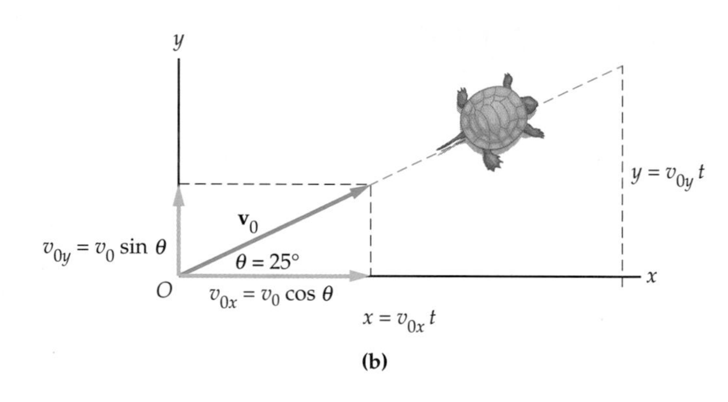

▲ **FIGURE 4–1 Constant velocity**
A turtle walks from the origin with a speed of $v_0 = 0.26$ m/s. **(a)** In a time t the turtle moves a straight-line distance $d = v_0 t$; thus the x and y displacements are $x = d \cos \theta, y = d \sin \theta$. **(b)** Equivalently, the turtle's x and y components of velocity are $v_{0x} = v_0 \cos \theta$ and $v_{0y} = v_0 \sin \theta$; hence $x = v_{0x} t$ and $y = v_{0y} t$.

First, note that the turtle moves in a straight line a distance

$$d = v_0 t = (0.26 \text{ m/s})(5.0 \text{ s}) = 1.3 \text{ m}$$

as indicated in Figure 4–1(a). From the definitions of sine and cosine given in the previous chapter, we see that

$$x = d \cos 25° = 1.2 \text{ m}$$

$$y = d \sin 25° = 0.55 \text{ m}$$

An alternative way to approach this problem is to treat the x and y motions separately. First, we determine the speed of the turtle in each direction. Referring to Figure 4–1(b), we see that the x component of velocity is

$$v_{0x} = v_0 \cos 25° = 0.24 \text{ m/s}$$

and the y component is

$$v_{0y} = v_0 \sin 25° = 0.11 \text{ m/s}$$

Next, we find the distance traveled by the turtle in the x and y directions by multiplying the speed in each direction by the time;

$$x = v_{0x} t = (0.24 \text{ m/s})(5.0 \text{ s}) = 1.2 \text{ m}$$

and

$$y = v_{0y} t = (0.11 \text{ m/s})(5.0 \text{ s}) = 0.55 \text{ m}$$

This is in agreement with our previous results. To summarize, we can think of the turtle's actual motion as a combination of separate x and y motions.

In general, the turtle might start at a position $x = x_0$ and $y = y_0$ at time $t = 0$. In this case, we have

$$x = x_0 + v_{0x}t \quad \text{4–1}$$

and

$$y = y_0 + v_{0y}t \quad \text{4–2}$$

as the x and y equations of motion.

Compare these equations with Equation 2–11, $x = x_0 + v_0t + \frac{1}{2}at^2$, which gives position as a function of time in one dimension. When acceleration is zero, as it is for the turtle, Equation 2–11 reduces to $x = x_0 + v_0t$. Replacing v_0 with the x component of the velocity, v_{0x}, yields Equation 4–1. Similarly, replacing each x in Equation 4–1 with y converts it to Equation 4–2, the y equation of motion.

A situation illustrating the use of Equations 4–1 and 4–2 is given in the next Example.

▲ To analyze the trajectory of this eagle, it is useful to consider the horizontal and vertical components of its motion separately. (See Example 4–1.)

EXAMPLE 4–1 The Eagle Descends

An eagle perched on a tree limb 19.5 m above the water spots a fish swimming near the surface. The eagle pushes off from the branch and descends toward the water. By adjusting its body in flight, the eagle maintains a constant speed of 3.10 m/s at an angle of 20.0° below the horizontal. **(a)** How long does it take for the eagle to reach the water? **(b)** How far has the eagle traveled in the horizontal direction when it reaches the water?

Picture the Problem

We set up our coordinate system so that the eagle starts at $x_0 = 0$ and $y_0 = h = 19.5$ m. The water level is $y = 0$.

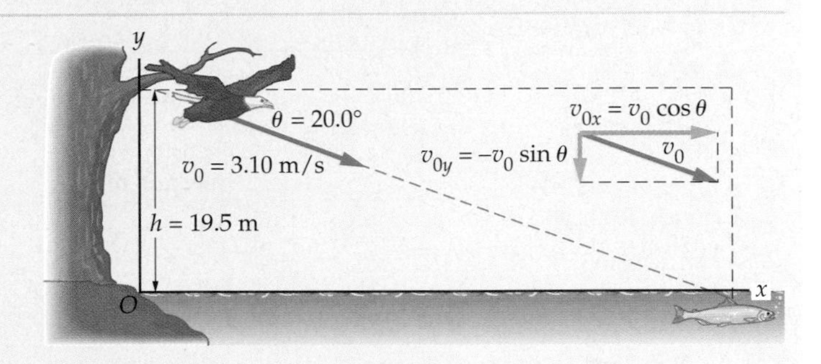

Strategy

As usual in such problems, it is best to treat the eagle's flight as a combination of separate x and y motions. Since we are given the speed of the eagle and the angle at which it descends, we can find the x and y components of its velocity. We then use the y equation of motion, $y = y_0 + v_{0y}t$, to find the time t when the eagle reaches the water. Finally, we use this value of t in the x equation of motion, $x = x_0 + v_{0x}t$, to find the horizontal distance the bird travels.

Solution

Part (a)

1. Begin by determining v_{0x} and v_{0y}:

$$v_{0x} = v_0 \cos\theta = (3.10\text{ m/s})\cos 20.0° = 2.91\text{ m/s}$$
$$v_{0y} = -v_0 \sin\theta = -(3.10\text{ m/s})\sin 20.0° = -1.06\text{ m/s}$$

2. Now, set $y = 0$ in $y = y_0 + v_{0y}t$ and solve for t:

$$y = y_0 + v_{0y}t = h + v_{0y}t = 0$$
$$t = -\frac{h}{v_{0y}} = -\frac{19.5\text{ m}}{(-1.06\text{ m/s})} = 18.4\text{ s}$$

Part (b)

3. Substitute $t = 18.4$ s into $x = x_0 + v_{0x}t$ to find x:

$$x = x_0 + v_{0x}t = 0 + (2.91\text{ m/s})(18.4\text{ s}) = 53.5\text{ m}$$

Insight

Notice how the two minus signs in step 2 combine to give a positive time.

Practice Problem

What is the location of the eagle 2.00 s after it takes flight? [**Answer:** $x = 5.82$ m, $y = 17.4$ m]

Some related homework problems: Problem 1, Problem 2

Constant Acceleration

To study motion with constant acceleration in two dimensions we repeat what was done in one dimension in Chapter 2, but with separate equations for both x and y. For example, to obtain x as a function of time we start with $x = x_0 + v_0t + \frac{1}{2}at^2$

(Equation 2–11), and replace both v_0 and a with the corresponding x components, v_{0x} and a_x. This gives

$$x = x_0 + v_{0x}t + \tfrac{1}{2}a_xt^2 \quad \text{4–3(a)}$$

To obtain y as a function of time we write y in place of x in Equation 4–3(a):

$$y = y_0 + v_{0y}t + \tfrac{1}{2}a_yt^2 \quad \text{4–3(b)}$$

These are the position versus time equations of motion for two dimensions. (Of course, in three dimensions we would simply replace x with z in Equation 4–3(a) to obtain z as a function of time.)

The same approach gives velocity as a function of time. Start with Equation 2-7, $v = v_0 + at$, and write it in terms of x and y components. This yields

$$v_x = v_{0x} + a_xt \quad \text{4–4(a)}$$

$$v_y = v_{0y} + a_yt \quad \text{4–4(b)}$$

Note that we simply repeat everything we did for one dimension, only now with separate equations for the x and y components.

Finally, we can write $v^2 = v_0^2 + 2a\Delta x$ in terms of components as well:

$$v_x^2 = v_{0x}^2 + 2a_x\Delta x \quad \text{4–5(a)}$$

$$v_y^2 = v_{0y}^2 + 2a_y\Delta y \quad \text{4–5(b)}$$

The following table summarizes these results:

TABLE 4–1 Constant-Acceleration Equations of Motion

Position as a function of time	Velocity as a function of time	Velocity as a function of position
$x = x_0 + v_{0x}t + \tfrac{1}{2}a_xt^2$	$v_x = v_{0x} + a_xt$	$v_x^2 = v_{0x}^2 + 2a_x\Delta x$
$y = y_0 + v_{0y}t + \tfrac{1}{2}a_yt^2$	$v_y = v_{0y} + a_yt$	$v_y^2 = v_{0y}^2 + 2a_y\Delta y$

We apply these equations throughout the rest of this chapter.

EXAMPLE 4–2 A Hummer Accelerates

A hummingbird is flying in such a way that it is initially moving vertically with a speed of 4.6 m/s and accelerating horizontally at 11 m/s^2. Assuming the bird's acceleration remains constant for the time interval of interest, find the horizontal and vertical distance through which it moves in 0.55 s.

Picture the Problem

In our sketch we have placed the origin of a two-dimensional coordinate system at the location of the hummingbird at the initial time, $t = 0$. In addition, we have chosen the initial direction of motion to be in the positive y direction, and the direction of acceleration to be in the positive x direction. As a result, it follows that $x_0 = y_0 = 0$, $v_{0x} = 0$, $v_{0y} = 4.6$ m/s, $a_x = 11$ m/s^2, and $a_y = 0$. As the hummingbird moves upward its x component of velocity increases, resulting in a curved path as shown.

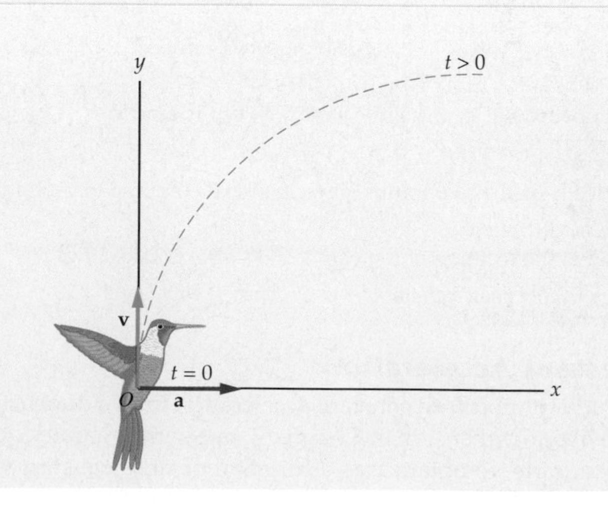

Strategy

Since we want to relate position and time, we find the horizontal position of the hummingbird using $x = x_0 + v_{0x}t + \tfrac{1}{2}a_xt^2$, and the vertical position using $y = y_0 + v_{0y}t + \tfrac{1}{2}a_yt^2$.

Solution

1. Use $x = x_0 + v_{0x}t + \frac{1}{2}a_xt^2$ to find x at $t = 0.55$ s: $x = x_0 + v_{0x}t + \frac{1}{2}a_xt^2 = \frac{1}{2}(11\ \text{m/s}^2)(0.55\ \text{s})^2 = 1.7\ \text{m}$
2. Use $y = y_0 + v_{0y}t + \frac{1}{2}a_yt^2$ to find y at $t = 0.55$ s: $y = y_0 + v_{0y}t + \frac{1}{2}a_yt^2 = (4.6\ \text{m/s})(0.55\ \text{s}) = 2.5\ \text{m}$

Insight

In 0.55 s the hummingbird moves 1.7 m horizontally and 2.5 m vertically.

Practice Problem

How much time is required for the hummingbird to move 2.0 m horizontally from its initial position? [**Answer:** $t = 0.60$ s]

Some related homework problems: Problem 3, Problem 4

4–2 Projectile Motion: Basic Equations

We now apply the independence of horizontal and vertical motions to projectiles. Just what do we mean by a projectile? Well, a **projectile** is an object that is thrown, kicked, batted, or otherwise launched into motion and then allowed to follow a path determined solely by the influence of gravity. As you might expect, this covers a wide variety of physical systems.

In studying projectile motion we make the following assumptions:

- air resistance is ignored
- the acceleration of gravity is constant, downward, and has a magnitude equal to $g = 9.81\ \text{m/s}^2$
- the Earth's rotation is ignored

Air resistance can be significant when a projectile moves with relatively high speed or if it encounters a strong wind. In many everyday situations, however, like tossing a ball to a friend or dropping a book, air resistance is relatively insignificant. As for the acceleration of gravity, $g = 9.81\ \text{m/s}^2$, this value varies slightly from place to place on the earth's surface and decreases with increasing altitude. In addition, the rotation of the earth can be significant when considering projectiles that cover great distances. Little error is made in ignoring the variation of g or the rotation of the earth, however, in the examples of projectile motion considered in this chapter.

Let's incorporate these assumptions into the equations of motion given in the previous section. Suppose, as in **Figure 4–2**, that the x axis is horizontal and the y axis is vertical, with the positive direction upward. Since downward is the negative direction, it follows that

$$a_y = -9.81\ \text{m/s}^2 = -g$$

Gravity causes no acceleration in the x direction. Thus, the x component of acceleration is zero:

$$a_x = 0$$

With these acceleration components substituted into the fundamental constant-acceleration equations of motion (Table 4–1) we find:

Projectile Motion

$$x = x_0 + v_{0x}t \qquad v_x = v_{0x} \qquad v_x^2 = v_{0x}^2$$
$$y = y_0 + v_{0y}t - \tfrac{1}{2}gt^2 \qquad v_y = v_{0y} - gt \qquad v_y^2 = v_{0y}^2 - 2g\Delta y \qquad \textbf{4–6}$$

Note that in these expressions the positive y direction is upward and the quantity g is positive.

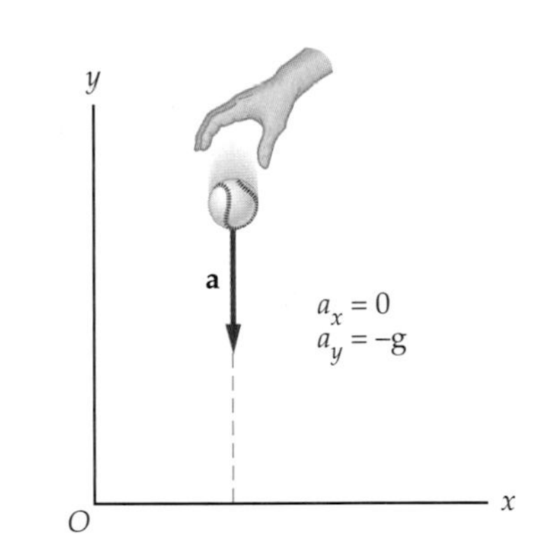

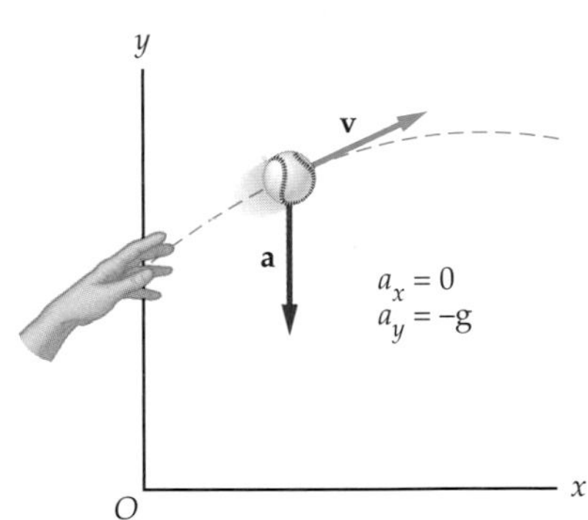

▲ **FIGURE 4–2 Acceleration in free fall**
All objects in free fall have acceleration components $a_x = 0$ and $a_y = -g$ when the coordinate system is chosen as shown here. This is true regardless of whether the object is dropped, thrown, kicked, etc.

PROBLEM SOLVING NOTE
Acceleration of a Projectile

When the x axis is chosen to be horizontal and the y axis points vertically upward, it follows that the acceleration of an ideal projectile is $a_x = 0$ and $a_y = -g$.

▲ In the multiple-exposure photo at left, a ball is projected upward from a moving cart. The ball retains its initial horizontal velocity; as a result, it follows a parabolic path and remains directly above the cart at all times. When the ball lands, it falls back into the cart, just as it would if the cart had been at rest. (In this sequence, the exposures were made at equal time intervals with light of different colors, making it easier to follow the relative motion of the ball and the cart.) In the photo at right, the pilot ejection seat of a jet fighter is being ground-tested. Here too, the horizontal and vertical motions are independent; thus, the test dummy is still almost directly above the cockpit from which it was ejected. Note, however, that air resistance is beginning to reduce the dummy's horizontal velocity. Eventually, it will fall far behind the speeding plane.

▶ **FIGURE 4–3 Independence of vertical and horizontal motions**

When you drop a ball while walking or running, it appears to you to drop straight down from the point where you released it. To a person at rest, the ball follows a curved path that combines horizontal and vertical motions.

A simple demonstration illustrates the independence of horizontal and vertical motions in projectile motion. First, while standing still, drop a rubber ball to the floor and catch it on the rebound. Note that the ball goes straight down, lands near your feet, and returns almost to the level of your hand in about a second.

Next, walk with constant speed before dropping the ball, then observe its motion carefully. To you, its motion looks the same as before: It goes straight down, lands near your feet, bounces straight back up, and returns in about one second. This is illustrated in Figure 4–3. The fact that you were moving in the horizontal direction the whole time had no effect on the ball's vertical motion—the motions were independent.

To an observer who sees you walking by, the ball follows a curved path, as shown. The precise shape of this curved path is determined in the next section.

4–3 Zero Launch Angle

A special case of some interest is a projectile launched horizontally, so that the angle between the initial velocity and the horizontal is $\theta = 0$. We devote this section to a brief look at this type of motion.

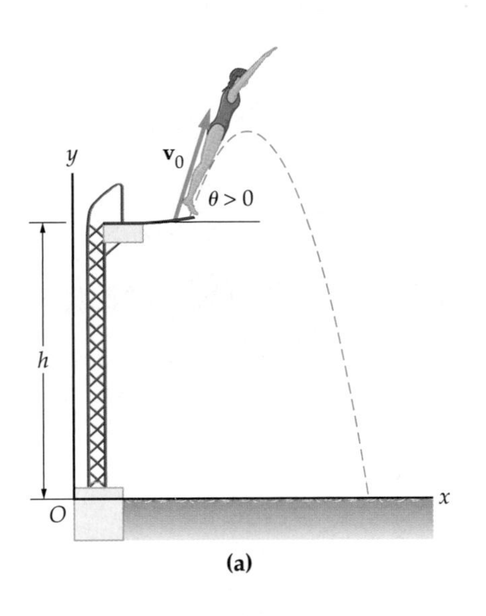

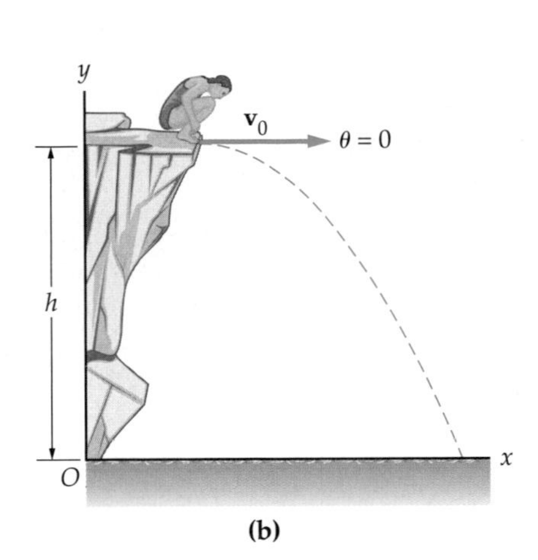

◀ **FIGURE 4–4 Launch angle of a projectile**

(a) A projectile launched at an angle above the horizontal, $\theta > 0$. A launch below the horizontal would correspond to $\theta < 0$. **(b)** A projectile launched horizontally, $\theta = 0$. In this section we consider $\theta = 0$. The next section deals with $\theta \neq 0$.

Equations of Motion

Suppose you are walking with a speed v_0 when you release a ball from a height h, as discussed in the previous section. If we choose ground level to be $y = 0$ and the release point to be directly above the origin, the initial position of the ball is given by

$$x_0 = 0$$

and

$$y_0 = h$$

This is illustrated in Figure 4–3.

The initial velocity is horizontal, corresponding to $\theta = 0$ in **Figure 4–4**. As a result, the x component of the initial velocity is simply the initial speed,

$$v_{0x} = v_0 \cos 0° = v_0$$

and the y component of the initial velocity is zero,

$$v_{0y} = v_0 \sin 0° = 0$$

Substituting these results into our basic equations for projectile motion (Equation 4–6) gives the following results for zero launch angle ($\theta = 0$):

$$x = v_0 t \qquad v_x = v_0 = \text{constant} \qquad v_x^2 = v_0^2 = \text{constant}$$
$$y = h - \tfrac{1}{2}gt^2 \qquad v_y = -gt \qquad v_y^2 = -2g\Delta y \qquad \textbf{4–7}$$

Note that the x component of velocity remains the same for all time and that the y component steadily decreases with time. As a result, x increases linearly with time, and y decreases with a t^2 dependence. Snapshots of this motion at equal time intervals are shown in **Figure 4–5**.

The launch point of a projectile determines x_0 and y_0. The initial velocity of a projectile determines v_{0x} and v_{0y}.

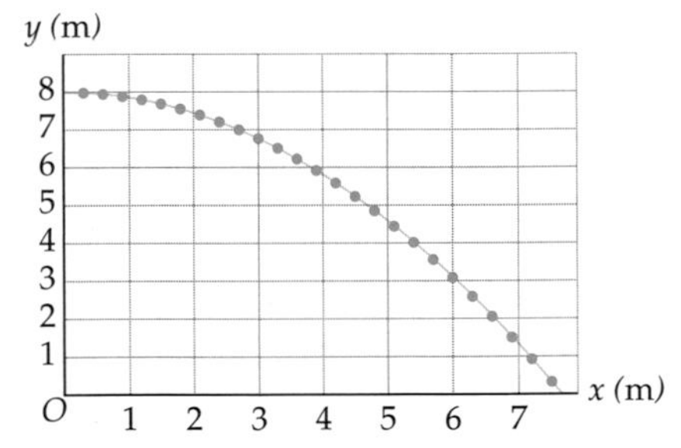

▲ **FIGURE 4–5 Trajectory of a projectile launched horizontally**

In this plot, the projectile was launched from a height of 8.0 m with an initial speed of 6.0 m/s. The positions shown in the plot correspond to the times $t = 0.05$ s, 0.10 s, 0.15 s, ...

EXAMPLE 4–3 Dropping a Ball

A person walking with a speed of 1.30 m/s releases a ball from a height of 1.25 m above the ground. Given that $x_0 = 0$ and $y_0 = h = 1.25$ m, find x and y for **(a)** $t = 0.250$ s and **(b)** $t = 0.500$ s. **(c)** Find the velocity, speed, and direction of motion of the ball at $t = 0.500$ s.

continued on the following page

continued from the previous page

Picture the Problem

The ball starts at $x_0 = 0$ and $y_0 = h = 1.25$ m. It accelerates downward in the y direction and moves with constant speed in the x direction.

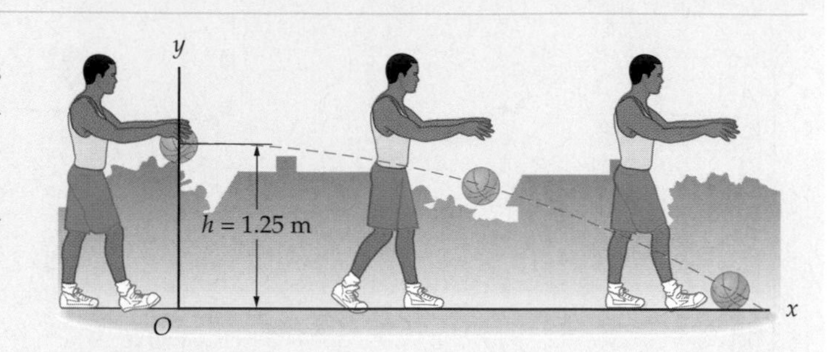

Strategy

The x and y positions are given by $x = v_0 t$ and $y = h - \frac{1}{2}gt^2$, respectively. We simply substitute time into these expressions. Similarly, the velocity components are $v_x = v_0$ and $v_y = -gt$.

Solution

Part (a)

1. Substitute $t = 0.250$ s into the x and y equations of motion:

$$x = v_0 t = (1.30 \text{ m/s})(0.250 \text{ s}) = 0.325 \text{ m}$$
$$y = h - \tfrac{1}{2}gt^2$$
$$= 1.25 \text{ m} - \tfrac{1}{2}(9.81 \text{ m/s}^2)(0.250 \text{ s})^2 = 0.943 \text{ m}$$

Part (b)

2. Substitute $t = 0.500$ s into the x and y equations of motion:

$$x = v_0 t = (1.30 \text{ m/s})(0.500 \text{ s}) = 0.650 \text{ m}$$

Note that the ball is only about an inch above the ground at this time:

$$y = h - \tfrac{1}{2}gt^2$$
$$= 1.25 \text{ m} - \tfrac{1}{2}(9.81 \text{ m/s}^2)(0.500 \text{ s})^2 = 0.0238 \text{ m}$$

Part (c)

3. First, calculate the x and y components of the velocity at $t = 0.500$ s using $v_x = v_0$ and $v_y = -gt$:

$$v_x = v_0 = 1.30 \text{ m/s}$$
$$v_y = -gt = -(9.81 \text{ m/s}^2)(0.500 \text{ s}) = -4.91 \text{ m/s}$$

4. Use these components to determine $\mathbf{v}$, v, and θ:

$$\mathbf{v} = (1.30 \text{ m/s})\hat{\mathbf{x}} + (-4.91 \text{ m/s})\hat{\mathbf{y}}$$
$$v = \sqrt{v_x^2 + v_y^2}$$
$$= \sqrt{(1.30 \text{ m/s})^2 + (-4.91 \text{ m/s})^2} = 5.08 \text{ m/s}$$
$$\theta = \tan^{-1}\frac{v_y}{v_x} = \tan^{-1}\frac{(-4.91 \text{ m/s})}{1.30 \text{ m/s}} = -75.2°$$

Insight

Note that the x position of the ball does not depend on the acceleration of gravity, g, and that its y position does not depend on the initial horizontal speed of the ball, v_0.

Practice Problem

How long does it take for the ball to land? [**Answer:** Referring to the results of part **(b)**, it is clear that the time of landing is slightly greater than 0.500 s. Setting $y = 0$ gives a precise answer; $t = \sqrt{2h/g} = 0.505$ s.]

Some related homework problems: Problem 12, Problem 13, Problem 17

CONCEPTUAL CHECKPOINT 4–1

Two youngsters dive off an overhang into a lake. Diver 1 drops straight down, diver 2 runs off the cliff with an initial horizontal speed v_0. Is the splashdown speed of diver 2 **(a)** greater than, **(b)** less than, or **(c)** equal to the splashdown speed of diver 1?

Reasoning and Discussion

Note that neither diver has an initial y component of velocity, and that they both fall with the same vertical acceleration—the acceleration of gravity. Therefore, the two divers fall for the same amount of time, and their y components of velocity are the same at splashdown. Since diver 2 also has a nonzero x component of velocity, unlike diver 1, the speed of diver 2 is greater.

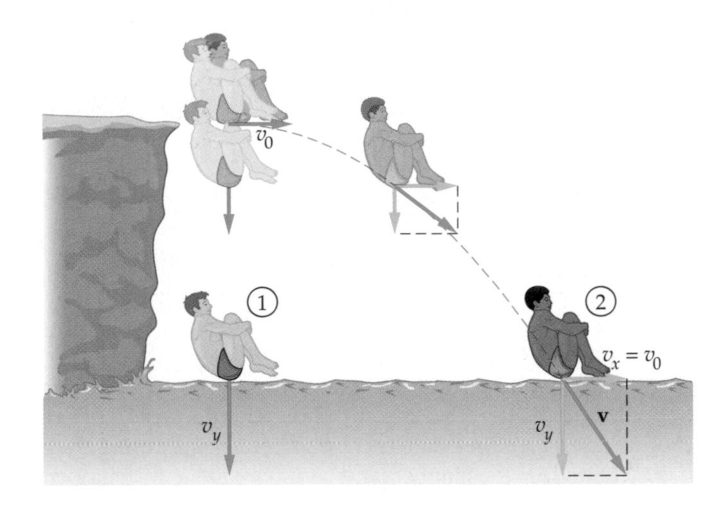

Answer:
(a) The speed of diver 2 is greater than that of diver 1.

Parabolic Path

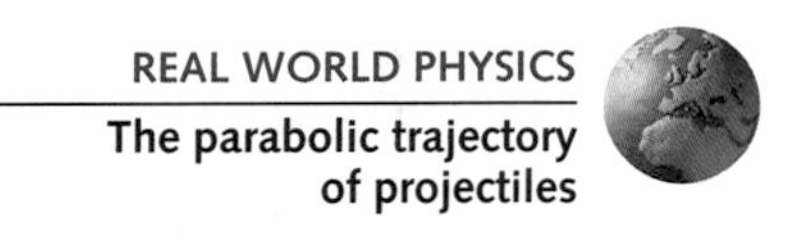

Just what is the shape of the curved path followed by a projectile launched horizontally? This can be found by combining $x = v_0t$ and $y = h - \frac{1}{2}gt^2$. First, solve for time using the x equation. This gives

$$t = x/v_0$$

Next, substitute this result into the y equation to eliminate t:

$$y = h - \tfrac{1}{2}g\left(\frac{x}{v_0}\right)^2 = h - \left(\frac{g}{2v_0^2}\right)x^2 \qquad \textbf{4–8}$$

Note that y has the form

$$y = a + bx^2$$

where $a = h =$ constant and $b = -g/2v_0^2 =$ constant. This is the equation of a parabola that curves downward, a characteristic shape in projectile motion.

Landing Site

Where does a projectile land if it is launched horizontally with a speed v_0 from a height h?

The most direct way to answer this question is to set $y = 0$ in Equation 4–8, since $y = 0$ corresponds to ground level. This gives

$$0 = h - \left(\frac{g}{2v_0^2}\right)x^2$$

Solving for x yields the landing site:

$$x = v_0\sqrt{\frac{2h}{g}} \qquad \textbf{4–9}$$

Note that we have chosen the positive sign for the square root since the projectile was launched in the positive x direction, and hence lands at a positive value of x.

A useful alternative approach is to find the time of landing and then substitute this time into $x = v_0t$. The landing time is found by setting $y = 0$ in $y = h - \frac{1}{2}gt^2$. This gives

$$0 = h - \tfrac{1}{2}gt^2 \quad \text{or} \quad t = \sqrt{\frac{2h}{g}}$$

▲ Lava bombs (top) and fountain jets (bottom) trace out parabolic paths, as is typical in projectile motion. The trajectories are only slightly altered by air resistance.

PROBLEM SOLVING NOTE
Use Independence of Motion

Projectile problems can be solved by breaking the problem into its x and y components, and then solving for the motion of each component separately.

Substituting this result into the x equation of motion, we again find

$$x = v_0\sqrt{\frac{2h}{g}}$$

Note that x is simply the constant speed in the x direction times the time of fall, as one would expect.

EXAMPLE 4–4 Jumping a Crevasse

A mountain climber encounters a crevasse in an ice field. The opposite side of the crevasse is 2.75 m lower, and is separated horizontally by a distance of 4.10 m. To cross the crevasse, the climber gets a running start and jumps in the horizontal direction. **(a)** What is the minimum speed needed by the climber to safely cross the crevasse? If, instead, the climber's speed is 6.00 m/s, **(b)** where does the climber land, and **(c)** what is the climber's speed on landing?

Picture the Problem

The mountain climber jumps from $y_0 = h = 2.75$ m and $x_0 = 0$. The landing site for part **(a)** is $y = 0$ and $x = w = 4.10$ m. Note that the y position of the climber decreases by h, thus $\Delta y = -h = -2.75$ m.

Strategy

(a) Given $x = 4.10$ m, we can use $x = v_0\sqrt{2h/g}$ to solve for v_0.

(b) Similarly, we can substitute $v_0 = 6.00$ m/s in $x = v_0\sqrt{2h/g}$ to find x.

(c) We already know v_x, and we can calculate v_y using $v_y^2 = -2g\Delta y$ (Equation 4-7). With the velocity components known, we can use the Pythagorean theorem to find the speed.

Solution

Part (a)

1. Solve $x = v_0\sqrt{2h/g}$ to obtain an expression for v_0:

$$x = v_0\sqrt{\frac{2h}{g}} \quad \text{or} \quad v_0 = x\sqrt{\frac{g}{2h}}$$

2. Substitute numerical values in this expression:

$$v_0 = x\sqrt{\frac{g}{2h}} = (4.10\text{ m})\sqrt{\frac{9.81\text{ m/s}^2}{2(2.75\text{ m})}} = 5.48\text{ m/s}$$

Part (b)

3. Substitute $v_0 = 6.00$ m/s in $x = v_0\sqrt{2h/g}$:

$$x = v_0\sqrt{\frac{2h}{g}} = (6.00\text{ m/s})\sqrt{\frac{2(2.75\text{ m})}{9.81\text{ m/s}^2}} = 4.49\text{ m}$$

Part (c)

4. Use the fact that the x component of velocity does not change to determine v_x, and use $v_y{}^2 = -2g\Delta y$ to determine v_y. For v_y, note that we choose the minus sign for the square root, because the climber is moving downward:

$$v_x = v_0 = 6.00\text{ m/s}$$
$$v_y = \pm\sqrt{-2g\Delta y}$$
$$= -\sqrt{-2(9.81\text{ m/s}^2)(-2.75\text{ m})} = -7.35\text{ m/s}$$

5. Use the Pythagorean theorem to determine the speed:

$$v = \sqrt{v_x^2 + v_y^2}$$
$$= \sqrt{(6.00\text{ m/s})^2 + (-7.35\text{ m/s})^2} = 9.49\text{ m/s}$$

Insight

The minimum speed needed to safely cross the crevasse is 5.48 m/s. If the initial horizontal speed is 6.00 m/s, the climber will land 4.49 m − 4.10 m = 0.39 m beyond the edge of the crevasse with a speed of 9.49 m/s.

Practice Problem

(a) When the climber's speed is the minimum needed to cross the crevasse, $v_0 = 5.48$ m/s, how long is the climber in the air? **(b)** How long is the climber in the air when $v_0 = 6.00$ m/s? **[Answer: (a)** $t = x/v_0 = (4.10 \text{ m})/(5.48 \text{ m/s}) = 0.748$ s. **(b)** $t = x/v_0 = t = x/v_0 = (4.49 \text{ m})(6.00 \text{ m/s}) = 0.748$ s. The times are the same! The answer to both parts is simply the time needed to fall through a height h; $t = \sqrt{2h/g} = 0.748$ s.]

Some related homework problems: Problem 8, Problem 9, Problem 14

CONCEPTUAL CHECKPOINT 4–2

If the height h is increased in the previous example but the width w remains the same, does the minimum speed needed to cross the crevasse **(a)** increase, **(b)** decrease, or **(c)** stay the same?

Reasoning and Discussion

If the height is greater, the time of fall is also greater. Since the climber is in the air for a greater time, the horizontal distance covered for a given initial speed is also greater. Thus, if the width of the crevasse is the same, a lower initial speed allows for a safe crossing.

Answer

(b) The minimum speed decreases.

4–4 General Launch Angle

We now consider the case of a projectile launched at an arbitrary angle with respect to the horizontal. To simplify the resulting equations, we take the launch site to be at the origin.

Figure 4–6 (a) shows a projectile launched with an initial speed v_0 at an angle θ above the horizontal. Since the projectile starts at the origin, the initial x and y positions are zero:

$$x_0 = y_0 = 0$$

The components of the initial velocity are determined as indicated in **Figure 4–6 (b)**:

$$v_{0x} = v_0 \cos \theta$$

and

$$v_{0y} = v_0 \sin \theta$$

As a quick check, note that if $\theta = 0$ then $v_{0x} = v_0$ and $v_{0y} = 0$. Similarly, if $\theta = 90°$ we find $v_{0x} = 0$ and $v_{0y} = v_0$. These checks are depicted in **Figure 4–6 (c)**.

Substituting these results into the projectile-motion equations from Section 4–2 (Equation 4–6) yields the following results for a general launch angle ($\theta \neq 0$):

$$x = (v_0 \cos \theta)t \qquad v_x = v_{0x} \cos \theta \qquad v_x^2 = v_{0x}^2 \cos^2 \theta$$

$$y = (v_0 \sin \theta)t - \tfrac{1}{2}gt^2 \qquad v_y = v_{0y} \sin \theta - gt \qquad v_y^2 = v_{0y}^2 \sin^2 \theta - 2g\Delta y \qquad \textbf{4–10}$$

In the next two exercises we calculate a projectile's position and velocity for three equally spaced times.

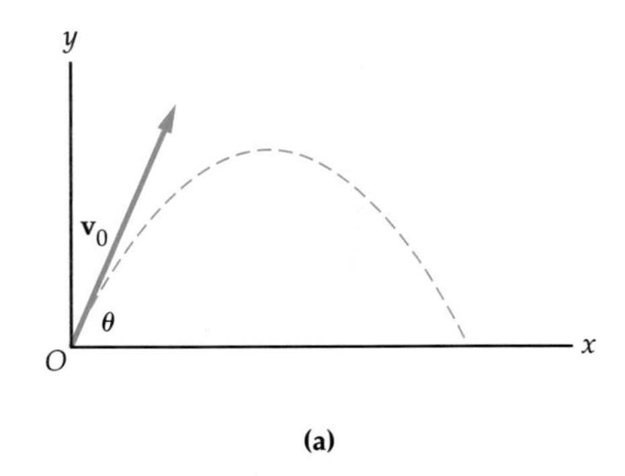

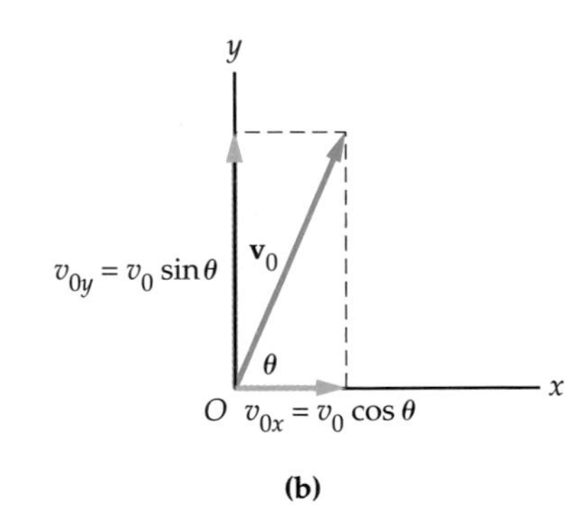

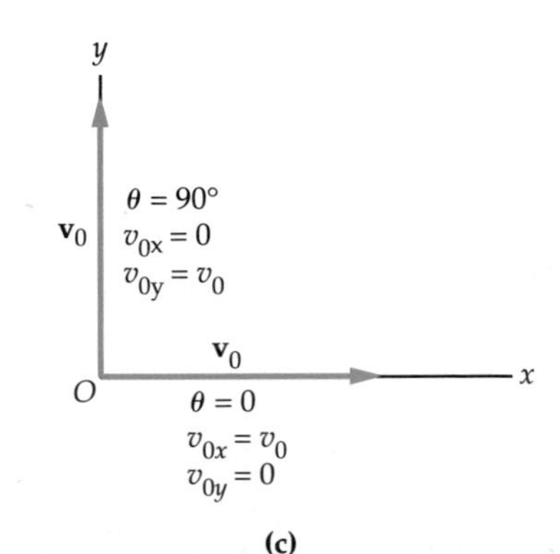

▲ **FIGURE 4–6 Projectile with arbitrary launch angle**

(a) A projectile launched from the origin at an angle θ above the horizontal. **(b)** The x and y components of the initial velocity. **(c)** Velocity components in the limits $\theta = 0$ and $\theta = 90°$.

EXERCISE 4–1

A projectile is launched from the origin with an initial speed of 20.0 m/s at an angle of 35.0° above the horizontal. Find the x and y positions of the projectile at times **(a)** $t = 0.500$ s, **(b)** $t = 1.00$ s, and **(c)** $t = 1.50$ s.

Solution

(a) $x = 8.19$ m, $y = 4.51$ m, **(b)** $x = 16.4$ m, $y = 6.57$ m, **(c)** $x = 24.6$ m, $y = 6.17$ m. Note that x increases steadily; y increases, then decreases.

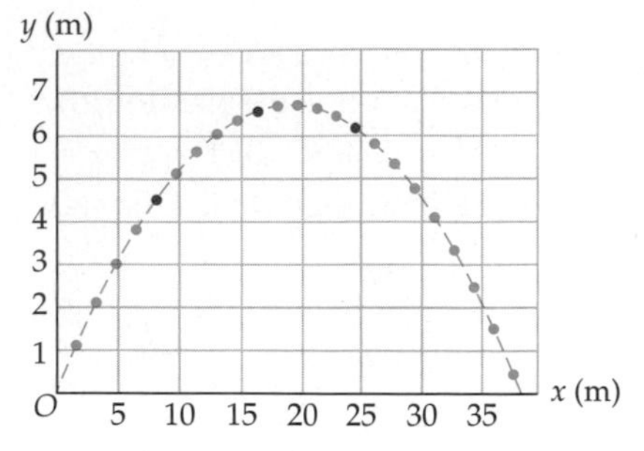

▲ **FIGURE 4–7 Snapshots of a trajectory**

This plot shows a projectile launched from the origin with an initial speed of 20.0 m/s at an angle of 35.0° above the horizontal. The positions shown in the plot correspond to the times $t = 0.1$ s, 0.2 s, 0.3 s, ... Red dots mark the positions considered in Exercises 4-1 and 4-2.

EXERCISE 4–2

Referring to the projectile in Exercise 4-1, find v_x and v_y at times **(a)** $t = 0.500$ s, **(a)** $t = 1.00$ s, and **(b)** $t = 1.50$ s.

Solution

(a) $v_x = 16.4$ m/s, $v_y = 6.57$ m/s, **(b)** $v_x = 16.4$ m/s, $v_y = 1.66$ m/s, **(c)** $v_x = 16.4$ m/s, $v_y = -3.24$ m/s.
Note that v_x is constant; v_y decreases steadily.

Figure 4–7 shows the projectile referred to in the previous exercises for a series of times spaced by 0.10 s. Note that the points in Figure 4–7 are not evenly spaced in terms of position, even though they are evenly spaced in time. In fact, the points bunch closer together at the top of the trajectory, showing that a comparatively large fraction of the flight time is spent near the highest point. This is why it seems that a basketball player soaring toward a slam dunk, or a ballerina performing a grand jeté, is "hanging" in air.

▲ "Hanging" in air near the peak of a jump requires no special knack—in fact, it's an unavoidable consequence of the laws of physics. This phenomenon, which makes big leapers (such as deer and dancers) look particularly graceful, can also make life more dangerous for salmon fighting their way upstream to spawn.

EXAMPLE 4–5 A Rough Shot

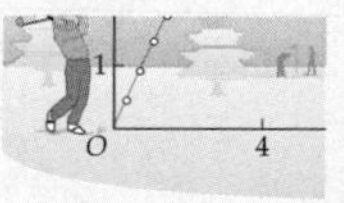

Chipping from the rough, a golfer sends the ball over a 3.00 m-high tree that is 14.0 m away. The ball lands at the same level from which it was struck after traveling a horizontal distance of 17.8 m—on the green, of course. **(a)** If the ball left the club 54.0° above the horizontal, and landed on the green 2.24 s later, what was its initial speed? **(b)** How high was the ball when it passed over the tree?

Picture the Problem

Our sketch shows the ball taking flight from the origin and arcing over the tree. The individual points correspond to equal time intervals.

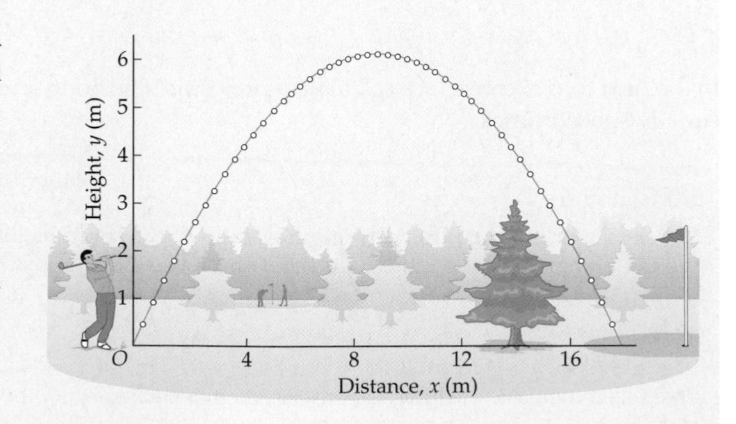

Strategy

(a) Since the projectile moves with constant speed in the x direction, the x component of velocity is simply horizontal distance divided by time. Knowing v_x and θ, we can find v_0 from $v_x = v_0 \cos\theta$.

(b) We can use $x = (v_0 \cos\theta)t$ to find the time when the ball is at $x = 14.0$ m. Substituting this time into $y = (v_0 \sin\theta)t - \frac{1}{2}gt^2$ gives the height.

Solution

1. **(a)** Divide the horizontal distance, d, by the time of flight, t, to obtain v_x:

$$v_x = \frac{d}{t} = \frac{17.8 \text{ m}}{2.24 \text{ s}} = 7.95 \text{ m/s}$$

2. Use $v_x = v_0 \cos\theta$ to find v_0, the initial speed:

$$v_x = v_0 \cos\theta \quad \text{or}$$
$$v_0 = \frac{v_x}{\cos\theta} = \frac{7.95 \text{ m/s}}{\cos 54.0^\circ} = 13.5 \text{ m/s}$$

3. **(b)** Use $x = (v_0 \cos\theta)t$ to find the time when $x = 14.0$ m:

$$x = (v_0 \cos\theta)t \quad \text{or}$$
$$t = \frac{x}{v_0 \cos\theta} = \frac{14.0 \text{ m}}{7.95 \text{ m/s}} = 1.76 \text{ s}$$

4. Evaluate $y = (v_0 \sin\theta)t - \frac{1}{2}gt^2$ at the time found in Step 3:

$$y = (v_0 \sin\theta)t - \tfrac{1}{2}gt^2$$
$$= [(13.5 \text{ m/s})\sin 54.0^\circ](1.76 \text{ s})$$
$$-\tfrac{1}{2}(9.81 \text{ m/s}^2)(1.76 \text{ s})^2 = 4.03 \text{ m}$$

Insight

The ball clears the top of the tree by 1.03 m and lands on the green 50.0° s later.

Practice Problem

What is the speed and direction of the ball when it passes over the tree? [**Answer:** The speed is $v = \sqrt{v_x^2 + v_y^2}$. To find the direction of the ball, note that $v_x = 7.95$ m/s and $v_y = v_0 \sin\theta - gt = -6.34$ m/s; thus $v = 10.2$ m/s and $\theta = \tan^{-1}(v_y/v_x) = -38.6^\circ$.]

Some related homework problems: Problem 25, Problem 33

ACTIVE EXAMPLE 4–1 An Elevated Green

A golfer hits a ball with an initial speed of 30.0 m/s at an angle of 50.0° above the horizontal. The ball lands on a green that is 5.00 m above the level where the ball was struck.

(a) How long is the ball in the air?
(b) How far has the ball traveled in the horizontal direction when it lands?
(c) What is the speed and direction of motion of the ball just before it lands?

Solution

Part (a)

1. Let $y = (v_0 \sin\theta)t - \frac{1}{2}gt^2 = 5.00$ m and solve for t: $t = 0.229$ s, 4.46 s
2. When $t = 0.229$ s the ball is moving upward, when $t = 4.46$ s the ball is on the way down. Choose the later time: $t = 4.46$ s

Part (b)

3. Substitute $t = 4.46$ s into $x = (v_0 \cos\theta)t$: $x = 86.0$ m

Part (c)

4. Use $v_x = v_0 \cos\theta$ to calculate v_x: $v_x = 19.3$ m/s
5. Substitute $t = 4.46$ s into $v_y = v_0 \sin\theta - gt$ to find v_y: $v_y = -20.8$ m/s
6. Calculate v and θ: $v = 28.4$ m/s, $\theta = -47.1^\circ$

The next Example presents a classic situation where two projectiles collide. One projectile is launched from the origin, and thus its equations of motion are given by Equation 4–10. The second projectile is simply dropped from a height, which is a special case of the equations of motion in Equation 4–7 with $v_0 = 0$.

EXAMPLE 4–6 A Leap of Faith

A trained dolphin leaps from the water with an initial speed of 12.0 m/s. It jumps directly toward a ball held by the trainer a horizontal distance of 5.50 m away and a vertical distance of 4.10 m above the water. In the absence of gravity the dolphin would move in a straight line to the ball and catch it, but because of gravity the dolphin follows a parabolic path well below the ball's initial position, as shown. If the trainer releases the ball the instant the dolphin leaves the water, show that the dolphin and the falling ball meet.

Picture the Problem

In our sketch we have the dolphin leaping from the water at the origin with an angle above the horizontal of $\theta = \tan^{-1}(h/d)$. The initial position of the ball is $x_0 = d = 5.50$ m and $y_0 = h = 4.10$ m, and its initial velocity is zero. The ball drops straight down with the acceleration of gravity.

Strategy

We want to show that when the dolphin is at $x = d$, its height above the water is the same as the height of the ball above the water. To do this we first find the time when the dolphin is at $x = d$, then calculate y for the dolphin at this time. Next, we calculate y of the ball at the same time, and then check to see if they are equal.

Since the ball drops from rest from a height h, its y equation of motion is $y = h - \frac{1}{2}gt^2$, as in Equation 4–7 in Section 4–3.

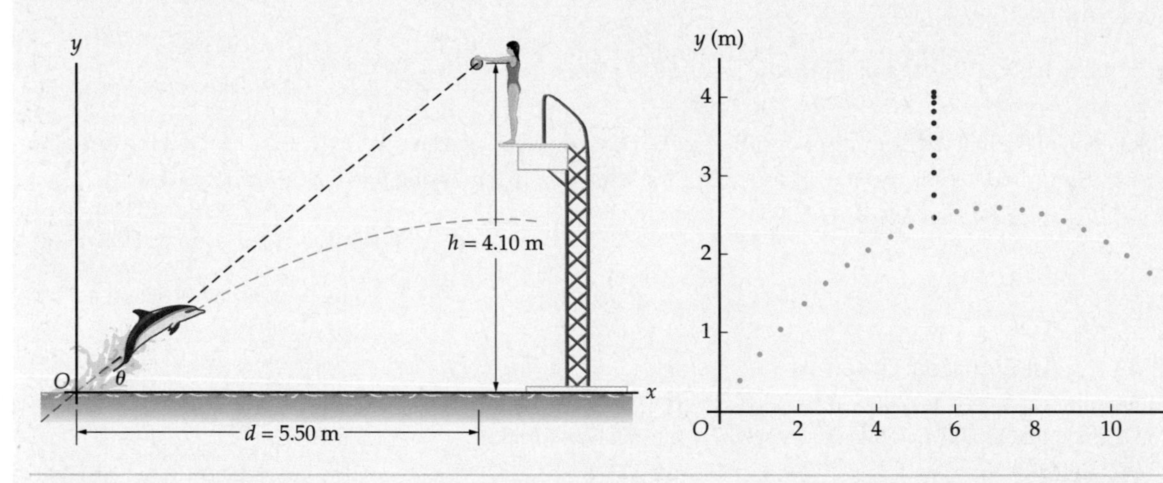

Solution

1. Calculate the angle at which the dolphin leaves the water:

$$\theta = \tan^{-1}\left(\frac{h}{d}\right) = \tan^{-1}\left(\frac{4.10 \text{ m}}{5.50 \text{ m}}\right) = 36.7°$$

2. Use this angle and the initial speed to find the time t when the x position of the dolphin, x_d, is equal to 5.50 m. The x equation of motion is $x_d = (v_0 \cos\theta)t$:

$$x_d = (v_0 \cos\theta)t = [(12.0 \text{ m/s})\cos 36.7°]t = (9.62 \text{ m/s})t$$
$$= 5.50 \text{ m}$$
$$t = \frac{5.50 \text{ m}}{9.62 \text{ m/s}} = 0.572 \text{ s}$$

3. Evaluate the y position of the dolphin, y_d, at $t = 0.572$ s. The y equation of motion is $y_d = (v_0 \sin\theta)t - \frac{1}{2}gt^2$:

$$y_d = (v_0 \sin\theta)t - \tfrac{1}{2}gt^2$$
$$= [(12.0 \text{ m/s})\sin 36.7°](0.572 \text{ s})$$
$$-\tfrac{1}{2}(9.81 \text{ m/s}^2)(0.572 \text{ s})^2$$
$$= 4.10 \text{ m} - 1.60 \text{ m} = 2.50 \text{ m}$$

4. Finally, evaluate the y position of the ball, y_b, at $t = 0.572$ s. The ball's equation of motion is $y_b = h - \frac{1}{2}gt^2$:

$$y_b = h - \tfrac{1}{2}gt^2 = 4.10 \text{ m} - \tfrac{1}{2}(9.81 \text{ m/s}^2)(0.572 \text{ s})^2$$
$$= 4.10 \text{ m} - 1.60 \text{ m} = 2.50 \text{ m}$$

Insight

In the absence of gravity, both the dolphin and the ball would be at $y = 4.10$ m at $t = 0.572$ s. Because of gravity, the dolphin and the ball fall below their zero-gravity positions—and by the same amount, 1.60 m.

Practice Problem

At what height does the dolphin catch the ball if it leaves the water with an initial speed is 8.00 m/s? [**Answer:** $y_d = y_b = 0.498$ m. If the dolphin's initial speed is less than 7.50 m/s it hits the water before catching the ball.]

Some related homework problems: Problem 25, Problem 34

*4–5 Projectile Motion: Key Characteristics

We conclude this chapter with a brief look at some additional characteristics of projectile motion that are both interesting and useful. In all cases our results follow as a direct consequence of the fundamental kinematic equations (Equation 4–10) describing projectile motion.

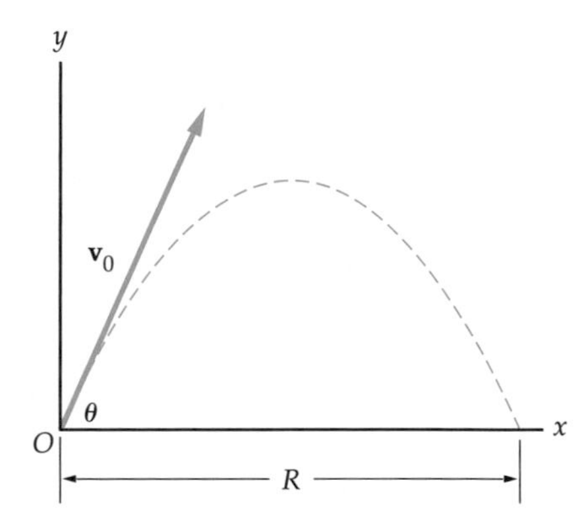

▲ **FIGURE 4–8 Range**
The range R of a projectile is the horizontal distance it travels.

Range

The **range**, R, of a projectile is the horizontal distance it travels before landing. We consider the case shown in Figure 4–8, where the initial and final elevations are the same ($y = 0$). One way to obtain the range, then, is as follows: (i) Find the time when the projectile lands by setting $y = 0$ in the expression $y = (v_0 \sin \theta)t - \frac{1}{2}gt^2$; (ii) Substitute the time found in (i) into the x equation of motion.

Carrying out the first part of the calculation yields the following:

$$(v_0 \sin \theta)t - \tfrac{1}{2}gt^2 = 0 \qquad \text{or} \qquad (v_0 \sin \theta)t = \tfrac{1}{2}gt^2$$

Clearly, $t = 0$ is a solution to this equation—corresponding to the initial condition—but the solution we seek is a time that is greater than zero. We can find the desired time by dividing both sides of the equation by t. This gives

$$(v_0 \sin \theta) = \tfrac{1}{2}gt$$

or

$$t = \left(\frac{2v_0}{g}\right)\sin \theta \qquad \textbf{4–11}$$

This is the time when the projectile lands.

Now, substitute this time into $x = (v_0 \cos \theta)t$ to find the value of x when the projectile lands:

$$x = (v_0 \cos \theta)t = (v_0 \cos \theta)\left(\frac{2v_0}{g}\right)\sin \theta = \left(\frac{2v_0^2}{g}\right)\sin \theta \cos \theta$$

This value of x is the range, R, thus

$$R = \left(\frac{2v_0^2}{g}\right)\sin \theta \cos \theta$$

Using the trigonometric identity $\sin 2\theta = 2 \sin \theta \cos \theta$, as given in Appendix A, we can write this more compactly as follows:

$$R = \left(\frac{v_0^2}{g}\right)\sin 2\theta \qquad \textit{(same initial and final elevation)} \qquad \textbf{4–12}$$

ACTIVE EXAMPLE 4–2 Kickoff!

A football game begins with a kickoff in which the ball travels a horizontal distance of 45 yd. If the ball was kicked at an angle of 40.0° above the horizontal, what was its initial speed?

Solution

1. Solve Equation 4–12 for the initial speed v_0: $\quad v_0 = \sqrt{gR/\sin 2\theta}$
2. Convert the range to meters: $\quad R = 41\ \text{m}$
3. Substitute numerical values: $\quad v_0 = 20\ \text{m/s}$

Insight

Note that we choose the positive square root in step 1 because we are interested only in the *speed* of the ball, which is always positive.

Note that R depends inversely on the acceleration of gravity, g—thus the smaller g, the larger the range. For example, a projectile launched on the Moon, where the acceleration of gravity is only about 1/6 that on Earth, travels about six times as far as it would on the Earth. It was for this reason that astronaut Alan Shepard simply couldn't resist the temptation of bringing a golf club and ball with him on the third lunar landing mission in 1971. He ambled out onto the Fra Mauro Highlands and became the first person to hit a tee shot on the Moon. His distance was undoubtedly respectable—unfortunately, his ball landed in a sand trap.

Now, what launch angle gives the greatest range? From Equation 4–12 we see that R varies with angle as $\sin 2\theta$; thus R is largest when $\sin 2\theta$ is largest—that is, when $\sin 2\theta = 1$. Since $\sin 90° = 1$ it follows that $\theta = 45°$ gives the maximum range. Thus

$$R_{\text{max}} = \frac{v_0^2}{g} \qquad \text{4–13}$$

As expected, the range (Equation 4–12) and maximum range (Equation 4–13) depend strongly on the initial speed of the projectile—they are both proportional to v_0^2.

Note that these results are specifically for the case where a projectile lands at the same level from which it was launched. If a projectile lands at a higher level, for example, the launch angle that gives maximum range is greater than 45°, and if it lands at a lower level the angle for maximum range is less that 45°.

Finally, the range given here applies only to the ideal case of no air resistance. In cases where air resistance is significant, as in the flight of a rapidly moving golf ball, for example, the overall range of the ball is reduced. In addition, the maximum range occurs for a launch angle less than 45° (**Figure 4–9**). The reason is that with a smaller launch angle the golf ball is in the air for less time, giving air resistance less time to affect its flight.

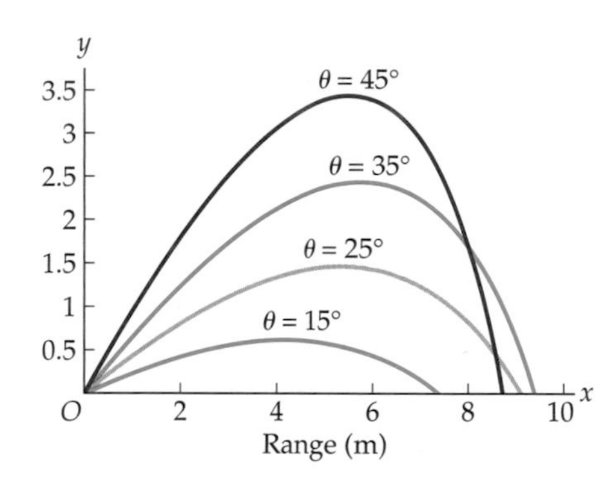

▲ **FIGURE 4–9 Projectiles with air resistance**

Projectiles with the same initial speed but different launch angles showing the effects of air resistance. Notice that the maximum range occurs for a launch angle less than 45°, and that the projectiles return to the ground at a steeper angle than the launch angle.

Symmetry in Projectile Motion

There are many striking symmetries in projectile motion, beginning with the graceful symmetry of the parabola itself. As a first example, recall that earlier in this section, in Equation 4–11, we found the time when a projectile lands:

$$t = \left(\frac{2v_0}{g}\right)\sin\theta$$

Now, by symmetry, the time it takes a projectile to reach its highest point (in the absence of air resistance) should be just half this time. After all, the projectile moves in the x direction with constant speed, and the highest point—by symmetry—occurs at $x = \frac{1}{2}R$.

This all seems reasonable, but is there another way to check? Well, at the highest point the projectile is moving horizontally, thus its y component of velocity is zero. Let's find the time when $v_y = 0$ and compare with the time to land:

$$v_y = v_{0y} - gt = v_0 \sin\theta - gt = 0$$

$$t = \left(\frac{v_0}{g}\right)\sin\theta \qquad \text{4–14}$$

As expected from symmetry, the time at the highest point is one-half the time at landing.

There is another interesting symmetry concerning speed. Recall that when a projectile is launched its y component of velocity is $v_y = v_0 \sin\theta$. When the projectile lands, at time $t = (2v_0/g)\sin\theta$, its y component of velocity is

$$v_y = v_0 \sin\theta - gt = v_0 \sin\theta - g\left(\frac{2v_0}{g}\right)\sin\theta = -v_0 \sin\theta$$

▲ To be successful, a juggler must master the behavior of projectile motion. Physicist Richard Feynman shows that just knowing the appropriate equations is not enough; one must also practice. In this sense, learning to juggle is similar to learning to solve physics problems.

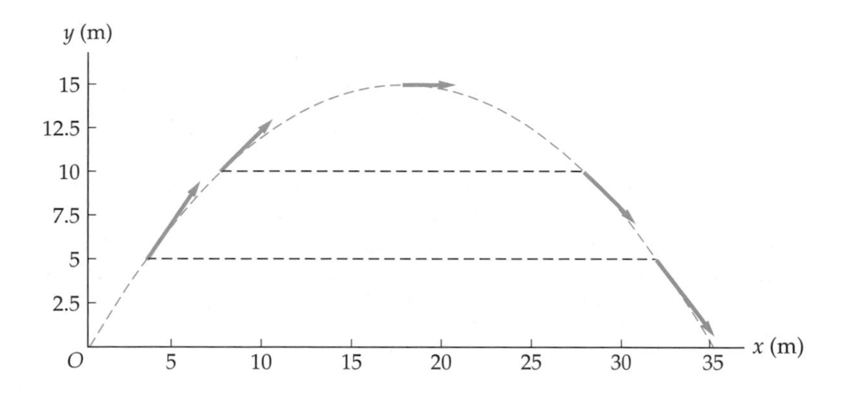

◀ **FIGURE 4–10 Velocity vectors for a projectile launched at the origin**
At a given height the speed (length of velocity vector) is the same on the way up as on the way down. The direction of motion on the way up is above the horizontal by the same amount that it is below the horizontal on the way down.

This is exactly the opposite of the y component of the velocity when it was launched. Since the x component of velocity is always the same, it follows that when the projectile lands, its speed, $v = \sqrt{v_x^2 + v_y^2}$, is the same as when it was launched—as one might expect from symmetry.

The velocities are different, however, since the direction of motion is different at launch and landing. Even so, there is still a symmetry—the initial velocity is *above* the horizontal by the angle θ; the landing velocity is *below* the horizontal by the same angle θ.

So far, these results have referred to launching and landing, which both occur at $y = 0$. The same symmetry extends to any level, though. That is, at a given height the speed of a projectile is the same on the way up as on the way down. In addition, the angle of the velocity above the horizontal on the way up is the same as the angle below the horizontal on the way down. This is illustrated in **Figure 4–10** and in the next Conceptual Checkpoint.

CONCEPTUAL CHECKPOINT 4–3

You and a friend stand on a snow-covered roof. You both throw snowballs with the same initial speed, but in different directions. You throw your snowball downward, at 40° *below* the horizontal; your friend throws her snowball upward, at 40° *above* the horizontal. When the snowballs land on the ground, is the speed of your snowball **(a)** greater than, **(b)** less than, or **(c)** the same as the speed of your friend's snowball?

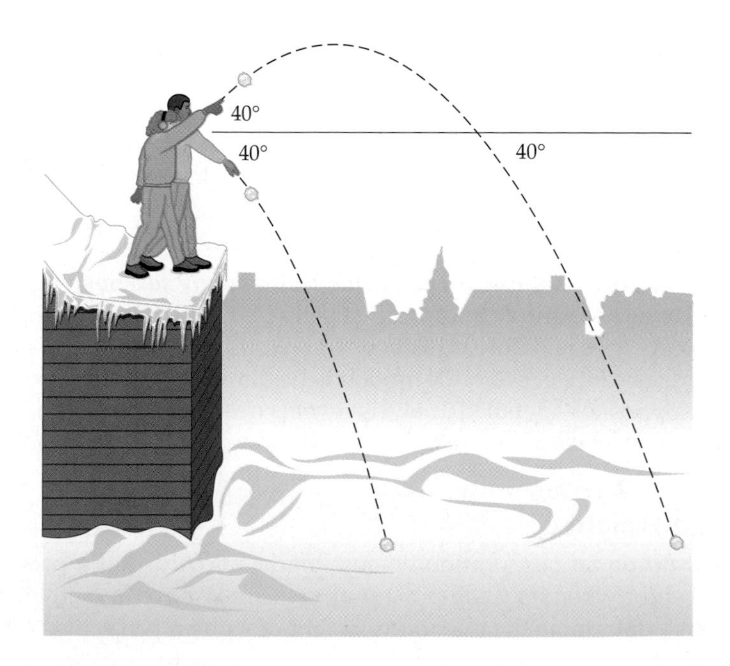

Reasoning and Discussion

One consequence of symmetry in projectile motion is that when your friend's snowball returns to the level of the throw, its speed will be the same as the initial speed. In addition, it will be moving downward, at 40° below the horizontal. From that point on its motion is the same as that of your snowball; thus it lands with the same speed.

What if you throw your snowball horizontally? Or suppose you throw it straight down? In either case, the final speed is unchanged! In fact, for a given initial speed, the speed on landing simply doesn't depend on the direction in which you throw the ball. This is shown in homework problems 29 and 60. We return to this point in Chapter 8 when we discuss potential energy and energy conservation.

Answer

(c) The snowballs have the same speed.

As our final example of symmetry, consider the range R. A plot of R versus launch angle θ is shown in **Figure 4–11** for $v_0 = 20$ m/s. Note that in the absence of air resistance, R is greatest at $\theta = 45°$, as pointed out previously. In addition, we can see from the figure that the range for angles equally above or below 45° is the same. For example, if air resistence is negligible, the range for $\theta = 30°$ is the same as the range for $\theta = 60°$.

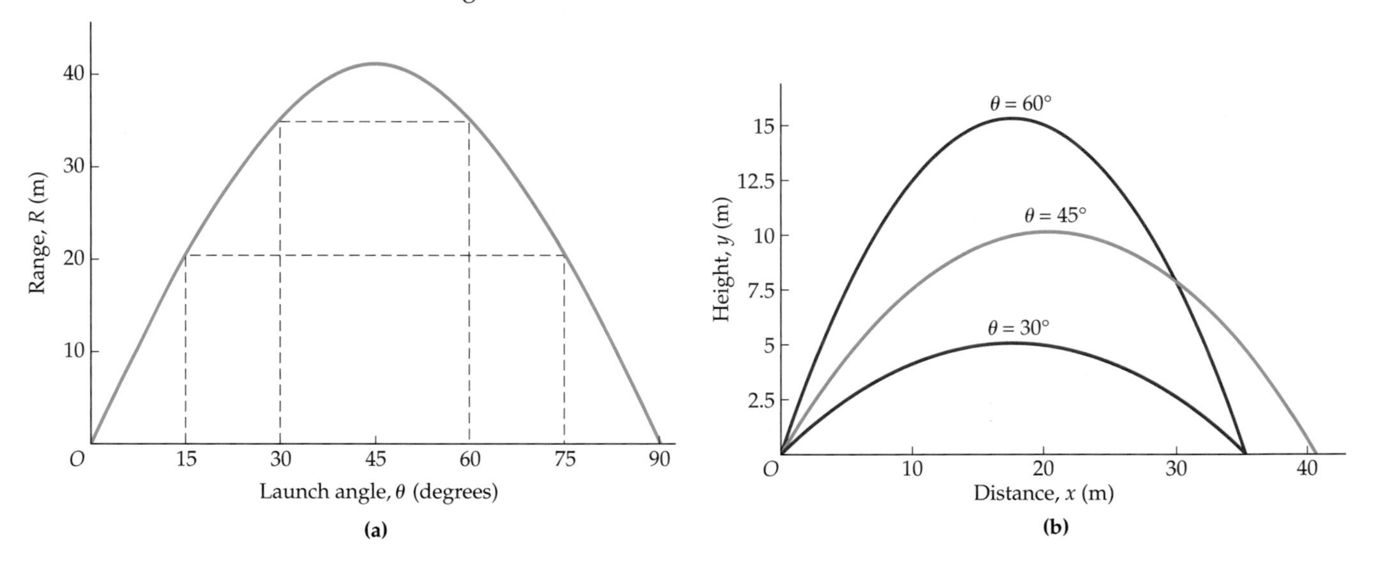

▲ **FIGURE 4–11 Range and launch angle in the absence of air resistance**

(a) A plot of range versus launch angle for a projectile launched with an initial speed of 20 m/s. Note that the maximum range occurs at $\theta = 45°$. Launch angles equally greater than or less than 45°, such as 30° and 60°, give the same range. **(b)** Trajectories of projectiles with initial speeds of 20 m/s and launch angles of 60°, 45°, and 30°. The projectiles with launch angles of 30° and 60° land at the same location.

Symmetries such as these are just some of the many reasons why physicists find physics to be "beautiful" and "aesthetically pleasing." Discovering such patterns and symmetries in nature is really what physics is all about. A physicist does not consider the beauty of projectile motion to be diminished by analyzing it in detail. Just the opposite—detailed analysis reveals deeper, more subtle, and sometimes unexpected levels of beauty.

Maximum Height

Let's follow up on an observation we just made; namely, that a projectile is at maximum height when its y component of velocity is zero. In fact, let's use this observation to determine the maximum height of a given projectile. We do so by

(i) finding the time when $v_y = 0$, and (ii) substituting this time into the y-versus-time expression $y = (v_0 \sin\theta)t - \frac{1}{2}gt^2$.

Part one of the calculation has already been done in Equation 4–14, with the result that $t = (v_0/g)\sin\theta$. Now for part two—substituting t into $y = (v_0 \sin\theta)t - \frac{1}{2}gt^2$. This yields

$$y = (v_0 \sin\theta)\left(\frac{v_0 \sin\theta}{g}\right) - \frac{1}{2}g\left(\frac{v_0 \sin\theta}{g}\right)^2 = \frac{v_0^2 \sin^2\theta}{2g}$$

Thus, the maximum height, y_{max}, of a projectile launched with a speed v_0 at an angle θ above the x axis is

$$y_{max} = \frac{v_0^2 \sin^2\theta}{2g} = \frac{(v_0 \sin\theta)^2}{2g} \qquad \text{4–15}$$

If the projectile is launched straight up—that is with $\theta = 90°$—it reaches the height

$$h = \frac{v_0^2}{2g}$$

We can check this result using the one-dimensional kinematics of Chapter 2. For example, if an object is thrown straight upward with an initial speed v_0, and this object accelerates downward with the acceleration of gravity, $a = -g$, it comes to rest after covering a distance, Δy, given by $0 = v_0^2 + 2(-g)\Delta y$, or

$$\Delta y = \frac{v_0^2}{2g}$$

This is an example of the type of internal consistency that characterizes the entire field of physics.

▲ An archerfish would have trouble procuring its lunch without an instinctive grasp of projectile motion.

EXAMPLE 4–7 What a Shot!

The archerfish hunts by dislodging an unsuspecting insect from its resting place with a stream of water expelled from the fish's mouth. Suppose the archerfish squirts water with an initial speed of 2.30 m/s at a beetle on a leaf 3.00 cm above the water's surface. **(a)** If the fish aims in such a way that the stream of water is moving horizontally when it hits the beetle, what is the launch angle? **(b)** How much time does the beetle have to react? **(c)** What is the horizontal distance d between the fish and the beetle when the water is launched?

Picture the Problem

Our sketch shows the fish squirting water from the origin, and the beetle at $x = d$, $y = h = 3.00$ cm. The water starts off with a speed $v_0 = 2.30$ m/s at an angle θ above the horizontal. Note that the water is moving horizontally when it reaches the beetle.

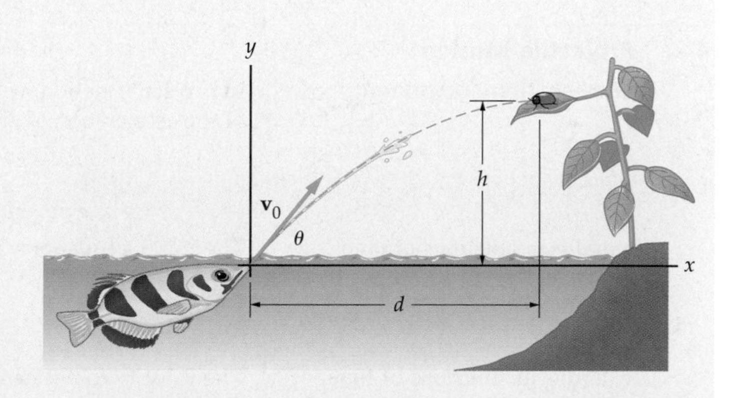

Strategy

(a) Since the stream of water is moving horizontally when it reaches the beetle, it is at the top of its parabolic trajectory, as can be seen in Figure 4–10. Thus, the water is at its maximum height (Equation 4–15). It follows that we can set $y_{max} = (v_0 \sin\theta)^2/2g$ equal to $h = 3.00$ cm and solve for θ.

(b) The water reaches the beetle at the top of its trajectory, where v_y is zero. Thus, we set $v_y = 0$, using $v_y = v_0 \sin\theta - gt$, and solve for t.

(c) Since we know θ and t we can find the distance with $x = (v_0 \cos\theta)t$.

continued on the following page

continued from the previous page

Solution

Part (a)

1. Let $y_{max} = h$ and solve for $\sin\theta$: $$y_{max} = \frac{(v_0 \sin\theta)^2}{2g} = h \quad \text{or} \quad \sin\theta = \frac{1}{v_0}\sqrt{2gh}$$

2. Substitute numerical values and solve for θ: $$\sin\theta = \frac{1}{(2.30 \text{ m/s})}\sqrt{2(9.81 \text{ m/s}^2)(0.0300 \text{ m})} = 0.334$$ $$\theta = \sin^{-1}(0.334) = 19.5°$$

Part (b)

3. Let $v_y = v_0 \sin\theta - gt = 0$ and solve for t: $$v_y = v_{0y} - gt = v_0 \sin\theta - gt = 0$$ $$t = \frac{v_0 \sin\theta}{g} = \frac{(2.30 \text{ m/s})\sin 19.5°}{9.81 \text{ m/s}^2} = 0.0783 \text{ s}$$

Part (c)

4. We can find the horizontal distance d using x as a function of time, $x = (v_0 \cos\theta)t$: $$x = (v_0 \cos\theta)t$$ $$d = [(2.30 \text{ m/s})\cos 19.5°](0.0783 \text{ s}) = 0.170 \text{ m}$$

Insight

We have found that the fish must aim 19.5° above the horizontal. For comparison, what is the straight-line angle from the fish to the beetle? This angle is $\tan^{-1}(0.0300/0.170) = 10.0°$. Thus, the fish cannot aim directly at its prey if it wants to have a meal. Also, note that the time for the water to reach the beetle is the same as the time it takes the beetle to drop a distance h; $t = \sqrt{2h/g} = 0.0783$ s.

Practice Problem

How far does the stream of water go if it happens to miss the beetle? [**Answer:** By symmetry, the distance d is half the range. Thus the stream of water travels a distance $R = 2d = 0.340$ m]

Some related homework problems: Problem 63, Problem 64

Chapter Summary

	Topic	Remarks and Relevant Equations	
4–1	**Motion in Two Dimensions**		
	independence of motion	Components of motion in the x and y directions can be treated independently of one another. Thus, two-dimensional motion with constant acceleration is described by the same kinematic equations derived in Chapter 2, only now written in terms of x and y components.	
4–2	**Projectile Motion**		
	acceleration components	In projectile motion, with the x axis horizontal and the y axis upward, the components of the acceleration of gravity are $a_x = 0$, $a_y = -g$	
	x and y as functions of time	The x and y equations of motion are $x = x_0 + v_{0x}t$, $y = y_0 + v_{0y}t - \frac{1}{2}gt^2$	4–6
	v_x and v_y as functions of time	The velocity components vary with time as follows: $v_x = v_{0x}$, $v_y = v_{0y} - gt$	4–6
	v_x and v_y as functions of displacement	v_x and v_y vary with displacement as $v_x^2 = v_{0x}^2$, $v_y^2 = v_{0y}^2 - 2g\Delta y$	4–6

4–3 Zero Launch Angle

equations of motion

A projectile launched horizontally from $x_0 = 0$, $y_0 = h$ with an initial speed v_0 has the following equations of motion:

$$x = v_0 t \qquad v_x = v_0 \qquad v_x^2 = v_0^2$$
$$y = h - \tfrac{1}{2}gt^2 \qquad v_y = -gt \qquad v_y^2 = -2g\Delta y \qquad \text{4–7}$$

parabolic path

The path followed by a projectile launched horizontally with an initial speed v_0 is described by

$$y = h - \left(\frac{g}{2v_0^2}\right)x^2 \qquad \text{4–8}$$

This path is a parabola.

landing site

The landing site of a projectile launched horizontally is

$$x = v_0\sqrt{\frac{2h}{g}} \qquad \text{4–9}$$

In this expression, v_0 is the initial speed and h is the initial height. Note that this result is simply the speed in the x direction multiplied by the time of fall.

4–4 General Launch Angle

launch from the origin

The equations of motion for a launch from the origin with an initial speed v_0 at an angle of θ with respect to the horizontal are

$$x = (v_0 \cos\theta)t \qquad v_x = v_0 \cos\theta \qquad v_x^2 = v_0^2 \cos^2\theta$$
$$y = (v_0 \sin\theta)t - \tfrac{1}{2}gt^2 \qquad v_y = v_0 \sin\theta - gt \qquad v_y^2 = v_0^2 \sin^2\theta - 2g\Delta y$$

***4–5 Projectile Motion: Key Characteristics**

range

The range of a projectile launched from the origin with an initial speed v_0 and a launch angle θ is

$$R = \left(\frac{v_0^2}{g}\right)\sin 2\theta \qquad \text{4–12}$$

This expression applies only to projectiles that land at the same level from which they were launched.

symmetry

Projectile motion exhibits many symmetries. For example, the speed of a projectile depends only on its height, and not on whether it is moving upward or downward.

maximum height

The maximum height of a projectile above its launch site is

$$y_{max} = \frac{v_0^2 \sin^2\theta}{2g} \qquad \text{4–15}$$

In this equation, v_0 is the initial speed and θ is the launch angle.

Problem-Solving Summary

Type of Problem	Relevant Physical Concepts	Related Examples
Study two-dimensional motion with constant acceleration.	Motion in the x and y directions are independent. This is the basis for the equations of motion given in Table 4-1. Note that these equations are the same as the kinematic equations of Chapter 2, only written in terms of x and y components.	Examples 4–1, 4–2
Find the location and velocity of a projectile launched horizontally.	When a projectile is launched horizontally with a speed v_0 its initial velocity components are $v_{0x} = v_0$ and $v_{0y} = 0$. Make these substitutions in the equations of projectile motion given in Equation 4–6.	Examples 4–3, 4–4 Conceptual Checkpoints 4–1, 4–2

Find the location and velocity of a projectile launched with an arbitrary launch angle.	If a projectile is launched at an angle θ, its initial velocity components are $v_{0x} = v_0 \cos\theta$ and $v_{0y} = v_0 \sin\theta$. Make these substitutions in the equations of projectile motion given in Equation 4–6.	Examples 4–5, 4–6, 4–7 Active Examples 4–1, 4–2

Conceptual Questions

1. What is the acceleration of a projectile when it reaches its highest point? What is its acceleration just before and just after reaching this point?
2. A projectile is launched with an initial speed of v_0 at an angle θ above the horizontal. It lands at the same level from which it was launched. What was its average velocity between launch and landing?
3. A projectile is launched from level ground. When it lands, its direction of motion has rotated clockwise through 60°. What was the launch angle?
4. Is it possible for the velocity of a projectile to be at right angles to its acceleration? If so, give an example.
5. A skateboarder, coasting with a constant velocity, jumps straight up. When she lands, does she come down (i) behind the skateboard, (ii) ahead of the skateboard, or (iii) on the skateboard? Explain.
6. Projectiles for which air resistance is nonnegligible, such as a bullet fired from a rifle, have maximum range when the launch angle is (i) greater than, (ii) less than, or (iii) equal to 45°? Explain.
7. A certain projectile is launched with an initial speed v_0. At its highest point its speed is $\frac{1}{2}v_0$. What was the launch angle?
8. Two projectiles are launched from the same point at the same angle above the horizontal. Projectile 1 reaches a maximum height twice that of projectile 2. What is the ratio of the initial speed of projectile 1 to the initial speed of projectile 2?
9. Two divers run horizontally off the end of a diving board. Diver 2 runs with twice the speed of diver 1. When the divers hit the water, the horizontal distance covered by diver 2 is (i) the same as, (ii) twice as much as, or (iii) four times the horizontal distance of diver 1?
10. Two divers run horizontally off the end of a diving board. When the divers hit the water, diver 2 has covered twice the horizontal distance covered by diver 1. Is the initial speed of diver 2 (i) the same as, (ii) twice as much as, or (iii) four times the initial speed of diver 1?
11. Three projectiles (a, b, and c) are launched with the same initial speed but with different launch angles, as shown in **Figure 4–12**. List the projectiles in order of increasing **(a)** horizontal component of initial velocity and **(b)** time of flight.

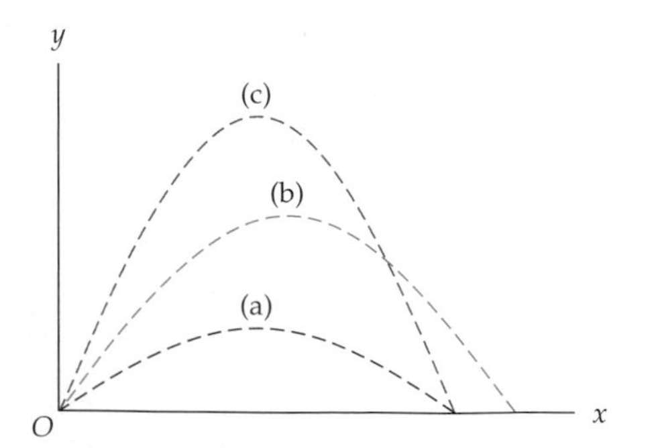

▲ **FIGURE 4–12** Question 11

12. Three projectiles (a, b, and c) are launched with different initial speeds so that they reach the same maximum height, as shown in **Figure 4–13**. List the projectiles in order of increasing **(a)** speed and **(b)** time of flight.

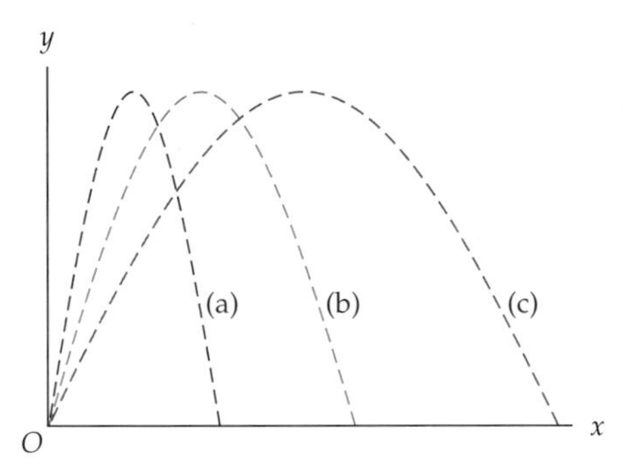

▲ **FIGURE 4–13** Question 12

13. The penguin shown in the accompanying photo is about to land on an ice floe. Just before it lands, is its speed greater than, less than, or the same as when it left the water? Explain.

▲ This penguin behaves much like a projectile from the time it leaves the water until it touches down on the ice. (Question 13.)

14. Driving down the highway you find yourself behind a heavily loaded tomato truck. You follow close behind the truck, keeping the same speed. Suddenly a tomato falls from the back of the truck. Will the tomato hit your car or land on the road, assuming you continue moving with the same speed and direction?

15. A child rides on a pony walking with constant velocity. The boy leans over to one side and a scoop of ice cream falls from his ice cream cone. Describe the path of the scoop of ice cream as seen by **(a)** the child and **(b)** his parents standing on the ground nearby.

16. A person flips a coin into the air and it lands on the ground a few feet away. If the person were to repeat the coin flip on an elevator rising with constant speed, would the coin's time of flight be (i) greater than, (ii) less than, or (iii) the same as when the person was at rest? Explain.

17. Child 1 throws a snowball horizontally from the top of a roof; child 2 throws a snowball straight down. Once in flight, is the acceleration of snowball 2 (i) greater than, (ii) equal to, or (iii) less than the acceleration of snowball 1? Explain.

18. A projectile is launched from the origin of a coordinate system where the positive x axis points horizontally to the right and the positive y axis points vertically upward. What was the projectile's launch angle with respect to the x axis if at its highest point its direction of motion has rotated **(a)** clockwise through 50° or **(b)** counterclockwise through 30°?

Problems

Note: **IP** *denotes an integrated conceptual/quantitative problem.* **BIO** *identifies problems of biological or medical interest. Blue bullets (•, ••, •••) are used to indicate the level of difficulty of each problem. Air resistance should be ignored in the problems for this chapter, unless specifically stated otherwise.*

Section 4–1 Motion in Two Dimensions

1. • A sailboat runs before the wind with a constant speed of 3.5 m/s in a direction 35° north of west. How far **(a)** west and **(b)** north has the sailboat traveled in 30 min?

2. • As you walk to class with a constant speed of 1.60 m/s you are moving in a direction that is 15.0° north of east. How long does it take you to move **(a)** 20.0 m east or **(b)** 30.0 m north?

3. • Starting from rest, a car accelerates at 2.0 m/s^2 up a hill that is inclined 5.5° above the horizontal. How far **(a)** horizontally and **(b)** vertically has the car traveled in 10 s?

4. •• A particle passes through the origin with a speed of 6.2 m/s in the positive y direction. If the particle accelerates in the negative x direction at 4.4 m/s^2, **(a)** what are its x and y positions after 5.0 s? **(b)** What are v_x and v_y at this time?

5. •• An electron in a cathode-ray tube is traveling horizontally at 2.10×10^9 cm/s when deflection plates give it an upward acceleration of 5.30×10^{17} cm/s^2. **(a)** How long does it take for the electron to cover a horizontal distance of 6.20 cm? **(b)** What is its vertical displacement during this time?

6. •• Two canoeists start paddling at the same time and head toward a small island in a lake, as shown in **Figure 4–14**. Canoeist 1 paddles with a speed of 1.35 m/s at an angle of 45° north of east. Canoeist 2 starts on the opposite shore of the lake, a distance of 1.5 km due east of canoeist 1. **(a)** In what direction relative to north must canoeist 2 paddle to reach the island? **(b)** What speed must canoeist 2 have if the two canoes are to arrive at the island at the same time?

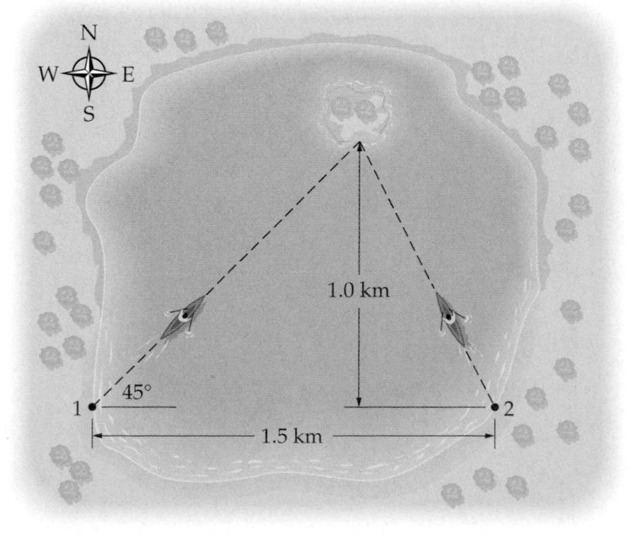

▲ **FIGURE 4–14** Problem 6

Section 4–3 Zero Launch Angle

7. • An archer shoots an arrow horizontally at a target 15 m away. The arrow is aimed directly at the center of the target, but it hits 52 cm lower. What was the initial speed of the arrow?

8. • The great, gray-green, greasy Zambezi River flows over Victoria Falls in south central Africa. The falls are approximately 108 m high. If the river is flowing horizontally at 3.60 m/s just before going over the falls, what is the speed of the water when it hits the bottom? Assume the water is in freefall as it drops.

9. • A diver runs horizontally off the end of a diving board with an initial speed of 1.75 m/s. If the diving board is 3.00 m above the water, what is the diver's speed just before she enters the water?

10. • An astronaut on the planet Zircon tosses a rock horizontally with a speed of 6.75 m/s. The rock falls through a vertical distance of 1.20 m and lands a horizontal distance of 8.95 m from the astronaut. What is the acceleration of gravity on Zircon?

11. •• **IP** Pitcher's mounds are raised to compensate for the vertical drop of the ball as it travels 18 m to the catcher. **(a)** If a pitch is thrown horizontally with an initial speed of 32 m/s, how far does it drop by the time it reaches the catcher? **(b)** If the speed of the pitch is increased, does the drop distance increase, decrease, or stay the same? Explain. **(c)** If this baseball game were to be played on the moon, would the drop distance increase, decrease, or stay the same? Explain.

12. •• Playing shortstop, you pick up a ground ball and throw it to second base. The ball is thrown horizontally, with a speed of 22 m/s, directly toward point A (**Figure 4–15**). When the ball reaches the second baseman 0.45 s later, it is caught at point B. **(a)** How far were you from the second baseman? **(b)** What is the distance of vertical drop, AB?

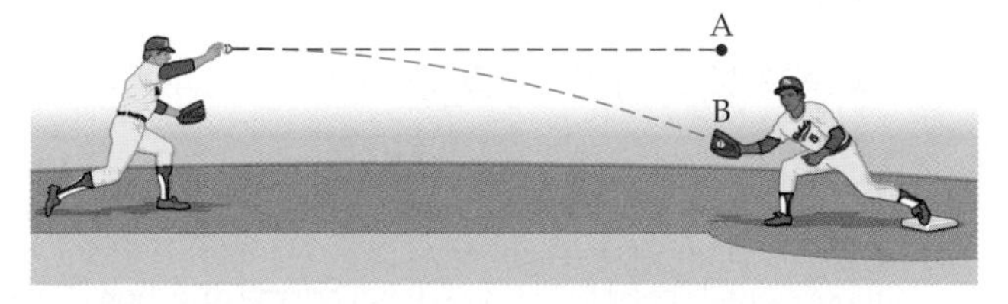

▲ **FIGURE 4–15** Problem 12

13. •• **IP** A crow is flying horizontally with a constant speed of 2.70 m/s when it releases a clam from its beak (**Figure 4–16**). The clam lands on the rocky beach 2.10 s later. Just before the clam lands, what is **(a)** its horizontal component of velocity, and **(b)** its vertical component of velocity? **(c)** How would your answers to parts **(a)** and **(b)** change if the speed of the crow were increased? Explain.

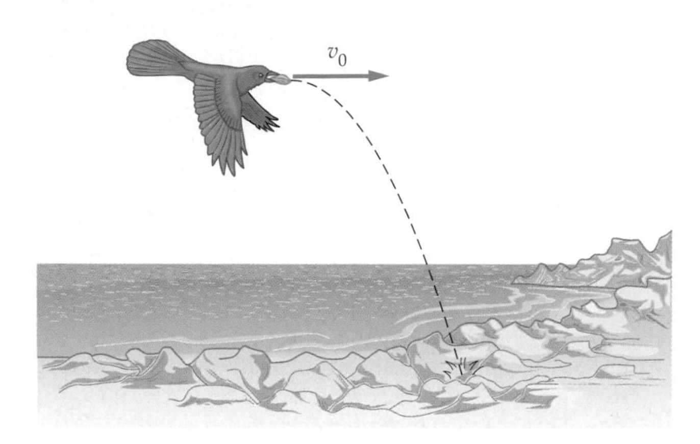

▲ **FIGURE 4–16** Problem 13

14. •• A mountain climber jumps a 3.0-m wide crevasse by leaping horizontally with a speed of 8.0 m/s. **(a)** If the climber's direction of motion on landing is −45°, what is the height difference between the two sides of the crevasse? **(b)** Where does the climber land?

15. •• **IP** A sparrow flying horizontally with a speed of 1.80 m/s folds its wings and begins to drop in freefall. **(a)** How far does the sparrow fall after traveling a horizontal distance of 0.500 m? **(b)** If the sparrow's initial speed is increased, does the distance of fall increase, decrease, or stay the same?

16. •• In Denver, children bring their old jack-o-lanterns to the top of a tower and compete for accuracy in hitting a target on the ground (**Figure 4–17**). Suppose that the tower is 9.0 m high, and that the bulls-eye is a horizontal distance of 3.5 m from the launch point. If the pumpkin is thrown horizontally, what is the launch speed needed to hit the bulls-eye?

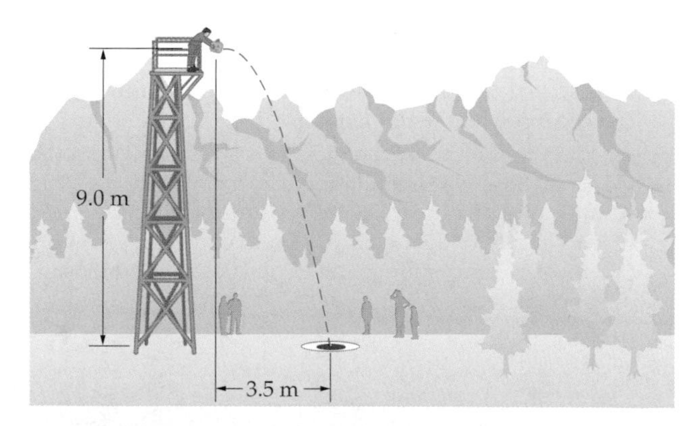

▲ **FIGURE 4–17** Problems 16 and 17

17. •• If, in the previous problem, a jack-o-lantern is given an initial horizontal speed of 3.3 m/s, what are the direction and magnitude of its velocity **(a)** 0.75 s later and **(b)** just before it lands?

18. •• Fairgoers ride a Ferris wheel with a radius of 5.00 m (**Figure 4–18**). The wheel completes one revolution every 32.0 s. **(a)** What is the average speed of a rider on this Ferris wheel? **(b)** If a rider accidentally drops a stuffed animal at the top of the wheel, where does it land relative to the base of the ride?

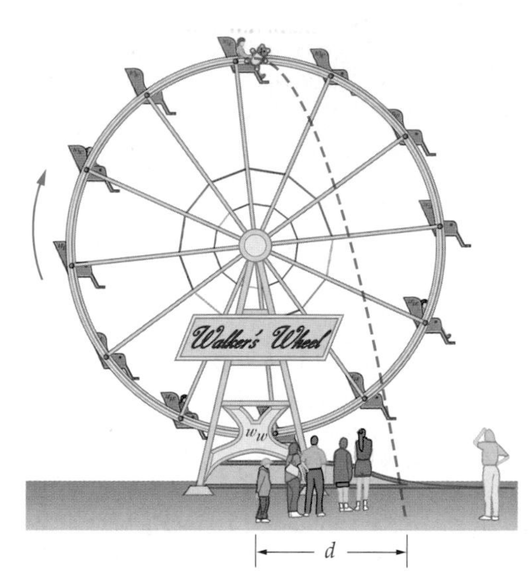

▲ **FIGURE 4–18** Problems 18 and 36

19. •• **IP** A swimmer runs horizontally off a diving board with a speed of 2.50 m/s, and hits the water a horizontal distance of 1.96 m from the end of the board. **(a)** How high above the water was the diving board? **(b)** If the swimmer runs off the board with a reduced speed, does it take more, less, or the same time to reach the water?

20. •• On August 25, 1894, Chicago catcher William Schriver caught a baseball thrown from the top of the Washington Monument (555 ft, 898 steps.) **(a)** If the ball was thrown horizontally from the top of the 555-foot monument with a speed of 5.00 m/s, where did it land? **(b)** What was the ball's speed and direction of motion when caught?

21. ••• A basketball is thrown horizontally with an initial speed of 4.0 m/s (**Figure 4–19**). A straight line drawn from the release point to the landing point makes an angle of 30.0° with the horizontal. What was the release height?

▲ **FIGURE 4–19** Problem 21

22. ••• **IP** A ball rolls off a table and falls 0.75 m to the floor, landing with a speed of 4.0 m/s. **(a)** What is the acceleration of the ball just before it strikes the ground? **(b)** What was the initial speed of the ball? **(c)** What initial speed must the ball have if it is to land with a speed of 5.0 m/s?

Section 4–4 General Launch Angle

23. • A second baseman tosses the ball to the first baseman, who catches it at the same level from which it was thrown. The throw is made with an initial speed of 17.0 m/s at an angle of 35.0° above the horizontal. **(a)** What is the horizontal component of the ball's velocity just before it is caught? **(b)** How long is the ball in the air?

24. • Referring to the previous problem, what is the y component of the ball's velocity and its direction of motion just before it is caught?

25. • A cork shoots out of a champagne bottle at an angle of 40.0° above the horizontal. If the cork travels a horizontal distance of 1.50 m in 1.25 s, what was its initial speed?

26. • A soccer ball is kicked with a speed of 9.50 m/s at an angle of 25.0° above the horizontal. If the ball lands at the same level from which it was kicked, how long was it in the air?

27. •• In a game of basketball, a forward makes a bounce pass to the center. The ball is thrown with an initial speed of 4.3 m/s at an angle of 15° below the horizontal. It is released 0.80 m above the floor. What horizontal distance does the ball cover before bouncing?

28. •• Repeat the previous problem for a bounce pass in which the ball is thrown 15° *above* the horizontal?

29. •• **IP** Snowballs are thrown with a speed of 13 m/s from a roof 7.0 m above the ground. Snowball A is thrown straight downward; snowball B is thrown in a direction 25° above the horizontal. **(a)** When the snowballs land, is the speed of A greater than, less than, or the same as the speed of B? Explain. **(b)** Verify your answer to part **(a)** by calculating the landing speed of both snowballs.

30. •• In the previous problem, find the direction of motion of the two snowballs when they land.

31. •• A golfer gives a ball a maximum initial speed of 30 m/s. **(a)** What is the longest possible hole in one for this golfer? Neglect any distance the ball might roll on the green, and assume that the tee and the green are at the same level. **(b)** What is the minimum speed of the ball during this hole-in-one shot?

32. •• What is the highest tree the ball in the previous problem could clear on its way to the hole in one?

33. •• The "hang time" of a punt is measured to be 4.50 s. If the ball was kicked at an angle of 63.0° above the horizontal and was caught at the same level from which it was kicked, what was its initial speed?

34. •• In a friendly game of handball, you hit the ball essentially at ground level and send it toward the wall with a speed of 13 m/s at an angle of 20° above the horizontal. **(a)** How long does it take for the ball to reach the wall if it is 5.2 m away? **(b)** How high is the ball when it hits the wall?

35. •• **IP** In the previous problem, **(a)** what are the magnitude and direction of the ball's velocity when it strikes the wall? **(b)** Has the ball reached the highest point of its trajectory at this time? Explain.

36. •• A passenger on the Ferris wheel described in Problem 18 drops his keys when he is on the way up and at an angle of 45° to the vertical. Where do the keys land relative to the base of the ride?

37. •• On a hot summer day a young girl swings on a rope above the local swimming hole (**Figure 4–20**). When she lets go of the rope her initial velocity is 2.25 m/s at an angle of 35.0° above the horizontal. If she is in flight for 1.60 s, how high above the water was she when she let go of the rope?

▲ **FIGURE 4–20** Problem 37

38. ••• A certain projectile is launched with an initial speed v_0. At its highest point its speed is $v_0/4$. What was the launch angle?

*Section 4–5 Projectile Motion: Key Characteristics

39. • In Delaware, a post-Halloween tradition is "pumpkin chunkin," in which contestants build cannons to launch pumpkins and compete for the greatest distance. In this contest, pumpkins are sometimes projected for as much as a quarter of a mile. What is the minimum initial speed needed for such a shot?

40. • A dolphin jumps with an initial velocity of 12.0 m/s at an angle of 40.0° above the horizontal. The dolphin passes through the center of a hoop before returning to the water. If the dolphin is moving horizontally when it goes through the hoop, how high above the water is the center of the hoop?

41. • Referring to Problem 33, how far did the football travel in the horizontal direction?

42. • A golf ball is struck with a five iron on level ground. It lands 90.0 m away 4.20 s later. What was the direction and magnitude of the initial velocity?

43. • Babe Didrikson holds the world record for the longest baseball throw (296 ft) by a woman. For the following questions, assume that the ball is thrown at an angle of 45.0° above the horizontal, that it travels a horizontal distance of 296 ft, and is caught at the same level from which it was thrown. **(a)** What is the ball's initial speed? **(b)** How long is the ball in the air?

44. •• **IP** A soccer ball is kicked with an initial speed of 10.2 m/s in a direction 25.0° above the horizontal. Find the magnitude and direction of its velocity **(a)** 0.250 s and **(b)** 0.500 s after being kicked. **(c)** Is the ball at its greatest height before or after 0.500 s? Explain.

45. •• A second soccer ball is kicked with the same initial speed as in the previous problem. After 0.750 s it is at its highest point. What was its initial direction of motion?

46. •• **IP** A golfer tees off on level ground, giving the ball an initial speed of 42.0 m/s and an initial direction of 35.0° above the horizontal. **(a)** How far from the golfer does the

ball land? **(b)** The next golfer in the group hits a ball with the same initial speed but at an angle above the horizontal that is greater than 45.0°. If the second ball travels the same horizontal distance as the first ball, what was its initial direction of motion? Explain.

47. •• **IP** Astronomers have discovered several volcanoes on Io, a moon of Jupiter. One of them, named Loki, ejects lava to a maximum height of 2.00×10^5 m. **(a)** What is the initial speed of the lava? (The acceleration of gravity on Io is 1.80 m/s^2.) **(b)** If this volcano were on Earth, would the maximum height of the ejected lava be greater than, less than, or the same as on Io? Explain.

▲ A volcano on Io, the innermost moon of Jupiter, displays the characteristic features of projectile motion. (Problem 47)

General Problems

48. • A train moving with a constant velocity of 27 m/s travels 150 m north in 10.0 s. **(a)** Find the direction of the train's motion relative to north. **(b)** How far west has the train traveled in this time?

49. • Find the y component of the hummingbird's velocity in Example 4-2 at the time $t = 0.55$ s.

50. • A racket ball is struck in such a way that it leaves the racket with a speed of 5.60 m/s in the horizontal direction. When the ball hits the court, it is a horizontal distance of 1.85 m from the racket. Find the height of the racket ball when it left the racket.

51. •• A hot air balloon rises from the ground at the rate of 2.0 m/s. A champagne bottle is opened to celebrate takeoff, expelling the cork with a speed of 5.0 m/s. When opened, the bottle is pointing horizontally and is 6.0 m above the ground. **(a)** What is the initial velocity of the cork, as seen by an observer on the ground? **(b)** Determine the maximum height above the ground attained by the cork. **(c)** How long does the cork remain in the air?

52. •• Repeat the previous problem, this time assuming that the balloon is *descending* with a speed of 2.0 m/s.

53. •• **IP** A soccer ball is kicked from the ground with an initial speed of 12.0 m/s. After 0.250 s its speed is 11.3 m/s. **(a)** Give a Strategy that will allow you to calculate the ball's initial direction of motion. **(b)** Use your strategy to find the initial direction.

54. •• A particle leaves the origin with an initial velocity $\mathbf{v} = (2.40\text{ m/s})\hat{\mathbf{x}}$, and moves with constant acceleration $\mathbf{a} = (-1.90\text{ m/s}^2)\hat{\mathbf{x}} + (3.20\text{ m/s}^2)\hat{\mathbf{y}}$. **(a)** How far does the particle move in the x direction before turning around? **(b)** What is the particle's velocity at this time? **(c)** Plot the particle's position at $t = 0.500$ s, 1.00 s, 1.50 s, and 2.00 s. Use these results to sketch position versus time for the particle.

55. •• When the dried up seed pod of a scotch broom plant bursts open, it shoots out a seed with an initial velocity of 2.7 m/s at an angle of 60.0° above the horizontal (**Figure 4–21**). If the seed pod is 1.0 m above the ground, **(a)** how long does it take for the seed to land? **(b)** What horizontal distance does it cover during its flight?

▲ **FIGURE 4–21** Problems 55 and 56

56. •• Referring to the previous problem, a second seed shoots out from the pod with the same speed but with a direction of motion 30° below the horizontal. **(a)** How long does it take for the second seed to land? **(b)** What horizontal distance does it cover during its flight?

57. •• A shot-putter throws the shot with an initial speed of 3.5 m/s from a height of 5.0 ft above the ground. What is the range of the shot if the launch angle is **(a)** 20°, **(b)** 30°, or **(c)** 40°?

58. •• A ball thrown straight upward returns to its original level in 2.50 s. A second ball is thrown at an angle of 40.0° above the horizontal. What is the initial speed of the second ball if it also returns to its original level in 2.50 s?

59. •• The men's world record for the shot put, 23.12 m, was set by Randy Barnes of the United States on May 20, 1990. If the shot was launched from 6.00 ft above the ground at an initial angle of 42.0°, what was its initial speed?

60. •• Referring to Conceptual Checkpoint 4–3, suppose the two snowballs are thrown from an elevation of 15 m with an initial speed of 12 m/s. What is the speed of each ball when it is 5.0 m above the ground?

61. •• **IP** A hockey puck just clears the 2.00-m high boards on its way out of the rink. The base of the boards is 20.2 m from the point where the puck is launched. **(a)** Given the launch angle of the puck, θ outline a strategy that you can use to find its initial speed, v_0. **(b)** Use your strategy to find v_0 for $\theta = 15.0°$.

62. •• Referring to Active Example 4–2, suppose the ball is punted from an initial height of 1.00 m. What is the initial speed of the ball in this case?

63. ••• As discussed in Example 4–7, the archerfish hunts by dislodging an unsuspecting insect from its resting place with a stream of water expelled from the fish's mouth. Suppose the archerfish squirts water with a speed of 2.0 m/s at an angle of 50.0° above the horizontal, and aims for a beetle on a leaf 3.0 cm above the water's surface. **(a)** At what horizontal distance from the beetle should the archerfish fire if it is to hit its target in the least time? **(b)** How much time will the beetle have to react?

64. ••• **(a)** What is the greatest horizontal distance from which the archerfish can hit the beetle, assuming the same squirt speed and direction as in the previous problem? **(b)** How much time does the beetle have to react in this case?

65. ••• Find the launch angle for which the range and maximum height of a projectile are the same.

66. ••• A mountain climber jumps a crevasse of width W by leaping horizontally with speed v_0. **(a)** If the height difference between the two sides of the crevasse is h, what is the minimum value of v_0 for the climber to land safely on the other side? **(b)** In this case, what is the climber's direction of motion on landing?

▲ Leaping a crevasse (Problems 66 and 70)

67. ••• Prove that the landing speed of a projectile is independent of launch angle for a given height of launch.

68. ••• Prove that the maximum height of a projectile, H, divided by the range of the projectile, R, satisfies the relation $H/R = \frac{1}{4}\tan\theta$.

69. ••• A projectile fired from $y = 0$ with initial speed v_0 and initial angle θ lands on a different level, $y = h$. Show that the time of flight of the projectile is

$$T = \tfrac{1}{2}T_0\left(1 + \sqrt{1 - \frac{h}{H}}\right)$$

where T_0 is the time of flight for $h = 0$ and H is the maximum height of the projectile.

70. ••• A mountain climber jumps a crevasse by leaping horizontally with speed v_0. If the climber's direction of motion on landing is θ below the horizontal, what is the height difference h between the two sides of the crevasse?

5 Newton's Laws of Motion

Most people know from experience that to accelerate an object, you must exert a force on it. But forces always come in pairs. Thus, each skater exerts a force on the other, and these forces are equal in magnitude but opposite in direction. This chapter introduces Newton's laws, the basis for our understanding of force and motion.

We are all subject to Newton's laws of motion, whether we know it or not. You can't move your body, drive a car, or toss a ball in a way that violates his rules. In short, our very existence is constrained and regulated by these three fundamental statements concerning matter and its motion.

Yet Newton's laws are surprisingly simple, especially when you consider that they apply equally well to galaxies, planets, comets, and yes, even apples falling from trees. In this chapter we present the three laws of Newton, and we show how they can be applied to everyday situations. Using them, we go beyond a simple description of motion, as in kinematics, to a study of the *causes* of motion, referred to as **dynamics**.

With the advent of Newtonian dynamics in 1687, science finally became quantitative and predictive. Edmund Halley, inspired by Newton's laws, used them to predict the return of the comet that today bears his name. In all of recorded history, no one had ever before predicted the appearance of a comet; in fact, they were generally regarded as supernatural apparitions. Though Halley didn't live to see his comet's return, his correct prediction illustrated the power of Newton's laws in a most dramatic and memorable way.

Today, we still recognize Newton's laws as the indispensable foundation for all of physics. It would be nice to say that these laws are the complete story when it comes to analyzing motion, but that is not the case. In the early part of this century, physicists discovered that Newton's laws must be modified for objects moving at speeds near the speed of light and for objects comparable in size to atoms. In the world of everyday experience, however, Newton's laws still reign supreme.

5–1 Force and Mass

A **force**, simply put, is a push or a pull. When you push on a box to slide it across the floor, for example, or pull on the handle of a wagon to give a child a ride, you are exerting a force. Similarly, when you hold this book in your hand you exert an upward force to oppose the downward pull of gravity. If you set the book on a table, the table exerts the same upward force you exerted a moment before. Forces are truly all around us.

Now, when you push or pull on something, there are two quantities that characterize the force you are exerting. The first is the strength or **magnitude** of your force; the second is the **direction** in which you are pushing or pulling. Because a force is determined by both a magnitude and a direction, it is a vector. We consider the vector properties of forces in more detail in Section 5–6.

In general, an object has several forces acting on it at any given time. In the previous example, a book at rest on a table experiences a downward force due to gravity and an upward force due to the table. If you push the book across the table, it also experiences a horizontal force due to your push. The total force exerted on the book is the vector sum of the individual forces acting on it.

A second key ingredient in Newton's laws is the **mass** of an object, which is a measure of how difficult it is to change its velocity—to start an object moving if it is at rest, to bring it to rest if it is moving, or to change its direction of motion. For example, if you throw a baseball or catch one thrown to you, the force required is not too great. But if you want to start a car moving or to stop one that is coming at you, the force involved is much greater. It follows that the mass of a car is greater than the mass of a baseball.

In agreement with everyday usage, mass can also be thought of as a measure of the quantity of matter in an object. Thus, it is clear that the mass of an automobile, for example, is much greater than the mass of a baseball, but much less than the mass of the earth. We measure mass in units of kilograms (kg), where one kilogram is defined as the mass of a standard cylinder of platinum-iridium, as discussed in Chapter 1. A list of typical masses is given in Table 5–1.

These properties of force and mass are developed in detail in the next three sections.

TABLE 5–1
Typical masses in kilograms (kg)

Earth	5.97×10^{24}
Space Shuttle	2,000,000
Blue whale (largest animal on earth)	178,000
Whale shark (largest fish)	18,000
Elephant (largest land animal)	5400
Automobile	1200
Human (adult)	70
Gallon of milk	3.6
Baseball	0.145
Honeybee	0.00015
Bacterium	10^{-15}

5–2 Newton's First Law of Motion

If you've ever stood in line at an airport, pushing your bags forward a few feet at a time, you know that as soon as you stop pushing the bags, they stop moving. Observations such as this often lead to the erroneous conclusion that a force is required for an object to move. In fact, according to Newton's First Law of Motion, a force is required only to *change* an object's motion.

What is missing in this analysis is the force of friction between the bags and the floor. When you stop pushing the bags, it is not true that they stop moving because they no longer have a force acting on them. On the contrary, there is a rather large *frictional force* between the bags and the floor. It is this force that causes the bags to come to rest.

To see how motion is affected by reducing friction, imagine that you slide into second base during a baseball game. You won't slide very far before stopping. On the other hand, if you slide with the same initial speed on a sheet of ice—where the friction is much less than on a ball field—you slide considerably farther. If you could reduce the friction more, you would slide even farther.

In the classroom, air tracks allow us to observe motion with practically no friction. An example of such a device is shown in **Figure 5–1**. Note that air is blown through small holes in the track, creating a cushion of air for a small "cart" to ride on. A cart placed at rest on a level track remains at rest—unless you push on it to get it started.

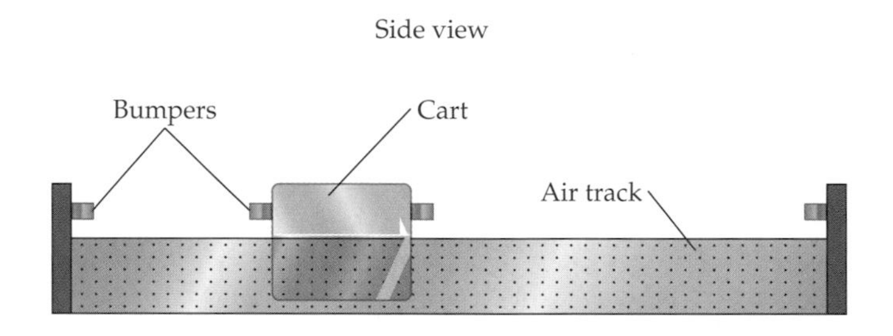

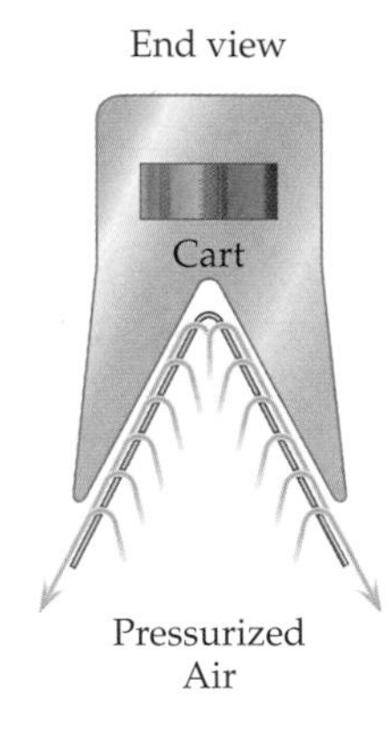

▶ **FIGURE 5–1 The air track**
An air track provides a cushion of air on which a cart can ride with virtually no friction.

▲ An air track provides a nearly frictionless environment for experiments involving linear motion.

Once set in motion, the cart glides along with constant velocity—constant speed in a straight line—until it hits a bumper at the end of the track. The bumper exerts a force on the cart, causing it to change its direction of motion. After bouncing off the bumper the cart again moves with constant velocity. If the track could be extended to infinite length, and could be made perfectly frictionless, the cart would simply keep moving with constant velocity forever.

Newton's first law of motion summarizes these observations in the following statements:

Newton's First Law

An object at rest remains at rest as long as no net force acts on it.
An object moving with constant velocity continues to move with the same speed *and* in the same direction as long as no net force acts on it.

Notice the recurring phrase, "no net force", in these statements. It is important to realize that this can mean one of two things: (i) no force acts on the object; or (ii) forces act on the object, but they sum to zero. We shall see examples of the second possibility later in this chapter and again in the next chapter.

Newton's first law is also known as the **law of inertia**, which is appropriate since the literal meaning of the word inertia is "laziness." Speaking loosely, we can say that matter is "lazy," in that it won't change its motion unless forced to do so. If an object is at rest, it won't start moving on its own. If an object is already moving with constant velocity, it won't alter its speed *or* direction, unless a force causes the change.

According to Newton's first law, being at rest and moving with constant velocity are actually equivalent. To see this, imagine two observers: one is in a train moving with constant velocity; the second is standing next to the tracks, at rest on the ground. The observer in the train places an ice cube on a dinner tray. From that person's point of view, the ice cube has no net force acting on it and it is at rest on the tray. It obeys the first law. From the point of view of the observer on the ground, the ice cube has no net force on it and it moves with constant velocity. This also agrees with the first law. Thus Newton's first law holds for both observers: They both see an ice cube with zero net force moving with constant velocity—it's just that for the first observer the constant velocity happens to be zero.

In this example, we say that each observer is in an **inertial frame of reference**; that is, a frame of reference in which the law of inertia holds. In general, if one frame is an inertial frame of reference, then any frame of reference that moves with constant velocity relative to that frame is also an inertial frame of reference. Thus, if an object moves with constant velocity in one inertial frame, it is always possible to find another inertial frame in which the object is at rest. It is in this sense that there really isn't any difference between being at rest and moving with constant velocity. It's all relative—relative to the frame of reference the object is viewed from.

This gives us a more compact statement of the first law:

If the net force on an object is zero, its velocity is constant.

As an example of a frame of reference that is not inertial, imagine that the train carrying the first observer suddenly comes to a halt. From the point of view of that observer, there is still no net force on the ice cube. However, because of the rapid braking, the ice cube flies off the tray. In fact, the ice cube simply continues to move forward with the same constant velocity while the *train* comes to rest. To the observer on the train, it appears that the ice cube has accelerated forward, even though no force acts on it, which is in violation of Newton's first law.

In general, any frame that accelerates relative to an inertial frame is a noninertial frame. The surface of the earth accelerates slightly, due to its rotational and orbital motions, but since the acceleration is so small, it may be considered an excellent approximation to an inertial frame of reference. Unless specifically stated otherwise, we will always consider the surface of the earth to be an inertial frame.

5-3 Newton's Second Law of Motion

To hold an object in your hand, you have to exert an upward force to oppose, or "balance," the force of gravity. If you suddenly remove your hand so that the only force acting on the object is gravity, it accelerates downward, as discussed in Chapter 2. This is one example of Newton's second law, which states, basically, that unbalanced forces cause accelerations.

To explore this in more detail, consider a spring scale of the type used to weigh fish. The scale gives a reading of the force, F, exerted by the spring contained within it. If we hang one weight from the scale it gives a reading that we will call F_1. If two identical weights are attached, the scale reads $F_2 = 2F_1$, as indicated in **Figure 5-2**. With these two forces marked on the scale, we are ready to perform some force experiments.

First, attach the scale to an air-track cart, as in **Figure 5-3**. If we pull with a force F_1, we observe that the cart accelerates at the rate a_1. If we now pull with a force

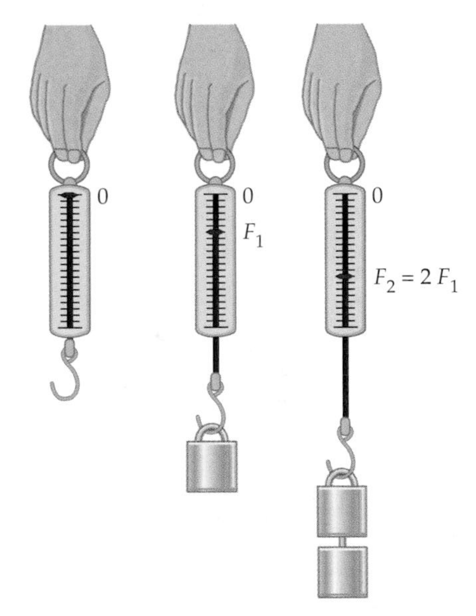

▲ **FIGURE 5-2 Calibrating a "force meter"**
With two weights, the force exerted by the scale is twice the force exerted when only a single weight is attached.

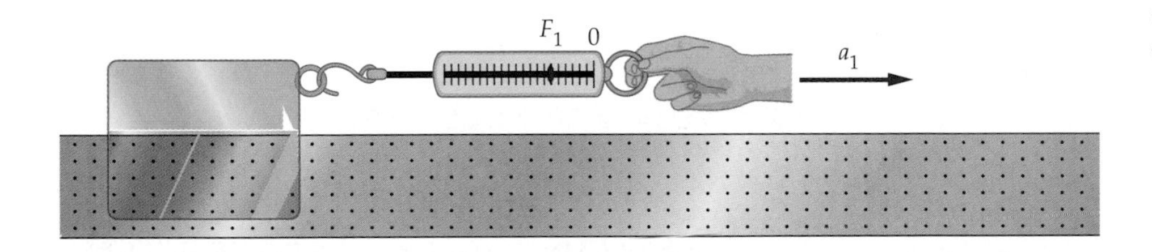

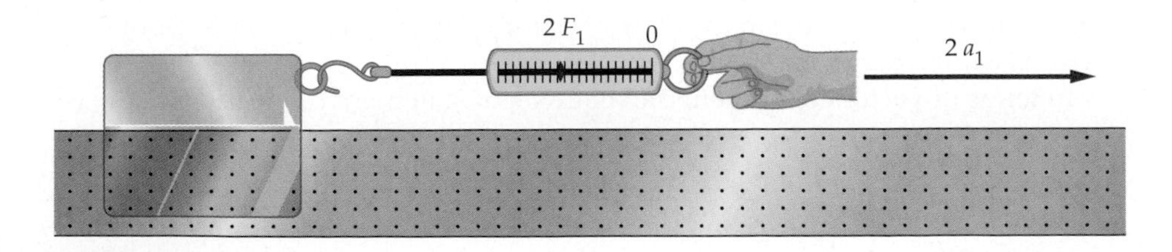

◀ **FIGURE 5-3 Acceleration is proportional to force**
The spring calibrated in Figure 5-2 is used to accelerate a mass on a "frictionless" air track. If the force is doubled, the acceleration is also doubled.

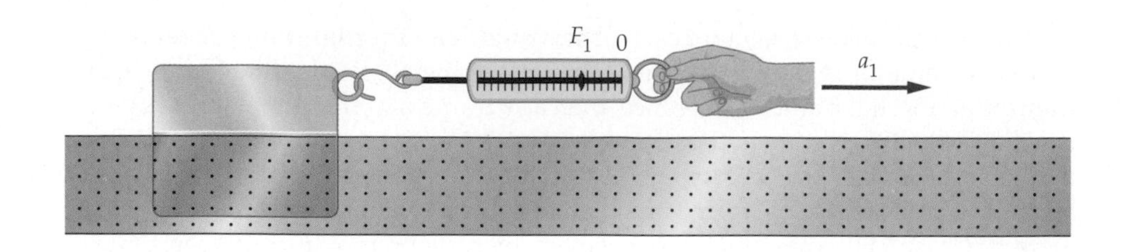

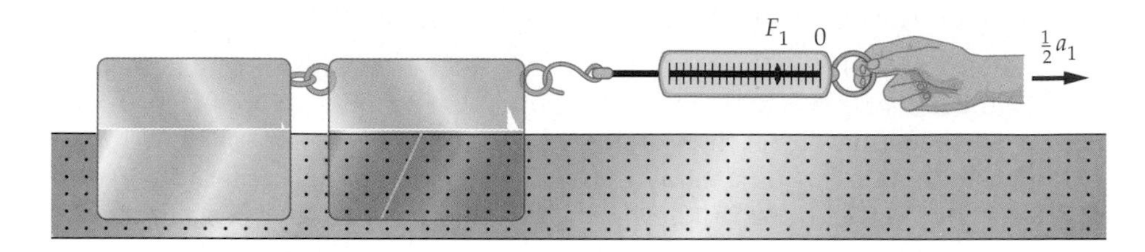

▶ FIGURE 5–4 Acceleration is inversely proportional to mass
If the mass of an object is doubled but the force remains the same, the acceleration is halved.

$F_2 = 2F_1$, the acceleration we observe is $a_2 = 2a_1$. Thus, the acceleration is proportional to the force—the greater the force, the greater the acceleration.

Second, instead of doubling the force, let's double the mass of the cart by connecting two together, as in **Figure 5–4**. In this case, if we pull with a force F_1 we find an acceleration equal to $\frac{1}{2}a_1$. Thus, the acceleration is inversely proportional to mass—the greater the mass, the less the acceleration.

Combining these results, we find that in this simple case—with just one force in just one direction—the acceleration is given by

$$a = \frac{F}{m}$$

Rearranging the equation yields the form of Newton's law that is perhaps best known, $F = ma$.

In general, there may be several forces acting on a given mass, and these forces may be in different directions. Thus, we replace F with the sum of the force vectors acting on a mass:

$$\text{sum of force vectors} = \mathbf{F}_{\text{net}} = \sum \mathbf{F}$$

The notation, $\Sigma\mathbf{F}$, which uses the Greek letter sigma (Σ), is read "sum **F**." Recalling that acceleration is also a vector, we arrive at the formal statement of Newton's second law of motion:

Newton's Second Law

$$\mathbf{a} = \frac{\sum \mathbf{F}}{m} \quad \text{or} \quad \sum \mathbf{F} = m\mathbf{a} \qquad 5\text{–}1$$

In words:

An object of mass m has an acceleration, **a**, equal to the net force, $\Sigma\mathbf{F}$, divided by m.

One should note that Newton's laws cannot be derived from anything more basic. In fact, this is what we mean by a law of nature. Their validity comes from comparisons with experiment.

In terms of vector components, an equivalent statement of the second law is:

$$\sum F_x = ma_x \quad \sum F_y = ma_y \quad \sum F_z = ma_z \qquad 5\text{–}2$$

▲ Even though the tugboat exerts a large force on this ship, the ship's acceleration is small. This is because the acceleration of an object is inversely proportional to its mass, and the mass of the ship is enormous. The force exerted on the unfortunate hockey player is much smaller. The resulting acceleration is much larger, however, due to the relatively small mass of the player compared to that of the ship.

Note that Newton's second law holds independently for each coordinate direction. This component form of the second law is particularly useful when solving problems.

Let's pause for a moment to consider an important special case of the second law. Suppose an object has zero net force acting upon it. Stated mathematically, we can write this as:

$$\sum \mathbf{F} = 0$$

Now, according to Newton's second law, we conclude that the acceleration of this object must be zero:

$$\mathbf{a} = \frac{\sum \mathbf{F}}{m} = \frac{0}{m} = 0$$

But if an object's acceleration is zero, its velocity must be constant. In other words, if the net force on an object is zero, the object moves with constant velocity. This is Newton's first law. Thus we see that Newton's first and second laws are consistent with one another.

Forces are measured in units called, appropriately enough, the newton (N). In particular, one newton is defined as the force required to give one kilogram of mass an acceleration of 1 m/s^2. Thus,

$$1\text{ N} = (1\text{ kg})(1\text{ m/s}^2) = 1\text{ kg}\cdot\text{m/s}^2 \tag{5–3}$$

In everyday terms, a newton is roughly a quarter of a pound. Note that a force in newtons divided by a mass in kilograms has the units of acceleration:

$$\frac{1\text{ N}}{1\text{ kg}} = \frac{1\text{ kg}\cdot\text{m/s}^2}{1\text{ kg}} = 1\text{ m/s}^2 \tag{5–4}$$

Other common units for force are presented in Table 5–2. Typical forces and their magnitudes in newtons are listed in Table 5–3.

TABLE 5–2 Units of mass, acceleration, and force

System of units	mass	acceleration	force
SI	kilogram (kg)	m/s^2	newton (N)
cgs	gram (g)	cm/s^2	dyne (dyn)
British	slug	ft/s^2	pound (lb)

(*Note:* $1\text{ N} = 10^5$ dyne = 0.225 lb.)

TABLE 5–3 Typical forces in newtons (N)

Saturn V rocket	33,800,000
Main engines of shuttle	2,000,000
Pulling force of locomotive	100,000
Thrust of jet engine	75,000
Force to accelerate a car	7000
Weight of adult human	1000
Weight of an apple	1
Weight of a rose	0.1
Weight of an ant	0.001

EXERCISE 5–1

The net force acting on a car is 540 N. If the car's acceleration is 0.39 m/s^2, what is its mass?

Solution

Since the net force and the acceleration are always in the same direction, we can replace the vectors in Equation 5–1 with magnitudes. Solving $\Sigma F = ma$ for the mass yields

$$m = \frac{\sum F}{a} = \frac{540\text{ N}}{0.39\text{ m/s}^2} = 1400\text{ kg}$$

The following Conceptual Checkpoint presents a situation in which both Newton's first and second laws play an important role.

CONCEPTUAL CHECKPOINT 5–1

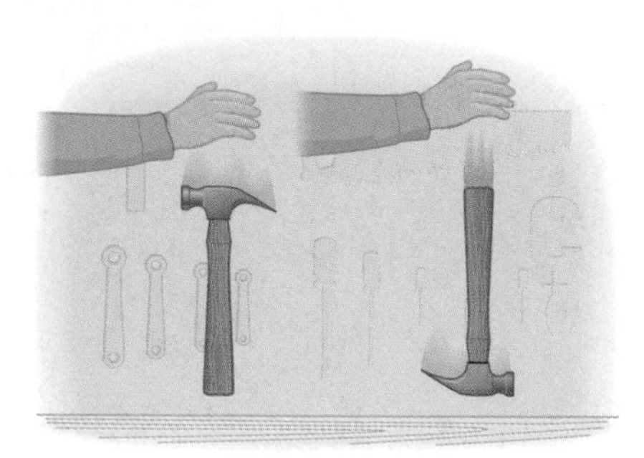

The metal head of a hammer is loose. To tighten it, you drop the hammer down onto a table. Should you (**a**) drop the hammer with the handle end down, (**b**) drop the hammer with the head end down, or (**c**) do you get the same result either way?

Reasoning and Discussion

It might seem that since the same hammer hits against the same table in either case, there shouldn't be a difference. Actually, there is.

In case (**a**) the handle of the hammer comes to rest when it hits the table, but the head continues downward until a force acts on it to bring it to rest. The force that acts on it is supplied by the handle, which results in the head being wedged more tightly onto the handle. Since the metal head is heavy, the force wedging it onto the handle is great. In case (**b**) the head of the hammer comes to rest, but the handle continues to move until a force brings it to rest. The handle is lighter than the head, however; thus the force acting on it is less, resulting in less tightening.

Answer:

(**a**) Drop the hammer with the handle end down.

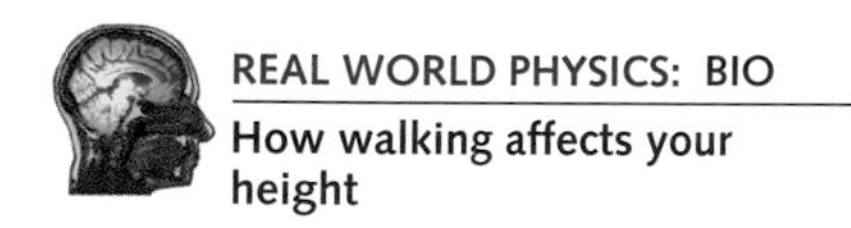

A similar effect occurs when you walk—with each step you take you tamp your head down onto your spine, as when dropping a hammer handle end down. This causes you to grow shorter during the day! Try it. Measure your height first thing in the morning, then again before going to bed. If you're like many people, you'll find that you have shrunk by an inch or so during the day.

Free-Body Diagrams

When solving problems involving forces and Newton's second law, it is essential to begin by making a sketch that indicates *each and every external force* acting on a given object. This type of sketch is referred to as a **free-body diagram**. If we are concerned only with nonrotational motion, as is the case in this and the next chapter, we treat the object of interest as a point particle and apply each of the forces acting on the object to that point, as **Figure 5–5** shows. Once the forces are drawn, we choose a coordinate system and resolve each force into components. At this point, Newton's second law can be applied to each coordinate direction separately.

External forces acting on an object fall into two main classes: (i) Forces at the point of contact with another object, and (ii) forces exerted by an external agent, such as gravity.

For example, in Figure 5–5 there are three external forces acting on the chair. One is the force **F** exerted by the person. In addition, gravity exerts a downward force, **W**, which is simply the weight of the chair. Finally, the floor exerts an upward force on the chair that prevents it from falling toward the center of the earth. This force is referred to as the *normal force*, **N**, because it is perpendicular (that is, normal) to the surface of the floor. We will consider the weight and the normal force in greater detail in Sections 5–6 and 5–7, respectively.

We can summarize the steps involved in constructing a free-body diagram as follows:

Sketch the Forces
Identify and sketch all of the external forces acting on an object.

Isolate the Object of Interest
Replace the object with a point particle of the same mass. Apply each of the forces acting on the object to that point.

Choose a Convenient Coordinate System
Any coordinate system will work; however, if the object moves in a known direction, it is often convenient to pick that direction for one of the coordinate axes. Otherwise, it is reasonable to choose a coordinate system that aligns with one or more of the forces acting on the object.

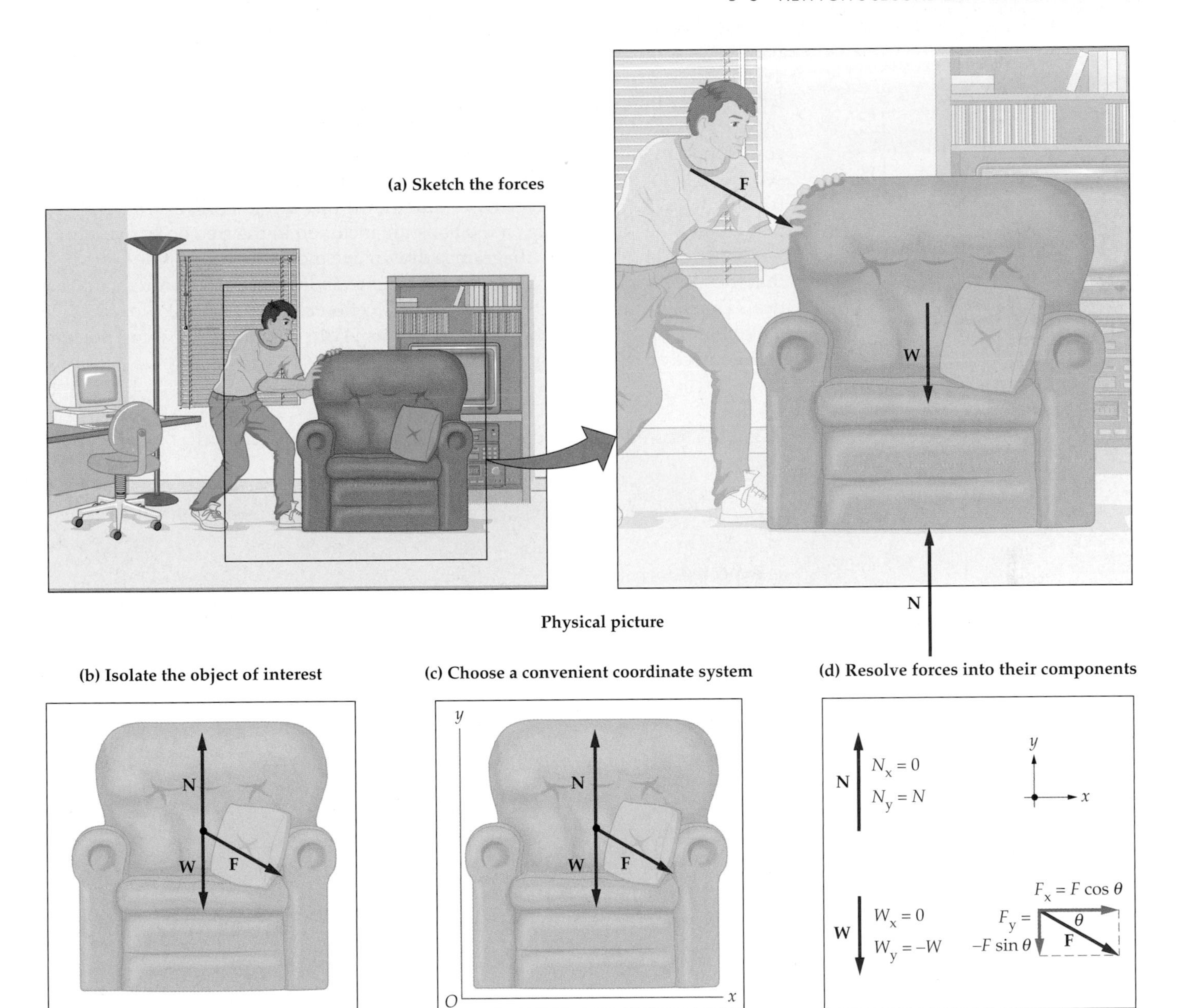

▲ FIGURE 5–5 Constructing and using a free-body diagram
(a) Sketch all of the external forces acting on an object of interest. **(b)** Isolate the object and treat it as a point particle. **(c)** Choose a convenient coordinate system. **(d)** Resolve each of the forces into components using the coordinate system of part **(c)**.

Resolve the Forces into Components
Determine the components of each force in the free-body diagram.

Apply Newton's Second Law to each Coordinate Direction
Analyze motion in each coordinate direction using the component form of Newton's second law, as given in Equation 5–2.

These basic steps are illustrated in Figure 5–5. Note that the figures in this chapter use the labels "Physical picture" to indicate a sketch of the physical situation and "Free-body diagram" to indicate a free-body sketch.

We begin by applying this procedure to a simple one-dimensional example, saving two-dimensional systems for Section 5–5. Suppose, for instance, that you hold a book at rest in your hand. What is the magnitude of the upward force that your

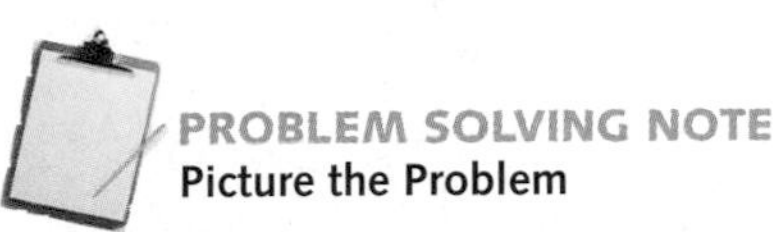

PROBLEM SOLVING NOTE
Picture the Problem

In problems involving Newton's laws it is important to begin with a free-body diagram and to identify all the external forces that act on an object. Once these forces are identified and resolved into their components, Newton's laws can be applied in a straightforward way. It is crucial, however, that only external forces acting on the object be included, and that none of the external forces be omitted.

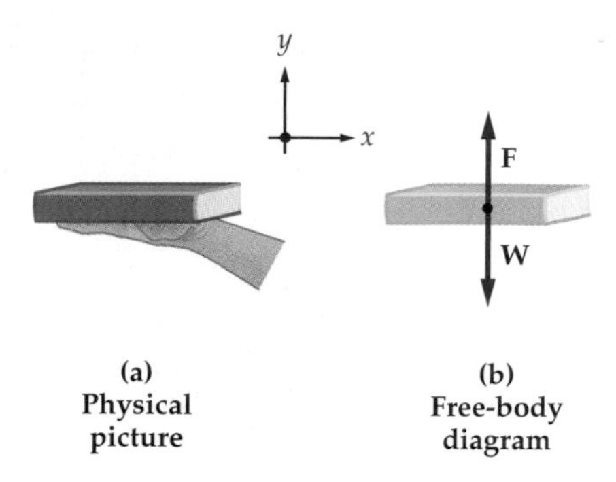

▲ FIGURE 5–6 A book supported in a person's hand
(a) The physical situation. **(b)** The free-body diagram for the book, showing the two external forces acting on it. We also indicate our choice for a coordinate system.

hand must exert in order to keep the book at rest? From everyday experience, we expect that the upward force must be equal in magnitude to the weight of the book, but let's see how this result can be obtained directly from Newton's second law.

We begin with a sketch of the physical situation, as shown in **Figure 5–6 (a)**. The corresponding free-body diagram, in Figure 5–6 (b), shows just the book, represented by a point, and the forces acting on it. Note that two forces act on the book: (i) the downward force of gravity, **W**, and (ii) the upward force, **F**, exerted by your hand. Only the forces acting *on* the book are included in the free-body diagram.

Now that the free-body diagram is drawn, we indicate a coordinate system so that the forces can be resolved into components. In this case all the forces are vertical. Thus we draw a y axis in the vertical direction in **Figure 5–6 (b)**. Note that we have chosen upward to be the positive direction. With this choice, the y components of the forces are $F_y = F$ and $W_y = -W$. It follows that

$$\sum F_y = F - W$$

Using the y component of the second law ($\sum F_y = ma_y$) we find

$$F - W = ma_y$$

Since the book remains at rest, its acceleration is zero. Thus, $a_y = 0$, which gives

$$F - W = ma_y = 0 \qquad \text{or} \qquad F = W$$

as expected.

Next, we consider a situation where the net force acting on an object is nonzero, meaning that its acceleration is also nonzero.

EXAMPLE 5–1 Three Forces

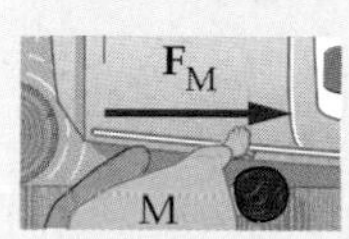

Moe, Larry, and Curly push on a 752-kg boat that floats next to a dock. They each exert an 80.5-N force parallel to the dock. **(a)** What is the acceleration of the boat if they all push in the same direction? Give both direction and magnitude. **(b)** What is the magnitude and direction of the boat's acceleration if Larry and Curly push in the opposite direction to Moe's push?

Picture the Problem

In our diagram we indicate the three relevant forces acting on the boat: $\mathbf{F}_M$, $\mathbf{F}_L$, and $\mathbf{F}_C$. Note that we have chosen the positive x direction to the right, in the direction that all three push for part **(a)**.

Strategy

Since we know the mass of the boat and the forces acting on it, we can find the acceleration using $\Sigma F_x = ma_x$.

Physical pictures

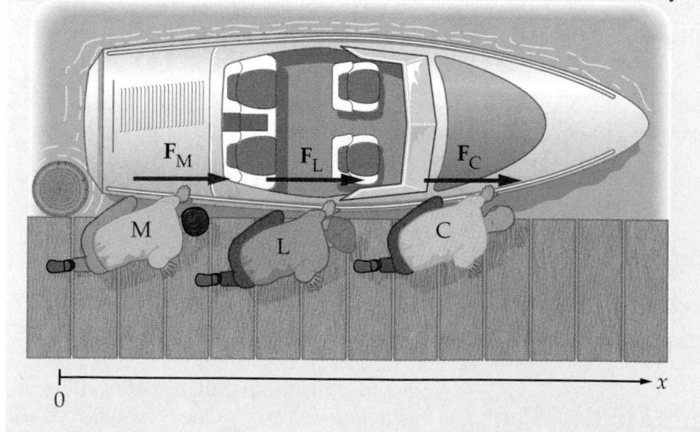

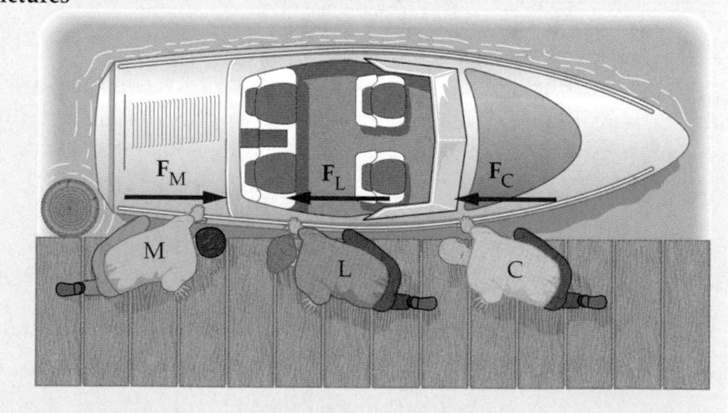

Free-body diagrams

F_M, F_L, F_C

(a)

F_L, F_M, F_C

(b)

Solution

Part (a)

1. Write out the x component for each of the three forces: $F_{M,x} = F_{L,x} = F_{C,x} = 80.5\text{ N}$

2. Sum the x components of force and set equal to ma_x: $\sum F_x = F_{M,x} + F_{L,x} + F_{C,x} = 241.5\text{ N} = ma_x$

3. Divide by the mass to find a_x. Since a_x is positive, the acceleration is to the right, as expected: $a_x = \dfrac{\sum F_x}{m} = \dfrac{241.5\text{ N}}{752\text{ kg}} = 0.321\text{ m/s}^2$

Part (b)

4. Again, start by writing the x component for each force:
$$F_{M,x} = 80.5\text{ N}$$
$$F_{L,x} = F_{C,x} = -80.5\text{ N}$$

5. Sum the x components of force and set equal to ma_x:
$$\sum F_x = F_{M,x} + F_{L,x} + F_{C,x}$$
$$= 80.5\text{ N} - 80.5\text{ N} - 80.5\text{ N} = -80.5\text{ N} = ma_x$$

6. Solve for a_x. In this case a_x is negative, indicating an acceleration to the left: $a_x = \dfrac{\sum F_x}{m} = \dfrac{-80.5\text{ N}}{752\text{ kg}} = -0.107\text{ m/s}^2$

Insight

Even though this problem is one dimensional, it is important to think of it in terms of vector components. For example, when we sum the x components of the forces, we are careful to use the appropriate signs—just as we always do when dealing with vectors.

Practice Problem

If Moe, Larry, and Curly all push to the right with 85.0-N forces, and the boat accelerates at 0.530 m/s^2, what is its mass? [**Answer:** 481 kg]

Some related homework problems: Problem 1, Problem 3

In some problems, we are given information that allows us to calculate an object's acceleration using the kinematic equations of Chapters 2 and 3. Once the acceleration is known, the second law can be used to find the net force that caused the acceleration.

For example, suppose that an astronaut uses a jet pack to push a satellite toward the space shuttle. These jet packs, which are known to NASA as Manned Maneuvering Units, or MMUs, are basically small "one-person rockets" strapped to the back of an astronaut's spacesuit. An MMU contains pressurized nitrogen gas that can be released through varying combinations of 24 nozzles spaced around the unit, producing a force of about 10 pounds. The MMUs contain enough propellant for a six-hour EVA (extra-vehicular activity).

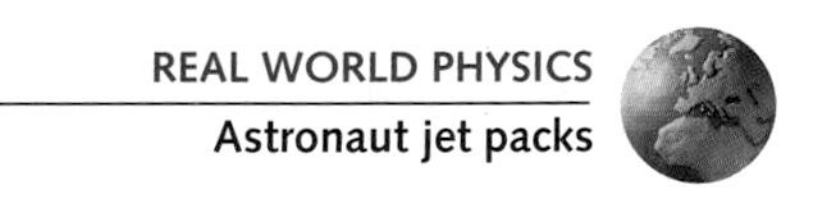

We show the physical situation in **Figure 5–7 (a)**, where an astronaut pushes on a 655-kg satellite. The corresponding free-body diagram for the satellite is shown in **Figure 5–7 (b)**. Note that we have chosen the x axis to point in the direction of the push. Now, if the satellite starts at rest and moves 0.675 m after 5.00 seconds of pushing, what is the force, F, exerted on it by the astronaut?

(a) Physical picture

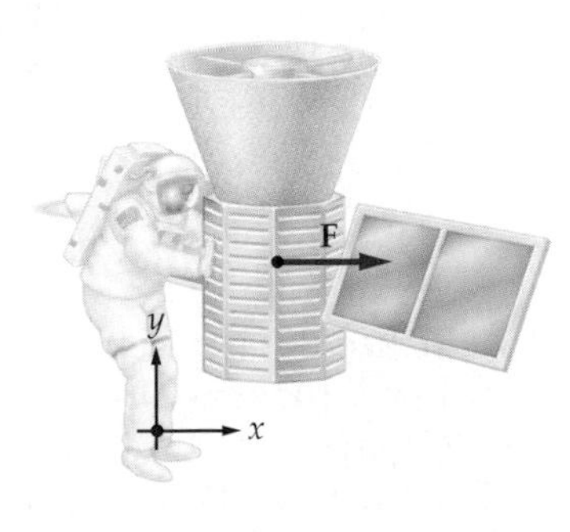

(b) Free-body diagram

◀ FIGURE 5–7 An astronaut using a jet pack to push a satellite
(a) The physical situation. **(b)** The free-body diagram for the satellite. Only one force acts on the satellite, and it is in the positive x direction.

▲ A technician inspects the landing gear of an airliner in a test of Foamcrete, a solid paving material that is just soft enough to collapse under the weight of an airliner. A plane that has run off the runway will slow safely to a stop as its wheels plow through the crumbling Foamcrete.

Clearly, we would like to use Newton's second law (basically, $\mathbf{F} = m\mathbf{a}$) to find the force, but we know only the mass of the satellite, not its acceleration. We can find the acceleration, however, by using the kinematic equation relating position to time; $x = x_0 + v_{0x}t + \frac{1}{2}a_x t^2$. We can choose the initial position of the satellite to be $x_0 = 0$, and we are given that it starts at rest, thus $v_{0x} = 0$. Hence,

$$x = \tfrac{1}{2}a_x t^2$$

Since we know the distance covered in a given time, we can solve for the acceleration:

$$a_x = \frac{2x}{t^2} = \frac{2(0.675\ \text{m})}{(5.00\ \text{s})^2} = 0.0540\ \text{m/s}^2$$

Now that kinematics has provided the acceleration, we use the x component of the second law to find the force. Only one force acts on the satellite, and its x component is F; thus,

$$\sum F_x = F = ma_x$$

$$F = ma_x = (655\ \text{kg})(0.0540\ \text{m/s}^2) = 35.4\ \text{N}$$

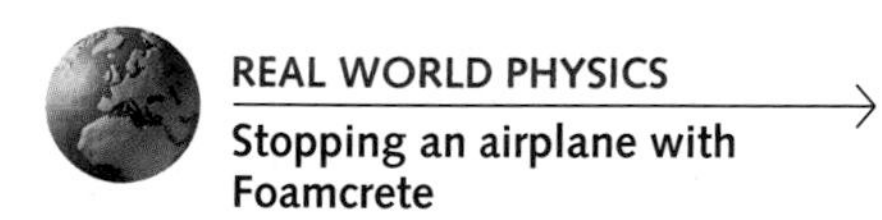

REAL WORLD PHYSICS

Stopping an airplane with Foamcrete

This force corresponds to a push of about 8 lb.

Another problem where we use kinematics to find the acceleration is presented in the following Active Example.

ACTIVE EXAMPLE 5–1 Foamcrete

Foamcrete is a substance designed to stop an airplane that has run off the end of a runway, without causing injury to passengers. It is solid enough to support a car, but crumbles under the weight of a large airplane. By crumbling, it slows the plane to a safe stop. For example, suppose a 747 jetliner with a mass of 1.75×10^5 kg and an initial speed of 26.8 m/s is slowed to a stop in 122 m. What is the magnitude of the retarding force **F** exerted by the Foamcrete on the plane?

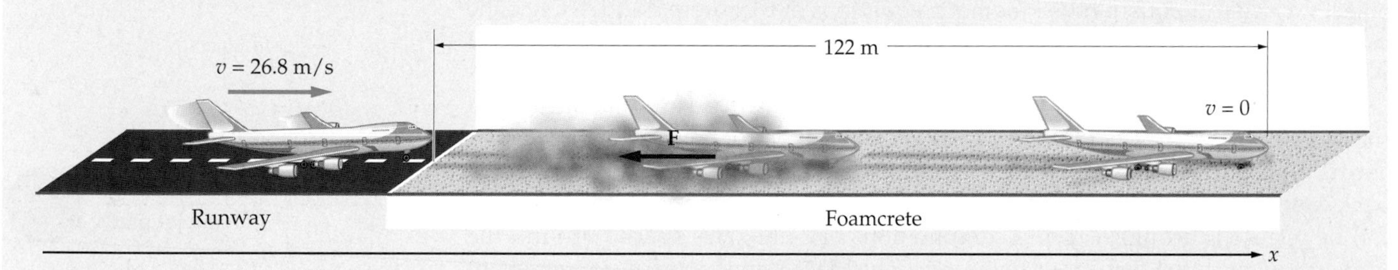

Solution

1. Use $v^2 = v_0^2 + 2a_x\Delta x$ to find the plane's acceleration: $a_x = -2.94\ \text{m/s}^2$
2. Sum the forces in the x direction. Let F represent the magnitude of the force **F**: $\Sigma F_x = -F$
3. Set the sum of forces equal to mass times acceleration: $-F = ma_x$
4. Solve for the magnitude of the force, F: $F = -ma_x = 5.15 \times 10^5\ \text{N}$

Note again the care we take with the signs. The plane's acceleration is negative, hence the net force acting on it, **F**, is in the negative x direction. On the other hand, the magnitude of the force, F, is positive, as is always the case for magnitudes.

Finally, we end this section with an estimation problem.

EXAMPLE 5–2 Pitch Man

A pitcher throws a 0.15-kg baseball, accelerating it from rest to a speed of about 90 mi/h. Estimate the force exerted by the pitcher on the ball.

Picture the Problem

We choose the x axis to point in the direction of the pitch. Also indicated in the sketch is the distance over which the pitcher accelerates the ball, Δx. Since we are interested only in the pitch, and not in the subsequent motion of the ball, we ignore the effects of gravity.

Strategy

We know the mass so we can find the force with $F_x = ma_x$ if we can estimate the acceleration. To find the acceleration, we estimate the distance, Δx, then find a_x using $v^2 = v_0^2 + 2a_x\Delta x$. From the sketch, a reasonable estimate for Δx is about 2.0 m. As for the speeds, we know that $v_0 = 0$ and $v \approx 90$ mi/h.

Solution

1. Starting with the fact that 60 mi/h = 1 mi/min, perform a rough "back-of-the-envelope" conversion of 90 mi/h to meters per second:

$$v \approx 90 \text{ mi/h} = \frac{1.5 \text{ mi}}{\text{min}} \approx \frac{2400 \text{ m}}{60 \text{ s}} = 40 \text{ m/s}$$

2. Solve $v^2 = v_0^2 + 2a_x\Delta x$ for the acceleration, a_x. Use the estimates $\Delta x \approx 2.0$ m and $v \approx 40$ m/s:

$$a_x = \frac{v^2 - v_0^2}{2\Delta x} \approx \frac{(40 \text{ m/s})^2 - 0}{2(2.0 \text{ m})} = 400 \text{ m/s}^2$$

3. Find the corresponding force with $F_x = ma_x$:

$$F_x = ma_x \approx (0.15 \text{ kg})(400 \text{ m/s}^2) = 60 \text{ N} \approx 10 \text{ lb}$$

Insight

This is a sizable force, especially when you consider that the ball itself weighs only about 1/3 lb. Thus, the pitcher exerts a force on the ball that is about 30 times greater than the force exerted by Earth's gravity. It follows that ignoring gravity during the pitch is a reasonable approximation.

Practice Problem

What is the approximate speed of the pitch if the force exerted by the pitcher is $\frac{1}{2}(60 \text{ N}) = 30$ N? [**Answer:** 30 m/s or 60 mi/h]

Some related homework problems: Problem 4, Problem 6

Another way to find the acceleration is to estimate the amount of time it takes to make the pitch. However, since the pitch is delivered so quickly—about 1/10 s—estimating the time would be more difficult than estimating the distance Δx.

5–4 Newton's Third Law of Motion

Nature never produces just one force at a time; *forces always come in pairs*. In addition, the forces in a pair, which always act on *different objects*, are equal in magnitude and opposite in direction. This is Newton's third law of motion.

Newton's Third Law

For every force that acts on an object, there is a reaction force acting on a different object that is equal in magnitude and opposite in direction.

In a somewhat more specific form:

If object 1 exerts a force **F** on object 2, then object 2 exerts a force −**F** on object 1.

This law, more commonly known by its abbreviated form, "for every action there is an equal and opposite reaction," completes Newton's laws of motion.

Figure 5–8 illustrates some action-reaction pairs. Notice that there is always a reaction force, whether the action force pushes on something hard to move, like a refrigerator, or on something that moves with no friction, like an air-track cart. In

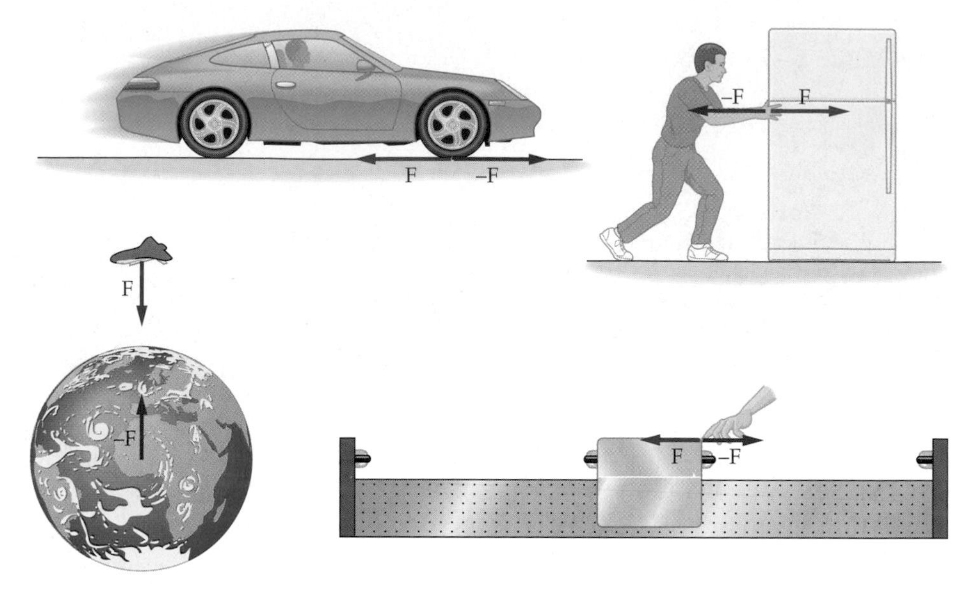

▶ **FIGURE 5–8 Examples of action-reaction force pairs**

▲ BASE jumpers are parachutists who descend from unlikely places: *b*uildings, *a*ntenna towers, *s*pans (bridges), or *e*arth (mountain cliffs). This BASE jumper is accelerating downward due to the attractive gravitational force of the Earth. By Newton's third law, however, the jumper also attracts the earth with a force that is equal in magnitude and opposite direction. As a result the Earth accelerates upward as the jumper accelerates downward. The acceleration of the Earth is imperceptibly small because its mass is many orders of magnitude greater than the mass of the jumper.

some cases, the reaction force tends to be overlooked, as when the Earth exerts a *downward* gravitational force on the space shuttle, and the shuttle exerts an equal and opposite *upward* gravitational force on the Earth. Still, the reaction force always exists.

Another important aspect of the third law is that the action-reaction forces always act on *different* objects. This, again, is illustrated in Figure 5–8. Thus, in drawing a free-body diagram, only one of the action-reaction pair would be drawn for a given object. The other force in the pair would appear in the free-body diagram of a different object. As a result, *the two forces do not cancel.*

For example, consider a car accelerating from rest, as in Figure 5–8. As the car's engine turns the wheels, the tires exert a force on the road. By the third law, the road exerts an equal and opposite force on the car's tires. It is this second force—which acts on the car through its tires—that propels the car forward. The force exerted by the tires on the road does not accelerate the car.

Since the action-reaction forces act on different objects, they generally produce very different accelerations. This is the case in the next example.

EXAMPLE 5–3 Tippy Canoe

Two groups of canoeists meet in the middle of a lake. After a brief visit, a person in canoe 1 pushes on canoe 2 with a force of 46 N to separate the canoes. If the mass of canoe 1 and its occupants is $m_1 = 150$ kg, and the mass of canoe 2 and its occupants is $m_2 = 250$ kg, **(a)** find the acceleration the push gives to each canoe. **(b)** What is the separation of the canoes after 1.2 s of pushing?

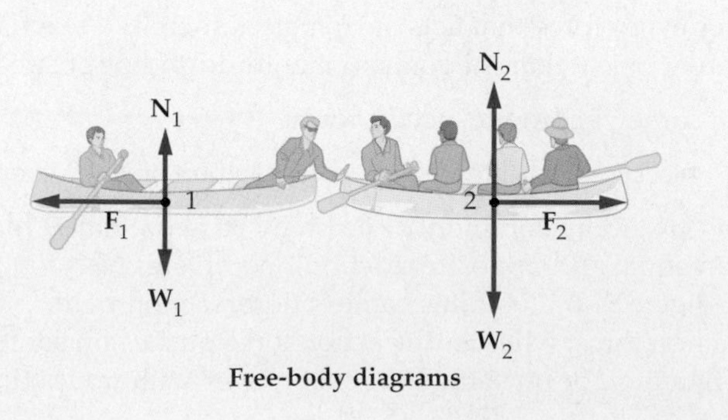

Free-body diagrams

Picture the Problem

We have chosen the positive x direction to point from canoe 1 to canoe 2. With this choice, the force exerted on canoe 2 is $\mathbf{F}_2 = (+46\text{ N})\hat{\mathbf{x}}$. By Newton's third law, the force exerted on the person in canoe 1, and thus on canoe 1 itself if the person is firmly seated, is $\mathbf{F}_1 = (-46\text{ N})\hat{\mathbf{x}}$. For convenience, we have placed the origin at the point where the canoes touch.

Strategy

(a) We can find the acceleration of each canoe by solving $\sum F_x = ma_x$ for a_x. **(b)** The kinematic equation relating position to time, $x = x_0 + v_{0x}t + \frac{1}{2}a_x t^2$, can then be used to find the displacement of each canoe.

Solution

Part (a)

1. Use Newton's second law to find the acceleration of canoe 2:

$$a_{2,x} = \frac{\sum F_{2,x}}{m_2} = \frac{46\text{ N}}{250\text{ kg}} = 0.18\text{ m/s}^2$$

2. Do the same calculation for canoe 1. Note that the acceleration of canoe 1 is in the negative direction:

$$a_{1,x} = \frac{\sum F_{1,x}}{m_1} = \frac{-46\text{ N}}{150\text{ kg}} = -0.31\text{ m/s}^2$$

Part (b)

3. Use $x = x_0 + v_{0x}t + \frac{1}{2}a_x t^2$ to find the position of canoe 2 at $t = 1.2$ s. From the problem statement, we know the canoes start at the origin ($x_0 = 0$), and at rest ($v_{0x} = 0$):

$$x_2 = \tfrac{1}{2}a_{2,x}t^2 = \tfrac{1}{2}(0.18\text{ m/s}^2)(1.2\text{ s})^2 = 0.13\text{ m}$$

4. Repeat the calculation for canoe 1:

$$x_1 = \tfrac{1}{2}a_{1,x}t^2 = \tfrac{1}{2}(-0.31\text{ m/s}^2)(1.2\text{ s})^2 = -0.22\text{ m}$$

5. Subtract the two positions to find the separation of the canoes:

$$x_2 - x_1 = 0.13\text{ m} - (-0.22\text{ m}) = 0.35\text{ m}$$

Insight

The same magnitude of force acts on each canoe, hence the lighter one has the greater acceleration and the greater displacement. If the heavier canoe were replaced by a large ship of great mass, both vessels would still accelerate as a result of the push. However, the acceleration of the large ship would be so small as to be practically imperceptible. In this case, it would appear as if only the canoe moved, whereas, in principle, both vessels move.

Practice Problem

If the mass of canoe 2 is increased, does its acceleration increase, decrease, or stay the same? Check your answer by calculating the acceleration for the case where canoe 2 is replaced by a 25,000-kg ship. [**Answer:** The acceleration will decrease. In this case, $a = 0.0018\text{ m/s}^2$.]

Some related homework problems: Problem 14, Problem 15

When objects are touching one another, the action-reaction forces are often referred to as **contact forces**. The behavior of contact forces is explored in the following Conceptual Checkpoint.

CONCEPTUAL CHECKPOINT 5–2

Two boxes—one large and heavy, the other small and light—rest on a smooth, level floor. You push with a force **F** on either the small box or the large box. Is the contact force between the two boxes **(a)** the same in either case, **(b)** larger when you push on the large box, or **(c)** larger when you push on the small box?

Reasoning and Discussion

Since the same force pushes on the boxes, you might think the force of contact is the same in both cases. It is not. What we can conclude, however, is that the boxes have the same acceleration in either case—the same net force acts on the same total mass, so the same acceleration, a, results.

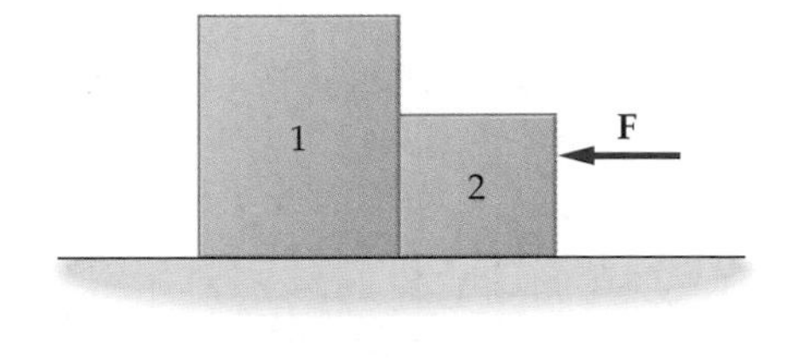

To find the contact force between the boxes we focus our attention on each box individually, and note that *Newton's second law must be satisfied for each of the boxes, just as it is for the entire two-box system.* For example, when the external force is applied to the small box, the only force acting on the *large* box (mass m_1) is the contact force; hence, the contact force must have a magnitude equal to $m_1 a$. In the second case, the only force acting on the *small* box (mass m_2) is the contact force, and so the magnitude of the contact force is $m_2 a$. Since m_1 is greater than m_2, it follows that the force of contact is larger when you push on the small box ($m_1 a$) than when you push on the large box ($m_2 a$).

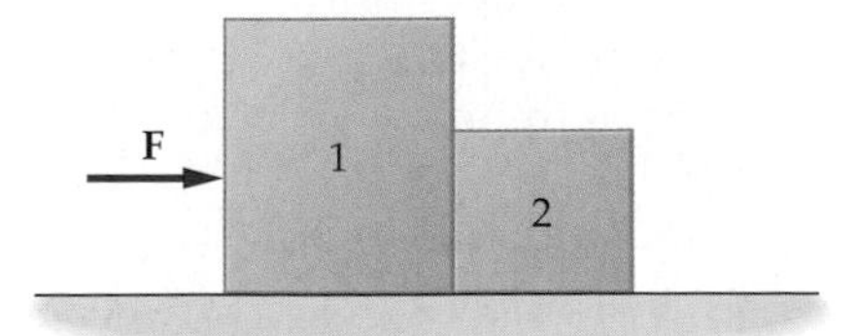

Answer:

(c) The contact force is larger when you push on the small box.

In the next example, we calculate a numerical value for the contact force.

EXAMPLE 5–4 When Push Comes to Shove

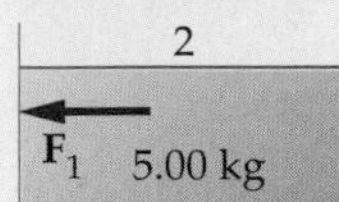

A box of mass $m_1 = 10.0$ kg rests on a smooth, horizontal floor next to a box of mass $m_2 = 5.00$ kg. If you push on box 1 with a horizontal force of magnitude $F = 20.0$ N, **(a)** what is the acceleration of the boxes? **(b)** What is the force of contact between the boxes?

Picture the Problem

We choose the x axis to be horizontal and pointing to the right. Thus, $\mathbf{F} = (20.0\text{ N})\hat{\mathbf{x}}$. The contact forces are labeled as follows: $\mathbf{F}_1$ is the contact force exerted on box 1; $\mathbf{F}_2$ is the contact force exerted on box 2. The contact forces have the same magnitude, f, but point in opposite directions. With our coordinate system, we have $\mathbf{F}_1 = -f\hat{\mathbf{x}}$ and $\mathbf{F}_2 = f\hat{\mathbf{x}}$.

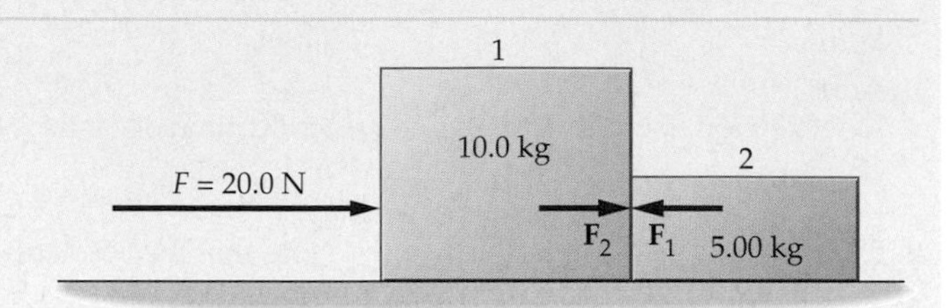

Physical picture

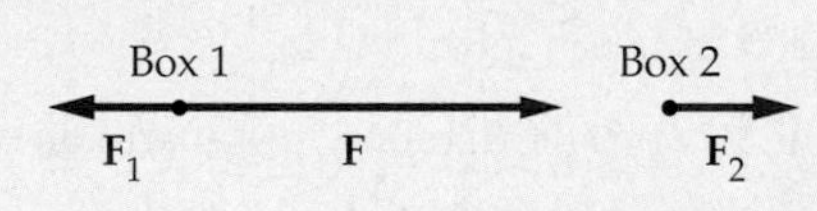

Free-body diagrams

Strategy

(a) Since the two boxes are in contact, they have the same acceleration. We find this acceleration by dividing the net horizontal force by the total mass of the two boxes. **(b)** Now let's consider the system consisting solely of box 2. The mass in this case is 5.00 kg, and the only horizontal force acting on the system is $\mathbf{F}_2$. Thus, we can find f, the magnitude of $\mathbf{F}_2$, by requiring that box 2 have the acceleration found in part (a).

Solution

Part (a)

1. Find the net horizontal force acting on the two boxes. Note that $\mathbf{F}_1$ and $\mathbf{F}_2$ are equal in magnitude but opposite in direction. Hence, they sum to zero; $\mathbf{F}_1 + \mathbf{F}_2 = 0$:

$$\sum_{\substack{\text{both}\\\text{boxes}}} F_x = F = 20.0\text{ N}$$

2. Divide the net force by the total mass, $m_1 + m_2$, to find the acceleration of the boxes:

$$a_x = \frac{\sum F_x}{m_1 + m_2}$$

$$= \frac{20.0\text{ N}}{(10.0\text{ kg} + 5.00\text{ kg})} = \frac{20.0\text{ N}}{15.0\text{ kg}} = 1.33\text{ m/s}^2$$

Part (b)

3. Find the net horizontal force acting on box 2, and set it equal to the mass of box 2 times its acceleration:

$$\sum_{\text{box 2}} F_x = F_{2,x} = f = m_2 a_x$$

4. Determine the magnitude of the contact force, f, by substituting numerical values for m_2 and a_x:

$$f = m_2 a_x = (5.00\text{ kg})(1.33\text{ m/s}^2) = 6.67\text{ N}$$

Insight

Since the net horizontal force acting on box 1 is $F - f = 20.0\text{ N} - 6.67\text{ N} = 13.3\text{ N}$, it follows that its acceleration is $(13.3\text{ N})/(10.0\text{ kg}) = 1.33\text{ m/s}^2$. Thus, as expected, box 1 and box 2 have precisely the same acceleration.

If box 2 were not present, the 20.0-N force acting on box 1 would give it an acceleration of 2.00 m/s^2. As it is, the contact force between the boxes slows box 1 so that its acceleration is less than 2.00 m/s^2, and accelerates box 2 so that its acceleration is greater than zero. The precise value of the contact force is simply the value that gives both boxes the same acceleration.

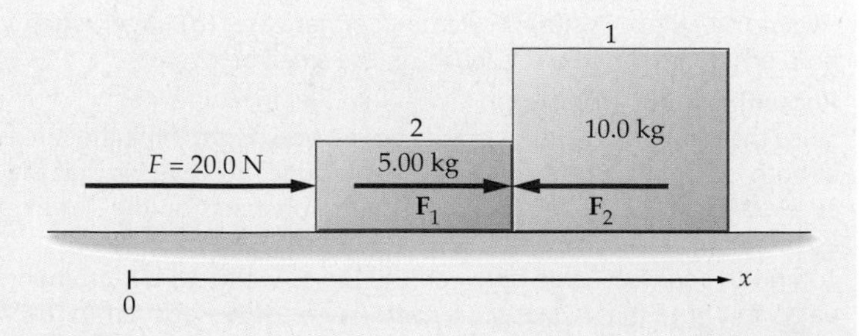

Practice Problem

Suppose the relative positions of the boxes are reversed so that **F** pushes on the small box. **(a)** What is the acceleration of the boxes? **(b)** What is the force of contact between the boxes? **(c)** What is the total force acting on box 2? [**Answer: (a)** 1.33 m/s^2, **(b)** 13.3 N, **(c)** 6.67 N]

Some related homework problems: Problem 16, Problem 17

5–5 The Vector Nature of Forces: Forces in Two Dimensions

When we presented Newton's second law in Section 5–3, we said that an object's acceleration is equal to the net force acting on it divided by its mass. For example, if only a single force acts on an object, its acceleration is found to be in the same direction as the force. If more than one force acts on an object, experiments show that its acceleration is in the direction of the vector sum of the forces. Thus forces are indeed vectors, and they exhibit all the vector properties discussed in Chapter 3.

The mass of an object, on the other hand, is simply a positive number with no associated direction. It represents the amount of matter in an object.

As an example of the vector nature of forces, suppose two astronauts are using jet packs to push a 940-kg satellite toward the space shuttle, as shown in **Figure 5–9**. With the coordinate system indicated in the figure, astronaut 1 pushes in the positive x direction and astronaut 2 pushes in a direction 52° above the x axis. If astronaut 1 pushes with a force of magnitude $F_1 = 26$ N and astronaut 2 pushes with a force of magnitude $F_2 = 41$ N, what are the magnitude and direction of the satellite's acceleration?

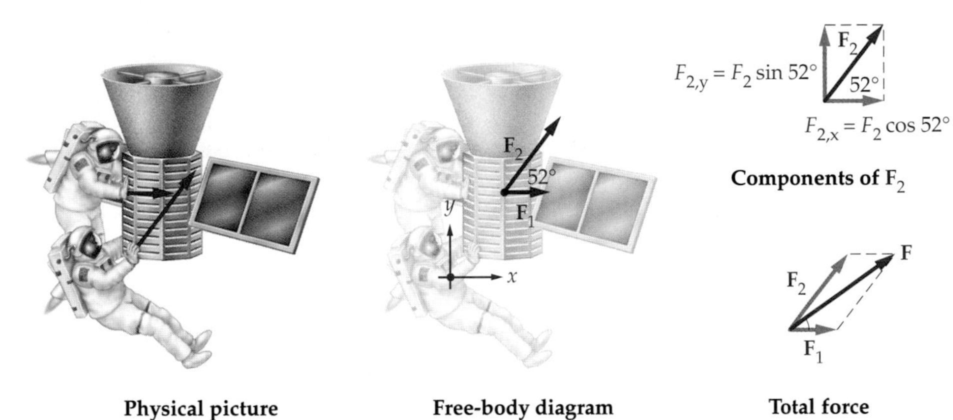

◀ **FIGURE 5–9 Two astronauts pushing a satellite with forces that differ in magnitude and direction**
The acceleration of the satellite can be found by calculating a_x and a_y separately.

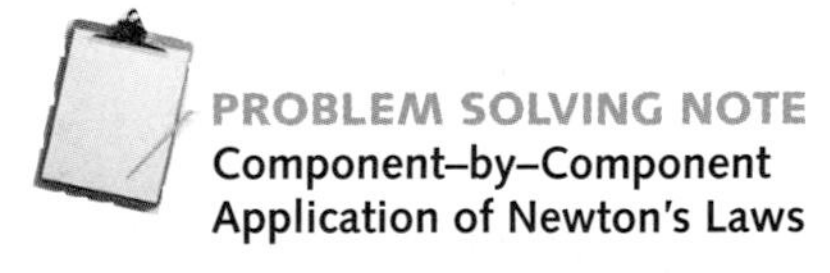

PROBLEM SOLVING NOTE
Component-by-Component Application of Newton's Laws

Newton's laws can be applied to each coordinate direction independently of the others. Therefore, when drawing a free-body diagram, be sure to include a coordinate system. Once the forces are resolved into their x and y components, the second law can be solved for each component separately. Working in a component-by-component fashion is the systematic way of using Newton's laws.

The easiest way to solve a problem like this is to treat each coordinate direction independently of the other, just as we did many times when studying two-dimensional kinematics in Chapter 4. Thus, we first resolve each force into its x and y components. Referring to Figure 5–9, we see that for the x direction

$$F_{1,x} = F_1$$
$$F_{2,x} = F_2 \cos 52°$$

For the y direction

$$F_{1,y} = 0$$
$$F_{2,y} = F_2 \sin 52°$$

Next, we find the acceleration in the x direction by using the x component of Newton's second law:

$$\sum F_x = ma_x$$

Applied to this system, we have

$$\sum F_x = F_{1,x} + F_{2,x} = F_1 + F_2 \cos 52° = 26\text{ N} + (41\text{ N})\cos 52° = 51\text{ N}$$
$$= ma_x$$

Solving for the acceleration yields

$$a_x = \frac{\sum F_x}{m} = \frac{51\text{ N}}{940\text{ kg}} = 0.054\text{ m/s}^2$$

Similarly, in the y direction we start with

$$\sum F_y = ma_y$$

This gives

$$\sum F_y = F_{1,y} + F_{2,y} = 0 + F_2 \sin 52° = (41 \text{ N})\sin 52° = 32 \text{ N}$$
$$= ma_y$$

As a result, the y component of acceleration is:

$$a_y = \frac{\sum F_y}{m} = \frac{32 \text{ N}}{940 \text{ kg}} = 0.034 \text{ m/s}^2$$

Thus, the satellite accelerates in both the x and the y directions. Its total acceleration has a magnitude of

$$a = \sqrt{a_x^2 + a_y^2} = \sqrt{(0.054 \text{ m/s}^2)^2 + (0.034 \text{ m/s}^2)^2} = 0.064 \text{ m/s}^2$$

From Figure 5–9 we expect the total acceleration to be in a direction above the x axis, but at an angle less than 52°. Straightforward calculation yields

$$\theta = \tan^{-1}\left(\frac{a_y}{a_x}\right) = \tan^{-1}\left(\frac{0.034 \text{ m/s}^2}{0.054 \text{ m/s}^2}\right) = \tan^{-1}(0.63) = 32°$$

This is the same direction as the total force in Figure 5-9, as expected.

The following Example and Active Example give further practice with resolving force vectors and using Newton's second law in component form.

EXAMPLE 5–5 Jack and Jill

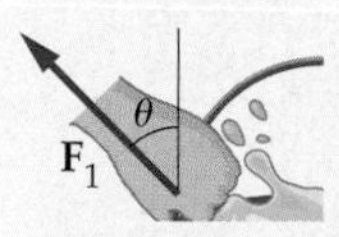

Jack and Jill lift upward on a 1.3-kg pail of water, with Jack exerting a force $\mathbf{F}_1$ of magnitude 7.0 N and Jill exerting a force $\mathbf{F}_2$ of magnitude 11 N. Jill's force is exerted at an angle of 28° with the vertical, as shown below. At what angle θ with respect to the vertical should Jack exert his force if the pail is to accelerate straight upward?

Picture the Problem

Our physical picture and free-body diagram show the pail and the three forces acting on it, as well as the angles relative to the vertical. In the panels at the right, we show the x and y components of the forces $\mathbf{F}_1$ and $\mathbf{F}_2$.

Strategy

We want the acceleration to be purely vertical. This means that the x component of acceleration must be zero, $a_x = 0$. For a_x to be zero it is necessary that the sum of forces in the x direction be zero, $\Sigma F_x = 0$. Since the x component of $\mathbf{F}_1$ depends on the angle θ, the equation $\Sigma F_x = 0$ can be used to find θ.

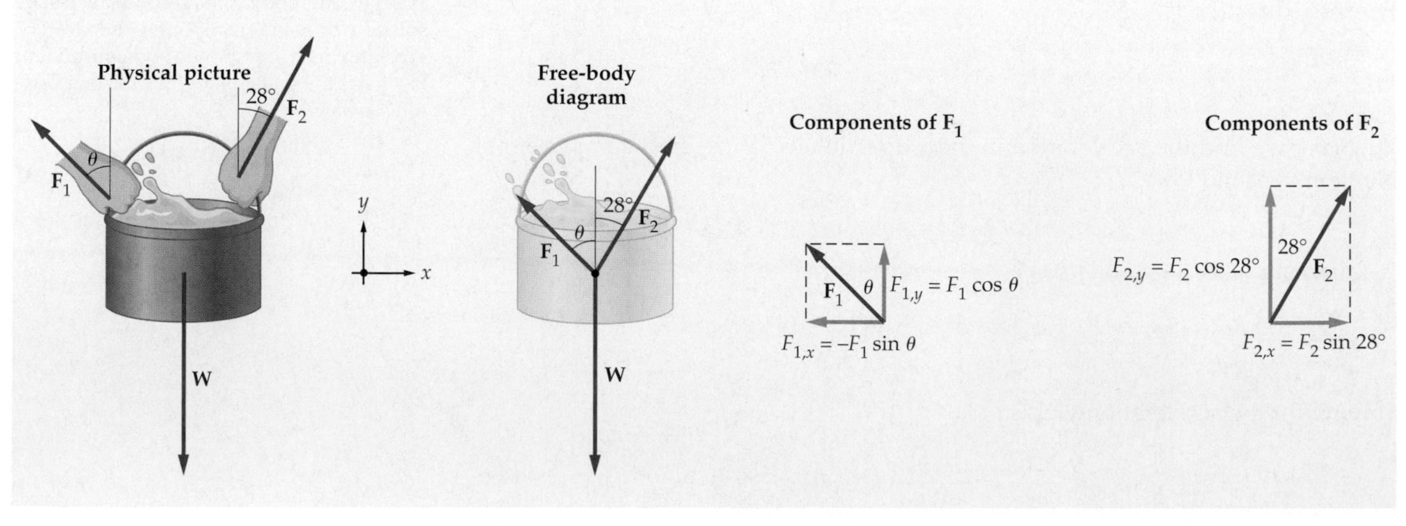

Solution

1. Begin by writing out the x components of each force. Note that **W** has no x component, and that the x component of $\mathbf{F}_1$ points in the negative x direction: $F_{1,x} = -F_1 \sin\theta \quad F_{2,x} = F_2 \sin 28° \quad W_x = 0$
2. Sum the x components of force and set equal to zero: $\sum F_x = -F_1 \sin\theta + F_2 \sin 28° + 0 = ma_x = 0$ or $F_1 \sin\theta = F_2 \sin 28°$

   Note that θ is the only unknown in this equation.
3. Solve for $\sin\theta$ and then for θ: $\sin\theta = \dfrac{F_2 \sin 28°}{F_1} = \dfrac{(11\text{ N})\sin 28°}{7.0\text{ N}} = 0.74$

   $\theta = \sin^{-1}(0.74) = 48°$

Insight

Note that only the y components of $\mathbf{F}_1$ and $\mathbf{F}_2$ contribute to the vertical acceleration of the pail; the x components cancel, leading to zero acceleration in the x direction.

Practice Problem

At what angle must Jack exert his force for the pail to accelerate straight upward if **(a)** $\mathbf{F}_2$ is at an angle of 19° with the vertical or **(b)** $\mathbf{F}_2$ is at an angle of 35° with the vertical? [**Answer: (a)** 31°, **(b)** 64°]

Some related homework problems: Problem 22, Problem 26

ACTIVE EXAMPLE 5–2 Mush

A 4.60-kg sled is pulled across a smooth ice surface. The force acting on the sled is of magnitude 6.20 N and points in a direction 35.0° above the horizontal. If the sled starts at rest, how fast is it going after being pulled for 1.15 s?

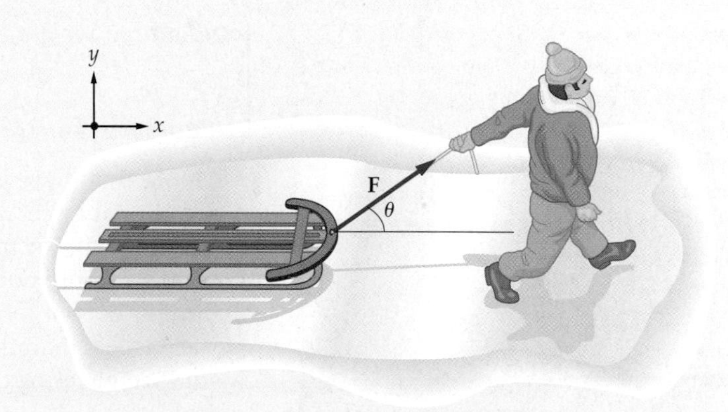

Solution

1. Find the x component of **F**: $F_x = 5.08\text{ N}$
2. Apply Newton's second law to the x direction: $\sum F_x = F_x = ma_x$
3. Solve for the x component of acceleration: $a_x = 1.10\text{ m/s}^2$
4. Use $v_x = v_{0x} + a_x t$ to find the speed of the sled: $v_x = 1.27\text{ m/s}$

Insight

Note that the y component of **F** has no effect on the acceleration of the sled.

5–6 Weight

When you step onto a scale to weigh yourself, the scale gives a measurement of the pull of Earth's gravity. This is your weight, W. Similarly, the weight of any object on the Earth's surface is simply the gravitational force exerted on it by the Earth.

- The weight, W, of an object on the Earth's surface is the gravitational force exerted on it by the Earth.

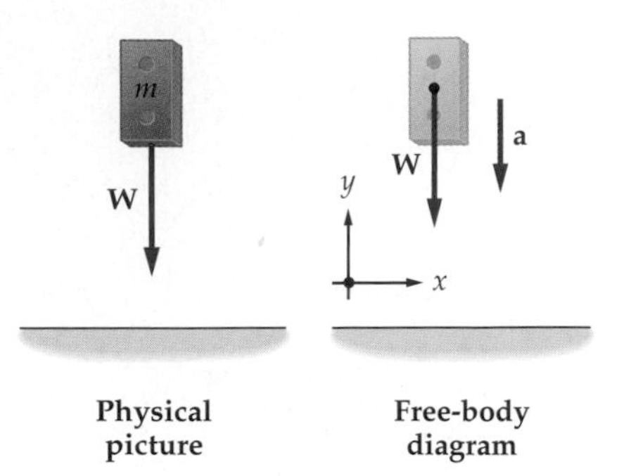

▲ **FIGURE 5–10 Weight and mass**
A brick of mass m has only one force acting on it in free fall—its weight, **W**. The resulting acceleration has a magnitude $a = g$; hence $W = mg$.

As we know from everyday experience, the greater the mass of an object, the greater its weight. For example, if you put a brick on a scale and weigh it, you might get a reading of 9.0 N. If you put a second, identical brick on the scale—which doubles the mass—you will find a weight of 2(9.0 N) = 18 N. Clearly, there must be a simple connection between weight, W, and mass, m.

To see exactly what this connection is, consider taking one of the bricks just mentioned and letting it drop in free fall. As indicated in **Figure 5–10**, the only force acting on the brick is its weight W, which is downward. If we choose upward to be the positive direction, we have

$$\sum F_y = -W$$

In addition, we know from Chapter 2 that the brick moves downward with an acceleration of $g = 9.81 \text{ m/s}^2$ regardless of its mass. Thus,

$$a_y = -g$$

Using these results in Newton's second law

$$\sum F_y = ma_y$$

we find

$$-W = -mg$$

Therefore, the weight of an object of mass m is $W = mg$:

Definition: Weight, *W*

$$W = mg \qquad \textbf{5–5}$$

SI unit: newton, N

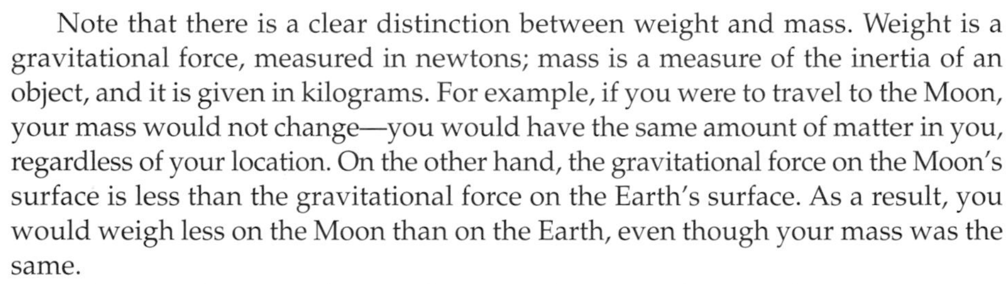

Note that there is a clear distinction between weight and mass. Weight is a gravitational force, measured in newtons; mass is a measure of the inertia of an object, and it is given in kilograms. For example, if you were to travel to the Moon, your mass would not change—you would have the same amount of matter in you, regardless of your location. On the other hand, the gravitational force on the Moon's surface is less than the gravitational force on the Earth's surface. As a result, you would weigh less on the Moon than on the Earth, even though your mass was the same.

To be specific, on Earth a 81.0-kg person has a weight given by

$$W_{\text{Earth}} = mg_{\text{Earth}} = (81.0 \text{ kg})(9.81 \text{ m/s}^2) = 795 \text{ N}$$

In contrast, the same person on the Moon, where the acceleration of gravity is 1.62 m/s^2, weighs only

$$W_{\text{Moon}} = mg_{\text{Moon}} = (81.0 \text{ kg})(1.62 \text{ m/s}^2) = 131 \text{ N}$$

This is roughly one-sixth the weight on Earth. If, sometime in the future, there is a Lunar Olympics, the Moon's low gravity would be a boon for pole-vaulters, gymnasts, and others.

▲ At the moment this picture was taken, the climber's acceleration was zero because the net force acting on him was zero. In particular, the upward forces exerted on the climber by the rocks exactly cancel the downward force of gravity.

Finally, since weight is a force—which is a vector quantity—it has both a magnitude and a direction. Its magnitude, of course, is mg, and its direction is simply the direction of gravitational acceleration. Thus, if $\mathbf{g}$ denotes a vector of magnitude g, pointing in the direction of free-fall acceleration, the weight of an object can be written in vector form as follows:

$$\mathbf{W} = m\mathbf{g}$$

We use the weight vector and its magnitude, mg, in the next example.

EXAMPLE 5–6 Where's the Fire?

The fire alarm goes off, and a 97-kg fireman slides 3.0 meters down a pole to the ground floor. Suppose the fireman starts from rest, slides with constant acceleration, and reaches the ground floor in 1.2 s. What was the upward force **F** exerted by the pole on the fireman?

Picture the Problem

Our sketch shows the fireman sliding down the pole and the two forces acting on him: the upward force exerted by the pole, **F**, and the downward force of gravity, **W**. Note that we choose the positive y direction to be upward, and $y = 0$ to be at ground level.

Strategy

The basic idea in approaching this problem is to apply Newton's second law to the y direction; $\Sigma F_y = ma_y$. The acceleration is not given directly, but we can find it using the kinematic equation $y = y_0 + v_{0y}t + \frac{1}{2}a_y t^2$. Substituting the result for a_y into Newton's second law, along with $W = mg$, allows us to solve for the unknown force, F.

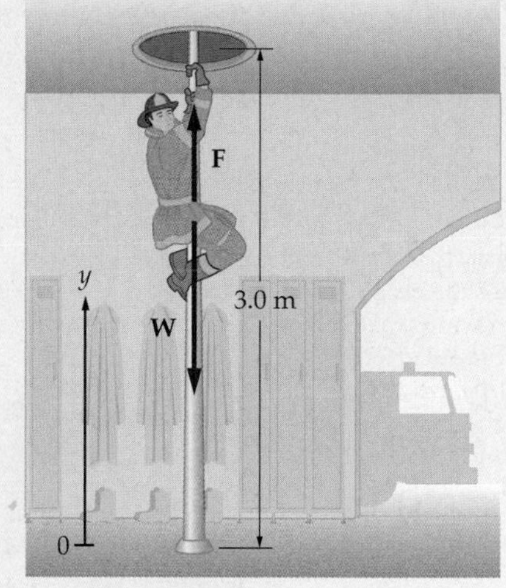

Physical picture

F

W

Free-body diagram

Solution

1. Solve $y = y_0 + v_{0y}t + \frac{1}{2}a_y t^2$ for a_y, using the fact that $v_{0y} = 0$:

$$y = y_0 + v_{0y}t + \tfrac{1}{2}a_y t^2 = y_0 + \tfrac{1}{2}a_y t^2$$

$$a_y = \frac{2(y - y_0)}{t^2}$$

2. Substitute $y = 0$, $y_0 = 3.0$ m, and $t = 1.2$ s to find the acceleration:

$$a_y = \frac{2(0 - 3.0\text{ m})}{(1.2\text{ s})^2} = -4.2\text{ m/s}^2$$

3. Sum the forces in the y direction:

$$\sum F_y = F - mg$$

4. Set the sum of the forces equal to mass times acceleration:

$$F - mg = ma_y$$

5. Solve for F, the force exerted by the pole:

$$F = mg + ma_y = m(g + a_y)$$
$$= (97\text{ kg})(9.81\text{ m/s}^2 - 4.2\text{ m/s}^2) = 544\text{ N}$$

Insight

How is it that the pole exerts a force on the fireman? Well, by wrapping his arms and legs around the pole as he slides, the fireman exerts a downward force on the pole. By Newton's third law, the pole exerts an upward force of equal magnitude on the fireman. These forces are due to friction, which we shall study in detail in the next chapter.

Practice Problem

What is the fireman's acceleration if the force exerted on him by the pole is 550 N? [**Answer:** $a_y = -4.1\text{ m/s}^2$]

Some related homework problems: Problem 29, Problem 33

Apparent Weight

We have all had the experience of riding in an elevator and feeling either heavy or light, depending on its motion. For example, when an elevator moving downward comes to rest by accelerating upward, we feel heavier. On the other hand, we feel lighter when an elevator moving upward comes to rest by accelerating downward. In short, the motion of an elevator can give rise to an **apparent weight** that differs from our true weight. Why?

The reason is that our sensation of weight in this case is due to the force exerted on our feet by the floor of the elevator. If this force is greater than our weight, mg, we feel heavy; if it is less than mg we feel light.

As an example, imagine you are in an elevator that is moving upward with an acceleration a, as indicated in **Figure 5–11**. Two forces act on you: (i) your weight, W, acting downward; and (ii) the upward normal force exerted on your feet by the floor of the elevator. Let's call the second force W_a, since it represents your apparent weight. We can find W_a by applying Newton's second law to the vertical direction.

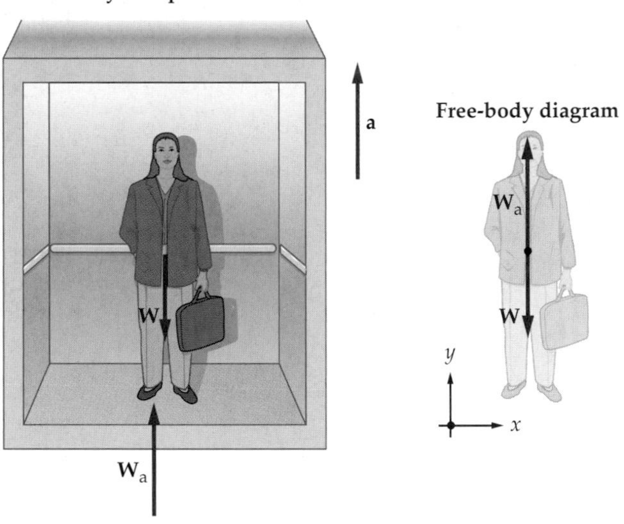

▶ FIGURE 5–11 Apparent Weight
A person rides in an elevator that is accelerating upward. Because the acceleration is upward, the net force must also be upward. As a result, the force exerted on the person by the floor of the elevator, $\mathbf{W}_a$, must be greater than the person's weight, $\mathbf{W}$. This means that the person feels heavier than normal.

To be specific, the sum of the forces acting on you is

$$\sum F_y = W_a - W$$

By Newton's second law, this sum must equal ma_y. Since $a_y = a$, we find

$$W_a - W = ma$$

Solving for the apparent weight, W_a, yields

$$\begin{aligned} W_a &= W + ma \\ &= mg + ma = m(g + a) \end{aligned} \qquad \textbf{5–6}$$

Note that W_a is greater than your weight, mg, and hence you feel heavier. In fact, your apparent weight is precisely what it would be if you were suddenly "transported" to a planet where the acceleration of gravity is $g + a$ instead of g.

On the other hand, if the elevator accelerates downward, so that $a_y = -a$, your apparent weight is found by simply replacing a with $-a$ in Equation 5–6:

$$\begin{aligned} W_a &= W - ma \\ &= mg - ma = m(g - a) \end{aligned} \qquad \textbf{5–7}$$

In this case you feel lighter than normal.

We explore these results in the next example, where we consider weighing a fish on a scale. The reading on the scale is equal to the upward force it exerts on an object. Thus, the upward force exerted by the scale is the apparent weight, W_a.

EXAMPLE 5–7 How Much Does the Salmon Weigh?

A 5.0-kg salmon is weighed by hanging it from a fish scale attached to the ceiling of an elevator. What is the apparent weight of the salmon, $\mathbf{W}_a$, if the elevator **(a)** is at rest, **(b)** moves with an upward acceleration of 2.5 m/s^2 or **(c)** moves with a downward acceleration of 3.2 m/s^2?

Picture the Problem
The free-body diagram for the salmon shows the weight of the salmon, $\mathbf{W}$, and the force exerted by the scale, $\mathbf{W}_a$. Note that upward is the positive direction.

Strategy
We know the weight, $W = mg$, and the acceleration, a. To find the apparent weight, W_a, we use $\sum \mathbf{F}_y = ma_y$. **(a)** Set $a_y = 0$. **(b)** Set $a_y = 2.5$ m/s^2. **(c)** Set $a_y = -3.2$ m/s^2.

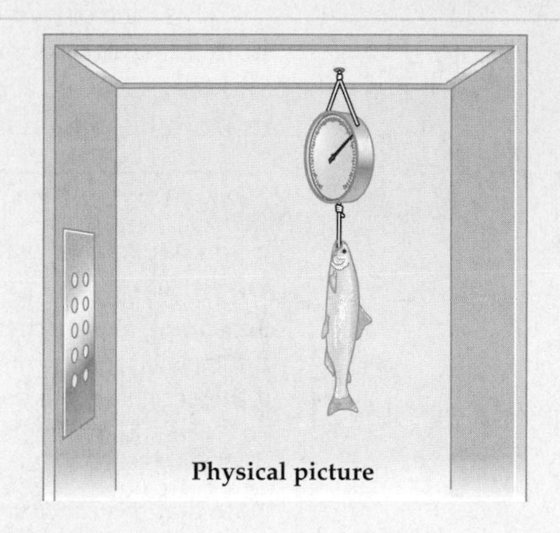

Physical picture

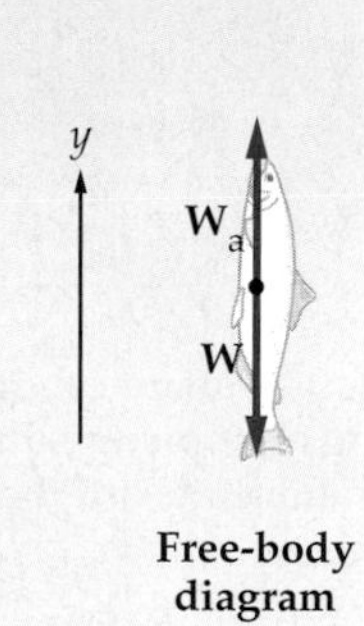

Free-body diagram

Solution

Part (a)

1. Sum the y component of the forces and set equal to mass times the y component of acceleration, with $a_y = 0$: $\sum F_y = W_a - W = ma_y = 0$
2. Solve for W_a: $W_a = W = mg = (5.0\text{ kg})(9.81\text{ m/s}^2) = 49\text{ N}$

Part (b)

3. Again, sum the forces and set equal to mass times acceleration, this time with $a_y = a = 2.5$ m/s^2: $\sum F_y = W_a - W = ma_y = ma$
4. Solve for W_a: $W_a = W + ma = mg + ma = 49\text{ N} + (5.0\text{ kg})(2.5\text{ m/s}^2) = 62\text{ N}$

Part (c)

5. Finally, sum the forces and set equal to mass times acceleration, with $a_y = -a = -3.2$ m/s^2: $\sum F_y = W_a - W = ma_y = -ma$
6. Solve for W_a: $W_a = W - ma = mg - ma = 49\text{ N} - (5.0\text{ kg})(3.2\text{ m/s}^2) = 33\text{ N}$

Insight
When the salmon is at rest, or moving with constant velocity, its acceleration is zero and the apparent weight is equal to the actual weight, mg. In part **(b)** the apparent weight is greater than the actual weight because the scale must exert an upward force capable not only of supporting the salmon, but of accelerating it upward as well. In part **(c)** the apparent weight is less than the actual weight. In this case the net force acting on the salmon is downward, hence its acceleration is downward.

Practice Problem
What is the elevator's acceleration if the scale gives a reading of **(a)** 55 N or **(b)** 45 N? [**Answer: (a)** 1.2 m/s^2, **(b)** -0.80 m/s^2]

Some related homework problems: Problem 31, Problem 32

Let's return for a moment to Equation 5–7:

$$W_a = m(g - a)$$

This result indicates that a person feels lighter than normal when riding in an elevator with a downward acceleration a. In particular, if the elevator's downward acceleration is g—that is, if the elevator is in free fall—it follows that $W_a = m(g - g) = 0$. Thus, a person feels "weightless" (zero apparent weight) in a freely falling elevator!

NASA uses this effect when training astronauts. Trainees are sent aloft in a KC-135 airplane affectionately known as the "vomit comet" (since many trainees experience nausea along with the weightlessness). To generate an experience of

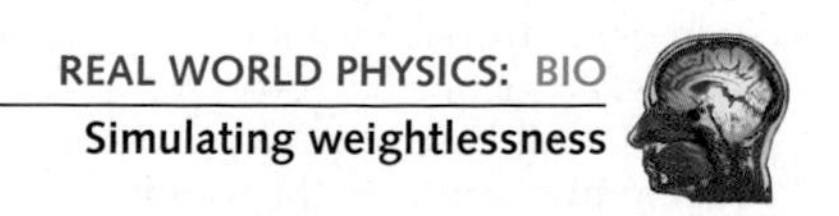

▲ Astronaut candidates pose for a floating class picture during weightlessness training aboard the "vomit comet."

weightlessness, the plane flies on a parabolic path—the same path followed by a projectile in free fall. Each round of weightlessness lasts about half a minute, after which the plane pulls up to regain altitude and start the cycle again. On a typical flight, trainees experience about 40 cycles of weightlessness. Many scenes in the movie *Apollo 13* were shot in 30-second takes aboard the vomit comet.

This idea of free-fall weightlessness applies to more than just the vomit comet. In fact, astronauts in orbit experience weightlessness for the same reason—they and their craft are actually in free fall. As we shall see in detail in Chapter 12 (Gravity), orbital motion is just a special case of free fall.

CONCEPTUAL CHECKPOINT 5–3

If you ride in an elevator moving upward with constant speed, is your apparent weight **(a)** the same as, **(b)** greater than, or **(c)** less than mg?

Reasoning and Discussion

If the elevator is moving in a straight line with constant speed, its acceleration is zero. Now, if the acceleration is zero, the net force must also be zero. Hence, the upward force exerted by the floor of the elevator, W_a, must equal the downward force of gravity, mg. As a result, your apparent weight is equal to mg.

Note that this conclusion agrees with Equations 5–6 and 5–7, with $a = 0$.

Answer:

(a) Your apparent weight is the same as mg.

5–7 Normal Forces

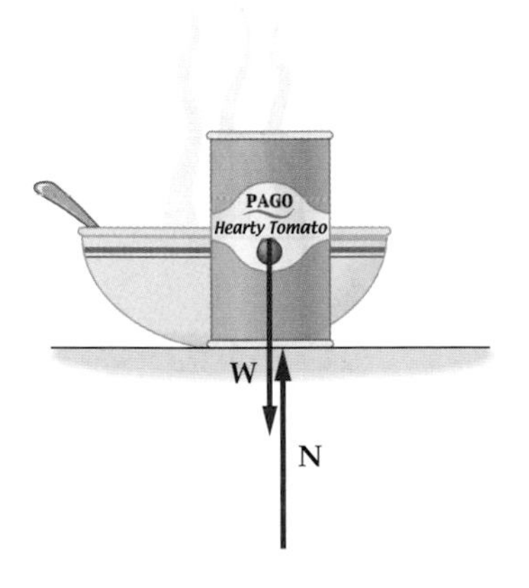

Physical picture

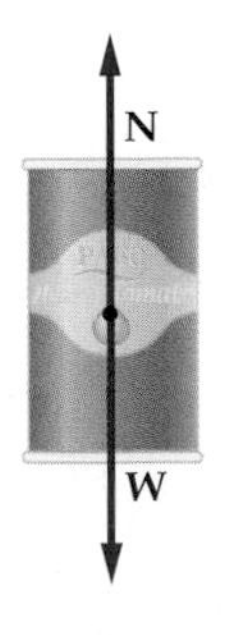

Free-body diagram

▲ FIGURE 5–12 The normal force may equal the weight
A can of soup rests on a kitchen counter, which exerts a normal force, **N**, to support it. In this case, the normal force is equal in magnitude to the weight, $\mathbf{W} = m\mathbf{g}$, and opposite in direction.

As you get ready for lunch, you take a can of soup from the cupboard and place it on the kitchen counter. The can is now at rest, which means that its acceleration is zero, so the net force acting on it is also zero. Thus, you know that the downward force of gravity is being opposed by an upward force exerted by the counter, as shown in **Figure 5–12**. As we have mentioned before, this force is referred to as the **normal force, N**. The reason the force is called "normal" is that it is *perpendicular to the surface*, and in mathematical terms, normal simply means perpendicular.

The origin of the normal force is the interaction between atoms in a solid that act to maintain its shape. When the can of soup is placed on the countertop, for example, it causes an imperceptibly small compression of the surface of the counter. This is similar to compressing a spring, and just like a spring, the countertop exerts a force to oppose the compression. Therefore, the greater the weight placed on the countertop, the greater the normal force it exerts to oppose being compressed.

In the example of the soup can and the countertop, the magnitude of the normal force is equal to the weight of the can. This is a special case, however. In general, the normal force may be greater than or less than the weight of an object.

To see how this can come about, consider pulling a 12.0-kg suitcase across a smooth floor by exerting a force, **F**, at an angle θ above the horizontal. The weight of the suitcase is $mg = (12.0\text{ kg})(9.81\text{ m/s}^2) = 118\text{ N}$. The normal force will have a magnitude less than this, however, because the force **F** has an upward component that supports part of the suitcase's weight. To be specific, suppose that **F** has a magnitude of 45.0 N and that $\theta = 20.0°$. What is the normal force exerted by the floor on the suitcase?

The situation is illustrated in **Figure 5–13**, where we show the three forces acting on the suitcase: (i) the weight of the suitcase, **W**, (ii) the force **F**, and (iii) the normal force, **N**. We also indicate a typical coordinate system in the figure, with the x axis horizontal and the y axis vertical. Now, the key to solving a problem like this is to realize that since the suitcase does not move in the y direction, its y component of acceleration is zero; that is, $a_y = 0$. It follows, from Newton's second law, that the sum of the y components of force must also equal zero; that is, $\Sigma F_y = ma_y = 0$. Using this condition, we can solve for the one force that is unknown, **N**.

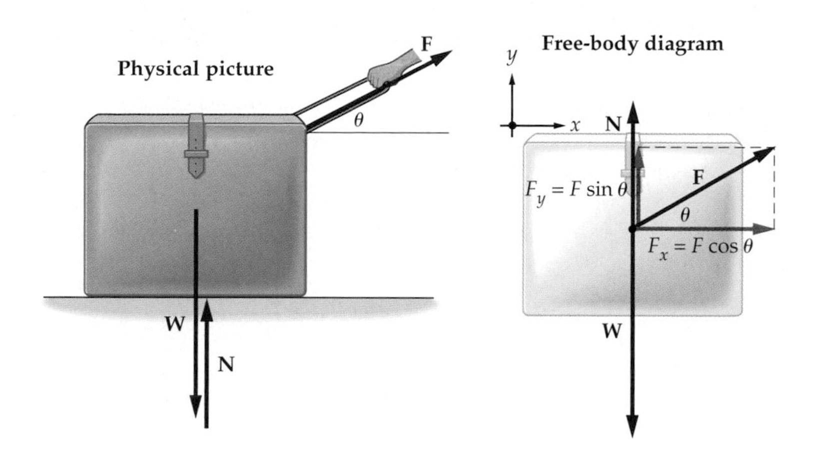

◀ FIGURE 5–13 The normal force may differ from the weight
A suitcase pulled across the floor by an applied force of magnitude F, directed at an angle θ above the horizontal. As a result of the upward component of **F**, the normal force **N** will have a magnitude less than the weight of the suitcase.

To find **N**, then, we start by writing out the y component of each force. For the weight we have $W_y = -mg = -118$ N; for the applied force, **F**, the y component is $F_y = F \sin 20.0° = (45.0 \text{ N})\sin 20.0° = 15.4$ N; and finally, the y component of the normal force is $N_y = N$. Setting the sum of the y components of force equal to zero yields

$$\sum F_y = W_y + F_y + N_y = -mg + F \sin 20.0° + N = 0$$

Solving for N gives

$$N = mg - F \sin 20.0° = 118 \text{ N} - 15.4 \text{ N} = 103 \text{ N}$$

In vector form,

$$\mathbf{N} = N_y\hat{\mathbf{y}} = (103 \text{ N})\hat{\mathbf{y}}$$

Thus, as mentioned, the normal force has a magnitude less than $mg = 118$ N because the y component of **F**, $F_y = F \sin 20.0°$, supports part of the weight. In the following example, however, the applied forces cause the normal force to be greater than the weight.

EXAMPLE 5–8 Ice Block

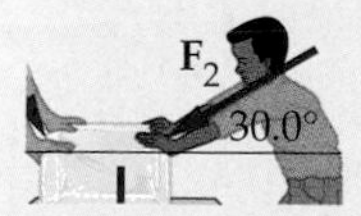

A 6.0-kg block of ice is acted on by two forces, $\mathbf{F}_1$ and $\mathbf{F}_2$, as shown in the diagram. If the magnitudes of the forces are $F_1 = 13$ N and $F_2 = 11$ N, find **(a)** the acceleration of the ice and **(b)** the normal force exerted on it by the table.

Picture the Problem

The diagram shows our choice of coordinate system, as well as all the forces acting on the block of ice. Note that $\mathbf{F}_1$ has a positive x component and a negative y component; $\mathbf{F}_2$ has negative x and y components.

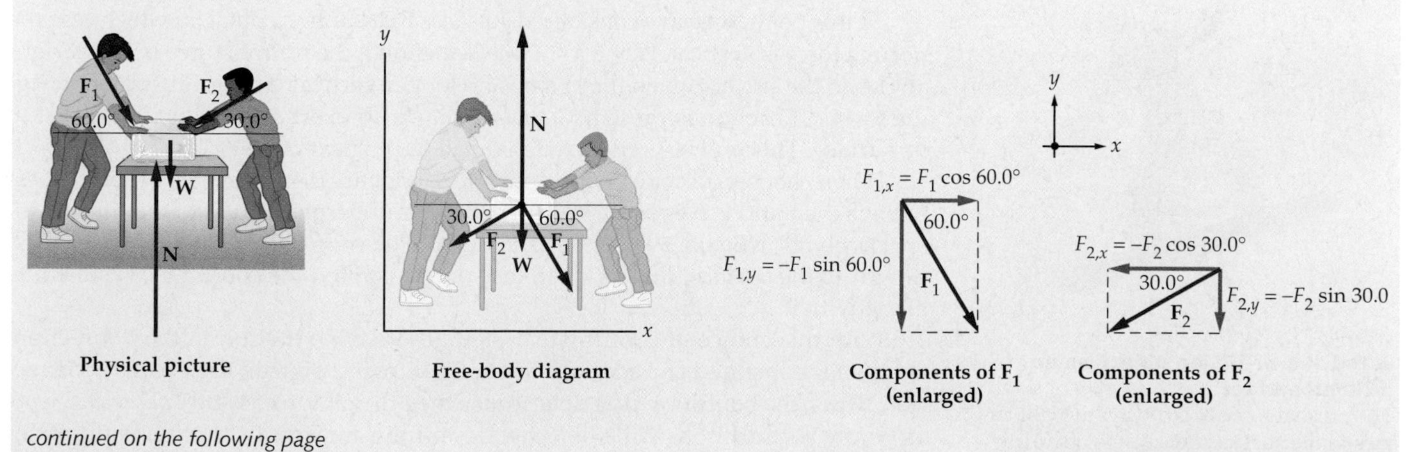

continued on the following page

continued from the previous page

Strategy

(a) The block can accelerate only in the horizontal direction; thus we find the acceleration by solving $\Sigma F_x = ma_x$ for a_x. **(b)** The acceleration in the y direction is zero. Hence, we can find the normal force **N** by setting $\Sigma F_y = ma_y = 0$.

Solution

Part (a)

1. Write out the x component of each force:

$$F_{1,x} = F_1 \cos 60.0° = (13\text{ N})\cos 60.0° = 6.5\text{ N}$$
$$F_{2,x} = -F_2 \cos 30.0° = -(11\text{ N})\cos 30.0° = -9.5\text{ N}$$
$$N_x = 0 \qquad W_x = 0$$

2. Sum the x components of force:

$$\sum F_x = F_{1,x} + F_{2,x} + N_x + W_x = 6.5\text{ N} - 9.5\text{ N} + 0 + 0 = -3.0\text{ N}$$

3. Divide by the mass to obtain the acceleration:

$$a_x = \frac{\sum F_x}{m} = \frac{-3.0\text{ N}}{6.0\text{ kg}} = -0.50\text{ m/s}^2$$

Part (b)

4. Write out the y component of each force:
   The only force we don't know is the normal. We represent its magnitude by N:

$$F_{1,y} = -F_1 \sin 60° = -(13\text{ N})\sin 60.0° = -11\text{ N}$$
$$F_{2,y} = -F_2 \sin 30° = -(11\text{ N})\sin 30.0° = -5.5\text{ N}$$
$$N_y = N \qquad W_y = -W = -mg$$

5. Sum the y components of force:

$$\sum F_y = F_{1,y} + F_{2,y} + N_y + W_y = -11\text{ N} - 5.5\text{ N} + N - mg$$

6. Set this sum equal to 0 since the acceleration in the y direction is zero, and solve for N:

$$-11\text{ N} - 5.5\text{ N} + N - mg = 0$$
or
$$N = 11\text{ N} + 5.5\text{ N} + mg = 17\text{ N} + (6.0\text{ kg})(9.81\text{ m/s}^2) = 76\text{ N}$$

Finally, we write the normal force in vector form:

$$\mathbf{N} = (76\text{ N})\hat{\mathbf{y}}$$

Insight

The block accelerates to the left, even though the force acting to the right, $\mathbf{F}_1$, has a greater magnitude than the force acting to the left, $\mathbf{F}_2$. Also, note that the normal force is greater in magnitude than the weight, $mg = 59$ N.

Practice Problem

At what angle must $\mathbf{F}_2$ be applied if the block of ice is to have zero acceleration? [**Answer:** $a_x = 0$ implies $F_1 \cos 60.0° = F_2 \cos \theta$. Thus, $\theta = 54°$.]

Some related homework problems: Problem 37, Problem 43

▲ FIGURE 5–14 An object on an inclined surface
The normal force **N** is always at right angles to the surface; hence, it is not always in the vertical direction.

To this point, we have considered surfaces that are horizontal, in which case the normal force is vertical. When a surface is inclined, the normal force is still at right angles to the surface, even though it is no longer vertical. This is illustrated in **Figure 5–14**. (If friction is present, a surface may also exert a force that is parallel to its surface. This will be considered in detail in the next chapter.)

When choosing a coordinate system for an inclined surface, it is generally best to have the x and y axes of the system parallel and perpendicular to the surface, respectively, as in **Figure 5–15**. One can imagine the coordinate system to be "bolted down" to the surface, so that when the surface is tilted the coordinate system tilts along with it.

With this choice of coordinate system, there is no motion in the y direction, even on the inclined surface, and the normal force points in the positive y direction. Thus, the condition that determines the normal force is still $\Sigma F_y = ma_y = 0$, as before. In addition, if the object slides on the surface, its motion is purely in the x direction.

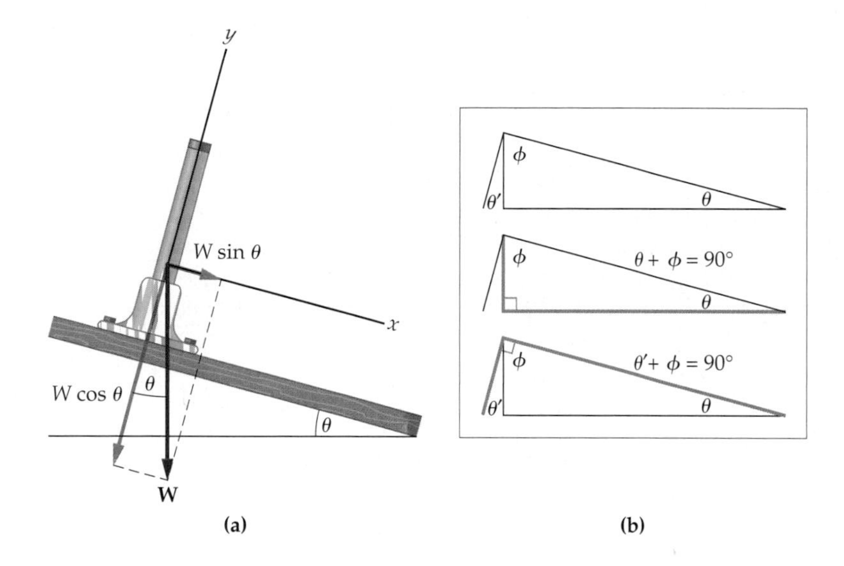

◀ FIGURE 5-15 Components of the weight on an inclined surface
Whenever a surface is tilted by an angle θ, the weight **W** makes the same angle θ with respect to the negative y axis. This is proven at right, where we show that $\theta + \phi = 90°$, and that $\theta' + \phi = 90°$. From these results it follows that $\theta' = \theta$. The component of the weight perpendicular to the surface is $W_y = -W\cos\theta$; the component parallel to the surface is $W_x = W\sin\theta$.

Finally, if the surface is inclined by an angle θ, note that the weight—which is still vertically downward—is at the same angle θ with respect to the negative y axis, as shown in Figure 5–15. As a result, the x and y components of the weight are

$$W_x = W\sin\theta = mg\sin\theta \qquad \textbf{5–8}$$

and

$$W_y = -W\cos\theta = -mg\cos\theta \qquad \textbf{5–9}$$

Let's quickly check a couple limits for these results. First, if $\theta = 0$ the surface is horizontal, and we find $W_x = 0$, $W_y = -mg$, as expected. Second, if $\theta = 90°$ the surface is vertical, hence the weight is parallel to the surface, pointing in the positive x direction. In this case, $W_x = mg$ and $W_y = 0$.

The next example shows how to use the weight components to find the acceleration of an object on an inclined surface.

EXAMPLE 5–9 Toboggan to the Bottom

A child of mass m rides on a toboggan down a slick, ice-covered hill inclined at an angle θ with respect to the horizontal. **(a)** What is the acceleration of the child? **(b)** What is the normal force exerted on the child by the toboggan?

Picture the Problem

We choose the x axis to be parallel to the slope, with the positive direction pointing downhill. Similarly, we choose the y axis to be perpendicular to the slope, and pointing up and to the right. With these choices, the x component of **W** is positive, and its y component is negative.

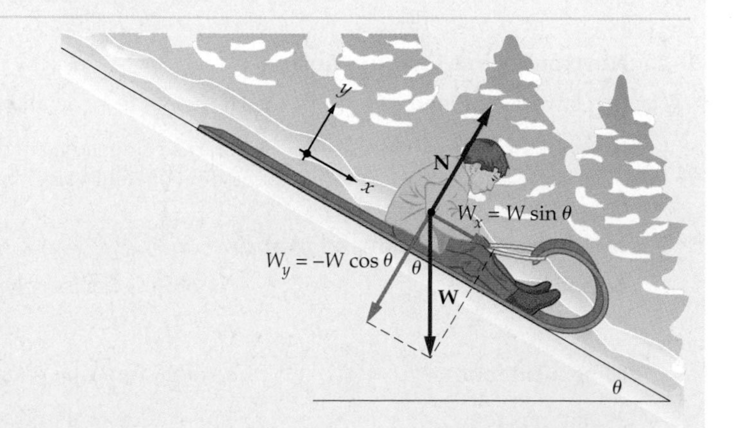

Strategy

Note that only two forces act on the child: (i) the weight, **W** and (ii) the normal force, **N**. **(a)** We find the child's acceleration by solving $\sum F_x = ma_x$ for a_x. **(b)** There is no motion in the y direction, hence the y component of acceleration is zero. Therefore, we can find the normal force by setting $\sum F_y = ma_y = 0$.

Solution

Part (a)

1. Write out the x component of the forces acting on the child: $\quad N_x = 0 \qquad W_x = W\sin\theta = mg\sin\theta$

continued on the following page

continued from the previous page

2. Sum the x components of the forces: $\sum F_x = N_x + W_x = mg \sin \theta$

3. Divide by the mass m to find the acceleration in the x direction: $a_x = \frac{\sum F_x}{m} = \frac{mg \sin \theta}{m} = g \sin \theta$

Part (b)

4. Write out the y components of the forces acting on the child: $N_y = N \quad W_y = -W \cos \theta = -mg \cos \theta$

5. Sum the y components of the forces and set the sum equal to zero, since $a_y = 0$: $\sum F_y = N_y + W_y = N - mg \cos \theta = ma_y = 0$

6. Solve for the magnitude of the normal force, N: $N - mg \cos \theta = 0$ or $N = mg \cos \theta$
   Write the normal force in vector form: $\mathbf{N} = (mg \cos \theta)\hat{\mathbf{y}}$

Insight

Note that for θ between 0 and 90° the acceleration of the child is less than the acceleration of gravity. This is due to the fact that only a *component* of the weight is causing the acceleration.

Let's check some special cases of our general result, $a_x = g \sin \theta$. First, let $\theta = 0$. In this case, we find zero acceleration; $a_x = g \sin 0 = 0$. This makes sense because with $\theta = 0$ the hill is actually level, and we don't expect an acceleration. Second, let $\theta = 90°$. In this case, the hill is vertical, and the toboggan should drop straight down in free fall. This also agrees with our general result; $a_x = g \sin 90° = g$.

Practice Problem

What is the child's acceleration if its mass is doubled to $2m$? [**Answer:** The acceleration is still $a_x = g \sin \theta$. As in free fall, the acceleration produced by gravity is independent of mass.]

Some related homework problems: Problem 38, Problem 42

Chapter Summary

	Topic	Remarks and Relevant Equations	
5–1	**Force and Mass**		
	force	A push or a pull.	
	mass	A measure of how much an object resists a change in its motion.	
5–2	**Newton's First Law of Motion**		
	first law (law of inertia)	If the net force on an object is zero, its velocity is constant.	
	inertial frame of reference	Frame of reference in which the first law holds. All inertial frames of reference move with constant velocity relative to one another.	
5–3	**Newton's Second Law of Motion**		
	second law	An object of mass m has an acceleration $\mathbf{a}$ given by the net force $\Sigma\mathbf{F}$ divided by m. That is	
		$\mathbf{a} = \Sigma\mathbf{F}/m$	5–1
	component form	$a_x = \Sigma F_x/m \quad a_y = \Sigma F_y/m \quad a_z = \Sigma F_z/m$	5–2
	SI unit: newton	$1\text{ N} = 1\text{ kg}\cdot\text{m/s}^2$	5–3
	free-body diagram	A sketch showing all external forces acting on an object.	
5–4	**Newton's Third Law of Motion**		
	third law	For every force that acts on an object, there is a reaction force acting on a different object that is equal in magnitude and opposite in direction.	

contact forces	Action-reaction pair of forces produced by physical contact of two objects.
5–5 The Vector Nature of Forces: Forces in Two Dimensions	Forces are vectors. Newton's second law can be applied to each component of force separately and independently.
5–6 Weight	Gravitational force exerted by the earth on an object. On another planet, or other astronomical body, the weight would be the gravitational force exerted by that body on the object. On the surface of the earth the weight, W, of an object of mass m is $W = mg$ **5–5**
apparent weight	Force felt from contact with the floor or a scale in an accelerating system. For example, the sensation of feeling heavier or lighter in an accelerating elevator.
5–7 Normal Forces	Force exerted by a surface that is *perpendicular* to the surface. The normal force is equal to the weight of an object only in special cases. In general, the normal force is more or less than the object's weight.

Problem-Solving Summary

Type of Calculation	Relevant Physical Concepts	Related Examples
Find the acceleration of an object.	Solve Newton's second law for each component of the acceleration; that is, $a_x = \Sigma F_x/m$ and $a_y = \Sigma F_y/m$.	Examples 5–1, 5–3, 5–4, 5–5, 5–8, 5–9 Active Example 5–2
Solve problems involving action-reaction forces.	Apply Newton's third law, being careful to note that the action-reaction forces act on different objects.	Examples 5–3, 5–4
Find the normal force exerted on an object.	Since there is no acceleration in the normal direction, set the sum of the normal components of force equal to zero.	Examples 5–8, 5–9

Conceptual Questions

1. Driving down the road you hit the brakes suddenly. As a result, your body moves toward the front of the car. Explain, using Newton's laws.
2. When a dog gets wet, it shakes its body from head to tail to shed the water. Explain, in terms of Newton's first law, why this works.

▲ A dog uses the principle of inertia to shake water from its coat. (Question 2)

3. When a plane accelerates down the runway during takeoff, passengers feel as if they are being pushed back into their seats. Why?
4. You've probably seen pictures of someone pulling a tablecloth out from under glasses, plates, and silverware set out for a formal dinner. Perhaps you've even tried it yourself. Using Newton's laws of motion, explain how this stunt works.
5. A young girl slides down a rope. As she slides faster and faster she tightens her grip, increasing the force exerted on her by the rope. What happens when this force is equal in magnitude to her weight? Explain.
6. A drag-racing car accelerates forward because of the force exerted on it by the road. Why, then, does it need an engine? Explain.
7. A block of mass m hangs from a string attached to a ceiling, as shown in **Figure 5-16**. An identical string hangs down from the

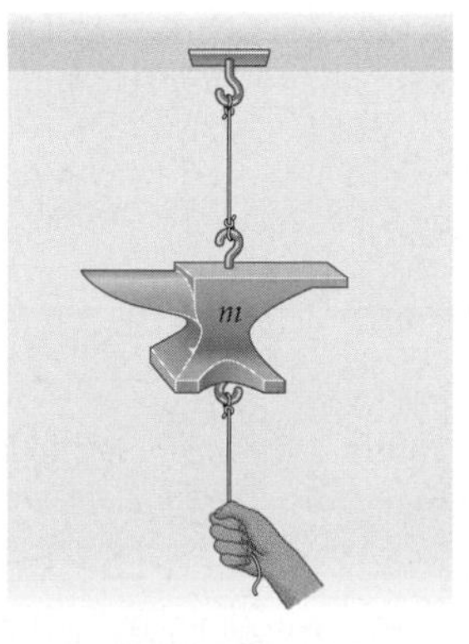

▲ **FIGURE 5-16** Question 7

bottom of the block. Which string breaks if **(a)** the lower string is pulled with a slowly increasing force or **(b)** the lower string is jerked rapidly downward? Explain.

8. An astronaut on a space walk discovers that his jet pack no longer works, leaving him stranded 50 m from the spacecraft. If the jet pack is removable, explain how the astronaut can still use it to return to the ship.
9. Two astronauts on a space walk decide to take a break and play catch with a baseball. Describe what happens as the game of catch progresses.
10. What are the action-reaction forces when a baseball bat hits a fast ball? What is the effect of each force?
11. As you read this, you are most likely sitting quietly in a chair. Can you conclude, therefore, that you are at rest? Explain.
12. A Cadillac bumps into a Volkswagen. Is the force exerted by the Cadillac on the Volkswagen greater than, less than, or equal to the force exerted by the Volkswagen on the Cadillac? Explain.
13. In **Figure 5–17** Wilbur asks Mr. Ed, the talking horse, to pull a cart. Mr. Ed replies that he would like to, but the laws of nature just won't allow it. According to Newton's third law, he says, if he pulls on the wagon it pulls back on him with an equal force. Clearly, then, the net force is zero and the wagon will stay put. How should Wilbur answer the clever horse?

▲ **FIGURE 5-17** Question 13

14. You are dribbling a basketball, ready to make your move to the hoop. What produces the force that causes the ball to return to your hand with each dribble?
15. Is it possible for an object to be in motion and yet have zero net force acting on it? Explain.
16. You jump out of an airplane and open your parachute after a brief period of free fall. To decelerate your fall, must the force exerted on you by the parachute be less than, equal to, or greater than your weight? Explain.
17. The force exerted by gravity on a whole brick is greater than the force exerted by gravity on half a brick. Why, then, doesn't a whole brick fall faster than half a brick? Explain.
18. A whole brick has more mass than half a brick, thus the whole brick is harder to accelerate. Why doesn't a whole brick fall more slowly than half a brick? Explain.
19. Suppose you jump from the cliffs of Acapulco and perform a perfect swan dive. As you fall, you exert an upward force on the Earth equal in magnitude to the downward force the Earth exerts on you. Why, then, does it seem that you are the one doing all the accelerating? Since the forces are the same, why aren't the accelerations?
20. You are told that an object is at rest and that a single force acts on it. Is this possible? Explain.
21. A friend tells you that since his car is at rest, there are no forces acting on it. How would you reply?
22. You drop two objects from the same height at the same time. Object 1 has a greater mass than object 2. If the upward force due to air resistance is the same for both objects, which one reaches the ground first?
23. Riding in an elevator moving upward with constant speed, you begin a game of darts. Do you have to adjust your aim, compared to the way you play darts normally? Explain.
24. Riding in an elevator moving with a constant upward acceleration, you begin a game of darts. Do you have to adjust your aim, compared to the way you play darts normally? Explain.
25. Since all objects are "weightless" for an astronaut in orbit, is it possible for astronauts to tell whether an object is heavy or light? Explain.
26. A bird cage, with a parrot inside, hangs from a scale. The parrot decides to hop (without flapping) from one perch to another. What can you say about the reading on the scale **(a)** when the parrot jumps, **(b)** when the parrot is in the air, and **(c)** when the parrot lands on the second perch? Assume that the scale responds rapidly so that it gives an accurate reading at all times.
27. If you step off a high board and drop to the water below, you plunge into the water without injury. On the other hand, if you were to drop the same distance onto solid ground, you might break a leg. Use Newton's laws to explain the difference.
28. To clean a rug, you can hang it from a clothesline and beat it with a tennis racket. Use Newton's laws to explain why beating the rug should have a cleaning effect.
29. Is it possible for an object to be moving in one direction while the net force is in another direction? Explain.
30. You hold your hands out in front of you and catch a bowling ball dropped from a foot above your hands. No problem. If, instead, you were to hold your hands about half an inch above the floor, and try the same catch, you would have a painful experience. Why? (Don't try this at home!)
31. Give the direction of the net force acting on each of the following objects. If the net force is zero, state "zero." **(a)** A car accelerating northward from a stop light. **(b)** A car traveling southward and slowing down. **(c)** A car traveling westward with constant speed. **(d)** A skydiver parachuting downward with constant speed. **(e)** A baseball during its flight from pitcher to catcher (ignoring air resistance).
32. Is it possible for the net force on an object to be at right angles to its direction of motion? Explain.
33. On a trip to Hawaii, you notice that your plane is flying with constant speed in a straight line. What can you conclude about the net force acting on the plane? Explain.
34. Since a bucket of water is "weightless" in space, would it hurt to kick the bucket? Explain.

Problems

Note: **IP** *denotes an integrated conceptual/quantitative problem.* **BIO** *identifies problems of biological or medical interest. Blue bullets (•, ••, •••) are used to indicate the level of difficulty of each problem.*

Section 5–3 Newton's Second Law of Motion

1. • On a planet far, far away, an astronaut picks up a rock. The rock has a mass of 5.00 kg, and on this particular planet its weight is 40.0 N. If the astronaut exerts an upward force of 46.2 N on the rock, what is its acceleration?

2. • In a grocery store, you push a 14.5-kg shopping cart with a force of 12.0 N. If the cart starts at rest, how far does it move in 3.00 s?

3. • You are pulling your little sister on her sled across an icy (frictionless) surface. When you exert a constant horizontal force of 110 N, the sled has an acceleration of 2.5 m/s^2. If the sled has a mass of 7.0 kg, what is the mass of your little sister?

4. • A 0.53-kg billiard ball is given a speed of 12 m/s during a time interval of 4.0 ms. What force acted on the ball?

5. • A 92-kg water skier floating in a lake is pulled from rest to a speed of 12 m/s in a distance of 25 m. What is the net force exerted on the skier, assuming his acceleration is constant?

6. •• **IP** A 42.0-kg parachutist lands moving straight downward with a speed of 3.85 m/s. **(a)** If the parachutist comes to rest with constant acceleration over a distance of 0.750 m, what force does the ground exert on her? **(b)** If the parachutist comes to rest over a shorter distance, is the force exerted by the ground greater than, less than, or the same as in part (a)?

7. •• **IP** In baseball, a pitcher can accelerate a 0.15-kg ball from rest to 90 mi/h in a distance of a meter and a half. **(a)** What is the average force exerted on the ball during the pitch? **(b)** If the mass of the ball is increased, is the force required of the pitcher increased, decreased, or unchanged?

8. •• A catcher stops a 92 mi/h pitch in his glove, bringing it to rest in 0.15 m. If the force exerted by the catcher is 803 N, what is the mass of the ball?

9. •• Driving home from school one day you spot a ball rolling out into the street (**Figure 5–18**). You brake for 1.20 s, slowing your 950-kg car from 16.0 m/s to 9.50 m/s. **(a)** What was the average force exerted on your car during braking? **(b)** How far did you travel while braking?

▲ **FIGURE 5-18** Problem 9

10. •• A 747 jetliner lands and begins to slow to a stop as it moves along the runway. If its mass is 3.50×10^5 kg, its speed is 27.0 m/s, and the net braking force is 4.30×10^5 N, **(a)** what is its speed 7.50 s later? **(b)** How far has it traveled in this time?

11. •• **IP** A drag racer crosses the finish line doing 210 mi/h, and promptly deploys her drag chute (the small parachute used for braking). **(a)** What force must the drag chute exert on the 870-kg car to slow it to 40.0 mi/h in a distance of 170 m? **(b)** Describe the Strategy you used to solve part (a).

Section 5–4 Newton's Third Law of Motion

12. • You hold a brick at rest in your hand. **(a)** How many forces act on the brick? **(b)** Identify these forces. **(c)** Are these forces equal and opposite? **(d)** Are these forces an action-reaction pair?

13. • Referring to the previous problem, you are now accelerating the brick upward. **(a)** How many forces act on the brick? **(b)** Identify these forces. **(c)** Are these forces equal and opposite? **(d)** Are these forces an action-reaction pair?

14. •• On vacation, your 1300-kg car pulls a 540-kg trailer away from a stop light with an acceleration of 1.90 m/s^2. **(a)** What is the net force exerted by the car on the trailer? **(b)** What force does the trailer exert on the car? **(c)** What is the net force acting on the car?

15. •• **IP** A 54-kg parent and a 18-kg child meet at the center of an ice rink. They place their hands together and push (**Figure 5–19**). **(a)** Is the force experienced by the child more than, less than, or the same as the force experienced by the parent? **(b)** Is the acceleration of the child more than, less than, or the same as the acceleration of the parent? Explain. **(c)** If the acceleration of the child is 2.6 m/s^2 in magnitude, what is the magnitude of the parent's acceleration?

▲ **FIGURE 5-19** Problem 15

16. •• A force of magnitude 7.50 N pushes three boxes with masses $m_1 = 1.30$ kg, $m_2 = 3.20$ kg, and $m_3 = 4.90$ kg, as shown in **Figure 5–20**. Find the contact force between **(a)** boxes 1 and 2, and **(b)** between boxes 2 and 3.

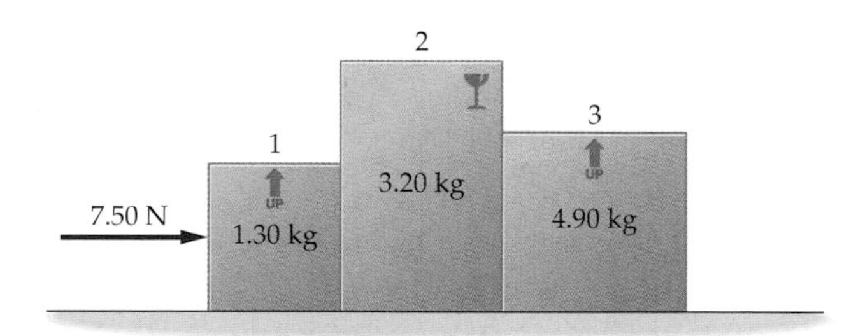

▲ **FIGURE 5-20** Problem 16

17. •• A force of magnitude 7.50 N pushes three boxes with masses $m_1 = 1.30$ kg, $m_2 = 3.20$ kg, and $m_3 = 4.90$ kg, as shown in **Figure 5–21**. Find the contact force between **(a)** boxes 1 and 2, and **(b)** between boxes 2 and 3.

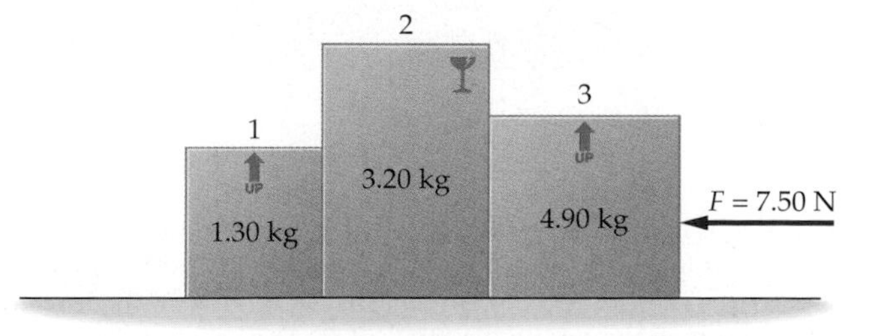

▲ **FIGURE 5-21** Problem 17

18. •• **IP** Two boxes sit side-by-side on a smooth horizontal surface. The lighter box has a mass of 5.2 kg, the heavier box has a mass of 7.4 kg. **(a)** Find the contact force between these boxes when a horizontal force of 5.0 N is applied to the light box. **(b)** If the 5.0-N force is applied to the heavy box instead, is the contact force between the boxes the same as, greater than, or less than the contact force in part (a)? Explain. **(c)** Verify your answer to part (b) by calculating the contact force in this case.

Section 5–5 The Vector Nature of Forces

19. • A farm tractor tows a 4400-kg trailer up a 21° incline at a steady speed of 3.0 m/s. What force does the tractor exert on the trailer? (Ignore friction.)

20. • A surfer "hangs ten," and accelerates down the sloping face of a wave. If the surfer's acceleration is 3.50 m/s^2 and friction can be ignored, what is the angle at which the face of the wave is inclined above the horizontal?

21. • A shopper pushes a 7.5-kg shopping cart up a 13° incline, as shown in **Figure 5–22**. Find the horizontal force, **F**, needed to give the cart an acceleration of 1.41 m/s^2.

▲ **FIGURE 5-22** Problem 21

22. • Two crewmen pull a boat through a lock, as shown in **Figure 5–23**. One crewman pulls with a force of 130 N at an angle of 34° relative to the forward direction of the boat. The second crewman, on the opposite side of the lock, pulls at an angle of 45°. With what force should the second crewman pull so that the net force of the two crewmen is in the forward direction?

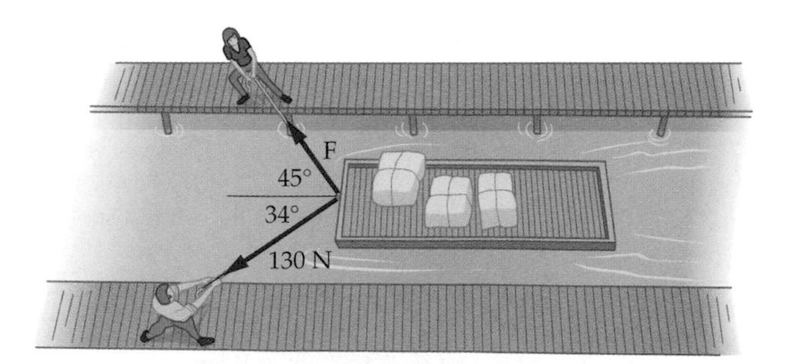

▲ **FIGURE 5-23** Problem 22

23. •• To give a 17-kg child a ride, two teenagers pull on a 3.4-kg sled with ropes, as indicated in **Figure 5–24**. Both teenagers pull with a force of 55 N at an angle of 35° relative to the forward direction, which is the direction of motion. In addition, the snow exerts a retarding force on the sled that points opposite to the direction of motion, and has a magnitude of 57 N. Find the acceleration of the sled and child.

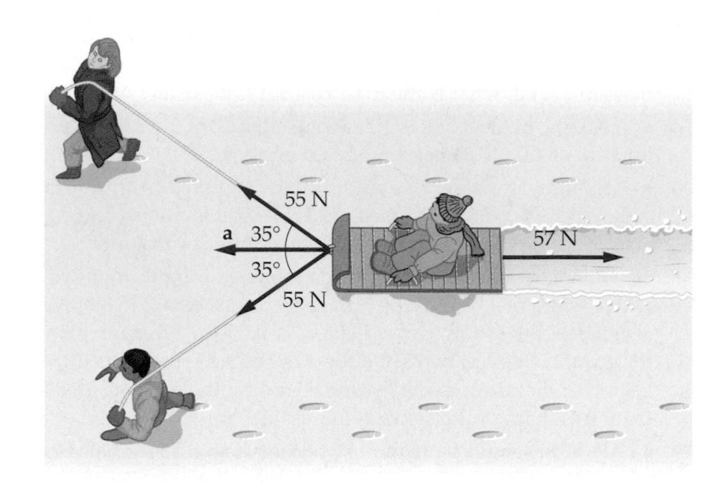

▲ **FIGURE 5-24** Problem 23

24. •• **IP BIO** Before practicing his routine on the rings, a 67-kg gymnast hangs motionless, with one hand grasping each ring and his feet touching the ground. Both arms make an angle of 24° with the vertical. **(a)** If the force exerted by the rings on each arm has a magnitude of 290 N, what is the magnitude of the force exerted by the floor on his feet? **(b)** If the angle his arms make with the vertical had been greater that 24°, would the force exerted by the floor be greater than or less than the value found in part (a)? Explain.

25. •• **IP** A 65-kg skier speeds down a trail, as shown in **Figure 5–25**. The surface is smooth and inclined at an angle of 22° with the horizontal. **(a)** Find the direction and magnitude of the net force acting on the skier. **(b)** Does the net force exerted on the skier increase, decrease, or stay the same as the slope becomes steeper? Explain.

▲ **FIGURE 5-25** Problem 25

26. •• An object acted on by three forces moves with constant velocity. One force acting on the object is in the positive x direction and has a magnitude of 6.5 N; a second force has a magnitude of 4.4 N and points in the negative y direction. Find the direction and magnitude of the third force acting on the object.

27. •• A train is traveling up a 3.0° incline at a speed of 3.35 m/s when the last car breaks free and begins to coast. How long does it take for the last car to come to rest?

28. •• In the previous problem, how far did the last car of the train move before stopping momentarily?

Section 5–6 Weight

29. • You pull upward on a stuffed suitcase with a force of 120 N, and it accelerates upward at $0.825\ m/s^2$. What is **(a)** the mass and **(b)** the weight of the suitcase?

30. • List three common objects that have a weight of approximately 1 N.

31. • Suppose a rocket launches with an acceleration of $32\ m/s^2$. What is the apparent weight of an 85-kg astronaut aboard this rocket?

32. • At the bow of a ship on a stormy sea, a crewman conducts an experiment by standing on a bathroom scale. In calm waters, the scale reads 180 lb. During the storm, the crewman finds a maximum reading of 225 lb and a minimum reading of 138 lb. Find **(a)** the maximum upward acceleration and **(b)** the maximum downward acceleration experienced by the crewman.

33. •• **IP** A 1.21-g samara—the winged fruit of a maple tree—falls toward the ground with a constant speed of 1.1 m/s (**Figure 5–26**). **(a)** What is the force of air resistance exerted on the samara? **(b)** If the constant speed of descent is greater than 1.1 m/s, is the force of air resistance greater than, less than, or the same as in part (a)? Explain.

▲ **FIGURE 5-26** Problem 33

34. •• When you weigh yourself on good old *terra firma* (solid ground), your weight is 160 lb. In an elevator your apparent weight is 142 lb. What are the direction and magnitude of the elevator's acceleration?

35. •• **IP** As part of a physics experiment, you stand on a bathroom scale in an elevator. Though your normal weight is 610 N, the scale at the moment reads 730 N. **(a)** Is the acceleration of the elevator upward, downward, or zero? Explain. **(b)** Calculate the magnitude of the elevator's acceleration.

36. ••• When you lift a bowling ball with a force of 82 N, the ball accelerates upward with an acceleration a. If you lift with a force of 92 N, the ball's acceleration is $2a$. Find **(a)** the weight of the bowling ball, and **(b)** the acceleration a.

Section 5–7 Normal Forces

37. • A 23-kg suitcase is being pulled by a handle that is at an angle of 25° above the horizontal. If the normal force exerted on the suitcase is 180 N, what is the force F applied to the handle?

38. • **(a)** Draw a free-body diagram for the skier in Problem 25. **(b)** Determine the normal force acting on the skier.

39. • A 9.0-kg child sits in a 2.3-kg high chair. **(a)** Draw a free-body diagram for the child, and find the normal force exerted by the chair on the child. **(b)** Draw a free-body diagram for the chair, and find the normal force exerted by the floor on the chair.

40. •• **Figure 5–27** shows the normal force as a function of the angle θ for the suitcase shown in Figure 5–13. Determine the magnitude of the force **F** for each of the three curves shown in Figure 5–27. Give your answer in terms of the weight of the suitcase, mg.

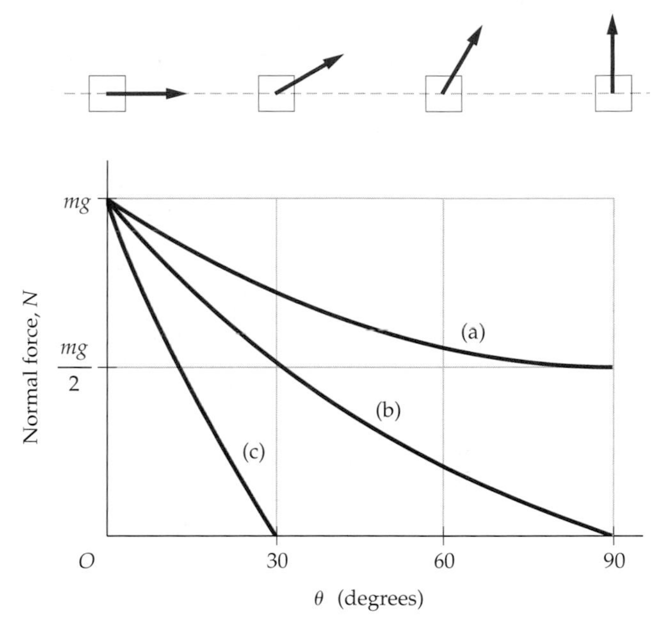

▲ **FIGURE 5-27** Problem 40

41. •• A 5.0-kg bag of potatoes sits on the bottom of a stationary shopping cart. Sketch a free-body diagram for the bag of potatoes. Now suppose the cart moves with a constant velocity. How does this affect your free-body diagram?

42. •• **IP** **(a)** Find the normal force exerted on a 2.5–kg book resting on a surface inclined at 28° above the horizontal. **(b)** If the angle of the incline is reduced, do you expect the normal force to increase, decrease, or stay the same? Explain.

43. •• **IP** A gardener mows a lawn with an old-fashioned push mower. The handle of the mower makes an angle of 32° with the surface of the lawn. **(a)** If a 249 N force is applied along the handle of the 21-kg mower, what is the normal force exerted by the lawn on the mower? **(b)** If the angle between the surface of the lawn and the handle of the mower is increased, does the normal force exerted by the lawn increase, decrease, or stay the same?

44. ••• An ant walks slowly away from the top of a bowling ball, as shown in **Figure 5–28**. If the ant starts to slip when the normal force on its feet drops below one-half its weight, at what angle θ does slipping begin?

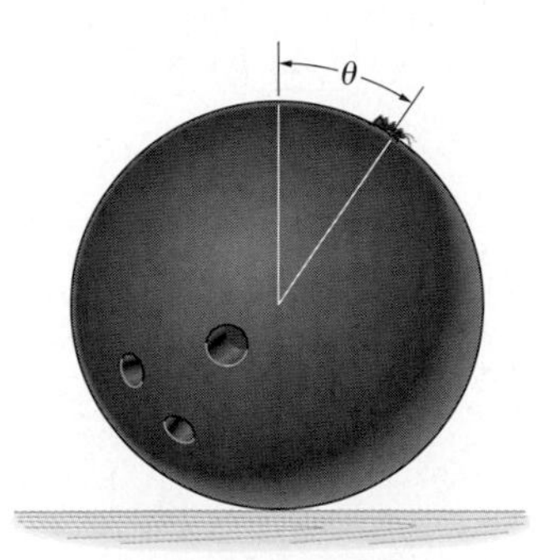

▲ **FIGURE 5-28** Problem 44

General Problems

45. • In a tennis serve, a 0.070-kg ball can be accelerated from rest to 35 m/s over a distance of 0.80 m. Find the average force exerted by the racket on the ball during the serve.

46. • A 40.0-kg swimmer with an initial speed of 1.75 m/s decides to coast until she comes to rest. If she slows with constant acceleration and stops after coasting 2.00 m, what was the force exerted on her by the water?

47. •• At the local grocery store, you push a 14.5-kg shopping cart. You stop for a moment to add a bag of dog food to your cart. With a force of 12.0 N, you accelerate the cart from rest through a distance of 2.29 m in 3.00 s. What was the mass of the dog food?

48. •• On an aircraft carrier, a jet can be catapulted from 0 to 150 mi/h in 2.00 s. If the average force exerted by the catapult is 5.10×10^6 N what is the mass of the jet?

▲ A jet takes off from the flight deck of an aircraft carrier. (Problem 48)

49. •• **IP** An archer shoots a 0.010-kg arrow at a target with a speed of 43 m/s. When it hits the target, it penetrates to a depth of 0.050 m. **(a)** What was the average force exerted by the target on the arrow? **(b)** If the mass of the arrow is doubled, and the force exerted by the target remains the same how does the penetration depth change? Explain.

50. •• **(a)** Draw a free-body diagram for the shopping cart in Problem 21. **(b)** Determine the force F required for the cart to move up the incline with constant speed. **(c)** What is the normal force acting on the cart in this case?

51. •• A 0.15–kg baseball is moving at 43 m/s toward home plate when it is hit with the bat. The ball is in contact with the bat for 0.0020 s, and leaves the bat at 25 m/s directly away from home plate. Determine the average force exerted by the bat on the ball.

52. •• Your groceries are in a bag with paper handles. The handles will tear off if a force greater than 52 N is applied to them. What is the greatest mass of groceries that can be lifted safely with this bag, given that the bag is raised **(a)** with constant speed, or **(b)** with an acceleration of 1.32 m/s^2?

53. •• **IP** While waiting at the airport for your flight to leave, you observe some of the jets as they take off. With your watch you find that it takes about 35 seconds for a plane to go from rest to takeoff speed. In addition, you estimate that the distance required is about 1.0 km. **(a)** If the mass of a jet is 1.70×10^5 kg, what force is needed for takeoff? **(b)** Describe the strategy you used to solve part (a).

54. ••• **IP** Responding to an alarm, a 92-kg fireman slides down a pole to the ground floor, 3.2 m below. The fireman starts at rest and lands with a speed of 4.1 m/s. **(a)** Find the force exerted on the fireman by the pole. **(b)** If the landing speed is half that in part (a), is the force exerted on the fireman doubled? Explain. **(c)** Find the force exerted on the fireman when the landing speed is 2.05 m/s.

55. ••• For a birthday gift, you and some friends take a hot air balloon ride. One friend is late, so the balloon floats a couple of feet off the ground as you wait. Before this person arrives the combined weight of the basket and people is 1220 kg, and the balloon is neutrally buoyant. When the late arrival climbs up into the basket, the balloon begins to accelerate downward at 0.56 m/s^2. What was the mass of the last person to climb aboard?

56. ••• A baseball of mass m and initial speed v strikes a catcher's mitt. If the mitt moves a distance Δx as it brings the ball to rest, what is the average force it exerts on the ball?

57. ••• When two people push in the same direction on an object of mass m they cause an acceleration a_1. When they push in opposite directions, the acceleration of the object is a_2. Determine the force exerted by each of the two people in terms of m, a_1, and a_2.

58. ••• **BIO** Seatbelts provide two main advantages in a car accident: (i) they keep you from being thrown from the car, and (ii) they reduce the force that acts on you during the collision to survivable levels. The second benefit can be illustrated by comparing the net force exerted on the driver of a car in a head-on collision with and without a seatbelt. **(a)** A driver wearing a seatbelt decelerates at the same rate as the car itself. Since modern cars have a "crumple zone" built into the front of the car, the car will decelerate over a distance of roughly 1.0 m. Find the net force acting on a 65-kg driver who is decelerated from 18 m/s to rest in a distance of 1.0 m. **(b)** A driver who does not wear a seatbelt continues to move forward with a speed of 18 m/s (due to inertia) until something solid is encountered. In this case, the driver comes to rest in a much shorter distance—perhaps only a centimeter. Find the net force acting on a 65-kg driver who is decelerated from 18 m/s to rest in 1.0 cm.

6 Applications of Newton's Laws

In Chapter 5, we considered Newton's laws in a simplified, idealized world. In the real world, however, we often have to deal with complicating factors: springs that exert varying forces, ropes and pulleys that transmit force between objects and sometimes change its direction. Above all, objects in the real world are subject to frictional forces. In this chapter we see how Newton's laws apply to such situations.

Newton's laws of motion can be applied to an immense variety of systems, a sampling of which were discussed in Chapter 5. In this chapter we extend our discussion of Newton's laws by introducing new types of forces and by considering new classes of systems. In addition, we show how Newton's laws can be applied to objects moving in a circular path.

6–1 Frictional Forces

In Chapter 5 we always assumed that surfaces were smooth and that objects could slide without resistance to their motion. No surface is perfectly smooth, however. When viewed on the atomic level, even the "smoothest" surface is actually rough and jagged, as indicated in **Figure 6–1**. To

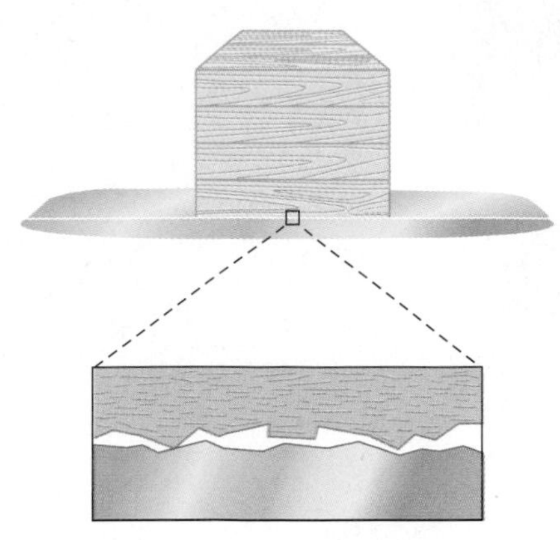

▲ **FIGURE 6–1 The origin of friction** Even "smooth" surfaces have irregularities when viewed at the microscopic level. This type of roughness contributes to friction.

▲ Friction plays an important role in almost everything we do. Sometimes it is desirable to reduce friction; in other cases we want as much friction as possible. For example, it is more fun to ride on a water slide (upper) if the friction is low. Similarly, an engine operates more efficiently when it is oiled. When running, however we need friction to help us speed up, slow down, and make turns. The sole of this running shoe (lower), like a car tire, is designed to maximize friction.

slide one such surface across another requires a force large enough to overcome the resistance of microscopic hills and valleys bumping together. This is the origin of the force we call **friction**.

We often think of friction as something that should be reduced, or even eliminated if possible. For example, roughly 20 percent of the gasoline you buy does nothing but overcome friction within your car's engine. Clearly, reducing that friction would be most desirable.

On the other hand, friction can be helpful—even indispensable—in other situations. Suppose, for example, that you are standing still and then decide to begin walking forward. The force that accelerates you is the force of friction between your shoes and the ground. We simply couldn't walk or run without friction—it's hard enough when friction is merely reduced, as on an icy sidewalk. Similarly, starting or stopping a car, or even turning a corner, all require friction. Friction is an important and common feature of everyday life.

Since friction is caused by the random, microscopic irregularities of a surface, and since it is greatly affected by other factors such as the presence of lubricants, there is no simple "law of nature" for friction. There are, however, some very useful rules of thumb that give us rather accurate, approximate results for calculating frictional forces. In what follows, we describe these rules of thumb for the two types of friction most commonly used in this text—kinetic friction and static friction.

Kinetic Friction

As its name implies, kinetic friction is the friction encountered when surfaces slide against one another with a finite relative speed. The force generated by this friction, which will be designated with the symbol f_k, acts to oppose the sliding motion at the point of contact between the surfaces.

A series of simple experiments illustrates the main characteristics of kinetic friction. First, imagine attaching a spring scale to a rough object, like a brick, and pulling it across a table, as shown in **Figure 6–2**. If the brick moves with constant velocity, Newton's second law tells us that the net force on the brick must be zero. Hence, the force read on the scale, F, has the same magnitude as the force of kinetic friction, f_k. Now, if we repeat the experiment, but this time put a second brick on top of the first, we find that the force needed to pull the brick with constant velocity is doubled, to $2F$.

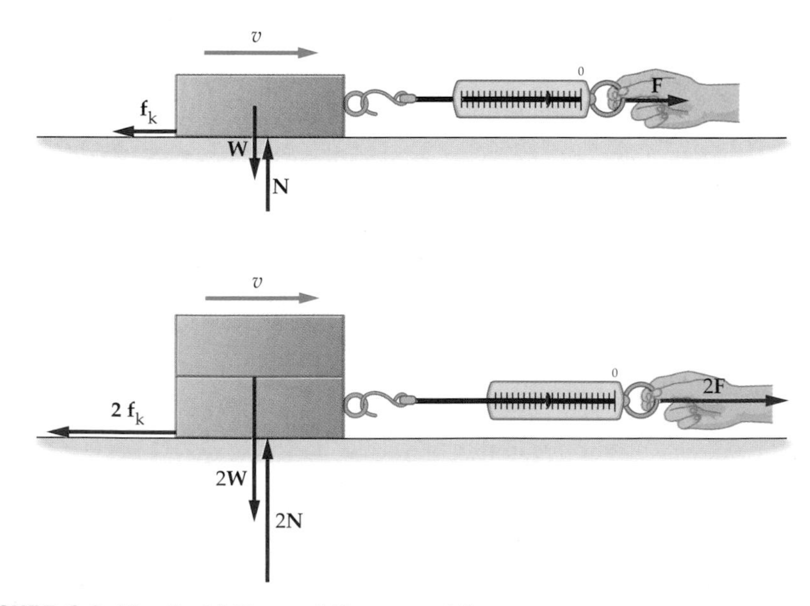

▲ **FIGURE 6–2 Kinetic friction and the normal force**
In the top part of the figure, a force F is required to pull the brick with constant speed v. Thus the force of kinetic friction is $f_k = F$. In the bottom part of the figure, the normal force has been doubled, and so has the force of kinetic friction, to $f_k = 2F$.

From this experiment we see that when we double the normal force—by stacking up two bricks, for example—the force of kinetic friction is also doubled. In general, the force of kinetic friction is found to be proportional to the magnitude of the normal force, N. Stated mathematically, this observation can be written as follows:

$$f_k = \mu_k N \qquad \textbf{6–1}$$

The constant of proportionality, μ_k (pronounced "mew sub k"), is referred to as the **coefficient of kinetic friction**.

Since f_k and N are both forces, and hence have the same units, we see that μ_k is a dimensionless number. Typical values for μ_k range between 0 and 1, as indicated in Table 6–1. The interpretation of μ_k is simple: If $\mu_k = 0.1$, for example, the force of kinetic friction is one-tenth of the normal force. Simply put, the greater μ_k the greater the friction; the smaller μ_k the smaller the friction.

TABLE 6–1 Typical Coefficients of Friction

Materials	Kinetic, μ_k	Static, μ_s
Rubber on concrete (dry)	0.80	0.90
Steel on steel	0.57	0.74
Glass on glass	0.40	0.94
Wood on leather	0.40	0.50
Copper on steel	0.36	0.53
Rubber on concrete (wet)	0.25	0.30
Steel on ice	0.06	0.10
Waxed ski on snow	0.05	0.10
Teflon on Teflon	0.04	0.04

As we know from everyday experience, the force of kinetic friction tends to oppose motion, as shown in Figure 6–2. Thus, $f_k = \mu_k N$ is not a vector equation, because N is perpendicular to the direction of motion. When doing calculations with the force of kinetic friction, we use $f_k = \mu_k N$ to find its magnitude, and we draw its direction so that it is opposite to the direction of motion.

There are two more friction experiments of particular interest. First, suppose that when we pull a brick, we initially pull it at the speed v, then later at the speed $2v$. What forces do we measure? It turns out that the force of kinetic friction is approximately the same in each case—it certainly does not double when we double the speed. Second, let's try standing the brick on end, so that it has a smaller area in contact with the table. If this smaller area is half the previous area, is the force halved? No, the force remains essentially the same, regardless of the area of contact.

We summarize these observations with the following three rules of thumb for kinetic friction:

Rules of Thumb for Kinetic Friction

The force of kinetic friction between two surfaces is:

1. Proportional to the magnitude of the normal force, N, between the surfaces,

$$f_k = \mu_k N$$

2. Independent of the relative speed of the surfaces.
3. Independent of the area of contact between the surfaces.

Again, these rules are useful and fairly accurate, though they are still only approximate. For simplicity, when we do calculations involving kinetic friction in this text, we will use these rules as if they were exact.

Before we show how to use f_k in calculations, we should make a comment regarding rule 3. This rule often seems rather surprising and counterintuitive. How is it that a larger area of contact doesn't produce a larger force? One way to think about this is to consider that when the area of contact is large, the normal force is spread out over a large area, giving a small force per area, F/A. As a result, the microscopic hills and valleys are not pressed too deeply against one another. On the other hand, if the area is small the normal force is concentrated in a small region, which presses the surfaces together more firmly, due to the large force per area. The net effect is roughly the same.

Now, let's consider a commonly encountered situation in which kinetic friction plays a decisive role.

EXAMPLE 6–1 Pass the Salt—Please

Someone at the other end of the table asks you to pass the salt. Feeling quite dashing, you slide the 50.0-g salt shaker in their direction, giving it an initial speed of 1.15 m/s. If the shaker comes to rest in 0.840 m, what is the coefficient of kinetic friction between the shaker and the table?

Picture the Problem

We choose the positive x direction to be the direction of motion, and the positive y direction to be upward. Two forces act in the y direction; the shaker's weight, **W**, and the normal force, **N**. Only one force acts in the x direction; the force of kinetic friction, $\mathbf{f}_k = (-\mu_k N)\hat{\mathbf{x}}$.

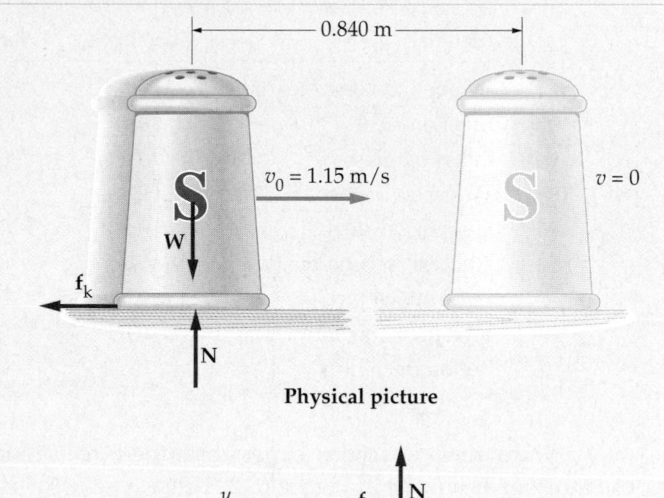

Physical picture

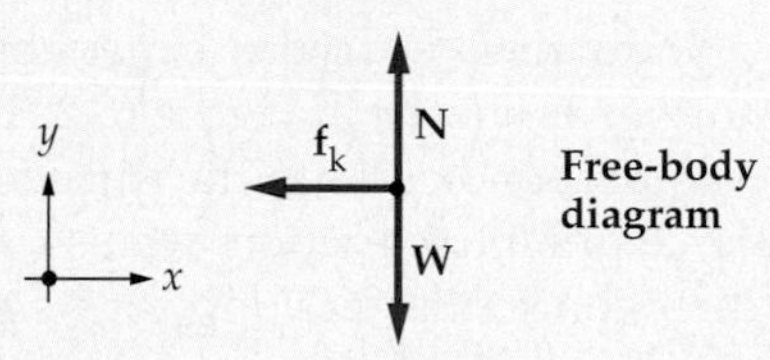

Free-body diagram

Strategy

Since the frictional force has a magnitude of $f_k = \mu_k N$, it follows that $\mu_k = f_k/N$. Therefore, we need to find the magnitudes of the frictional force, f_k, and the normal force, N.

To find f_k we set $\Sigma F_x = ma_x$, and find a_x with the kinematic equation $v_x^2 = v_{0x}^2 + 2a_x\Delta x$.

To find N we set $a_y = 0$ (since there is no motion in the y direction) and solve for N using $\Sigma F_y = ma_y = 0$.

Solution

1. Set $\Sigma F_x = ma_x$ to find f_k in terms of a_x:

$$\sum F_x = -f_k = ma_x \quad \text{or} \quad f_k = -ma_x$$

2. Determine a_x by using the kinematic equation relating velocity to position, $v_x^2 = v_{0x}^2 + 2a_x\Delta x$:

$$v_x^2 = v_{0x}^2 + 2a_x\Delta x$$

$$a_x = \frac{v_x^2 - v_{0x}^2}{2\Delta x} = \frac{0 - (1.15 \text{ m/s})^2}{2(0.840 \text{ m})} = -0.787 \text{ m/s}^2$$

3. Set $\Sigma F_y = ma_y = 0$ to find the normal force, N:

$$\sum F_y = N + (-W) = ma_y = 0 \quad \text{or} \quad N = W = mg$$

4. Substitute $N = mg$ and $f_k = -ma_x$ (with $a_x = -0.787 \text{ m/s}^2$) into $\mu_k = f_k/N$ to find μ_k:

$$\mu_k = \frac{f_k}{N} = \frac{-ma_x}{mg} = \frac{-a_x}{g} = \frac{-(-0.787 \text{ m/s}^2)}{9.81 \text{ m/s}^2} = 0.0802$$

Insight

Note that m canceled in line 4, so our final result is independent of the shaker's mass. For example, if we were to slide a shaker with twice the mass, but with the same initial speed, it would slide the same distance. It is unlikely this independence would have been apparent if we had worked the problem numerically rather than symbolically.

Practice Problem

Given the same initial speed and a coefficient of kinetic friction equal to 0.120, what is **(a)** the acceleration of the shaker, and **(b)** the distance it slides? [**Answer: (a)** -1.18 m/s^2, **(b)** 0.560 m]

Some related homework problems: Problem 1, Problem 12

In the next Example we consider a system that is inclined at an angle θ relative to the horizontal. To be very clear about how we handle the force vectors in such a case, we begin by resolving each vector into its x and y components.

EXAMPLE 6–2 Making a Big Splash

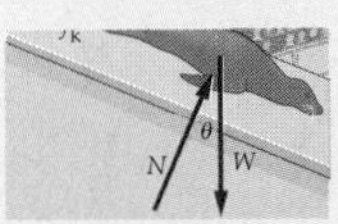

A trained sea lion slides from rest down a 3.0-m-long ramp into a pool of water. If the ramp is inclined at an angle of 23° above the horizontal and the coefficient of kinetic friction between the sea lion and the ramp is 0.26, how long does it take for the sea lion to make a splash in the pool?

Picture the Problem

As is usual with inclined surfaces, we choose the x axis to be parallel to the surface and the y axis to be perpendicular to it. In our sketch the sea lion slides in the positive x direction, and there is no motion in the y direction (hence $a_y = 0$).

Strategy

We can use the kinematic equation relating position to time, $x = x_0 + v_{0x}t + \frac{1}{2}a_x t^2$, to find the time of the sea lion's slide. It will be necessary, however, to first determine the acceleration of the sea lion in the x direction, a_x.

To find a_x we apply Newton's second law to the sea lion. First, we can find N by setting $\Sigma F_y = ma_y$ equal to zero (since $a_y = 0$). It is important to start by finding N because we need it to find the force of kinetic friction, $f_k = \mu_k N$. Using f_k in the sum of forces in the x direction, $\Sigma F_x = ma_x$, allows us to solve for a_x and, finally, for the time.

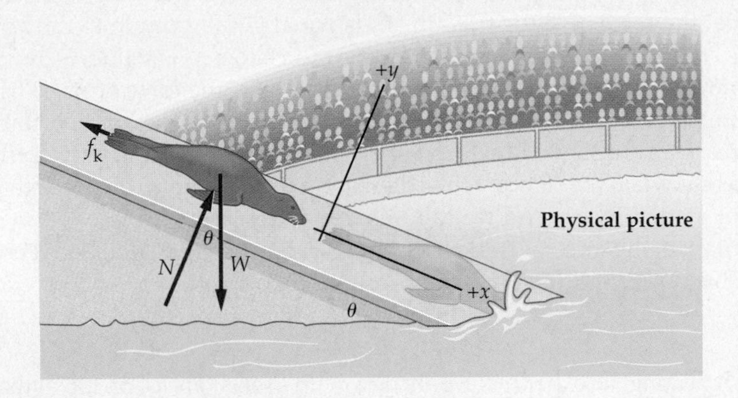

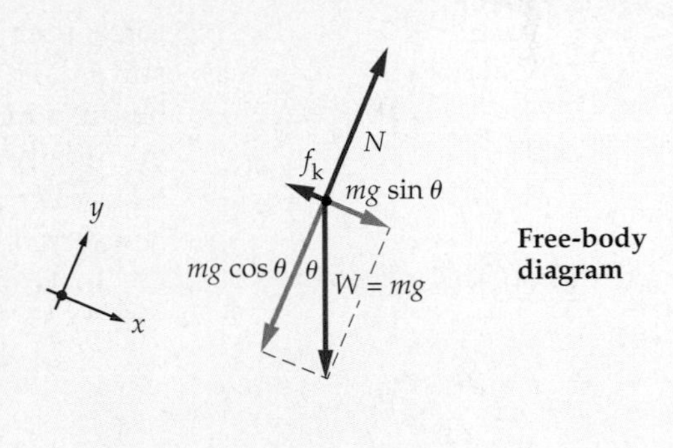

Solution

1. We begin by resolving each of the three force vectors into x and y components:

$$N_x = 0 \qquad N_y = N$$
$$f_{k,x} = -f_k = -\mu_k N \qquad f_{k,y} = 0$$
$$W_x = mg\sin\theta \qquad W_y = -mg\cos\theta$$

2. Set $\Sigma F_y = ma_y = 0$ to find N:

$$\sum F_y = N - mg\cos\theta = ma_y = 0$$
$$N = mg\cos\theta$$

3. Next, set $\Sigma F_x = ma_x$:
Note that the mass cancels in this equation:

$$\sum F_x = mg\sin\theta - \mu_k N$$
$$= mg\sin\theta - \mu_k mg\cos\theta = ma_x$$

4. Solve for the acceleration in the x direction, a_x:

$$a_x = g(\sin\theta - \mu_k\cos\theta)$$
$$= (9.81\ \text{m/s}^2)[\sin 23° - (0.26)\cos 23°]$$
$$= 1.5\ \text{m/s}^2$$

5. Use $x = x_0 + v_{0x}t + \frac{1}{2}a_x t^2$ to find the time when the sea lion reaches the bottom. We choose $x_0 = 0$, and we are given that $v_{0x} = 0$, hence we set $x = \frac{1}{2}a_x t^2 = 3.0$ m and solve for t:

$$x = \tfrac{1}{2}a_x t^2$$
$$t = \sqrt{\frac{2x}{a_x}} = \sqrt{\frac{2(3.0\ \text{m})}{1.5\ \text{m/s}^2}} = 2.0\ \text{s}$$

Insight

Note that we don't need the sea lion's mass in order to find the time. On the other hand, if we wanted the magnitude of the force of kinetic friction, the mass would be needed.

continued on the following page

continued from the previous page

Practice Problem

How long would it take the sea lion to reach the water if there were no friction in this system? [**Answer:** 1.3 s]

Some related homework problems: Problem 8, Problem 54

PROBLEM SOLVING NOTE

Choice of Coordinate System: Incline

On an incline, align one axis (x) parallel to the surface, and the other axis (y) perpendicular to the surface. That way the motion is in the x direction. Since no motion occurs in the y direction, we know that $a_y = 0$.

Static Friction

Static friction tends to keep two surfaces from moving relative to one another. It, like kinetic friction, is due to the microscopic irregularities of surfaces that are in contact. In fact, static friction is typically stronger than kinetic friction because when surfaces are in static contact, their microscopic hills and valleys can nestle down deeply into one another, thus forming a strong connection between the surfaces. In kinetic friction, the surfaces bounce along and don't become as deeply enmeshed.

As we did with kinetic friction, let's use the results of some simple experiments to determine the rules of thumb for static friction. We start with a brick at rest on a table, with no horizontal force pulling on it, as in **Figure 6–3**. Of course, in this case the force of static friction is zero; no force is needed to keep the brick from sliding.

Next, attach a spring scale to the brick and pull with a small force of magnitude F_1, a force small enough that the brick doesn't move. Since the brick is still at rest, it follows that the force of static friction, f_s, is equal in magnitude to the applied force; that is, $f_s = F_1$. Now, increase the applied force to a new value, F_2, which is still small enough that the brick stays at rest. In this case, the force of static friction has also increased so that $f_s = F_2$. If we continue increasing the applied force we eventually reach a value beyond which the brick starts to move and kinetic friction takes over, as shown in the figure. Thus, there is an upper limit to the force that can be exerted by static friction, and we call this upper limit $f_{s,max}$.

To summarize, the force of static friction, f_s, can have any value between zero and $f_{s,max}$. This can be written mathematically as follows:

$$0 \leq f_s \leq f_{s,max} \qquad 6\text{–}2$$

Imagine repeating the experiment, only now with a second brick on top of the first. This doubles the normal force and it also doubles the maximum force of static friction. Thus, the maximum force is proportional to the magnitude of the normal force, or

$$f_{s,max} = \mu_s N \qquad 6\text{–}3$$

▶ **FIGURE 6–3 Static friction**
As the force applied to an object increases, so does the force of static friction—up to a certain point. Beyond this maximum value, static friction can no longer hold the object, and it begins to slide. Now kinetic friction takes over.

The constant of proportionality is called μ_s (pronounced "mew sub s"), the **coefficient of static friction**. Note that μ_s, like μ_k, is dimensionless. Typical values are given in Table 6–1. In most cases, μ_s is greater than μ_k, indicating that the force of static friction is greater than the force of kinetic friction, as mentioned.

Finally, two additional comments regarding the nature of static friction: (i) Experiments show that static friction, like kinetic friction, is independent of the area of contact. (ii) The force of static friction is not in the direction of the normal force, thus $f_{s,max} = \mu_s N$ is not a vector relation. The direction of f_s is parallel to the surface of contact, and opposite to the direction the object would move if there were no friction.

These observations are summarized in the following rules of thumb:

Rules of Thumb for Static Friction

The force of static friction between two surfaces has the following properties:

1. It takes on any value between zero and the maximum possible force of static friction, $f_{s,max} = \mu_s N$:

$$0 \leq f_s \leq \mu_s N$$

2. It is independent of the area of contact between the surfaces.
3. It is parallel to the surface of contact, and in the direction that opposes relative motion.

Next, we consider a practical method of determining the coefficient of static friction. As with the last example, we begin by resolving all relevant force vectors into their x and y components.

▲ The coefficient of static friction between two surfaces depends on many factors, including whether the surfaces are dry or wet. On the desert floor of Death Valley, in California, occasional rains can reduce the friction between rocks and the sandy ground to such an extent that strong winds can move the rocks over considerable distances. This results in linear "rock trails," which record the direction of the winds at different times.

EXAMPLE 6–3 Slightly Tilted

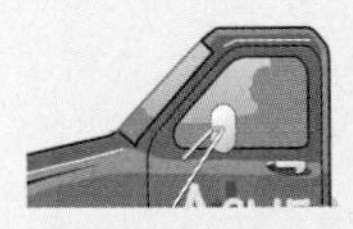

A dump truck slowly tilts its bed upward to dispose of a 95.0-kg crate. For small angles of tilt the crate stays put, but when the tilt angle exceeds 23.2° the crate begins to slide. What is the coefficient of static friction between the bed of the truck and the crate?

Picture the Problem

We align our coordinate system with the incline, and choose the positive x direction to be down slope. Note that three forces act on the crate: the weight, **W**; the normal force, **N**; and the force of static friction, $\mathbf{f}_s$.

Strategy

When the crate is on the verge of slipping, but has not yet slipped, its acceleration is zero in both the x and y directions. In addition, "verge of slipping" means that the magnitude of the static friction is at its maximum value, $f_s = f_{s,max} = \mu_s N$. Thus, we set $\Sigma F_y = ma_y = 0$ to find N, then use $\Sigma F_x = ma_x = 0$ to find μ_s.

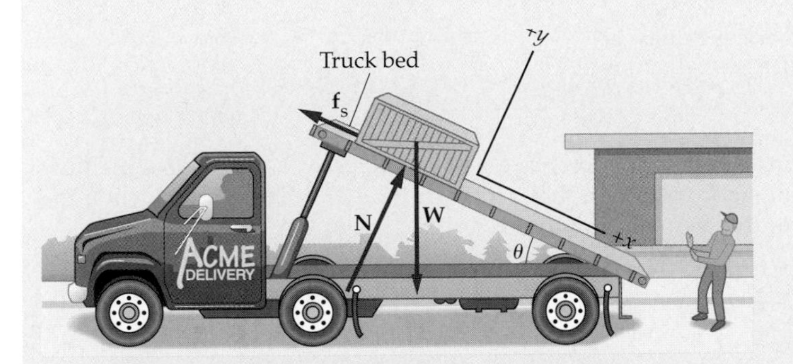

Physical picture

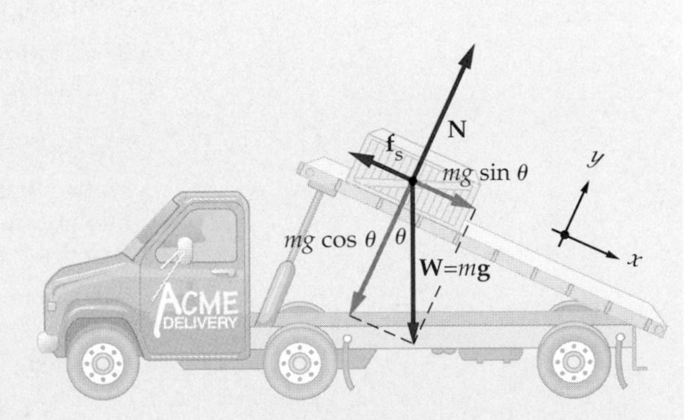

Free-body diagram

continued on the following page

continued from the previous page

Solution

1. Resolve the three force vectors acting on the crate into x and y components:

$$N_x = 0 \qquad N_y = N$$
$$f_{s,x} = -f_{s,\max} = -\mu_s N \qquad f_{s,y} = 0$$
$$W_x = mg \sin\theta \qquad W_y = -mg \cos\theta$$

2. Set $\Sigma F_y = ma_y = 0$, since $a_y = 0$:

$$\sum F_y = N_y + f_{s,y} + W_y = N + 0 - mg\cos\theta = ma_y = 0$$

Solve for the normal force, N:

$$N = mg\cos\theta$$

3. Set $\Sigma F_x = ma_x = 0$, since the crate is at rest, and use the result for N obtained in step 2:

$$\sum F_x = N_x + f_{s,x} + W_x = ma_x = 0$$
$$= 0 - \mu_s N + mg\sin\theta$$
$$= 0 - \mu_s mg\cos\theta + mg\sin\theta$$

4. Solve the expression for the coefficient of static friction, μ_s:

$$\mu_s mg\cos\theta = mg\sin\theta$$
$$\mu_s = \frac{mg\sin\theta}{mg\cos\theta} = \tan\theta = \tan 23.2° = 0.429$$

Insight

In general, if an object is on the verge of slipping when the surface on which it rests is tilted at an angle θ_c, the coefficient of static friction between the object and the surface is $\mu_s = \tan\theta_c$. Note that this result is independent of the mass of the object.

Practice Problem

Find the magnitude of the force of static friction acting on the crate. [**Answer:** $f_{s,\max} = \mu_s N = 367$ N]

Some related homework problems: Problem 9, Problem 64

Recall that static friction can have magnitudes less than its maximum possible value. This point is emphasized in the following Active Example.

ACTIVE EXAMPLE 6–1 Less Tilted

In the previous example, what is the magnitude of the force of static friction acting on the crate when the truck bed is tilted at an angle of 20.0°?

Solution

1. Sum the x components of force acting on the crate: $\Sigma F_x = 0 - f_s + mg\sin\theta$
2. Set this sum equal to zero (since $a_x = 0$) and solve for the magnitude of the static friction force, f_s: $f_s = mg\sin\theta$
3. Substitute numerical values, including $\theta = 20.0°$: $f_s = 319$ N

Insight

Notice that the force of static friction has a magnitude (319 N) that is less than the maximum value of static friction in this case, $f_{s,\max} = \mu_s N = 367$ N, as given in the Practice Problem of the previous example.

Also, note that if we substitute $\theta = 23.2°$ into the expression for static friction given in step 2, $f_s = mg\sin\theta$, we find $f_s = 367$ N, in agreement with $f_{s,\max} = \mu_s N$, as expected.

Finally, friction often enters into problems dealing with vehicles with rolling wheels. In Conceptual Checkpoint 6–1, we consider which type of friction is appropriate in such cases.

CONCEPTUAL CHECKPOINT 6–1

A car drives with its tires rolling freely. Is the friction between the tires and the road **(a)** kinetic or **(b)** static?

Reasoning and Discussion

A reasonable-sounding answer is that because the car is moving, the friction between its tires and the road must be kinetic friction—but this is not the case.

Actually, the friction is static because the bottom of the tire is in static contact with the road. To understand this, watch your feet as you walk. Even though you are moving, each foot is in static contact with the ground once you step down on it. Your foot doesn't move again until you lift it up and move it forward for the next step. A tire can be thought of as a succession of feet arranged in a circle, each of which is momentarily in static contact with the ground.

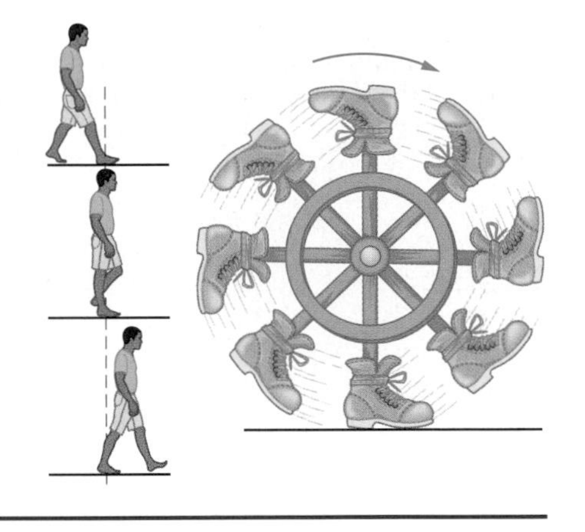

Answer:

(b) The friction between the tires and the road is static friction.

To summarize, if a car skids, the friction acting on it is kinetic; if its wheels are rolling, the friction is static. Since static friction is generally greater than kinetic friction, it follows that a car can be stopped in less distance if its wheels are rolling (static friction) than if its wheels are locked up (kinetic friction). This is the idea behind the antilock braking systems (ABS) that are available on many newer cars. When the brakes are applied in a car with ABS, an electronic rotation sensor at each wheel detects whether the wheel is about to start skidding. To prevent skidding, a small computer automatically begins to modulate the hydraulic pressure in the brake lines in short bursts, causing the brakes to release and then reapply in rapid succession. This allows the wheels to continue rotating, even in an emergency stop, and for static friction to determine the stopping distance. **Figure 6–4** shows a comparison of braking distances for cars with and without ABS. An added benefit of ABS is that a driver is better able to steer and control a braking car if its wheels are rotating.

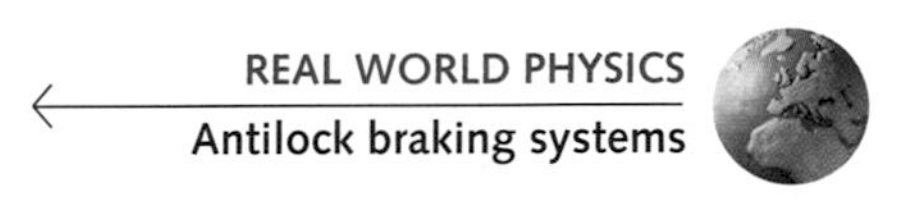

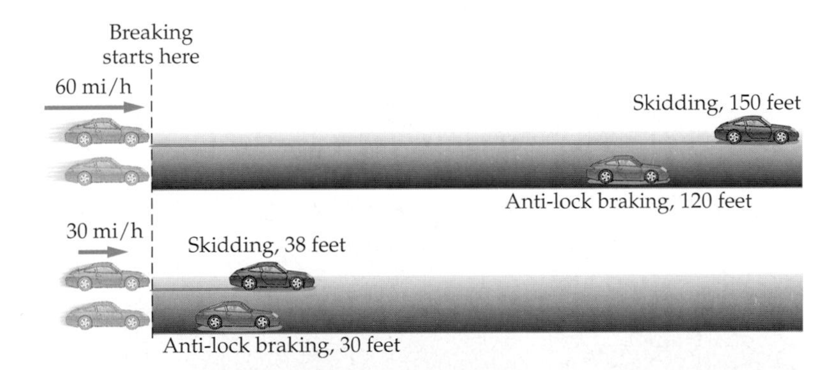

◀ **FIGURE 6–4 Stopping distance with and without ABS**

◀ The angle that the sloping sides of a sand pile (left) make with the horizontal is determined by the coefficient of static friction between grains of sand, in much the same way that static friction determines the angle at which the crate in Example 6–3 begins to slide. The same basic mechanism determines the angle of the cone-shaped mass of rock debris at the base of a cliff, known as a talus slope (right).

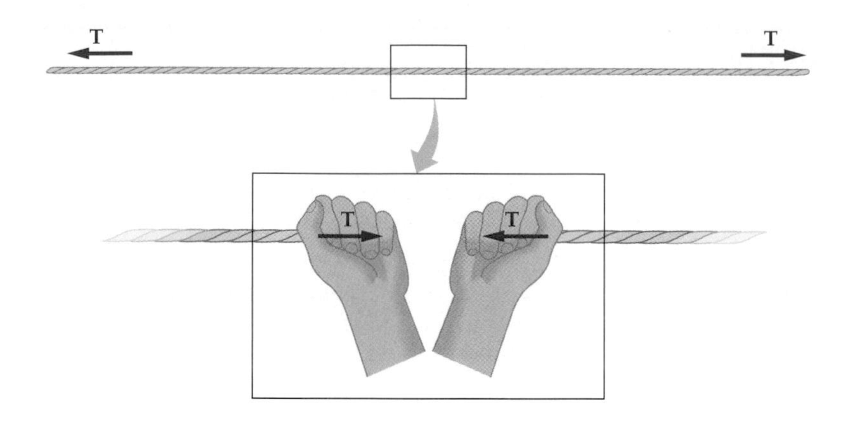

▶ **FIGURE 6–5 Tension in a string**
A string, pulled from either end, has a tension, T. If the string were to be cut at any point, the force required to hold the ends together is T.

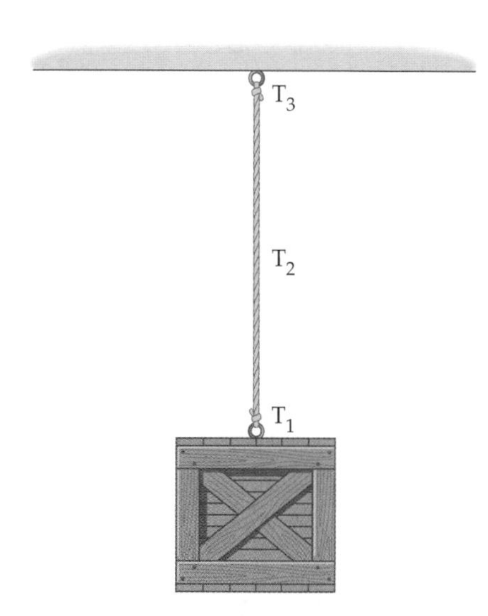

▲ **FIGURE 6–6 Tension in a heavy rope**
Because of the weight of the rope, the tension is noticeably different at points 1, 2, and 3. As the rope becomes lighter, however, the difference in tension decreases. In the limit of a rope of zero mass, the tension is the same throughout the rope.

6–2 Strings and Springs

A common way to exert a force on an object is to pull on it with a string, a rope, a cable, or a wire. Similarly, you can push or pull on an object if you attach it to a spring. In this section we discuss the basic features of strings and springs and how they transmit forces.

Strings and Tension

Imagine picking up a light string and holding it with one end in each hand. If you pull to the right with your right hand and to the left with your left hand, the string becomes taut. In such a case, we say that there is a **tension** in the string. To be more specific, if you cut the string at some point, the tension T is the force pulling the string apart, as illustrated in **Figure 6–5**. Note that at any given point, the tension pulls equally to the right and to the left.

As an example, consider a rope that is attached to the ceiling at one end, and to a box with a weight of 105 N at the other end, as shown in **Figure 6–6**. In addition, suppose the rope is uniform, and that it has a total weight of 2.00 N. What is the tension in the rope (i) where it attaches to the box, (ii) at its midpoint, and (iii) where it attaches to the ceiling?

First, the rope holds the box at rest, thus the tension where the rope attaches to the box is simply the weight of the box, $T_1 = 105$ N. At the midpoint of the rope, the tension supports the weight of the box, plus the weight of half the rope. Thus, $T_2 = 105\text{ N} + \frac{1}{2}(2.00\text{ N}) = 106\text{ N}$. Similarly, at the ceiling the tension supports the box plus all of the rope, giving a tension of $T_3 = 107$ N. Note that the tension pulls down on the ceiling but pulls up on the box.

▶ The rope supporting this climber is under tension, but that tension is not uniform along its length. Most of the tension is a reaction to the the weight of the climber, but as we move upward, the weight of the rope itself makes an increasing contribution.

From this discussion, we can see that the tension in the rope changes slightly from top to bottom because of the mass of the rope. If the rope had less mass, the difference in tension between its two ends would also be less. In particular, if the rope's mass were to be vanishingly small, the difference in tension would vanish as well. In this text, we will assume that all ropes, strings, wires, and so on are practically massless—unless specifically stated otherwise—and, hence, that the tension is the same throughout their length.

Pulleys are often used to redirect a force exerted by a string, as indicated in **Figure 6–7**. In the ideal case, a pulley has no mass, and no friction in its bearings. Thus, *an ideal pulley simply changes the direction of the tension in a string, without changing its magnitude*. An application of pulleys in a traction device is considered in the next Example.

EXAMPLE 6–4 A Bad Break: Setting a Broken Leg with Traction

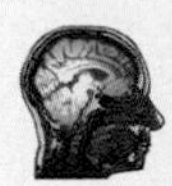

Real World Physics: Bio

A traction device employing three pulleys is applied to a broken leg, as shown in the sketch. The middle pulley is attached to the sole of the foot, and a mass m supplies the tension in the ropes. Find the value of the mass m if the net force exerted on the sole of the foot is to be 165 N.

Picture the Problem

Our sketch shows the physical picture as well as the free-body diagram for the middle pulley. Notice that the rope makes an angle of 40.0° with the horizontal on either side of the middle pulley.

Strategy

We begin by noting that the rope supports the hanging mass m. As a result, the tension in the rope, T, must be equal in magnitude to the weight of the mass; $T = mg$.

Next, the pulleys simply change the direction of the tension without changing its magnitude. Therefore, the tension in each segment of the rope is the same. The net force exerted on the sole of the foot is the sum of the tension T at 40.0° above the horizontal plus the tension T at 40.0° below the horizontal. We will calculate the net force component by component.

Once we calculate the net force acting on the foot, we set it equal to 165 N and solve for the tension T. Finally, we find the mass using the relation $T = mg$.

40.0°
40.0°
m

Physical picture

y
x
T
$T \sin 40.0°$
$T \cos 40.0°$
40.0°
40.0°
$T \sin 40.0°$
$T \cos 40.0°$
T

Free-body diagram for middle pulley

Solution

1. First, consider the tension that acts upward on the middle pulley. Resolve this tension into x and y components:

$$T_x = T \cos 40.0° \qquad T_y = T \sin 40.0°$$

2. Next, consider the tension that acts downward on the middle pulley. Resolve this tension into x and y components. Note the minus sign in the y component:

$$T_x = T \cos 40.0° \qquad T_y = -T \sin 40.0°$$

3. Sum the x and y components of force acting on the foot. We see that the net force acts only in the x direction, as one might expect from symmetry:

$$\sum F_x = T \cos 40.0° + T \cos 40.0° = 2T \cos 40.0°$$
$$\sum F_y = T \sin 40.0° - T \sin 40.0° = 0$$

4. Step 3 shows that the net force acting on the foot is $2T \cos 40.0°$. Set this force equal to 165 N and solve for T:

$$2T \cos 40.0° = 165 \text{ N}$$
$$T = \frac{165 \text{ N}}{2 \cos 40.0°} = 108 \text{ N}$$

5. Solve for the mass, m, using $T = mg$:

$$T = mg$$
$$m = \frac{T}{g} = \frac{108 \text{ N}}{9.81 \text{ m/s}^2} = 11.0 \text{ kg}$$

continued on following page

continued from previous page

Insight

Notice that this pulley arrangement "magnifies the force" in the sense that a 108 N weight attached to the rope produces a 165 N force exerted on the foot. In addition, the net force on the foot produces an opposing force in the leg that acts in the direction of the head (a cephalad force), as desired to set a broken leg.

Practice Problem

(a) Would the required mass m increase or decrease if the angles in this device were changed from 40.0° to 30.0°? **(b)** Find the mass m for an angle of 30.0°. [**Answer: (a)** The required mass m will decrease. **(b)** 9.71 kg]

Some related homework problems: Problem 16, Problem 19, Problem 29

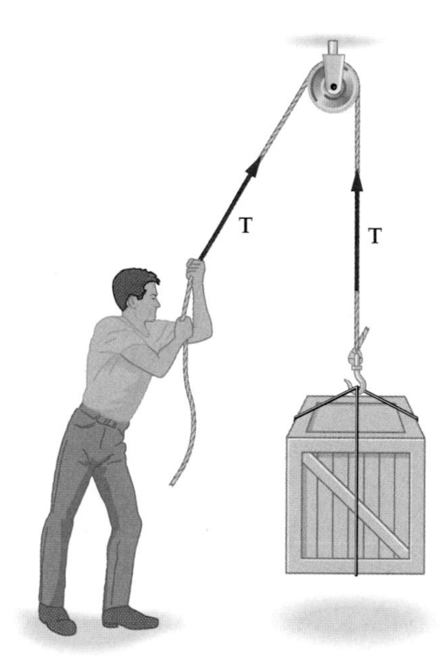

▲ **FIGURE 6–7 A pulley changes the direction of a tension**

In an ideal string, the tension has the same magnitude, T, throughout its length. A pulley can serve to redirect the string, however, so that the tension acts in a different direction.

CONCEPTUAL CHECKPOINT 6–2

The scale at left reads 9.81 N. Is the reading of the scale at right **(a)** greater than 9.81 N, **(b)** equal to 9.81 N, or **(c)** less than 9.81 N?

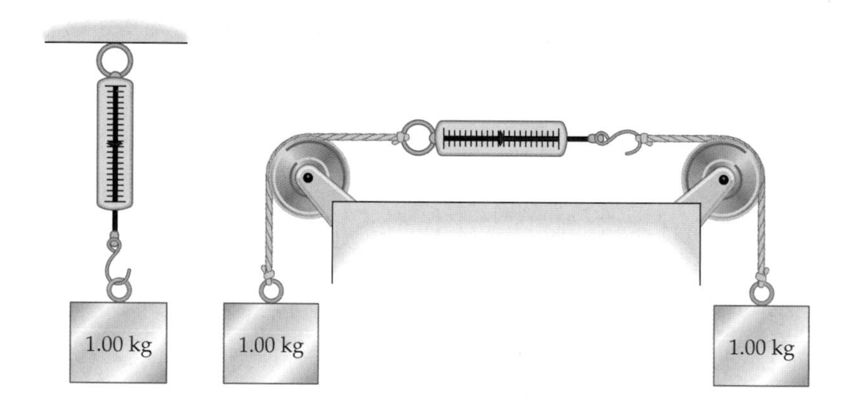

Reasoning and Discussion

Since a pulley simply changes the direction of the tension in a string without changing its magnitude, it is clear that the scale at left above reads the same as the scale shown in the figure below.

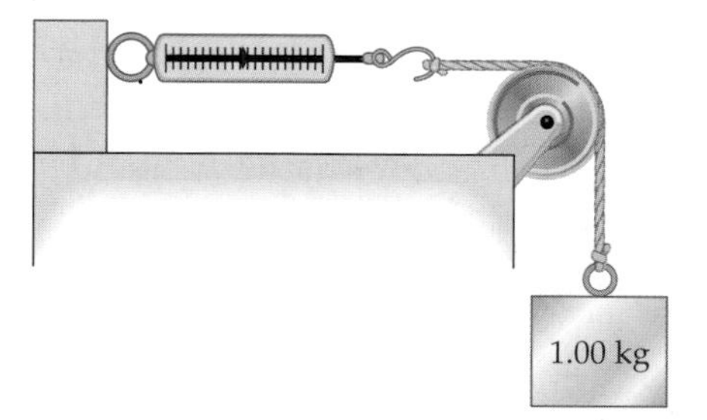

There is no difference, however, between attaching the top end of the scale to something rigid and attaching it to another 1.00-kg hanging mass. In either case, the fact that the scale is at rest means that a force of 9.81 N must be exerted to the left on the top of the scale to balance the 9.81-N force exerted on the lower end of the scale. As a result, the two scales read the same.

Answer:

(b) The reading of the scale at right is equal to 9.81 N.

Springs and Hooke's Law

Suppose you take a spring of length L, as shown in **Figure 6–8 (a)**, and attach it to a block. If you pull on the spring, causing it to stretch to a length $L + x$, the spring pulls on the block with a force of magnitude F. If you increase the length of the spring to $L + 2x$, the force exerted by the spring increases to $2F$. Similarly, if you compress the spring to a length $L - x$, the spring pushes on the block with a force of magnitude F, where F is the same force given previously. As you might expect, compression to a length $L - 2x$ results in a push of magnitude $2F$.

As a result of these experiments, we can say that a spring exerts a force that is proportional to the amount, x, by which it is stretched or compressed. Thus, if F is the magnitude of the spring force, we can say that

$$F = kx$$

In this expression, k is a constant of proportionality, referred to simply as the **force constant**. Since F has units of newtons and x has units of meters, it follows that k has units of newtons per meter, or N/m. The larger the value of k, the stiffer the spring.

To be more precise, consider the spring shown in **Figure 6–8 (b)**. Note that we have placed the origin of the x axis at the equilibrium length of the spring. Now, if we stretch the spring so that the end of the spring is at a positive value of x $(x > 0)$, we find that the spring exerts a force of magnitude kx in the negative x direction. Thus, the spring force (which has only an x component) can be written as

$$F_x = -kx$$

Similarly, consider compressing the spring so that its end is at a negative value of x $(x < 0)$. In this case, the force exerted by the spring is of magnitude kx, and points in the positive x direction, as is shown in Figure 6–8(b). Again, we can write the spring force as

$$F_x = -kx$$

To see that this is correct—that is, that F_x is positive in this case—recall that x is negative, which means that $(-x)$ is positive.

This result for the force of a spring is known as Hooke's law, after Robert Hooke (1635–1703). It is really just a good rule of thumb rather than a law of nature. Clearly, it can't work for any amount of stretching. For example, we know that if we stretch a spring far enough it will be permanently deformed, and will never return to its original length. Still, for small stretches or compressions, Hooke's law is quite accurate.

Rules of Thumb for Springs (Hooke's Law)

A spring stretched or compressed by the amount x from its equilibrium length exerts a force given by

$$F_x = -kx \qquad 6\text{–}4$$

In this text, we consider only **ideal springs**—that is, springs that are massless, and that are assumed to obey Hooke's law exactly.

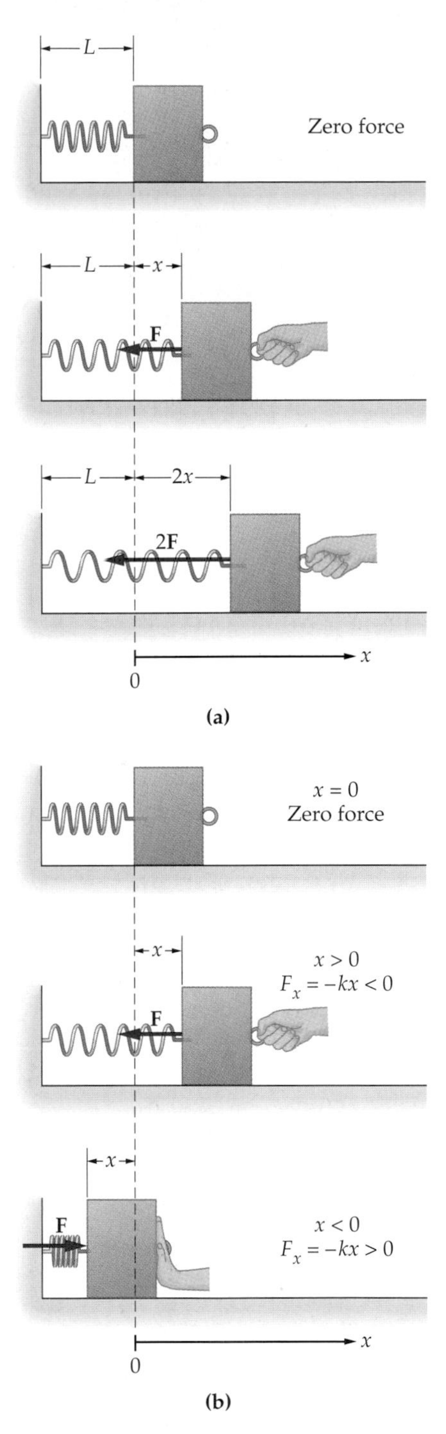

▲ **FIGURE 6–8 Spring Forces**
When dealing with a spring, it is convenient to choose the origin at the equilibrium position. In the case shown above, the force is strictly in the x direction, and is given by $F_x = -kx$. Note that the minus sign means that the force is opposite to the displacement; that is, the force is restoring.

EXERCISE 6–1

A 1.50-kg object hangs motionless from a spring with a force constant of $k = 250$ N/m. How far is the spring stretched from its equilibrium length?

Solution

Setting the magnitude of the upward spring force equal to the weight yields $kx = mg$, or

$$x = \frac{mg}{k} = \frac{(1.50 \text{ kg})(9.81 \text{ m/s}^2)}{250 \text{ N/m}} = 0.0589 \text{ m}$$

Since the stretch of a spring and the force it exerts are proportional, we can now see how a spring scale operates. In particular, pulling on the two ends of a scale stretches the spring inside it by an amount proportional to the applied force. Once the scale is calibrated, by stretching the spring with a known, or reference, force, we can use it to measure other unknown forces.

6–3 Translational Equilibrium

When we say that an object is in **translational equilibrium**, we mean that the net force acting on it is zero:

$$\sum \mathbf{F} = 0 \qquad \textbf{6–5}$$

From Newton's second law, this is equivalent to saying that the object's acceleration is zero.

Later, in Chapters 10 and 11, we will study objects that have both rotational and linear motions. In such cases, rotational equilibrium will be as important as translation equilibrium. For now, however, when we say equilibrium, we simply mean translational equilibrium.

As a first example, consider the situation illustrated in **Figure 6–9**. Here we see a person lifting a bucket of water from a well by pulling down on a rope that passes over a pulley. If the bucket's mass is m, and it is rising with constant speed v, what is the tension T_1 in the rope attached to the bucket? In addition, what is the tension T_2 in the wire that supports the pulley?

To answer these questions, we first note that both the bucket and the pulley are in equilibrium; that is, they both have zero acceleration. As a result, the net force on each of them must be zero.

Let's start with the bucket. Its free-body diagram is shown in Figure 6–9, where we see that just two forces act on the bucket: (i) its weight $W = mg$ downward, and (ii) the tension in the rope, T_1 upward. If we take upward to be the positive direction, we can say that for the bucket

$$T_1 - mg = 0$$

Physical picture

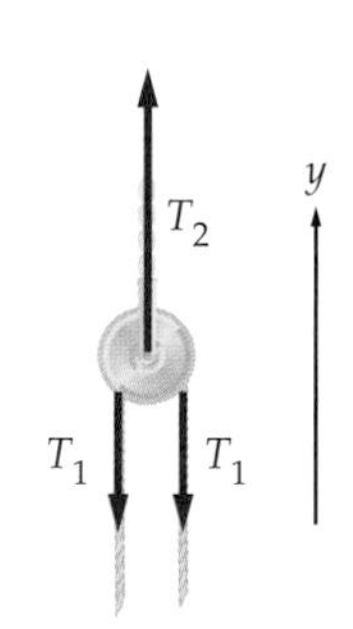

Free-body diagram for pulley

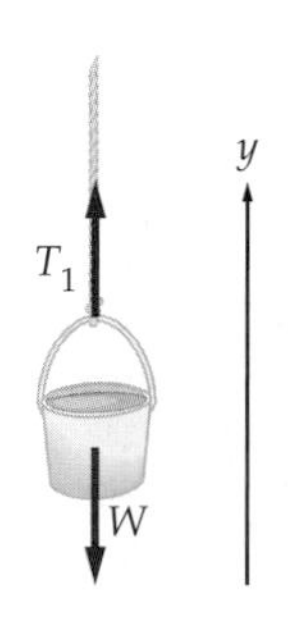

Free-body diagram for bucket

▶ **FIGURE 6–9 Raising a bucket**
A person lifts a bucket of water from the bottom of a well with a constant speed, v. Because the speed is constant, the net force acting on the bucket must be zero.

Thus, the tension in the rope is $T_1 = mg$. Note that this is also the force the person must exert downward on the rope, as expected.

Next, we consider the pulley. In its free-body diagram, also shown in Figure 6–9, we show the three forces that act on it: (i) the tension in the wire, T_2 upward, (ii) the tension in the part of the rope leading to the bucket, T_1 downward, and (iii) the tension in the part of the rope leading to the person, T_1 downward. Note that we don't include the weight of the pulley since we consider it to be ideal; that is, massless and frictionless. If we again take upward to be positive, the statement that the net force on the pulley is zero can be written

$$T_2 - T_1 - T_1 = 0$$

Thus, the tension in the wire is $T_2 = 2T_1 = 2mg$, twice the weight of the bucket of water!

In the next Conceptual Checkpoint we consider a slight variation of this situation.

CONCEPTUAL CHECKPOINT 6–3

A person hoists a bucket of water from a well and holds the rope, keeping the bucket at rest, as at left. A short time later, the person ties the rope to the bucket so that the rope holds the bucket in place, as at right. In this case, is the tension in the rope **(a)** greater than, **(b)** less than, or **(c)** equal to the tension in the first case?

Reasoning and Discussion

In the first case (left), the only upward force exerted on the bucket is the tension in the rope. Since the bucket is at rest, the tension must be equal in magnitude to the weight of the bucket. In the second case (right), the two ends of the rope exert equal upward forces on the bucket, hence the tension in the rope is only half the weight of the bucket. To see this more clearly, imagine cutting the bucket in half so that each end of the rope supports half the weight, as shown schematically in the following diagram.

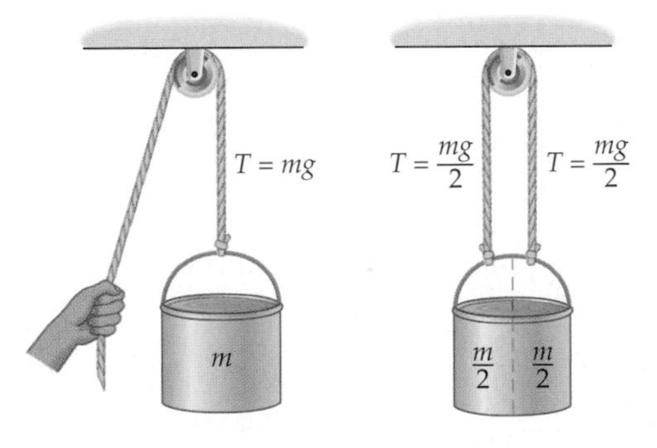

Answer:

(b) The tension in the second case is less than in the first.

In the next two Examples, we consider systems where forces act at various angles with respect to one another. Hence, our first step is to resolve the relevant vectors into their x and y components.

EXAMPLE 6–5 Suspended Vegetation

To hang a 6.20-kg pot of flowers, a gardener uses two wires—one attached horizontally to a wall, the other sloping upward at an angle of $\theta = 40.0°$ and attached to the ceiling. Find the tension in each wire.

Picture the Problem

We choose a typical coordinate system. Notice that $\mathbf{T}_1$ is in the positive x direction, $\mathbf{W}$ is in the negative y direction, and $\mathbf{T}_2$ has a negative x component and a positive y component.

Strategy

The pot is at rest, therefore the net force acting on it is zero. As a result, we can say that (i) $\Sigma F_x = 0$ and (ii) $\Sigma F_y = 0$. These two conditions allow us to determine the magnitude of the two tensions, T_1 and T_2.

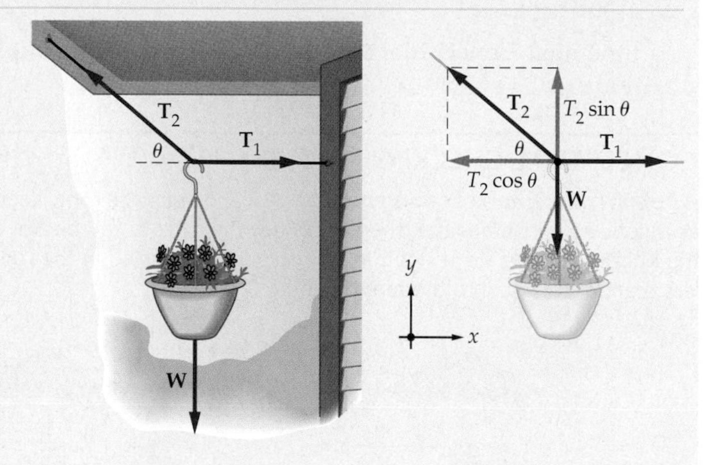

Solution

1. First, resolve each of the forces acting on the pot into x and y components:

$$T_{1,x} = T_1 \qquad T_{1,y} = 0$$
$$T_{2,x} = -T_2 \cos\theta \qquad T_{2,y} = T_2 \sin\theta$$
$$W_x = 0 \qquad W_y = -mg$$

2. Now, set $\Sigma F_x = 0$. Note that this condition gives a relation between T_1 and T_2:

$$\sum F_x = T_{1,x} + T_{2,x} + W_x = T_1 + (-T_2 \cos\theta) + 0 = 0$$
$$T_1 = T_2 \cos\theta$$

3. Next, set $\Sigma F_y = 0$. This time, the resulting condition determines T_2 in terms of the weight, mg:

$$\sum F_y = T_{1,y} + T_{2,y} + W_y = 0 + T_2 \sin\theta + (-mg) = 0$$
$$T_2 \sin\theta = mg$$

4. Use the relation obtained in step 3 to find T_2:

$$T_2 = \frac{mg}{\sin\theta} = \frac{(6.20\text{ kg})(9.81\text{ m/s}^2)}{\sin 40.0°} = 94.6\text{ N}$$

5. Finally, use the connection between the two tensions (obtained from $\Sigma F_x = 0$) to find T_1:

$$T_1 = T_2 \cos\theta = (94.6\text{ N})\cos 40.0° = 72.5\text{ N}$$

Insight

Even though two wires suspend the pot, they both have tensions greater than the pot's weight, $mg = 60.8$ N.

Practice Problem

Find T_1 and T_2 if the second wire slopes upward at the angle **(a)** $\theta = 20°$, **(b)** $\theta = 60.0°$, or **(c)** $\theta = 90.0°$. [**Answer: (a)** $T_1 = 167$ N, $T_2 = 178$ N **(b)** $T_1 = 35.1$ N, $T_2 = 70.2$ N **(c)** $T_1 = 0$, $T_2 = mg = 60.8$ N]

Some related homework problems: Problem 27, Problem 30

ACTIVE EXAMPLE 6–2 Low-tech Laundry

A 1.84-kg bag of clothes pins hangs in the middle of a clothesline, causing it to sag by an angle $\theta = 3.50°$. Find the tension, T, in the clothesline.

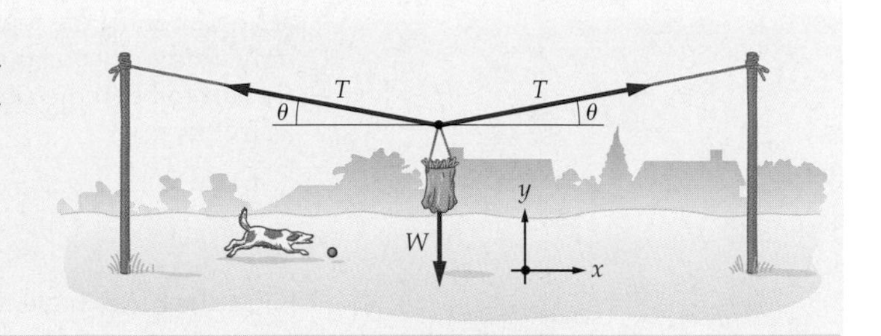

Solution

1. Find the y component for each tension: $T_y = T \sin \theta$
2. Find the y component of the weight: $W_y = -mg$
3. Set $\Sigma F_y = 0$: $T \sin \theta + T \sin \theta - mg = 0$
4. Solve for T: $T = mg/(2 \sin \theta) = 148\ N$

Insight

Note that we only considered the y components of force in our calculation. This is because forces in the x direction automatically balance, due to the symmetry of the system.

At 148 N, the tension in the clothesline is quite large, especially when you consider that the weight of the clothespin bag itself is only 18.1 N. The reason for such a large value is that the vertical component of the two tensions is $2T \sin \theta$, which, for $\theta = 3.50°$, is $(0.122)T$. If $(0.122)T$ is to equal the weight of the bag, it is clear that T must be roughly eight times the bag's weight.

If you and a friend were to pull on the two ends of the clothesline, in an attempt to straighten it out, you would find that no matter how hard you pulled, the line would still sag. You may be able to reduce θ to quite a small value, but as you do so the corresponding tension increases rapidly. In principle, it would take an infinite force to completely straighten the line and reduce θ to zero.

On the other hand, if θ were 90°, so that the two halves of the clothesline were vertical, the tension would be $T = mg/(2 \sin 90°) = mg/2$. In this case, each side of the line supports half the weight of the bag, as expected.

▲ Like the bag of clothespins in Active Example 6–2, this mountain climber is in static equilibrium. Since the ropes suspending the climber are nearly horizontal the tension in them is significantly greater than the climber's weight.

6–4 Connected Objects

Interesting applications of Newton's laws arise when we consider accelerating objects that are tied together. Suppose, for example, that a force of magnitude F pulls two boxes—connected by a string—along a frictionless surface, as in **Figure 6–10**. In such a case, the string has a certain tension, T, and the two boxes have the same acceleration, a. Given the masses of the boxes and the applied force F, we would like to determine both the tension in the string and the acceleration of the boxes.

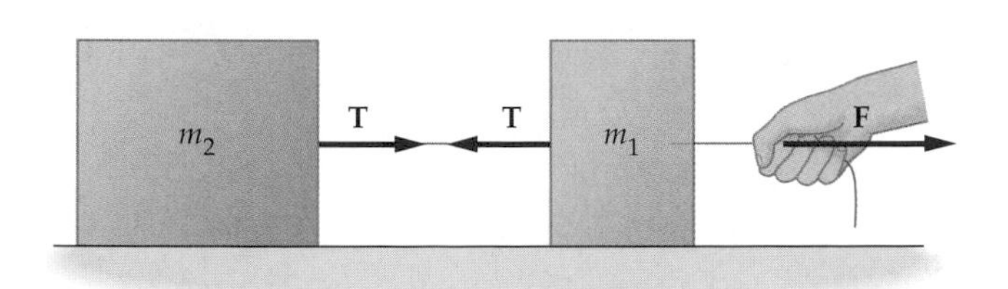

Physical picture

Box 2   Box 1

T   T   F

x

Free-body diagrams
(horizontal components only)

▲ **FIGURE 6–10 Two boxes connected by a string**

The string forces the two boxes to have the same acceleration. This physical connection results in a mathematical connection, as shown in Equation 6–6. Note that in this case we treat each box as a separate system.

First, sketch the free-body diagram for each box. Box 1 has two horizontal forces acting on it: (i) the tension T to the left, and (ii) the force F to the right. Box 2 has only a single horizontal force, the tension T to the right. If we take the positive direction to be to the right, Newton's second law for the two boxes can be written as follows:

$$\begin{aligned} F - T &= m_1 a_1 = m_1 a \qquad \text{box 1} \\ T &= m_2 a_2 = m_2 a \qquad \text{box 2} \end{aligned} \tag{6-6}$$

Since the boxes have the same acceleration, a, we have set $a_1 = a_2 = a$.

Next, we can eliminate the tension T by adding the two equations:

$$\begin{aligned} F - T &= m_1 a \\ T &= m_2 a \\ \hline F &= (m_1 + m_2)a \end{aligned}$$

With this result, it is straightforward to solve for the acceleration in terms of the applied force F:

$$a = \frac{F}{m_1 + m_2} \tag{6-7}$$

Finally, substitute this expression for a into either of the second law equations to find the tension. The algebra is simpler if we use the equation for box 2. We find

$$T = m_2 a = \left(\frac{m_2}{m_1 + m_2}\right)F \tag{6-8}$$

It is left as an exercise to show that the equation for box 1 gives the same expression for T.

A second way to approach this problem is to treat both boxes together as a single system with a mass $m_1 + m_2$, as shown in **Figure 6–11**. The only *external* horizontal force acting on this system is the applied force **F**—the two tension forces are now *internal to the system*, and internal forces are not included when applying Newton's second law. As a result, the horizontal acceleration is simply $F/(m_1 + m_2)$, as given in Equation 6–7. This is certainly a quick way to find the acceleration a, but to find the tension T we must still use one of the relations given in Equation 6–6.

In general, we are always free to choose the "system" any way we like—we can choose any individual object, as when we considered box 1 and box 2 separately, or we can choose all the objects together. The important point is that Newton's second law is equally valid no matter what choice we make for the system, as long as we remember to include only forces *external to that system* in the corresponding free-body diagram.

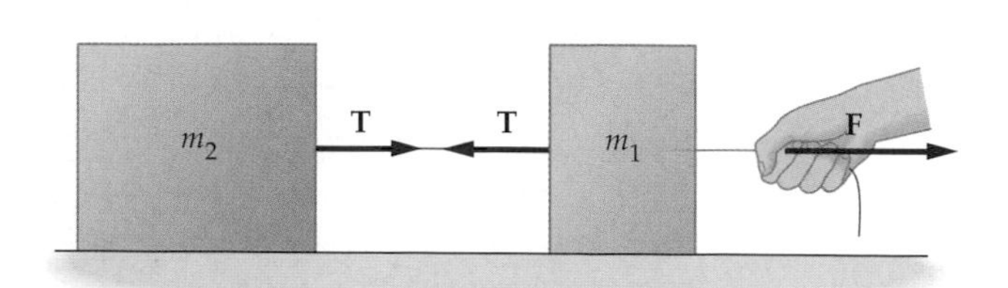

Physical picture

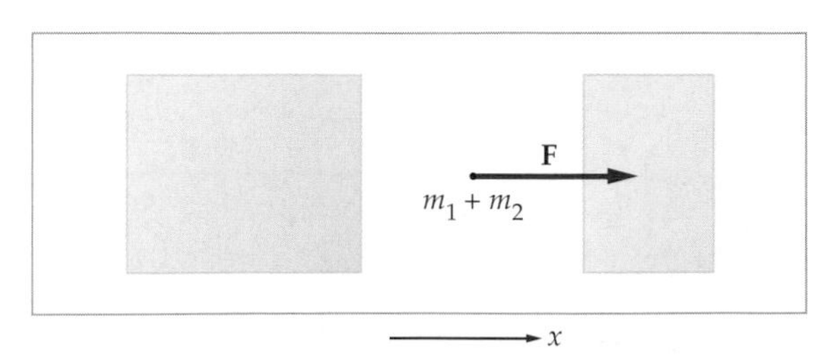

Free-body diagram
(horizontal components only)

▲ **FIGURE 6–11 Two boxes, one system**

In this case we consider the two boxes together as a single system of mass $m_1 + m_2$. The only horizontal force acting on this system is **F**; hence the horizontal acceleration of the system is $a = F/(m_1 + m_2)$, in agreement with Equation 6–7.

CONCEPTUAL CHECKPOINT 6–4

Two masses, m_1 and m_2, are connected by a string that passes over a pulley. Mass m_1 slides without friction on a horizontal tabletop, and mass m_2 falls vertically downward. Both masses move with a constant acceleration of magnitude a. Is the tension in the string **(a)** greater than, **(b)** equal to, or **(c)** less than m_2g?

Reasoning and Discussion

First, note that m_2 accelerates downward, which means that the net force acting on it is downward. Only two forces act on m_2, however: the tension in the string (upward), and its weight (downward). Since the net force is downward, the tension in the string must be less than the weight, m_2g.

A common misconception is that since m_2 has to pull m_1 behind it, the tension in the string must be greater than m_2g. Certainly, attaching the string to m_1 has an effect on the tension. If the string were not attached, for example, its tension would be zero. Hence, m_2 pulling on m_1 increases the tension to a value greater than zero, though still less than m_2g.

Answer:

(c) The tension in the string is less than m_2g.

In the next Example, we verify the qualitative conclusions given in the Conceptual Checkpoint with a detailed calculation. But first, a note about choosing a coordinate system for a problem such as this. Rather than apply the same coordinate system to both masses, it is useful to take into consideration the fact that a pulley simply changes the direction of the tension in a string. With this in mind, we choose a set of axes that "follow the motion" of the string, so that both masses accelerate in the positive x direction with accelerations of equal magnitude. Example 6–6 illustrates the use of this type of coordinate system.

PROBLEM SOLVING NOTE
Choice of Coordinate System: Connected Objects

If two objects are connected by a string passing over a pulley, let the coordinate system follow the direction of the string. With this choice, both objects have accelerations of the same magnitude and in the same coordinate direction.

EXAMPLE 6–6 Connected Blocks

A block of mass m_1 slides on a frictionless tabletop. It is connected to a string that passes over a pulley and suspends a mass m_2. Find the acceleration of the masses and the tension in the string.

Picture the Problem

Our coordinate system follows the motion of the string so that both masses move in the positive x direction. Since the masses are connected, their accelerations have the same magnitude. Thus, $a_{1,x} = a_{2,x} = a$.

Strategy

Applying Newton's second law to the two masses yields the following relations: For mass 1, $\Sigma F_{1,x} = T = m_1a_{1,x} = m_1a$ and for mass 2, $\Sigma F_{2,x} = m_2g - T = m_2a_{2,x} = m_2a$. These two equations can be solved for the two unknowns, a and T.

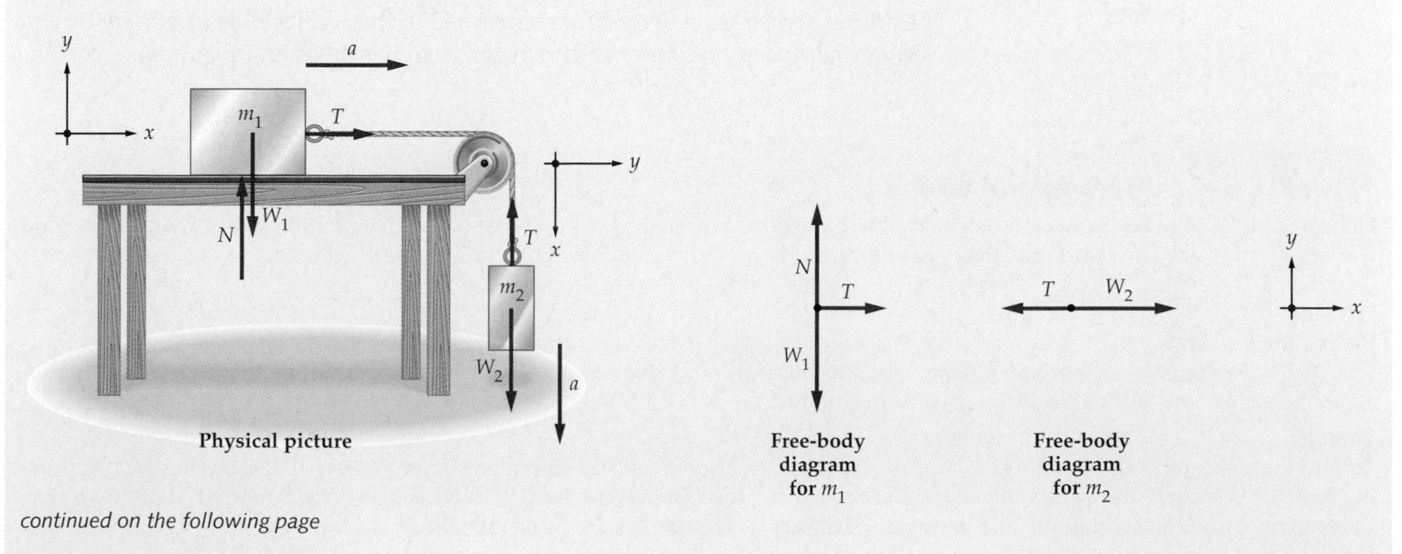

continued on the following page

continued from the previous page

Solution

1. First, write $\Sigma F_{1,x} = m_1 a$. Note that the only force acting on m_1 in the x direction is T:

$$\sum F_{1,x} = T = m_1 a$$
$$T = m_1 a$$

2. Next, write $\Sigma F_{2,x} = m_2 a$. In this case, two forces act in the x direction: $W_2 = m_2 g$ (positive direction) and T (negative direction):

$$\sum F_{2,x} = m_2 g - T = m_2 a$$
$$m_2 g - T = m_2 a$$

3. Sum the two relations obtained to eliminate T:

$$T = m_1 a$$
$$\underline{m_2 g - T = m_2 a}$$
$$m_2 g = (m_1 + m_2) a$$

4. Solve for a:

$$a = \left(\frac{m_2}{m_1 + m_2}\right) g$$

5. Substitute a into the first relation ($T = m_1 a$) to find T:

$$T = m_1 a = \left(\frac{m_1 m_2}{m_1 + m_2}\right) g$$

Insight

We could just as well have determined T using $m_2 g - T = m_2 a$, though the algebra is a bit messier. Also, note that $a = 0$ if $m_2 = 0$, and that $a = g$ if $m_1 = 0$, as expected. Similarly, $T = 0$ if either m_1 or m_2 is zero. This type of check, where you connect equations with physical situations, is one of the best ways to increase your understanding of physics.

Practice Problem

Find the tension for the case $m_1 = 1.50$ kg and $m_2 = 0.750$ kg, and compare the tension to $m_2 g$. [**Answer:** $a = 3.27\ \text{m/s}^2$, $T = 4.91\ \text{N} < m_2 g = 7.36\ \text{N}$]

Some related homework problems: Problem 34, Problem 38

Conceptual Checkpoint 6–4 shows that the tension in the string should be less than $m_2 g$. Let's rewrite our solution for T to show that this is indeed the case. From Example 6–6 we have

$$T = \left(\frac{m_1 m_2}{m_1 + m_2}\right) g = \left(\frac{m_1}{m_1 + m_2}\right) m_2 g$$

Since the ratio $m_1/(m_1 + m_2)$ is always less than 1 (as long as m_2 is nonzero), it follows that $T < m_2 g$, as expected.

We conclude this section with a classic system that can be used to measure the acceleration of gravity. It is referred to as Atwood's machine, and it is basically two blocks of different mass connected by a string that passes over a pulley. The resulting acceleration of the blocks is related to the acceleration of gravity by a relatively simple expression, which we derive in the following Example.

EXAMPLE 6–7 Atwood's Machine

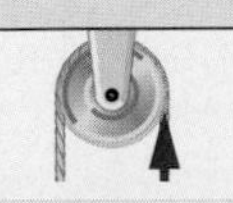

Atwood's machine consists of two masses connected by a string that passes over a pulley, as shown below. Find the acceleration of the masses for general m_1 and m_2, and evaluate for the case $m_1 = 3.1$ kg, $m_2 = 4.4$ kg.

Picture the Problem

Our sketch shows Atwood's machine, along with our choice of coordinate directions for the two blocks. Note that both blocks accelerate in the x direction with accelerations of equal magnitude.

Strategy

To find the acceleration of the blocks we follow the same strategy given in the previous Example. In particular, we start by applying Newton's second law to each block individually, using the fact that $a_{1,x} = a_{2,x} = a$. This gives two equations, both involving the tension T and the acceleration a. Eliminating T allows us to solve for the acceleration.

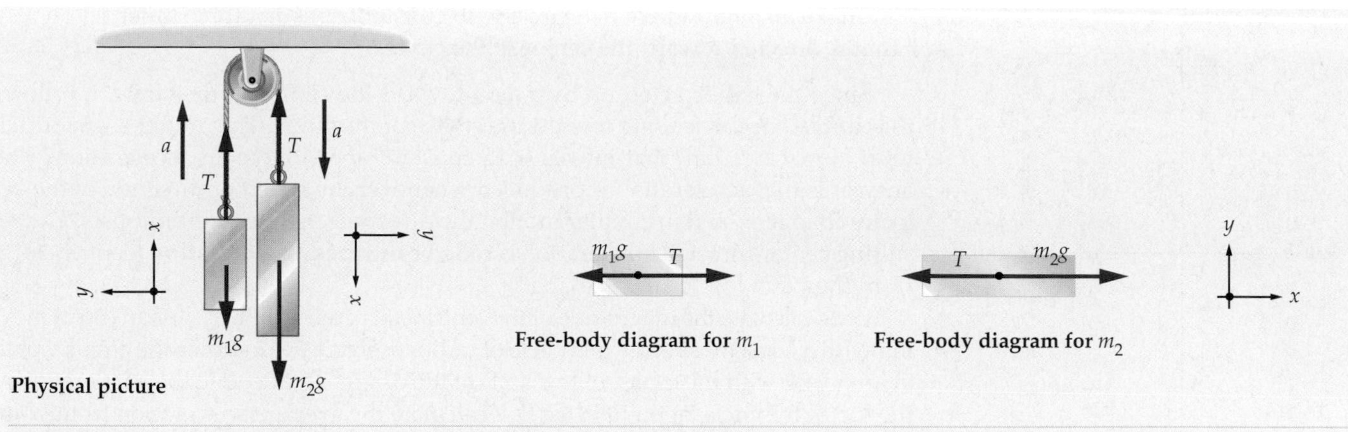

Solution

1. Begin by writing out the expression $\Sigma F_{1,x} = m_1 a$. Note that two forces act in the x direction: T (positive direction) and $m_1 g$ (negative direction):

$$\sum F_{1,x} = T - m_1 g = m_1 a$$

2. Next, write out $\Sigma F_{2,x} = m_2 a$. The two forces acting in the x direction in this case are $m_2 g$ (positive direction) and T (negative direction):

$$\sum F_{2,x} = m_2 g - T = m_2 a$$

3. Sum the two relations obtained above to eliminate T:

$$\begin{aligned} T - m_1 g &= m_1 a \\ m_2 g - T &= m_2 a \\ \hline (m_2 - m_1)g &= (m_1 + m_2)a \end{aligned}$$

4. Solve for a:

$$a = \left(\frac{m_2 - m_1}{m_1 + m_2}\right)g$$

5. To evaluate the acceleration, substitute numerical values for the masses and for g:

$$a = \left(\frac{m_2 - m_1}{m_1 + m_2}\right)g = \left(\frac{4.4\text{ kg} - 3.1\text{ kg}}{3.1\text{ kg} + 4.4\text{ kg}}\right)(9.81\text{ m/s}^2) = 1.7\text{ m/s}^2$$

Insight

Since m_2 is greater than m_1, we find that the acceleration is positive, meaning that the masses accelerate in the positive x direction. On the other hand, if m_1 were greater than m_2, we would find that a is negative, indicating that the masses accelerate in the negative x direction, as expected.

Practice Problem

If m_1 is increased by a small amount, does the acceleration of the blocks increase, decrease, or stay the same? Check your answer by evaluating the acceleration for $m_1 = 3.3$ kg. [**Answer:** If m_1 is increased only slightly the acceleration will decrease. For $m_1 = 3.3$ kg we find $a = 1.4\text{ m/s}^2$.]

Some related homework problems: Problem 38, Problem 40

6–5 Circular Motion

According to Newton's second law, if no force acts on an object, it will move with constant speed in a constant direction. A force is required to change the speed, the direction, or both. For example, if you drive a car with constant speed on a circular track, the direction of the car's motion changes continuously. A force must act on the car to cause this change in direction. We would like to know two things about a force that causes circular motion: (i) what is its direction, and (ii) what is its magnitude?

First, let's consider the direction of the force. Imagine swinging a ball tied to a string in a circle about your head, as shown in **Figure 6–12**. As you swing the ball you feel a tension in the string pulling outward. Of course, on the other end of the string, where it attaches to the ball, the tension pulls inward, toward the center of the circle. Thus, the force the ball feels is a force that is always directed toward the center of the circle.

▲ **FIGURE 6–12 Swinging a ball in a circle**
The tension in the string pulls outward on the person's hand and pulls inward on the ball.

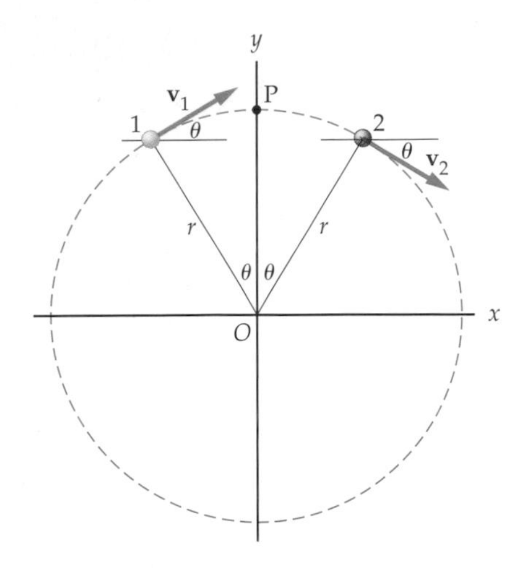

▲ **FIGURE 6–13 A particle moving in a circular path centered on the origin.** The speed of the particle is constant.

To make an object move in a circle with constant speed, a force must act on it that is directed toward the center of the circle.

Since the ball is acted on by a *force* toward the center of the circle, it follows that it must be *accelerating* toward the center of the circle. This might seem odd at first: How can a ball that moves with constant speed have an acceleration? The answer is that acceleration is produced whenever the speed or direction of the velocity changes—and in circular motion the direction changes continuously. The resulting center-directed acceleration is called **centripetal acceleration** (centripetal is from the Latin for "center seeking").

Let's calculate the magnitude of the centripetal acceleration, a_{cp}, for an object moving with a constant speed v in a circle of radius r. **Figure 6–13** shows the circular path of an object, with the center of the circle at the origin. To calculate the acceleration at the top of the circle, at point P, we first calculate the average acceleration from point 1 to point 2:

$$\mathbf{a}_{av} = \frac{\Delta \mathbf{v}}{\Delta t} = \frac{\mathbf{v}_2 - \mathbf{v}_1}{\Delta t} \qquad \text{6–9}$$

The instantaneous acceleration at P is the limit of $\mathbf{a}_{av}$ as points 1 and 2 move closer to P.

Referring to the figure, we see that $\mathbf{v}_1$ is at an angle θ above the horizontal, and $\mathbf{v}_2$ is at an angle θ below the horizontal. Both $\mathbf{v}_1$ and $\mathbf{v}_2$ have a magnitude v. Therefore, we can write the two velocities in vector form as follows:

$$\mathbf{v}_1 = (v \cos \theta)\hat{\mathbf{x}} + (v \sin \theta)\hat{\mathbf{y}}$$
$$\mathbf{v}_2 = (v \cos \theta)\hat{\mathbf{x}} + (-v \sin \theta)\hat{\mathbf{y}}$$

Substituting these results into $\mathbf{a}_{av}$ gives

$$\mathbf{a}_{av} = \frac{\mathbf{v}_2 - \mathbf{v}_1}{\Delta t} = \frac{-2v \sin \theta}{\Delta t}\hat{\mathbf{y}} \qquad \text{6–10}$$

TABLE 6–2
$\frac{\sin \theta}{\theta}$ for Values of θ Approaching Zero

θ, radians	$\frac{\sin \theta}{\theta}$
1.00	0.841
0.500	0.959
0.250	0.990
0.125	0.997
0.0625	0.999

Note that $\mathbf{a}_{av}$ points in the negative y direction, which, at point P, is toward the center of the circle.

To complete the calculation we need Δt, the time it takes the object to go from point 1 to point 2. Since the object's speed is v, and the distance from point 1 to point 2 is $d = r(2\theta)$ where θ is measured in radians (see Chapter 1 for a discussion of radians and degrees), we find

$$\Delta t = \frac{d}{v} = \frac{2r\theta}{v} \qquad \text{6–11}$$

Combining this result for Δt with the previous result for $\mathbf{a}_{av}$ gives

$$\mathbf{a}_{av} = \frac{-2v \sin \theta}{(2r\theta/v)}\hat{\mathbf{y}} = -\frac{v^2}{r}\left(\frac{\sin \theta}{\theta}\right)\hat{\mathbf{y}} \qquad \text{6–12}$$

To find $\mathbf{a}$ at the point P we let points 1 and 2 approach P, which means letting θ go to zero. Table 6–2 shows that as θ goes to zero ($\theta \rightarrow 0$) the ratio $(\sin \theta)/\theta$ goes to 1:

$$\frac{\sin \theta}{\theta} \xrightarrow[\text{as } \theta \to 0]{} 1$$

Finally, then, the acceleration at point P is

$$\mathbf{a} = -\frac{v^2}{r}\hat{\mathbf{y}} = -a_{cp}\,\hat{\mathbf{y}} \qquad \text{6–13}$$

As mentioned, the direction of the acceleration is toward the center of the circle, and now we see that its magnitude is

$$a_{cp} = \frac{v^2}{r} \qquad \text{6–14}$$

We can summarize these results as follows:

▲ The people enjoying this carnival ride are expecting a centripetal acceleration of roughly 10 m/s^2 directed inward, toward the axis of rotation. The force needed to produce this acceleration, which keeps the riders moving in a circular path, is provided by the horizontal component of the tension in the chains.

- When an object moves in a circle of radius r with constant speed v, its centripetal acceleration is $a_{cp} = v^2/r$.
- A force must be applied to an object to give it circular motion. For an object of mass m, the force must have a magnitude

$$f_{cp} = ma_{cp} = m\frac{v^2}{r} \quad \textbf{6–15}$$

and must be directed toward the center of the circle.

Note that the **centripetal force**, f_{cp}, can be produced in any number of ways. For example, f_{cp} might be the tension in a string, as in the example with the ball, or it might be due to friction between tires and the road, as when a car turns a corner. In addition, f_{cp} could be the force of gravity causing a satellite, or the Moon, to orbit the Earth. Thus, f_{cp} is a force that must be present to cause circular motion, but the specific cause of f_{cp} varies from system to system.

We now show how these results for centripetal force and centripetal acceleration can be applied in practice.

PROBLEM SOLVING NOTE
Choice of Coordinate System: Circular Motion

In circular motion, it is convenient to choose the coordinate system so that one axis points toward the center of the circle. Then, we know that the acceleration in that direction must be $a_{cp} = v^2/r$.

EXAMPLE 6–8 Rounding a Corner

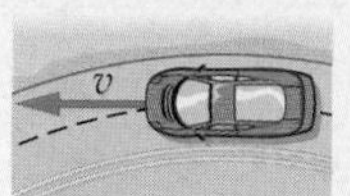

A 1200-kg car rounds a corner of radius $r = 45$ m. If the coefficient of static friction between the tires and the road is $\mu_s = 0.82$, what is the greatest speed the car can have in the corner without skidding?

Picture the Problem

In the first sketch we show a bird's-eye view of the car as it moves along its circular path. The next sketch shows the car moving directly toward the observer. We also indicate the three forces acting on the car: gravity, **W**; the normal force, **N**; and the force of static friction, $\mathbf{f}_s$. Note that we choose the x direction toward the center of the circular path, and the y axis is vertical.

Strategy

In this system, the force of static friction provides the centripetal force required for the car to move in a circular path. If the car moves faster, more centripetal force (i.e., more friction) is required. Thus, the greatest speed for the car corresponds to the maximum static friction, $f_s = \mu_s N$. Hence, if we set $\mu_s N$ equal to the centripetal force, $ma_{cp} = mv^2/r$, we can solve for v.

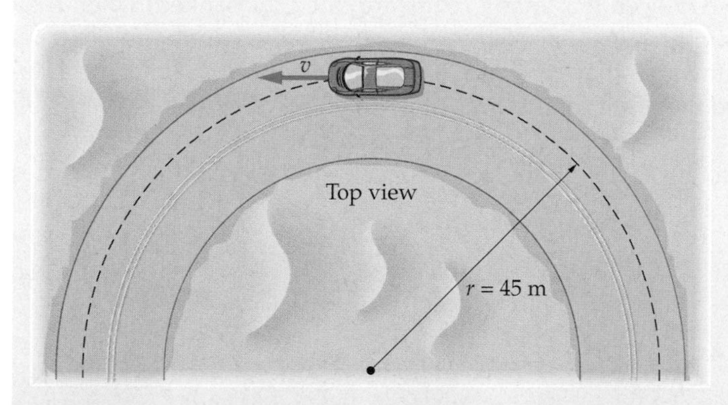

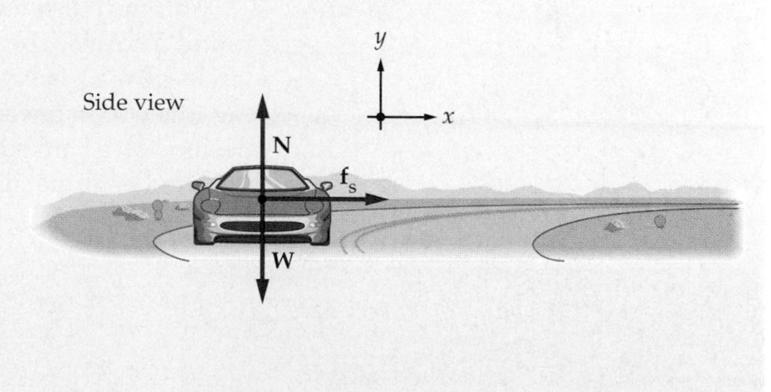

Solution

1. Sum the x components of force to relate the force of static friction to the centripetal acceleration of the car: $\quad \sum F_x = f_s = ma_x$

   Since the car moves in a circular path, with the center of the circle in the x direction, it follows that $a_x = a_{cp} = v^2/r$. Make this substitution, along with $f_s = \mu_s N$ for the force of static friction: $\quad \mu_s N = ma_{cp} = m\dfrac{v^2}{r}$

2. Next, set the sum of the y components of force equal to zero (since $a_y = 0$) to find the normal force, N: $\quad \sum F_y = N - W = ma_y = 0$

   Solve for the normal force: $\quad N = W = mg$

continued on the following page

continued from the previous page

3. Substitute $N = mg$ in step 1 and solve for v. Notice that the mass of the car cancels:

$$\mu_s mg = m\frac{v^2}{r}$$

$$v = \sqrt{\mu_s rg}$$

4. Substitute numerical values to determine v:

$$v = \sqrt{(0.82)(45\text{ m})(9.81\text{ m/s}^2)} = 19\text{ m/s}$$

Insight

Note that the maximum speed is less if the radius is smaller (tighter corner) or if μ_s is smaller (slick road). The mass of the vehicle, however, is irrelevant.

Practice Problem

Suppose the situation described in this Example takes place on the Moon, where the acceleration of gravity is less than it is on Earth. If a lunar rover goes around this same corner, is its maximum speed greater than, less than, or the same as the result found in step 4? To check your answer, find the maximum speed for a lunar rover when it rounds a corner with $r = 45$ m and $\mu_s = 0.82$. (On the Moon, $g = 1.62\text{ m/s}^2$.) [**Answer:** The maximum speed will be less. On the Moon we find $v = 7.7$ m/s.]

Some related homework problems: Problem 41, Problem 43, Problem 47

If you try to round a corner too rapidly you may experience a skid; that is, your car may begin to slide sideways across the road. A common bit of road wisdom is that you should turn in the direction of the skid to regain control—which, to most people, sounds counterintutitive. The advice is sound, however. Suppose, for example, that you are turning to the left and begin to skid to the right. If you turn more sharply to the left to try to correct for the skid, you simply reduce the turning radius of your car, r. The result is that the centripetal acceleration, v^2/r, becomes larger, and an even larger force would be required from the road to make the turn. The tendency to skid would therefore be increased. On the other hand, if you turn slightly to the right when you start to skid, you *increase* your turning radius and the centripetal acceleration decreases. In this case your car may stop skidding, and you can then regain control of your vehicle.

You may also have noticed that many roads are tilted, or banked, when they round a corner. The same type of banking is observed on many automobile racetracks as well. Next time you drive around a banked curve, notice that the banking tilts you in toward the center of the circular path you are following. This is by design. On a banked curve, the normal force exerted by the road contributes to the required centripetal force. If the tilt angle is just right, the normal force provides

▲ The steeply banked track at the Talledega Speedway in Alabama (left) helps to keep the rapidly moving cars from skidding off along a tangential path. Even when there is no solid roadway, however, banking can still help—airplanes bank when making turns (center) to keep from "skidding" sideways. Banking is beneficial in another way as well. Occupants of cars on a banked roadway or of a banking airplane feel no sideways force when the banking angle is just right, so turns become a safer and more comfortable experience. For this reason, some trains use hydraulic suspension systems to bank when rounding corners (right), even though the tracks themselves are level.

all of the centripetal force so that the car can negotiate the curve even if there is no friction between its tires and the road. The next Example determines the optimum banking angle for a given speed and given radius of turn.

EXAMPLE 6–9 Bank on It

If a roadway is banked at the proper angle, a car can round a corner without any assistance from friction between the tires and the road. Find the appropriate banking angle for a 900-kg car traveling at 20.5 m/s in a turn of radius 85.0 m.

Picture the Problem

A car on a properly banked roadway is acted on by only two forces: its weight, **W**, and the normal force, **N**—there is no friction in this case. Note that we choose the y axis to be vertical and the x axis to point toward the center of the circular path.

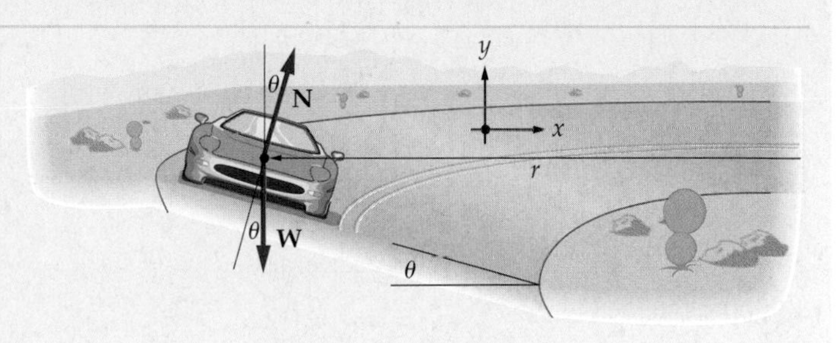

Strategy

In order for the car to move in a circular path, there must be a force acting on it in the positive x direction. Since the weight **W** has no x component, it follows that the normal force **N** must supply the needed centripetal force. Thus, we find N by setting $\Sigma F_y = ma_y = 0$, since there is no motion in the y direction. Then we use N in $\Sigma F_x = ma_x = mv^2/r$ to find the angle θ.

Solution

1. Start by determining N from the condition $\Sigma F_y = 0$:

$$\sum F_y = N\cos\theta - W = 0$$
$$N = \frac{W}{\cos\theta} = \frac{mg}{\cos\theta}$$

2. Next, set $\Sigma F_x = mv^2/r$:

$$\sum F_x = N\sin\theta$$
$$= ma_x = ma_{\text{cp}} = m\frac{v^2}{r}$$

3. Substitute $N = mg/\cos\theta$ (from $\Sigma F_y = 0$) and solve for θ, using the fact that $\sin\theta/\cos\theta = \tan\theta$. Notice that, once again, the mass of the car cancels:

$$N\sin\theta = \frac{mg}{\cos\theta}\sin\theta = m\frac{v^2}{r}$$
$$\tan\theta = \frac{v^2}{gr} \qquad \theta = \tan^{-1}\left(\frac{v^2}{gr}\right)$$

4. Substitute numerical values to determine θ:

$$\theta = \tan^{-1}\left[\frac{(20.5\text{ m/s})^2}{(9.81\text{ m/s}^2)(85.0\text{ m})}\right] = 26.7°$$

Insight

The banking angle increases with increasing speed and with decreasing radius of turn, as expected.

Practice Problem

A turn of radius 65 m is banked at 30.0°. What speed should a car have in order to make the turn with no assistance from friction? [**Answer:** $v = 19$ m/s]

Some related homework problems: Problem 50, Problem 84

If you've ever driven through a dip in the road, you know that you feel momentarily heavier near the bottom of the dip. This change in apparent weight is due to the approximately circular motion of the car, as we show next.

ACTIVE EXAMPLE 6–3 Don't Bottom Out

While driving along a country lane with a constant speed of 17.0 m/s, you encounter a dip in the road. The dip can be approximated as a circular arc, with a radius of 65.0 m. What is the normal force exerted by a car seat on a 80.0-kg passenger when the car is at the bottom of the dip?

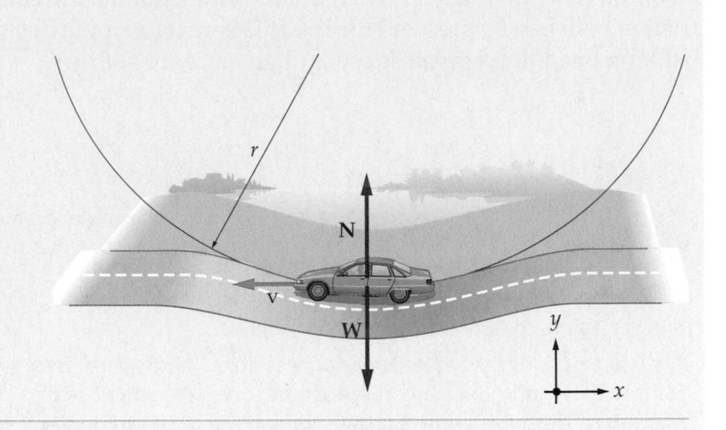

Solution

1. Write $\Sigma F_y = ma_y$ for the passenger: $N - mg = ma_y$
2. Replace a_y with the centripetal acceleration: $a_y = v^2/r$
3. Solve for N: $N = mg + mv^2/r$
4. Substitute numerical values: $N = 1140 \text{ N}$

Insight

At the bottom of the dip the normal force is greater than the weight of the passenger, since it must also supply the centripetal force. As a result, the passenger feels heavier than usual. In this case, the 80.0-kg passenger feels as if his mass has increased by 45 percent, to 116 kg!

A similar calculation can be applied to a car going over the top of a bump. In that case, circular motion results in a reduced apparent weight.

REAL WORLD PHYSICS: BIO Centrifuges and ultracentrifuges

Finally, we determine the acceleration produced in a **centrifuge**, a common device in biological and medical laboratories that uses large centripetal accelerations to perform such tasks as separating red and white blood cells from serum. A simplified top view of a centrifuge is shown in **Figure 6–14**.

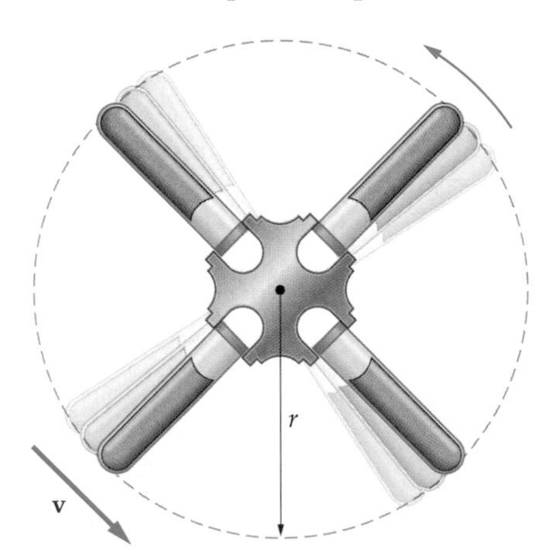

▲ **FIGURE 6–14 Simplified top view of a centrifuge in operation**

▲ A laboratory centrifuge of the kind commonly used to separate blood components.

EXERCISE 6–2

The centrifuge in Figure 6–14 rotates at a rate that gives the bottom of the test tube a linear speed of 89.3 m/s. If the bottom of the test tube is 8.50 cm from the axis of rotation, what is the centripetal acceleration experienced there?

Solution

Applying the relation $a_{cp} = v^2/r$ yields

$$a_{cp} = \frac{v^2}{r} = \frac{(89.3\ \text{m/s})^2}{0.0850\ \text{m}} = 93{,}800\ \text{m/s}^2 = 9560\ g$$

In this expression, g is the acceleration of gravity, 9.81 m/s^2.

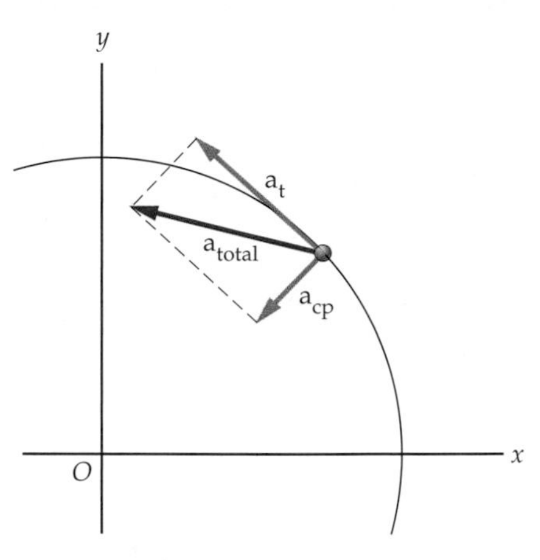

▲ **FIGURE 6–15 A particle moving in a circular path with tangential acceleration**

In this case, the particle's speed is increasing at the rate given by a_t.

Thus, a centrifuge can produce centripetal accelerations that are many thousand times greater than the acceleration of gravity. In fact, devices referred to as **ultracentrifuges** can produce accelerations as great as 1 million g. Even in the relatively modest case considered in Exercise 6–2 the forces involved in a centrifuge can be quite significant. For example, if the contents of the test tube have a mass of 12.0 g, the centripetal force that must be exerted by the bottom of the tube is (0.0120 kg)(9560 g) = 1130 N, or about 250 lbs!

Finally, an object moving in a circular path may increase or decrease its speed. In such a case, the object has both an acceleration tangential to its path that changes its speed, $\mathbf{a}_t$, plus a centripetal acceleration perpendicular to its path, $\mathbf{a}_{cp}$, that changes its direction of motion. Such a situation is illustrated in Figure 6–15. The total acceleration of the object is the vector sum of $\mathbf{a}_t$ and $\mathbf{a}_{cp}$. We will explore this case more fully in Chapter 10.

Chapter Summary

	Topic	Remarks and Relevant Equations
6–1	**Frictional Forces**	Frictional forces are due to the microscopic roughness of surfaces in contact. As a rule of thumb, friction is independent of the area of contact and independent of the relative speed of the surfaces.
	kinetic friction	Friction experienced by surfaces that are in contact and moving relative to one another. The force of kinetic friction is given by
		$f_k = \mu_k N$ **6–1**
		In this expression, μ_k is the coefficient of kinetic friction and N is the magnitude of the normal force.
	static friction	Friction experienced by surfaces that are in static contact. The maximum force of static friction is given by
		$f_{s,\max} = \mu_s N$ **6–3**
		In this expression, μ_s is the coefficient of static friction and N is the magnitude of the normal force. The force of static friction can have any magnitude between zero and its maximum value.
6–2	**Strings and Springs**	Strings and springs provide a common way of exerting forces on objects. Ideal strings and springs are massless.
	tension	The force transmitted through a string. The tension is the same throughout the length of an ideal string.
	Hooke's law	The force exerted by an ideal spring stretched by the amount x is
		$F_x = -kx$ **6–4**
		In words, the force exerted by a spring is proportional to the amount of stretch or compression, and is in the opposite direction.

6–3 Translational Equilibrium

An object is in translational equilibrium if the net force acting on it is zero. Equivalently, an object is in equilibrium if it has zero acceleration.

6–4 Connected Objects

Connected objects are linked physically, and hence they are linked mathematically as well. For example, objects connected by strings have the same magnitude of acceleration.

6–5 Circular Motion

An object moving with speed v in a circle of radius r has an acceleration directed toward the center of the circle of magnitude v^2/r. This is referred to as the centripetal acceleration, a_{cp}.

If the object has a mass m, the force required for the circular motion is

$$f_{cp} = ma_{cp} = mv^2/r. \quad \text{6–15}$$

Problem-Solving Summary

Type of Calculation	Relevant Physical Concepts	Related Examples
Find the acceleration when kinetic friction is present.	First, find the magnitude of the normal force, N. The corresponding kinetic friction has a magnitude of $f_k = \mu_k N$ and points opposite to the direction of motion. Include this force with the others when applying Newton's second law.	Examples 6–1, 6–2
Solve problems involving static friction.	Start by finding the magnitude of the normal force, N. The corresponding static friction has a magnitude between zero and $\mu_s N$. Its direction opposes motion.	Example 6–3 Active Example 6–1
Find the acceleration and the tension for masses connected by a string.	Apply Newton's second law to each mass separately. This generates two equations, which can be solved for the two unknowns, a and T.	Examples 6–6, 6–7
Solve problems involving circular motion.	Set up the coordinate system so that one axis points to the center of the circle. When applying Newton's second law to that direction, set the acceleration equal to $a_{cp} = v^2/r$.	Examples 6–8, 6–9 Active Example 6–3

Conceptual Questions

1. A clothesline always sags a little, even if nothing hangs from it. Explain.
2. In the *Jurassic Park* sequel, *The Lost World*, a man tries to keep a large vehicle from going over a cliff by connecting a cable from his Jeep to the vehicle. The man then puts the Jeep in gear and spins the rear wheels. Do you expect that spinning the tires will increase the force exerted by the Jeep on the vehicle? Why or why not?
3. An object moves on a flat surface with an acceleration of constant magnitude. If the acceleration is always perpendicular to the direction of motion, what is the shape of the object's path? Does the object's speed change? Explain.
4. In a car with rear-wheel drive, the maximum acceleration is often less than the maximum deceleration. Why?
5. A train typically requires a much greater distance to come to rest, for a given initial speed, than does a car. Why?
6. Give some everyday examples of situations where friction is beneficial.
7. At the local farm you buy a flat of strawberries and place them on the backseat of the car. On the way home you begin to brake as you approach a stop sign. At first the strawberries stay put, but as you brake a bit harder, they begin to slide off the seat. Explain.
8. It is possible to spin a bucket of water in a vertical circle, and yet have none of the water spill when the bucket is upside down. How would you explain this to members of your family?
9. When a traffic accident is investigated, it is common for the length of the skid marks to be measured. How could this information be used to estimate the initial speed of the vehicle that left the skid marks?
10. If you weigh yourself at the equator you get a smaller value than if you weigh yourself at one of the poles. Why?
11. Water sprays off a rapidly turning bicycle wheel. Why?
12. Can an object be in equilibrium if it is moving? Explain.
13. Referring to Example 6–7, suppose both m_1 and m_2 are increased by the same amount. In this case, does the acceleration of the blocks increase, decrease, or stay the same? Explain.
14. Suppose you stand on a bathroom scale and get a reading of 900 N. In principle, would the scale read more, less, or the same if the Earth did not rotate?
15. In a dramatic circus act, a motorcyclist drives his bike around the inside of a vertical circle. How is this possible, considering that the motorcycle is upside down at the top of the circle?
16. The gravitational attraction of the Earth is only slightly less at the altitude of an orbiting spacecraft than it is on the Earth's surface. Why is it, then, that astronauts feel weightless?
17. A popular carnival ride has passengers stand with their backs against the inside wall of a cylinder. As the cylinder begins to spin, the passengers feel as if they are being pushed against the wall. Explain.

18. Referring to the previous question, after the cylinder reaches operating speed, the floor is lowered away, leaving the passengers "stuck" to the wall. Explain.
19. To stop a car in the minimum possible distance, is it wise to lock up the wheels and screech to a halt, or brake just to the verge of locking so that the wheels continue to turn? Explain.
20. Discuss the physics involved in the spin cycle of a washing machine. In particular, how is circular motion related to the removal of water from the clothes?
21. Your car is stuck on an icy side street. Some students on their way to class see your predicament and help out by sitting on the trunk of your car to increase its traction. Why does this help?
22. A certain spring has a force constant k. If this spring is cut in half, does the resulting half-spring have a force constant equal to k, greater than k, or less than k? Explain.
23. The gas pedal and the brake pedal are capable of causing a car to accelerate. Can the steering wheel also produce an acceleration? Explain.
24. In the movie *2001: A Space Odyssey*, a rotating space station provides "artificial gravity" for its inhabitants. How does this work?

▲ The rotating space station from the movie *2001: A Space Odyssey* (Question 24)

25. A car drives with constant speed around a circular track. **(a)** Is its velocity constant? **(b)** Is the magnitude of its acceleration constant? **(c)** Is the direction of its acceleration constant?
26. When rounding a corner on a bicycle or a motorcycle, the driver leans inward, toward the center of the circle. Why?

Problems

Note: **IP** *denotes an integrated conceptual/quantitative problem.* **BIO** *identifies problems of biological or medical interest. Blue bullets (•, ••, •••) are used to indicate the level of difficulty of each problem.*

Section 6–1 Frictional Forces

1. • A baseball player dives into third base with an initial speed of 7.90 m/s. If the coefficient of kinetic friction between the player and the ground is 0.41, how far does the player slide before coming to rest?
2. • A child goes down a playground slide with an acceleration of 1.05 m/s^2. Find the coefficient of kinetic friction between the child and the slide if the slide is inclined at an angle of 35.0° below the horizontal.
3. • Hopping into your Porsche, you floor it and accelerate at 12 m/s^2 without spinning the tires. Determine the minimum coefficient of static friction between the tires and the road needed to make this possible.
4. • When you push a 1.80-kg book resting on a tabletop it takes 2.25 N to start the book sliding. Once it is sliding, however, it takes only 1.50 N to keep the book moving with constant speed. What are the coefficients of static and kinetic friction between the book and the tabletop?
5. • In the previous problem, what is the frictional force exerted on the book when you push on it with a force of 0.75 N?
6. •• **IP** A tie is laid out on a table, with a fraction of its length hanging over the edge. Initially, the tie is at rest. **(a)** If the fraction hanging from the table is increased, the tie eventually slides to the ground. Explain. **(b)** What is the coefficient of static friction between the tie and the table if the tie begins to slide when one-fourth of its length hangs over the edge?
7. •• To move a large crate across a rough floor, you push down on it at an angle of 21°, as shown in **Figure 6–16**. Find the force necessary to start the crate moving, given that the mass of the crate is 32 kg and the coefficient of static friction between the crate and the floor is 0.57.

▲ **FIGURE 6–16** Problem 7

8. •• In the previous problem, find the acceleration of the crate if the applied force is 330 N and the coefficient of kinetic friction is 0.45.
9. •• A 45-kg crate is placed on an inclined ramp. When the angle the ramp makes with the horizontal is increased to 23° the crate begins to slide downward. **(a)** What is the coefficient of static friction between the crate and the ramp? **(b)** At what angle does the crate begin to slide if its mass is doubled?
10. •• **IP** A 95-kg sprinter wishes to accelerate from rest to a speed of 12 m/s in a distance of 20 m. **(a)** What coefficient of static friction is required between the sprinter's shoes and the track? **(b)** Explain the strategy used to find the answer to part (a).
11. •• A person places a cup of coffee on the roof of her car while she dashes back into the house for a forgotten item. When she returns to the car she hops in and takes off with the coffee cup still on the roof. **(a)** If the coefficient of static friction between the

coffee cup and the roof of the car is 0.24, what is the maximum acceleration the car can have without causing the cup to slide? Ignore the effects of air resistance. **(b)** What is the smallest amount of time in which the person can accelerate the car from rest to 15 m/s and still keep the coffee cup on the roof?

12. ••• **IP** The coefficient of kinetic friction between the tires of your car and the roadway is μ. **(a)** If your initial speed is v and you lock your tires during braking, how far do you skid? Give your answer in terms of v, μ, and m, the mass of your car. **(b)** If you double your speed, what happens to the stopping distance? **(c)** What is the stopping distance for a truck with twice the mass of your car, assuming the same initial speed and coefficient of kinetic friction?

Section 6–2 Strings and Springs

13. • Pulling down on a rope, you lift a 4.50-kg bucket of water from a well with an acceleration of 2.10 m/s^2. What is the tension in the rope?
14. • When a 9.00-kg mass is placed on top of a vertical spring, the spring compresses 4.50 cm. Find the force constant of the spring.
15. • A 120-kg box is loaded into the trunk of a car. If the height of the car's bumper decreases by 12 cm, what is the force constant of its rear suspension?
16. • A 50.0-kg person takes a nap in a backyard hammock. Both ropes supporting the hammock are at an angle of 15.0° above the horizontal. Find the tension in the ropes.
17. • **IP** A backpack full of books weighing 52.0 N rests on a table in a physics laboratory classroom. A spring with a force constant of 150 N/m is attached to the backpack and pulled horizontally, as indicated in **Figure 6–17**. **(a)** If the spring is pulled until it stretches 2.00 cm and the pack remains at rest, what is the force of friction exerted on the backpack by the table? **(b)** Does your answer to part (a) change if the mass of the backpack is doubled? Explain.

▲ **FIGURE 6–17** Problems 17 and 18

18. • If the spring in Figure 6–17 stretches by 2.50 cm before the 52.0-N backpack begins to slip, what is the coefficient of static friction between the backpack and the table?
19. •• The pulley system in **Figure 6–18** is used to lift a 52-kg crate. Note that a chain connects the upper pulley to the ceiling and a second chain connects the lower pulley to the crate. Assuming the masses of the chains, pulleys, and ropes are negligible, determine **(a)** the force **F** required to lift the crate with constant speed and **(b)** the tension in each chain.

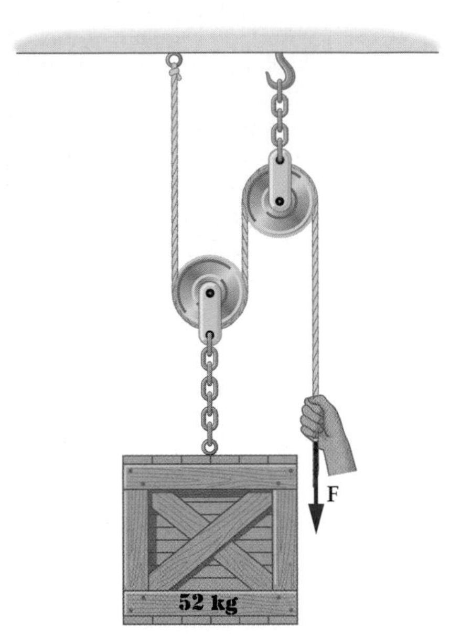

▲ **FIGURE 6–18** Problem 19

20. •• In the previous problem, determine the force and the tension in the chains if the crate is lifted with an acceleration of 0.23 m/s^2.
21. •• **IP** A spring with a force constant of 120 N/m is used to push a 0.27-kg block of wood against a wall, as shown in **Figure 6–19**. **(a)** Find the minimum compression of the spring needed to keep the block from falling, given that the coefficient of static friction between the block and the wall is 0.46. **(b)** Does your answer to part (a) change if the mass of the block of wood is doubled? Explain.

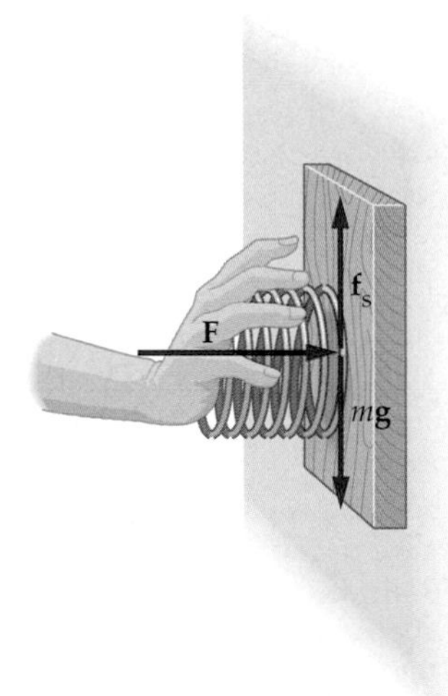

▲ **FIGURE 6–19** Problem 21

22. •• **IP** Your friend's 10.0 g graduation tassel hangs on a string from his rear-view mirror. **(a)** When he accelerates from a stop light, the tassel moves backward toward the rear of the car. Explain. **(b)** If the tassel hangs at an angle of 5.00° relative to the vertical, what is the acceleration of the car?
23. •• In the previous problem, find the tension in the string holding the tassel.

24. •• **IP** Illinois Jones is being pulled from a snake pit with a rope that breaks if the tension in it exceeds 750 N. **(a)** If Illinois Jones has a mass of 70.0 kg and the snake pit is 3.40 m deep, what is the minimum time necessary to pull our intrepid explorer from the pit? **(b)** Explain why the rope breaks if Jones is pulled from the pit in less time than that calculated.

25. •• **IP** The equilibrium length of a certain spring with a force constant of $k = 250$ N/m is 0.2 m. **(a)** What force is required to stretch this spring to twice its equilibrium length? **(b)** Is the force required to compress the spring to half its length the same as in part (a)? Explain.

26. •• **IP** A picture hangs on the wall suspended by two strings, as shown in **Figure 6–20**. The tension in string 1 is 1.7 N. **(a)** Is the tension in string 2 greater than, less than, or equal to 1.7 N? Explain. **(b)** Verify your answer to part (a) by calculating the tension in string 2. **(c)** What is the weight of the picture?

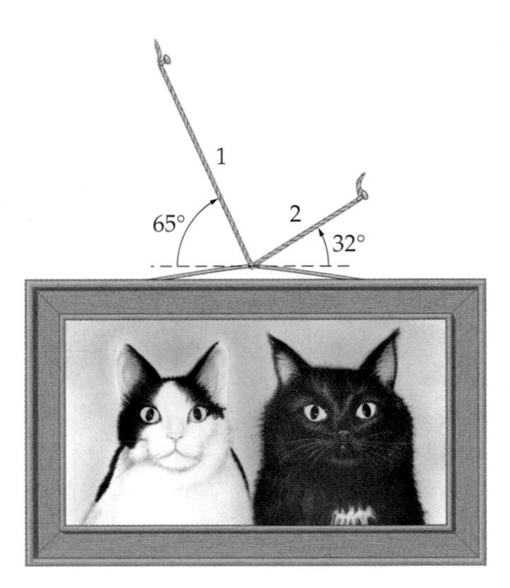

▲ **FIGURE 6–20** Problem 26

Section 6–3 Translational Equilibrium

27. • Pulling the string back with a force of 25.0 lb, an archer prepares to shoot an arrow. If the archer pulls in the center of the string, and the angle between the two halves is 145°, what is the tension in the string?

28. • In **Figure 6–21** we see two blocks connected by a string and tied to a wall. The mass of the lower block is 1.0 kg; the mass of the upper block is 2.0 kg. Given that the angle of the incline is 31°, find the tensions in **(a)** the string connecting the two blocks and **(b)** the string that is tied to the wall.

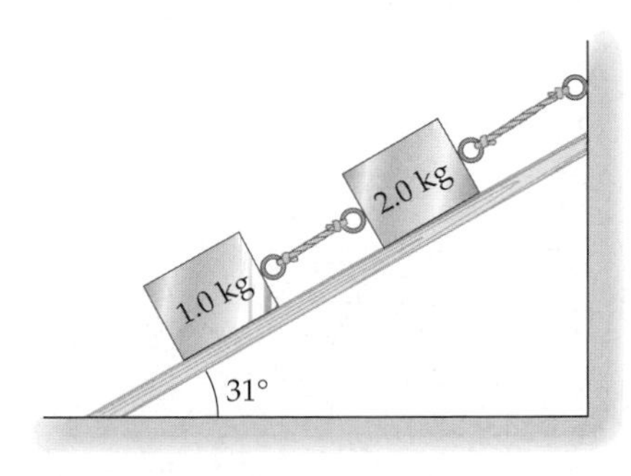

▲ **FIGURE 6–21** Problem 28

29. • BIO After a skiing accident your leg is in a cast and supported in a traction device, as shown in **Figure 6–22**. Find the magnitude of the force **F** exerted by the leg on the small pulley. (By Newton's third law, the small pulley exerts an equal and opposite force on the leg.) Let the mass m be 2.50 kg.

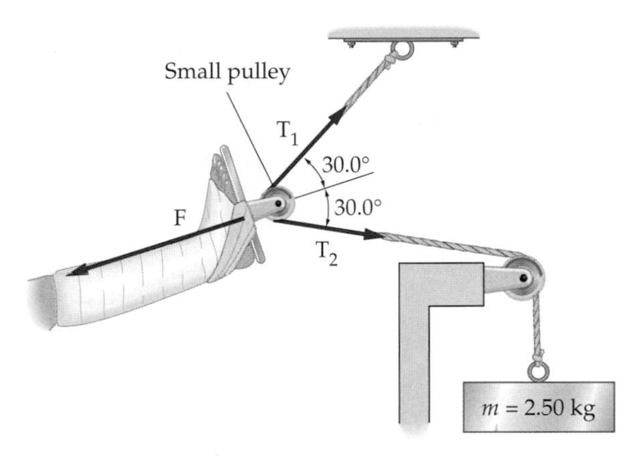

▲ **FIGURE 6–22** Problem 29

30. • Two blocks are connected by a string, as shown in **Figure 6–23**. The smooth inclined surface makes an angle of 42° with the horizontal, and the block on the incline has a mass of 6.7 kg. Find the mass of the hanging block that will cause the system to be in equilibrium.

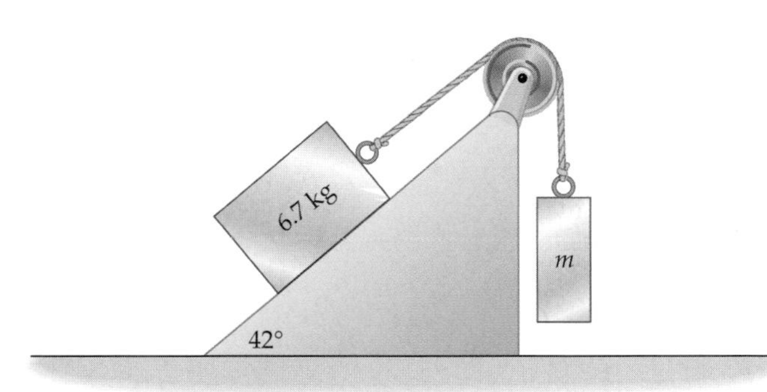

▲ **FIGURE 6–23** Problem 30

31. •• A 0.15-kg ball is placed in a shallow wedge with an opening angle of 120°, as shown in **Figure 6–24**. For each contact point between the wedge and the ball, determine the force exerted on the ball. Assume the system is frictionless.

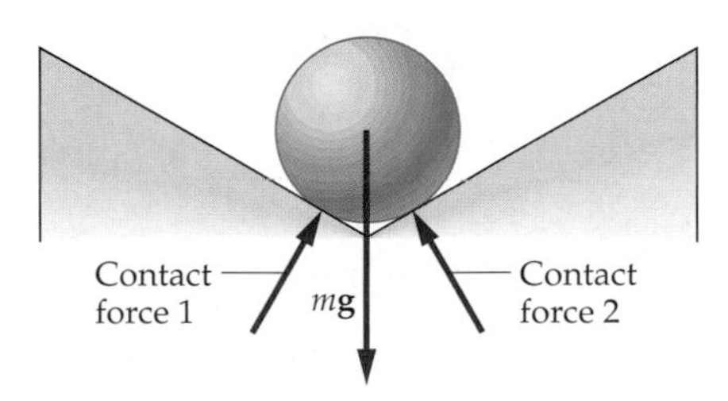

▲ **FIGURE 6–24** Problem 31

32. •• **IP** You want to nail a 1.3-kg board onto the wall of a barn. To position the board before nailing, you push it against the wall with a horizontal force **F** to keep it from sliding to the

ground.(**Figure 6–25**) **(a)** If the coefficient of static friction between the board and the wall is 0.86, what is the least force you can apply and still hold the board in place? **(b)** What happens to the force of static friction if you push against the wall with a force greater than that found in part (a)?

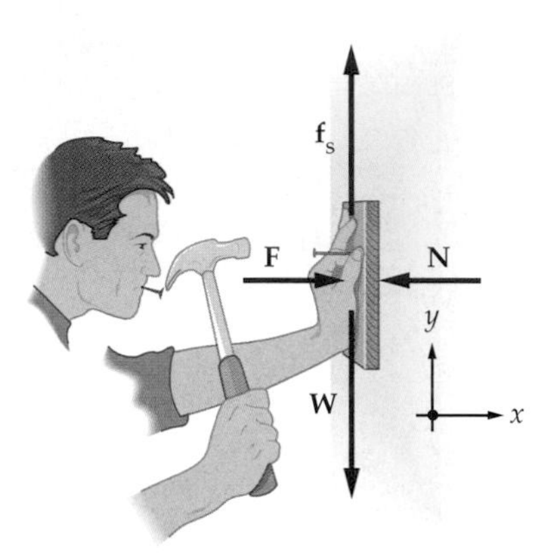

▲ **FIGURE 6–25** Problem 32

33. ••• **BIO** To immobilize a fractured femur (the thigh bone), doctors often utilize the Russell traction system illustrated in **Figure 6–26**. Notice that one force is applied directly to the knee, $\mathbf{F}_1$, while two other forces, $\mathbf{F}_2$ and $\mathbf{F}_3$, are applied to the foot. The latter two forces combine to give a force $\mathbf{F}_2 + \mathbf{F}_3$ that is transmitted through the lower leg to the knee. The result is that the knee experiences the total force $\mathbf{F}_{\text{total}} = \mathbf{F}_1 + \mathbf{F}_2 + \mathbf{F}_3$. The goal of this traction system is to have $\mathbf{F}_{\text{total}}$ directly in line with the fractured femur. Find **(a)** the angle θ required to produce this alignment of $\mathbf{F}_{\text{total}}$ and **(b)** the magnitude of the force, $\mathbf{F}_{\text{total}}$, that is applied to the femur in this case.

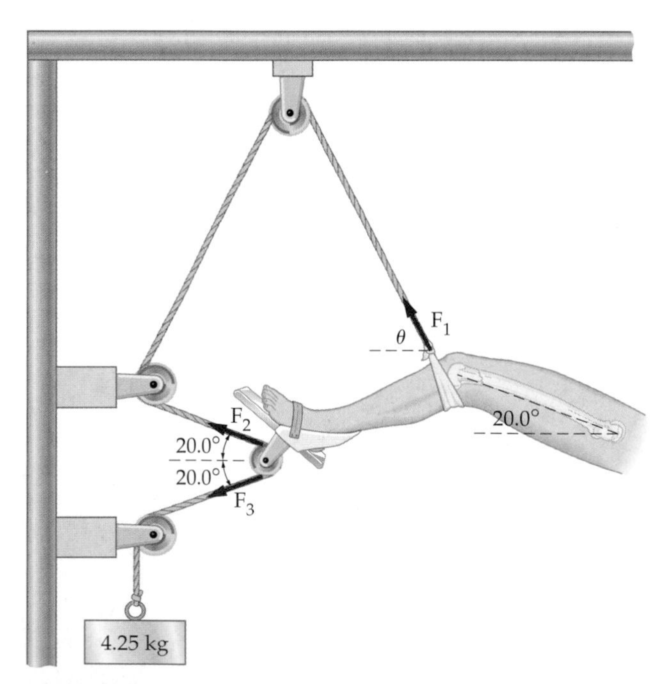

▲ **FIGURE 6–26** Problem 33

Section 6–4 Connected Objects

34. • Find the acceleration of the masses shown in **Figure 6–27**, given that $m_1 = 1.0$ kg, $m_2 = 2.0$ kg, and $m_3 = 3.0$ kg.

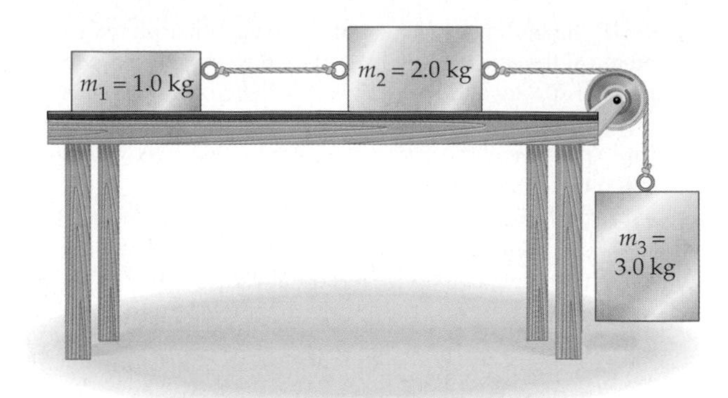

▲ **FIGURE 6–27** Problem 34

35. • Two blocks are connected by a string, as shown in **Figure 6–28**. The smooth inclined surface makes an angle of 35° with the horizontal, and the block on the incline has a mass of 5.7 kg. The mass of the hanging block is 3.2 kg. Find the direction and magnitude of the hanging block's acceleration.

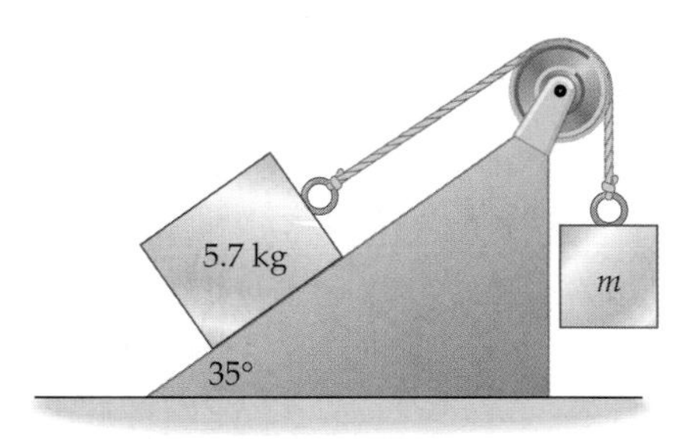

▲ **FIGURE 6–28** Problem 35

36. • Referring to the previous problem, find the direction and magnitude of the hanging block's acceleration if its mass is 4.2 kg.

37. •• Find the tension in each of the strings in Figure 6–27, given that $m_1 = 1.0$ kg, $m_2 = 2.0$ kg, and $m_3 = 3.0$ kg.

38. •• **IP** A 3.50-kg block on a smooth tabletop is attached by a string to a hanging block of mass 2.80 kg, as shown in **Figure 6–29**. The blocks are released from rest and allowed to move freely. **(a)** Is the tension in the string greater than, less than, or equal to the weight of the hanging mass? **(b)** Find the acceleration of the blocks and the tension in the string.

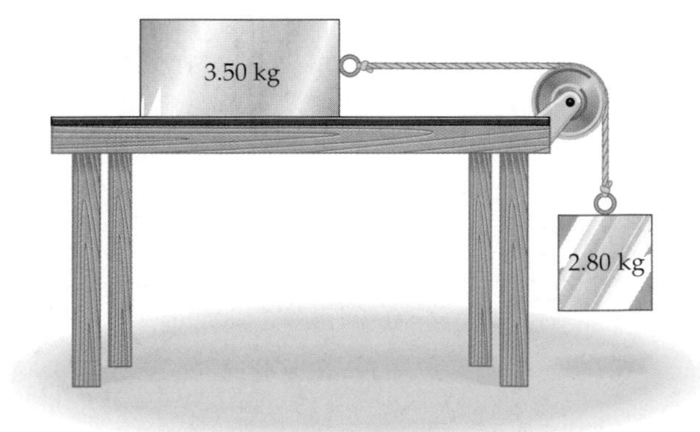

▲ **FIGURE 6–29** Problem 38

39. •• **IP** A 6.4-N force pulls horizontally on a 1.5-kg block that slides on a smooth, horizontal surface. This block is con-

nected by a horizontal string to a second block of mass $m_2 = 0.93$ kg on the same surface. **(a)** What is the acceleration of the blocks? **(b)** What is the tension in the string? **(c)** If the mass of block 1 is increased, does the tension in the string increase, decrease, or stay the same?

40. ••• Two buckets of sand hang from opposite ends of a rope that passes over a pulley. One bucket is full and weighs 110 N, the other is only partly filled and weighs 63 N. **(a)** Initially, you hold onto the lighter bucket to keep it from moving. What is the tension in the rope? **(b)** You release the bucket and the heavier one descends. What is the tension in the rope now? **(c)** Eventually the heavier bucket lands and the two buckets come to rest. What is the tension in the rope now?

Section 6–5 Circular Motion

41. • When you take your 1100-kg car out for a spin, you go around a corner of radius 55 m with a speed of 15 m/s. Assuming your car doesn't skid, what is the force exerted on it by static friction?

42. • Find the linear speed that the bottom of a test tube must have in a centrifuge if the centripetal acceleration there is to be 52,000 times the acceleration of gravity. The distance from the axis of rotation to the bottom of the test tube is 7.5 cm.

43. • **BIO** To test the effects of high acceleration on the human body, the National Aeronautics and Space Administration (NASA) has constructed a large centrifuge at the Manned Spacecraft Center in Houston. In this device, astronauts are placed in a capsule that moves in a circular path with a radius of 15 m. If the astronauts in this centrifuge experience a centripetal acceleration nine times that of gravity, what is the linear speed of the capsule?

44. • A car goes around a curve on a road that is banked at an angle of 30.0°. Even though the road is slick, the car will stay on the road without any friction between its tires and the road when its speed is 24.0 m/s. What is the radius of the curve?

45. •• Jill of the Jungle swings on a vine 6.5 m long. What is the tension in the vine if Jill, whose mass is 61 kg, is moving at 2.4 m/s when the vine is vertical?

46. •• **IP** In the previous problem, how does the tension in the vine change if Jill's speed is doubled? How does the tension change if her mass is doubled instead? Explain.

47. •• **IP (a)** As you ride on a Ferris wheel your apparent weight is different at the top and at the bottom. Explain. **(b)** Calculate your apparent weight at the top and bottom of a Ferris wheel, given that the radius of the wheel is 7.2 m, it completes on revolution every 28 s, and your mass is 55 kg.

▲ A Ferris Wheel (Problem 47)

48. •• Driving in your car with a constant speed of 12 m/s, you encounter a bump in the road that has a circular cross section, as indicated in **Figure 6–30**. If the radius of curvature of the bump is 35 m, find the apparent weight of a 67-kg person in your car as you pass over the top of the bump.

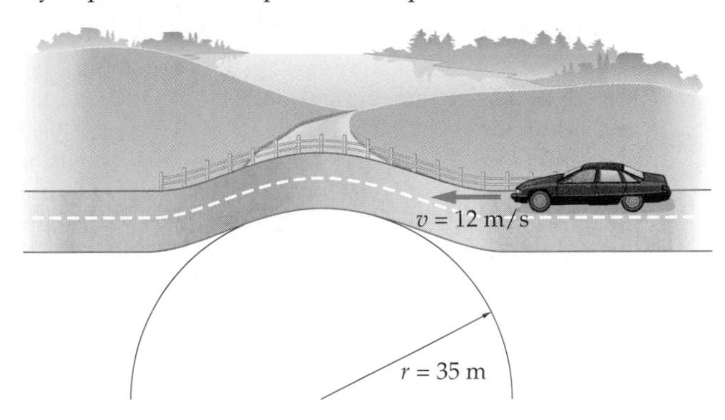

▲ **FIGURE 6–30** Problem 48

49. •• Referring to the previous problem, at what speed must you go over the bump if people in your car are to feel "weightless?"

50. •• **IP** You swing a 5.0-kg bucket of water in a vertical circle of radius 1.2 m. **(a)** What speed must the bucket have if it is to complete the circle without spilling any water? **(b)** How does your answer depend on the mass of the bucket?

General Problems

51. • **BIO** A skateboard accident leaves your leg in a cast and supported by a traction device, as in Figure 6–22. Find the mass m that will cause the net force exerted by the small pulley on the leg to have a magnitude of 32 N.

52. • **BIO** Find the centripetal acceleration at the top of a test tube in a centrifuge, given that it is 4.2 cm from the axis of rotation and that its linear speed is 77 m/s.

53. • Find the coefficient of kinetic friction between a 3.85-kg block and the horizontal surface on which it rests if a 85 N/m spring must be stretched by 6.20 cm to pull it with constant speed. Assume that the spring pulls in the horizontal direction.

54. • A child goes down a playground slide that is inclined at an angle of 25.0° below the horizontal. Find the acceleration of the child given that the coefficient of kinetic friction between the child and the slide is 0.42.

55. • When a block is placed on top of a vertical spring, the spring compresses 3.75 cm. Find the mass of the block, given that the force constant of the spring is 1800 N/m.

56. •• A force of 9.4 N pulls horizontally on a 1.1-kg block that slides on a rough, horizontal surface. This block is connected by a horizontal string to a second block of mass $m_2 = 1.92$ kg on the same surface. The coefficient of kinetic friction is $\mu_k = 0.24$ for both blocks. **(a)** What is the acceleration of the blocks? **(b)** What is the tension in the string?

57. •• You swing a 3.25-kg bucket of water in a vertical circle of radius 0.950 m. At the top of the circle the speed of the bucket is 2.23 m/s; at the bottom of the circle its speed is 2.74 m/s. Find the tension in the rope tied to the bucket at **(a)** the top and **(b)** the bottom of the circle.

58. •• A 12-g coin slides upward on a surface that is inclined at an angle of 15° above the horizontal. The coefficient of kinetic friction between the coin and the surface is 0.23; the coefficient of static friction is 0.31. Find the magnitude and direction of the force of friction **(a)** when the coin is sliding and **(b)** after it comes to rest.

59. •• In the previous problem, the angle of the incline is increased to 25°. Find the magnitude and direction of the force of friction when the coin is **(a)** sliding upward and **(b)** sliding downward.

60. •• A physics textbook weighing 22 N rests on a table. The coefficient of static friction between the book and the table is $\mu_s = 0.60$, and the coefficient of kinetic friction is $\mu_k = 0.40$. You push horizontally on the book with a gradually increasing and then decreasing force. For each value of the applied force listed in the following table, give the magnitude of the force of friction and state whether the book is accelerating, decelerating, at rest, or moving with constant speed.

applied force	friction force	motion
0		
5 N		
10 N		
15 N		
10 N		
8 N		
5 N		

61. •• A ball of mass m is placed in a wedge with an opening angle of 90°, as shown in **Figure 6–31**. For each contact point, determine the force exerted by the frictionless wall of the wedge on the ball.

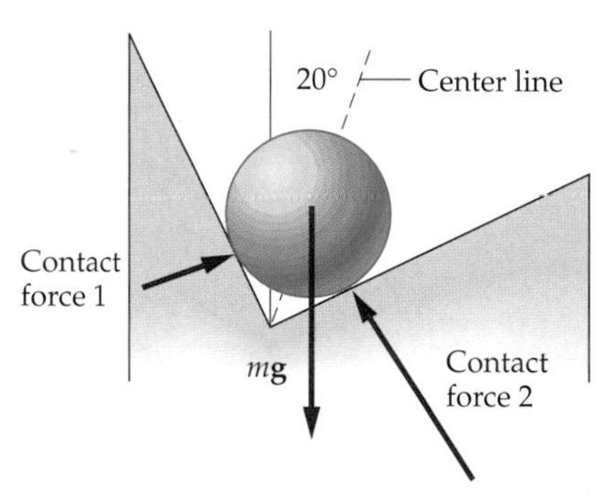

▲ **FIGURE 6–31** Problem 61

62. •• A 2.0-kg box rests on a plank that is inclined at an angle of 65° above the horizontal. The upper end of the box is attached to a spring with a force constant of 18 N/m, as shown in **Figure 6–32**. If the coefficient of static friction between the box and the plank is 0.22, what is the maximum amount the spring can be stretched and the box remain at rest?

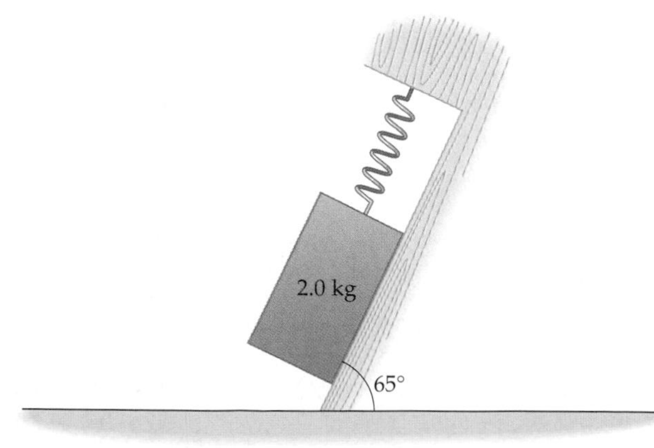

▲ **FIGURE 6–32** Problem 62

63. •• **IP** The system shown in **Figure 6–33** is in equilibrium. **(a)** Find the frictional force exerted on block A given that the mass of block A is 8.50 kg, the mass of block B is 2.25 kg, and the coefficient of static friction between block A and the surface on which it rests is 0.320. **(b)** If the mass of block A is doubled, does the frictional force exerted on it increase, decrease, or stay the same? Explain.

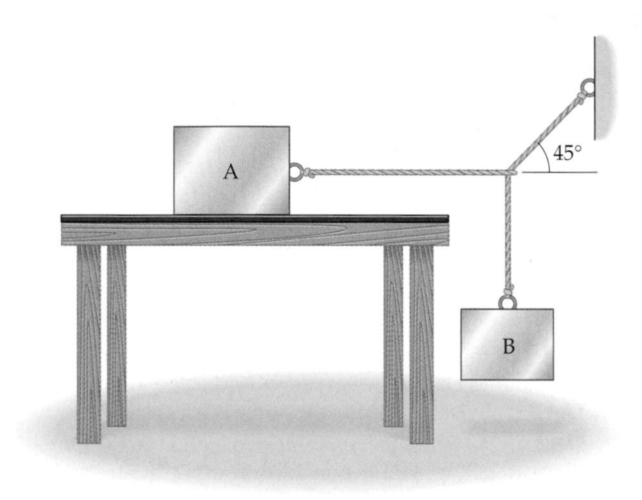

▲ **FIGURE 6–33** Problem 63

64. •• In part (a) of the previous problem, what is the maximum mass block B can have and the system still remain in equilibrium?

65. •• A 0.075-kg toy airplane is tied to the ceiling with a string. When the airplane's motor is started it moves with a constant speed of 1.21 m/s in a horizontal circle of radius 0.44 m, as illustrated in **Figure 6–34**. Find the angle the string makes with the vertical and the tension in the string.

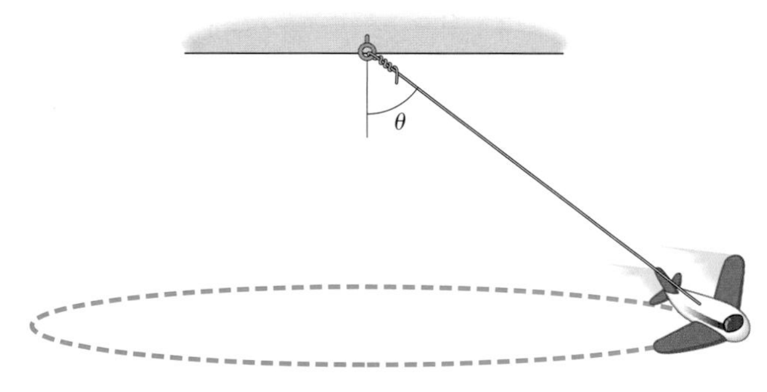

▲ **FIGURE 6–34** Problem 65

66. •• A tugboat tows a barge at constant speed with a 14,000-kg cable, as shown in **Figure 6–35**. If the angle the cable makes with the horizontal where it attaches to the barge and the tugboat is 22°, find the force exerted on the barge in the forward direction.

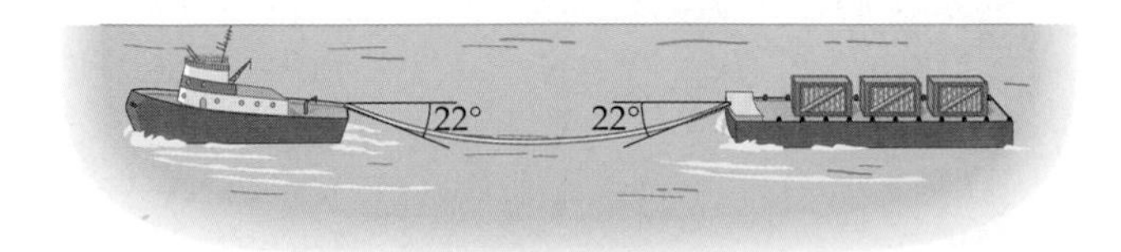

▲ **FIGURE 6–35** Problem 66

67. •• **IP** Two blocks, stacked one on top of the other, slide on a frictionless, horizontal surface (**Figure 6–36**). The surface between the two blocks is rough, however, with a coefficient of static friction equal to 0.47. **(a)** If a horizontal force F is applied to the 5.0-kg bottom block, what is the maximum value F can have before the 2.0-kg top block begins to slip? **(b)** If the mass of the top block is increased, does the maximum value of F increase, decrease, or stay the same? Explain.

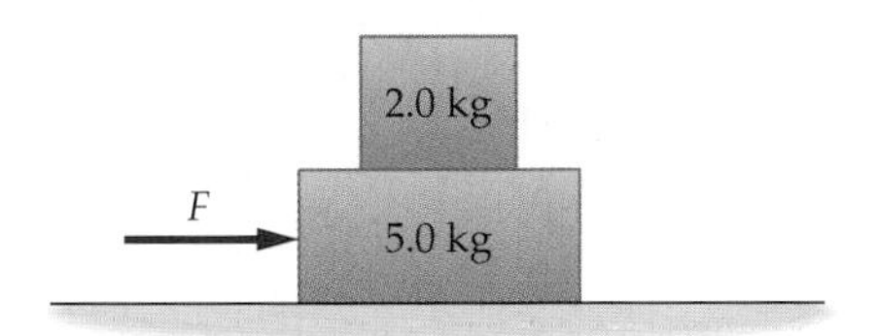

▲ **FIGURE 6–36** Problem 67

68. •• Find the coefficient of kinetic friction between a 4.5-kg block and the horizontal surface on which it rests if an 85 N/m spring must be stretched by 2.1 cm to pull it with constant speed. Assume that the spring pulls in a direction 13° above the horizontal.

69. •• **IP** In a daring rescue by helicopter, two men with a combined mass of 172 kg are lifted to safety. **(a)** If the helicopter lifts the men straight up with constant acceleration, is the tension in the rescue cable greater than, less than, or equal to the combined weight of the men? Explain. **(b)** Determine the tension in the cable if the men are lifted with an acceleration of 1.10 m/s^2.

70. •• At the airport, you pull a 15-kg suitcase across the floor with a strap that is at an angle of 45° above the horizontal. Find the normal force and the tension in the strap, given that the suitcase moves with constant speed and that the coefficient of kinetic friction between the suitcase and the floor is 0.36.

71. •• **IP** A light spring with a force constant of 13 N/m is connected to a wall and to a 3.0-kg toy bulldozer (**Figure 6–37**). When the electric motor in the bulldozer is turned on, it stretches the spring for a distance of 2.0 m before its tires begin to slip. **(a)** Which coefficient of friction (static or kinetic) can be determined from this information? Explain. **(b)** What is the numerical value of this coefficient of friction?

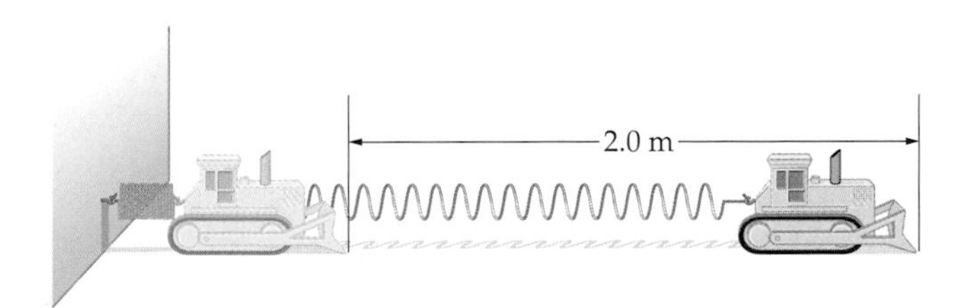

▲ **FIGURE 6–37** Problem 71

72. •• **IP** A 0.10-g spider hangs from the middle of the first thread of its future web. The thread makes an angle of 5.2° with the horizontal on both sides of the spider. **(a)** What is the tension in the thread? **(b)** If the angle made by the thread had been less than 5.2°, would its tension have been greater than, less than, or the same as in part (a)? Explain.

73. •• Find the acceleration that the cart in **Figure 6–38** must have in order for the cereal box not to fall. Assume that the coefficient of static friction between the cart and the box is 0.34.

74. •• **IP** The tension in a violin string is 2.7 N. When pushed down against the neck of the violin, the string makes an angle of 4.1° with the horizontal. **(a)** With what force must you push down on the string to bring it into contact with the neck? **(b)** If the angle were less than 4.1°, would the required force be greater than, less than, or the same as in part (a)? Explain.

▲ **FIGURE 6–38** Problem 73

75. •• **IP** A pair of fuzzy dice hangs from a string attached to your rear-view mirror. As you turn a corner with a radius of 95 m and a constant speed of 25 mi/h, what angle will the dice make with the vertical? Why is it unnecessary to give the mass of the dice?

76. •• Referring to Problem 16, find the tension in the two ropes supporting the hammock if one is at an angle of 15° above the horizontal and the other is at an angle of 35° above the horizontal.

77. •• As your plane circles an airport, it moves in a horizontal circle of radius 2100 m with a speed of 380 km/h. If the lift of the airplane's wings is perpendicular to the wings, at what angle should the plane be banked?

78. •• A child sits on a rotating merry-go-round, 2.1 m from its center. If the speed of the child is 1.9 m/s, what is the minimum coefficient of static friction between the child and the merry-go-round that will prevent the child from slipping?

79. ••• A wood block of mass m rests on a larger wood block of mass M that rests on a wooden table. The coefficient of static friction between all surfaces is the same, μ. What is the minimum horizontal force, F, applied to the lower block that will cause it to slide out from under the upper block?

80. ••• Find the tension in each of the two strings shown in Figure 6–27 for general values of the masses. Your answer should be in terms of m_1, m_2, m_3, and g.

81. ••• The coefficient of static friction between a rope and the table on which it rests is μ_s. Find the fraction of the rope that can hang over the edge of the table before it begins to slip.

82. ••• A hockey puck of mass m is attached to a string that passes through a hole in the center of a table, as shown in **Figure 6–39**. The hockey puck moves in a circle of radius r. Tied to the other end of the string, and hanging vertically beneath the table, is a mass M. Assuming the tabletop is perfectly smooth, what speed must the hockey puck have if the mass M is to remain at rest?

83. ••• To move a crate of mass m across a rough floor, you push down on it at an angle θ, as shown in Figure 6–16. **(a)** Find the force necessary to start the crate moving, given that the coefficient of static friction between the crate and the floor is μ_s. **(b)** Show that it is impossible to move the crate, no matter how great the force, if the coefficient of static friction is greater than or equal to $1/\tan\theta$.

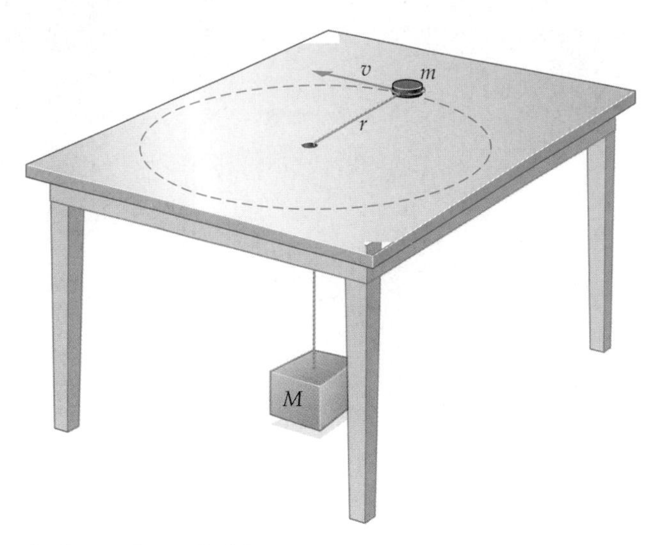

▲ **Figure 6–39** Problem 82

84. ••• **IP** A popular ride at amusement parks is illustrated in **Figure 6–40**. In this ride, people sit in a swing that is suspended from a long, rotating arm. Riders are at a distance of 12 m from the axis of rotation and move with a speed of 25 mi/h. **(a)** Find the centripetal acceleration of the riders. **(b)** Find the angle θ the supporting wires make with the vertical. **(c)** Notice that the swings in Figure 6–40 are at the same angle to the vertical, regardless of the weight of the rider. Explain.

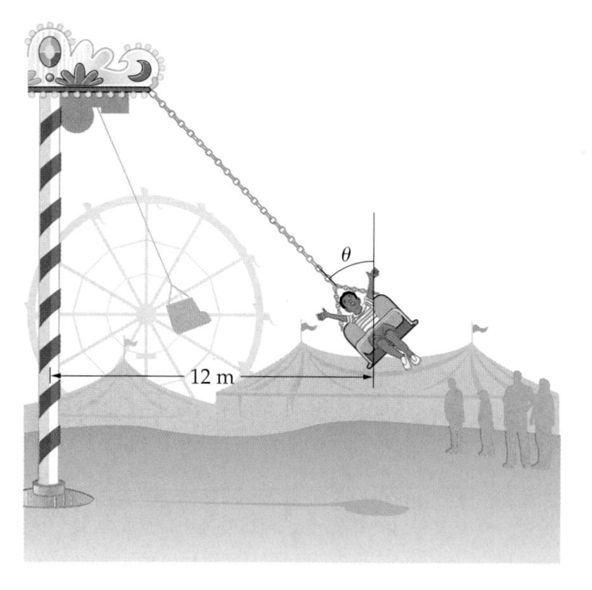

▲ **FIGURE 6–40** Problem 84

85. ••• A box is placed on a conveyor belt that moves with a constant speed of 1.30 m/s. The coefficient of kinetic friction between the box and the belt is 0.830. **(a)** How long does it take before the box stops sliding relative to the belt? **(b)** How far has the box moved in this time?

86. ••• You push a box along the floor against a constant force of friction. When you push with a horizontal force of 75 N the acceleration of the box is 0.50 m/s^2; when you increase the force to 81 N the acceleration is 0.75 m/s^2. Find the mass of the box and the coefficient of kinetic friction between the box and the floor.

87. ••• A mountain climber of mass m hangs onto a rope to keep from sliding down a smooth, ice-covered slope (**Figure 6–41**). Find a formula for the tension in the rope when the slope is inclined at an angle θ above the horizontal. Check your results in the limits $\theta = 0$ and $\theta = 90°$.

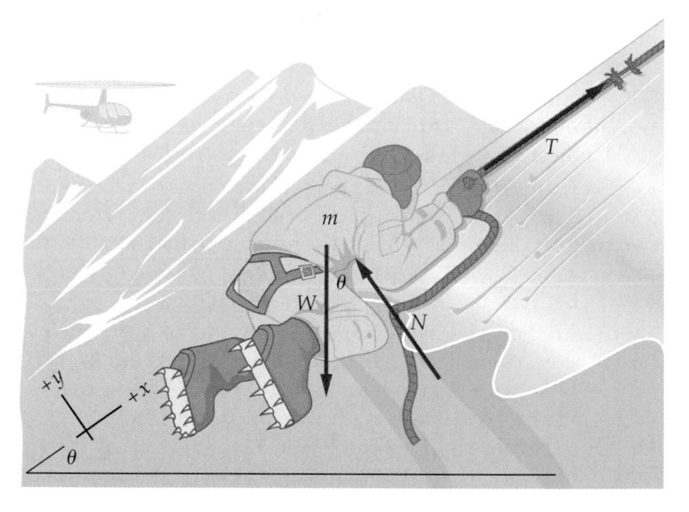

▲ **FIGURE 6–41** Problem 87

7 Work and Kinetic Energy

We all know intuitively that motion, energy, and work are somehow related. For example, the chemical energy stored in our muscles enables us to do work, as when we throw a javelin. The work done on the javelin appears as kinetic energy—the energy of motion—and when the javelin strikes an object, its kinetic energy can in turn do work. In this chapter we'll define more precisely the concepts of work, kinetic energy, and power, and explore the physical relationships among them.

The concept of force is one of the foundations of physics, as we have seen in the previous two chapters. Equally important, though less obvious, is the idea that a force times the distance through which it acts is also an important physical quantity. We refer to this quantity as the *work* done by a force.

Now, we all know what work means in everyday life: We get up in the morning and go to work, or we "work up a sweat" as we hike a mountain trail. Later in the day we eat lunch, which gives us the "energy" to continue working or to continue our hike. In this chapter we give a precise physical definition of work, and show how it is related to another important physical quantity–the energy of motion, or *kinetic energy*. When these concepts are extended in the next chapter, we are lead to the rather sweeping observation that the total amount of energy in the universe remains constant at all times.

7–1 Work Done by a Constant Force

In this section we define work–in the physics sense of the word–and apply our definition to a variety of physical situations. We start with the simplest case; namely, the work done when force and displacement are in the same direction. Later in the section we generalize our definition to include cases where the force and displacement are in arbitrary directions. We conclude with a discussion of the work done on an object when it is acted on by more than one force.

Force in the Direction of Displacement

When we push a shopping cart in a store or pull a suitcase through an airport, we do work. The greater the force, the greater the work; the greater the distance, the greater the work. These simple ideas form the basis for our definition of work.

To be specific, suppose we push a box with a constant force **F**, as shown in Figure 7–1. If we move the box *in the direction of* **F** through a distance d, the **work** W we have done is Fd:

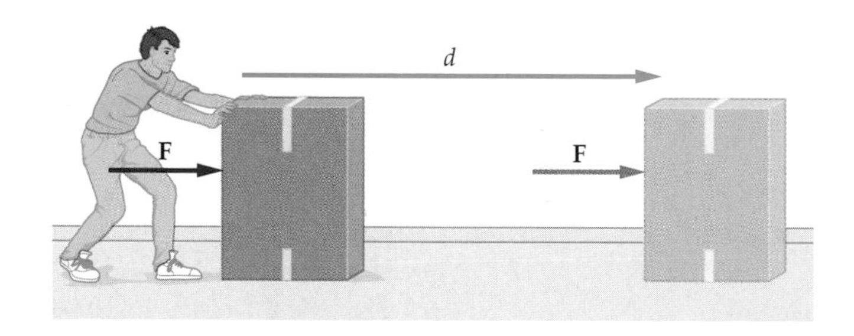

▶ **FIGURE 7–1 Work: force in the direction of motion**
A force **F** pushes a box through a distance d. In this case the force is in the direction of motion, and the work done by the force is $W = Fd$.

Definition of Work, *W*, When Force Is in the Direction of Displacement

$$W = Fd \qquad \text{7–1}$$

SI unit: newton-meter (N·m) = joule, J

Note that work is the product of two magnitudes, and hence it is a scalar.

The dimensions of work are newtons (force) times meters (distance), or N·m. This combination of dimensions is called the **joule** (rhymes with school, as commonly pronounced) in honor of James Prescott Joule (1818–1889), a dedicated physicist who is said to have conducted physics experiments even while on his honeymoon. We define a joule as follows:

Definition of the joule, J

$$1 \text{ joule} = 1 \text{ J} = 1 \text{ N}\cdot\text{m} = 1 \ (\text{kg}\cdot\text{m/s}^2)\cdot\text{m} = 1 \text{ kg}\cdot\text{m}^2/\text{s}^2 \qquad \text{7–2}$$

For example, suppose you exert a force of 8.20 N on the box in Figure 7–1 and move it in the direction of the force through a distance of 3.00 m. The work you have done is

$$W = Fd = (8.20 \text{ N})(3.00 \text{ m}) = 24.6 \text{ N}\cdot\text{m} = 24.6 \text{ J}$$

Similarly, if you do 5.00 J of work to lift a book through a vertical distance of 0.750 m, the force you exerted on the book is

$$F = \frac{W}{d} = \frac{5.00 \text{ J}}{0.750 \text{ m}} = \frac{5.00 \text{ N}\cdot\text{m}}{0.750 \text{ m}} = 6.67 \text{ N}$$

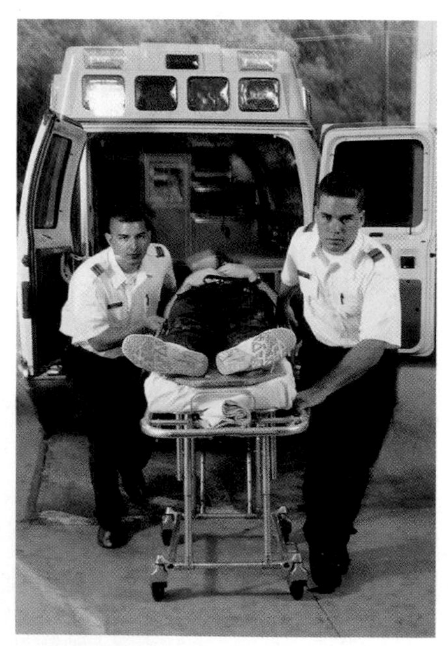

▲ The work done on the gurney by these paramedics is directly proportional to the force they exert on it.

EXERCISE 7–1

One species of Darwin's finch, *magnirostris*, can exert a force of 205 N with its beak as it cracks open a Tribulus seed case. If its beak moves through a distance of 0.40 cm during this operation, how much work does the finch do to get the seed?

Solution

$$W = Fd = (205 \text{ N})(0.0040 \text{ m}) = 0.82 \text{ J}$$

Just how much work is a joule, anyway? Well, you do one joule of work when you lift a gallon of milk through a height of about an inch, or lift an apple a meter. One joule of work lights a 100-watt lightbulb for 0.01 seconds or heats a glass of water 0.00125 degrees Celsius. Clearly, a joule is a modest amount of work in everyday terms. Additional examples of work are listed in Table 7–1.

TABLE 7–1 Typical Values of Work

Activity	Equivalent work (J)	Activity	Equivalent work (J)
Annual U. S. energy use	8×10^{19}	Lighting a 100-W bulb for 1 minute	6×10^{3}
Mt. St. Helens eruption	10^{18}	Lifting a baseball two feet	0.9
Thunderstorm	10^{15}	Heartbeat	0.5
Burning one gallon of gas	10^{8}	Turning page of a book	10^{-3}
Human food intake/day	10^{7}	Hop of a flea	10^{-7}
Burning one lump of coal	10^{6}	Breaking a bond in DNA	10^{-20}
Melting an ice cube	10^{4}		

EXAMPLE 7–1 Heading for the ER

An intern pushes a 72-kg patient on a 15-kg gurney, producing an acceleration of 0.60 m/s^2. How much work does the intern do by pushing the patient and gurney through a distance of 2.5 m? Assume the gurney moves without friction.

Picture the Problem

Our sketch shows the situation for this problem. Note that the force and the displacement are in the same direction.

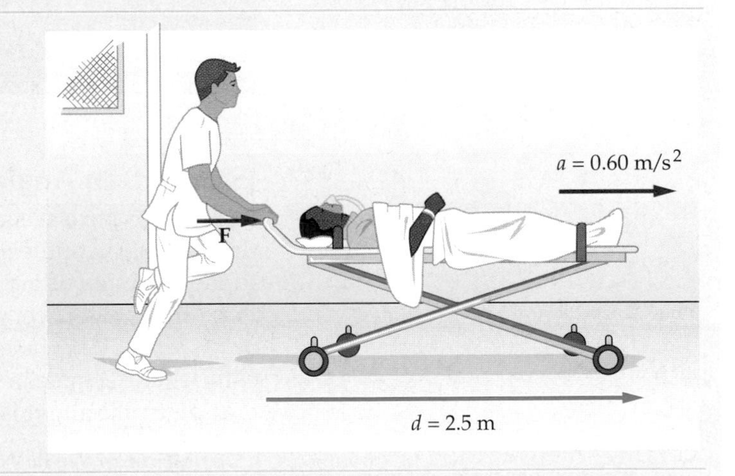

Strategy

We are not given the magnitude of the force F, so we cannot apply Equation 7–1 directly. However, we are given the mass and acceleration of the patient and gurney, from which we can calculate the force with $F = ma$. The work done by the intern is then $W = Fd$.

Solution

1. First, find the force F exerted by the intern: $F = ma = (72\text{ kg} + 15\text{ kg})(0.60\text{ m/s}^2) = 52\text{ N}$

2. The work done by the intern, W, is the force times the distance: $W = Fd = (52\text{ N})(2.5\text{ m}) = 130\text{ J}$

Insight

One might wonder whether the work done by the intern depends on the speed of the gurney. The answer is no. The work done on an object, $W = Fd$, doesn't depend on whether the object moves through the distance d quickly or slowly. What does depend on the speed of the gurney is the *rate* at which work is done, as we discuss in detail in Section 7–4.

Practice Problem

If the total mass of the gurney plus patient is halved and the acceleration is doubled, does the work done by the intern increase, decrease, or remain the same? [**Answer:** The work remains the same.]

Some related homework problems: Problem 1, Problem 2

Before moving on, let's note an interesting point about our definition of work. It's clear from Equation 7–1 that *the work W is zero if the distance d is zero*—

and this is true regardless of how great the force might be. For example, if you push against a solid wall you do no work on it, even though you may become tired from your efforts. Similarly, if you stand in one place holding a 50-pound suitcase in your hand you do no work on the suitcase. The fact that we become tired when we push against a wall or hold a heavy object is due to the repeated contraction and expansion of individual cells within our muscles. Thus, even when we are "at rest," our muscles are doing mechanical work on the microscopic level.

▶ The weightlifter at right does more work in raising 150 kilograms above her head than Atlas, who is supporting the entire world. Why?

Force at an Angle to the Displacement

In **Figure 7–2** we see a person pulling a suitcase on a level surface with a strap that makes an angle θ with the horizontal—in this case the force is at an angle to the direction of motion. How do we calculate the work now? Well, instead of force times distance, we say that work is the *component* of force in the *direction* of displacement times the magnitude of the displacement. In Figure 7–2, the component of force in the direction of displacement is $F \cos \theta$ and the magnitude of the displacement is d. Therefore, the work is $F \cos \theta$ times d:

Definition of Work When the Angle between the Force and Displacement Is θ

$$W = (F \cos \theta)d = Fd \cos \theta \qquad \text{7–3}$$

SI unit: joule, J

▶ **FIGURE 7–2 Work: force at an angle to direction of motion**

A person pulls a suitcase with a strap at an angle θ to the direction of motion. The component of force in the direction of motion is $F \cos \theta$, and the work done by the person is $W = (F \cos \theta)d$.

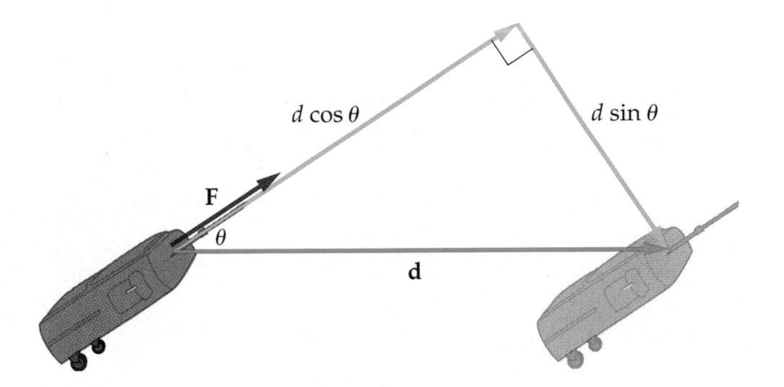

◀ **FIGURE 7–3 Force at an angle to direction of motion: another look**
The displacement of the suitcase in Figure 7–2 is equivalent to a displacement of magnitude $d \cos\theta$ in the direction of the force **F** plus a displacement of magnitude $d \sin\theta$ perpendicular to the force. Only the displacement parallel to the force results in nonzero work, hence the total work done is $F(d \cos\theta)$ as expected.

Of course, in the case where the force is in the direction of motion, the angle θ is zero; then $W = Fd \cos 0° = Fd \cdot 1 = Fd$, in agreement with Equation 7–1.

Equally interesting is a situation in which the force and the displacement are at right angles to one another. In this case $\theta = 90°$ and the work done by the force F is zero; $W = Fd \cos 90° = 0$.

This result leads naturally to an alternative way to think about the expression $W = Fd \cos\theta$. In **Figure 7–3** we show the displacement and the force for the suitcase in Figure 7–2. Notice that the displacement is equivalent to a displacement in the direction of the force of magnitude $(d \cos\theta)$ *plus* a displacement at right angles to the force of magnitude $(d \sin\theta)$. Since the displacement at right angles to the force corresponds to zero work and the displacement in the direction of the force corresponds to a work $W = F(d \cos\theta)$, it follows that the work done in this case is $Fd \cos\theta$, as given in Equation 7–3. Thus, the work can be thought of in the following two *equivalent* ways:

The work done by a force is:

(i) *The component of force in the direction of displacement* times *the magnitude of the displacement*.

(ii) *The component of displacement in the direction of the force* times *the magnitude of the force*.

In either case, the mathematical expression for the work is exactly the same.

EXAMPLE 7–2 Slip Sliding Away

A 75.0-kg person slides a distance of 5.00 m on a straight water slide, dropping through a vertical height of 2.50 m. How much work does gravity do on the person?

Picture the Problem
Note that the force of gravity $m\mathbf{g}$, and displacement **d** are at an angle θ relative to one another, where θ is the angle the slide makes with the vertical.

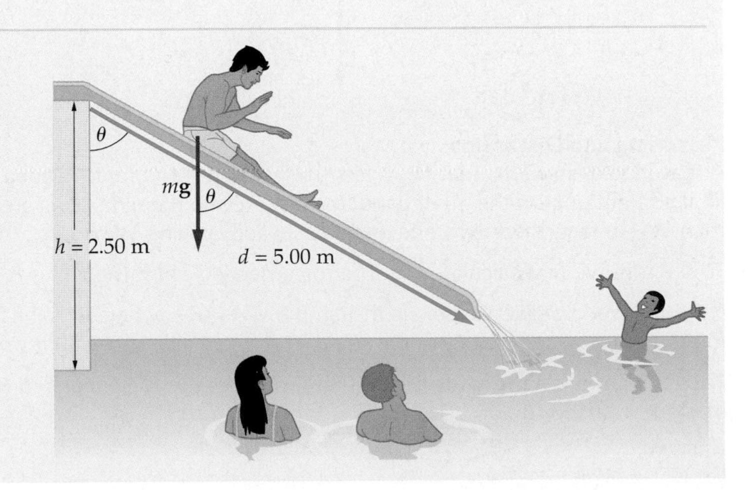

Strategy
By definition, the work done by gravity in this case is $W = (F \cos\theta)d$, where $F = mg$ and θ is the angle between $m\mathbf{g}$ and **d**. We are not given θ in the problem statement, but from the right triangle that forms the slide we see that $\cos\theta = h/d$, where h is the vertical height.

continued on the following page

continued from the previous page

Solution

1. First, find the component of $\mathbf{F} = m\mathbf{g}$ in the direction of motion:

$$F\cos\theta = (mg)\left(\frac{h}{d}\right)$$
$$= (75.0\text{ kg})(9.81\text{ m/s}^2)\left(\frac{2.50\text{ m}}{5.00\text{ m}}\right) = 368\text{ N}$$

2. Multiply by distance to find the work:

$$W = (F\cos\theta)d = (368\text{ N})(5.00\text{ m}) = 1840\text{ J}$$

3. Alternatively, cancel d algebraically before substituting numerical values:

$$W = Fd\cos\theta = (mg)(d)\left(\frac{h}{d}\right)$$
$$= mgh = (75.0\text{ kg})(9.81\text{ m/s}^2)(2.50\text{ m}) = 1840\text{ J}$$

Insight

The work is simply $W = mgh$, exactly the same as if the person had fallen straight down through the height h.

Notice that working the problem symbolically, as in step 3, results in two distinct advantages. First, it makes for a simpler expression for the work. Second, and more importantly, it shows that the distance d cancels, and hence the work depends on the height h but not on d. Such a result is not apparent when we work solely with numbers, as in steps 1 and 2.

Practice Problem

What is the height h if the work done by gravity is 2010 J? [**Answer:** $h = 2.73$ m]

Some related homework problems: Problem 8, Problem 9

Next, we present a Conceptual Checkpoint that compares the work required to move an object along two different paths.

CONCEPTUAL CHECKPOINT 7–1

You want to load a box into the back of a truck. One way is to lift it straight up through a height h, as shown, doing a work W_1. Alternatively, you can slide the box up a loading ramp a distance L, doing a work W_2. Assuming the box slides on the ramp without friction, which of the following is correct: **(a)** $W_1 < W_2$, **(b)** $W_1 = W_2$, **(c)** $W_1 > W_2$?

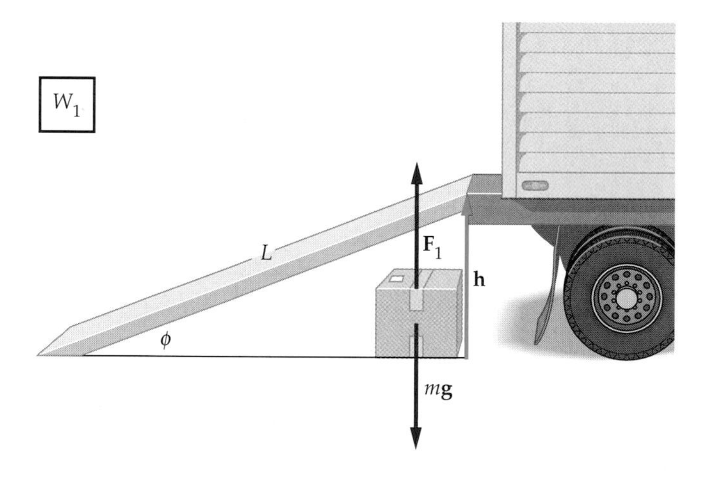

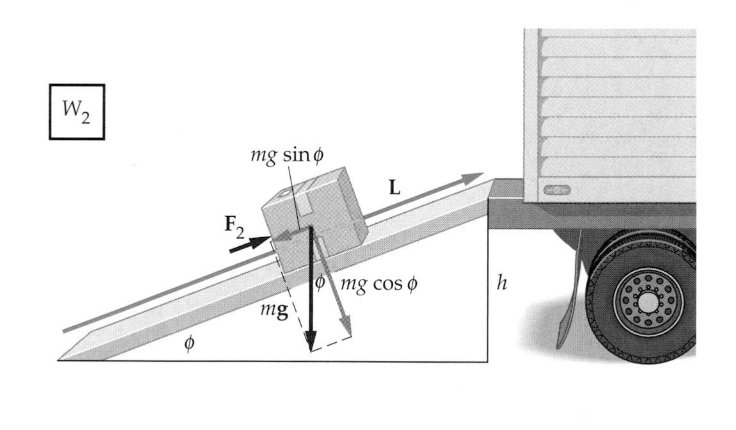

Reasoning and Discussion

A reasonable answer is that W_2 is less than W_1 since the force needed to slide the box up the ramp, F_2, is less than the force needed to lift it straight up. On the other hand, the distance up the ramp, L, is greater than the vertical distance, h, so perhaps W_2 should be greater than W_1. In fact, these two effects cancel exactly, giving $W_1 = W_2$.

To see this, we first calculate W_1. The force needed to lift the box is $F_1 = mg$, and the height is h, therefore $W_1 = mgh$.

Next, the work to slide the box up the ramp is $W_2 = F_2L$, where F_2 is the force required to push against the tangential component of gravity. In the figure we see that $F_2 = mg\sin\phi$. The figure also shows that $\sin\phi = h/L$; thus $W_2 = (mg\sin\phi)L = (mg)(h/L)L = mgh = W_1$.

Thus the ramp is a useful device, in that it reduces the force required to move the box upward. Even so, it doesn't decrease the amount of work we need to do.

Answer:

(b) $W_1 = W_2$

Negative Work and Total Work

Work depends on the angle between the force, **F**, and the displacement (or direction of motion), **d**. This dependence gives rise to three distinct possibilities, as shown in **Figure 7–4**:

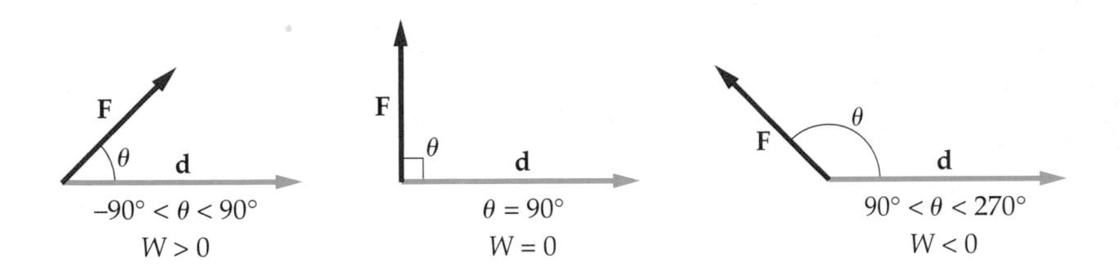

◀ FIGURE 7–4 Positive, negative, and zero work

Work is positive when the force is in the same general direction as the displacement and is negative if the force is generally opposite to the displacement. Zero work is done if the force is at right angles to the displacement.

(i) Work is positive if the force has a component in the direction of motion ($-90° < \theta < 90°$).

(ii) Work is zero if the force has no component in the direction of motion ($\theta = \pm 90°$).

(iii) Work is negative if the force has a component opposite to the direction of motion ($90° < \theta < 270°$).

Thus, whenever we calculate work we must be careful about its sign, and not just assume it to be positive.

When more than one force acts on an object, the total work is the sum of the work done by each force separately. Thus, if force $\mathbf{F}_1$ does work W_1, force $\mathbf{F}_2$ does work W_2, and so on, the total work is

$$W_{\text{total}} = W_1 + W_2 + W_3 + \cdots = \sum_i W_i \qquad \textbf{7–4}$$

Equivalently, the total work can be calculated by first performing a vector sum of all the forces acting on an object to obtain $\mathbf{F}_{\text{total}}$ and then using our basic definition of work:

$$W_{\text{total}} = (F_{\text{total}} \cos \theta)d = F_{\text{total}} d \cos \theta \qquad \textbf{7–5}$$

where θ is the angle between $\mathbf{F}_{\text{total}}$ and the displacement **d**. In the next two Examples we calculate the total work in each of these ways.

PROBLEM SOLVING NOTE
Be Careful About the Angle θ

In calculating $W = (F \cos \theta)d$, be sure that the angle you use in the cosine is the angle between the force and the direction of motion. Sometimes θ may be used to label a different angle in a given problem. For example, θ is often used to label the angle of a slope, in which case it may have nothing to do with the angle between the force and the displacement. To summarize: Just because an angle is labeled θ doesn't mean it's automatically the right one to use in the work formula.

EXAMPLE 7–3 A Coasting Car

A car of mass m coasts down a hill inclined at an angle ϕ below the horizontal. The car is acted on by three forces: (i) the normal force **N** exerted by the road, (ii) a force due to air resistance, $\mathbf{F}_{\text{air}}$, and (iii) the force of gravity, $m\mathbf{g}$. Find the total work done on the car as it travels a distance d along the road.

Picture the Problem
Since ϕ is the angle of the slope, it is also the angle between $m\mathbf{g}$ and the downward normal direction, as was shown in Section 5-7.

Strategy
For each force we calculate the work using $W = Fd \cos \theta$, where θ is the angle between that particular force and the displacement **d**. The total work is the sum of the work done by each of the three forces.

continued on the following page

continued from the previous page

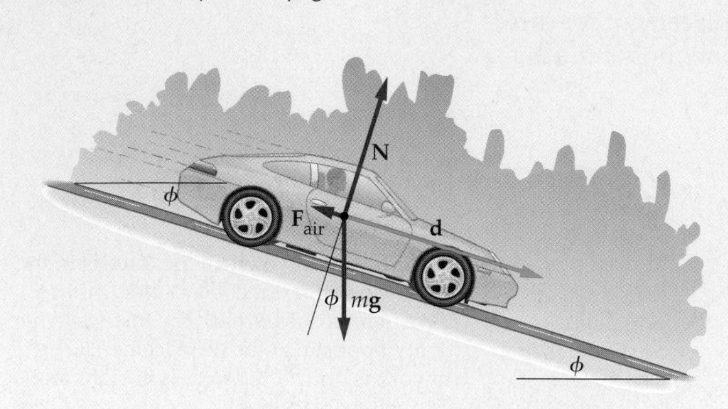

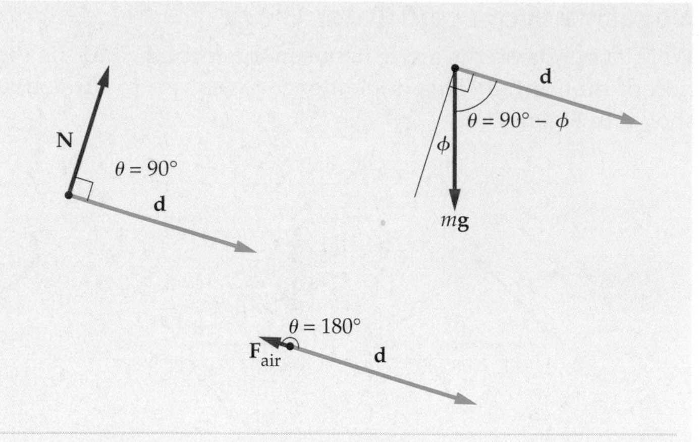

Solution

1. We start with the normal force, **N**. From the figure we see that $\theta = 90°$ for this force:

$$W_N = Nd\cos\theta = Nd\cos 90° = Nd(0) = 0$$

2. For the force of air resistance, $\theta = 180°$:

$$W_{air} = F_{air}d\cos 180° = F_{air}d(-1) = -F_{air}d$$

3. For gravity the angle θ is $\theta = 90° - \phi$, as indicated in the figure. Recall that $\cos(90° - \phi) = \sin\phi$ (see Appendix A):

$$W_{mg} = mgd\cos(90° - \phi) = mgd\sin\phi$$

4. The total work is the sum of the individual works:

$$W_{total} = W_N + W_{air} + W_{mg} = 0 - F_{air}d + mgd\sin\phi$$

Insight

The normal force is perpendicular to the motion of the car, and thus does no work. Air resistance points in a direction that opposes the motion, so it does negative work. On the other hand, gravity has a component in the direction of motion; therefore its work is positive.

Practice Problem

Calculate the total work done on a 1550-kg car as it coasts 20.4 m down a hill with $\phi = 5.00°$. Let the force due to air resistance be 15.0 N. **[Answer:** $W_{total} = W_N + W_{air} + W_{mg} = 0 - F_{air}d + mgd\sin\phi = 0 - 306\text{ J} + 2.67 \times 10^4\text{ J} = 2.64 \times 10^4\text{ J}$]

Some related homework problems: Problem 12, Problem 56

Next, we sum the forces acting on the car to find F_{total}, then calculate the total work using $W_{total} = F_{total}d\cos\theta$.

EXAMPLE 7–4 A Coasting Car II

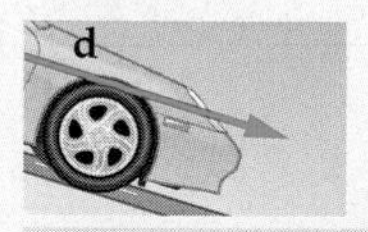

Consider the car described in Example 7–3. Calculate the total work done on the car using $W_{total} = F_{total}d\cos\theta$.

Picture the Problem

There is no acceleration in the y direction, which means that the total force in that direction must be zero. As a result, the total force acting on the car is in the x direction.

Strategy

We begin by finding the x component of each force vector, and then summing them to find the total force acting on the car. As can be seen from the figure, the total force points in the positive x direction; that is, in the same direction as the displacement. Therefore, the angle θ in $W = F_{total}d\cos\theta$ is zero.

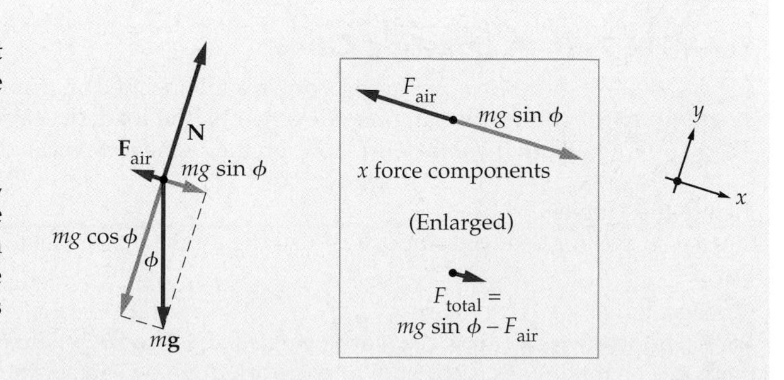

Solution

1. Referring to the figure above, we see that the *magnitude* of the total force is $mg\sin\phi$ minus F_{air}:

$$F_{total} = mg\sin\phi - F_{air}$$

2. The *direction* of $\mathbf{F}_{\text{total}}$ is the same as the direction of **d**, thus $\theta = 0°$. We can now calculate W_{total}:

$$W_{\text{total}} = F_{\text{total}}d\cos\theta = (mg\sin\phi - F_{\text{air}})d\cos 0°$$
$$= mgd\sin\phi - F_{\text{air}}d$$

Insight

Note that we were careful to calculate both the magnitude and the direction of the total force. The magnitude (which is always positive) gives F_{total} and the direction gives θ, allowing us to use $W_{\text{total}} = F_{\text{total}}d\cos\theta$.

Practice Problem

The total work done on a 1620-kg car as it coasts 25.0 m down a hill with $\phi = 6.00°$ is $W_{\text{total}} = 3.75 \times 10^4$ J. Find the magnitude of the force due to air resistance. [**Answer:** $F_{\text{air}}d = -W_{\text{total}} + mgd\sin\phi = 4030$ J, thus $F_{\text{air}} = (4030\text{ J})/d = 161$ N]

Some related homework problems: Problem 12, Problem 56

The full significance of positive versus negative work is seen in the next section, where we relate the work done on an object to the change in its speed.

7–2 Kinetic Energy and the Work-Energy Theorem

Suppose you drop an apple. As it falls, gravity does positive work on it, as indicated in Figure 7–5, and its speed increases. If you toss the apple upward gravity does negative work, and the apple slows down. In general, whenever the total work done on an object is positive its speed increases; when the total work is negative its speed decreases. In this section we derive an important result, the **work-energy theorem**, which makes this connection between work and change in speed precise.

◀ **FIGURE 7–5 Gravitational work** The work done by gravity on a freely falling apple moving downward is positive, and its speed increases. In contrast, gravity does negative work on a freely falling apple moving upward, and in this case its speed decreases.

To begin, consider an apple of mass m falling through the air, and suppose that two forces act on the apple—gravity, $m\mathbf{g}$, and the average force of air resistance, $\mathbf{F}_{\text{air}}$. The total force acting on the apple, $\mathbf{F}_{\text{total}}$, gives the apple a constant downward acceleration of magnitude

$$a = F_{\text{total}}/m$$

Since the total force is downward and the motion is downward, the work done on the apple is positive.

Now, suppose the initial speed of the apple is v_i, and that after falling a distance d its speed increases to v_f. The apple falls with constant acceleration a, hence constant-acceleration kinematics (Equation 2-12) gives

$$v_f^2 = v_i^2 + 2ad$$

or, with a slight rearrangement,

$$2ad = v_f^2 - v_i^2$$

Next, substitute $a = F_{total}/m$ into this equation:

$$2\left(\frac{F_{total}}{m}\right)d = v_f^2 - v_i^2$$

Multiplying both sides by m and dividing by 2 yields

$$F_{total}d = \tfrac{1}{2}mv_f^2 - \tfrac{1}{2}mv_i^2$$

where $F_{total}d$ is simply the total work done on the apple. Thus we find

$$W_{total} = \tfrac{1}{2}mv_f^2 - \tfrac{1}{2}mv_i^2$$

showing that total work is directly related to change in speed, as just mentioned. Note that $W_{total} > 0$ means $v_f > v_i$, $W_{total} < 0$ means $v_f < v_i$, and $W_{total} = 0$ implies that $v_f = v_i$.

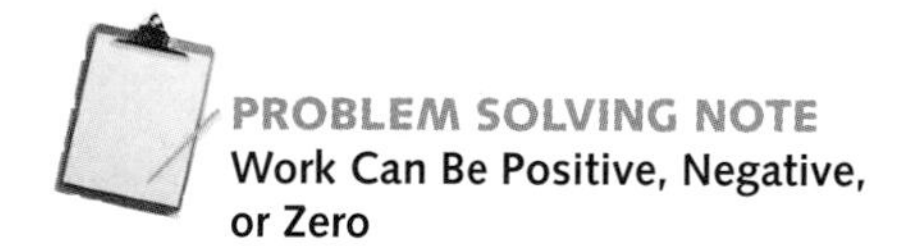

PROBLEM SOLVING NOTE
Work Can Be Positive, Negative, or Zero

When you calculate work be sure to keep track of whether it is positive or negative. The distinction is important, since positive work increases speed whereas negative work decreases speed. Zero work, of course, has no effect on speed.

The quantity $\frac{1}{2}mv^2$ in the last Equation 7–8 has a special significance in physics, as we shall see. We call it the **kinetic energy**, K:

Definition of Kinetic Energy, *K*

$$K = \tfrac{1}{2}mv^2 \qquad \textbf{7–6}$$

SI unit: $kg \cdot m^2/s^2$ = joule, J

We measure kinetic energy in joules, the same units as work, and both kinetic energy and work are scalars. Unlike work, however, the kinetic energy is never negative. Instead, K is always greater than or equal to zero, independent of the direction of motion or the direction of any forces.

To get a feeling for typical values of the kinetic energy, consider your kinetic energy when jogging. Assuming a mass of about 62 kg and a speed of 2.5 m/s, your kinetic energy is $K = \frac{1}{2}(62\text{ kg})(2.5\text{ m/s})^2 = 190\text{ J}$. Additional examples of kinetic energy are given in Table 7–2.

TABLE 7–2
Typical Kinetic Energies

Source	Approximate kinetic energy (joules)
Jet aircraft at 500 mph	10^9
Car at 60 mph	10^6
Home-run baseball	10^3
Person at walking speed	50
Housefly in flight	10^{-3}

EXERCISE 7–2

A truck moving at 15 m/s has a kinetic energy of 1.4×10^5 J. What is the mass of the truck?

Solution

$K = \frac{1}{2}mv^2$; therefore $m = 2K/v^2 = 1200$ kg.

In terms of kinetic energy, the work-energy theorem can be stated as follows:

Work-Energy Theorem

The total work done on an object is equal to the change in its kinetic energy:

$$W_{total} = \Delta K = \tfrac{1}{2}mv_f^2 - \tfrac{1}{2}mv_i^2 \qquad \textbf{7–7}$$

Though we have derived the work-energy theorem for a force that is constant in direction and magnitude it is valid for any force, as can be shown using the methods of calculus. Thus, the work-energy theorem is completely general, making it one of the more important and fundamental results in physics.

PROBLEM SOLVING NOTE
Starts from Rest Means $v_i = 0$

A problem statement that uses a phrase like "starts from rest" or "is raised from rest" is telling you that $v_i = 0$.

EXERCISE 7–3

How much work is required for a 74-kg sprinter to accelerate from rest to 2.2 m/s?

Solution

Since $v_i = 0$, we have $W = \frac{1}{2}mv_f^2 - \frac{1}{2}mv_i^2 = \frac{1}{2}mv_f^2 = 180\ \text{J}$.

We now present a variety of examples showing how the work-energy theorem is used in practical situations.

EXAMPLE 7–5 Hit the Books

A 4.1-kg box of books is lifted vertically from rest a distance of 1.6 m by an upward applied force of 60.0 N. Find **(a)** the work done by the applied force, **(b)** the work done by gravity, and **(c)** the final speed of the box.

Picture the Problem

Our sketch shows the direction of motion and the direction of the two forces, $\mathbf{F}_{app}$ and $m\mathbf{g}$, that act on the box.

Strategy

The applied force is in the direction of motion, so the work it does, W_{app}, is positive. Gravity is opposite in direction to the motion, thus its work, W_g, is negative. The total work is the sum of W_{app} and W_g, and the final speed of the box is found by applying the work-energy theorem, $W_{total} = \Delta K$.

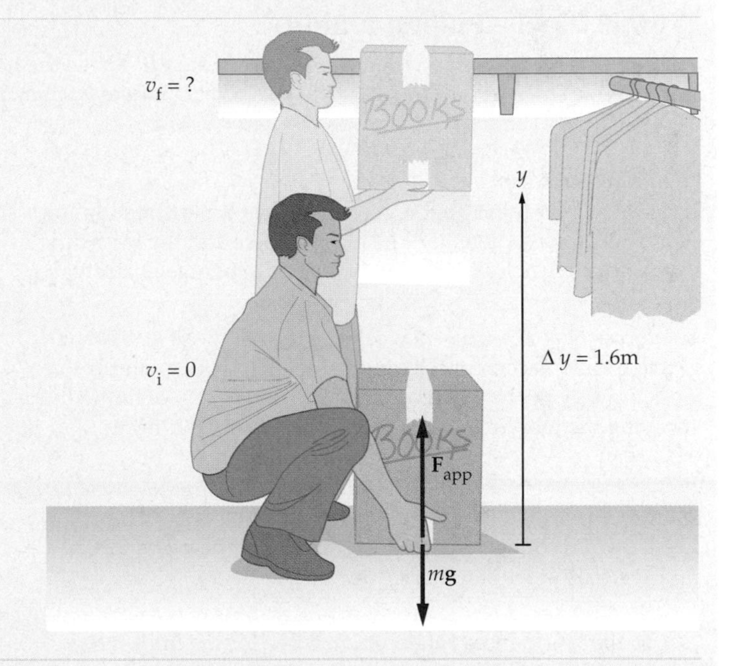

Solution

Part (a)

1. First we find the work done by the applied force. In this case, $\theta = 0°$ and the displacement is $\Delta y = 1.6$ m:

$$W_{app} = F_{app} \cos 0° \ \Delta y = (60.0\ \text{N})(1)(1.6\ \text{m}) = 96\ \text{J}$$

Part (b)

2. Next, we calculate the work done by gravity. The displacement is $\Delta y = 1.6$ m, as before, but now $\theta = 180°$:

$$W_g = mg \cos 180° \ \Delta y = (4.1\ \text{kg})(9.81\ \text{m/s}^2)(-1)(1.6\ \text{m}) = -64\ \text{J}$$

Part (c)

3. The total work done on the box, W_{total}, is the sum of W_{app} and W_g:

$$W_{total} = W_{app} + W_g = 96\ \text{J} - 64\ \text{J} = 32\ \text{J}$$

4. To find the final speed, v_f, we apply the work-energy theorem. Recall that the box started at rest, thus $v_i = 0$:

$$W_{total} = \tfrac{1}{2}mv_f^2 - \tfrac{1}{2}mv_i^2 = \tfrac{1}{2}mv_f^2$$

$$v_f = \sqrt{\frac{2W_{total}}{m}} = \sqrt{\frac{2(32\ \text{J})}{4.1\ \text{kg}}} = 3.9\ \text{m/s}$$

continued on the following page

continued from the previous page

Insight

As a check on our result, we can find v_f in a completely different way. First, calculate the acceleration of the box, with the result $a = (F_{app} - mg)/m = 4.8\ m/s^2$. Next, use this result in the kinematic equation $v^2 = v_0^2 + 2a\Delta y$. With $v_0 = 0$ and $\Delta y = 1.6$ m we find $v = 3.9$ m/s, in agreement with the results using the work-energy theorem.

Practice Problem

If the box is lifted only a quarter of the distance, is the final speed 1/8, 1/4, or 1/2 of the value found in step 4? Calculate v_f in this case as a check on your answer. [**Answer:** Since work depends linearly on Δy, and v_f depends on the square root of the work, it follows that the final speed is $\sqrt{1/4} = \frac{1}{2}$ the value in step 4. Letting $\Delta y = (1.6\ m)/4 = 0.40$ m we find $v_f = \frac{1}{2}(3.9\ m/s)$.]

Some related homework problems: Problem 13, Problem 16, Problem 17

In the previous example the initial speed was zero. This is not always the case, of course. The next example illustrates how to use the work-energy theorem when the initial velocity is nonzero.

EXAMPLE 7–6 Pulling a Sled

A boy exerts a force of 11.0 N at 29.0° above the horizontal on a 6.40-kg sled. Find the work done by the boy and the final speed of the sled after it moves 2.00 m, assuming the sled starts with an initial speed of 0.500 m/s and slides horizontally without friction.

Picture the Problem

The sketch shows the direction of motion and the directions of each of the forces. Note that the normal force and the force due to gravity are vertical, whereas the displacement is horizontal.

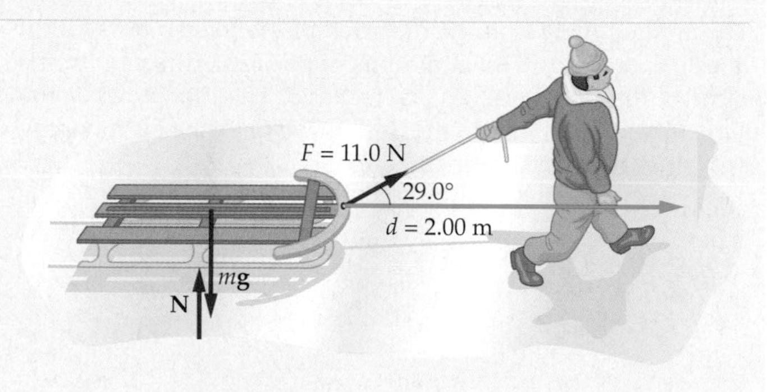

Strategy

The forces **N** and $m\mathbf{g}$ do no work because they are at right angles to the displacement. Therefore, the total work is simply the work done by the boy. After calculating this work, we find v_f by applying the work-energy theorem with $v_i = 0.500$ m/s.

Solution

1. The work done by the boy is $(F\cos\theta)d$, where $\theta = 29.0°$. This is also the total work done on the sled:

$$W_{boy} = (F\cos\theta)d$$
$$= (11.0\ N)(\cos 29.0°)(2.00\ m) = 19.2\ J = W_{total}$$

2. Use the work-energy theorem to solve for the final speed:

$$W_{total} = \Delta K = \tfrac{1}{2}mv_f^2 - \tfrac{1}{2}mv_i^2$$
$$\tfrac{1}{2}mv_f^2 = W_{total} + \tfrac{1}{2}mv_i^2$$
$$v_f = \sqrt{\frac{2W_{total}}{m} + v_i^2}$$

3. Substitute numerical values to get the final answer:

$$v_f = \sqrt{\frac{2(19.2\ J)}{6.40\ kg} + (0.500\ m/s)^2}$$
$$= 2.50\ m/s$$

Insight

If the sled had started from rest, instead of with an initial speed of 0.500 m/s, would its final speed be 2.50 m/s − 0.500 m/s = 2.00 m/s? No. If the initial speed is zero, then $v_f = \sqrt{\frac{2W_{total}}{m}} = \sqrt{\frac{2(19.2\ J)}{6.40\ kg}} = 2.45$ m/s. Why don't the speeds add and subtract in a straightforward way? The reason is that the work-energy theorem depends on the *square* of the speeds rather than on v_i and v_f directly.

Practice Problem
Suppose the sled starts with a speed of 0.500 m/s and has a final speed of 2.50 m/s after the boy pulls it through a distance of 3.00 m. What force did the boy exert on the sled? [**Answer:** $F = W_{\text{total}}/(d \cos \theta) = \Delta K/(d \cos \theta) = 7.32$ N]

Some related homework problems: Problem 20, Problem 46

The final speeds in the previous Examples could have been found using Newton's laws and the constant-acceleration kinematics of Chapter 2, as indicated in the Insight following Example 7–5. The work-energy theorem provides an alternative method of calculation that is often much easier to apply than Newton's laws. We return to this point in the next chapter.

CONCEPTUAL CHECKPOINT 7–2

To accelerate a certain car from rest to the speed v requires the work W_1. The work needed to accelerate the car from v to $2v$ is W_2. Which of the following is correct? **(a)** $W_2 = W_1$, **(b)** $W_2 = 2W_1$, **(c)** $W_2 = 3W_1$?

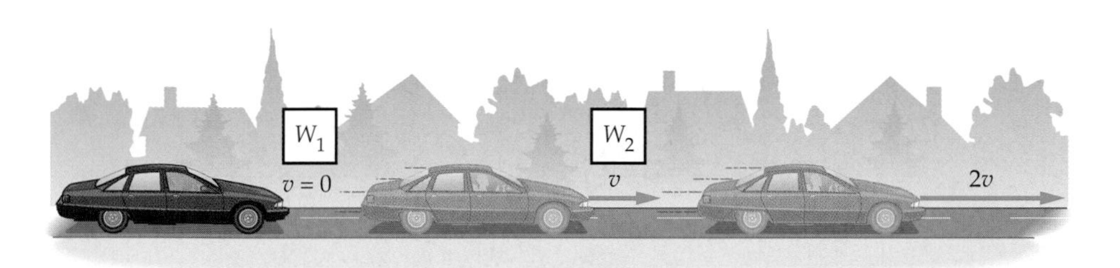

Be Careful About Linear Reasoning

Though some relations are linear—if you *double* the mass you *double* the kinetic energy—others are not. For example, if you *double* the speed you *quadruple* the kinetic energy. Be careful not to jump to conclusions based on linear reasoning.

Reasoning and Discussion
A common mistake is to reason that since we increase the speed by the same amount in each case, the work required is the same. It is not, and the reason is that work depends on the speed squared rather than on the speed itself.

To see how this works, first calculate W_1, the work needed to go from rest to a speed v. From the work-energy theorem, with $v_i = 0$, and $v_f = v$, we find $W_1 = \frac{1}{2}mv_f^2 - \frac{1}{2}mv_i^2 = \frac{1}{2}mv^2$. Similarly, the work needed to go from rest, $v_i = 0$, to a speed $v_f = 2v$ is simply $\frac{1}{2}m(2v)^2 = 4(\frac{1}{2}mv^2) = 4W_1$. Therefore, the work needed to increase the speed from v to $2v$ is the difference; $W_2 = 4W_1 - W_1 = 3W_1$.

Answer:
(c) $W_2 = 3W_1$

7–3 Work Done by a Variable Force

Thus far we have calculated work only for constant forces, yet most forces in nature vary with position. For example, the force exerted by a spring depends on how far the spring is stretched, and the force of gravity between planets depends on their separation. In this section we show how to calculate the work for a force that varies with position.

First, let's review briefly the case of a constant force, and develop a graphical interpretation of work. **Figure 7–6** shows a constant force plotted versus position, x. If the force moves an object a distance d, from x_1 to x_2, the work it does is $W = Fd = F(x_2 - x_1)$. Referring to the figure, we see that the work is equal to the shaded area[1] under the force line.

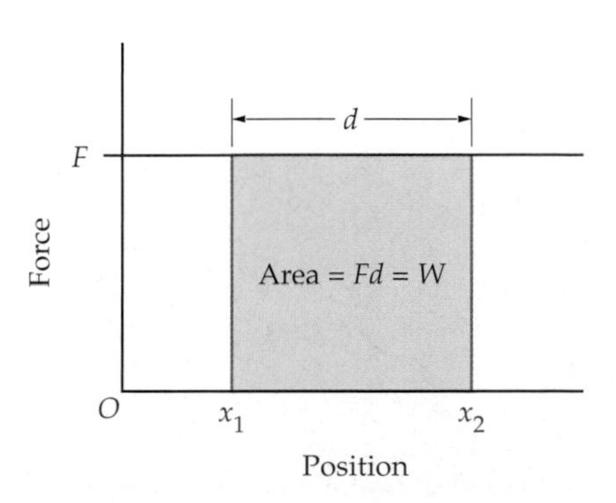

▲ **FIGURE 7–6 Graphical representation of work by constant force**
A constant force F acting through a distance d does a work $W = Fd$. Note that Fd is also equal to the shaded area between the force line and the x axis.

[1]Usually, area has the dimensions of (length) × (length), or length^2. In this case, the vertical axis is force and the horizontal axis is distance, therefore the dimensions of area are (force) × (distance), which in SI units is N·m = J.

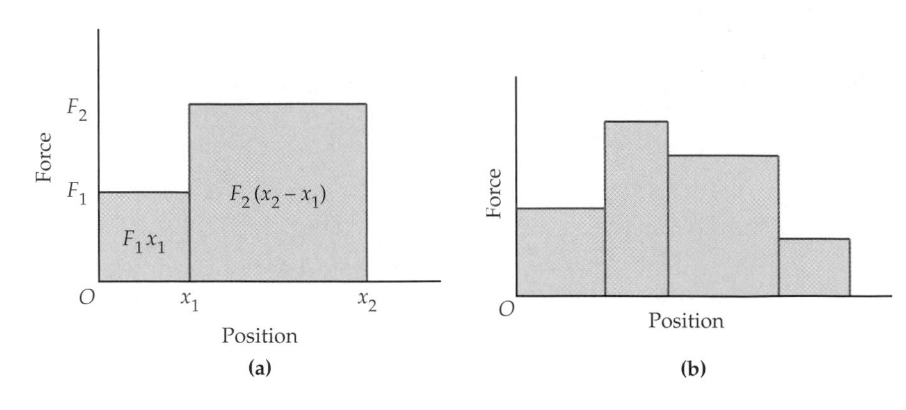

▶ **FIGURE 7–7 Work done by a non-constant force**
(a) A force with a value F_1 from 0 to x_1 and a value F_2 from x_1 to x_2 does the work $W = F_1x_1 + F_2(x_2 - x_1)$. This is simply the area of the two shaded rectangles. **(b)** If a force takes on a number of different values, the work it does is still the total area between the force lines and the x axis, just as in part (a).

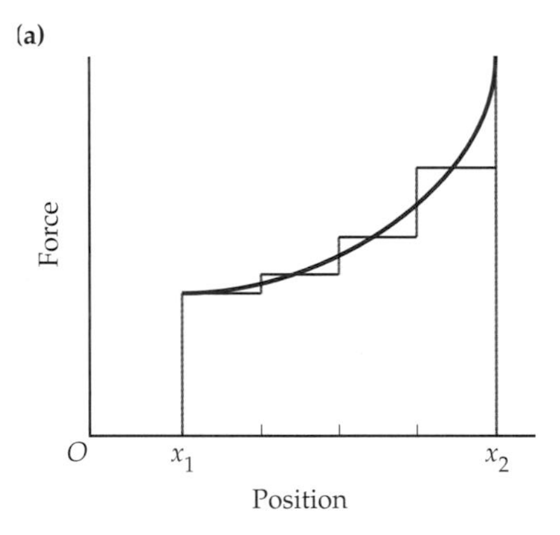

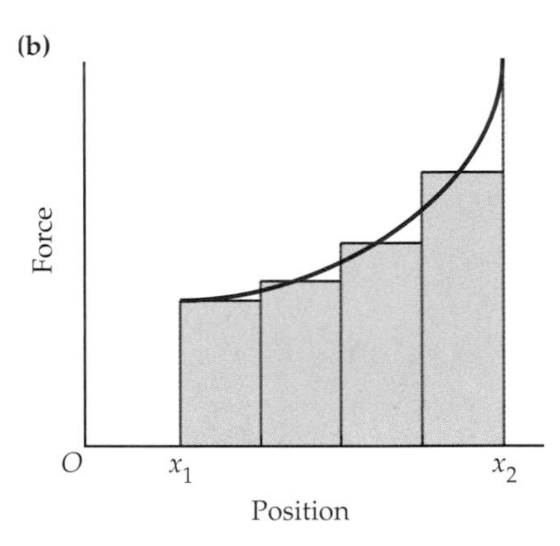

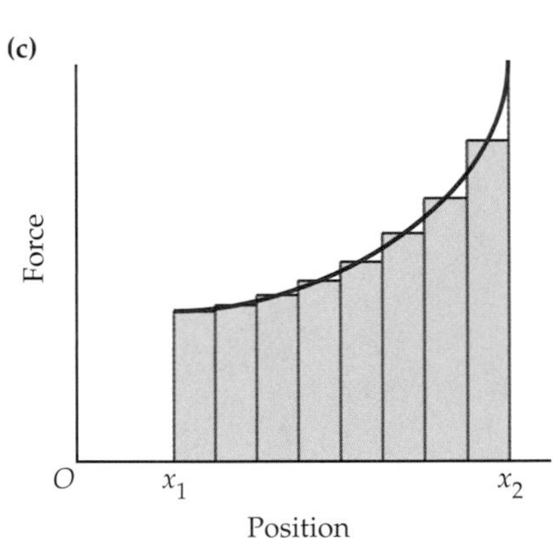

▲ **FIGURE 7–8 Work done by a continuously varying force**
(a) A continuously varying force can be approximated by a series of constant values that follow the shape of the curve. **(b)** The work done by the continuous force is approximately equal to the area of the small rectangles corresponding to the constant values of force shown in part **(a)**. **(c)** In the limit of an infinite number of vanishingly small rectangles, we see that the work done by the force is equal to the area between the force curve and the x axis.

Next, consider a force that has the value F_1 from $x = 0$ to $x = x_1$ and a different value F_2 from $x = x_1$ to $x = x_2$, as in **Figure 7–7 (a)**. The work in this case is the sum of the works done by F_1 and F_2. Therefore, $W = F_1x_1 + F_2(x_2 - x_1)$ which, again, is the area between the force lines and the x axis. Clearly, this type of calculation can be extended to a force with any number of different values, as indicated in **Figure 7–7 (b)**.

If a force varies continuously with position we can approximate it with a series of constant values that follow the shape of the curve, as shown in **Figure 7–8 (a)**. It follows that the work done by the continuous force is approximately equal to the area of the corresponding rectangles, as **Figure 7–8 (b)** shows. The approximation can be made better by using more rectangles, as illustrated in **Figure 7–8 (c)**. In the limit of an infinite number of vanishingly small rectangles the area of the rectangles becomes identical to the area under the force curve. Hence this area is the work done by the continuous force. To summarize:

> The work done by a force in moving an object from x_1 and x_2 is equal to the corresponding area between the force curve and the x axis.

▲ The restoring force exerted by a spring is typically proportional to the distance it is stretched or compressed. This makes springs useful for absorbing bumps and smoothing out the ride of many vehicles, including railroad cars.

A case of particular interest is that of a spring. Since the force exerted by a spring is $F = -kx$ (Section 6-2), it follows that the force we must exert to hold it at the position x is $+kx$. This is illustrated in **Figure 7–9**, where we also show that the corresponding force curve is a straight line extending from the origin. Therefore, the work we do in stretching a spring from $x = 0$ (equilibrium) to the general position x is the shaded, triangular area shown in **Figure 7–10**. This area is equal to $\frac{1}{2}$(base)(height), where in this case the base is x and the height is kx. As a result,

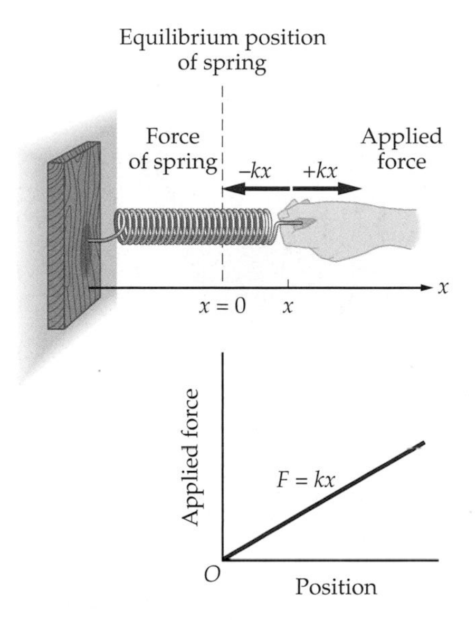

▲ **FIGURE 7–9 Stretching a spring**
The force we must exert on a spring to stretch it a distance x is $+kx$. Thus, applied force versus position for a spring is a straight line of slope k.

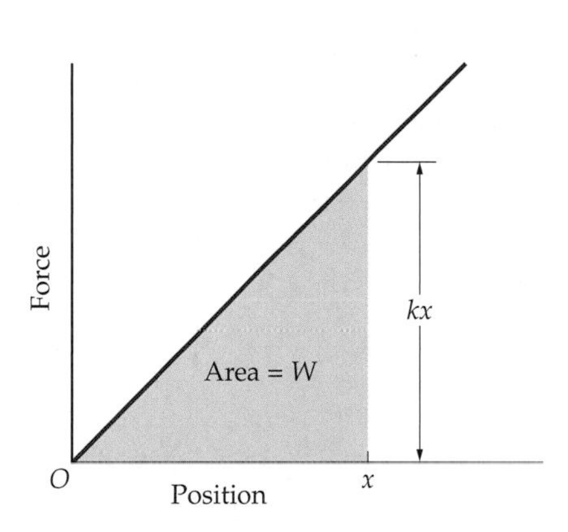

▲ **FIGURE 7–10 Work needed to stretch a spring a distance *x***
The work done is equal to the shaded area, which is a right triangle. The area of the triangle is $\frac{1}{2}(x)(kx) = \frac{1}{2}kx^2$.

the work is $\frac{1}{2}(x)(kx) = \frac{1}{2}kx^2$. Similar reasoning shows that the work needed to compress a spring a distance x is also $\frac{1}{2}kx^2$. Therefore,

Work To Stretch or Compress a Spring a Distance *x* from Equilibrium,

$$W = \frac{1}{2}kx^2 \qquad \text{7–8}$$

SI unit: joule, J

We can get a feeling for the amount of work required to compress a typical spring in the following Exercise.

EXERCISE 7–4

The spring in a pinball launcher has a force constant of 405 N/m. How much work is required to compress the spring a distance of 3.00 cm?

Solution

$$W = \frac{1}{2}kx^2 = \frac{1}{2}(405\text{ N/m})(0.0300\text{ m})^2 = 0.182\text{ J}$$

Note that the work done in compressing or expanding a spring varies with the second power of the distance. The consequences of this dependence are explored in the next Example.

EXAMPLE 7–7 Stretching the Slinky Dog

In the chase scene from the movie *Toy Story*, the slinky dog is stretched a considerable distance. **(a)** If the work required to stretch the slinky dog one meter is $W = 2$ J, what is the force constant for the slinky? **(b)** How much work is required to stretch the slinky dog from 1 m to 2 m?

continued on the following page

continued from the previous page

Picture the Problem

Our sketch shows the slinky dog being stretched to 1 m, then from 1 m to 2 m.

Strategy

(a) Given $W = 2$ J and $x = 1$ m, we can find the force constant k using $W = \frac{1}{2}kx^2$. **(b)** To find the work required to stretch from $x = 1$ m to $x = 2$ m, $W_{1\to2}$, we calculate the work to stretch from $x = 0$ to $x = 2$ m, $W_{0\to2}$, and subtract the work needed to stretch from $x = 0$ to $x = 1$ m, $W_{0\to1}$. (Note that we *cannot* simply assume the work to go from $x = 1$ m to $x = 2$ m is the same as the work to go from $x = 0$ to $x = 1$ m.)

Solution

Part (a)

1. Solve $W = \frac{1}{2}kx^2$ for the force constant k:

$$k = \frac{2W}{x^2} = \frac{2(2\text{ J})}{(1\text{ m})^2} = 4\text{ N/m}$$

Part (b)

2. First, calculate the work needed to stretch the dog from $x = 0$ to $x = 2$ m:

$$W_{0\to2} = \tfrac{1}{2}kx^2 = \tfrac{1}{2}(4\text{ N/m})(2\text{ m})^2 = 8\text{ J}$$

3. Subtract from this result the work to stretch from $x = 0$ to $x = 1$ m, which the problem statement gives as 2 J:

$$W_{1\to2} = W_{0\to2} - W_{0\to1} = 8\text{ J} - 2\text{ J} = 6\text{ J}$$

Insight

Our results show that more energy is needed to stretch the spring the second meter than to stretch it the first meter. Why? The reason is that the force of the spring increases with distance. Thus, the average force over the second meter is greater than the average force over the first meter. In fact, we can see from the adjacent figure that the average force between 1 m and 2 m (6 N) is three times the average force between 0 and 1 m (2 N). It follows, then, that the work required for the second meter is three times the work required for the first meter.

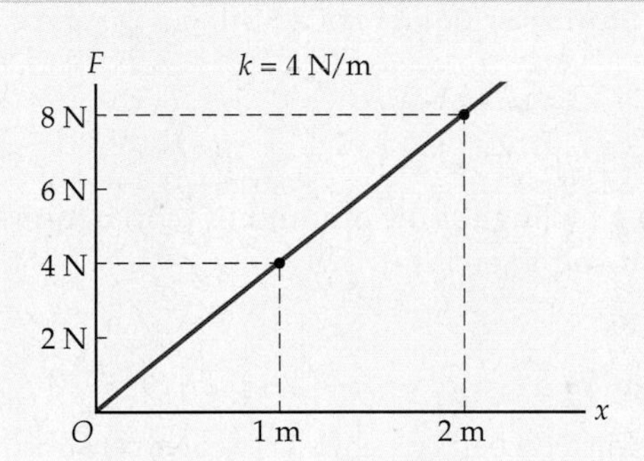

Practice Problem

How much stretch from equilibrium does 0.5 J of work produce in the slinky dog? [**Answer:** $x = 0.5$ m]

Some related homework problems: Problem 23, Problem 28

An equivalent way to calculate the work for a variable force is to multiply the average force, F_{av}, by the distance, d:

$$W = F_{av}\, d \qquad \textbf{7–9}$$

For a spring that is stretched a distance x from equilibrium the force varies linearly from 0 to kx. Thus, the average force is $F_{av} = \frac{1}{2}kx$, as indicated in **Figure 7–11**. Therefore, the work is

$$W = (\tfrac{1}{2}kx)(x) = \tfrac{1}{2}kx^2$$

As expected, our result agrees with Equation 7–8.

Finally, when you stretch or compress a spring the work you do is always positive. The work done *by* a spring, however, may be either positive or negative, depending on the situation. For example, consider a block on a smooth, horizontal surface, as shown in Figure 7–12. In part (a) the spring is expanding and pushing on the block. Since the force exerted by the spring is in the same direction as the block's motion, the spring does positive work. According to the work-energy theorem, the speed of the block will increase, as one would expect.

On the other hand, the block in Figure 7–12 (b) is compressing a spring. As a result, the spring's force is opposite in direction to the motion of the block, which means that the work done by the spring is negative. In particular, if the block compresses the spring by an amount x, the work done by the spring is $W = -\frac{1}{2}kx^2$. The block comes to rest when the negative work done by the spring is equal in magnitude to the block's initial kinetic energy: $W = \Delta K = K_f - K_i = 0 - K_i = -K_i$. We apply this result in Active Example 7–1.

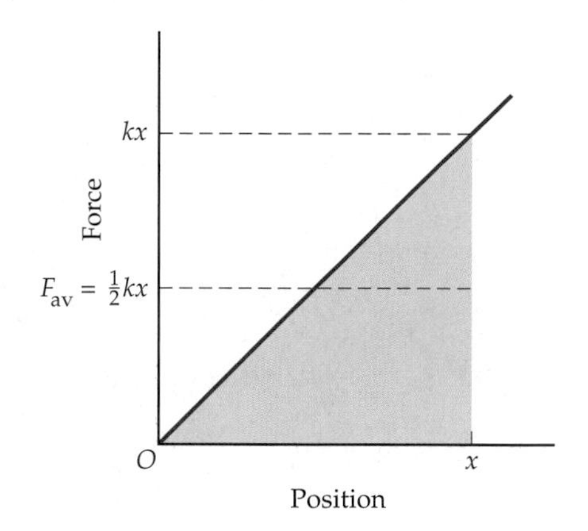

▲ **FIGURE 7–11 Work done in stretching a spring**

The average force of a spring from $x = 0$ to x is $F_{av} = \frac{1}{2}kx$, and the work done is $W = F_{av}d = (\frac{1}{2}kx)(x) = \frac{1}{2}kx^2$.

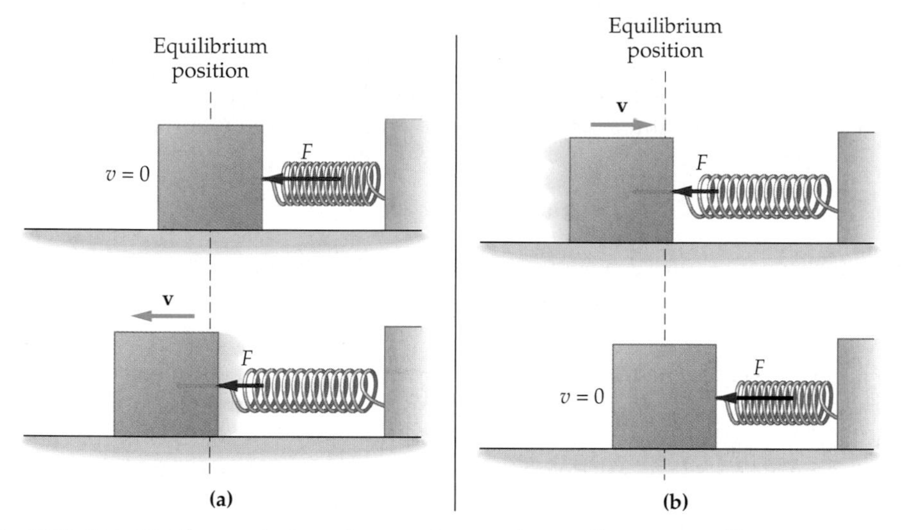

▲ **FIGURE 7–12 The work done by a spring can be positive or negative**

(a) In this case the spring pushes in the same direction as the block's motion; hence the work it does is positive, and the kinetic energy of the block increases. **(b)** Here the spring force is opposite to the direction of the block's motion. As a result, the work done by the spring is negative, and the block's kinetic energy decreases.

ACTIVE EXAMPLE 7–1 A Block Compresses a Spring

Suppose the block in Figure 7–12 (b) has a mass of 1.5 kg and moves with an initial speed of 2.2 m/s. Find the compression of the spring, whose force constant is 475 N/m, when the block momentarily comes to rest.

Solution

1. Calculate the initial and final kinetic energies of the block: $K_i = 3.6\text{ J}, K_f = 0$
2. Calculate the change in kinetic energy of the block: $\Delta K = -3.6\text{ J}$
3. Set the work done by the spring equal to the change in kinetic energy of the block: $-\frac{1}{2}kx^2 = \Delta K = -3.6\text{ J}$
4. Solve for the compression, x, and substitute numerical values: $x = 0.12\text{ m}$

Insight

After the block comes to rest the spring expands back to its equilibrium position. During this expansion the force exerted by the spring does *positive* work in the amount $W = \frac{1}{2}kx^2$. As a result, the block leaves the spring with the same speed it had initially.

7–4 Power

Power is a measure of how *quickly* work is done. To be precise, suppose the work W is performed in the time t. The average power delivered during this time is defined as follows:

Definition of Average Power, P

$$P = \frac{W}{t} \tag{7–10}$$

SI unit: J/s = watt, W

For simplicity of notation we drop the usual subscript av for an average quantity, and simply understand that the power P refers to an average power unless stated otherwise.

Note that the dimensions of power are joules (work) per second (time). We define one joule per second to be a watt (W), after James Watt (1736–1819), the Scottish engineer and inventor who played a key role in the development of practical steam engines:

$$1\ \text{watt} = 1\ \text{W} = 1\ \text{J/s} \tag{7–11}$$

Another common unit of power is horsepower (hp). It is defined as follows:

$$1\ \text{horsepower} = 1\ \text{hp} = 746\ \text{W} \tag{7–12}$$

TABLE 7–3 Typical Values of Power

Source	Approximate power (W)
Hoover Dam	1.34×10^9
Car moving at 40 mph	7×10^4
Home stove	1.2×10^4
Sunlight falling on one square meter	1380
Refrigerator	615
Television	200
Person walking up stairs	150

To get a feel for the magnitude of the watt and the horsepower, consider the power you might generate when walking up a flight of stairs. Suppose, for example, that an 80.0-kg person walks up a flight of stairs in 20.0 s, and that the altitude gain is 12.0 ft (3.66 m). Referring to Example 7–2 and Conceptual Checkpoint 7–1, we find that the work done by the person is $W = mgh = (80.0\ \text{kg})(9.81\ \text{m/s}^2)(3.66\ \text{m}) = 2870\ \text{J}$. To find the power we simply divide by the time; $P = W/t = (2870\ \text{J})/(20.0\ \text{s}) = 144\ \text{W} = 0.193\ \text{hp}$. Thus, a leisurely stroll up the stairs requires about 1/5 hp or 150 W. Similarly, the power produced by a sprinter bolting out of the starting blocks is about 1 hp, and the greatest power most people can produce for sustained periods of time is roughly 1/3 to 1/2 hp. (A horse, actually, can produce only about 2/3 hp for sustained periods.) Further examples of power are given in Table 7–3.

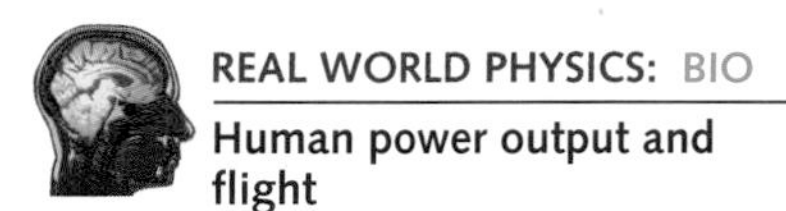

REAL WORLD PHYSICS: BIO
Human power output and flight

Human-powered flight is a feat just barely within our capabilities, since the most efficient human-powered airplanes require a steady power output of about 1/3 hp. On August 23, 1977, the Gossamer Condor, designed by Paul MacCready and flown by Bryan Allen, became the first human-powered airplane to complete a prescribed one-mile, figure-eight course and claim the Kremer Prize. Allen, an accomplished bicycle racer, used bicycle-like pedals to spin the propeller. Controlling the slow-moving craft while pedaling at full power was no easy task. Allen also piloted the Gossamer Albatross, which in 1979 became the first (and so far the only) human-powered aircraft to fly across the English Channel. This 22.25-mile flight, from Folkestone, England to Cap Gris-Nez, France, took 2 hours 49 minutes and required a total energy output roughly equivalent to climbing to the top of the Empire State Building 10 times.

▲ The Gossamer Albatross on its record-breaking flight across the English Channel in 1979. On two occasions the aircraft actually touched the surface of the water, but the pilot was able to maintain control and complete the 22.25-mile flight.

Power output is also an important factor in the performance of a car. For example, suppose it takes a certain amount of work, W, to accelerate a car from 0 to 60 mi/h. If the average power provided by the engine is P, then according to Equation 7–10 the amount of time required to reach 60 mi/h is $t = W/P$. Clearly, the greater the power P, the less the time required to accelerate. Thus, in a loose way of speaking, we can say that the power of a car is a measure of "how fast it can go fast."

EXAMPLE 7–8 Passing Fancy

To pass a slow-moving truck you want your fancy, 1.30×10^3-kg car to accelerate from 13.4 m/s (30.0 mph) to 17.9 m/s (40.0 mph) in 3.00 s. What is the minimum power required for this pass?

Picture the Problem

Our sketch shows the car accelerating from an initial speed of $v_i = 13.4$ m/s to a final speed of $v_f = 17.9$ m/s. We assume the road is level.

Strategy

Power is work divided by time, and work is equal to the change in kinetic energy as the car accelerates. Thus, $W = \Delta K$ and $P = \Delta K/t$.

Solution

1. First, calculate the change in kinetic energy:

$$\Delta K = \tfrac{1}{2}mv_f^2 - \tfrac{1}{2}mv_i^2 = \tfrac{1}{2}(1.30 \times 10^3\ \text{kg})(17.9\ \text{m/s})^2 - \tfrac{1}{2}(1.30 \times 10^3\ \text{kg})(13.4\ \text{m/s})^2$$
$$= 9.16 \times 10^4\ \text{J}$$

2. Divide by time to find the minimum power. (The actual power would have to be greater to overcome frictional losses):

$$P = \frac{W}{t} = \frac{\Delta K}{t} = \frac{9.16 \times 10^4\ \text{J}}{3.00\ \text{s}} = 3.05 \times 10^4\ \text{W} = 40.9\ \text{hp}$$

Insight

Suppose that your fancy car continues to produce the same 3.05×10^4 W of power as it accelerates from $v = 17.9$ m/s (40.0 mph) to $v = 22.4$ m/s (50.0 mph). Is the time required more than, less than, or equal to 3.00 s? *Answer*: It will take more than 3.00 s. The reason is that ΔK is greater for a change in speed from 40.0 mph to 50.0 mph than for a change in speed from 30.0 mph to 40.0 mph, since K depends on speed squared. Since ΔK is greater, the time $t = \Delta K/P$ is also greater.

Practice Problem

Find the time required to accelerate from 40.0 mph to 50.0 mph with 3.05×10^4 W of power. [**Answer:** First, $\Delta K = 1.17 \times 10^5$ J. Second, $P = \Delta K/t$ can be solved for time to give $t = \Delta K/P$. Thus, $t = 3.84$ s.]

Some related homework problems: Problem 33, Problem 44

Finally, consider a system in which a car, or some other object, is moving with a constant speed v. For example, a car might be traveling uphill on a road inclined at an angle θ above the horizontal. To maintain a constant speed, the engine must exert a constant force F to overcome friction, gravity, and air resistance, as indicated in **Figure 7–13**. Now, as the car travels a distance d, the work done by the engine is $W = Fd$, and the power it delivers is

$$P = \frac{W}{t} = \frac{Fd}{t}$$

Since the car has a constant speed v, the distance it covers in the time t is $d = vt$, thus

$$P = \frac{Fd}{t} = \frac{F(vt)}{t} = Fv \qquad \textbf{7–13}$$

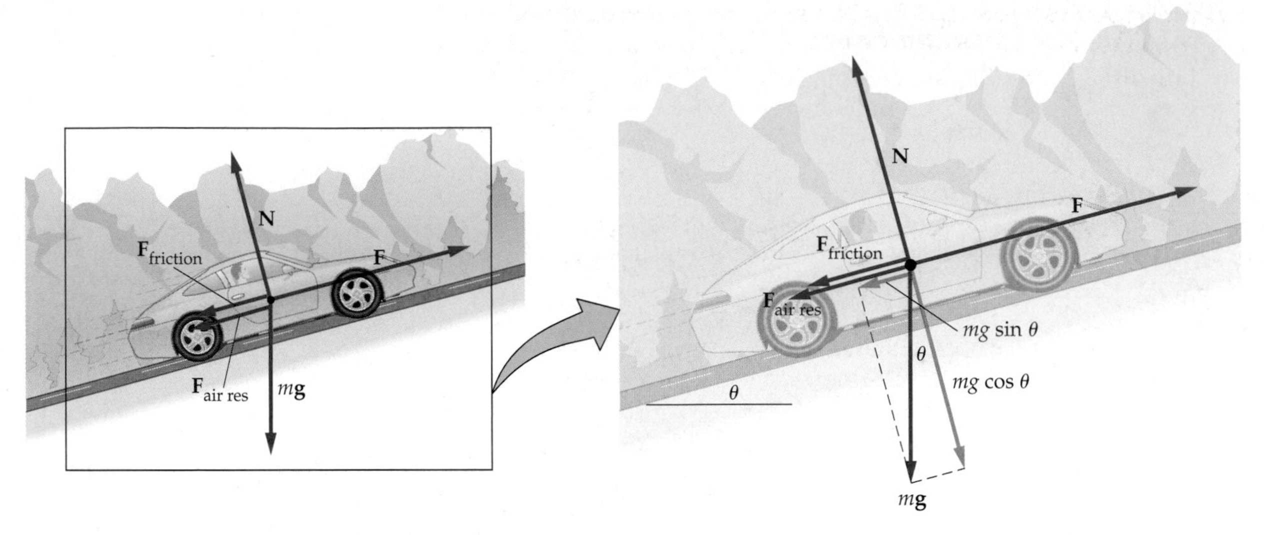

▲ **FIGURE 7–13 Driving up a hill**
A car traveling uphill at constant speed requires a constant force F, of magnitude $mg \sin \theta + F_{\text{air res}} + F_{\text{friction}}$, applied in the direction of motion.

This expression can be applied to cases where the force and speed vary, as long as we replace F and v with average values.

ACTIVE EXAMPLE 7–2 **Maximum Speed**

It takes a force of 1280 N to keep a 1500-kg car moving with constant speed up a slope of 5.00°. If the engine delivers 50.0 hp to the drive wheels, what is the maximum speed of the car?

Solution

1. Convert the power of 50.0 hp to watts: $P = 3.73 \times 10^4\ \text{W}$
2. Solve Equation 7–13 for the speed v: $v = P/F$
3. Substitute numerical values for the power and force: $v = 29.1\ \text{m/s}$

Insight
Thus, the maximum speed of the car on this slope is approximately 65 mi/h.

Chapter Summary

	Topic	Remarks and Relevant Equations	
7–1	**Work Done by a Constant Force**	A force exerted through a distance performs mechanical work.	
	force in direction of motion	In this, the simplest case, work is force times distance: $W = Fd$	7–1
	force at an angle θ to motion	Work is the component of force in the direction of motion, $F \cos \theta$, times distance, d: $W = (F \cos \theta)d = Fd \cos \theta$	7–3
	negative and total work	Work is negative if the force opposes the motion; that is, if $90° < \theta < 270°$.	
		If more than one force does work, the total work is the sum of the works done by each force separately: $W_{\text{total}} = W_1 + W_2 + W_3 + \cdots$	7–4

	Equivalently, sum the forces first to find F_{total}, then	
	$$W_{total} = (F_{total} \cos \theta)d = F_{total} d \cos \theta$$	7–5
units	The SI unit of work and energy is the joule, J:	
	$$1 \text{ J} = \text{N} \cdot \text{m}$$	7–2
7–2 Work-Energy Theorem and Kinetic Energy	Total work is equal to the change in kinetic energy:	
	$$W_{total} = \Delta K = \tfrac{1}{2}mv_f^2 - \tfrac{1}{2}mv_i^2$$	7–7
	Note: To apply this theorem correctly it is always necessary to use the *total* work.	
	Kinetic energy is one-half mass times speed squared:	
	$$K = \tfrac{1}{2}mv^2$$	7–6
	It follows that the kinetic energy is always positive or zero.	
7–3 Work Done by a Variable Force	Work is equal to the area between the force curve and the displacement on the x axis. For the case of a spring force, the work is	
	$$W = \tfrac{1}{2}kx^2$$	7–8
7–4 Power	Average power is work divided by the time required to do the work:	
	$$P = \frac{W}{t}$$	7–10
	Equivalently, power is force times speed:	
	$$P = Fv$$	7–13
units	The SI unit of power is the watt, W:	
	$$1 \text{ W} = 1 \text{ J/s}$$	7–11
	$$746 \text{ W} = 1 \text{ hp}$$	7–12

Problem-Solving Summary

Type of Calculation	Relevant Physical Concepts	Related Examples
Find the work done by a constant force.	Work is defined as force times distance, $W = Fd$, when F is in the direction of motion. Use $W = (F \cos \theta)d$ when there is an angle θ between the force and the direction of motion.	Examples 7–1 through 7–6
Calculate the change in speed.	The change in kinetic energy is given by the work-energy theorem, $W_{total} = \Delta K$. From this the change in speed can be found by recalling that $K = \frac{1}{2}mv^2$. Be sure W_{total} is the total work and that it has the correct sign.	Examples 7–5, 7–6
Calculate the power.	Find the work done, then divide by time: $P = W/t$. Alternatively, find the force, then multiply by the speed: $P = Fv$.	Example 7–8 Active Example 7–2

Conceptual Questions

1. Is it possible to do work on an object that remains at rest?
2. True or false: Only the total force acting on an object can do work.
3. True or false: A force that is always perpendicular to the velocity of a particle does no work on the particle.
4. A pendulum bob swings back and forth along a circular path. Does the tension in the string do any work on the bob? Does gravity do work on the bob? Explain.
5. To get out of bed in the morning, do you have to do work?
6. The net work done on a certain object is zero. What can you say about its speed?
7. A catcher stops a 90 mph pitch. Has the catcher done **(a)** positive work, **(b)** negative work, or **(c)** zero work on the ball? Explain.
8. Give an example of a frictional force doing negative work.
9. Give an example of a frictional force doing positive work.

10. A certain amount of work W_0 is required to accelerate a car from rest to a speed v. How much work is required to accelerate the car from rest to $v/2$?

11. The work W_0 accelerates a car from 0 to 50 km/h. How much work is needed to accelerate the car from 50 km/h to 150 km/h?

12. By what factor does the kinetic energy of a car change when its speed is tripled?

13. A package rests on the floor of an elevator that is rising with constant speed. The elevator exerts an upward normal force on the package, and hence does positive work on it. Why doesn't the kinetic energy of the package increase?

14. A ski boat moves with constant velocity. Is the net force acting on the boat doing work? Explain.

15. A youngster rides on a skateboard with a speed of 2 m/s. After a force acts on the youngster, her speed is 3 m/s. Was the work done by the force positive, negative, or zero?

16. Is it possible for a slow-moving elephant to have more kinetic energy than a fast-moving antelope? Explain.

17. Car 1 has twice the mass of car 2, but they both have the same kinetic energy. How do their speeds compare?

18. An object moves with constant velocity. Is it safe to conclude that no force acts on the object? Why, or why not?

19. A block of mass m and speed v collides with a spring, compressing it a distance Δx. How does Δx change if **(a)** v is doubled, or **(b)** m is doubled?

20. How does the work required to stretch a spring 2 cm compare with the work required to stretch it 1 cm?

21. Force F_1 does 5 J of work in 10 seconds. Force F_2 does 3 J of work in 5 seconds. Which force produces the greater power?

22. Engine 1 does twice the work of engine 2. Is it correct to conclude that engine 1 produces twice as much power as engine 2? Explain.

23. Engine 1 produces twice the power of engine 2. Is it correct to conclude that engine 1 does twice as much work as engine 2? Explain.

Problems

Note: ***IP*** *denotes an integrated conceptual/quantitative problem.* ***BIO*** *identifies problems of biological or medical interest. Blue bullets (•, ••, •••) are used to indicate the level of difficulty of each problem.*

Section 7.1 Work Done by a Constant Force

1. • A farmhand pushes a 23-kg bale of hay 3.0 m across the floor of a barn. If she exerts a horizontal force of 80.0 N on the hay, how much work has she done?

2. • Children in a tree house lift a small dog in a basket 4.5 m up to their house. If it takes 216 J of work to do this, what is the combined mass of the dog and basket?

3. • Early one October you go to a pumpkin patch to select your Halloween pumpkin. You lift the 3.2-kg pumpkin to a height of 1.2 m, then carry it 50.0 m (on level ground) to the check-out stand. **(a)** Calculate the work you do on the pumpkin as you lift it from the ground. **(b)** How much work do you do on the pumpkin as you carry it from the field?

4. • The coefficient of kinetic friction between a suitcase and the floor is 0.26. If the suitcase has a mass of 70.0 kg, how far can it be pushed across the level floor with 640 J of work?

5. •• You pick up a 3.6-kg can of paint from the ground and lift it to a height of 2.1 m. **(a)** How much work do you do on the can of paint? **(b)** You hold the can stationary for half a minute, waiting for a friend on a ladder to take it. How much work do you do during this time? **(c)** Your friend decides against the paint, so you lower it back to the ground. How much work do you do on the can as you lower it?

6. •• **IP** A tow rope, parallel to the water, pulls a water skier with constant velocity for a distance of 55 m before the skier falls. The tension in the rope is 105 N. **(a)** Is the work done on the skier by the rope positive, negative, or zero? Explain. **(b)** Calculate the work done by the rope on the skier.

7. •• **IP** In the situation described in the previous problem, **(a)** is the work done on the boat by the rope positive, negative, or zero? Explain. **(b)** Calculate the work done by the rope on the boat.

8. •• A child pulls a friend in a little red wagon with constant speed. If the child pulls with a force of 16 N for 10.0 m, and the handle of the wagon is inclined at an angle of 25° above the horizontal, how much work does the child do on the wagon?

9. •• A 55-kg packing crate is pulled with constant speed across a rough floor with a rope that is at an angle of 40.0° above the horizontal. If the tension in the rope is 125 N, how much work is done on the crate to move it 5.0 m?

10. •• **IP** To clean a floor, a janitor pushes on a mop handle with a force of 50.0 N. **(a)** If the mop handle is at an angle of 55° above the horizontal, how much work is required to push the mop 0.50 m? **(b)** If the angle the mop handle makes with the horizontal is increased to 65°, does the work done by the janitor increase, decrease, or stay the same? Explain.

11. •• A small plane tows a glider at constant speed and altitude. If the plane does 2.00×10^5 J of work to tow the glider 145 m and the tension in the tow rope is 2560 N, what is the angle between the tow rope and the horizontal?

12. •• Water skiers often ride to one side of the center line of a boat, as shown in **Figure 7–14**. In this case, the ski boat is traveling at 15 m/s and the tension in the rope is 75 N. If the boat does 3500 J of work on the skier in 50.0 m, what is the angle θ between the tow rope and the center line of the boat?

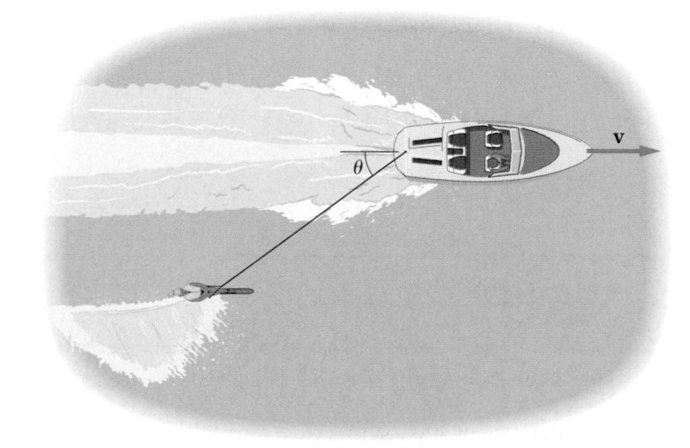

▲ **FIGURE 7–14** Problems 12 and 56

Section 7.2 Kinetic Energy and the Work-Energy Theorem

13. • How much work is needed for a 65-kg runner to accelerate from rest to 10.0 m/s?

14. • When Skylab reentered the Earth's atmosphere on July 11, 1979, it broke into a myriad of pieces. One of the largest fragments was a 1770-kg lead-lined film vault, and it landed with an estimated speed of 120 m/s. What was the kinetic energy of the film vault when it landed?

15. • **IP** A 10.0-g bullet has a speed of 1.20 km/s. **(a)** What is its kinetic energy in joules? **(b)** What is the bullet's kinetic energy if its speed is halved? **(c)** If its speed is doubled?

16. •• **IP** A 0.1-kg pine cone falls 12 m to the ground, where it lands with a speed of 13 m/s. **(a)** With what speed would the pine cone have landed if there had been no air resistance? **(b)** Did air resistance do positive work, negative work, or zero work on the pine cone? Explain.

17. •• In the previous problem, **(a)** how much work was done on the pine cone by air resistance? **(b)** What was the average force of air resistance exerted on the pine cone?

18. •• At $t = 1.0$ s, a 0.40-kg object is falling with a speed of 6.0 m/s. At $t = 2.0$ s, it has a kinetic energy of 25 J. **(a)** What is the kinetic energy of the object at $t = 1.0$ s? **(b)** What is the speed of the object at $t = 2.0$ s? **(c)** How much work was done on the object between $t = 1.0$ s and $t = 2.0$ s?

19. •• After hitting a long fly ball that goes over the right fielder's head and lands in the outfield, the batter decides to keep going past second base and try for third base. The 60.0-kg player begins sliding 3.00 m from the base with a speed of 4.2 m/s. If the player comes to rest at third base, **(a)** how much work was done on the player by friction? **(b)** What was the coefficient of kinetic friction between the player and the ground?

20. •• **IP** A 1300-kg car coasts on a horizontal road with a speed of 18 m/s. After crossing an unpaved, sandy stretch of road 30.0 m long its speed decreases to 15 m/s. **(a)** Was the net work done on the car positive, negative, or zero? Explain. **(b)** Find the magnitude of the average net force on the car in the sandy section.

21. •• **IP** **(a)** In the previous problem, the car's speed decreased by 3.0 m/s as it coasted across a sandy section of road 30.0 m long. If the sandy portion of the road had been only 15.0 m long, would the car's speed have decreased by 1.5 m/s, more than 1.5 m/s, or less than 1.5 m/s? Explain. **(b)** Calculate the change in speed in this case.

22. •• A 65-kg bicyclist rides his 10.0-kg bicycle with a speed of 12 m/s. **(a)** How much work must be done by the brakes to bring the bike and rider to a stop? **(b)** How far does the bicycle travel if it takes 4.0 s to come to rest? **(c)** What is the magnitude of the braking force?

Section 7.3 Work Done by a Variable Force

23. • A spring with a force constant of 2.5×10^4 N/m is initially at its equilibrium length. **(a)** How much work must you do to stretch the spring 0.050 m? **(b)** How much work must you do to compress it 0.050 m?

24. • A 1.2-kg block is held against a spring of force constant 1.0×10^4 N/m, compressing it a distance of 0.15 m. How fast is the block moving after it is released and the spring pushes it away?

25. • Initially sliding with a speed of 2.1 m/s, a 1.8-kg block collides with a spring and compresses it 0.35 m before coming to rest. What is the force constant of the spring?

26. • The force shown in **Figure 7–15** moves an object from $x = 0$ to $x = 0.75$ m. How much work is done by the force?

27. • An object is acted on by the force shown in **Figure 7–16**. What is the final position of the object if its initial position is $x = 0.50$ m and the work done on it is equal to **(a)** 0.12 J or **(b)** -0.29 J?

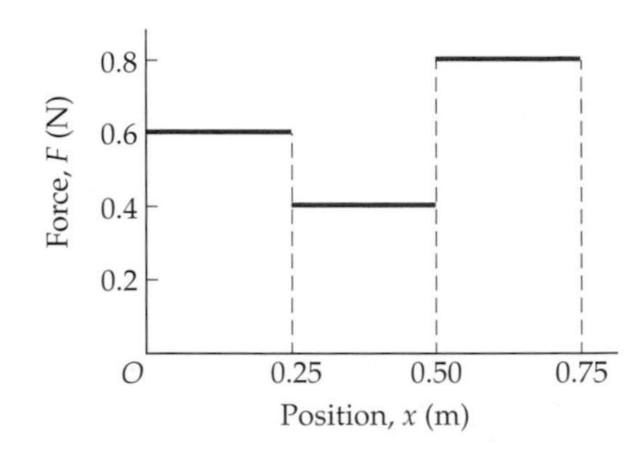

▲ **FIGURE 7–15** Problem 26

28. •• To compress spring 1 by 0.20 m takes 150 J of work. Stretching spring 2 by 0.30 m requires 210 J of work. Which spring is stiffer?

29. •• **IP** It takes 130 J of work to compress a certain spring 0.10 m. **(a)** What is the force constant of this spring? **(b)** To compress the spring an additional 0.10 m, does it take 130 J, more than 130 J, or less than 130 J? Verify your answer with a calculation.

30. •• The force shown in Figure 7–16 acts on a 1.7-kg object whose initial speed is 0.44 m/s and initial position is $x = 0.27$ m. **(a)** Find the speed of the object when it is at the location $x = 0.99$ m. **(b)** At what location would the object's speed be 0.16 m/s?

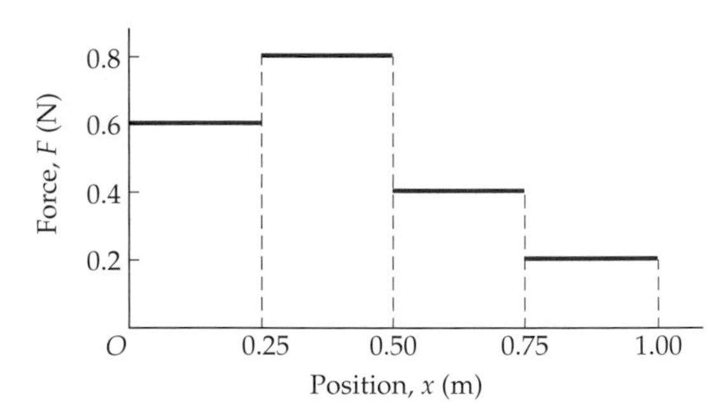

▲ **FIGURE 7–16** Problems 27 and 30

31. ••• A block is acted on by a force that varies as $(2.0 \times 10^4 \text{ N/m})x$ for $0 \le x \le 0.21$ m, and then remains constant at 4200 N for larger x. How much work does the force do on the block in moving it **(a)** from $x = 0$ to $x = 0.30$ m, or **(b)** from $x = 0.10$ m to $x = 0.40$ m?

32. ••• To compress spring 1 a certain distance takes a work W. To stretch spring 2 twice that distance takes a work $W/2$. If the force constant of spring 1 is k, what is the force constant of spring 2?

Section 7.4 Power

33. • What is the average power needed to accelerate a 750-kg car from 0 to 65 mi/h in 6.0 seconds? Assume that all forms of frictional losses can be ignored.

34. • **BIO** A record was set for stair climbing when a man ran up the 1600 steps of the Empire State Building in 10 minutes and 59 seconds. If the height gain of each step was 0.20 m, and the man's mass was 70.0 kg, what was his average power output during the climb? Give your answer in both watts and horsepower.

35. • How many joules are in a kilowatt-hour?

36. • Calculate the power output of a 1.0-g fly as it walks straight up a window pane at 2.5 cm/s.

37. • An ice cube is placed in a microwave oven. Suppose the oven delivers 105 W of power to the ice cube and that it takes 32,000 J to melt it. How long does it take for the ice cube to melt?

38. • You raise a bucket of water from the bottom of a deep well. If your power output is 108 W, and the mass of the bucket and the water in it is 5.00 kg, with what speed can you raise the bucket? Ignore the weight of the rope.

39. •• In order to keep a leaking ship from sinking, it is necessary to pump 10.0 lb of water each second from below deck up a height of 2.00 m and over the side. What is the minimum horsepower motor that can be used to save the ship?

40. •• **IP** A kayaker paddles with a power output of 50.0 W to maintain a steady speed of 1.50 m/s. **(a)** Calculate the resistive force exerted by the water on the kayak. **(b)** If the kayaker doubles her power output, and the resistive force due to the water remains the same, by what factor does the kayaker's speed change?

41. •• **BIO** Human-powered aircraft require a pilot to pedal, as in a bicycle, and produce a sustained power output of about 0.30 hp. The Gossamer Albatross flew across the English Channel on June 12, 1979 in 2 h 49 min. **(a)** How much energy did the pilot expend during the flight? **(b)** How many Snickers candy bars (280 Cal per bar) would the pilot have to consume to be "fueled up" for the flight?

(Note: The nutritional calorie, 1 Cal, is equivalent to 1000 calories (1000 cal) as defined in physics. In addition, the conversion factor between calories and joules is as follows: 1 Cal = 1000 cal = 1 kcal = 4186 J.)

42. •• **IP** A grandfather clock is powered by the descent of a 4.00-kg weight. **(a)** If the weight descends through a distance of 0.750 m in 3.00 days, how much power does it deliver to the clock? **(b)** To increase the power delivered to the clock, should the time it takes for the mass to descend be increased or decreased? Explain.

43. •• **BIO** Estimate the power you produce in running up a flight of stairs. Give your answer in horsepower.

44. ••• **IP** A certain car can accelerate from rest to the speed v in T seconds. If the power output of the car remains constant, **(a)** how long does it take for the car to accelerate from v to $2v$? **(b)** How fast is the car moving at $2T$ seconds after starting?

45. ••• **BIO** The force of water resistance on a swimming whale can be approximated by $F = bv$, where v is the swimming speed and b is a constant with units N · s/m. Find the maximum speed of the whale, assuming it can produce a power P.

General Problems

46. • Coasting along at 7.50 m/s, a 60.0-kg bicyclist on a 7.00-kg bicycle encounters a small hill. If the speed of the bicyclist is 6.00 m/s at the top of the hill, how much work was done on the bicyclist and her bike?

47. • A Mountain bar has a mass of 0.045 kg and a calorie rating of 210 Cal. What speed would this candy bar have if its kinetic energy were equal to its metabolic energy. (See the note following Problem 41.)

48. • A small motor runs a lift that raises a load of bricks weighing 830 N to a height of 12 m in 25 s. Assuming that the bricks are lifted with constant speed, what is the minimum power the motor must produce?

49. • You push a 75-kg box across a floor where the coefficient of kinetic friction is $\mu_k = 0.65$. The force you exert is horizontal. **(a)** How much power is needed to push the box at a speed of 0.50 m/s? **(b)** How much work do you do if you push the box for 35 s?

50. •• After a tornado, a 0.40-g straw was found embedded 2.6 cm into the trunk of a tree. If the average force exerted on the straw by the tree was 65 N, what was the speed of the straw when it hit the tree?

51. •• **IP** A hockey puck slows from 45 m/s to 44 m/s as it slides 25 m across the ice. **(a)** What is the coefficient of kinetic friction between the ice and the puck? **(b)** If the puck slides another 25 m, is its speed reduced to 43 m/s, more than 43 m/s, or less than 43 m/s? Explain.

52. •• **IP** The coefficient of kinetic friction between a suitcase and the floor is 0.240. **(a)** If the suitcase has a mass of 50.0 kg, how much work does it take to push the suitcase horizontally with constant speed a distance of 2.60 m across the floor? **(b)** If the suitcase speeds up as you push it, is the work you have done more than, less than, or the same as in part (a)?

53. •• You throw a glove straight upward to celebrate a victory. Its initial kinetic energy is K and it reaches a maximum height h. What is the kinetic energy of the glove when it is at the height $h/2$?

54. •• The force shown in **Figure 7–17** acts on an object that moves along the x axis. How much work is done by the force as the object moves from **(a)** $x = 0$ to $x = 2.0$ m, **(b)** $x = 1.0$ m to $x = 4.0$ m, and **(c)** $x = 2.0$ m to $x = 3.0$ m?

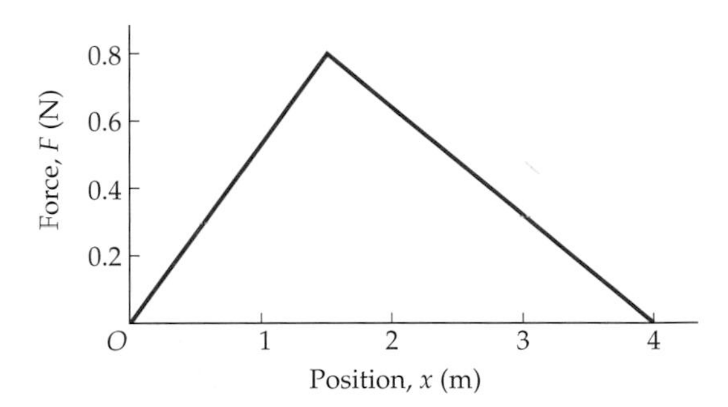

▲ **FIGURE 7–17** Problem 54

55. •• A juggling ball of mass m is thrown straight upward from an initial height h with an initial speed v_0. How much work has gravity done on the ball **(a)** when it reaches its greatest height, h_{max}, and **(b)** when it reaches ground level? **(c)** Find an expression for the kinetic energy of the ball as it lands.

56. •• The water skier in Figure 7–14 is at an angle of 35° with respect to the center line of the boat, and is being pulled at a constant speed of 14 m/s. If the tension in the tow rope is 90.0 N, **(a)** how much work does the rope do on the skier in 10.0 s? **(b)** How much work does the resistive force of water do on the skier in the same time?

57. •• A 6.0-kg block rests on a rough, horizontal surface with a coefficient of static friction equal to 0.45. A spring with a force constant $k = 1500$ N/m is attached to one side of the block. How much work does it take to stretch the spring to the point where the block is on the verge of moving?

58. •• Car 1, with mass m and power P, can accelerate from 0 to 60.0 mi/h in 7.00 s. Car 2 has mass $2m$, and can accelerate from 0 to 50.0 mi/h in 7.00 s. What is the power of car 2?

59. •• Calculate the power output of a 1.2-g beetle as it walks up a window pane at 2.3 cm/s. The beetle walks on a path that is at 25° to the vertical, as illustrated in **Figure 7–18**.

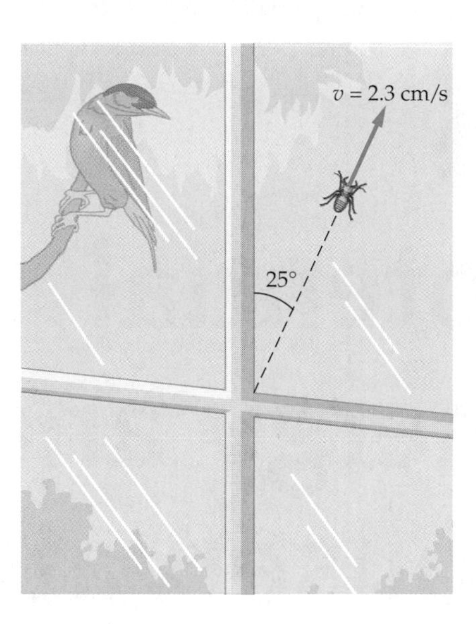

▲ **FIGURE 7–18** Problem 59

60. •• The motor of a ski boat produces a power of 36,600 W to maintain a constant speed of 14.0 m/s. To pull a water skier at the same constant speed the motor must produce a power of 37,800 W. What is the tension in the rope pulling the skier?

61. •• To make a batch of cookies you mix half a bag of chocolate chips into a bowl of cookie dough, exerting a 25-N force on the stirring spoon. Assume that your force is always in the direction of motion of the spoon. **(a)** What power is needed to move the spoon at a speed of 0.24 m/s? **(b)** How much work do you do if you stir the mixture for 1.5 min?

62. •• A pitcher accelerates a 0.14-kg hardball from rest to 40.0 m/s. **(a)** How much work does the pitcher do on the ball? **(b)** If the ball is in contact with the pitcher's hand for 0.050 s, what is the pitcher's power output during the pitch?

63. •• A catapult launcher on an aircraft carrier accelerates a jet from rest to 72 m/s. The work done by the catapult during the launch is 7.6×10^7 J. **(a)** What is the mass of the jet? **(b)** If the jet is in contact with the catapult for 2.0 s, what is the power output of the catapult?

64. •• On October 9, 1992, a 27-pound meteorite struck a car in Peekskill, NY, creating a dent about 22 cm deep. If the initial speed of the meteorite was 550 m/s, what was the average force exerted on the meteorite by the car?

▲ An interplanetary fender-bender (Problem 64)

65. ••• **BIO** A pigeon in flight experiences a force of air resistance given by $F = bv^2$, where v is the flight speed and b is a constant with units $\text{N}\cdot\text{s}^2/\text{m}^2$. What is the maximum speed of the pigeon if its power output is P?

66. ••• **IP** A block of mass m sliding with an initial speed v_0 collides with a spring and compresses it a distance x. **(a)** Find the force constant of the spring. **(b)** If the experiment is repeated, this time with an initial speed of $2v_0$, how far is the spring compressed?

67. ••• Two springs, with force constants k_1 and k_2, are connected in parallel, as shown in **Figure 7–19**. How much work is needed to stretch this system a distance x?

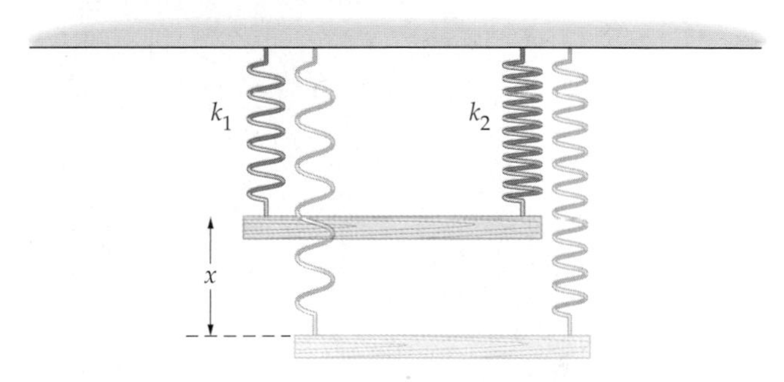

▲ **FIGURE 7–19** Problem 67

68. ••• Two springs, with force constants k_1 and k_2, are connected in series, as shown in **Figure 7–20**. How much work is needed to stretch this system a distance x?

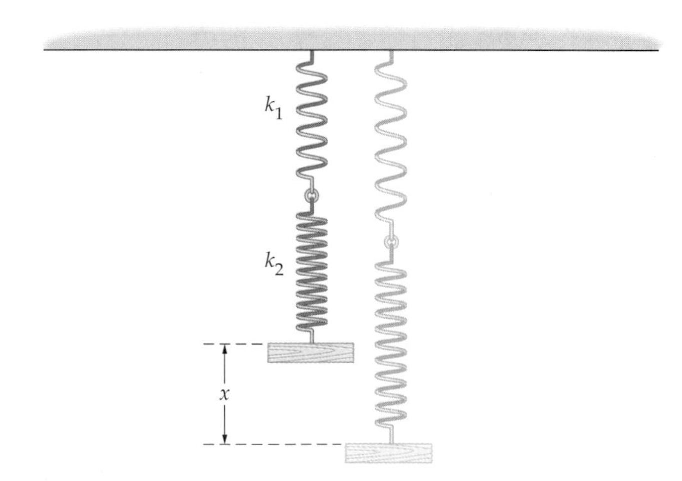

▲ **FIGURE 7–20** Problem 68

69. ••• A block rests on a horizontal, frictionless surface. A string is attached to the block, and is pulled with a force of 45.0 N at an angle θ above the horizontal, as shown in **Figure 7–21**. After the block is pulled through a distance of 1.50 m its speed is 2.60 m/s, and 50.0 J of work has been done on it. **(a)** What is the angle θ? **(b)** What is the mass of the block?

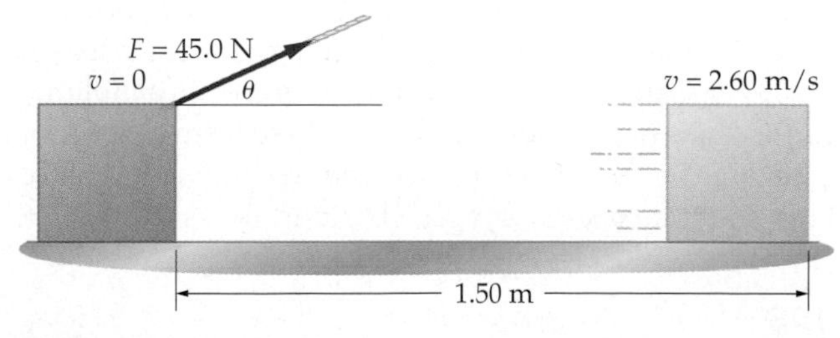

▲ **FIGURE 7–21** Problem 69

8 Potential Energy and Conservative Forces

Probably everyone has seen a skateboarder zoom up a ramp, slow, hang motionless in midair for an instant, and then start to descend, picking up speed on the way. At the top of his trajectory, where has his kinetic energy gone? And how does it reappear as he descends? Trying to answer such questions, we will find that there are other kinds of energy besides those considered in the last chapter.

One of the greatest accomplishments of physics is the concept of energy and its conservation. To realize, for example, that there is an important physical quantity that we can neither see nor touch is an impressive leap of the imagination. Even more astonishing, however, is the discovery that energy comes in a multitude of forms, and that the sum total of all these forms of energy is a constant. The universe, in short, has a certain amount of energy, and that energy simply ebbs and flows from one form to another, with the total amount remaining fixed.

In this chapter we focus on the conservation of energy, the first "conservation law" to be studied in this text. Though only a handful of conservation laws are known, they are all of central importance in physics. Not only do they give deep insight into the workings of nature, they are also practical tools in problem solving. As we shall see in this chapter, many problems that would be difficult to solve using Newton's laws can be solved with ease using the principle of energy conservation.

8–1 Conservative and Nonconservative Forces

In physics, we classify forces according to whether they are *conservative* or *non-conservative*. The key distinction is that when a **conservative force** acts, the work it does is stored in the form of energy that can be released at a later time. In this section we sharpen this distinction and explore some examples of conservative and nonconservative forces.

Perhaps the simplest case of a conservative force is gravity. Imagine lifting a box of mass m from the floor to a height h, as in **Figure 8–1**. To lift the box with constant speed, the force you must exert against gravity is mg. Since the upward distance is h, the work you do on the box is $W = mgh$. If you now release the box and allow it to drop back to the floor, gravity does the same work, $W = mgh$, and in the process gives the box an equivalent amount of kinetic energy.

Contrast this with the force of kinetic friction, which is nonconservative. To slide a box of mass m across the floor with constant speed, as shown in **Figure 8–2**, you must exert a force of magnitude $\mu_k N = \mu_k mg$. After sliding the box a distance d, the work you have done is $W = \mu_k mgd$. In this case, when you release the box it simply stays put—friction does no work on it after you let go. Thus, the work done by a **nonconservative force** cannot be recovered later as kinetic energy; instead, it is converted to other forms of energy, such as a slight warming of the floor and box in our example.

The differences between conservative and nonconservative forces are even more apparent if we consider moving an object around a closed path. Consider, for example, the path shown in **Figure 8–3**. If we move a box of mass m along this path, the total work done by gravity is the sum of the work done on each segment of the path; that is $W_{total} = W_{AB} + W_{BC} + W_{CD} + W_{DA}$. The work done by gravity from A to B and from C to D is zero since the force is at right angles to the dis-

◀ **FIGURE 8–1 Work against gravity**
Lifting a box against gravity with constant speed takes a work mgh. When the box is released, gravity does the same work on the box as it falls. Gravity is a conservative force.

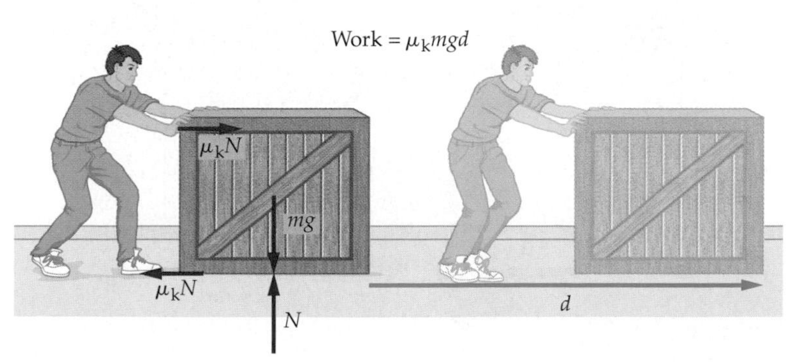

◀ **FIGURE 8–2 Work against friction**
Pushing a box with constant speed against friction takes a work $\mu_k mgd$. When the box is released it quickly comes to rest and friction does no further work. Friction is a nonconservative force.

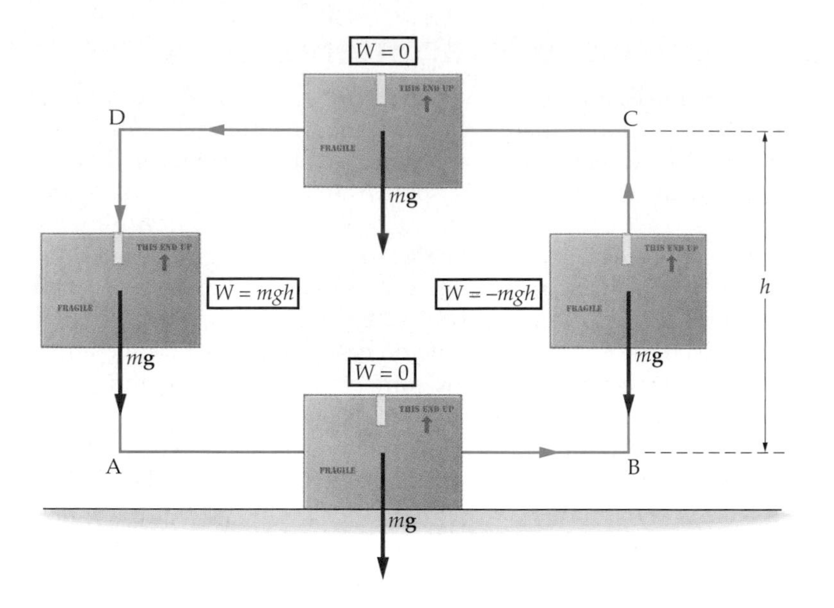

▶ FIGURE 8–3 Work done by gravity on a closed path
Gravity does no work on the two horizontal segments of the path. On the two vertical segments, the amounts of work done are equal in magnitude but opposite in sign. Thus, the total work done by gravity on a closed path is zero.

placement on these segments. Thus $W_{AB} = W_{CD} = 0$. On the segment from B to C gravity does negative work (displacement and force are in opposite directions), but it does positive work from D to A (displacement and force are in the same direction). Hence, $W_{BC} = -mgh$ and $W_{DA} = mgh$. As a result, the total work done by gravity is zero:

$$W_{\text{total}} = 0 + (-mgh) + 0 + mgh = 0$$

With friction, the results are quite different. If we move the box around the closed path shown in **Figure 8–4**, the total work done by friction does not vanish. In fact, friction does the negative work $W = -\mu_k mgd$ on each segment. Therefore, the total work done by kinetic friction is

$$W_{\text{total}} = (-\mu_k mgd) + (-\mu_k mgd) + (-\mu_k mgd) + (-\mu_k mgd) = -4\mu_k mgd$$

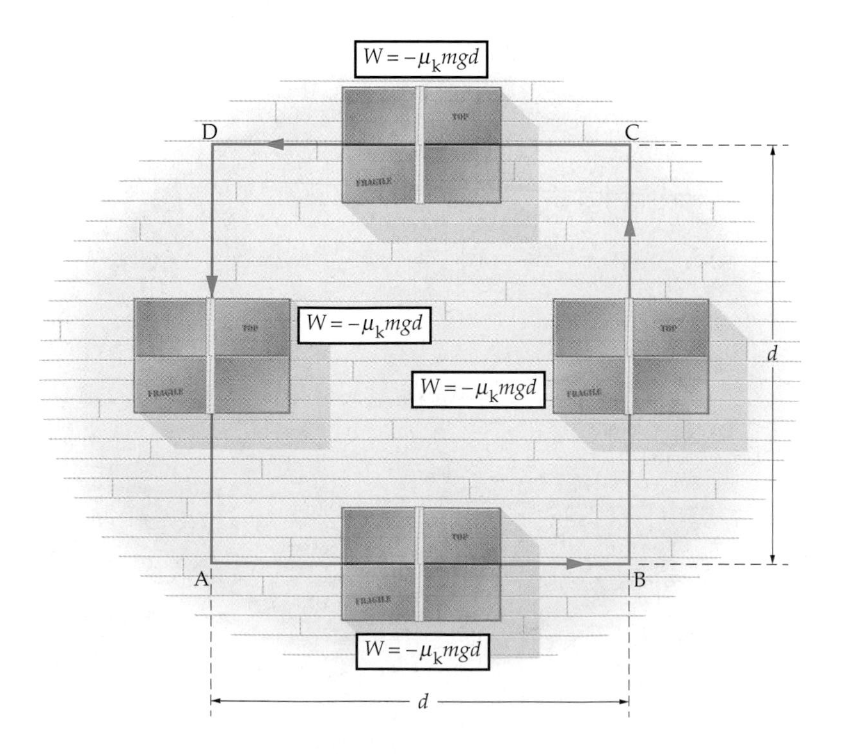

▶ FIGURE 8–4 Work done by friction on a closed path
The work done by friction when an object moves through a distance d is $-\mu_k mgd$. Thus, the total work done by friction on a closed path is not zero. In this case it is equal to $-4\mu_k mgd$.

◀ **FIGURE 8–5 Gravity is a conservative force**

If frictional forces can be ignored, a roller coaster car will have the same speed at points A and D, since they are at the same height. Hence, after any complete circuit of the track the speed of the car returns to its initial value. It follows that the change in kinetic energy is zero for a complete circuit, and therefore the work done by gravity is also zero.

These results lead to the following definition of a conservative force:

Conservative Force: Definition 1

A conservative force does zero total work on a closed path.

A roller coaster provides a good illustration of this definition. If a car on a roller coaster has a speed v at point A in **Figure 8–5**, it speeds up as it drops to point B, slows down near point C, and so on. When the car returns to its original height, at point D, it will again have the speed v, as long as friction and other nonconservative forces can be neglected. Similarly, if the car completes a circuit of the track and returns to point A, it will again have the speed v. Hence, a car's kinetic energy is unchanged ($\Delta K = 0$) after *any* complete circuit of the track. From the work energy theorem, $W_{\text{net}} = \Delta K$, it follows that the work done by gravity is zero for the closed path of the car, as expected for a conservative force.

This property of conservative forces has interesting consequences. For instance, consider the closed paths shown in **Figure 8–6**. On each of these paths, we know that the work done by a conservative force is zero. Thus, it follows from paths 1 and 2 that $W_{\text{total}} = W_1 + W_2 = 0$, or

$$W_2 = -W_1$$

Similarly, using paths 1 and 3 we have $W_{\text{total}} = W_1 + W_3 = 0$, or

$$W_3 = -W_1$$

As a result, we see that the work done on path 3 is the same as the work done on path 2:

$$W_3 = W_2$$

But paths 2 and 3 are arbitrary, as long as they start at point A and end at point B. This leads to an equivalent definition of a conservative force:

Conservative Force: Definition 2

The work done by a conservative force in going from point A to point B is *independent of the path* from A to B.

This definition is given an explicit check in Example 8–1.

Table 8–1 summarizes the different kinds of conservative and nonconservative forces we have encountered thus far in this text.

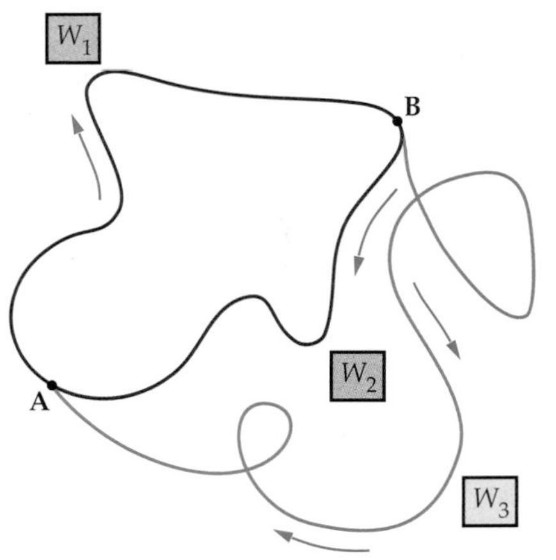

▲ **FIGURE 8–6 The work done by a conservative force is zero on any closed path**

$W_1 + W_2 = 0$, or $W_2 = -W_1$. Similarly, $W_1 + W_3 = 0$, or $W_3 = -W_1$. Note that $W_3 = W_2$ since they are both equal to $-W_1$; hence the work done in going from A to B is independent of the path.

TABLE 8–1
Conservative and Nonconservative Forces

Force	Section
Conservative forces	
Gravity	5–6
Spring force	6–2
Nonconservative forces	
Friction	6–1
Tension in a rope, cable, etc.	6–2
Forces exerted by a motor	7–4
Forces exerted by muscles	5–3

EXAMPLE 8–1 Different Paths, Different Forces

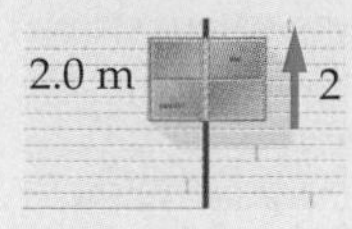

(a) A 4.57-kg box is moved with constant speed from A to B along the two paths shown at left below. Calculate the work done by gravity on each of these paths. **(b)** The same box is pushed across a floor from A to B along path 1 and path 2 at right below. If the coefficient of kinetic friction between the box and the surface is $\mu_k = 0.63$, how much work is done by friction along each path?

Picture the Problem

The two paths are shown in **(a)** and **(b)**, along with the relevant dimensions.

Strategy

To calculate the work for each path, we break it down into segments. Path 1 is made up of two segments, path 2 has four segments.

(a) For gravity, the work is zero on horizontal segments. On vertical segments, the work done by gravity is positive when motion is downward and negative when motion is upward.

(b) The work done by kinetic friction is negative on all segments of both paths.

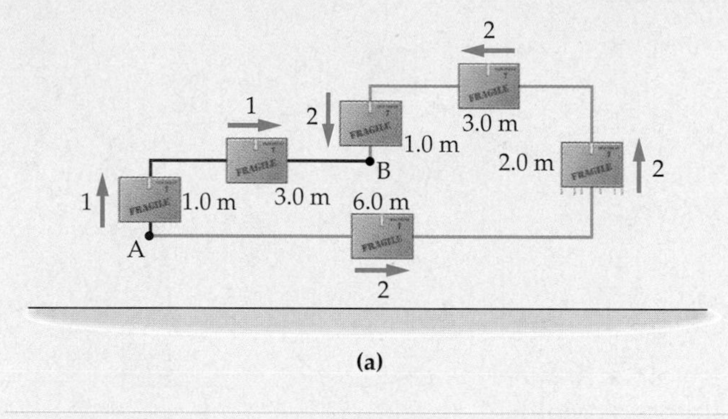

(a)

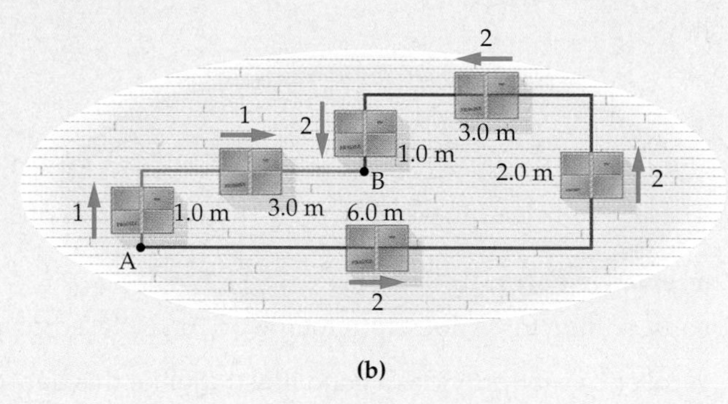

(b)

Solution

Part (a)

1. Using $W = Fd = mgy$, calculate the work done by gravity along the two segments of path 1:

$$W_1 = -(4.57\text{ kg})(9.81\text{ m/s}^2)(1.0\text{ m}) + 0 = -45\text{ J}$$

2. In the same way, calculate the work done by gravity along the four segments of path 2:

$$W_2 = 0 - (4.57\text{ kg})(9.81\text{ m/s}^2)(2.0\text{ m}) + 0 + (4.57\text{ kg})(9.81\text{ m/s}^2)(1.0\text{ m}) = -45\text{ J}$$

Part (b)

3. Using $F = \mu_k N$, calculate the work done by kinetic friction along the two segments of path 1:

$$W_1 = -(0.63)(4.57\text{ kg})(9.81\text{ m/s}^2)(1.0\text{ m}) - (0.63)(4.57\text{ kg})(9.81\text{ m/s}^2)(3.0\text{ m}) = -110\text{ J}$$

4. Similarly, calculate the work done by kinetic friction along the four segments of path 2:

$$W_2 = -(0.63)(4.57\text{ kg})(9.81\text{ m/s}^2)(6.0\text{ m}) - (0.63)(4.57\text{ kg})(9.81\text{ m/s}^2)(2.0\text{ m}) - (0.63)(4.57\text{ kg})(9.81\text{ m/s}^2)(3.0\text{ m}) - (0.63)(4.57\text{ kg})(9.81\text{ m/s}^2)(1.0\text{ m}) = -340\text{ J}$$

Insight

As expected, the conservative force of gravity gives the same work in going from A to B, regardless of the path. The work done by kinetic friction, however, is greater on the path of greater length.

Practice Problem

The work done by gravity when the box moves from point B to a point C is 140 J. Is point C above or below point B? What is the vertical distance between points B and C? [**Answer:** Point C is 3.1 m below point B.]

Some related homework problems: Problem 1, Problem 2

8–2 Potential Energy and the Work Done by Conservative Forces

Work must be done to lift a bowling ball from the floor to a shelf. Once on the shelf, the bowling ball has zero kinetic energy, just as it did on the floor. Even so, the work done in lifting the ball has not been lost. If the ball is allowed to fall from the

▲ Because gravity is a conservative force, the work done against gravity in lifting these logs (left) can, in principle, all be recovered. If the logs are released, for example, they will acquire an amount of kinetic energy exactly equal to the work done to lift them and to the gravitational potential energy that they gained in being lifted. Friction, by contrast is a nonconservative force. Some of the work done by this spinning grindstone (right) goes into removing material from the object being ground, while the rest is transformed into sound energy and (especially) heat. Most of this work can never be recovered as kinetic energy.

shelf, gravity does the same amount of work on it as you did to lift it in the first place. As a result, the work you did is "recovered" in the form of kinetic energy. Thus we say that when the ball is lifted to a new position there is an increase in **potential energy**, U, and that this potential energy can be converted to kinetic energy when the ball falls.

In a sense, potential energy is a storage system for energy. When we increase the separation between the ball and the ground, the work we do is stored in the form of an increased potential energy. Not only that, but the storage system is perfect, in the sense that the energy is never lost, as long as the separation remains the same. The ball can rest on the shelf for a million years, and still, when it falls, it gains the same amount of kinetic energy.

Work done against friction, however, is not "stored" as potential energy. Instead, it is dissipated into other forms of energy such as heat or sound. The same is true of other nonconservative forces. Only conservative forces have the potential-energy storage system.

Before proceeding, we should point out an interesting difference between kinetic and potential energy. Kinetic energy is given by the expression $K = \frac{1}{2}mv^2$, no matter what force might be involved. On the other hand, each different conservative force has a different expression for its potential energy. To see how this comes about, we turn now to a precise definition of potential energy.

Potential Energy, *U*

When a conservative force does an amount of work W_c (the subscript c stands for conservative), the corresponding potential energy U is changed, according to the following definition:

Definition of Potential Energy, *U*

$$W_c = U_i - U_f = -(U_f - U_i) = -\Delta U \qquad \text{8–1}$$

SI unit: joule, J

In words, the work done by a conservative force is equal to the negative of the change in potential energy. For example, when an object falls, gravity does *positive*

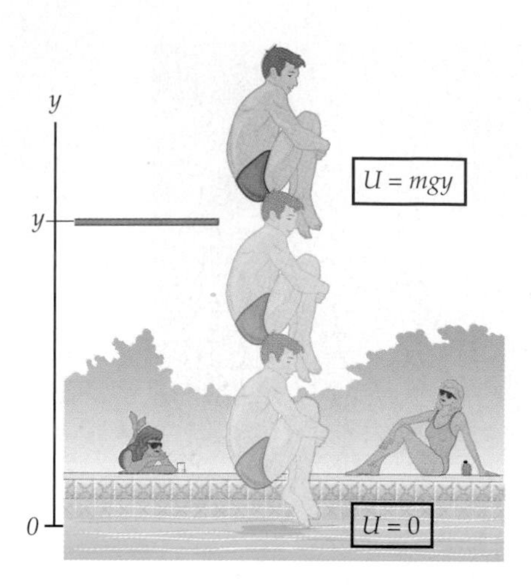

▲ **FIGURE 8–7 Gravitational Potential Energy**
A person drops from a diving board into a swimming pool. The diving board is at the height y, and the surface of the water is at $y = 0$. We choose the gravitational potential energy to be zero at $y = 0$; hence the potential energy is mgy at the diving board.

work on it and its potential energy *decreases*. Similarly, when an object is lifted, gravity does *negative* work and the potential energy *increases*.

Note that since work is a scalar with units of joules, the same is true of potential energy. In addition, our definition determines only the *difference* in potential energy between two points, not the actual value of the potential energy. Hence, we are free to choose the place where the potential energy is zero ($U = 0$) in much the same way we are free to choose the location of the origin in a coordinate system.

Gravity

Let's apply our definition of potential energy to the force of gravity near the Earth's surface. Suppose a person of mass m drops a distance y from a diving board into a pool, as shown in **Figure 8–7**. As the person drops, gravity does the work

$$W_c = mgy$$

Applying the definition given in Equation 8–1, the corresponding change in potential energy is

$$-\Delta U = U_i - U_f = W_c = mgy$$

In this expression, U_i is the potential energy when the diver is on the board, and U_f is the potential energy when the diver enters the water. Rearranging slightly, we have

$$U_i = mgy + U_f \qquad \textbf{8–2}$$

PROBLEM SOLVING NOTE
Zero of Potential Energy

When working potential energy problems it is important to make a definite choice for the location where the potential energy is to be set equal to zero. Any location can be chosen, but once the choice is made it must be used consistently.

Note that U_i is greater than U_f.

As mentioned above, we are free to choose $U = 0$ anywhere we like; only the difference in U is important. For example, if you slip and fall to the ground, you hit with the same thud whether you fall in Denver (altitude 1 mile) or in Honolulu (at sea level). It's the difference in height that matters, not the height itself. (The acceleration of gravity does vary slightly with altitude, as we shall see, but the difference is small enough to be unimportant in this case.) The only point to be careful about when choosing a location for $U = 0$ is to be consistent with the choice once it is made.

In general, we choose $U = 0$ in a convenient location. In Figure 8–7, a reasonable place for $U = 0$ is the surface of the water, where $y = 0$; that is, $U_f = 0$. Then, Equation 8–2 becomes $U_i = mgy$. If we omit the subscript on U_i, letting U stand for the potential energy at the arbitrary height y, we have

Gravitational Potential Energy (Near Earth's Surface)

$$U = mgy \qquad \textbf{8–3}$$

Note that the gravitational potential energy depends only on the height, y, and is independent of horizontal position.

▲ The $U = 0$ level for the gravitational potential energy of this system can be assigned to the point where the diver starts his dive or to the water level. Regardless of the choice, however, his kinetic energy when he strikes the water will be exactly equal to the difference in gravitational potential energy between these two points.

EXERCISE 8–1

Find the gravitational potential energy of a 65-kg person on a 3.0-m-high diving board. Let $U = 0$ be at water level.

Solution

Substituting $m = 65\,\text{kg}$ and $y = 3.0\,\text{m}$ in Equation 8–3 yields

$$U = mgy = (65\text{ kg})(9.81\text{ m/s}^2)(3.0\text{ m}) = 1900\text{ J}$$

The next Example considers the change in gravitational potential energy of a mountain climber, given different choices for the location of $U = 0$.

EXAMPLE 8–2 Pikes Peak or Bust

An 82.0-kg mountain climber is in the final stage of the ascent of 4301-m-high Pikes Peak. What is the change in gravitational potential energy as the climber gains the last 100.0 m of altitude? Let $U = 0$ be **(a)** at sea level or **(b)** at the top of the peak.

Picture the Problem

Our sketch shows the mountain climber and the last 100.0 m of altitude to be climbed. We choose a coordinate system with the y axis upward and the x axis to the right.

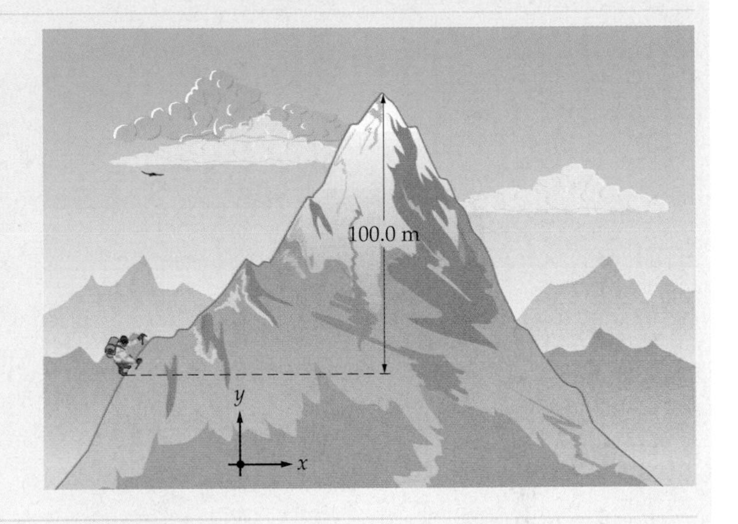

Strategy

The gravitational potential energy of the Earth-climber system depends only on the height y; the path followed in gaining the last 100.0 m of altitude is unimportant. The change in potential energy is $\Delta U = U_f - U_i = mgy_f - mgy_i$, where y_f is the altitude of the peak and y_i is 100.0 m less than y_f.

Solution

Part (a)

1. Calculate ΔU with $y_f = 4301$ m and $y_i = 4201$ m:

$$\Delta U = mgy_f - mgy_i = (82.0 \text{ kg})(9.81 \text{ m/s}^2)(4301 \text{ m}) - (82.0 \text{ kg})(9.81 \text{ m/s}^2)(4201 \text{ m}) = 80{,}400 \text{ J}$$

Part (b)

2. Calculate ΔU with $y_f = 0$ and $y_i = -100.0$ m:

$$\Delta U = mgy_f - mgy_i = (82.0 \text{ kg})(9.81 \text{ m/s}^2)(0) - (82.0 \text{ kg})(9.81 \text{ m/s}^2)(-100.0 \text{ m}) = 80{,}400 \text{ J}$$

Insight

As expected, the *change* in gravitational potential energy does not depend on where we choose $U = 0$.

Practice Problem

At what altitude is the climber's gravitational potential energy 1.00×10^5 J less than at the summit? [**Answer:** 4180 m]

Some related homework problems: Problem 5, Problem 10

A single item of food can be converted into a surprisingly large amount of potential energy. This is shown for the case of a candy bar in Example 8–3.

EXAMPLE 8–3 Converting Food Energy to Mechanical Energy

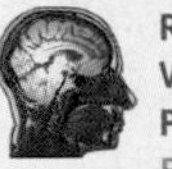

Real World Physics: Bio

A candy bar called the Mountain Bar has a calorie content of 210.0 Cal = 210.0 kcal, which is equivalent to an energy of 8.791×10^5 J. If an 82.0-kg mountain climber eats a Mountain Bar and magically converts it all to potential energy, what gain of altitude would be possible?

Picture the Problem

We show the mountain climber eating the candy bar at $y = 0$. The gain in altitude is indicated by $y = h$.

Strategy

The mountain climber's initial gravitational potential energy is $U = 0$; the final potential energy is $U = mgh$. To find the altitude gain, set $U = mgh$ equal to the energy provided by the candy bar, 8.791×10^5 J, and solve for h.

continued on the following page

continued from the previous page

Solution

1. Solve $U = mgh$ for h:

$$U = mgh$$
$$h = \frac{U}{mg}$$

2. Substitute numerical values, with $U = 8.791 \times 10^5$ J:

$$h = \frac{U}{mg} = \frac{8.791 \times 10^5\ \text{J}}{(82.0\ \text{kg})(9.81\ \text{m/s}^2)} = 1090\ \text{m}$$

Insight

This is over two-thirds of a mile in elevation. Even if we take into account the fact that metabolic efficiency is only about 70 percent, the height would still be 765 m, or nearly half a mile. It is remarkable how much our bodies can do with so little.

Practice Problem

If the mass of the mountain climber is increased—by adding more items to the backpack, for example—does the possible elevation gain increase, decrease, or stay the same? Calculate the elevation gain for a climber with a mass of 90.0 kg. [**Answer:** The altitude gain will decrease. For $m = 90.0$ kg we find $h = 996$ m.]

Some related homework problems: Problem 5, Problem 10

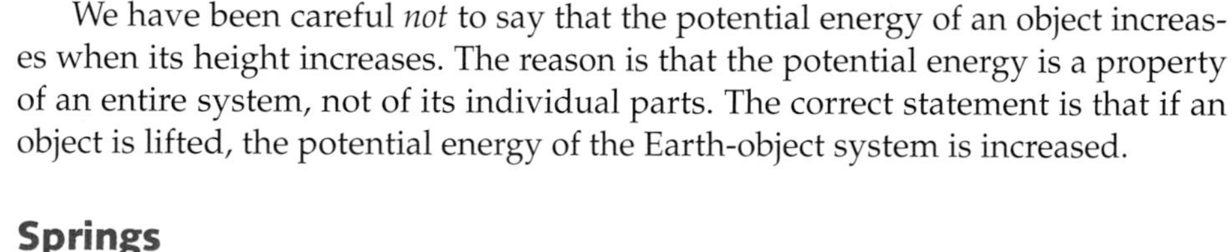

We have been careful *not* to say that the potential energy of an object increases when its height increases. The reason is that the potential energy is a property of an entire system, not of its individual parts. The correct statement is that if an object is lifted, the potential energy of the Earth-object system is increased.

▲ Because the spring force is a conservative force, the springs in these pogo sticks function as energy storage devices. They convert kinetic energy to potential energy when they are compressed, then quickly give it back.

Springs

Consider a spring that is stretched from its equilibrium position a distance x. According to Equation 6-11, the work required to cause this stretch is $W = \frac{1}{2}kx^2$. Therefore, if the spring is released—and allowed to move from the stretched position back to the equilibrium position—it will do the same work, $\frac{1}{2}kx^2$. From our definition of potential energy, then, we see that

$$W_c = \tfrac{1}{2}kx^2 = U_i - U_f \qquad \text{8–4}$$

Note that in this case U_f is the potential energy when the spring is at $x = 0$ (equilibrium position), and U_i is the potential energy when the spring is stretched by the amount x.

A convenient choice for $U = 0$ is the equilibrium position of the spring. With this choice we have $U_f = 0$, and Equation 8–4 becomes $U_i = \frac{1}{2}kx^2$. Omitting the subscript i, so that U represents the potential energy of the spring for an arbitrary amount of stretch x, we have

Spring Potential Energy

$$U = \tfrac{1}{2}kx^2 \qquad 8\text{–}5$$

Since U depends on x^2, which is positive even if x is negative, the spring potential energy is always greater than or equal to zero. Thus, a spring's potential energy increases whenever it is displaced from equilibrium.

EXERCISE 8–2

Find the potential energy of a spring with force constant $k = 680$ N/m if it is **(a)** stretched by 5.00 cm or **(b)** compressed by 7.00 cm.

Solution

Substituting $x = 0.0500$ m and $x = -0.0700$ m yields

(a) $U = \frac{1}{2}(680 \text{ N/m})(0.0500 \text{ m})^2 = 0.85 \text{ J}$,

(b) $U = \frac{1}{2}(680 \text{ N/m})(-0.0700 \text{ m})^2 = 1.7 \text{ J}$.

Finally, comparing Equation 8–3 to Equation 8–5, we see that the potential energy for gravity and for a spring are given by different expressions. As mentioned, each conservative force has its own potential energy.

EXAMPLE 8–4 Compressed Energy

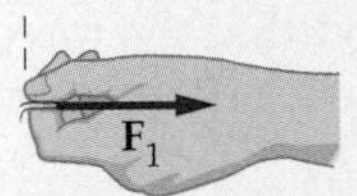

When a force of 120.0 N is applied to a certain spring, it causes a stretch of 2.25 cm. What is the potential energy of this spring when it is compressed by 3.50 cm?

Picture the Problem

The top sketch shows the spring stretched by the force $F_1 = 120.0$ N. The lower sketch shows the spring compressed by a second force, F_2, causing a compression of 3.50 cm.

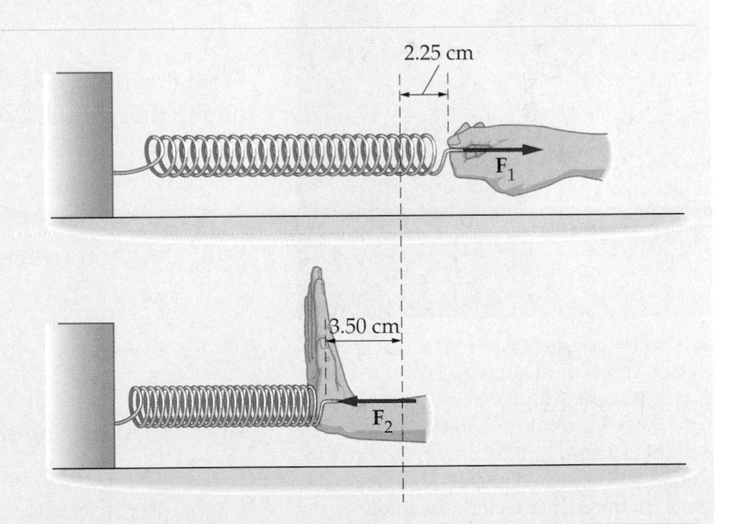

Strategy

From the first piece of information—a certain force causes a certain stretch—we can calculate the force constant using $F = kx$. Once we know k, we find the potential energy with $U = \frac{1}{2}kx^2$.

Solution

1. Solve $F = kx$ for the spring constant, k:

$$F = kx$$

$$k = \frac{F}{x} = \frac{120.0 \text{ N}}{0.0225 \text{ m}} = 5330 \text{ N/m}$$

2. Substitute $k = 5330$ N/m and $x = -0.0350$ m into the potential energy expression, $U = \frac{1}{2}kx^2$:

$$U = \tfrac{1}{2}kx^2 = \tfrac{1}{2}(5330 \text{ N/m})(-0.0350 \text{ m})^2 = 3.26 \text{ J}$$

Insight

Note that the potential energy of the spring is the same whether it is compressed by 3.50 cm or stretched by the same amount.

continued on the following page

continued from the previous page

Practice Problem
What stretch is necessary for this spring to have a potential energy of 5.00 J? [**Answer:** 4.33 cm]
Some related homework problems: Problem 7, Problem 9

8–3 Conservation of Mechanical Energy

In this section, we show how potential energy can be used as a powerful tool in solving a variety of problems, and in gaining greater insight into the workings of physical systems. To do so, we begin by defining the **mechanical energy**, E, as the sum of the potential and kinetic energies of an object:

$$E = U + K \quad \text{8–6}$$

The significance of mechanical energy is that it is **conserved** in systems involving only conservative forces. By conserved, we mean that its value never changes; that is, $E =$ constant. (In situations where nonconservative forces are involved the mechanical energy can change, as when friction causes warming by converting mechanical energy to thermal energy. When *all* possible forms of energy are considered, energy is always found to be conserved.)

To show that E is conserved for conservative forces, we start with the work-energy theorem from Chapter 7:

$$W_{\text{total}} = \Delta K = K_f - K_i$$

Suppose for a moment that the system has only a single force, and that the force is conservative. If this is the case, then the total work, W_{total}, is the work done by the conservative force, W_c:

$$W_{\text{total}} = W_c$$

From the definition of potential energy we know that $W_c = -\Delta U = U_i - U_f$. Combining these results, we have

$$W_{\text{total}} = W_c$$

$$K_f - K_i = U_i - U_f$$

With a slight rearrangement we find

$$U_f + K_f = U_i + K_i$$

or

$$E_f = E_i$$

Since the initial and final points can be chosen arbitrarily, it follows that E is conserved:

$$E = \text{constant}$$

If the system has more than one conservative force, the only change to these results is to replace U with the sum of potential energies for each force.

To summarize:

Conservation of Mechanical Energy

In systems with conservative forces only, the mechanical energy E is conserved; that is, $E = U + K =$ constant.

In terms of physical systems, conservation of mechanical energy means that energy can be converted between potential and kinetic forms, but that the sum remains the same. As an example, in the roller coaster shown in Figure 8–5, the gravitational potential energy decreases as the car approaches point B, and as it does, the car's kinetic energy increases by the same amount. From a practical point

▲ A roller coaster (top) illustrates the conservation of mechanical energy. With every descent, gravitational potential energy is converted into kinetic energy: with every rise, kinetic energy is converted back into gravitational potential energy. If friction is neglected, the total mechanical energy of the car remains constant. The same principle is exploited at a pumped-storage facility, such as this one at the Mormon Flat Dam in Phoenix, Arizona (bottom). When surplus electrical power is available, it is used to pump water uphill into the reservoir. This process, in effect, stores electrical energy as gravitational potential energy. When power demand is high, the stored water is allowed to flow back downhill through the electrical generators in the dam, converting the gravitational energy to kinetic energy and the kinetic energy to electrical energy.

of view, conservation of mechanical energy means that many physics problems can be solved by what amounts to simple bookkeeping.

For example, consider a key chain of mass m that is dropped to the floor from a height h, as illustrated in **Figure 8–8**. The question is, how fast are the keys moving just before they land? We know how to solve this problem using Newton's laws and kinematics, but now let's see how energy conservation can be used instead.

First, note that the only force acting on the keys is gravity—ignoring air resistance, of course—and that gravity is a conservative force. As a result, we can say that $E = U + K$ is constant during the entire time the keys are falling. To solve the problem, then, we pick two points on the motion of the keys, say i and f in Figure 8–8, and we set the mechanical energy equal at these points:

$$E_\mathrm{i} = E_\mathrm{f} \qquad \textbf{8–7}$$

Writing this out in terms of potential and kinetic energies, we have

$$U_\mathrm{i} + K_\mathrm{i} = U_\mathrm{f} + K_\mathrm{f} \qquad \textbf{8–8}$$

This one equation—which is nothing but bookkeeping—can be used to solve for the one unknown, the final speed.

To be specific, in Figure 8–8 (a) we choose $y = 0$ at ground level, which means that $U_\mathrm{i} = mgh$. In addition, the fact that the keys are released from rest means that $K_\mathrm{i} = 0$. Similarly, at point f—just before hitting the ground—the energy is all kinetic, and the potential energy is zero; that is, $U_\mathrm{f} = 0$, $K_\mathrm{f} = \frac{1}{2}mv^2$. Substituting these values into Equation 8–8, we find

$$mgh + 0 = 0 + \tfrac{1}{2}mv^2$$

Canceling m and solving for v yields the same result we get with kinematics:

$$v = \sqrt{2gh}$$

Suppose, instead, that we had chosen $y = 0$ to be at the release point of the keys, as in Figure 8–8 (b), so that the keys land at $y = -h$. Now, when the keys are released, we have $U_\mathrm{i} = 0$ and $K_\mathrm{i} = 0$, and when they land $U_\mathrm{f} = -mgh$ and $K_\mathrm{f} = \frac{1}{2}mv^2$. Substituting these results in $U_\mathrm{i} + K_\mathrm{i} = U_\mathrm{f} + K_\mathrm{f}$ yields

$$0 + 0 = -mgh + \tfrac{1}{2}mv^2$$

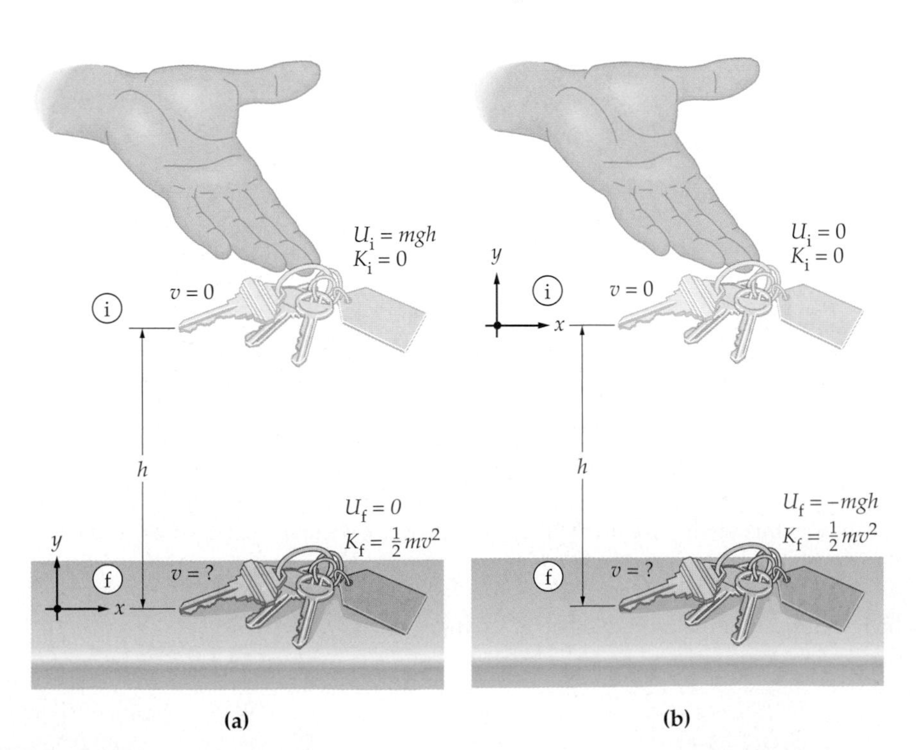

◀ **FIGURE 8–8 Solving a kinematics problem using conservation of energy**

(a) A set of keys falls to the floor. Ignoring frictional forces, we know that the mechanical energy at points i and f must be equal; $E_\mathrm{i} = E_\mathrm{f}$. Using this condition, we can find the speed of the keys just before they land. **(b)** The same physical situation as in part (a), except this time we have chosen $y = 0$ to be at the point where the keys are dropped. As before, we set $E_\mathrm{i} = E_\mathrm{f}$ to find the speed of the keys just before they land. The result is the same.

Solving for v gives the same result:

$$v = \sqrt{2gh}$$

Thus, as expected, changing the zero level has no effect on the physical results.

EXAMPLE 8–5 Graduation Fling

At the end of a graduation ceremony, graduates fling their caps into the air. Suppose a 0.120-kg cap is thrown straight upward with an initial speed of 7.85 m/s, and that frictional forces can be ignored. **(a)** Use kinematics to find the speed of the cap when it is 1.18 m above the release point. **(b)** Show that the mechanical energy at the release point is the same as the mechanical energy 1.18 m above the release point.

Picture the Problem

In our sketch we choose $y = 0$ to be at the level where the cap is released. We also designate the release point as i (initial) and the point at which $y = 1.18$ m as f (final).

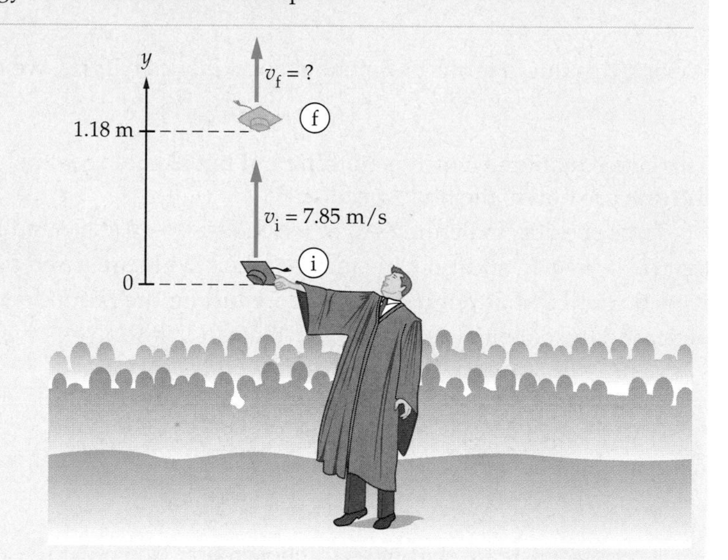

Strategy

(a) Since we want to relate velocity to position, we use $v_y^2 = v_{0y}^2 + 2a_y\Delta y$ (Section 2–5). In this case, $v_{0y} = 7.85$ m/s, $\Delta y = 1.18$ m, and $a_y = -g$. Substituting these values gives v_y.

(b) At each point we simply calculate $E = U + K$, with $U = mgy$ and $K = \frac{1}{2}mv^2$.

Solution

Part (a)

1. Use kinematics to solve for v_y:

$$v_y^2 = v_{0y}^2 + 2a_y\Delta y$$
$$v_y = \pm\sqrt{v_{0y}^2 + 2a_y\Delta y}$$

2. Substitute $v_{0y} = 7.85$ m/s, $\Delta y = 1.18$ m, and $a_y = -g$ to find v_y. Choose the plus sign, since we are interested only in the speed:

$$v_y = \sqrt{v_{0y}^2 + 2a_y\Delta y}$$
$$= \sqrt{(7.85 \text{ m/s})^2 - 2(9.81 \text{ m/s}^2)(1.18 \text{ m})} = 6.20 \text{ m/s}$$

Part (b)

3. Calculate E_i. At this point $y_i = 0$ and $v_i = 7.85$ m/s:

$$E_i = U_i + K_i = mgy_i + \tfrac{1}{2}mv_i^2$$
$$= 0 + \tfrac{1}{2}(0.120 \text{ kg})(7.85 \text{ m/s})^2 = 3.70 \text{ J}$$

4. Calculate E_f. At this point $y_f = 1.18$ m and $v_f = 6.20$ m/s:

$$E_f = U_f + K_f = mgy_f + \tfrac{1}{2}mv_f^2$$
$$= (0.120 \text{ kg})(9.81 \text{ m/s}^2)(1.18 \text{ m}) +$$
$$\tfrac{1}{2}(0.120 \text{ kg})(6.20 \text{ m/s})^2 = 1.39 \text{ J} + 2.31 \text{ J} = 3.70 \text{ J}$$

Insight

As expected, E_f is equal to E_i. In the remaining Examples in this section we turn this process around; we start with $E_f = E_i$, and use this relation to find a final speed or a final height.

Practice Problem

Use energy conservation to find the height at which the speed of the cap is 5.00 m/s. [**Answer:** 1.87 m]

Some related homework problems: Problem 14, Problem 15

◀ **FIGURE 8–9 Speed is independent of path**

If the speed of the cap is v_i at the height y_i, its speed is v_f at the height y_f, independent of the path between the two heights. This assumes, of course, that frictional forces can be neglected.

An interesting extension of this Example is shown in **Figure 8–9**. In this case, we are given that the speed of the cap is v_i at the height y_i, and we would like to know its speed v_f when it is at the height y_f.

To find v_f, we apply energy conservation to the points i and f:

$$U_i + K_i = U_f + K_f$$

Writing out U and K specifically for these two points yields the following:

$$mgy_i + \tfrac{1}{2}mv_i^2 = mgy_f + \tfrac{1}{2}mv_f^2$$

As before, we cancel m and solve for the unknown, v_f:

$$v_f = \sqrt{v_i^2 + 2g(y_i - y_f)}$$

This result is in agreement with the kinematic equation, $v_y^2 = v_{0y}^2 + 2a_y\Delta y$.

Note that v_f depends only on y_i and y_f, not on the path connecting them. This is because conservative forces such as gravity do work that is path independent. What this means physically is that the cap has the same speed v_f at the height y_f, whether it goes straight upward or follows some other trajectory, as in Figure 8–9. All that matters is the height difference.

PROBLEM SOLVING NOTE
Conservative Systems

A convenient approach to problems involving energy conservation is to first sketch the system, and then label the initial and final points with i and f, respectively. To apply energy conservation, write out the energy at these two points and set $E_i = E_f$.

EXAMPLE 8–6 Catching a Home Run

In the bottom of the ninth inning a player hits a 0.15-kg baseball over the outfield fence. The ball leaves the bat with a speed of 36 m/s, and a fan in the bleachers catches it 7.2 m above the point where it was hit. Assuming frictional forces can be ignored, find **(a)** the kinetic energy of the ball when it is caught and **(b)** its speed when caught.

Picture the Problem

The sketch shows the ball's trajectory. We label the hit point i and the catch point f. At point i we choose $y_i = 0$; at point f, then, $y_f = h$.

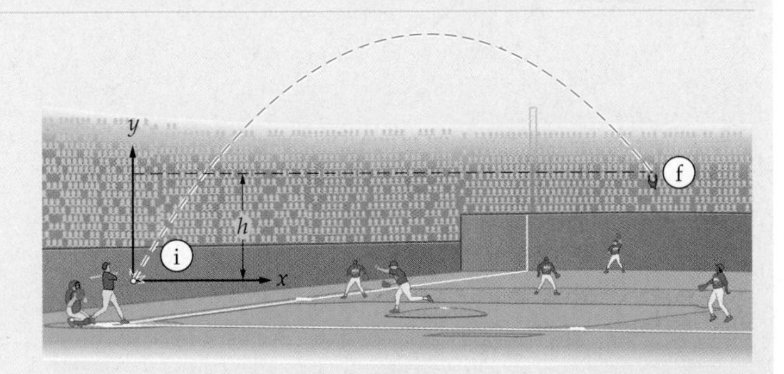

Strategy

(a) Use the fact that the initial mechanical energy is equal to the final mechanical energy, $U_i + K_i = U_f + K_f$, to find K_f.

(b) Once K_f is determined, use $K_f = \tfrac{1}{2}mv_f^2$ to find v_f.

continued on the following page

continued from the previous page

Solution

Part (a)

1. Begin by writing U and K for point i:

$$U_i = 0$$

$$K_i = \tfrac{1}{2}mv_i^2 = \tfrac{1}{2}(0.15\text{ kg})(36\text{ m/s})^2 = 97\text{ J}$$

2. Next, write U and K for point f:

$$U_f = mgh = (0.15\text{ kg})(9.81\text{ m/s}^2)(7.2\text{ m}) = 11\text{ J}$$

$$K_f = \tfrac{1}{2}mv_f^2$$

3. Set $U_i + K_i$ equal to $U_f + K_f$ and solve for K_f:

$$0 + 97\text{ J} = 11\text{ J} + K_f$$
$$K_f = 97\text{ J} - 11\text{ J} = 86\text{ J}$$

Part (b)

4. Use $K_f = \tfrac{1}{2}mv_f^2$ to find v_f:

$$K_f = \tfrac{1}{2}mv_f^2$$

$$v_f = \sqrt{\frac{2K_f}{m}} = \sqrt{\frac{2(86\text{ J})}{0.15\text{ kg}}} = 34\text{ m/s}$$

Insight

To find the speed when the ball was caught we needed the height of point f, but we don't need to know any details about the ball's trajectory. For example, it is not necessary to know the angle at which the ball leaves the bat, or the maximum height reached by the ball.

Practice Problem

If the mass of the ball were increased, would the catch speed be greater than, less than, or the same as the value we just found? [**Answer:** The same. U and K depend on mass in the same way, hence the mass cancels.]

Some related homework problems: Problem 13, Problem 14

The connection between height difference and speeds is explored further in the following Conceptual Checkpoint and Example.

CONCEPTUAL CHECKPOINT 8–1

Swimmers at a water park can enter a pool using one of two frictionless slides of equal height. Slide 1 approaches the water with a uniform slope; slide 2 dips rapidly at first, then levels out. Is the speed, v_2, at the bottom of slide 2 **(a)** greater than, **(b)** less than, or **(c)** the same as the speed v_1 at the bottom of slide 1?

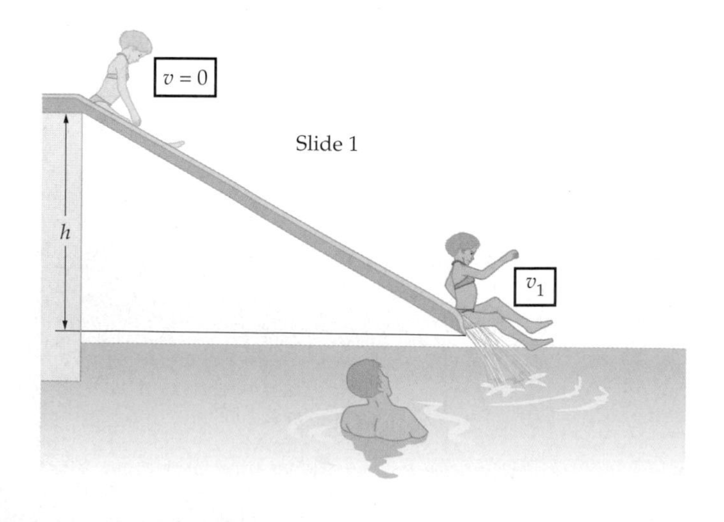

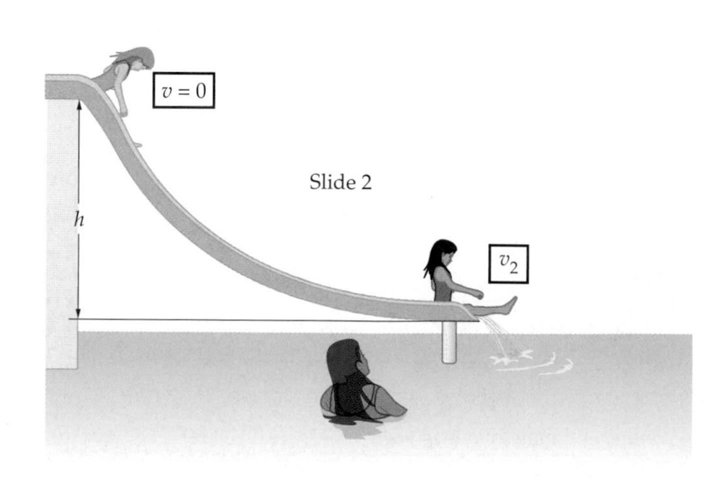

Reasoning and Discussion
In both cases, the same amount of potential energy, *mgh*, is converted to kinetic energy. Since the conversion of gravitational potential energy to kinetic energy is the *only* energy transaction taking place, it follows that the speed is the same for each slide.

Interestingly, although the final speeds are the same, the time required to reach the water is less for slide 2. The reason is that swimmer 2 reaches a high speed early and maintains it, whereas the speed of swimmer 1 increases slowly and steadily.

Answer:
(c) The speeds are the same.

EXAMPLE 8–7 Skateboard Exit Ramp

A 55-kg skateboarder enters a ramp moving horizontally with a speed of 6.5 m/s, and leaves the ramp moving vertically with a speed of 4.1 m/s. Find the height of the ramp, assuming no energy loss to frictional forces.

Picture the Problem
We choose $y = 0$ to be the level of the bottom of the ramp, thus the gravitational potential energy is zero there. Point i indicates the skateboarder entering the ramp; point f is the top of the ramp.

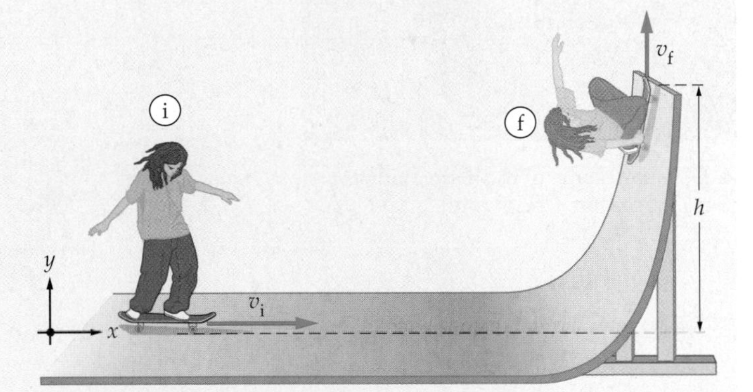

Strategy
To find h, simply set the initial energy, $E_i = U_i + K_i$, equal to the final energy, $E_f = U_f + K_f$.

Solution

1. Write expressions for U_i and K_i: $U_i = mg \cdot 0 = 0$ $\quad K_i = \frac{1}{2}mv_i^2$

2. Write expressions for U_f and K_f: $U_f = mgh$ $\quad K_f = \frac{1}{2}mv_f^2$

3. Set $E_i = U_i + K_i$ equal to $E_f = U_f + K_f$: $\frac{1}{2}mv_i^2 = mgh + \frac{1}{2}mv_f^2$

4. Solve for h. Note that m cancels: $mgh = \frac{1}{2}mv_i^2 - \frac{1}{2}mv_f^2$

$$h = \frac{v_i^2 - v_f^2}{2g}$$

5. Substitute numerical values:

$$h = \frac{(6.5\ \text{m/s})^2 - (4.1\ \text{m/s})^2}{2(9.81\ \text{m/s}^2)} = 1.3\ \text{m}$$

Insight
Note that our value for h is independent of the shape of the ramp—it is equally valid for one with the shape shown here, or one that simply inclines upward at a constant angle. In addition, the height does not depend on the person's mass, as we see in step 4.

continued on the following page

continued from the previous page

Practice Problem

What is the skateboarder's maximum height above the bottom of the ramp? [**Answer:** 2.2 m]

Some related homework problems: Problem 13, Problem 17

▲ Does the shape of the slide matter? (See Conceptual Checkpoint 8–1.)

It is interesting to express the equation in line 3 from Example 8–7 in words. First, the left side of the equation is the initial kinetic energy of the skateboarder, $\frac{1}{2}mv_i^2$. This is the initial energy content of the system. At point f the system still has the same amount of energy, only now part of it, mgh, is in the form of gravitational potential energy. The remainder is the final kinetic energy, $\frac{1}{2}mv_f^2$.

Conceptual Checkpoint 8–2 considers the effect of a slight change in the initial speed of an object.

CONCEPTUAL CHECKPOINT 8–2

A snowboarder coasts on a smooth track that rises from one level to another. If the snowboarder's initial speed is 4 m/s, the snowboarder just makes it to the upper level and comes to rest. With a slightly greater initial speed of 5 m/s, the snowboarder is still moving to the right on the upper level. Is the snowboarder's final speed in this case **(a)** 1 m/s, **(b)** 2 m/s, or **(c)** 3 m/s?

Reasoning and Discussion

A plausible-sounding answer is that since the initial speed is greater by 1 m/s in the second case, the final speed should be greater by 1 m/s as well. Therefore, the answer should be 0 + 1 m/s = 1 m/s. This is incorrect, however.

As surprising as it may seem, an increase in the initial speed from 4 m/s to 5 m/s results in an increase in the final speed from 0 to 3 m/s. This is due to the fact that kinetic energy depends on v^2 rather than v; thus, it is the difference in v^2 that counts. In this case, the initial value of v^2 increases from 16 m^2/s^2 to 25 m^2/s^2, for a total increase of 25 m^2/s^2 − 16 m^2/s^2 = 9 m^2/s^2. The final value of v^2 must increase by the same amount, 9 m^2/s^2 = $(3\ m/s)^2$. As a result, the final speed is 3 m/s.

Answer:

(c) The final speed of the snowboarder in the second case is 3 m/s.

Let's check the results of the previous Conceptual Checkpoint with a specific numerical example. Suppose the snowboarder has a mass of 74.0 kg. It follows that in the first case the initial kinetic energy is $K_i = \frac{1}{2}(74.0\text{ kg})(4.0\text{ m/s})^2 = 592\text{ J}$. At the top of the hill all of this kinetic energy is converted to gravitational potential energy, mgh.

In the second case, the initial speed of the snowboarder is 5.0 m/s; thus, the initial kinetic energy is $K_i = \frac{1}{2}(74.0\text{ kg})(5.0\text{ m/s})^2 = 925\text{ J}$. When the snowboarder reaches the top of the hill, 592 J of this kinetic energy is converted to gravitational potential energy, leaving the snowboarder with a final kinetic energy of $925\text{ J} - 592\text{ J} = 333\text{ J}$. The corresponding speed is given by

$$\frac{1}{2}mv^2 = 333\text{ J}$$

$$v = \sqrt{\frac{2(333\text{ J})}{m}} = \sqrt{\frac{2(333\text{ J})}{74.0\text{ kg}}} = \sqrt{9.0\text{ m}^2/\text{s}^2} = 3.0\text{ m/s}$$

Thus, as expected, the snowboarder in the second case has a final speed of 3.0 m/s.

We conclude this section with two Examples involving springs.

EXAMPLE 8–8 Spring Time

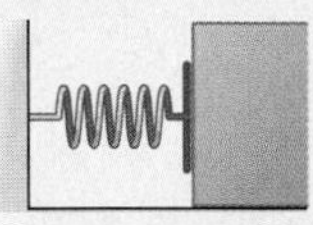

A 1.70-kg block slides on a horizontal, frictionless surface until it encounters a spring with a force constant of 955 N/m. The block comes to rest after compressing the spring a distance of 4.60 cm. Find the initial speed of the block. (Ignore air resistance and any energy lost when the block collides with the spring.)

Picture the Problem

Point i refers to times before the block makes contact with the spring; point f refers to the situation where the block has come to rest, and the spring is compressed.

We can choose the center of the block to be the $y = 0$ level. With this choice, the gravitational potential energy of the system is zero at all times.

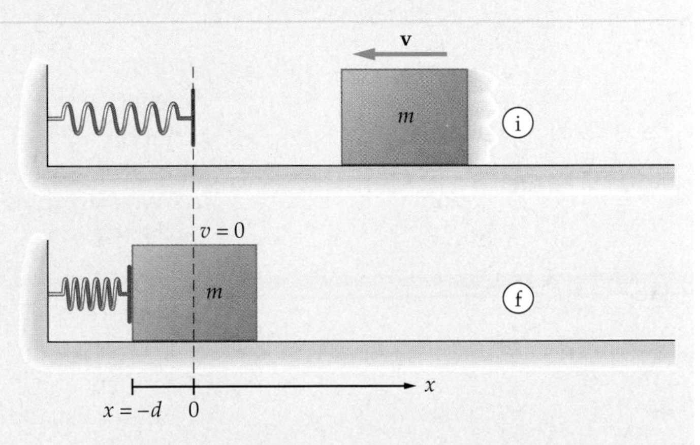

Strategy

Set E_i equal to E_f to find the one unknown, v.

Solution

1. Write expressions for U_i and K_i. For U, we consider only the potential energy of the spring, $U = \frac{1}{2}kx^2$:
$$U_i = \frac{1}{2}k \cdot 0^2 = 0 \qquad K_i = \frac{1}{2}mv^2$$

2. Do the same for U_f and K_f:
$$U_f = \frac{1}{2}k(-d)^2 = \frac{1}{2}kd^2 \qquad K_f = \frac{1}{2}m \cdot 0^2 = 0$$

3. Set $E_i = U_i + K_i$ equal to $E_f = U_f + K_f$ and solve for v:
$$\frac{1}{2}mv^2 = \frac{1}{2}kd^2$$
$$v = d\sqrt{\frac{k}{m}}$$

4. Substitute numerical values:
$$v = d\sqrt{\frac{k}{m}} = (0.0460\text{ m})\sqrt{\frac{955\text{ N/m}}{1.70\text{ kg}}} = 1.09\text{ m/s}$$

continued on the following page

continued from the previous page

Insight

After the block comes to rest, the spring expands again, converting its potential energy back into the kinetic energy of the block. When the block leaves the spring, moving to the right, its speed is once again 1.09 m/s.

Practice Problem

What is the compression distance, d, if the block's initial speed is 0.500 m/s? [**Answer:** 2.11 cm]

Some related homework problems: Problem 16, Problem 18

ACTIVE EXAMPLE 8–1 A Vertical Spring

Suppose the spring and block in Example 8–8 are oriented vertically, as shown here. Initially, the spring is compressed 4.60 cm and the block is at rest. When the block is released it accelerates upward. Find the speed of the block when the spring has returned to its equilibrium position.

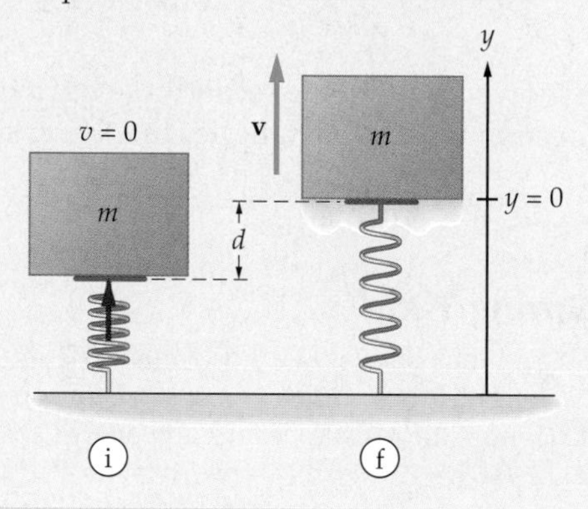

Solution

1. Write an expression for the initial mechanical energy E_i: $E_i = U_i + K_i = -mgd + \frac{1}{2}kd^2 + 0$

2. Write an expression for the final mechanical energy E_f: $E_f = U_f + K_f = 0 + 0 + \frac{1}{2}mv^2$

3. Set E_i equal to E_f and solve for v: $-mgd + \frac{1}{2}kd^2 = \frac{1}{2}mv^2$
$v = \sqrt{kd^2/m - 2gd}$

4. Substitute numerical values: $v = 0.535$ m/s

Insight

In this system, part of the initial potential energy of the spring ($\frac{1}{2}kd^2$) goes into increasing the gravitational potential energy of the block (mgd). The remainder of the initial energy, $\frac{1}{2}kd^2 - mgd$, is converted into the block's kinetic energy.

8–4 Work Done by Nonconservative Forces

Nonconservative forces change the amount of mechanical energy in a system. They might decrease the mechanical energy by converting it to thermal energy, or increase it by converting muscular work to kinetic or potential energy. In some systems, both types of processes occur at the same time.

To see the connection between the work done by a nonconservative force, W_{nc}, and the mechanical energy, E, we return once more to the work-energy theorem, which says that the *total* work is equal to the change in kinetic energy:

$$W_{total} = \Delta K$$

Suppose, for instance, that a system has one conservative and one nonconservative force. In this case, the total work is the sum of the conservative work W_c and the nonconservative work W_{nc}:

$$W_{total} = W_c + W_{nc}$$

Recalling that conservative work is related to the change in potential energy by the definition given in Equation 8–1, $W_c = -\Delta U$, we have

$$W_{total} = -\Delta U + W_{nc} = \Delta K$$

Solving this relation for the nonconservative work yields

$$W_{nc} = \Delta U + \Delta K$$

Finally, since the total mechanical energy is $E = U + K$, it follows that the *change* in mechanical energy is $\Delta E = \Delta U + \Delta K$. As a result, the nonconservative work is simply the change in mechanical energy:

$$W_{nc} = \Delta E \qquad \text{8–9}$$

If more than one nonconservative force acts, we simply add the nonconservative work done by each such force to obtain W_{nc}.

At this point it may be useful to collect the three "working relationships" that have been introduced in the last two chapters:

$$W_{total} = \Delta K$$
$$W_c = -\Delta U \qquad \text{8–10}$$
$$W_{nc} = \Delta E$$

Note that positive nonconservative work increases the total mechanical energy of a system, while negative nonconservative work decreases the mechanical energy—and converts it to other forms. In the next Example, for instance, part of the initial mechanical energy of a leaf is converted to heat and other forms of energy by air resistance as it falls to the ground.

▲ The kinetic energy of this hydraulic shovel is being used to do nonconservative work—reducing a concrete structure to rubble.

EXAMPLE 8–9 A Leaf Falls in the Forest

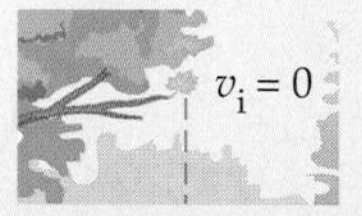

Deep in the forest, a 17-g leaf falls from a tree and drops straight to the ground. If its initial height was 5.3 m, and its speed on landing was 1.3 m/s, how much nonconservative work was done on the leaf?

Picture the Problem

The leaf drops from rest at a height $y = h$, and lands with a speed v at $y = 0$. These two points are labeled i and f, respectively.

Strategy

To begin, calculate the initial mechanical energy, E_i, and the final mechanical energy, E_f. Once these energies have been determined, the nonconservative work is $W_{nc} = \Delta E = E_f - E_i$.

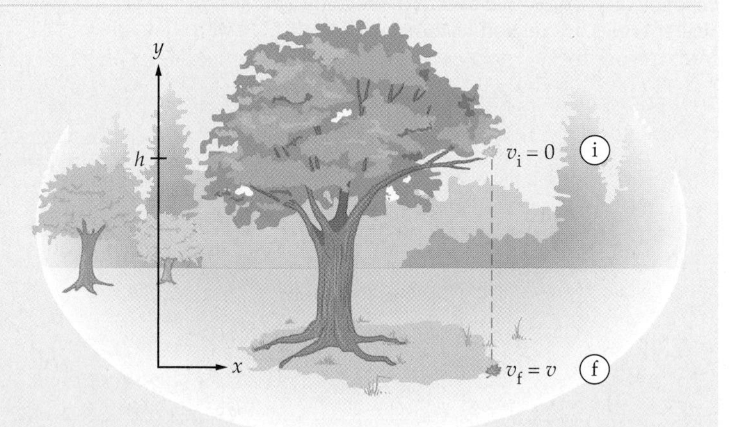

continued on the following page

continued from the previous page

Solution

1. Evaluate U_i, K_i, and E_i:

$$U_i = mgh = (0.017\text{ kg})(9.81\text{ m/s}^2)(5.3\text{ m}) = 0.88\text{ J}$$
$$K_i = \tfrac{1}{2}m \cdot 0^2 = 0$$
$$E_i = U_i + K_i = 0.88\text{ J}$$

2. Next, evaluate U_f, K_f, and E_f.

$$U_f = mg \cdot 0 = 0$$
$$K_f = \tfrac{1}{2}mv^2 = \tfrac{1}{2}(0.017\text{ kg})(1.3\text{ m/s})^2 = 0.014\text{ J}$$
$$E_f = U_f + K_f = 0.014\text{ J}$$

3. Use $W_{nc} = \Delta E$ to find the nonconservative work:

$$W_{nc} = \Delta E = E_f - E_i = 0.014\text{ J} - 0.88\text{ J} = -0.87\text{ J}$$

Insight

Note that most of the initial mechanical energy is dissipated as the leaf falls. The small amount that remains (only about 1.5%) appears as the kinetic energy of the leaf just before it lands. If a cherry had fallen from the tree, it would have struck the ground with a considerably greater speed—perhaps 5 times the speed of the leaf. In that case, the percentage of the initial potential energy remaining as kinetic energy would have been $5^2 = 25$ times greater than the percentage retained by the leaf.

Practice Problem

What was the average nonconservative force exerted on the leaf as it fell? [**Answer:** $W_{nc} = -Fh$, $F = -W_{nc}/h = 0.16$ N, upward]

Some related homework problems: Problem 22, Problem 23

In the following Active Example, we use a knowledge of the nonconservative work to find the depth at which a diver comes to rest.

ACTIVE EXAMPLE 8–2 Take a Dive

A 95.0-kg diver steps off a diving board and drops into the water, 3.00 m below. At some depth d below the water's surface the diver comes to rest. If the nonconservative work done on the diver is $W_{nc} = -5120$ J, what is the depth, d?

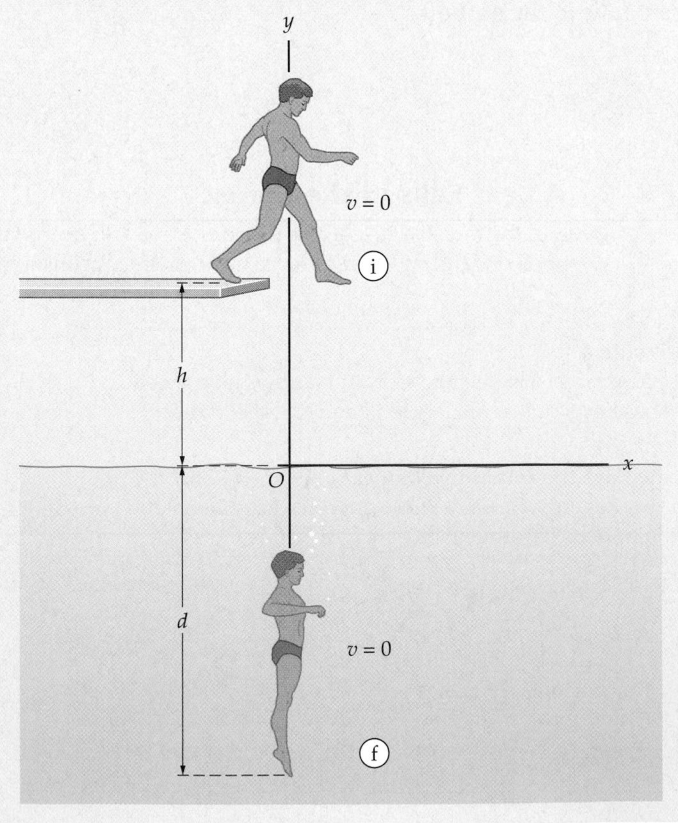

PROBLEM SOLVING NOTE
Nonconservative Systems

Start by sketching the system and labeling the initial and final points with i and f, respectively. The initial and final mechanical energies are related to the nonconservative work by $W_{nc} = E_f - E_i$.

Solution

1. Write the initial mechanical energy, E_i: $E_i = mgh + 0 = mgh$
2. Write the final mechanical energy, E_f: $E_f = mg(-d) + 0 = -mgd$
3. Set W_{nc} equal to ΔE: $W_{nc} = \Delta E = E_f - E_i = -mgd - mgh$
4. Solve for d: $d = -(W_{nc} + mgh)/mg$
5. Substitute numerical values: $d = 2.49$ m

Insight

Another way to write line 3 is $E_f = E_i + W_{nc}$. In words, this equation says that the final mechanical energy is the initial mechanical energy plus the nonconservative work done on the system. In this case, $W_{nc} < 0$; hence the final mechanical energy is less than the initial mechanical energy.

We now present a Conceptual Checkpoint that further examines the relationship between nonconservative work and distance.

CONCEPTUAL CHECKPOINT 8–3

A golfer badly misjudges a putt, sending the ball only one-quarter the distance to the hole. The original putt gave the ball an initial speed of v_0. If the force of resistance due to the grass is constant, would an initial speed of **(a)** $2v_0$, **(b)** $3v_0$, or **(c)** $4v_0$ be needed to get the ball to the hole from its original position?

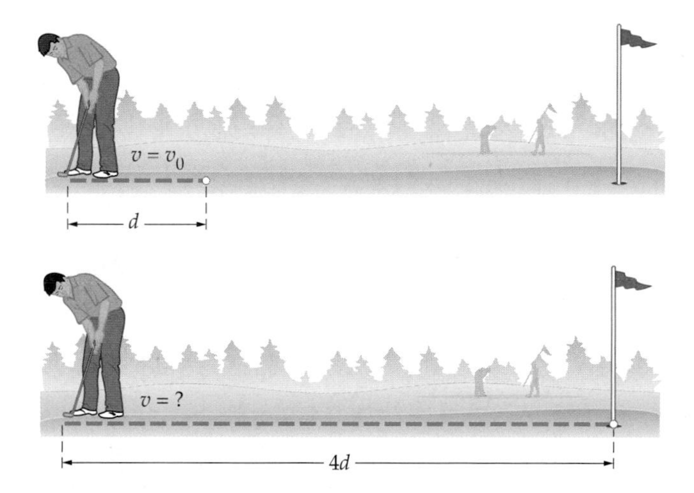

Reasoning and Discussion

In the original putt, the ball started with a kinetic energy of $\frac{1}{2}mv_0^2$ and came to rest in the distance d. The kinetic energy was dissipated by the nonconservative force due to grass resistance, F, which does the work $W_{nc} = -Fd$. Since the change in mechanical energy is $\Delta E = 0 - \frac{1}{2}mv_0^2 = -\frac{1}{2}mv_0^2$, it follows from $W_{nc} = \Delta E$ that $Fd = \frac{1}{2}mv_0^2$. Therefore, to go four times the distance, $4d$, we need to give the ball four times as much kinetic energy. Noting that kinetic energy is proportional to v^2, we see that the initial speed need only be doubled.

Answer:

(a) The initial speed should be doubled to $2v_0$.

A common example of a nonconservative force is kinetic friction. In the next Example, we show how to include the effects of friction in a system that also includes kinetic energy and gravitational potential energy.

EXAMPLE 8–10 Landing With a Thud

A block of mass $m_1 = 2.40$ kg is connected to a second block of mass $m_2 = 1.80$ kg, as shown here. When the blocks are released from rest they move through a distance $d = 0.500$ m, at which point m_2 hits the floor. Given that the coefficient of kinetic friction between m_1 and the horizontal surface is $\mu_k = 0.450$, find the speed of the blocks just before m_2 lands.

Picture the Problem

We choose $y = 0$ to be at floor level; therefore, the gravitational potential energy of m_2 is zero when it lands. The potential energy of m_1 doesn't change during this process; it is always m_1gh. Thus, it isn't necessary to know the value of h. Note that we label the beginning and ending points with i and f, respectively.

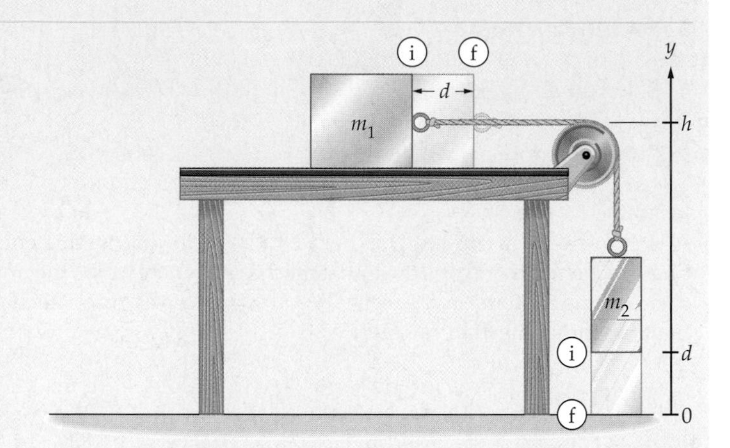

Strategy

Since a nonconservative force (friction) is doing work in this system, we use $W_{nc} = \Delta E = E_f - E_i$. Thus, we must calculate not only the mechanical energies, E_i and E_f, but also the nonconservative work, W_{nc}. Note that E_f can be written in terms of the unknown speed of the blocks just before m_2 lands. Therefore, we can set W_{nc} equal to ΔE and solve for the final speed.

Solution

1. Evaluate U_i, K_i, and E_i. Be sure to include contributions from both masses:

$$U_i = m_1gh + m_2gd$$

$$K_i = \tfrac{1}{2}m_1 \cdot 0^2 + \tfrac{1}{2}m_2 \cdot 0^2 = 0$$

$$E_i = U_i + K_i = m_1gh + m_2gd$$

2. Next, evaluate U_f, K_f, and E_f.

$$U_f = m_1gh + 0$$

$$K_f = \tfrac{1}{2}m_1v^2 + \tfrac{1}{2}m_2v^2$$

Note that E_f depends on the unknown speed, v:

$$E_f = U_f + K_f = m_1gh + \tfrac{1}{2}m_1v^2 + \tfrac{1}{2}m_2v^2$$

3. Calculate the nonconservative work, W_{nc}. Recall that the force of friction is $f_k = \mu_k N = \mu_k m_1 g$, and that it points opposite to the displacement of distance d:

$$W_{nc} = -f_k d = -\mu_k m_1 gd$$

4. Set W_{nc} equal to $\Delta E = E_f - E_i$. Notice that m_1gh cancels because it occurs in both E_i and E_f.

$$W_{nc} = E_f - E_i$$

$$-\mu_k m_1 gd = \tfrac{1}{2}m_1v^2 + \tfrac{1}{2}m_2v^2 - m_2 gd$$

5. Solve for v:

$$v = \sqrt{\frac{2(m_2 - \mu_k m_1)gd}{m_1 + m_2}}$$

6. Substitute numerical values:

$$v = \sqrt{\frac{2[1.80\text{ kg} - (0.450)(2.40\text{ kg})](9.81\text{ m/s}^2)(0.500\text{ m})}{1.80\text{ kg} + 2.40\text{ kg}}}$$

$$= 1.30\text{ m/s}$$

Insight

Note that line 4 can be rearranged as follows: $\tfrac{1}{2}m_1v^2 + \tfrac{1}{2}m_2v^2 = m_2gd - \mu_k m_1 gd$. Translating this to words, we can say that the final kinetic energy of the blocks is equal to the initial gravitational potential energy of m_2, minus the energy dissipated by friction.

Practice Problem

Find the coefficient of kinetic friction if the final speed of the blocks is 0.950 m/s. [**Answer:** $\mu_k = 0.589$]

Some related homework problems: Problem 27, Problem 28

Finally, we present an Active Example for the common situation of a system in which two different nonconservative works are done.

ACTIVE EXAMPLE 8–3 Marathon Man

An 80.0-kg jogger starts from rest and runs uphill into a stiff breeze. At the top of the hill the jogger has done a work $W_{nc1} = +1.80 \times 10^4$ J, air resistance has done a work $W_{nc2} = -4420$ J, and the jogger's speed is 3.50 m/s. Find the height of the hill.

Solution

1. Write the initial mechanical energy, E_i: $E_i = U_i + K_i = 0 + 0 = 0$
2. Write the final mechanical energy, E_f: $E_f = U_f + K_f = mgh + \frac{1}{2}mv^2$
3. Set W_{nc} equal to ΔE: $W_{nc} = \Delta E = mgh + \frac{1}{2}mv^2$
4. Use $W_{nc} = \Delta E$ to solve for h: $h = (W_{nc} - \frac{1}{2}mv^2)/mg$
5. Calculate the total nonconservative work: $W_{nc} = W_{nc1} + W_{nc2} = 13{,}600$ J
6. Substitute numerical values to determine h: $h = 16.7$ m

Insight

As usual when dealing with energy calculations, our final result is independent of the shape of the hill.

▲ Highways that descend steeply are often provided with escape ramps that enable truck drivers whose brakes fail to bring their rigs to a safe stop. These ramps provide a perfect illustration of the conservation of energy. From a physics point of view, the driver's problem is to get rid of an enormous amount of kinetic energy in the safest possible way. The ramps run uphill, so some of the kinetic energy is simply converted back into gravitational potential energy (just as in a roller coaster). In addition, the ramps are typically surfaced with sand or gravel, allowing much of the initial kinetic energy to be dissipated by friction into other forms of energy, such as sound and heat.

8–5 Potential Energy Curves and Equipotentials

Figure 8–10 shows a metal ball rolling on a roller coaster-like track. Initially the ball is at rest at point A. Since the height at A is $y = h$, the ball's initial mechanical energy is $E_0 = mgh$. If friction and other nonconservative forces can be ignored, the ball's mechanical energy remains fixed at E_0 throughout its motion. Thus,

$$E = U + K = E_0$$

As the ball moves, its potential energy falls and rises in the same way as the track. After all, the gravitational potential energy, $U = mgy$, is directly proportional to the height of the track, y. In a sense, then, the track itself represents a graph of the corresponding potential energy.

This is shown explicitly in **Figure 8–11**, where we plot energy on the vertical axis and x on the horizontal axis. The potential energy U looks just like the track in Figure 8–10. In addition, we plot a horizontal line at the value E_0, indicating the constant energy of the ball. Since the potential energy plus the kinetic energy must always add up to E_0, it follows that K is the amount of energy

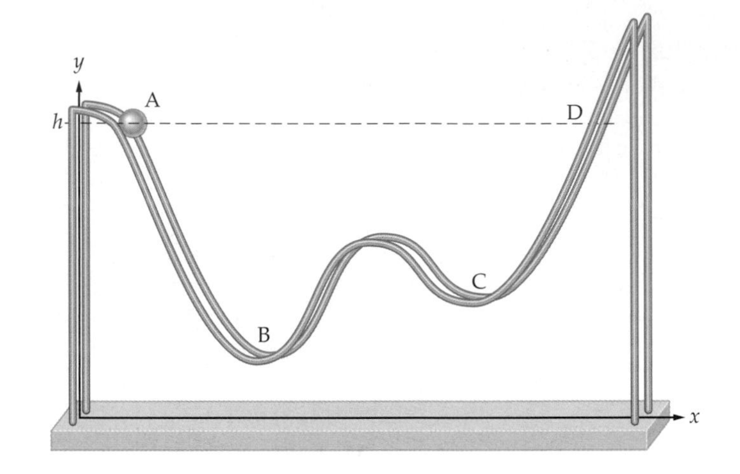

▶ **FIGURE 8–10 A ball rolling on a frictionless track**

The ball starts at A, where $y = h$, with zero speed. Its greatest speed occurs at B. At D, where $y = h$ again, its speed returns to zero.

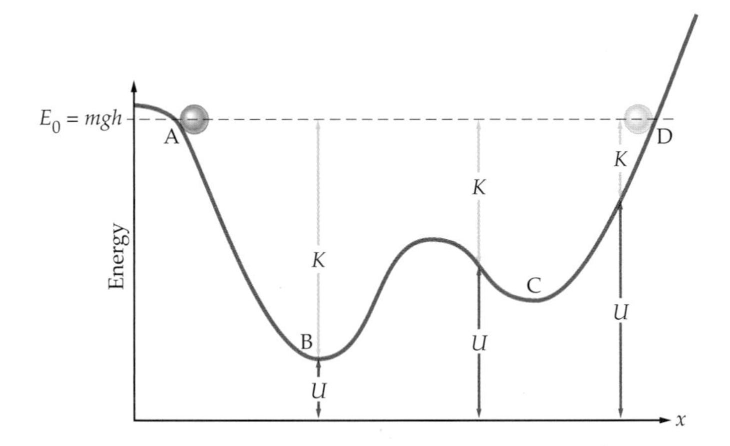

▶ **FIGURE 8–11 Gravitational potential energy versus position for the track shown in Figure 8–10**

The shape of the potential energy curve is exactly the same as the shape of the track. In this case, the total mechanical energy is fixed at its initial value, $E_0 = U + K = mgh$. Because the height of the curve is U, by definition, it follows that K is the distance from the curve up to the dashed line at $E_0 = mgh$. Note that K is largest at B. In addition, K vanishes at A and D, which are turning points of the motion.

from the potential energy curve up to the horizontal line at E_0. This is also shown in Figure 8–11.

Examining an energy plot like Figure 8–11 gives a great deal of information about the motion of an object. For example, at point B the potential energy has its lowest value, and thus the kinetic energy is greatest there. At point C the potential energy has increased, indicating a corresponding decrease in kinetic energy. As the ball continues to the right, the potential energy increases until, at point D, it is again equal to the total energy, E_0. At this point the kinetic energy is zero, and the ball comes to rest momentarily. It then "turns around" and begins to move to the left, eventually returning to point A where it again stops, changes direction, and begins a new cycle. Points A and D, then, are referred to as **turning points** of the motion.

Turning points are also seen in the motion of a mass on a spring, as indicated in **Figure 8–12**. Figure 8–12 (a) shows a mass pulled to the position $x = A$, and released from rest; Figure 8–12 (b) shows the potential energy of the system, $U = \frac{1}{2}kx^2$. Starting the system this way gives it an initial energy $E_0 = \frac{1}{2}kA^2$, shown by the horizontal line in Figure 8–12 (b). As the mass moves to the left its speed increases, reaching a maximum where the potential energy is lowest, at $x = 0$. If no nonconservative forces act, the mass continues to $x = -A$, where it stops momentarily before returning to $x = A$. This type of **oscillatory motion** will be studied in detail in Chapter 13.

The next Example uses a potential-energy curve to find the speed of an object at a given value of x.

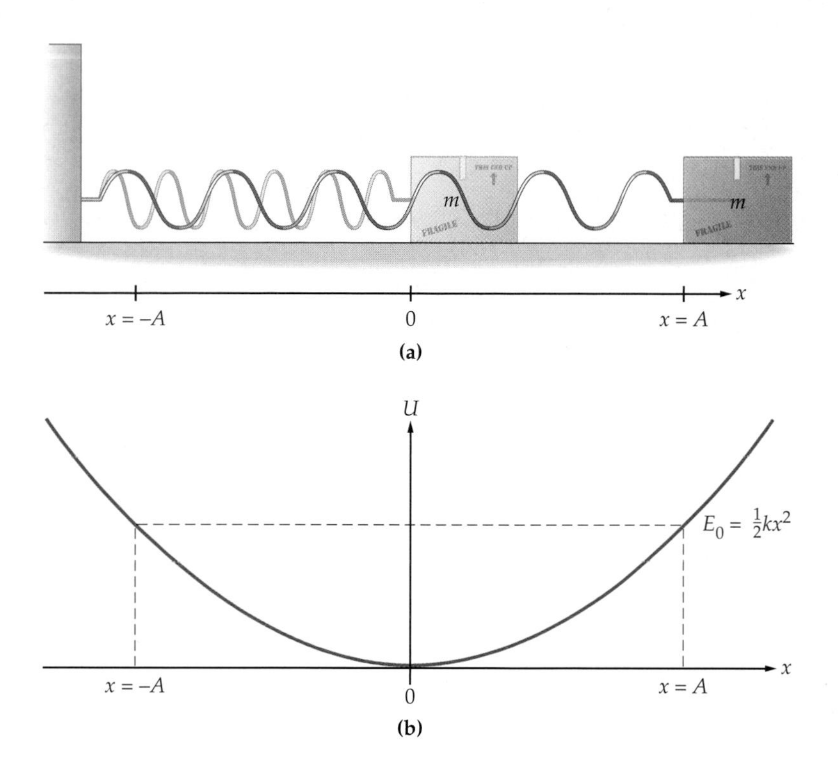

◀ **FIGURE 8–12 A mass on a spring**
(a) A spring is stretched by an amount A, giving it a potential energy of $U = \frac{1}{2}kA^2$. **(b)** The potential energy curve, $U = \frac{1}{2}kx^2$, for the spring in (a). Because the mass starts at rest, its initial mechanical energy is $E_0 = \frac{1}{2}kA^2$. The mass oscillates between $x = A$ and $x = -A$.

EXAMPLE 8–11 A Potential Problem

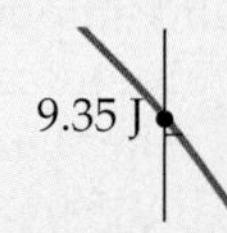

A 1.60-kg object in a conservative system moves along the x axis, where the potential energy is as shown. A physical example would be a bead sliding on a wire with the shape of the potential energy curve. If the object's speed at $x = 0$ is 2.30 m/s, what is its speed at $x = 2.00$ m?

Picture the Problem

The plot shows U as a function of x. The values of U at $x = 0$ and $x = 2.00$ m are 9.35 J and 4.15 J, respectively.

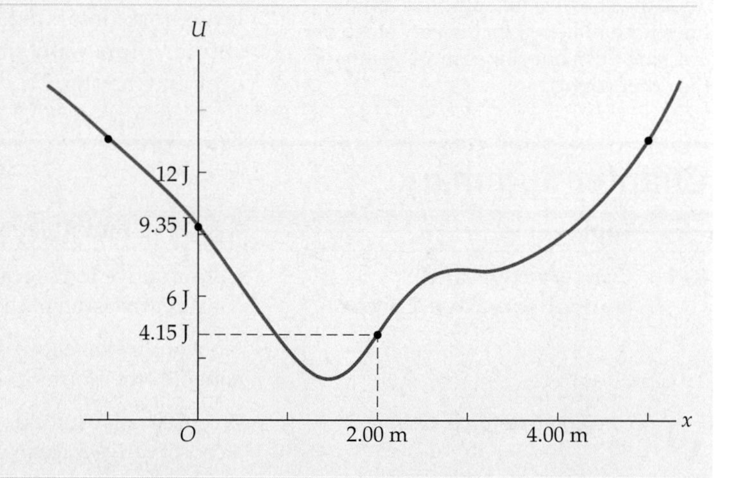

Strategy

Since mechanical energy is conserved, we know that the total energy at $x = 0$ $(U_i + K_i)$ is equal to the total energy at $x = 2.00$ m $(U_f + K_f)$.

The problem statement gives U_i, and since we also know the speed at $x = 0$ we can use $K = \frac{1}{2}mv^2$ to calculate the corresponding kinetic energy, K_i. At $x = 2.00$ m we know the potential energy, U_f; hence we can use $U_i + K_i = U_f + K_f$ to solve for K_f. Once the final kinetic energy is known, it is possible to solve for the final speed by once again using $K = \frac{1}{2}mv^2$.

Solution

1. Evaluate U_i, K_i, and E_i at $x = 0$:

$$U_i = 9.35\text{ J}$$
$$K_i = \tfrac{1}{2}mv_i^2 = \tfrac{1}{2}(1.60\text{ kg})(2.30\text{ m/s})^2 = 4.23\text{ J}$$
$$E_i = U_i + K_i = 9.35\text{ J} + 4.23\text{ J} = 13.6\text{ J}$$

2. Write expressions for U_f, K_f, and E_f at $x = 2.00$ m:

$$U_f = 4.15\text{ J}$$
$$K_f = \tfrac{1}{2}mv_f^2$$
$$E_f = U_f + K_f = 4.15\text{ J} + \tfrac{1}{2}mv_f^2$$

3. Set E_f equal to E_i:

$$4.15\text{ J} + \tfrac{1}{2}mv_f^2 = 13.6\text{ J}$$

Solve for v_f.

$$v_f = \sqrt{\frac{2(13.6\text{ J} - 4.15\text{ J})}{m}}$$

continued on the following page

continued from the previous page

4. Substitute the numerical value of the object's mass:

$$v_f = \sqrt{\frac{2(13.6\text{ J} - 4.15\text{ J})}{1.60\text{ kg}}} = 3.44\text{ m/s}$$

Insight

As we see in step 1, the total mechanical energy of the system is 13.6 J. This means that turning points for this object occur at values of x where $U = 13.6$ J.

Practice Problem

Using the graph provided, estimate the location of the turning points for this object. [**Answer:** $x = -1.00$ m and $x = 5.00$ m]

Some related homework problems: Problem 36, Problem 37

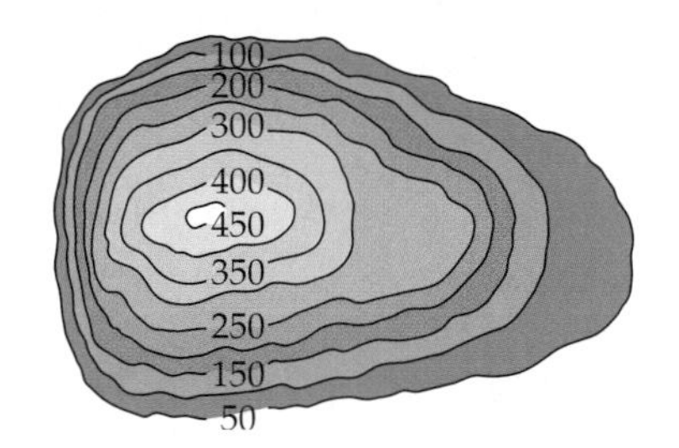

▲ **FIGURE 8–13 A contour map**
The map shows equal-altitude contour lines for a hill (top) that is very steep on one side (left) but slopes more gently on the other (right).

Oscillatory motion between turning points is also observed in molecules. If the energy of oscillation is relatively small, as is usual at room temperature, the atoms in a molecule simply vibrate back and forth—like masses connected by a spring. As long as no energy is gained or lost, the molecular oscillations continue unchanged. On the other hand, if the energy of the molecule is increased by heating, or some other mechanism, the molecule will eventually dissociate—fly apart—as the atoms move to infinite separation.

In some cases, a two-dimensional plot of potential energy contours is useful. For instance, **Figure 8–13** shows a contour map of a hill. Each contour corresponds to a given altitude, and hence, to a given value of the gravitational potential energy. In general, lines corresponding to constant values of potential energy are called **equipotentials**. Since the altitude changes by equal amounts from one contour to the next, it follows that when gravitational equipotentials are packed close together, the corresponding terrain is steep. On the other hand, when the equipotentials are widely spaced, the ground is nearly flat, since a large horizontal distance is required for a given change in altitude. We shall see similar plots with similar interpretations when we study electric potential energy in Chapter 21.

Chapter Summary

	Topic	Remarks and Relevant Equations
8–1	**Conservative and Nonconservative Forces**	Conservative forces conserve the mechanical energy of a system. Thus, in a conservative system the total mechanical energy remains constant.
		Nonconservative forces convert mechanical energy into other forms of energy, or convert other forms of energy into mechanical energy.
	conservative force, definition	A conservative force does zero total work on a closed path. In addition, the work done by a conservative force in going from point A to point B is *independent of the path* from A to B.
	examples of conservative forces	Gravity, spring.
	nonconservative force, definition	The work done by a nonconservative force on a closed path is nonzero. The work is also path dependent.
	examples of nonconservative forces	Friction, air resistance, tension in ropes and cables, forces exerted by muscles and motors.
8–2	**Potential Energy and the Work Done by Conservative Forces**	Potential energy, U, can "store" energy in a system. Energy in the form of potential energy can be converted to kinetic or other forms of energy.
	potential energy, definition	The work done by a conservative force is minus the change in potential energy: $W_c = -\Delta U = U_i - U_f$ **8–1**
	zero level	Any location can be chosen for $U = 0$. Once the choice is made, however, it must be used consistently.

	gravity	Choosing $y = 0$ to be the zero level near Earth's surface, $U = mgy$.	**8–3**
	spring	Choosing $x = 0$ (the equilibrium position) to be the zero level, $U = \frac{1}{2}kx^2$.	**8–5**
8–3	**Conservation of Mechanical Energy**	Mechanical energy, E, is conserved in systems with conservative forces only.	
	mechanical energy, definition	Mechanical energy is the sum of the potential and kinetic energies of a system: $$E = U + K$$	**8–6**
8–4	**Work Done by Nonconservative Forces**	Nonconservative forces can change the mechanical energy of a system.	
	change in mechanical energy	The work done by a nonconservative force is equal to the change in the mechanical energy of a system: $$W_{nc} = \Delta E = E_f - E_i$$	**8–9**
8–5	**Potential Energy Curves and Equipotentials**	A potential energy curve plots U as a function of position. An equipotential plot shows contours corresponding to constant values of U.	
	turning points	Turning points occur where an object stops momentarily before reversing direction. At turning points the kinetic energy is zero.	
	oscillatory motion	An object moving back and forth between two turning points is said to have oscillatory motion.	

Problem-Solving Summary

Type of Calculation	Relevant Physical Concepts	Related Examples
Calculate the gravitational or potential energy.	The potential energy for gravity is $U = mgy$; the potential energy for a spring is $U = \frac{1}{2}kx^2$.	Examples 8–2, 8–3, 8–4
Apply energy conservation in a system involving gravity.	Choose a horizontal level for $y = 0$, then use $U = mgy$.	Examples 8–6, 8–7 Active Example 8–1
Apply energy conservation in a system involving a spring.	Use $U = \frac{1}{2}kx^2$, where x measures the expansion or compression of the spring from its equilibrium position.	Example 8–8, Active Example 8–1
Find the nonconservative work done on a system.	Calculate the initial energy, E_i, and the final energy, E_f. Then use $W_{nc} = \Delta E = E_f - E_i$.	Examples 8–9, 8–10, Active Examples 8–2, 8–3

Conceptual Questions

1. Is it possible for the kinetic energy of an object to be negative? Is it possible for the gravitational potential energy of an object to be negative?
2. An avalanche occurs when a mass of snow slides down a steep mountain slope. Discuss the energy conversions responsible for water vapor rising to form clouds, falling as snow on a mountain, and then sliding down a slope as an avalanche.
3. Taking a leap of faith, a bungee jumper steps off a platform and falls until the cord brings her to rest. Suppose you analyze this system, choosing $y = 0$ either at the platform level or at the ground level. Will your answers agree or disagree on the following quantities: **(a)** The jumper's initial and final potential energy; **(b)** the jumper's change in potential energy?
4. A ball dropped to the floor rises to a height less than its original height. Discuss some of the energy conversions that take place in a system like this.
5. Is the change in gravitational potential energy of a stone that falls to the ground different from that of a stone that is thrown to the ground? Explain.
6. If the stretch of a spring is doubled, the force it exerts is also doubled. By what factor does the potential energy increase?
7. A mass is attached to a vertical spring. This causes the spring to stretch and the mass to move downward. Does the potential energy of the spring increase or decrease? Does the gravitational potential energy of the mass increase or decrease?
8. When a mass is placed on a vertical spring, the spring compresses and the mass moves downward. Analyze this system in terms of mechanical energy.
9. You and a friend both solve a problem involving a skier going down a slope. When comparing solutions, you notice that you have chosen different levels for $y = 0$. Will your answers agree or

disagree on the following quantities: **(a)** The skier's potential energy; **(b)** the skier's change in potential energy; **(c)** the skier's kinetic energy?

10. If a spring is stretched so far that it is permanently deformed, its force is no longer conservative. Why?

11. A block slides on a frictionless, horizontal surface with a speed v until it encounters a spring. The block compresses the spring a certain distance, d. When the spring expands again, what is the speed it gives to the block?

12. When an object is thrown upward to a person on a roof, at what point is the object's **(a)** kinetic energy and **(b)** potential energy a maximum? At what locations do these energies have their minimum values?

13. You throw a ball upward and let it fall to the ground. Your friend drops an identical ball straight down to the ground from the same height. Compare the change in gravitational potential energy for the two balls. Compare the change in kinetic energies.

14. It is a law of nature that the total energy of the universe is conserved. What do politicians mean, then, when they urge "energy conservation"?

15. Discuss the various energy conversions that occur during a pole vault.

▲ How many energy conversions can you identify? (Question 15)

16. Discuss the various energy conversions that occur during a swan dive from a diving board.

17. For an object moving along the x axis, the potential energy of the system is shown in **Figure 8–14**. At what point(s) is the speed of the object the same as it is at **(a)** point A, and **(b)** point B?

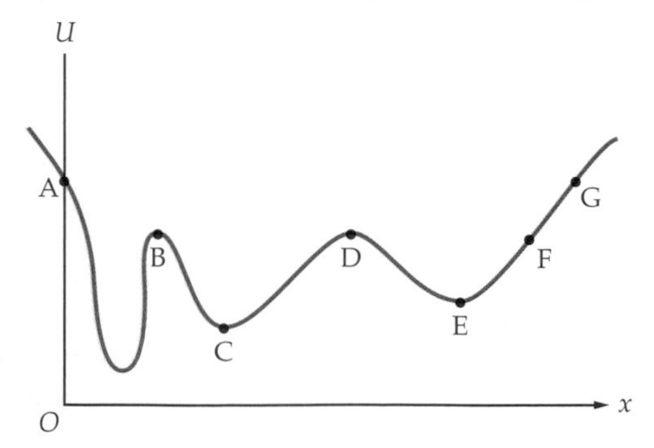

▲ **FIGURE 8–14** Question 17

18. You drive your bicycle down a hill and stop at the bottom. What happened to the potential energy you and your bicycle had at the top of the hill?

19. When a ball is thrown upward, its mechanical energy, $E = mgy + \frac{1}{2}mv^2$, is constant with time if air resistance can be ignored. How does E vary with time if air resistance cannot be ignored?

20. When a ball is thrown upward, it spends the same amount of time on the way up as on the way down—as long as air resistance can be ignored. If air resistance is taken into account, is the time on the way down (i) the same as, (ii) greater than, or (iii) less than the time on the way up? Explain.

21. A leaf falls to the ground with constant speed. Is $U_i + K_i$ greater than, less than, or the same as $U_f + K_f$? Explain.

22. On reentry, the space shuttle's heat tiles become extremely hot. What is the source of the energy that causes this heating?

23. Suppose the situation discussed in Conceptual Checkpoint 8–2 were to be repeated on a planet with a smaller acceleration of gravity. Would the height of the hill be (i) greater than, (ii) the same as, or (iii) less than the height on Earth? In the second case, where $v_i = 5\text{ m/s}$, do you expect the final speed to be (i) greater than, (ii) the same as, or (iii) less than the final speed on Earth?

24. If the force on an object is zero, does that mean the potential energy of the system is zero? If the potential energy of a system is zero, is the force zero?

25. You coast up a hill on your bicycle with decreasing speed. Your friend pedals up the hill with constant speed. In which case is mechanical energy conserved? Explain.

26. A toy frog consists of a suction cup and a spring. When the suction cup is pressed against a smooth surface the frog is held down. When the suction cup lets go, the frog leaps into the air. Discuss the behavior of the frog in terms of energy conversions.

27. Discuss the nature of the work done by the equipment shown in this photo. What types of forces are involved?

▲ Conservative or nonconservative? (Question 27)

Problems

Note: **IP** *denotes an integrated conceptual/quantitative problem.* **BIO** *identifies problems of biological or medical interest. Blue bullets (•, ••, •••) are used to indicate the level of difficulty of each problem.*

Section 8–1 Conservative and Nonconservative Forces

1. • Calculate the work done by gravity as a 2.6-kg object is moved from point A to point B in **Figure 8–15** along paths 1, 2, and 3.

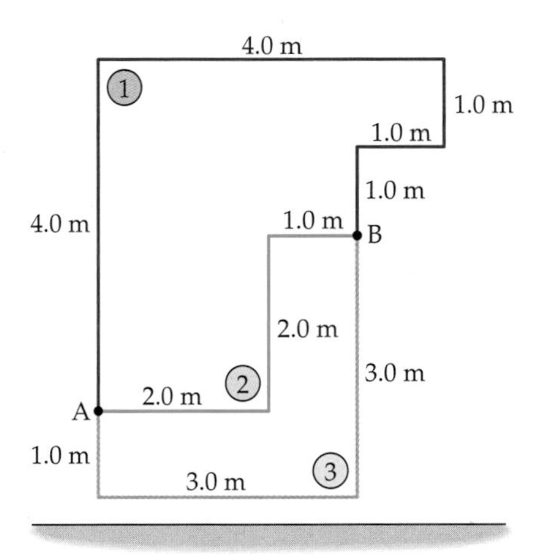

▲ **FIGURE 8–15** Problem 1

2. • Calculate the work done by friction as a 2.6-kg box is slid along a floor from point A to point B in **Figure 8–16** along paths 1, 2, and 3. Assume that the coefficient of kinetic friction between the box and the floor is 0.23.

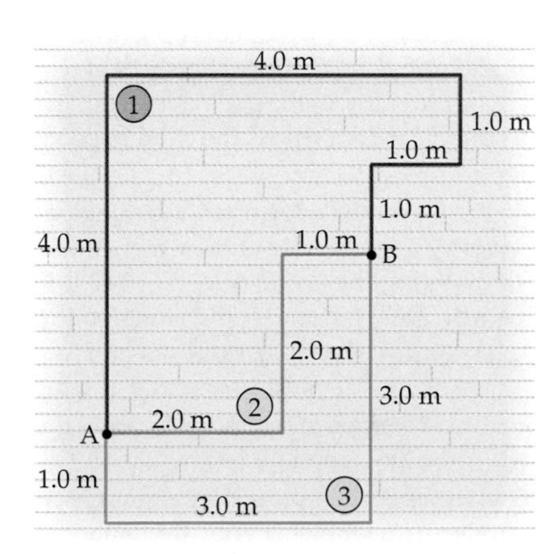

▲ **FIGURE 8–16** Problem 2

3. • **IP** A 4.1-kg block is attached to a spring with a force constant of 550 N/m, as shown in **Figure 8–17**. **(a)** Find the work done by the spring on the block as the block moves from A to B along paths 1 and 2. **(b)** How do your results depend on the mass of the block?

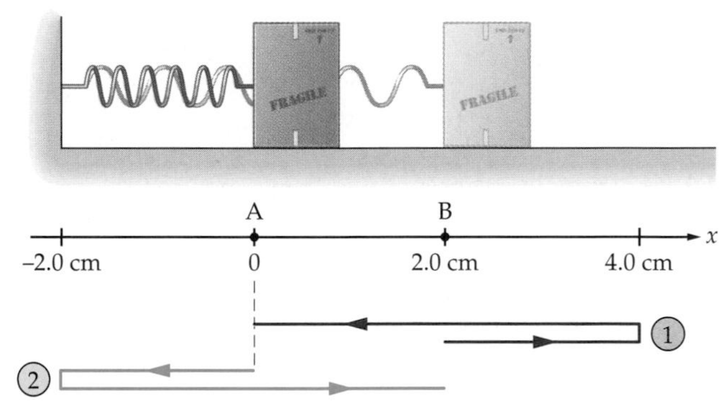

▲ **FIGURE 8–17** Problem 3

4. • **IP (a)** Calculate the work done by gravity as a 5.2-kg object is moved from A to B in **Figure 8–18** along paths 1 and 2. **(b)** How do your results depend on the mass of the block?

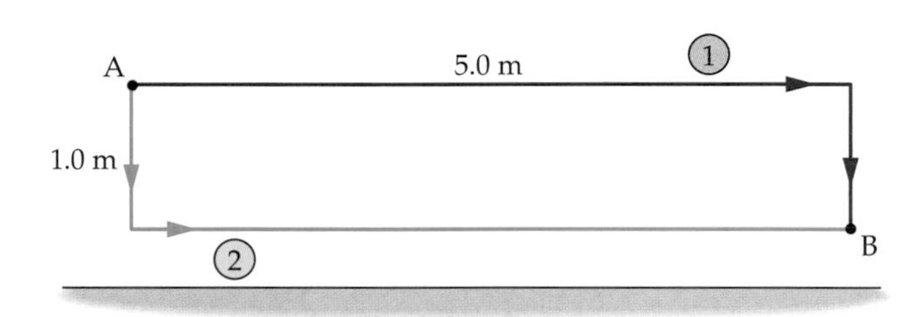

▲ **FIGURE 8–18** Problem 4

Section 8–2 Potential Energy and the Work Done by Conservative Forces

5. • Find the gravitational potential energy of an 80.0-kg person standing atop Mt. Everest, at an altitude of 8848 m. Use sea level as the location for $y = 0$.

6. • As an Acapulco cliff diver drops to the water from a height of 40.0 m, his gravitational potential energy decreases by 25,000 J. How much does the diver weigh?

7. •• Pushing on the pump of a soap dispenser compresses a small spring. When the spring is compressed 0.50 cm, the spring potential energy is 0.0025 J. What compression is required for the spring potential energy to equal 0.0084 J?

8. •• **IP** A vertical spring stores 0.962 J in spring potential energy when a 3.0-kg mass is suspended from it. **(a)** By what factor does the spring potential energy change if the mass attached to the spring is doubled? **(b)** Verify your answer to part (a) by calculating the spring potential energy when a 6.0-kg mass is attached to the spring.

9. •• If 30.0 J of work are required to stretch a spring from a 4.00-cm elongation to a 5.00-cm elongation, how much work is needed to stretch it from a 5.00-cm to a 6.00-cm elongation?

10. •• A 0.33-kg pendulum bob is attached to a string 1.2 m long. What is the change in the bob's gravitational potential energy as it swings from point A to point B in **Figure 8–19**?

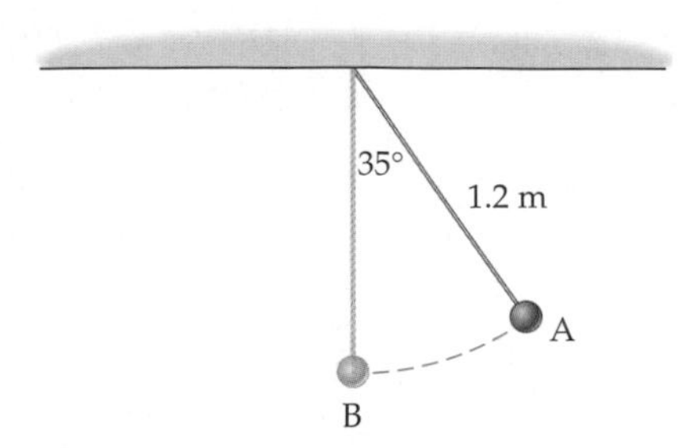

▲ **FIGURE 8–19** Problem 10

Section 8–3 Conservation of Mechanical Energy

11. • At an amusement park, a swimmer uses a water slide to enter the main pool. If the swimmer starts at rest, slides without friction, and falls through a vertical height of 2.61 m, what is her speed at the bottom of the slide?

12. • In the previous problem, find the swimmer's speed at the bottom of the slide if she starts with an initial speed of 0.840 m/s.

13. • **IP** A player passes a 0.600-kg basketball downcourt for a fast break. The ball leaves the player's hands with a speed of 8.30 m/s and slows down to 7.10 m/s at its highest point. **(a)** Ignoring air resistance, how high above the release point is the ball when it is at its maximum height? **(b)** How would doubling the ball's mass affect the result in part (a)?

14. •• **IP** In a tennis match, a player wins a point by hitting the ball sharply to the ground on the opponent's side of the net. **(a)** If the ball bounces upward from the ground with a speed of 16 m/s and is caught by a fan with a speed of 12 m/s, how high above the court is the fan? Ignore air resistance. **(b)** Explain why it is not necessary to know the mass of the tennis ball.

15. •• A 0.21-kg apple falls from a tree to the ground, 4.0 m below. Ignoring air resistance, determine the apple's gravitational potential energy, U, kinetic energy, K, and total mechanical energy, E, when its height above the ground is each of the following: 4.0 m, 3.0 m, 2.0 m, 1.0 m, and 0 m. Take ground level to be $y = 0$.

16. •• **IP** A 2.7-kg block slides with a speed of 1.1 m/s on a frictionless, horizontal surface until it encounters a spring. **(a)** If the block compresses the spring 6.0 cm before coming to rest, what is the force constant of the spring? **(b)** What initial speed should the block have if it is to compress the spring by 1.5 cm?

17. •• A 0.26-kg rock is thrown vertically upward from the top of a cliff that is 32 m high. When it hits the ground at the base of the cliff the rock has a speed of 29 m/s. Assuming that air resistance can be ignored, find **(a)** the initial speed of the rock and **(b)** the greatest height of the rock as measured from the base of the cliff.

18. •• A 1.60-kg block slides with a speed of 0.950 m/s on a frictionless, horizontal surface until it encounters a spring with a force constant of 902 N/m. The block comes to rest after compressing the spring 4.00 cm. Find the spring potential energy, U, the kinetic energy of the block, K, and the total mechanical energy of the system, E, for the following compressions: 0 cm, 1.00 cm, 2.00 cm, 3.00 cm, 4.00 cm.

19. •• A 5.00-kg rock is dropped and allowed to fall freely. Find the initial kinetic energy, the final kinetic energy, and the change in kinetic energy for **(a)** the first two meters of fall and **(b)** the second two meters of fall.

20. ••• The two masses in the Atwood's machine shown in **Figure 8–20** are initially at rest at the same height. After they are released, the large mass, m_2, falls through a height h and hits the floor, and the small mass, m_1, rises through a height h. **(a)** Find the speed of the masses just before m_2 lands, giving your answer in terms of m_1, m_2, g, and h. Assume the ropes and pulley have negligible mass and that friction can be ignored. **(b)** Evaluate your answer to part (a) for the case where h = 1.2 m, m_1 = 3.7 kg, and m_2 = 4.1 kg.

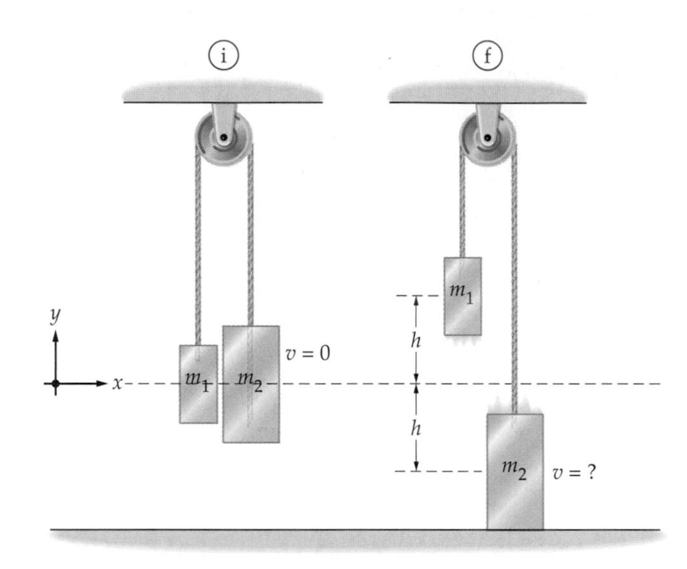

▲ **FIGURE 8–20** Problems 20, 21, and 54

21. ••• In the previous problem, suppose that m_2 has an initial upward speed of 0.20 m/s. How high does m_2 rise above its initial position before momentarily coming to rest, given that m_1 = 3.7 kg and m_2 = 4.1 kg?

Section 8–4 Work Done by Nonconservative Forces

22. • Catching a wave, a 72-kg surfer starts with a speed of 1.3 m/s, drops through a height of 1.75 m, and ends with a speed of 8.2 m/s. How much nonconservative work was done on the surfer?

23. • At a playground, an 18-kg child plays on a slide that drops through a height of 2.2 m. The child starts at rest at the top of the slide. On the way down, the slide does a nonconservative work of −373 J on the child. What is the child's speed at the bottom of the slide?

24. • Starting at rest at the edge of a swimming pool, a 72.0-kg athlete swims along the surface of the water and reaches a speed of 1.20 m/s by doing the work W_{nc1} = +160 J. Find the nonconservative work, W_{nc2}, done by the water on the athlete .

25. • A 17,000-kg airplane lands with a speed of 82 m/s on a stationary aircraft carrier deck that is 115 m long. Find the work done by nonconservative forces in stopping the plane.

26. • **IP** The driver of a 1100-kg car moving at 17 m/s brakes quickly to 12 m/s when he spots a local garage sale. **(a)** Find the change in the car's kinetic energy. **(b)** Explain where the "missing" kinetic energy has gone.

27. • Suppose the system in Example 8–10 starts with m_2 moving downward with a speed of 1.3 m/s. What speed do the masses have just before m_2 lands?

28. •• A 30.0-kg seal at an amusement park slides down a ramp into the pool below. The top of the ramp is 1.50 m higher than the surface of the water and the ramp is inclined at an angle of 30.0° above the horizontal. If the seal reaches the water with a speed of 4.90 m/s, what is **(a)** the coefficient of kinetic friction between the seal and the ramp and **(b)** the work done by kinetic friction?

29. •• A 1.75-kg rock is released from rest at the surface of a pond 1.00 m deep. As the rock falls, a constant upward force of 4.10 N is exerted on it by water resistance. Calculate the nonconservative work, W_{nc}, done by water resistance on the rock, the gravitational potential energy of the system, U, the kinetic energy of the rock, K, and the total mechanical energy of the system, E, for the following depths below the water's surface: $d = 0.00$ m, 0.500 m, 1.00 m. Let $y = 0$ be at the bottom of the pond.

30. •• A 1300-kg car drives up a 17.0-m hill. During the drive, two nonconservative forces do work on the car: (i) the force of friction, and (ii) the force generated by the car's engine. The work done by friction is -3.31×10^5 J; the work done by the engine is $+6.34 \times 10^5$ J. Find the change in the car's kinetic energy from the bottom of the hill to the top of the hill.

31. •• **IP** An 81.0-kg in-line skater does +3500 J of nonconservative work by pushing against the ground with his skates. In addition, friction does −710 J of nonconservative work on the skater. The skater's initial and final speeds are 2.50 m/s and 1.60 m/s, respectively. **(a)** Has the skater gone uphill, downhill, or remained at the same level? Explain. **(b)** Calculate the change in height of the skater.

32. •• In Example 8–10, suppose the two masses start from rest and are moving with a speed of 2.05 m/s just before m_2 hits the floor. If the coefficient of kinetic friction is $\mu_k = 0.350$, what is the distance of travel, d, for the masses?

33. •• **IP** A 15,800-kg truck is moving at 12.0 m/s when it starts down a 6.00° incline in the Canadian Rockies. At the start of the descent the driver notices that the altitude is 1630 m. When she reaches an altitude of 1440 m her speed is 29.0 m/s. Find the change in the truck's **(a)** gravitational potential energy and **(b)** kinetic energy. **(c)** Is the total mechanical energy of the truck conserved? Explain.

34. ••• A 1.80-kg block slides on a rough, horizontal surface. The block hits a spring with a speed of 2.00 m/s and compresses it a distance of 11.0 cm before coming to rest. If the coefficient of kinetic friction between the block and the surface is $\mu_k = 0.560$, what is the force constant of the spring?

Section 8–5 Potential Energy Curves and Equipotentials

35. • **Figure 8–21** shows a potential energy curve as a function of x. Describe the subsequent motion of an object that starts at rest at point A.

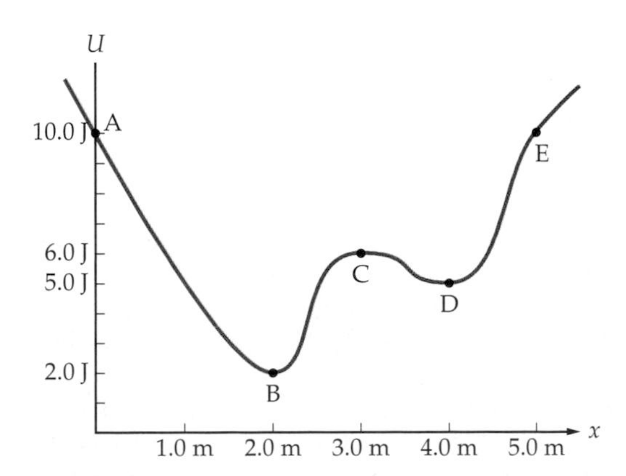

▲ **FIGURE 8–21** Problems 35, 36, 37 and 40

36. • An object moves along the x axis, subject to the potential energy shown in Figure 8–21. The object has a mass of 1.1 kg, and starts at rest at point A. **(a)** What is the object's speed at point B? At point C? At point D? **(b)** What are the turning points for this object?

37. • A 1.3 kg object moves along the x axis, subject to the potential energy shown in Figure 8–21. If the object's speed at point C is 1.65 m/s, what are the approximate locations of its turning points?

38. • A 23-kg child swings back and forth on a swing suspended by 2.0-m-long ropes. Plot the gravitational potential energy of this system as a function of the angle the ropes make with the vertical, assuming the potential energy is zero when the ropes are vertical. Consider angles up to 90° on either side of the vertical.

39. •• Find the turning-point angles in the previous problem if the child has a speed of 0.89 m/s when the ropes are vertical. Indicate the turning points on a plot of the system's potential energy.

40. •• The potential energy of a particle moving along the x axis is shown in Figure 8–21. When the particle is at $x = 1.0$ m it has 3.0 J of kinetic energy. **(a)** What is the particle's total mechanical energy? **(b)** What is its range of motion?

41. ••• A mass m is attached to a spring of force constant k. Find the turning points for a mass with a maximum speed v_{max}.

General Problems

42. •• **IP** A sled slides without friction down a small, ice-covered hill. If the sled starts from rest at the top of the hill, its speed at the bottom is 8.50 m/s. **(a)** On a second run, the sled starts with a speed of 1.50 m/s at the top. When it reaches the bottom of the hill, is its speed 10.0 m/s, more than 10.0 m/s, or less than 10.0 m/s? Explain. **(b)** Find the speed of the sled at the bottom of the hill after the second run.

43. •• In the previous problem, what is the height of the hill?

44. •• A 61-kg skier encounters a dip in the snow's surface that has a circular cross section with a radius of curvature of 12 m. If the skier's speed at point A in **Figure 8–22** is 8.0 m/s, what is the normal force exerted by the snow on the skier at point B? Ignore frictional forces.

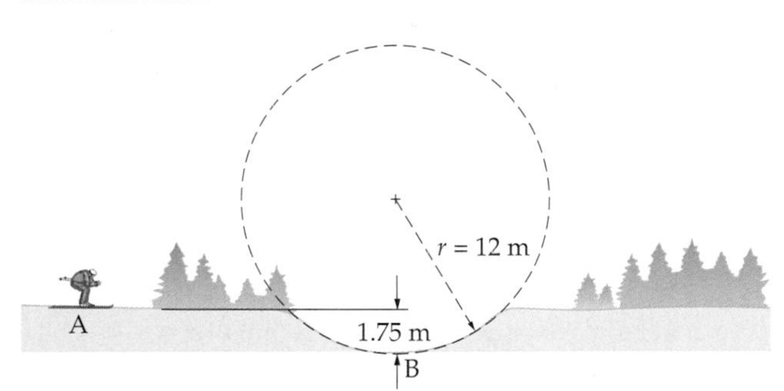

▲ **FIGURE 8–22** Problem 44

45. •• The spring in a clothespin is compressed 0.40 cm, storing energy in the form of spring potential energy. If the spring is compressed twice as far, the spring potential energy increases by 0.0046 J. What is the force constant, k, for this spring?

46. •• In a circus act, a 60.0-kg trapeze artist starts from rest with the 5.00-m trapeze rope horizontal. What is the tension in the rope when it is vertical?

47. •• An 865-kg airplane starts at rest on an airport runway at sea level. **(a)** What is the change in mechanical energy of the airplane if it climbs to a cruising altitude of 2420 m and maintains a constant speed of 90.0 m/s? **(b)** What cruising speed would the plane need at this altitude if its increase in kinetic energy is to equal its increase in potential energy?

48. •• **IP** At the local playground a child on a swing has a speed of 2.12 m/s when the swing is at its lowest point. **(a)** To what maximum vertical height does the child rise, assuming he sits still

and "coasts"? **(b)** How do your results change if the initial speed of the child is halved?

49. •• The water slide shown in **Figure 8–23** ends at a height of 1.50 m above the pool. If the person starts from rest at point A and lands in the water at point B, what is the height h of the water slide? (Assume the water slide is frictionless.)

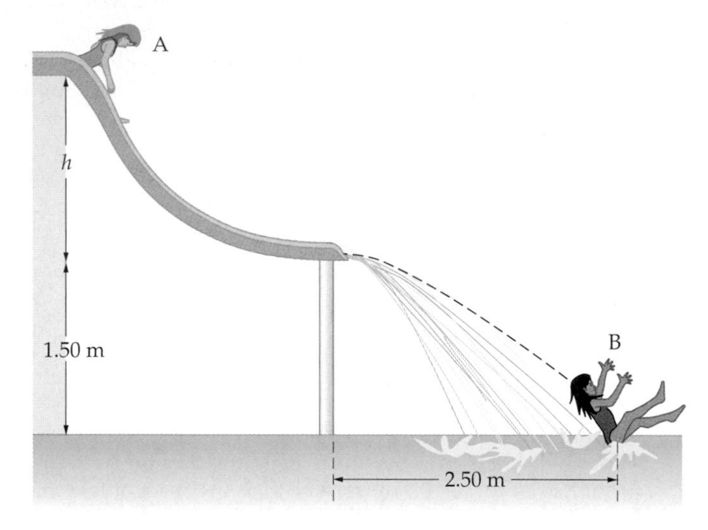

▲ **FIGURE 8–23** Problems 49 and 50

50. •• If the height of the water slide in Figure 8–23 is $h = 3.2$ m and the person's initial speed at point A is 0.54 m/s, at what location does the swimmer splash down in the pool?

51. •• **IP** A person is to be released from rest on a swing pulled away from the vertical by an angle of 20.0°. The two frayed ropes of the swing are 2.75 m long, and will break if the tension in either of them exceeds 355 N. **(a)** What is the maximum weight the person can have and not break the ropes? **(b)** If the person is released at an angle greater than 20.0°, does the maximum weight increase, decrease, or stay the same? Explain.

52. •• **IP** A car is coasting without friction toward a hill of height h and radius of curvature r. **(a)** What initial speed, v_0, will result in the car's wheels just losing contact with the roadway as the car crests the hill? **(b)** What happens if the initial speed of the car is greater than the value found in part (a)?

53. •• A skateboarder starts at point A in **Figure 8–24** and rises to a height of 2.64 m above the top of the ramp at point B. What was the skateboarder's initial speed at point A?

54. •• In the Atwood's machine of Problem 20, the mass m_2 remains at rest once it hits the floor, but the mass m_1 continues moving upward. How much higher does m_1 go after m_2 has landed? Give your answer for the case $h = 1.2$ m, $m_1 = 3.7$ kg, and $m_2 = 4.1$ kg.

55. •• An 8.70-kg block slides with an initial speed of 1.66 m/s up a ramp inclined at an angle of 27.4° with the horizontal. The coefficient of kinetic friction between the block and the ramp is 0.62. Use energy conservation to find the distance the block slides before coming to rest.

56. •• Repeat the previous problem for the case of an 8.70-kg block sliding down the ramp, with an initial speed of 1.66 m/s.

57. •• Jeff of the Jungle swings on a 7.00-m vine that initially makes an angle of 35.0° with the vertical. If Jeff starts at rest and has a mass of 73.0 kg, what is the tension in the vine at the lowest point of the swing?

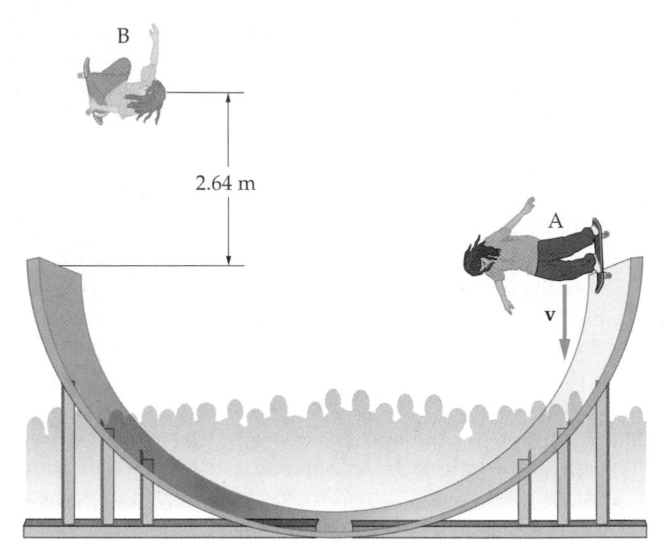

▲ **FIGURE 8–24** Problems 53 and 58

58. •• A skateboard track has the form of a circular arc with a 4.00 m radius, extending to an angle of 90.0° relative to the vertical on either side of the lowest point, as shown in Figure 8–24. A 57.0-kg skateboarder starts from rest at the top of the circular arc. What is the normal force exerted on the skateboarder at the bottom of the circular arc?

59. •• **IP** A 1.3-kg mass is attached to a spring of force constant 510 N/m. **(a)** Find the turning points for a mass with a maximum speed of 1.6 m/s. **(b)** By what factor do the turning-point positions change if the force constant of the spring is doubled?

60. •• A 1.9-kg block slides down a frictionless ramp, as shown in **Figure 8–25**. The top of the ramp is 1.5 m above the ground; the bottom of the ramp is 0.25 m above the ground. The block leaves the ramp moving horizontally, and lands a horizontal distance d away. Find the distance d.

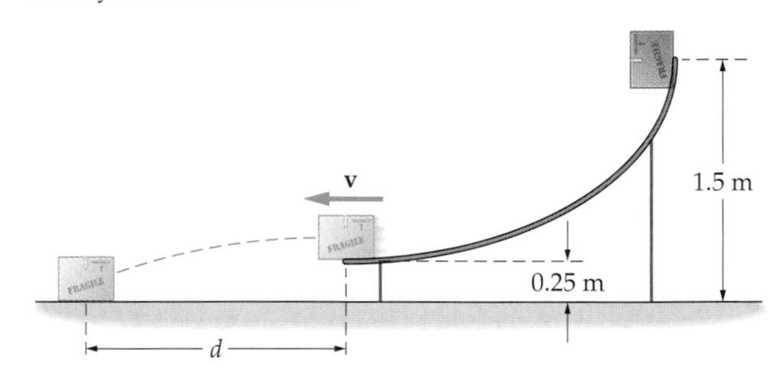

▲ **FIGURE 8–25** Problem 60

61. •• A 1.4-kg block is pushed up against a spring whose force constant is 650 N/m, compressing it 6.1 cm. When the block is released it moves horizontally without friction until it launches off the edge of a table 0.65 m above the ground. Find the horizontal distance, d, covered by the block during its fall to the ground.

62. ••• A trapeze artist of mass m swings on a rope of length L. Initially, the trapeze artist is at rest, and the rope makes an angle of 30° with the vertical. Find the tension in the rope when it is vertical. Explain why your result depends on L in the way it does.

63. ••• A mass m is attached to a string of length L and released from rest at the point A in **Figure 8–26**. Show that the tension in the string when the mass just reaches point B is $3mg$, independent of the length l.

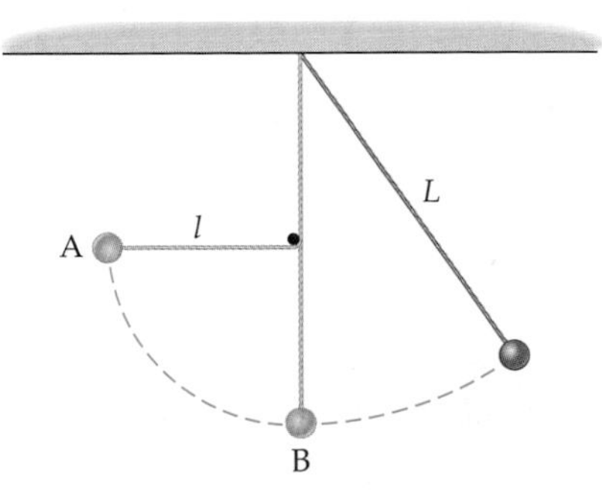

▲ **FIGURE 8–26** Problems 63 and 64

64. •• Referring to Figure 8–26, suppose that $L = 0.75$ cm and $l = 0.43$ cm. Find the maximum angle the string makes with the vertical when the mass swings as far to the right as it can.

65. ••• An ice cube is placed on top of an overturned spherical bowl of radius r, as indicated in **Figure 8–27**. If the ice cube slides downward from rest at the top of the bowl, at what angle θ does it separate from the bowl? In other words, at what angle does the normal force between the ice cube and the bowl go to zero?

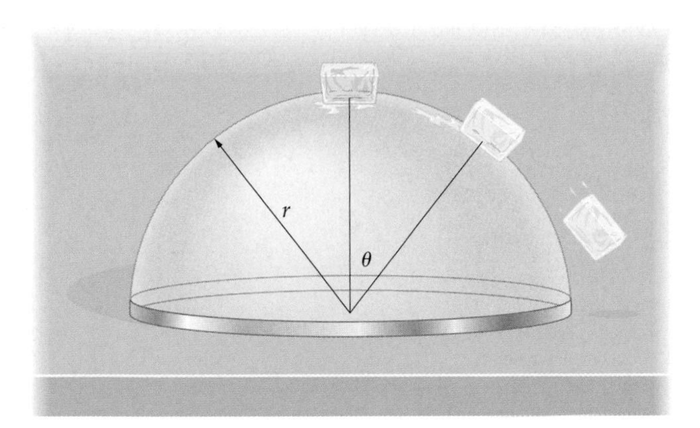

▲ **FIGURE 8–27** Problem 65

66. ••• The two blocks shown in **Figure 8–28** are moving with an initial speed v. If the system is frictionless, find the distance d the blocks travel before coming to rest. (Let $U = 0$ at the initial position of block 2.)

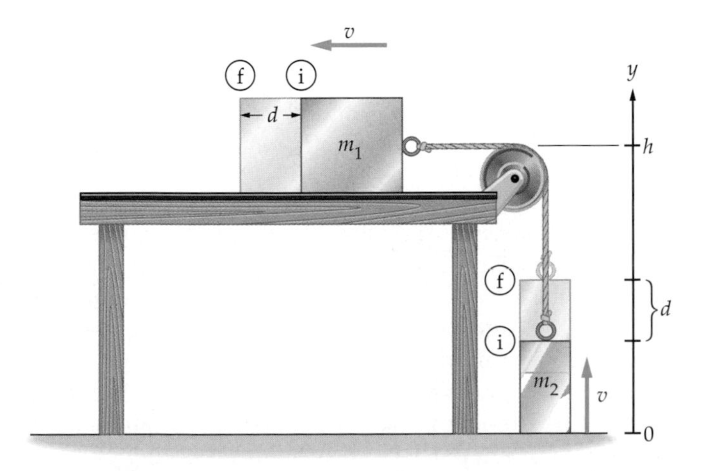

▲ **FIGURE 8–28** Problems 66 and 67

67. •• Referring to the previous problem, what initial speed is required if the blocks $m_1 = 3.1$ kg and $m_2 = 1.4$ kg are to travel a distance of 3.5 cm before coming to rest?

68. ••• **IP** **(a)** A block of mass m slides from rest on a frictionless loop-the-loop track, as shown in **Figure 8–29**. What is the minimum release height, h, required for the block to maintain contact with the track at all times? Give your answer in terms of the radius of the loop, r. **(b)** Explain why the release height obtained in part (a) is independent of the block's mass.

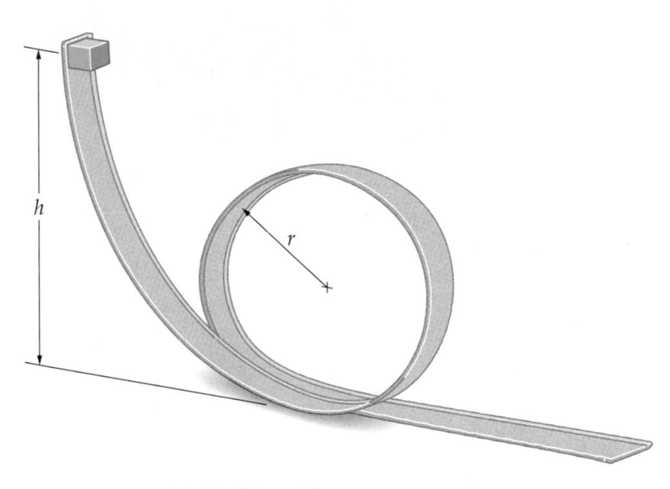

▲ **FIGURE 8–29** Problem 68

69. ••• **Figure 8–30** shows a 1.50-kg block at rest on a ramp of height h. When the block is released, it reaches the bottom of the ramp and moves across a surface that is frictionless except for one section of width 10.0 cm that has a coefficient of kinetic friction $\mu_k = 0.640$. Find h such that the block's speed after crossing the rough patch is 3.50 m/s.

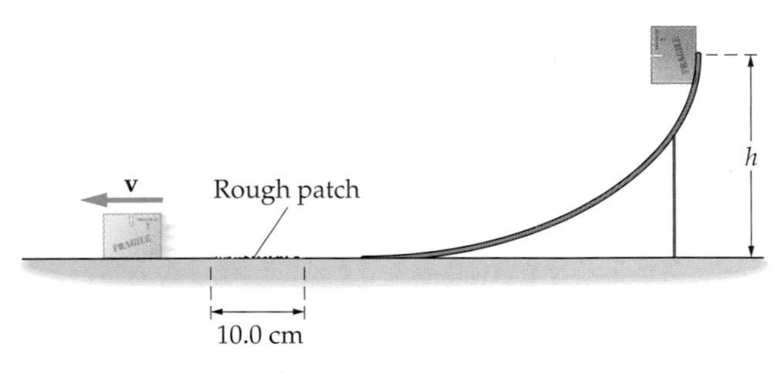

▲ **FIGURE 8–30** Problem 69

70. ••• In **Figure 8–31** a 1.2-kg block is held at rest against a spring with a force constant $k = 730$ N/m. Initially, the spring is compressed a distance d. When the block is released it slides across a surface that is frictionless, except for a section of width 5.0 cm that has a coefficient of kinetic friction $\mu_k = 0.44$. Find d such that the block's speed after crossing the rough patch is 2.3 m/s.

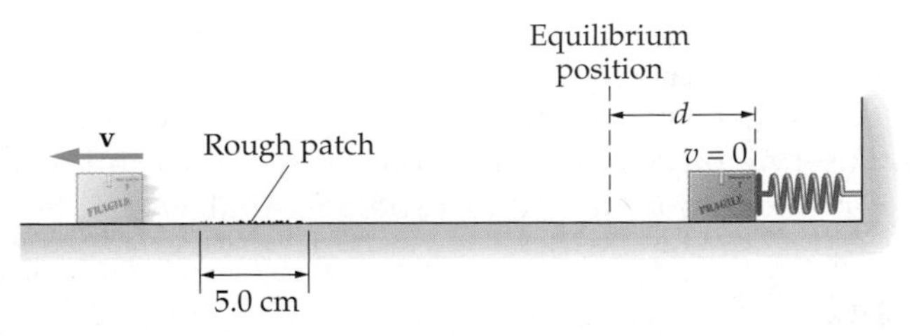

▲ **FIGURE 8–31** Problem 70

9 Linear Momentum and Collisions

A swift kick provides the impetus to send this soccer ball flying. In this chapter, we show how to analyze collisions between various objects—such as a foot and a soccer ball—in terms of Newton's laws and a new physical quantity called the momentum.

Conservation laws play a central role in physics. In this chapter we introduce the concept of *momentum* and show that it, like energy, is a conserved quantity. Nothing we can do, in fact nothing that can occur in nature, can change the total energy or the total momentum of the universe.

As with conservation of energy, we shall see that the conservation of momentum provides a powerful way of approaching a variety of problems that would be extremely difficult to solve using Newton's laws directly. In particular, problems involving the collision of two or more objects—such as a baseball bat striking a ball, or one car bumping into another at an intersection—are especially well suited to a momentum approach. Finally, we introduce the concept of the *center of mass* and show that it allows us to extend many of the results that have been obtained for point particles to systems involving more realistic objects.

9–1 Linear Momentum

Imagine for a moment that you are sitting at rest on a skateboard that can roll without friction on a smooth surface. If you catch a heavy, slow-moving ball tossed to you by a friend, you begin to move. If, on the other hand, your friend tosses you a light, yet fast-moving ball, the net effect may be the same—that is, catching a lightweight ball moving fast enough will cause you to move with the same speed as when you caught the heavy ball.

In physics, the previous observations are made precise by defining a quantity called the **linear momentum, p**, which is defined as the product of the mass m and velocity $\mathbf{v}$ of an object:

Definition of linear momentum, p

$$\mathbf{p} = m\mathbf{v} \qquad 9\text{–}1$$

SI unit: kg·m/s

In our example, if the heavy ball has twice the mass of the light ball, but the light ball has twice the speed of the heavy ball, the momenta of the two balls are equal. We can see from Equation 9–1 that the units of linear momentum are simply the units of mass times the units of velocity: kg·m/s. There is no special shorthand name given to this combination of units.

It is important to note that a constant *linear* momentum $\mathbf{p}$ is the momentum of an object of mass m *moving in a straight line* with a velocity $\mathbf{v}$. In Chapter 11 we introduce a similar quantity to describe the momentum of an object that rotates. This momentum will be referred to as the *angular momentum*. In general, when we simply say momentum, we are referring to the linear momentum $\mathbf{p}$. We will always specify angular momentum when referring to the momentum associated with rotation.

Since the velocity $\mathbf{v}$ is a vector with both a magnitude and a direction, so too is the momentum, $\mathbf{p} = m\mathbf{v}$. The next exercise gives some feeling for the *magnitude* of the momentum, $p = mv$, for everyday objects.

EXERCISE 9–1

(a) A 1180-kg car drives along a city street at 30.0 miles per hour (13.4 m/s). What is the magnitude of the car's momentum? **(b)** A major-league pitcher can give a 0.142-kg baseball a speed of 101 mi/h (45.1 m/s). Find the magnitude of the baseball's momentum.

Solution

(a) Using $p = mv$, we find

$$p_c = m_c v_c = (1180 \text{ kg})(13.4 \text{ m/s}) = 15{,}800 \text{ kg}\cdot\text{m/s}$$

(b) Similarly,

$$p_b = m_b v_b = (0.142 \text{ kg})(45.1 \text{ m/s}) = 6.40 \text{ kg}\cdot\text{m/s}$$

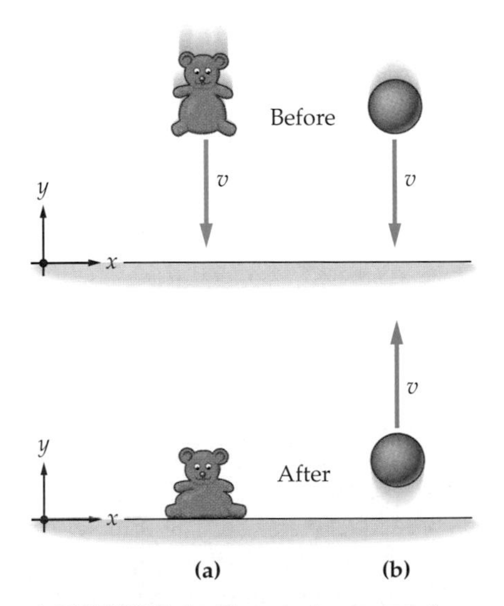

▲ **FIGURE 9–1 Change in momentum**
A beanbag bear and a rubber ball, with the same mass m and the same downward speed v, hit the floor. **(a)** The beanbag bear comes to rest on hitting the floor. Its change in momentum is mv upward. **(b)** The rubber ball bounces upward with a speed v. Its change in momentum is $2mv$ upward.

As an illustration of the vector nature of momentum, consider the situations shown in **Figures 9–1 (a)** and **(b)**. In Figure 9–1 (a), a 0.10-kg beanbag bear is dropped to the floor, where it hits with a speed of 4.0 m/s and sticks. In Figure 9–1 (b) a 0.10-kg rubber ball also hits the floor with a speed of 4.0 m/s, but in this case the ball bounces upward off the floor. Assuming an ideal rubber ball, its initial upward speed is 4.0 m/s. Now the question in each case is, "What is the change in momentum?"

To approach the problem systematically, we introduce a coordinate system as shown in Figure 9–1. With this choice, we can see that neither object has momentum in the x direction; thus we need only consider the y component of momentum,

Be sure to draw a coordinate system for momentum problems, even if the problem is only one-dimensional. It is important to use the coordinate system to assign the correct sign to velocities and momenta in the system.

p_y. The problem, therefore, is one-dimensional; still, we must be careful about the sign of p_y.

We begin with the beanbag. Just before hitting the floor, it moves downward (that is, in the negative y direction) with a speed of $v = 4.0$ m/s. Letting m stand for the mass of the beanbag, we find that the initial momentum is

$$p_{y,\mathrm{i}} = m(-v)$$

After landing on the floor the beanbag is at rest; hence, its final momentum is zero:

$$p_{y,\mathrm{f}} = m(0) = 0$$

Therefore the change in momentum is

$$\begin{aligned} \Delta p_y &= p_{y,\mathrm{f}} - p_{y,\mathrm{i}} = 0 - m(-v) = mv \\ &= (0.10 \text{ kg})(4.0 \text{ m/s}) = 0.40 \text{ kg}\cdot\text{m/s} \end{aligned}$$

Note that the change in momentum is positive—that is, in the upward direction. This makes sense because before the bag landed it had a negative (downward) momentum in the y direction. In order to increase the momentum from a negative value to zero, it is necessary to add a positive (upward) momentum.

Next, consider the rubber ball in Figure 9–1 (b). Before bouncing, its momentum is

$$p_{y,\mathrm{i}} = m(-v)$$

the same as for the beanbag. After bouncing, when the ball is moving in the upward (positive) direction, its momentum is

$$p_{y,\mathrm{f}} = mv$$

As a result, the change in momentum for the rubber ball is

$$\begin{aligned} \Delta p_y &= p_{y,\mathrm{f}} - p_{y,\mathrm{i}} = mv - m(-v) = 2mv \\ &= 2(0.10 \text{ kg})(4.0 \text{ m/s}) = 0.80 \text{ kg}\cdot\text{m/s} \end{aligned}$$

This is *twice* the change in momentum of the beanbag! The reason is that in this case, the momentum in the y direction must first be increased from $-mv$ to 0, then increased again from 0 to mv. For the beanbag, the change was merely from $-mv$ to 0.

Note how important it is to be careful about the vector nature of the momentum and to use the correct sign for p_y. Otherwise, we might have concluded—erroneously—that the rubber ball had zero change in momentum, since the *magnitude* of its momentum was unchanged by the bounce. In fact, its momentum does change due to the change in its *direction* of motion.

One additional point: Since momentum is a vector, the total momentum of a system of objects is the *vector* sum of the momentum of each object separately. That is,

$$\mathbf{p}_{\text{total}} = \mathbf{p}_1 + \mathbf{p}_2 + \mathbf{p}_3 + \ldots \qquad \textbf{9–2}$$

This is illustrated for the case of three objects in the following Example.

EXAMPLE 9–1 Duck, Duck, Goose: Adding Momenta

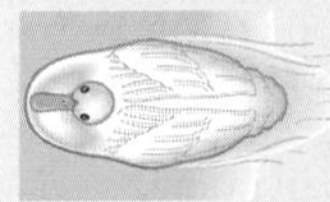

At a city park, a person throws some bread into a duck pond. Two 4.00-kg ducks and a 9.00-kg goose paddle rapidly toward the bread, as shown in the figure. If the ducks swim at 1.10 m/s, and the goose swims with a speed of 1.30 m/s, find the magnitude and direction of the total momentum of the three birds.

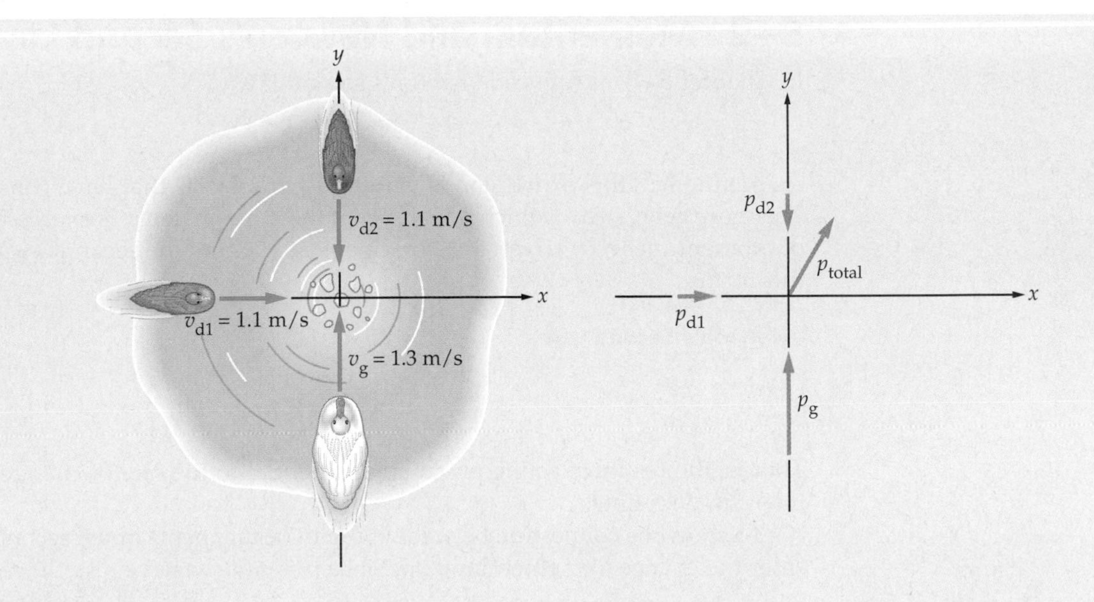

Picture the Problem

In our sketch we place the origin where the bread floats on the water. Note that duck 1 swims in the positive x direction, duck 2 swims in the negative y direction, and the goose swims in the positive y direction.

Strategy

Write the momentum of each bird as a vector, using unit vectors in the x and y directions. Next, sum the vectors component by component. Finally, use the components of the total momentum to calculate its magnitude and direction.

Solution

1. Use x and y unit vectors to express the momentum of each bird in vector form:

$$\mathbf{p}_{d1} = m_d v_d \hat{\mathbf{x}} = (4.00\text{ kg})(1.10\text{ m/s})\hat{\mathbf{x}} = (4.40\text{ kg}\cdot\text{m/s})\hat{\mathbf{x}}$$

$$\mathbf{p}_{d2} = -m_d v_d \hat{\mathbf{y}} = -(4.00\text{ kg})(1.10\text{ m/s})\hat{\mathbf{y}} = -(4.40\text{ kg}\cdot\text{m/s})\hat{\mathbf{y}}$$

$$\mathbf{p}_{g} = m_g v_g \hat{\mathbf{y}} = (9.00\text{ kg})(1.30\text{ m/s})\hat{\mathbf{y}} = (11.7\text{ kg}\cdot\text{m/s})\hat{\mathbf{y}}$$

2. Sum the momentum vectors to obtain the total momentum:

$$\mathbf{p}_{total} = \mathbf{p}_{d1} + \mathbf{p}_{d2} + \mathbf{p}_{g} = (4.40\text{ kg}\cdot\text{m/s})\hat{\mathbf{x}} + [-4.40\text{ kg}\cdot\text{m/s} + 11.7\text{ kg}\cdot\text{m/s}]\hat{\mathbf{y}} = (4.40\text{ kg}\cdot\text{m/s})\hat{\mathbf{x}} + (7.30\text{ kg}\cdot\text{m/s})\hat{\mathbf{y}}$$

3. Calculate the magnitude of the total momentum:

$$p_{total} = \sqrt{p_{total,x}^2 + p_{total,y}^2} = \sqrt{(4.40\text{ kg}\cdot\text{m/s})^2 + (7.30\text{ kg}\cdot\text{m/s})^2} = 8.52\text{ kg}\cdot\text{m/s}$$

4. Calculate the direction of the total momentum:

$$\theta = \tan^{-1}\left(\frac{p_{total,y}}{p_{total,x}}\right) = \tan^{-1}\left(\frac{7.30\text{ kg}\cdot\text{m/s}}{4.40\text{ kg}\cdot\text{m/s}}\right) = 58.9°$$

Insight

Note that the momentum of each bird depends only on its mass and velocity; it is independent of the bird's location.

Practice Problem

Should the speed of the goose be increased or decreased if the total momentum of the three birds is to point in the positive x direction? Verify your answer by calculating the required speed. [**Answer:** The goose's speed must be decreased. Setting the momentum of the goose equal to minus the momentum of duck 2 yields $v_g = 0.489$ m/s.]

Some related homework problems: Problem 1, Problem 2, Problem 3

9–2 Momentum and Newton's Second Law

In Chapter 5 we introduced Newton's second law:

$$\sum \mathbf{F} = m\mathbf{a}$$

As mentioned, this expression is valid only for objects that have constant mass. The more general law, which holds even if the mass changes, is expressed in terms of momentum. In fact, Newton's original statement of the second law was in just this form:

Newton's Second Law

$$\sum \mathbf{F} = \frac{\Delta \mathbf{p}}{\Delta t} \qquad 9\text{–}3$$

That is, the net force acting on an object is equal to the rate of change of its momentum with time.

To show the connection between these two statements of the second law, consider the change in momentum, $\Delta\mathbf{p}$. Since $\mathbf{p} = m\mathbf{v}$, we have

$$\Delta\mathbf{p} = \mathbf{p}_{\mathrm{f}} - \mathbf{p}_{\mathrm{i}} = m_{\mathrm{f}}\mathbf{v}_{\mathrm{f}} - m_{\mathrm{i}}\mathbf{v}_{\mathrm{i}}$$

However, if the mass is constant, so that $m_{\mathrm{f}} = m_{\mathrm{i}} = m$, it follows that the change in momentum is simply m times $\Delta\mathbf{v}$:

$$\Delta\mathbf{p} = m_{\mathrm{f}}\mathbf{v}_{\mathrm{f}} - m_{\mathrm{i}}\mathbf{v}_{\mathrm{i}} = m(\mathbf{v}_{\mathrm{f}} - \mathbf{v}_{\mathrm{i}}) = m\Delta\mathbf{v}$$

As a result, Newton's second law, for objects of constant mass, can be written as follows:

$$\sum \mathbf{F} = \frac{\Delta\mathbf{p}}{\Delta t} = m\frac{\Delta\mathbf{v}}{\Delta t}$$

Finally, recall that acceleration is the rate of change of velocity with time,

$$\mathbf{a} = \frac{\Delta\mathbf{v}}{\Delta t}$$

Therefore, we can write Equation 9–3 as

$$\sum \mathbf{F} = \frac{\Delta\mathbf{p}}{\Delta t} = m\mathbf{a} \qquad 9\text{–}4$$

Hence, the two statements are equivalent if the mass is constant.

It should be noted, however, that $\sum\mathbf{F} = \Delta\mathbf{p}/\Delta t$ is the general form of Newton's second law, and that it is valid no matter how the mass may vary. In the remainder of this chapter we use this form of the second law to investigate the connections between forces and changes in momentum.

9–3 Impulse

The pitcher delivers a fast ball, the batter takes a swing, and with a crack of the bat the ball that was approaching home plate at 90 mi/h is now heading toward the pitcher at 60 mi/h. In the language of physics, we say that the bat has delivered an **impulse**, **I**, to the ball.

During the brief time the ball and bat are in contact—perhaps as little as a thousandth of a second—the force between them rises rapidly to a large value, as shown in **Figure 9–2**, then falls back to zero as the ball takes flight. It would be difficult, of course, to describe every detail of the way the force varies with time. In-

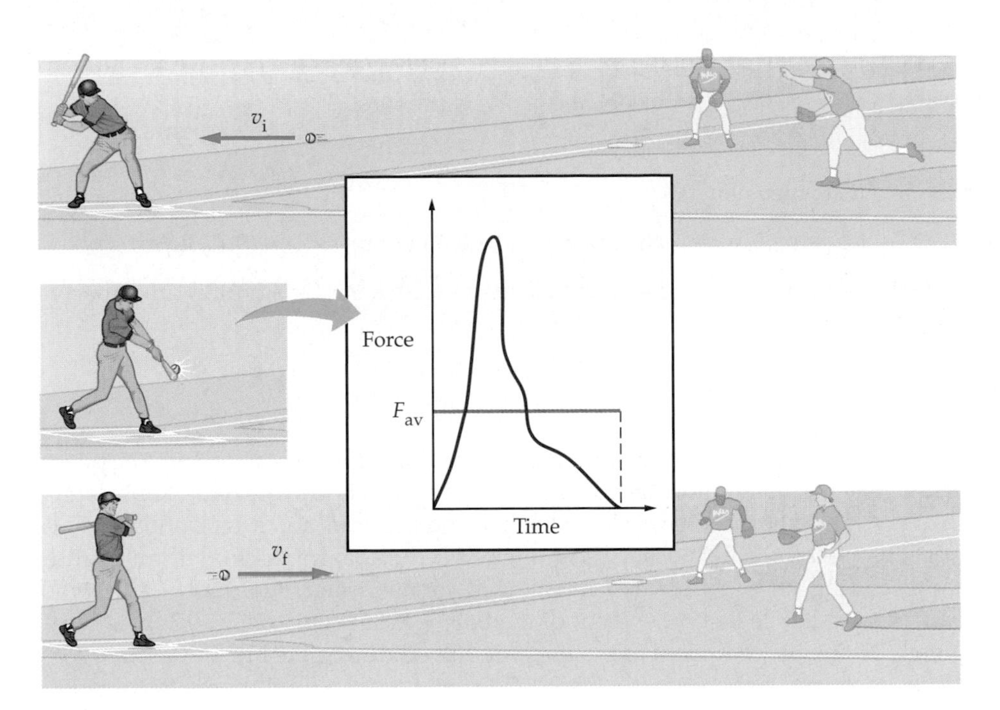

◀ **FIGURE 9–2 The average force during a collision**
The force between two objects that collide, as when a bat hits a baseball, rises rapidly to very large values, then drops again to zero in a matter of milliseconds. Rather than try to describe the complex behavior of the force, we focus on its average value, F_{av}. Note that the area under the F_{av} rectangle is the same as the area under the actual force curve.

stead, we focus on the average force exerted by the bat, $\mathbf{F}_{av}$, which is also shown in Figure 9–2, and define the impulse to be $\mathbf{F}_{av}$ times the length of time, Δt, that the ball and bat are in contact.

Definition of impulse, I

$$\mathbf{I} = \mathbf{F}_{av}\Delta t \qquad 9\text{–}5$$

SI unit: N·s = kg·m/s

Note that impulse is a vector and that it points in the same direction as the average force. In addition, its units are $\text{N}\cdot\text{s} = (\text{kg}\cdot\text{m/s}^2)\cdot\text{s} = \text{kg}\cdot\text{m/s}$, the same as the units of momentum.

It is no accident that impulse and momentum have the same units. In fact, rearranging Newton's second law, Equation 9–3, we see that the average force times Δt is simply the change in momentum of the ball due to the bat:

$$\mathbf{F}_{av} = \frac{\Delta \mathbf{p}}{\Delta t}$$

$$\mathbf{F}_{av}\Delta t = \Delta \mathbf{p}$$

Hence, in general, impulse is just the change in momentum:

$$\mathbf{I} = \mathbf{F}_{av}\Delta \text{t} = \Delta \mathbf{p} \qquad 9\text{–}6$$

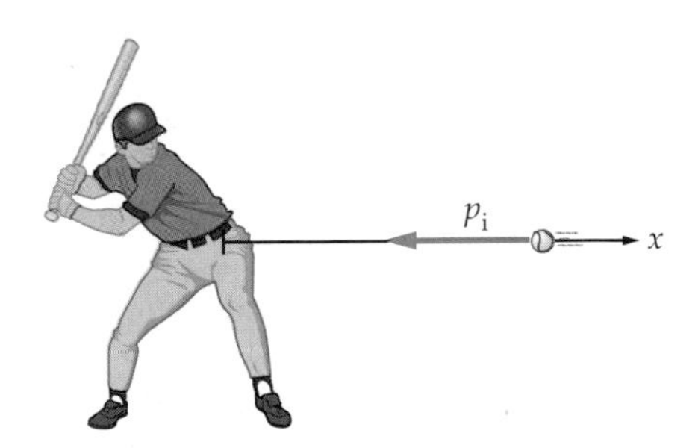

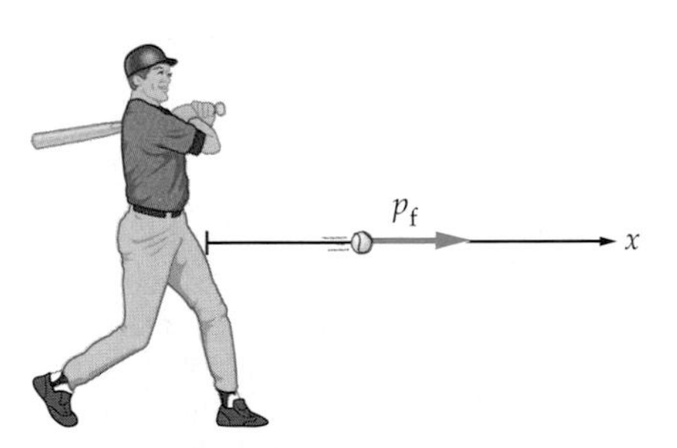

▲ **FIGURE 9–3 Hitting a baseball**
A batter hits a ball, sending it back toward the mound.

For example, if we know the impulse delivered to an object—that is, its change in momentum—and the time interval during which the change occurs, we can find the average force that caused the impulse.

As an example, let's calculate the impulse given to the baseball considered at the beginning of this section, as well as the average force between the ball and the bat. First, set up a coordinate system with the positive x axis pointing from home plate toward the pitcher's mound, as indicated in **Figure 9–3**. If the ball's mass is 0.145 kg, its initial momentum, which is in the negative x direction, is

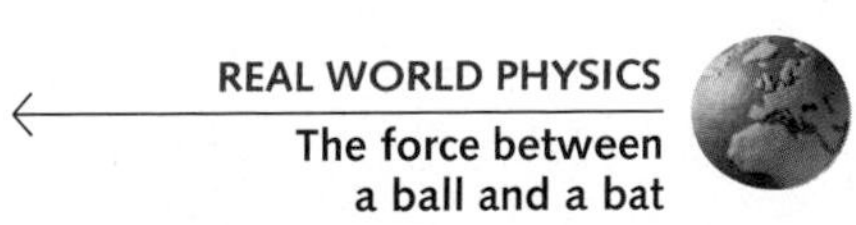

$$\mathbf{p}_i = -mv_i\hat{\mathbf{x}} = -(0.145\text{ kg})(90.0\text{ mi/h})\left(\frac{0.447\text{ m/s}}{1\text{ mi/h}}\right)\hat{\mathbf{x}} = -(5.83\text{ kg}\cdot\text{m/s})\hat{\mathbf{x}}$$

▲ When a golf ball is struck by a club (top), an enormous force (thousands of newtons) acts for a very short period of time—perhaps only a few ms. During this time, the ball is dramatically deformed by the impact. To keep the same thing from happening to a pole vaulter, who must fall nearly 20 feet after clearing the bar (bottom), a deeply padded landing area is provided. The change in the pole vaulter's momentum as he is brought to a stop, $mv = F\Delta t$, is the same whether he lands on a mat or on concrete. However, the padding is very yielding, greatly prolonging the time Δt during which he is in contact with the mat. The corresponding force on the vaulter is thus markedly decreased.

Immediately after the hit, the ball's final momentum is in the positive x direction:

$$\mathbf{p}_\mathrm{f} = mv_\mathrm{f}\hat{\mathbf{x}} = (0.145\ \mathrm{kg})(60.0\ \mathrm{mi/h})\left(\frac{0.447\ \mathrm{m/s}}{1\ \mathrm{mi/h}}\right)\hat{\mathbf{x}} = (3.89\ \mathrm{kg\cdot m/s})\hat{\mathbf{x}}$$

The impulse, then, is

$$\mathbf{I} = \Delta\mathbf{p} = \mathbf{p}_\mathrm{f} - \mathbf{p}_\mathrm{i} = [3.89\ \mathrm{kg\cdot m/s} - (-5.83\ \mathrm{kg\cdot m/s})]\hat{\mathbf{x}} = (9.72\ \mathrm{kg\cdot m/s})\hat{\mathbf{x}}$$

If the ball and bat are in contact for 1.20 ms $= 1.20 \times 10^{-3}$ s, a typical time, the average force is

$$\mathbf{F}_\mathrm{av} = \frac{\Delta\mathbf{p}}{\Delta t} = \frac{\mathbf{I}}{\Delta t} = \frac{(9.72\ \mathrm{kg\cdot m/s})\hat{\mathbf{x}}}{1.20 \times 10^{-3}\ \mathrm{s}} = (8.10 \times 10^3\ \mathrm{N})\hat{\mathbf{x}}$$

Note that the average force is in the positive x direction; that is, toward the pitcher, as expected. In addition, the magnitude of the average force is remarkably large. In everyday units, the force between the ball and the bat is about 1800 pounds. This explains why the ball is observed in high-speed photographs to deform significantly during a hit—the force is so large that, for an instant, it partially flattens the ball. Finally, notice that the weight of the ball, which is only about 0.3 lb, is completely negligible compared to the forces involved during the hit.

In problems that are strictly one-dimensional we can drop the vector notation when dealing with impulse. However, we must still be careful about the signs of the various quantities in the system. This is illustrated in the following Active Example.

ACTIVE EXAMPLE 9–1 The Natural

A 0.144-kg baseball is moving toward home plate with a speed of 43.0 m/s when it is bunted (hit softly). The bat exerts an average force of 6.50×10^3 N on the ball for 1.30 ms. The average force is directed toward the pitcher. What is the final speed of the ball?

Solution

1. Relate change in momentum to impulse (Equation 9–5): $\Delta p = p_\mathrm{f} - p_\mathrm{i} = I = F_\mathrm{av}\Delta t$
2. Solve for the final momentum: $p_\mathrm{f} = F_\mathrm{av}\Delta t + p_\mathrm{i}$
3. Calculate the initial momentum: $p_\mathrm{i} = -6.19\ \mathrm{kg\cdot m/s}$
4. Calculate the impulse: $I = F_\mathrm{av}\Delta t = 8.45\ \mathrm{kg\cdot m/s}$
5. Use these results to find the final momentum: $p_\mathrm{f} = 2.26\ \mathrm{kg\cdot m/s}$
6. Divide by the mass to find the final velocity: $v_\mathrm{f} = p_\mathrm{f}/m = 15.7\ \mathrm{m/s}$

Insight

As in Figure 9–3, we have chosen the positive direction to be toward the pitcher. Hence, the initial momentum, p_i, is negative. The impulse, which is directed toward the pitcher, is positive.

We saw in Section 9–1 that the change in momentum is different for an object that hits something and sticks compared with an object that hits and bounces off. This means that the impulse, and hence the force, is different in the two cases. We explore this in the following Conceptual Checkpoint.

CONCEPTUAL CHECKPOINT 9–1

A person stands under an umbrella during a rain shower. A few minutes later the raindrops turn to hail—though the number of "drops" hitting the umbrella per time and their speed remains the same. Is the force required to hold the umbrella in the hail **(a)** the same as, **(b)** more than, or **(c)** less than the force required in the rain?

▲ Most bats can take off simply by dropping from their perch on a branch or the ceiling of a cave, but vampire bats like this one must leap from the ground to become airborne. They do so by rocking forward onto their front limbs and then pushing off, using the extremely strong pectoral muscles that are also their main source of power in flight. Pushing downward on the ground, a bat experiences an upward reaction force exerted on it by the ground, with a corresponding impulse sufficient to propel it upward a considerable distance. In fact, a vampire bat can launch itself 1 m or more into the air in a mere 30 ms.

Reasoning and Discussion
When raindrops strike the umbrella they tend to splatter and run off; when hailstones hit the umbrella they bounce back upward. As a result, the change in momentum is greater for the hail—just as the change in momentum is greater for a rubber ball bouncing off the floor than it is for a beanbag landing on the floor. Hence, the impulse and the force is greater with hail.

Answer:
(b) The force is greater in the hail.

We conclude this section with an additional calculation involving impulse.

EXAMPLE 9–2 Jumping for Joy

After winning a prize on a game show, a 72-kg contestant jumps for joy. **(a)** If the jump results in an upward speed of 2.1 m/s, what is the impulse experienced by the contestant? **(b)** Before the jump, the floor exerts an upward force of mg on the contestant. What additional average upward force does the floor exert if the contestant pushes down on it for 0.36 s during the jump?

Picture the Problem
The sketch shows that the motion is purely one-dimensional. We choose the upward direction to be positive, hence the final momentum is positive.

Strategy
(a) The impulse is simply the change in momentum. We know the initial and final speeds, plus the mass of the contestant; hence the change in momentum, Δp, can be calculated using $p = mv$.
(b) The magnitude of the additional average force is $\Delta p/\Delta t$, where Δt is given as 0.36 s.

continued on the following page

continued from the previous page

Solution

Part (a)

1. Write an expression for the impulse, noting that $v_i = 0$: $I = \Delta p = p_f - p_i = mv_f$

2. Substitute numerical values: $I = mv_f = (72 \text{ kg})(2.1 \text{ m/s}) = 150 \text{ kg}\cdot\text{m/s}$

Part (b)

3. Express the average force in terms of the impulse I and the time interval Δt: $F_{av} = \dfrac{I}{\Delta t} = \dfrac{150 \text{ kg}\cdot\text{m/s}}{0.36 \text{ s}} = 420 \text{ kg}\cdot\text{m/s}^2 = 420 \text{ N}$

Insight

The additional average force exerted by the floor is rather large; in fact, 420 N is approximately 95 lb, or about 60% of the contestant's weight. Thus, the total upward force exerted by the floor is $mg + 420 \text{ N} = 710 \text{ N} + 420 \text{ N}$, which is about 250 lb. The contestant, of course, exerts the same force downward. Fortunately, the contestant only needs to exert that force for about a third of a second.

Practice Problem

Suppose the contestant lands with a speed of 2.1 m/s and comes to rest in 0.25 s. What is the magnitude of the average force exerted by the floor during landing? [**Answer:** $mg + 600 \text{ N} \sim 290 \text{ lb}$]

Some related homework problems: Problem 7, Problem 8

9–4 Conservation of Linear Momentum

Recall that the net force acting on an object is equal to the rate of change of its momentum

$$\sum \mathbf{F} = \frac{\Delta \mathbf{p}}{\Delta t}$$

Rearranging this expression, we find that the change in momentum during a time interval Δt is

$$\Delta \mathbf{p} = \left(\sum \mathbf{F}\right)\Delta t \qquad 9\text{–}7$$

Clearly, then, if the net force acting on an object is zero,

$$\sum \mathbf{F} = 0$$

its change in momentum is also zero

$$\Delta \mathbf{p} = \left(\sum \mathbf{F}\right)\Delta t = 0$$

Writing the change of momentum in terms of its initial and final values, we have

$$\Delta \mathbf{p} = \mathbf{p}_f - \mathbf{p}_i = 0$$

or

$$\mathbf{p}_f = \mathbf{p}_i \qquad 9\text{–}8$$

Since the momentum does not change in this case, we say that it is **conserved**. To summarize:

Conservation of Momentum

If the net force acting on an object is zero, its momentum is conserved; that is, $\mathbf{p}_f = \mathbf{p}_i$.

Note that in some cases the force may be zero in one direction and nonzero in another. For example, an object in free fall has a nonzero y component of force, $F_y \neq 0$, but no force in the x direction, $F_x = 0$. As a result, the object's y component of momentum changes with time while its x component of momentum remains constant. Thus, in applying momentum conservation, we must remember that both the force and the momentum are vector quantities and that the momentum conservation principle applies separately to each coordinate direction.

Thus far, our discussion has referred to the forces acting on a single object. Next, we consider a system composed of more than one object.

Internal Versus External Forces

The net force acting on a system of objects is the sum of forces applied from outside the system (external forces, $\mathbf{F}_{\text{ext}}$) and forces acting between objects within the system (internal forces, $\mathbf{F}_{\text{int}}$.) Thus, we can write

$$\mathbf{F}_{\text{net}} = \sum \mathbf{F} = \sum \mathbf{F}_{\text{ext}} + \sum \mathbf{F}_{\text{int}}$$

As we shall see, internal and external forces play very *different* roles in terms of how they affect the momentum of a system.

To illustrate the distinction, consider the case of two canoes floating at rest next to one another on a lake, as described in Example 5–3 and shown in **Figure 9–4**. In this case, let's consider the "system" to be the two canoes and the people inside them. When a person in canoe 1 pushes on canoe 2, a force $\mathbf{F}_2$ is exerted on canoe 2. By Newton's third law, an equal and opposite force, $\mathbf{F}_1 = -\mathbf{F}_2$, is exerted on the person in canoe 1. Note that $\mathbf{F}_1$ and $\mathbf{F}_2$ are internal forces, since they act between objects in the system. In addition, note that they sum to zero:

$$\mathbf{F}_1 + \mathbf{F}_2 = (-\mathbf{F}_2) + \mathbf{F}_2 = 0$$

This is a special case, of course, but it demonstrates the following general principles:

- Internal forces, like all forces, always occur in action-reaction pairs.
- Since the forces in action-reaction pairs are equal and opposite—due to Newton's third law—internal forces must *always* sum to zero:

$$\sum \mathbf{F}_{\text{int}} = 0$$

Since the internal forces always cancel, the net force acting on a system of objects is simply the sum of the *external* forces acting on it:

$$\mathbf{F}_{\text{net}} = \sum \mathbf{F}_{\text{ext}} + \sum \mathbf{F}_{\text{int}} = \sum \mathbf{F}_{\text{ext}}$$

The external forces, on the other hand, may or may not sum to zero—it all depends on the particular situation. For example, if the system consists of the two canoes in Figure 9–4, the external forces are the weights of the people and the canoes acting downward, and the upward, normal force exerted by the water to keep the canoes afloat. These forces sum to zero, and there is no acceleration in the vertical direction. In the next few sections we consider a variety of systems in which the external forces either sum to zero, or are so small that they can be ignored. Later, in Section 9–7, we consider situations where the external forces do not sum to zero, and hence must be taken into account.

◀ **FIGURE 9–4 Separating two canoes**
A system comprised of two canoes and their occupants. The forces $\mathbf{F}_1$ and $\mathbf{F}_2$ are internal to the system. They sum to zero.

▶ If the astronaut in this photo pushes on the satellite, the satellite exerts an equal but opposite force on him, in accordance with Newton's third law. If we are calculating the change in the astronaut's momentum, we must take this force into account. However, if we define the system to be the astronaut *and* the satellite, the forces between them are internal to the system. Whatever affect they may have on the astronaut or the satellite individually, they do not affect the momentum of the system as a whole. Whether a particular force counts as internal or external thus depends entirely on where we draw the boundaries of the system.

Finally, how do external and internal forces affect the momentum of a system? To see the connection, first note that Newton's second law gives the change in the net momentum for a given time interval Δt:

$$\Delta \mathbf{p}_{\text{net}} = \mathbf{F}_{\text{net}} \Delta t$$

Since the internal forces cancel, the change in the net momentum is directly related to the net *external* force:

$$\Delta \mathbf{p}_{\text{net}} = \left(\sum \mathbf{F}_{\text{ext}}\right) \Delta t \qquad 9\text{–}9$$

Therefore, the key distinction between internal and external forces is the following:

Conservation of momentum for a system of objects

Internal forces have absolutely no affect on the net momentum of a system.

- If the *net external* force acting on a system is zero, its net momentum is conserved. That is,

$$\mathbf{p}_{1,\text{f}} + \mathbf{p}_{2,\text{f}} + \mathbf{p}_{3,\text{f}} + \ldots = \mathbf{p}_{1,\text{i}} + \mathbf{p}_{2,\text{i}} + \mathbf{p}_{3,\text{i}} + \ldots$$

It is important to note that these statements apply only to the *net* momentum of a system, not to the momentum of each individual object. For example, suppose a system consists of two objects, 1 and 2, and that the net external force acting on the system is zero. As a result, the net momentum must remain constant:

$$\mathbf{p}_{\text{net}} = \mathbf{p}_1 + \mathbf{p}_2 = \text{constant}$$

This does not mean, however, that $\mathbf{p}_1$ and $\mathbf{p}_2$ are constant. All we can say is that the *sum* of $\mathbf{p}_1$ and $\mathbf{p}_2$ does not change.

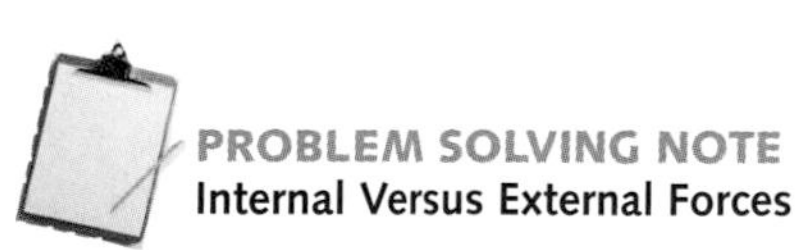

PROBLEM SOLVING NOTE
Internal Versus External Forces

It is important to keep in mind that internal forces cannot change the momentum of a system—only a net external force can do that.

As a specific example, consider the case of the two canoes floating on a lake, as described previously. Initially the momentum of the system is zero, since the canoes are at rest. After a person pushes the canoes apart, they are both moving, and hence both have nonzero momentum. Thus, the momentum of each canoe has changed. On the other hand, since the net external force acting on the system is zero, the sum of the canoes' momenta must still vanish. We show this in the next example.

EXAMPLE 9–3 Tippy Canoe: Comparing Velocity and Momentum

Two groups of canoeists meet in the middle of a lake. After a brief visit, a person in canoe 1 pushes on canoe 2 with a force of 46 N to separate the canoes. If the mass of canoe 1 and its occupants is 130 kg, and the mass of canoe 2 and its occupants is 250 kg, find the momentum of each canoe after 1.20 s of pushing.

Picture the Problem
We choose the positive x direction to point from canoe 1 to canoe 2. With this choice, the force exerted on canoe 2 is $\mathbf{F}_2 = (46\text{ N})\hat{\mathbf{x}}$ and the force exerted on canoe 1 is $\mathbf{F}_1 = (-46\text{ N})\hat{\mathbf{x}}$.

Strategy
First, we find the acceleration of each canoe using $a = F/m$. Next, we use $v = v_0 + at$ to find the velocity at time t. Note that the canoes start at rest, hence $v_0 = 0$. Finally, the momentum can be calculated using $p = mv$.

Solution

1. Use Newton's second law to find the acceleration of canoe 2:
$$a_{2,x} = \frac{\sum F_{2,x}}{m_2} = \frac{46\text{ N}}{250\text{ kg}} = 0.18\text{ m/s}^2$$

2. Do the same calculation for canoe 1. Note that the acceleration of canoe 1 is in the negative direction:
$$a_{1,x} = \frac{\sum F_{1,x}}{m_1} = \frac{-46\text{ N}}{130\text{ kg}} = -0.35\text{ m/s}^2$$

3. Calculate the velocity of each canoe at $t = 1.20$ s:
$$v_{1,x} = a_{1,x}t = (-0.35\text{ m/s}^2)(1.20\text{ s}) = -0.42\text{ m/s}$$
$$v_{2,x} = a_{2,x}t = (0.18\text{ m/s}^2)(1.20\text{ s}) = 0.22\text{ m/s}$$

4. Calculate the momentum of each canoe at $t = 1.20$ s:
$$p_{1,x} = m_1 v_{1,x} = (130\text{ kg})(-0.42\text{ m/s}) = -55\text{ kg}\cdot\text{m/s}$$
$$p_{2,x} = m_2 v_{2,x} = (250\text{ kg})(0.22\text{ m/s}) = 55\text{ kg}\cdot\text{m/s}$$

Insight
Note that the sum of the momenta of the two canoes is zero. This is just what one would expect: The canoes start at rest with zero momentum, there is zero net external force acting on the system, hence the final momentum must also be zero. The final velocities *do not* add to zero; it is momentum ($m\mathbf{v}$) that is conserved, not velocity ($\mathbf{v}$).

Practice Problem
What are the final momenta if the canoes are pushed apart with a force of 56 N? [**Answer:** $p_{1,x} = -67\text{ kg}\cdot\text{m/s}$, $p_{2,x} = 67\text{ kg}\cdot\text{m/s}$]

Some related homework problems: Problem 14, Problem 15

In a situation like that described in Example 9–3, the person in canoe 1 pushes canoe 2 away. At the same time, canoe 1 begins to move in the opposite direction. This is referred to as **recoil**. It is essentially the same as the recoil one experiences when firing a gun or when turning on a strong stream of water.

A particularly interesting example of recoil involves the human body. Perhaps you have noticed, when resting quietly in a rocking or reclining chair, that the chair wobbles back and forth slightly about once a second. The reason for this movement is that each time your heart pumps blood in one direction (from the atria to the ventricles, then to the aorta and pulmonary arteries, and so on) your body recoils in the opposite direction. Since the recoil depends on the force exerted by your heart on the blood and the volume of blood expelled from the heart with each beat, it is possible to gain valuable medical information regarding the health of your heart by analyzing the recoil it produces.

The medical instrument that employs the physical principle of recoil is called the *ballistocardiograph*. It is a completely noninvasive technology that simply requires the patient to sit comfortably in a chair fitted with sensitive force sensors under the seat and behind the back. Sophisticated bathroom scales also utilize this technology. A ballistocardigraphic (BCG) scale detects the recoil vibrations of the

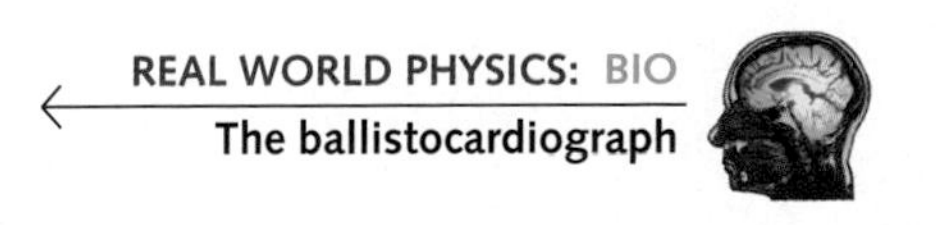

body as a person stands on the scale. This allows the BCG scale to display not only the person's body weight but his or her heart rate as well.

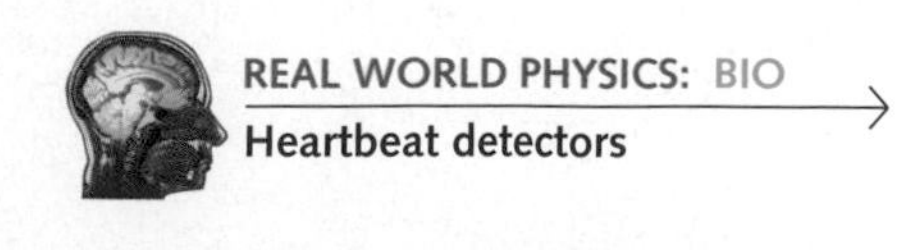

A more dramatic application of heartbeat recoil is currently being used at the Riverbend Maximum Security Institution in Tennessee. The only successful breakout from this prison occurred when four inmates hid in a secret compartment in a delivery truck that was leaving the facility. The institution now uses a heartbeat recoil detector that would have foiled this escape. Vehicles leaving the prison must stop at a checkpoint where a small motion detector is attached to it with a suction cup. Any persons hidden in the vehicle will reveal their presence by the very beating of their hearts. These heartbeat detectors have proved to be 100 percent effective, even though the recoil of the heart may displace a large truck by only a few millionths of an inch. Similar systems are being used at other high-security installations and border crossings.

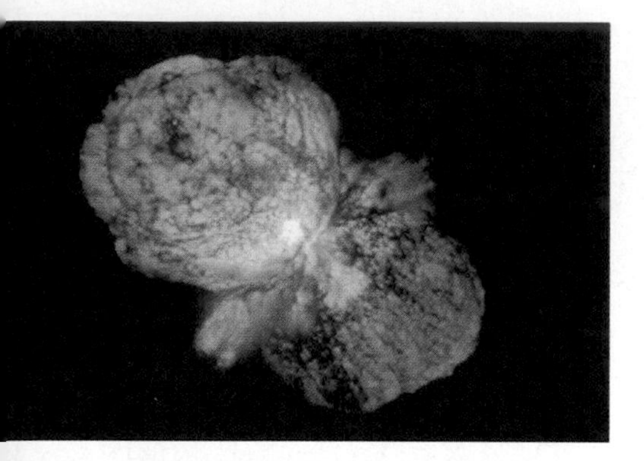

▲ This Hubble Space Telescope photograph shows the aftermath of a violent explosion of the star Eta Carinae. The explosion, which was observed on Earth in 1841 and briefly made Eta Carinae the second brightest star in the sky, produced two bright lobes of matter spewing outward in opposite directions. In this photograph, these lobes have expanded to about the size of our solar system. The momentum of the star before the explosion must be the same as the total momentum of the star and the bright lobes after the explosion. Since the lobes are roughly symmetric and move in opposite directions, their net momentum is essentially zero. Thus, we conclude that the momentum of the star itself was virtually unchanged by the explosion.

CONCEPTUAL CHECKPOINT 9–2

In Example 9–3, the final momentum of the system (consisting of the two canoes and their occupants) is equal to the initial momentum of the system. Is the final kinetic energy **(a)** equal to, **(b)** less than, or **(c)** greater than the initial kinetic energy?

Reasoning and Discussion

The final momentum of the two canoes is zero because one canoe has a positive momentum and the other has a negative momentum of the same magnitude. The two momenta, then, sum to zero. Kinetic energy, which is $\frac{1}{2}mv^2$, cannot be negative, hence no such cancellation is possible. Both canoes have positive kinetic energies, hence the final kinetic energy is greater than the initial kinetic energy, which is zero.

Where does the increase in kinetic energy come from? It comes from the muscular work done by the person who pushes the canoes apart.

Answer:

(c) K_f is greater than K_i.

A special case of some interest is the universe. Since there is nothing external to the universe—by definition—it follows that the net external force acting on it is zero. Therefore, its net momentum is conserved. No matter what happens—a comet collides with the Earth, a star explodes and becomes a supernova, a black hole swallows part of a galaxy—the total momentum of the universe simply cannot change. A particularly vivid illustration of momentum conservation is provided by the exploding star Eta Carinae. As can be seen in the Hubble Space Telescope photograph, jets of material are moving away from the star in opposite directions, just like the canoes moving apart from one another in Example 9–3.

Conservation of momentum also applies to the more everyday situation described in the next Active Example.

ACTIVE EXAMPLE 9–2 Bee on a Stick

A honeybee lands on one end of a floating 4.75-g Popsicle stick. After sitting at rest for a moment, it runs toward the other end with a speed of 3.80 cm/s relative to the still water. The stick moves in the opposite direction at 0.120 cm/s. What is the mass of the bee?

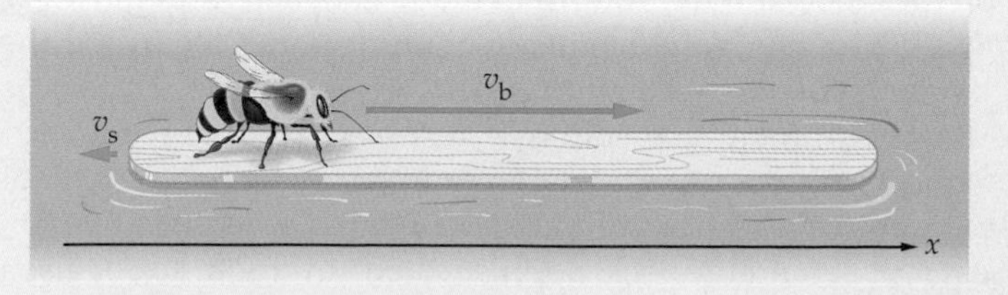

Solution

1. Set the total momentum of the system equal to zero:	$p_b + p_s = 0$
2. Solve for the momentum of the bee:	$p_b = -p_s = m_b v_b$
3. Calculate the momentum of the stick:	$p_s = -m_s v_s = -0.570\ \text{g} \cdot \text{cm/s}$
4. Calculate the momentum of the bee:	$p_b = m_b v_b = -p_s = 0.570\ \text{g} \cdot \text{cm/s}$
5. Divide by the bee's velocity to find its mass:	$m_b = p_b / v_b = 0.150\ \text{g}$

Insight

Since only internal forces are at work while the bee walks on the stick, the system's total momentum must remain zero.

9–5 Inelastic Collisions

We now turn our attention to **collisions**. By a collision we mean a situation in which two objects strike one another, and in which the net external force is either zero or negligibly small. For example, if two train cars roll along on a level track and hit one another, this is a collision. In this case, the net external force—the weight downward and the normal force exerted by the tracks upward—is zero. As a result, the momentum of the two-car system is conserved.

Another example of a collision is a baseball being struck by a bat. In this case, the external forces are not zero because the weight of the ball is not balanced by any other force. However, as we have seen in Section 9–3, the forces exerted during the hit are much larger than the weight of the ball or the bat. Hence, to a good approximation, we may neglect the external forces (the weight of the ball and bat) in this case, and say that the momentum of the ball-bat system is conserved.

Now it may seem surprising at first, but the fact that the momentum of a system is conserved during a collision does not necessarily mean that the system's kinetic energy is conserved. In fact most, or even all, of a system's kinetic energy may be converted to other forms during a collision while, at the same time, not one bit of momentum is lost. This shall be explored in detail in this section.

In general, collisions are categorized according to what happens to the kinetic energy of the system. There are basically two possibilities. After a collision, the final kinetic energy, K_f, is either equal to the initial kinetic energy, K_i, or it is not. If $K_f = K_i$, the collision is said to be **elastic**. We shall consider elastic collisions in the next section.

On the other hand, the kinetic energy may change during a collision. Usually it decreases due to losses associated with sound, heat, and deformation. Sometimes it increases, if the collision sets off an explosion, for instance. In any event, collisions in which the kinetic energy is not conserved are referred to as **inelastic**:

Inelastic Collisions

In an inelastic collision, the momentum of a system is conserved,

$$\mathbf{p}_f = \mathbf{p}_i$$

but its kinetic energy is not,

$$K_f \neq K_i$$

Finally, in the special case where objects stick together after the collision, we say that the collision is **completely inelastic**.

▶ In both elastic and inelastic collisions, momentum is conserved. The same is not true of kinetic energy, however. In the largely inelastic collision at left, much of the hockey players' initial kinetic energy is transformed into work: rearranging the players' anatomies and shattering the glass of the rink. In the highly elastic collision at right, the ball rebounds with very little diminution of its kinetic energy (though a little energy is lost as sound and heat).

Completely Inelastic Collisions

When objects stick together after colliding, the collision is completely inelastic.

In completely inelastic collisions, the maximum amount of kinetic energy is lost.

Inelastic Collisions in One Dimension

Consider a system of two identical train cars of mass m on a smooth, level track. One car is at rest initially while the other moves toward it with a speed v_0, as shown in **Figure 9–5**. When the cars collide the coupling mechanism latches, causing the cars to stick together and move as a unit. What is the speed of the cars after the collision?

To answer this question, we begin by considering the general case of a completely inelastic collision, and then specialize to the specific case of these two train cars. In general, suppose that two masses, m_1 and m_2, have initial velocities $v_{1,i}$ and $v_{2,i}$ respectively. The initial momentum of the system is

$$p_i = m_1 v_{1,i} + m_2 v_{2,i}$$

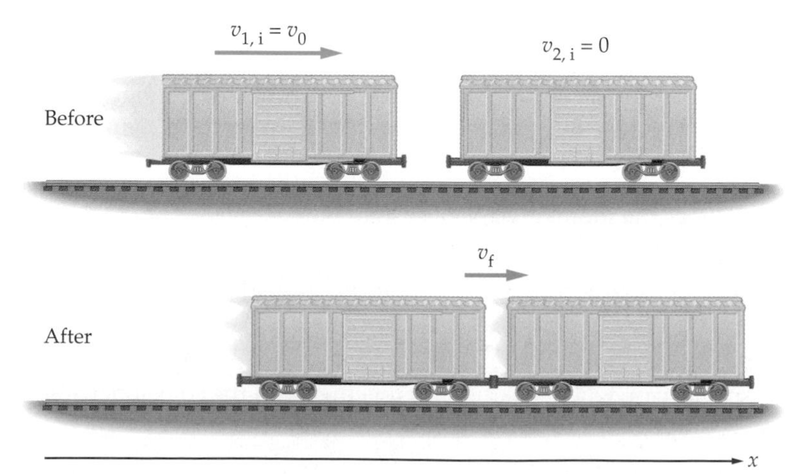

▶ FIGURE 9–5 Railroad cars collide and stick together

A moving train car collides with an identical car that is stationary. After the collision, the cars stick together and move with the same speed.

After the collision, the objects move together with a common velocity v_f. Therefore, the final momentum is

$$p_f = (m_1 + m_2)v_f$$

Equating the initial and final momenta yields

$$v_f = \frac{m_1 v_{1,i} + m_2 v_{2,i}}{m_1 + m_2} \qquad \text{9–10}$$

We can apply this general result to the case of the two railroad cars by noting that $m_1 = m_2 = m$, $v_{1,i} = v_0$, and $v_{2,i} = 0$. Thus, the final velocity is

$$v_f = \frac{mv_0 + m \cdot 0}{m + m} = \frac{m}{2m}v_0 = \tfrac{1}{2}v_0 \qquad \text{9–11}$$

As one might have guessed, the final speed is one-half the initial speed.

EXERCISE 9–2

A 1200-kg car moving at 2.5 m/s is struck in the rear by a 2600-kg truck moving at 6.2 m/s. If the vehicles stick together after the collision, what is their speed immediately after colliding? (Assume external forces may be ignored.)

Solution

Applying Equation 9–10 yields $v_f = 5.0$ m/s.

PROBLEM SOLVING NOTE
Momentum Versus Energy Conservation

Be sure to distinguish between momentum conservation and energy conservation. A common error is to assume that kinetic energy is conserved just because the momentum is conserved.

During the collision of the railroad cars, some of the initial kinetic energy is converted to other forms. Some propagates away as sound, some is converted to heat, some creates permanent deformations in the metal of the latching mechanism. The precise amount of kinetic energy that is lost is addressed in the following Conceptual Checkpoint.

CONCEPTUAL CHECKPOINT 9–3

A railroad car of mass m and speed v collides and sticks to an identical railroad car that is initially at rest. After the collision, is the kinetic energy of the system **(a)** 1/2, **(b)** 1/3, or **(c)** 1/4 of its initial kinetic energy?

Reasoning and Discussion

Before the collision, the kinetic energy of the system is

$$K_i = \tfrac{1}{2}mv^2$$

After the collision, the mass doubles and the speed is halved. Hence, the final kinetic energy is

$$K_f = \tfrac{1}{2}(2m)\left(\frac{v}{2}\right)^2 = \tfrac{1}{2}\left(\tfrac{1}{2}mv^2\right) = \tfrac{1}{2}K_i$$

Therefore, one-half of the initial kinetic energy is converted to other forms of energy.

An equivalent way to arrive at this conclusion is to express the kinetic energy in terms of the momentum, $p = mv$:

$$K = \tfrac{1}{2}mv^2 = \tfrac{1}{2}\left(\frac{m^2v^2}{m}\right) = \frac{p^2}{2m}$$

Since the momentum is the same before and after the collision, the fact that the mass doubles means the kinetic energy is halved.

Answer:

(a) The final kinetic energy is one-half the initial kinetic energy.

Note that we know the precise amount of kinetic energy that was lost, even though we don't know just how much went into sound, how much went into heat, and so on. It is not necessary to know all of those details to determine how much kinetic energy was lost.

We also know how much momentum was lost—none.

EXAMPLE 9–4 Goal-Line Stand

On a touchdown attempt, a 95.0-kg running back runs toward the end zone at 3.75 m/s. A 111-kg linebacker moving at 4.10 m/s meets the runner in a head-on collision. If the two players stick together, **(a)** what is their velocity immediately after the collision? **(b)** What are the initial and final kinetic energies of the system?

Picture the Problem

In our sketch, we let subscript 1 refer to the running back, who carries the ball, and subscript 2 refer to the linebacker, who will make the tackle. The direction of the running back's initial motion is taken to be positive.

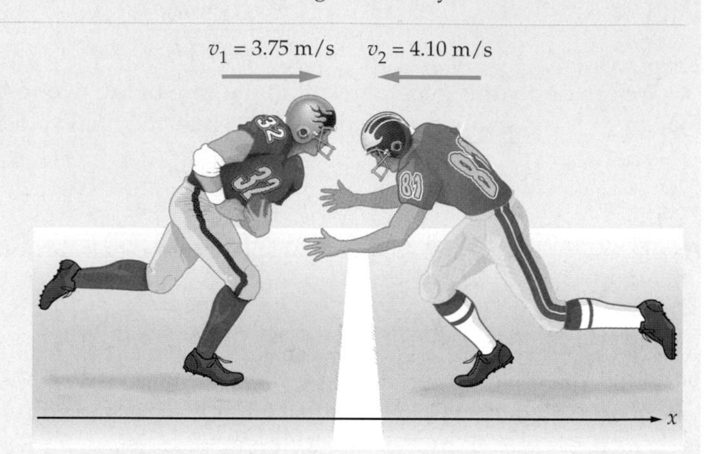

Strategy

(a) The final velocity is found by applying momentum conservation, just as was done to derive Equation 9–10.

(b) The kinetic energies are calculated with $\frac{1}{2}mv^2$.

Solution

Part (a)

1. Set the initial momentum equal to the final momentum:

$$m_1 v_1 + m_2 v_2 = (m_1 + m_2)v_f$$

2. Solve for the final velocity and substitute numerical values, being careful to use the appropriate signs:

$$v_f = \frac{m_1 v_1 + m_2 v_2}{m_1 + m_2}$$
$$= \frac{(95.0 \text{ kg})(3.75 \text{ m/s}) + (111 \text{ kg})(-4.10 \text{ m/s})}{95.0 \text{ kg} + 111 \text{ kg}}$$
$$= -0.480 \text{ m/s}$$

Part (b)

3. Calculate the initial kinetic energy of the two players:

$$K_i = \tfrac{1}{2}m_1 v_1^2 + \tfrac{1}{2}m_2 v_2^2$$
$$= \tfrac{1}{2}(95.0 \text{ kg})(3.75 \text{ m/s})^2 + \tfrac{1}{2}(111 \text{ kg})(-4.10 \text{ m/s})^2$$
$$= 1600 \text{ J}$$

4. Calculate the final kinetic energy of the players, noting that they both move with the same velocity after the collision:

$$K_f = \tfrac{1}{2}(m_1 + m_2)v_f^2$$
$$= \tfrac{1}{2}(95.0 \text{ kg} + 111 \text{ kg})(-0.480 \text{ m/s})^2 = 23.7 \text{ J}$$

Insight

After the collision, the two players are moving in the negative direction; that is, away from the end zone. This is because the linebacker had more negative momentum than the running back had positive momentum.

As for the kinetic energy, of the original 1600 J, only 23.7 J is left after the collision. This means that over 98% of the original kinetic energy is lost to other forms. Even so, *none* of the momentum is lost.

Practice Problem

If the final speed of the two players is to be zero, should the speed of the running back be increased or decreased? Check your answer by calculating the required speed for the running back. [**Answer:** The running back's speed should be increased to 4.79 m/s.]

Some related homework problems: Problem 21, Problem 28

EXAMPLE 9–5 Ballistic Pendulum

Real World Physics

In a ballistic pendulum, an object of mass m is fired with an initial speed v_0 at the bob of a pendulum. The bob has a mass M, and is suspended by a rod of negligible mass. After the collision, the object and the bob stick together and swing through an arc, eventually gaining a height h. Find the height h in terms of m, M, v_0, and g.

Picture the Problem
Our sketch shows the physical setup of a ballistic pendulum. Immediately after the collision, the bob and object move together with a new speed, v_f, which is determined by momentum conservation.

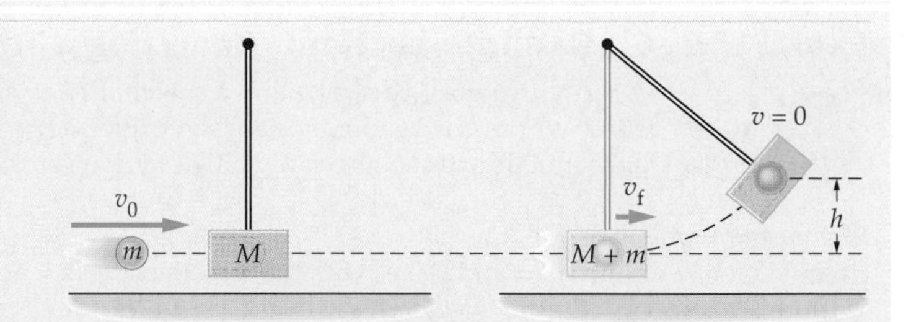

Strategy
There are two distinct physical processes at work in the ballistic pendulum. The first is a completely inelastic collision between the bob and the object. Momentum is conserved during this collision, but kinetic energy is not. After the collision, the remaining kinetic energy is converted into gravitational potential energy, which determines how high the bob and object will rise.

Solution

1. Set the momentum just before the bob-object collision equal to the momentum just after the collision. Let v_f be the speed just after the collision:
$$mv_0 = (M + m)v_f$$

2. Solve for the speed just after the collision, v_f:
$$v_f = \left(\frac{m}{M + m}\right)v_0$$

3. Calculate the kinetic energy just after the collision:
$$K_f = \tfrac{1}{2}(M + m)v_f^2 = \tfrac{1}{2}(M + m)\left(\frac{m}{M + m}\right)^2 v_0^2$$
$$= \tfrac{1}{2}mv_0^2\left(\frac{m}{M + m}\right)$$

4. Set the kinetic energy after the collision equal to the gravitational potential energy at the height h:
$$\tfrac{1}{2}mv_0^2\left(\frac{m}{M + m}\right) = (M + m)gh$$

5. Solve for the height, h:
$$h = \left(\frac{m}{M + m}\right)^2\left(\frac{v_0^2}{2g}\right)$$

Insight
A ballistic pendulum is often used to measure the speed of a rapidly moving object, such as a bullet. If a bullet were shot straight up it would rise to the height $v_0^2/2g$, which can be thousands of feet. On the other hand, if a bullet of mass m is fired into a ballistic pendulum, in which M is much greater than m, the bullet reaches only a small fraction of this height. Thus, the ballistic pendulum makes for a more convenient and practical measurement.

Practice Problem
A 7.00-g bullet is fired into a ballistic pendulum whose bob has a mass of 0.950 kg. If the bob rises to a height of 0.220 m, what was the initial speed of the bullet? [**Answer:** v_0 = 284 m/s. If this bullet were fired straight up, it would rise 4.11 km ≈ 13,000 ft.]

Some related homework problems: Problem 24, Problem 26

Inelastic Collisions in Two Dimensions

Next we consider collisions in two dimensions, where we must conserve the momentum component by component. To do this, we set up a coordinate system and resolve the initial momentum into x and y components. Next, we demand that the final momentum have precisely the same x and y components as the initial momentum. That is,

$$p_{x,i} = p_{x,f}$$

and

$$p_{y,i} = p_{y,f}$$

The following Example shows how to carry out such a calculation in a practical situation.

PROBLEM SOLVING NOTE
Sketch the System Before and After the Collision

In problems involving collisions, it is useful to draw the system before and after the collision. Be sure to label the relevant masses, velocities, and angles.

EXAMPLE 9–6 Bad Intersection: Analyzing a Traffic Accident

Real World Physics

A car with a mass of 950 kg and a speed of 16 m/s approaches an intersection, as shown. A 1300-kg minivan traveling at 21 m/s is heading for the same intersection. The car and minivan collide and stick together. Find the speed and direction of the wrecked vehicles just after the collision, assuming external forces can be ignored.

Picture the Problem

In the sketch, we align the x and y axes with the crossing streets. With this choice, $\mathbf{v}_1$ (the car's velocity) is in the positive x direction, and $\mathbf{v}_2$ (the minivan's velocity) is in the positive y direction. In addition, the problem statement indicates that $m_1 = 950$ kg and $m_2 = 1300$ kg.

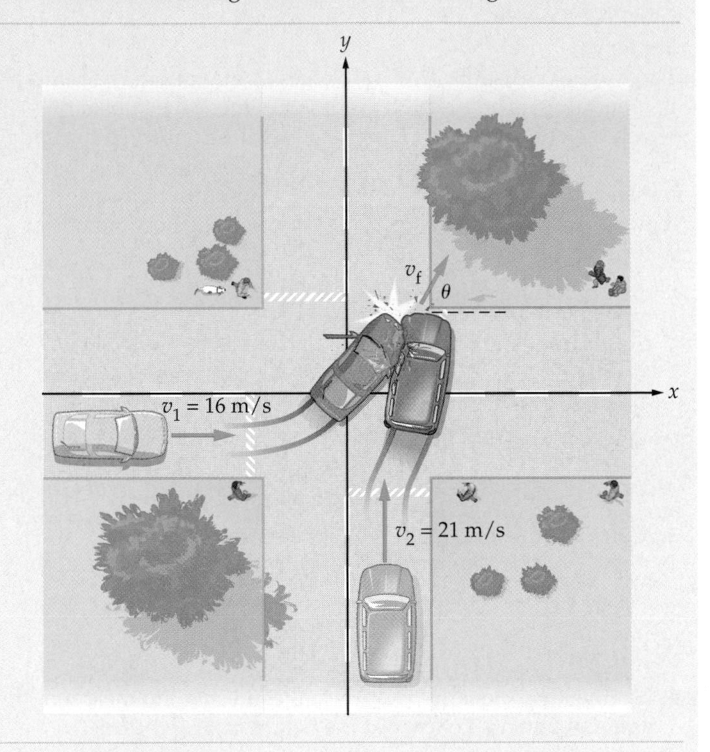

Strategy

Since external forces can be ignored, the total momentum of the system must be conserved during the collision. This is really two conditions: (i) the x component of momentum is conserved, and (ii) the y component of momentum is conserved. These two conditions determine the two unknowns: the final speed, v_f, and the final direction, θ.

Solution

1. Set the initial x component of momentum equal to the final x component of momentum:

$$m_1 v_1 = (m_1 + m_2) v_f \cos\theta$$

2. Do the same for the y component of momentum:

$$m_2 v_2 = (m_1 + m_2) v_f \sin\theta$$

3. Divide the y momentum equation by the x momentum equation. This eliminates v_f, giving an equation involving θ alone:

$$\frac{m_2 v_2}{m_1 v_1} = \frac{(m_1 + m_2) v_f \sin\theta}{(m_1 + m_2) v_f \cos\theta} = \frac{\sin\theta}{\cos\theta} = \tan\theta$$

4. Solve for θ:

$$\theta = \tan^{-1}\left(\frac{m_2 v_2}{m_1 v_1}\right) = \tan^{-1}\left[\frac{(1300\text{ kg})(21\text{ m/s})}{(950\text{ kg})(16\text{ m/s})}\right]$$
$$= \tan^{-1}(1.8) = 61^\circ$$

5. The final speed can be found using either the x or the y momentum equation. Here we use the x equation:

$$v_f = \frac{m_1 v_1}{(m_1 + m_2)\cos\theta}$$
$$= \frac{(950\text{ kg})(16\text{ m/s})}{(950\text{ kg} + 1300\text{ kg})\cos 61^\circ} = 14\text{ m/s}$$

Insight

As a check, you should verify that the y momentum equation gives the same value for v_f.

Practice Problem

Suppose the speed and direction immediately after the collision are known to be $v_f = 12.5$ m/s and $\theta = 42^\circ$, respectively. Find the initial speed of each car. [**Answer:** $v_1 = 22$ m/s, $v_2 = 14$ m/s]

Some related homework problems: Problem 22, Problem 23

9–6 Elastic Collisions

In this section we consider collisions in which both momentum and kinetic energy are conserved. As mentioned in the previous section, such collisions are referred to as elastic:

Elastic Collisions

In an elastic collision, momentum and kinetic energy are conserved. That is,

$\mathbf{p}_f = \mathbf{p}_i$

and

$K_f = K_i$

Most collisions in everyday life are rather poor approximations to being elastic—usually there is a significant amount of energy converted to other forms. However, the collision of objects that bounce off one another with little deformation, like billiard balls for example, provides a reasonably good approximation to an elastic collision. In the subatomic world, on the other hand, elastic collisions are common. Elastic collisions, then, are not merely an ideal that is approached but never attained—they are constantly taking place in nature.

Elastic Collisions in One Dimension

Consider a head-on collision of two carts on an air track, as pictured in **Figure 9–6**. The carts are provided with bumpers that give an elastic bounce when the carts collide. Let's suppose that initially cart 1 is moving to the right with a speed v_0 toward cart 2, which is at rest. If the masses of the carts are m_1 and m_2, respectively, then momentum conservation can be written as follows:

$$m_1 v_0 = m_1 v_{1,f} + m_2 v_{2,f}$$

In this expression, $v_{1,f}$ and $v_{2,f}$ are the final velocities of the two carts. Note that we say velocities, not speeds, since it is possible for cart 1 to reverse direction, in which case $v_{1,f}$ would be negative.

Next, the fact that this is an elastic collision means the final velocities must also satisfy energy conservation:

$$\tfrac{1}{2}m_1 v_0^2 = \tfrac{1}{2}m_1 v_{1,f}^2 + \tfrac{1}{2}m_2 v_{2,f}^2$$

Thus, we now have two equations for the two unknowns, $v_{1,f}$ and $v_{2,f}$. Straightforward algebra yields the following results:

$$v_{1,f} = \left(\frac{m_1 - m_2}{m_1 + m_2}\right)v_0$$

$$v_{2,f} = \left(\frac{2m_1}{m_1 + m_2}\right)v_0 \qquad \text{9–12}$$

Note that the final velocity of cart 1 can be positive, negative, or zero, depending on whether m_1 is greater than, less than, or equal to m_2, respectively. The final velocity of cart 2, however, is always positive.

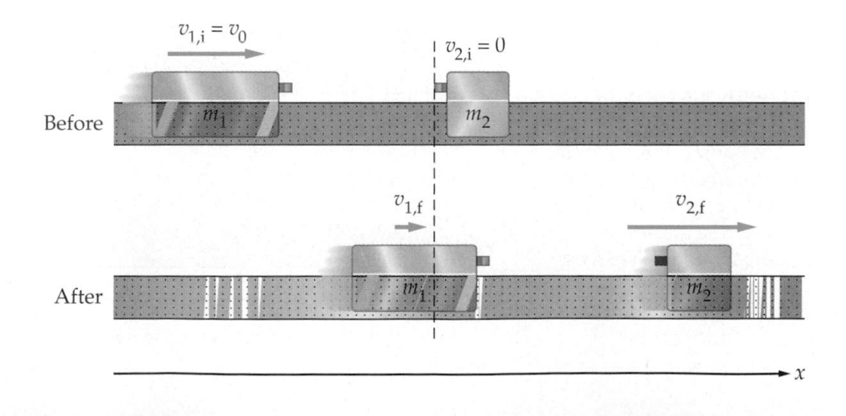

◀ **FIGURE 9–6 An elastic collision between two air carts**

In the case pictured, $v_{1,f}$ is to the right (positive), which means that m_1 is greater than m_2. In fact, we have chosen $m_1 = 2m_2$ for this plot; therefore, $v_{1,f} = v_0/3$ and $v_{2,f} = 4v_0/3$ as given by Equations 9-12. If m_1 were less than m_2, cart 1 would bounce back toward the left, meaning that $v_{1,f}$ would be negative.

EXERCISE 9–3

At an amusement park, a 96.0-kg bumper car moving with a speed of 1.24 m/s bounces elastically off a 135-kg bumper car at rest. Find the final velocities of the cars.

Solution

Using Equations 9–12, we find the final velocities to be $v_{1,f} = -0.209$ m/s and $v_{2,f} = 1.03$ m/s. Note that the direction of travel of car 1 has been reversed.

Let's check a few special cases of our results. First, consider the case where the two carts have equal masses, $m_1 = m_2 = m$. Substituting into Equations 9–12, we find

$$v_{1,f} = \frac{m - m}{m + m} v_0 = 0$$

and

$$v_{2,f} = \frac{2m}{m + m} v_0 = v_0$$

Thus, after the collision, the cart that was moving with velocity v_0 is now at rest, and the cart that was at rest is now moving with velocity v_0. In effect, the carts have "exchanged" velocities. This case is illustrated in **Figure 9–7 (a)**.

Next, suppose that m_2 is much greater than m_1, or, equivalently, that m_1 approaches zero. Returning to Equations 9–12, and setting $m_1 = 0$, we find

$$v_{1,f} = \frac{0 - m_2}{0 + m_2} v_0 = \frac{-m_2}{m_2} v_0 = -v_0$$

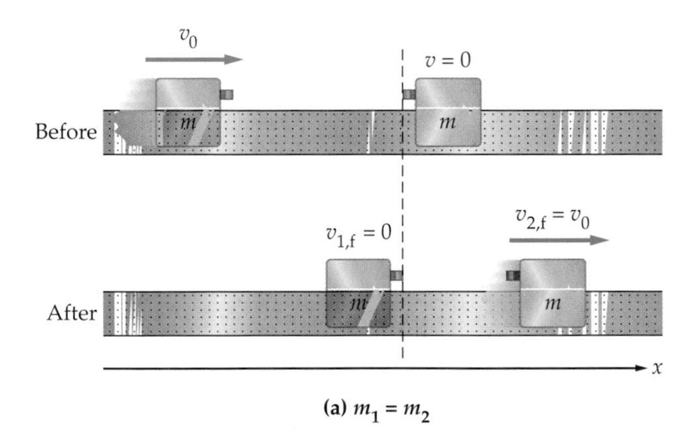

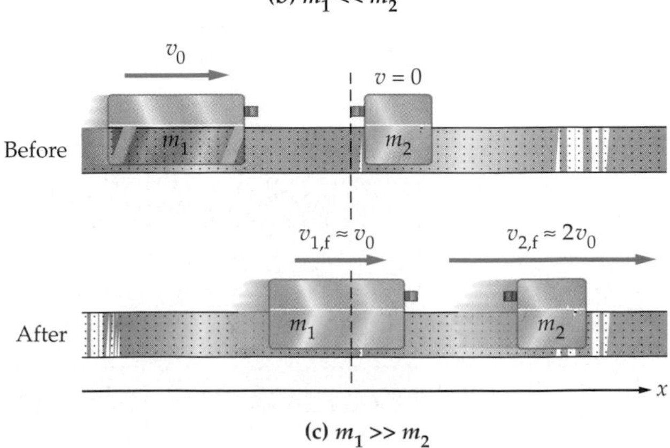

▶ **FIGURE 9–7 Elastic collisions between air carts of various masses**

(a) Carts of equal mass exchange velocites when they collide. **(b)** When a light cart collides with a stationary, heavy cart its direction of motion is reversed. Its speed is practically unchanged. **(c)** When a heavy cart collides with a stationary, light cart it continues to move in the same direction with essentially the same speed. The light cart moves off with a speed that is roughly twice the initial speed of the heavy cart.

and

$$v_{2,f} = \frac{2 \cdot 0}{0 + m_2} v_0 = 0$$

Physically, we interpret these results as follows: A very light cart collides with a heavy cart that is at rest. The heavy cart hardly budges, but the light cart is reflected, heading *backward* (remember the minus sign in $-v_0$) with the same speed it had initially. For example, if you throw a ball against a wall, the wall is the very heavy object, and the ball is the light object. The ball bounces back with the same speed it had initially (assuming an ideal elastic collision). We show a case in which m_1 is much less than m_2 in **Figure 9–7 (b)**.

Finally, what happens when m_1 is much greater than m_2? To check this limit we can set m_2 equal to zero. We consider the results in the following Conceptual Checkpoint.

CONCEPTUAL CHECKPOINT 9–4

A hover fly is happily maintaining a fixed position about 10 ft above the ground when an elephant charges out of the bush and collides with it. The fly bounces elastically off the forehead of the elephant. If the initial speed of the elephant is v_0, is the speed of the fly after the collision equal to **(a)** v_0, **(b)** $1.5v_0$, or **(c)** $2v_0$?

Reasoning and Discussion

We can use Equation 9–12 to find the final speeds of the fly and the elephant. First, let m_1 be the mass of the elephant, and m_2 be the mass of the fly. Clearly, m_2 is vanishingly small compared with m_1, hence we can evaluate Equation 9–12 in the limit $m_2 \to 0$. This yields

$$v_{1,f} = \frac{m_1 - m_2}{m_1 + m_2} v_0 \underset{m_2 \to 0}{\longrightarrow} \frac{m_1}{m_1} v_0 = v_0$$

and

$$v_{2,f} = \frac{2m_1}{m_1 + m_2} v_0 \underset{m_2 \to 0}{\longrightarrow} \frac{2m_1}{m_1} v_0 = 2v_0$$

As expected, the speed of the elephant is unaffected. The fly, however, rebounds with twice the speed of the elephant. **Figure 9–7 (c)** illustrates this case with air carts.

Answer:

(c) The speed of the fly is $2v_0$.

Note that after the collision the fly is separating from the elephant with the speed $2v_0 - v_0 = v_0$. Before the collision the elephant was approaching the fly with the same speed, v_0. This is a special case of the following general result:

> The speed of separation after a head-on, elastic collision is always equal to the speed of approach before the collision.

Elastic Collisions in Two Dimensions

In a two-dimensional elastic collision, if we are given the final speed and direction of one of the objects, we can find the speed and direction of the other object using energy conservation and momentum conservation. For example, consider the collision of two 7.0-kg curling stones, as depicted in **Figure 9–8**. One stone is at rest initially, the other approaches with a speed $v_{1,i} = 1.5$ m/s. The collision is not head-on, and after the collision, stone 1 moves with a speed of $v_{1,f} = 0.61$ m/s in a direction 66° away from the initial line of motion. What is the speed and direction of stone 2?

▲ The apparatus shown here illustrates some of the basic features of elastic collisions between objects of equal mass. The device consists of five identical metal balls suspended by strings. When the end ball is pulled out to the side and then released so as to fall back and strike the second ball, it creates a rapid succession of elastic collisions among the balls. In each collision, one ball comes to rest while the next one begins to move with the original speed, just as with the air carts in Figure 9–7 (a) When the collisions reach the other end of the apparatus, the last ball swings out to the same height from which the first ball was released.

If two balls are pulled out and released, two balls swing out at the other side, and so on. To see why this must be so, imagine that the two balls swing in with a speed v and a single ball swings out at the other side with a speed v'. What value must v' have (a) to conserve momentum, and (b) to conserve kinetic energy? Since the required speed is $v' = 2v$ for (a) and $v' = \sqrt{2}\, v$ for (b), it follows that it is not possible to conserve both momentum and kinetic energy with two balls swinging in and one ball swinging out.

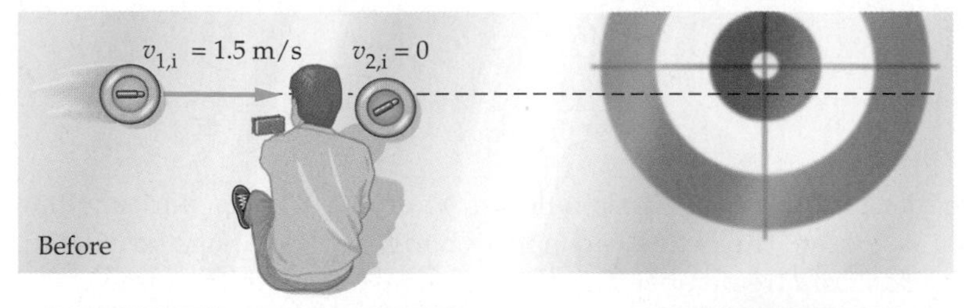

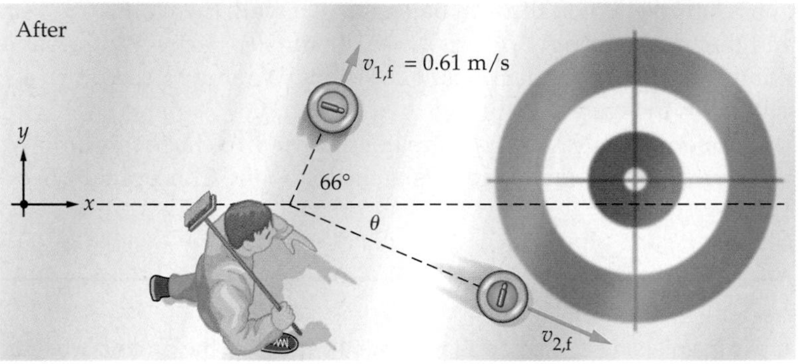

▶ FIGURE 9–8 Two curling stones undergo an elastic collision

The speed of curling stone 2 after this collision can be determined using energy conservation; its direction of motion can be found using momentum conservation in either the x or the y direction.

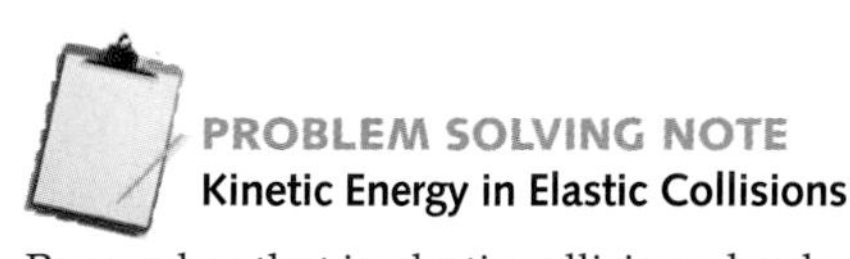

PROBLEM SOLVING NOTE

Kinetic Energy in Elastic Collisions

Remember that in elastic collisions, by definition, the kinetic energy is conserved.

First, let's find the speed of stone 2. The easiest way to do this is to simply require that the final kinetic energy be equal to the initial kinetic energy. Initially, the kinetic energy is

$$K_{\mathrm{i}} = \tfrac{1}{2}m_1 v_{1,\mathrm{i}}^2 = \tfrac{1}{2}(7.0\text{ kg})(1.5\text{ m/s})^2 = 7.9\text{ J}$$

After the collision stone 1 has a speed of 0.61 m/s and stone 2 has the speed $v_{2,\mathrm{f}}$. Hence, the final kinetic energy is

$$K_{\mathrm{f}} = \tfrac{1}{2}m_1 v_{1,\mathrm{f}}^2 + \tfrac{1}{2}m_2 v_{2,\mathrm{f}}^2 = \tfrac{1}{2}(7.0\text{ kg})(0.61\text{ m/s})^2 + \tfrac{1}{2}m_2 v_{2,\mathrm{f}}^2$$
$$= 1.3\text{ J} + \tfrac{1}{2}m_2 v_{2,\mathrm{f}}^2 = K_{\mathrm{i}}$$

Solving for the speed of stone 2, we find

$$v_{2,\mathrm{f}} = 1.4\text{ m/s}$$

Next, we can find the direction of motion of stone 2 by requiring that the momentum be conserved. For example, initially there is no momentum in the y direction. This must be true after the collision as well. Hence, we have the following condition:

$$0 = m_1 v_{1,\mathrm{f}} \sin 66° - m_2 v_{2,\mathrm{f}} \sin\theta$$

Solving for the angle θ we find

$$\theta = 23°$$

As a final check, compare the initial and final x component of momentum. Initially, we have

$$p_{x,\mathrm{i}} = m_1 v_{1,\mathrm{i}} = (7.0\text{ kg})(1.5\text{ m/s}) = 11\text{ kg}\cdot\text{m/s}$$

Following the collision, the x component of momentum is

$$p_{x,\mathrm{f}} = m_1 v_{1,\mathrm{f}} \cos 66° + m_2 v_{2,\mathrm{f}} \cos 23°$$
$$= (7.0\text{ kg})(0.61\text{ m/s}) \cos 66° + (7.0\text{ kg})(1.4\text{ m/s}) \cos 23°$$
$$= 11\text{ kg}\cdot\text{m/s}$$

As expected, the momentum is unchanged.

EXAMPLE 9–7 Two Fruits in Two Dimensions: Analyzing an Elastic Collision

Two astronauts on opposite ends of a space ship are comparing lunches. One has an apple, the other has an orange. They decide to trade. Astronaut 1 tosses the 0.130-kg apple toward astronaut 2 with a speed of 1.11 m/s. The 0.160-kg orange is tossed from astronaut 2 to astronaut 1 with a speed of 1.21 m/s. Unfortunately, the fruits collide, sending the orange off with a speed of 1.16 m/s at an angle of 42.0° with respect to its original direction of motion. Find the final speed and direction of the apple, assuming an elastic collision. Give the apple's direction relative to its original direction of motion.

Picture the Problem

In our sketch we refer to the apple as object 1 and to the orange as object 2. We also choose the positive x direction to be the initial direction of motion of the apple. Thus, initially the apple has a positive x component of momentum. The initial x component of momentum of the orange is negative. There is no momentum in the y direction before the collision.

Strategy

As described in the text, we first find the speed of the apple by demanding that the initial and final kinetic energies be the same. Next, we find the angle θ by conserving momentum in either the x or the y direction—the results are the same whichever direction is chosen.

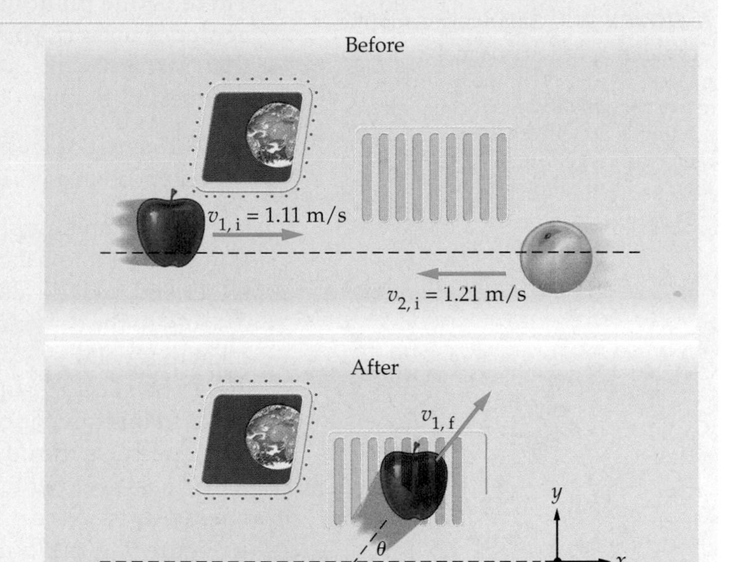

Solution

1. Calculate the initial kinetic energy of the system:

$$K_i = \tfrac{1}{2}m_1v_{1,i}^2 + \tfrac{1}{2}m_2v_{2,i}^2$$
$$= \tfrac{1}{2}(0.130\text{ kg})(1.11\text{ m/s})^2 + \tfrac{1}{2}(0.160\text{ kg})(1.21\text{ m/s})^2$$
$$= 0.197\text{ J}$$

2. Calculate the final kinetic energy of the system in terms of $v_{1,f}$:

$$K_f = \tfrac{1}{2}m_1v_{1,f}^2 + \tfrac{1}{2}m_2v_{2,f}^2$$
$$= \tfrac{1}{2}(0.130\text{ kg})v_{1,f}^2 + \tfrac{1}{2}(0.160\text{ kg})(1.16\text{ m/s})^2$$
$$= \tfrac{1}{2}(0.130\text{ kg})v_{1,f}^2 + 0.108\text{ J}$$

3. Set $K_f = K_i$ to find $v_{1,f}$:

$$v_{1,f} = \sqrt{\frac{2(0.197\text{ J} - 0.108\text{ J})}{0.130\text{ kg}}} = 1.17\text{ m/s}$$

4. Set the final y component of momentum equal to zero to determine the angle, θ:

$$0 = m_1v_{1,f}\sin\theta - m_2v_{2,f}\sin 42.0°$$

Solve for $\sin\theta$:

$$\sin\theta = \frac{m_2v_{2,f}\sin 42.0°}{m_1v_{1,f}}$$

5. Substitute numerical values:

$$\sin\theta = \frac{(0.160\text{ kg})(1.16\text{ m/s})\sin 42.0°}{(0.130\text{ kg})(1.17\text{ m/s})} = 0.817$$

$$\theta = \sin^{-1}(0.817) = 54.8°$$

Insight

The x momentum equation gives the same value for θ, as expected.

Practice Problem

Suppose that after the collision the apple moves in the positive y direction with a speed of 1.27 m/s. What is the final speed and direction of the orange in this case? [**Answer:** The orange moves with a speed of 1.08 m/s in a direction of 73.4° below the negative x axis.]

Some related homework problems: Problem 31, Problem 71

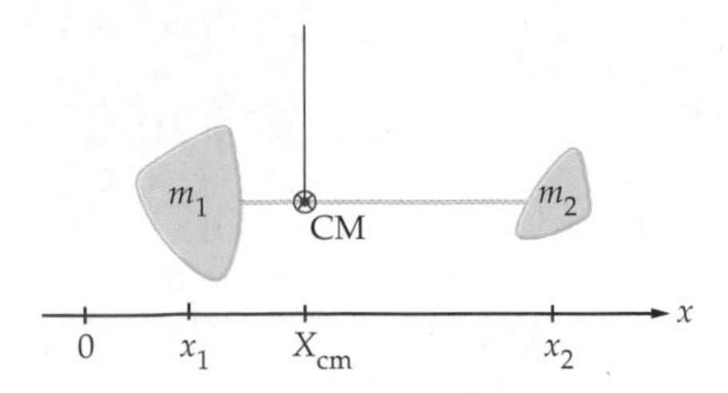

▲ **FIGURE 9–9 Balancing a mobile**
Consider a portion of a mobile with masses m_1 and m_2 at the locations x_1 and x_2, respectively. The object balances when a string is attached at the center of mass. Since the center of mass is closer to m_1 than to m_2, it follows that m_1 is greater than m_2.

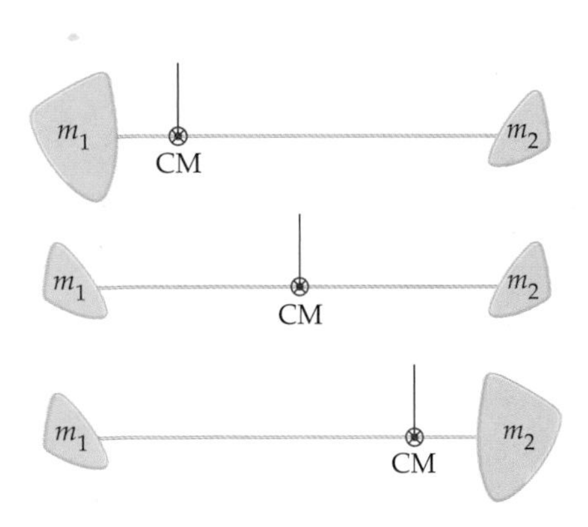

▲ **FIGURE 9–10 The center of mass of two objects**
The center of mass is closest to the larger mass, or equidistant between the masses if they are equal.

▲ Mobiles like *Antennae with Red and Blue Dots* by Alexander Calder illustrate the concept of center of mass with artistic flair. Each arm of the mobile is in balance because it is suspended at its center of mass.

9–7 Center of Mass

In this section we introduce the concept of the center of mass. We begin by defining its location for a given system of masses. Next we consider the motion of the center of mass, and show how it is related to the net external force acting on the system.

Location of the Center of Mass

There is one point in any system of objects that has special significance—the **center of mass (CM)**. One of the reasons the center of mass is so special is the fact that, in many ways, a system behaves as if all of its mass were concentrated there. As a result, a system can be balanced at its center of mass:

> The center of mass of a system of masses is the point where the system can be balanced in a uniform gravitational field.

For example, suppose you are making a mobile. At one stage in its construction you want to balance a light rod with objects of mass m_1 and m_2 connected to either end, as indicated in **Figure 9–9**. To make the rod balance you should attach a string to the center of mass of the system, just as if all its mass were concentrated at that point.

To be specific, suppose the two objects connected to the rod have the same mass. In this case the center of mass is at the midpoint of the rod, since this is where it balances. On the other hand, if one object has more mass than the other, the center of mass is closer to the heavier object, as indicated in **Figure 9–10**. In general, if a mass m_1 is on the x axis at the position x_1, and a mass m_2 is at the position x_2, as in Figure 9–9, the location of the center of mass, X_{cm}, is defined as follows:

Center of Mass for Two Objects

$$X_{cm} = \frac{m_1x_1 + m_2x_2}{m_1 + m_2} = \frac{m_1x_1 + m_2x_2}{M} \qquad \text{9–13}$$

Note that we have used $M = m_1 + m_2$ for the total mass of the two objects.

To see that this definition of X_{cm} agrees with our expectations, consider first the case where the masses are equal; $m_1 = m_2 = m$. In this case, $M = m_1 + m_2 = 2m$, and $X_{cm} = (mx_1 + mx_2)/2m = \frac{1}{2}(x_1 + x_2)$. Thus, as expected, if two masses are equal their center of mass is halfway between them. On the other hand, if m_1 is significantly greater than m_2, it follows that $M = m_1 + m_2 \sim m_1$ and $m_1x_1 + m_2x_2 \sim m_1x_1$, since m_2 can be ignored in comparison to m_1. As a result, we find that $X_{cm} \sim m_1x_1/m_1 = x_1$; that is, the center of mass is essentially at the location of the extremely heavy mass, m_1. In general, as one mass becomes larger than the other, the center of mass moves closer to the larger mass.

EXERCISE 9–4

Suppose the masses in Figure 9–9 are separated by 0.500 m, and that $m_1 = 0.260$ kg and $m_2 = 0.170$ kg. What is the distance from m_1 to the center of mass of the system?

Solution

Letting $x_1 = 0$ and $x_2 = 0.500$ m in Figure 9–9, we have

$$X_{cm} = \frac{m_1x_1 + m_2x_2}{m_1 + m_2} = \frac{(0.260 \text{ kg}) \cdot 0 + (0.170 \text{ kg})(0.500 \text{ m})}{0.260 \text{ kg} + 0.170 \text{ kg}} = 0.198 \text{ m}$$

Thus, the center of mass is closer to m_1 (the larger mass) than to m_2.

To extend the definition of X_{cm} to more general situations, first consider a system that contains many objects, not just two. In that case, X_{cm} is the sum of m times x for each object, divided by the total mass of the system, M. If, in addition, the objects in the system are not in a line, but are distributed in two dimensions, the center of mass will have both an x coordinate, X_{cm}, and a y coordinate, Y_{cm}. As one would expect, Y_{cm} is simply the sum of m times y for each object, divided by M. Thus, the x coordinate of the center of mass is

X Coordinate of the Center of Mass

$$X_{cm} = \frac{m_1x_1 + m_2x_2 + \ldots}{m_1 + m_2 + \ldots} = \frac{\sum mx}{M} \qquad 9\text{–}14$$

Similarly, the y coordinate of the center of mass is

Y Coordinate of the Center of Mass

$$Y_{cm} = \frac{m_1y_1 + m_2y_2 + \ldots}{m_1 + m_2 + \ldots} = \frac{\sum my}{M} \qquad 9\text{–}15$$

In systems with a continuous, uniform distribution of mass, the center of mass is at the geometric center of the object, as illustrated in Figure 9–11. Note that it is common for the center of mass to be located in a position where no mass exists, as in a donut, where the center of mass is precisely in the center of the hole.

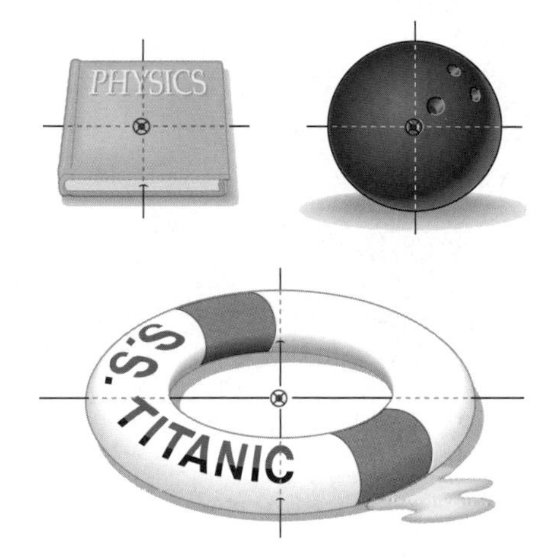

▲ **FIGURE 9–11 Locating the center of mass**

In an object of continuous, uniform mass distribution, the center of mass is located at the geometric center of the object. In some cases, this means that the center of mass is not located within the object.

EXAMPLE 9–8 Center of Mass of the Arm

Real World Physics: Bio

A person's arm is held with the upper arm vertical, the lower arm and hand horizontal. Find the center of mass of the arm in this configuration, given the following data: The upper arm has a mass of 2.5 kg and a center of mass 18 cm above the elbow; the lower arm has a mass of 1.6 kg and a center of mass 12 cm to the right of the elbow; the hand has a mass of 0.64 kg and a center of mass 0.40 m to the right of the elbow.

Picture the Problem

We place the origin at the elbow, with the x and y axes pointing along the lower and upper arms, respectively. The center of mass of each part of the arm is indicated by an x; the center of mass of the entire arm is at the point labeled CM.

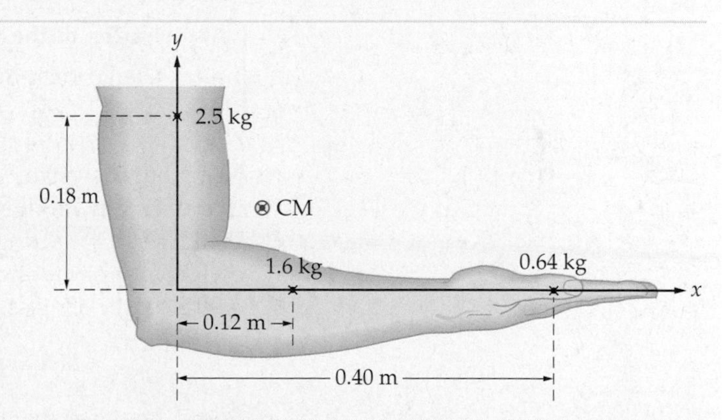

Strategy

Using the information given in the problem statement, we can treat the arm as a system of three point masses placed as follows: 2.5 kg at (0, 18 cm); 1.6 kg at (12 cm, 0); 0.64 kg at (40 cm, 0). We substitute these masses and locations into Equations 9–14 and 9–15 to find the x and y coordinates of the center of mass, respectively.

Solution

1. Calculate the x coordinate of the center of mass:

$$X_{cm} = \frac{(2.5\text{ kg})(0) + (1.6\text{ kg})(12\text{ cm}) + (0.64\text{ kg})(40\text{ cm})}{2.5\text{ kg} + 1.6\text{ kg} + 0.64\text{ kg}} = 9.5\text{ cm}$$

2. Do the same calculation for the y coordinate of the center of mass:

$$Y_{cm} = \frac{(2.5\text{ kg})(18\text{ cm}) + (1.6\text{ kg})(0) + (0.64\text{ kg})(0)}{2.5\text{ kg} + 1.6\text{ kg} + 0.64\text{ kg}} = 9.5\text{ cm}$$

Insight

As is often the case, the center of mass is in a location where no mass exists.

Practice Problem

Suppose a 1.2-kg ball is placed in the center of the hand; that is, 0.40 m to the right of the elbow. **(a)** Does X_{cm} increase, decrease, or stay the same? **(b)** Does Y_{cm} increase, decrease, or stay the same? **(c)** Check your answers to parts (a) and (b) by finding the center of mass of the arm-ball system. [**Answer:** (a) increases; (b) decreases; (c) $X_{cm} = 16$ cm, $Y_{cm} = 7.6$ cm]

Some related homework problems: Problem 38, Problem 40

Motion of the Center of Mass

Another reason the center of mass is of such importance is that its motion often displays a remarkable simplicity when compared with the motion of other parts of a system. To analyze this motion, we consider both the velocity and the acceleration of the center of mass. Each of these quantities is defined in complete analogy with the definition of the center of mass itself.

For example, to find the velocity of the center of mass we first multiply the mass of each object in a system, m, by its velocity, $\mathbf{v}$, to give $m_1\mathbf{v}_1$, $m_2\mathbf{v}_2$, and so on. Next, we add all these products together, $m_1\mathbf{v}_1 + m_2\mathbf{v}_2 + \ldots$, and divide by the total mass, $M = m_1 + m_2 + \ldots$. The result, by definition, is the velocity of the center of mass, $\mathbf{V}_{cm}$:

Velocity of the Center of Mass

$$\mathbf{V}_{cm} = \frac{m_1\mathbf{v}_1 + m_2\mathbf{v}_2 + \ldots}{m_1 + m_2 + \ldots} = \frac{\sum m\mathbf{v}}{M} \qquad \text{9–16}$$

Comparing with Equation 9–14, we see that $\mathbf{V}_{cm}$ is the same as X_{cm} with each position x replaced with a velocity vector $\mathbf{v}$. In addition, note that the total mass of the system, M, times the velocity of the center of mass, $\mathbf{V}_{cm}$, is simply the total momentum of the system:

$$M\mathbf{V}_{cm} = m_1\mathbf{v}_1 + m_2\mathbf{v}_2 + \ldots = \mathbf{p}_1 + \mathbf{p}_2 + \ldots = \mathbf{p}_{total}$$

To gain more information on how the center of mass moves, we next consider its acceleration, $\mathbf{A}_{cm}$. As expected by analogy with $\mathbf{V}_{cm}$, the acceleration of the center of mass is defined as follows:

Acceleration of the Center of Mass

$$\mathbf{A}_{cm} = \frac{m_1\mathbf{a}_1 + m_2\mathbf{a}_2 + \ldots}{m_1 + m_2 + \ldots} = \frac{\sum m\mathbf{a}}{M} \qquad \text{9–17}$$

Note that the vector $\mathbf{A}_{cm}$ contains terms like $m_1\mathbf{a}_1$, $m_2\mathbf{a}_2$, and so on, for each object in the system. From Newton's second law, however, we know that $m_1\mathbf{a}_1$, is simply $\mathbf{F}_1$, the net force acting on mass 1. The same conclusion applies to each of the masses. Therefore, we find that the total mass of the system, M, times the acceleration of the center of mass, $\mathbf{A}_{cm}$, is simply the total force acting on the system:

$$M\mathbf{A}_{cm} = m_1\mathbf{a}_1 + m_2\mathbf{a}_2 + \ldots = \mathbf{F}_1 + \mathbf{F}_2 + \ldots = \mathbf{F}_{total}$$

Recall, however, that the total force acting on a system is the same as the net external force, $\mathbf{F}_{net,ext}$, since the internal forces cancel. Therefore, $M\mathbf{A}_{cm}$ is the net external force acting on the system:

Newton's Second Law for a System of Particles

$$M\mathbf{A}_{cm} = \mathbf{F}_{net,ext} \qquad \text{9–18}$$

Zero Net External Force For systems in which $\mathbf{F}_{net,ext}$ is zero, it follows that the acceleration of the center of mass is zero. Hence, if the center of mass is initially at rest, it remains at rest. Similarly, if the center of mass is moving initially, it continues to move with the same velocity. For example, in a collision between two air track carts, the velocity of each cart changes as a result of the collision. The velocity of the center of mass of the two carts, however, is the same before and after the collision. We explore cases in which $\mathbf{F}_{net,ext} = 0$ in the following Example and Active Example.

EXAMPLE 9–9 Crash of the Air Carts

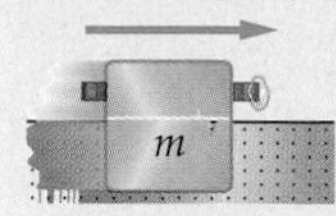

An air cart of mass m and speed v_0 moves toward a second, identical air cart that is at rest. When the carts collide they stick together and move as one. Find the velocity of the center of mass of this system **(a)** before and **(b)** after the carts collide.

Picture the Problem

We choose the positive direction to be the direction of motion. Note that the carts have wads of putty on their bumpers so they will stick when they collide.

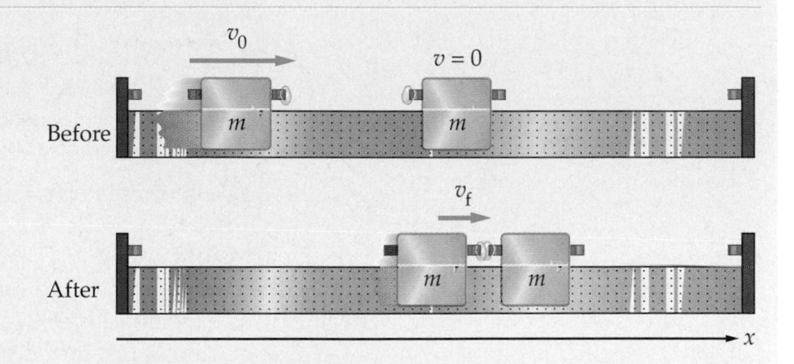

Strategy

(a) We can find the velocity of the center of mass by applying Equation 9–16 to the case of just two masses; $V_{cm} = (m_1 v_1 + m_2 v_2)/M$. In this case, $v_1 = v_0$, $v_2 = 0$, and $m_1 = m_2 = m$.

(b) After the collision the two masses have the same velocity, v_f, which is given by momentum conservation (Equations 9–10 and 9–11). Hence, $V_{cm} = (m_1 v_f + m_2 v_f)/M$.

Solution

Part (a)

1. Use $V_{cm} = (m_1 v_1 + m_2 v_2)/M$ to find the velocity of the center of mass before the collision:

$$V_{cm} = \frac{m_1 v_1 + m_2 v_2}{m_1 + m_2} = \frac{m v_0 + m \cdot 0}{m + m} = \tfrac{1}{2} v_0$$

Part (b)

2. Use momentum conservation, as in Equation 9–11, to find the speed of the carts after the collision:

$$m v_0 = m v_f + m v_f$$
$$v_f = \tfrac{1}{2} v_0$$

3. Calculate the velocity of the center of mass of the two carts:

$$V_{cm} = \frac{m_1 v_1 + m_2 v_2}{m_1 + m_2} = \frac{m v_f + m v_f}{m + m} = v_f = \tfrac{1}{2} v_0$$

Insight

As expected, the velocity of each cart changes as a result of the collision, while the center of mass continues to move with the same velocity. This is illustrated here, where we show a sequence of equal-time snapshots of the system before and after the collision.

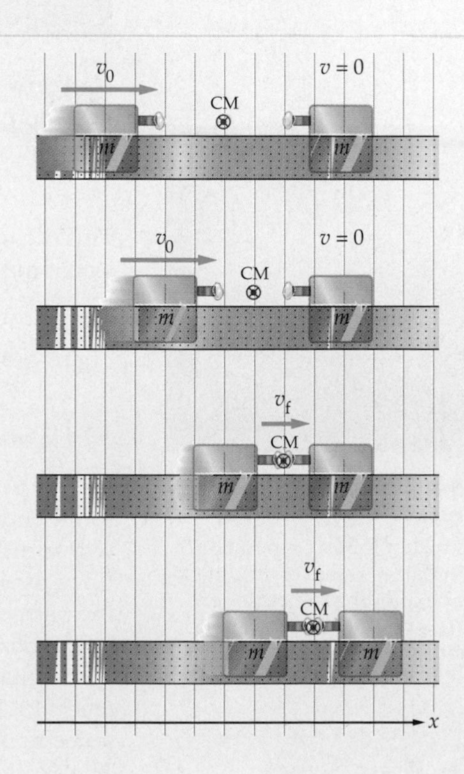

Practice Problem

If the mass of the cart that is moving initially is doubled to $2m$, does the velocity of the center of mass increase, decrease, or stay the same? Verify your answer by calculating the velocity of the center of mass in this case. [**Answer:** The velocity of the center of mass increases. We found that $V_{cm} = 2v_0/3$, both before and after the collision.]

Some related homework problems: Problem 45, Problem 67

ACTIVE EXAMPLE 9–3 Return of the Bee on a Stick

In Active Example 9–2 we found that as a 0.150-g bee runs with a speed of 3.80 cm/s in one direction, the 4.75-g Popsicle stick on which it floats moves with a speed of 0.120 cm/s in the opposite direction. Find the velocity of the center of mass of the bee and the stick.

Solution

1. Write the velocity of the bee, using the correct sign: $v_b = 3.80\text{ cm/s}$
2. Write the velocity of the stick, using the correct sign: $v_s = -0.120\text{ cm/s}$
3. Use these velocities to calculate V_{cm}: $V_{cm} = (m_b v_b + m_s v_s)/(m_b + m_s) = 0$

Insight

V_{cm} is zero, hence the center of mass stays at rest as the bee and the stick move. This is as expected, since the net external force is zero for this system, and the bee and stick started at rest initially.

▶ **FIGURE 9–12 Motion of the center of mass**

At first glance the motion of this thrown hammer seems complex. However, if you draw a curve through the successive positions of the center of mass (marked with a black dot), you will find that it follows a parabolic path, just as if the hammer were a projectile with all of its mass concentrated at that one spot.

Nonzero Net External Force Recall that Newton's second law, as expressed in Equation 9–18, states that the acceleration of the center of mass is related to the net external force as follows:

$$M\mathbf{A}_{cm} = \mathbf{F}_{net,ext}$$

This is completely analogous to the relationship between the acceleration of an object of mass m and the net force $\mathbf{F}_{net}$ applied to it:

$$m\mathbf{a} = \mathbf{F}_{net}$$

Therefore, when $\mathbf{F}_{net,ext}$ is nonzero we can conclude the following:

> The center of mass of a system accelerates precisely as if it were a point particle of mass M acted on by the force $\mathbf{F}_{net,ext}$.

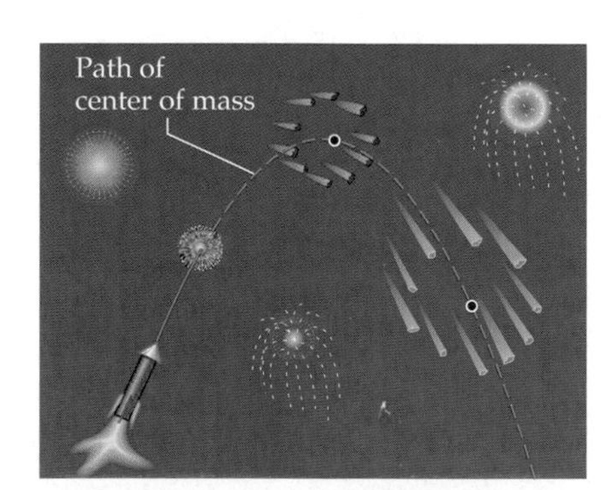

▲ **FIGURE 9–13 Center of mass of an exploding rocket**

A fireworks rocket follows a parabolic path, ignoring air resistance, until it explodes. After exploding, its center of mass continues on the same parabolic path until some of the fragments start to land.

For this reason, the motion of the center of mass can be quite simple compared to the motion of its constituent parts. For example, a hammer tossed into the air with a rotation is shown in **Figure 9–12**. The motion of one part of the hammer, the tip of the handle, let's say, follows a complicated path in space. On the other hand, the path of the center of mass is a simple parabola, precisely the same path that a point mass would follow.

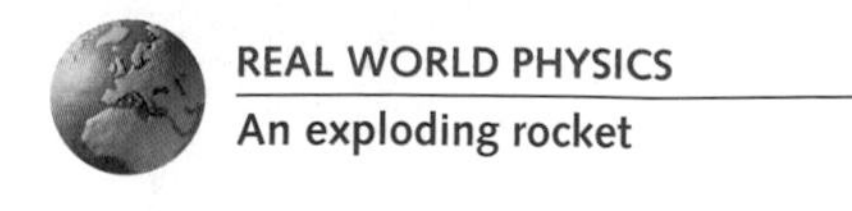

REAL WORLD PHYSICS
An exploding rocket

Similarly, consider a fireworks rocket launched into the sky, as illustrated in **Figure 9–13**. The center of mass of the rocket follows a parabolic path, ignoring air resistance. At some point in its path it explodes into numerous individual pieces. The explosion is due to internal forces, however, which must therefore sum to zero. Hence, the net external force acting on the pieces of the rocket is the same before, during, and after the explosion. As a result, the center of mass has a constant downward acceleration, and continues to follow the original parabolic path. It is only when an additional external force acts on the system, as when one of the pieces of the rocket hits the ground, that the path of the center of mass changes.

To see how to apply $M\mathbf{A}_{cm} = \mathbf{F}_{net,ext}$, consider the system shown in **Figure 9–14**. Here we see a box of mass m_1, inside of which is a ball of mass m_2 suspended from a light string. The entire system rests on a scale reading its weight. The scale exerts an upward force on the box of magnitude F_s. Initially, of course, $F_s = (m_1 + m_2)g$.

Now, suppose the string breaks, allowing the ball to fall with constant acceleration g toward the bottom of the box. What is the reading on the scale while the ball falls? We can guess that the answer should be simply $m_1 g$, the weight of the box alone, but let's analyze the problem from the point of view of the center of mass.

Taking upward as the positive direction, the net external force acting on the box and the ball is

$$F_{net,ext} = F_s - m_1 g - m_2 g$$

The acceleration of the center of mass is

$$A_{cm} = \frac{m_1 \cdot 0 - m_2 g}{M} = -\frac{m_2}{M} g$$

Setting $MA_{cm} = F_{net,ext}$ yields

$$MA_{cm} = M\left(-\frac{m_2}{M}\right)g = -m_2 g = F_{net,ext} = F - m_1 g - m_2 g$$

Finally, canceling the term $-m_2 g$ and solving for the weight read by the scale, F_s, we find, as expected, that

$$F_s = m_1 g$$

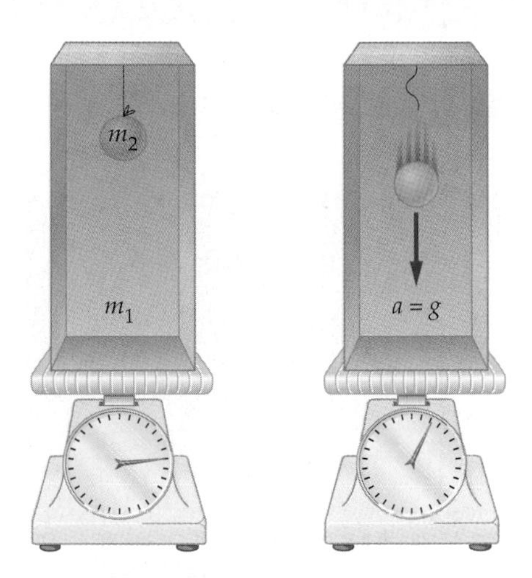

▲ **FIGURE 9–14 Weight and acceleration of the center of mass**
A box with a ball suspended from a string is weighed on a scale. The scale reads the weight of the box and the ball. When the string breaks and the ball falls with the acceleration of gravity, the scale reads only the weight of the box.

*9–8 Systems With Changing Mass: Rocket Propulsion

We close this chapter by considering systems in which the mass can change. A rocket, for example, changes its mass as it operates its engines because it ejects part of the fuel as it burns. The burning process is produced by internal forces, hence the total momentum of the rocket and its fuel remains constant.

Consider, then, a rocket in outer space, far from any large, massive objects. When the rocket's engine is fired, it expels a certain mass of fuel out the back with a speed v. If the mass of the ejected fuel is Δm, then the momentum of the ejected fuel has a magnitude equal to $(\Delta m)v$. Since the total momentum of the system must still be zero, the rocket acquires an equivalent amount of momentum in the forward direction. Hence, the momentum increase of the rocket is

$$\Delta p = (\Delta m)v$$

If the mass of fuel Δm is ejected in the time Δt, the force exerted on the rocket is the change in its momentum divided by the time interval (Equation 9–3); that is

$$F = \frac{\Delta p}{\Delta t} = \left(\frac{\Delta m}{\Delta t}\right)v$$

The force exerted on the rocket by the ejected fuel is referred to as the **thrust**. Thus, the thrust of a rocket is

Thrust

$$\text{thrust} = \left(\frac{\Delta m}{\Delta t}\right)v \qquad \textbf{9–19}$$

SI unit: newton, N

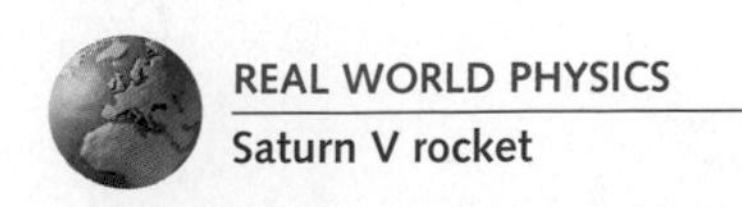

By $\Delta m/\Delta t$, we simply mean the amount of mass per time coming out of the rocket. For example, on the Saturn V rocket, the one used on the manned missions to the Moon, the main engines eject fuel at the rate of 13,800 kg/s with a speed of 2440 m/s. As a result, the thrust produced by these engines is

$$\text{thrust} = \left(\frac{\Delta m}{\Delta t}\right)v = (13{,}800 \text{ kg/s})(2440 \text{ m/s}) = 33.7 \times 10^6 \text{ N}$$

Since this is about 7.60 million pounds, and the weight of the rocket at liftoff is only 6.30 million pounds $= 28.0 \times 10^6$ N, the thrust is sufficient to launch the rocket and give it an upward acceleration. In fact, the initial net force acting on the rocket is

$$F_{\text{net}} = \text{thrust} - mg = 33.7 \times 10^6 \text{ N} - 28.0 \times 10^6 \text{ N} = 5.7 \times 10^6 \text{ N}$$

The rocket's initial weight is $W = mg = 28.0 \times 10^6$ N, hence its initial mass is $m = W/g = 2.85 \times 10^6$ kg. Therefore, the rocket lifts off with an upward acceleration of

$$a = \frac{F_{\text{net}}}{m} = \frac{5.7 \times 10^6 \text{ N}}{2.85 \times 10^6 \text{ kg}} = 2.0 \text{ m/s}^2 \approx 0.20g$$

This is a rather gentle acceleration. The gentleness lasts only a matter of seconds, however, since the decreasing mass of the rocket results in an increasing acceleration.

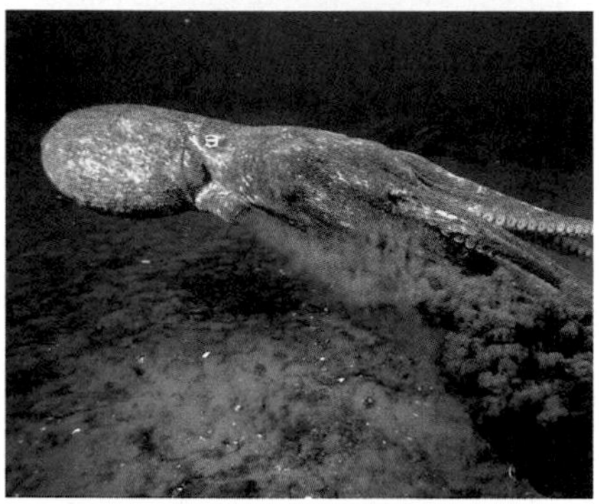

▲ A rocket (top) makes use of the principle of conservation of momentum: mass (the products of explosive burning of fuel) is ejected at high speed in one direction, causing the rocket to move in the opposite direction. The same method of propulsion has evolved in octopi (bottom) and some other animals. When danger threatens and a quick escape is needed, powerful muscles contract to create a jet of water that propels the animal to safety. (A smokescreen of ink provides additional security.)

EXERCISE 9–5

The ascent stage of the lunar lander was designed to produce 15,500 N of thrust at liftoff. If the speed of the ejected fuel is 2500 m/s, what is the rate at which the fuel must be burned?

Solution

The rate of fuel consumption is

$$\frac{\Delta m}{\Delta t} = \frac{\text{thrust}}{v} = \frac{15{,}500 \text{ N}}{2500 \text{ m/s}} = 6.2 \text{ kg/s}$$

A common question regarding rockets is: "How can a rocket accelerate in outer space when it has nothing to push against?" The answer is that rockets, in effect, push against their own fuel. The situation is similar to firing a gun. When a bullet is ejected by the internal combustion of the gunpowder, the person firing the gun feels a recoil. If the person were in space, or standing on a frictionless surface, the recoil would give him or her a speed in the direction opposite to the bullet. The burning of a rocket engine provides a continuous recoil, almost as if the rocket were firing a steady stream of bullets out the back.

Chapter Summary

	Topic	Remarks and Relevant Equations
9–1	**Linear Momentum**	The linear momentum of an object of mass m moving with velocity $\mathbf{v}$ is $\mathbf{p} = m\mathbf{v}$ (9–1)
	momentum is a vector	Linear momentum is a vector, pointing in the same direction as the velocity vector, $\mathbf{v}$.
	momentum of a system of objects	In a system of several objects, the total linear momentum is the vector sum of the individual momenta: $\mathbf{p}_{\text{total}} = \mathbf{p}_1 + \mathbf{p}_2 + \mathbf{p}_3 + \dots$ (9–2)

9–2 Newton's Second Law

In terms of momentum, Newton's second law is

$$\sum \mathbf{F} = \frac{\Delta \mathbf{p}}{\Delta t} \qquad 9\text{–}3$$

That is, the net force acting on an object is equal to the rate of change of its momentum.

constant mass

For cases in which the mass is constant, Newton's second law reduces to the familiar form

$$\sum \mathbf{F} = m\mathbf{a} \qquad 9\text{–}4$$

9–3 Impulse

The impulse delivered to an object by an average force $\mathbf{F}_{av}$ acting for a time Δt is

$$\mathbf{I} = \mathbf{F}_{av}\Delta t \qquad 9\text{–}5$$

impulse is a vector

Impulse is a vector, proportional to the force vector.

impulse and momentum

By Newton's second law, the impulse delivered to an object is equal to the change in its momentum:

$$\mathbf{I} = \mathbf{F}_{av}\Delta t = \Delta \mathbf{p} \qquad 9\text{–}6$$

magnitude of the impulse and force

Since an impulse is often delivered in a very short time interval, the average force can be large.

9–4 Conservation of Linear Momentum

The momentum of an object is conserved (remains constant) if the net force acting on it is zero.

internal/external forces

In a system of objects, internal forces always sum to zero. The net force acting on a system of objects, then, is the sum of the external forces.

conservation of momentum in a system

In a system of objects, the net momentum is conserved if the net external force acting on the system is zero.

9–5 Inelastic Collisions

In collisions, we assume that external forces either sum to zero, or are small enough to be ignored. Hence, momentum is conserved in all collisions.

inelastic collision

In an inelastic collision, the final kinetic energy is different from the initial kinetic energy. The kinetic energy is usually less after a collision, but it can also be more than the initial kinetic energy.

completely inelastic collision

A collision in which objects hit and stick together is referred to as completely inelastic.

collisions in one dimension

A one dimensional collision occurs along a line, which we can choose to be the x axis. After the collision, the x component of momentum is equal to the x component of momentum before the collision; that is, the x component of momentum is conserved.

If two objects, of mass m_1 and m_2 and with initial velocities $v_{1,i}$ and $v_{2,i}$, collide and stick, the final velocity is

$$v_f = \frac{m_1 v_{1,i} + m_2 v_{2,i}}{m_1 + m_2} \qquad 9\text{–}10$$

collisions in two dimensions

In a two-dimensional collision, there are two separate momentum relations to be satisfied: (i) the x component of momentum is conserved, and (ii) the y component of momentum is conserved.

9–6 Elastic Collisions

In collisions, we assume that external forces either sum to zero or are small enough to be ignored. Hence, momentum is conserved in all collisions.

elastic collision

In an elastic collision, the final kinetic energy is equal to the initial kinetic energy.

collisions in one dimension

In an elastic collision in one dimension where mass m_1 is moving with an initial velocity v_0, and mass m_2 is initially at rest, the velocities of the masses after the collision are:

$$v_{1,f} = \frac{m_1 - m_2}{m_1 + m_2} v_0$$

and

$$v_{2,f} = \frac{2m_1}{m_1 + m_2} v_0 \qquad 9\text{–}12$$

collisions in two dimensions

In elastic collisions in two dimensions, three separate conditions are satisfied: (i) kinetic energy is conserved, (ii) the x component of momentum is conserved, and (iii) the y component of momentum is conserved.

9–7 Center of Mass

The location of the center of mass of a two-dimensional system of objects is defined as follows:

$$X_{cm} = \frac{m_1x_1 + m_2x_2 + \ldots}{m_1 + m_2 + \ldots} = \frac{\sum mx}{M} \quad \text{9–14}$$

and

$$Y_{cm} = \frac{m_1y_1 + m_2y_2 + \ldots}{m_1 + m_2 + \ldots} = \frac{\sum my}{M} \quad \text{9–15}$$

motion of the center of mass

The velocity of the center of mass is

$$\mathbf{V}_{cm} = \frac{m_1\mathbf{v}_1 + m_2\mathbf{v}_2 + \ldots}{m_1 + m_2 + \ldots} = \frac{\sum m\mathbf{v}}{M} \quad \text{9–16}$$

Note that $M\mathbf{V}_{cm} = m_1\mathbf{v}_1 + m_2\mathbf{v}_2 + \ldots = \mathbf{P}_{total}$. If a system's momentum is conserved, its center of mass has constant velocity.

Similarly, the acceleration of the center of mass is

$$\mathbf{A}_{cm} = \frac{m_1\mathbf{a}_1 + m_2\mathbf{a}_2 + \ldots}{m_1 + m_2 + \ldots} = \frac{\sum m\mathbf{a}}{M} \quad \text{9–17}$$

Note that $M\mathbf{A}_{cm} = m_1\mathbf{a}_1 + m_2\mathbf{a}_2 + \ldots =$ (net external force). That is,

$$M\mathbf{A}_{cm} = \mathbf{F}_{net,ext} \quad \text{9–18}$$

The center of mass accelerates as if the net external force acted on a single object of mass $M = m_1 + m_2 + \ldots$.

***9–8 Systems With Changing Mass: Rocket Propulsion**

The mass of a rocket changes because its engines expel fuel when they are fired. If fuel is expelled with the speed v and at the rate $\Delta m/\Delta t$, the thrust experienced by the rocket is

$$\text{thrust} = \left(\frac{\Delta m}{\Delta t}\right)v \quad \text{9–19}$$

Problem-Solving Summary

Type of Calculation	Relevant Physical Concepts	Related Examples
Calculate the momentum of a system.	Each object in a system has a momentum of magnitude mv that points in the direction of its velocity vector. The total momentum is the vector sum of the individual momenta.	Example 9–1
Relate force and time to the impulse.	The impulse acting on a system is the average force, F_{av}, times the time interval, Δt.	Example 9–2 Active Example 1
Apply momentum conservation.	Momentum is conserved when the net external force acting on a system is zero.	Examples 9–3, 9–4, 9–5, 9–6, 9–7 Active Example 9–2
Find the center of mass.	The location of the center of mass is given by Equations 9–14 and 9–15.	Example 9–8
Determine the motion of the center of mass.	The center of mass moves the same as if it were a point particle of mass M (the total mass of the system) acted on by the net external force, $\mathbf{F}_{net,ext}$.	Example 9–9 Active Example 9–3

Conceptual Questions

1. If you drop your keys, their momentum increases as they fall. Why is the momentum of the keys not conserved? Does this mean that the momentum of the universe increases as the keys fall?
2. By what factor does an object's kinetic energy change if its speed is doubled? By what factor does its momentum change?
3. Two objects are known to have the same momentum. Do these two objects necessarily have the same kinetic energy? Explain, and give a specific example.
4. Two objects are known to have the same kinetic energy. Do these two objects necessarily have the same momentum? Explain, and give a specific example.
5. A system of particles is known to have zero kinetic energy. What can you say about the momentum of the system?
6. A system of particles is known to have zero momentum. Does it follow that the kinetic energy of the system is also zero?
7. A block of wood is struck by a bullet. Is the block more likely to be knocked over if the bullet is metal and embeds itself in the wood, or if the bullet is rubber and bounces off the wood? Explain.
8. As you approach a stoplight you apply the brakes and bring your car to rest. What happened to your car's initial momentum?
9. Is the impulse produced by a larger force always greater than the impulse produced by a smaller force? Explain, and give a specific example.
10. Your car rolls slowly in a parking lot and bangs into the metal base of a light pole. In terms of safety, is it be better for your collision with the light pole to be elastic or inelastic? Explain.
11. A net force of 200 N acts on a 100-kg boulder, and a force of the same magnitude acts on a 110-g pebble. Is the rate of change of the boulder's momentum greater than, less than, or equal to the rate of change of the pebble's momentum? Explain.
12. Referring to the previous question, is the rate of change in velocity for the boulder greater than, less than, or equal to the rate of change of velocity for the pebble? Explain.
13. On a calm day you connect an electric fan to a battery on your sailboat and generate a breeze. Can the wind produced by the fan be used to power the sailboat? Explain.
14. In the previous question, can you use the wind generated by the fan to move a boat that has no sail? Explain why or why not.
15. Crash statistics show that it is safer to be riding in a heavy car in an accident than in a light car. Explain in terms of physical principles.
16. Two cars collide at an intersection. If the cars do not stick together, can we conclude that their collision was elastic? Explain.
17. An object at rest on a frictionless surface is struck by a second object. Is it possible that after this collision both objects are at rest? Explain.
18. In the previous question, is it possible for one of the two objects to be at rest after the collision? Explain.
19. Can two objects have a collision in which all the initial kinetic energy of the system is lost? Explain, and give a specific example.
20. You tee up a golf ball and drive it down the fairway. When the ball leaves the tee is its speed the same as, greater than, or less than the speed of the golf club? Explain.
21. A friend tosses a ball of mass m to you with a speed v. When you catch the ball you feel a noticeable sting in your hand. If you now catch a ball of mass $2m$ and speed $v/2$, is the sting you feel greater than, less than, or the same as that felt when you caught the first ball? Explain, being careful to distinguish between momentum and kinetic energy.
22. In an inelastic collision it is possible for objects to lose most of their kinetic energy, while at the same time losing none of their momentum. Explain how this can happen.
23. At the instant a bullet is fired from a gun, the bullet and the gun have equal and opposite momenta. Which object—the bullet or the gun—has the greater kinetic energy? Explain. How does your answer apply to the observation that it is safe to hold a gun while it is fired, whereas the bullet is deadly?
24. Suppose you throw a rubber ball at a charging elephant (not a good idea). When the ball bounces back toward you, is its speed greater than, less than, or equal to the speed with which you threw it? Explain.
25. Two objects undergo an elastic collision. Is the kinetic energy of each object the same before and after the collision? Is the momentum of each object the same before and after the collision? Explain.
26. In the "Fosbury flop" method of high jumping, named for the track and field star Dick Fosbury, an athlete's center of mass may pass under the bar while the athlete's body passes over the bar. Explain how this is possible.

▲ The "Fosbury flop" (Question 26)

27. Lifting one foot into the air, you balance on the other foot. What can you say about the location of your center of mass?
28. A stalactite in a cave has drops of water falling from it to the cave floor below. The drops are equally spaced in time and come in rapid succession, so that at any given moment there are many drops in midair. Where is the center of mass of these midair drops? In particular, is the center of mass higher than, lower than, or at the halfway distance between the tip of the stalactite and the cave floor?
29. Is the center of mass of a baseball bat midway between the two ends of the bat, nearer the thick end, or nearer the thin end? Explain.
30. An hourglass is turned over and the sand is allowed to pour from the upper half of the glass to the lower half. If the hourglass is resting on a scale, and its total mass is M, describe the reading on the scale as the sand runs to the bottom.

31. A pencil standing upright on its eraser end falls over and lands on a table. As the pencil falls, its eraser does not slip. The following questions refer to the contact force exerted on the pencil by the table. **(a)** During the pencil's fall, is the x component of the contact force positive, negative, or zero? Explain. **(b)** Is the y component of the contact force greater than, less than, or equal to the weight of the pencil? Explain.

32. A juggler performs a series of tricks with three bowling balls while standing on a bathroom scale. Is the average reading of the scale greater than, less than, or equal to the weight of the juggler plus the weight of the three balls? Explain.

33. In the classic movie "The Spirit of St. Louis," Jimmy Stewart portrays Charles Lindbergh on his history-making transatlantic flight. Lindbergh is concerned about the weight of his plane. As he flies over Newfoundland he notices a fly on the dashboard. Speaking to the fly, he wonders aloud, "Does the plane weigh less if you fly inside it as it's flying? Now that's an interesting question." What do you think?

Problems

Note: **IP** *denotes an integrated conceptual/quantitative problem.* **BIO** *identifies problems of biological or medical interest. Blue bullets (•, ••, •••) are used to indicate the level of difficulty of each problem.*

Section 9–1 Linear Momentum

1. • Referring to Exercise 9–1, what speed must the baseball have if its momentum is to be equal in magnitude to that of the car? Give your result in miles per hour.

2. • Find the total momentum of the birds in Example 9–1 if the goose reverses direction.

3. •• A 20.0-kg dog is running northward at 2.50 m/s, while a 5.00-kg cat is running eastward at 3.00 m/s. Their 70.0-kg owner has the same momentum as the two pets taken together. Find the direction and magnitude of the owner's velocity.

4. •• **IP** Two air track carts move toward one another on an air track. Cart 1 has a mass of 0.45 kg and a speed of 1.1 m/s. Cart 2 has a mass of 0.65 kg. **(a)** What speed must cart 2 have if the total momentum of the system is to be zero? **(b)** Since the momentum of the system is zero, does it follow that the kinetic energy of the system is also zero? **(c)** Verify your answer to part (b) by calculating the system's kinetic energy.

5. •• A 0.150-kg baseball is dropped from rest. If the magnitude of the baseball's momentum is 0.680 kg·m/s just before it lands on the ground, from what height was it dropped?

6. •• **IP** A 220-g ball falls vertically downward, hitting the floor with a speed of 2.5 m/s and rebounding upward with a speed of 2.0 m/s. **(a)** Find the magnitude of the change in the ball's momentum. **(b)** Find the change in the magnitude of the ball's momentum. **(c)** Which of the two quantities calculated in parts (a) and (b) is more directly related to the net force acting on the ball during its collision with the floor? Explain.

Section 9–3 Impulse

7. • In a typical golf swing, the club is in contact with the ball for about 0.0010 s. If the 45-g ball acquires a speed of 65 m/s, estimate the magnitude of the force exerted by the club on the ball.

8. • Find the magnitude of the impulse delivered to a soccer ball when a player kicks it with a force of 1250 N. Assume that the player's foot is in contact with the ball for 6.20×10^{-3} s.

9. • When spiking a volleyball, a player changes the velocity of the ball from 4.5 m/s to −23 m/s along a certain direction. If the impulse delivered to the ball by the player is −9 kg·m/s, what is the mass of the volleyball?

10. • A 0.50-kg croquet ball is initially at rest on the grass. When the ball is struck by a mallet, the average force exerted on it is 230 N. If the ball's speed after being struck is 3.2 m/s, how long was the mallet in contact with the ball?

11. •• To make a bounce pass, a player throws a 0.60-kg basketball toward the floor. The ball hits the floor with a speed of 5.4 m/s at an angle of 65° to the vertical. If the ball rebounds with the same speed and angle, what was the impulse delivered to it by the floor?

12. •• **IP** A 15.0-g marble is dropped from rest onto the floor 1.44 m below. **(a)** If the marble bounces straight upward to a height of 0.640 m, what is the magnitude and direction of the impulse delivered to the marble by the floor? **(b)** If the marble had bounced to a greater height, would the impulse delivered to it have been greater or less than the impulse found in part (a)? Explain.

13. •• A 0.14-kg baseball moves horizontally with a speed of 35 m/s toward a bat. After striking the bat the ball moves vertically upward with half its initial speed. Find the direction and magnitude of the impulse delivered to the ball by the bat.

Section 9–4 Conservation of Momentum

14. • In a situation similar to Example 9–3, suppose the speeds of the two canoes after they are pushed apart are 0.52 m/s for canoe 1 and 0.44 m/s for canoe 2. If the mass of canoe 1 is 340 kg, what is the mass of canoe 2?

15. • Two ice skaters stand at rest in the center of an ice rink. When they push off against one another the 45-kg skater acquires a speed of 0.62 m/s. If the speed of the other skater is 0.89 m/s, what is this skater's mass?

16. • Suppose the bee in Active Example 9–2 has a mass of 0.175 g. If the bee walks with a speed of 1.41 cm/s relative to the still water, what is the speed of the 4.75-g stick relative to the water?

17. •• An object initially at rest breaks into two pieces as the result of an explosion. One piece has twice the kinetic energy of the other piece. What is the ratio of the masses of the two pieces? Which piece has the larger mass?

18. •• A 97-kg astronaut and a 1100-kg satellite are at rest relative to the space shuttle. The astronaut pushes on the satellite, giving it a speed of 0.13 m/s directly away from the shuttle. Seven-and-a-half seconds later the astronaut comes into contact with the shuttle. What was the initial distance from the shuttle to the astronaut?

19. •• **IP** An 85-kg lumberjack stands at one end of a 380-kg floating log, as shown in **Figure 9–15**. Both the log and the lumberjack are at rest initially. **(a)** If the lumberjack now trots toward the other end of the log with a speed of 2.7 m/s relative to the log, what is the lumberjack's speed relative to the shore? Ignore friction between the log and the water. **(b)** If the mass of the log had been greater, would the lumberjack's speed relative to the shore be greater than, less than, or the same as in part (a)? Explain. **(c)** Check your an-

swer to part (b) by calculating the lumberjack's speed relative to the shore for the case of a 450-kg log.

▲ **FIGURE 9–15** Problem 19

20. ••• A plate drops onto a smooth floor and shatters into three pieces of equal mass. Two of the pieces go off with equal speeds v at right angles to one another. Find the speed and direction of the third piece.

Section 9–5 Inelastic Collisions

21. • A cart moves with a speed v on a frictionless air track and collides with an identical cart that is stationary. If the two carts stick together after the collision, what is the final kinetic energy of the system?

22. • Suppose the car in Example 9–6 has an initial speed of 20.0 m/s, and that the direction of the wreckage after the collision is 40.0° above the x axis. What is the initial speed of the minivan and the final speed of the wreckage?

23. • Two 75.0-kg hockey players skating at 5.50 m/s collide and stick together. If the angle between their initial directions was 120°, what is their velocity after the collision?

24. •• **IP** **(a)** Referring to Exercise 9–2, is the final kinetic energy of the car and truck together greater than, less than, or equal to the sum of the initial kinetic energies of the car and truck separately? Explain. **(b)** Verify your answer to part (a) by calculating the initial and final kinetic energies of the system.

25. •• **IP** A bullet with a mass of 4.0 g and a speed of 650 m/s is fired at a block of wood with a mass of 0.095 kg. The block rests on a frictionless surface, and is thin enough that the bullet passes completely through it. Immediately after the bullet exits the block, the speed of the block is 23 m/s. **(a)** What is the speed of the bullet when it exits the block? **(b)** Is the final kinetic energy of this system equal to, less than, or greater than the initial kinetic energy? Explain. **(c)** Verify your answer to part (b) by calculating the initial and final kinetic energies of the system.

26. •• **IP** A 0.470-kg block of wood hangs from the ceiling by a string, and a 0.0700-kg wad of putty is thrown straight upward, striking the bottom of the block with a speed of 5.60 m/s. The wad of putty sticks to the block. **(a)** Is the mechanical energy of this system conserved? **(b)** How high does the putty-block system rise above the original position of the block?

27. •• A 0.430-kg block is attached to a horizontal spring that is at its equilibrium length, and whose force constant is 20.0 N/m. The block rests on a frictionless surface. A 0.0500-kg wad of putty is thrown horizontally at the block, hitting it with a speed of 2.30 m/s and sticking. How far does the putty-block system compress the spring?

28. ••• Two objects moving with a speed v travel in opposite directions in a straight line. The objects stick together when they collide, and move with a speed of $v/4$ after the collision. **(a)** What is the ratio of the final kinetic energy of the system to the initial kinetic energy? **(b)** What is the ratio of the mass of the more massive object to the mass of the less massive object?

Section 9–6 Elastic Collisions

29. • A 722-kg car stopped at an intersection is rear-ended by a 1620-kg truck moving with a speed of 14.5 m/s. If the car was in neutral and its brakes were off, so that the collision is approximately elastic, find the final speed of both vehicles after the collision.

30. • The collision between a hammer and a nail can be considered to be approximately elastic. Estimate the kinetic energy acquired by a 10-g nail when it is struck by a 550-g hammer moving with a speed of 4.5 m/s.

31. •• In the apple-orange collision in Example 9–7, suppose the final velocity of the orange is 1.03 m/s in the negative y direction. What is the final speed and direction of the apple in this case?

32. •• In a nuclear reactor, neutrons released by nuclear fission must be slowed down before they can trigger additional reactions in other nuclei. To see what sort of material is most effective in slowing (or moderating) a neutron, calculate the ratio of a neutron's final kinetic energy to its initial kinetic energy, K_f/K_i, for a head-on, elastic collision with each of the following stationary target particles. (*Note:* The mass of a neutron is $m = 1.009\ u$, where the atomic mass unit, u, is defined as follows: $1\ u = 1.66 \times 10^{-27}$ kg.) **(a)** An electron ($M = 5.49 \times 10^{-4}\ u$). **(b)** A proton ($M = 1.007\ u$). **(c)** The nucleus of a lead atom ($M = 207.2\ u$).

33. •• **IP** A charging bull elephant with a mass of 5400 kg comes directly toward you with a speed of 4.30 m/s. You toss a 0.150-kg rubber ball at the elephant with a speed of 8.11 m/s. **(a)** When the ball bounces back toward you, what is its speed? **(b)** How do you account for the fact that the ball's kinetic energy has increased?

Section 9–7 Center of Mass

34. • Find the x coordinate of the center of mass of the bricks shown in **Figure 9–16**.

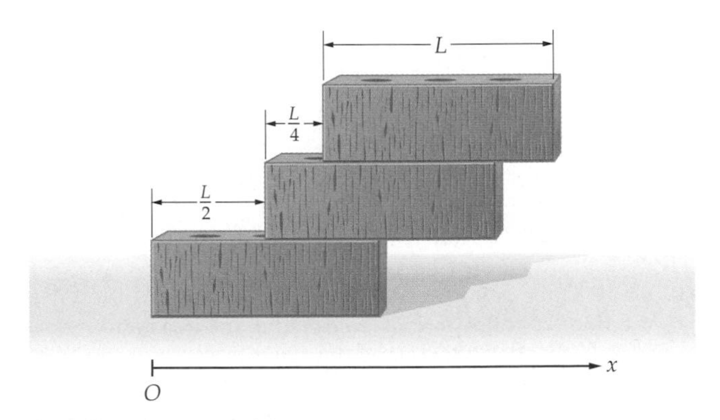

▲ **FIGURE 9–16** Problem 34

35. • You are holding a shopping basket at the grocery store with two 0.55-kg cartons of cereal at one end of the basket. The basket is 0.75-m long. Where should you place a half gallon of milk (1.8 kg) so that the center of mass of your groceries is at the center of the basket?

36. • The Earth has a mass of 5.98×10^{24} kg, the Moon has a mass of 7.35×10^{22} kg, and their center-to-center distance is 3.85×10^{8} m. How far from the center of the Earth is the Earth-Moon center of mass? Is the Earth-Moon center of mass above or below the surface of the earth? By what distance?

37. •• A cardboard box is in the shape of a cube with each side of length L. If the top of the box is missing, where is the center of mass of the open box?

38. •• The location of the center of mass of the partially eaten, 12-inch diameter pizza shown in **Figure 9–17** is $X_{cm} = -1.2$ in. and $Y_{cm} = -1.2$ in. Assuming the pizza to be uniform, find the center of mass of the portion of the pizza above the x axis (that is, the portion of the pizza in the second quadrant).

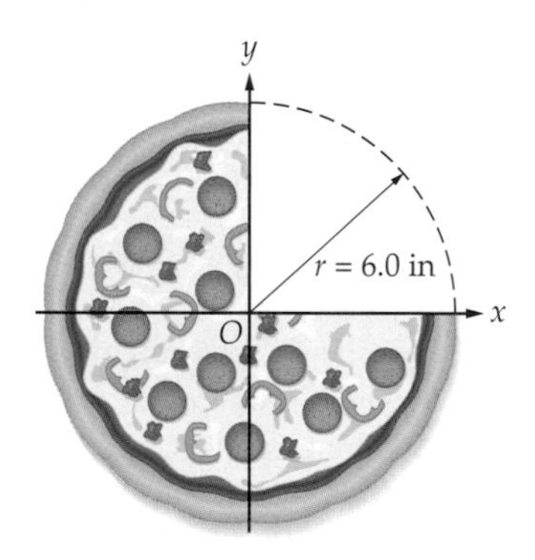

▲ **FIGURE 9–17** Problem 38

39. •• Sulfur dioxide (SO_2) consists of two oxygen atoms (each of mass 16 u, where u is defined in Problem 32) and a single sulfur atom (of mass 32 u). The center-to-center distance between the sulfur atom and either of the oxygen atoms is 0.143 nm, and the angle formed by the three atoms is 120°, as shown in **Figure 9–18**. Find the x and y coordinates of the center of mass of this molecule.

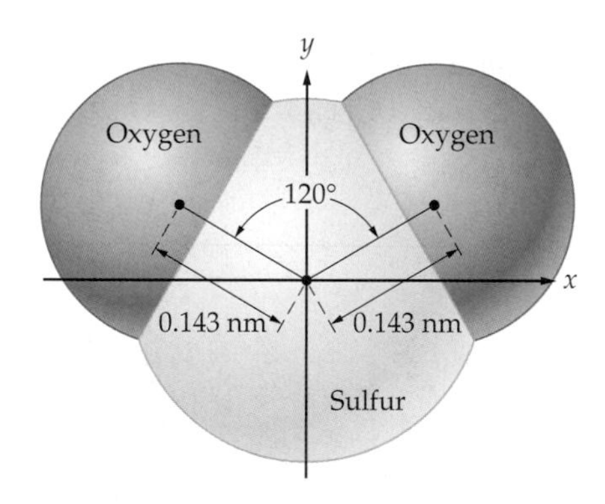

▲ **FIGURE 9–18** Problem 39

40. •• **IP** Three uniform meter sticks, each of mass M, are placed on the floor as follows: stick 1 lies along the y axis from $y = 0$ to $y = 1.0$ m, stick 2 lies along the x axis from $x = 0$ to $x = 1.0$ m, stick 3 lies along the x axis from $x = 1.0$ m to $x = 2.0$ m. **(a)** Find the location of the center of mass of the meter sticks. **(b)** How would the location of the center of mass be affected if the mass of the meter sticks were doubled?

41. •• A 0.604-kg rope 2.00 meters long lies on a floor. You grasp one end of the rope and begin lifting it upward with a constant speed of 0.910 m/s. Find the position and velocity of the rope's center of mass from the time you begin lifting the rope to the time the last piece of rope lifts off the floor. Plot your results. (Assume the rope occupies negligible volume directly below the point where it is being lifted.)

42. •• Repeat the previous problem, this time lowering the rope onto a floor instead of lifting it.

43. •• Consider the system shown in **Figure 9–19**. Assume that after the string breaks the ball falls through the liquid with constant speed. If the mass of the bucket *and* the liquid is 1.20 kg, and the mass of the ball is 0.150 kg, what is the reading on the scale **(a)** before and **(b)** after the string breaks?

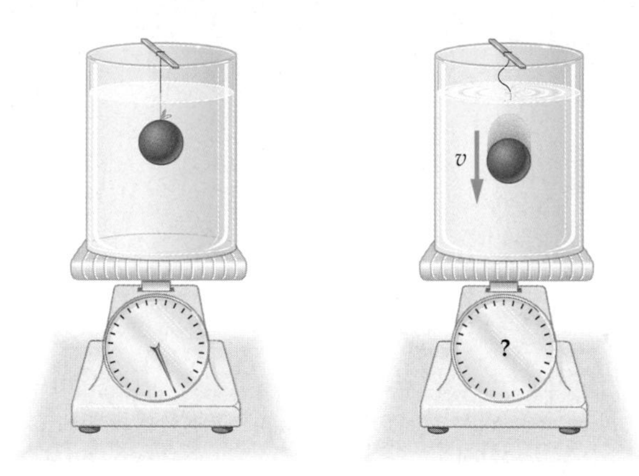

▲ **FIGURE 9–19** Problems 43 and 63

44. •• A cooking pot and the water it contains has a mass of 2.80 kg. This pan is placed on a scale, and a 0.0450-kg egg is dropped into the water. The egg falls toward the bottom of the pan with an acceleration equal to half the acceleration of gravity. Find **(a)** the acceleration of the center of mass of this system and **(b)** the reading on the scale. **(c)** What is the reading on the scale after the egg comes to rest on the bottom of the pot?

45. •• Consider a collision of air carts in a system similar to the one described in Example 9–9. Suppose the air cart to the left has a mass of 0.750 kg and an initial speed of 0.455 m/s. The cart to the right is initially at rest, and has a mass of 0.275 kg. Find the velocity of the center of mass **(a)** before and **(b)** after the carts collide and stick together. **(c)** Find the kinetic energy of the system before and after the collision.

46. ••• A metal block of mass m is attached to the ceiling by a spring. Connected to the bottom of this block is a string that supports a second block of the same mass m, as shown in **Figure 9–20**.

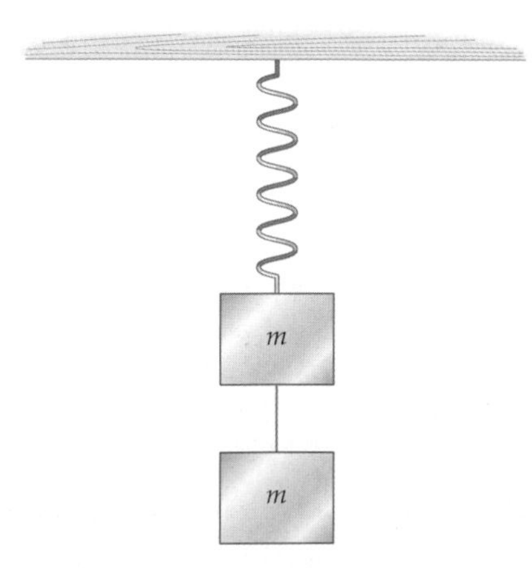

▲ **FIGURE 9–20** Problem 46

The string connecting the two blocks is now cut. **(a)** What is the net force acting on the two-block system immediately after the string is cut? **(b)** What is the acceleration of the center of mass of the two-block system immediately after the string is cut?

*Section 9–8 Systems With Changing Mass: Rocket Propulsion

47. • During a rescue operation, a 5500-kg helicopter hovers above a fixed point. The helicopter blades send air downward with a speed of 60.0 m/s. What mass of air must pass through the blades every second to produce enough thrust for the helicopter to hover?

▲ The powerful downdraft from this helicopter's blades creates a circular wave pattern in the water below. The thrust resulting from this downdraft is sufficient to support the weight of the helicopter. (Problem 47)

48. • A child sits in a wagon with a pile of 0.50-kg rocks. If she can throw each rock with a speed of 11 m/s relative to the ground, how many rocks must she throw per minute to maintain a constant average speed against a 3.4-N force of friction?

49. • A 57.0-kg person holding two 0.850-kg bricks stands on a 2.10-kg skateboard. Initially, the skateboard and the person are at rest. The person now throws the two bricks at the same time so that their speed relative to the person is 18.0 m/s. What is the recoil speed of the person and the skateboard relative to the ground, assuming the skateboard moves without friction?

50. •• In the previous problem, calculate the final speed of the person and the skateboard relative to the ground if the person throws the bricks one at a time. Assume that each brick is thrown with a speed of 18.0 m/s relative to the person.

51. •• A 0.540-kg bucket rests on a scale. Into this bucket you pour sand at the constant rate of 56.0 g/s. If the sand lands in the bucket with a speed of 3.20 m/s, **(a)** what is the reading of the scale when there is 0.750 kg of sand in the bucket? **(b)** What is the weight of the bucket and 0.750 kg of sand?

52. •• **IP** Holding a long rope by its upper end, you lower it onto a scale. The rope has a mass of 0.13 kg per meter of length, and is lowered onto the scale at the constant rate of 1.4 m/s. **(a)** Calculate the thrust exerted by the rope as it lands on the scale. **(b)** When 0.25-kg of rope is on the scale (corresponding to a weight of 2.5 N), does the scale read 2.5 N, more than 2.5 N, or less than 2.5 N? Explain. **(c)** Check your answer to part (b) by calculating the reading on the scale at this time.

General Problems

53. • A 70.0-kg tourist climbs the stairs to the top of the Washington Monument, which is 555 ft high. How far does the Earth move in the opposite direction as the tourist climbs?

54. • A car moving with an initial speed v collides with a second stationary car that is one-half as massive. After the collision the first car moves in the same direction as before with a speed $v/3$. **(a)** Find the final speed of the second car. **(b)** Is this collision elastic or inelastic?

55. •• A 1.30-kg block of wood sits at the edge of a table, 0.750 m above the floor. A 0.0100-kg bullet moving horizontally with a speed of 725 m/s embeds itself within the block. What horizontal distance does the block cover before hitting the ground?

56. •• **IP** The carton of eggs shown in **Figure 9–21** is filled with a dozen eggs, each of mass m. Initially, the center of mass of the eggs is at the center of the carton. **(a)** Does the location of the center of mass of the eggs change more if egg 1 is removed or if egg 2 is removed? Explain. **(b)** Find the center of mass of the eggs when egg 1 is removed. **(c)** Find the center of mass of the eggs if egg 2 is removed instead.

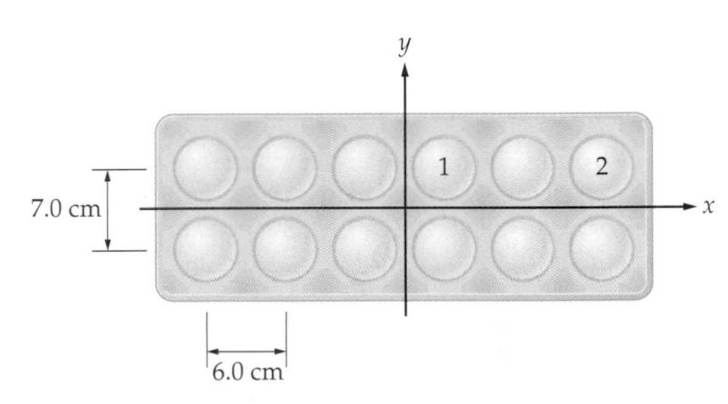

▲ **FIGURE 9–21** Problem 56

57. •• During a severe storm in Palm Beach, Fla., in January 1999, 31 inches of rain fell in a period of nine hours. Assuming that the raindrops hit the ground with a speed of 10 m/s, estimate the average force exerted on a square meter of ground during this storm. (Note: One cubic meter of water has a mass of 1000 kg.)

58. •• An apple that weighs 3.0 N falls vertically downward from rest for 1.5 s. **(a)** What is the change in the apple's momentum per second? **(b)** What is the total change in its momentum during the 1.5-second fall?

59. •• To balance a 35.5-kg automobile tire and wheel, a mechanic must place a 50.2-g lead weight 25.0 cm from the center of the wheel. When the wheel is balanced, its center of mass is exactly at the center of the wheel. How far from the center of the wheel was its center of mass before the lead weight was added?

60. •• A hoop of mass M and radius R rests on a smooth, level surface. The inside of the hoop has ridges on either side, so that it forms a track on which a ball can roll, as indicated in **Figure 9–22**. If a ball of mass $2M$ and radius $r = R/4$ is released as shown, the system rocks back and forth until it comes to rest with the ball at the bottom of the hoop. When the ball comes to rest, what is the x coordinate of its center?

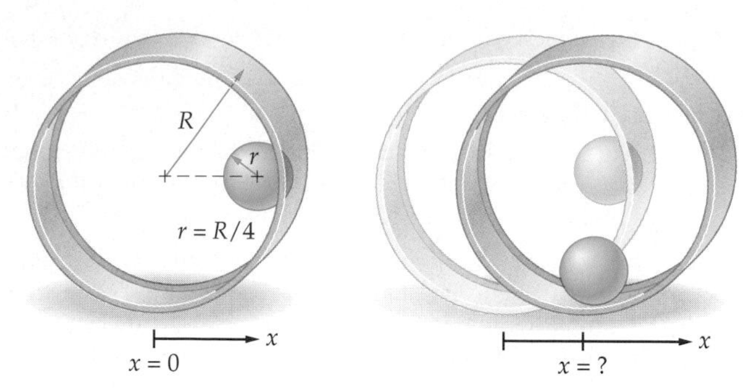

▲ **FIGURE 9–22** Problem 60

61. •• **IP** A 63-kg canoeist stands in the middle of her 22-kg canoe. The canoe is 3.0 m long, and the end that is closest to land is 2.5 m from the shore. The canoeist now walks toward the shore until she comes to the end of the canoe. **(a)** When the canoeist stops at the end of her canoe, is her distance from the shore equal to, greater than, or less than 2.5 m? Explain. **(b)** Verify your answer to part (a) by calculating the distance from the canoeist to shore.

62. •• In the previous problem, suppose the canoeist is 3.4 m from shore when she reaches the end of her canoe. What is the canoe's mass?

63. •• Referring to Problem 43, find the reading on the scale **(a)** before and **(b)** after the string breaks, assuming the ball falls through the liquid with an acceleration equal to one-quarter the acceleration of gravity.

64. •• A young hockey player stands at rest on the ice holding a 1.1-kg helmet. The player tosses the helmet with a speed of 6.2 m/s in a direction 13° above the horizontal, and recoils with a speed of 0.25 m/s. Find the mass of the hockey player.

65. •• Find the center of mass of a water molecule, referring to **Figure 9–23** for the relevant angles and distances. The mass of a hydrogen atom is 1.0 u, and the mass of an oxygen atom is 16 u, where u is the atomic mass unit. Use the oxygen atom as the origin of your coordinate system.

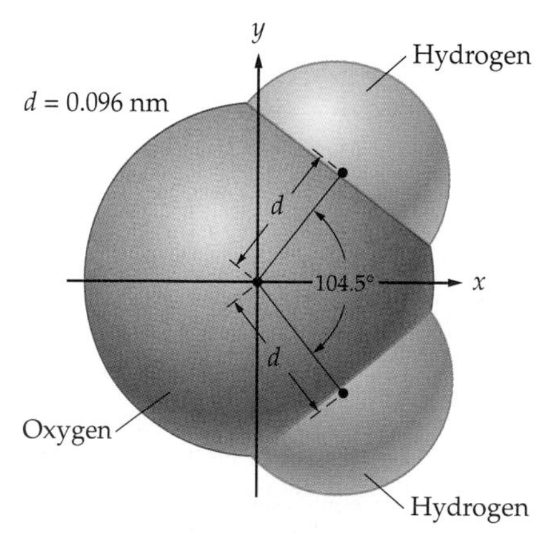

▲ **FIGURE 9–23** Problem 65

66. •• A long uniform rope with a mass of 0.105 kg per meter lies on the ground. You grab one end of the rope and lift it at the constant rate of 1.25 m/s. Calculate the upward force you must exert at the moment when the top end of the rope is 0.500 m above the ground.

67. •• Suppose the air carts in Example 9–9 are both moving to the right initially. The cart to the left has a mass m and an initial speed v_0; the cart to the right has an initial speed $v_0/2$. If the center of mass of this system moves to the right with a speed $2v_0/3$, what is the mass of the cart on the right?

68. ••• Consider a one-dimensional, head-on elastic collision. One object has a mass m_1 and an initial velocity v_1; the other has a mass m_2 and an initial velocity v_2. Use momentum conservation and energy conservation to show that the final velocities of the two masses are

$$v_{1,\mathrm{f}} = \left(\frac{m_1 - m_2}{m_1 + m_2}\right)v_1 + \left(\frac{2m_2}{m_1 + m_2}\right)v_2$$

$$v_{2,\mathrm{f}} = \left(\frac{2m_1}{m_1 + m_2}\right)v_1 + \left(\frac{m_2 - m_1}{m_1 + m_2}\right)v_2$$

69. ••• Two objects with masses m_1 and m_2 and initial speeds $v_{1,\mathrm{i}}$ and $v_{2,\mathrm{i}}$ move along a straight line and collide elastically. Assuming that the objects move along the same straight line after the collision, show that their relative speeds are unchanged; that is, that $v_{1,\mathrm{i}} - v_{2,\mathrm{i}} = v_{2,\mathrm{f}} - v_{1,\mathrm{f}}$. (You can use the results given in Problem 68.)

70. ••• An air track cart of mass m is pushed against a spring attached to the end of the track and released. The resulting speed of the cart is v. Now the same spring is placed between two identical carts, each of mass m. The spring is compressed by the same amount as before and the two carts are released simultaneously. Find the final speeds of the carts in terms of the speed v.

71. ••• An object of mass m undergoes an elastic collision with an identical object that is at rest. The collision is not head-on. Show that the angle between the velocities of the two objects after the collision is 90°.

72. ••• On a cold winter's morning a child sits on a sled resting on smooth ice. When the 9.50-kg sled is pulled with a horizontal force of 40.0 N it begins to move with an acceleration of 2.32 m/s^2. The 21.0-kg child accelerates too, but with a smaller acceleration than that of the sled. Thus, the child moves forward relative to the ice, but slides backward relative to the sled. Find the acceleration of the child relative to the ice.

73. ••• Two objects of equal mass and initial speed v stick together as the result of a completely inelastic collision. After the collision, the two objects as a group move off with a speed $v/3$. What was the angle between the initial directions of the objects?

74. ••• Two small rubber balls are dropped from rest at a height h above a hard floor. When the balls are released, the lighter ball (with mass m) is directly above the heavier ball (with mass M). Assume the heavier ball reaches the floor first and bounces elastically; thus, when the balls collide, the ball of mass M is moving upward with a speed v and the ball of mass m is moving downward with essentially the same speed. In terms of h, find the height to which the ball of mass m rises after the collision. (Use the results given in Problem 68.)

10 Rotational Kinematics and Energy

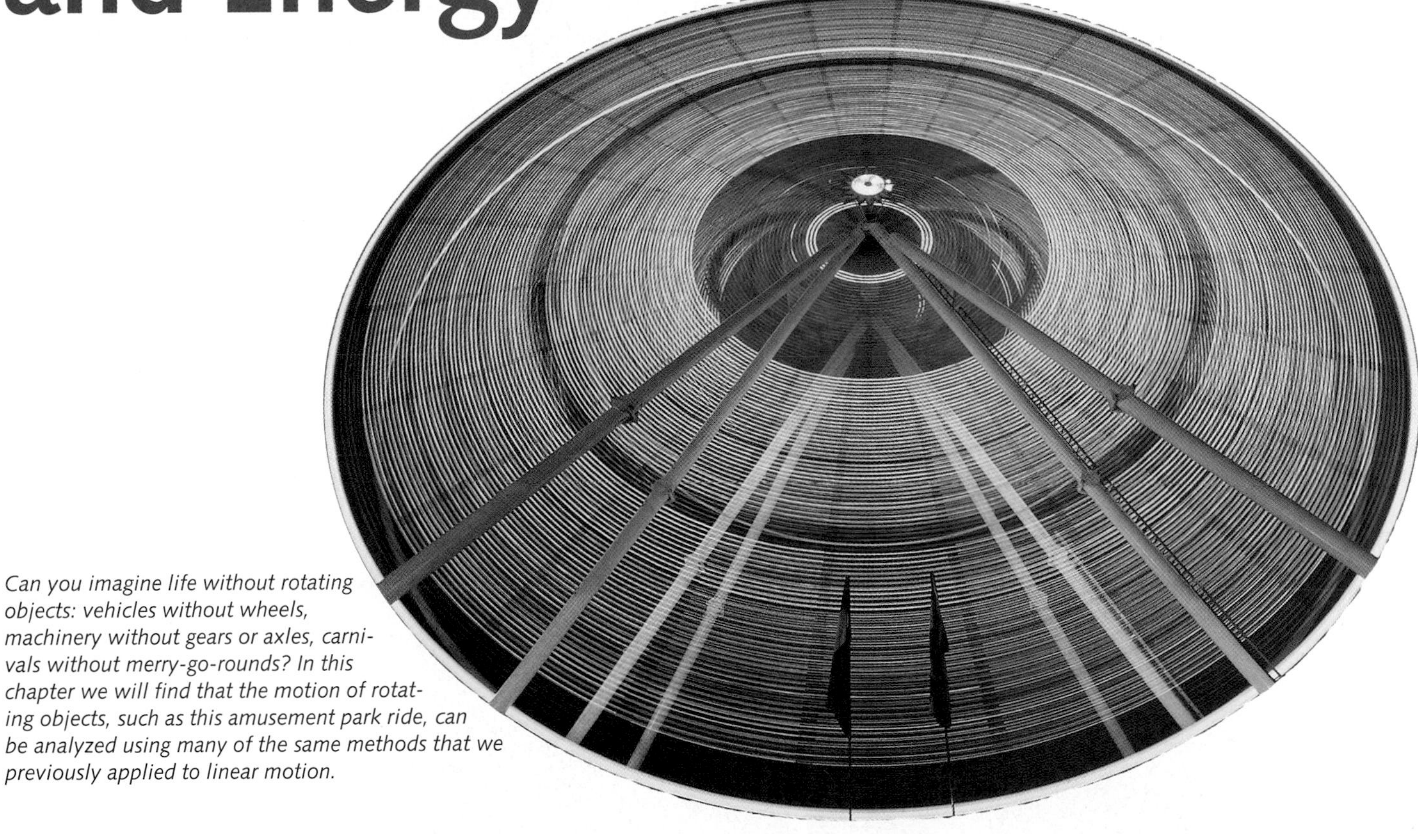

Can you imagine life without rotating objects: vehicles without wheels, machinery without gears or axles, carnivals without merry-go-rounds? In this chapter we will find that the motion of rotating objects, such as this amusement park ride, can be analyzed using many of the same methods that we previously applied to linear motion.

It is certainly no exaggeration to say that rotation is a part of everyday life. After all, we live on a planet that rotates about its axis once a day and that revolves about the Sun once a year. The apparent motion of the Sun across the sky, for example, is actually the result of the Earth's rotational motion. In addition, engines that power cars and trucks have moving parts that rotate quite rapidly, as do CDs, CD-ROMs, and DVDs, not to mention the tumbling, rotating molecules in the air we breathe. Thus, a study of rotation yields results that apply to a great variety of natural phenomena.

In this chapter, then, we study various aspects of rotational motion. As we do, we shall make extensive use of the close analogies that exist between rotational and linear motion. In fact, many of the results derived in earlier chapters can be applied to rotation by simply replacing linear quantities with their rotational counterparts.

10–1 Angular Position, Velocity, and Acceleration

To describe the motion of an object moving in a straight line, it is useful to establish a coordinate system with a definite origin and positive direction. In terms of this coordinate system we can measure the object's position, velocity, and acceleration.

Similarly, to describe rotational motion we define "angular" quantities that are analogous to the linear position, velocity, and acceleration. These angular quantities form the basis of our study of rotation. We begin by defining the most basic angular quantity—the angular position.

Angular Position, θ

Consider a bicycle wheel that is free to rotate about its axle, as shown in Figure 10–11. Suppose that there is a small spot of red paint on the tire, and we want to describe the rotational motion of the spot. The **angular position** of the spot is defined to be the angle, θ, that a line from the axle to the spot makes with a reference line, as indicated in Figure 10–1.

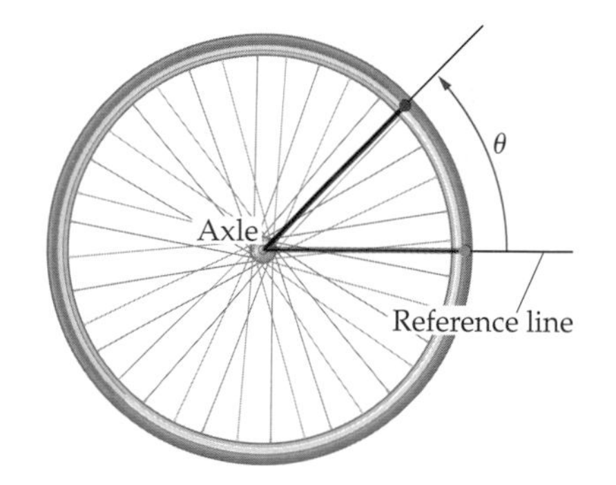

▲ **FIGURE 10–1 Angular position**
The angular position, θ, of a spot of paint on a bicycle wheel. The reference line, where $\theta = 0$, is drawn horizontal here but can be chosen in any direction.

Definition of Angular Position, θ

$$\theta = \text{angle measured from reference line} \qquad \text{10–1}$$

SI unit: radian (rad), which is dimensionless

The reference line simply defines $\theta = 0$; it is analogous to the origin in a linear coordinate system. The reference line begins at the axis of rotation, and may be chosen to point in any direction—just as an origin may be placed anywhere along a coordinate axis. Once chosen, however, the reference line must be used consistently.

Note that the spot of paint in Figure 10–1 is rotated counterclockwise from the reference line by the angle θ. By convention, we say that this angle is positive. Similarly, clockwise rotations correspond to negative angles.

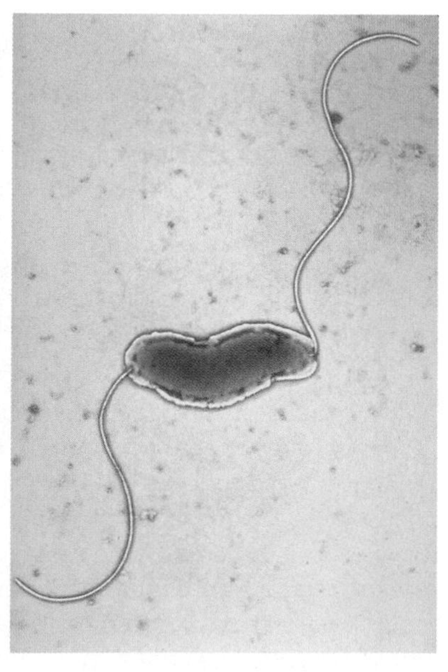

▲ Rotational motion is everywhere in our universe, on every scale of length and time. A galaxy like the one at left may take millions of years to complete a single rotation about its center, while the skater in the middle photo spins several times in a second. The bacterium at right moves in a corkscrew path by rapidly twirling its flagella (the fine projections at either end of the cell) like whips.

Sign Convention for Angular Position

By convention:

$\theta > 0$ counterclockwise rotation from reference line

$\theta < 0$ clockwise rotation from reference line

Now that we have established a reference line (for $\theta = 0$), and a positive direction of rotation (counterclockwise), we must choose units in which to measure angles. Common units are degrees (°) and revolutions (rev), where one revolution—that is, going completely around a circle—corresponds to 360°:

$$1 \text{ rev} = 360°$$

The most convenient units for scientific calculations, however, are radians. A **radian** (rad) is defined as follows:

> A radian is the angle for which the arc length on a circle of radius r is equal to the radius of the circle.

This definition is useful because it establishes a particularly simple relationship between an angle measured in radians and the corresponding arc length, as illustrated in **Figure 10–2**. For example, it follows from our definition that for an angle of one radian, the arc length s is equal to the radius; $s = r$. Similarly, an angle of two radians corresponds to an arc length of two radii, $s = 2r$, and so on. Thus, the arc length s for an arbitrary angle θ measured in radians is given by the following relation:

$$s = r\theta \qquad \textbf{10–2}$$

This simple and straightforward relation does not hold for degrees or revolutions—additional conversion factors would be needed.

In one complete revolution, the arc length is the circumference of a circle, $C = 2\pi r$. Comparing with $s = r\theta$, we see that a complete revolution corresponds to 2π radians:

$$1 \text{ rev} = 360° = 2\pi \text{ rad}$$

Equivalently,

$$1 \text{ rad} = 57.3°$$

One final note on the units for angles: Radians, as well as degrees and revolutions, are dimensionless. In the relation $s = r\theta$, for example, the arc length and the radius both have SI units of meters. For the equation to be dimensionally consistent, it is necessary that θ have no dimensions. Still, if an angle θ is, let's say, three radians, we will write it as $\theta = 3$ rad to remind us of the angular units being used.

Angular Velocity, ω

As the bicycle wheel in Figure 10–1 rotates, the angular position of the spot of red paint changes. This is illustrated in **Figure 10–3**. The angular displacement of the spot, $\Delta\theta$, is

$$\Delta\theta = \theta_f - \theta_i$$

If we divide the angular displacement by the time, Δt, during which the displacement occurs, the result is the **average angular velocity**, ω_{av}.

Definition of Average Angular Velocity, ω_{av}

$$\omega_{av} = \frac{\Delta\theta}{\Delta t} \qquad \textbf{10–3}$$

SI unit: radian per second (rad/s) = s^{-1}

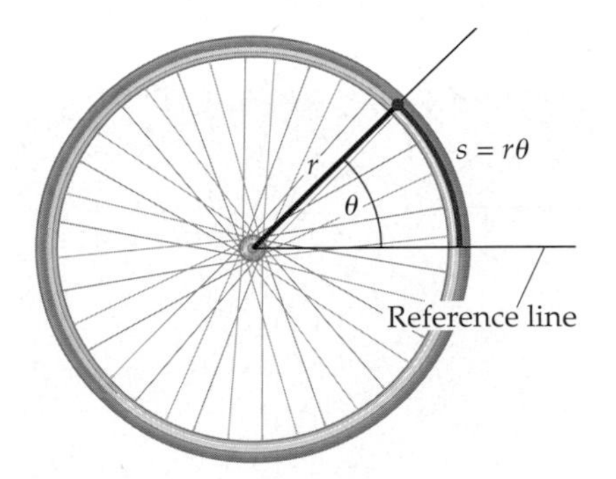

▲ **FIGURE 10–2 Arc length**
The arc length, s, from the reference line to the spot of paint is given by $s = r\theta$, if the angular position θ is measured in radians.

PROBLEM SOLVING NOTE
Radians

Remember to measure angles in radians when using the relation $s = r\theta$.

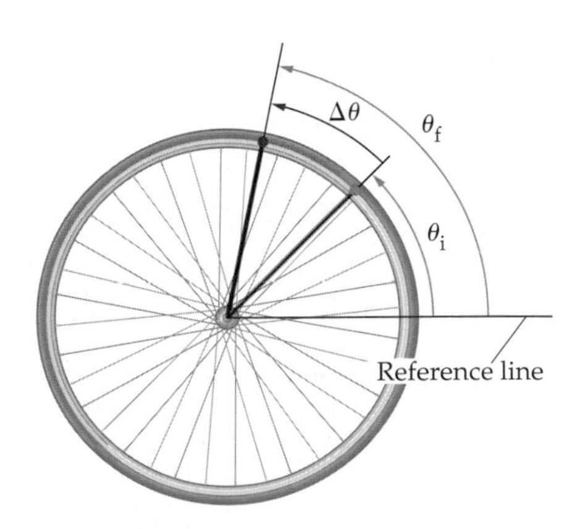

▲ **FIGURE 10–3 Angular displacement**
As the wheel rotates, the spot of paint undergoes an angular displacement, $\Delta\theta = \theta_f - \theta_i$.

▲ Star trails provide a clear illustration of the relationship between angle, arc, and radius in circular motion. The stars, of course, do not actually move like this, but because of the Earth's rotation they appear to follow circular paths across the sky each night, with Polaris, the North Star, very near the axis of rotation. This photo was made by opening the camera shutter for an extended period of time. Notice that each star moves through the same angle in the course of the exposure. However, the farther a star is from the axis of rotation, the longer the arc it traces out in a given period of time. (Can you estimate the length of the exposure?)

This is analogous to the definition of the average linear velocity $v_{av} = \Delta x/\Delta t$. Note that the units of linear velocity are m/s, whereas the units of angular velocity are rad/s.

In addition to the average angular velocity, we can define an **instantaneous angular velocity** as the limit of ω_{av} as the time interval, Δt, approaches zero. The instantaneous angular velocity, then, is

Definition of Instantaneous Angular Velocity, ω

$$\omega = \lim_{\Delta t \to 0} \frac{\Delta\theta}{\Delta t} \qquad \text{10–4}$$

SI unit: $\text{rad/s} = \text{s}^{-1}$

Generally, we shall refer to the instantaneous angular velocity simply as the angular velocity.

Note that we call ω the angular velocity, not the angular speed. The reason is that ω can be positive or negative, depending on the sense of rotation. For example, if the red paint spot rotates in the counterclockwise sense, the angular position, θ, increases. As a result, $\Delta\theta$ is positive and therefore, so is ω. Similarly, clockwise rotation corresponds to a negative $\Delta\theta$ and hence a negative ω.

Sign Convention for Angular Velocity

By convention:

$\omega > 0$ counterclockwise rotation

$\omega < 0$ clockwise rotation

The sign convention for angular velocity is illustrated in **Figure 10–4.** In analogy with linear motion, the magnitude of the angular velocity is the **angular speed**.

In the following exercise we utilize the definitions and conversion factors presented so far in this section.

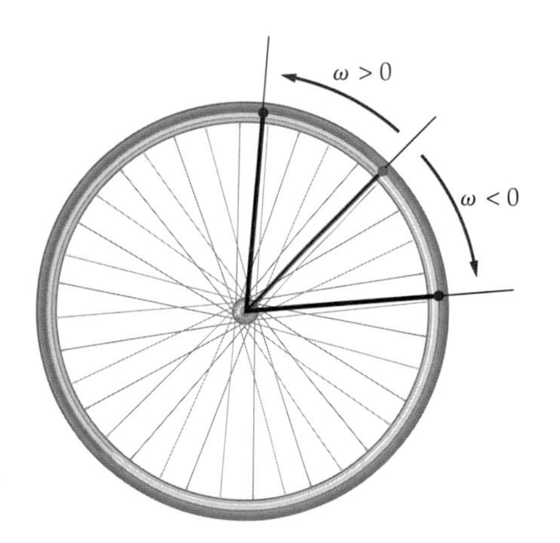

▲ **FIGURE 10–4 Angular speed and velocity**

Counterclockwise rotation is defined to correspond to a positive angular velocity, ω. Similarly, clockwise rotation corresponds to a negative angular velocity. The magnitude of the angular velocity is referred to as the angular speed.

EXERCISE 10–1

(a) An old phonograph record rotates clockwise at $33\frac{1}{3}$ rpm (revolutions per minute). What is its angular velocity in rad/s? **(b)** If a CD rotates at 22.0 rad/s, what is its angular speed in rpm?

Solution

(a) Convert from rpm to rad/s, and note that clockwise rotation corresponds to a negative angular velocity:

$$\omega = -33\tfrac{1}{3}\text{ rpm} = \left(-33\tfrac{1}{3}\frac{\text{rev}}{\text{min}}\right)\left(\frac{2\pi\text{ rad}}{1\text{ rev}}\right)\left(\frac{1\text{ min}}{60\text{ s}}\right) = -3.49\text{ rad/s}$$

(b) Converting angular speed from rad/s to rpm gives

$$\omega = \left(22.0\frac{\text{rad}}{\text{s}}\right)\left(\frac{1\text{ rev}}{2\pi\text{ rad}}\right)\left(\frac{60\text{ s}}{1\text{ min}}\right) = 210\frac{\text{rev}}{\text{min}} = 210\text{ rpm}$$

Note that the same symbol, ω, is used for both angular velocity and angular speed in the Exercise. Which quantity is meant in a given situation will be clear from the context in which it is used.

As a simple application of angular velocity, consider the following question: An object rotates with a constant angular velocity, ω. How much time, T, is required for it to complete one full revolution?

To solve this problem, note that since ω is constant, the instantaneous angular velocity is equal to the average angular velocity. That is,

$$\omega = \omega_{av} = \frac{\Delta\theta}{\Delta t}$$

In one revolution, we know that $\Delta\theta = 2\pi$ and $\Delta t = T$. Therefore,

$$\omega = \frac{\Delta\theta}{\Delta t} = \frac{2\pi}{T}$$

Finally, solving for T we find

$$T = \frac{2\pi}{\omega}$$

The time to complete one revolution, T, is referred to as the **period**.

Definition of Period, T

$$T = \frac{2\pi}{\omega} \qquad \textbf{10–5}$$

SI unit: second, s

EXERCISE 10–2

Find the period of a record that is rotating at 45 rpm.

Solution

To apply $T = 2\pi/\omega$ we must first express ω in terms of rad/s:

$$45 \text{ rpm} = \left(45\frac{\text{rev}}{\text{min}}\right)\left(\frac{2\pi \text{ rad}}{1 \text{ rev}}\right)\left(\frac{1 \text{ min}}{60 \text{ s}}\right) = 4.7 \text{ rad/s}$$

Now we can calculate the period:

$$T = \frac{2\pi}{\omega} = \frac{2\pi \text{ rad}}{4.7 \text{ rad/s}} = 1.3 \text{ s}$$

Angular Acceleration, α

If the angular velocity of the rotating bicycle wheel increases or decreases with time, we say that the wheel experiences an **angular acceleration**, α. The average angular acceleration is the change in angular velocity, $\Delta\omega$, in a given interval of time, Δt:

Definition of Average Angular Acceleration, α_{av}

$$\alpha_{av} = \frac{\Delta\omega}{\Delta t} \qquad \textbf{10–6}$$

SI unit: radian per second per second $(\text{rad/s}^2) = \text{s}^{-2}$

Note that the SI units of α are rad/s^2, which, since rad is dimensionless, is simply s^{-2}.

As expected, the instantaneous angular acceleration is the limit of α_{av} as the time interval, Δt, approaches zero:

Definition of Instantaneous Angular Acceleration, α

$$\alpha = \lim_{\Delta t \to 0} \frac{\Delta\omega}{\Delta t} \qquad \textbf{10–7}$$

SI unit: $\text{rad/s}^2 = \text{s}^{-2}$

When referring to the instantaneous angular acceleration, we will usually just say angular acceleration.

The sign of the angular acceleration is determined by whether the change in angular velocity is positive or negative. For example, if ω is becoming more positive, so that ω_f is greater than ω_i, it follows that α is positive. Similarly, if ω is becoming

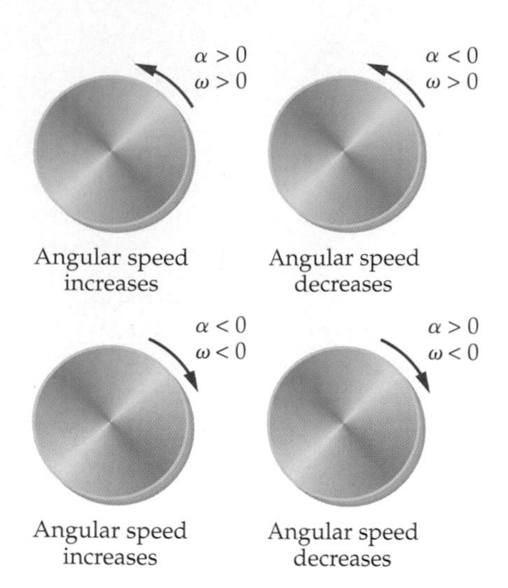

▲ **FIGURE 10–5 Angular acceleration**
When angular velocity and acceleration have the same sign, the angular speed increases. When angular velocity and angular acceleration have opposite signs, the angular speed decreases.

more negative, so that ω_f is less than ω_i, it follows that α is negative. Therefore, if ω and α have the same sign, the speed of rotation is increasing. If ω and α have opposite signs, the speed of rotation is decreasing. This is illustrated in **Figure 10–5**.

EXERCISE 10–3

As the wind dies, a windmill that was rotating at 2.1 rad/s begins to slow down with a constant angular acceleration of 0.45 rad/s^2. How long does it take for the windmill to come to a complete stop?

Solution

If we choose the initial angular velocity to be positive, the angular acceleration is negative, corresponding to a deceleration. Hence, Equation 10–6 gives

$$\Delta t = \frac{\Delta\omega}{\alpha_{av}} = \frac{\omega_f - \omega_i}{\alpha} = \frac{0 - 2.1 \text{ rad/s}}{-0.45 \text{ rad/s}^2} = 4.7 \text{ s}$$

10–2 Rotational Kinematics

Just as the kinematics of Chapter 2 described linear motion, rotational kinematics describes rotational motion. In this section, as in Chapter 2, we concentrate on the important special case of constant acceleration.

As an example of a system with constant angular acceleration, consider the pulley shown in **Figure 10–6**. Wrapped around the circumference of the pulley is a string, with a mass attached to its free end. When the mass is released, the pulley begins to rotate—slowly at first, but then faster and faster. As we shall see in Chapter 11, the pulley is accelerating with constant angular acceleration.

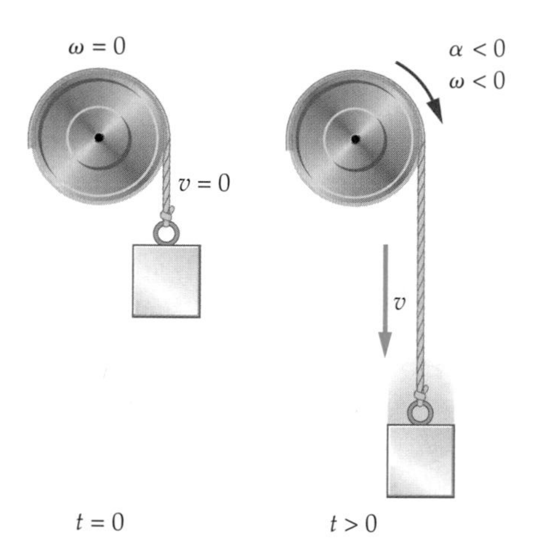

▲ **FIGURE 10–6 A pulley with constant angular acceleration**
A mass is attached to a string wrapped around a pulley. As the mass falls, it causes the pulley to increase its angular speed with a constant angular acceleration.

Since α is constant, it follows that the average and instantaneous angular accelerations are equal. Hence,

$$\alpha = \alpha_{av} = \frac{\Delta\omega}{\Delta t}$$

Suppose the pulley starts with the initial angular velocity ω_0 at time $t = 0$, and that at the later time t its angular velocity is ω. Substituting these values into the preceding expression for α yields

$$\alpha = \frac{\Delta\omega}{\Delta t} = \frac{\omega - \omega_0}{t - 0} = \frac{\omega - \omega_0}{t}$$

Rearranging, we see that the angular velocity, ω, varies with time as follows:

$$\omega = \omega_0 + \alpha t \qquad \textbf{10–8}$$

EXERCISE 10–4

If the angular velocity of the pulley in Figure 10–6 is -8.4 rad/s at a given time, and its angular acceleration is -2.8 rad/s^2, what is the angular velocity of the pulley 1.5 s later?

Solution

The angular velocity, ω, is found by applying Equation 10–8:

$$\omega = \omega_0 + \alpha t = -8.4 \text{ rad/s} + (-2.8 \text{ rad/s}^2)(1.5 \text{ s}) = -13 \text{ rad/s}$$

Note that the angular speed has increased, as expected, since ω and α have the same sign.

Note the close analogy between Equation 10–8 for angular velocity and the corresponding relation for linear velocity, Equation 2–7:

$$v = v_0 + at$$

Clearly, the equation for angular velocity can be obtained from our previous equation for linear velocity by replacing v with ω and replacing a with α. This type of analogy between linear and angular quantities can be most useful both in deriving angular equations—by starting with linear equations and using analogies—and in obtaining a better physical understanding of angular systems. Several linear-to-angular analogs are listed in the following table.

Linear Quantity	Angular Quantity
x	θ
v	ω
a	α

Using these analogies, we can rewrite all the kinematic equations in Chapter 2 in angular form. The following Table gives both the linear kinematic equations and their angular counterparts.

Linear Equation (a = constant)		Angular Equation (α = constant)	
$v = v_0 + at$	2–7	$\omega = \omega_0 + \alpha t$	10–8
$x = x_0 + \frac{1}{2}(v_0 + v)t$	2–10	$\theta = \theta_0 + \frac{1}{2}(\omega_0 + \omega)t$	10–9
$x = x_0 + v_0 t + \frac{1}{2}at^2$	2–11	$\theta = \theta_0 + \omega_0 t + \frac{1}{2}\alpha t^2$	10–10
$v^2 = v_0^2 + 2a(x - x_0)$	2–12	$\omega^2 = \omega_0^2 + 2\alpha(\theta - \theta_0)$	10–11

In solving kinematic problems involving rotation, we apply these angular equations in the same way that the linear equations were applied in Chapter 2. In a sense, then, this material is a review—since the mathematics is essentially the same. The only difference comes in the physical interpretation of the results. We will emphasize the rotational interpretations throughout the chapter.

Using analogies between linear and angular quantities often helps when solving problems involving rotational kinematics.

EXAMPLE 10–1 Thrown for a Curve

To throw a curve ball, a pitcher gives the ball an initial angular speed of 36.0 rad/s. When the catcher gloves the ball 0.595 s later, its angular speed has decreased (due to air resistance) to 34.2 rad/s. **(a)** What is the ball's angular acceleration, assuming it to be constant? **(b)** How many revolutions does the ball make before being caught?

Picture the Problem

We choose the ball's initial direction of rotation to be positive. As a result, the angular acceleration will be negative. We can also identify the initial angular velocity to be $\omega_0 = 36.0$ rad/s, and the final angular velocity to be $\omega = 34.2$ rad/s.

Strategy

(a) To relate angular velocity to time, we use $\omega = \omega_0 + \alpha t$. This can be solved for α.

(b) To relate angle to time we use $\theta = \theta_0 + \omega_0 t + \frac{1}{2}\alpha t^2$. The angular displacement of the ball is $\theta - \theta_0$.

Solution

Part (a)

1. Solve $\omega = \omega_0 + \alpha t$ for the angular acceleration, α:

$$\omega = \omega_0 + \alpha t$$
$$\alpha = \frac{\omega - \omega_0}{t}$$

2. Substitute numerical values to find α:

$$\alpha = \frac{\omega - \omega_0}{t} = \frac{34.2 \text{ rad/s} - 36.0 \text{ rad/s}}{0.595 \text{ s}} = -3.03 \text{ rad/s}^2$$

continued on the following page

continued from previous page

Part (b)

3. Use $\theta = \theta_0 + \omega_0 t + \frac{1}{2}\alpha t^2$ to calculate the angular displacement of the ball:

$$\theta - \theta_0 = \omega_0 t + \tfrac{1}{2}\alpha t^2$$
$$= (36.0 \text{ rad/s})(0.595 \text{ s}) + \tfrac{1}{2}(-3.03 \text{ rad/s}^2)(0.595 \text{ s})^2$$
$$= 20.9 \text{ rad}$$

4. Convert the angular displacement to revolutions:

$$\theta - \theta_0 = 20.9 \text{ rad} = 20.9 \text{ rad}\left(\frac{1 \text{ rev}}{2\pi \text{ rad}}\right) = 3.33 \text{ rev}$$

Insight

The ball rotates through three-and-one-third revolutions during its time in flight.

Practice Problem

(a) What is the angular velocity of the ball 0.500 s after it is thrown? (b) What is the ball's angular velocity after it completes its first full revolution? [**Answer:** (a) Use $\omega = \omega_0 + \alpha t$ to find $\omega = 34.5$ rad/s. (b) Use $\omega^2 = \omega_0^2 + 2\alpha(\theta - \theta_0)$ to find $\omega = 35.5$ rad/s.]

Some related homework problems: Problem 15, Problem 18

EXAMPLE 10–2 Wheel of Misfortune

On a certain game show, contestants spin a wheel when it is their turn. One contestant gives the wheel an initial angular speed of 3.40 rad/s. It then rotates through one-and-one-quarter revolutions and comes to rest on the BANKRUPT space. **(a)** Find the angular acceleration of the wheel, assuming it to be constant. **(b)** How long does it take for the wheel to come to rest?

Picture the Problem

We choose the initial angular velocity to be positive, and indicate it with a counterclockwise rotation in our sketch. Since the wheel slows to a stop, the angular acceleration must be negative; that is, in the clockwise direction. After a rotation of 1.25 rev the wheel will read BANKRUPT.

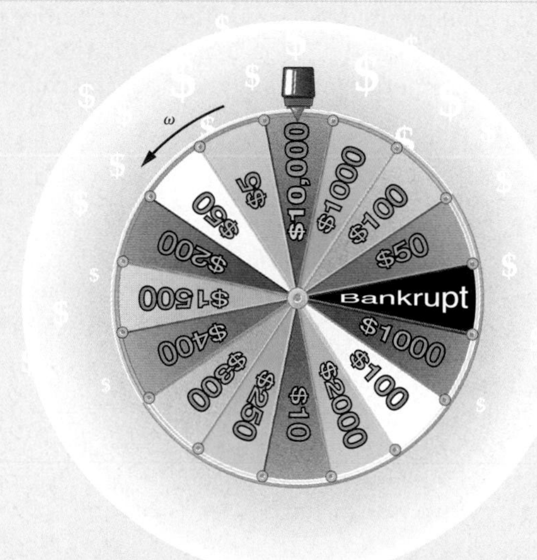

Strategy

(a) We are given the initial angular velocity, $\omega_0 = +3.40$ rad/s, the final angular velocity, $\omega = 0$ (the wheel comes to rest), and the angular displacement, $\theta - \theta_0 = 1.25$ rev. We can find the angular acceleration using $\omega^2 = \omega_0^2 + 2\alpha(\theta - \theta_0)$.

(b) Knowing the angular velocity and acceleration, we can find the time with $\omega = \omega_0 + \alpha t$.

Solution

Part (a)

1. Solve $\omega^2 = \omega_0^2 + 2\alpha(\theta - \theta_0)$ for the angular acceleration, α:

$$\omega^2 = \omega_0^2 + 2\alpha(\theta - \theta_0)$$
$$\alpha = \frac{\omega^2 - \omega_0^2}{2(\theta - \theta_0)}$$

2. Convert $\theta - \theta_0 = 1.25$ rev to radians:

$$\theta - \theta_0 = 1.25 \text{ rev} = 1.25 \text{ rev}\left(\frac{2\pi \text{ rad}}{1 \text{ rev}}\right) = 7.85 \text{ rad}$$

3. Substitute numerical values to find α:

$$\alpha = \frac{\omega^2 - \omega_0^2}{2(\theta - \theta_0)} = \frac{0 - (3.40 \text{ rad/s})^2}{2(7.85 \text{ rad})} = -0.736 \text{ rad/s}^2$$

Part (b)

4. Solve $\omega = \omega_0 + \alpha t$ for the time, t:

$$\omega = \omega_0 + \alpha t$$
$$t = \frac{\omega - \omega_0}{\alpha}$$

5. Substitute numerical values to find t:

$$t = \frac{\omega - \omega_0}{\alpha} = \frac{0 - 3.40\text{ rad/s}}{(-0.736\text{ rad/s}^2)} = 4.62\text{ s}$$

Insight

Note that it was not necessary to define a reference line; that is, a direction for $\theta = 0$. All we need to know is the angular displacement, $\theta - \theta_0$, not the individual angles θ and θ_0.

Practice Problem

What is the angular speed of the wheel after one complete revolution? [**Answer:** $\omega = 1.52$ rad/s]

Some related homework problems: Problem 14, Problem 16

Finally, we consider a pulley that is rotating in such a way that initially it is lifting a mass with speed v. Gravity acting on the mass causes it and the pulley to slow and momentarily come to rest.

ACTIVE EXAMPLE 10–1 Time To Rest

A pulley rotating in the counterclockwise direction is attached to a mass suspended from a string. The mass causes the pulley's angular velocity to decrease with a constant angular acceleration $\alpha = -2.10\text{ rad/s}^2$. **(a)** If the pulley's initial angular velocity is $\omega_0 = 5.40$ rad/s, how long does it take for the pulley to come to rest? **(b)** Through what angle does the pulley turn during this time?

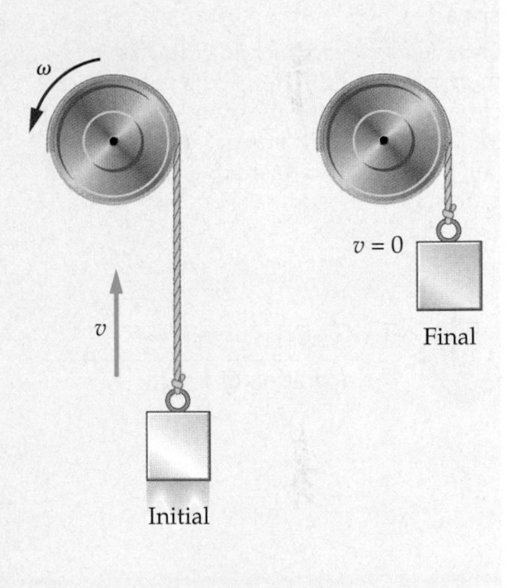

Solution

1. **(a)** Relate angular velocity to time: $\omega = \omega_0 + \alpha t$
2. Solve for the time, t: $t = (\omega - \omega_0)/\alpha$
3. Substitute numerical values: $t = 2.57$ s
4. **(b)** Use $\theta = \theta_0 + \omega_0 t + \frac{1}{2}\alpha t^2$ to solve for $\theta - \theta_0$: $\theta - \theta_0 = \omega_0 t + \frac{1}{2}\alpha t^2 = 6.94$ rad
5. Alternatively, use $\omega^2 = \omega_0{}^2 + 2\alpha(\theta - \theta_0)$: $\theta - \theta_0 = (\omega^2 - \omega_0^2)/2\alpha = 6.94$ rad

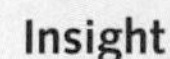

Insight

After the pulley comes to rest, it immediately begins to rotate in the clockwise direction as the mass falls. The pulley's angular *acceleration* is constant—it has the same value before the pulley stops, when it stops, and after it begins rotating in the opposite direction.

10–3 Connections Between Linear and Rotational Quantities

At a local county fair a child rides on a merry-go-round. The ride completes one circuit every $T = 7.50$ s. Therefore, the angular velocity of the child, from Equation 10–5, is

$$\omega = \frac{2\pi}{T} = \frac{2\pi}{7.50\text{ s}} = 0.838\text{ rad/s}$$

The path followed by the child is circular, with the center of the circle at the axis of rotation of the merry-go-round. In addition, at any instant of time the child is moving in a direction that is *tangential* to the circular path, as **Figure 10–7** shows. What is the tangential speed, v_t, of the child? In other words, what is the speed of the wind in the child's face?

We can find the child's tangential speed by dividing the circumference of the circular path, $2\pi r$, by the time required to complete one circuit, T. Thus,

$$v_t = \frac{2\pi r}{T} = r\left(\frac{2\pi}{T}\right)$$

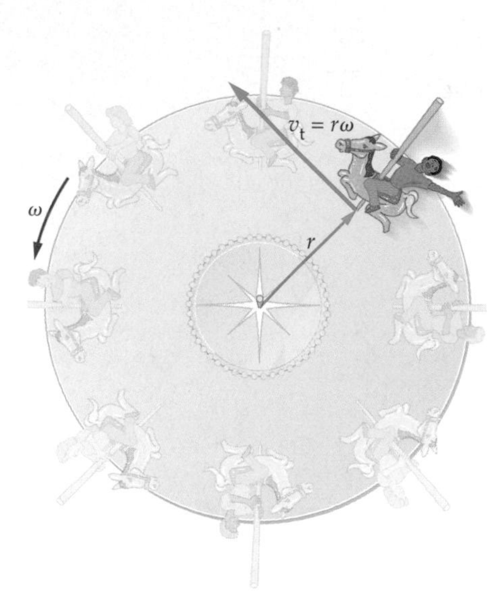

▲ **FIGURE 10–7 Angular and linear speed**
Overhead view of a child riding on a merry-go-round. The child's path is a circle centered on the merry-go-round's axis of rotation. At any given time the child is moving tangential to the circular path with a speed $v_t = r\omega$.

Since $2\pi/T$ is simply ω, we can express the tangential speed as follows:

Tangential Speed of a Rotating Object

$$v_t = r\omega \qquad \text{10–12}$$

SI unit: m/s

Note that ω must be given in rad/s for this relation to be valid.

In the case of the merry-go-round, if the radius of the child's circular path is $r = 4.25$ m, the tangential speed is $v_t = r\omega = (4.25 \text{ m})(0.838 \text{ rad/s}) = 3.56$ m/s. When it is clear that we are referring to the tangential speed, we will often drop the subscript t, and simply write $v = r\omega$.

An interesting application of the relation between linear and angular speeds is provided in the operation of a compact disk (CD). As you know, a CD is played by shining a laser beam onto the disk and then converting the reflected light into sound. For proper operation, however, the linear speed of the disk where the laser beam shines on it must be maintained at the constant value of 1.25 m/s. As the CD is played, the laser beam scans the disk in a spiral track from near the center outward to the rim. In order to maintain the required linear speed, the angular speed of the disk must decrease as the beam scans outward. The required angular speeds are determined in the following Exercise.

EXERCISE 10–5

Find the angular speed a CD must have to give a linear speed of 1.25 m/s when the laser beam shines on the disk **(a)** 2.50 cm and **(b)** 6.00 cm from its center.

Solution

(a) Using $v = 1.25$ m/s and $r = 0.0250$ m in Equation 10–12, we find

$$\omega = \frac{v}{r} = \frac{1.25 \text{ m/s}}{0.0250 \text{ m}} = 50.0 \text{ rad/s} = 477 \text{ rpm}$$

(b) Similarly, with $r = 0.0600$ m we find

$$\omega = \frac{v}{r} = \frac{1.25 \text{ m/s}}{0.0600 \text{ m}} = 20.8 \text{ rad/s} = 199 \text{ rpm}$$

Thus, a CD slows from about 500 rpm to roughly 200 rpm as it plays.

How do the angular and tangential speed of an object vary from one point to another? We explore this question in the following Conceptual Checkpoint.

CONCEPTUAL CHECKPOINT 10–1

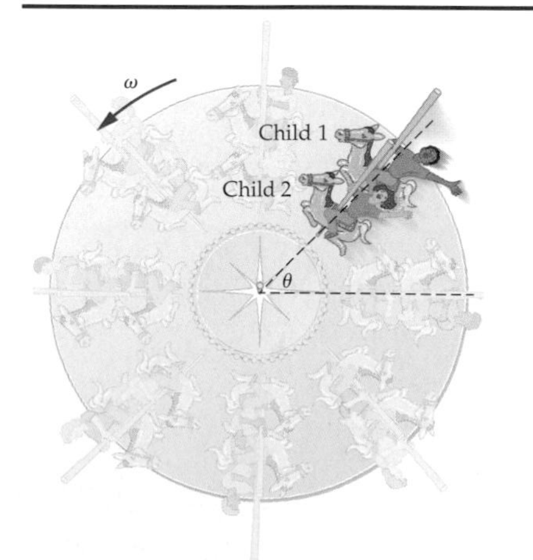

Two children ride on a merry-go-round, with child 1 at a greater distance from the axis of rotation than child 2. Is the angular speed of child 1 **(a)** greater than, **(b)** less than, or **(c)** the same as the angular speed of child 2?

Reasoning and Discussion

At any given time, the angle θ for child 1 is the same as the angle for child 2, as shown. Therefore, when the angle for child 1 has gone through 2π, for example, so has the angle for child 2. As a result, they have the same angular speed. In fact, *each and every point on the merry-go-round has exactly the same angular speed.*

The tangential speeds are different, however. Child 1 has the greater tangential speed since he travels around a larger circle in the same time that child 2 travels around a smaller circle. This is in agreement with $v = r\omega$, since child 1 has the larger radius.

Answer:

(c) The angular speeds are the same.

▲ In the photo at left, two plastic letter "E"s have been placed on a rotating turntable at different distances from the axis of rotation. The stretching and blurring of the image of the outermost letter clearly shows that it is moving faster than the letter closer to the axis. Similarly, the boy near the rim of this playground merry-go-round is moving faster than the girl near the hub.

Since the children on the merry-go-round move in a circular path, they experience a centripetal acceleration, a_{cp} (Section 6–5). The centripetal acceleration is always directed toward the axis of rotation, and has a magnitude given by

$$a_{cp} = \frac{v^2}{r}$$

Since the speed v in this expression is the tangential speed, $v = v_t = r\omega$, the centripetal acceleration in terms of ω is

$$a_{cp} = \frac{(r\omega)^2}{r}$$

Canceling one power of r, we have

Centripetal Acceleration of a Rotating Object

$$a_{cp} = r\omega^2 \qquad \textbf{10–13}$$

SI unit: $\mathrm{m/s^2}$

If the radius of a child's circular path on the merry-go-round is 4.25 m, and the angular speed of the ride is 0.838 rad/s, the centripetal acceleration of the child is $a_{cp} = r\omega^2 = (4.25\ \mathrm{m})(0.838\ \mathrm{rad/s})^2 = 2.98\ \mathrm{m/s^2}$.

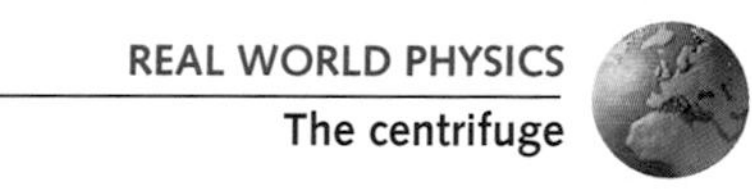

Though the centripetal acceleration of a merry-go-round is typically only a fraction of the acceleration of gravity, rotating devices referred to as **centrifuges** can produce centripetal accelerations many times greater than gravity. For example, the world's most powerful research centrifuge, operated by the U. S. Army Corps of Engineers, can subject 2.2-ton payloads to accelerations as high as 350 g (350 times greater than the acceleration of gravity.) This centrifuge is used to study earthquake engineering and dam erosion. The Air Force uses centrifuges to subject prospective jet pilots to the accelerations they will experience during rapid flight maneuvers, and in the future NASA may even use a human-powered centrifuge for gravity studies aboard the International Space Station.

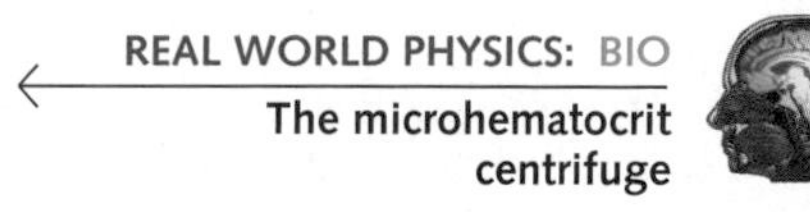

The centrifuges most commonly encountered in everyday life are those found in virtually every medical laboratory in the world. These devices, which can produce centripetal accelerations in excess of 13,000 g, are used to separate blood cells from blood plasma. The ratio of the packed cell volume to the total blood volume gives the *hematocrit value,* which is a useful clinical indicator of blood quality. In the next Example we consider the operation of a *microhematocrit centrifuge,* which measures the hematocrit value of a small (micro) sample of blood.

▲ The large centrifuge shown at left, at the Gagarin Cosmonaut Training Center, is used to train Russian cosmonauts for space missions. This device, which rotates at 36 rpm, can produce a centripetal acceleration of over 290 m/s^2, 30 times the acceleration of gravity. The device at right is a microhematocrit centrifuge, used to separate blood cells from plasma. The volume of red blood cells in a given quantity of whole blood is a major factor in determining the oxygen-carrying capacity of the blood, an important clinical indicator.

EXAMPLE 10–3 The Microhematocrit

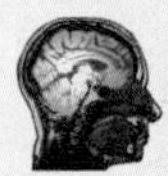

Real World Physics: Bio

In a microhematocrit centrifuge, small samples of blood are placed in heparinized capillary tubes (heparin is an anticoagulant). The tubes are rotated at 11,500 rpm, with the bottom of the tubes 9.07 cm from the axis of rotation. **(a)** Find the linear speed of the bottom of the tubes. **(b)** What is the centripetal acceleration at the bottom of the tubes?

Picture the Problem

Our sketch shows a top view of the centrifuge, with the capillary tubes rotating at 11,500 rpm. Notice that the bottom of the tubes move in a circular path of radius 9.07 cm.

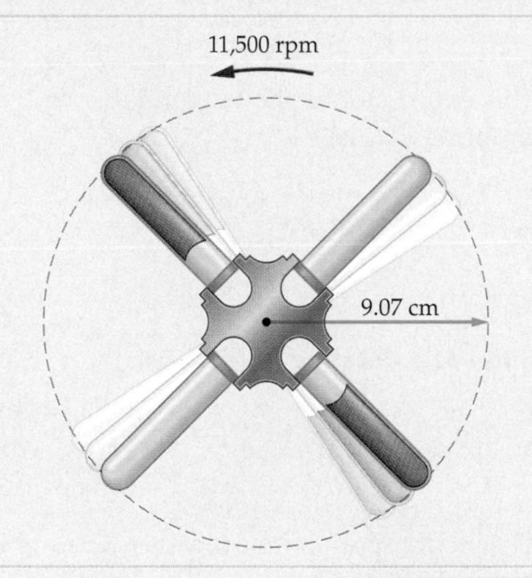

Strategy

(a) Linear and angular speeds are related by $v = r\omega$. Once we convert the angular speed to rad/s we can use this relation to determine v.

(b) The centripetal acceleration is $a_{cp} = r\omega^2$. Using ω from part **(a)** yields the desired result.

Solution

Part (a)

1. Convert the angular speed, ω, to radians per second:

$$\omega = (11{,}500 \text{ rev/min})\left(\frac{2\pi \text{ rad}}{1 \text{ rev}}\right)\left(\frac{1\text{min}}{60 \text{ s}}\right) = 1200 \text{ rad/s}$$

2. Use $v = r\omega$ to calculate the linear speed:

$$v = r\omega = (0.0907 \text{ m})(1200 \text{ rad/s}) = 109 \text{ m/s}$$

Part (b)

3. Calculate the centripetal acceleration using $a_{cp} = r\omega^2$:

$$a_{cp} = r\omega^2 = (0.0907 \text{ m})(1200 \text{ rad/s})^2 = 131{,}000 \text{ m/s}^2$$

4. As a check, calculate the centripetal acceleration using $a_{cp} = v^2/r$:

$$a_{cp} = \frac{v^2}{r} = \frac{(109 \text{ m/s})^2}{0.0907 \text{ m}} = 131{,}000 \text{ m/s}^2$$

Insight

Note that every point on a tube has the same angular speed. As a result, points near the top of a tube have smaller linear speeds and centripetal accelerations than do points near the bottom of a tube. In this case, the bottoms of the tubes experience a centripetal acceleration of 13,400 g.

Practice Problem

What angular speed must this centrifuge have if the centripetal acceleration at the bottom of the tubes is to be 98,100 m/s^2 ($\approx$10,000 g)? [**Answer:** $\omega = \sqrt{a_{cp}/r} = 1040$ rad/s $= 9940$ rpm]

Some related homework problems: Problem 25, Problem 28

If the angular speed of the merry-go-round in Conceptual Checkpoint 10–1 changes, the tangential speed of the children changes as well. It follows, then, that the children will experience a tangential acceleration, a_t. We can determine a_t by considering the relation $v_t = r\omega$. If ω changes by the amount $\Delta\omega$, with r remaining constant, the corresponding change in tangential speed is

$$\Delta v_t = r\Delta\omega$$

If this change in ω occurs in the time Δt, the tangential acceleration is

$$a_t = \frac{\Delta v_t}{\Delta t} = r\frac{\Delta\omega}{\Delta t}$$

Since $\Delta\omega/\Delta t$ is the angular acceleration, α, we find that

Tangential Acceleration of a Rotating Object

$$a_t = r\alpha \qquad \textbf{10–14}$$

SI unit: m/s^2

As with the tangential speed, we will often drop the subscript t in a_t when no confusion will arise.

In general, the children on the merry-go-round may experience both tangential and centripetal accelerations at the same time. Recall that a_t is due to a changing tangential speed, and that a_{cp} is caused by a changing direction of motion, even if the tangential speed remains constant. To summarize:

Tangential Versus Centripetal Acceleration

$a_t = r\alpha$	due to changing speed
$a_{cp} = r\omega^2$	due to changing direction of motion

As an example, suppose an object is rotating with a constant angular acceleration, α. In this case, the tangential acceleration, $a_t = r\alpha$, is constant in magnitude. On the other hand, the centripetal acceleration, $a_{cp} = r\omega^2$, changes with time since the angular speed changes.

In cases in which both the centripetal and tangential accelerations are present, the total acceleration is the vector sum of the two, as indicated in **Figure 10–8**. Note that a_t and a_{cp} are at right angles to one another, hence the total acceleration is given by the Pythagorean theorem:

$$a = \sqrt{a_t^2 + a_{cp}^2}$$

The direction of the total acceleration, measured relative to the tangential direction, is

$$\phi = \tan^{-1}\left(\frac{a_{cp}}{a_t}\right)$$

This angle is shown in Figure 10–8.

▲ FIGURE 10–8 Centripetal and tangential acceleration
If the angular speed of the merry-go-round is increased, the child will experience two accelerations: (i) a tangential acceleration, $\mathbf{a}_t$, and (ii) a centripetal acceleration, $\mathbf{a}_{cp}$. The child's total acceleration, $\mathbf{a}$, is the vector sum of $\mathbf{a}_t$ and $\mathbf{a}_{cp}$.

ACTIVE EXAMPLE 10–2 The Centrifuge Accelerates

Suppose the centrifuge in Example 10–3 is just starting up, and that it has an angular speed of 8.00 rad/s and an angular acceleration of 95.0 rad/s^2. **(a)** What is the magnitude of the centripetal, tangential, and total acceleration of the bottom of a tube? **(b)** What angle does the total acceleration make with the direction of motion?

continued on the following page

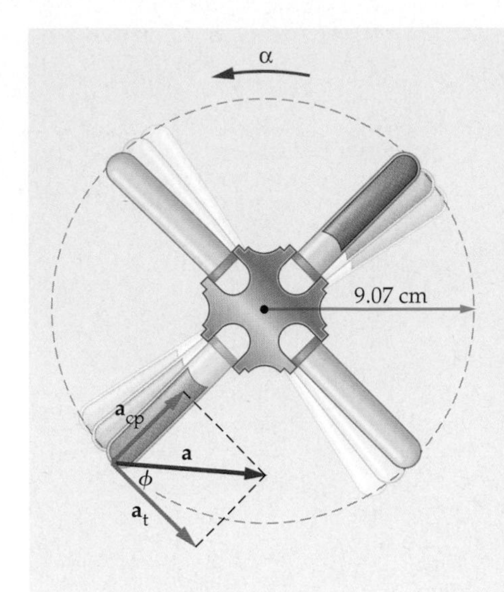

continued from the previous page

Solution

Part (a)

1. Calculate the centripetal acceleration: $a_{\mathrm{cp}} = r\omega^2 = 5.80 \text{ m/s}^2$
2. Calculate the tangential acceleration: $a_{\mathrm{t}} = r\alpha = 8.62 \text{ m/s}^2$
3. Find the magnitude of the total acceleration: $a = \sqrt{a_{\mathrm{cp}}^2 + a_{\mathrm{t}}^2} = 10.4 \text{ m/s}^2$

Part (b)

4. Find the angle ϕ for the total acceleration: $\phi = \tan^{-1}(a_{\mathrm{cp}}/a_{\mathrm{t}}) = 33.9°$

Insight

Every point on a tube has the same angular speed *and* the same angular acceleration.

10–4 Rolling Motion

We began this chapter with a bicycle wheel rotating about its axle. In that case, the axle was at rest and every point on the wheel, such as the spot of red paint, moved in a circular path about the axle. We would like to consider a different situation now. Suppose the bicycle wheel is rolling freely, as indicated in **Figure 10–9**, with no slipping between the tire and the ground. The wheel still rotates about the axle, but the axle itself is moving in a straight line. As a result, the motion of the wheel is a combination of both rotational motion and linear (or **translational**) motion.

▶ **FIGURE 10–9 Rolling without slipping**

A wheel of radius r rolling without slipping. During one complete revolution, the center of the wheel moves forward through a distance $2\pi r$.

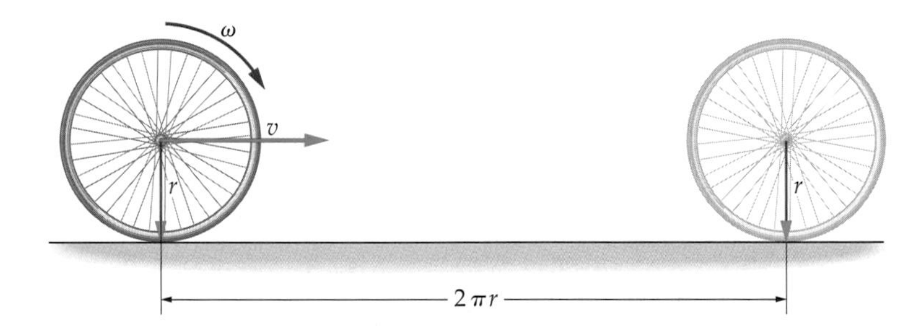

To see the connection between the wheel's rotational and translational motions, we show one full rotation of the wheel in Figure 10–9. During this rotation, the axle translates forward through a distance equal to the circumference of the wheel, $2\pi r$. Since the time required for one rotation is the period, T, the translational speed of the axle is

$$v = \frac{2\pi r}{T}$$

Recalling that $\omega = 2\pi/T$, we find

$$v = r\omega = v_{\mathrm{t}} \qquad \textbf{10–15}$$

Hence, the translational speed of the axle is equal to the tangential speed of a point on the rim of a wheel spinning with angular speed ω.

A rolling object, then, combines rotational motion with angular speed ω, and translation motion with linear speed $v = r\omega$, where r is the radius of the object. Let's consider these two motions one at a time. First, in **Figure 10–10 (a)** we show pure rotational motion with angular speed ω. In this case, the axle is at rest, and points at

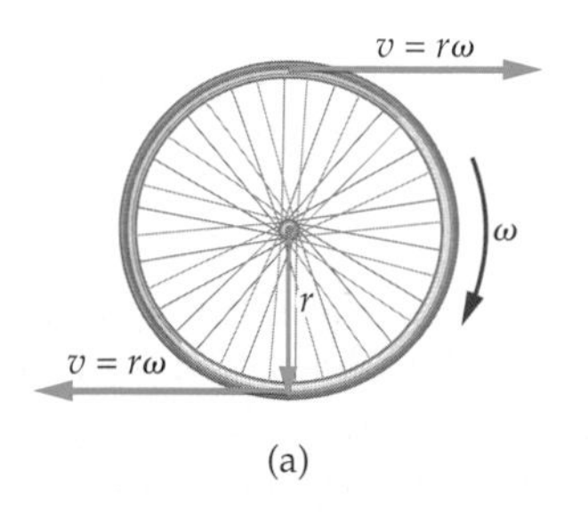

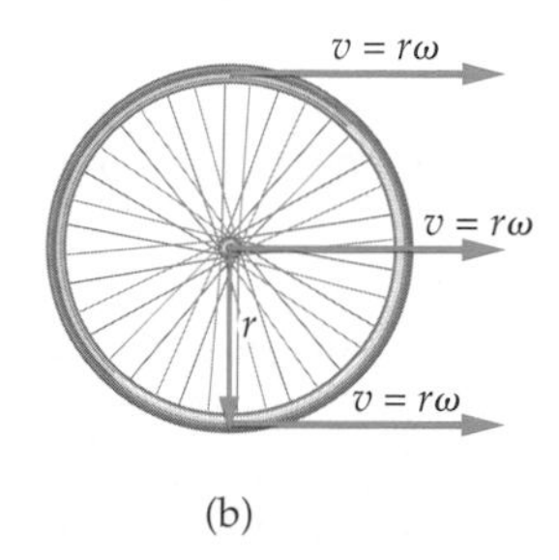

◀ **FIGURE 10–10 Rotational and translational motions of a wheel**

(a) In pure rotational motion, the velocities at the top and bottom of the wheel are in opposite directions. **(b)** In pure translational motion, each point on the wheel moves with the same speed in the same direction.

the top and bottom of the wheel have tangential velocities that are equal in magnitude, $v = r\omega$, but point in opposite directions.

Next, we consider translational motion with speed $v = r\omega$. This is illustrated in **Figure 10–10 (b)**, where we see that each point on the wheel moves in the same direction with the same speed. If this were the only motion the wheel had, it would be skidding across the ground, instead of rolling without slipping.

Finally, we combine these two motions by simply adding the velocity vectors in Figures 10–10 (a) and (b). The result is shown in **Figure 10–11**. At the top of the wheel the two velocity vectors are in the same direction, so they sum to give a speed of $2v$. At the axle, the velocity vectors sum to give a speed v. Finally, at the bottom of the wheel, the velocity vectors from rotation and translation have equal magnitude, but are in opposite directions. As a result, these velocities cancel, giving a speed of zero where the wheel is in contact with the ground.

The fact that the bottom of the wheel is instantaneously at rest, so that it is in static contact with the ground, is precisely what is meant by "rolling without slipping." Thus, a wheel that rolls without slipping is just like the situation when you are walking—even though your body as a whole moves forward, the soles of your shoes are momentarily at rest every time you place them on the ground. This point was discussed in detail in Conceptual Checkpoint 6–1.

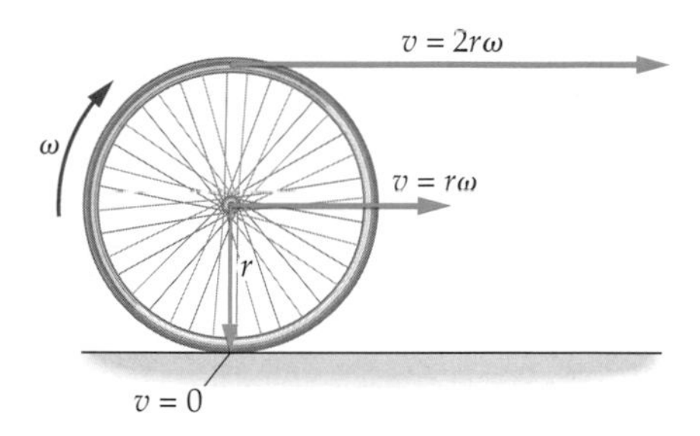

▲ **FIGURE 10–11 Velocities in rolling motion**

In a wheel that rolls without slipping, the point in contact with the ground is instantaneously at rest. The center of the wheel moves forward with the speed $v = r\omega$, and the top of the wheel moves forward with twice that speed, $v = 2r\omega$.

EXERCISE 10–6

A car with tires of radius 32 cm drives on the highway at 55 mph. **(a)** What is the angular speed of the tires? **(b)** What is the linear speed of the top of the tires?

Solution

(a) Using Equation 10–15 we find

$$\omega = \frac{v}{r} = \frac{(55 \text{ mph})\left(\dfrac{0.447 \text{ m/s}}{1 \text{ mph}}\right)}{0.32 \text{ m}} = 77 \text{ rad/s}$$

This is about 12 revolutions per second. **(b)** The top of the tires have a speed of $2v = 110$ mph.

10–5 Rotational Kinetic Energy and the Moment of Inertia

An object in motion has kinetic energy, whether that motion is translational, rotational, or a combination of the two. In translational motion, for example, the kinetic energy of a mass m moving with a speed v is $K = \frac{1}{2}mv^2$. We cannot use this expression for a rotating object, however, because the speed v of each particle within a rotating object varies with its distance r from the axis of rotation, as we have seen in Equation 10–12. Thus, there is no unique value of v for an entire rotating object. On the other hand, there *is* a unique value of ω the angular speed, that applies to all particles in the object.

To see how the kinetic energy of a rotating object depends on its angular speed, we start with a particularly simple system consisting of a rod of length r and negligible mass rotating about one end with an angular speed ω. Attached to the other

▲ This photograph of a rolling wheel gives a visual indication of the speed of its various parts. The bottom of the wheel is at rest at any instant, so the image there is sharp. The top of the wheel has the greatest speed, and the image there shows the most blurring. (Compare Figure 10-11.)

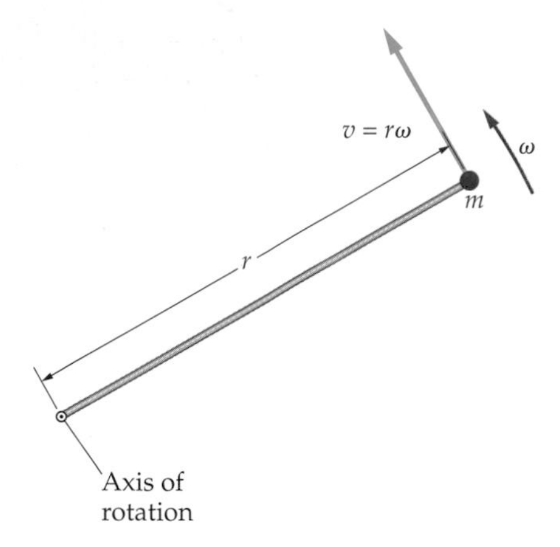

▲ **FIGURE 10–12 Kinetic energy of a rotating object**

As this rod rotates about the axis of rotation with an angular speed ω, the mass has a speed of $v = r\omega$. It follows that the kinetic energy of the mass is $K = \frac{1}{2}mv^2 = \frac{1}{2}(mr^2)\omega^2$.

end of the rod is a point mass m, as **Figure 10–12** shows. To find the kinetic energy of the mass, recall that its linear speed is $v = r\omega$ (Equation 10–12). Therefore, the translational kinetic energy of the mass m is

$$K = \tfrac{1}{2}mv^2 = \tfrac{1}{2}m(r\omega)^2 = \tfrac{1}{2}(mr^2)\omega^2 \qquad \textbf{10–16}$$

Notice that the kinetic energy of the mass depends not only on the angular speed squared (analogous to the way the translational kinetic energy depends on the linear speed squared), but also on the radius squared—that is, the kinetic energy depends on the *distribution* of mass in the rotating object. To be specific, mass near the axis of rotation contributes little to the kinetic energy since its speed ($v = r\omega$) is small. On the other hand, the farther a mass is from the axis of rotation the greater its speed v for a given angular velocity, and thus the greater its kinetic energy.

You have probably noticed that the kinetic energy in Equation 10–16 is similar in form to the translational kinetic energy. Instead of $\frac{1}{2}(m)v^2$ we now have $\frac{1}{2}(mr^2)\omega^2$. Clearly, then, the quantity mr^2 plays the role of the mass for the rotating object. This "rotational mass" is given a special name in physics: the **moment of inertia,** I. Thus, in general, the kinetic energy of an object rotating with an angular speed ω can be written as:

Rotational Kinetic Energy

$$K = \tfrac{1}{2}I\omega^2 \qquad \textbf{10–17}$$

The greater the moment of inertia, the greater an object's rotational kinetic energy. As we have just seen, in the special case of a point mass m a distance r from the axis of rotation the moment of inertia is simply $I = mr^2$.

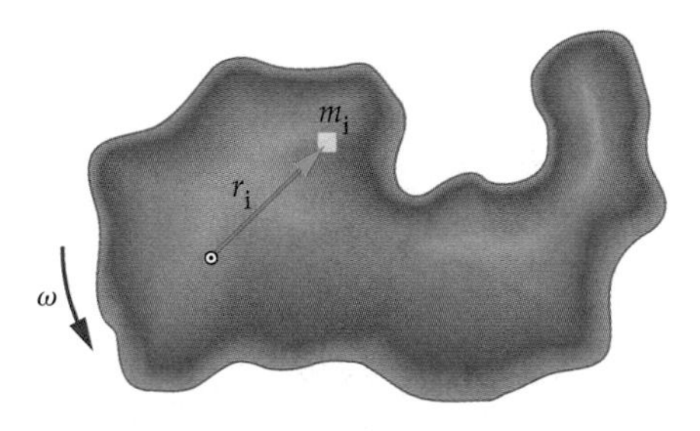

▲ **FIGURE 10–13 Kinetic energy of a rotating object of arbitrary shape**

To calculate the kinetic energy of an object of arbitrary shape as it rotates about an axis with angular speed ω, imagine dividing it into small mass elements, m_i. The total kinetic energy of the object is the sum of the kinetic energies of all the mass elements.

We now show how to find the moment of inertia for an object of arbitrary, fixed shape, as in **Figure 10–13**. Suppose, for example, that this object rotates about the axis indicated in the figure with an angular speed ω. To calculate the kinetic energy of the object, we first imagine dividing it into a collection of small mass elements, m_i. We then calculate the kinetic energy of each element and sum over all elements. This extends to a large number of mass elements what we did for the single mass m.

Following this plan, the total kinetic energy of an arbitrary rotating object is

$$K = \sum(\tfrac{1}{2}m_i v_i^2)$$

In this expression, m_i is the mass of one of the small mass elements and v_i is its speed. If m_i is at the radius r_i from the axis of rotation, as indicated in Figure 10–13, its speed is $v_i = r_i\omega$. Note that it is not necessary to write a separate angular speed, ω_i, for each element, because all mass elements of the object have exactly the same angular speed, ω. Therefore,

$$K = \sum(\tfrac{1}{2}m_i r_i^2\omega^2) = \tfrac{1}{2}\left(\sum m_i r_i^2\right)\omega^2$$

Now, in analogy with our results for the single mass, we can define the moment of inertia, I, as follows:

Definition of Moment of Inertia, I

$$I = \sum m_i r_i^2 \qquad \textbf{10–18}$$

SI unit: $kg \cdot m^2$

The precise value of I for a given object depends on its distribution of mass. A simple example of this dependence is given in the following Exercise.

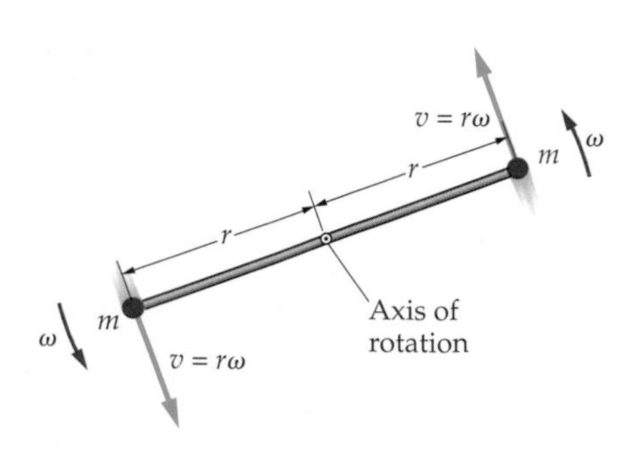

▲ **FIGURE 10–14 A dumbbell-shaped object rotating about its center.**

EXERCISE 10–7

Use the general definition of the moment of inertia, as given in Equation 10–18, to find the moment of inertia for the dumbbell-shaped object shown in **Figure 10–14**. Note that the axis of rotation goes through the center of the object and points out of the page. In addition, assume that the masses may be treated as point masses.

Solution

Referring to Figure 10–14 we see that $m_1 = m_2 = m$ and $r_1 = r_2 = r$. Therefore, the moment of inertia is

$$I = \sum m_i r_i^2 = m_1 r_1^2 + m_2 r_2^2 = mr^2 + mr^2 = 2mr^2$$

The connection between rotational kinetic energy and the moment of inertia is explored in more detail in the following Example.

EXAMPLE 10–4 Nose to the Grindstone

A grindstone with a radius of 0.610 m is being used to sharpen an ax. If the linear speed of the stone relative to the ax is 1.50 m/s, and the stone's rotational kinetic energy is 13.0 J, what is its moment of inertia?

Picture the Problem

In our sketch, we show that the tangential speed at the rim of the grindstone is $v = 1.50$ m/s. The radius of the grindstone is $r = 0.610$ m.

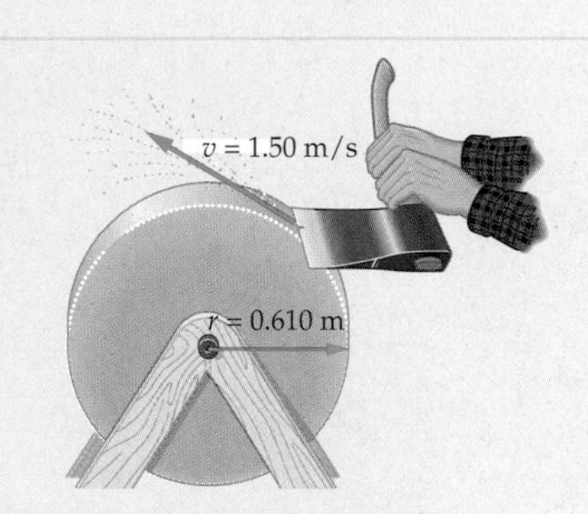

Strategy

Recall that rotational kinetic energy and moment of inertia are related by $K = \frac{1}{2}I\omega^2$; thus $I = 2K/\omega^2$. We are not given ω, but we can find it from the connection between linear and angular speed, $v = r\omega$. Thus, we begin by finding ω. We then use ω, along with the kinetic energy K, to find I.

Solution

1. Find the angular speed of the grindstone: $\omega = \dfrac{v}{r} = \dfrac{1.50 \text{ m/s}}{0.610 \text{ m}} = 2.46 \text{ rad/s}$

2. Solve for the moment of inertia in terms of kinetic energy: $K = \frac{1}{2}I\omega^2$, or $I = \dfrac{2K}{\omega^2}$

3. Substitute numerical values for K and ω: $I = \dfrac{2K}{\omega^2} = \dfrac{2(13.0 \text{ J})}{(2.46 \text{ rad/s})^2} = 4.30 \text{ J}\cdot\text{s}^2 = 4.30 \text{ kg}\cdot\text{m}^2$

Insight

In this calculation we found I by relating it to the rotational kinetic energy of the grindstone. Later in the section we show how to calculate the moment of inertia of a disk given its radius and mass.

Practice Problem

When the ax is pressed firmly against the grindstone for sharpening, the angular speed of the grindstone decreases. If the rotational kinetic energy of the grindstone is cut in half to 6.50 J, what is its angular speed? [**Answer:** The moment of inertia is unchanged; it depends only on the size, shape, and mass of the grindstone. Hence, $\omega = \sqrt{2K/I} = 1.74$ rad/s.]

Some related homework problems: Problem 43, Problem 44

We return now to the dependence of the moment of inertia on the particular shape, or mass distribution, of an object. Suppose, for example, that a mass M is formed into the shape of a *hoop* of radius R. In addition, consider the case where the axis of rotation is perpendicular to the plane of the hoop and passes through its center, as shown in **Figure 10–15**. This is similar to a bicycle wheel rotating about its axle, if one ignores the spokes. In terms of small mass elements, we can write the moment of inertia as

$$I = \sum m_i r_i^2$$

Each mass element of the hoop, however, is at the same radius R from the axis of rotation; that is, $r_i = R$. Hence, the moment of inertia in this case is

$$I = \sum m_i r_i^2 = \sum m_i R^2 = \left(\sum m_i\right) R^2$$

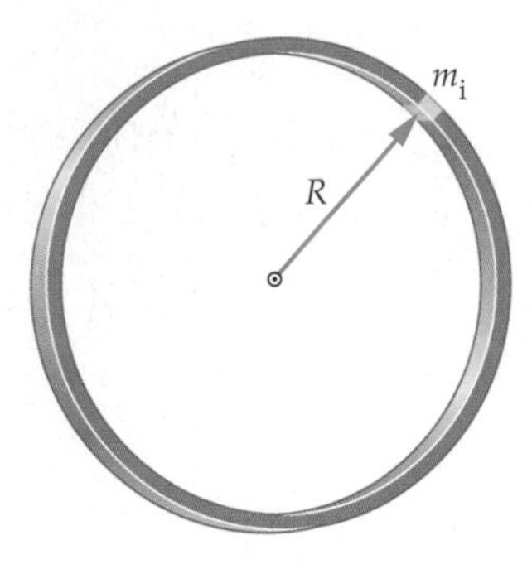

▲ FIGURE 10–15 The moment of inertia of a hoop

Consider a hoop of mass M and radius R. Each small mass element is at the same distance, R, from the center of the hoop. The moment of inertia in this case is $I = MR^2$.

Clearly, the sum of all the elementary masses is simply the total mass of the hoop, $\Sigma m_i = M$. Therefore, the moment of inertia of a hoop of mass M and radius R is

$$I = MR^2 \text{ (hoop)}$$

In contrast, if the same mass, M, is formed into a uniform *disk* of the same radius, R, the moment of inertia is different. To see this, note that it is no longer true that $r_i = R$ for all mass elements. In fact, most of the mass elements are closer to the axis of rotation than was the case for the hoop, as indicated in **Figure 10–16**. Thus, since the r_i are generally less than R, the moment of inertia will be smaller for the disk than for the hoop. A detailed calculation, summing over all mass elements, yields the following result:

$$I = \tfrac{1}{2}MR^2 \text{ (disk)}$$

As expected, I is less for the disk than for the hoop.

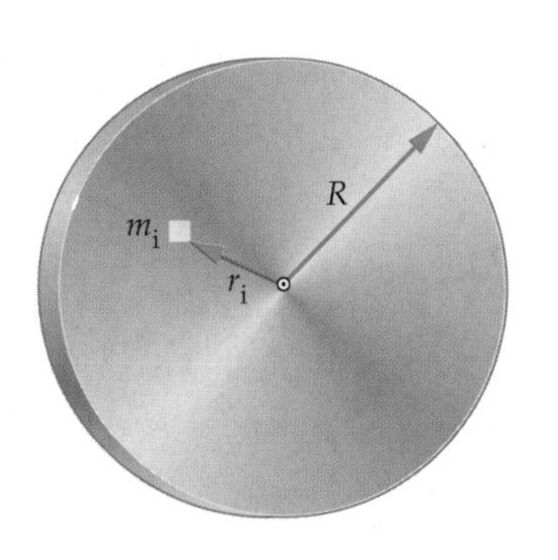

▲ FIGURE 10–16 The moment of inertia of a disk

Consider a disk of mass M and radius R. Mass elements for the disk are at distances from the center ranging from 0 to R. The moment of inertia in this case is $I = \tfrac{1}{2}MR^2$.

EXERCISE 10–8

If the grindstone in Example 10–4 is a uniform disk, what is its mass?

Solution

Applying the preceding equation yields

$$M = \frac{2I}{R^2} = \frac{2(4.30 \text{ kg}\cdot\text{m}^2)}{(0.610 \text{ m})^2} = 23.1 \text{ kg}$$

Thus, the grindstone has a weight of roughly 51 lb.

Table 10–1 collects moments of inertia for a variety of objects. Note that in all cases the moment of inertia is of the form $I = (\text{constant})MR^2$. It is only the constant in front of MR^2 that changes from one object to another.

Note also that objects of the same general shape but with different mass distributions—such as solid and hollow spheres—have different moments of inertia. In particular, the hollow sphere has a larger I than the solid sphere, for the same reason that the hoop's moment of inertia is greater than the disk's—more of its mass is at a greater distance from the axis of rotation. Thus, I is a measure of both the shape *and* the mass distribution of an object.

TABLE 10–1 Moments of Inertia for Uniform, Rigid Objects of Various Shapes

Hoop or cylindrical shell
$I = MR^2$

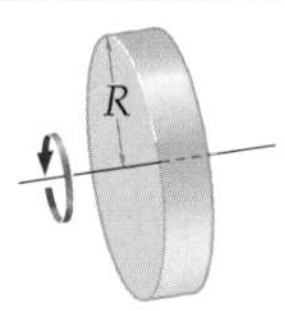

Disk or solid cylinder
$I = \frac{1}{2}MR^2$

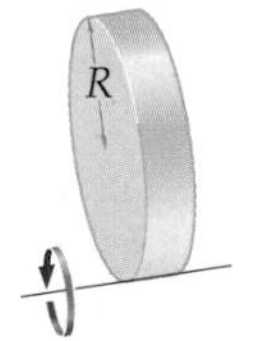

Disk or solid cylinder (axis at rim)
$I = \frac{3}{2}MR^2$

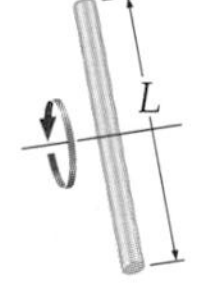

Long thin rod (axis through midpoint)
$I = \frac{1}{12}ML^2$

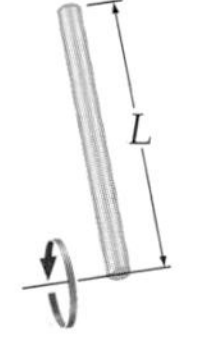

Long thin rod (axis at one end)
$I = \frac{1}{3}ML^2$

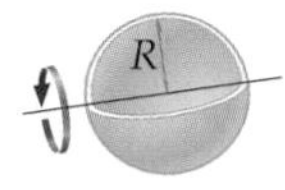

Hollow sphere
$I = \frac{2}{3}MR^2$

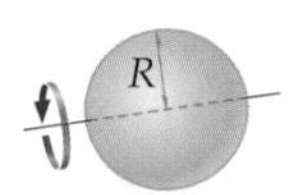

Solid sphere
$I = \frac{2}{5}MR^2$

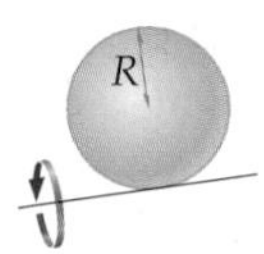

Solid sphere (axis at rim)
$I = \frac{7}{5}MR^2$

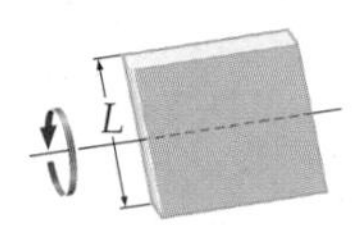

Solid plate (axis through center, in plane of plate)
$I = \frac{1}{12}ML^2$

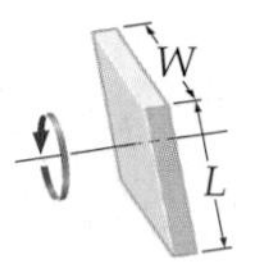

Solid plate (axis perpendicular to plane of plate)
$I = \frac{1}{12}M(L^2 + W^2)$

Consider, for example, the moment of inertia of the Earth. If the Earth were a uniform sphere of mass M_E and radius R_E, its moment of inertia would be $\frac{2}{5}M_E R_E^2 = 0.4M_E R_E^2$. In fact, the Earth's moment of inertia is only $0.331M_E R_E^2$, considerably less than for a uniform sphere. This is due to the fact that the Earth is not homogeneous, but instead has a dense inner core surrounded by a less dense outer core and an even less dense mantle. The resulting concentration of mass near its axis of rotation gives the Earth a much smaller moment of inertia than it would have if its mass were uniformly distributed.

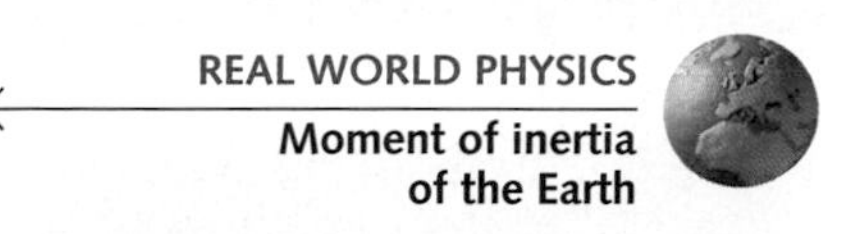

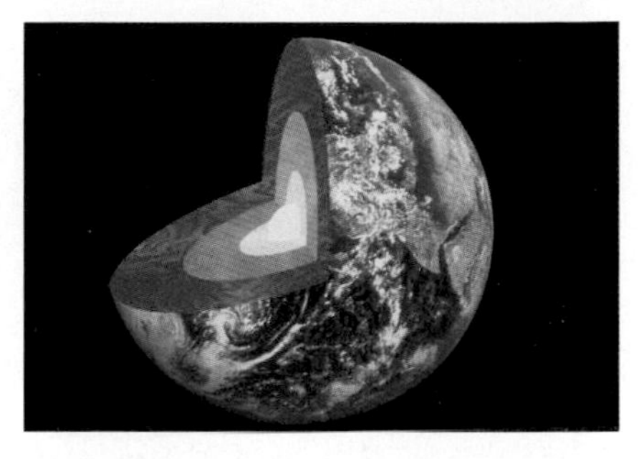

▲ The distribution of mass in the Earth is not uniform. Dense materials, like iron and nickel, have concentrated near the center, while less dense materials, like silicon and aluminum, have risen to the surface. This concentration of mass near the axis of rotation lowers the Earth's moment of inertia.

On the other hand, if the polar ice caps were to melt and release their water into the oceans, the Earth's moment of inertia would increase. This is because mass that had been near the axis of rotation (in the polar ice) would now be distributed more or less uniformly around the Earth (in the oceans). With more of the Earth's mass at greater distances from the axis of rotation, the moment of inertia would increase. If such an event were to occur, not only would the moment of inertia increase, but the length of the day would increase as well. We shall discuss the reasons for this in the next chapter.

The moment of inertia of an object also depends on the location and orientation of the axis of rotation. If the axis of rotation is moved, all of the r_i change, leading to a different result for I. This is investigated for the dumbbell system in the following Conceptual Checkpoint.

CONCEPTUAL CHECKPOINT 10–2

If the dumbbell-shaped object in Figure 10–14 is rotated about one end, is its moment of inertia **(a)** more than, **(b)** less than, or **(c)** the same as the moment of inertia about its center? As before, assume that the masses can be treated as point masses.

Reasoning and Discussion

As we saw in Exercise 10–7, the moment of inertia about the center of the dumbbell is $I = 2mR^2$. When the axis is at one end, that mass is at the radius $r = 0$, and the other mass is at $r = 2R$. Therefore, the moment of inertia is

$$I = \sum m_i r_i^2 = m \cdot 0 + m(2R)^2 = 4mR^2$$

Thus, the moment of inertia doubles when the axis of rotation is moved from the center to one end.

The reason I increases is that the moment of inertia depends on the radius squared. Hence, even small increases in r can cause significant increases in I. By moving the axis to one end, the radius to the other mass is increased to its greatest possible value. As a result, I increases.

Answer:

(a) The moment of inertia is greater about one end than about the center.

Finally, we summarize in the accompanying table the similarities between the translational kinetic energy, $K = \frac{1}{2}mv^2$, and the rotational kinetic energy, $K = \frac{1}{2}I\omega^2$. As expected, we see that the linear speed, v, has been replaced with the angular speed, ω. In addition, note that the mass m has been replaced with the moment of inertia I.

Linear Quantity	Angular Quantity
v	ω
m	I
$\frac{1}{2}mv^2$	$\frac{1}{2}I\omega^2$

As suggested by these analogies, the moment of inertia I plays the same role in rotational motion that mass plays in translational motion. For example, the larger I the more resistant an object is to any change in its angular velocity—an object with a large I is difficult to start rotating, and once it is rotating it is difficult to stop. We shall see further applications of this analogy in the next chapter when we consider angular momentum.

10–6 Conservation of Energy

In this section, we consider the mechanical energy of objects that roll without slipping, and show how to apply energy conservation to such systems. In addition, we consider objects that rotate as a string or rope unwinds: for example, a pulley with a string wrapped around its circumference, or a yo-yo with a string wrapped around its axle. As long as the unwinding process and the rolling motion occur without slipping, the two situations are basically the same—at least as far as energy considerations are concerned.

To apply energy conservation to rolling objects, we first need to determine the kinetic energy of rolling motion. In Section 10–4 we saw that rolling motion is a combination of rotation and translation. It follows, then, that the kinetic energy of a rolling object is simply the sum of its translational kinetic energy, $\frac{1}{2}mv^2$, and its rotational kinetic energy, $\frac{1}{2}I\omega^2$:

Kinetic Energy Of Rolling Motion

$$K = \tfrac{1}{2}mv^2 + \tfrac{1}{2}I\omega^2 \qquad \text{10–19}$$

Note that I in this expression is the moment of inertia about the center of the rolling object.

We can simplify the expression for the kinetic energy of a rolling object by using the fact that linear and angular speeds are related. In fact, recall that $v = r\omega$ (Equation 10–12), which can be rewritten as $\omega = v/r$. Substituting this into our expression for the rolling kinetic energy yields

Kinetic Energy of Rolling Motion

$$K = \tfrac{1}{2}mv^2 + \tfrac{1}{2}I\left(\frac{v}{r}\right)^2 = \tfrac{1}{2}mv^2\left(1 + \frac{I}{mr^2}\right) \qquad \text{10–20}$$

Since $I = (\text{constant})mr^2$, the last term in Equation 10–20 is a constant that depends on the shape and mass distribution of the rolling object.

A special case of some interest is the point particle. In this case, by definition, all of the mass is at a single point. Therefore, $r = 0$, and hence $I = 0$. Substituting $I = 0$ in either Equation 10–19 or Equation 10–20 yields $K = \frac{1}{2}mv^2$, as expected.

Next, we apply Equations 10–19 and 10–20 to a disk that rolls with no slipping.

EXAMPLE 10–5 Like a Rolling Disk

A 1.20-kg disk with a radius of 10.0 cm rolls without slipping. If the linear speed of the disk is 1.41 m/s, find **(a)** the translational kinetic energy, **(b)** the rotational kinetic energy, and **(c)** the total kinetic energy of the disk.

Picture the Problem

Since the disk rolls without slipping, the angular speed and the linear speed are related by $v = r\omega$.

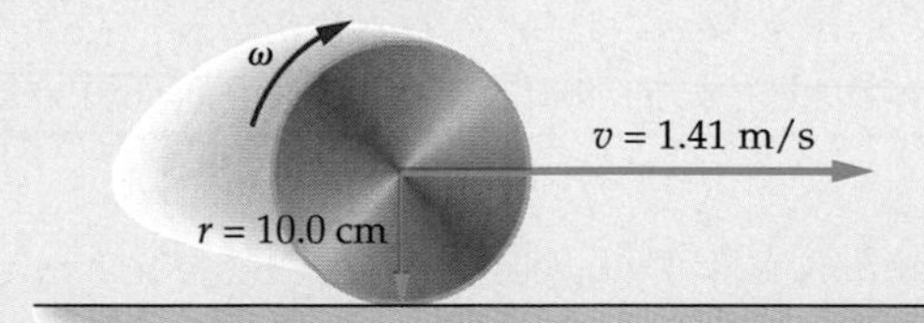

Strategy

We calculate each contribution to the kinetic energy separately. The linear kinetic energy, of course, is simply $\frac{1}{2}mv^2$. For the rotational kinetic energy, $\frac{1}{2}I\omega^2$, we must use the fact that the moment of inertia for a disk is $I = \frac{1}{2}mr^2$. Finally, since the disk rolls without slipping, its angular speed is $\omega = v/r$.

Solution

Part (a)

1. Calculate the translational kinetic energy, $\frac{1}{2}mv^2$:

$$\tfrac{1}{2}mv^2 = \tfrac{1}{2}(1.20\ \text{kg})(1.41\ \text{m/s})^2 = 1.19\ \text{J}$$

Part (b)

2. Calculate the rotational kinetic energy symbolically, using $I = \frac{1}{2}mr^2$ and $\omega = v/r$:

$$\tfrac{1}{2}I\omega^2 = \tfrac{1}{2}(\tfrac{1}{2}mr^2)\left(\frac{v}{r}\right)^2 = \tfrac{1}{2}(\tfrac{1}{2}mv^2)$$

3. Substitute the numerical value for $\frac{1}{2}mv^2$ (the translational kinetic energy) obtained in step 1:

$$\tfrac{1}{2}I\omega^2 = \tfrac{1}{2}(1.19\ \text{J}) = 0.595\ \text{J}$$

Part (c)

4. Sum the kinetic energies obtained in Parts **(a)** and **(b)**:

$$K = 1.19\text{ J} + 0.595\text{ J} = 1.79\text{ J}$$

5. Note that the same result is obtained using Equation 10–20:

$$K = \tfrac{1}{2}mv^2\left(1 + \frac{I}{mr^2}\right) = \tfrac{1}{2}mv^2\left(1 + \frac{1}{2}\right)$$
$$= \tfrac{3}{2}(\tfrac{1}{2}mv^2) = \tfrac{3}{2}(1.19\text{ J}) = 1.79\text{ J}$$

Insight

The symbolic result in step 2 shows that the rotational kinetic energy of a uniform disk rolling without slipping is precisely one-half the disk's translational kinetic energy. Thus, 2/3 of the disk's kinetic energy is translational, 1/3 rotational. This result is independent of the disk's radius, as we can see by the cancellation of the radius r in step 2.

To understand this cancellation, note that a larger disk has a larger moment of inertia, since it has mass farther from the axis of rotation. On the other hand, the larger disk also has a smaller angular speed, since the angular speed is inversely proportional to the radius: $\omega = v/r$. These two effects cancel, giving the same rotational kinetic energy for uniform disks of any radius—provided their linear speed is the same.

Practice Problem

Repeat this problem for the case of a rolling, hollow sphere. [**Answer:** (a) 1.19 J, (b) 0.793 J, (c) 1.98 J]

Some related homework problems: Problem 45, Problem 48

CONCEPTUAL CHECKPOINT 10–3

A solid sphere and a hollow sphere of the same mass and radius roll without slipping at the same speed. Is the kinetic energy of the solid sphere **(a)** more than, **(b)** less than, or **(c)** the same as the kinetic energy of the hollow sphere?

Reasoning and Discussion

Both spheres have the same translational kinetic energy since they have the same mass and speed. The rotational kinetic energy, however, is proportional to the moment of inertia. Since the hollow sphere has the greater moment of inertia, it has the greater kinetic energy.

Answer:

(b) The solid sphere has less kinetic energy than the hollow sphere.

Now that we can calculate the kinetic energy of rolling motion, we show how to apply it to energy conservation. For example, consider an object of mass m, radius r, and moment of inertia I at the top of a ramp, as shown in **Figure 10–17**. The object is released from rest and allowed to roll to the bottom, a vertical height h below the starting point. What is the object's speed on reaching the bottom?

The simplest way to solve this problem is to use energy conservation. To do so, we set the initial mechanical energy at the top (i) equal to the final mechanical energy at the bottom (f). That is,

$$K_i + U_i = K_f + U_f$$

Since we are dealing with rolling motion, the kinetic energy is

$$K = \tfrac{1}{2}mv^2\left(1 + \frac{I}{mr^2}\right)$$

The potential energy is simply that due to the uniform gravitational field. Therefore,

$$U = mgy$$

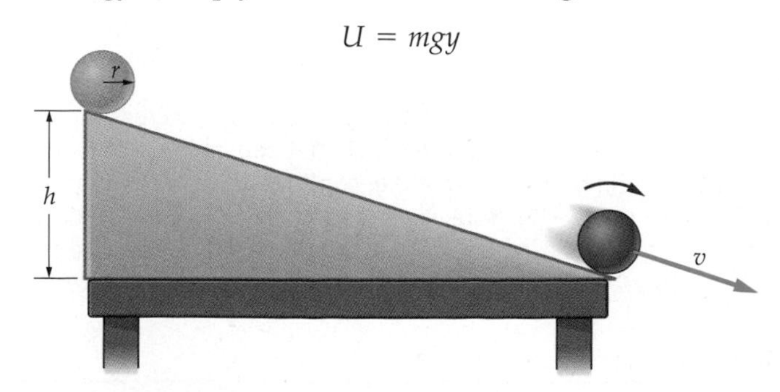

◀ **FIGURE 10–17 An object rolls down an incline**

An object starts at rest at the top of an inclined plane and rolls without slipping to the bottom. The speed of the object at the bottom depends on its moment of inertia—a larger moment of inertia results in a lower speed.

With $y = h$ at the top of the ramp and the object starting at rest, we have

$$K_i + U_i = 0 + mgh = mgh$$

Similarly, with $y = 0$ at the bottom of the ramp and the object rolling with a speed v, we find

$$K_f + U_f = \tfrac{1}{2}mv^2\left(1 + \frac{I}{mr^2}\right) + 0 = \tfrac{1}{2}mv^2\left(1 + \frac{I}{mr^2}\right)$$

Setting the initial and final energies equal yields

$$mgh = \tfrac{1}{2}mv^2\left(1 + \frac{I}{mr^2}\right)$$

Solving for v, we find

$$v = \sqrt{\frac{2gh}{1 + \dfrac{I}{mr^2}}}$$

Let's quickly check one special case: namely, $I = 0$. With this substitution, we find

$$v = \sqrt{2gh}$$

This is the speed an object would have after falling straight down with no rotation through a distance h. Thus, setting $I = 0$ means there is no rotational kinetic energy, and hence the result is the same as for a point particle. As I becomes larger, the speed at the bottom of the ramp is smaller.

CONCEPTUAL CHECKPOINT 10–4

A disk and a hoop of the same mass and radius are released at the same time at the top of an inclined plane. Does the disk reach the bottom of the plane **(a)** before, **(b)** after, or **(c)** at the same time as the hoop?

Reasoning and Discussion

As we have just seen, the larger the moment of inertia, I, the smaller the speed, v. Hence the object with the larger moment of inertia (the hoop in this case) loses the race to the bottom, because its speed is less than the speed of the disk at any given height.

Another way to think about this is to recall that both objects have the same mechanical energy to begin with, namely, mgh. For the hoop, more of this initial potential energy goes into rotational kinetic energy, since the hoop has the larger moment of inertia; therefore less energy is left for translational motion. As a result, the hoop moves more slowly and loses the race.

Answer:

(a) The disk wins the race by reaching the bottom before the hoop.

In the next Conceptual Checkpoint, we consider the effects of a surface that changes from nonslip to frictionless.

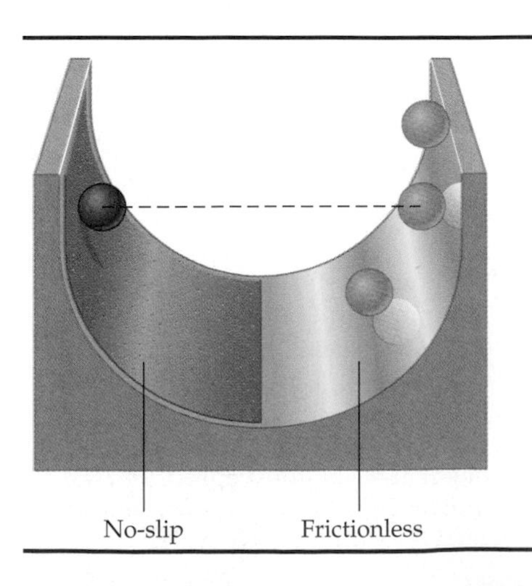

CONCEPTUAL CHECKPOINT 10–5

A ball is released from rest on a no-slip surface, as shown. After reaching its lowest point, the ball begins to rise again, this time on a frictionless surface. When the ball reaches its maximum height on the frictionless surface, is it **(a)** at a greater height, **(b)** at a lesser height, or **(c)** at the same height as when it was released?

Reasoning and Discussion

As the ball descends on the no-slip surface it begins to rotate, increasing its angular speed until it reaches the lowest point of the surface. When it begins to rise again, there is no friction to slow the rotational motion; thus the ball continues to rotate with the same angular speed it had at its lowest point. Therefore, some of the ball's initial gravitational potential energy remains in the form of rotational kinetic energy. As a result, less energy is available to be converted back into gravitational potential energy, and the height is less.

Answer:

(b) The height on the frictionless side is less.

We can also apply energy conservation to the case of a pulley, or similar object, with a string that winds or unwinds without slipping. In such cases, the relation $v = r\omega$ is valid and we can follow the same methods applied to an object that rolls without slipping.

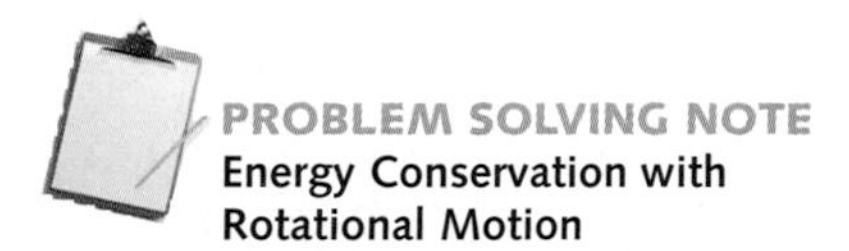

When applying energy conservation to a system with rotational motion be sure to include the rotational kinetic energy, $\frac{1}{2}I\omega^2$.

EXAMPLE 10–6 Spinning Wheel

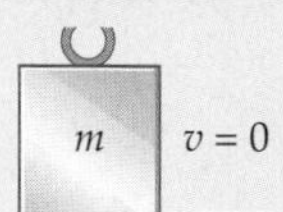

A block of mass m is attached to a string that is wrapped around the circumference of a wheel of radius R and moment of inertia I. The wheel rotates freely about its axis and the string wraps around its circumference without slipping. Initially the wheel rotates with an angular speed ω, causing the block to rise with a linear speed v. To what height does the block rise before coming to rest? Give a symbolic answer.

Picture the Problem

We choose the origin of the y axis to be at the initial height of the block. When the block comes to rest, it is at the height $y = h$.

Strategy

The problem statement gives two key pieces of information. First, the string wraps onto the disk without slipping; therefore, $v = R\omega$. Second, the wheel rotates freely, which means that the mechanical energy of the system is conserved. Thus, at the height h the initial kinetic energy of the system has been converted to gravitational potential energy. This condition can be used to find h.

Before we continue, note that the mechanical energy of the system includes the following contributions: (i) linear kinetic energy for the block, (ii) rotational kinetic energy for the wheel, and (iii) gravitational potential energy for the block. We do not include the gravitational potential energy of the wheel because its height does not change.

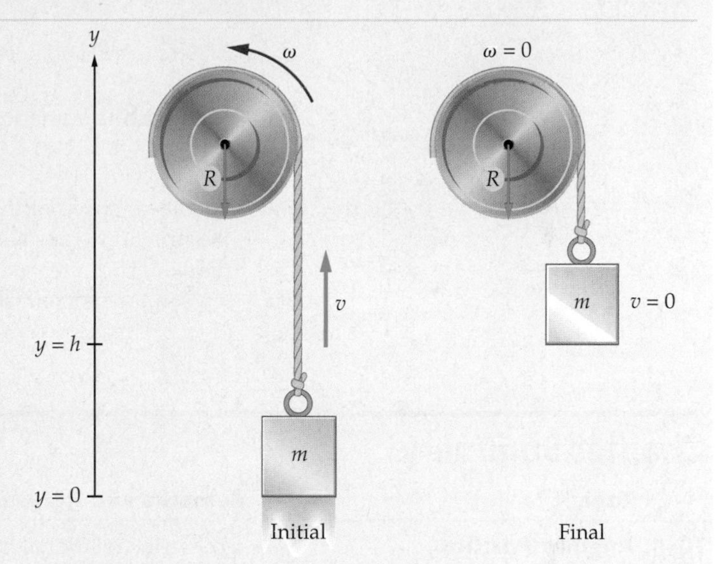

Solution

1. Write an expression for the initial mechanical energy of the system, E_i, including all three contributions mentioned in the Strategy:

$$E_i = \tfrac{1}{2}mv^2 + \tfrac{1}{2}I\omega^2 + mgy = \tfrac{1}{2}mv^2 + \tfrac{1}{2}I\left(\frac{v}{R}\right)^2 + 0$$

2. Write an expression for the final mechanical energy of the system, E_f:

$$E_f = \tfrac{1}{2}mv^2 + \tfrac{1}{2}I\omega^2 + mgy = 0 + 0 + mgh$$

3. Set the initial and final mechanical energies equal to one another, $E_i = E_f$:

$$E_i = \tfrac{1}{2}mv^2 + \tfrac{1}{2}I\left(\frac{v}{R}\right)^2 = \tfrac{1}{2}mv^2\left(1 + \frac{I}{mR^2}\right) = E_f = mgh$$

4. Solve for the height, h:

$$h = \left(\frac{v^2}{2g}\right)\left(1 + \frac{I}{mR^2}\right)$$

Insight

If the block were moving upward with speed v on its own—not attached to anything—it would rise to the height $h = v^2/2g$. We recover this result if $I = 0$, since in that case it is as if the wheel were not there. If the wheel is there, and I is nonzero, the block rises to a height greater than $v^2/2g$. The reason is that the wheel has kinetic energy, in addition to the kinetic energy of the block, and the sum of these kinetic energies must be converted to gravitational potential energy before the block and the wheel stop moving. With more kinetic energy to start with, the height must be greater as well.

Practice Problem

If the wheel has a mass m, and is shaped like a disk, what is the height h? [**Answer:** In this case, $I = \frac{1}{2}mR^2$. Thus, $h = (3/2)(v^2/2g)$.]

Some related homework problems: Problem 51, Problem 54, Problem 57

The situation with a yo-yo is similar, as we see in the next Active Example.

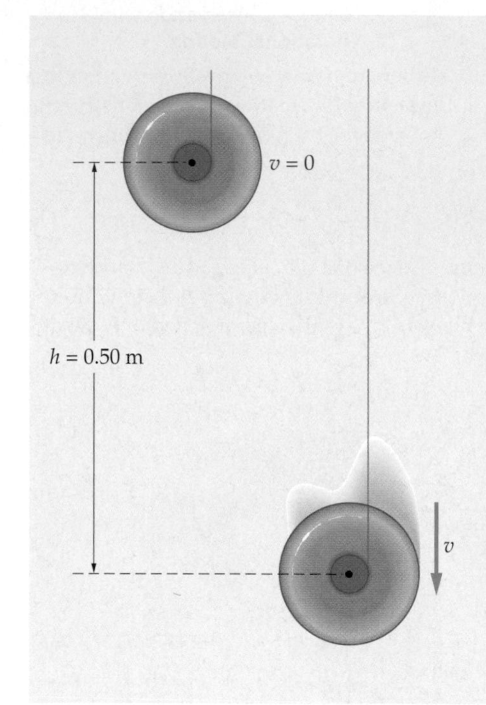

ACTIVE EXAMPLE 10–3 Yo-Yo Man

Yo-Yo man releases a yo-yo from rest and allows it to drop, as he keeps the top end of the string stationary. The mass of the yo-yo is 0.056 kg, its moment of inertia is $2.9 \times 10^{-5}\ \text{kg}\cdot\text{m}^2$, and the radius of the axle the string wraps around is 0.0064 m. What is the linear speed, v, of the yo-yo after it has dropped through a height $h = 0.50$ m?

Solution

1. Write the initial energy of the system: $E_i = mgh$
2. Write the final energy of the system: $E_f = \frac{1}{2}mv^2(1 + I/mr^2)$
3. Set $E_f = E_i$ and solve for v: $v = \sqrt{2gh/(1 + I/mr^2)}$
4. Substitute numerical values: $v = 0.85$ m/s

Insight

The linear speed of the yo-yo is $v = r\omega$, where r is the radius of the axle from which the string unwraps without slipping. Therefore, the r in the term I/mr^2 is the radius of the axle. The outer radius of the yo-yo affects its moment of inertia, but since I is given to us in the problem statement, the outer radius is not pertinent.

Chapter Summary

Topic	Remarks and Relevant Equations
10–1 Angular Position, Velocity, and Acceleration	To describe rotational motion, rotational analogs of position, velocity, and acceleration are defined.

angular position

Angular position, θ, is the angle measured from an arbitrary reference line.

$$\theta \text{ (in radians)} = \text{arc length/radius} = s/r \qquad \textbf{10–2}$$

angular velocity

Angular velocity, ω, is the rate of change of angular position. The average angular velocity is

$$\omega_{av} = \frac{\Delta\theta}{\Delta t} \qquad \textbf{10–3}$$

The instantaneous angular velocity is the limit of ω_{av} as Δt approaches zero:

$$\omega = \lim_{\Delta t \to 0} \frac{\Delta\theta}{\Delta t} \qquad \textbf{10–4}$$

angular acceleration

Angular acceleration, α, is the rate of change of angular velocity. The average angular acceleration is

$$\alpha_{av} = \frac{\Delta\omega}{\Delta t} \qquad \textbf{10–6}$$

The instantaneous angular acceleration is the limit of α_{av} as Δt approaches zero:

$$\alpha = \lim_{\Delta t \to 0} \frac{\Delta\omega}{\Delta t} \qquad \textbf{10–7}$$

period of rotation

The period, T, is the time required to complete one full rotation. If the angular velocity is constant, T is related to ω as follows:

$$T = \frac{2\pi}{\omega} \qquad \textbf{10–5}$$

sign convention

Counterclockwise rotations are positive, clockwise rotations are negative.

10–2 Rotational Kinematics

Rotational kinematics is the description of angular motion, in the same way that linear kinematics describes linear motion. In both cases, we assume constant acceleration.

linear/angular analogs

Rotational kinematics is related to linear kinematics by the following linear/angular analogies:

Linear Quantity	Angular Quantity
x	θ
v	ω
a	α

kinematic equations (constant acceleration)

The equations of rotational kinematics are the same as the equations of linear kinematics, with the substitutions indicated by the linear-angular analogies:

Linear Equation		Angular Equation	
$v = v_0 + at$	2–7	$\omega = \omega_0 + \alpha t$	10–8
$x = x_0 + \frac{1}{2}(v_0 + v)t$	2–10	$\theta = \theta_0 + \frac{1}{2}(\omega_0 + \omega)t$	10–9
$x = x_0 + v_0 t + \frac{1}{2}at^2$	2–11	$\theta = \theta_0 + \omega_0 t + \frac{1}{2}\alpha t^2$	10–10
$v^2 = v_0^2 + 2a(x - x_0)$	2–12	$\omega^2 = \omega_0^2 + 2\alpha(\theta - \theta_0)$	10–11

10–3 Connections Between Linear and Rotational Quantities

A point on a rotating object follows a circular path. At any instant of time, the point is moving in a direction tangential to the circle, with a linear speed and acceleration. The linear speed and acceleration are related to the angular speed and acceleration.

tangential speed

The tangential speed, v_t, of a point on a rotating object is

$$v_t = r\omega \qquad 10\text{–}12$$

centripetal acceleration

The centripetal acceleration, a_{cp}, of a point on a rotating object is

$$a_{cp} = r\omega^2 \qquad 10\text{–}13$$

Centripetal acceleration is due to a change in direction of motion.

tangential acceleration

The tangential acceleration, a_t, of a point on a rotating object is

$$a_t = r\alpha \qquad 10\text{–}14$$

Tangential acceleration is due to a change in speed.

total acceleration

The total acceleration of a rotating object is the vector sum of its tangential and centripetal accelerations.

10–4 Rolling Motion

Rolling motion is a combination of translational and rotational motions. An object of radius r, rolling without slipping, translates with linear speed v and rotates with angular speed

$$\omega = v/r \qquad 10\text{–}15$$

10–5 Rotational Kinetic Energy and the Moment of Inertia

Rotating objects have kinetic energy, just as objects in linear motion have kinetic energy.

rotational kinetic energy

The kinetic energy of a rotating object is

$$K = \tfrac{1}{2}I\omega^2 \qquad 10\text{–}17$$

The quantity I is the moment of inertia.

moment of inertia, discrete masses

The moment of inertia, I, of a collection of masses, m_i, at distances r_i from the axis of rotation is

$$I = \sum m_i r_i^2 \qquad 10\text{–}18$$

moment of inertia, continuous distribution of mass

In a continuous object, the moment of inertia is calculated by dividing the object into a collection of small mass elements and summing $m_i r_i^2$ for each element. Results for a variety of continuous objects are collected in Table 10–1 on p. 288.

linear/angular analog

The moment of inertia is the rotational analog to mass in linear systems. In particular, an object with a large moment of inertia is hard to start rotating and hard to stop rotating.

10–6 Energy Conservation

Energy conservation can be applied to a variety of rotational systems in the same way that it is applied to translational systems.

kinetic energy of rolling motion

The kinetic energy of an object that rolls without slipping is

$$K = \tfrac{1}{2}mv^2 + \tfrac{1}{2}I\omega^2 \qquad \text{10–19}$$

Since rolling without slipping implies that $\omega = v/r$, the kinetic energy can be written as follows:

$$K = \tfrac{1}{2}mv^2 + \tfrac{1}{2}I\left(\frac{v}{r}\right)^2 = \tfrac{1}{2}mv^2\left(1 + \frac{I}{mr^2}\right) \qquad \text{10–20}$$

energy conservation

Conservation of mechanical energy is a statement that the initial kinetic plus potential energy is equal to the final kinetic plus potential energy: $K_i + U_i = K_f + U_f$.
By taking into account both rotational and translational kinetic energy, energy conservation can be applied in the same way as was done for linear systems.

Problem-Solving Summary

Type of Problem	Relevant Physical Concepts	Related Examples
Rotational kinematics with constant angular acceleration.	Rotational kinematics is completely analogous to the linear kinematics studied in Chapter 2. Angular problems are solved in the same way as the corresponding linear problems.	Example 10–1, Example 10–2 Active Example 10–1
Relate linear and angular motion.	Linear speed and angular speed are related by $v = r\omega$. Similarly, linear and angular acceleration are related by $a = r\alpha$. The centripetal acceleration of an object in circular motion is $a_{cp} = r\omega^2$.	Example 10–3 Active Example 10–2
Find the rotational kinetic energy of an object.	Rotational kinetic energy is given by $K = \tfrac{1}{2}I\omega^2$. The moment of inertia, I, plays the same role in rotational motion as the mass in linear motion.	Example 10–4, Example 10–5
Apply energy conservation to a rotational system.	To use energy conservation in a system with rotational motion it is necessary to include the kinetic energy of rotation as one of the forms of energy.	Example 10–6 Active Example 10–3

Conceptual Questions

1. An object at rest begins to rotate with a constant angular acceleration. If this object rotates through an angle θ in the time t, through what angle did it rotate in the time $t/2$?
2. If the angular speed of the object in the previous question is ω after the time t, what was its angular speed at the time $t/2$?
3. A rigid object rotates about an axis. Do all points on the object have the same angular speed? Do all points on the object have the same linear speed?
4. Two children, Jason and Betsy, ride on the same merry-go-round. Jason is a distance R from the axis of rotation; Betsy is a distance $2R$ from the axis. How do the periods of rotation compare for Jason and Betsy? Explain.
5. Referring to the previous question, what is (a) the ratio of Jason's angular speed to Betsy's angular speed, (b) the ratio of Jason's linear speed to Betsy's linear speed, and (c) the ratio of Jason's centripetal acceleration to Betsy's centripetal acceleration?
6. Can you drive your car so as to have a nonzero centripetal acceleration while your tangential acceleration is zero? Give an example.
7. Is it possible to drive your car in a circular path in such a way that you have a nonzero tangential acceleration at the same time your centripetal acceleration is zero? Explain.
8. The fact that the Earth rotates gives people in New York a linear speed of about 750 mi/h. Where should you stand on the Earth to have the smallest possible linear speed?

9. When you ride on a Ferris wheel that is operating at constant angular speed, is your linear velocity constant? Is your angular velocity constant? Is your linear acceleration constant? Is your angular acceleration constant? Explain.
10. The world's tallest buildings are the Petronas Towers in Kuala Lumpur, which rise to a height of 452 m (1,483 ft). When standing on the top floor of one of these buildings, is your linear speed due to the Earth's rotation greater than, less than, or the same as when you stand on the ground floor? Explain.
11. In the previous question, how does your angular speed on the top floor compare to your angular speed on the ground floor?
12. Suppose a bicycle wheel is rotated about an axis through its rim and parallel to its axle. Is its moment of inertia greater than, less than, or the same as when it rotates about its axle? Explain.
13. Is it possible to change the rotational kinetic energy of an object without changing its translational kinetic energy?
14. The minute and hour hands of a clock have a common axis of rotation and equal mass. The minute hand is long and thin; the hour hand is short and thick. Which hand has the greatest moment of inertia?
15. Tons of dust and small particles rain down onto the Earth from space everyday. As a result, does the Earth's moment of inertia increase, decrease, or stay the same? Explain.
16. Why should changing the axis of rotation of an object change its moment of inertia, given that its shape and mass remain the same?
17. As you drive down the highway, the top of your tires are moving with a speed v. What is the reading on your speedometer?
18. Two spheres have identical radii and masses. How might you tell which of these spheres is hollow and which is solid?
19. One way to tell whether an egg is raw or hard boiled—without cracking it open—is to place it on a kitchen counter and give it a spin. If you do this to two eggs, one raw the other hard boiled, you will find that one spins considerably longer than the other. Which egg is this, and why?
20. When the Hoover Dam was completed and filled with water, how was the moment of inertia of the earth affected?
21. Place two quarters on a table with their rims touching, as shown in **Figure 10–18**. While holding one quarter fixed, roll the other one—without slipping—around the circumference of the fixed quarter until it has completed one round trip. How many revolutions has the rolling quarter made?

▲ **FIGURE 10–18** Question 21

22. At the grocery store you pick up a can of beef broth and a can of chunky beef stew. The cans are identical in diameter and weight. Rolling both of them down the aisle with the same initial speed, you notice that the can of chunky stew rolls much farther than the can of broth. Why?
23. The L-shaped object in **Figure 10–19** can be rotated in one of the following three ways: (i) about the x axis; (ii) about the y axis; and (iii) about the z axis (which passes through the origin perpendicular to the plane of the figure.) In which of these cases is the object's moment of inertia greatest? In which case is it least? Explain.

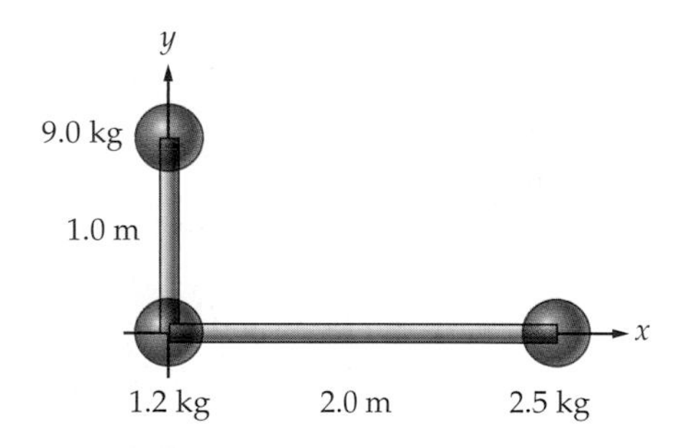

▲ **FIGURE 10–19** Question 23

Problems

Note: **IP** *denotes an integrated conceptual/quantitative problem.* **BIO** *identifies problems of biological or medical interest. Blue bullets (•, ••, •••) are used to indicate the level of difficulty of each problem.*

Section 10–1 Angular Position, Velocity, and Acceleration

1. • The following angles are given in degrees. Convert them to radians: 30°, 45°, 90°, 180°.
2. • The following angles are given in radians. Convert them to degrees: $\pi/6$, 0.70π, 1.5π, 5π.
3. • Find the angular speed of **(a)** the minute hand and **(b)** the hour hand of the famous clock in London, England, that rings the bell known as Big Ben.
4. • Express the angular velocity of the second hand on a clock in the following units: **(a)** rev/hr and **(b)** deg/min and **(c)** rad/s.
5. • List the following in order of increasing angular speed: an automobile tire rotating at 2.00×10^3 deg/s, an electric drill rotating at 400.0 rev/min, and an airplane propeller rotating at 40.0 rad/s.
6. • A spot of paint on a bicycle tire moves in a circular path of radius 0.35 m. When the spot has traveled a linear distance of 1.25 m, through what angle has the tire rotated? Give your answer in radians.
7. • One of the most studied objects in the sky is the Crab nebula, the remains of a supernova explosion observed by the Chinese in 1054. In 1968 it was discovered that a pulsar—a rapidly rotating neutron star that emits a pulse of radio waves with each revolution—lies near the center of the Crab nebula. The period of this pulsar is 33 ms. What is the angular speed (in rad/s) of the Crab nebula pulsar?

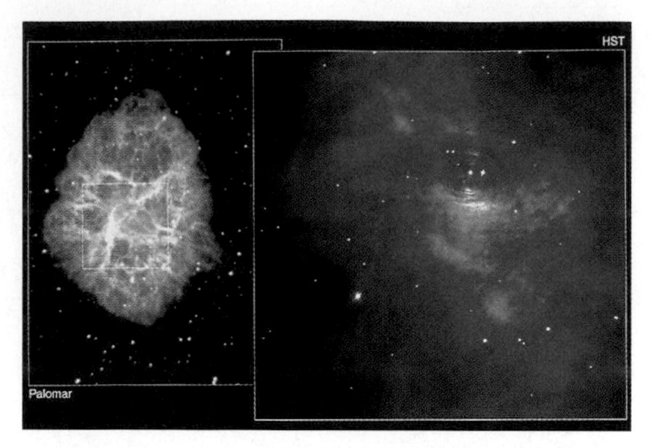

▲ The photo at left is a true-color visible light image of the Crab nebula. In the false-color breakout, the pulsar can be seen as the left member of the pair of stars just above the center of the frame. (Problems 7 and 78)

▲ When the little hand is on the 3 and the big hand is on the 12 . . . . (Problem 17)

8. • Find the angular speed of the Earth as it spins about its axis. Give your result in rad/s.

9. • What is the angular speed (in rev/min) of the Earth as it orbits about the Sun?

10. •• **IP** A 3.5-inch floppy disk in a computer rotates with a period of 2.00×10^{-1} s. What is **(a)** the angular speed of the disk and **(b)** the linear speed of a point on the rim of the disk? **(c)** Does a point near the center of the disk have an angular speed that is greater than, less than, or the same as the angular speed found in part (a)? Explain. (Note: A 3.5-inch floppy disk is 3.5 inches in diameter.)

11. •• During a solar eclipse the Moon moves directly across the face of the Sun, as it appears in the sky. Given that the angle subtended by the Sun is about 0.50°, how long does it take for the Moon to go from just starting to block the Sun to just clearing the Sun? Assume that the Sun is motionless during this time and that the period of the Moon is 27 days.

Section 10–2 Rotational Kinematics

12. • In Active Example 10–1, how long does it take before the angular velocity of the pulley is equal to −5.0 rad/s?

13. • In Example 10–2, through what angle has the wheel turned when its angular speed is 2.0 rad/s?

14. •• After fixing a flat tire on a bicycle you give the wheel a spin. **(a)** If its initial angular speed was 6.15 rad/s and it rotated 13.2 revolutions before coming to rest, what was its average angular acceleration? **(b)** How long did the wheel rotate?

15. •• **IP** A ceiling fan is rotating at 0.50 rev/s. When turned off it slows uniformly to a stop in 12 s. **(a)** How many revolutions does it make in this time? **(b)** Using the result from part (a), find the number of revolutions the fan must make for its speed to decrease from 0.50 rev/s to 0.25 rev/s.

16. •• A discus thrower starts from rest and begins to rotate with a constant angular acceleration of 2.2 rad/s^2. **(a)** How many revolutions does it take for the discus thrower's angular speed to reach 6.3 rad/s? **(b)** How long does this take?

17. •• At 3:00 the hour hand and the minute hand of a clock point in directions that are 90° apart. What is the first time after 3:00 that the angle between the two hands has decreased to 45°?

18. •• **BIO** A centrifuge is a common laboratory instrument that separates components of differing densities in solution. This is accomplished by spinning a sample around in a circle with a large angular speed. Suppose that after a centrifuge in a medical laboratory is turned off it continues to rotate with a constant angular deceleration for 11 s before coming to rest. **(a)** If its initial angular speed was 3750 rpm, what is the magnitude of its angular deceleration? **(b)** How many revolutions did the centrifuge complete after being turned off?

19. •• The Earth's rate of rotation is constantly decreasing, causing the day to increase in duration. In the year 2000 the Earth takes about 0.548 s longer to complete 365 revolutions than it did in the year 1900. What is the average angular acceleration of the Earth?

20. •• **IP** A compact disk (CD) speeds up uniformly from rest to 310 rpm in 3.0 s. **(a)** Describe a strategy that would allow you to calculate the number of revolutions the CD makes in this time. **(b)** Use your strategy to find the number of revolutions.

21. •• When a carpenter shuts off his circular saw the 10.0-inch diameter blade slows from 5900 rpm to zero in 2.0 s. **(a)** What is the angular acceleration of the blade? **(b)** What is the distance traveled by a point on the rim of the blade during the 2-second deceleration? **(c)** What is the magnitude of the displacement of a point on the rim of the blade during the 2-second deceleration?

22. •• A dentist's drill can accelerate with a constant angular acceleration of 750 rad/s^2. **(a)** How long does it take for the drill to accelerate from rest to an angular speed of 2.0×10^5 rpm? **(b)** How many revolutions does the drill make in this time?

Section 10–3 Connections Between Linear and Rotational Quantities

23. •• The hour hand on a certain clock is 8.2 cm long. Find the tangential speed of the tip of this hand.

24. • Two children ride on the merry-go-round shown in Conceptual Checkpoint 10–1. Child 1 is 2.0 m from the axis of rotation, and child 1 is 1.5 m from the axis. If the merry-go-round completes one revolution every 4.5 s, find **(a)** the angular speed and **(b)** the linear speed of each child.

25. • Jeff of the Jungle swings on a vine that is 7.20 m long (**Figure 10–20**). At the bottom of the swing, just before hitting a tree, Jeff's linear speed is 8.50 m/s. **(a)** Find Jeff's angular speed at this time. **(b)** What centripetal acceleration does Jeff experience?

▲ **FIGURE 10–20** Problems 25 and 26

26. •• Suppose, in Problem 25, that at some point in his swing Jeff of the Jungle has an angular speed of 0.850 rad/s and an angular acceleration of 0.620 rad/s^2. Find the magnitude of his centripetal, tangential, and total accelerations, and the angle his total acceleration makes with respect to the tangential direction of motion.

27. •• A compact disk, which has a diameter of 12.0 cm, speeds up uniformly from zero to 4.0 rev/s in 3.0 s. What is the tangential acceleration of a point on the outer rim of the disk at the moment when its angular speed is **(a)** 2.0 rev/s and **(b)** 3.0 rev/s?

28. •• When the compact disk in the previous problem is rotating at 4.0 rev/s what is **(a)** the linear speed and **(b)** the centripetal acceleration of a point on its outer rim?

29. •• **IP** As Tony the fisherman reels in a "big one" he turns the spool on his fishing reel at the rate of 3.0 complete revolutions every second (**Figure 10–21**). **(a)** If the radius of the reel is 3.7 cm, what is the linear speed of the fishing line as it is reeled in? **(b)** How would your answer to part (a) change if the radius of the reel were doubled?

▲ **FIGURE 10–21** Problem 29

30. •• A Ferris wheel with a radius of 9.5 m rotates at a constant rate, completing one revolution every 32 s. Find the direction and magnitude of a passenger's acceleration when at **(a)** the top and **(b)** the bottom of the wheel.

31. •• Suppose the Ferris wheel in the previous problem begins to decelerate at the rate of 0.22 rad/s^2 when the passenger is at the top of the wheel. Find the direction and magnitude of the passenger's acceleration at that time.

32. •• A person swings a 0.52-kg tether ball tied to a 5.0-m rope in an approximately horizontal circle. If the maximum tension the rope can withstand before breaking is 10.0 N, what is the maximum angular speed of the ball?

33. •• To polish a filling, a dentist attaches a sanding disk with a radius of 3.2 mm to the drill. **(a)** When the drill is operated at 2.15×10^4 rad/s, what is the tangential speed of the rim of the disk? **(b)** What period of rotation must the disk have if the tangential speed of its rim is to be 280 m/s?

34. •• In the previous problem, suppose the disk has an angular acceleration of 230 rad/s^2 when its angular speed is 640 rad/s. Find both the translational and centripetal acceleration of a point on the rim of the disk.

35. •• The Bohr model of the hydrogen atom pictures the electron as a tiny particle moving in a circular orbit about a stationary proton. In the lowest-energy orbit the distance from the proton to the electron is 5.29×10^{-11} m, and the linear speed of the electron is 2.18×10^6 m/s. **(a)** What is the angular velocity of the electron? **(b)** How many orbits about the proton does it make each second? **(c)** What is the electron's centripetal acceleration?

36. ••• A wheel of radius R starts from rest and accelerates with a constant angular acceleration α about a fixed axis. At what time t will the centripetal and tangential accelerations of a point on the rim have the same magnitude?

Section 10–4 Rolling Motion

37. • The tires on a car have a radius of 31 cm. What is the angular speed of these tires when the car is driven at 25 m/s?

38. • A child pedals a tricycle, giving the driving wheel an angular speed of 0.373 rev/s (**Figure 10–22**, p. 300). If the diameter of the wheel is 0.52 m, what is the child's linear speed?

39. • A soccer ball, which has a circumference of 70.0 cm, rolls 12 yards in 3.45 s. What was the average angular speed of the ball during this time?

40. •• As you drive down the road at 17 m/s you press on the gas pedal and speed up with a uniform acceleration of 1.12 m/s^2 for 0.65 s. If the tires on your car have a radius of 33 cm, what is their angular displacement during this period of acceleration?

41. •• **IP** A bicycle coasts downhill and accelerates from rest to a linear speed of 8.90 m/s in 12.2 s. **(a)** If the bicycle's tires have a radius of 36 cm, what is their angular acceleration? **(b)** If the radius of the tires had been smaller, would their angular acceleration be greater than or less than the result found in part (a)?

▲ **FIGURE 10–22** Problem 38

Section 10–5 Rotational Kinetic Energy and the Moment of Inertia

42. • The moment of inertia of a 0.98-kg bicycle wheel rotating about its center is $0.13\ \text{kg}\cdot\text{m}^2$. What is the radius of this wheel, assuming the weight of the spokes can be ignored?

43. • What is the kinetic energy of the grindstone in Example 10–4 if it completes one revolution every 4.20 s?

44. • An electric fan spinning with an angular speed of 12 rad/s has a kinetic energy of 4.1 J. What is the moment of inertia of the fan?

45. • Repeat Example 10–5 for the case of a rolling hoop of the same mass and radius.

46. •• **IP** A 3.0-g CD with a radius of 6.0 cm rotates with an angular speed of 20.0 rad/s. **(a)** What is its kinetic energy? **(b)** What angular speed must the CD have if its kinetic energy is to be doubled?

47. •• When a pitcher throws a curve ball, the ball is given a fairly rapid spin. If a 0.15-kg baseball with a radius of 3.7 cm is thrown with a linear speed of 40.0 m/s and an angular speed of 41 rad/s, how much of its kinetic energy is translational and how much is rotational? Assume the ball is a uniform, solid sphere.

48. •• **IP** A basketball rolls along the floor with a constant linear speed v. **(a)** Find the fraction of its total kinetic energy that is in the form of rotational kinetic energy about the center of the ball. **(b)** If the linear speed of the ball is doubled to $2v$, does your answer to part (a) increase, decrease, or stay the same? Explain.

49. •• Referring to Problem 19, find the rate at which the rotational kinetic energy of the Earth is decreasing. The Earth has a moment of inertia of $0.331\,M_E R_E$, where $R_E = 6.38 \times 10^6$ m and $M_E = 5.97 \times 10^{24}$ kg. Give your answer in watts and horsepower.

50. •• A lawn mower has a flat, rod-shaped steel blade that rotates about its center. The mass of the blade is 0.58 kg and its length is 0.56 m. **(a)** What is the rotational energy of the blade at its operating angular speed of 3500 rpm? **(b)** If all of the rotational kinetic energy of the blade could be converted to gravitational potential energy, to what height would the blade rise?

Section 10–6 Conservation of Energy

51. • Suppose the block in Example 10–6 has a mass of 2.1 kg and an initial upward speed of 0.33 m/s. Find the moment of inertia of the wheel if its radius is 8.0 cm and the block rises to a height of 7.4 cm before momentarily coming to rest.

52. • Through what height must the yo-yo in Active Example 10–3 fall for its linear speed to be 0.50 m/s?

53. •• Calculate the speeds of **(a)** the disk and **(b)** the hoop at the bottom of the inclined plane in Conceptual Checkpoint 10–4 if the height of the incline is 1.1 m.

54. •• **IP** The two masses ($m_1 = 5.0$ kg and $m_2 = 3.0$ kg) in the Atwood's machine shown in **Figure 10–23** are released from rest, with m_1 at a height of 0.75 m above the floor. When m_1 hits the ground its speed is 0.22 m/s. Assuming that the pulley is a uniform disk with a radius of 12 cm, **(a)** outline a strategy that allows you to find the mass of the pulley. **(b)** Implement the strategy given in part (a) and determine the pulley's mass.

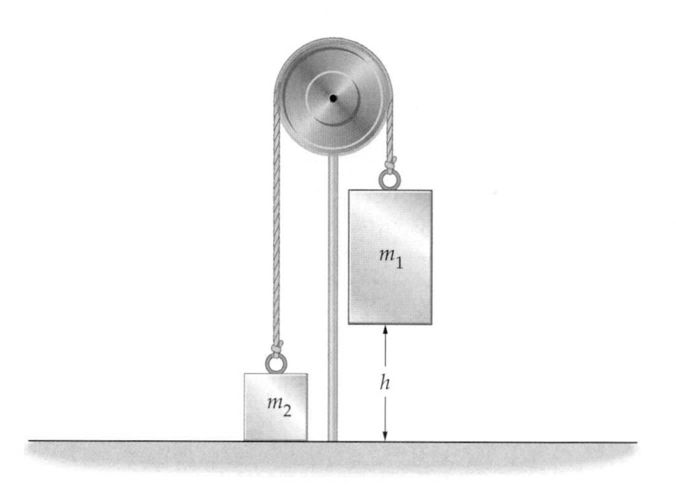

▲ **FIGURE 10–23** Problem 54

55. •• In Conceptual Checkpoint 10–5, assume that the ball is a solid sphere of radius 2.8 cm and mass 0.10 kg. If the ball is released from rest at a height of 0.75 m above the bottom of the track on the no-slip side, how high does it rise on the frictionless side?

56. •• After you pick up a spare, your bowling ball rolls without slipping back toward the ball rack with a linear speed of 2.85 m/s (**Figure 10–24**). To reach the rack, the ball rolls up a ramp that gives the ball a 0.53-m vertical rise. What is the speed of the ball when it reaches the top of the ramp?

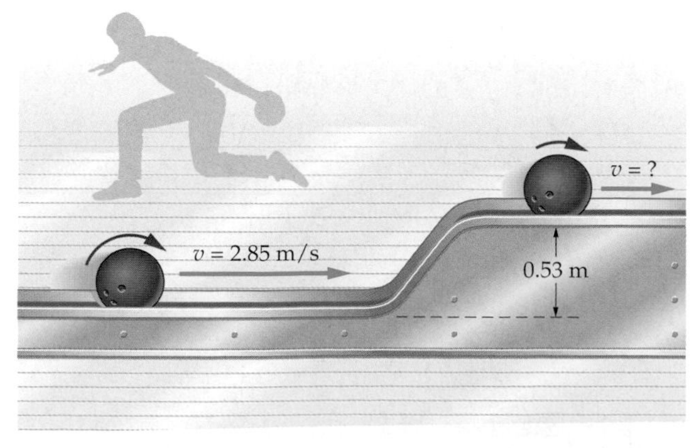

▲ **FIGURE 10–24** Problem 56

57. •• **IP** A 1.3-kg block is tied to a string that is wrapped around the rim of a pulley of radius 7.2 cm. The block is released from rest. **(a)** Assuming the pulley is a uniform disk with a mass of 0.31 kg, find the speed of the block after it has fallen through a height of 0.50 m. **(b)** If a small lead weight is attached near the rim of the pulley and this experiment is repeated, will the speed of the block increase, decrease, or stay the same? Explain.

58. •• After doing some exercises on the floor, you are lying on your back with one leg pointing straight up. If you allow your leg to fall freely until it hits the floor (**Figure 10–25**), what is the tangential speed of your foot just before it lands? Assume the leg can be treated as a uniform rod 0.95-m long that pivots freely about the hip.

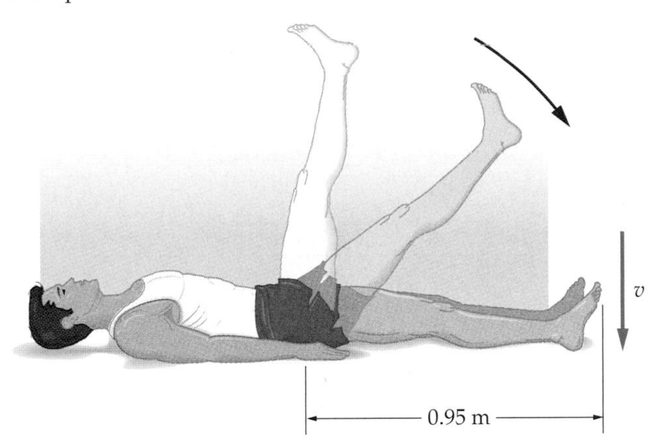

▲ **FIGURE 10–25** Problem 58

59. ••• A 2.0-kg cylinder (radius = 0.10 m, length = 0.50 m) is released from rest at the top of a ramp and allowed to roll without slipping. The ramp is 0.75 m high and 5.0 m long. When the cylinder reaches the bottom of the ramp, what is **(a)** its total kinetic energy, **(b)** its rotational kinetic energy, and **(c)** its translational kinetic energy?

60. ••• A 2.0-kg solid sphere (radius = 0.10 m) is released from rest at the top of a ramp and allowed to roll without slipping. The ramp is 0.75 m high and 5.0 m long. When the sphere reaches the bottom of the ramp, what is **(a)** its total kinetic energy, **(b)** its rotational kinetic energy, and **(c)** its translational kinetic energy?

General Problems

61. • A diver completes $2\frac{1}{2}$ somersaults during a 2.1-s dive. What was the diver's average angular speed during the dive?

62. • What linear speed must a 0.050-kg hula hoop have if its total kinetic energy is to be 0.10 J? Assume the hoop rolls on the ground without slipping.

63. • A pilot performing a horizontal turn will lose consciousness if she experiences a centripetal acceleration greater than 7.00 times the acceleration of gravity. What is the minimum radius turn she can make without losing consciousness if her plane is flying with a constant speed of 250 m/s?

64. • The accompanying double-exposure photograph illustrates a method for determining the speed of a BB. First, note the circular disk in the upper part of the photo. This disk rotates with a constant angular speed of 50.4 revolutions per second. A single white radial line drawn on the disk is seen in two locations in the double exposure. Below the disk are two bright images of a BB taken during the two exposures. Use the information given here and in the photo to estimate the speed of the BB.

▲ Speeding BB and spinning wheel. (Problems 64 and 65)

65. •• Referring to the previous problem, **(a)** estimate the linear speed of a point on the rim of the rotating disk. **(b)** By comparing the arc length between the two white lines to the distance covered by the BB, estimate the speed of the BB. **(c)** What radius must the disk have for the linear speed of a point on its rim to be the same as the speed of the BB? **(d)** Suppose a 1.0-g lump of putty is stuck to the rim of the disk. What centripetal force is required to hold the putty in place?

66. •• A mathematically inclined friend emails you the following instructions: "Meet me in the cafeteria the first time after 2 p.m. today that the hands of a clock point in the same direction." When should you meet your friend?

67. •• In the previous problem, suppose the instructions are to meet your friend the first time after 2 p.m. that the hands of a clock point in opposite directions. What time should you meet your friend?

68. •• **IP** A potter's wheel of radius 6.5 cm rotates with a period of 0.55 s. What is **(a)** the linear speed and **(b)** the centripetal acceleration of a small lump of clay on the rim of the wheel? **(c)** How do your answers to parts (a) and (b) change if the period of rotation is doubled?

69. •• A diver runs horizontally off the end of a diving tower 3.0 m above the surface of water. During her fall she rotates with an average angular speed of 2.4 rad/s. How many revolutions has she made when she hits the water?

70. •• A rubber ball with a radius of 4.0 cm rolls along the horizontal surface of a table with a constant linear speed v. When the ball rolls off the edge of the table it falls 0.86 m to the floor below. If the ball completes 0.77 revolutions during its fall, what was its linear speed v?

71. •• A college campus features a large fountain surrounded by a circular pool. Two students start at the northernmost point of the pool and begin walking around it in opposite directions. **(a)** If the angular speed of the student walking in the clockwise direction (as viewed from above) is 0.050 rad/s and the angular speed of the other student is 0.037 rad/s, how long does it take before they meet? **(b)** At what angle relative to north do they meet?

72. •• **IP** A yo-yo moves downward until it reaches the end of its string, where it "sleeps." As it sleeps—that is, spins in place—its angular speed decreases from 35 rad/s to 25 rad/s. During this time it completes 120 revolutions. **(a)** How long did it take for the yo-yo to slow from 35 rad/s to 25 rad/s? **(b)** How long

does it take for the yo-yo to slow from 25 rad/s to 15 rad/s? Assume a constant angular acceleration as the yo-yo sleeps.

73. •• **IP** **(a)** An automobile with tires of radius 32 cm accelerates from 0 to 40.0 mph in 7.1 s. Find the angular acceleration of the tires. **(b)** How does your answer to part (a) change if the radius of the tires is halved?

74. •• **IP** In Problems 59 and 60 we considered a cylinder and a solid sphere, respectively, rolling down a ramp. **(a)** Which object do you expect to have the greater speed at the bottom of the ramp? **(b)** Verify your answer to part (a) by calculating the speed of the cylinder and the sphere when they reach the bottom of the ramp.

75. •• A centrifuge (Problem 18) with an angular speed of 6050 rpm produces a maximum centripetal acceleration equal to 6835 g (that is, 6835 times the acceleration of gravity). **(a)** What is the diameter of this centrifuge? **(b)** What force must the bottom of the sample holder exert on a 15-g sample under these conditions?

76. •• Yomega ("The yo-yo with a brain") is constructed with a clever clutch mechanism in its axle that allows it to rotate freely and "sleep" when its angular speed is greater than a certain critical value. When the yo-yo's angular speed falls below this value the clutch engages, causing the yo-yo to climb the string to the user's hand. If the moment of inertia of the yo-yo is $7.4 \times 10^{-5}\ \text{kg}\cdot\text{m}^2$, its mass is 0.11 kg, and the string is 1.0 m long, what is the critical value of the angular speed?

▲ A brain, or just a clutch? (Problem 76)

77. •• The rotor in a centrifuge has an initial angular speed of 430 rad/s. After 8.2 s of constant angular acceleration its angular speed has increased to 550 rad/s. During this time, what was **(a)** the angular acceleration of the rotor and **(b)** the angle through which it turned?

78. ••• The pulsar in the Crab nebula (Problem 7) was created by a supernova explosion that was observed on Earth in A.D. 1054. Its current period of rotation (33 ms) is observed to be increasing by 1.26×10^{-5} seconds per year. **(a)** What is the angular acceleration of the pulsar in rad/s^2? **(b)** Assuming the angular acceleration of the pulsar to be constant, how many years will it take for the pulsar to slow to a stop? **(c)** Under the same assumption, what was the period of the pulsar when it was created?

79. ••• A wooden plank rests on two soup cans laid on their sides. Each can has a diameter of 6.5 cm. The plank is 3.0 m long and, initially, one can is placed 1.0 m inward from either end of the plank, as **Figure 10–26** shows. The plank is now pulled 1.0 m to the right, and the cans roll without slipping. **(a)** How far does the center of each can move? **(b)** How many rotations does each can make?

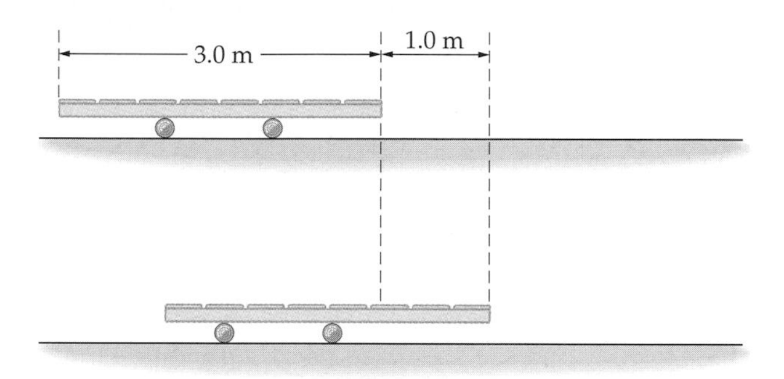

▲ **FIGURE 10–26** Problem 79

80. ••• A person rides on a 12-m diameter Ferris wheel that rotates at the constant rate of 8.1 rpm. Calculate the magnitude and direction of the force that the seat exerts on a 65-kg person when he is **(a)** at the top of the wheel, **(b)** at the bottom of the wheel, and **(c)** halfway up the wheel.

81. ••• A 0.17-m diameter solid sphere released from rest rolls down a ramp, dropping through a vertical height of 0.61 m. The ball leaves the bottom of the ramp, which is 1.22 m above the floor, moving horizontally (**Figure 10–27**). **(a)** What distance d does the ball move in the horizontal direction before landing? **(b)** How many revolutions does it complete during its fall?

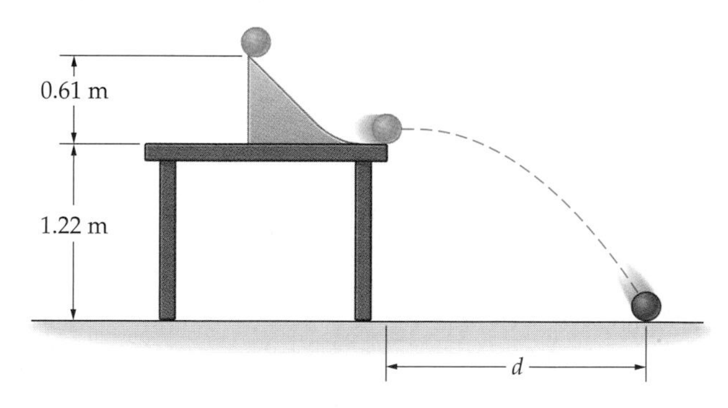

▲ **FIGURE 10–27** Problem 81

11 Rotational Dynamics and Static Equilibrium

Equilibrium isn't just a matter forces—a lot depends on where the forces are applied. (Imagine trying to support these five dancers with a force at the shins rather than at the waist.) In this chapter we will see how a new concept, that of torque, helps us to determine when a system is in equilibrium—and what happens if it is not.

In the previous chapter we learned how to describe uniformly accelerated rotational motion, but we did not discuss how a given angular acceleration is caused by a given force. The connection between forces and angular acceleration is the focus of this chapter.

We begin by defining a quantity that is the rotational equivalent of force. This quantity is called the *torque*. Although torque may not be as familiar a term as force, your muscles are exerting torques on your body at this very moment. In fact, every time you raise an arm, extend a finger, or stretch a leg, you exert torques to carry out these motions. Thus, our ability to move from place to place, or to hold our body still, is intimately related to our ability to exert precisely controlled torques on our limbs.

We also introduce the notion of *angular momentum* in this chapter, and show that it is related to torque in essentially the same way that linear momentum is related to force. As a result, it follows that angular momentum is conserved when the net external torque acting on a system is zero. Thus, conservation of angular momentum joins conservation of energy and conservation of linear momentum as one of the fundamental principles on which all physics is based.

▲ The long handle of this wrench enables the user to produce a large torque without having to exert a very great force.

11–1 Torque

Suppose you want to loosen a nut by rotating it counterclockwise with a wrench. If you have ever used a wrench in this way, you probably know that the nut is more likely to turn if you apply your force as far from the nut as possible, as indicated in **Figure 11–1 (a)**. Applying a force near the nut would not be very effective—you could still get the nut to turn, but it would require considerably more effort! Similarly, it is much easier to open a revolving door if you push far from the axis of rotation, as indicated in **Figure 11–1 (b)**. Clearly, then, the tendency for a force to cause a rotation increases with the distance, r, from the axis of rotation to the force. As a result, it is useful to define a quantity called the **torque**, τ, that takes into account both the magnitude of the force, F, *and* the distance from the axis of rotation, r:

Definition of Torque, τ, for a Tangential Force

$$\tau = rF \qquad \text{11–1}$$

SI unit: N·m

Note that the torque increases with both the force and the distance.

Equation 11–1 is valid only when the applied force is *tangential* to a circle of radius r centered on the axis of rotation, as indicated in Figure 11–1. The more general case is considered next. First, we use Equation 11–1 to determine how much force is needed to open a swinging door, depending on where we apply the force.

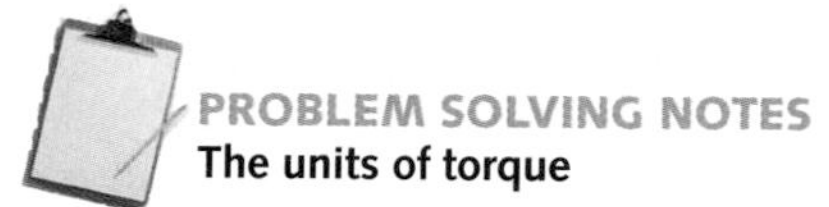

PROBLEM SOLVING NOTES
The units of torque

Note that the units of torque are N·m, the same as the units of work. Though their units are the same, torque, τ, and work, W, represent different physical quantities and should not be confused with one another.

EXERCISE 11–1

To open the door in Figure 11–1 a tangential force F is applied at a distance r from the axis of rotation. If the minimum torque required to open the door is 3.1 N·m, what force must be applied if r is **(a)** 0.35 m, or **(b)** 0.94 m?

Solution

(a) Setting $\tau = rF = 3.1\ \text{N}\cdot\text{m}$, we find that the required force is

$$F = \frac{\tau}{r} = \frac{3.1\ \text{N}\cdot\text{m}}{0.35\ \text{m}} = 8.9\ \text{N}$$

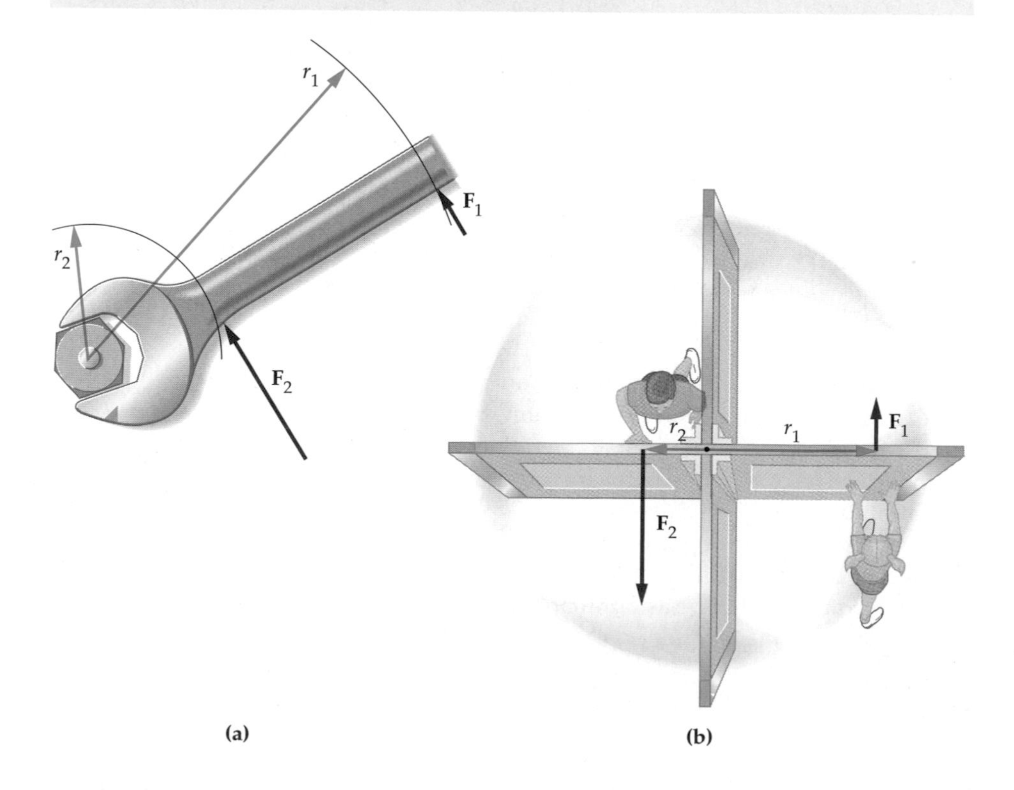

▶ **FIGURE 11–1 Applying a torque**
(a) When a wrench is used to loosen a nut, less force is required if it is applied far from the nut. **(b)** Similarly, less force is required to open a revolving door if it is applied far from the axis of rotation.

(b) Repeat the calculation, this time with $r = 0.94$ m:

$$F = \frac{\tau}{r} = \frac{3.1\ \mathrm{N \cdot m}}{0.94\ \mathrm{m}} = 3.3\ \mathrm{N}$$

As expected, the required force is less when it is applied farther from the hinges.

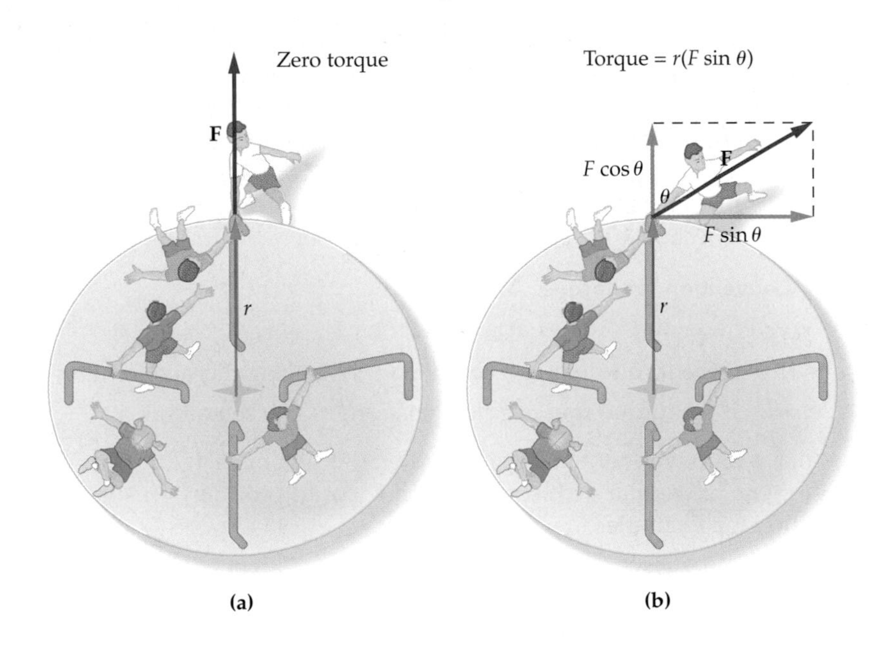

◀ **FIGURE 11–2 Only the tangential component of a force causes a torque**
(a) A radial force causes no rotation. In this case, the force **F** is opposed by an equal and opposite force exerted by the axle of the merry-go-round. The merry-go-round does not rotate. **(b)** A force applied at an angle θ with respect to the radial direction. The radial component of this force, $F\cos\theta$, causes no rotation; the tangential component, $F\sin\theta$, can cause a rotation.

To this point, we have considered tangential forces only. What happens if you exert a force in a direction that is not tangential? Suppose, for example, that you pull on a playground merry-go-round in a direction that is radial—that is, along a line that extends through the axis of rotation—as in Figure 11–2 (a). In this case, your force has no tendency to cause a rotation. Instead, the axle of the merry-go-round simply exerts an equal and opposite force, and the merry-go-round remains at rest. Similarly, if you were to push or pull in a radial direction on a swinging door it would not rotate. We conclude that a *radial force produces zero torque*.

On the other hand, what if your force is at an angle θ relative to a radial line, as shown in Figure 11–2 (b)? To analyze this case, we first resolve the force vector **F** into radial and tangential components. Referring to the figure, we see that the radial component has a magnitude of $F\cos\theta$, and the tangential component has a magnitude of $F\sin\theta$. Since it is the tangential component alone that causes rotation, we define the torque to have a magnitude of $r(F\sin\theta)$. That is,

General Definition of Torque, τ

$$\tau = r(F\sin\theta) \qquad \text{11–2}$$

SI units: N·m

As a quick check, note that a radial force corresponds to $\theta = 0$. In this case, $\tau = r(F\sin 0) = 0$, as expected. If the force is tangential, however, it follows that $\theta = \pi/2$. This gives $\tau = r(F\sin\pi/2) = rF$, in agreement with Equation 11–1.

An equivalent way to define the torque is in terms of the **moment arm**, $r_\perp$. The idea here is to extend a line through the force vector, as in Figure 11–3, and then draw a second line from the axis of rotation perpendicular to the line of the force. The perpendicular distance from the axis of rotation to the line of the force is defined to be $r_\perp$. From the figure, we see that

$$r_\perp = r\sin\theta$$

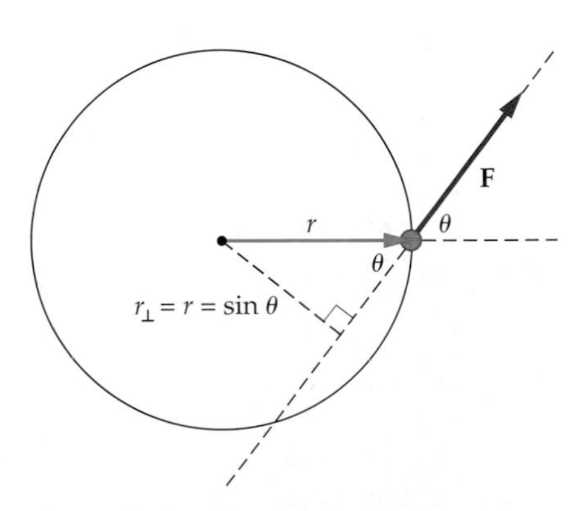

▲ **FIGURE 11–3 The moment arm**
To find the moment arm, $r_\perp$, for a given force, first extend a line through the force vector. Next, drop a perpendicular line from the axis of rotation to the line of the force. The perpendicular distance is $r_\perp = r\sin\theta$.

▲ The net torque on the wheel of this ship is the sum of the torques exerted by the two helmsmen. (If they want the ship to turn to the right, should they be exerting a positive or a negative torque?)

In addition, we note that a simple rearrangement of the torque expression in Equation 11–2 yields

$$\tau = r(F \sin\theta) = (r \sin\theta)F$$

Thus, the torque can be written as the moment arm times the force:

$$\tau = r_{\perp}F \qquad \text{11–3}$$

Just as a force applied to an object gives rise to a linear acceleration, a torque applied to an object gives rise to an angular acceleration. For example, if a torque acts on an object at rest, the object will begin to rotate; if a torque acts on a rotating object, the object's angular velocity will change. In fact, the greater the torque applied to an object, the greater its angular acceleration, as we shall see in the next section. For this reason, the sign of the torque is determined by the same convention used in Section 10–1 for angular acceleration:

Sign Convention for Torque

By convention, if a torque τ acts alone, then

$\tau > 0$ if the torque causes a counterclockwise angular acceleration

$\tau < 0$ if the torque causes a clockwise angular acceleration

PROBLEM SOLVING NOTES
The Sign of Torques

The sign of a torque is determined by the direction of rotation it would cause if it were the only torque acting in the system.

In a system with more than one torque, the sign of each torque is determined by the type of angular acceleration *it alone* would produce. This is illustrated in the following Example.

EXAMPLE 11–1 Torques to the Left and Torques to the Right

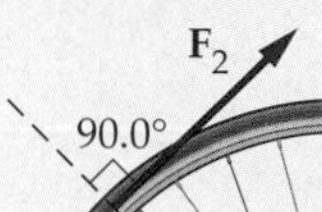

Two forces act on a wheel, as shown below. The wheel is free to rotate without friction, has a radius of 0.42 m, and is initially at rest. Given that $F_1 = 12$ N and $F_2 = 9.5$ N, find **(a)** the torque caused by $\mathbf{F}_1$ and **(b)** the torque caused by $\mathbf{F}_2$. **(c)** In which direction does the wheel turn as a result of these two forces?

Picture the Problem

The sketch shows that both forces are applied at the distance $r = 0.42$ m from the axis of rotation. $\mathbf{F}_1$ is at an angle of 50° relative to the radial direction, $\mathbf{F}_2$ is tangential.

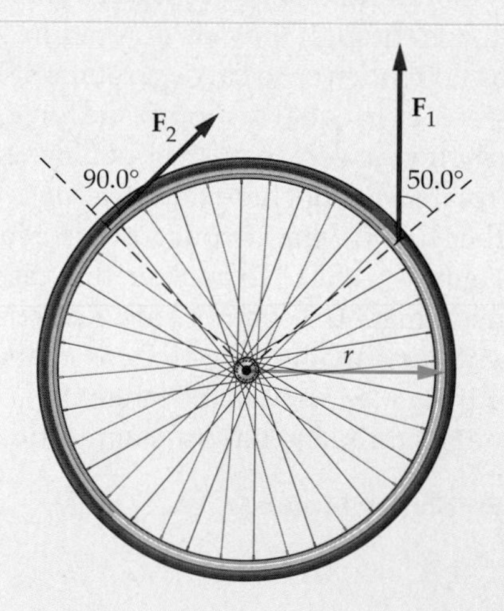

Strategy

For each force, we find the magnitude of the corresponding torque using $\tau = rF \sin\theta$. As for the signs of the torques, we must consider the angular acceleration each force alone would cause. $\mathbf{F}_1$ acting alone would cause the wheel to accelerate counterclockwise, hence its torque is positive. $\mathbf{F}_2$ would accelerate the wheel clockwise if it acted alone, hence its torque is negative. If the sum of the two torques is positive, the wheel accelerates counterclockwise; if the sum of the two torques is negative, the wheel accelerates clockwise.

Solution

Part (a)

1. Use Equation 11–2 to calculate the torque due to $\mathbf{F}_1$. Recall that this torque is positive: $\tau_1 = rF_1 \sin 50° = (0.42\text{ m})(12\text{ N}) \sin 50° = 3.9\text{ N}\cdot\text{m}$

Part (b)

2. Similarly, calculate the torque due to $\mathbf{F}_2$. Recall that this torque is negative: $\tau_2 = -rF_2 \sin 90° = -(0.42\text{ m})(9.5\text{ N}) = -4.0\text{ N}\cdot\text{m}$

Part (c)

3. Sum the torques from parts **(a)** and **(b)** to find the net torque: $\tau_{\text{net}} = \tau_1 + \tau_2 = 3.9\text{ N}\cdot\text{m} - 4.0\text{ N}\cdot\text{m} = -0.1\text{ N}\cdot\text{m}$

Insight
Since the net torque is negative, the wheel accelerates clockwise. Thus, even though $\mathbf{F}_2$ is the smaller force, it has the greater effect in determining the wheel's direction of acceleration. This is because $\mathbf{F}_2$ is applied tangentially, whereas $\mathbf{F}_1$ is applied in a direction that is partially radial.

Practice Problem
What magnitude of $\mathbf{F}_2$ would yield zero net torque on the wheel? [**Answer:** $F_2 = 9.3$ N]

Some related homework problems: Problem 1, Problem 3

11–2 Torque and Angular Acceleration

In the previous section we indicated that a torque causes a change in the rotational motion of an object. To be more precise, a single torque, τ, acting on an object causes the object to have an angular acceleration, α. In this section we develop the specific relationship between τ and α.

Consider, for example, a small object of mass m connected to an axis of rotation by a light rod of length r, as in **Figure 11–4**. If a tangential force of magnitude F is applied to the mass, it will move with an acceleration given by Newton's second law:

$$a = \frac{F}{m}$$

From Chapter 10, Equation 10–14, we know that the linear and angular accelerations are related by

$$\alpha = \frac{a}{r}$$

Combining these results yields the following expression for the angular acceleration:

$$\alpha = \frac{a}{r} = \frac{F}{mr}$$

Finally, multiplying both numerator and denominator by r gives

$$\alpha = \left(\frac{r}{r}\right)\frac{F}{mr} = \frac{rF}{mr^2}$$

Now this last result is rather interesting, since the numerator and denominator have simple interpretations. First, the numerator is the torque, $\tau = rF$, for the case of a tangential force (Equation 11–1). Second, the denominator is the moment of inertia of a single mass m rotating at a radius r; that is, $I = mr^2$. Therefore, we find that

$$\alpha = \frac{rF}{mr^2} = \frac{\tau}{I}$$

Rewriting this, we have the rotational version of Newton's second law:

Newton's Second Law for Rotational Motion

$$\tau = I\alpha \qquad \text{11–4}$$

Thus, once we calculate the torque, as described in the previous section, we can find the angular acceleration of a system using $\tau = I\alpha$.

The relationship $\tau = I\alpha$ was derived for the special case of a tangential force and a single mass rotating at a radius r. However, the result is completely general. In addition, for a system with more than one torque, the relation $\tau = I\alpha$ is replaced with $\tau_{\text{net}} = I\alpha$, where τ_{net} is the net torque acting on the system.

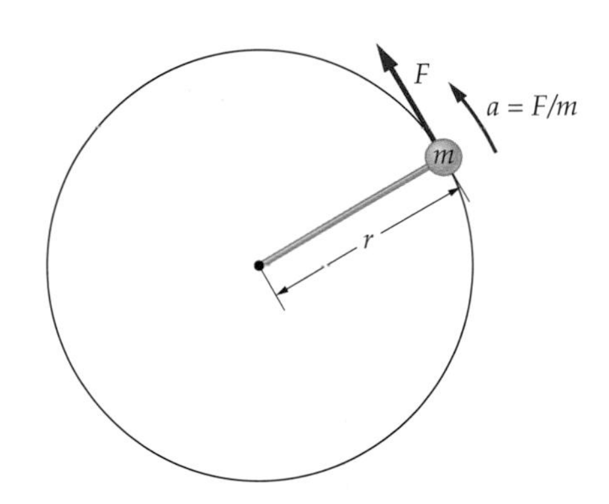

▲ **FIGURE 11–4 Torque and angular acceleration**

A tangential force **F** applied to a mass m gives it a linear acceleration of magnitude $a = F/m$. The corresponding angular acceleration is $\alpha = \tau/I$, where $\tau = rF$ and $I = mr^2$.

EXERCISE 11–2

A light rope wrapped around a disk-shaped pulley is pulled with a force of 0.53 N. Find the angular acceleration of the pulley, given that its mass is 1.3 kg and its radius is 0.11 m.

Solution

The torque applied to the disk is

$$\tau = rF = (0.11\ \text{m})(0.53\ \text{N}) = 5.8 \times 10^{-2}\ \text{N}\cdot\text{m}$$

Since the pulley is a disk, its moment of inertia is given by

$$I = \tfrac{1}{2}mr^2 = \tfrac{1}{2}(1.3\ \text{kg})(0.11\ \text{m})^2 = 7.9 \times 10^{-3}\ \text{kg}\cdot\text{m}^2$$

Thus, the angular acceleration of the pulley is

$$\alpha = \frac{\tau}{I} = \frac{5.8 \times 10^{-2}\ \text{N}\cdot\text{m}}{7.9 \times 10^{-3}\ \text{kg}\cdot\text{m}^2} = 7.3\ \text{rad/s}^2$$

It is easy to remember the rotational version of Newtons' second law, $\tau = I\alpha$, by using analogies between rotational and linear quantities. We have already seen that I is the analog of m, and that α is the analog of a. Similarly, τ, which causes an angular acceleration, is the analog of F, which causes a linear acceleration. To summarize:

Linear Quantity	Angular Quantity
m	I
a	α
F	τ

Thus, just as $F = ma$ describes linear motion, $\tau = I\alpha$ describes rotational motion.

EXAMPLE 11–2 A Fish Takes the Line

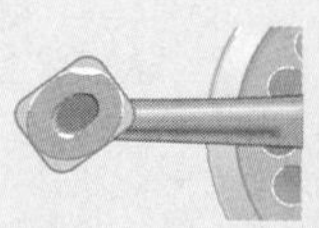

A fisherman is dozing when a fish takes the line and pulls it with a tension T. The spool of the fishing reel is at rest initially and rotates without friction (since the fisherman left the drag off) as the fish pulls for a time t. If the radius of the spool is R, and its moment of inertia is I, find **(a)** the angular displacement of the spool, **(b)** the length of line pulled from the spool, and **(c)** the final angular speed of the spool.

Picture the Problem

Our sketch shows the fishing line being pulled tangentially from the spool with a tension T. Since the radius of the spool is R, the torque produced by the line is $\tau = RT$. Also, note that as the spool rotates through an angle $\Delta\theta$, the line moves through a linear distance $\Delta x = R\,\Delta\theta$. Finally, the spool starts at rest, hence $\omega_0 = 0$.

Strategy

This is basically an angular kinematics problem, as in Chapter 10, but in this case the angular acceleration must first be calculated using $\alpha = \tau/I$. Once α is known, we can find the angular displacement, $\Delta\theta$, using $\theta = \theta_0 + \omega_0 t + \tfrac{1}{2}\alpha t^2$. Similarly, we can find the angular speed of the spool, ω, using $\omega = \omega_0 + \alpha t$.

Solution

1. Calculate the torque acting on the spool. Note that $\theta = 90°$, since the pull is tangential. The radius is $r = R$, and the force applied to the reel is the tension in the line, T: $\quad \tau = rF\sin\theta = RT\sin 90° = RT$
2. Using the result just obtained for the torque, find the angular acceleration of the reel: $\quad \alpha = \dfrac{\tau}{I} = \dfrac{RT}{I}$

Part (a)

3. Calculate the angular displacement $\Delta\theta = \theta - \theta_0$:

$$\Delta\theta = \theta - \theta_0 = \omega_0 t + \tfrac{1}{2}\alpha t^2 = \tfrac{1}{2}\alpha t^2 = \left(\frac{RT}{2I}\right)t^2$$

Part (b)

4. Calculate the length of line pulled from the spool with $\Delta x = R\,\Delta\theta$:

$$\Delta x = R\,\Delta\theta = \left(\frac{R^2T}{2I}\right)t^2$$

Part (c)

5. Use $\omega = \omega_0 + \alpha t$ to find the final angular speed:

$$\omega = \omega_0 + \alpha t = \left(\frac{RT}{I}\right)t$$

Insight

Note that the final angular speed can also be obtained using $\omega^2 = \omega_0^2 + 2\alpha\Delta\theta$.

Practice Problem

How fast is the line moving at time t? [**Answer:** $v = R\omega = (R^2T/I)t$]

Some related homework problems: Problem 7, Problem 14

CONCEPTUAL CHECKPOINT 11–1

The rotating systems shown below differ only in that the two identical movable masses are positioned either far from the axis of rotation (left), or near the axis of rotation (right). If the hanging blocks are released simultaneously from rest, is it observed that **(a)** the mass at left lands first, **(b)** the mass at right lands first, or **(c)** both masses land at the same time?

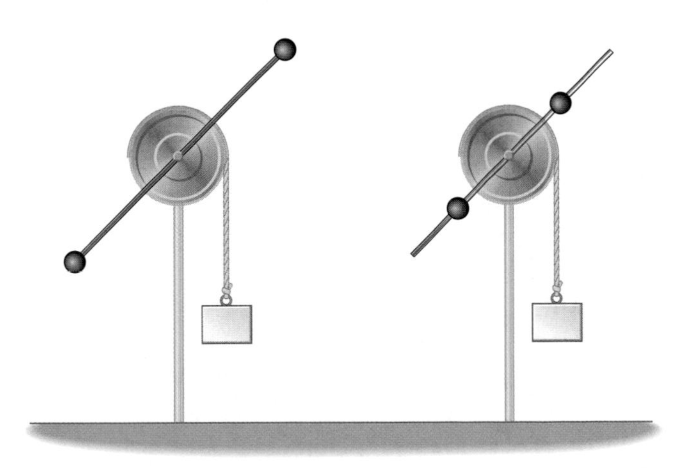

Reasoning and Discussion

The net external torque, supplied by the hanging blocks, is the same for each of these systems. However, the moment of inertia of the system at right is less than that of the system at left because the movable masses are closer to the axis of rotation. Since the angular acceleration is inversely proportional to the moment of inertia ($\alpha = \tau_{net}/I$), the system at right has the greater angular acceleration, and it wins the race.

Answer:

(b) The mass at right lands first.

EXAMPLE 11–3 Drop It

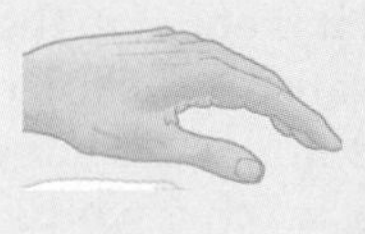

A person holds his outstretched arm at rest in a horizontal position. The mass of the arm is m and its length is 0.740 m. When the person releases his arm, allowing it to drop freely, it begins to rotate about the shoulder joint. Find **(a)** the initial angular acceleration of the arm, and **(b)** the initial linear acceleration of the man's hand. (Hint: In calculating the torque, assume the mass of the arm is concentrated at its center. In calculating the angular acceleration, use the moment of inertia of a uniform rod of length L; $I = \tfrac{1}{3}mL^2$.)

continued on the following page

continued from the previous page

Picture the Problem

The arm is initially horizontal and at rest. When released, it rotates downward about the shoulder joint. The force of gravity, mg, acts at a distance of $(0.740 \text{ m})/2 = 0.370$ m from the shoulder.

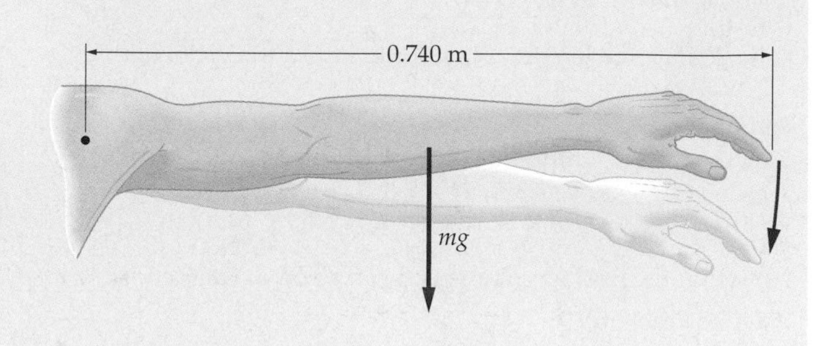

Strategy

The angular acceleration, α, can be found using $\tau = I\alpha$. In this case, the initial torque is $\tau = mg(L/2)$, where $L = 0.740$ m, and the moment of inertia is $I = \frac{1}{3}mL^2$.

Once the initial angular acceleration is found, the corresponding linear acceleration is obtained from $a = r\alpha$.

Solution

Part (a)

1. Use $\tau = I\alpha$ to find the angular acceleration, α:

$$\alpha = \frac{\tau}{I}$$

2. Write expressions for the initial torque, τ, and the moment of inertia, I:

$$\tau = mg\frac{L}{2}$$
$$I = \tfrac{1}{3}mL^2$$

3. Substitute τ and I into the expression for the angular acceleration. Note that the mass of the arm cancels:

$$\alpha = \frac{\tau}{I} = \frac{mgL/2}{mL^2/3} = \frac{3g}{2L}$$

4. Substitute numerical values:

$$\alpha = \frac{3g}{2L} = \frac{3(9.81 \text{ m/s}^2)}{2(0.740 \text{ m})} = 19.9 \text{ rad/s}^2$$

Part (b)

5. Use $a = r\alpha$ to calculate the linear acceleration at the man's hand, a distance $r = L$ from the shoulder:

$$a = L\alpha = L\left(\frac{3g}{2L}\right) = \tfrac{3}{2}g = 14.7 \text{ m/s}^2$$

Insight

Note that the linear acceleration of the hand is 1.50 times greater than the acceleration of gravity, regardless of the mass of the arm. This can be demonstrated with the following simple experiment: Hold your arm straight out with a pen resting on your hand. Now, relax your deltoid muscles, and let your arm rotate freely downward about your shoulder joint. Notice that as your arm falls downward your hand moves more rapidly than the pen, which appears to "lift off" your hand. The pen drops with the acceleration of gravity, which is clearly less than the acceleration of the hand. This effect can be seen clearly in the photo below.

Practice Problem

At what distance from the shoulder is the initial linear acceleration of the arm equal to the acceleration of gravity? [**Answer:** Set $a = r\alpha$ equal to g. This gives $r = 2L/3 = 0.493$ m.]

Some related homework problems: Problem 10, Problem 12

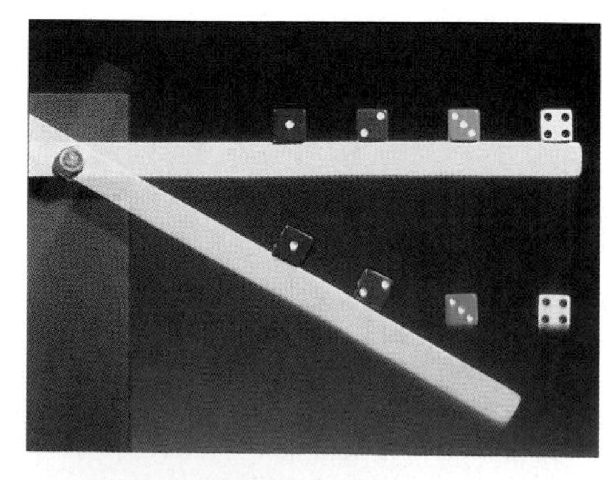

As a rod of length L rotates freely about one end, points farther from the axle than $2L/3$ have an acceleration greater than g (see the Practice Problem for Example 11–3). Thus, the rod falls out from under the last two dice.

11–3 Zero Torque and Static Equilibrium

The parents of a young girl are supporting her on a long, lightweight plank, as illustrated in **Figure 11–5**. If the mass of the child is m, the upward forces exerted by the parents must sum to mg; that is,

$$F_1 + F_2 = mg$$

This condition ensures that the net force acting on the plank is zero. It *does not*, however, guarantee that the plank remains at rest.

To see why, imagine for a moment that the parent on the right lets go of the plank and that the parent on the left increases his or her force until it is equal to the weight of the child. In this case, $F_1 = mg$ and $F_2 = 0$, which clearly satisfies the force equation we have just written. Since the right end of the plank is no longer supported, however, it drops toward the ground while the left end rises. In other words, the plank rotates in a clockwise sense. For the plank to remain completely at rest, with no translation or rotation, we must impose the following *two* conditions: First, the net force acting on the plank must be zero, so that there is no translational ac-

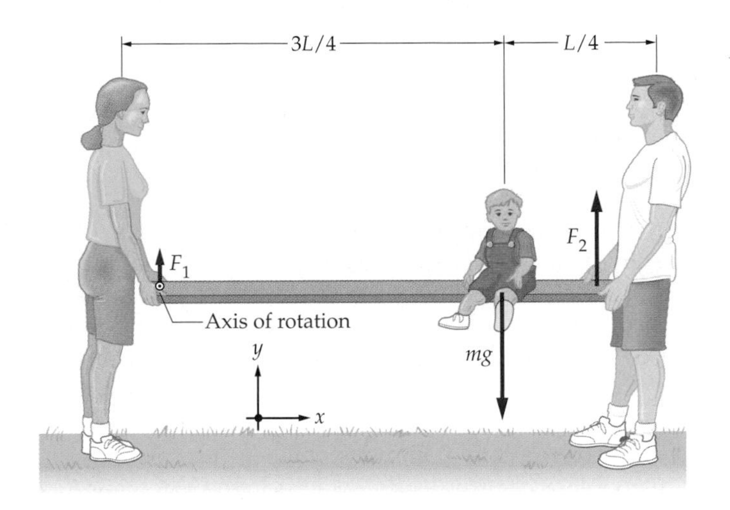

◀ **FIGURE 11–5 Forces required for static equilibrium**

Two parents support a child on a lightweight plank of length L. For the calculation described in the text we choose the axis of rotation to be the left end of the plank.

celeration. Second, the net torque acting on the plank must also be zero, so that there is no rotational acceleration. If both of these conditions are met, an extended object, like the plank, will remain at rest if it starts at rest. To summarize:

Conditions for Static Equilibrium

For an extended object to be in static equilibrium, the following two conditions must be met:

(i) The net force acting on the object must be zero,

$$\sum F_x = 0, \ \sum F_y = 0 \qquad \textbf{11–5}$$

(ii) The net torque acting on the object must be zero,

$$\sum \tau = 0 \qquad \textbf{11–6}$$

Let's apply these conditions to the plank that supports the child. First, we consider the forces acting on the plank, with upward chosen as the positive direction, as in Figure 11–5. Setting the net force equal to zero yields

$$F_1 + F_2 - mg = 0 \qquad \textbf{11–5}$$

Clearly, this agrees with the force equation we wrote down earlier.

Next, we apply the torque condition. To do so, we must first choose an axis of rotation. For example, we might take the left end of the plank to be the axis, as in Figure 11–5. With this choice, we see that the force F_1 exerts zero torque, since it acts directly through the axis of rotation. On the other hand, F_2 acts at the far end of the plank, a distance L from the axis. In addition, F_2 would cause a counterclockwise (positive) rotation if it acted alone, as we can see in Figure 11–5. Therefore, the torque due to F_2 is

$$\tau_2 = F_2 L$$

Finally, the weight of the child, mg, acts at a distance of $3L/4$ from the axis, and would cause a clockwise (negative) rotation if it acted alone. Hence, its torque is negative:

$$\tau_{mg} = -mg(\tfrac{3}{4}L)$$

Setting the net torque equal to zero, then, yields the following condition:

$$F_2 L - mg(\tfrac{3}{4}L) = 0$$

This torque condition, along with the force condition in $F_1 + F_2 - mg = 0$, can be used to determine the two unknowns, F_1 and F_2. For example, we can begin by canceling L in the torque equation to find F_2:

$$F_2 = \tfrac{3}{4}mg$$

Substituting this result into the force condition gives

$$F_1 + \tfrac{3}{4}mg - mg = 0$$

Therefore, F_1 is

$$F_1 = \tfrac{1}{4}mg$$

PROBLEM SOLVING NOTES
Axis of Rotation

Any point in a system may be used as the axis of rotation when calculating torque. It is generally best, however, to choose an axis that gives zero torque for at least one of the unknown forces in the system. Such a choice simplifies the algebra needed to solve for the forces.

These two forces support the plank, *and* keep it from rotating. As one might expect, the force nearest the child is greatest.

Our choice of the left end of the plank as the axis of rotation was completely arbitrary. In fact, if an object is in static equilibrium the net torque acting on it is zero, regardless of the location of the axis of rotation. Hence, we are free to choose an axis of rotation that is most convenient for a given problem. In general, it is useful to pick the axis to be at the location of one of the unknown forces. This eliminates that force from the torque condition, and simplifies the remaining algebra. We consider an alternative choice for the axis of rotation in the following Active Example.

ACTIVE EXAMPLE 11–1 Axis on the Right

A child of mass m is supported on a light plank by her parents, who exert the forces F_1 and F_2 as indicated. Find the forces required to keep the plank in static equilibrium. Use the right end of the plank as the axis of rotation.

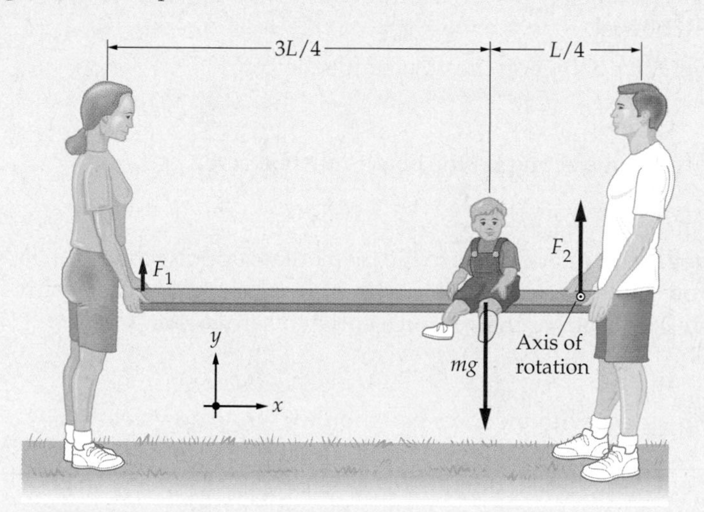

Solution

1. Set the net force acting on the plank equal to zero: $F_1 + F_2 - mg = 0$
2. Set the net torque acting on the plank equal to zero: $-F_1(L) + mg(\tfrac{1}{4}L) = 0$
3. Note that the torque condition involves only one of the two unknowns, F_1. Use this condition to solve for F_1: $F_1 = \tfrac{1}{4}mg$
4. Substitute F_1 into the force condition to solve for F_2: $F_2 = mg - \tfrac{1}{4}mg = \tfrac{3}{4}mg$

Insight

As expected, the results are identical to those obtained previously. Note that in this case the torque produced by the child would cause a counterclockwise rotation, hence it is positive. Thus, the magnitude *and* sign of the torque produced by a given force depend on the location chosen for the axis of rotation.

A third choice for the axis of rotation is considered in Homework Problem 18. As expected, all three choices give the same results.

In the next Example we show that the forces supporting a person or other object sometimes act in different directions. To emphasize the direction of the forces we solve the Example in terms of the components of the relevant forces.

EXAMPLE 11–4 Taking the Plunge

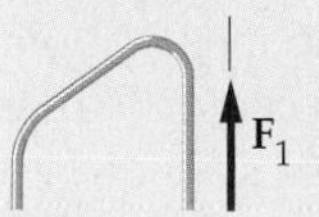

A 5.00-m long diving board of negligible mass is supported by two pillars. One pillar is at the left end of the diving board, as shown below, the other is 1.50 m away. Find the forces exerted by the pillars when a 90.0-kg diver stands at the far end of the board.

Picture the Problem

We choose upward to be the positive direction for the forces. When calculating torques, we use the left end of the diving board as the axis of rotation. Note that $\mathbf{F}_2$ would cause a counterclockwise rotation if it acted alone, so its torque is positive. On the other hand, $m\mathbf{g}$ would cause a clockwise rotation, so its torque is negative.

Strategy

As usual in this type of problem, we use the conditions of (i) zero net force and (ii) zero net torque to determine the unknown forces, $\mathbf{F}_1$ and $\mathbf{F}_2$. In this system all forces act in the positive or negative y direction; thus we need only set the net y component of force equal to zero.

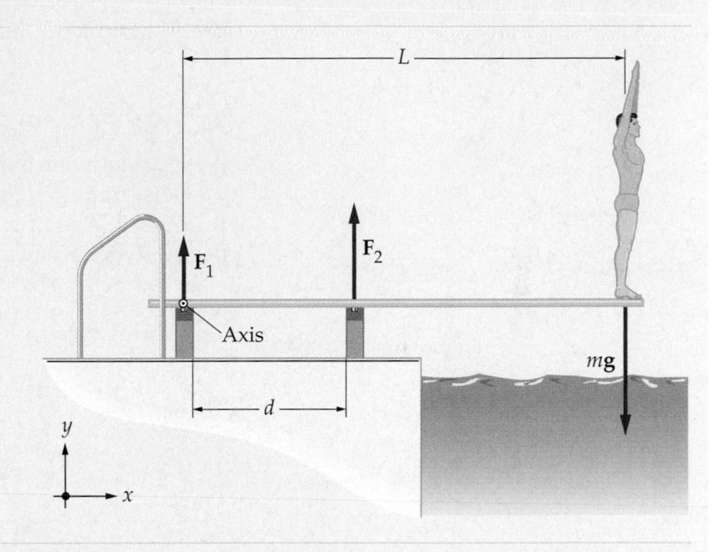

Solution

1. Set the net y component of force acting on the diving board equal to zero:

$$\sum F_y = F_{1,y} + F_{2,y} - mg = 0$$

2. Calculate the torque due to each force, using the left end of the board as the axis of rotation. Note that each force is at right angles to the radius, and that $\mathbf{F}_1$ goes directly through the axis of rotation:

$$\tau_1 = F_{1,y}(0) = 0$$
$$\tau_2 = F_{2,y}(d)$$
$$\tau_3 = -mg(L)$$

3. Set the net torque acting on the diving board equal to zero:

$$\sum \tau = F_{1,y}(0) + F_{2,y}(d) - mg(L) = 0$$

4. Solve the torque equation for the force $F_{2,y}$:

$$F_{2,y} = mg(L/d)$$
$$= (90.0\text{ kg})(9.81\text{ m/s}^2)(5.00\text{ m}/1.50\text{ m}) = 2940\text{ N}$$

5. Use the force equation to determine $F_{1,y}$:

$$F_{1,y} = mg - F_{2,y}$$
$$= (90.0\text{ kg})(9.81\text{ m/s}^2) - 2940\text{ N} = -2060\text{ N}$$

Insight

The first point to notice about our solution is that $F_{1,y}$ is negative, which means that $\mathbf{F}_1$ is actually directed *downward*, as shown here. To see why, imagine for a moment that the board is no longer connected to the first pillar. In this case, the board would rotate clockwise about the second pillar, and the left end of the board would move upward. Thus, a downward force is required on the left end of the board to hold it in place.

The second point is that both forces have magnitudes that are considerably larger than the diver's weight, $mg = 883$ N. In particular, the first pillar must pull downward with a force of 2.33 mg, while the second pillar pushes upward with a force of 3.33 mg.

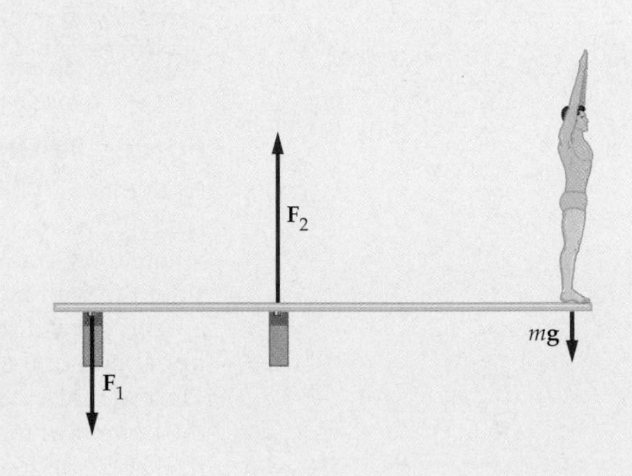

continued on the following page

continued from the previous page

Practice Problem
Find the forces exerted by the pillars when the diver is 1.00 m from its right end. [**Answer:** $F_{1,y} = -1470$ N, $F_{2,y} = 2350$ N]
Some related homework problems: Problem 20, Problem 26

To this point we have ignored the mass of the plank holding the child and the diving board holding the swimmer, since they were described as lightweight. If we want to consider the torque exerted by an extended object of finite mass, however, we can simply treat it as if all its mass were concentrated at its center of mass, as was done in similar situations in Section 9-7. We consider such a system in the next Active Example.

ACTIVE EXAMPLE 11–2 **Walking the Plank**

A cat walks along a uniform plank that is 4.00 m long and has a mass of 7.00 kg. The plank is supported by two sawhorses, one 0.500 m from the left end of the board and the other 1.50 m from its right end. When the cat reaches the right end, the plank just begins to tip. What is the mass of the cat?

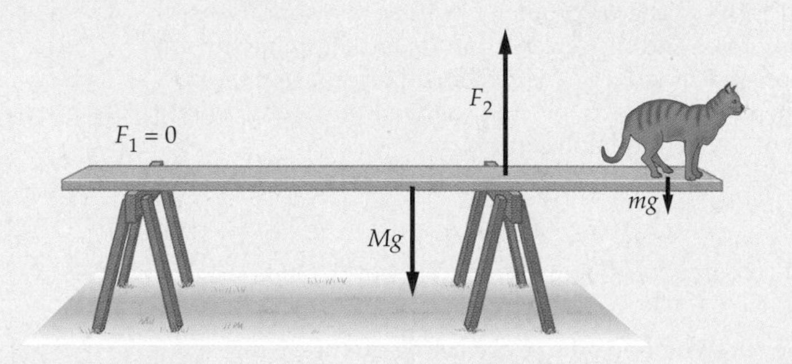

Solution

1. Since the board is just beginning to tip, there is no weight on the left sawhorse: $F_1 = 0$
2. Calculate the torque about the right sawhorse: $Mg(0.500 \text{ m}) - mg(1.50 \text{ m}) = 0$
3. Solve the torque equation for the mass of the cat, m: $m = 0.333M = 2.33$ kg

Insight
Note that we did not include a torque for the left sawhorse, since F_1 is zero. As an exercise, you might try repeating the calculation with the axis of rotation at the left sawhorse, or at the center of mass of the plank.

Forces With Both Vertical and Horizontal Components

Note that all of the previous examples have dealt with forces that point either directly upward, or directly downward. We now consider a more general situation, where forces may have both vertical and horizontal components. For example, consider the wall-mounted lamp (sconce) shown in **Figure 11–6**. The sconce consists of a light, curved rod that is bolted to the wall at its lower end. Suspended from the upper end of the rod, a horizontal distance h from the wall, is the lamp of mass m. The rod is also connected to the wall by a horizontal wire a vertical distance v above the bottom of the rod. We would like to know the tension in the wire, and the vertical and horizontal components of the force exerted by the bolt on the rod.

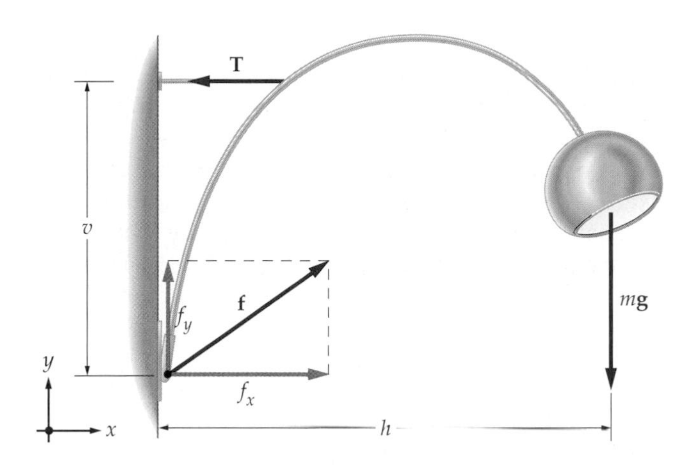

◀ FIGURE 11–6 A lamp in static equilibrium

A wall-mounted lamp of mass m is suspended from a light, curved rod. The bottom of the rod is bolted to the wall. The rod is also connected to the wall by a horizontal wire a vertical distance v above the bottom of the rod.

To solve this problem, we apply the same conditions as before: the net force and the net torque must be zero. In this case, however, forces may have both horizontal and vertical components. Thus, the condition of zero net force is really two separate conditions: (i) zero net force in the horizontal direction; and (ii) zero net force in the vertical direction. These two conditions plus (iii) zero net torque, allow for a full solution of the problem.

We begin with the torque condition. A convenient choice for the axis of rotation is the bottom end of the rod, since this eliminates one of the unknown forces (f). With this choice we can readily calculate the torques acting on the rod by using the moment arm expression for the torque, $\tau = r_{\perp}F$ (Equation 11–3). We find

$$\sum \tau = T(v) - mg(h) = 0$$

This relation can be solved immediately for the tension, giving

$$T = mg(h/v)$$

Note that the tension is increased if the wire is connected closer to the bottom of the rod; that is, if v is reduced.

Next, we apply the force conditions. First, we sum the y components of all the forces, and set the sum equal to zero:

$$\sum F_y = f_y - mg = 0$$

Thus, the vertical component of the force exerted by the bolt simply supports the weight of the lamp

$$f_y = mg$$

Finally, we sum the x components of the forces and set that sum equal to zero:

$$\sum F_x = f_x - T = 0$$

Clearly, the x component of the force exerted by the bolt is of the same magnitude as the tension, but points in the opposite direction:

$$f_x = T = mg(h/v)$$

The bolt, then, pushes upward on the rod to support the lamp, and at the same time pushes to the right to keep the rod from rotating.

For example, suppose the lamp in Figure 11–6 has a mass of 2.00 kg, and that $v = 12.0$ cm and $h = 15.0$ cm. In this case, we find the following forces:

$$T = mg(h/v) = (2.00\text{ kg})(9.81\text{ m/s}^2)(15.0\text{ cm})/(12.0\text{ cm}) = 24.5\text{ N}$$

$$f_x = T = 24.5\text{ N}$$

$$f_y = mg = (2.00\text{ kg})(9.81\text{ m/s}^2) = 19.6\text{ N}$$

▲ The chains that support this sign maintain it in a state of translational and rotational equilibrium. The forces in the chains are most easily analyzed by resolving them into vertical and horizontal components and applying the conditions for equilibrium. In particular, the net vertical force, the net horizontal force, and the net torque must all be zero.

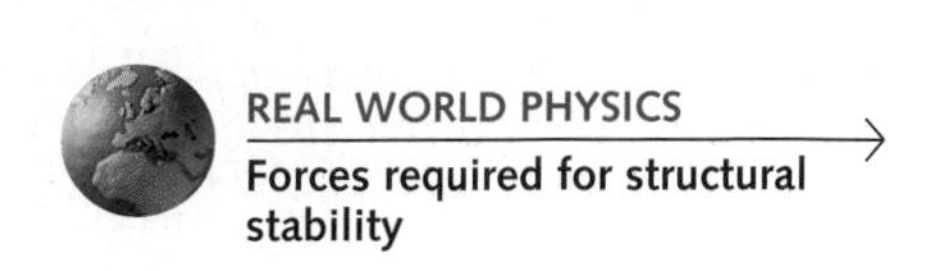

Note that f_x and T are greater than the weight, mg, of the lamp. Just as we found with the diving board in Example 11–4, the forces required of structural elements can be greater than the weight of the object to be supported—an important consideration when designing a structure like a bridge or an airplane. The same effect occurs in the human body. We find in homework Problem 19, for example, that the force exerted by the biceps to support a baseball in the hand is several times larger the baseball's weight. Similar conclusions apply to muscles throughout the body.

We consider another system in which forces have both vertical and horizontal components in the following Example.

EXAMPLE 11–5 Arm in a Sling

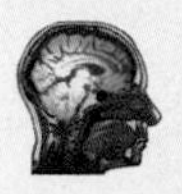

Real World Physics: Bio

A hiker who has broken his forearm rigs a temporary sling using a cord stretching from his shoulder to his hand. The cord holds the forearm level and makes an angle of 40.0° with the horizontal where it attaches to the hand. Considering the forearm and hand to be uniform, with a mass of 1.31 kg and a length of 0.300 m, find **(a)** the tension in the cord and **(b)** the horizontal and vertical components of the force, **f**, exerted by the humerus (the bone of the upper arm) on the radius and ulna (the bones of the forearm).

Picture the Problem

In the sketch, we use the typical conventions for the positive x and y directions. In addition, since the forearm and hand are uniform we indicate the weight, mg, as acting at its center. The length of the forearm and hand is $L = 0.300$ m.

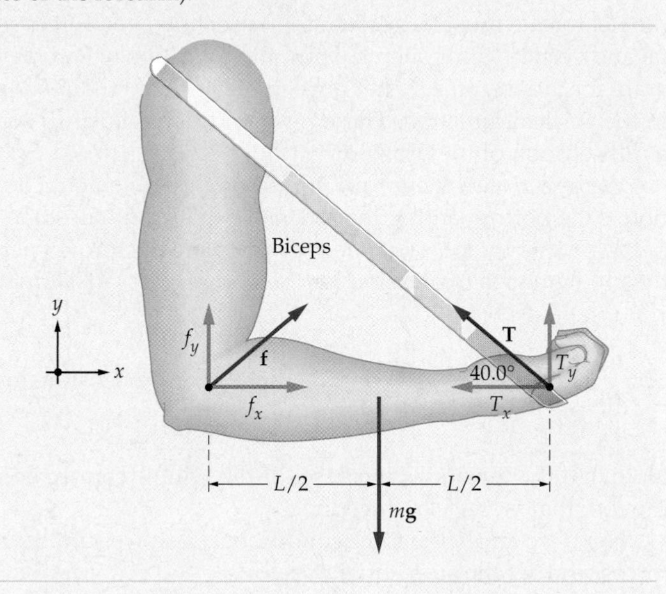

Strategy

In this system there are three unknowns: T, f_x, and f_y. These unknowns can be determined using the following three conditions: (i) net torque equals zero; (ii) net x component of force equals zero; and (iii) net y component of force equals zero.

We start with the torque condition, using the elbow joint as the axis of rotation. This choice of axis eliminates f, and gives a direct solution for the tension T. Next, we use T and the two force conditions to determine f_x and f_y.

Solution

Part (a)

1. Calculate the torque about the elbow joint. Note that f causes zero torque, mg causes a negative torque, and the vertical component of T causes a positive torque. The horizontal component of T produces no torque, since it is on a line with the axis:

$$\sum \tau = (T \sin 40.0°)L - mg(L/2) = 0$$

2. Solve the torque condition for the tension, T:

$$T = \frac{mg}{2 \sin 40.0°} = \frac{(1.31 \text{ kg})(9.81 \text{ m/s}^2)}{2 \sin 40.0°} = 10.0 \text{ N}$$

Part (b)

3. Set the sum of the x components of force equal to zero, and solve for f_x:

$$\sum F_x = f_x - T \cos 40.0° = 0$$
$$f_x = T \cos 40.0° = (10.0 \text{ N}) \cos 40.0° = 7.66 \text{ N}$$

4. Set the sum of the y components of force equal to zero, and solve for f_y:

$$\sum F_y = f_y - mg + T \sin 40.0° = 0$$
$$f_y = mg - T \sin 40.0°$$
$$= (1.31 \text{ kg})(9.81 \text{ m/s}^2) - (10.0 \text{ N}) \sin 40.0° = 6.43 \text{ N}$$

Insight

It is not necessary to determine T_x and T_y separately, since we know the direction of the cord. In particular, it is clear from our sketch that the components of **T** are $T_x = -T \cos \theta = -7.66$ N and $T_y = T \sin \theta = 6.43$ N.

Practice Problem

Suppose the forearm and hand are nonuniform, and that the center of mass is located at a distance of $L/4$ from the elbow joint. What are T, f_x, and f_y in this case? [**Answer:** $T = 5.00$ N, $f_x = 3.83$ N, $f_y = 9.64$ N]

Some related homework problems: Problem 27, Problem 79

ACTIVE EXAMPLE 11–3 Don't Walk under the Ladder

An 85-kg person stands on a lightweight ladder, as shown. The floor is rough; hence it exerts both a normal force, f_1, and a frictional force, f_2, on the ladder. The wall, on the other hand, is frictionless; it exerts only a normal force, f_3. Using the dimensions given in the figure, find the magnitudes of f_1, f_2, and f_3.

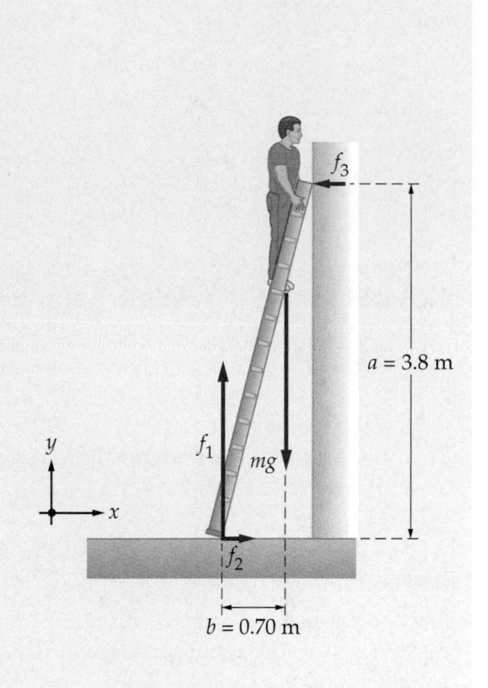

Solution

1. Set the net torque acting on the ladder equal to zero. Use the bottom of the ladder as the axis: $f_3(a) - mg(b) = 0$
2. Solve for f_3: $f_3 = mg(b/a) = 150\text{ N}$
3. Sum the x components of force and set equal to zero: $f_2 - f_3 = 0$
4. Solve for f_2: $f_2 = f_3 = 150\text{ N}$
5. Sum the y components of force and set equal to zero: $f_1 - mg = 0$
6. Solve for f_1: $f_1 = mg = 830\text{ N}$

Insight

If the floor is quite smooth, the ladder might slip—it depends on whether the coefficient of static friction is great enough to provide the needed force $f_2 = 150$ N. In this case, the normal force exerted by the floor is 830 N. Therefore, if the coefficient of static friction is greater than 0.18 [since $0.18(830\text{ N}) = 150\text{ N}$], the ladder will stay put. Ladders often have rubberized pads on the bottom in order to increase the static friction, and hence increase the safety of the ladder.

11–4 Center of Mass and Balance

Suppose you decide to construct a mobile. To begin, you tie a thread to a light rod, as in **Figure 11–7**. Note that the rod extends a distance x_1 to the left of the thread and a distance x_2 to the right. At the left end of the rod you attach an object of mass m_1. What mass, m_2, should be attached to the right end if the rod is to be balanced?

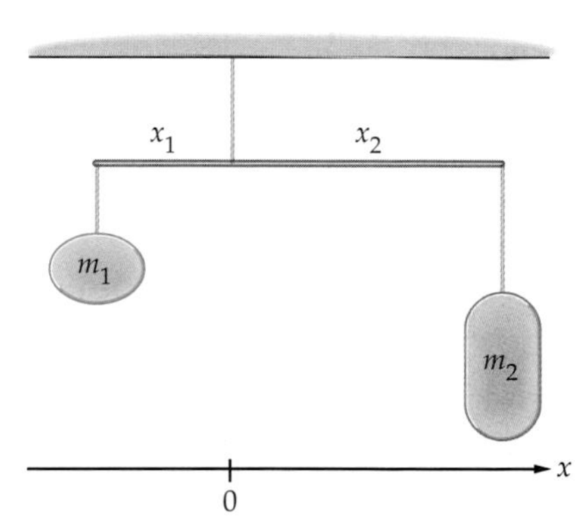

▲ **FIGURE 11–7 Zero torque and balance**

One section of a mobile. The rod is balanced when the net torque acting on it is zero. This is equivalent to having the center of mass directly under the suspension point.

From the discussions in the previous sections, it is clear that if the rod is to be in static equilibrium (balanced), the net torque acting on it must be zero. Taking the point where the thread is tied to the rod as the axis of rotation, this zero-torque condition can be written as:

$$m_1g(x_1) - m_2g(x_2) = 0$$

Canceling g and rearranging, we find

$$m_1x_1 = m_2x_2 \qquad \text{11–7}$$

This gives the following result for m_2:

$$m_2 = m_1(x_1/x_2)$$

For example, if $x_2 = 2x_1$, it follows that m_2 should be one-half of m_1.

Let's now consider a slightly different question: Where is the center of mass of m_1 and m_2? Choosing the origin of the x axis to be at the location of the thread, as indicated in Figure 11–7, we can use the definition of the center of mass, Equation 9-13, to find x_{cm}:

$$x_{cm} = \frac{m_1(-x_1) + m_2(x_2)}{m_1 + m_2} = -\left(\frac{m_1x_1 - m_2x_2}{m_1 + m_2}\right)$$

Referring to the zero-torque condition in Equation 11–7, we see that $m_1x_1 - m_2x_2 = 0$; hence the center of mass is at the origin:

$$x_{cm} = 0$$

This is precisely where the string is attached. We conclude, then, that the rod balances when its center of mass is at the point from which it is suspended. This is a general result.

Let's apply this result to the case of the mobile shown in the next Example.

EXAMPLE 11–6 A Well-Balanced Meal

As a grade-school project, students construct a mobile representing some of the major food groups. Their completed artwork is shown below. Find the masses m_1, m_2, and m_3 that are required for a perfectly balanced mobile.

Picture the Problem

The dimensions of the horizontal rods, and the given masses, are indicated in the sketch.

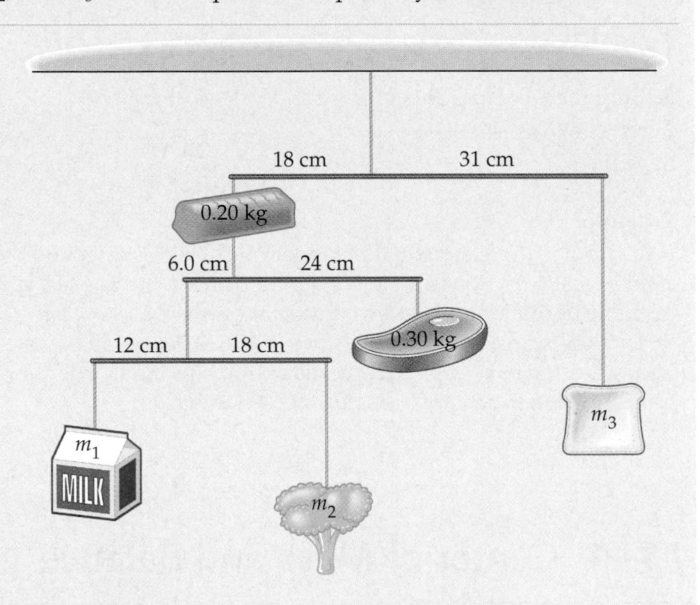

Strategy

We can find all three unknown masses by repeatedly applying the condition for balance, $m_1x_1 = m_2x_2$.

First, we apply the balance condition to m_1 and m_2, with the distances $x_1 = 12$ cm and $x_2 = 18$ cm. This gives a relation between m_1 and m_2.

To get a second relation between m_1 and m_2, we apply the balance condition again at the next higher level of the mobile. That is, the mass $(m_1 + m_2)$ at the distance 6.0 cm must balance the mass 0.30 kg at the distance 24 cm. These two conditions determine m_1 and m_2.

To find m_3 we again apply the balance condition, this time with the mass $(m_1 + m_2 + 0.30 \text{ kg} + 0.20 \text{ kg})$ at the distance 18 cm, and the mass m_3 at the distance 31 cm.

Solution

1. Apply the balance condition to m_1 and m_2:

$$m_1(12 \text{ cm}) = m_2(18 \text{ cm})$$
$$m_1 = (1.5)m_2$$

2. Apply the balance condition to the next level up in the mobile. Solve for the sum, $m_1 + m_2$:

$$(m_1 + m_2)(6.0 \text{ cm}) = (0.30 \text{ kg})(24 \text{ cm})$$
$$m_1 + m_2 = \frac{(0.30 \text{ kg})(24 \text{ cm})}{6.0 \text{ cm}} = 1.2 \text{ kg}$$

3. Substitute $m_1 = (1.5)m_2$ into $m_1 + m_2 = 1.2$ kg to find m_2:

$$(1.5)m_2 + m_2 = (2.5)m_2 = 1.2 \text{ kg}$$
$$m_2 = 1.2 \text{ kg}/2.5 = 0.48 \text{ kg}$$

4. Use $m_1 = (1.5)m_2$ to find m_1:

$$m_1 = (1.5)m_2 = (1.5)0.48 \text{ kg} = 0.72 \text{ kg}$$

5. Apply the balance condition to the top level of the mobile:

$$(0.72 \text{ kg} + 0.48 \text{ kg} + 0.30 \text{ kg} + 0.20 \text{ kg})(18 \text{ cm}) = m_3(31 \text{ cm})$$

6. Solve for m_3:

$$m_3 = \frac{(1.7 \text{ kg})(18 \text{ cm})}{31 \text{ cm}} = 0.99 \text{ kg}$$

Insight

With the values for m_1, m_2, and m_3 found above, the mobile balances at every level. In particular, the center of mass of the entire mobile is directly below the suspension point.

Practice Problem

Find m_1, m_2, and m_3 if the 0.30 kg mass is replaced with a 0.40 kg mass. [**Answer:** $m_1 = 0.96$ kg, $m_2 = 0.64$ kg, $m_3 = 1.3$ kg]

Some related homework problems: Problem 36, Problem 37

In general, if you allow an arbitrarily shaped object to hang freely, its center of mass is directly below the suspension point. To see why, note that when the center of mass is directly below the suspension point, the torque due to gravity is zero, since the force of gravity goes right through the axis of rotation. This is shown in **Figure 11–8 (a)**. If the object is rotated slightly, as in **Figure 11–8 (b)**, the force of gravity is not in line with the axis of rotation—hence gravity produces a torque. This torque tends to rotate the object, bringing the center of mass back under the suspension point.

For example, suppose you cut a piece of wood into the shape of the United States, as shown in **Figure 11–9**, drill a small hole in it and hang it from the point A. The result is that the center of mass lies somewhere on the line aa'. Similarly, if a second hole is drilled at point B, we find that the center of mass lies somewhere on the line bb'. The only point that is on both the line aa' *and* the line bb' is the point x, near Smith Center, Kansas, which marks the location of the center of mass of the United States.

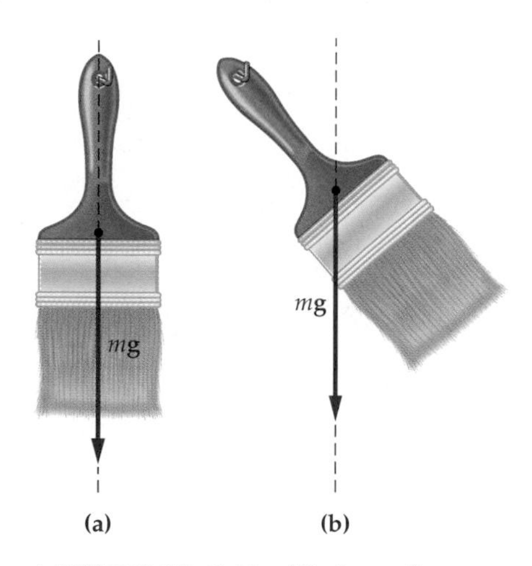

▲ **FIGURE 11–8 Equilibrium of a suspended object**
(a) If an object's center of mass is directly below the suspension point, its weight creates zero torque, and the object is in equilibrium. **(b)** When an object is rotated, so that the center of mass is no longer directly below the suspension point, the object's weight creates a torque. The torque tends to rotate the object to bring the center of mass under the suspension point.

CONCEPTUAL CHECKPOINT 11–2

A croquet mallet balances when suspended from its center of mass, as indicated in the drawing at left. If you cut the mallet in two at its center of mass, as in the drawing at right, how do the masses of the two pieces compare? **(a)** The masses are equal; **(b)** the piece with the head of the mallet has the greater mass; or **(c)** the piece with the head of the mallet has the smaller mass.

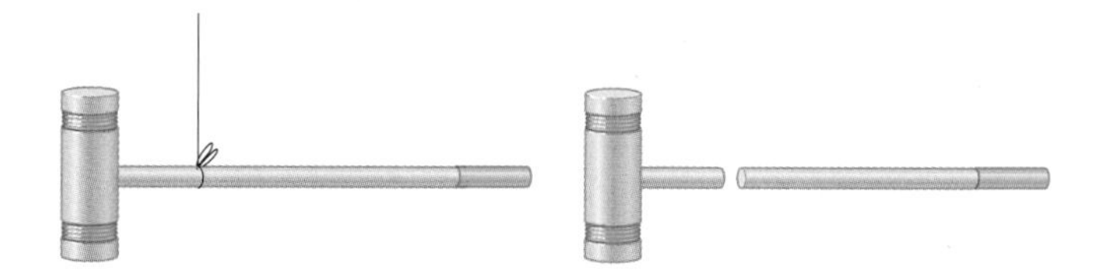

Reasoning and Discussion

The mallet balances because the torques due to the two pieces are of equal magnitude. The piece with the head of the mallet extends a smaller distance from the point of suspension than does the other piece, hence its mass must be greater; that is, a large mass at a small distance creates the same torque as a small mass at a large distance.

Answer:

(b) The piece with the head of the mallet has more mass.

▲ In this scene from the movie *Mission Impossible*, Tom Cruise is attempting to download top-secret computer files without setting off the elaborate security system in the room. To accomplish this nearly impossible mission he is suspended from the ceiling, since touching the floor would immediately give away his presence. To remain in equilibrium above the floor as he works, he must carefully adjust the position of his arms and legs to keep his center of mass directly below the suspension point.

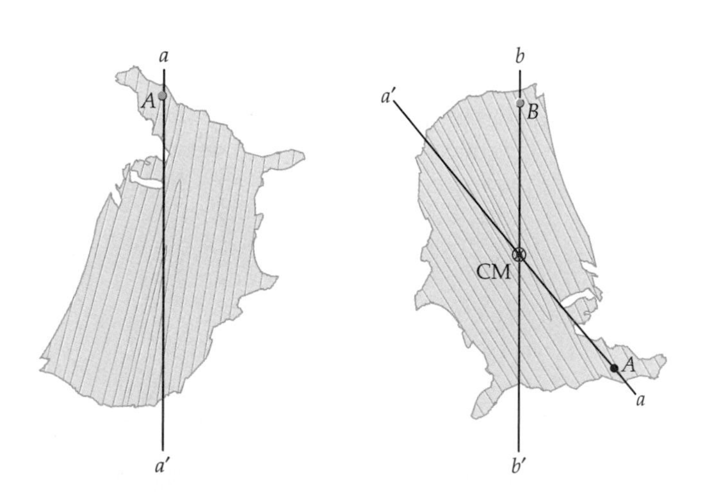

▲ **FIGURE 11–9 The geometric center of the United States**
To find the center of mass of an irregularly shaped object, such as the United States, suspend it from two or more points. The center of mass lies on a vertical line extending downward from the suspension point. The intersection of these vertical lines gives the precise location of the center of mass.

Similar considerations apply to an object that is at rest on a surface, as opposed to being suspended from a point. In such a case, the object is in equilibrium as long as its center of mass is directly above the base on which it is supported. For example, when you stand upright with normal posture your feet provide a base of support, and your center of mass is above a point roughly halfway between your feet. If you lift your right foot from the floor—without changing your posture—you will begin to lose your balance and tip over. The reason is that your center of mass is no longer above the base of support, which is now your left foot. To balance on your left foot you must lean slightly in that direction so as to position your center of mass directly above the foot. This principle applies to everything from a performer in a high-wire act to one of the "balancing rocks" that are a familiar sight in the desert southwest. In homework Problem 37 we apply this condition for stability to a stack of books on the edge of a table.

▶ (Left) Although it looks precarious, this rock in Arches National Park, Utah, has probably been balancing above the desert for many thousands of years. It will remain secure on its perch as long as its center of mass lies above its base of support. (Right) Although his knowledge may be based more on experience than on physics, the man on this ladder knows what he must do to keep from falling. By extending his leg backward as he leans forward, he keeps his center of mass safely positioned over the foot that supports him.

11–5 Dynamic Applications of Torque

In this section we focus on applications of Newton's second law for rotation. For example, consider a disk-shaped pulley of radius R and mass M with a string wrapped around its circumference, as in Figure 11–10 (a). Hanging from the string is a mass m. When the mass is released, it accelerates downward and the pulley begins to rotate. If the pulley rotates without friction, and the string unwraps without slipping, what is the acceleration of the mass and the tension in the string?

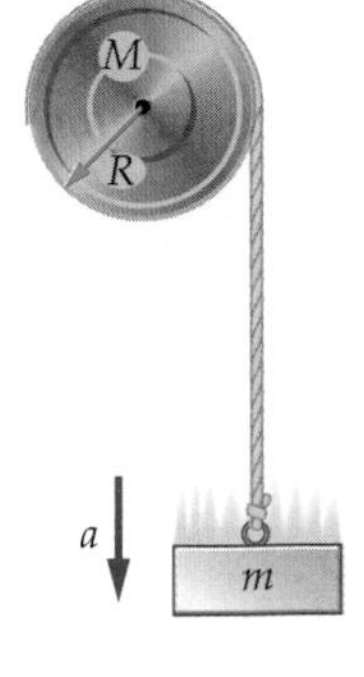

(a) Physical picture

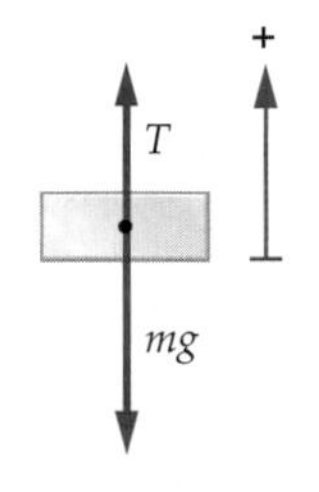

(b) Free-body diagram for mass

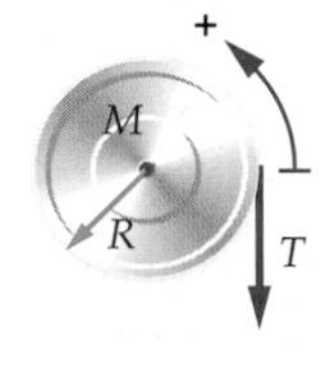

(c) Free-body diagram for pulley

▶ **FIGURE 11–10 A mass suspended from a pulley**

A mass m hangs from a string wrapped around the circumference of a disk-shaped pulley of radius R and mass M. When the mass is released, it accelerates downward. Positive directions of motion for the system are shown in parts **(b)** and **(c)**.

At first it may seem that since the pulley rotates freely, the mass will simply fall with the acceleration of gravity. But remember, the pulley has a nonzero moment of inertia, $I > 0$, which means that it resists any change in its rotational motion. In order for the pulley to rotate, the string must pull downward on it. This means that the string also pulls upward on the mass m with a tension T. As a result, the net downward force on m is less than mg, and thus its acceleration is less than g.

To solve for the acceleration of the mass, we must apply Newton's second law to both the linear motion of the mass *and* the rotational motion of the pulley. The first step is to define a consistent choice of positive directions for the two motions. In Figure 11–10 (a) we note that when the pulley rotates counterclockwise, the mass moves upward. Thus, we choose counterclockwise to be positive for the pulley, and upward to be positive for the mass.

With our positive directions established, we proceed to apply Newton's second law. Referring to the free-body diagram for the mass, shown in **Figure 11–10 (b)**, we see that

$$T - mg = ma \qquad \textbf{11–8}$$

Similarly, the free-body diagram for the pulley is shown in **Figure 11–10 (c)**. Note that the tension in the string, T, exerts a tangential force on the pulley at a distance R from the axis of rotation. This produces a torque of magnitude TR. Since the tension tends to cause a clockwise rotation, it follows that the torque is negative; thus, $\tau = -TR$. As a result, Newton's second law for the pulley gives

$$-TR = I\alpha \qquad \textbf{11–9}$$

Now, these two statements of Newton's second law are related by the fact that the string unwraps without slipping. As was discussed in Chapter 10, when a string unwraps without slipping, the angular and linear accelerations are related by

$$\alpha = \frac{a}{R}$$

Using this relation in Equation 11–9 we have

$$-TR = I\frac{a}{R}$$

or, dividing by R,

$$T = -I\frac{a}{R^2}$$

Substituting this result into Equation 11–8 yields

$$-I\frac{a}{R^2} - mg = ma$$

Finally, dividing by m and rearranging yields the acceleration, a:

$$a = -\frac{g}{\left(1 + \dfrac{I}{mR^2}\right)} \qquad \textbf{11–10}$$

Let's briefly check our solution for a. First, note that a is negative. This is to be expected since the mass accelerates downward, which is the negative direction. Second, if the moment of inertia were zero, $I = 0$, or if the mass m were infinite, $m \to \infty$, the mass would fall with the acceleration of gravity, $a = -g$. When I is greater than zero and m is finite, however, the acceleration of the mass has a magnitude less than g. In fact, in the limit of an infinite moment of inertia, $I \to \infty$, the acceleration vanishes.

A further example of using Newton's laws to relate linear and rotational motions is given next.

EXAMPLE 11–7 The Pulley Matters

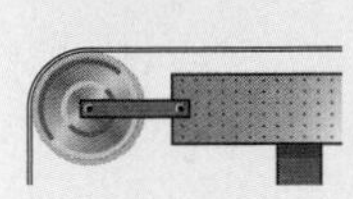

A 0.31-kg cart on a horizontal air track is attached to a string. The string passes over a disk-shaped pulley of mass 0.080 kg and radius 0.012 m and is pulled vertically downward with a constant force of 1.1 N. Find **(a)** the tension in the string between the pulley and the cart, and **(b)** the acceleration of the cart.

Picture the Problem

The system is shown below. We label the mass of the cart with M, the mass of the pulley with m, and the radius of the pulley with r. The applied downward force creates a tension $T_1 = 1.1$ N in the vertical portion of the string. The horizontal portion of the string, from the pulley to the cart, has a tension T_2. If the pulley had zero mass, these two tensions would be equal. In this case, however, T_2 will have a different value than T_1.

We also show free-body diagrams for the pulley and the cart. The positive direction of rotation is counterclockwise, and the corresponding positive direction of motion for the cart is to the left.

Strategy

The two unknowns, T_2 and a, can be found by applying Newton's second law to both the pulley and the cart. This gives two equations for two unknowns.

In applying Newton's second law to the pulley, note that since the pulley is a disk, it follows that $I = \frac{1}{2}mr^2$. Also, since the string is not said to slip as it rotates the pulley, we can assume that the angular and linear accelerations are related by $\alpha = a/r$.

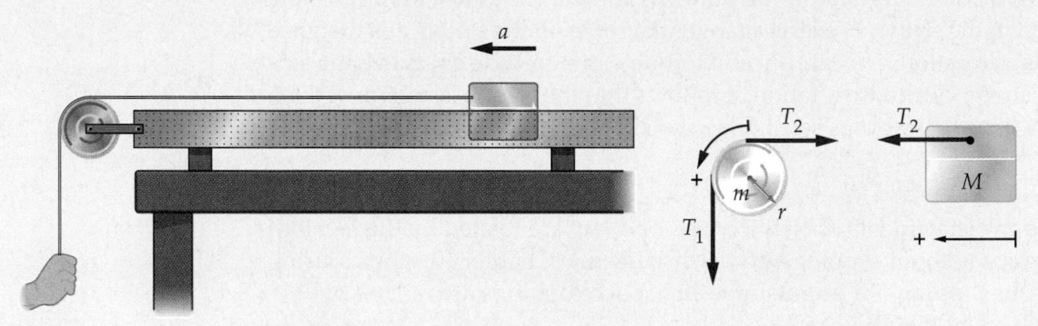

Physical picture **Free-body diagrams**

Solution

Part (a)

1. Apply Newton's second law to the cart:

$$T_2 = Ma$$

2. Apply Newton's second law to the pulley. Note that T_1 causes a positive torque, and T_2 causes a negative torque. In addition, use the relation $\alpha = a/r$.

$$\Sigma\tau = I\alpha$$

$$rT_1 - rT_2 = \left(\tfrac{1}{2}mr^2\right)\left(\frac{a}{r}\right) = \tfrac{1}{2}mra$$

3. Use the cart equation, $T_2 = Ma$, to eliminate a in the pulley equation:

$$a = \frac{T_2}{M}$$

$$rT_1 - rT_2 = \tfrac{1}{2}mr\left(\frac{T_2}{M}\right)$$

4. Cancel r and solve for T_2:

$$T_2 = \frac{T_1}{1 + m/2M} = \frac{1.1\text{ N}}{1 + 0.080\text{ kg}/[2(0.31\text{ kg})]} = 0.97\text{ N}$$

Part (b)

5. Use $T_2 = Ma$ to find the acceleration:

$$a = \frac{T_2}{M} = \frac{0.97\text{ N}}{0.31\text{ kg}} = 3.1\text{ m/s}^2$$

Insight

Note that T_2 is less than T_1. As a result, the net torque acting on the pulley is in the counterclockwise direction, causing a rotation in that direction, as expected. If the mass of the pulley were zero ($m = 0$), the two tensions would be equal, and the acceleration of the cart would be $T_1/M = 3.5\text{ m/s}^2$.

Practice Problem

What applied force is necessary to give the cart an acceleration of 2.2 m/s^2? [**Answer:** $T_1 = T_2(1 + m/2M) = (Ma)(1 + m/2M) = 0.77$ N]

Some related homework problems: Problem 42, Problem 43

11–6 Angular Momentum

When an object of mass m moves with a speed v in a straight line, we say that it has a linear momentum, $p = mv$. When the same object moves with an angular speed ω along the circumference of a circle of radius r, as in **Figure 11–11**, we say that it has an **angular momentum**, L. The magnitude of L is given by replacing m and v in the expression for p with their angular analogs I and ω (Section 10–5). Thus, we define the angular momentum as follows:

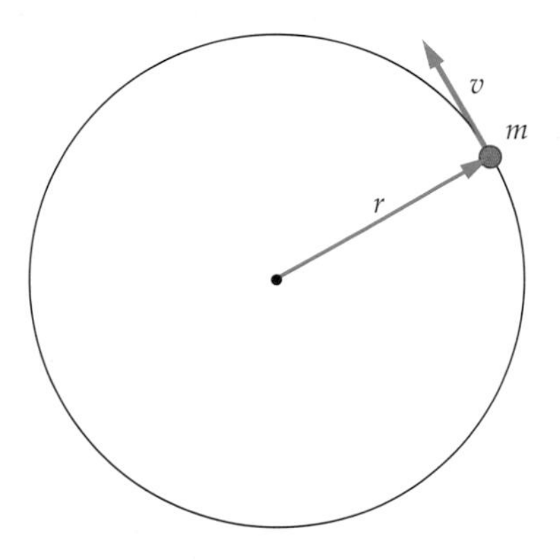

▲ FIGURE 11–11 The angular momentum of circular motion

A particle of mass m, moving in a circle of radius r with a speed v. This particle has an angular momentum of magnitude $L = rmv$.

Definition of the Angular Momentum, L

$$L = I\omega \qquad \textbf{11–11}$$

SI unit: $\text{kg}\cdot\text{m}^2/\text{s}$

This expression applies to any object undergoing angular motion, whether it is a point mass moving in a circle, as in Figure 11–11, or a rotating hoop, disk, or other object.

Returning for a moment to the case of a point mass m moving in a circle of radius r, recall that the moment of inertia in this case is $I = mr^2$ (Equation 10-18). In addition, the linear speed of the mass is $v = r\omega$ (Equation 10-12). Combining these results we find

$$L = I\omega = (mr^2)(v/r) = rmv$$

Noting that mv is the linear momentum p, we find that the angular momentum of a point mass can be written in the following form:

$$L = rmv = rp \qquad \textbf{11–12}$$

It is important to recall that this expression applies specifically to a point particle moving along the circumference of a circle.

More generally, a point object may be moving at an angle θ with respect to a radial line, as indicated in **Figure 11–12 (a)**. In this case, it is only the tangential component of the momentum, $p \sin\theta = mv \sin\theta$, that contributes to the angular momentum, just as the tangential component of the force, $F \sin\theta$, is all that contributes to the torque. Thus, the magnitude of the angular momentum for a point particle is defined as:

Angular Momentum, L, for a Point Particle

$$L = rp \sin\theta = rmv \sin\theta \qquad \textbf{11–13}$$

SI unit: $\text{kg}\cdot\text{m}^2/\text{s}$

Note that if the particle moves in a circular path the angle θ is 90° and the angular momentum is $L = rmv$, in agreement with Equation 11–12. On the other hand, if the object moves radially, so that $\theta = 0$, the angular momentum is zero; $L = rmv \sin 0 = 0°$.

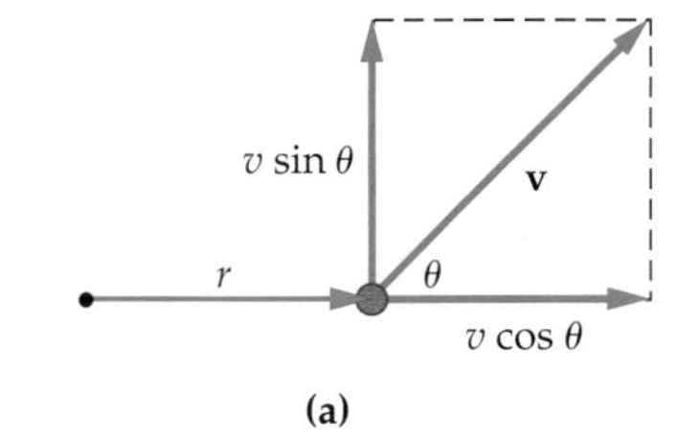

(a)

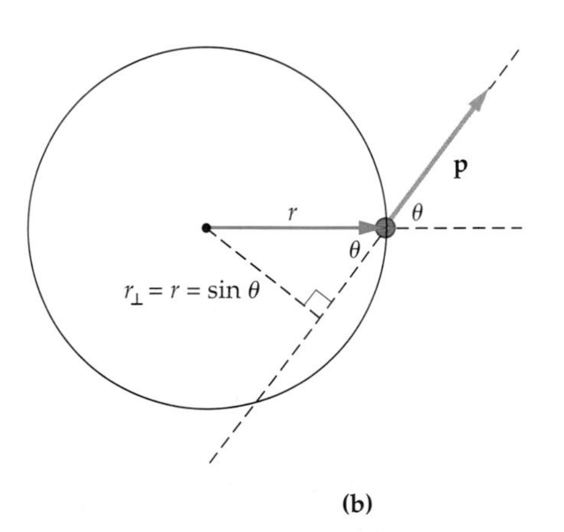

(b)

▲ FIGURE 11–12 The angular momentum of non-tangential motion

(a) When a particle moves at an angle θ with respect to the radial direction, only the tangential component of velocity, $v \sin\theta$, contributes to the angular momentum. In the case shown here, the particle's angular momentum has a magnitude given by $L = rmv \sin\theta$. **(b)** The angular momentum of an object can also be defined in terms of the moment arm, $r_\perp$. Since $r_\perp = r \sin\theta$, it follows that $L = rmv \sin\theta = r_\perp mv$. Note the similarity between this figure and Figure 11–3.

EXERCISE 11–3

Find the angular momentum of **(a)** a 0.13-kg Frisbee (considered to be a uniform disk of radius 7.5 cm) spinning with an angular speed of 1.15 rad/s and **(b)** a 95-kg person running with a speed of 5.1 m/s on a circular track of radius 25 m.

Solution

(a) Recalling that $I = \frac{1}{2}mR^2$ for a uniform disk (Table 10–1), we have

$$L = I\omega$$
$$= \left(\tfrac{1}{2}mR^2\right)\omega = \tfrac{1}{2}(0.13\ \text{kg})(0.075\ \text{m})^2(1.15\ \text{rad/s}) = 4.2 \times 10^{-4}\ \text{kg}\cdot\text{m}^2/\text{s}$$

(b) Treating the person as a particle of mass m we find

$$L = rmv = (25\ \text{m})(95\ \text{kg})(5.1\ \text{m/s}) = 12{,}000\ \text{kg}\cdot\text{m}^2/\text{s}$$

An alternative definition of the angular momentum uses the moment arm, $r_\perp$, as was done for the torque in Equation 11–3. To apply this definition, start by extending a line through the momentum vector, **p**, as in **Figure 11–12 (b)**. Next, draw a line from the axis of rotation perpendicular to the line through **p**. The perpendicular distance from the axis of rotation to the line of **p** is the moment arm. From the figure we see that $r_\perp = r \sin\theta$. Hence, from Equation 11–13, the angular momentum is

$$L = r_\perp p = r_\perp mv$$

If an object moves in a circle of radius r the moment arm is $r_\perp = r$ and the angular momentum reduces to our earlier result, $L = rp$.

CONCEPTUAL CHECKPOINT 11–3

Does an object moving in a straight line have non-zero angular momentum **(a)** always, **(b)** sometimes, or **(c)** never?

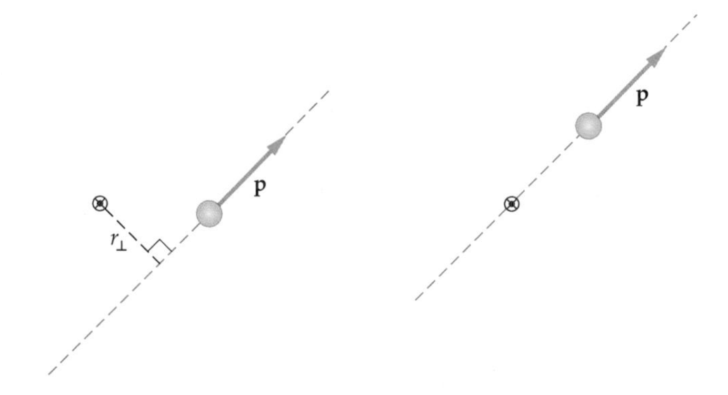

Reasoning and Discussion

The answer is sometimes, because it depends on the choice of the axis of rotation. If the axis of rotation is not on the line drawn through the momentum vector, as in the sketch at left, the moment arm is nonzero, and therefore $L = r_\perp p$ is also nonzero. If the axis of rotation is on the line of motion, as in the sketch at right, the moment arm is zero; hence the linear momentum is radial and L vanishes.

Answer:

(b) An object moving in a straight line may or may not have angular momentum, depending on the location of the axis of rotation.

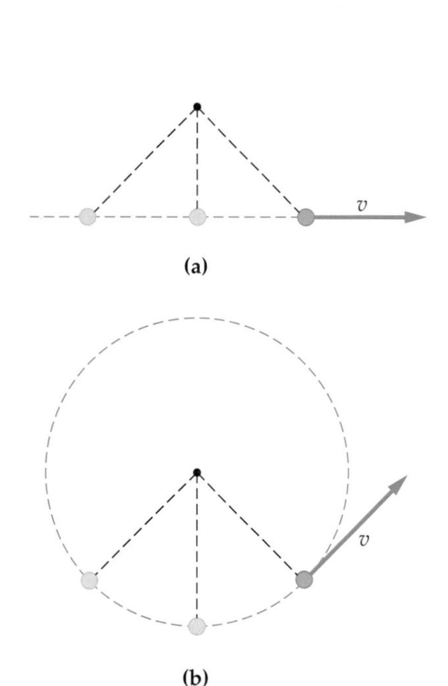

▲ **FIGURE 11–13 Angular momentum in linear and circular motion**

An object moving in **(a)** a straight line and **(b)** a circular path. In both cases, the angular position increases with time, hence the angular momentum is positive.

Note that an object moving with a momentum p in a straight line that does not go through the axis of rotation has an *angular* position that changes with time. This is illustrated in **Figure 11–13 (a)**. It is for this reason that such an object is said to have an *angular* momentum.

The sign of L is determined by whether the angle to a given object is increasing or decreasing with time. For example, the object moving counterclockwise in a circular path in **Figure 11–13 (b)** has a positive angular momentum, since θ is increasing with time. Similarly, the object in Figure 11–13 **(a)** also has an angle θ that increases with time, hence its angular momentum is positive as well. On the other hand, if these objects were to have their direction of motion reversed, they would have angles that decrease with time and their angular momenta would be negative.

EXAMPLE 11–8 Jump on

Running with a speed of 4.10 m/s, a 23.2-kg child heads toward the rim of a merry-go-round, as shown. If the radius of the merry-go-round is 2.00 m, and the child moves in the direction shown, what is the child's angular momentum with respect to the center of the merry-go-round?

Picture the Problem

Our sketch shows the child approaching the rim of the merry-go-round. Since the line of motion of the child does not go through the center of the merry-go-round, the child has nonzero angular momentum with respect to that point.

Strategy

The child moves at an angle of 135° with respect to the radial line, as shown in the sketch. Therefore, the angular momentum is given by $L = rmv \sin \theta$, with $\theta = 135°$.

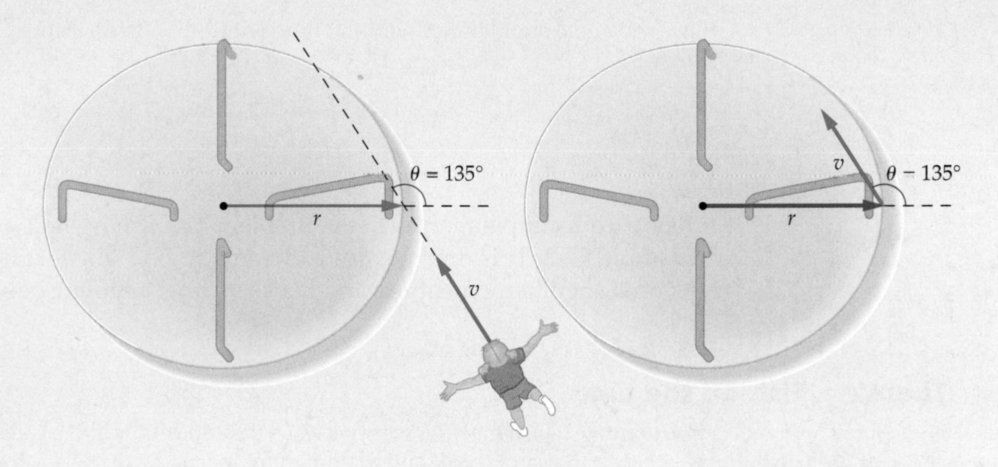

Solution

1. Evaluate $L = rmv \sin \theta$:

$$L = rmv \sin \theta = (2.00 \text{ m})(23.2 \text{ kg})(4.10 \text{ m/s}) \sin 135° = 135 \text{ kg} \cdot \text{m}^2/\text{s}$$

Insight

When the child lands on the merry-go-round, she will transfer angular momentum to the merry-go-round, causing it to rotate about its center. This will be discussed in more detail in the next section.

Practice Problem

For what angle relative to the radial line does the child have a maximum angular momentum? What is the angular momentum in this case? [**Answer:** $\theta = 90°$, for which $L = rmv = 190 \text{ kg} \cdot \text{m}^2/\text{s}$]

Some related homework problems: Problem 49, Problem 50, Problem 51

Next, we consider the rate of change of angular momentum with time. Since the moment of inertia is a constant—as long as the mass and shape of the object remain unchanged—the change in L in a time interval Δt is

$$\frac{\Delta L}{\Delta t} = I\frac{\Delta \omega}{\Delta t}$$

Recall, however, that $\Delta\omega/\Delta t$ is the angular acceleration, α. Therefore, we have

$$\frac{\Delta L}{\Delta t} = I\alpha$$

Since $I\alpha$ is the torque, it follows that Newton's second law for rotational motion can be written as

Newton's Second Law for Rotational Motion

$$\tau = I\alpha = \frac{\Delta L}{\Delta t} \qquad \textbf{11–14}$$

Clearly, this is the rotational analog of $\mathbf{F} = \Delta\mathbf{p}/\Delta t$.

EXERCISE 11–4

In a light wind, a windmill experiences a constant torque of 255 N · m. If the windmill is initially at rest, what is its angular momentum 2.00 s later?

Solution

Solve Equation 11–14 for the change in angular momentum:

$$\Delta L = L_f - L_i = \tau \Delta t$$

Since the initial angular momentum of the windmill is zero, its final angular momentum is

$$L_f = \tau\Delta t = (255\ \text{N}\cdot\text{m})(2.00\ \text{s}) = 510\ \text{kg}\cdot\text{m}^2/\text{s}$$

The next Example returns to the problem of a fishing reel, considered earlier in Example 11–2. This time we use Newton's second law for rotation to find the reel's final angular momentum, and hence its final angular speed.

EXAMPLE 11–9 There's a Fish on the Line

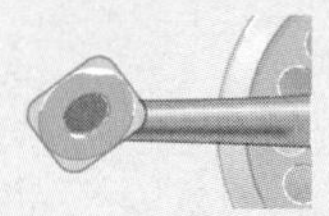

The spool of a fishing reel is free to rotate (since the drag was not set), when suddenly a fish pulls on the line with a tension T. If the spool, initially at rest, has a radius R and a moment of inertia I, what is its angular speed after a time t?

Picture the Problem

In the sketch we see that the fishing line exerts a tangential force T at a radius R from the center of the spool. Hence, the torque exerted on the spool is RT.

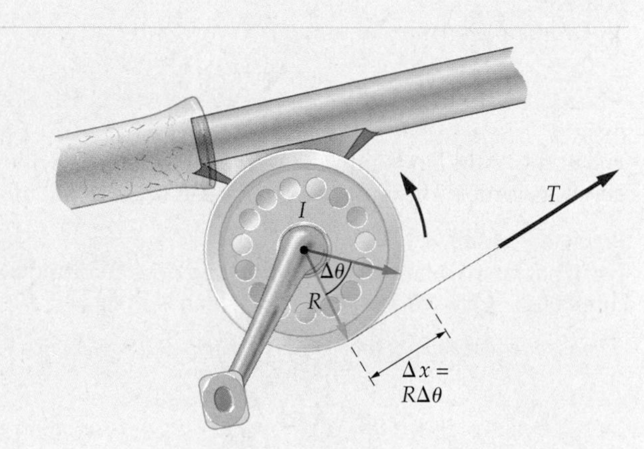

Strategy

Knowing the torque and the elapsed time, $\Delta t = t$, we can find the change in angular momentum using $\Delta L = \tau\,\Delta t$. Next, we can relate angular momentum to angular speed using $L = I\omega$. Combining these relations yields the final angular speed.

Solution

1. Calculate the torque acting on the spool: $\tau = RT$
2. Find the change in angular momentum: $\Delta L = L_f - L_i = \tau\Delta t = RTt$
3. Find the final angular momentum, using the fact that the initial angular momentum is zero: $L_f = RTt$
4. Relate the angular momentum to the angular speed: $L_f = I\omega_f$
5. Solve for the final angular speed: $I\omega_f = RTt$

$$\omega_f = \left(\frac{RT}{I}\right)t$$

Insight

Note that the angular speed obtained using angular momentum agrees with the result obtained in Example 11–2 using angular kinematics.

Practice Problem

If the initial angular speed of the reel is ω_0, what is its angular speed after a time t? [**Answer:** $\omega_f = \omega_0 + (RT/I)t$]

Some related homework problems: Problem 52, Problem 53

11–7 Conservation of Angular Momentum

When an ice skater goes into a spin and pulls her arms inward to speed up, she probably doesn't think about angular momentum. Neither does a diver, who springs into the air and folds his body to speed his rotation. Most people, in fact, are not aware that the actions of these athletes are governed by the same basic laws of physics that cause a collapsing star to spin faster as it becomes a rapidly rotating pulsar. Yet in all these cases, as we shall see, **conservation of angular momentum** is at work.

To see the origin of angular momentum conservation, consider an object with an initial angular momentum L_i acted on by a single torque τ. After a period of time, Δt, the object's angular momentum changes in accordance with Newton's second law:

$$\tau = \frac{\Delta L}{\Delta t}$$

Solving for ΔL, we find

$$\Delta L = L_f - L_i = \tau \, \Delta t$$

Thus, the final angular momentum of the object is

$$L_f = L_i + \tau \, \Delta t$$

If the torque acting on the object is zero, $\tau = 0$, it follows that the initial and final angular momenta are equal—that is, the angular momentum is conserved:

$$L_f = L_i \qquad (\text{if } \tau = 0)$$

Angular momentum is also conserved in systems acted on by more than one torque, provided that the *net external torque* is zero. The reason that internal torques can be ignored is that, just as internal forces come in equal and opposite pairs that cancel, so too do internal torques. As a result, the internal torques in a system sum to zero, and the net torque acting on it is simply the net external torque. Thus, for a general system, angular momentum is conserved if $\tau_{\text{net, ext}}$ is zero:

Conservation of Angular Momentum

$$L_f = L_i \qquad (\text{if } \tau_{\text{net,ext}} = 0) \qquad \textbf{11–15}$$

As an illustration of angular momentum conservation, we consider the case of a student rotating on a piano stool in the next Example. Notice how a change in moment of inertia results in a change in angular speed.

▲ Once she has launched herself into space, this diver is essentially a projectile. However, the principle of conservation of angular momentum allows her to control the rotational part of her motion. By curling her body up into a tight "tuck" she decreases her moment of inertia, thereby increasing the speed of her spin. To slow down for an elegant entry into the water, she will extend her body, increasing her moment of inertia.

EXAMPLE 11–10 Going for a Spin

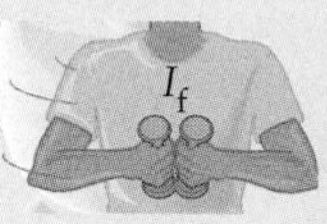

For a classroom demonstration, a student sits on a piano stool holding a sizable mass in each hand. Initially, the student holds his arms outstretched and spins about the axis of the stool with an angular speed of 3.74 rad/s. The moment of inertia in this case is 5.33 kg·m². While still spinning, the student pulls his arms in to his chest, reducing the moment of inertia to 1.60 kg·m². What is the student's angular speed now?

Picture the Problem

The initial and final configurations of the student are shown in the sketch. Clearly, the mass distribution in the final configuration will give a smaller moment of inertia.

Strategy

Ignoring friction in the axis of the stool, since none was mentioned, we conclude that no external torques act on the system. As a result, the angular moment is conserved. Therefore, setting the initial angular momentum, $L_i = I_i\omega_i$, equal to the final angular momentum, $L_f = I_f\omega_f$, yields the final angular speed.

continued on the following page

continued from previous page

Solution

1. Apply angular momentum conservation to this system:
$$L_i = L_f$$
$$I_i\omega_i = I_f\omega_f$$

2. Solve for the final angular speed, ω_f:
$$\omega_f = \left(\frac{I_i}{I_f}\right)\omega_i$$

3. Substitute numerical values:
$$\omega_f = \left(\frac{5.33\ \text{kg}\cdot\text{m}^2}{1.60\ \text{kg}\cdot\text{m}^2}\right)(3.74\ \text{rad/s}) - 12.5\ \text{rad/s}$$

Insight

Initially the student completes one revolution roughly every two seconds. After pulling the weights in, the student's rotation rate has increased to almost two revolutions a second—quite a dizzying pace.

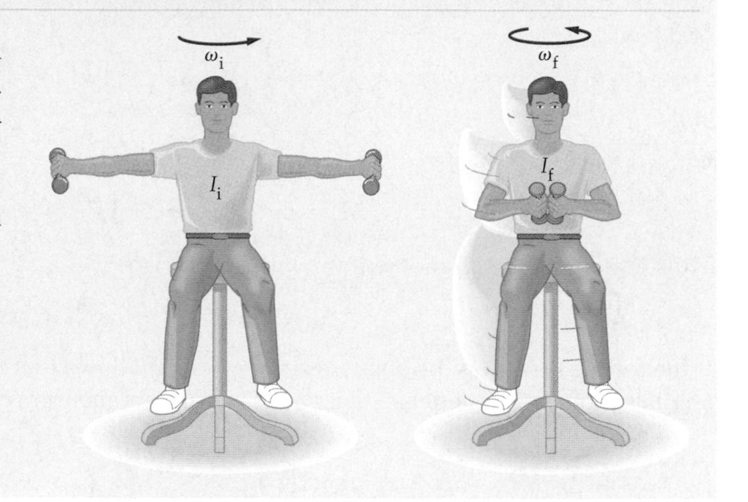

Practice Problem

What moment of inertia would be required to give a final spin rate of 10.0 rad/s? **[Answer:** $I_f = (\omega_i/\omega_f)I_i = 1.99\ \text{kg}\cdot\text{m}^2$]

Some related homework problems: Problem 55, Problem 64

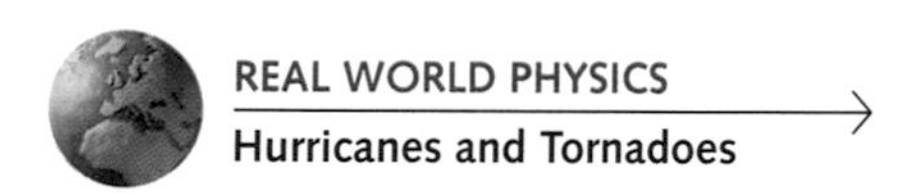

An increasing angular speed, as experienced by the student in Example 11–10, can be observed in nature as well. For example, a hurricane draws circulating air in at ground level toward its "eye," where it then rises to an altitude of 10 miles or more. As air moves inward toward the axis of rotation its angular speed increases, just as the masses held by the student speed up when they are pulled in-

▲ This 1992 satellite photo of Hurricane Andrew (left), one of the largest and most devastating hurricanes of recent decades, clearly suggests the rotating structure of the storm. The violence of the hurricane winds can be attributed in large part to conservation of angular momentum: as air is sucked in closer to the eye of the storm, its rotational velocity increases. The same principle, operating on a smaller scale, explains the tremendous destructive power of tornadoes. The tornado shown at right passed through downtown Miami on May 12, 1997.

ward. For example, if the wind has a speed of only 3.0 mph at a distance of 300 miles from the center of the hurricane, it would speed up to 150 mph when it comes to within 6.0 miles of the center. Of course, this analysis ignores friction, which would certainly decrease the wind speed. Still, the basic principle—that a decreasing distance from the axis of rotation implies an increasing speed—applies to both the student and the hurricane. Similar behavior is observed in tornadoes and waterspouts.

Another example of conservation of angular momentum occurs in stellar explosions. On occasion a star will explode, sending a portion of its material out into space. After the explosion, the star collapses to a fraction of its original size, speeding up its rotation in the process. If the mass of the star is greater than 1.44 times the mass of the Sun, the collapse can continue until a *neutron star* is formed, with a radius of only about 10 to 20 km. Neutron stars have incredibly high densities; in fact, if you could bring a teaspoonful of neutron star material to the Earth, it would weigh about 100 million tons! In addition, neutron stars produce powerful beams of x-rays and other radiation that sweep across the sky like a gigantic lighthouse beam as the star rotates. On the Earth we see pulses of radiation from these rotating beams, one for each revolution of the star. These "pulsating stars," or *pulsars*, typically have periods ranging from about 2 ms to nearly one second. The Crab Nebula (see Homework Problems 7 and 78 in Chapter 10) is a famous example of such a system. The dependence of angular speed on radius for a collapsing star is considered in Active Example 11–4.

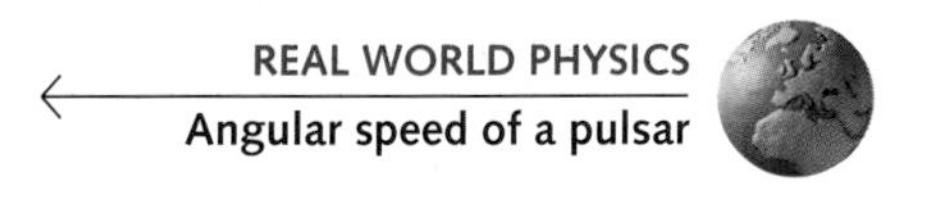

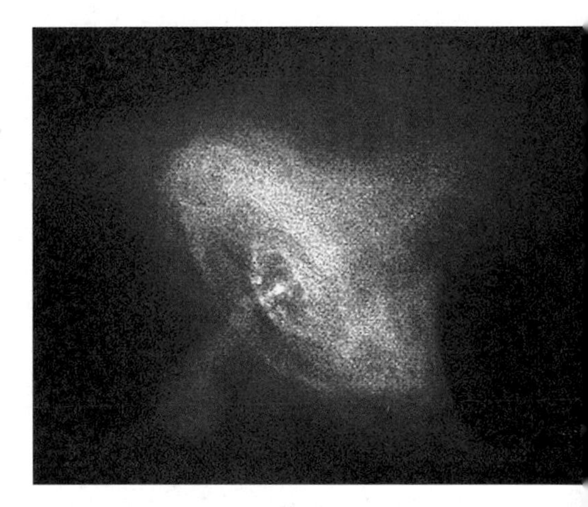

▲ Among the fastest rotating objects known in nature are pulsars: stars that have collapsed to a tiny fraction of their original size. Since all the angular momentum of a star must be conserved when it collapses, the dramatic decrease in radius is accompanied by a correspondingly great increase in rotational speed. The Crab nebula pulsar, the remains of a star whose explosion was observed on Earth nearly 1000 years ago, spins at about 30 rev/s. This X-ray photograph shows rings and jets of high-energy particles flying outward from the whirling neutron star at the center.

ACTIVE EXAMPLE 11–4 A Stellar Performance

A star of radius $R = 2.3 \times 10^8$ m rotates with an angular speed $\omega = 2.4 \times 10^{-6}$ rad/s. If this star collapses to a radius of 20.0 km, find its final angular speed. (Treat the star as if it were a uniform sphere, and assume that no mass is lost as the star collapses.)

Solution

1. Apply conservation of angular momentum: $I_i\omega_i = I_f\omega_f$
2. Write expressions for the initial and final moments of inertia: $I_i = \frac{2}{5}MR_i^2$, and $I_f = \frac{2}{5}MR_f^2$
3. Solve for the final angular speed: $\omega_f = (I_i/I_f)\omega_i = (R_i^2/R_f^2)\omega_i$
4. Substitute numerical values: $\omega_f = 320$ rad/s

Insight

The final angular speed corresponds to a period of about 20 ms, a typical period for pulsars. Since 320 rad/s is roughly 3000 rpm, it follows that a pulsar, which has the mass of a star, rotates as fast as the engine in a racing car.

Note that if the student in Example 11–10 were to stretch his arms back out again, the resulting *increase* in the moment of inertia would cause a *decrease* in his angular speed. The same effect might apply to the Earth one day. For example, a melting of the polar ice caps would lead to an increase in the Earth's moment of inertia (as we saw in Chapter 10) and thus, by angular momentum conservation, the angular speed of the Earth would decrease. This would mean that more time would be required for the Earth to complete a revolution about its axis of rotation; that is, the day would lengthen.

Since angular momentum is conserved in the systems we have studied so far, it is natural to ask whether the energy is conserved as well. We consider this question in the next Conceptual Checkpoint.

CONCEPTUAL CHECKPOINT 11–4

A skater pulls in her arms, decreasing her moment of inertia by a factor of two, and doubling her angular speed. Is her final kinetic energy **(a)** equal to, **(b)** greater than, or **(c)** less than her initial kinetic energy?

Reasoning and Discussion

Let's calculate the initial and final kinetic energies, and compare them. The initial kinetic energy is

$$K_i = \tfrac{1}{2}I_i\omega_i^2$$

After pulling in her arms, the skater has half the moment of inertia and twice the angular speed. Hence, her final kinetic energy is

$$K_f = \tfrac{1}{2}I_f\omega_f^2 = \tfrac{1}{2}(I_i/2)(2\omega_i)^2 = 2(\tfrac{1}{2}I_i\omega_i^2) = 2K_i$$

Thus, the fact that K depends on the square of ω leads to an increase in the kinetic energy. The source of this additional energy is the work done by the muscles in the skater's arms as she pulls them in to her body.

Answer:

(b) The skater's kinetic energy increases.

Rotational Collisions

In the not-too-distant past, a person would play music by placing a record on a rotating turntable. Suppose, for example, that a turntable with a moment of inertia I_t is rotating freely with an initial angular speed ω_0. A record, with a moment of inertia I_r and initially at rest, is dropped straight down onto the rotating turntable, as in **Figure 11–14**. When the record lands, frictional forces between it and the turntable cause the record to speed up and the turntable to slow down, until they both have the same angular speed. Since only internal forces are involved during this process, it follows that the system's angular momentum is conserved. We can think of this event, then, as a "rotational collision."

Before the collision, the angular momentum of the system is

$$L_i = I_t\omega_0$$

After the collision, when both the record and the turntable are rotating with the angular speed ω_f, the system's angular momentum is

$$L_f = I_t\omega_f + I_r\omega_f$$

Setting $L_f = L_i$ yields the final angular speed:

$$\omega_f = \left(\frac{I_t}{I_t + I_r}\right)\omega_0 \qquad \text{11–16}$$

Since this collision is completely inelastic, we expect the final kinetic energy to be less than the initial kinetic energy.

We conclude this section with a somewhat different example of a rotational collision. The physical principles involved are precisely the same, however.

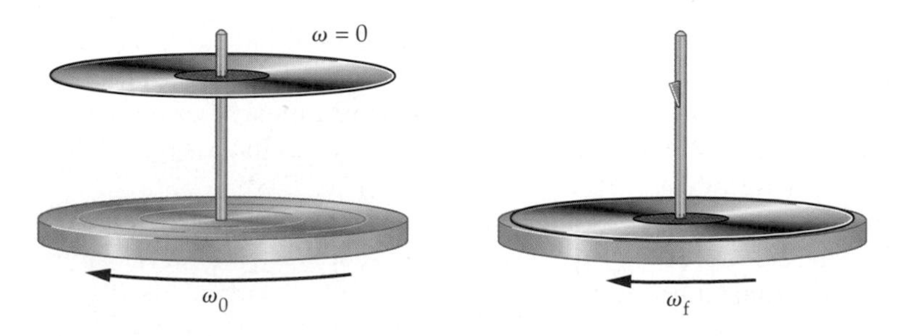

▶ FIGURE 11–14 A rotational collision

A nonrotating record dropped onto a rotating turntable is an example of a "rotational collision." Since only internal forces are involved during the collision, the final angular momentum is equal to the initial angular momentum.

ACTIVE EXAMPLE 11–5 Run and Jump

A 34.0-kg child runs with a speed of 2.80 m/s tangential to the rim of a stationary merry-go-round. The merry-go-round has a moment of inertia of 510 kg · m^2 and a radius of 2.31 m. When the child jumps onto the merry-go-round the entire system begins to rotate. What is the angular speed of the system?

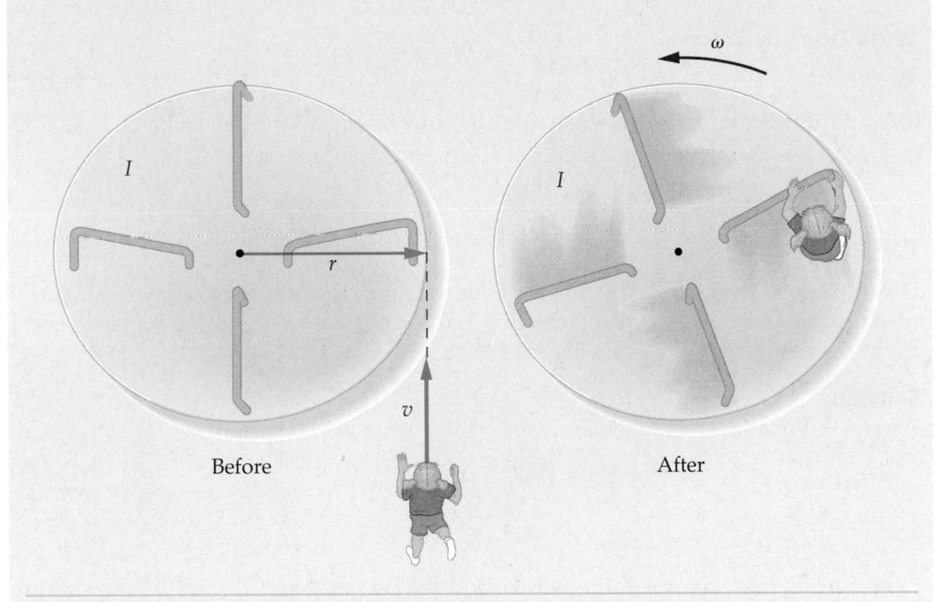

Solution

1. Write the initial angular momentum of the child: $L_i = rmv$
2. Write the final angular momentum of the system: $L_f = (I + mr^2)\omega$
3. Set $L_f = L_i$ and solve for the angular speed: $\omega = rmv/(I + mr^2)$
4. Substitute numerical values: $\omega = 0.318$ rad/s

Insight

If the moment of inertia of the merry-go-round had been zero, $I = 0$, the angular speed would be $\omega = v/r$. This means that the linear speed of the child, $r\omega = v$, is unchanged. If $I > 0$, however, the linear speed of the child is decreased. In this particular case, the child's linear speed after the collision is only $v = r\omega = 0.735$ m/s.

The initial and final kinetic energies of the system in Active Example 11–5 are considered in Homework Problem 56.

11–8 Rotational Work

Just as a force acting through a distance performs work on an object, so too does a torque acting through an angular displacement. To see this, consider again the fishing line pulled from a reel. If the line is pulled with a force F for a distance Δx, as in **Figure 11–15**, the work done on the reel is

$$W = F\,\Delta x$$

Now, since the line is unwinding without slipping, it follows that the linear displacement of the line, Δx, is related to the angular displacement of the reel, $\Delta\theta$, by the following relation:

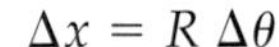

$$\Delta x = R\,\Delta\theta$$

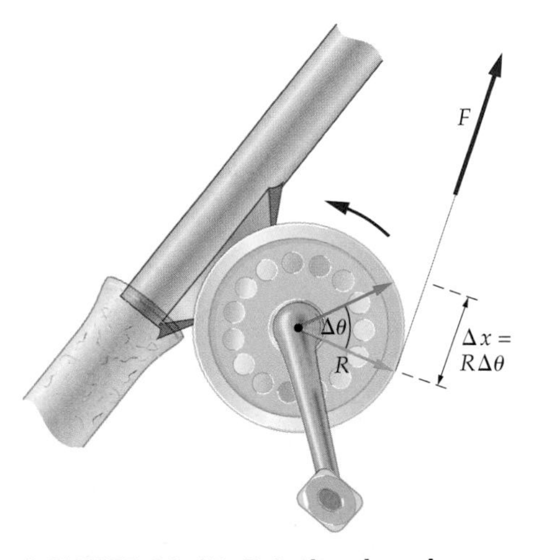

▲ **FIGURE 11–15 Rotational work**
A force F pulling a length of line Δx from a fishing reel does a work $W = F\,\Delta x$. In terms of torque and angular displacement, the work can be expressed as $W = \tau\,\Delta\theta$.

In this equation, R is the radius of the reel, and $\Delta\theta$ is measured in radians. Thus, the work can be written as

$$W = F\,\Delta x = FR\,\Delta\theta$$

Finally, the torque exerted on the reel by the line is $\tau = RF$, hence the work done on the reel is simply torque times angular displacement:

Work Done by Torque

$$W = \tau\,\Delta\theta \qquad \text{11–17}$$

Note again the analogies between angular and linear quantities in $W = F\,\Delta x$ and $W = \tau\,\Delta\theta$. As usual, τ is the analog of F, and θ is the analog of x.

EXERCISE 11–5

It takes a good deal of effort to make homemade ice cream. If the torque required to turn the handle on the ice cream maker is 5.00 N·m, how much work is expended on each complete revolution of the handle?

Solution

Applying Equation 11–17 yields

$$W = \tau\,\Delta\theta = (5.00\ \text{N}\cdot\text{m})(2\pi\ \text{rad}) = 31.4\ \text{J}$$

As we saw in Chapter 7, the net work done on an object is equal to the change in its kinetic energy. This is the work-energy theorem:

$$W = \Delta K = K_f - K_i \qquad \text{11–18}$$

The work-energy theorem applies regardless of whether the work is done by a force acting through a distance or a torque acting through an angle.

Thus, in the case of the fishing reel, we can say that

$$W = K_f - K_i = \tau\,\Delta\theta$$

In addition, the kinetic energy of the reel is $\frac{1}{2}I\omega^2$ (Equation 10–17). Thus, if the reel starts at rest, for example, and a torque $\tau = RT$ rotates it through the angle $\Delta\theta$, the final angular speed of the reel is given by

$$\tau\,\Delta\theta = K_f = \tfrac{1}{2}I\omega_f^2$$

Solving for ω_f, we find

$$\omega_f = \sqrt{\frac{2\tau\,\Delta\theta}{I}} = \sqrt{\frac{2RT\,\Delta\theta}{I}}$$

Now the angular displacement, $\Delta\theta$, that occurs in the time t has already been calculated in Example 11–2. The result is

$$\Delta\theta = \left(\frac{RT}{2I}\right)t^2$$

Substituting this into the expression for the final angular speed we find

$$\omega_f = \left(\frac{RT}{I}\right)t$$

This agrees with the results of both Example 11–2 and Example 11–9, obtained by different means.

Example 11–11 Pull a Fast One

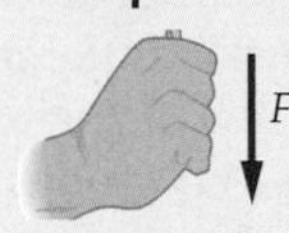

A light rope tied around the circumference of a disk-shaped pulley is pulled downward with a constant force of 1.30 N. The pulley rotates freely, has a mass of 0.950 kg, and a radius of 0.150 m. **(a)** How much work is done on the pulley as it rotates through half a revolution? **(b)** If the pulley starts at rest, what is its angular speed after half a revolution?

Picture the Problem

The force exerted by the rope is tangential, and at a distance r from the axis of rotation. Hence, it exerts a torque $\tau = rF$. The pulley itself is disk-shaped; thus its moment of inertia is $I = \frac{1}{2}mr^2$. Finally, the disk rotates through half a revolution, therefore $\Delta\theta = \pi$.

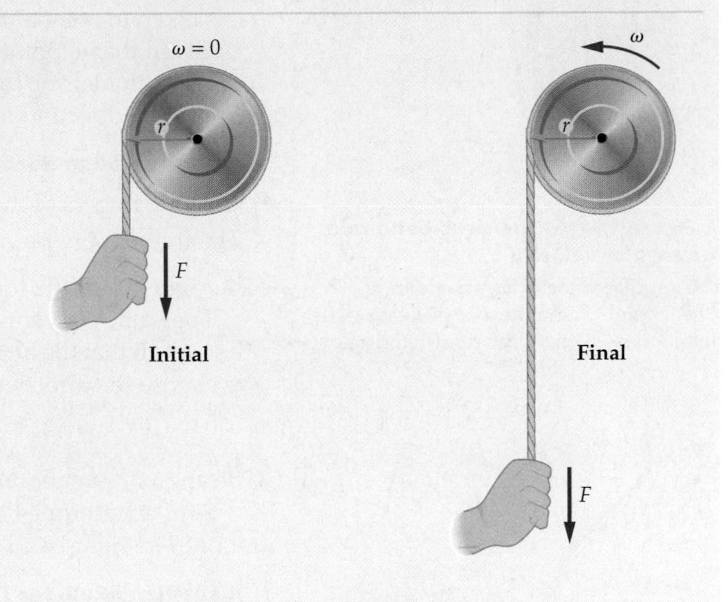

Strategy

(a) The work can be calculated using $W = \tau\,\Delta\theta$, as derived in Equation 11–17. In this case, the torque is $\tau = rF$, since the force F acts tangentially at the distance r. In addition, since the angular displacement of one revolution is 2π radians, the angular displacement of this pulley is $\Delta\theta = \pi$ rad.

(b) To find the final angular speed, ω_f, we set the work done by the torque equal to the change in the pulley's kinetic energy; $W = \Delta K = \frac{1}{2}I\omega_f^2 - \frac{1}{2}I\omega_i^2$. The work, W, has been calculated in part (a), and since the pulley starts at rest, we know that $\omega_i = 0$.

Solution

Part (a)

1. Calculate the work done on the pulley:

$$W = \tau\,\Delta\theta = rF\,\Delta\theta = (0.150\text{ m})(1.30\text{ N})(\pi\text{ rad}) = 0.613\text{ J}$$

Part (b)

2. Use the work-energy theorem, Equation 11–18, to solve for the final angular speed. Use the fact that $\omega_i = 0$:

$$W = K_f - K_i = \tfrac{1}{2}I\omega_f^2 - \tfrac{1}{2}I\omega_i^2 = \tfrac{1}{2}I\omega_f^2$$

$$\omega_f = \sqrt{\frac{2W}{I}}$$

3. Calculate the moment of inertia of the pulley:

$$I = \tfrac{1}{2}mr^2 = \tfrac{1}{2}(0.950\text{ kg})(0.150\text{ m})^2 = 0.0107\text{ kg}\cdot\text{m}^2$$

4. Substitute numerical values to determine the final angular speed, ω_f:

$$\omega_f = \sqrt{\frac{2W}{I}} = \sqrt{\frac{2(0.613\text{ J})}{(0.0107\text{ kg}\cdot\text{m}^2)}} = 10.7\text{ rad/s}$$

Insight

Note that the work done on the pulley can also be thought of as force times distance. That is, the force F acts through a distance of half the circumference of the pulley, $(2\pi r)/2 = \pi r$. Thus, the work is $W = F(\pi r) = 0.613$ J, in agreement with the rotational calculation of part (a).

Practice Problem

Find the final angular speed of the pulley if its initial angular speed is 5.00 rad/s. [**Answer:** $\omega_f = \sqrt{\omega_i^2 + 2W/I} = 11.8$ rad/s]

Some related homework problems: Problem 66, Problem 68

*11–9 The Vector Nature of Rotational Motion

We have mentioned many times that the angular velocity is a vector, and that we must be careful to use the proper sign for ω. But if the angular velocity is a vector, what is its direction?

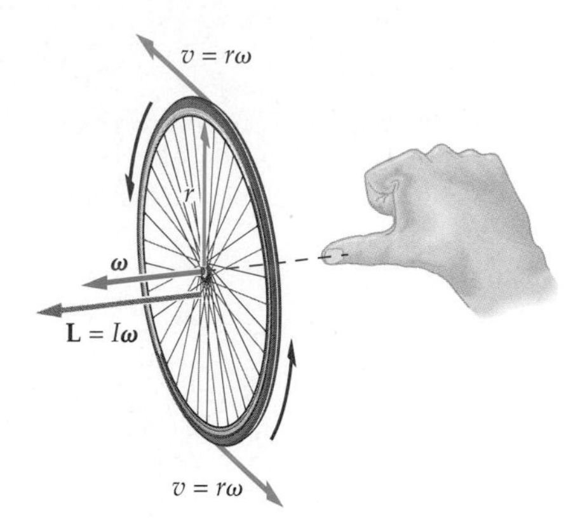

▲ **Figure 11–16 The right-hand rule for angular velocity**

The angular velocity, ω, of a rotating wheel points along the axis of rotation. Its direction is given by the right-hand rule.

To address this question, consider the rotating wheel shown in **Figure 11–16**. Each point on the rim of this wheel has a velocity vector pointing in a different direction in the plane of rotation. Since different parts of the wheel move in different directions, how can we assign a single direction to the angular velocity vector, $\boldsymbol{\omega}$? The answer is that there is only one direction that remains fixed as the wheel rotates; the direction of the axis of rotation. By definition, then, the angular velocity vector, $\boldsymbol{\omega}$, is taken to point along the axis of rotation.

Given that $\boldsymbol{\omega}$ points along the axis of rotation, we must still decide whether it points to the left or to the right in Figure 11–16. The convention we use for assigning the direction of $\boldsymbol{\omega}$ is referred to as the right-hand rule:

Right-Hand Rule for the Angular Velocity, $\boldsymbol{\omega}$

Curl the fingers of the right hand in the direction of rotation.

The thumb now points in the direction of the angular velocity, $\boldsymbol{\omega}$.

The right-hand rule for $\boldsymbol{\omega}$ is illustrated in Figure 11–16.

The same convention for direction applies to the angular momentum vector. First, recall that the angular momentum has a magnitude given by $L = I\omega$. Hence, we choose the direction of $\mathbf{L}$ to be the same as the direction of $\boldsymbol{\omega}$. That is

$$\mathbf{L} = I\boldsymbol{\omega} \qquad \mathbf{11\text{–}19}$$

The angular momentum vector is also illustrated in Figure 11–16.

Similarly, torque is a vector, and it too is defined to point along the axis of rotation. The right-hand rule for torque is similar to that for angular velocity:

Right-Hand Rule for Torque, $\boldsymbol{\tau}$

Curl the fingers of the right hand in the direction of rotation that this torque would cause if it acted alone.

The thumb now points in the direction of the torque vector, $\boldsymbol{\tau}$.

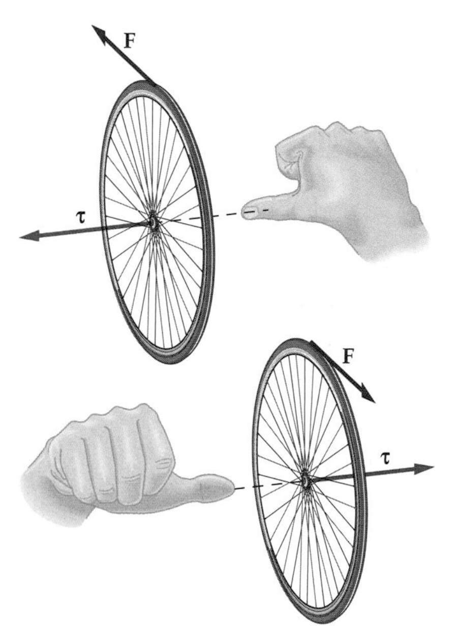

▲ **FIGURE 11–17 The right-hand rule for torque**

Examples of torque vectors obtained using the right-hand rule.

Examples of torque vectors are given in **Figure 11–17**.

As an example of torque and angular momentum vectors, consider the spinning bicycle wheel shown in **Figure 11–18**. The angular momentum vector for the wheel points to the left, along the axis of rotation. If a person pushes on the rim of the wheel in the direction indicated, the resulting torque is also to the left, as shown in the figure. If this torque lasts for a time Δt the angular momentum changes by the amount

$$\Delta\mathbf{L} = \boldsymbol{\tau}\,\Delta t$$

Adding $\Delta\mathbf{L}$ to the original angular momentum $\mathbf{L}_i$ yields the final angular momentum, $\mathbf{L}_f$, shown in Figure 11–18. Since $\mathbf{L}_f$ is in the same direction as $\mathbf{L}_i$, but with a greater magnitude, it follows that the wheel is spinning in the same direction as before, only faster. This is to be expected, considering the direction of the person's push on the wheel.

On the other hand, if a person pushes on the wheel in the opposite direction, the torque vector points to the right. As a result, $\Delta\mathbf{L}$ points to the right as well.

▶ Children have always been fascinated by tops—but not only children. The physicists in the photo at right, Wolfgang Pauli and Niels Bohr, seem as delighted by a spinning top as any child. Their contributions to modern physics, discussed in Chapter 30, helped to show that subatomic particles, the ultimate constituents of matter, have a property (now referred to as "spin") that is in some ways analogous to the rotation of a top or gyroscope.

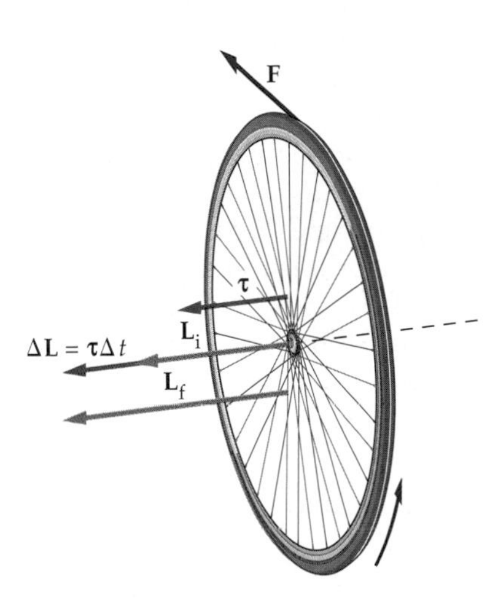

▲ Figure 11–18 Torque and angular momentum vectors
A tangential push on the spinning wheel in the direction shown causes a torque to the left. As a result, the angular momentum increases. Hence, the wheel spins faster, as expected.

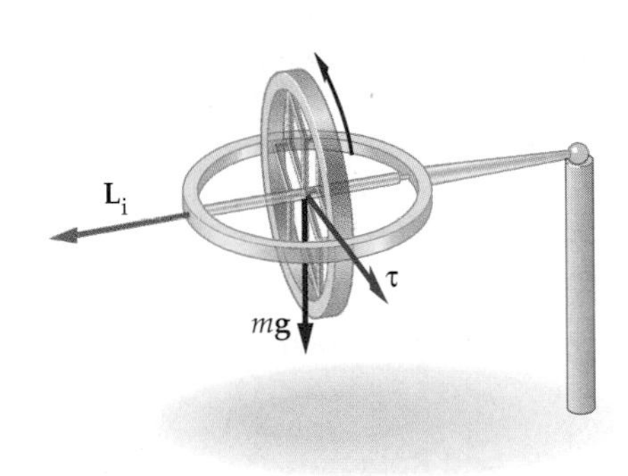

▲ FIGURE 11–19 The torque exerted on a gyroscope
A spinning gyroscope has an initial angular momentum to the left. The torque due to gravity is out of the page.

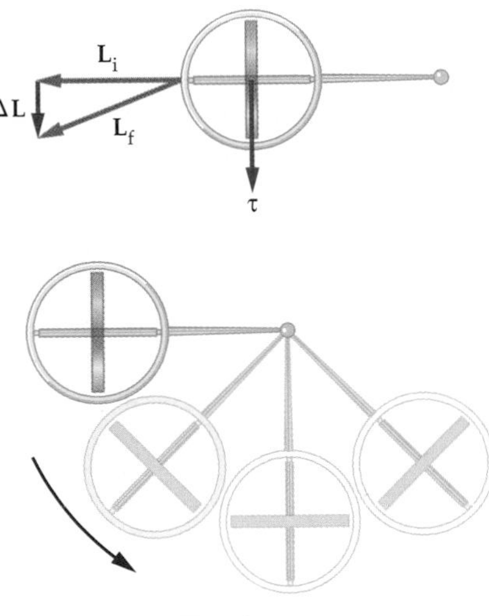

▲ FIGURE 11–20 Precession of a gyroscope
The gyroscope as viewed from above. In a time Δt the angular momentum changes by the amount $\Delta \mathbf{L} = \boldsymbol{\tau}\,\Delta t$. This causes the angular momentum vector, and hence the gyroscope as a whole, to rotate in a counterclockwise direction.

When we add $\Delta \mathbf{L}$ and $\mathbf{L}_i$ to obtain the final angular momentum, $\mathbf{L}_f$, we find that it has the same direction as $\mathbf{L}_i$, but a smaller magnitude. Hence, we conclude that the wheel spins more slowly, as one would expect.

Finally, a case of considerable interest is when the torque and angular momentum vectors are at right angles to one another. The classic example of such a system is the **gyroscope**. To begin, consider a gyroscope whose axis of rotation is horizontal, as in **Figure 11–19**. If the gyroscope were to be released with no spin it would simply fall, rotating counterclockwise downward about its point of support. Curling the fingers of the right hand in the counterclockwise direction we see that the thumb, and hence the torque due to gravity, points out of the page.

Next, imagine the gyroscope to be spinning rapidly—as would normally be the case—with its angular momentum pointing to the left in Figure 11–19. If the gyroscope is released now, it doesn't fall as before, even though the torque is the same. To see what happens instead, consider the change in angular momentum, $\Delta \mathbf{L}$, caused by the torque, $\boldsymbol{\tau}$, acting for a small interval of time. As shown in **Figure 11–20**, the small change, $\Delta \mathbf{L}$, is at right angles to $\mathbf{L}_i$, hence the final angular momentum, $\mathbf{L}_f$, is essentially the same length as $\mathbf{L}_i$, but pointing in a direction slightly out of the page. With each small interval of time, the angular momentum vector continues to change in direction so that, viewed from above as in Figure 11–20, the gyroscope as a whole rotates in a counterclockwise sense around its support point. This type of motion, where the axis of rotation changes direction with time, is referred to as **precession**.

Because of its spinning motion about its rotational axis, the Earth may be considered as one rather large gyroscope. Gravitational forces exerted on the Earth by the Sun and the Moon subject it to external torques that cause its rotational axis to precess. At the moment, the rotational axis of the Earth points toward Polaris, the "North Star," which remains almost fixed in position in time-lapse photographs while the other stars move in circular paths about it. In a few hundred years, however, Polaris will also move in a circular path in the sky because the Earth's axis of rotation will point in a different direction. After 26,000 years the Earth will complete one full cycle of precession, and Polaris will again be the pole star.

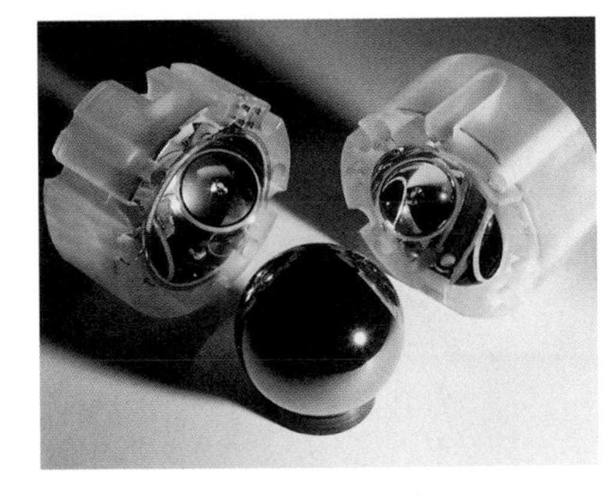

▲ The 1.5-inch fused quartz sphere shown here is no ordinary ball. In fact, it is the most perfect sphere ever manufactured. If the Earth were this smooth, the change in elevation from the deepest ocean trench to the highest mountain peak would be only 16 feet. Such precision is required because this sphere is designed to serve as the rotor for an extremely sensitive gryoscope. The device, a million times more sensitive than those used in the best inertial navigation systems, will orbit the Earth as part of an experiment to test predictions of Einstein's theory of general relativity.

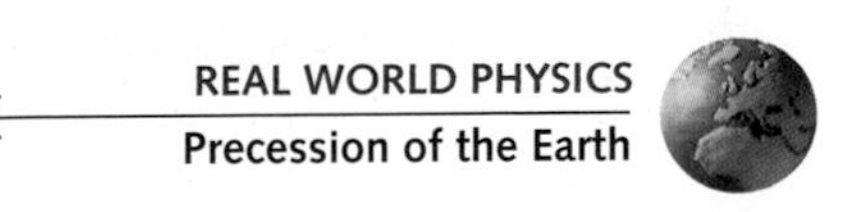

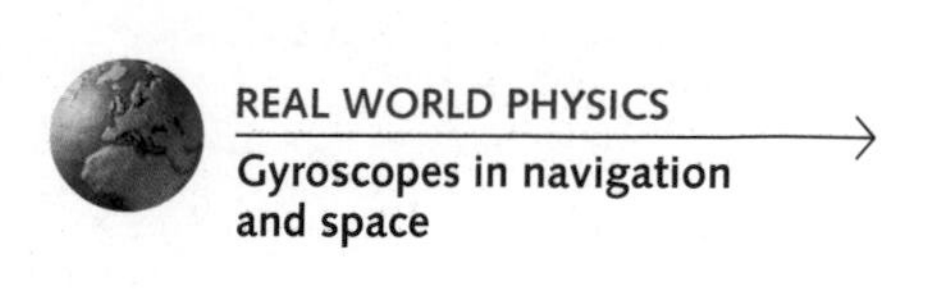

On a smaller scale, gyroscopes are used in the navigational systems of a variety of vehicles. In such applications, the rapidly spinning wheel of a gyroscope is mounted on nearly frictionless bearings so that it is practically free from external torques. If no torque acts on the gyroscope, its angular momentum vector remains unchanged both in magnitude and—here is the important point—in direction. With the axis of its gyroscope always pointing in the same, known direction, it is possible for a vehicle to maintain a desired direction of motion relative to the gyroscope's reference direction.

On the Hubble Space Telescope, for example, six gyroscopes are used for pointing and stability, though it can operate with only three working gyroscopes if necessary. Gyroscopes are also being used to test Einstein's theory of general relativity. In the Gravity Probe experiment, scheduled for launch in September of 2001, a highly sensitive gyroscope, using the world's roundest sphere as its rotor, will search for the minute deviations in its axis of rotation that are predicted by the theory.

Chapter Summary

Topic	Remarks and Relevant Equations
11–1 Torque	A force applied so as to cause an angular acceleration is said to exert a torque, τ.
tangential force	A force is tangential if it is tangent to a circle centered on the axis of rotation.
torque due to a tangential force	A tangential force F applied at a distance r from the axis of rotation produces a torque $\tau = rF$ **11–1**
torque for a general force	A force exerted at an angle θ with respect to the radial direction, and applied at a distance r from the axis of rotation, produces the torque $\tau = rF \sin\theta$ **11–2**
11–2 Torque and Angular Acceleration	A single torque applied to an object gives it an angular acceleration.
Newton's second law for rotation	The connection between torque and angular acceleration is $\tau = I\alpha$ **11–4** In this expression, I is the moment of inertia about the axis of rotation and α is the angular acceleration about this axis.
rotational/translational analogies	Torque is analogous to force, the moment of inertia is analogous to mass, and the angular acceleration is analogous to linear acceleration. Therefore, the rotational analog of $F = ma$ is $\tau = I\alpha$
11–3 Zero Torque and Static Equilibrium	The conditions for an object to be in static equilibrium are that the total force and the total torque acting on the object must be zero: $\sum F_x = 0,\quad \sum F_y = 0,\quad \sum\tau = 0$
11–4 Center of Mass and Balance	An object balances when it is supported at its center of mass.
11–5 Dynamic Applications of Torque	Newton's second law can be applied to rotational systems in a way that is completely analogous to its application to linear systems.
systems involving both rotational and linear elements	In a system with both rotational and linear motion—such as a string passing over a pulley and attached to a mass—Newton's second law must be applied separately to the rotational and linear motions of the system. Connections between the two motions, such as $\alpha = a/r$, can be used to solve for all the accelerations in the system.

11–6 Angular Momentum

A moving object has angular momentum, as long as its direction of motion does not extend through the axis of rotation.

tangential motion

An object of mass m moving tangentially with a speed v at a distance r from the axis of rotation has an angular momentum, L, given by

$$L = rmv \quad \text{11–12}$$

general motion

If an object of mass m is a distance r from the axis of rotation, and moves with a speed v at an angle θ with respect to the radial direction, its angular momentum is

$$L = rmv \sin\theta \quad \text{11–13}$$

angular momentum and angular speed

Angular momentum can be expressed in terms of angular speed and the moment of inertia as follows:

$$L = I\omega \quad \text{11–11}$$

This is the rotational analog of $p = mv$.

Newton's second law

Newton's second law can be expressed in terms of the rate of change of the angular momentum:

$$\tau = \Delta L/\Delta t \quad \text{11–14}$$

This is the rotational analog of $F = \Delta p/\Delta t$.

11–7 Conservation of Angular Momentum

If the net external torque acting on a system is zero, its angular momentum is conserved:

$$L_f = L_i$$

rotational collisions

Systems in which two rotational objects come into contact can be thought of in terms of a "rotational collision." In such a case, the total angular momentum of the system is conserved.

11–8 Rotational Work

work done by a torque

A torque acting through an angle does work, just as does a force acting through a distance. A torque τ acting through an angle $\Delta\theta$ does a work W given by

$$W = \tau\Delta\theta \quad \text{11–17}$$

work-energy theorem

The work-energy theorem is

$$W = \Delta K = K_f - K_i \quad \text{11–18}$$

This theorem applies whether the work is done by a force or by a torque. In the linear case the kinetic energy is $\frac{1}{2}mv^2$; in the rotational case, the kinetic energy is $K = \frac{1}{2}I\omega^2$: (Equation 10–17).

***11–9 The Vector Nature of Rotational Motion**

Rotational quantities have directions that point along the axis of rotation. The precise direction is given by the right-hand rule.

right-hand rule

If the fingers of the *right hand* are curled in the direction of rotation, the thumb points in the direction of the rotational quantity in question. This rule applies to the angular velocity vector, $\boldsymbol{\omega}$, the angular acceleration vector, $\boldsymbol{\alpha}$, the angular momentum vector, $\mathbf{L}$, and the torque vector, $\boldsymbol{\tau}$.

Problem-Solving Summary

Type of Problem	Relevant Physical Concepts	Related Examples
Find the torque exerted on a system.	The torque exerted by a tangential force a distance r from the axis of rotation is $\tau = rF$. If the force is at an angle θ to the radial direction the torque is $\tau = rF \sin\theta$.	Example 11–1
Determine the angular acceleration of a system.	First, calculate the torque exerted on the system. Next, find the angular acceleration using Newton's second law as applied to rotation; namely, $\tau = I\alpha$.	Examples 11–2, 11–3

Find the forces required for static equilibrium.	Static equilibrium requires that both the net force and the net torque acting on a system be zero.	Examples 11–4, 11–5, 11–6 Active Examples 11–1, 11–2, 11–3
Find the final angular momentum of a system.	A torque changes the angular momentum L of a system with time as follows; $\tau = \Delta L/\Delta t$. If no net torque acts on a system its angular momentum is conserved.	Examples 11–8, 11–9, 11–10 Active Examples 11–4, 11–5
Calculate the mechanical work done by a torque, and relate it to an object's rotational kinetic energy.	The work done by a torque τ rotating an object through an angle $\Delta\theta$ is $W = \tau\,\Delta\theta$. Once the work is known, the kinetic energy of the object can be found using the work-energy theorem, $W = \Delta K$.	Example 11–11

Conceptual Questions

1. Two forces produce the same torque. Does it follow that they have the same magnitude? Explain.
2. A car pitches down in front when the brakes are applied sharply. Explain this observation in terms of torques.
3. A tightrope walker uses a long pole to aid in balancing. Why?
4. When a motorcycle accelerates rapidly from a stop it sometimes "pops a wheelie"; that is, its front wheel may lift off the ground. Explain this behavior in terms of torques.
5. Is the tension in the string in each of the two cases shown in Conceptual Checkpoint 11–1 greater than, equal to, or less than the weight of the mass attached to the string? Explain.
6. Suppose your body is rotated about an axis that runs along your spine, or one that passes through your hips. In which case is the moment of inertia greatest? In which case will a given torque produce the greatest angular acceleration? Explain.
7. Give an example of a system in which the net torque is zero but the net force is nonzero.
8. The net force acting on a given system is zero. Is the net torque on the system necessarily zero as well?
9. Is the normal force exerted by the ground the same for all four tires on your car? Explain.
10. In Active Example 11–3, is the ladder more or less likely to slip as the person climbs higher?
11. Give some examples of everyday objects that are in static equilibrium.
12. Give some everyday examples of objects that are not in static equilibrium.
13. Can an object have both zero translational acceleration and nonzero rotational acceleration? Give an example to support your answer.
14. An empty drinking glass sits on a table. Give a qualitative description of the location of the center of mass of the glass and its contents as the glass is slowly filled with water.
15. Stars form when a large, rotating cloud of gas collapses. What happens to the angular speed of the gas cloud as it collapses?
16. What purpose does the tail rotor on a helicopter serve?
17. A puck on a horizontal, frictionless surface is attached to a string that wraps around a pole of finite radius, as shown in **Figure 11–21**. What happens to the linear speed of the puck during this process? What happens to its angular speed? Its angular momentum?

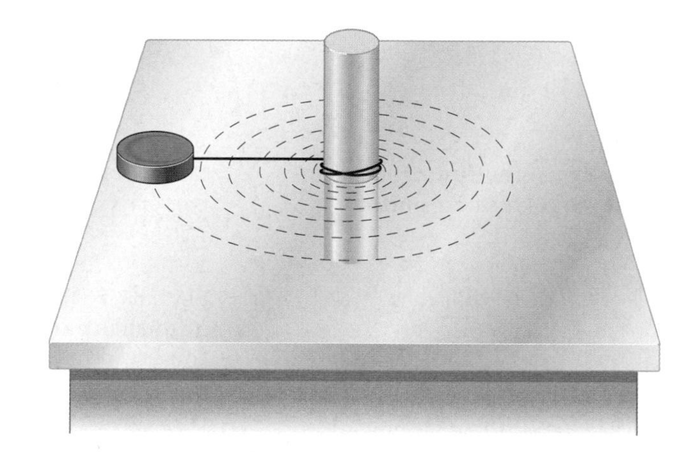

▲ **FIGURE 11–21** Question 17

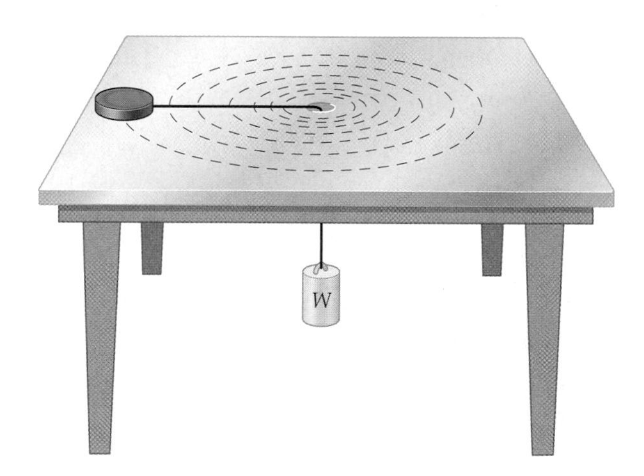

▲ **FIGURE 11–22** Question 18

18. A puck on a horizontal, frictionless surface is attached to a string that passes through a hole in the surface, as shown in **Figure 11–22**. As the puck rotates about the hole, the string is pulled downward, bringing the puck closer to the hole. During this process, what happens to the puck's linear speed? Its angular speed? Its angular momentum?
19. As the student in Example 11–10 spins with arms outstretched on the piano stool, he suddenly lets go of the weights and they drop to the floor. What happens to his angular speed?

20. Can the angular momentum of an object be changed without changing its linear momentum?
21. A uniform disk stands upright on a piece of paper on a tabletop. If the paper is pulled horizontally to the right, in which direction does the disk rotate? In which direction does the center of the disk move?
22. You have two eggs, one hard boiled the other raw. To determine which is which, you spin the eggs on the kitchen counter. Which egg spins faster?
23. A beetle sits near the rim of a turntable that rotates without friction about a vertical axis. What happens to the angular speed of the turntable as the beetle walks toward the axis of rotation?
24. Suppose a diver springs into the air with no initial angular velocity. Can the diver begin to rotate by folding into a tucked position? Explain.
25. If the Earth were to magically expand, doubling its radius while keeping its mass the same, would the length of the day increase, decrease, or stay the same? Explain.
26. A beetle sits near the rim of a turntable that is at rest, but is free to rotate about a vertical axis. What happens if the beetle begins to walk around the perimeter of the turntable? Does the beetle move relative to the ground? Consider the limits of a very massive turntable and a very light turntable.
27. A meter stick has a large mass attached to it at the 20 cm mark. If the 0 cm end of the meter stick is placed on the ground, and the stick is allowed to fall, it lands in a time T. If the 100 cm end of the stick is placed on the ground instead, is the time for the stick to hit the ground equal to, greater than, or less than T? Explain.
28. Is it more difficult to do sit ups with your hands behind your head, or with your arms outstretched in front of you? Explain in terms of torque and moment of inertia.
29. A disk and a hoop of equal radius and mass have a string wrapped around their circumferences. Hanging from the string, halfway between the disk and the hoop, is a block of mass m, as shown in **Figure 11–23**. The disk and the hoop are free to rotate about their centers. When the block is allowed to fall, does it stay on the center line, move toward the right, or move toward the left? Explain.

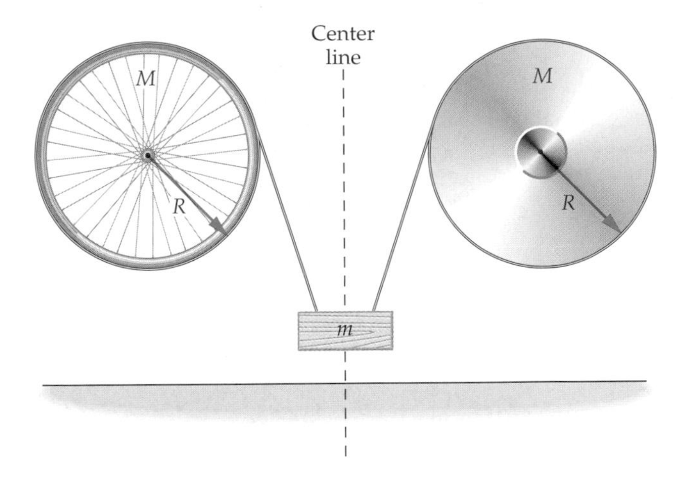

▲ **FIGURE 11–23** Question 29

30. Two spheres of equal mass and radius are rolling across the floor with the same speed. One sphere is solid, the other is hollow. Which sphere is harder to stop? Why?
31. The L-shaped object in Figure 11–27 (see Problem 12) can be rotated in one of the following three ways: (i) about the x axis; (ii) about the y axis; and (iii) about the z axis (which passes through the origin perpendicular to the plane of the figure.) If a torque τ is applied to this object, in which of the above three cases is the resulting angular acceleration greatest? In which case is it least? Explain.

Problems

Note: **IP** *denotes an integrated conceptual/quantitative problem.* **BIO** *identifies problems of biological or medical interest. Blue bullets (•, ••, •••) are used to indicate the level of difficulty of each problem.*

Section 11–1 Torque

1. • To tighten a spark plug, it is recommended that a torque of 15 N · m be applied. If a mechanic tightens the spark plug with a wrench that is 25-cm long, what is the minimum force necessary to create the desired torque?
2. • The gardening tool shown in **Figure 11–24** is used to pull weeds. If a 1.23-N · m torque is required to pull a given weed, what force did the weed exert on the tool?
3. • A 1.41-kg bowling trophy is held at arm's length, a distance of 0.600 m from the shoulder joint. What torque does the trophy exert about the shoulder if the arm is **(a)** horizontal, or **(b)** at an angle of 20.0° below the horizontal?
4. • A person slowly lowers a 3.3-kg crab trap over the side of a dock, as shown in **Figure 11–25**. What torque does the trap exert about the person's shoulder?
5. •• **IP BIO** A person holds a 1.42-N baseball in his hand, a distance of 34.0 cm from the elbow joint, as shown in **Figure 11–26**. The biceps, attached at a distance of 2.75 cm from the elbow, exert an upward force of 12.6 N on the forearm. Consider the forearm and hand to be a uniform rod with a mass of 1.20 kg. **(a)** Calculate the net torque acting on the forearm and hand.

▲ **FIGURE 11–24** Problem 2

▲ **FIGURE 11–25** Problem 4

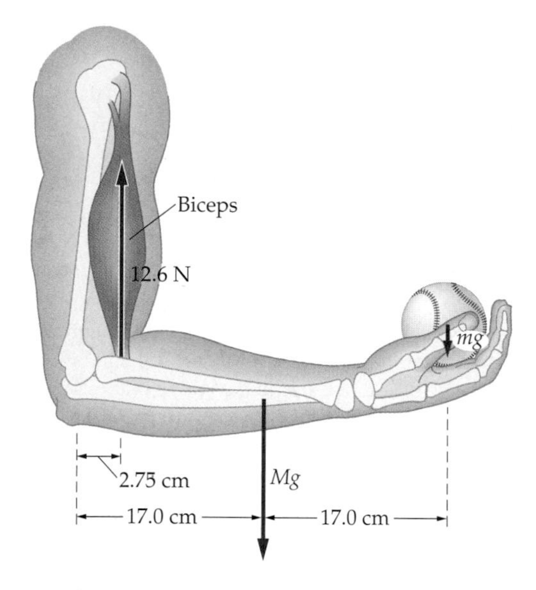

▲ **FIGURE 11–26** Problem 5

Use the elbow joint as the axis of rotation. **(b)** If the net torque obtained in part **(a)** is nonzero, in which direction will the forearm and hand rotate? **(c)** Would the net torque exerted on the forearm and hand increase or decrease if the biceps were attached farther from the elbow joint?

6. •• At the local playground, a 16-kg child sits on the end of a horizontal teeter totter, 1.5 m from the pivot point. On the other side of the pivot an adult pushes straight down on the teeter totter with a force of 95 N. In which direction does the teeter totter rotate if the adult applies the force at a distance of **(a)** 3.0 m, **(b)** 2.5 m, or **(c)** 2.0 m from the pivot?

Section 11–2 Torque and Angular Acceleration

7. • A torque of 0.97 N·m is applied to a bicycle wheel of radius 35 cm and mass 0.75 kg. Treating the wheel as a hoop, find its angular acceleration.

8. • When a ceiling fan rotating with an angular speed of 2.55 rad/s is turned off, a frictional torque of 0.220 N·m slows it to a stop in 5.75 s. What is the moment of inertia of the fan?

9. • When the play button is pressed, a CD accelerates uniformly from rest to 450 rev/min in 3.0 revolutions. If the CD has a radius of 12 cm, and a mass of 17 g, what is the torque exerted on it?

10. •• A person holds a ladder horizontally at its center. Treating the ladder as a uniform rod of length 3.05 m and mass 7.60 kg, find the torque the person must exert on the ladder to give it an angular acceleration of 0.322 rad/s^2.

11. •• **IP** A wheel on a game show is given an initial angular speed of 1.22 rad/s. It comes to rest after rotating through 3/4 of a turn. **(a)** Find the average torque exerted on the wheel given that it is a disk of radius 0.71 m and mass 6.4 kg. **(b)** If the mass of the wheel is doubled and its radius is halved, will the angle through which it rotates before coming to rest increase, decrease, or stay the same? Explain. (Assume that the average torque is unchanged.)

12. •• The L-shaped object in **Figure 11–27** consists of three masses connected by light rods. What torque must be applied to this object to give it an angular acceleration of 1.20 rad/s^2 if it is rotated about **(a)** the x axis, **(b)** the y axis, or **(c)** the z axis (which is through the origin and perpendicular to the page)?

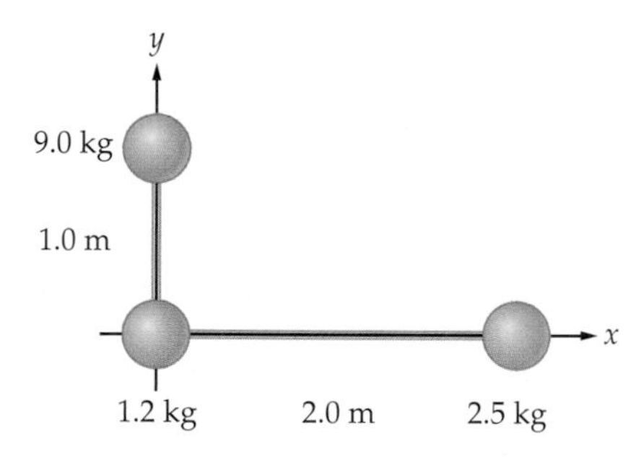

▲ **FIGURE 11–27** Problem 12

13. •• What is the angular acceleration of the rectangular object in **Figure 11–28** if a torque of 13 N·m is applied about **(a)** the x axis, **(b)** the y axis, or **(c)** the z axis (which is through the origin and perpendicular to the page)?

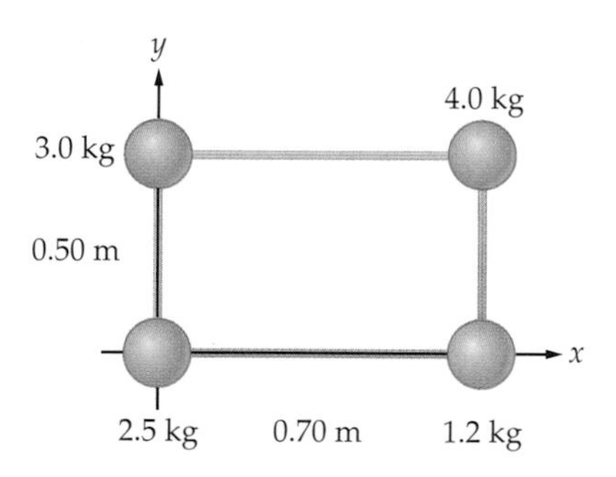

▲ **FIGURE 11–28** Problem 13

14. •• A fish takes the bait and pulls on the line with a force of 2.1 N. The fishing reel, which rotates without friction, is a cylinder of radius 0.055 m and mass 0.84 kg. **(a)** What is the angular acceleration of the fishing reel? **(b)** How much line does the fish pull from the reel in 0.25 s?

15. •• Repeat the previous problem, only now assume the reel has a friction clutch that exerts a restraining torque of 0.047 N·m.

Section 11–3 Zero Torque and Static Equilibrium

16. • A string that passes over a pulley has a 0.351-kg mass attached to one end and a 0.615-kg mass attached to the other end. The pulley, which is a disk of radius 9.50 cm, has friction in its axle. What is the magnitude of the frictional torque that must be exerted by the axle if the system is to be in static equilibrium?

17. • To loosen the lid on a jar of jam 8.9 cm in diameter a torque of 9.2 N · m must be applied to the circumference of the lid. If a jar wrench whose handle extends 15 cm from the center of the jar is attached to the lid, what is the minimum force required to open the jar?

18. • Consider the system in Active Example 11–1, this time with the axis of rotation at the location of the child. Write out both the condition for zero net force and the condition for zero net torque. Solve for the two forces.

19. • **(a)** Referring to the person holding a baseball in Problem 5, determine the force that must be exerted by the biceps for the system to be in static equilibrium. **(b)** Compare the force exerted by the biceps to the total weight of the forearm, hand, and baseball. Which is greater?

20. •• To determine the location of his center of mass, a physics student lies on a lightweight plank supported by two scales 2.50 m apart, as indicated in **Figure 11–29**. If the left scale reads 290 N, and the right scale reads 122 N, find **(a)** the student's mass and **(b)** the distance from the student's head to his center of mass.

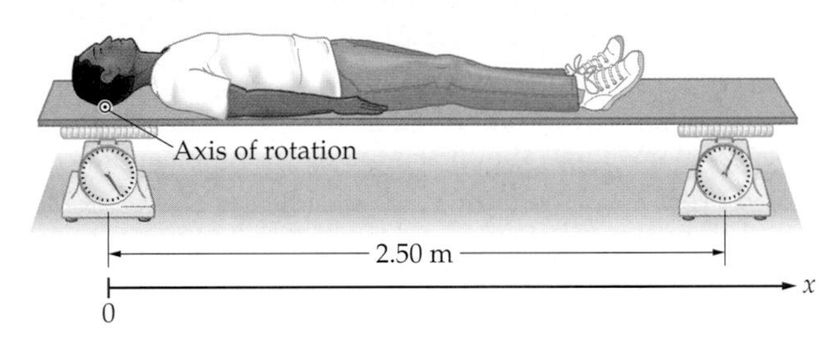

▲ **FIGURE 11–29** Problem 20

21. •• **BIO** A set of fossilized triceratops footprints show that the front and rear feet were 3.2 m apart. The rear footprints are observed to be twice as deep as the front footprints. Assuming that the rear feet pressed down on the ground with twice the force exerted by the front feet, find the horizontal distance from the rear feet to the triceratops's center of mass.

22. •• **IP** A school yard teeter totter with a total length of 5.2 m and a mass of 36 kg is pivoted at its center. A 18-kg child sits on one end of the teeter totter. **(a)** Where should a parent push with a force of 210 N in order to hold the teeter totter level? **(b)** Where should the parent push with a force of 310 N? **(c)** How would your answers to parts (a) and (b) change if the mass of the teeter totter were doubled? Explain.

23. •• A 0.110-kg remote control 21.0 cm long rests on a table, as shown in **Figure 11–30**, with a length L overhanging its edge. To operate the power button on this remote requires a force of 0.365 N. How far can the remote control extend beyond the edge of the table and still not tip over when you press the power button? Assume the mass of the remote is distributed uniformly, and that the power button is on the end of the remote overhanging the table.

24. •• **IP** A 0.23-kg meter stick is held perpendicular to a wall by a 2.5-m string going from the wall to the far end of the stick. **(a)** Find the tension in the string. **(b)** If a shorter string is used, will its tension be greater than, less than, or the same as that found in part (a)? **(c)** Find the tension in a 2.0-m string.

25. •• Repeat Example 11–4, this time with a uniform diving board that weighs 66 N.

26. •• Babe Ruth steps to the plate and casually points to left center field to indicate the location of his next home run. The mighty Babe holds his bat across his shoulder, with one hand holding the small end of the bat. The bat is horizontal, and the distance from the small end of the bat to the shoulder is 22.5 cm. If the bat has a mass of 1.10 kg, and has a center of mass that is 67.0 cm from the small end of the bat, find the magnitude and direction of the force exerted by **(a)** the hand and **(b)** the shoulder.

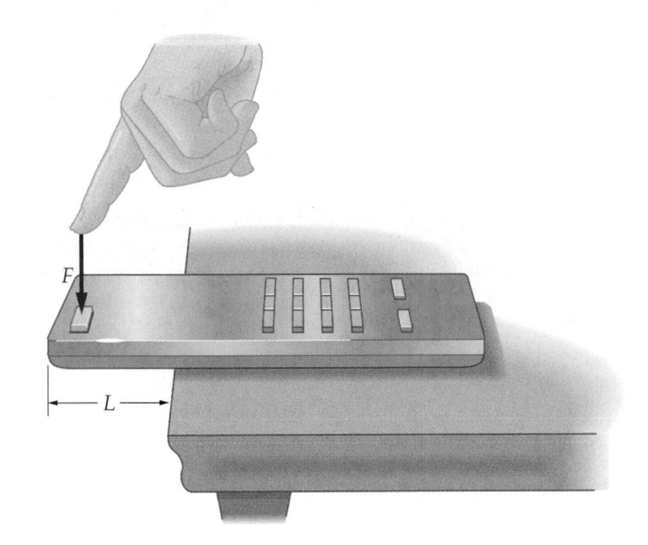

▲ **FIGURE 11–30** Problem 23

27. •• A uniform metal rod, with a mass of 3.1-kg and a length of 1.2 m, is attached to a wall by a hinge at its base. A horizontal wire bolted to the wall 0.51 m above the base of the rod holds the rod at an angle of 25° above the horizontal. **(a)** Find the tension in the wire. **(b)** Find the horizontal and vertical components of the force exerted on the rod by the hinge.

28. •• **IP** In the previous problem, suppose the wire is shortened, so that the rod now makes an angle of 35° with the horizontal. The wire is horizontal, as before. **(a)** Do you expect the tension in the wire to increase, decrease, or stay the same as a result of its new length? Explain. **(b)** Calculate the tension in the wire.

29. •• Repeat Active Example 11–3, this time with a uniform 7.2-kg ladder that is 4.0 m long.

30. •• A rigid, vertical rod of negligible mass is connected to the floor by an axle through its lower end, as shown in **Figure 11–31**. The rod also has a wire connected between its top end and the floor. If a horizontal force F is applied at the midpoint of the rod, find **(a)** the tension in the wire, and **(b)** the horizontal and vertical components of force exerted by the bolt on the rod.

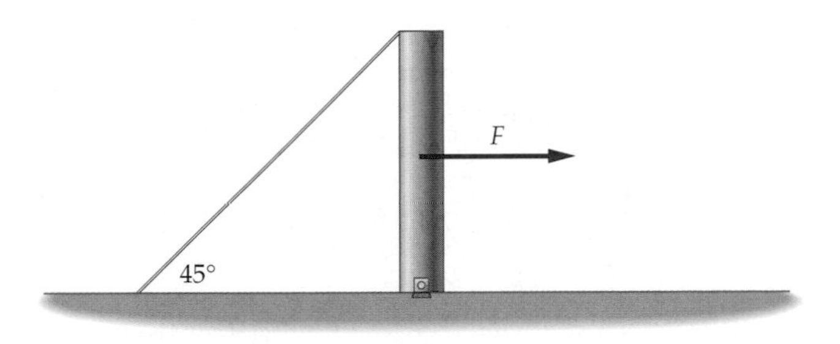

▲ **FIGURE 11–31** Problem 30

31. •• A 0.214-kg meter stick rests in contact with a frictionless bowling ball and a rough floor. The bowling ball has a diameter of 21.6 cm, and the angle the meter stick makes with the horizontal is 30.0°. Find the force exerted on the meter stick by **(a)** the bowling ball and **(b)** the floor.

32. ••• **IP** A uniform crate with a mass of 14.2 kg rests on a floor with a coefficient of static friction equal to 0.551. The crate is a uniform cube with sides 1.21 m in length. **(a)** What horizontal force applied to the top of the crate will initiate tipping? **(b)** If the horizontal force is applied halfway to the top of the crate it will begin to slip before it tips. Explain.

33. ••• In the previous problem, **(a)** what is the minimum height where the force F can be applied so that the crate begins to tip before sliding? **(b)** What is the magnitude of the force in this case?

Section 11–4 Center of Mass and Balance

34. • A hand-held shopping basket 62.0 cm long has a 1.81-kg carton of milk at one end, and a 0.722-kg box of cereal at the other end. Where should a 1.80-kg container of orange juice be placed so that the basket balances at its center?

35. • If the cat in Active Example 11–2 has a mass of 2.5 kg, how close to the right end of the two-by-four can it walk before the board begins to tip?

36. •• A 0.24-kg meter stick balances at its center. If a necklace is suspended from one end of the stick, the balance point moves 8.5 cm toward that end. What is the mass of the necklace?

37. •• Three identical, uniform books of length L are stacked one on top the other. Find the maximum overhang distance d in **Figure 11–32** such that the books do not fall over.

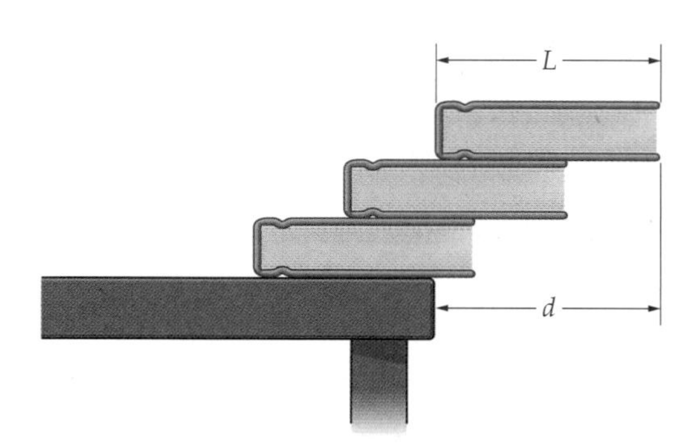

▲ **FIGURE 11–32** Problem 37

38. •• A baseball bat balances 71.1 cm from one end. If an 0.560-kg glove is attached to that end, the balance point moves 24.7 cm toward the glove. Find the mass of the bat.

Section 11–5 Dynamic Applications of Torque

39. •• A 2.85-kg bucket is attached to a disk-shaped pulley of radius 0.121 m and mass 0.742 kg. If the bucket is allowed to fall, **(a)** what is its linear acceleration? **(b)** What is the angular acceleration of the pulley? **(c)** How far does the bucket drop in 1.50 s?

40. •• **IP** In the previous problem, **(a)** is the tension in the rope greater than, less than, or equal to the weight of the bucket? **(b)** Calculate the tension in the rope.

41. •• A child exerts a tangential 40.0-N force on the rim of a disk-shaped merry-go-round with a radius of 2.40 m. If the merry-go-round starts at rest and acquires an angular speed of 0.0870 rev/s in 3.50 s, what is its mass?

42. •• **IP** You pull downward with a force of 25 N on a rope that passes over a disk-shaped pulley of mass 1.3 kg and radius 0.075 m. The other end of the rope is attached to a 0.67-kg mass. **(a)** Is the tension in the rope the same on both sides of the pulley? If not, which side has the largest tension? **(b)** Find the tension in the rope on both sides of the pulley.

43. •• Referring to the previous problem, find the linear acceleration of the 0.67-kg mass.

44. ••• A uniform meter stick of mass M has a half-filled can of fruit juice of mass m attached to one end. The meter stick and the can balance at a point 20.0 cm from the end of the stick where the can is attached. When the balanced stick-can system is suspended from a scale, the reading on the scale is 2.54 N. Find the mass of **(a)** the meter stick and **(b)** the can of juice.

45. ••• An Atwood's machine consists of two masses, m_1 and m_2, connected by a string that passes over a pulley. If the pulley is a disk of radius R and mass M, find the acceleration of the masses.

Section 11–6 Angular Momentum

46. • Calculate the angular momentum of the Earth about its own axis, due to its daily rotation. Assume that the Earth is a uniform sphere.

47. • A 0.17-kg record with a radius of 15 cm rotates with an angular speed of 33 1/3 rpm. Find the angular momentum of the record.

48. • In the previous problem, a 1.1-g fly lands on the rim of the record. What is the fly's angular momentum?

49. • Jogger 1 in **Figure 11–33** has a mass of 70.1 kg and runs in a straight line with a speed of 3.35 m/s. **(a)** What is the jogger's linear momentum? **(b)** What is the jogger's angular momentum with respect to the origin, O?

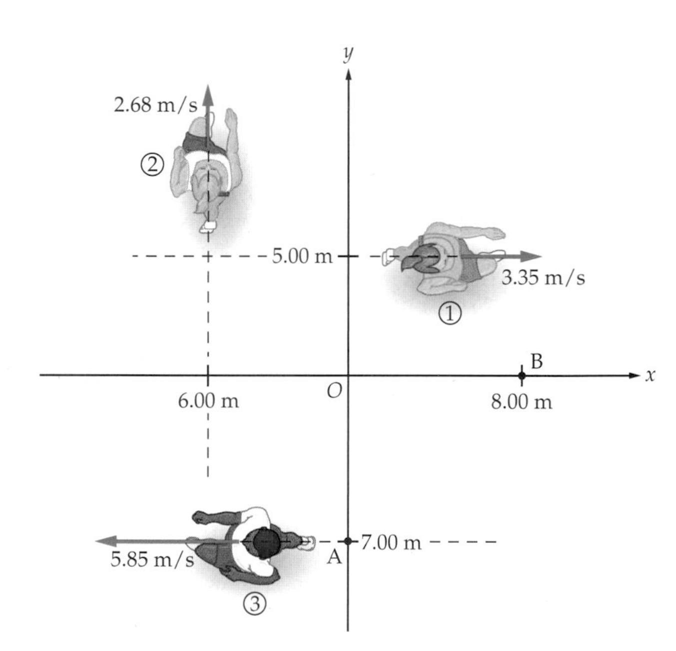

▲ **FIGURE 11–33** Problems 49, 50, 51

50. • Repeat the previous problem for the case of jogger 2, whose speed is 2.68 m/s and mass is 56.4 kg.

51. •• What is the angular momentum of jogger 3 in Figure 11–33 with respect to **(a)** point A, **(b)** point B, and **(c)** the origin, O? Jogger 3 has a mass of 62.2 kg and a speed of 5.85 m/s.

52. •• A torque of 0.12 N · m is applied to an egg beater. **(a)** If the egg beater starts at rest, what is its angular momentum after 0.50 s? **(b)** If the momentum of inertia of the egg beater is 2.5×10^{-3} kg · m^2, what is its angular speed after 0.50 s?

53. •• A windmill has an initial angular momentum of 8500 kg · m^2/s. The wind picks up, and 5.66 s later the windmill's angular mo-

mentum is 9700 $kg \cdot m^2/s$. What was the torque acting on the windmill, assuming it was constant during this time?

54. •• Two gerbils run in place with a linear speed of 0.45 m/s on an exercise wheel that is shaped like a hoop. Find the angular momentum of the system if each gerbil has a mass of 0.33 kg and the exercise wheel has a radius of 9.5 cm and a mass of 5.0 g.

Section 11–7 Conservation of Angular Momentum

55. • As an ice skater begins a spin, his angular speed is 3.28 rad/s. After pulling in his arms, his angular speed increases to 5.72 rad/s. Find the ratio of the skater's final moment of inertia to his initial moment of inertia.

56. • Calculate both the initial and the final kinetic energy of the system described in Active Example 11–5.

57. • A diver tucks her body in midflight, decreasing her moment of inertia by a factor of two. By what factor does her angular speed change?

58. •• **IP** In the previous problem, **(a)** does the diver's kinetic energy increase, decrease, or stay the same? **(b)** Calculate the ratio of the final kinetic energy to the initial kinetic energy for the diver.

59. •• A disk-shaped merry-go-round of radius 2.63 m and mass 155 kg rotates freely with an angular speed of 0.641 rev/s. A 59.4-kg person running tangential to the rim of the merry-go-round at 3.41 m/s jumps onto its rim and holds on. Before jumping on the merry-go-round, the person was moving in the same direction as the merry-go-round's rim. What is the final angular speed of the merry-go-round?

60. •• **IP** In the previous problem, **(a)** does the kinetic energy of the system increase, decrease, or stay the same when the person jumps on the merry-go-round? **(b)** Calculate the initial and final kinetic energies for this system.

61. •• A student sits at rest on a piano stool that can rotate without friction. The moment of inertia of the student-stool system is 4.1 $kg \cdot m^2$. A second student tosses a 1.5-kg mass with a speed of 2.7 m/s to the student on the stool, who catches it at a distance of 0.40 m from the axis of rotation. What is the resulting angular speed of the student and the stool?

62. •• **IP** Referring to the previous problem, **(a)** does the kinetic energy of the mass-student-stool system increase, decrease, or stay the same as the mass is caught? **(b)** Calculate the initial and final kinetic energy of the system.

63. •• **IP** A turntable with a moment of inertia of $5.4 \times 10^{-3}\ kg \cdot m^2$ rotates freely with an angular speed of 33 1/3 rpm. Riding on the rim of the turntable, 15 cm from the center, is a 1.3-g cricket. **(a)** If the cricket walks to the center of the turntable, will the turntable rotate faster, slower, or at the same rate? Explain. **(b)** Calculate the angular speed of the turntable when the cricket reaches the center.

64. •• A student on a piano stool rotates freely with an angular speed of 2.95 rev/s. The student holds a 1.25-kg mass in each outstretched arm, 0.759 m from the axis of rotation. The combined moment of inertia of the student and the stool, ignoring the two masses, is 5.43 $kg \cdot m^2$, a value that remains constant. **(a)** As the student pulls his arms inward, his angular speed increases to 3.54 rev/s. How far are the masses from the axis of rotation at this time, considering the masses to be points? **(b)** Calculate the initial and final kinetic energy of the system.

65. ••• A child of mass m stands at rest near the rim of a stationary merry-go-round of radius R and moment of inertia I. The child now begins to walk around the circumference of the merry-go-round with a tangential speed v with respect to the merry-go-round's surface. **(a)** What is the child's speed with respect to the ground. **(b)** Check your result in the limits $I \to 0$ and $I \to \infty$.

Section 11–8 Rotational Work

66. • How much work must be done to accelerate a baton from rest to an angular speed of 7.9 rad/s about its center. Consider the baton to be a uniform rod of length 0.52 m and mass 0.56 kg.

67. • Turning a doorknob through 1/4 of a revolution requires 0.12 J of work. What is the torque required to turn the doorknob?

68. • A person exerts a tangential force of 36.1 N on the rim of a disk-shaped merry-go-round of radius 2.74 m and mass 167 kg. If the merry-go-round starts at rest, what is its angular speed after the person has rotated it through an angle of 60.0°?

69. • To prepare homemade ice cream a crank must be turned with a torque of 3.3 $N \cdot m$. How much work is required for each complete turn of the crank?

70. • After getting a drink of water, a hamster jumps onto an exercise wheel for a run. A few seconds later the hamster is running in place with a speed of 1.4 m/s. Find the work done by the hamster to get the exercise wheel moving, assuming it is a hoop of radius 0.13 m and mass 6.5 g.

71. •• The L-shaped object in Figure 11–27 consists of three masses connected by light rods. Find the work that must be done on this object to accelerate it from rest to an angular speed of 2.75 rad/s about **(a)** the x-axis, **(b)** the y-axis, and **(c)** an axis through the origin and perpendicular to the page.

72. •• The rectangular object in Figure 11–28 consists of four masses connected by light rods. Find the work that must be done on this object to accelerate it from rest to an angular speed of 2.75 rad/s about **(a)** the x-axis, **(b)** the y-axis, and **(c)** an axis through the origin and perpendicular to the page.

73. •• A circular saw blade accelerates from rest to an angular speed of 3600 rpm in 6.30 revolutions. **(a)** Find the torque exerted on the saw blade, assuming it is a disk of radius 12.2 cm and mass 0.855 kg. **(b)** What is the angular speed of the saw blade after 3.15 revolutions?

General Problems

74. •• A 64.0-kg person stands on a lightweight diving board supported by two pillars, one at the end of the board, the other 1.10-m away. The pillar at the end of the board exerts a downward force of 828 N. **(a)** How far from that pillar is the person standing? **(b)** Find the force exerted by the second pillar.

75. •• A 47.0-kg uniform rod 4.25 m long is attached to a wall with a hinge at one end. The rod is held in a horizontal position by a wire attached to its other end. The wire makes an angle of 30.0° with the horizontal, and is bolted to the wall directly above the hinge. If the wire can withstand a maximum tension of 1400 N before breaking, how far from the wall can a 68.0-kg person sit without breaking the wire?

76. •• **IP** A puck attached to a string moves in a circular path on a frictionless surface, as shown in Figure 11–22. Initially, the speed of the puck is v and the radius of the circle is r. If the string passes through a hole in the surface, and is pulled downward until the radius of the circular path is $r/2$, **(a)** does the speed of the puck increase, decrease, or stay the same? **(b)** Calculate the final speed of the puck.

77. •• **IP** You hold a uniform, 25-g pen horizontal with your thumb pushing down on one end and your index finger pushing upward 3.5 cm from your thumb. The pen is 14 cm long. **(a)** Which

of these two forces is greater in magnitude? **(b)** Find the two forces.

78. •• In Active Example 11–3, suppose the ladder is uniform, 4.0 m long and weighs 60.0 N. Find the forces exerted on the ladder when the person is **(a)** halfway up the ladder and **(b)** three-fourths of the way up the ladder.

79. •• When you arrive at Duke's Dude Ranch, you are greeted by the large wooden sign shown in **Figure 11–34**. The left end of the sign is held in place by a bolt, the right end is tied to a rope that makes an angle of 20.0° with the horizontal. If the sign is uniform, 3.20 m long, and has a mass of 16.0 kg, what is **(a)** the tension in the rope, and **(b)** the horizontal and vertical components of the force, **F**, exerted by the bolt?

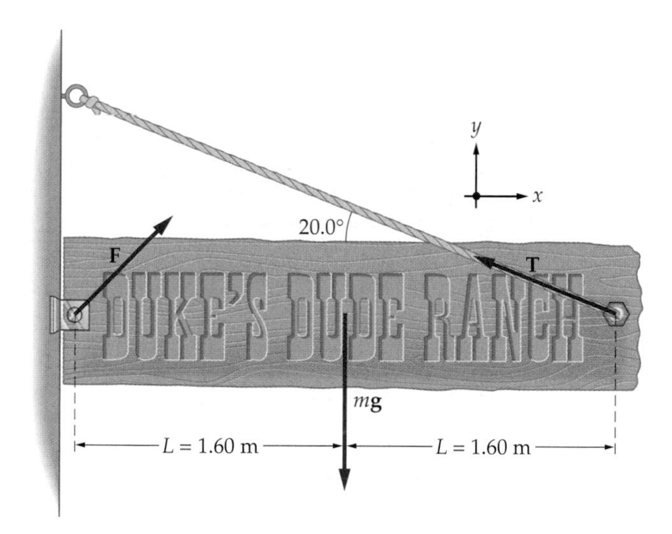

▲ **FIGURE 11–34** Problem 79

80. •• Two students carry a couch up a flight of stairs. The couch is 2.20 m long, makes an angle of 30.0° with the horizontal, and weighs 180 N. The upper student exerts a horizontal force on the top end of the couch; the lower student holds the bottom end of the couch. Find the magnitude of the force exerted by each student.

81. •• In Example 11–4, find F_1 and F_2 as a function of the distance, x, of the swimmer from the left end of the diving board. Assume that the diving board is uniform and has a mass of 85 kg.

82. •• Referring to Problem 37, what is the maximum overhang for a stack of four books?

83. •• **IP** Suppose partial melting of the polar ice caps increases the moment of inertia of the Earth from $0.331\ M_ER_E^2$ to $0.332\ M_ER_E^2$. **(a)** Would the length of a day (the time required for the Earth to complete one revolution about its axis) increase or decrease? Explain. **(b)** Calculate the change in the length of a day. Give your answer in seconds.

84. ••• In Problem 30, assume that the rod has a mass of M and that its bottom end simply rests on the floor, and is held in place by static friction. If the coefficient of static friction is μ_s, find the maximum force F that can be applied to the rod at its midpoint before it slips.

85. ••• In the previous problem, suppose the rod has a mass of 2.3 kg and the coefficient of static friction is 1/7. **(a)** Find the greatest force F that can be applied at the midpoint of the rod without causing it to slip. **(b)** Show that if F is applied 1/8 of the way from the top of the rod it will never slip at all, no matter how large the force F.

86. ••• A cylinder of mass m and radius r has a string wrapped around its circumference. The upper end of the string is held fixed, and the cylinder is allowed to fall. Show that its linear acceleration is $(2/3)g$.

87. ••• Repeat the previous problem, replacing the cylinder with a solid sphere. Show that its linear acceleration is $(5/7)g$.

88. ••• A mass M is attached to a rope that passes over a disk-shaped pulley of mass m and radius r. On the other side of the pulley, the rope is pulled downward with a force F. Find **(a)** the acceleration of the mass, and **(b)** the tension in the rope on both sides of the pulley. **(c)** Check your results in the limits $m \to 0$ and $m \to \infty$.

89. ••• A wheel of radius R is at rest against a step of height $3R/4$. Find the minimum horizontal force that must be exerted on the axle to make the wheel start to rise up over the step.

90. ••• A 0.205-kg yo-yo has an outer radius 5.50 times greater than the radius of its axle. The yo-yo is in equilibrium if a mass m is suspended from its outer edge, as shown in **Figure 11–35**. Find the tension in the two strings, T_1 and T_2, and the mass m.

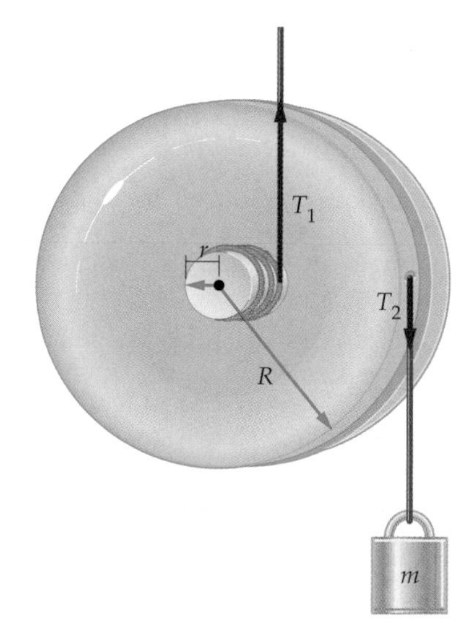

▲ **FIGURE 11–35** Problem 90

91. ••• Consider a system of four uniform bricks of length L stacked as shown in **Figure 11–36**. What is the maximum distance, x, that the middle bricks can be displaced outward before they begin to tip?

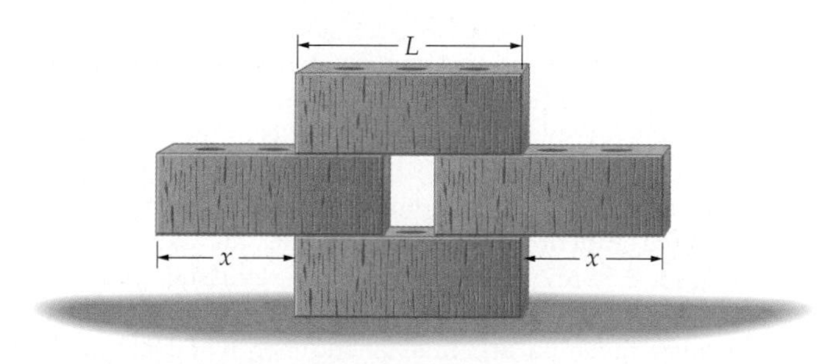

▲ **FIGURE 11–36** Problem 91

12 Gravity

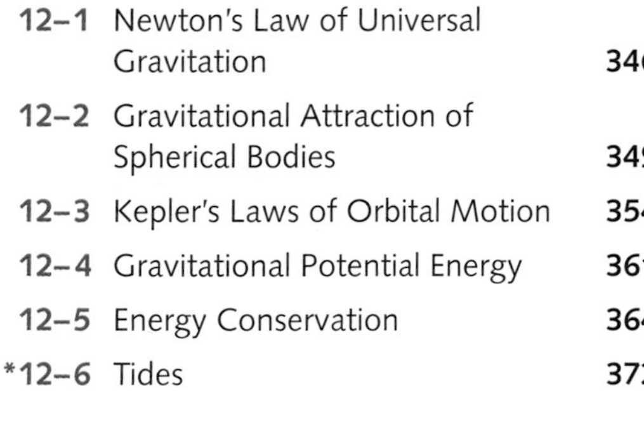

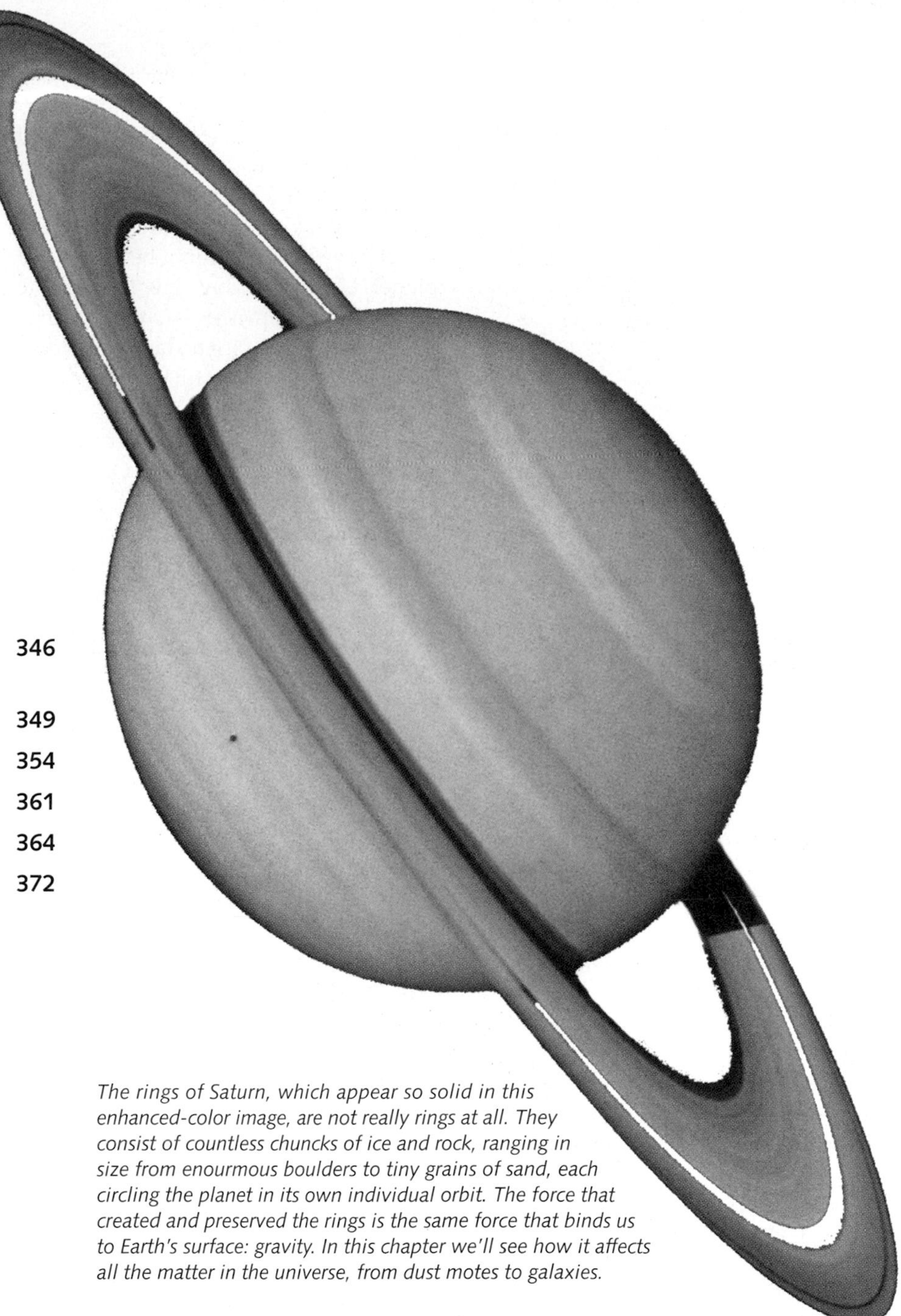

The rings of Saturn, which appear so solid in this enhanced-color image, are not really rings at all. They consist of countless chuncks of ice and rock, ranging in size from enourmous boulders to tiny grains of sand, each circling the planet in its own individual orbit. The force that created and preserved the rings is the same force that binds us to Earth's surface: gravity. In this chapter we'll see how it affects all the matter in the universe, from dust motes to galaxies.

The study of gravity has always been a central theme in physics, from Galileo's early experiments on free fall in the seventeenth century to Einstein's general theory of relativity in the early years of the twentieth century, and Stephen Hawking's work on black holes in recent years. Perhaps the grandest milestone in this endeavor, however, was the discovery by Newton of the **universal law of gravitation**. With just one simple equation to describe the force of gravity, Newton was able to determine the orbits of planets, moons, and comets, and to explain such earthly phenomena as the tides and the fall of an apple.

Before Newton's work, it was generally thought that the heavens were quite separate from the Earth, and that they obeyed their own "heavenly" laws. Newton showed, on the contrary, that the same law of gravity that operates on the surface of the Earth applies to the Moon and to other astronomical objects. As a result of Newton's efforts, physics expanded its realm of applicability to natural phenomena throughout the universe.

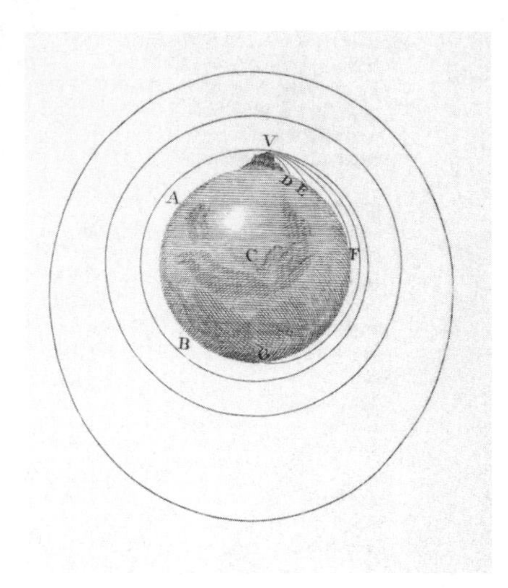

▲ In this illustration from his great work, the *Principia*, published in 1687, Newton presents a "thought experiment" to show the connection between free fall and orbital motion. Imagine throwing a projectile horizontally from the top of a mountain. The greater initial speed of the projectile, the farther it travels in free fall before striking the ground. In the absence of air resistance, a great enough initial speed could result in the projectile circling the Earth and returning to its starting point. Thus, an object orbiting the Earth is actually in free fall—it simply has a large horizontal speed.

So successful was Newton's law of gravitation that Edmond Halley (1656–1742) was able to use it to predict the return of the comet that today bears his name. Though he did not live to see its return in 1758, the fact that the comet did reappear when predicted was an event unprecedented in human history. Roughly a hundred years later, Newton's theory of gravity scored an even more impressive success. Astronomers observing the planet Uranus noticed small deviations in its orbit, which they thought might be due to the gravitational tug of a previously unknown planet. Using Newton's law to calculate the predicted position of the new planet—now called Neptune—it was found on the very first night of observations, September 23, 1846. The fact that Neptune was precisely where the law of gravitation said it should be still stands as one of the most astounding triumphs in the history of science.

Today, Newton's law of gravitation is used to determine the orbits that take spacecraft from the Earth to various destinations within our solar system and beyond. Appropriately enough, spacecraft were even sent to view Halley's comet at close range in 1986. In addition, the law allows us to calculate with pinpoint accuracy the time of solar eclipses and other astronomical events in the distant past and remote future. This incredibly powerful and precise law of nature is the subject of this chapter.

12–1 Newton's Law of Universal Gravitation

It's ironic, but the first fundamental force of nature to be recognized as such, **gravity**, is also the weakest of the fundamental forces. Still, it is the force most apparent to us in our everyday lives, and the force responsible for the motion of the Moon, the Earth, and the planets. Yet the connection between falling objects on Earth and planets moving in their orbits was not known before Newton.

The flash of insight that came to Newton—whether it was due to seeing an apple fall to the ground or not—is simply this: The force causing an apple to accelerate downward is the same force causing the Moon to move in a circular path around the Earth. To put it another way, Newton was the first to realize that the Moon is *constantly falling* toward the Earth, and that it falls for the same reason that an apple falls. This is illustrated in a classic drawing due to Newton, shown here.

To be specific, in the case of the apple the motion is linear, as it accelerates downward toward the center of the Earth. In the case of the Moon the motion is circular with constant speed. As discussed in Section 6–5, an object in uniform circular motion accelerates toward the center of the circle. It follows, therefore, that the Moon *also* accelerates toward the center of the Earth. In fact, the force responsible for the Moon's centripetal acceleration is the Earth's gravitational attraction, the same force responsible for the fall of the apple.

To describe the force of gravity, Newton proposed the following simple law:

Newton's Law of Universal Gravitation

The force of gravity between any two point objects of mass m_1 and m_2 is attractive and of magnitude

$$F = G\frac{m_1 m_2}{r^2} \qquad \textbf{12–1}$$

In this expression, r is the distance between the masses and G is a constant referred to as the **universal gravitation constant**. Its value is

$$G = 6.67 \times 10^{-11}\ \mathrm{N \cdot m^2/kg^2} \qquad \textbf{12–2}$$

The force is directed along the line connecting the masses, as indicated in **Figure 12–1**.

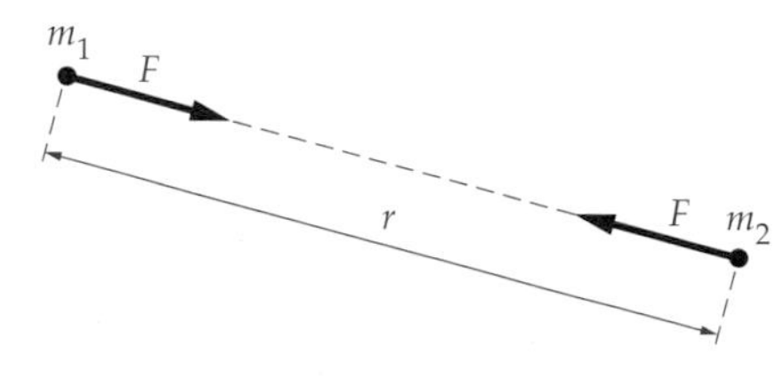

▲ **FIGURE 12–1 Gravitational force between point masses**

Two point masses, m_1 and m_2, separated by a distance r exert equal and opposite attractive forces on one another. The magnitude of the forces, F, is given by Equation 12–1.

Note that each mass experiences a force of the same magnitude, $F = Gm_1m_2/r^2$, but acting in opposite directions. That is, the force of gravity between two objects forms an action-reaction pair.

According to Newton's law, all objects in the universe attract all other objects in the universe by way of the gravitational interaction. It is in this sense that the force law is termed "universal." Thus, the net gravitational force acting on you is due not only to the planet on which you stand, which is certainly responsible for the majority of the net force, but also to people nearby, planets, and even stars in far-off galaxies. In short, everything in the universe "feels" everything else, thanks to gravity.

The fact that G is such a small number means that the force of gravity between objects of human proportions is imperceptibly small. This is shown in the following Exercise.

EXERCISE 12–1

A man takes his dog for a walk on a deserted beach. Treating people and dogs as point objects for the moment, find the force of gravity between the 105-kg man and his 11.2-kg dog when they are separated by a distance of **(a)** 1.00 m and **(b)** 10.0 m.

Solution

(a) Substituting numerical values into Equation 12–1 yields

$$F = G\frac{m_1m_2}{r^2} = (6.67 \times 10^{-11}\ \text{N}\cdot\text{m}^2/\text{kg}^2)\frac{(105\ \text{kg})(11.2\ \text{kg})}{(1.00\ \text{m})^2} = 7.84 \times 10^{-8}\ \text{N}$$

(b) Repeating the calculation for $r = 10.0$ m gives

$$F = G\frac{m_1m_2}{r^2} = (6.67 \times 10^{-11}\text{N}\cdot\text{m}^2/\text{kg}^2)\frac{(105\ \text{kg})(11.2\ \text{kg})}{(10.0\ \text{m})^2} = 7.84 \times 10^{-10}\ \text{N}$$

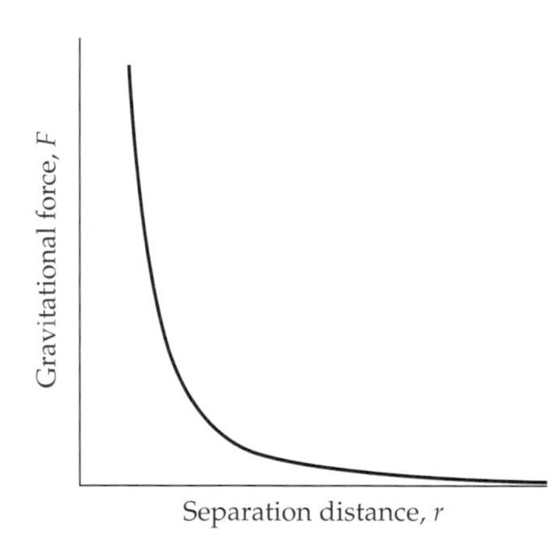

▲ **FIGURE 12–2 Dependence of the gravitational force on separation distance, r**

Note that the gravitational force decreases rapidly with distance, but never completely vanishes. We say that gravity is a force of infinite range.

The forces found in this Exercise are imperceptibly small. In comparison, the force exerted by the Earth on the man is 1030 N and the force exerted on the dog is 110 N—these forces are several orders of magnitude greater than the force between the man and the dog. In general, gravitational forces are significant only when large masses, such as the Earth or the Moon, are involved.

Exercise 12–1 also illustrates how rapidly the force of gravity decreases with distance. In particular, since F varies as $1/r^2$, it is said to have an **inverse square dependence** on distance. Thus, for example, an increase in distance by a factor of 10 results in a decrease in the force by a factor of $10^2 = 100$. A plot of the force of gravity versus distance is given in **Figure 12–2**. Note that even though the force diminishes rapidly with distance, it never completely vanishes; thus, we say that gravity is a force of infinite range.

Note also that the force of gravity between two masses depends on the product of the masses, m_1 times m_2. With this type of dependence, it follows that if either mass is doubled, the force of gravity is doubled as well. This would not be the case, for example, if the force of gravity depended on the sum of the masses, $m_1 + m_2$.

Finally, if a given mass is acted on by gravitational interactions with a number of other masses, the net force acting on it is simply the vector sum of each of the forces individually. This property of gravity is referred to as **superposition**. As an example, superposition implies that the net gravitational force exerted on you at this moment is the vector sum of the force exerted by the Earth, plus the force exerted by the Moon, plus the force exerted by the Sun, and so on. The following example illustrates superposition.

PROBLEM SOLVING NOTE
Net Gravitational Force

To find the net gravitational force acting on an object one should (i) resolve each of the forces acting on the object into components and (ii) add the forces component by component.

EXAMPLE 12–1 How Much Force Is With You?

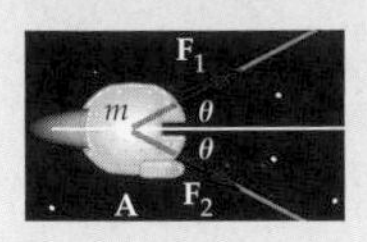

As part of a daring rescue attempt, the Millennium Eagle passes between a pair of twin asteroids, as shown. If the mass of the spaceship is 2.50×10^7 kg and the mass of each asteroid is 3.50×10^{18} kg, find the net gravitational force exerted on the Millennium Eagle **(a)** when it is at location A and **(b)** when it is at location B. Treat the spaceship and the asteroids as if they were point objects.

Picture the Problem

The sketch shows the spaceship as it follows a path between the twin asteroids. The relevant distances and masses are indicated, as are the two points of interest, A and B.

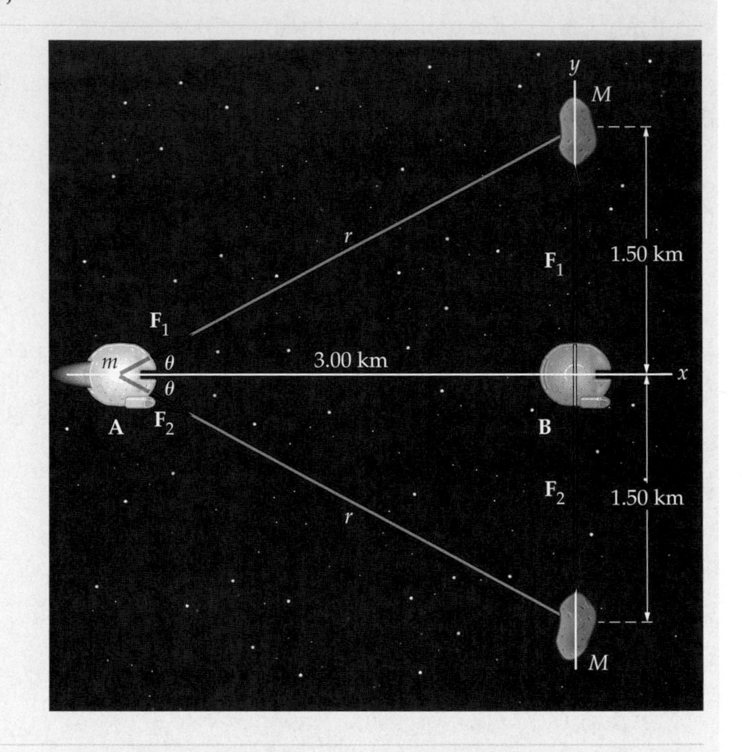

Strategy

To find the net gravitational force exerted on the spaceship we first determine the magnitude of the force exerted on it by each asteroid. This is done by using Equation 12–1 and the distances given in the sketch. Next, we resolve these forces into x and y components. Finally, we sum the force components to find the net force.

Solution

Part (a)

1. Use the Pythagorean theorem to find the distance r from point A to each asteroid. Also, refer to the sketch to find the angle between $\mathbf{F}_1$ and the x-axis. The angle between $\mathbf{F}_2$ and the x-axis has the same magnitude but the opposite sign:

$$r = \sqrt{(3.00 \times 10^3\ \text{m})^2 + (1.50 \times 10^3\ \text{m})^2} = 3350\ \text{m}$$

$$\theta = \tan^{-1}\left(\frac{1.50 \times 10^3\ \text{m}}{3.00 \times 10^3\ \text{m}}\right) = \tan^{-1}(0.500) = 26.6°$$

2. Use r and Equation 12–1 to calculate the magnitude of the forces $\mathbf{F}_1$ and $\mathbf{F}_2$ at point A:

$$F_1 = F_2 = G\frac{mM}{r^2}$$

$$= (6.67 \times 10^{-11}\ \text{N} \cdot \text{m}^2/\text{kg}^2)\frac{(2.50 \times 10^7\ \text{kg})(3.50 \times 10^{18}\ \text{kg})}{(3350\ \text{m})^2}$$

$$= 5.20 \times 10^8\ \text{N}$$

3. Use the value of θ found in step 1 to calculate the x and y components of $\mathbf{F}_1$ and $\mathbf{F}_2$:

$$F_{1,x} = F_1 \cos\theta = (5.19 \times 10^8\ \text{N})\cos 26.6° = 4.65 \times 10^8\ \text{N}$$

$$F_{1,y} = F_1 \sin\theta = (5.19 \times 10^8\ \text{N})\sin 26.6° = 2.33 \times 10^8\ \text{N}$$

$$F_{2,x} = F_2 \cos\theta = (5.19 \times 10^8\ \text{N})\cos(-26.6°) = 4.65 \times 10^8\ \text{N}$$

$$F_{2,y} = F_2 \sin\theta = (5.19 \times 10^8\ \text{N})\sin(-26.6°) = -2.33 \times 10^8\ \text{N}$$

4. Add the components of $\mathbf{F}_1$ and $\mathbf{F}_2$ to find the components of the net force, $\mathbf{F}$:

$$F_x = F_{1,x} + F_{2,x} = 9.30 \times 10^8\ \text{N}$$

$$F_y = F_{1,y} + F_{2,y} = 0$$

Part (b)

5. Use Equation 12–1 to find the magnitude of the forces exerted on the spaceship by the asteroids at location B:

$$F_1 = F_2 = \frac{GmM}{r^2}$$

$$= (6.67 \times 10^{-11}\ \text{N} \cdot \text{m}^2/\text{kg}^2)\frac{(2.50 \times 10^7\ \text{kg})(3.50 \times 10^{18}\ \text{kg})}{(1.50 \times 10^3\ \text{m})^2}$$

$$= 2.59 \times 10^9\ \text{N}$$

6. Use the fact that $\mathbf{F}_1$ and $\mathbf{F}_2$ have equal magnitudes and point in opposite directions to determine the net force, **F**, acting on the spaceship:

$$\mathbf{F} = \mathbf{F}_1 + \mathbf{F}_2 = 0$$

Insight

We find that the net force at location A is in the positive x direction, as one would expect by symmetry. At location B the net force is zero since the attractive forces exerted by the two asteroids are equal and opposite, and thus cancel.

Practice Problem

Find the net gravitational force acting on the spaceship when it is at the location $x = 5.00 \times 10^3$ m, $y = 0$. [**Answer:** 4.10×10^8 N in the negative x direction.]

Some related homework problems: Problem 8, Problem 10, Problem 11

12–2 Gravitational Attraction of Spherical Bodies

Newton's law of gravity applies to point objects. How, then, do we calculate the force of gravity for an object of finite size? In general, the approach is to divide the finite object into a collection of small mass elements, then use superposition and the methods of calculus to determine the net gravitational force. For an arbitrary shape, this calculation can be quite difficult. For objects with a uniform spherical shape, however, the final result is remarkably simple, as was shown by Newton. (In fact, Newton invented the calculus to do this very calculation.)

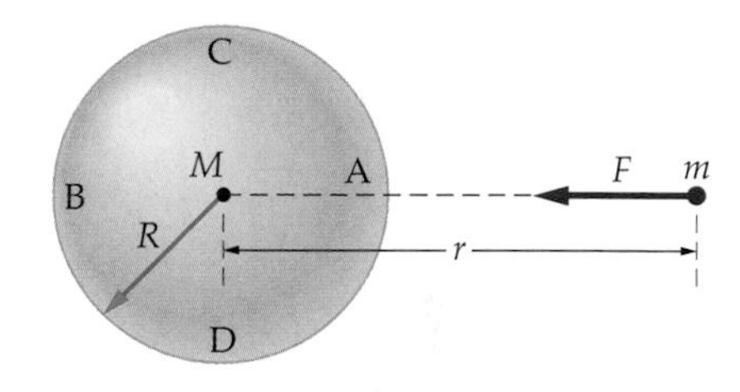

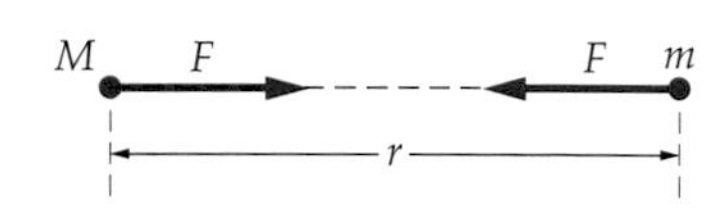

▲ **FIGURE 12–3 Gravitational force between a point mass and a sphere**
The force is the same as if all the mass of the sphere were concentrated at its center.

Uniform Sphere

Consider a uniform sphere of radius R and mass M, as in **Figure 12–3**. A point object of mass m is brought near the sphere, though still outside it at a distance r from its center. The object experiences a relatively strong attraction from mass near the point A, and a weaker attraction from mass near point B. In both cases the force is along the line connecting the mass m and the center of the sphere; that is, along the x-axis. In addition, mass at the points C and D exert a net force that is also along the x-axis—just as in the case of the twin asteroids in Example 12–1. Thus, the symmetry of the sphere guarantees that the net force it exerts on m is directed toward its center. The magnitude of the force exerted by the sphere must be calculated with the methods of calculus–which Newton invented and then applied to this problem. As a result of his calculations, Newton was able to show that **the net force exerted by the sphere on the mass m is the same as if all the mass of the sphere were concentrated at its center**. That is, the force between the mass m and the sphere of mass M is simply

$$F = G\frac{mM}{r^2} \qquad \textbf{12–3}$$

Let's apply this result to the case of a mass m on the surface of the Earth. If the mass of the Earth is M_E, and its radius is R_E, it follows that the force exerted on m by the Earth is

$$F = G\frac{mM_E}{R_E^2} = m\left(\frac{GM_E}{R_E^2}\right)$$

We also know, however, that the gravitational force experienced by a mass m on the Earth's surface is simply $F = mg$, where g is the acceleration due to gravity. Therefore, we see that

$$m\left(\frac{GM_E}{R_E^2}\right) = mg$$

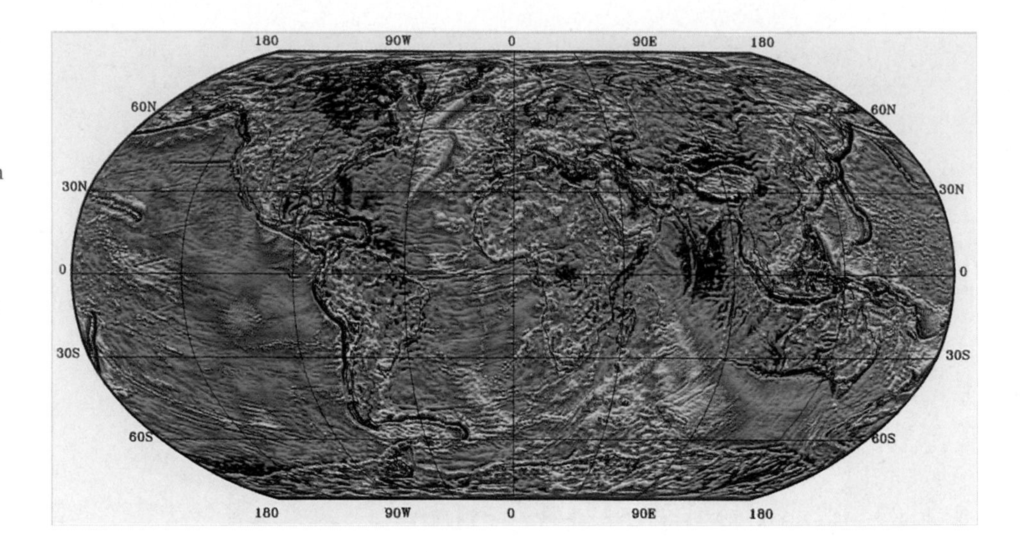

▶ This global model of the Earth's gravitational strength was constructed from a combination of surface gravity measurements and satellite tracking data. It shows how the acceleration of gravity varies from the value at an idealized "sea level" that takes into account the Earth's nonspherical shape. (The Earth is somewhat flattened at the poles—its radius is greatest at the equator.) Gravity is strongest in the red areas and weakest in the dark blue areas.

or

$$g = \frac{GM_E}{R_E^2} = 9.81 \text{ m/s}^2 \qquad \text{12–4}$$

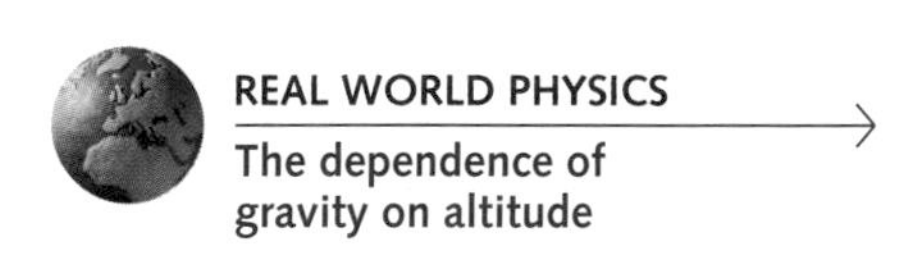

REAL WORLD PHYSICS
The dependence of gravity on altitude

This result can be extended to objects above the Earth's surface, and hence farther from the center of the Earth, as we show in the next Example.

EXAMPLE 12–2 On a Mountain Top

If you climb to the top of Mt. Everest, you will be about 5.50 mi above sea level. What is the acceleration due to gravity at this altitude?

Picture the Problem

At the top of the mountain your distance from the center of the Earth is $r = R_E + h$, where $h = 5.50$ mi is the altitude.

Strategy

First, use $F = GmM_E/r^2$ to find the force due to gravity on the mountain top. Then, set $F = mg_h$ to find the acceleration g_h at the height h.

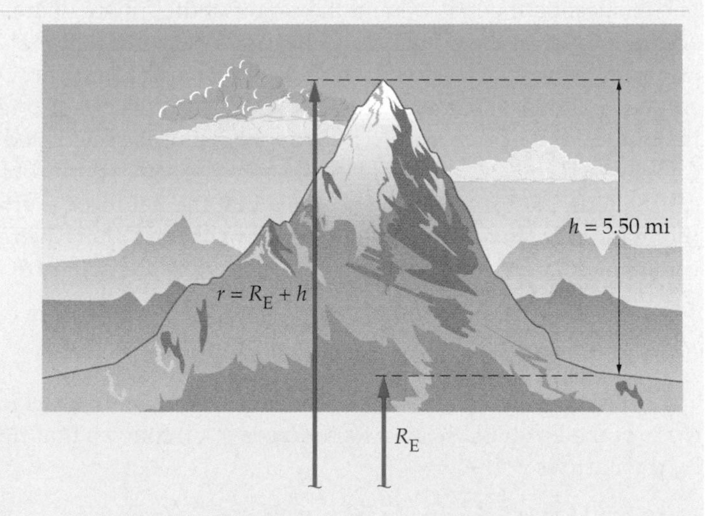

Solution

1. Calculate the force F due to gravity at a height h above the Earth's surface:

$$F = G\frac{mM_E}{(R_E + h)^2}$$

2. Set F equal to mg_h and solve for g_h:

$$F = G\frac{mM_E}{(R_E + h)^2} = mg_h$$

$$g_h = G\frac{M_E}{(R_E + h)^2}$$

3. Factor out R_E^2 from the denominator, and use the fact that $GM_E/R_E^2 = g$:

$$g_h = \left(\frac{GM_E}{R_E^2}\right)\frac{1}{\left(1+\dfrac{h}{R_E}\right)^2} = \frac{g}{\left(1+\dfrac{h}{R_E}\right)^2}$$

4. Substitute numerical values, with $h = 5.50$ mi $= (5.50$ mi$)(1609$ m/mi$) = 8850$ m, and $R_E = 6.37 \times 10^6$ m:

$$g_h = \frac{g}{\left(1+\dfrac{h}{R_E}\right)^2} = \frac{9.81\text{ m/s}^2}{\left(1+\dfrac{8850\text{ m}}{6.37\times 10^6\text{ m}}\right)^2} = 9.78\text{ m/s}^2$$

Insight

As expected, the acceleration of gravity is less as one moves farther from the center of the Earth. Thus, if you were to climb to the top of Mt. Everest you would lose weight not only because of the physical exertion required for the climb, but also because of the reduced gravity. In particular, a person with a mass of 60 kg (about 130 lbs) would lose about half a pound of weight just by standing on the summit of the mountain.

A plot of g_h as a function of h is shown in **Figure 12-4 (a)**. The plot indicates the altitude of Mt. Everest and the orbit of the space shuttle. **Figure 12-4 (b)** shows g_h out to the orbit of communications and weather satellites, which orbit at an altitude of roughly 22,300 mi.

Practice Problem

Find the acceleration of gravity at the altitude of the Space Shuttle's orbit, 250 km above the Earth's surface. [**Answer:** $g_h = 9.08\text{ m/s}^2$]

Some related homework problems: Problem 14, Problem 16

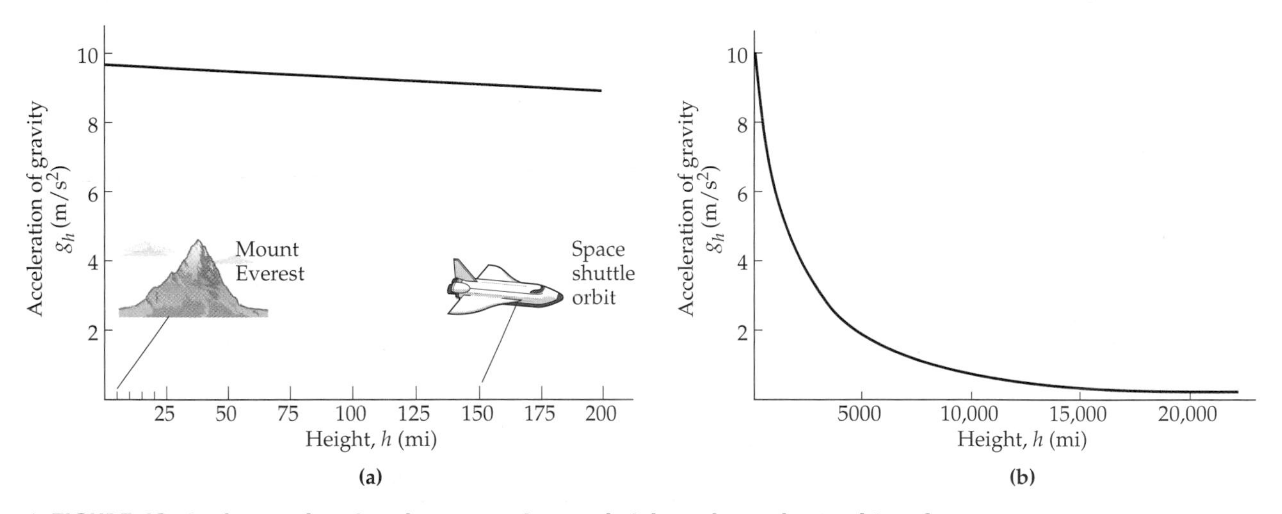

▲ **FIGURE 12–4 The acceleration due to gravity at a height h above the Earth's suface**

(a) In this plot, the peak of Mt. Everest is at about $h = 5.50$ mi, and the Space Shuttle orbit is at roughly $h = 150$ mi. **(b)** The decrease in the acceleration of gravity from the surface of the Earth to an altitude of about 25,000 mi. The orbit of geosynchronous satellites—ones that orbit above a fixed point on the Earth—is at roughly $h = 22{,}300$ mi.

Equation 12–4 can be used to calculate the acceleration of gravity on other objects in the solar system besides the Earth. For example, to calculate the acceleration of gravity on the Moon, g_m, we simply use the mass and radius of the Moon in Equation 12–4. Once g_m is known, the weight of an object of mass m on the Moon is found by using $W_m = mg_m$.

EXERCISE 12–2

(a) Find the acceleration of gravity on the surface of the Moon.

(b) The lunar rover had a mass of 220 kg. What was its weight on the Earth and on the Moon ($M_m = 7.35 \times 10^{22}$ kg)?

Solution

(a) For the Moon, the acceleration of gravity is

$$g_m = \frac{GM_m}{R_m^2} = \frac{(6.67 \times 10^{-11}\ \mathrm{N \cdot m^2/kg^2})(7.35 \times 10^{22}\ \mathrm{kg})}{(1.74 \times 10^6\ \mathrm{m})^2} = 1.62\ \mathrm{m/s^2}$$

This is about 1/6 the acceleration of gravity on the Earth.

(b) On the Earth, the rover's weight was

$$W = mg = (220\ \mathrm{kg})(9.81\ \mathrm{m/s^2}) = 2160\ \mathrm{N}$$

On the Moon, its weight was

$$W = mg_m = (220\ \mathrm{kg})(1.62\ \mathrm{m/s^2}) = 356\ \mathrm{N}$$

As expected, this is roughly 1/6 its Earth weight.

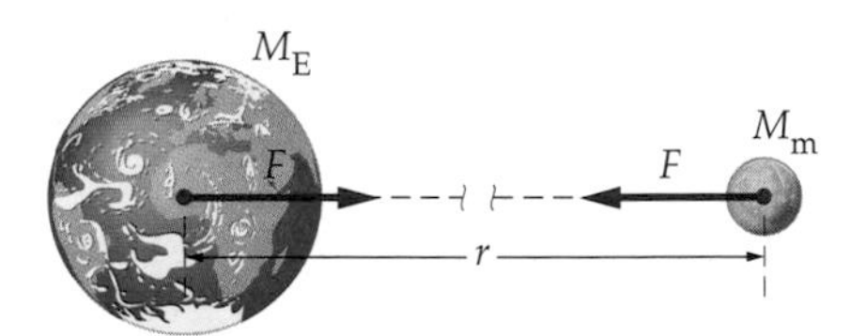

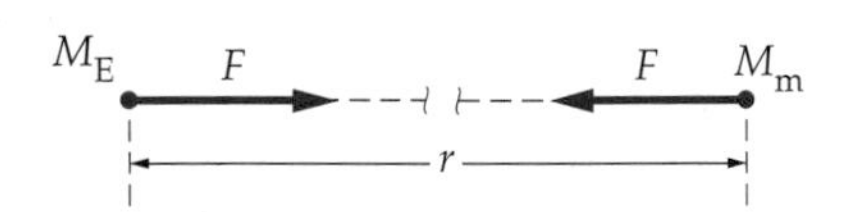

▲ **FIGURE 12–5 Gravitational force between the Earth and Moon**

The force is the same as if both the Earth and the Moon were point masses. (The sizes of the Earth and Moon are in correct proportion in this figure, but the separation between the two should be much greater than that shown here. In reality, it is about 30 times the diameter of the Earth, and so would be about 2 ft on this scale.)

The replacement of a sphere with a point mass at its center can be applied to many physical systems. For example, the force of gravity between two spheres of finite size is the same as if *both* were replaced by point masses. Thus, the gravitational force between the Earth, with mass M_E, and the Moon, with mass M_m, is

$$F = G\frac{M_E M_m}{r^2}$$

The distance r in this expression is the center-to-center distance between the Earth and the Moon, as shown in **Figure 12–5**. It follows, then, that in many calculations involving the solar system, Moons and planets can be treated as point objects.

Weighing the Earth

The British physicist Henry Cavendish performed an experiment in 1798 that is often referred to as "weighing the Earth." What he did, in fact, was measure the value of the universal gravitation constant, G, that appears in Newton's law of gravity. As we have pointed out before, G is a very small number, hence a sensitive experiment is needed for its measurement. It is because of this experimental difficulty that G was not measured until more than 100 years after Newton published the law of gravitation.

In the Cavendish experiment, illustrated in **Figure 12–6**, two masses m are suspended from a thin thread. Near each suspended mass is a large, stationary mass M, as shown. Each suspended mass is attracted by the force of gravity toward the large mass near it; hence the rod holding the suspended masses tends to rotate and twist the thread. The angle through which the thread twists can be measured by bouncing a beam of light from a mirror attached to the thread. If the force required to twist the thread through a given angle is known (from previous experiments), a measurement of the twist angle gives the magnitude of the force of gravity. Fi-

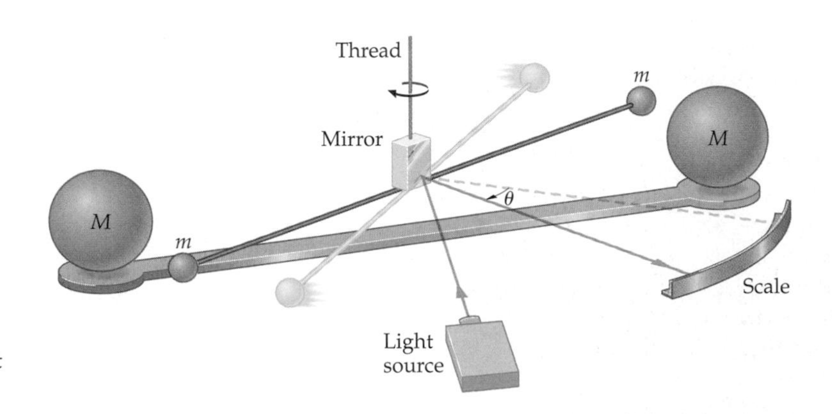

▶ **FIGURE 12–6 The Cavendish experiment**

The gravitational attraction between the masses m and M causes the rod and the suspending thread to twist. Measurement of the twist angle allows for a direct measurement of the gravitational force.

nally, knowing the masses m and M, and the distance between their centers, r, we can use Equation 12–1 to solve for G. Cavendish found $6.754 \times 10^{-11}\ \text{N}\cdot\text{m}^2/\text{kg}^2$, in good agreement with the currently accepted value given in Equation 12–2.

To see why Cavendish is said to have weighed the Earth, recall that the force of gravity on the surface of the Earth, mg, can be written as follows:

$$mg = G\frac{mM_E}{R_E^2}$$

Canceling m and solving for M_E yields

$$M_E = \frac{gR_E^2}{G} \qquad \text{12–5}$$

Before the Cavendish experiment, the quantities g and R_E were known from direct measurement, but G had yet to be determined. When Cavendish measured G, he didn't actually "weigh" the Earth, of course. Instead, he calculated its mass, M_E.

▲ (Top) The weak lunar gravity permits astronauts, even encumbered by their massive space suits, to bound over the Moon's surface. The low gravitational pull, only about one-sixth that of Earth, is a consequence not only of the Moon's smaller size but also of its lower average density. (Bottom) The Sojourner rover vehicle, brought to Mars by the Pathfinder mission of 1997, exploring the Martian surface. The gravity of Mars is greater than that of the Moon but still only about 38 percent that of Earth.

EXERCISE 12–3

Use $M_E = gR_E^2/G$ to calculate the mass of the Earth.

Solution

Substituting numerical values, we find

$$M_E = \frac{gR_E^2}{G} = \frac{(9.81\ \text{m/s}^2)(6.37 \times 10^6\ \text{m})^2}{6.67 \times 10^{-11}\ \text{N}\cdot\text{m}^2/\text{kg}^2} = 5.97 \times 10^{24}\ \text{kg}$$

As soon as Cavendish determined the mass of the Earth, geologists were able to use the result to calculate its average density; that is, its average mass per volume. Assuming a spherical Earth of radius R_E, its total volume is

$$V_E = \tfrac{4}{3}\pi R_E^3 = \tfrac{4}{3}\pi(6.37 \times 10^6\ \text{m})^3 = 1.08 \times 10^{21}\ \text{m}^3$$

Dividing this into the total mass yields the average density, ρ:

$$\rho = \frac{M_E}{V_E} = \frac{5.97 \times 10^{24}\ \text{kg}}{1.08 \times 10^{21}\ \text{m}^3} = 5510\ \text{kg/m}^3 = 5.51\ \text{g/cm}^3$$

This is an interesting result because typical rocks found near the surface of the Earth, such as granite, have a density of only about $3.00\ \text{g/cm}^3$. We conclude, then, that the interior of the Earth must have a greater density than its surface. In fact, by analyzing the propagation of seismic waves around the world, we now know that the Earth has a rather complex interior structure, including a solid inner core with a density of about $15.0\ \text{g/cm}^3$ (see Section 10–5).

A similar calculation for the Moon yields an average density of about $3.33\ \text{g/cm}^3$, essentially the same as the density of the lunar rocks brought back during the Apollo program. Hence, it is likely that the Moon does not have an internal structure similar to that of the Earth.

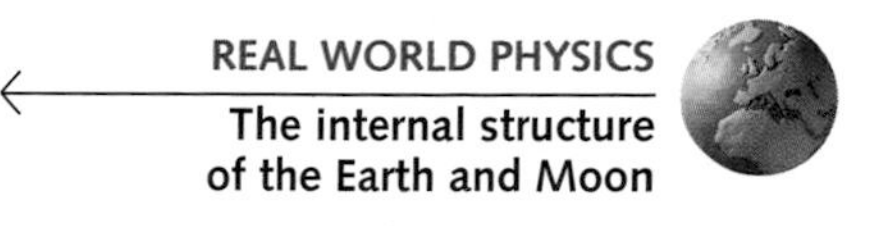

Since G is a universal constant—with the same value everywhere in the universe—it can be used to calculate the mass of other bodies in the solar system as well. This is illustrated in the following Example.

EXAMPLE 12–3 Mars Attracts!

Find the mass of Mars, given that its radius is 3.39×10^6 m, and that the acceleration of gravity on its surface is $3.73\ \text{m/s}^2$.

continued on the following page

continued from the previous page

Picture the Problem

We use the subscript M to denote quantities referring to Mars. Thus, $g_M = 3.73 \text{ m/s}^2$ and $R_M = 3.39 \times 10^6 \text{ m}$.

Strategy

Since the acceleration of gravity is g_M on the surface of Mars, it follows that the force of gravity on an object of mass m is $F = mg_M$. This force is also given by Newton's law of gravity; that is, $F = GmM_M/R_M^2$. Setting these expressions for the force equal to one another yields the mass of Mars, M_M.

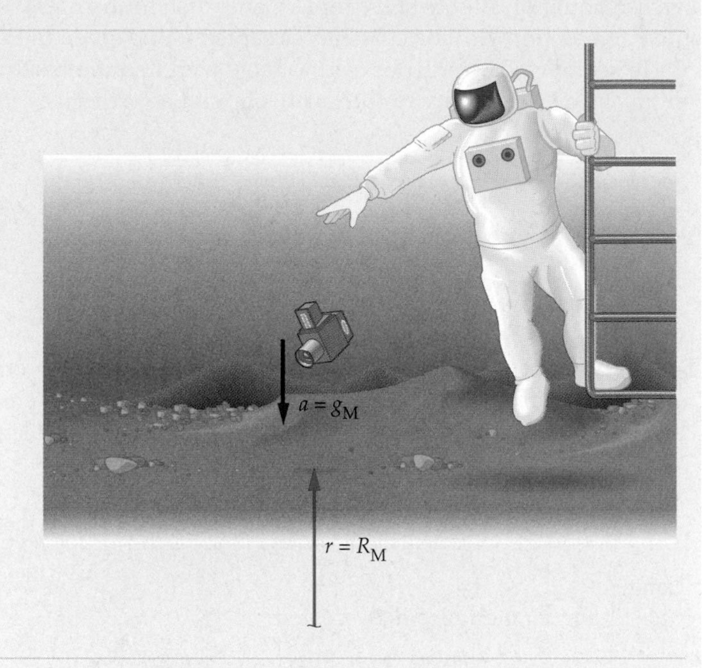

Solution

1. Set mg_M equal to GmM_M/R_M^2:

$$mg_M = G\frac{mM_M}{R_M^2}$$

2. Cancel m and solve for the mass of Mars:

$$M_M = \frac{g_M R_M^2}{G}$$

3. Substitute numerical values:

$$M_M = \frac{(3.73 \text{ m/s}^2)(3.39 \times 10^6 \text{ m})^2}{6.67 \times 10^{-11} \text{ N}\cdot\text{m}^2/\text{kg}^2} = 6.43 \times 10^{23} \text{ kg}$$

Insight

The important point here is that G applies equally well to Mars as to the Earth.

Practice Problem

If the radius of Mars were reduced to 3.00×10^6 m, with its mass remaining the same, would the acceleration of gravity on Mars increase, decrease, or stay the same? Check your answer by calculating the acceleration of gravity for this case. [**Answer:** The acceleration of gravity increases to 4.77 m/s^2.]

Some related homework problems: Problem 19, Problem 20

12–3 Kepler's Laws of Orbital Motion

If you go outside each clear night and observe the position of Mars with respect to the stars, you will find that its motion is rather complex. Instead of moving on a simple, curved path, it occasionally reverses direction (this is known as *retrograde motion*). A few months later it reverses direction yet again and resumes its original direction of motion. Other planets exhibit similar odd behavior.

The Danish astronomer Tycho Brahe (1546–1601) followed the paths of the planets, and Mars in particular, for many years, even though the telescope had not yet been invented. He used, instead, an elaborate sighting device to plot the precise position of the planets. Brahe was joined in his work by Johannes Kepler (1571–1630) in 1600, and after Brahe's death, Kepler inherited his astronomical observations.

Kepler made good use of Brahe's life work, extracting from his carefully collected data the three laws of orbital motion we know today as Kepler's laws. These laws make it clear that the Sun and the planets do not orbit the Earth, as Ptolemy claimed, but rather that the Earth, along with the other planets, orbit the Sun, as proposed by Copernicus (1473–1543).

Why the planets obey Kepler's laws no one knew—not even Kepler—until Newton considered the problem decades after Kepler's death. Newton was able to show that each of Kepler's laws follows as a direct consequence of the universal law of gravitation. In the remainder of this section we consider Kepler's three laws one at a time, and point out the connection between them and the law of gravitation.

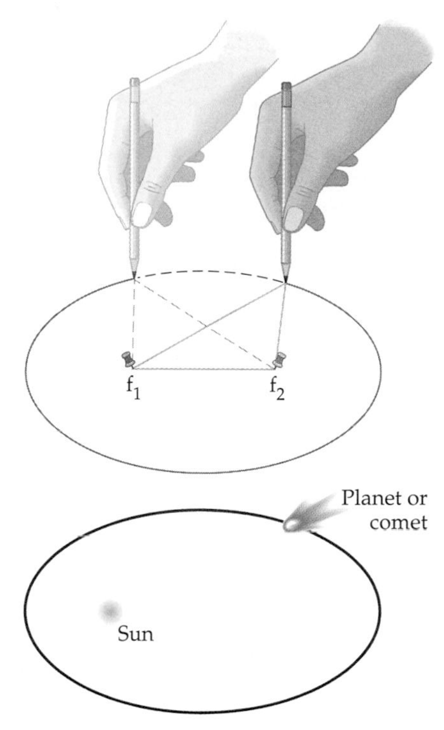

▲ **FIGURE 12–7 Drawing an ellipse**
To draw an ellipse, put two tacks in a piece of cardboard. The tacks define the "foci" of the ellipse. Now, connect a length of string to the two tacks, and use a pencil and the string to sketch out a smooth, closed curve, as shown. This closed curve is an ellipse. In a planetary orbit a planet follows an elliptical path, with the sun at one focus. Nothing is at the other focus.

Kepler's First Law

Kepler tried long and hard to find a circular orbit around the Sun that would match Brahe's observations of Mars. After all, up to that time everyone from Ptolemy to Copernicus believed that celestial objects moved in circular paths of one sort or another. Though the orbit of Mars was exasperatingly close to being circular, the small differences between a circular path and the experimental observations just could not be ignored. Eventually, after a great deal of hard work and disappointment over the loss of circular orbits, Kepler discovered that Mars followed an orbit that was elliptical rather than circular. The same applied to the other planets. This observation became Kepler's first law:

Planets follow elliptical orbits, with the Sun at one focus of the ellipse.

This is a fine example of the scientific method in action. Though Kepler expected and wanted to find circular orbits, he would not allow himself to ignore the data. If Brahe's observations had not been so accurate, Kepler probably would have chalked up the small differences between the data and a circular orbit to error. As it was, he had to discard a treasured—but incorrect—theory, and move on to an unexpected, but ultimately correct, view of nature.

Kepler's first law is illustrated in **Figure 12–7**, along with a definition of an ellipse in terms of its two foci. In the case where the two foci merge, as in **Figure 12–8**, the ellipse reduces to a circle. Thus, a circular orbit *is* allowed by Kepler's first law, but only as a special case.

Newton was able to show that, because the force of gravity decreases with distance as $1/r^2$, closed orbits must have the form of ellipses or circles, as stated in Kepler's first law. He also showed that orbits that are not closed—say the orbit of a comet that passes by the Sun once and then leaves the solar system—are either parabolic or hyperbolic.

Kepler's Second Law

When Kepler plotted the position of a planet on its elliptical orbit, indicating at each position the time the planet was there, he made an interesting observation. First, draw a line from the Sun to a planet at a given time. Then a certain time later–perhaps a month–draw a line again from the Sun to the new position of the planet. The result is that the planet has "swept out" a wedge-shaped area, as

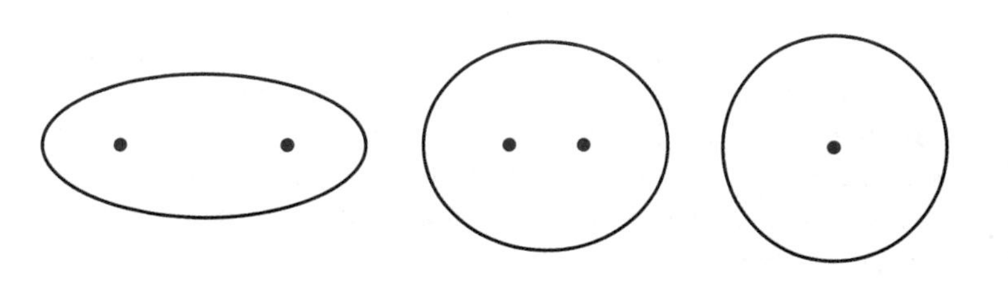

◀ **FIGURE 12–8 The circle as a special case of the ellipse**
As the two foci of an ellipse approach one another the ellipse becomes more circular. In the limit that the foci merge, the ellipse becomes a circle.

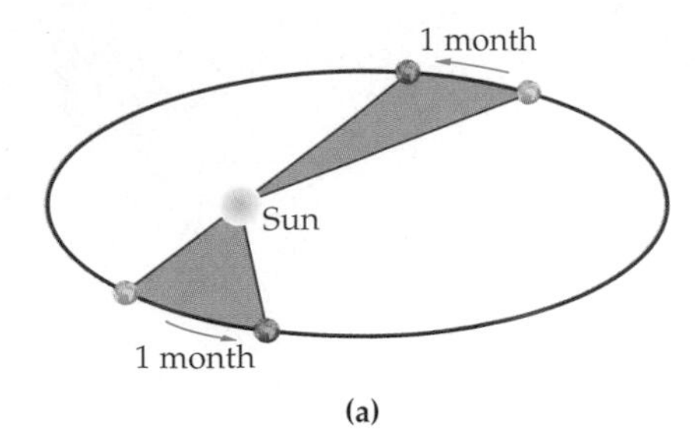

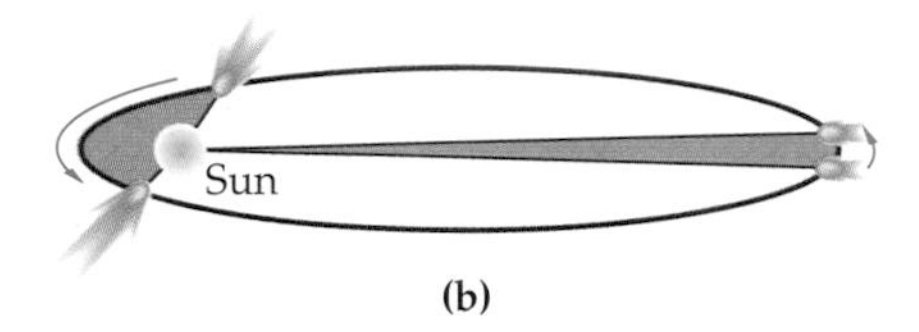

▲ **FIGURE 12–9 Kepler's second law**
(a) The second law states that a planet sweeps out equal areas in equal times. **(b)** In a highly eliptical orbit, the long thin area is equal to the broad, fan-shaped area.

indicated in **Figure 12–9 (a)**. If this procedure is repeated, when the planet is on a different part of its orbit, another wedge-shaped area is generated. Kepler's observation was that the area of these two wedges are equal:

> As a planet moves in its orbit, it sweeps out an equal amount of area in an equal amount of time.

Kepler's second law follows from the fact that the force of gravity on a planet is directly toward the Sun. As a result, gravity exerts zero torque about the Sun, which means that the angular momentum of a planet in its orbit must be conserved. As Newton showed, conservation of angular momentum is equivalent to the equal-area law stated by Kepler.

CONCEPTUAL CHECKPOINT 12–1

The Earth's orbit is slightly elliptical. In fact, the Earth is closer to the Sun during the northern hemisphere winter than it is during the summer. Is the speed of the Earth during winter **(a)** greater than, **(b)** less than, or **(c)** the same as its speed during summer?

Reasoning and Discussion

According to Kepler's second law, the area swept out by the Earth per month is the same in winter as it is in summer. In winter, however, the radius from the Sun to the Earth is less than it is in summer. Therefore, if this smaller radius is to sweep out the same area, the Earth must move more rapidly.

Answer:

(a) The speed of the Earth is greater during the winter.

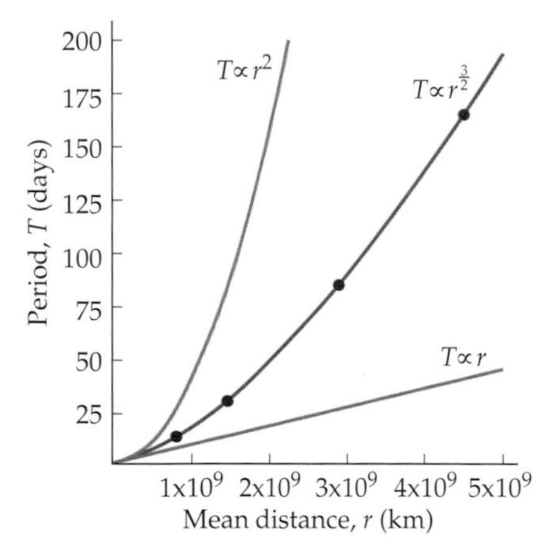

▲ **FIGURE 12–10 Kepler's third law and some near misses**

These plots represent three possible mathematical relationships between period of revolution, T (in days), and mean distance from the Sun, r (in kilometers). The lower curve shows $T = (\text{constant})r$, the upper curve is $T = (\text{constant})r^2$. The middle curve, which fits the data, is $T = (\text{constant})r^{3/2}$. This is Kepler's third law.

Though we have stated the first two laws in terms of planets, they apply equally well to any object orbiting the Sun. For example, a comet might follow a highly elliptical orbit, as in **Figure 12–9 (b)**. When it is near the Sun it moves very quickly, for the reason discussed in Conceptual Checkpoint 12–1, sweeping out a broad wedge-shaped area in a month's time. Later in its orbit, the comet is far from the Sun and moving slowly. In this case, the area it sweeps out in a month is a long, thin wedge. Still, the two wedges have equal areas.

Kepler's Third Law

Finally, Kepler studied the relation between the mean distance of a planet from the Sun, r, and its period—that is, the time, T, it takes for the planet to complete one orbit. **Figure 12–10** shows a plot of period versus distance for the planets of the solar system. Kepler tried to "fit" these results to a simple dependence between T and r. If he tried a linear fit, that is T proportional to r, (the bottom curve in Figure 12–10), he found that the period did not increase rapidly enough with distance. On the other hand, if he tried T proportional to r^2, (the top curve in Figure 12–10), the period increased too rapidly. Splitting the difference, and trying T proportional to $r^{3/2}$, yields a good fit, (the middle curve in Figure 12–10). This is Kepler's third law:

> The period, T, of a planet increases as its mean distance from the Sun, r, raised to the 3/2 power. That is,
>
> $$T = (\text{constant})r^{3/2} \qquad \text{12–6}$$

It is straightforward to derive this result for the special case of a circular orbit. Consider, then, a planet orbiting the Sun at a distance r, as in **Figure 12–11**. Since the planet moves in a circular path, a centripetal force must act on it, as we saw in Section 6-5. In addition, this force must be directed toward the center of the circle; that is, toward the Sun. It is as if you were to swing a ball on the end of a string in a circle above your head, as in Figure 6–12 (p. 157). In order for the ball to move in a circular path, you have to exert a force on the ball toward the center of the circular path. This force is exerted through the string. In the case of a planet orbiting the Sun, the centripetal force is provided by the force of gravity between the Sun and the planet.

If the planet has a mass m, and the Sun has a mass M_s, the force of gravity between them is

$$F = G\frac{mM_s}{r^2}$$

Now, this force creates the centripetal acceleration of the planet, a_{cp}, which, according to Equation 6–14, is

$$a_{cp} = \frac{v^2}{r}$$

Thus, the centripetal force necessary for the planet to orbit is ma_{cp}:

$$F = ma_{cp} = m\frac{v^2}{r}$$

Since the speed of the planet, v, is the circumference of the orbit, $2\pi r$, divided by the time to complete an orbit, T, we have

$$F = m\frac{v^2}{r} = m\frac{(2\pi r/T)^2}{r} = \frac{4\pi^2 rm}{T^2}$$

Setting the centripetal force equal to the force of gravity yields

$$\frac{4\pi^2 rm}{T^2} = G\frac{mM_s}{r^2}$$

Eliminating m and rearranging, we find

$$T^2 = \frac{4\pi^2}{GM_s}r^3$$

or

$$T = \left(\frac{2\pi}{\sqrt{GM_s}}\right)r^{3/2} = (\text{constant})r^{3/2} \qquad \text{12–7}$$

As predicted by Kepler, T is proportional to $r^{3/2}$.

Deriving Kepler's third law by using Newton's law of gravitation has allowed us to calculate the constant that multiplies $r^{3/2}$. Note that the constant depends on the mass of the Sun; that is, *T depends on the mass being orbited*. It does not depend on the mass of the planet orbiting the Sun, however, as long as the planet's mass is much less than the mass of the Sun. As a result, Equation 12–7 applies equally to all the planets.

This result can also be applied to the case of a moon or a satellite (an artificial moon) orbiting a planet. To do so, we simply note that it is the planet that is being orbited, not the Sun. Hence, to apply Equation 12–7, we simply replace the mass of the Sun, M_s, with the mass of the appropriate planet.

As an example, let's calculate the mass of Jupiter. One of the four moons of Jupiter discovered by Galileo is Io, which completes one orbit every 42 hr 27 min $= 1.53 \times 10^5$ s. Given that the average distance from the center of Jupiter to Io is 4.22×10^8 m, we can find the mass of Jupiter as follows:

$$T = \left(\frac{2\pi}{\sqrt{GM_J}}\right)r^{3/2}$$

$$M_J = \frac{4\pi^2 r^3}{GT^2} = \frac{4\pi^2(4.22 \times 10^8\ \text{m})^3}{(6.67 \times 10^{-11}\ \text{N}\cdot\text{m}^2/\text{kg}^2)(1.53 \times 10^5\ \text{s})^2} = 1.90 \times 10^{27}\ \text{kg}$$

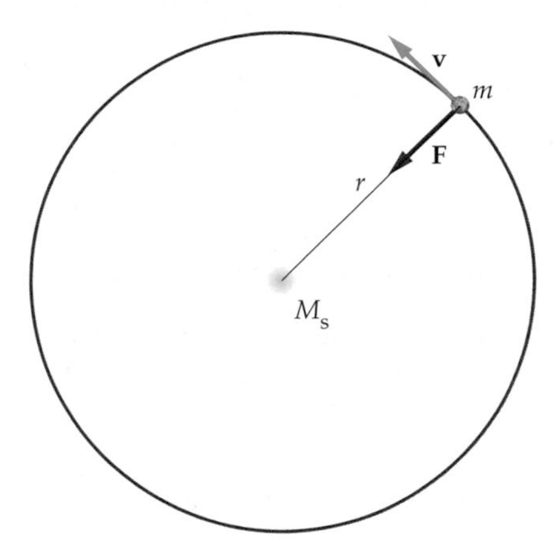

▲ **FIGURE 12–11 Centripetal force on a planet in orbit**

As a planet revolves about the Sun in a circular orbit of radius r, the force of gravity between it and the Sun, $F = GmM_s/r^2$, provides the required centripetal force.

PROBLEM SOLVING NOTE

The Mass in Kepler's Third Law

When applying Kepler's third law, recall that the mass in Equation 12–7, M_s, refers to the mass of the object being orbited. Thus, the third law can be applied to satellites of any object, as long as M_s is replaced by the orbited mass.

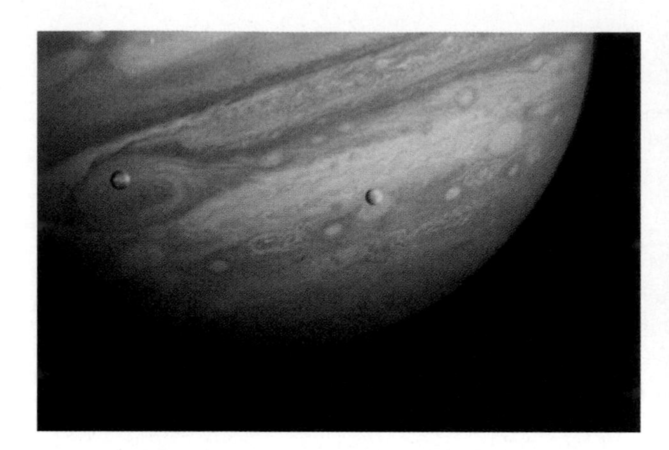

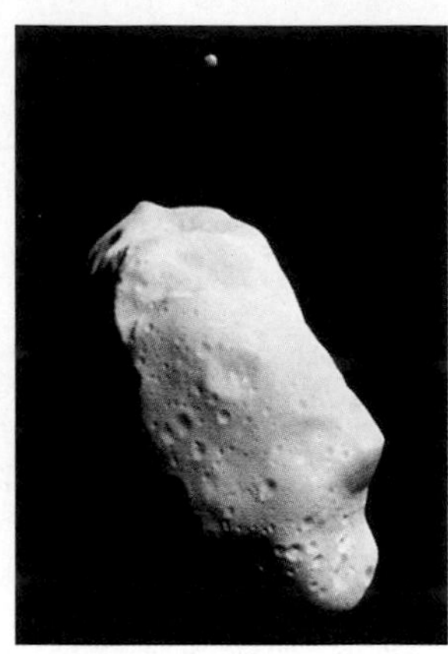

▶ Kepler's laws of orbital motion apply to planetary satellites as well as planets. Jupiter, the largest planet in the solar system, has at least 16 Moons, all of which travel in elliptical orbits that obey Kepler's laws. (The moons in the photo at left, passing in front of Jupiter, are Io and Europa, two of the four largest Jovian satellites discovered by Galileo in 1609.) Even some asteroids have been found to have their own satellites. The large cratered object in the photo at right is Ida, an asteroid some 56 km long; its miniature companion at the top of the photo is Dactyl, about 1.5 km in diameter. Like all gravitationally bound bodies, Ida and Dactyl orbit their common center of mass.

EXAMPLE 12–4 The Sun and Mercury

Real World Physics

The Earth revolves around the Sun once a year at an average distance of 1.50×10^{11} m. **(a)** Use this information to calculate the mass of the Sun. **(b)** Find the period of revolution for the planet Mercury, whose average distance from the Sun is 5.79×10^{10} m.

Picture the Problem

Our sketch shows the orbits of Mercury, Venus, and the Earth in correct proportion. In addition, each of these orbits is slightly elliptical, though the deviation from circularity is too small for the eye to see.

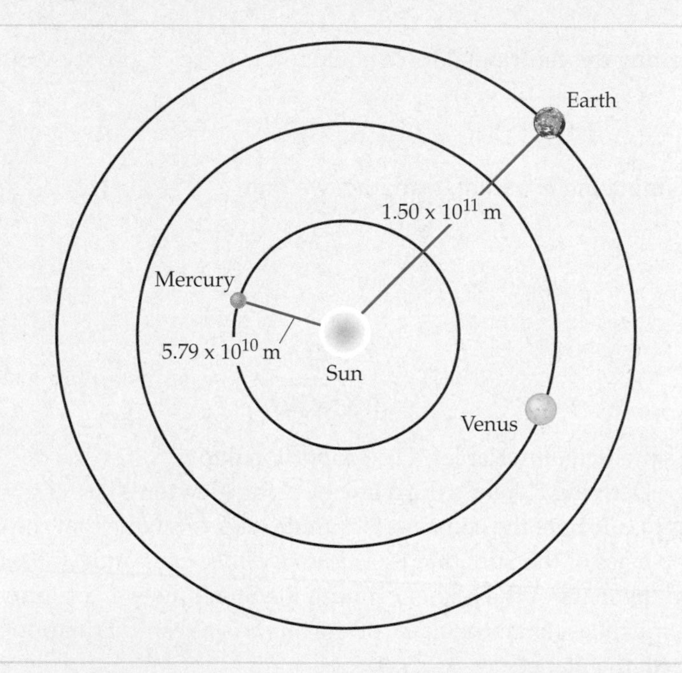

Strategy

(a) To find the mass of the Sun, we solve Equation 12–7 for M_S. Note that the period, $T = 1$ yr, must be converted to seconds before we evaluate the formula.

(b) The period of Mercury is found by substituting $r = 5.79 \times 10^{10}$ m in Equation 12–7.

Solution

Part (a)

1. Solve Equation 12–7 for the mass of the Sun:

$$T = \left(\frac{2\pi}{\sqrt{GM_S}}\right) r^{3/2}$$

$$M_S = \frac{4\pi^2 r^3}{GT^2}$$

2. Calculate the period of the Earth in seconds:

$$T = 1\text{ yr}\left(\frac{365.24\text{ days}}{1\text{ yr}}\right)\left(\frac{24\text{ hr}}{1\text{ day}}\right)\left(\frac{3600\text{ s}}{1\text{ hr}}\right) = 3.16 \times 10^7\text{ s}$$

3. Substitute numerical values in the expression for the mass of the Sun obtained in step 1:

$$M_S = \frac{4\pi^2 r^3}{GT^2}$$

$$= \frac{4\pi^2 (1.50 \times 10^{11}\text{ m})^3}{(6.67 \times 10^{-11}\text{ N}\cdot\text{m}^2/\text{kg}^2)(3.16 \times 10^7\text{ s})^2}$$

$$= 2.00 \times 10^{30}\text{ kg}$$

Part (b)

4. Substitute $r = 5.79 \times 10^{10}$ m into Equation 12–7. In addition, use the mass of the Sun obtained in part (a):

$$T = \left(\frac{2\pi}{\sqrt{GM_S}}\right) r^{3/2}$$
$$= \left(\frac{2\pi}{\sqrt{(6.67 \times 10^{-11}\ \mathrm{N \cdot m^2/kg^2})(2.00 \times 10^{30}\ \mathrm{kg})}}\right) \times (5.79 \times 10^{10}\ \mathrm{m})^{3/2}$$
$$= 7.58 \times 10^6\ \mathrm{s} = 0.240\ \mathrm{yr} = 87.5\ \mathrm{days}$$

Insight

As expected, Mercury, with its smaller orbital radius, has a shorter year than the Earth.

Practice Problem

Venus orbits the Sun with a period of 1.94×10^7 s. What is its average distance from the Sun? [**Answer:** $r = 1.08 \times 10^{11}$ m]

Some related homework problems: Problem 21, Problem 25

A *geosynchronous satellite* is one that orbits above the equator with a period equal to one day. From the Earth, such a satellite appears to be in the same location in the sky at all times, making it particularly useful for applications such as communications and weather forecasting. From Kepler's third law, we know that a satellite has a period of one day only if its orbital radius has a particular value. We determine this value in the following Active Example.

ACTIVE EXAMPLE 12–1 Geosynchronous Satellites

Find the altitude above the Earth's surface where a satellite orbits with a period of one day ($R_E = 6.37 \times 10^6$ m, $M_E = 5.97 \times 10^{24}$ kg, $T = 1$ day $= 8.64 \times 10^4$ s).

Solution

1. Rewrite Equation 12–7, using the mass of the Earth in place of the mass of the Sun: $T = \left(2\pi/\sqrt{GM_E}\right) r^{3/2}$
2. Solve for the radius, r: $r = (T/2\pi)^{2/3}(GM_E)^{1/3}$
3. Substitute numerical values: $r = 4.22 \times 10^7$ m
4. Subtract the radius of the Earth to find the altitude: $r - R_E = 3.58 \times 10^7$ m

Insight

Thus, *all* geosynchronous satellites orbit 3.58×10^7 m $\approx$ 22,300 mi above our heads.

Not all spacecraft are placed in geosynchronous orbits, however. The U. S. Space Shuttle, for example, orbits at an altitude of about 150 mi. At that altitude, it takes less than an hour and a half to complete one orbit. The Russian space station, Mir, and the International Space Station, currently under construction, orbit at similar altitudes.

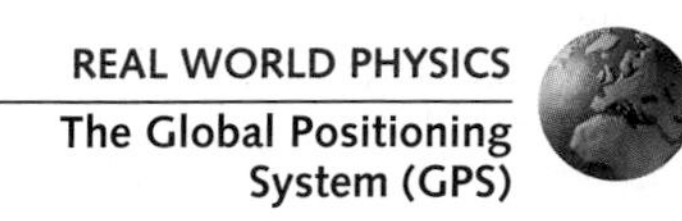

The 24 satellites of the Global Positioning System (GPS) are also in relatively low orbits. These satellites, which have an average altitude of 12,550 mi and orbit the Earth every 12 hours, are used to provide a precise determination of an observer's position anywhere on Earth. The operating principle of the GPS is illustrated in **Figure 12–12**. Imagine, for example, that satellite 2 emits a radio signal at a particular time (all GPS satellites carry atomic clocks on board). This signal travels away from the satellite with the speed of light (see Chapter 25) and is detected a short time later by an observer's GPS receiver. Multiplying the time delay by the speed of light gives the distance of the receiver from satellite 2. Thus, in our example, the observer must

▲ Many weather and communications satellites are placed in geosynchronous orbits that allow them to remain 'stationary' in the sky—that is, fixed over one point on the Earth's equator. Since the Earth rotates, the period of such a satellite must exactly match that of the Earth. The altitude needed for such an orbit is about 36,000 km (see Active Example 12–1). Other satellites, such as those used in the Global Positioning System (GPS), as well as the Russian space station Mir, the Hubble Space Telescope, and the American Space Shuttles, operate at much lower altitudes—typically just a few hundred miles. The photo at left shows the communications satellite Intelsat VI just prior to its capture by astronauts of the Space Shuttle Endeavour. A launch failure had left the satellite stranded in low orbit. The astronauts snared the satellite (right) and fitted it with a new engine that boosted it to its geosynchronous orbit, where it is still in operation today.

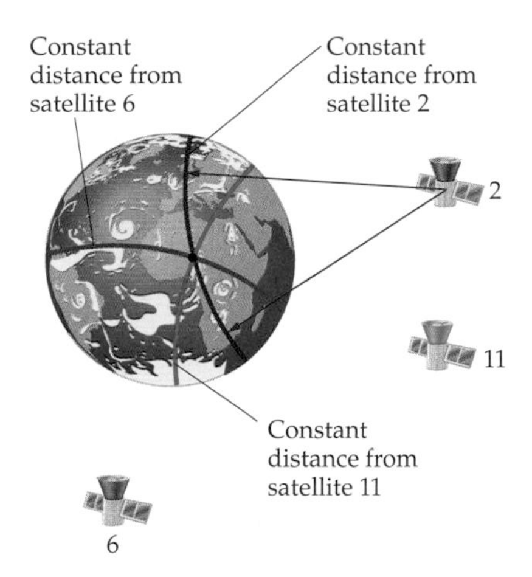

▲ FIGURE 12–12 The Global Positioning System

A system of 24 satellites in orbit about the Earth make it possible to determine a person's location with great accuracy. Measuring the distance of a person from satellite 2 places the person somewhere on the red circle. Similar measurements using satellite 11 place the person's position somewhere on the green circle, and further measurements can pinpoint the person's location.

lie somewhere on the red circle in Figure 12–12. Similar time delay measurements for signals from satellite 11 show that the observer is also somewhere on the green circle; hence the observer is either at the point shown in Figure 12–12, or at the second intersection of the red and green circles on the other side of the planet. Measurements from satellite 6 can resolve the ambiguity, and place the observer at the point shown in the figure. Measurements from additional satellites can even determine the observer's altitude. GPS receivers, which are used by hikers, boaters, and others who need to know their precise location, typically use signals from as many as 12 satellites. As currently operated, the GPS can give positions with an accuracy of 15 m to 100 m. In the future, the accuracy may be as high as 1 m.

Orbital Maneuvers

We now show how Kepler's laws can give insight into maneuvering a satellite in orbit. Suppose, for example, that you are piloting a satellite in a circular orbit, and you would like to move to a lower circular orbit. As you might expect, you should begin by using your rockets to decrease your speed—that is, fire the rockets that point in the forward direction so that their thrust (Section 9-8) is opposite to your direction of motion. The result of firing the decelerating rockets at a given point A in your original orbit is shown in **Figure 12–13 (a)**. Note that your new orbit is not a circle, as desired, but rather an ellipse. To produce a circular orbit you can simply fire the decelerating rockets once again at point B, on the opposite side of the Earth from point A. The net result of these two firings is that you now move in a circular orbit of smaller radius.

Similarly, to move to a larger orbit, you must fire your accelerating rockets twice. The first firing puts you into an elliptical orbit that moves farther from the Earth, as **Figure 12–13 (b)** shows. After the second firing you are again in a circular orbit. This simplest type of orbital transfer, requiring just two rocket burns, is referred to as a *Hohmann transfer.* The Hohmann transfer is the basic maneuver used to send spacecraft such as the Mars lander from Earth's orbit about the Sun to the orbit of Mars.

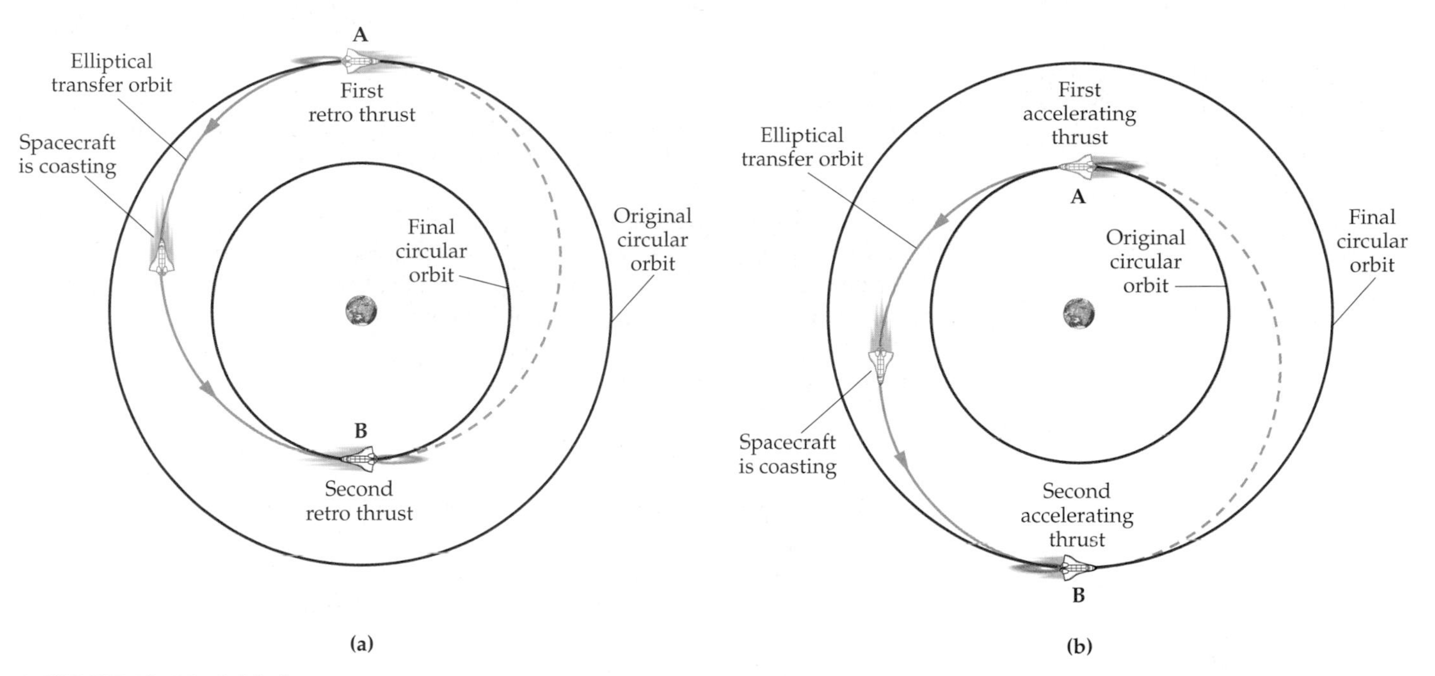

▲ **FIGURE 12–13 Orbital maneuvers**

(a) The radius of a satellite's orbit can be decreased by firing the decelerating rockets once at point A and again at point B. Between firings the satellite follows an elliptical orbit. The satellite speeds up as it falls inward toward the earth during this maneuver. For this reason its final speed in the new circular orbit is greater than its speed in the original orbit, even though the decelerating rockets have slowed it down twice. **(b)** The radius of a satellite's orbit can be increased by firing the accelerating rockets once at point A and again at point B. Between firings the satellite follows an elliptical orbit. The satellite slows down as it moves farther from the earth during this maneuver. For this reason its final speed in the new circular orbit is less than its speed in the original orbit, even though the accelerating rockets have sped it up twice.

CONCEPTUAL CHECKPOINT 12–2

As you pilot your spacecraft in a circular orbit about the Earth you notice the space station you want to dock with several miles ahead in the same orbit. To catch up with the space station, should you **(a)** fire your accelerating rockets or **(b)** fire your decelerating rockets?

Reasoning and Discussion

Since you want to catch up with something miles ahead, you must accelerate, right? Well, not in this case. Accelerating moves you into an elliptical orbit, as in Figure 12–13 (b), and with a second acceleration you can make your new orbit circular with a greater radius. Recall from Kepler's third law, however, that the larger the radius of an orbit the larger the period, as Equation 12–6 shows. Thus, on your new higher path you take longer to complete an orbit, so you fall farther behind the space station. The same is true even if you fire your rockets only once and stay on the elliptical orbit–it also has a longer period than the original orbit.

On the other hand, two decelerating burns will put you into a circular orbit of smaller radius, and thus smaller period. As a result you complete an orbit in less time than before, and catch up with the space station. After catching up you can perform two accelerating burns to move you back into the original orbit to dock.

Answer:

(b) You should fire your decelerating rockets.

12–4 Gravitational Potential Energy

In Chapter 8 we saw that the principle of conservation of energy can be used to solve a number of problems that would be difficult to handle with a straightforward application of Newton's laws of mechanics. Before we can apply energy conservation to astronomical situations, however, we must know the gravitational potential energy for a spherical object such as the Earth. Now you may be wondering, "Don't we already know the potential energy of gravity?" Well, in fact, in Chapter 8 we said that the gravitational potential energy a distance h above the Earth's surface is $U = mgh$. As was mentioned at the time, however, this result is valid only near the Earth's surface, where we can say that the acceleration of gravity, g, is approximately constant.

As the distance from the Earth increases we know that g decreases, as was shown in Example 12–2. It follows that mgh cannot be valid for arbitrary h. Indeed, it can be shown that the gravitational potential energy of a system consisting of a mass m a distance r from the center of the Earth is

$$U = -G\frac{mM_E}{r} \qquad \textbf{12–8}$$

A plot of $U = -GmM_E/r$ is presented in **Figure 12–14**. Note that U approaches zero as r approaches infinity. This is a common convention in astronomical systems. In fact, since only *differences* in potential energy matter, as was mentioned in Chapter 8, the choice of the reference point ($U = 0$) is completely arbitrary. When we considered systems that were near the Earth's surface, it was natural to let $U = 0$ at ground level. When we consider, instead, distances of astronomical scale, it is generally more convenient to choose the potential energy to be zero when objects are separated by an infinite distance.

EXERCISE 12–4

Use Equation 12–8 to find the gravitational potential energy of a 12.0-kg meteorite when it is **(a)** one Earth radius above the surface of the Earth, and **(b)** on the surface of the Earth.

Solution

(a) In this case, the distance from the center of the Earth is $2R_E$, thus

$$U = -G\frac{mM_E}{2R_E}$$

$$= -(6.67 \times 10^{-11}\,\text{N}\cdot\text{m}^2/\text{kg}^2)\frac{(12.0\text{ kg})(5.97 \times 10^{24}\text{ kg})}{2(6.37 \times 10^6\text{ m})} = -3.75 \times 10^8\text{ J}$$

(b) Now, the distance from the center of the Earth is R_E, therefore

$$U = -\frac{GmM_E}{R_E}$$

$$= -(6.67 \times 10^{-11}\,\text{N}\cdot\text{m}^2/\text{kg}^2)\frac{(12.0\text{ kg})(5.97 \times 10^{24}\text{ kg})}{6.37 \times 10^6\text{ m}} = -7.50 \times 10^8\text{ J}$$

Note that the potential energy in part (b) is twice what it was in part (a), since the distance from the center of the Earth to the meteorite has been halved.

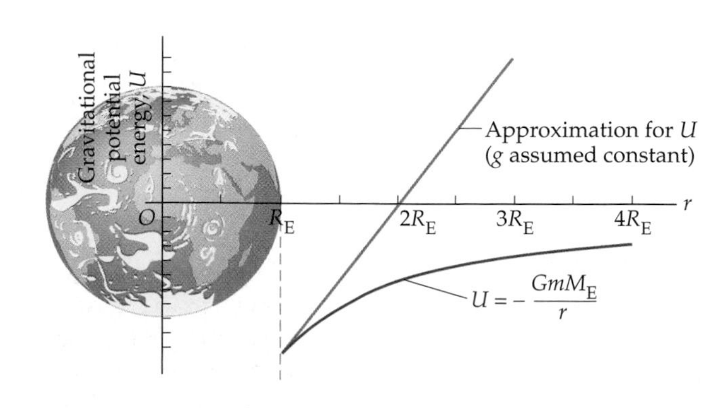

▶ FIGURE 12–14 Gravitational potential energy as a function of the distance r from the center of the Earth

The lower curve in this plot shows the gravitational potential energy, $U = -GmM_E/r$, for r greater than R_E. Near the Earth's surface, U is approximately linear, corresponding to the result $U = mgh$ given in Chapter 8.

At first glance, Equation 12–8 doesn't seem to bear any similarity to mgh, which we know to be valid near the surface of the Earth. Even so, there is a direct connection between these two expressions. Recall that when we say that the potential energy at a height h is mgh, what we mean is that when a mass m is raised from the ground to a height h the potential energy of the system increases by the amount mgh. Let's calculate the corresponding difference in potential energy using Equation 12–8.

First, at a height h above the surface of the Earth we have $r = R_E + h$; hence the potential energy there is

$$U = -G\frac{mM_E}{R_E + h}$$

On the surface of the Earth, where $r = R_E$, we have

$$U = -G\frac{mM_E}{R_E}$$

Thus, if we assume that the height h is much smaller than the radius of the Earth, R_E, it can be shown that the change in potential energy is given approximately by the following expression:

$$\Delta U = \left(-G\frac{mM_E}{R_E + h}\right) - \left(-G\frac{mM_E}{R_E}\right) \approx m\left[\frac{GM_E}{R_E^2}\right]h$$

The term in square brackets should look familiar—according to Equation 12–4 it is simply g. Hence, the increase in potential energy at the height h is

$$\Delta U = mgh$$

as expected.

The straight line in Figure 12–14 corresponds to the potential energy mgh. Near the Earth's surface, it is clear that mgh and $-GmM_E/r$ are in close agreement. For larger r, however, the fact that gravity is getting weaker means that the potential energy does not continue rising as rapidly as it would if gravity were of constant strength.

An important distinction between the potential energy, U, and the gravitational force, $\mathbf{F}$, is that the force is a vector, whereas the potential energy is a scalar—that is, U is simply a number. As a result:

> The total gravitational potential energy of a system of objects is the sum of the gravitational potential energies of each pair of objects separately.

Since U is not a vector, there are no x or y components to consider, as would be the case with a vector. Finally, the potential energy given in Equation 12–8 applies to a mass m and the Earth, with mass M_E. More generally, if two point masses, m_1 and m_2 are separated by a distance r, their gravitational potential energy is

Gravitational Potential Energy, *U*

$$U = -G\frac{m_1 m_2}{r} \qquad \textbf{12–9}$$

SI unit: joule, J

In the next Example we use this result, and the fact that U is a scalar, to find the total gravitational potential energy for a system of three point masses.

EXAMPLE 12–5 Simple Addition

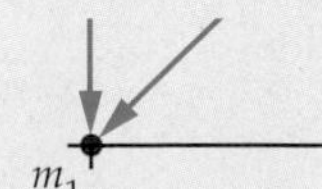

Three masses are positioned as follows: $m_1 = 2.5$ kg is at the origin; $m_2 = 0.75$ kg is at $x = 0$, $y = 1.25$ m; and $m_3 = 0.75$ kg is at $x = 1.25$ m and $y = 1.25$ m. Find the total gravitational potential energy of this system.

continued on the following page

continued from the previous page

Picture the Problem
The masses and their positions are shown in the sketch. The distances along the sides of the square are $r = 1.25$ m; the diagonal distance is $\sqrt{2}\,r$.

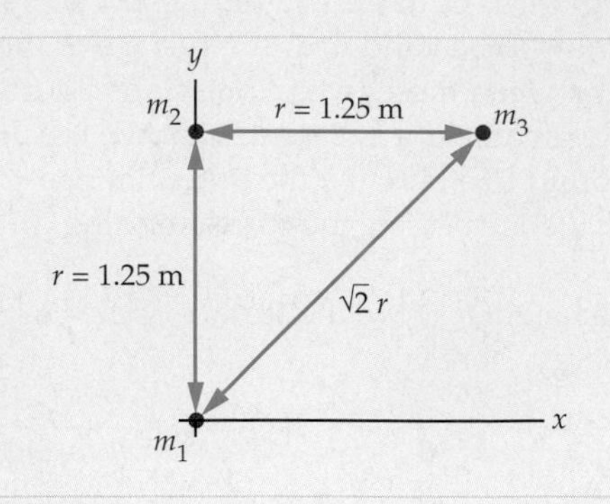

Strategy
The potential energy associated with each pair of masses is given by Equation 12–9. The total potential energy of the system is the sum of the potential energy for each of the three pairs of masses.

Solution

1. Use Equation 12–9 to calculate the potential energy for masses 1 and 2:

$$U_{12} = -G\frac{m_1m_2}{r_{12}}$$
$$= -(6.67 \times 10^{-11}\,\mathrm{N\cdot m^2/kg^2})\frac{(2.5\,\mathrm{kg})(0.75\,\mathrm{kg})}{(1.25\,\mathrm{m})}$$
$$= -1.0 \times 10^{-10}\,\mathrm{J}$$

2. Similarly, calculate the potential energy for masses 2 and 3:

$$U_{23} = -G\frac{m_2m_3}{r_{23}}$$
$$= -(6.67 \times 10^{-11}\,\mathrm{N\cdot m^2/kg^2})\frac{(0.75\,\mathrm{kg})(0.75\,\mathrm{kg})}{(1.25\,\mathrm{m})}$$
$$= -3.0 \times 10^{-11}\,\mathrm{J}$$

3. Do the same calculation for masses 1 and 3:

$$U_{13} = -G\frac{m_1m_3}{r_{13}}$$
$$= -(6.67 \times 10^{-11}\,\mathrm{N\cdot m^2/kg^2})\frac{(2.5\,\mathrm{kg})(0.75\,\mathrm{kg})}{\sqrt{2}(1.25\,\mathrm{m})}$$
$$= -7.1 \times 10^{-11}\,\mathrm{J}$$

4. The total potential energy is the sum of the three contributions calculated above:

$$U_{\text{total}} = U_{12} + U_{23} + U_{13}$$
$$= -1.0 \times 10^{-10}\,\mathrm{J} - 3.0 \times 10^{-11}\,\mathrm{J} - 7.1 \times 10^{-11}\,\mathrm{J}$$
$$= -2.0 \times 10^{-10}\,\mathrm{J}$$

Insight
Note that the total gravitational potential energy of this system $U_{\text{total}} = -2.0 \times 10^{-10}$ J is less than it would be if the separation of the masses were to approach infinity, in which case $U_{\text{total}} = 0$. The implications of this change in potential energy, in terms of energy conservation, are considered in the next section.

Practice Problem
If the distance $r = 1.25$ m is reduced by a factor of two to $r = 0.625$ m, does the potential energy of the system increase, decrease, or stay the same? Verify your answer by calculating the potential energy in this case. [**Answer:** The potential energy decreases; that is, it becomes more negative. We find $U = 2(-2.0 \times 10^{-10}\,\mathrm{J})$].

Some related homework problems: Problem 34, Problem 35

12–5 Energy Conservation

Now that we know the gravitational potential energy, U, at an arbitrary distance from a spherical object, we can apply energy conservation to astronomical situations in the same way we applied it to systems near the Earth's surface in Chapter 8. To be specific, the mechanical energy, E, of an object of mass m a distance r from the Earth is

$$E = K + U = \tfrac{1}{2}mv^2 - G\frac{mM_E}{r} \qquad \text{12–10}$$

Using energy conservation—that is, setting the initial mechanical energy equal to the final mechanical energy—we can answer questions such as the following: Suppose that an asteroid has zero speed infinitely far from the Earth. If this asteroid were to fall directly toward the Earth, what speed would it have when it strikes the Earth's surface?

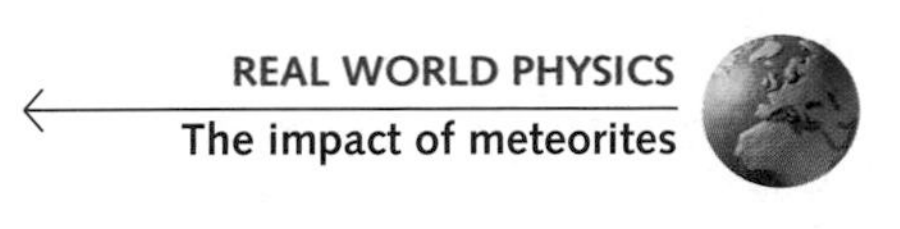

As you probably know, this is not an entirely academic question. Asteroids and comets, both large and small, have struck the Earth innumerable times during its history. In fact, a particularly large object appears to have struck the Earth on the Yucatan Peninsula in Mexico, near the town of Chicxulub, some 65 million years ago. Evidence suggests that this impact may have lead to the mass extinctions of the Cretaceous period during which the dinosaurs disappeared from the Earth. Unfortunately, such events are not limited to the distant past. For example, as recently as 25,000 years ago an iron asteroid tens of meters in diameter and shining 10,000 times brighter than the Sun (from atmospheric heating) slammed into the ground near Winslow, Ariz., forming the 1.2-km wide Barringer Meteor Crater. More recently yet, at sunrise on June 30, 1908, a relatively small stony asteroid streaked through the atmosphere and exploded at an altitude of several kilometers near the Tunguska River in Siberia. The energy released by the explosion was comparable to that of an H-bomb, and it flattened the forest for kilometers in all directions. One can only imagine the consequences if an event like this were to occur near a populated area. Finally, an uncomfortably close call occurred in the early evening of December 9, 1994, when an asteroid the size of a mountain passed the Earth at a distance only one-third the distance from the Earth to the Moon. Thus, though extremely unlikely, the scenarios depicted in movies such as *Armageddon* and *Deep Impact* are not completely unrealistic.

Returning to our original question, we can use energy conservation to determine the speed such an asteroid or comet might have when it hits the Earth. To begin, we assume the asteroid starts at rest, hence its initial kinetic energy is zero, $K_i = 0$. In addition, the initial potential energy of the system, U_i, is also zero, since $U = -GmM_E/r$ approaches zero as r approaches infinity. As a result, the total initial mechanical energy of the asteroid-Earth system is zero; $E_i = K_i + U_i = 0$. Since gravity is a conservative force (as discussed in Section 8-1), the total mechanical energy remains constant as the asteroid falls toward the Earth. Thus, as the asteroid moves closer to the Earth and U becomes increasingly negative, the kinetic energy K must become increasingly positive so that their sum, $U + K$, is always zero.

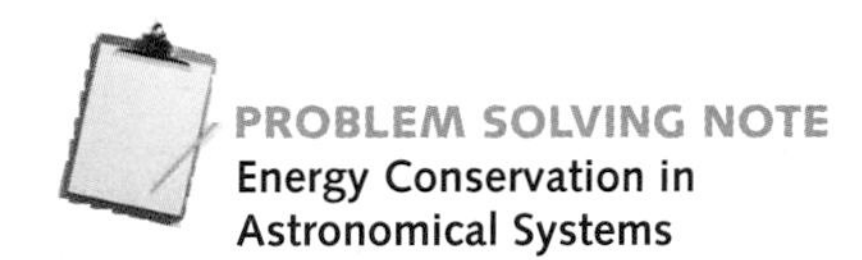

To apply conservation of energy to an object that moves far from the surface of a planet, one must use $U = -GmM/r$, where r is the distance from the center of the planet.

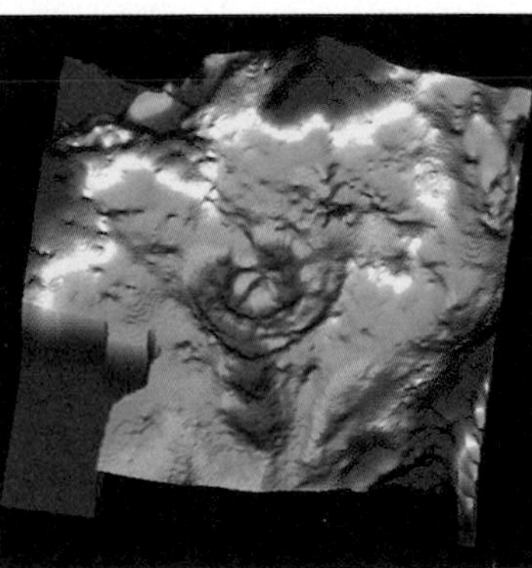

▲ Bodies from space have struck the Earth countless times in the past, and continue to do so on a regular basis. Most such objects are relatively small, ranging in size from grains of dust to fist-sized rocks, and burn up from friction as they pass through the atmosphere, creating the bright streaks that we know as meteors. But larger objects, including the occasional comet or asteroid, also cross our path from time to time, and some of these make it to the surface—often with very dramatic results. The crater at left, in Rotorua, New Zealand, must be of relatively recent origin (thousands rather than millions of years old), since erosion has not yet erased this scar on the Earth's surface.

The image at right is a false-color gravity anomaly map of the Chicxulub impact crater in Mexico. The object that struck here some 65 million years ago may have produced such far-reaching climatic disruption that the dinosaurs and many other species became extinct as a result. At the center of the crater the strength of gravity is lower than normal (blue) because of the presence of low-density rock: debris from the impact and sediments that have accumulated in the crater.

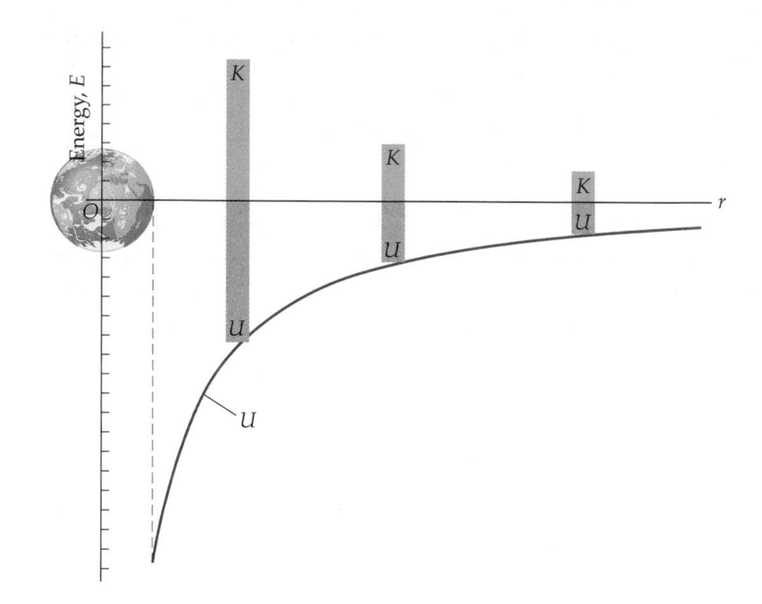

▶ FIGURE 12–15 Potential and kinetic energies of an object falling toward Earth

As an object with zero total energy moves closer to the Earth, its gravitational potential energy, U, becomes increasingly negative. In order for the total energy to remain zero, $E = U + K = 0$, it is necessary for the kinetic energy to become increasingly positive.

We now set the initial energy equal to the final energy to determine the final speed, v_f. Recalling that the final distance r is the radius of the Earth, R_E, we have

$$E_i = E_f$$

$$0 = \frac{1}{2}mv_f^2 - G\frac{mM_E}{R_E}$$

Solving for the final speed yields

$$v_f = \sqrt{\frac{2GM_E}{R_E}} \qquad \textbf{12–11}$$

Substituting numerical values into this expression gives

$$v_f = \sqrt{\frac{2GM_E}{R_E}} = \sqrt{\frac{2(6.67 \times 10^{-11}\,\mathrm{N \cdot m^2/kg^2})(5.97 \times 10^{24}\,\mathrm{kg})}{6.37 \times 10^6\,\mathrm{m}}} \qquad \textbf{12–12}$$

$$= 11{,}200\ \mathrm{m/s}\ (25{,}000\ \mathrm{mi/h})$$

Thus, a typical asteroid hits the Earth moving at about 7.0 mi/s—about 16 times faster than a rifle bullet! Note that this result is independent of the asteroid's mass.

To help visualize energy conservation in this system, we plot the gravitational potential energy U in **Figure 12–15**. Also indicated in the plot is the total energy, $E = 0$. Since $U + K$ must always equal zero, the value of K goes up as the value of U goes down. This is illustrated graphically in the figure with the help of several histograms.

Another way to think about this is to imagine a smooth wooden or plastic surface constructed to have the same shape as the plot of U shown in Figure 12–15. An object placed on this surface has a gravitational potential energy proportional to the height of the surface above a given reference level. Thus, if a small block is allowed to slide without friction on the surface, it will move downhill and speed up as it drops lower in elevation. That is, its kinetic energy will increase as the potential energy of the system decreases. This is completely analogous to the behavior of an asteroid as it "falls" toward the Earth.

A somewhat more elaborate plot showing the same physics is presented in **Figure 12–16**. The two-dimensional surface in this case represents the potential energy function U as one moves away from the Earth in any direction. In partic-

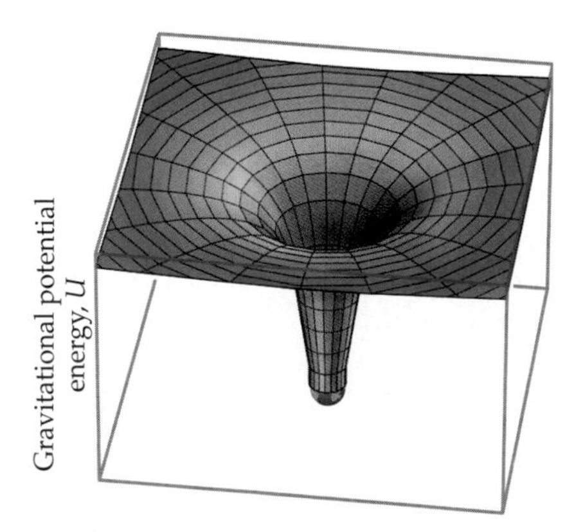

▲ FIGURE 12–16 A gravitational potential "well"

The illustration is a three-dimensional plot of the gravitational potential energy near an object such as the earth. An object approaching the earth speeds up as it "falls" into the gravitational potential well.

ular, the dependence of U on distance r along any radial line in Figure 12–16 is the same as the shape of U versus r in Figure 12–15. Because the potential energy drops downward in such a plot, this type of situation is often referred to as a "potential well." If a marble is allowed to roll on such a surface, its motion is similar in many ways to the motion of an object near the Earth. In fact, if the marble is started with the right initial velocity, it will roll in a circular or elliptical "orbit" for a long time before falling into the center of the well. (Eventually, of course, the well does swallow up the marble. Though the retarding force of rolling friction is quite small, it still causes the marble to descend into a lower and lower orbit—just as air resistance causes a satellite to descend lower and lower into the Earth's atmosphere until it finally burns up.)

EXAMPLE 12–6 Armageddon Rendezvous

In the movie *Armageddon*, a crew of hard-boiled oil drillers rendezvous with a menacing asteroid just as it passes the orbit of the Moon on its way toward Earth. Assuming the asteroid starts at rest infinitely far from the Earth, as in the previous discussion, find its speed when it passes the Moon's orbit. Assume the Moon orbits at a distance of $60R_E$ from the center of the Earth, and that its gravitational influence may be neglected.

Picture the Problem

Our sketch shows the Earth, the Moon, and the asteroid. The asteroid is moving directly toward the Earth with the final speed v_f.

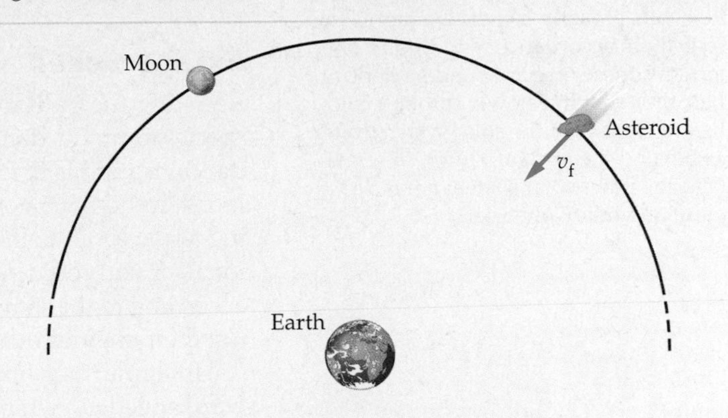

Strategy

The basic strategy is the same as that used to obtain the speed of an asteroid in Equation 12–12; namely, we set the initial energy equal to the final energy and solve for the final speed v_f. In this case, the final radius is $r = 60R_E$. As before, the initial energy is zero.

Solution

1. Set the initial energy of the system equal to its final energy:

$$E_i = E_f$$
$$0 = \tfrac{1}{2}mv_f^2 - G\frac{mM_E}{60R_E}$$

2. Solve for the final speed, v_f:

$$v_f = \sqrt{\frac{2GM_E}{60R_E}} = \frac{1}{\sqrt{60}}\left(\sqrt{\frac{2GM_E}{R_E}}\right)$$

3. Substitute numerical values, and use the value of v_f obtained in Equation 12–12:

$$v_f = \frac{1}{\sqrt{60}}(11{,}200 \text{ m/s}) = 1440 \text{ m/s} \sim 3200 \text{ mi/h}$$

Insight

Note that the majority of the asteroid's increase in speed occurs after it passes the Moon. The reason for this can be seen in the following plot of the gravitational potential energy, U.

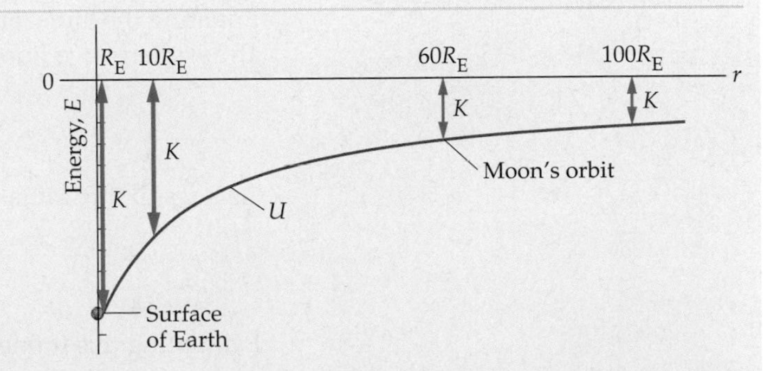

Note that U drops downward more and more rapidly as the Earth is approached. Thus, while there is relatively little increase in K from infinite distance to $r = 60R_E$, there is a substantially larger increase in K from $r = 60R_E$ to $r = R_E$.

Practice Problem

At what distance from the center of the Earth is the asteroid's speed equal to 3540 m/s? [**Answer:** $r = 6.37 \times 10^7$ m $= 10R_E$]

Some related homework problems: Problem 40, Problem 42

▲ Comet Hale-Bopp, one of the largest and brightest comets to visit our celestial neighborhood in recent decades, photographed in April 1997. While most of the planets and planetary satellites in the solar system have roughly circular orbits, the orbits of many comets are highly elliptical. In accordance with Kepler's second law, these objects spend most of their time moving slowly through cold, distant regions of the solar system (often far beyond the orbit of Pluto). Their visits to the inner solar system are infrequent and relatively brief.

ACTIVE EXAMPLE 12–2 Elliptical Orbit

A satellite in an elliptical orbit has a speed of 9.00 km/s when it is at its closest approach to the Earth (perigee). The satellite is 7.00×10^6 m from the center of the Earth at this time. When the satellite is at its greatest distance from the center of the Earth (apogee), its speed is 3.66 km/s. How far is the satellite from the center of the Earth at apogee? ($R_E = 6.37 \times 10^6$ m, $M_E = 5.97 \times 10^{24}$ kg)

Solution

1. Set the energy at perigee, E_1, equal to the energy at apogee, E_2: $\frac{1}{2}mv_1^2 - GmM_E/r_1 = \frac{1}{2}mv_2^2 - GmM_E/r_2$
2. Solve for $1/r_2$: $1/r_2 = 1/r_1 + (v_2^2 - v_1^2)/(2GM_E)$
3. Substitute numerical values: $1/r_2 = 5.80 \times 10^{-8}\ \text{m}^{-1}$
4. Invert to obtain r_2: $r_2 = 1.73 \times 10^7$ m

Insight

In this case, apogee is about 2.5 times farther from the center of the Earth than perigee.

Escape Speed

Resisting the pull of Earth's gravity has always held a fascination for the human species, from Daedalus and Icarus with their wings of feathers and wax, to Leonardo da Vinci and his flying machine, to the Montgolfier brothers and their hot air balloons. In his 1865 novel, *From the Earth to the Moon*, Jules Verne imagined launching a spacecraft to the Moon by firing it straight upward from a cannon. Not a bad idea—if you could survive the initial blast. Today, rockets fired into space operate according to the same basic idea, though they smooth out the initial blast by burning their engines over a period of several minutes.

Imagine, then, that you would like to give a spacecraft of mass m an initial upward speed, v_0, which is just enough to carry it to the Moon's orbit. The initial energy of the spacecraft is

$$E_0 = K_0 + U_0 = \tfrac{1}{2}mv_0^2 - G\frac{mM_E}{R_E}$$

At the Moon's orbit, the spacecraft comes to rest momentarily before falling back toward the Earth. Thus, since the radius of the Moon's orbit is approximately $60R_E$, the final energy of the spacecraft is

$$E_f = K_f + U_f = 0 - G\frac{mM_E}{60R_E}$$

Equating the initial and final energies (that is, ignoring air resistance in the Earth's atmosphere), we find

$$\tfrac{1}{2}mv_0^2 - G\frac{mM_E}{R_E} = -G\frac{mM_E}{60R_E}$$

As a result, the initial speed of the spacecraft must be

$$v_0 = \sqrt{\frac{2GM_E}{R_E}\left(1 - \frac{1}{60}\right)} = \sqrt{\frac{2GM_E}{R_E}\left(\frac{59}{60}\right)}$$

Evaluating this numerically we find the following rather large speed:

$$v_0 = 11{,}100\ \text{m/s} \approx 24{,}800\ \text{mi/h} \approx 7.00\ \text{mi/s}$$

Suppose that, instead of going to the Moon, you would like to just keep going, and escape the Earth altogether. What is the **escape speed**, v_e, that would be required of your rocket? In this case, we would like the rocket to be able to travel an infinite distance from the Earth before coming to rest. Hence, we can repeat the previous calculation, simply replacing $60R_E$ with ∞. This gives

$$E_0 = K_0 + U_0 = \tfrac{1}{2}mv_e^2 - G\frac{mM_E}{R_E}$$

for the initial energy, and

$$E_f = K_f + U_f = 0 - G\frac{mM_E}{\infty} = 0$$

for the final energy. Equating these energies yields

$$\tfrac{1}{2}mv_e^2 - G\frac{mM_E}{R_E} = 0$$

Therefore, the escape speed from the Earth is

$$v_e = \sqrt{\frac{2GM_E}{R_E}} = 11{,}200 \text{ m/s} \approx 25{,}000 \text{ mi/h} \qquad \textbf{12–13}$$

Note that the escape speed is precisely the same as the speed of the asteroid calculated in Equation 12–12. This is not surprising when you consider that an object launched from the Earth to infinity is just the reverse of an object falling from infinity to the Earth.

The result given in Equation 12–13 can be applied to other astronomical objects as well by simply replacing M_E and R_E with the appropriate mass and radius for that object.

EXERCISE 12–5

Calculate the escape speed for the Moon.

Solution

For the Moon we use $M_m = 7.35 \times 10^{22}$ kg and $R_m = 1.74 \times 10^6$ m. With these values, the escape speed is

$$v_e = \sqrt{\frac{2GM_m}{R_m}} = \sqrt{\frac{2(6.67 \times 10^{-11} \text{ N}\cdot\text{m}^2/\text{kg}^2)(7.35 \times 10^{22} \text{ kg})}{1.74 \times 10^6 \text{ m}}}$$

$$= 2370 \text{ m/s } (5320 \text{ mi/h})$$

The relatively low escape speed of the Moon means that it is much easier to launch a rocket into space from the Moon than from the Earth. For example, the tiny lunar module that blasted off from the Moon to return the astronauts to Earth could not have come close to escaping the Earth.

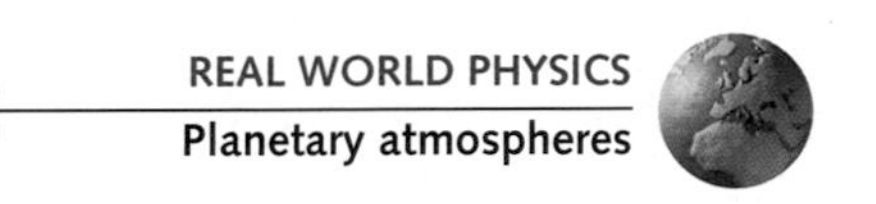

Similarly, the Moon's low escape speed is the reason it has no atmosphere. Even if one could magically supply the Moon with an atmosphere, it would soon evaporate into space because the individual molecules in the air move with speeds great enough to escape. On the Earth, however, where the escape speed is much higher, gravity can prevent the rapidly moving molecules from moving off into space. Even so, light molecules, like hydrogen and helium, move faster for a given temperature than the heavier molecules like nitrogen and oxygen, as we shall see in Chapter 17. For this reason, the Earth's atmosphere contains virtually no hydrogen or helium. (In fact, helium was first discovered on the Sun, as we point out in Chapter 31, hence its name.) Since a stable atmosphere is a likely requirement for the development of life, it follows that the escape speed is an important quantity when considering the possibility of life on other planets.

CONCEPTUAL CHECKPOINT 12–3

Is the escape speed for a 10-N rocket **(a)** equal to, **(b)** less than, or **(c)** greater than the escape speed for a 10,000-N rocket?

Reasoning and Discussion

The derivation of the escape speed in Equation 12–13 shows that the mass of the rocket, m, cancels. Hence, the escape speed is the same for all objects, regardless of their mass. On the other hand, the kinetic energy required to give the 10,000-N rocket the escape speed is 1000 times greater than the kinetic energy required for the 10-N rock.

Answer:

(a) Equal. The escape speed is independent of the mass that is escaping.

EXAMPLE 12–7 Half Escape

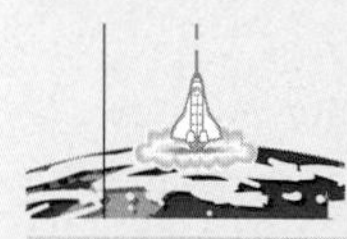

Suppose Jules Verne's cannon launches a rocket straight upward with an initial speed equal to half the escape speed. How far from the center of the Earth does this rocket travel before momentarily coming to rest? (Ignore air resistance in the Earth's atmosphere.)

Picture the Problem

The rocket is launched vertically from the surface of the Earth. It moves radially away from the Earth until it comes to rest at a distance r from the Earth's center.

Strategy

Since we ignore air resistance, the final energy of the rocket, E_f, must be equal to its initial energy, E_0. Setting these energies equal determines the point where the rocket comes to rest.

Solution

1. The initial speed, v_0, is one-half the escape speed. Use Equation 12–13 to write an expression for v_0:

$$v_0 = \tfrac{1}{2}v_e = \tfrac{1}{2}\sqrt{\frac{2GM_E}{R_E}} = \sqrt{\frac{GM_E}{2R_E}}$$

2. Write out the initial energy of the rocket, E_0:

$$E_0 = K_0 + U_0 = \tfrac{1}{2}mv_0^2 - \frac{GmM_E}{R_E}$$

$$= \tfrac{1}{2}m\left(\sqrt{\frac{GM_E}{2R_E}}\right)^2 - \frac{GmM_E}{R_E} = -\tfrac{3}{4}\frac{GmM_E}{R_E}$$

3. Write out the final energy of the rocket. Note that the rocket is a distance r from the center of the Earth when it comes to rest:

$$E_f = K_f + U_f = 0 - \frac{GmM_E}{r} = -\frac{GmM_E}{r}$$

4. Equate the initial and final energies:

$$-\tfrac{3}{4}\frac{GmM_E}{R_E} = -\frac{GmM_E}{r}$$

5. Solve the relation for r:

$$r = \tfrac{4}{3}R_E$$

Insight

An initial speed of v_e allows the rocket to go to infinity before stopping. If the rocket is launched with half that initial speed, however, it can only rise to a height of $4R_E/3 - R_E = R_E/3$ above the Earth's surface. Quite a dramatic difference.

Practice Problem

Find the rocket's maximum distance from the center of the Earth, r, if its launch speed is $3v_e/4$. [**Answer:** $r = 16R_E/7 = 2.29R_E$]

Some related homework problems: Problem 39, Problem 46

A plot of the speed of a rocket as a function of its distance from the center of the Earth is presented in **Figure 12–17** for a variety of initial speeds. Note that when the initial speed is less than the escape speed the rocket comes to rest momentarily at a finite distance, r. In particular, if the launch speed is $0.75\,v_e$, as in the Practice Problem of Example 12–7, the rocket's maximum distance from the center of the Earth is $2.29R_E$.

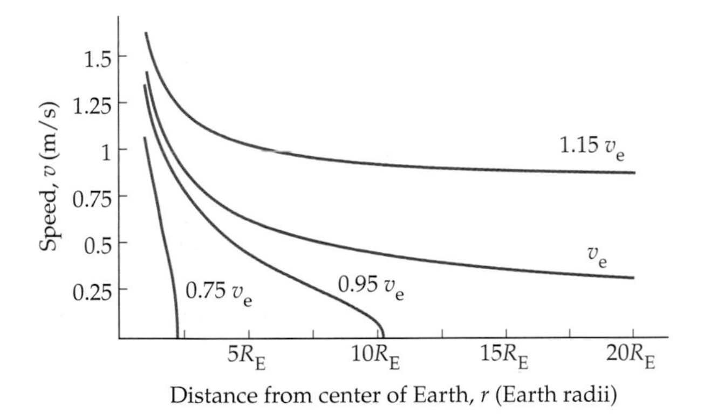

▲ **FIGURE 12–17 Speed of a rocket as a function of distance from the center of the Earth, r, for various vertical launch speeds**

The lower two curves show launch speeds that are less than the escape speed, v_e. In these cases the rocket comes to rest momentarily at a finite height above the Earth. The top curve corresponds to a launch speed greater than the escape speed—this rocket has finite speed at infinite distance. The final curve shows the speed of a rocket launched with the escape speed. It slows to zero speed as the distance approaches infinity.

Black Holes

As we can see from Equation 12–13, the escape speed of an object increases with increasing mass and decreasing radius. Thus, for example, if a massive star were to collapse to a relatively small size, its escape speed would become very large. According to Einstein's theory of general relativity, the escape speed of a compressed, massive star could even exceed the speed of light. In this case nothing—not even light—could escape from the star. For this reason, such objects are referred to as *black holes*. Anything entering a black hole would be making a one-way trip to an unknown destiny.

Since black holes cannot be seen directly, our evidence for their existence is indirect. However, we can predict that as matter is drawn toward a black hole it should become heated to the point where it would emit strong beams of x-rays before disappearing from view. X-ray beams matching these predictions have in fact been observed. These observations, coupled with a variety of others, give astronomers confidence that massive black holes reside at the core of many galaxies—including our own!

Finally, just as a black hole can bend a beam of light back on itself and prevent it from escaping, any massive object can bend light—at least a little. For example, light from distant stars is deflected as it passes by the Sun by 1.75 seconds of an arc (the size of a quarter at a distance of 1.8 miles). Light passing by an entire galaxy of stars or cluster of galaxies can be bent by significant amounts, however, as **Figure 12–18** indicates. This effect is referred to as *gravitational lensing*, since the galaxies act much like the lenses we will study in Chapter 26. Because of gravitational lensing, the images of very distant galaxies or quasars in deep-space astronomical photographs sometimes appear in duplicate, in quadruplicate, or even spread out into circular arcs.

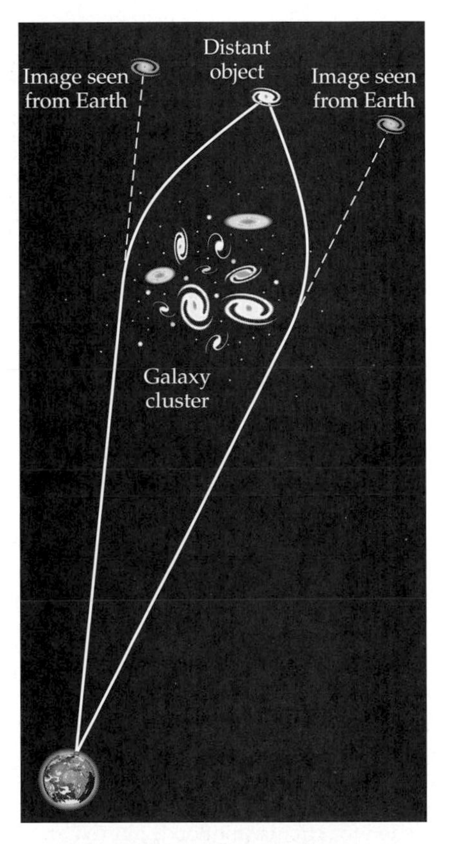

▲ **FIGURE 12–18 Gravitational lensing**

Astronomers often find that very distant objects seem to produce multiple images in their photographs. The cause is the gravitational attraction of intervening galaxies or clusters of galaxies, which are so massive that they can significantly bend the light from remote objects as it passes by them on its way to Earth.

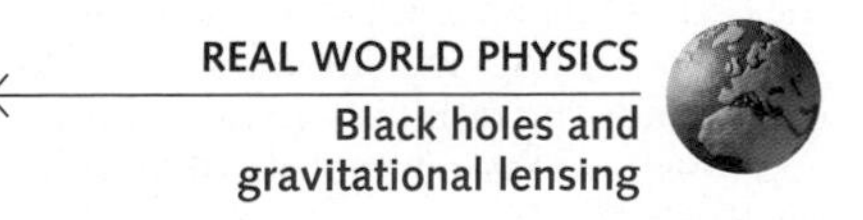

*12–6 Tides

The reason for the ocean tides that rise and fall twice a day was a perplexing and enduring mystery until Newton introduced his law of universal gravitation. Even Galileo, who made so many advances in physics and astronomy, could not explain the tides. However, with the understanding that a force is required to cause an object to move in a circular path, and that the force of gravity becomes weaker with distance, it is possible to describe the tides in detail. In this section we show how it can be done. In addition, we extend the basic idea of tides to several related phenomena.

To begin, consider the idealized situation shown in **Figure 12–19 (a)**. Here we see an object of finite size (a Moon or a planet, for example) orbiting a point mass. If all the mass of the object were concentrated at its center, the gravitational force exerted on it by the central mass would be precisely the amount needed to cause it to move in its circular path. Since the object is of finite size, however, the force exerted on various parts of it have different magnitudes. For example, points closer to the central mass experience a greater force than points farther away.

To see the effect of this variation in force, we use a dark red vector in Figure 12–19 (a) to indicate the force exerted by the central mass at three different points on the object. In addition, we use a light red vector to show the force that is required at each of these three points to cause a mass at that distance to orbit the central mass. Comparing these vectors, we see that the forces are identical at the center of the object—as expected. On the near side of the object, however, the force exerted by the central mass is larger than the force needed to hold the object in orbit, and on the far side the force due to the central mass is less than the force needed to hold the object in orbit. The result is that the near side of the object is pulled closer to the central mass and the far side tends to move farther from the central mass. This causes an egg-shaped deformation of the object, as indicated in Figure 12–19 (a).

REAL WORLD PHYSICS
Tides

Any two objects orbiting one another cause deformations of this type. For example, the Earth causes a deformation in the Moon, and the Moon causes a similar deformation in the Earth. In **Figure 12–19 (b)** we show the Earth and the waters of its ocean deformed into an egg shape. Since the waters in the oceans can flow, they deform much more than the underlying rocky surface of the Earth. As a result, the water level relative to the surface of the Earth is greater at the *tidal bulges* shown in the figure. As the Earth rotates about its axis, a person at a given location will observe two high tides and two low tides each day.

The Moon has no oceans, but the tidal bulges produced in it by the Earth are the reason we see only one side of the Moon. Specifically, the Earth exerts gravi-

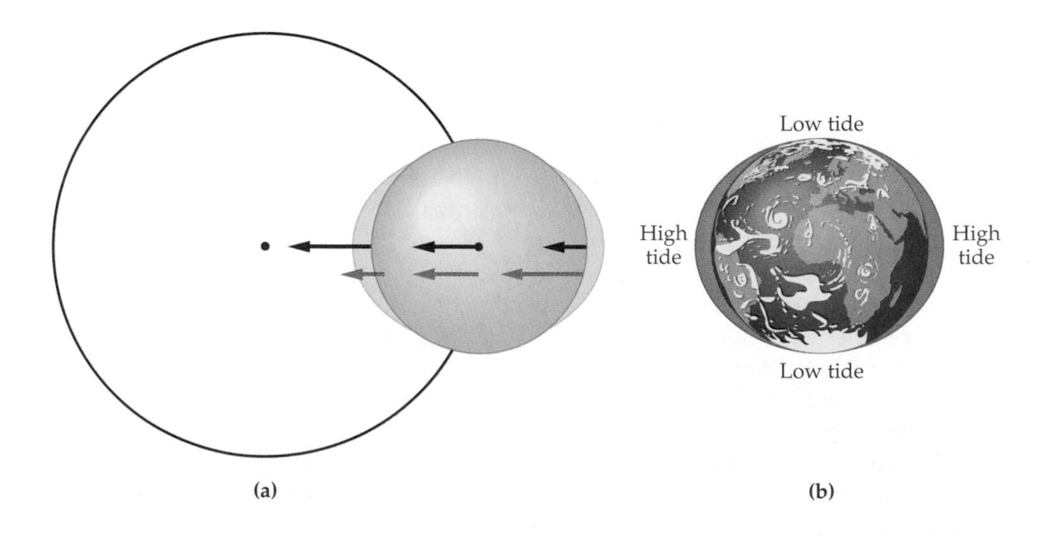

▶ FIGURE 12–19
(a) Tides are caused by a disparity between the gravitational force exerted at various points on a finite-sized object (dark red arrows) and centripetal force needed for circular motion (light red arrows). Note that the gravitational force decreases with distance, as expected. On the other hand, the centripetal force required to keep an object moving in a circular path *increases* with distance. On the near side, therefore, the gravitational force is stronger than required, and the object is stretched inward. On the far side, the gravitational force is weaker than required, and the object stretches outward. **(b)** On the Earth, the water in the oceans responds more to the deforming effects of tides than do the solid rocks of the land. The result is two high tides and two low tides on opposite sides of the Earth.

◀ Tides on Earth are caused chiefly by the Moon's gravitational pull, though at full and new moon, when the Moon and Sun are aligned, the Sun's gravity can enhance the effect. In some places on Earth, such as the Bay of Fundy between Maine and Nova Scotia, local topographic conditions produce abnormally large tides.

tational forces on the tidal bulges of the Moon, causing them to point directly toward the Earth. If the Moon were to rotate slightly away from this alignment the forces exerted by the Earth would cause a torque that would return the Moon to the original alignment. The net result is that the Moon's period of rotation about its axis is equal to its period of revolution about the Earth. This effect, known as **tidal locking**, is common among the various moons in the solar system.

A particularly interesting example of tidal locking is provided by Jupiter's moon Io, a site of intense volcanism (see the photo on p. 102). Io follows an elliptical orbit around Jupiter, and its tidal deformation is larger when it is closer to Jupiter than when it is farther away. As a result, each time Io orbits Jupiter it is squeezed into a greater deformation and then released. This continual flexing of Io causes its internal temperature to rise, just as a rubber ball gets warmer if you squeeze and release it in your hand over and over. It is this mechanism that is largely responsible for Io's ongoing volcanic activity.

In extreme cases, tidal deformation can become so large that an object is literally torn apart. Since tidal deformation increases as a moon moves closer to the planet it orbits, there is a limiting orbital radius—known as the **Roche limit**—inside of which this breakup occurs. A most spectacular example of this effect can be seen in the rings of Saturn, all of which exist well within the Roche limit. The small chunks of ice and other materials that make up the rings may be the remains of a moon that moved too close to Saturn and was destroyed by tidal forces. On the other hand, they may represent material that tidal forces prevented from aggregating to form a moon in the first place. In either case, this dramatic debris field will now never coalesce to form a moon—tidal effects will not allow such a process to occur.

Chapter Summary

	Topic	Remarks and Relevant Equations
12–1	**Newton's Law of Universal Gravitation**	The force of gravity between two point masses, m_1 and m_2, separated by a distance r is attractive and of magnitude $$F = G\frac{m_1 m_2}{r^2} \quad \text{12–1}$$ G is the universal gravitation constant, $$G = 6.67 \times 10^{-11}\ \text{N}\cdot\text{m}^2/\text{kg}^2 \quad \text{12–2}$$ Gravity exerts an action-reaction pair of forces on m_1 and m_2; that is, the force exerted by gravity on m_1 is equal in magnitude but opposite in direction to the force exerted on m_2.
	inverse square dependence	The force of gravity decreases with distance, r, as $1/r^2$. This is referred to as an inverse square dependence.
	superposition	If more than one mass exerts a gravitational force on a given object, the net force is simply the vector sum of each force individually.

12–2 Gravitational Attraction of Spherical Bodies

In calculating gravitational forces, spherical objects can be replaced by point masses.

uniform sphere

If a mass m is outside a uniform sphere of mass M, the gravitational force between m and the sphere is equivalent to the force exerted by a point mass M located at the center of the sphere.

acceleration of gravity

Replacing the Earth with a point mass at its center, we find that the acceleration of gravity on the surface of the Earth is

$$g = \frac{GM_E}{R_E^2} \qquad \text{12–4}$$

weighing the Earth

Cavendish was the first to determine the value of the universal gravitation constant G by direct experiment. Knowing G allows one to calculate the mass of the Earth:

$$M_E = \frac{gR_E^2}{G} \qquad \text{12–5}$$

12–3 Kepler's Laws of Orbital Motion

Kepler determined three laws that describe the motion of the planets in our solar system. Newton showed that Kepler's laws are a direct consequence of his law of universal gravitation.

Kepler's first law

The orbits of the planets are ellipses, with the Sun at one focus.

Kepler's second law

Planets sweep out equal area in equal time.

Kepler's third law

The period of a planet's orbit, T, is proportional to the 3/2 power of its average distance from the Sun, r:

$$T = \left(\frac{2\pi}{\sqrt{GM_s}}\right)r^{3/2} = (\text{constant})r^{3/2} \qquad \text{12–7}$$

12–4 Gravitational Potential Energy

The gravitational potential energy, U, between two point masses m_1 and m_2 separated by a distance r is

$$U = -G\frac{m_1 m_2}{r} \qquad \text{12–9}$$

zero level

The zero of the gravitational potential energy between two point masses is chosen to be at infinite separation of the two masses.

***U* is a scalar**

The gravitational potential energy, U, is a scalar. Therefore, the total potential energy for a group of objects is simply the numerical sum of the potential energy associated with each pair of masses.

12–5 Energy Conservation

With the gravitational potential energy given in Section 12–4, energy conservation can be applied to astronomical situations.

total mechanical energy

An object with mass m, speed v, and at a distance r from the center of the Earth has a total energy given by

$$E = K + U = \tfrac{1}{2}mv^2 - \frac{GmM_E}{r} \qquad \text{12–10}$$

escape speed

An object launched from the surface of the Earth with the escape speed, v_e, can move infinitely far from the Earth. In the limit of infinite separation, the object slows to zero speed.

The escape speed for the Earth is given by

$$v_e = \sqrt{\frac{2GM_E}{R_E}} \qquad \text{12–13}$$

Its numerical value is 11,200 m/s = 25,000 mi/h. A similar expression can be applied to other astronomical bodies.

***12–6 Tides**	Tides result from the variation of the gravitational force from one side of an astronomical object to the other side.
tidal locking	Tidal locking occurs when one astronomical object always points its tidal bulge at the object it orbits.
Roche limit	Tidal deformation increases as an astronomical object moves closer to the body it orbits. At the Roche limit, the tidal deformation is so great that it breaks the object into small pieces.

Problem-Solving Summary

Type of Problem	Relevant Physical Concepts	Related Examples
Find the force due to gravity.	The magnitude of the force is given by Newton's law of universal gravitation, $F = Gm_1m_2/r^2$. The direction of the force is attractive, and along the line connecting m_1 and m_2. If more than one force is involved, the net force is the vector sum of the individual forces.	Examples 12–1, 12–2, 12–3
Relate the period of a planet to the radius of its orbit and the mass of the body it orbits.	Use Kepler's third law, $T = (2\pi/\sqrt{GM})r^{3/2}$.	Example 12–4 Active Example 12–1
Determine the speed of an object at a particular location, given its initial speed and location.	Use energy conservation, with the gravitational potential energy given by $U = -Gm_1m_2/r$.	Examples 12–6, 12–7 Active Example 12–2

Conceptual Questions

1. It is often said that astronauts experience weightlessness because they are beyond the pull of Earth's gravity. Is this statement correct? Explain.
2. When a person passes you on the street you do not feel a gravitational tug. Explain.
3. Does the radius vector of Mars sweep out the same amount of area per time as that of the Earth? Why or why not?
4. If the Earth were to suddenly shrink to half its current diameter, with its mass remaining constant, would the escape speed increase, decrease, or stay the same?
5. On June 22, 1978, James Christy made the first observation of a moon orbiting Pluto. Until that time the mass of Pluto was not known, but with the discovery of its moon, Charon, its mass could be calculated with some accuracy. Explain.
6. When a communications satellite is placed in a geosynchronous orbit above the equator, it remains fixed over a given point on the ground. Is it possible to put a satellite into an orbit so that it remains fixed above the North Pole? Explain.
7. As a satellite goes through one complete orbit of the Earth, how much work does the Earth's gravitational force do on the satellite? Does it matter whether the orbit is circular or elliptical?
8. Is the amount of energy required to get a spacecraft from the Earth to the Moon greater than, less than, or the same as the energy required to go from the Moon to the Earth? Explain.
9. One day in the future you may take a pleasure cruise to the Moon. While there you might climb a lunar mountain and throw a rock horizontally from its summit. If, in principle, you could throw the rock fast enough, it might end up hitting you in the back. Explain.
10. Rockets are launched into space from Cape Canaveral in an easterly direction. Is there an advantage to launching to the east versus launching to the west? Explain.
11. If you light a candle on the Space Shuttle—which would not be a good idea—would it burn the same as on the Earth? Explain.
12. Skylab, the largest spacecraft ever to fall back to the Earth, met its end on July 11, 1979. The cause of Skylab's crash was the friction it experienced in the upper reaches of the Earth's atmosphere. As the radius of Skylab's orbit decreased, did its speed increase, decrease, or stay the same? Explain.
13. You weigh yourself on a scale inside an airplane that is flying with constant velocity at 20,000 ft. Is your weight greater than, less than, or the same as when you are on the surface of the Earth?
14. Does it take more energy to launch a rocket vertically to a height h, or to put a rocket into orbit at the height h? Explain.
15. You weigh yourself on a scale inside an airplane flying above the equator. Does your apparent weight depend on whether the plane is flying to the east or to the west? Explain.
16. Apollo astronauts orbiting the Moon at low altitude noticed occasional changes in their orbit that they attributed to localized concentrations of mass below the lunar surface. Just what effect would such "mascons" have on their orbit?
17. A satellite orbits the Earth in a circular orbit of radius r. At some point its rocket engine is fired in such a way that its speed increases rapidly by a small amount. **(a)** Describe the subsequent path of the satellite. Does the **(b)** apogee distance and **(c)** perigee distance increase, decrease, or stay the same? Explain.
18. Repeat the previous problem, only this time with the rocket engine of the satellite fired in such a way as to slow the satellite.

19. Is it possible for a satellite to have any desired speed in a given orbit? Explain.

20. The force exerted by the Sun on the Moon is more than twice the force exerted by the Earth on the Moon. Should the Moon be thought of as orbiting the Earth or the Sun? Explain.

21. The speed of the Earth in its orbit is greatest around January 4 and least around July 4. When is the Earth closest to the Sun? When is the Earth farthest from the Sun? Explain.

Problems

Note: **IP** *denotes an integrated conceptual/quantitative problem.* **BIO** *identifies problems of biological or medical interest. Blue bullets (•, ••, •••) are used to indicate the level of difficulty of each problem. Astronomical data needed for problems in this chapter can be found on the inside back cover and in Appendix C. Ignore air resistance.*

Section 12–1 Newton's Law of Universal Gravitation

1. • In each hand you hold a 0.10-kg apple. What is the gravitational force exerted by each apple on the other when their separation is **(a)** 0.25 m and **(b)** 0.50 m?

2. • A 6.1-kg bowling ball and a 7.2-kg bowling ball rest on a rack 0.75 m apart. **(a)** What is the force of gravity exerted on each of the balls by the other ball? **(b)** At what separation is the force of gravity between the balls equal to 2.0×10^{-9} N?

3. • A communications satellite with a mass of 250 kg is in a circular orbit about the Earth. The radius of the orbit is 35,000 km as measured from the center of the Earth. Calculate **(a)** the weight of the satellite on the surface of the Earth and **(b)** the gravitational force exerted on the satellite by the Earth when it is in orbit.

4. • The largest asteroid known, Ceres, has a mass of roughly 7.0×10^{20} kg. If Ceres passes within 10,000 m of the spaceship in which you are traveling, what force does it exert on you? Use an approximate value for your mass.

5. • In one hand you hold a 0.12-kg apple, in the other hand a 0.20-kg orange. The apple and orange are separated by 0.75 m. What is the magnitude of the force of gravity that **(a)** the orange exerts on the apple and **(b)** the apple exerts on the orange?

6. •• A spaceship travels from the Earth to the Moon along a line joining the center of the Earth and the center of the Moon. At what distance from the center of the Earth is the force due to the Earth of equal magnitude to the force due to the Moon?

7. •• At new Moon, the Earth, Moon and Sun are in a line, as indicated in **Figure 12–20**. Find the net gravitational force exerted on **(a)** the Earth, **(b)** the Moon, and **(c)** the Sun.

Sun Moon Earth

▲ **FIGURE 12–20** Problem 7

8. •• When the Earth, Moon, and Sun form a right triangle with the Earth located at the right angle, as shown in **Figure 12–21**, the Moon is approaching its third quarter. (The Earth is viewed here from above its north pole.) Find the magnitude and direction of the net force exerted on the Earth.

Moon

Sun Earth

▲ **FIGURE 12–21** Problems 8, 9, and 53

9. •• Repeat the previous problem, this time finding the magnitude and direction of the net force acting on the Sun.

10. •• **IP** Three 5.25-kg masses are at the corners of an equilateral triangle and located in space far from any other masses. **(a)** If the sides of the triangle are 1.75 m long, find the magnitude of the net force exerted on each of the three masses. **(b)** How does your answer to part (a) change if the sides of the triangle are doubled in length?

11. •• **IP** Four masses are positioned at the corners of a rectangle, as indicated in **Figure 12–22**. **(a)** Find the magnitude and direction of the net force acting on the 2.0-kg mass. **(b)** How do your answers to part **(a)** change (if at all) if all sides of the rectangle are doubled in length?

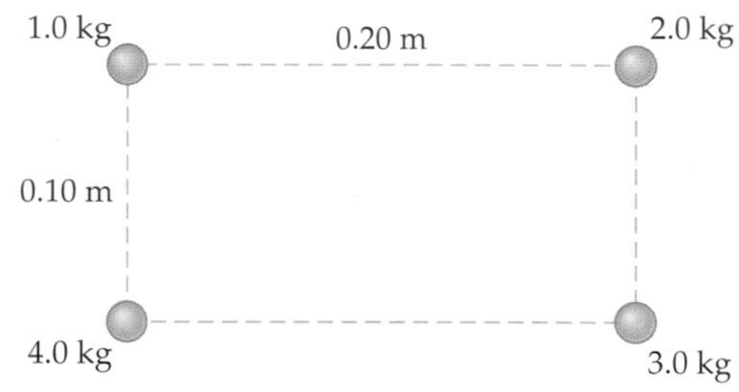

▲ **FIGURE 12–22** Problems 11 and 34

12. ••• Suppose that three astronomical objects (1, 2, and 3) are observed to lie on a line, and that the distance from object 1 to object 3 is D. Given that object 1 has four times the mass of object 3 and seven times the mass of object 2, find the distance between objects 1 and 2 for which the net force on object 2 is zero.

Section 12–2 Gravitational Attraction of Spherical Bodies

13. • Find the acceleration of gravity on the surface of **(a)** Mercury, and **(b)** Venus.

14. • At what altitude above the Earth's surface is the acceleration of gravity equal to $g/2$?

15. • Two 6.3-kg bowling balls, each with a radius of 0.11 m, are in contact with one another. What is the gravitational attraction between the bowling balls?

16. • What is the acceleration due to Earth's gravity at a distance from the center of the Earth equal to the orbital radius of the Moon?

17. • Titan is the largest moon of Saturn, and the only moon in the solar system known to have a substantial atmosphere. Find the acceleration of gravity on Titan's surface, given that its mass is 1.35×10^{23} kg and its radius is 2570 km.

18. •• **(a)** Find the distance from the center of the Earth at which a 3.0-kg object has a weight of 2.0 N. **(b)** If the object is released at

this location and allowed to fall toward the Earth, what is its initial acceleration?

19. •• The acceleration of gravity on the Moon's surface is known to be about 1/6 the acceleration of gravity on the Earth. Given that the radius of the Moon is roughly 1/4 that of the Earth, find the mass of the Moon in terms of the mass of the Earth.

20. •• **IP** Several volcanoes have been observed erupting on the surface of Jupiter's closest moon, Io. Suppose that material ejected from one of these volcanoes reaches a height of 5.00 km after being projected straight upward with an initial speed of 190 m/s. Given that the radius of Io is 3630 km, **(a)** outline a strategy that allows you to calculate the mass of Io. **(b)** Use your strategy to calculate Io's mass.

Section 12–3 Kepler's Laws of Orbital Motion

21. • On Apollo missions to the Moon, the command module orbited at an altitude of 110 km above the lunar surface. How long did it take for the command module to complete one orbit?

22. • Find the orbital speed of a satellite in a circular orbit 1500 km above the surface of the Earth.

23. • In July of 1999 a planet was reported to be orbiting the sun-like star Iota Horologii with a period of 320 days. Find the radius of the planet's orbit, assuming that Iota Horologii has the same mass as the Sun. (This planet is presumably similar to Jupiter, but it may have large, rocky moons that enjoy a pleasant climate.)

24. • Phobos, one of the moons of Mars, orbits at a distance of 9378 km from the center of the red planet. What is the period of Phobos?

25. • The largest moon in the solar system is Ganymede, a moon of Jupiter. Ganymede orbits at a distance of 1.07×10^9 m from the center of Jupiter with a period of about 6.18×10^5 s. Using this information, find the mass of Jupiter.

26. •• **IP** The asteroid 243 Ida has its own small moon, Dactyl. **(a)** Outline a strategy to find the mass of 243 Ida, given that the orbital radius of Dactyl is 90 km and its period is 19 hr. **(b)** Use your strategy to calculate the mass of 243 Ida.

27. •• A typical GPS (Global Positioning System) satellite orbits at an altitude of 2.0×10^7 m. Find **(a)** the orbital period, and **(b)** the orbital speed of such a satellite.

28. •• Compare the orbital speeds of satellites that orbit **(a)** one Earth radius above the surface of the Earth and **(b)** two Earth radii above the surface of the Earth.

29. •• Compare the orbital periods of satellites that orbit **(a)** one Earth radius above the surface of the Earth and **(b)** two Earth radii above the surface of the Earth.

30. •• **IP** The Martian moon Deimos has a period that is greater than the other Martian moon, Phobos. Both moons have approximately circular orbits. **(a)** Is Deimos closer to or farther from Mars than Phobos? Explain. **(b)** Calculate the distance from the center of Mars to Deimos given that its period is 1.10×10^5 s.

31. ••• Centauri A and Centauri B are binary stars with a separation of 3.45×10^{12} m and an orbital period of 2.52×10^9 s. Assuming the two stars are equally massive (which is approximately the case), determine their mass.

32. ••• Find the speed of Centauri A and Centauri B, using the information given in the previous problem.

Section 12–4 Gravitational Potential Energy

33. • The first artificial satellite to orbit the Earth was Sputnik I, launched October 4, 1957. The mass of Sputnik I was 83.5 kg, and its distances from the center of the Earth at apogee and perigee were 7330 km and 6610 km, respectively. Find the difference in gravitational potential energy for Sputnik I as it moved from apogee to perigee.

34. • Find the total gravitational potential energy of the system shown in Figure 12–22.

35. •• Calculate the gravitational potential energy of a 5.0-kg mass **(a)** on the surface of the Earth and **(b)** at an altitude of 1.0 km. **(c)** Take the difference between the results for parts (b) and (a), and compare with mgh, where $h = 1.0$ km.

36. •• Two 0.59-kg basketballs, each with a radius of 12 cm, are just touching. How much energy is required to separate the center of the basketballs by **(a)** 1.0 m and **(b)** 10 m? (Ignore any other gravitational interactions.)

37. •• Find the minimum kinetic energy needed for a 29,000-kg rocket to escape **(a)** the Moon or **(b)** the Earth.

Section 12–5 Energy Conservation

38. • Suppose one of the Global Positioning System satellites has a speed of 4.46 km/s at perigee and a speed of 3.64 km/s at apogee. If the distance from the center of the Earth to the satellite at perigee is 2.00×10^4 km, what is the corresponding distance at apogee?

39. • **BIO** Several meteorites found in Antarctica are believed to have come from Mars, including the famous ALH84001 meteorite that some believe contains fossils of ancient life on Mars. Meteorites from Mars are thought to get to Earth by being blasted off the Martian surface when a large object (such as an asteroid or a comet) crashes into the planet. What speed must a rock have to escape Mars?

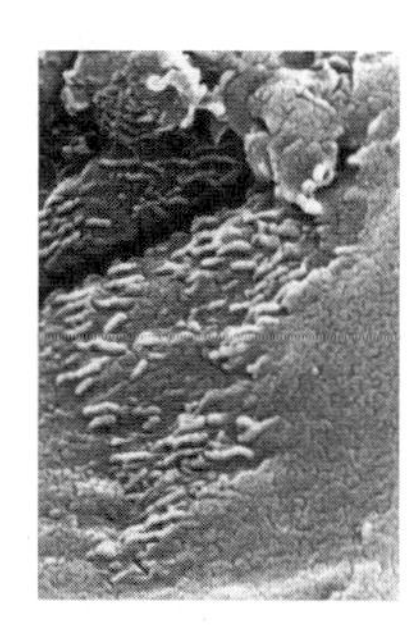

▲ The meteorite ALH84001(left), dislodged from the Martian surface by a tremendous impact, drifted through space for millions of years before falling to Earth in Antarctica about 13,000 years ago. The electron micrograph at right shows tubular structures within the meteorite; some scientists think they are traces of primitive, bacteria-like organisms that may have lived on Mars billions of years ago. (Problem 39)

40. • Referring to Example 12–1, if the Millennium Eagle is at rest at point A what is its speed at point B?

41. • What is the launch speed of a projectile that rises above the Earth to an altitude equal to one Earth radius?

42. • A projectile launched vertically from the surface of the Moon rises to an altitude of 2.75 km. What was the projectile's initial speed?

43. • Find the escape velocity for **(a)** Mercury and **(b)** Venus.

44. •• **IP** Halley's comet, which passes around the Sun every 76 yr, has an elliptical orbit. When closest to the Sun (perihelion) it is at a distance of 8.823×10^{10} m and moves with a speed of 54.6 km/s. The greatest distance between Halley's comet and the Sun (aphelion) is 6.152×10^{12} m. **(a)** Is the speed of Halley's comet greater than or less than 54.6 km/s when it is at aphelion? Explain. **(b)** Calculate its speed at aphelion.

45. •• On Apollo Moon missions, the lunar module would blast off from the Moon's surface and dock with the command module in lunar orbit. After docking, the lunar module would be jettisoned and allowed to crash back onto the lunar surface. Seismometers placed on the Moon's surface by the astronauts would then pick up the resulting seismic waves. Find the impact speed of the lunar module, given that it is jettisoned from an orbit 110 km above the lunar surface moving with a speed of 1630 m/s.

46. •• If a projectile is launched from Earth with a speed equal to the escape speed, how high above the Earth's surface is it when its speed is half the escape speed?

47. •• Suppose that a planet is discovered orbiting a distant star that has 10 times the mass of the Earth and one-tenth its radius. How does the escape speed on this planet compare with that of the Earth?

48. •• A projectile is launched vertically from the surface of the Moon with an initial speed of 1500 m/s. At what altitude is the projectile's speed one-half its initial value?

49. •• To what radius would the Sun have to be contracted for its escape speed to equal the speed of light? (Black holes have escape speeds greater than the speed of light; hence we see no light from them.)

50. •• Two baseballs, each with a mass of 0.145 kg, are separated by a distance of 250 m in outer space. If the balls are released from rest, what speed do they have when their separation has decreased to 150 m? Ignore the gravitational effects from any other objects.

51. ••• On Earth, a person can jump vertically and rise to the height h. What is the radius of the largest spherical asteroid from which this person could escape by jumping? Assume that each cubic meter of the asteroid has a mass of 3500 kg.

General Problems

52. •• An astronaut exploring a distant solar system lands on an unnamed planet with a radius of 3560 km. When the astronaut jumps upward with an initial speed of 3.00 m/s, she rises to a height of 0.570 m. What is the mass of the planet?

53. •• **IP** When the Moon is approaching its third quarter, the Earth, Moon, and Sun form a right triangle, as shown in Figure 12–21. Calculate the force exerted on the Moon by **(a)** the Earth and **(b)** the Sun. **(c)** Does it make more sense to think of the Moon as orbiting the Sun, with a small effect due to the Earth, or as orbiting the Earth, with a small effect due to the Sun?

54. •• An equilateral triangle 10.0 m on a side has a 1.00-kg mass at one corner, a 2.00-kg mass at another corner, and a 3.00-kg mass at the third corner (**Figure 12–23**). Find the magnitude and direction of the net force acting on the 1.00-kg mass.

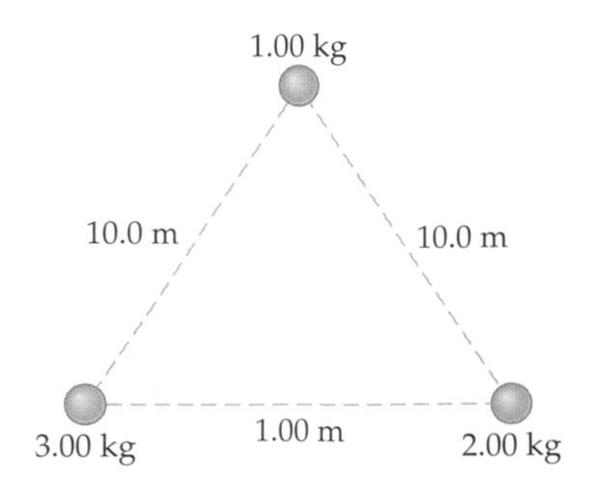

▲ **FIGURE 12–23** Problem 54

55. •• **IP** Suppose that a planet is discovered that has the same amount of mass in a given volume as the Earth, but has half its radius. **(a)** Is the acceleration of gravity on this planet more than, less than, or the same as the acceleration of gravity on the Earth? Explain. **(b)** Calculate the acceleration of gravity on this planet.

56. •• **IP** Suppose that a planet is discovered that has the same mass as the Earth, but half its radius. **(a)** Is the acceleration of gravity on this planet more than, less than, or the same as the acceleration of gravity on the Earth? Explain. **(b)** Calculate the acceleration of gravity on this planet.

57. •• Show that the speed of a satellite in a circular orbit a height h above the surface of the Earth is

$$v = \sqrt{\frac{GM_E}{R_E + h}}$$

58. •• In a binary star system, two stars orbit about their common center of mass, as shown in **Figure 12–24**. If $r_2 = 3r_1$, what is the ratio of the masses m_2/m_1 of the two stars?

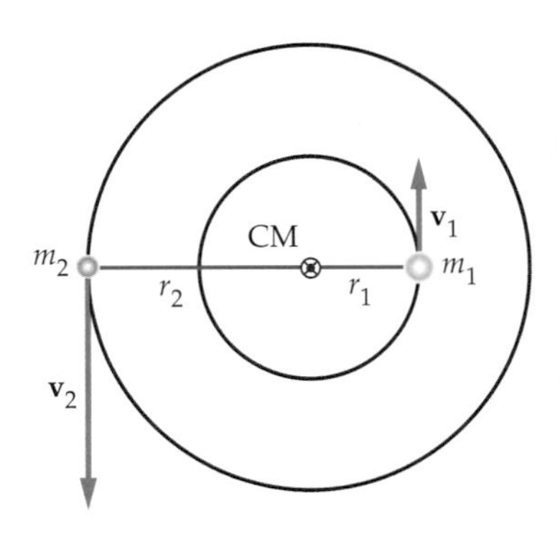

▲ **FIGURE 12–24** Problems 58 and 59

59. •• Find the orbital period of the binary star system described in the previous problem.

60. •• Using the results from Problem 44, find the angular momentum of Halley's comet **(a)** at perihelion and **(b)** at aphelion. (Take the mass of Halley's comet to be 9.8×10^{14} kg.)

61. •• At what altitude above the Moon's surface would a satellite orbiting the equator of the Moon remain fixed above a given spot?

62. •• **IP** A satellite is placed in Earth orbit 1,000 miles higher than the altitude of a geosynchronous satellite. Referring to Active Example 12–1, we see that the altitude of the satellite is 23,300 mi. **(a)** Is the period of this satellite greater than or less than 24 hours? **(b)** As viewed from the surface of the Earth, does the satellite move eastward or westward? Explain. **(c)** Find the period of this satellite.

63. •• Find the speed of the Millennium Eagle at point A in Example 12–1 if its speed at point B is 631 m/s.

64. •• Show that the force of gravity between the Moon and the Sun is always greater than the force of gravity between the Moon and the Earth.

65. •• The astronomical unit A.U. is defined as the mean distance from the Sun to the Earth (1 A.U. = 1.50×10^{11} m). Apply Kepler's

third law (Equation 12–7) to the solar system, and show that it can be written as

$$T = Cr^{3/2}$$

where the period T is measured in years, the distance r is measured in astronomical units, and the constant C is equal to $1\ \text{yr}/(\text{A.U.})^{3/2}$.

66. •• **(a)** Find the kinetic energy of a 1200-kg satellite in a circular orbit with a radius of 15,000 miles. **(b)** How much energy is required to move this satellite to a circular orbit with a radius of 25,000 miles?

67. •• **IP** On a typical mission, the Space Shuttle ($m = 2.00 \times 10^6$ kg) orbits at an altitude of 250 km above the Earth's surface. **(a)** Does the orbital speed of the shuttle depend on its mass? Explain. **(b)** Find the speed of the shuttle in its orbit. **(c)** How long does it take for the shuttle to complete one orbit of the Earth?

68. ••• Consider a spherical asteroid with a radius of 12 km and a mass of 3.45×10^{15} kg. **(a)** What is the acceleration of gravity on the surface of this asteroid? **(b)** Suppose the asteroid spins about an axis through its center, like the Earth, with an angular speed ω. What is the greatest value ω can have before loose rocks on the asteroid's equator begin to fly off the surface?

69. ••• Three identical stars, at the vertices of an equilateral triangle, orbit about their common center of mass (**Figure 12–25**). Find the period of this orbital motion in terms of the orbital radius, R, and the mass of each star, M.

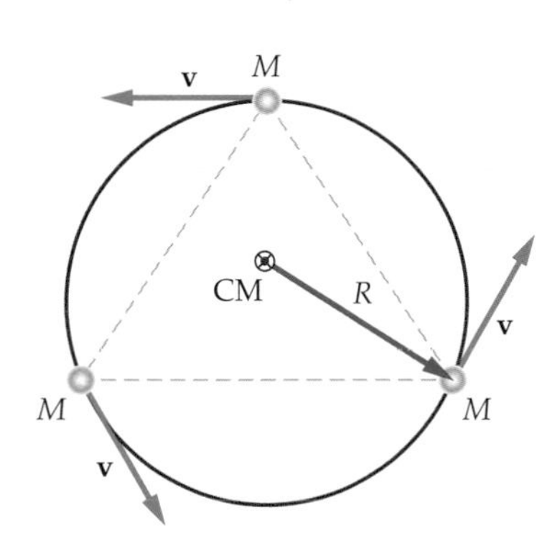

▲ **FIGURE 12–25** Problem 69

70. ••• Find an expression for the kinetic energy of a satellite of mass m in an orbit of radius r about a planet of mass M.

71. ••• Referring to Example 12–1, find the x component of the net force acting on the Millennium Eagle as a function of x. Plot your result, showing both negative and positive values of x.

72. ••• A dumbbell has a mass m on either end of a rod of length $2a$. The center of the dumbbell is a distance r from the center of the Earth, and the dumbbell is aligned radially. If $r \gg a$, show that the difference in the gravitational force exerted on the two masses by the Earth is approximately $4GmM_E a/r^3$. (Note: The difference in force causes a tension in the rod connecting the masses. We refer to this as a *tidal force*.) [*Hint*: Use the fact that $1/(r-a)^2 - 1/(r+a)^2 \sim 4a/r^3$ for $r \gg a$.]

73. ••• Referring to the previous problem, suppose the rod connecting the two masses m is removed. In this case, the only force between the two masses is their mutual gravitational attraction. In addition, suppose the masses are spheres of radius a and mass $\frac{4}{3}\pi a^3(917\ \text{kg/m}^3)$ that touch each other. **(a)** Write an expression for the gravitational force between the masses m. **(b)** Find the distance from the center of the Earth, r, for which the gravitational force found in part (a) is equal to the tidal force found in Problem 72. This distance is known as the *Roche limit*. **(c)** Calculate the Roche limit for the Earth and for Saturn. (The famous rings of Saturn are within the Roche limit for that planet. Thus, the innumerable small objects, composed mostly of ice, that make up the rings will never coalesce to form a moon.)

74. ••• A satellite orbits the Earth in an elliptical orbit. At perigee its distance from the center of the Earth is 22,500 km and its speed is 4280 m/s. At apogee its distance from the center of the Earth is 24,100 km and its speed is 3990 m/s. Using this information, calculate the mass of the Earth.

75. ••• Show that the total mechanical energy of an object of mass m orbiting the Earth is equal to (-1) times its kinetic energy. Does your result apply to an object orbiting the Sun? Explain.

76. ••• In a binary star system two stars orbit about their common center mass. Find the orbital period of such a system, given that the stars are separated by a distance d and have masses m and $2m$.

13 Oscillations About Equilibrium

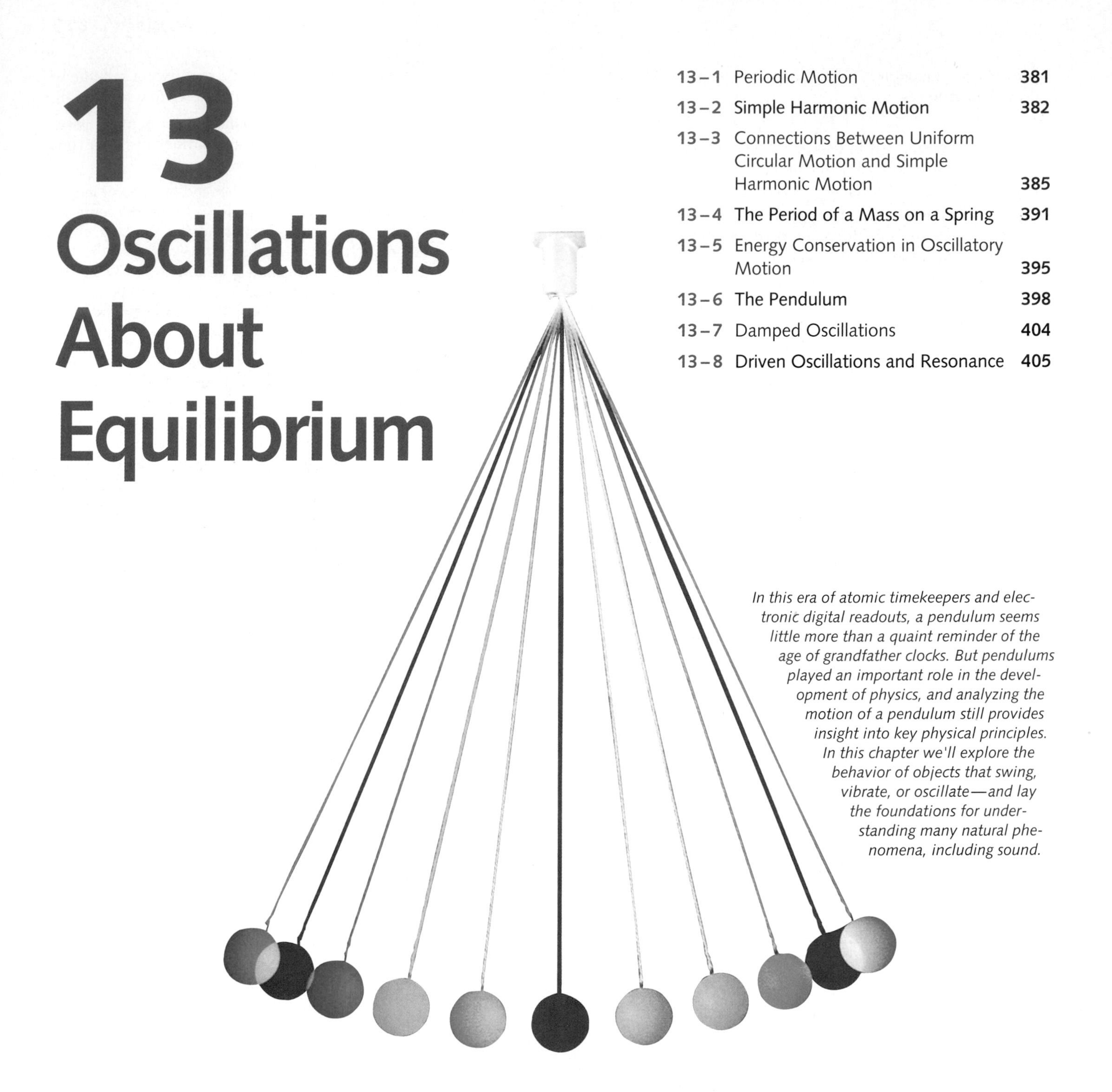

In this era of atomic timekeepers and electronic digital readouts, a pendulum seems little more than a quaint reminder of the age of grandfather clocks. But pendulums played an important role in the development of physics, and analyzing the motion of a pendulum still provides insight into key physical principles. In this chapter we'll explore the behavior of objects that swing, vibrate, or oscillate—and lay the foundations for understanding many natural phenomena, including sound.

In Chapter 11 we considered systems in static equilibrium. Such systems are seldom left undisturbed for very long, it seems, before they are displaced from equilibrium by a bump, a kick, or a nudge. When this happens to a system, it often results in **oscillations** back and forth from one side of the equilibrium position to the other.

The basic cause of oscillations is the fact that when an object is displaced from a position of stable equilibrium it experiences a *restoring force* that is directed back toward the equilibrium position. Thus, the restoring force accelerates the object in the direction of its initial position. When it reaches equilibrium the force acting on it is zero, but it doesn't come to rest. In moving back to equilibrium it has gained speed and momentum, hence its inertia carries it through the equilibrium position to the other side, where the restoring force is now in the opposite direction. The process repeats itself, leading to a series of oscillations.

Perhaps the most familiar oscillating system is the simple pendulum, like the ones that keep time in grandfather clocks. Modern digital wristwatches also use oscillators to keep time, but theirs are tiny quartz crystals. In fact, oscillating systems are found in nature over virtually all length scales, from water molecules that oscillate in a microwave oven, to planets that oscillate when struck by an asteroid, to the universe itself, which some think may oscillate in a series of "big bangs" followed by equally momentous "big crunches."

13–1 Periodic Motion

A motion that repeats itself over and over is referred to as **periodic motion**. The beating of your heart, the ticking of a clock, and the movement of a child on a swing are familiar examples. One of the key characteristics of a periodic system is the time required for the completion of one cycle of its repetitive motion. For example, the pendulum in a grandfather clock might take one second to swing from maximum displacement in one direction to maximum displacement in the opposite direction, or two seconds for a complete cycle of oscillation. In this case, we say that the **period**, T, of the pendulum is 2 s.

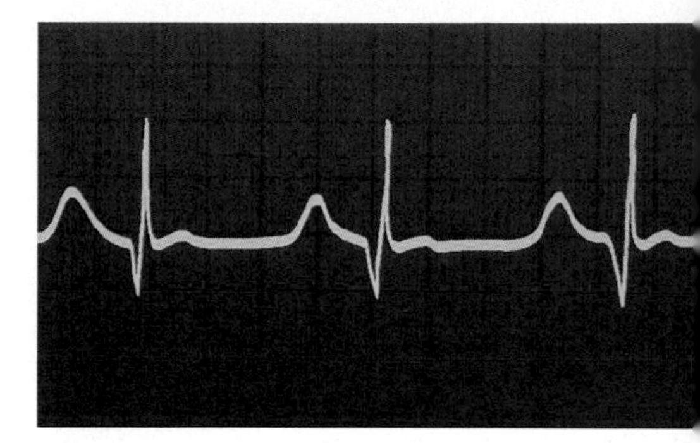

▲ Periodic phenomena are found everywhere in nature, from the movements of the heavenly bodies to the vibration of individual atoms. The trace of an electrocardiogram (ECG or EKG), as shown here, records the rhythmic electrical activity that accompanies the beating of our hearts.

Definition of Period, *T*

T = time required for one cycle of a periodic motion

SI unit: seconds/cycle = s

Note that a cycle (that is, an oscillation) is dimensionless.

Closely related to the period is another common measure of periodic motion, the **frequency**, f. The frequency of an oscillation is simply the number of oscillations per unit of time. Thus f tells us how frequently, or rapidly, an oscillation takes place—the higher the frequency, the more rapid the oscillations. By definition, the frequency is simply the inverse of the period, T:

Definition of Frequency, *f*

$$f = \frac{1}{T} \qquad \text{13–1}$$

SI unit: cycle/second = 1/s = s^{-1}

Note that if the period of an oscillation, T, is very small, corresponding to rapid oscillations, the corresponding frequency, $f = 1/T$, will be large, as expected.

A special unit has been introduced for the measurement of frequency. It is the **hertz** (Hz), named for the German physicist Heinrich Hertz (1857–1894), in honor of his pioneering studies of radio waves. By definition, one Hz is one cycle per second:

$$1\text{ Hz} = 1\text{ cycle/second}$$

High frequencies are often measured in kilohertz (kHz), where 1 kHz = 10^3 Hz, or megahertz (MHz), where 1 MHz = 10^6 Hz.

EXERCISE 13–1

If the processing speed of a personal computer is 525 MHz, how much time is required for one processing cycle?

Solution

We can use Equation 13–1 to solve for the processing period:

$$T = \frac{1}{f} = \frac{1}{525\text{ MHz}} = \frac{1}{5.25 \times 10^8\text{ cycles/s}} = 1.90 \times 10^{-9}\text{ s}$$

The high frequency of the computer corresponds to a very small period; less than 2 billionths of a second to complete one operation. We next consider a situation in which the frequency is considerably smaller.

EXERCISE 13–2

A tennis ball is hit back and forth between two players warming up for a match. If it takes 2.31 s for the ball to go from one player to the other, what is the period and frequency of the ball's motion?

Solution

The period of this motion is the time for the ball to complete one round trip. Therefore,

$$T = 2(2.31 \text{ s}) = 4.62 \text{ s}$$

$$f = \frac{1}{T} = \frac{1}{4.62 \text{ s}} = 0.216 \text{ Hz}$$

13–2 Simple Harmonic Motion

Periodic motion can take many forms, as illustrated by the output of an electrocardiogram or the variation of light from a distant star. There is one type of periodic motion, however, that is of particular importance. It is referred to as **simple harmonic motion**.

A classic example of simple harmonic motion is provided by the oscillations of a mass attached to a spring. (See Section 6–2 for a discussion of ideal springs and the forces they exert.) To be specific, consider an air-track cart of mass m attached to a spring of force constant k, as in **Figure 13–1**. When the spring is at its equilibrium length, which places the cart at $x = 0$, the cart remains at rest. If the cart is displaced from equilibrium by a distance x, the spring exerts a restoring force given by Hooke's law, $F = -kx$. In other words:

> A spring exerts a restoring force that is proportional to the displacement from equilibrium.

This is the key feature of a mass-spring system that leads to simple harmonic motion.

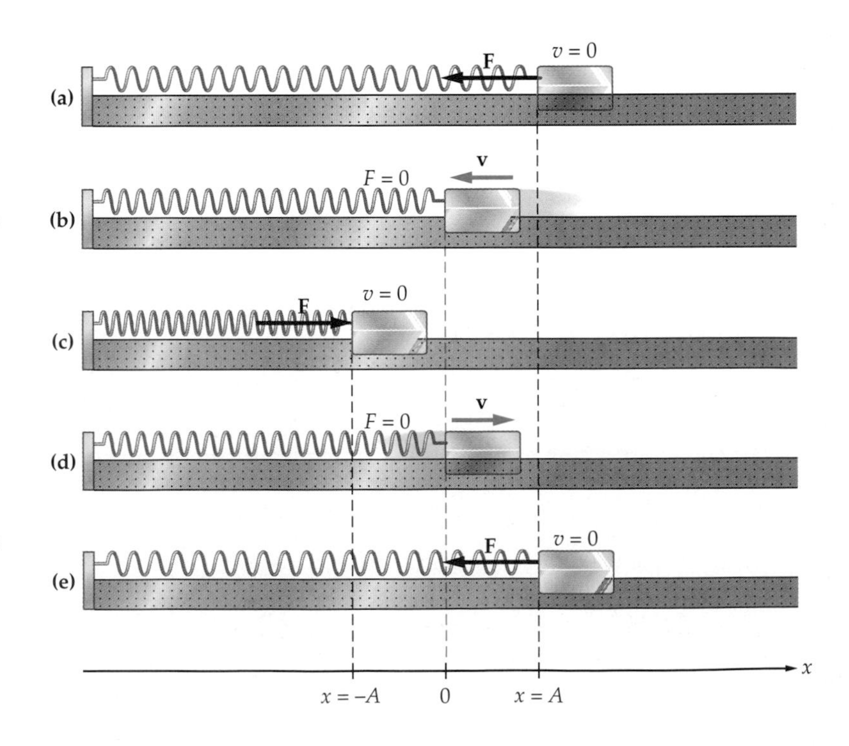

▶ FIGURE 13–1 A mass attached to a spring undergoes simple harmonic motion about $x = 0$

(a) The mass is at its maximum positive value of x. Its velocity is zero, and the force on it points to the left with maximum magnitude. **(b)** The mass is at the equilibrium position of the spring. Here the speed has its maximum value, and the force exerted by the spring is zero. **(c)** The mass is at its maximum displacement in the negative x direction. The velocity is zero here, and the force points to the right with maximum magnitude. **(d)** The mass is at the equilibrium position of the spring, with zero force acting on it and maximum speed. **(e)** The mass has completed one cycle of its oscillation about $x = 0$.

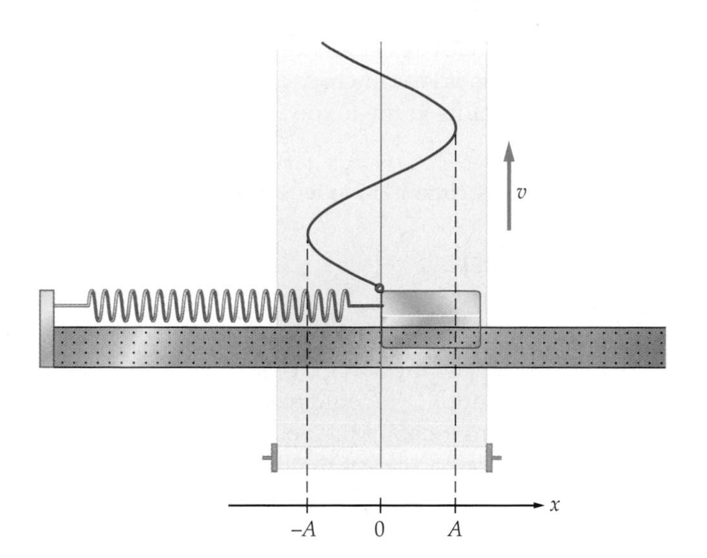

◀ **FIGURE 13–2 Displaying position versus time for simple harmonic motion**

As an air-track cart oscillates about its equilibrium position, a pen attached to it traces its motion onto a moving sheet of paper. This produces a "strip chart," showing that the cart's motion has the shape of a sine or a cosine.

Now, suppose the cart is released from rest at the location $x = A$. As indicated in Figure 13–1, the spring exerts a force on the cart to the left, causing the cart to accelerate toward the equilibrium position. When the cart reaches $x = 0$ the net force acting on it is zero. Its speed is not zero at this point, however, and so it continues to move to the left. As the cart compresses the spring it experiences a force to the right, causing it to decelerate and finally come to rest at $x = -A$. The spring continues to exert a force to the right, thus the cart immediately begins to move to the right until it comes to rest again at $x = A$, completing one oscillation in the time T.

If a pen is attached to the cart it can trace its motion on a strip of paper moving with constant speed, as indicated in **Figure 13–2**. On this "strip chart" we obtain a record of the cart's motion as a function of time. As we see in Figure 13–2, the motion of the cart looks like a sine or a cosine function.

Mathematical analysis, using the methods of calculus, shows that this is indeed the case; that is, the position of the cart as a function of time can be represented by a sine or a cosine function. The reason that either function works can be seen by considering **Figure 13–3**. If we take $t = 0$ to be at the point (1), for example, the position as a function of time starts at zero, just like a sine function; if we choose $t = 0$ to be at point (2), however, the position versus time starts at its maximum value, just like a cosine function. It is really the same mathematical function, differing only in the choice of starting point.

Returning to Figure 13–1, note that the position of the mass oscillates between $x = +A$ and $x = -A$. Since A represents the extreme displacement of the cart on either side of equilibrium, we refer to it as the **amplitude** of the motion. It follows that the amplitude is half the total range of motion. In addition, recall that the cart's

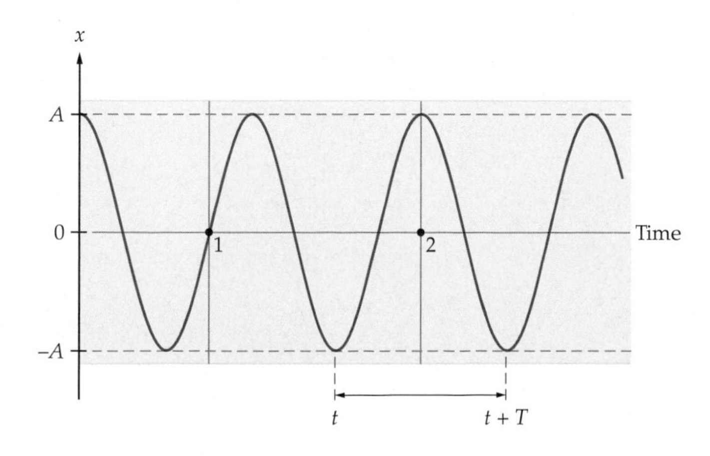

◀ **FIGURE 13–3 Simple harmonic motion as a sine or a cosine**

The strip chart from Figure 13–2. The cart oscillates from $x = +A$ to $x = -A$, completing one cycle in the time T. The function traced by the pen can be represented as a sine function if $t = 0$ is taken to be at point (1), where the function is equal to zero. The function can be represented by a cosine if $t = 0$ is taken to be at point (2), where the function has its maximum value.

motion repeats with a period T. As a result, the position of the cart is the same at the time $t + T$ as it is at the time t, as shown in Figure 13–3. Combining all these observations results in the following mathematical description of position versus time:

PROBLEM SOLVING NOTE
Using Radians

When evaluating Equation 13–2, be sure you have your calculator set to "radians" mode.

Position Versus Time in Simple Harmonic Motion

$$x = A\cos\left(\frac{2\pi}{T}t\right) \qquad 13\text{–}2$$

SI unit: m

This type of dependence on time—as a sine or a cosine—is characteristic of simple harmonic motion. We could have used the sine function just as well, of course. With this particular choice the position at $t = 0$ is $x = A\cos(0) = A$; thus, Equation 13–2 describes an object that has its maximum displacement at $t = 0$.

To see how Equation 13–2 works, recall that the cosine oscillates between $+1$ and -1. Therefore, $x = A\cos(2\pi t/T)$ will oscillate between $+A$ and $-A$, just as in the strip chart. Next, consider what happens if we replace the time t with the time $t + T$. This gives

$$x = A\cos\left(\frac{2\pi}{T}(t + T)\right)$$

$$= A\cos\left(\frac{2\pi}{T}t + \frac{2\pi}{T}T\right) = A\cos\left(\frac{2\pi}{T}t + 2\pi\right)$$

Finally, using the fact that $\cos(\theta + 2\pi) = \cos\theta$ for any angle θ, we can rewrite the last expression as follows:

$$x = A\cos\left(\frac{2\pi}{T}t\right)$$

Therefore, as expected, the position at time $t + T$ is precisely the same as the position at time t.

EXAMPLE 13–1 Spring Time

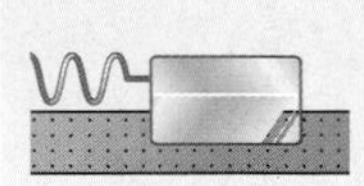

An air-track cart attached to a spring completes one oscillation every 2.4 s. At $t = 0$ the cart is released from rest at a distance of 0.10 m from its equilibrium position. What is the position of the cart at **(a)** 0.30 s, **(b)** 0.60 s, **(c)** 2.7 s, and **(d)** 3.0 s?

Picture the Problem

In our sketch the x-axis is chosen so that the origin is at the equilibrium position of the mass. In addition, we choose positive to be to the right. The cart is released at $x = 0.10$ m, hence the amplitude is $A = 0.10$ m.

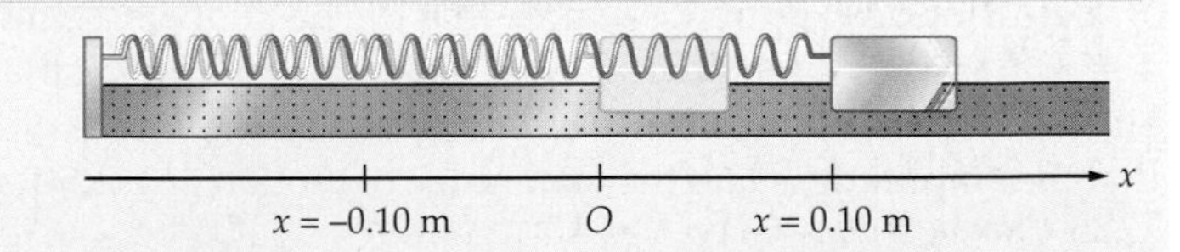

Strategy

Note that the period of oscillation, $T = 2.4$ s, is given in the problem statement. Thus, we can find the position of the cart by evaluating $x = A\cos(2\pi t/T)$ at the desired times, using $A = 0.10$ m.

Solution

Part (a)

1. Calculate x at the time $t = 0.30$ s:

$$x = A\cos\left(\frac{2\pi}{T}t\right) = (0.10\text{ m})\cos\left[\left(\frac{2\pi}{2.4\text{ s}}\right)(0.30\text{ s})\right]$$
$$= (0.10\text{ m})\cos(\pi/4) = 7.1\text{ cm}$$

Part (b)

2. Now, substitute $t = 0.60$ s:

$$x = A\cos\left(\frac{2\pi}{T}t\right) = (0.10\text{ m})\cos\left[\left(\frac{2\pi}{2.4\text{ s}}\right)(0.60\text{ s})\right]$$
$$= (0.10\text{ m})\cos(\pi/2) = 0$$

Part (c)

3. Repeat with $t = 2.7$ s:

$$x = A\cos\left(\frac{2\pi}{T}t\right) = (0.10\text{ m})\cos\left[\left(\frac{2\pi}{2.4\text{ s}}\right)(2.7\text{ s})\right]$$
$$= (0.10\text{ m})\cos(9\pi/4) = 7.1\text{ cm}$$

Part (d)

4. Repeat with $t = 3.0$ s:

$$x = A\cos\left(\frac{2\pi}{T}t\right) = (0.10\text{ m})\cos\left[\left(\frac{2\pi}{2.4\text{ s}}\right)(3.0\text{ s})\right]$$
$$= (0.10\text{ m})\cos(5\pi/2) = 0$$

Insight

Note that the results for parts **(c)** and **(d)** are the same as for parts **(a)** and **(b)**, respectively. This is because the times in **(c)** and **(d)** are greater than the corresponding times in **(a)** and **(b)** by one period; that is, 2.7 s = 0.30 s + 2.4 s and 3.0 s = 0.60 s + 2.4 s.

Practice Problem

What is the first time the cart is at the position $x = -5.0$ cm? [**Answer:** $t = T/3 = 0.80$ s]

Some related homework problems: Problem 8, Problem 13, Problem 14

As you might imagine, a pendulum oscillates about its stable equilibrium in much the same way as a mass on a spring. We will consider the pendulum in detail in Section 13–6. In the next section, however, we examine the close relationship between simple harmonic motion and uniform circular motion.

13–3 Connections Between Uniform Circular Motion and Simple Harmonic Motion

Imagine a turntable that rotates with a constant angular speed $\omega = 2\pi/T$, taking the time T to complete a revolution. At the rim of the turntable we place a small peg, as indicated in **Figure 13–4**. If we view the turntable from above, we see the peg undergoing uniform circular motion.

On the other hand, suppose we view the turntable from the side, so that the peg appears to move back and forth. Perhaps the easiest way to view this motion is to shine a light that casts a shadow of the peg on a screen, as shown in Figure 13–4. While the peg itself moves on a circular path, its shadow moves back and forth in a straight line.

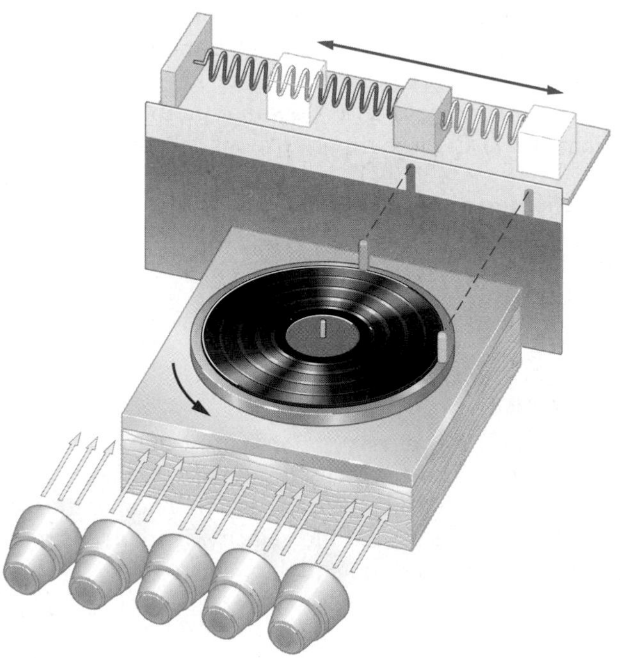

◀ FIGURE 13–4 The relationship between uniform circular motion and simple harmonic motion

A peg is placed at the rim of a turntable that rotates with constant angular velocity. Viewed from above, the peg exhibits uniform circular motion. If the peg is viewed from the side, however, it appears to move back and forth in a straight line, as we can see by shining a light to cast a shadow of the peg onto a screen. The shadow moves with simple harmonic motion. If we compare this motion with the behavior of a mass on a spring, moving with the same period as the turntable and an amplitude of motion equal to the radius of the turntable, we find that the mass and the shadow of the peg move together in simple harmonic motion.

To be specific, let the radius of the turntable be $r = A$, so that the shadow moves from $x = +A$ to $x = -A$. When the shadow is at $x = +A$, release a mass on a spring that is also at $x = +A$, so that the mass and the shadow start together. If we adjust the period of the mass so that it completes one oscillation in the same time T that the turntable completes one revolution, we find that the mass and the shadow move as one for all times. This is also illustrated in Figure 13–4. Since the mass undergoes simple harmonic motion, it follows that the shadow does so as well.

We now use this connection, plus our knowledge of circular motion, to obtain detailed results for the position, velocity, and acceleration of a particle undergoing simple harmonic motion.

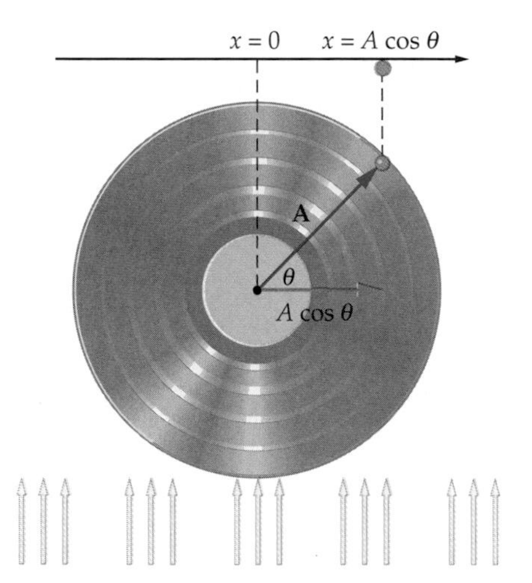

▲ **FIGURE 13–5 Position versus time in simple harmonic motion**

A peg rotates on the rim of a turntable of radius A. When the peg is at the angular position θ its shadow is at $x = A\cos\theta$. Note that $A\cos\theta$ is also the x component of the radius vector **A** from the center of the turntable to the peg.

Position

In **Figure 13–5** we show the peg at the angular position θ, where θ is measured relative to the x-axis. If the peg starts at $\theta = 0$ at $t = 0$, and the turntable rotates with a constant angular speed ω, we know from Equation 10-10 that the angular position of the peg is simply

$$\theta = \omega t \qquad 13\text{–}3$$

That is, the angular position increases linearly with time.

Now, imagine drawing a radius vector of length A to the position of the peg, as indicated in Figure 13–5. When we project the shadow of the peg onto the screen, the shadow is at the location $x = A\cos\theta$, which is the x component of the radius vector. Therefore, the position of the shadow as a function of time is

Position of the Shadow as a Function of Time

$$x = A\cos\theta = A\cos(\omega t) = A\cos\left(\frac{2\pi}{T}t\right) \qquad 13\text{–}4$$

SI unit: m

Note that we have used Equation 13–3 to express θ in terms of the time, t. Clearly, Equations 13–4 and 13–2 are identical, so the shadow does indeed exhibit simple harmonic motion, just like a mass on a spring.

For notational simplicity, we will often write the position of a mass on a spring in the form $x = A\cos(\omega t)$, which is more compact than $x = A\cos(2\pi t/T)$. When referring to a rotating turntable, ω is called the angular speed; when referring to simple harmonic motion, or other periodic motion, we have a slightly different name for ω. In these situations, ω is called the **angular frequency**:

Definition of Angular Frequency, ω

$$\omega = 2\pi f = \frac{2\pi}{T} \qquad 13\text{–}5$$

SI unit: $\text{rad/s} = \text{s}^{-1}$

Velocity

We can find the velocity of the shadow in the same way that we determined its position; first find the velocity of the peg, then take its x component. The result of this calculation will be the velocity as a function of time for simple harmonic motion.

To begin, recall that the velocity of an object in uniform circular motion of radius r has a magnitude equal to

$$v = r\omega$$

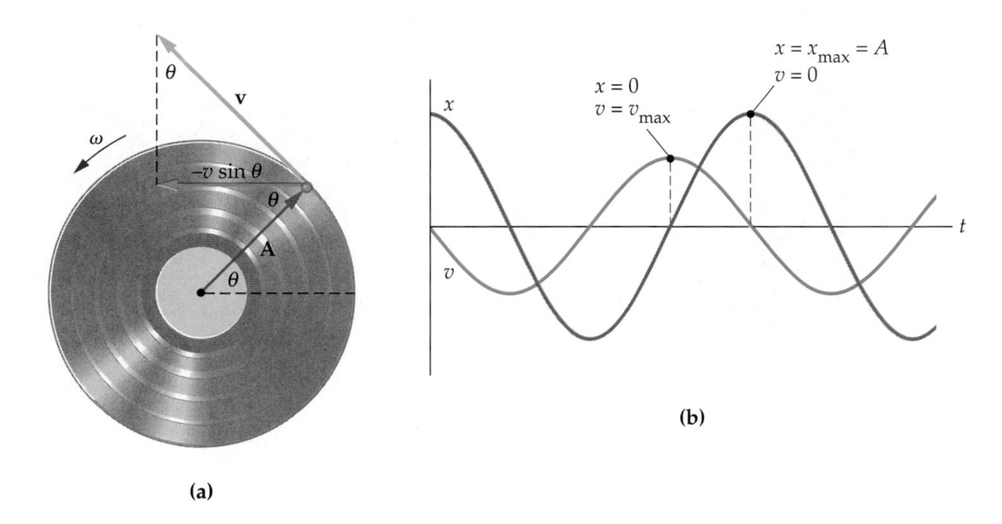

FIGURE 13–6 Velocity versus time in simple harmonic motion

(a) The velocity of a peg rotating on the rim of a turntable is tangential to its circular path. As a result, when the peg is at the angle θ, its velocity makes an angle of θ with the vertical. The x component of the velocity, then, is $-v \sin \theta$. **(b)** Position, x, and velocity, v, as a function of time for simple harmonic motion. The speed is greatest when the object passes through equilibrium, $x = 0$. On the other hand, the speed is zero when the position is greatest; that is, at the turning points. Finally, note that as x moves in the negative direction the velocity is negative. Similar remarks apply to the positive direction.

In addition, the velocity is tangential to the object's circular path, as indicated in **Figure 13–6 (a)**. Therefore, referring to the figure, we see that when the peg is at the angular position θ, the velocity vector makes an angle θ with the vertical. As a result, the x component of the velocity is $-v \sin \theta$. Combining these results, we find that the velocity of the peg, along the x-axis, is

$$v_x = -v \sin\theta = -r\omega \sin\theta$$

In what follows we shall drop the x subscript, since we know that the shadow and a mass on a spring move only along the x-axis. Recalling that $r = A$ and $\theta = \omega t$, we have

Velocity in Simple Harmonic Motion

$$v = -A\omega \sin(\omega t) \qquad \textbf{13–6}$$

SI unit: m/s

We plot x and v for simple harmonic motion in **Figure 13–6 (b)**. Note that when the displacement from equilibrium is a maximum, the velocity is zero. This is to be expected, since at $x = +A$ and $x = -A$ the object is momentarily at rest as it turns around. Not surprisingly, these points are referred to as **turning points** of the motion.

On the other hand, the speed is a maximum when the displacement from equilibrium is zero. Similarly, a mass on a spring is moving with its greatest speed as it goes through $x = 0$. From the expression $v = -A\omega \sin(\omega t)$, and the fact that the largest value of $\sin \theta$ is 1, we see that the maximum speed of the mass is

$$v_{max} = A\omega \qquad \textbf{13–7}$$

After the mass passes $x = 0$ it begins to either compress or to stretch the spring, and hence it slows down.

Acceleration

The acceleration of an object in uniform circular motion has a magnitude given by

$$a_{cp} = r\omega^2$$

In addition, the direction of the acceleration is toward the center of the circular path, as indicated in **Figure 13–7 (a)**. Thus, when the angular position of the peg is θ, the acceleration vector is at an angle θ below the x-axis, and its x component is $-a_{cp} \cos \theta$. Again setting $r = A$ and $\theta = \omega t$, we find

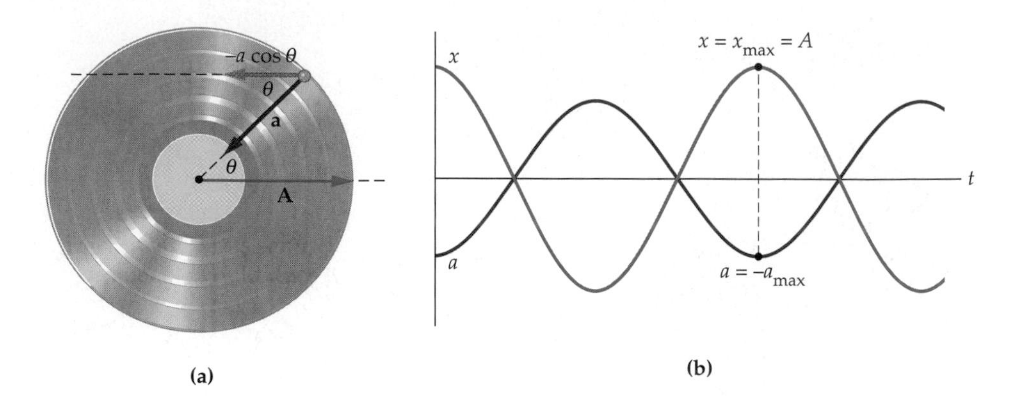

▶ FIGURE 13–7 Acceleration versus time in simple harmonic motion

(a) The acceleration of a peg on the rim of a uniformly rotating turntable is directed toward the center of the turntable. Hence, when the peg is at the angle θ, the acceleration makes an angle θ with the horizontal. The x component of the acceleration is $-a\cos\theta$. **(b)** Position, x, and acceleration, a, as a function of time for simple harmonic motion. Note that when the position has its greatest positive value, the acceleration has its greatest negative value.

Acceleration in Simple Harmonic Motion.

$$a = -A\omega^2 \cos(\omega t) \qquad \textbf{13–8}$$

SI unit: m/s^2

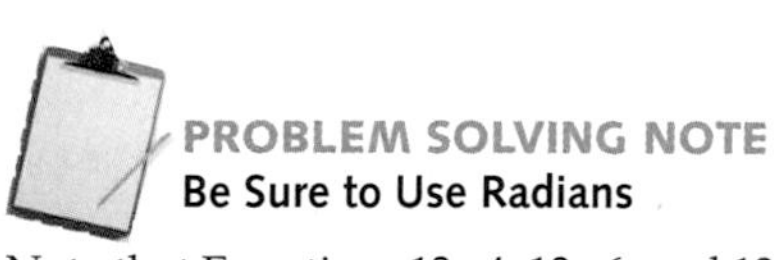

PROBLEM SOLVING NOTE

Be Sure to Use Radians

Note that Equations 13–4, 13–6, and 13–8 must all be evaluated in terms of radians

The position and acceleration for simple harmonic motion are plotted in **Figure 13–7 (b)**. Note that the acceleration and position vary with time in the same way, but with opposite signs. That is, when the position has its maximum *positive* value, the acceleration has its maximum *negative* value, and so on. After all, the restoring force of the spring is opposite to the position, hence the acceleration, $a = F/m$, must also be opposite to the position. In fact, comparing Equations 13–4 and 13–8 we see that the acceleration can be written as:

$$a = -\omega^2 x$$

Finally, since the largest value of x is the amplitude A, we see that the maximum acceleration is of magnitude

$$a_{max} = A\omega^2 \qquad \textbf{13–9}$$

We conclude this section with a few examples using position, velocity, and acceleration in simple harmonic motion.

EXAMPLE 13–2 Velocity and Acceleration

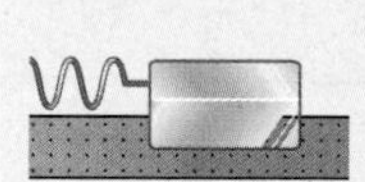

As in Example 13–1, an air-track cart attached to a spring completes one oscillation every 2.4 s. At $t = 0$ the cart is released from rest at a distance of 0.10 m from its equilibrium position. What is the velocity and acceleration of the cart at **(a)** 0.30 s and **(b)** 0.60 s?

Picture the Problem

As before, the x-axis is chosen so that the origin is at the equilibrium position of the mass, with the positive direction to the right.

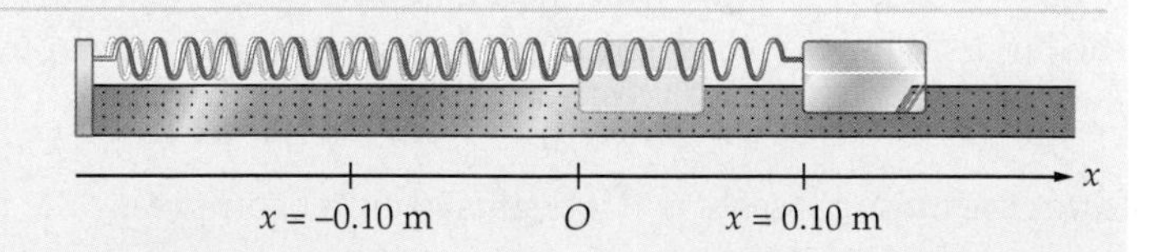

Strategy

After calculating the angular frequency, $\omega = 2\pi/T$, we simply substitute $t = 0.30$ s and $t = 0.60$ s into $v = -A\omega\sin(\omega t)$ and $a = -A\omega^2\cos(\omega t)$.

Solution

1. Calculate the angular frequency for this motion:

$$\omega = \frac{2\pi}{T} = \frac{2\pi}{(2.4\text{ s})} = 2.6\text{ rad/s}$$

Part (a)

2. Calculate v at the time $t = 0.30$ s:

$$v = -A\omega\sin(\omega t)$$
$$= -(0.10\text{ m})(2.6\text{ rad/s})\sin\left[\left(\frac{2\pi}{2.4\text{ s}}\right)(0.30\text{ s})\right]$$
$$= -(26\text{ cm/s})\sin(\pi/4) = -18\text{ cm/s}$$

3. Similarly, calculate a at $t = 0.30$ s:

$$a = -A\omega^2 \cos(\omega t)$$
$$= -(0.10\text{ m})(2.6\text{ rad/s})^2 \cos\left[\left(\frac{2\pi}{2.4\text{ s}}\right)(0.30\text{ s})\right]$$
$$= -(68\text{ cm/s}^2)\cos(\pi/4) = -48\text{ cm/s}^2$$

Part (b)

4. Calculate v at the time $t = 0.60$ s:

$$v = -A\omega \sin(\omega t)$$
$$= -(0.10\text{ m})(2.6\text{ rad/s})\sin\left[\left(\frac{2\pi}{2.4\text{ s}}\right)(0.60\text{ s})\right]$$
$$= -(26\text{ cm/s})\sin(\pi/2) = -26\text{ cm/s}$$

5. Similarly, calculate a at $t = 0.60$ s:

$$a = -A\omega^2 \cos(\omega t)$$
$$= -(0.10\text{ m})(2.6\text{ rad/s})^2 \cos\left[\left(\frac{2\pi}{2.4\text{ s}}\right)(0.60\text{ s})\right]$$
$$= -(68\text{ cm/s}^2)\cos(\pi/2) = 0$$

Insight

Note that the cart speeds up from $t = 0.30$ s to $t = 0.60$ s; in fact, the maximum speed of the cart, $v_{max} = A\omega = 26$ cm/s, occurs at $t = 0.60\text{ s} = T/4$. Referring to Example 13–1 we see that this is precisely the time when the cart is at the equilibrium position, $x = 0$. As expected, the acceleration is zero at this time.

Practice Problem

What is the first time the velocity of the cart is +26 cm/s? [**Answer:** $v = +26$ cm/s at $t = 3T/4 = 1.8$ s]

Some related homework problems: Problem 19, Problem 20

In problems like the preceding Example, it is often useful to express the time t in terms of the period, T. For example, if the period is $T = 2.4$ s it follows that $t = 0.60$ s is $T/4$. Thus, the angular frequency times the time is

$$\omega t = \left(\frac{2\pi}{T}\right)\left(\frac{T}{4}\right) = \pi/2$$

This result was used in steps 4 and 5 in Example 13–2. Using $\omega t = \pi/2$ we find that the position of the cart is

$$x = A\cos(\omega t) = A\cos(\pi/2) = 0$$

Similarly, the velocity of the cart is

$$v = -A\omega\sin(\omega t) = -A\omega\sin(\pi/2) = -A\omega$$

and its acceleration is

$$a = -A\omega^2\cos(\omega t) = -A\omega^2\cos(\pi/2) = 0$$

When expressed in this way, it is clear why x and a are zero, and v has its maximum negative value.

When evaluating the expression $x = A\cos(2\pi t/T)$ at the time t, it is often helpful to express t in terms of the period T.

EXAMPLE 13–3 Turbulence!

On December 29, 1997, a United Airlines flight from Tokyo to Honolulu was hit with severe turbulence 31 minutes after takeoff. Data from the airplane's "black box" indicated the 747 moved up and down with an amplitude of 30.0 m and a maximum acceleration of 1.8 g. Treating the up-down motion of the plane as simple harmonic, find **(a)** the time required for one complete oscillation and **(b)** the plane's maximum vertical speed.

continued on the following page

continued from the previous page

Picture the Problem

Our sketch shows the 747 moving up and down with an amplitude of 30.0 m relative to its normal horizontal flight path. We also indicate the time T for one oscillation.

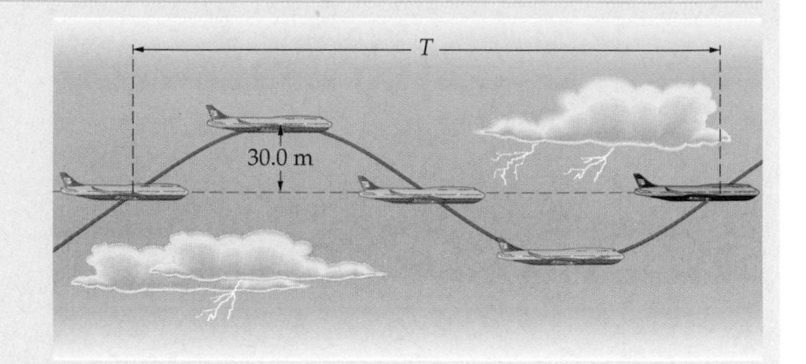

Strategy

(a) We know that the maximum acceleration of simple harmonic motion is $a_{max} = A\omega^2$ (Equation 13–9). Setting a_{max} equal to 1.8 g allows us to solve for ω. Finally, $\omega = 2\pi/T$ yields the period of the motion, T.

(b) The maximum speed is found using $v_{max} = A\omega$ (Equation 13–7).

Solution

Part (a)

1. Set a_{max} equal to 1.8 g: $a_{max} = A\omega^2 = 1.8\,g$

2. Solve for ω, and express in terms of T: $\omega = \sqrt{1.8\,g/A} = 2\pi/T$

3. Solve for T and substitute numerical values: $T = \dfrac{2\pi}{\sqrt{1.8\,g/A}} = \dfrac{2\pi}{\sqrt{1.8(9.81\text{ m/s}^2)/(30.0\text{ m})}} = 8.2\text{ s}$

Part (b)

4. Calculate the maximum speed using $v_{max} = A\omega = 2\pi A/T$: $v_{max} = A\omega = \dfrac{2\pi A}{T} = \dfrac{2\pi(30.0\text{ m})}{8.2\text{ s}} = 23\text{ m/s}$

Insight

We don't expect the up-down motion of the plane to be exactly simple harmonic, but this should be a reasonably good approximation.

Practice Problem

What amplitude of motion would result in a maximum acceleration of 0.5 g, everything else remaining the same?
[**Answer:** $A = 0.5\,g/\omega^2 = 0.5\,gT^2/4\pi^2 = 8.4$ m]

Some related homework problems: Problem 19, Problem 20

ACTIVE EXAMPLE 13–1 Bobbing for Apples

A red delicious apple floats in a barrel of water. If you lift the apple 2.00 cm above its floating level and release it, it bobs up and down with a period of $T = 0.750$ s. Assuming the motion is simple harmonic, find the position, velocity, and acceleration of the apple at the times **(a)** $T/4$ and **(b)** $T/2$.

Solution

1. Identify the amplitude of motion: $A = 2.00$ cm
2. Calculate the angular frequency: $\omega = 2\pi/T = 8.38$ rad/s

Part (a)

3. Evaluate $x = A\cos(\omega t)$ at $t = T/4$: $x = A\cos(\pi/2) = 0$
4. Evaluate $v = -A\omega\sin(\omega t)$ at $t = T/4$: $v = -A\omega\sin(\pi/2) = -A\omega = -16.8$ cm/s
5. Evaluate $a = -A\omega^2\cos(\omega t)$ at $t = T/4$: $a = -A\omega^2\cos(\pi/2) = 0$

Part (b)

6. Evaluate $x = A\cos(\omega t)$ at $t = T/2$: $x = A\cos(\pi) = -A = -2.00$ cm
7. Evaluate $v = -A\omega\sin(\omega t)$ at $t = T/2$: $v = -A\omega\sin(\pi) = 0$
8. Evaluate $a = -A\omega^2\cos(\omega t)$ at $t = T/2$: $a = -A\omega^2\cos(\pi) = A\omega^2 = 140\text{ cm/s}^2$

Insight

The acceleration in part **(b)** may seem rather large, but remember that $140\text{ cm/s}^2 = 1.40\text{ m/s}^2$, so the acceleration is only a fraction of the acceleration of gravity.

13–4 The Period of a Mass on a Spring

In this section we show how the period of a mass on a spring is related to the mass m, the force constant k, and the amplitude of motion, A. As a first step, note that the net force acting on the mass at the position x is

$$F = -kx$$

Now, since $F = ma$, it follows that

$$ma = -kx$$

Substituting the time dependence of x and a, as given in Equations 13–4 and 13–8 in the previous section, we find

$$m[-A\omega^2 \cos(\omega t)] = -k[A\cos(\omega t)]$$

Canceling $-A\cos(\omega t)$ from each side of the equation yields

$$\omega^2 = k/m$$

or

$$\omega = \sqrt{\frac{k}{m}} \qquad 13\text{–}10$$

Finally, since $\omega = 2\pi/T$, it follows that the period of a mass on a spring is

Period of a Mass on a Spring

$$T = 2\pi\sqrt{\frac{m}{k}} \qquad 13\text{–}11$$

SI unit: s

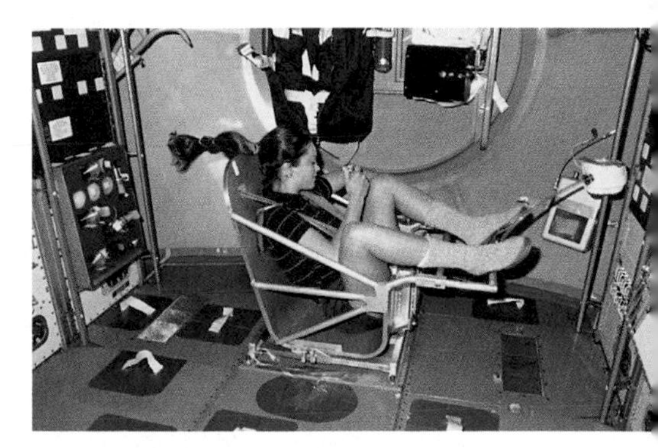

▲ Since astronauts are in free fall when orbiting the Earth, they behave as if they were "weightless." It is nevertheless possible to determine the mass of an astronaut by exploiting the properties of oscillatory motion. The chair into which astronaut Tamara Jernigan is strapped is attached to a spring. If the force constant of the spring is known, her mass can be determined simply by measuring the period with which she rocks back and forth. This instrument is known as a Body Mass Measurement Device (BMMD).

EXERCISE 13–3

When a 0.22-kg air-track cart is attached to a spring it oscillates with a period of 0.84 s. What is the force constant for this spring?

Solution

From Equation 13–11 we find

$$k = 4\pi^2 m/T^2 = 12\ \text{N/m}$$

As one might expect, the period increases with the mass and decreases with the spring's force constant. For example, a larger mass has greater inertia, and hence it takes longer for the mass to move back and forth through an oscillation. On the other hand, a larger value of the force constant, k, indicates a stiffer spring. Clearly, a mass on a stiff spring completes an oscillation in less time than one on a soft, squishy spring.

The relationship between mass and period given in Equation 13–11 is used by NASA to measure the mass of astronauts in orbit. Since astronauts in orbit are in free fall, as was discussed in Chapter 12, they are "weightless." As a result, they cannot simply step onto a bathroom scale to determine their mass. Thus, NASA has developed a device, known as the Body Mass Measurement Device (BMMD), to get around this problem. The BMMD is basically a spring attached to a chair, into which an astronaut is strapped. As the astronaut oscillates back and forth the period of oscillation is measured. Knowing the force constant k and the period of oscillation, T, the astronaut's mass can be

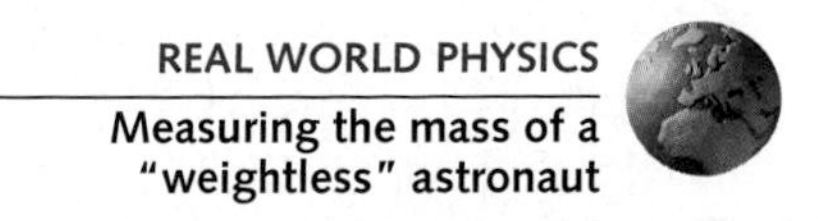

determined from Equation 13–11. The result is simply $m = kT^2/4\pi^2$. See Problem 58 for an application of this result.

Note that the period given in Equation 13–11 is independent of the amplitude, A, which canceled in the derivation of this equation. This might seem counterintuitive at first: Shouldn't it take more time for a mass to cover the greater distance implied by a larger amplitude? While it is true that a mass will cover a greater distance when the amplitude is increased, it is also true that a larger amplitude implies a larger force exerted by the spring. With a greater force acting on it, the mass moves more rapidly; in fact, the speed of the mass is increased just enough that it covers the greater distance in precisely the same time.

These relationships between the motion of a mass on a spring and the mass, the force constant, and the amplitude are summarized in **Figure 13–8**.

EXAMPLE 13–4 Spring into Motion

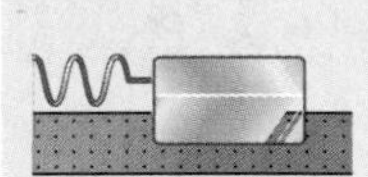

A 0.120-kg mass attached to a spring oscillates with an amplitude of 0.0750 m and a maximum speed of 0.524 m/s. Find **(a)** the force constant and **(b)** the period of motion.

Picture the Problem

The mass oscillates between $x = 0.0750$ m and $x = -0.0750$ m. Its maximum speed, when it is at $x = 0$, is $v_{max} = 0.524$ m/s.

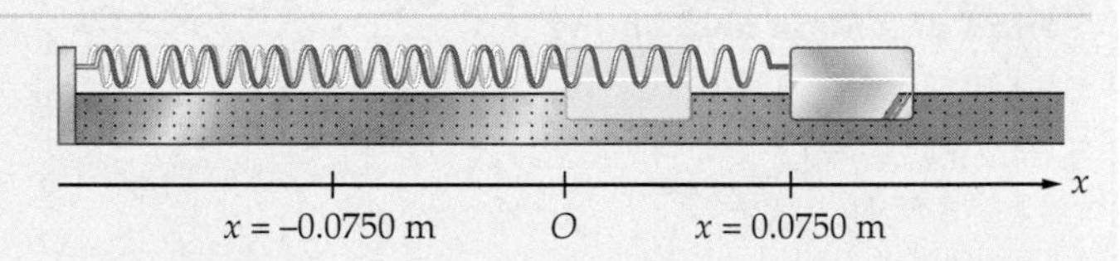

Strategy

(a) To find the force constant we first use the maximum speed, $v_{max} = A\omega$ (Equation 13–7), to determine the angular frequency ω. Once we know ω we can obtain the force constant with $\omega = \sqrt{k/m}$ (Equation 13–10).

(b) We can find the period from the angular frequency, using $\omega = 2\pi/T$. Alternatively, we can use the force constant and the mass in $T = 2\pi\sqrt{m/k}$ (Equation 13–11).

Solution

Part (a)

1. Calculate the angular frequency in terms of the maximum speed:

$$v_{max} = A\omega$$
$$\omega = \frac{v_{max}}{A} = \frac{0.524 \text{ m/s}}{0.0750 \text{ m}} = 6.99 \text{ rad/s}$$

2. Solve $\omega = \sqrt{k/m}$ for the force constant:

$$\omega = \sqrt{k/m}$$
$$k = m\omega^2 = (0.120 \text{ kg})(6.99 \text{ rad/s})^2 = 5.86 \text{ N/m}$$

Part (b)

3. Use $\omega = 2\pi/T$ to find the period:

$$\omega = 2\pi/T$$
$$T = \frac{2\pi}{\omega} = \frac{2\pi}{6.99 \text{ rad/s}} = 0.899 \text{ s}$$

4. Use $T = 2\pi\sqrt{m/k}$ to find the period:

$$T = 2\pi\sqrt{m/k} = 2\pi\sqrt{\frac{0.120 \text{ kg}}{5.86 \text{ N/m}}} = 0.899 \text{ s}$$

Insight

What would happen if we were to attach a larger mass to this same spring, and release it with the same amplitude? The answer is that the period would increase, the angular frequency would decrease, and hence the maximum speed would also decrease.

Practice Problem

What is the maximum acceleration of the mass described in this Example? [**Answer:** $a_{max} = A\omega^2 = 3.66 \text{ m/s}^2$]

Some related homework problems: Problem 27, Problem 28

ACTIVE EXAMPLE 13–2 Mass-Spring

When a 0.420-kg mass is attached to a spring, it oscillates with a period of 0.350 s. If a second mass, m_2, is attached to the same spring, it oscillates with a period of 0.700 s. Find **(a)** the force constant of the spring and **(b)** the mass m_2.

Solution

Part (a)

1. Let the initial mass and period be m_1 and T_1, respectively: $m_1 = 0.420$ kg, $T_1 = 0.350$ s
2. Use Equation 13–11 to write an expression for T_1: $T_1 = 2\pi\sqrt{m_1/k}$
3. Solve this expression for the force constant, k: $k = 4\pi^2 m_1/T_1^2 = 135$ N/m

Part (b)

4. Write an expression for T_2: $T_2 = 2\pi\sqrt{m_2/k}$
5. Solve for m_2: $m_2 = kT_2^2/4\pi^2 = 1.68$ kg

Insight

Note that if the period is to be doubled, the mass must be increased by a factor of 4, in accordance with Equation 13–11; in this case, from 0.420 kg to 1.68 kg.

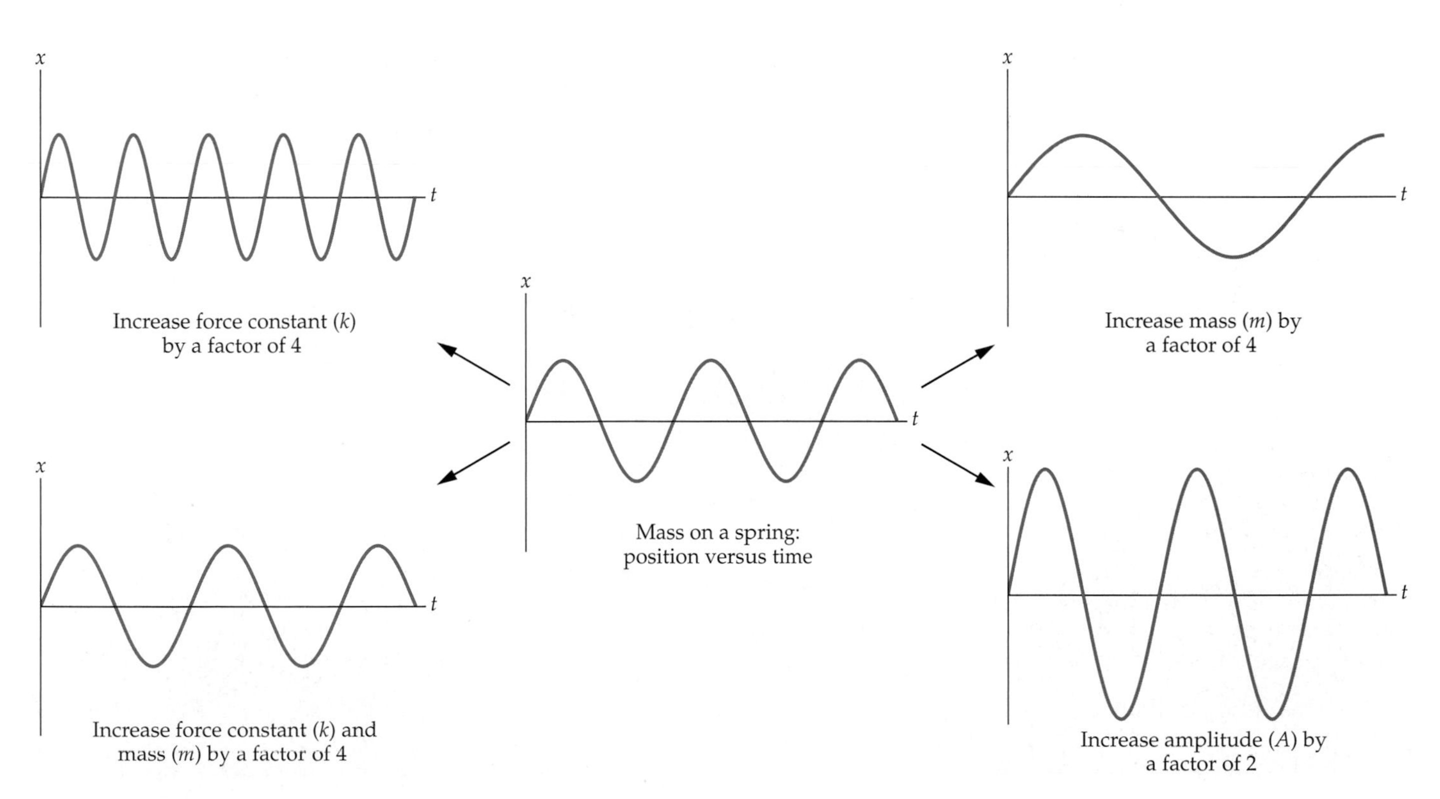

▲ **FIGURE 13–8 Factors affecting the motion of a mass on a spring**
Increasing the force constant k increases the frequency of motion; increasing the mass m decreases the frequency. If both m and k are increased by the same factor, the motion is unchanged. Changing the amplitude of motion has no effect on the frequency.

A Vertical Spring

To this point we have considered only springs that are horizontal, and that are therefore unstretched at their equilibrium position. In many cases, however, we may wish to consider a vertical spring, as in **Figure 13–9**.

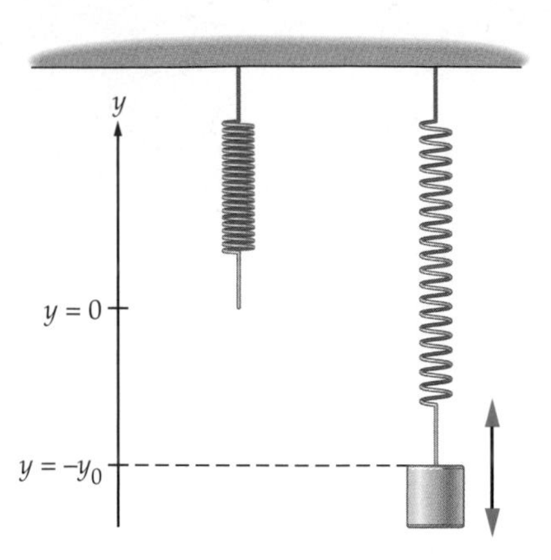

▲ **FIGURE 13–9 A mass on a vertical spring**
A mass stretches a vertical spring from its initial equilibrium at $y = 0$ to a new equilibrium at $y = -y_0 = -mg/k$. The mass executes simple harmonic motion about this new equilibrium.

▲ A ball attached to a vertical spring oscillates up and down with simple harmonic motion. If successive images taken at equal time intervals are displaced laterally, as in the sequence of photos at left, the ball appears to trace out a sinusoidal pattern (compare Figure 13–2). The bungee jumper at right will oscillate in a similar fashion, though friction and air resistance will reduce the amplitude of his bounces.

Now, when a mass m is attached to a vertical spring, it causes the spring to stretch. In fact, the vertical spring is in equilibrium when it exerts an upward force equal to the weight of the mass. That is, the spring stretches by an amount y_0 given by

$$ky_0 = mg$$

or

$$y_0 = mg/k \qquad \textbf{13–12}$$

Thus, a mass on a vertical spring oscillates about the equilibrium point y_0. In all other respects the oscillations are the same as for a horizontal spring. In particular, the motion is simple harmonic, and the period is given by Equation 13–11.

EXAMPLE 13–5 It's a Stretch

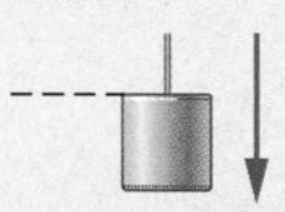

A 0.260-kg mass is attached to a vertical spring. When the mass is put into motion, its period is 1.12 s. How much does the mass stretch the spring when it is at rest?

Picture the Problem

We choose the vertical axis to have its origin at the unstretched position of the spring. Once the mass is attached, the spring stretches to the position $y = -y_0$. The mass oscillates about this point with a period of 1.12 s.

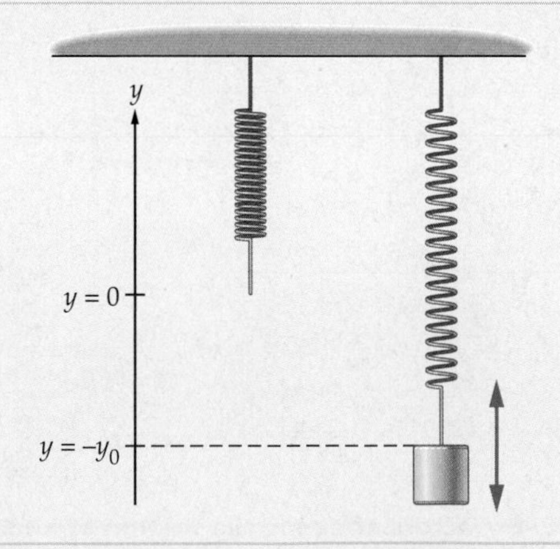

Strategy

In order to find the stretch of the spring, $y_0 = mg/k$, we need to know the force constant, k. We can find k from the period of oscillation; that is, from $T = 2\pi\sqrt{m/k}$.

Solution

1. Use the period $T = 2\pi\sqrt{m/k}$ to solve for the force constant:

$$T = 2\pi\sqrt{m/k}$$

$$k = \frac{4\pi^2 m}{T^2} = \frac{4\pi^2(0.260\text{ kg})}{(1.12\text{ s})^2} = 8.18\text{ kg/s}^2 = 8.18\text{ N/m}$$

2. Set the magnitude of the spring force, ky_0, equal to mg to solve for y_0:

$$ky_0 = mg$$

$$y_0 = mg/k = \frac{(0.260\text{ kg})(9.81\text{ m/s}^2)}{8.18\text{ N/m}} = 0.312\text{ m}$$

Insight

Note that even though gravity determines the amount by which the spring stretches when at rest, it plays no role in determining the period of oscillation. Thus, one would observe the same period on the moon, or even in orbit, as in the case of the Body Mass Measurement Device mentioned earlier in this section.

Practice Problem

A 0.170-kg mass stretches a vertical spring 0.250 m when at rest. What is its period when set into vertical motion? [**Answer:** $T = 1.00$ s]

Some related homework problems: Problem 29, Problem 32

CONCEPTUAL CHECKPOINT 13–1

When a mass m is attached to a vertical spring with a force constant k it oscillates with a period T. If the spring is cut in half and the same mass is attached to it, is the period of oscillation **(a)** greater than, **(b)** less than, or **(c)** equal to T?

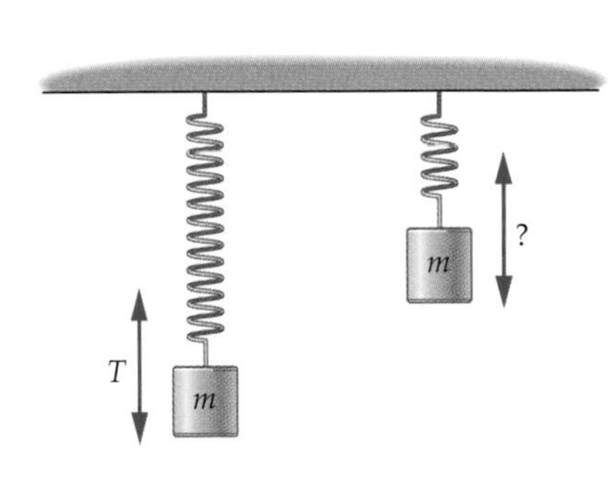

Reasoning and Discussion

The downward force exerted by the mass is of magnitude mg, as is the upward force exerted by the spring. Note that *each coil* of the spring experiences the *same force*, just as each point in a string experiences the same tension. Therefore, each coil elongates by the same amount, regardless of how many coils there are in a given spring. It follows, then, that the total elongation of the spring with half the number of coils is half the total elongation of the longer spring. Since half the elongation for the same applied force means a greater force constant, the half spring has a larger value of k—it is stiffer. As a result, its period of oscillation is less than the period of the full spring.

Answer:

(b) The period for the half spring is less than for the full spring.

13–5 Energy Conservation in Oscillatory Motion

In an ideal system with no friction or other nonconservative forces, the total energy is conserved. For example, the total energy E of a mass on a spring is the sum of its kinetic energy, $K = \frac{1}{2}mv^2$, and its potential energy, $U = \frac{1}{2}kx^2$. Therefore,

$$E = K + U = \tfrac{1}{2}mv^2 + \tfrac{1}{2}kx^2 \qquad \textbf{13–13}$$

Since E remains the same throughout the motion, it follows that there is a continual tradeoff between kinetic and potential energy.

This energy tradeoff is illustrated in **Figure 13–10**, where the horizontal line represents the total energy of the system, E, and the parabolic curve is the spring's potential energy, U. At any given value of x, the sum of U and K must equal E;

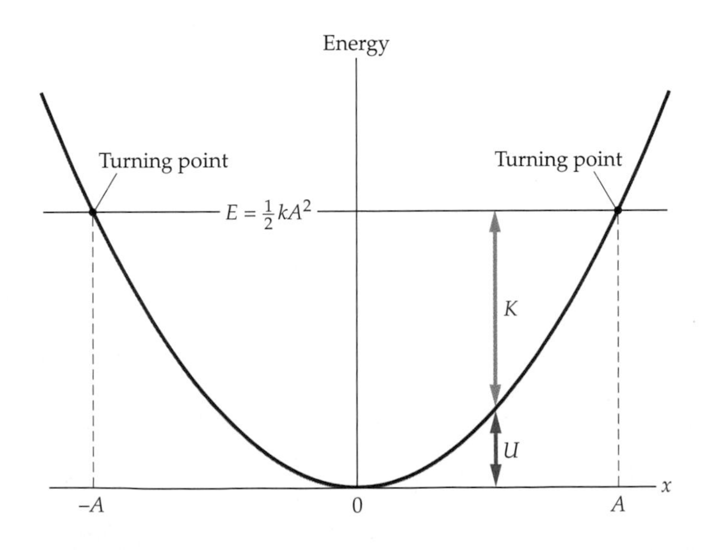

◀ **FIGURE 13–10 Energy as a function of position in simple harmonic motion**

The parabola represents the potential energy, U, of the spring. The horizontal line shows the total energy, E, of the system, which is constant, and the distance from the parabola to the horizontal line is the kinetic energy, K. Note that the kinetic energy vanishes at the turning points, $x = A$ and $x = -A$. At these points the energy is purely potential, thus the total energy of the system is $E = \frac{1}{2}kA^2$.

therefore, since U is the amount of energy from the axis to the parabola, K is the amount of energy from the parabola to the horizontal line. We can see, then, that the kinetic energy vanishes at the turning points, $x = +A$ and $x = -A$, as expected. On the other hand, the kinetic energy is greatest at $x = 0$ where the potential energy vanishes.

Since the mass oscillates back and forth with time, the kinetic and potential energies also change with time. For example, the potential energy is

$$U = \tfrac{1}{2}kx^2$$

Letting $x = A\cos(\omega t)$, we have

$$U = \tfrac{1}{2}kA^2\cos^2(\omega t) \qquad \textbf{13–14}$$

Clearly, the maximum value of U is

$$U_{max} = \tfrac{1}{2}kA^2$$

From Figure 13–10, we see that the maximum value of U is simply the total energy of the system, E. Therefore

$$E = U_{max} = \tfrac{1}{2}kA^2 \qquad \textbf{13–15}$$

This result leads to the following conclusion:

In simple harmonic motion, the total energy is proportional to the square of the amplitude of motion.

Similarly, the kinetic energy of the mass is

$$K = \tfrac{1}{2}mv^2$$

Letting $v = -A\omega\sin(\omega t)$ yields

$$K = \tfrac{1}{2}mA^2\omega^2\sin^2(\omega t)$$

It follows that the maximum kinetic energy is

$$K_{max} = \tfrac{1}{2}mA^2\omega^2 \qquad \textbf{13–16}$$

As noted, the maximum kinetic energy occurs when the potential energy is zero; hence the maximum kinetic energy must equal the total energy, E. That is,

$$E = U + K = U_{max} + 0 = 0 + K_{max}$$

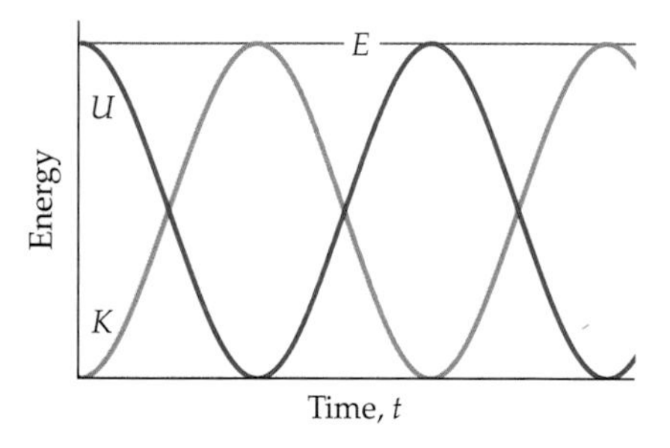

▲ **FIGURE 13–11 Energy as a function of time in simple harmonic motion**

The sum of the potential energy, U, and the kinetic energy, K, is equal to the (constant) total energy E at all times. Note that when one energy (U or K) has its maximum value the other energy is zero.

At first glance, Equation 13–16 doesn't seem to be the same as Equation 13–15. However, if we recall that $\omega^2 = k/m$ (Equation 13–10), we see that

$$K_{max} = \tfrac{1}{2}mA^2\omega^2 = \tfrac{1}{2}mA^2(k/m) = \tfrac{1}{2}kA^2$$

Therefore, $K_{max} = U_{max} = E$, as expected, and the kinetic energy as a function of time is

$$K = \tfrac{1}{2}kA^2\sin^2(\omega t) \qquad \textbf{13–17}$$

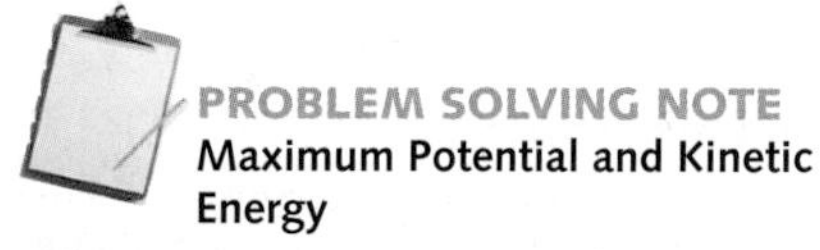

PROBLEM SOLVING NOTE
Maximum Potential and Kinetic Energy

The maximum potential energy of a mass-spring system is the same as the maximum kinetic energy of the mass. When the system has its maximum potential energy the kinetic energy of the mass is zero; when the mass has its maximum kinetic energy the potential energy of the system is zero.

K and U are plotted as functions of time in **Figure 13–11**. The horizontal line at the top is E, the sum of U and K at all times. This shows quite graphically the back and forth tradeoff of energy between kinetic and potential. Mathematically, we can see that the total energy E is constant as follows:

$$E = U + K = \tfrac{1}{2}kA^2\cos^2(\omega t) + \tfrac{1}{2}kA^2\sin^2(\omega t)$$
$$= \tfrac{1}{2}kA^2[\cos^2(\omega t) + \sin^2(\omega t)] = \tfrac{1}{2}kA^2$$

The last step follows from the fact that $\cos^2\theta + \sin^2\theta = 1$ for all θ.

EXAMPLE 13–6 **Stop the Block**

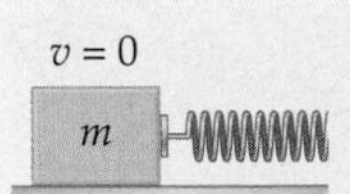

A 0.980-kg block slides on a frictionless, horizontal surface with a speed of 1.32 m/s. The block encounters an unstretched spring with a force constant of 245 N/m, as shown in the sketch. **(a)** How far is the spring compressed before the block comes to rest? **(b)** How long is the block in contact with the spring before it comes to rest?

Picture the Problem

Initially, the system's energy is entirely kinetic; the kinetic energy of the block. When the block comes to rest, its kinetic energy has been converted into the potential energy of the spring.

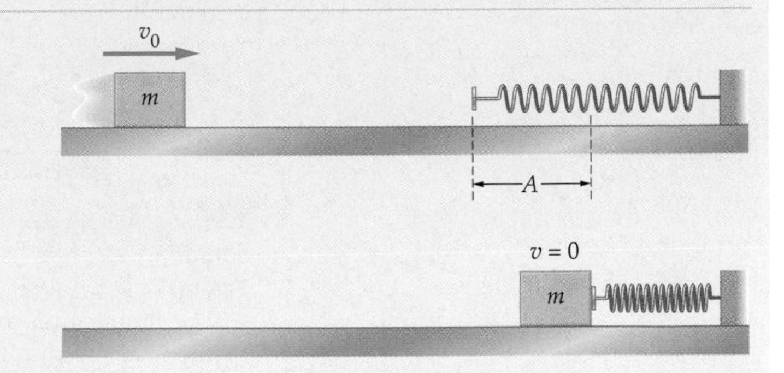

Strategy

(a) We can find the compression, A, by using energy conservation. We set the initial kinetic energy of the block, $\frac{1}{2}mv^2$, equal to the spring potential energy, $\frac{1}{2}kA^2$, and solve for A.

(b) If the mass were attached to the spring it would complete one oscillation in the time $T = 2\pi\sqrt{m/k}$. In moving from the equilibrium position of the spring to maximum compression, the mass has undergone $\frac{1}{4}$ of a cycle; thus the time is $T/4$.

Solution

Part (a)

1. Set the initial kinetic energy of the block equal to the spring potential energy:

$$\tfrac{1}{2}mv^2 = \tfrac{1}{2}kA^2$$

2. Solve for A, the maximum compression:

$$A = v\sqrt{m/k} = (1.32\ \text{m/s})\sqrt{\frac{0.980\ \text{kg}}{245\ \text{N/m}}} = 0.0835\ \text{m}$$

Part (b)

3. Calculate the period of one oscillation:

$$T = 2\pi\sqrt{m/k} = 2\pi\sqrt{\frac{0.980\ \text{kg}}{245\ \text{N/m}}} = 0.397\ \text{s}$$

4. Divide T by four, since the block has been in contact with the spring for 1/4 of an oscillation:

$$t = \tfrac{1}{4}T = \tfrac{1}{4}(0.397\ \text{s}) = 0.0993\ \text{s}$$

Insight

If the horizontal surface had been rough, some of the block's initial kinetic energy would have been converted to thermal energy. In this case, the maximum compression of the spring would be less than that just calculated.

Practice Problem

If the initial speed of the mass in this Example is increased, does the time required to bring it to rest increase, decrease, or stay the same? Check your answer by calculating the time for an initial speed of 1.50 m/s. [**Answer:** Increasing v increases the amplitude, A. The period is independent of amplitude, however. Thus, the time is the same, $t = 0.0993$ s.]

Some related homework problems: Problem 44, Problem 45

In the following Active Example we consider a bullet striking a block that is attached to a spring. As the bullet embeds itself in the block, the completely inelastic bullet-block collision dissipates some of the initial kinetic energy into thermal energy. Only the kinetic energy remaining after the collision is available for compressing the spring.

ACTIVE EXAMPLE 13–3 **Bullet-Block Collision**

A bullet of mass m embeds itself in a block of mass M, which is attached to a spring of force constant k. If the initial speed of the bullet is v_0, find **(a)** the maximum compression of the spring and **(b)** the time for the bullet-block system to come to rest.

continued on the following page

continued from the previous page

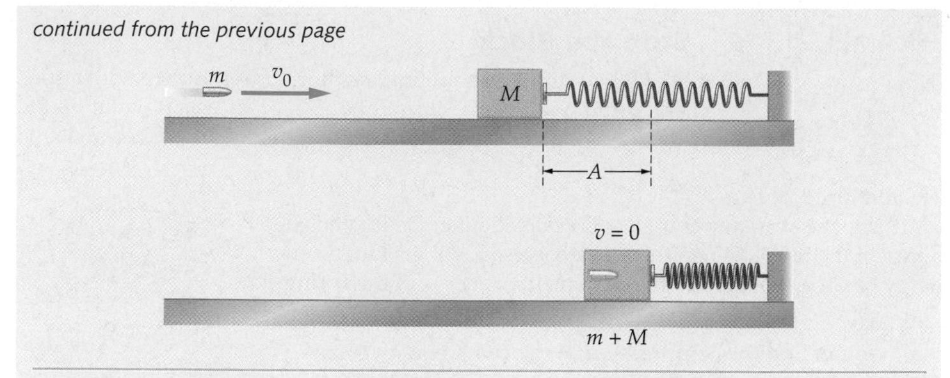

Solution

Part (a)

1. Use momentum conservation to find the final speed, v, of the bullet-block system: $mv_0 = (m + M)v$, $v = mv_0/(m + M)$
2. Find the kinetic energy of the bullet-block system after the collision: $\frac{1}{2}(m + M)v^2 = \frac{1}{2}m^2v_0^2/(m + M)$
3. Set this kinetic energy equal to $\frac{1}{2}kA^2$ to find the maximum compression of the spring: $A = mv_0/\sqrt{k(m + M)}$

Part (b)

4. As in the previous Example, the time to come to rest is one-quarter of a period: $t = T/4 = \frac{1}{2}\pi\sqrt{(m + M)/k}$

Insight

Note that the maximum compression depends directly on the initial speed of the bullet. Thus, a measurement of the compression could be used to determine the bullet's speed. On the other hand, the time for the bullet-block to come to rest is independent of the bullet's speed.

13–6 The Pendulum

One Sunday in 1583, as Galileo Galilei attended services in a cathedral in Pisa, Italy, he suddenly realized something interesting about the chandeliers hanging from the ceiling. Air currents circulating through the cathedral had set them in motion with small oscillations, and Galileo noticed that chandeliers of equal length oscillated with equal periods, even if their amplitudes were different. Indeed, as any given chandelier oscillated with decreasing amplitude its period remained constant. He verified this observation by timing the oscillations with his pulse!

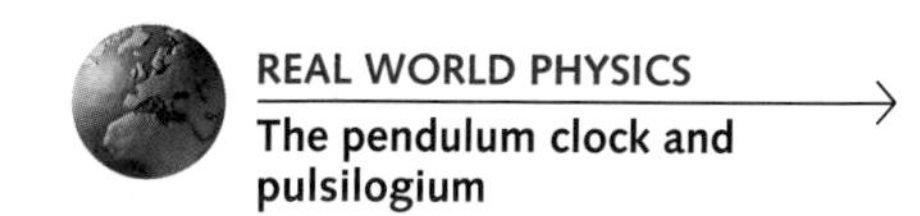

Galileo was much struck by this observation, and after rushing home he experimented with pendula constructed from different lengths of string and different weights. Continuing to use his pulse as a stopwatch, he observed that the period of a pendulum varies with its length, but is independent of the weight attached to the string. Thus, in one exhilarating afternoon, the young medical student Galileo discovered the key characteristics of a pendulum and launched himself on a new career in science. Later, he would go on to construct the first crude pendulum clock and a medical device, known as the pulsilogium, to measure a patient's pulse rate.

In modern terms, we would say that the chandeliers observed by Galileo were undergoing simple harmonic motion, as expected for small oscillations. As we know, the period of simple harmonic motion is independent of amplitude. The fact that the period is also independent of the mass is a special property of the pendulum, as we shall see next.

The Simple Pendulum

A simple pendulum consists of a mass m suspended by a light string or rod of length L. The pendulum has a stable equilibrium when the mass is directly below the suspension point, and oscillates about this position if displaced from it.

To understand the behavior of the pendulum, let's begin by considering the potential energy of the system. As shown in **Figure 13–12**, when the pendulum is at an angle θ with respect to the vertical, the mass m is above its lowest point by a vertical height $L(1 - \cos\theta)$. If we let the potential energy be zero at $\theta = 0$, the potential energy for general θ is

$$U = mgL(1 - \cos\theta) \qquad \textbf{13–18}$$

This function is plotted in **Figure 13–13**.

Note that the stable equilibrium of the pendulum corresponds to a minimum of the potential energy, as expected. Near this minimum, the shape of the potential energy curve is approximately the same as for a mass on a spring, as indicated in Figure 13–13. As a result, when a pendulum oscillates with small displacements from the vertical, its motion is virtually the same as the motion of a mass on a spring; that is, the pendulum exhibits simple harmonic motion.

Next, we consider the forces acting on the mass m. In **Figure 13–14** we show the force of gravity, $m\mathbf{g}$, and the tension force in the supporting string, $\mathbf{T}$. The tension acts in the radial direction, and supplies the force needed to keep the mass moving

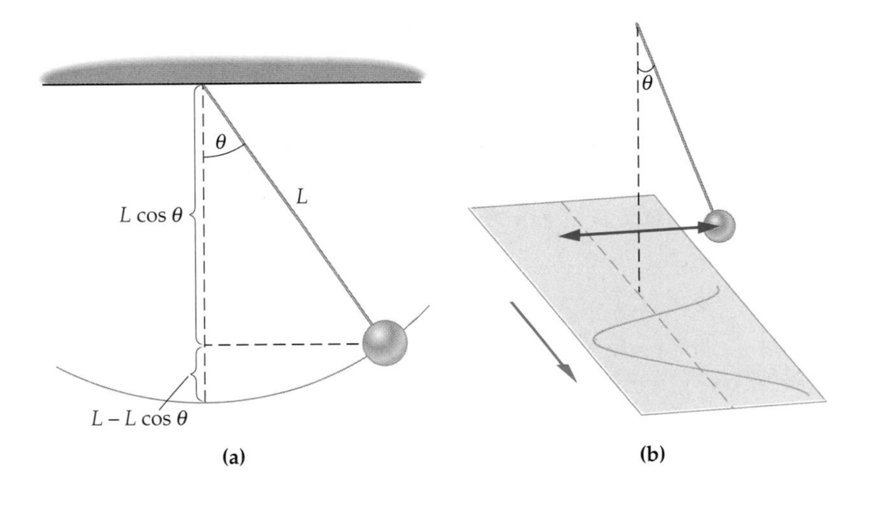

◀ **FIGURE 13–12 Motion of a pendulum**

(a) As a pendulum swings away from its equilibrium position it rises a vertical distance $L - L\cos\theta = L(1 - \cos\theta)$. **(b)** As a pendulum bob swings back and forth it leaks a trail of sand onto a moving sheet of paper, creating a "strip chart" similar to that produced by a mass on a spring in Figure 13–2. In particular, the angle of the pendulum with the vertical varies with time like a sine or a cosine.

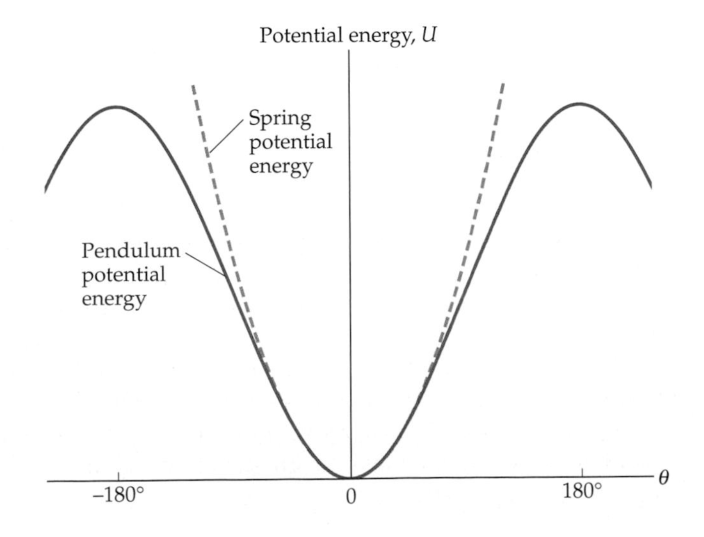

◀ **FIGURE 13–13 The potential energy of a simple pendulum**

As a simple pendulum swings away from the vertical by an angle θ its potential energy increases, as indicated by the solid curve. Near $\theta = 0$ the potential energy of the pendulum is essentially the same as that of a mass on a spring (dashed curve). Therefore, when a pendulum oscillates with small displacements from the vertical it exhibits simple harmonic motion—the same as a mass on a spring.

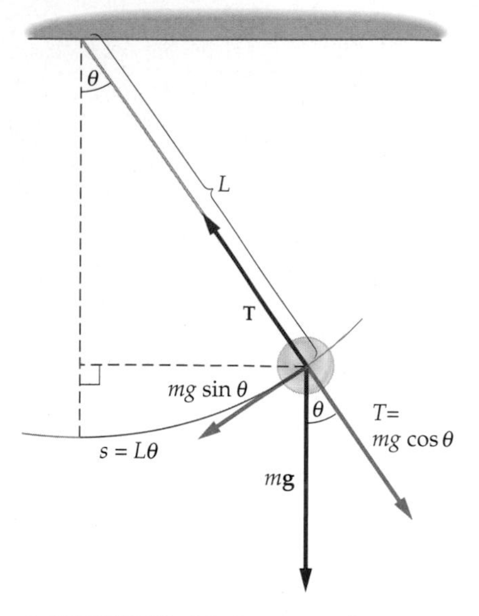

▲ **FIGURE 13–14 Forces acting on a pendulum bob**

When the bob is displaced by an angle θ from the vertical, the restoring force is the tangential component of the weight, $mg \sin \theta$.

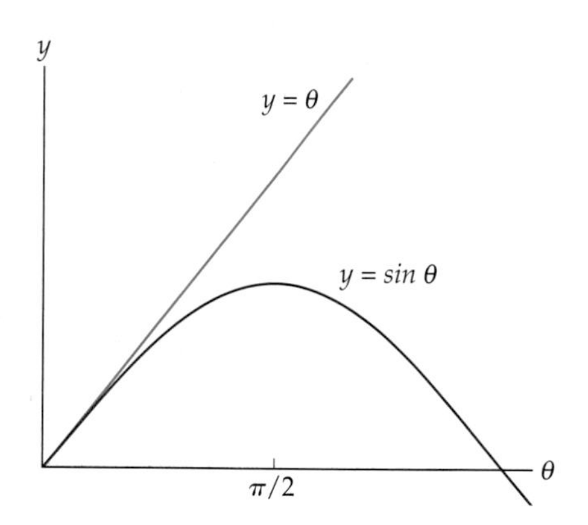

▲ **FIGURE 13–15 Relationship between sin θ and θ**

For small angles, $\sin \theta$ is approximately equal to θ. Thus, when considering small oscillations of a pendulum, we can replace $\sin \theta$ with θ. (See also Appendix A.)

along its circular path. The net tangential force acting on m, then, is simply the tangential component of its weight:

$$F = mg \sin\theta$$

The direction of the net tangential force is always toward the equilibrium point. Thus F is a restoring force, as expected.

Now, for small angles θ (measured in *radians*), the sine of θ is approximately equal to the angle itself. That is,

$$\sin\theta \approx \theta$$

This is illustrated in **Figure 13–15**. Notice that there is little difference between $\sin \theta$ and θ for angles smaller than about $\pi/8$. In addition, we see from Figure 13–14 that the arc length displacement of the mass from equilibrium is

$$s = L\theta$$

Equivalently,

$$\theta = s/L$$

Therefore, if the mass m is displaced from equilibrium by a small arc length s, the force it experiences is restoring and of magnitude

$$F = mg \sin\theta \approx mg\theta = (mg/L)s \qquad \textbf{13–19}$$

Note that the restoring force is proportional to the displacement, just as expected for simple harmonic motion.

Let's compare the pendulum to a mass on a spring. In the latter case, the restoring force has a magnitude given by

$$F = kx$$

The restoring force acting on the pendulum has precisely the same form, if we let $x = s$ and

$$k = mg/L$$

Therefore, the period of a pendulum is simply the period of a mass on a spring, $T = 2\pi\sqrt{m/k}$, with k replaced by mg/L:

$$T = 2\pi\sqrt{\frac{m}{k}} = 2\pi\sqrt{\frac{m}{(mg/L)}}$$

Canceling the mass m, we find

Period of a Pendulum (small amplitude)

$$T = 2\pi\sqrt{\frac{L}{g}} \qquad \textbf{13–20}$$

SI unit: s

This is the classic formula for the period of a pendulum. Note that T depends on the length of the pendulum, L, and on the acceleration of gravity, g. It is independent, however, of the mass m and the amplitude A, as noted by Galileo.

The mass does not appear in the expression for the period of a pendulum for the same reason that different masses free fall with the same acceleration. In particular, a large mass tends to move more slowly because of its large inertia; on the other hand, the larger a mass the greater the gravitational force acting on it. These two effects cancel in free fall, as well as in a pendulum.

EXERCISE 13–4

The pendulum in a grandfather clock is designed to take one second to swing in each direction; that is, 2.00 seconds for a complete period. Find the length of a pendulum with a period of 2.00 seconds.

Solution

Solve Equation 13–20 for L and substitute numerical values:

$$L = \frac{gT^2}{4\pi^2} = \frac{(9.81\ \text{m/s}^2)(2.00\ \text{s})^2}{4\pi^2} = 0.994\ \text{m}$$

CONCEPTUAL CHECKPOINT 13–2

If you look carefully at a grandfather clock, you will notice that the weight at the bottom of the pendulum can be moved up or down by turning a small screw. Suppose you have a grandfather clock at home that runs slow. Should you turn the adjusting screw so as to **(a)** raise the weight or **(b)** lower the weight?

Reasoning and Discussion

To make the clock run faster, we want it to go from tick to tick more rapidly; in other words, we want the period of the pendulum to be decreased. From Equation 13–20 we can see that shortening the pendulum—that is, decreasing L—decreases the period. Hence, the weight should be raised, which effectively shortens the pendulum.

Answer:

(a) The weight should be raised. This shortens the period and makes the clock run faster.

EXAMPLE 13–7 Drop Time

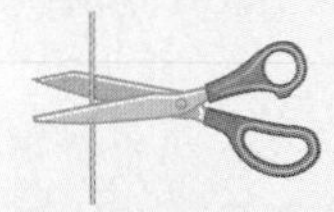

A pendulum is constructed from a string 0.627 m long attached to a mass of 0.250 kg. When set in motion, the pendulum completes one oscillation every 1.59 s. If the pendulum is held at rest and the string is cut, how long will it take for the mass to fall through a distance of 1.00 m?

Picture the Problem

The pendulum has a length L and a period T. When the string is cut the mass falls straight downward with an acceleration g through a distance of 1.00 m.

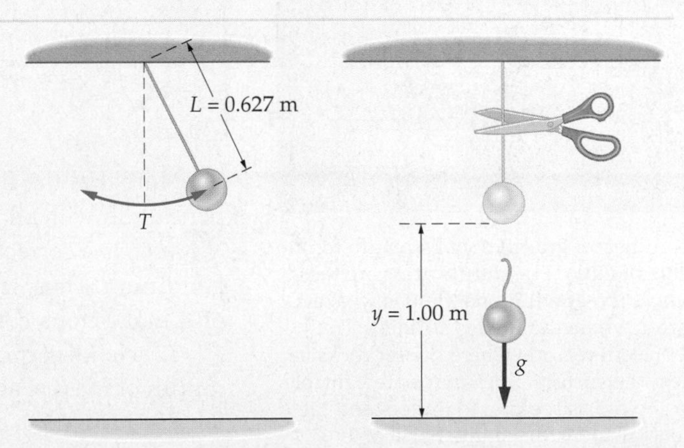

Strategy

At first it might seem that the period of oscillation and the time of fall are unrelated. Recall, however, that the period of a pendulum depends on both its length *and* the acceleration of gravity g *at the location of the pendulum.*

To solve this problem, then, we first use the period T to find the acceleration of gravity g. Once g is known, the time of fall is a straightforward kinematics problem (see Example 2-10).

Solution

1. Use the formula for the period of a pendulum to solve for the acceleration of gravity:

$$T = 2\pi\sqrt{L/g}$$
$$g = \frac{4\pi^2 L}{T^2}$$

2. Substitute numerical values to find g:

$$g = \frac{4\pi^2 L}{T^2} = \frac{4\pi^2(0.627\ \text{m})}{(1.59\ \text{s})^2} = 9.79\ \text{m/s}^2$$

3. Use kinematics to solve for the time to drop from rest through a distance y:

$$y = \tfrac{1}{2}gt^2$$
$$t = \sqrt{2y/g}$$

continued on the following page

continued from the previous page

4. Substitute $y = 1.00$ m and the value of g just found to find the time: $t = \sqrt{2y/g} = \sqrt{\dfrac{2(1.00 \text{ m})}{9.79 \text{ m/s}^2}} = 0.452 \text{ s}$

Insight

The fact that the acceleration of gravity g varies from place to place on the earth was mentioned in Chapter 2. In fact, "gravity maps," such as the one shown in the photo below, are valuable tools for geologists attempting to understand the underground properties of a given region. One instrument geologists use to make gravity maps is basically a very precise pendulum whose period can be accurately measured. Slight changes in the period from one location to another, or from one elevation to another, correspond to slight changes in g.

Practice Problem

If a mass falls 1.00 m in a time of 0.451 s, what is **(a)** the acceleration of gravity and **(b)** the period of a pendulum of length 0.500 m at that location? [**Answer:** (a) 9.83 m/s^2, (b) 1.42 s]

Some related homework problems: Problem 47, Problem 49

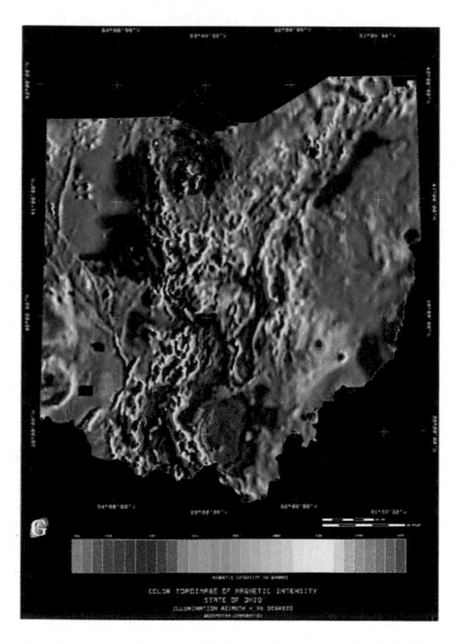

▲ A map of gravitational strength for the state of Ohio. The purple areas are those where the gravitational field is weakest. Areas where the field is strongest (red) represent regions where denser rocks lie near the surface. Such maps are valuable to geologists seeking to understand the geological history of the Earth or prospecting for subterranean resources such as oil and metallic ores.

*The Physical Pendulum

In the ideal version of a simple pendulum, a bob of mass m swings back and forth a distance L below the suspension point. All of the pendulum's mass is assumed to be concentrated in the bob, which is treated as a point mass. On the other hand, a **physical pendulum** is one in which the mass is not concentrated at a point, but instead is distributed over a finite volume. Examples are shown in **Figure 13–16**. Detailed mathematical analysis shows that if the moment of inertia (Chapter 11) of a physical pendulum about its axis of rotation is I, and the distance from the axis to the center of mass is ℓ, the period of the pendulum is given by the following:

Period of a Physical Pendulum

$$T = 2\pi\sqrt{\frac{\ell}{g}}\left(\sqrt{\frac{I}{m\ell^2}}\right) \qquad \text{13–21}$$

SI unit: s

Note that the first part of the expression, $2\pi\sqrt{\ell/g}$, is the period of a simple pendulum with all its mass concentrated at the center of mass. The second factor, $\sqrt{I/m\ell^2}$, is a correction that takes into account the size and shape of the physical pendulum. Thus, writing the period in this form, rather than canceling one power of ℓ, makes for a convenient comparison with the simple pendulum.

To see how Equation 13–21 works in practice, we first apply it to a simple pendulum of mass m and length L. In this case, the moment of inertia is $I = mL^2$, and the distance to the center of mass is $\ell = L$. As a result, the period is

$$T = 2\pi\sqrt{\frac{\ell}{g}}\left(\sqrt{\frac{I}{m\ell^2}}\right) = 2\pi\sqrt{\frac{L}{g}}\left(\sqrt{\frac{mL^2}{mL^2}}\right) = 2\pi\sqrt{\frac{L}{g}}$$

Thus, as expected, Equation 13–21 also applies to a simple pendulum.

Next, we apply Equation 13–21 to a nontrivial physical pendulum; namely, your leg. When you walk, your leg rotates about the hip joint much like a uniform rod pivoted about one end, as indicated in **Figure 13–17**. Thus, if we approximate your leg as a uniform rod of length L, its period can be found using Equation 13–21. Recall from Chapter 10 (see Table 10–1) that the moment of inertia of a rod of length L is

$$I = \tfrac{1}{3}mL^2$$

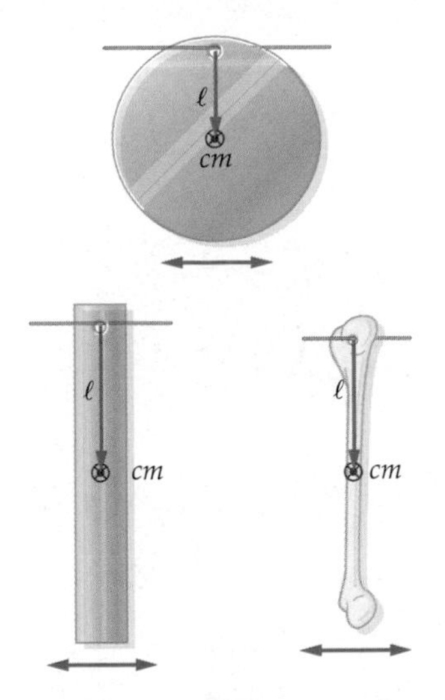

▲ FIGURE 13–16 Examples of physical pendulums

In each case, an object of definite size and shape oscillates about a given pivot point. The period of oscillation depends in detail on the location of the pivot point as well as on the distance ℓ from it to the center of mass.

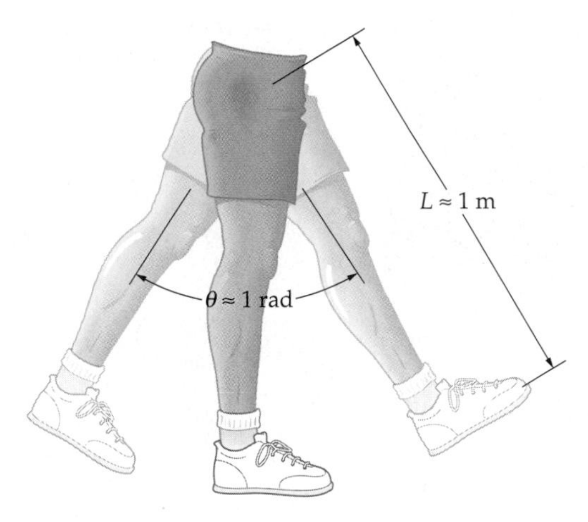

▲ FIGURE 13–17 The leg as a physical pendulum

As a person walks, each leg swings much like a physical pendulum. A reasonable approximation is to treat the leg as a uniform rod about 1 m in length.

Similarly, the center of mass of your leg is essentially at the center of your leg; thus,

$$\ell = \tfrac{1}{2}L$$

Combining these results, we find that the period of your leg is roughly

$$T = 2\pi\sqrt{\frac{\ell}{g}}\left(\sqrt{\frac{I}{m\ell^2}}\right) = 2\pi\sqrt{\frac{\frac{1}{2}L}{g}}\left(\sqrt{\frac{\frac{1}{3}mL^2}{m(\frac{1}{2}L)^2}}\right) = 2\pi\sqrt{\frac{L}{g}}\left(\sqrt{\frac{2}{3}}\right)$$

Given that a typical human leg is about a meter long ($L = 1.0$ m), we find that its period of oscillation about the hip is approximately $T = 1.6$ s.

The significance of this result is that the natural walking pace of humans and other animals is largely controlled by the swinging motion of their legs as physical pendula. In fact, a great deal of research in animal locomotion has focused on precisely this type of analysis. In the case just considered, suppose that the leg in Figure 13–17 swings through an angle of roughly 1.0 radian, as the foot moves from behind the hip to in front of the hip for the next step. The arc length through which the foot moves is $s = r\theta \approx (1.0\text{ m})(1.0\text{ rad}) = 1.0$ m. Since it takes half a period ($T/2 = 0.80$ s) for the foot to move from behind the hip to in front of it, the average speed of the foot is roughly

$$v = \frac{d}{t} \approx \frac{1.0\text{ m}}{0.80\text{ s}} \approx 1.3\text{ m/s}$$

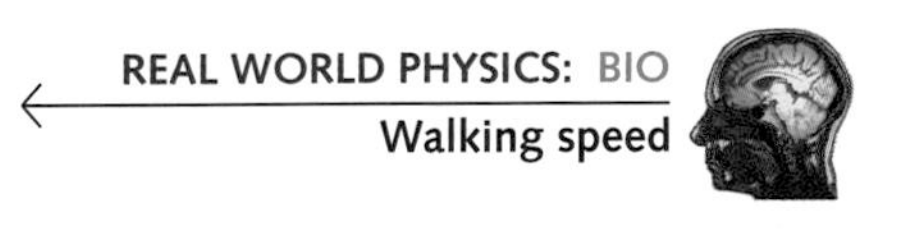

This is the typical speed of a person taking a brisk walk.

Returning to Equation 13–21, we can see that the smaller the moment of inertia I, the smaller the period. After all, an object with a small moment of inertia rotates quickly and easily. It therefore completes an oscillation in less time than an object with a larger moment of inertia. In the case of the human leg, the fact that its mass is distributed uniformly along its length—rather than concentrated at the far end—means that its moment of inertia and period are less than for a simple pendulum 1 m long.

When finding the period of a physical pendulum, recall that I is the moment of inertia about the pivot point and ℓ is the distance from the pivot point to the center of mass.

▼ When animals walk, the swinging movement of their legs can be approximated fairly well by treating the legs as physical pendulums. Such analysis has proved useful in analyzing the gaits of various creatures, from Beatles to elephants.

ACTIVE EXAMPLE 13–4 Christmas Ornament

A Christmas ornament is made from a hollow glass sphere of mass M and radius R. The ornament is suspended from a small hook near its surface. If the ornament is nudged slightly, what is its period of oscillation? (*Note*: The moment of inertia about the pivot point is $I = \frac{5}{3}mR^2$.)

Solution

1. Identify the distance from the axis to the center of mass: $\ell = R$
2. Substitute I and ℓ into Equation 13–21: $T = 2\pi\sqrt{\frac{R}{g}}\left(\sqrt{\frac{5}{3}}\right)$

Insight

The period is greater than that of a simple pendulum of length R.

13–7 Damped Oscillations

To this point we have restricted our considerations to oscillating systems in which no energy is gained or lost. In most physical systems, however, there is some loss of energy to friction, air resistance, or other nonconservative forces. As the energy of a system decreases, its amplitude of oscillation decreases as well, as expected from Equation 13–15. This type of motion is referred to as a **damped oscillation**.

In a typical situation, an oscillating mass may lose its energy to a force like air resistance that is proportional to the speed of the mass and opposite in direction. The force in such a case can be written as:

$$\mathbf{F} = -b\mathbf{v}$$

The constant b is referred to as the **damping constant**; it is a measure of the strength of the damping force.

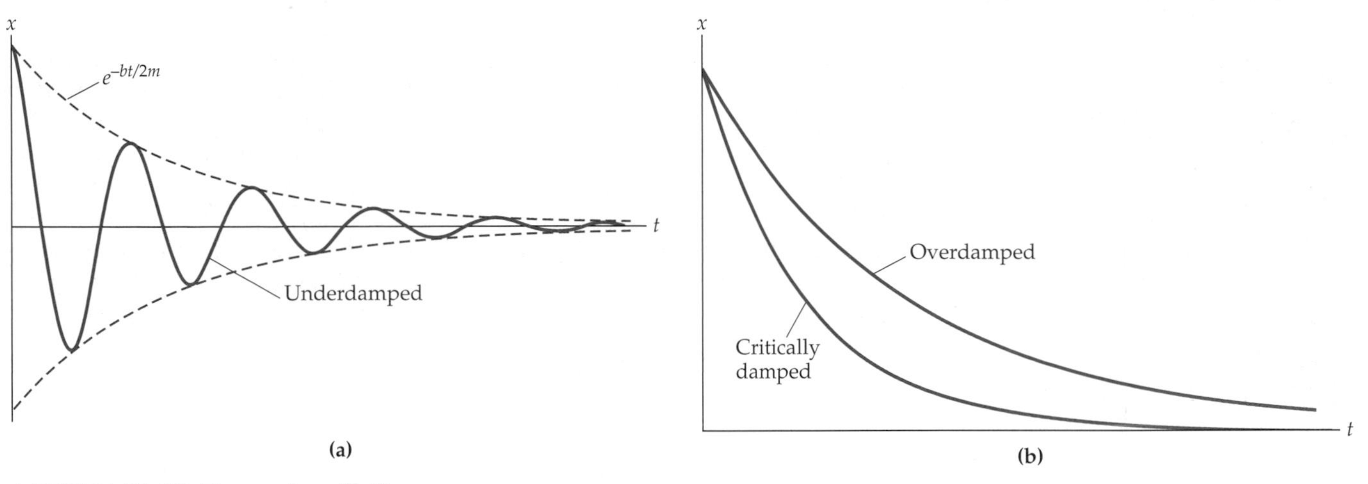

▲ **FIGURE 13–18 Damped oscillations**
In underdamped oscillations **(a)**, the position continues to oscillate as a function of time, but the amplitude of oscillation decreases exponentially. In critically damped and overdamped motion **(b)**, no oscillations occur. Instead, an object simply settles back to its equilibrium position without overshooting. Equilibrium is reached most rapidly in the critically damped case.

If the damping constant is small the system will continue to oscillate, but with a continuously decreasing amplitude. This type of motion, referred to as **underdamped**, is illustrated in **Figure 13–18 (a)**. In such cases, the amplitude decreases exponentially with time. Thus, if A_0 is the initial amplitude of an oscillating mass m, the amplitude at the time t is

$$A = A_0 e^{-bt/2m}$$

The exponential dependence of the amplitude is indicated by the dashed curve in the figure.

As the damping is increased, a point is reached where the system no longer oscillates, but instead simply relaxes back to the equilibrium position, as shown in **Figure 13–18 (b)**. A system with this type of behavior is said to be **critically damped**. If the damping is increased beyond this point the system is said to be **overdamped**. In this case, the system still returns to equilibrium without oscillating, but the time required is greater. This is also illustrated in the figure.

Some mechanical systems are designed to be near the condition for critical damping. If such a system is displaced from equilibrium it will return to equilibrium, without oscillating, in the shortest possible time. For example, shock absorbers are designed so that a car that has just hit a bump will return to equilibrium quickly, without a lot of up and down oscillations.

▲ Many people are familiar with shock absorbers in cars, but similar devices are also used in buildings to reduce the swaying caused by earthquakes. Like automotive shocks, these giant shock absorbers are "tuned" to provide critical damping for the natural vibration frequency of the building.

13–8 Driven Oscillations and Resonance

In the previous section we considered the effects of removing energy from an oscillating system. It is also possible, however, to increase the energy of a system, or to replace the energy lost to various forms of friction. This can be done by applying an external force that does positive work.

Suppose, for example, that you hold the end of a string from which a small weight is suspended, as in **Figure 13–19**. If the weight is set in motion and you hold your hand still, it will soon stop oscillating. If you move your hand back and forth in a horizontal direction, however, you can keep the weight oscillating indefinitely. The motion of your hand is said to be "driving" the weight, leading to **driven oscillations**.

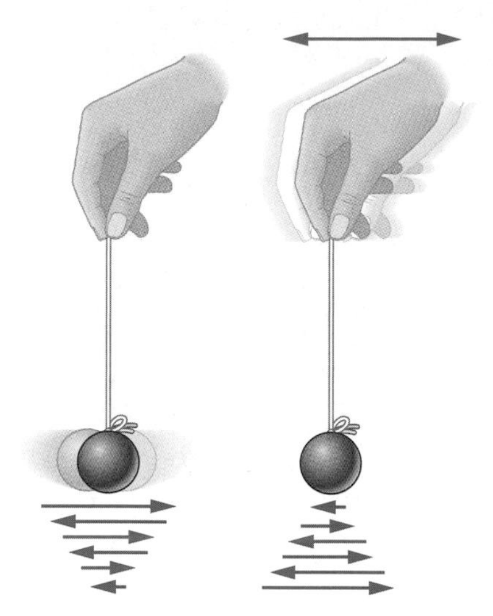

▲ FIGURE 13–19 Driven oscillations
If the support point of a pendulum is held still, its oscillations quickly die away. If the support point is oscillated back and forth, however, the pendulum will continue swinging. This is called "driving" the pendulum. If the driving frequency is close to the natural frequency of the pendulum—the frequency at which it oscillates when the support point is held still—its amplitude of motion can become quite large.

The response of the weight in this example depends on the frequency of your hand's back and forth motion, as you can readily verify for yourself. For instance, if you move your hand very slowly, the weight will simply track the motion of your hand. Similarly, if you oscillate your hand very rapidly, the weight will exhibit only small oscillations. Oscillating your hand at an intermediate frequency, however, can result in large amplitude oscillations for the weight.

Just what is an appropriate intermediate frequency? Well, to achieve a large response, your hand should drive the weight at the frequency at which it oscillates when not being driven. This is referred to as the **natural frequency**, f_0, of the system. For example, the natural frequency of a pendulum of length L is simply the inverse of its period:

$$f_0 = \frac{1}{T} = \frac{1}{2\pi}\sqrt{\frac{g}{L}}$$

Similarly, the natural frequency of a mass on a spring is

$$f_0 = \frac{1}{T} = \frac{1}{2\pi}\sqrt{\frac{k}{m}}$$

In general, driving any system at a frequency near its natural frequency results in large oscillations.

As an example, **Figure 13–20** shows a plot of amplitude, A, versus driving frequency, f, for a mass on a spring. Note the large amplitude for frequencies near f_0. This type of large response, due to frequency matching, is known as **resonance**, and the curve shown in Figure 13–20 is the **resonance curve**. Also shown in Figure 13–20 are resonance curves for various amounts of damping. As we can see, systems with small damping have a high, narrow peak on their resonance curve. This means that resonance is a large effect in these systems, and that it is very sensitive to frequency. On the other hand, systems with large damping have resonance peaks that are broad and low.

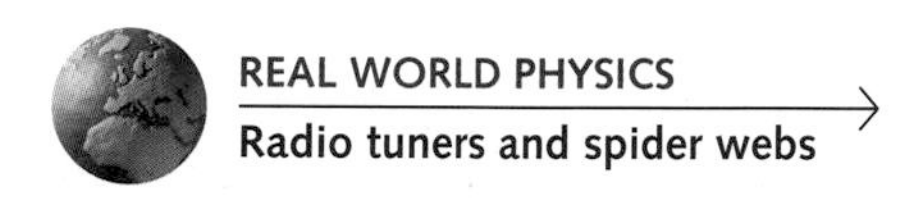

REAL WORLD PHYSICS
Radio tuners and spider webs

Resonance plays an important role in a variety of physical systems, from a pendulum to atoms in a laser to a tuner in a radio or TV. As we shall see in Chapter 24, for example, adjusting the tuning knob in a radio changes the resonance frequency of the electrical circuit in the tuner. When its resonance frequency matches the frequency being broadcast by a station (101 MHz, perhaps), that station is picked up. To change stations, we simply change the resonance frequency of the tuner to the frequency of another station. A good tuner will have little damping, thus stations that are even slightly "off resonance" will have small response, and hence will not be heard.

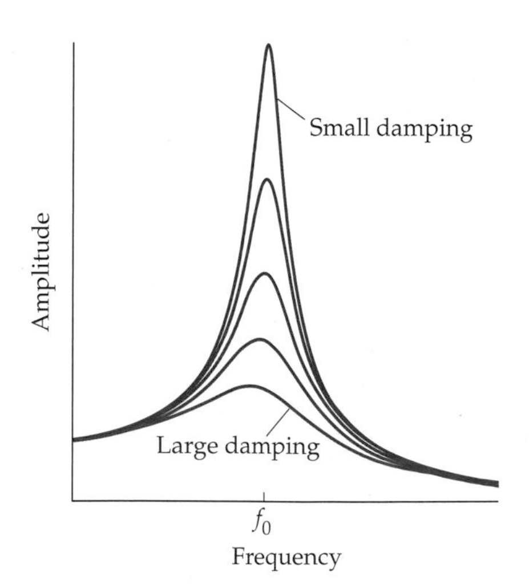

▲ FIGURE 13–20 Resonance curves for various amounts of damping
When the damping is small, the amplitude of oscillation can become very large for frequencies close to the natural frequency, f_0. When the damping is large, the amplitude has only a low, broad peak near the natural frequency.

Mechanical examples of resonance are all around us as well. In fact, you might want to try the following experiment the next time you notice a spider in its web. Move close to the web and hum, starting with a low pitch. Slowly increase the pitch of your humming, and soon you will notice the spider react excitedly. You have hit the resonance frequency of its web, causing it to vibrate and making the spider think he has snagged a lunch. If you continue to increase your humming frequency, the spider quiets down again, because you have gone past the resonance. Each individual spider web you encounter will resonate at a specific, and different, frequency.

Man-made structures can show resonance effects as well. One of the most dramatic and famous examples is the collapse of the Tacoma Narrows bridge in 1940. High winds through the narrows had often set the bridge into a gentle swaying motion, resulting in its being known by the affectionate nickname "Galloping Girdie." During one particular wind storm, however, the bridge experienced a resonance-like effect, and the amplitude of its swaying motion began to increase. Alarmed officials closed the bridge to traffic, and a short time later the swaying

▲ Anyone who has ever pushed a child on a swing (left) knows that the timing of the pushes is critical. If they are synchronized with the natural frequency of the swing, the amplitude can increase rapidly. This phenomenon of resonance can have dangerous consequences. In 1940, the Tacoma Narrows bridge (right), completed only four months earlier, collapsed when high winds set the bridge swaying at one of its resonant frequencies.

motion became so great that the bridge broke apart and fell into the waters below. Needless to say, bridges built since that time have been designed to prevent such catastrophic oscillations.

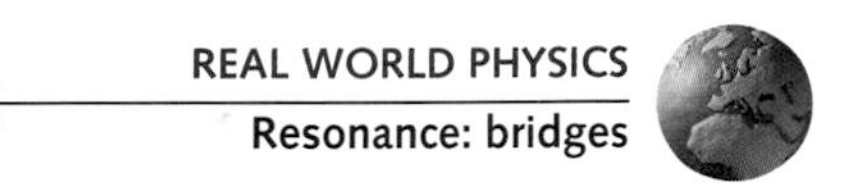

Chapter Summary

	Topic	Remarks and Relevant Equations	
13–1	**Periodic Motion**	Periodic motion repeats after a definite length of time.	
	period	The time, T, required for a motion to repeat is referred to as the period: T = time required for one cycle of a periodic motion	
	frequency	Frequency, f, is the inverse of the period: $f = \frac{1}{T}$	13–1
	angular frequency	Angular frequency, ω, is 2π times the frequency: $\omega = 2\pi f = \frac{2\pi}{T}$	13–5
		Rapid motion corresponds to a short period and a large frequency.	
13–2	**Simple Harmonic Motion**	A particular type of periodic motion is simple harmonic motion. A classic example of simple harmonic motion is the oscillation of a mass attached to a spring.	
	restoring force	Simple harmonic motion occurs when the restoring force is proportional to the displacement from equilibrium.	
	amplitude	The maximum displacement from equilibrium is referred to as the amplitude, A.	
	position versus time	The position, x, of an object undergoing simple harmonic motion varies with time as $\cos(\omega t)$: $x = A\cos\left(\frac{2\pi}{T}t\right) = A\cos(\omega t)$	13–2, 13–4

13–3 Connections Between Uniform Circular Motion and Simple Harmonic Motion

A close relationship exists between uniform circular motion and simple harmonic motion. In particular, circular motion viewed from the side, by projecting a shadow on a screen, for example, is simple harmonic. Using this connection, and our previous knowledge of circular motion, we can derive a number of results for simple harmonic motion.

velocity

The velocity as a function of time in simple harmonic motion is

$$v = -A\omega \sin(\omega t) \quad \text{13–6}$$

acceleration

The acceleration as a function of time in simple harmonic motion is

$$a = -A\omega^2 \cos(\omega t) \quad \text{13–8}$$

maximum speed and acceleration

The maximum speed of an object in simple harmonic motion is

$$v_{\text{max}} = A\omega \quad \text{13–7}$$

Its maximum acceleration is

$$a_{\text{max}} = A\omega^2 \quad \text{13–9}$$

13–4 The Period of a Mass on a Spring

An important special case of simple harmonic motion is a mass on a spring.

period

The period of a mass m attached to a spring of force constant k is

$$T = 2\pi\sqrt{\frac{m}{k}} \quad \text{13–11}$$

vertical spring

A mass attached to a vertical spring causes it to stretch to a new equilibrium position. Oscillations about this new equilibrium are simple harmonic, with a period given by the previous expression.

13–5 Energy Conservation in Oscillatory Motion

In an ideal oscillatory system the total energy remains constant. This means that the kinetic and potential energy of the system vary with time in such a way that their sum remains fixed.

total energy

The total energy in simple harmonic motion is proportional to the amplitude, A, squared. For a mass on a spring, the total energy, E, is

$$E = \tfrac{1}{2}kA^2 \quad \text{13–15}$$

potential energy as a function of time

For a mass on a spring, the potential energy varies with time as follows:

$$U = \tfrac{1}{2}kA^2 \cos^2(\omega t) \quad \text{13–14}$$

kinetic energy as a function of time

Similarly, the kinetic energy as a function of time is

$$K = \tfrac{1}{2}kA^2 \sin^2(\omega t) \quad \text{13–17}$$

13–6 The Pendulum

A pendulum oscillating with small amplitude also exhibits simple harmonic motion.

simple pendulum

A simple, or ideal, pendulum is one in which all the mass is concentrated at a single point a distance L below the suspension point.

period of a simple pendulum

The period of a simple pendulum of length L is

$$T = 2\pi\sqrt{\frac{L}{g}} \quad \text{13–20}$$

physical pendulum

In a physical pendulum, the mass is distributed throughout a finite volume. Thus, a physical pendulum has a definite shape, whereas a simple pendulum is characterized by a point mass.

period of a physical pendulum

The period of a physical pendulum is

$$T = 2\pi\sqrt{\frac{\ell}{g}}\left(\sqrt{\frac{I}{m\ell^2}}\right)$$

In this expression, I is the moment of inertia about the pivot point, and ℓ is the distance from the pivot point to the center of mass.

13–7 Damped Oscillations

Systems in which mechanical energy is lost to other forms, such as heat or sound, eventually come to rest at the equilibrium position. How they move as they come to rest depends on the amount of damping.

underdamping

In an underdamped case, a system of mass m and damping constant b continues to oscillate as its amplitude steadily decreases with time. The decrease in amplitude is exponential:

$$A = A_0 e^{-bt/2m}$$

critical damping

In a system with critical damping, no oscillations occur. The system simply relaxes back to equilibrium in the least possible time.

overdamping

An overdamped system also relaxes back to equilibrium with no oscillations. The relaxation occurs more slowly in this case than in critical damping.

13–8 Driven Oscillations and Resonance

If an oscillating system is driven by an external force, it is possible for energy to be added to the system. This added energy may simply replace energy lost to friction, or, in the case of resonance, it may result in oscillations of large amplitude and energy.

natural frequency

The natural frequency of an oscillating system is the frequency at which it oscillates when free from external disturbances. For example, if a mass is allowed to oscillate freely on a spring, or a pendulum swings freely from a fixed support, they are oscillating at their natural frequencies.

resonance

Resonance is the response of an oscillating system to a driving force of the appropriate frequency. In particular, if the driving force varies with time with roughly the same frequency as a system's natural frequency, the response will be large and the system is in resonance.

Problem-Solving Summary

Type of Problem	Relevant Physical Concepts	Related Examples
Find the position, velocity, and acceleration as a function of time for an object undergoing simple harmonic motion.	In simple harmonic motion, the position, velocity, and acceleration are all sinusoidal functions of time. In particular, $x = A\cos(\omega t)$, $v = -A\omega\sin(\omega t)$, and $a = -A\omega^2\cos(\omega t)$, where $\omega = 2\pi/T$.	Examples 13–1, 13–2 Active Example 13–1
Calculate the maximum speed and acceleration for simple harmonic motion.	The maximum speed is $v_{max} = A\omega$, and the maximum acceleration is $a_{max} = A\omega^2$.	Examples 13–3, 13–4
Relate the period of a mass on a spring to the mass and the force constant.	The period of a mass m attached to a spring of force constant k is $T = 2\pi\sqrt{m/k}$.	Examples 13–4, 13–5, 13–6 Active Examples 13–2, 13–3
Relate the period of a pendulum to its length and to the acceleration of gravity.	The period of a simple pendulum of length L is $T = 2\pi\sqrt{L/g}$. Note that the period is independent of the mass, but does depend on the acceleration of gravity.	Example 13–7
Find the period of a physical pendulum.	Identify the moment of inertia, I, about the pivot point, and the distance from the pivot point to the center of mass, ℓ. Then use $T = 2\pi\sqrt{\ell/g}\,(\sqrt{I/m\ell^2})$.	Active Example 13–4

Conceptual Questions

1. A basketball player dribbles a ball with a steady period of T seconds. Is the motion of the ball periodic? Is it simple harmonic? Explain.
2. Give an example of an oscillating system that is not a simple harmonic oscillator.

3. An air-track cart bounces back and forth between the two ends of an air track. Is this motion simple harmonic? Explain.
4. A mass moves in simple harmonic motion with amplitude A and period T. How long does it take for the mass to move a distance $2A$? A distance $3A$?
5. A mass moves in simple harmonic motion with amplitude A and period T. How far does it move in the time T? In the time $5T/2$?
6. A mass on a spring oscillates with simple harmonic motion of amplitude A. If the mass is doubled, but the amplitude remains the same, does the total energy (i) increase, (ii) decrease, or (iii) stay the same? Explain.
7. A mass on a spring oscillates with simple harmonic motion of amplitude A about the equilibrium position $x = 0$. Its maximum speed is v_{max} and its maximum acceleration is a_{max}. What is the speed and acceleration of the mass at $x = 0$? At $x = A$?
8. An object executes simple harmonic motion of amplitude A and period T. How long does it take for the object to travel a distance of $6A$?
9. If the amplitude of a simple harmonic oscillator is doubled, by what factor do the following quantities change: **(a)** angular frequency, **(b)** frequency, **(c)** period, **(d)** maximum speed, **(e)** maximum acceleration, **(f)** total mechanical energy?
10. If a mass m and a mass $2m$ oscillate on identical springs with identical amplitudes, they both have the same maximum kinetic energy. How can this be? Shouldn't the larger mass have more kinetic energy? Explain.
11. An object oscillating with simple harmonic motion completes a cycle in a time T. If the object's amplitude is doubled, the time required for one cycle is still T, even though the object covers twice the distance. How can this be? Explain.
12. An object moves with simple harmonic motion. If the period of motion is doubled, how do the following quantities change: **(a)** angular frequency, **(b)** frequency, **(c)** maximum speed, **(d)** maximum acceleration, **(e)** total mechanical energy.
13. If a mass m is attached to a given spring, its period of oscillation is T. If two such springs are connected end to end and the same mass m is attached, is its period (i) more than, (ii) less than, or (iii) the same as with a single spring? Explain.
14. An old car with worn out shock absorbers oscillates with a given frequency when it hits a speed bump. If the driver adds a passenger to the car, does its frequency of oscillation (i) increase, (ii) decrease, or (iii) remain the same?
15. A grandfather clock keeps correct time at sea level. If the clock is taken to the top of a nearby mountain, would you expect it to (i) keep correct time, (ii) run slow, or (iii) run fast? Explain.
16. The pendulum bob in Figure 13–12 leaks sand onto the strip chart. What effect does this loss of sand have on the period of the pendulum? Explain.
17. A pendulum of length L has a period T. How long must the pendulum be if its period is to be $2T$?
18. Considering the leg as a physical pendulum, is the period of a short person's leg (i) greater than, (ii) less than, or (iii) the same as the period of a tall person's leg? Explain.
19. Metronomes, such as the penguin shown in the photo, are useful devices for music students. If it is desired to have the metronome tick with a greater frequency, should the penguin's bow tie be moved upward or downward? Explain.
20. A pendulum of length L is suspended from the ceiling of an elevator. When the elevator is at rest, the period of the pendulum is T. Does the period increase, decrease, or remain the same when the elevator **(a)** moves upward with constant speed or **(b)** moves downward with constant speed? Explain.

▲ How do you like my tie? Question 19

21. A pendulum of length L is suspended from the ceiling of an elevator. When the elevator is at rest the period of the pendulum is T. Does the period increase, decrease, or remain the same when the elevator **(a)** moves upward with constant acceleration or **(b)** moves downward with constant acceleration? Explain.
22. A mass m is suspended from the ceiling of an elevator by a spring of force constant k. When the elevator is at rest, the period of the mass is T. Does the period increase, decrease, or remain the same when the elevator **(a)** moves upward with constant speed or **(b)** moves downward with constant speed? Explain.
23. A mass m is suspended from the ceiling of an elevator by a spring of force constant k. When the elevator is at rest, the period of the mass is T. Does the period increase, decrease, or remain the same when the elevator **(a)** moves upward with constant acceleration or **(b)** moves downward with constant acceleration? Explain.
24. The two blocks in **Figure 13–21** have the same mass. When set into oscillation, is the period of block 1 (i) greater than, (ii) less than, or (iii) the same as the period of block 2? Explain.

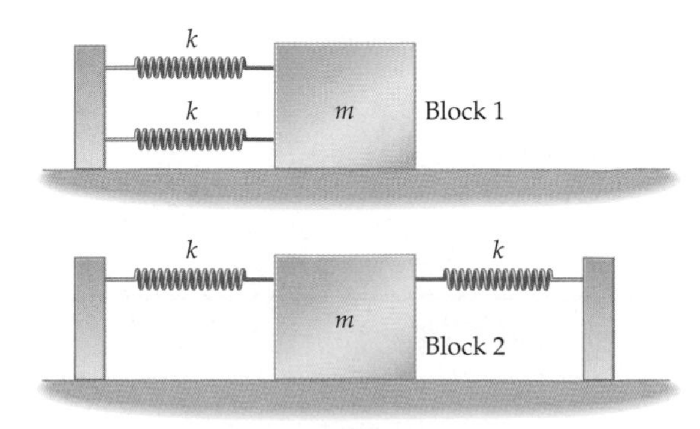

▲ **FIGURE 13–21** Question 24, Problem 28

25. A mass oscillates on a vertical spring with a period T. If this system is taken to the moon, does the period of oscillation (i) increase, (ii) decrease, or (iii) stay the same? Explain.

26. A pendulum oscillates with a period T. If this system is taken to the moon, does the period of oscillation (i) increase, (ii) decrease, or (iii) stay the same? Explain.

27. Soldiers on the march are often ordered to break cadence in their step when crossing a bridge. Why is this a good idea?

Problems

Note: **IP** *denotes an integrated conceptual/quantitative problem.* **BIO** *identifies problems of biological or medical interest. Blue bullets (•, ••, •••) are used to indicate the level of difficulty of each problem.*

Section 13–1 Periodic Motion

1. • A small cart on a 5.0-m long air track moves with a speed of 0.75 m/s. Bumpers at either end of the track cause the cart to reverse direction and maintain the same speed. Find the period and frequency of this motion.

2. • A person in a rocking chair completes 15 cycles in 23 s. What is the period and frequency of the rocking?

3. • While fishing for catfish, a fisherman suddenly notices that the bobber (a floating device) attached to his line is bobbing up and down with a frequency of 2.8 Hz. What is the period of the bobber's motion?

4. • If you dribble a basketball with a frequency of 1.88 Hz, how long does it take for you to complete 10 dribbles?

5. • You take your pulse and observe 74 heartbeats in a minute. What is the period and frequency of your heartbeat?

6. •• **IP** **(a)** Your heart beats with a frequency of 1.45 Hz. How many beats occur in a minute? **(b)** If the frequency of your heartbeat increases, will the number of beats in a minute increase, decrease, or stay the same? **(c)** How many beats occur in a minute if the frequency increases to 1.55 Hz?

7. •• You rev your car's engine to 2500 rpm (rev/min). **(a)** What is the period and frequency of the engine? **(b)** If you change the period of the engine to 0.034 s, how many rpms is it doing?

Section 13–2 Simple Harmonic Motion

8. • The position of a mass oscillating on a spring is given by $x = (5.2\text{ cm})\cos[2\pi t/(0.54\text{ s})]$. **(a)** What is the period of this motion? **(b)** When is the mass first at the position $x = 0$?

9. • The position of a mass oscillating on a spring is given by $x = (7.8\text{ cm})\cos[2\pi t/(0.78\text{ s})]$. **(a)** What is the frequency of this motion? **(b)** When is the mass first at the position $x = -7.8$ cm?

10. •• A mass oscillates on a spring with a period of 0.83 s and an amplitude of 6.4 cm. Write an equation giving x as a function of time, assuming that x starts at $x = A$ at time $t = 0$.

11. •• An atom in a molecule oscillates about its equilibrium position with a frequency of 3.00×10^{14} Hz and a maximum displacement of 3.00 nm. Write an expression for its position as a function of time.

12. •• A mass oscillates on a spring with a period T and an amplitude of 0.48 cm. Where is the mass at the times $t = 0, T/8, \ldots, T$? Plot your results with the vertical axis representing position and the horizontal axis representing time.

13. •• The position of a mass on a spring is given by $x = (6.5\text{ cm})\cos[2\pi t/(0.88\text{ s})]$. **(a)** What is the period of this motion? **(b)** Where is the mass at $t = 0.25$ s? **(c)** Show that the mass is at the same location at $0.25\text{ s} + T$ seconds as it is at 0.25 s.

14. •• A mass attached to a spring oscillates with a period of 2.16 s and an amplitude of 0.0320 m. If the mass starts at $x = 0.0320$ m at time $t = 0$, where is it at time $t = 6.37$ s?

15. •• An object moves with simple harmonic motion of period T and amplitude A. During one complete cycle, for what length of time is the position of the object greater than $A/2$?

16. •• An object moves with simple harmonic motion of period T and amplitude A. During one complete cycle, for what length of time is the speed of the object greater than $v_{max}/2$?

17. ••• An object executing simple harmonic motion has a maximum speed v_{max} and a maximum acceleration a_{max}. Find **(a)** the amplitude and **(b)** the period of this motion. Express your answers in terms of v_{max} and a_{max}.

Section 13–3 Connections Between Uniform Circular Motion and Simple Harmonic Motion

18. • A ball rolls on a circular track of radius 0.50 m with a constant angular speed of 1.3 rad/s. If the angular position of the ball at $t = 0$ is $\theta = 0$, find the x component of the ball's position at the times 2.5 s, 5.0 s, and 7.5 s.

19. • An object executing simple harmonic motion has a maximum speed of 3.3 m/s and a maximum acceleration of 0.65 m/s^2. Find **(a)** the amplitude and **(b)** the period of this motion.

20. • A child rocks back and forth on a porch swing with an amplitude of 0.204 m and a period of 2.80 s. Assuming the motion is approximately simple harmonic, find the child's maximum speed.

21. • A 30.0-g goldfinch lands on a slender branch, where it oscillates up and down with simple harmonic motion of amplitude 0.0412 m and period 1.85 s. What is the maximum acceleration of the finch? Express your answer as a fraction of the acceleration of gravity, g.

22. •• A peg on a turntable moves with a constant linear speed of 0.67 m/s in a circle of radius 0.45 m. The peg casts a shadow on a wall. Find the following quantities related to the motion of the shadow: **(a)** the period, **(b)** the amplitude, **(c)** the maximum speed, and **(d)** the maximum magnitude of the acceleration.

23. •• The pistons in an internal combustion engine undergo a motion that is approximately simple harmonic. If the amplitude of motion is 3.5 cm, and the engine runs at 1700 rev/min, find **(a)** the maximum acceleration of the pistons and **(b)** their maximum speed.

24. •• A 0.50-kg air cart is attached to a spring and allowed to oscillate. If the displacement of the air cart from equilibrium is $x = (10.0\text{ cm})\cos[(2.00\text{ s}^{-1})t + \pi]$, find **(a)** the maximum kinetic energy of the cart and **(b)** the maximum force exerted on it by the spring.

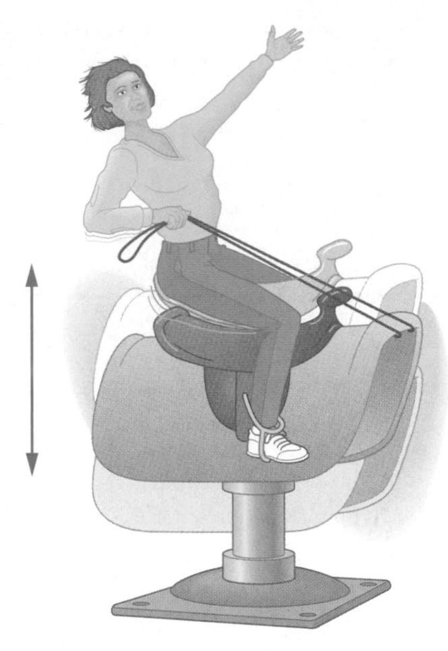

▲ **FIGURE 13–22** Problem 25

25. •• **IP** A person rides on a mechanical bucking horse (see **Figure 13–22**) that oscillates up and down with simple harmonic motion. The period of the bucking is 0.74 s, and the amplitude is slowly increasing. At a certain amplitude the rider must hang on to prevent separating from the mechanical horse. **(a)** Give a strategy that will allow you to calculate this amplitude. **(b)** Carry out your strategy and find the desired amplitude.

Section 13–4 The Period of a Mass on a Spring

26. • Show that the units of the quantity $\sqrt{k/m}$ are s^{-1}.

27. • A 0.21-kg mass attached to a spring undergoes simple harmonic motion with a period of 0.55 s. What is the force constant of the spring?

28. •• Find the periods of block 1 and block 2 in Figure 13–21, given that $k = 49.2$ N/m and $m = 1.25$ kg.

29. •• When a 0.50-kg mass is attached to a vertical spring, the spring stretches by 15 cm. How much mass must be attached to the spring to result in a 0.75 s period of oscillation?

30. •• A spring with a force constant of 65 N/m is attached to a 0.50-kg mass. Determine the following quantities for this system: **(a)** ω, **(b)** v, **(c)** T.

31. •• Two people with a combined mass of 125 kg hop into an old car with worn out shock absorbers. This causes the springs to compress by 8.00 cm. When the car hits a bump in the road it oscillates up and down with a period of 1.65 s. Find **(a)** the total load supported by the springs and **(b)** the mass of the car.

32. •• A 0.85-kg mass attached to a vertical spring of force constant 150 N/m oscillates with a maximum speed of 0.35 m/s. Find the following quantities related to the motion of the mass: **(a)** the period, **(b)** the amplitude, **(c)** the maximum magnitude of the acceleration.

33. •• When a 0.213–kg mass is attached to a vertical spring it causes the spring to stretch a distance d. If the mass is now displaced slightly from equilibrium, it is found to make 100 oscillations in 56.7 s. Find the stretch distance, d.

34. •• **IP** The springs of a 511-kg motorcycle have an effective force constant of 9130 N/m. **(a)** If a person sits on the motorcycle, does its period of oscillation increase, decrease, or stay the same? **(b)** By what percent and in what direction does the period of oscillation change when a 112-kg person rides the motorcycle?

35. ••• **IP** If a mass m is attached to a given spring, its period of oscillation is T. If two such springs are connected end to end, and the same mass m is attached, **(a)** is its period more than, less than, or the same as with a single spring? **(b)** Verify your answer to part (a) by calculating the new period, T', in terms of the old period T.

Section 13–5 Energy Conservation in Oscillatory Motion

36. • How much work is required to stretch a spring 0.175 m if its force constant is 9.87 N/m?

37. • A 0.321-kg mass is attached to a spring with a force constant of 12.3 N/m. If the mass is displaced 0.256 m from equilibrium and released, what is its speed when it is 0.128 m from equilibrium?

38. • Find the total mechanical energy of the system described in the previous problem.

39. •• A 1.2-kg mass attached to a spring oscillates with an amplitude of 7.8 cm and a frequency of 2.6 Hz. What is its energy of motion?

40. •• **IP** A 0.40-kg mass is attached to a spring with a force constant of 26 N/m and released from rest a distance of 3.2 cm from the equilibrium position of the spring. **(a)** Give a strategy that allows you to find the speed of the mass when it is halfway to the equilibrium position. **(b)** Use your strategy to find this speed.

41. •• **(a)** What is the maximum speed of the mass in the previous problem? **(b)** How far is the mass from the equilibrium position when its speed is half the maximum speed?

42. •• A bunch of grapes is placed in a spring scale at a supermarket. The grapes oscillate up and down with a period of 0.35 s, and the spring in the scale has a force constant of 650 N/m. What is **(a)** the mass and **(b)** the weight of the grapes?

43. •• What is the maximum speed of the grapes in the previous problem, if their amplitude of oscillation is 2.3 cm?

44. •• **IP** A 0.540-kg block slides on a frictionless, horizontal surface with a speed of 1.13 m/s. The block encounters an unstretched spring and compresses it 25 cm before coming to rest. **(a)** What is the force constant of this spring? **(b)** How long is the block in contact with the spring before it comes to rest? **(c)** If the force constant of the spring is increased, does the time required to stop the block increase, decrease, or stay the same? Explain.

45. •• A 10.0-g bullet embeds itself in a 0.500-kg block, which is attached to a spring of force constant 36.0 N/m. If the maximum compression of the spring is 1.50 cm, find **(a)** the initial speed of the bullet and **(b)** the time for the bullet-block system to come to rest.

Section 13–6 The Pendulum

46. • An observant fan at a baseball game notices that the radio commentators have lowered a microphone from their booth to just a few inches above the ground, as shown in **Figure 13–23**. The microphone is used to pick up sound from the field and from the fans. The fan also notices that the microphone is

▲ **FIGURE 13–23** Problem 46

slowly swinging back and forth like a simple pendulum. Using her digital watch, she finds that 10 complete oscillations take 60.0 s. How high above the field is the radio booth? (Assume the microphone and its cord can be treated as a simple pendulum.)

47. • A simple pendulum of length 2.5 m makes 5 complete swings in 16 s. What is the acceleration of gravity at the location of the pendulum?

48. • A large, simple pendulum is on display in the lobby of the United Nations building. If the pendulum is 10.0 m in length, what is the least amount of time it takes for the bob to swing from a position of maximum displacement to the equilibrium position of the pendulum? (Assume that the acceleration of gravity is $g = 9.81\ \text{m/s}^2$ at the UN building.)

49. • Find the length of a simple pendulum that has a period of 1.00 s. Assume that the acceleration of gravity is $g = 9.81\ \text{m/s}^2$.

50. •• **IP** If the pendulum in the previous problem were to be taken to the moon, where the acceleration of gravity is $g/6$, **(a)** would its period increase, decrease, or stay the same? **(b)** Check your result in part (a) by calculating the period of the pendulum on the moon.

***51.** •• A hula hoop hangs from a peg. Find the period of the hoop as it gently rocks back and forth on the peg. (For a hoop with axis at the rim $I = 2mR^2$, where R is the radius of the hoop.)

***52.** •• A fireman tosses his 0.82-kg hat onto a peg, where it oscillates as a physical pendulum (**Figure 13–24**). If the center of mass of the hat is 8.2 cm from the pivot point, and its period of oscillation is 0.75 s, what is the moment of inertia of the hat about the pivot point?

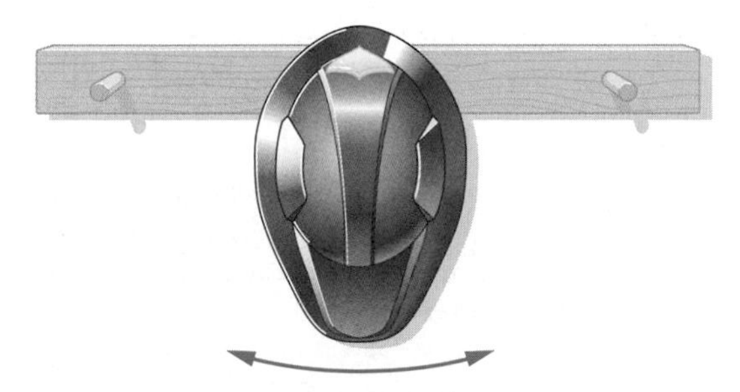

▲ **FIGURE 13–24** Problem 52

***53.** •• What must be the length of a simple pendulum if its period is to equal the period of a meter stick pivoted about one end?

***54.** •• On the construction site for a new skyscraper, a uniform beam of steel is suspended from one end. If the beam swings back and forth with a period of 2.00 s, what is its length?

***55.** •• **BIO** Find the period of a child's leg as it swings about the hip joint. Assume the leg is 0.50 m long and can be treated as a uniform rod. Estimate the child's walking speed.

56. ••• Suspended from the ceiling of an elevator is a pendulum of length L. What is the period of this pendulum if the elevator **(a)** accelerates upward with an acceleration a, or **(b)** accelerates downward with an acceleration $a < g$? Give your answer in terms of L, g, and a.

General Problems

57. • A 1.1-kg mass is attached to a spring with a force constant of 56 N/m. If the mass is released with a speed of 0.25 m/s at a distance of 8.4 cm from the equilibrium position of the spring, what is its speed when it is halfway to the equilibrium position?

58. • **BIO** An astronaut uses a Body Mass Measurement Device (BMMD) to determine her mass. What is the astronaut's mass, given that the force constant of the BMMD is 2600 N/m and the period of oscillation is 0.85 s? See the discussion on page 391 for more details on the BMMD.

59. • A typical atom in a solid might oscillate with a frequency of 10^{12} Hz and an amplitude of 0.10 angstrom (10^{-11} m). Find the maximum acceleration of the atom, and compare it with the acceleration of gravity.

60. •• A 1.24-g spider oscillates on its web, which has a damping constant of 4.30×10^{-5} kg/s. How long does it take for the spider's amplitude of oscillation to decrease by 10.0%?

61. •• An object undergoes simple harmonic motion with a period T and amplitude A. How long does it take the object to travel from $x = A$ to $x = A/2$?

***62.** •• Find the period of oscillation of a disk of mass 0.32 kg and radius 0.15 m if it is pivoted about a small hole drilled near its rim.

***63.** •• **IP** A small wad of putty of mass 8.5 g is placed at the lowest point of the disk in the previous problem. **(a)** Does the period of oscillation increase, decrease, or stay the same? **(b)** Check your answer to part (a) by calculating the new period of oscillation.

64. •• When a mass m is attached to a spring with a force constant 96 N/m, it stretches the spring by the amount 12 cm. Compare the period of this mass with the period of a simple pendulum of length 12 cm.

65. •• Calculate the ratio of the kinetic energy to the potential energy of a simple harmonic oscillator when its displacement is half its amplitude.

66. •• A 0.30-kg mass slides on a frictionless floor with a speed of 1.34 m/s. The mass strikes and compresses a spring with a force constant of 45 N/m. **(a)** How far does the mass travel after contacting the spring? **(b)** How long does it take for the spring to stop the mass?

67. •• A large rectangular barge floating on a lake oscillates up and down with a period of 4.5 s. Find the damping constant for the barge, given that its mass is 2.44×10^5 kg and that its amplitude of oscillation decreases by a factor of 2 in 5.0 minutes.

68. •• **Figure 13–25** shows a displacement versus time graph of the periodic motion of a 3.0-kg mass on a spring. How much energy is stored in this system?

69. •• A 0.86-kg crow lands on a slender branch and bobs up and down with a period of 1.1 s. An eagle flies up to the same branch, scaring the crow away, and lands. The eagle now bobs up and

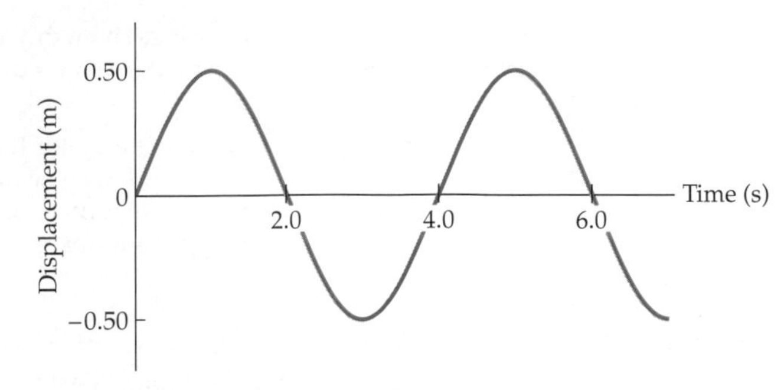

▲ **FIGURE 13–25** Problem 68

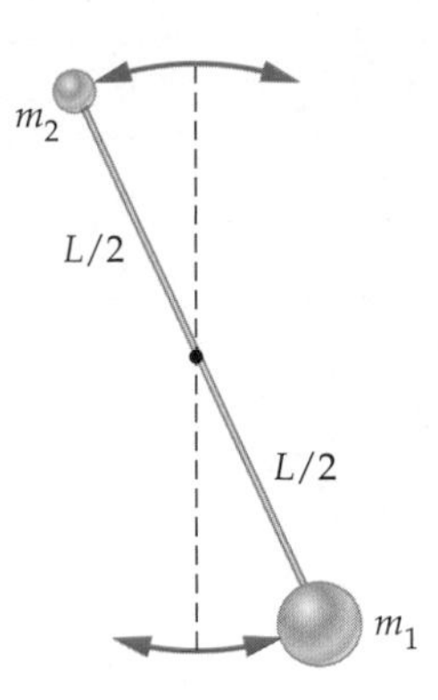

▲ **FIGURE 13–27** Problem 72

down with a period of 0.52 s. Treating the branch as an ideal spring, find **(a)** the effective force constant of the branch and **(b)** the mass of the eagle.

70. ••• Two identical masses connected by a light string are suspended from the bottom of a vertical spring and are undergoing simple harmonic vertical motion of amplitude A. At the instant when the acceleration of the masses is a maximum in the upward direction the spring breaks, allowing the lower mass to drop to the floor. Find the resulting amplitude of motion of the remaining mass.

71. ••• **IP** Consider the pendulum shown in **Figure 13–26**. Note that the pendulum's string is stopped by a peg when the bob swings to the left, but moves freely when the bob swings to the right. **(a)** Is the period of this pendulum greater than, less than, or the same as the period of the same pendulum without the peg?

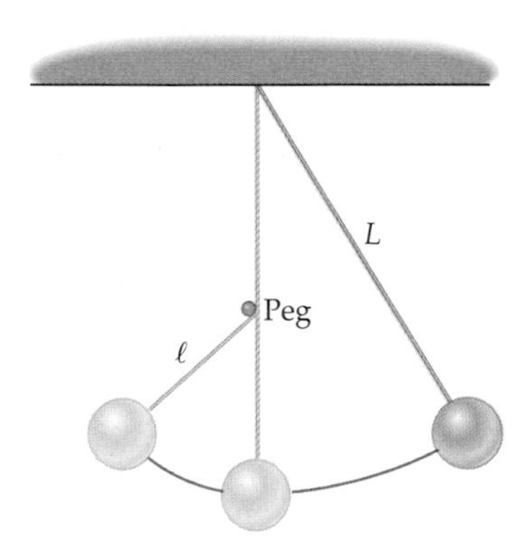

▲ **FIGURE 13–26** Problem 71

(b) Calculate the period of this pendulum in terms of L and ℓ. **(c)** Evaluate your result for $L = 1.0$ m and $\ell = 0.25$ m.

***72.** ••• A physical pendulum consists of a light rod of length L suspended in the middle. A large mass m_1 is attached to one end of the rod, and a lighter mass m_2 is attached to the other end, as illustrated in **Figure 13–27**. Find the period of oscillation for this pendulum.

73. ••• A vertical, hollow tube is connected to a speaker, which vibrates vertically with simple harmonic motion (**Figure 13–28**). The speaker operates with constant amplitude, A, but variable frequency, f. A slender pencil is placed inside the tube. At low frequency the pencil stays in contact with the speaker. At higher frequency the pencil begins to rattle. Find an expression for the frequency at which the rattling begins.

▲ **FIGURE 13–28** Problem 73

74. ••• When a mass m is attached to a spring with a force constant k it stretches the spring by the amount L. Calculate both the period of this mass and the period of a simple pendulum of length L.

75. ••• An object undergoes simple harmonic motion of amplitude A and angular frequency ω about the equilibrium point $x = 0$. Use energy conservation to show that the speed of the object at the general position x is given by the following expression:

$$v = \omega\sqrt{A^2 - x^2}$$

14 Waves and Sound

Have you ever wondered why a grand piano has this somewhat peculiar shape? It's not just tradition—there's also a physical reason, having to do with the way vibrating strings produce sound. But to understand this and other aspects of sound, it is first necessary to learn about waves in general—for sound, as we shall see, is merely a particular kind of wave, though one that has a special importance in our lives.

In the last chapter, we studied the behavior of an oscillator. Here, we consider the behavior of a series of oscillators that are connected to one another. Connecting oscillators leads to an assortment of new phenomena, including waves on a string, water waves, and sound. In this chapter, we focus our attention on the behavior of such waves, and in particular on the way they propagate, their speed of propagation, and their interactions with one another. Later, in Chapter 25, we shall see that light is also a type of wave, and that it displays many of the same phenomena exhibited by the waves considered in this chapter.

14–1 Types of Waves

Consider a group of swings in a playground swing set. We know that each swing by itself behaves like a simple pendulum; that is, like an oscillator. Now, let's connect the swings to one another. To be specific, suppose we tie a rope from the seat of the first swing to its neighbor, and then another rope from the second swing to the third swing, and so on. When the swings are at rest—in equilibrium—the connecting ropes have no effect. If you now sit in the first swing and begin oscillating—thus "disturbing" the equilibrium—the connecting ropes cause the other swings along the line to start oscillating as well. You have created a traveling disturbance.

In general, a disturbance that propagates from one place to another is referred to as a **wave**. Waves propagate with well-defined speeds determined by the properties of the material through which they travel. In addition, waves carry energy. For example, part of the energy you put into sound waves when you speak is carried to the ears of others, where some of the sound energy is converted into electrical energy carried by nerve impulses to the brain which, in turn, creates the sensation of hearing.

It is important to distinguish between the motion of the wave itself and the motion of the individual particles that make up the wave. Common examples include the waves that propagate through a field of wheat. The individual wheat stalks sway back and forth as a wave passes, but they do not change their location. Similarly, a "wave" at a ball game may propagate around the stadium more quickly than a person can run, but the individual people making up the wave simply stand and sit in one place. From these simple examples it is clear that waves can come in a variety of types. We discuss some of the more common types in this section. In addition, we show how the speed of a wave is related to some of its basic properties.

▲ A wave can be viewed as a disturbance that propagates through space. Although the wave itself moves steadily in one direction, the particles that create the wave do not share in this motion. Instead, they oscillate back and forth about their equilibrium positions. The water in an ocean wave, for example, moves mainly up and down—as it passes, you bob up and down with it rather than being carried onto the shore. Similarly, the people in a human "wave" at a ballpark simply stand or raise their arms in place—they do not travel around the stadium.

Transverse Waves

Perhaps the easiest type of wave to visualize is a wave on a string, as illustrated in **Figure 14–1**. To generate such a wave, start with a long string or rope tied to a wall at one end. Pull on the free end with your hand, producing a tension in the string, and then move your hand up and down. As you do so, a wave will travel along the string toward the wall. In fact, if your hand moves up and down with simple harmonic motion, the wave on the string will have the shape of a sine or a cosine; we refer to such a wave as a **harmonic wave**.

Note that the wave travels in the horizontal direction, even though your hand oscillates vertically about one spot. In fact, if you look at *any* point on the string, it too moves vertically up and down, with no horizontal motion at all. This is shown in **Figure 14–2**, where we see the location of an individual point on a string as a wave travels past. Notice, in particular, that the displacement of particles in a string is at right angles to the direction of propagation of the wave. A wave with this property is called a **transverse wave**:

> In a transverse wave, the displacement of individual particles is at *right angles* to the direction of propagation of the wave.

Other examples of transverse waves include light and radio waves. These will be discussed in detail in Chapter 25.

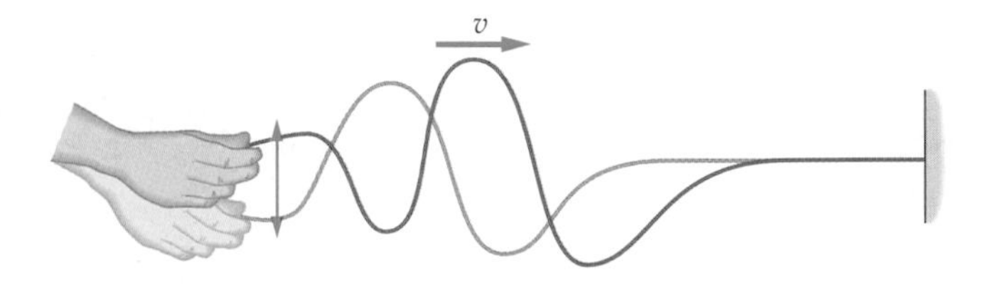

▶ **FIGURE 14–1 A wave on a string**

A wave on a string can be generated by vibrating one end of the string with an up-and-down motion.

Longitudinal Waves

Longitudinal waves differ from transverse waves in the way that particles in the wave move. In particular, a longitudinal wave is defined as follows:

> In a longitudinal wave, the displacement of individual particles is in the *same direction* as the direction of propagation of the wave.

The classic example of a longitudinal wave is sound. When you speak, for example, the vibrations in your vocal cords create a series of compressions and expansions (rarefactions) in the air. The same kind of situation occurs with a loudspeaker, as illustrated in **Figure 14–3**. Here we see a speaker diaphragm vibrating horizontally with simple harmonic motion. As it moves to the right it compresses the air momentarily; as it moves to the left it rarefies the air. A series of compressions and rarefactions then travel horizontally away from the loudspeaker with the speed of sound.

Figure 14–3 also indicates the motion of an individual particle in the air as a sound wave passes. Note that the particle moves back and forth horizontally; that is, in the same direction as the propagation of the wave. The particle does not travel with the wave—each individual particle simply oscillates about a given position in space.

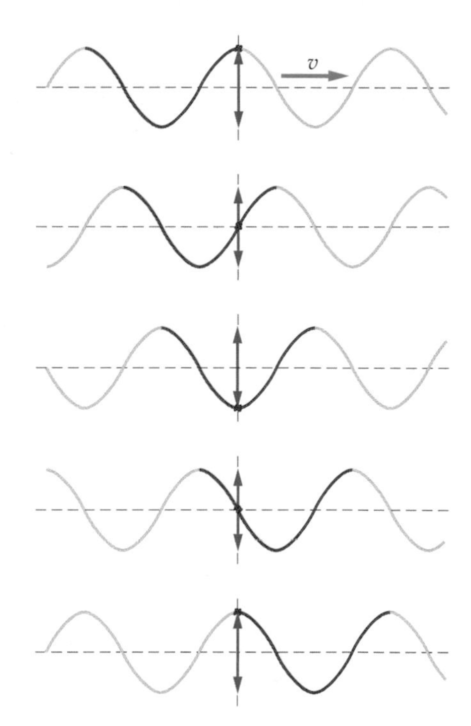

▲ **FIGURE 14–2 The motion of a wave on a string**

As a wave on a string moves horizontally, all points on the string vibrate in the vertical direction.

Water Waves

If a pebble is dropped into a pool of water, a series of concentric waves move away from the drop point. This is illustrated in **Figure 14–4**. To visualize the movement of the water as a wave travels by, place a small piece of cork into the water. As a wave passes, the motion of the cork will trace out the motion of the water itself, as indicated in **Figure 14–5**.

Notice that the cork moves in a roughly circular path, returning to approximately its starting point. Thus, each element of water moves both vertically and horizontally as the wave propagates by in the horizontal direction. In this sense, a water wave is a combination of both transverse and longitudinal waves. This makes the water wave more difficult to analyze. Hence, most of our results will refer to the simpler cases of purely transverse and purely longitudinal waves.

Wavelength, Frequency, and Speed

A simple wave can be thought of as a regular, rhythmic disturbance that propagates from one point to another, repeating itself both in *space* and in *time*. We now show that the repeat length and the repeat time of a wave are directly related to its speed of propagation.

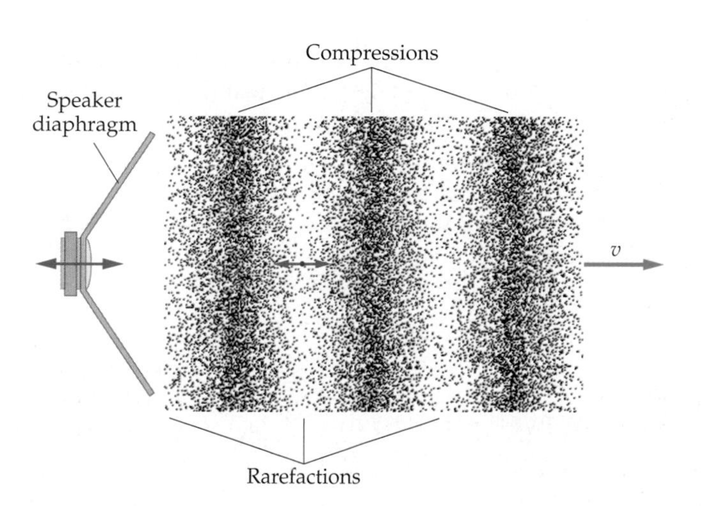

◀ **FIGURE 14–3 Sound produced by a speaker**

As the diaphragm of a speaker vibrates back and forth it alternately compresses and rarefies the surrounding air. These regions of high and low density propagate away from the speaker with the speed of sound.

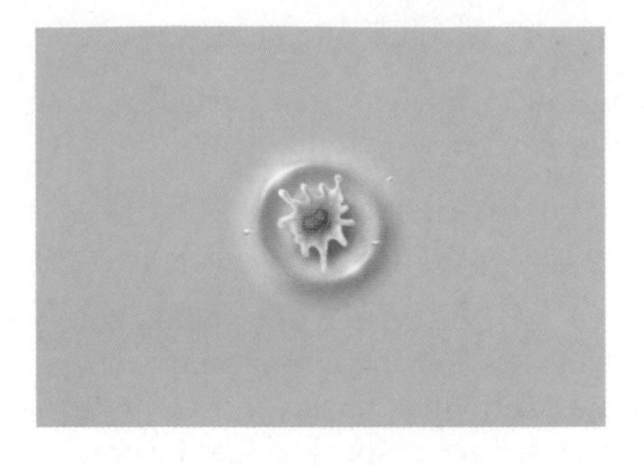

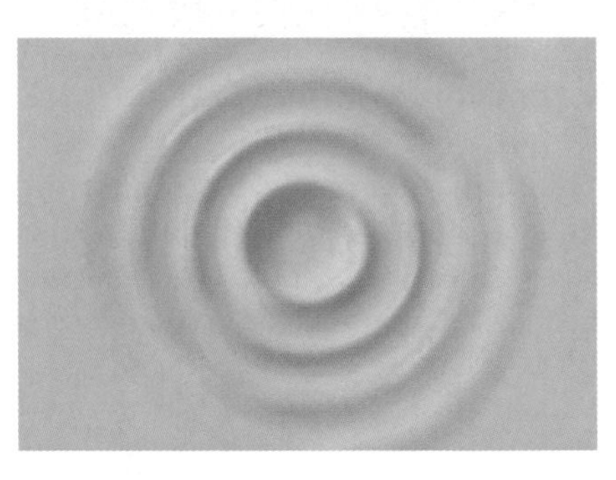

▲ **FIGURE 14–4 Water waves from a disturbance**

An isolated disturbance in a pool of water, caused by a pebble dropped into the water, creates waves that propagate symmetrically away from the disturbance. The crests and troughs form circles on the surface of the water as they move outward.

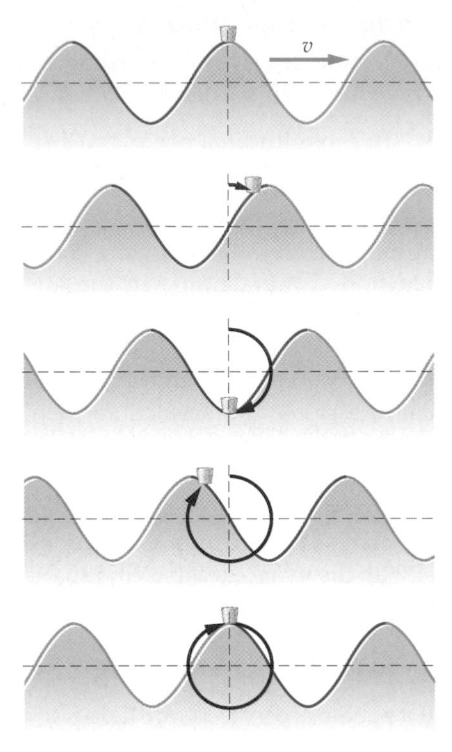

▲ **FIGURE 14–5 The motion of a water wave**

As a water wave passes a given point, a molecule (or a small piece of cork) moves in a roughly circular path. This means that the water molecules move both vertically and horizontally. In this sense, the water wave has characteristics of both transverse and longitudinal waves.

We begin by considering the snapshots of a wave shown in **Figure 14–6**. Points on the wave corresponding to maximum upward displacement are referred to as **crests**; points corresponding to maximum downward displacement are called **troughs**. The distance from one crest to the next, or from one trough to the next, is the repeat length—or **wavelength**, λ—of the wave.

Definition of Wavelength, λ

λ = distance over which a wave repeats

SI unit: m

Similarly, the repeat time—or **period**, T—of a wave is the time required for one wavelength to pass a given point, as illustrated in Figure 14–6. Closely related to the period of a wave is its **frequency**, f, which, as with oscillations, is defined by the relation $f = 1/T$.

Combining these observations, we see that a wave travels a distance λ in the time T. Applying the definition of speed—distance divided by time—it follows that the speed of a wave is

Speed of a Wave

$$v = \frac{\text{distance}}{\text{time}} = \frac{\lambda}{T} = \lambda f \qquad \text{14–1}$$

SI unit: m/s

This result applies to all waves.

EXERCISE 14–1

Sound waves travel in air with a speed of 343 m/s. The lowest frequency sound we can hear is 20.0 Hz; the highest frequency is 20.0 kHz. Find the wavelength of sound for frequencies of 20.0 Hz and 20.0 kHz.

Solution

Solve Equation 14–1 for λ:

$$\lambda = \frac{v}{f} = \frac{343\ \text{m/s}}{20.0\ \text{s}^{-1}} = 17.2\ \text{m}$$

$$\lambda = \frac{v}{f} = \frac{343\ \text{m/s}}{20{,}000\ \text{s}^{-1}} = 1.72\ \text{cm}$$

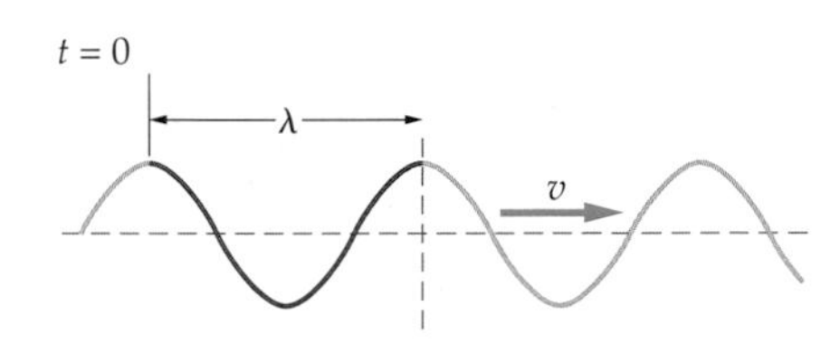

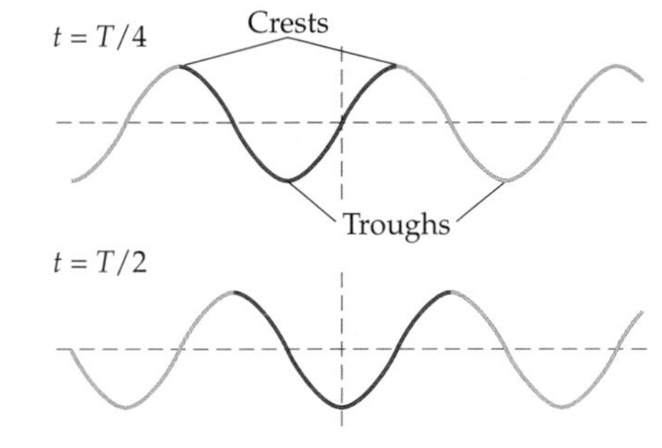

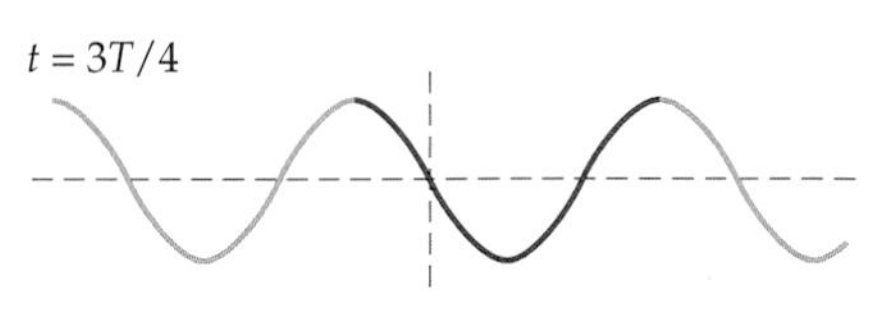

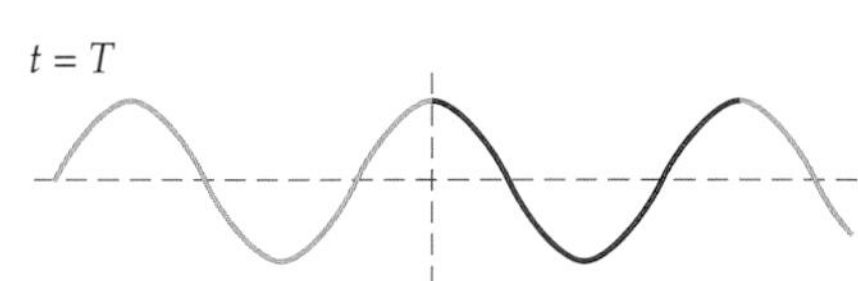

▲ **FIGURE 14–6 The speed of a wave**
A wave repeats over a distance equal to the wavelength, λ. The time necessary for a wave to move one wavelength is the period, T. Thus, the speed of a wave is $v = \lambda/T = \lambda f$.

14–2 Waves on a String

In this section we consider some of the basic properties of waves traveling on a string, a rope, a wire, or any similar linear medium.

The Speed of a Wave on a String

The speed of a wave is determined by the properties of the medium through which it propagates. In the case of a string of length L, there are two basic characteristics that determine the speed of a wave: (i) the tension in the string, and (ii) the mass of the string.

Let's begin with the tension, which is the force F transmitted through the string (we will use F for the tension rather than T to avoid confusion between the tension and the period). Clearly, there must be a tension in a string in order for it to propagate a wave. Imagine, for example, that a string lies on a smooth floor with both ends free. If you take one end into your hand and shake it, the portions of the string near your hand will oscillate slightly, but no wave will travel to the other end of the string. If someone else takes hold of the other end of the string and pulls enough to set up a tension, then any movement you make on your end will propagate to the other end. In fact, if the tension is increased—so that the string becomes less slack—waves will travel through the string more rapidly.

Next, we consider the mass m of the string. A heavy string responds slowly to a given disturbance because of its inertia. Thus, if you try sending a wave through a kite string or a large rope, both under the same tension, you will find that the wave in the rope travels more slowly. Thus, the heavier a rope or string the slower the speed of waves in it. Of course, the total mass of a string doesn't really matter; a longer string has more mass, but its other properties are basically the same. What is important is the mass of the string per length. We give this quantity the label μ:

Definition of Mass per Length, μ

μ = mass per length = m/L

SI unit: kg/m

Thus, we expect the speed v to increase with the tension F and decrease with the mass per length, μ. Assuming these are the only factors determining the speed of a wave on a string, we can obtain the dependence of v on F and μ using dimensional analysis (see Chapter 1, Section 3). First, we identify the dimensions of v, F, and μ:

$$[v] = \text{m/s}$$

$$[F] = \text{N} = \text{kg}\cdot\text{m/s}^2$$

$$[\mu] = \text{kg/m}$$

Next, we seek a combination of F and μ that has the dimensions of v; namely, m/s. Suppose, for example, that v depends on F to the power a and μ to the power b. Then, we have

$$v = F^a \mu^b$$

In terms of dimensions, this equation is

$$\text{m/s} = (\text{kg}\cdot\text{m/s}^2)^a(\text{kg/m})^b = \text{kg}^{a+b}\text{m}^{a-b}\text{s}^{-2a}$$

Comparing dimensions, we see that kg does not appear on the left side of the equation; therefore, we conclude that $a + b = 0$ so that kg does not appear on the right side of the equation. Hence, $a = -b$. Looking at the time dimension, s, we see that on the left we have s^{-1}; thus on the right side we must have $-2a = -1$, or $a = \frac{1}{2}$. It follows that $b = -a = -\frac{1}{2}$. This gives the following result:

Speed of a Wave on a String, *v*

$$v = \sqrt{\frac{F}{\mu}} \qquad \text{14–2}$$

SI unit: m/s

As expected, the speed increases with F and decreases with μ.

Dimensional analysis does not guarantee that this is the complete, final result; there could be a dimensionless factor like $\frac{1}{2}$ or 2π left unaccounted for. It turns out, however, that a complete analysis based on Newton's laws gives precisely the same result.

PROBLEM SOLVING NOTE
Mass Versus Mass-per-Length

To find the mass of a string, multiply its mass per length, μ, by its length L. That is, $m = \mu L$.

EXERCISE 14–2

A 5.0-m length of rope, with a mass of 0.52 kg, is pulled taut with a tension of 46 N. Find the speed of waves on the rope.

Solution

First, calculate the mass per length, μ:

$$\mu = m/L = (0.52\text{ kg})/(5.0\text{ m}) = 0.10\text{ kg/m}$$

Now, substitute μ and F into Equation 14–2:

$$v = \sqrt{\frac{F}{\mu}} = \sqrt{\frac{46\text{ N}}{0.10\text{ kg/m}}} = 21\text{ m/s}$$

EXAMPLE 14–1 A Wave on a Rope

A 12-m rope is pulled tight with a tension of 92 N. When one end of the rope is given a "thunk" it takes 0.45 s for the disturbance to propagate to the other end. What is the mass of the rope?

Picture the Problem

Our sketch shows the thunk propagating with a speed v from one end of the rope to the other.

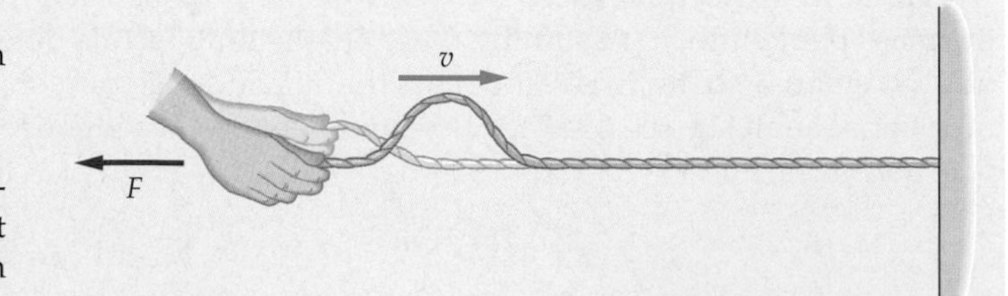

Strategy

We know that the speed of waves (disturbances) on a rope is determined by the tension and the mass per length. Thus, we first calculate the speed of the wave with the information given in the problem statement. Next, we solve for the mass per length, then multiply by the length to get the mass.

Solution

1. Calculate the speed of the thunk: $v = \frac{d}{t} = \frac{12\text{ m}}{0.45\text{ s}} = 27\text{ m/s}$

2. Use $v = \sqrt{F/\mu}$ to solve for the mass per length: $\mu = F/v^2$

3. Substitute numerical values for F and v: $\mu = \frac{F}{v^2} = \frac{92\text{ N}}{(27\text{ m/s})^2} = 0.13\text{ kg/m}$

4. Multiply μ by $L = 12$ m to find the mass: $m = \mu L = (0.13\text{ kg/m})(12\text{ m}) = 1.6\text{ kg}$

Insight

Note that the speed of a wave on this rope is about the same as the speed of a car on a highway.

Practice Problem

If the tension in this rope is doubled, how long will it take for the thunk to travel from one end to the other? [**Answer:** In this case the wave speed is $v = 38$ m/s; hence the time is $t = 0.32$ s.]

Some related homework problems: Problem 8, Problem 9

In the following Conceptual Checkpoint, we consider the speed of a wave on a vertical rope of finite mass.

CONCEPTUAL CHECKPOINT 14–1

A rope of length L and mass M hangs from a ceiling. If the bottom of the rope is given a gentle wiggle, a wave will travel to the top of the rope. As the wave travels upward does its speed **(a)** increase, **(b)** decrease, or **(c)** stay the same?

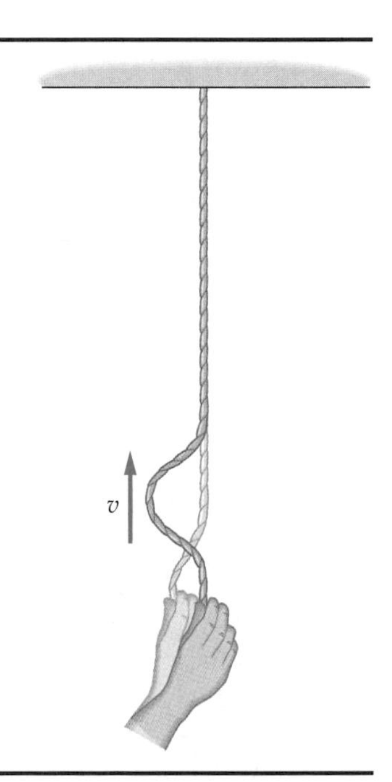

Solution

The speed of the wave is determined by the tension in the rope and its mass per length. The mass per length is the same from bottom to top, but not the tension. In particular, the tension at any point in the rope is equal to the weight of rope below that point. Thus, the tension increases from almost zero near the bottom to essentially Mg near the top. Since the tension increases with height, so too does the speed, according to Equation 14–2.

Answer:

(a) The speed increases.

Reflections

Thus far we have discussed only the situation in which a wave travels along a string; but at some point the wave must reach the end of the string. What happens then? Clearly, we expect the wave to be reflected, but the precise way in which the reflection occurs needs to be considered.

Suppose, for example, that the far end of a string is anchored firmly into a wall, as shown in **Figure 14–7**. If you give a flick to your end of the string you set up a wave "pulse" that travels toward the far end. When it reaches the end it exerts an upward force on the wall, trying to pull it up into the pulse. Since the end is tied down, however, the wall exerts an equal and opposite downward force to keep the end at rest. Thus, the wall exerts a downward force on the string that is just the *opposite* of the upward force you exerted when you created the pulse. As a result,

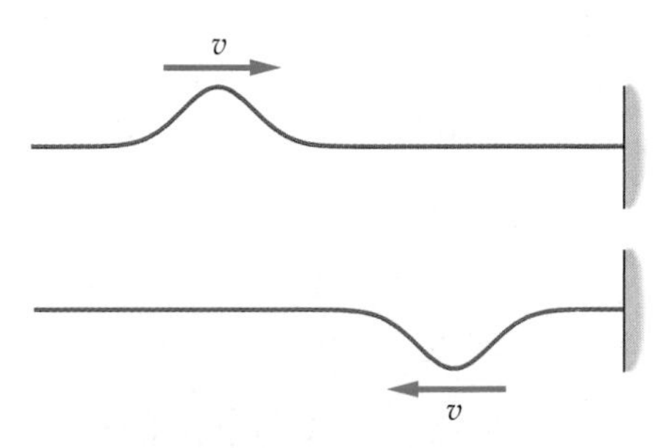

▲ **FIGURE 14–7 A reflected wave pulse: fixed end**
A wave pulse on a string is inverted when it reflects from an end that is tied down.

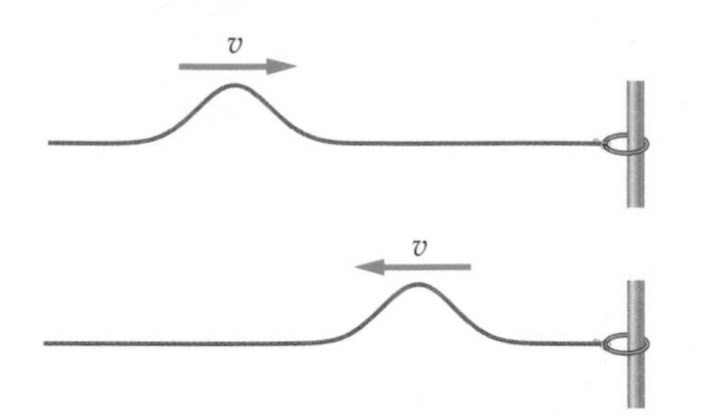

▲ **FIGURE 14–8 A reflected wave pulse: free end**

A wave pulse on a string whose end is free to move is reflected without inversion.

the reflection is an inverted, or upside-down, pulse, as indicated in Figure 14–7. We shall encounter this same type of inversion under reflection when we consider the reflection of light in Chapter 28.

Another way to tie off the end of the string is shown in **Figure 14–8**. In this case, the string is tied to a small ring that slides vertically without friction on a vertical pole. In this way, the string still has a tension in it, since it pulls on the ring, but it is also free to move up and down.

Consider a pulse moving along such a string, as in Figure 14–8. When the pulse reaches the end it lifts the ring upward and then lowers it back down. In fact, the pulse flicks the far end of the string in the *same* way that you flicked it when you created the pulse. Therefore, the far end of the string simply creates a new pulse, identical to the first, only traveling in the opposite direction. This is illustrated in the figure.

Thus, when waves reflect they may or may not be inverted, depending on how the reflection occurs.

*14–3 Harmonic Wave Functions

If a wave is generated by oscillating one end of a string with simple harmonic motion, the waves will have the shape of a sine or a cosine. This is shown in **Figure 14–9**, where the y direction denotes the vertical displacement of the string, and $y = 0$ corresponds to the flat string with no wave present. In what follows, we consider the mathematical formula that describes y as a function of time, t, and position, x, for such a harmonic wave.

First, note that the harmonic wave in Figure 14–9 repeats when x increases by an amount equal to the wavelength, λ. Thus, the dependence of the wave on x must be of the form

$$y(x) = A\cos\left(\frac{2\pi}{\lambda}x\right) \qquad \textbf{14–3}$$

To see that this is the correct dependence, note that replacing x with $x + \lambda$ gives the same value for y:

$$y(x+\lambda) = A\cos\left[\frac{2\pi}{\lambda}(x+\lambda)\right] = A\cos\left(\frac{2\pi}{\lambda}x + 2\pi\right) = A\cos\left(\frac{2\pi}{\lambda}x\right) = y(x)$$

It follows that Equation 14–3 describes a vertical displacement that repeats with a wavelength λ, as desired for a wave.

This is only part of the "wave function, " however, since we have not yet described the way the wave changes with time. This is illustrated in Figure 14–9, where we see a harmonic wave at time $t = 0$, $t = T/4$, $t = T/2$, $t = 3T/4$, and $t = T$. Note that the peak in the wave that was originally at $x = 0$ at $t = 0$ moves to $x = \lambda/4$, $x = \lambda/2$, $x = 3\lambda/4$, and $x = \lambda$ for the times just given. Thus, the position x of this peak can be written as follows:

$$x = \lambda\frac{t}{T}$$

Equivalently, we can say that the peak that was at $x = 0$ is now at the location given by

$$x - \lambda\frac{t}{T} = 0$$

Similarly, the peak that was originally at $x = \lambda$ at $t = 0$ is at the following position at the general time t:

$$x - \lambda\frac{t}{T} = \lambda$$

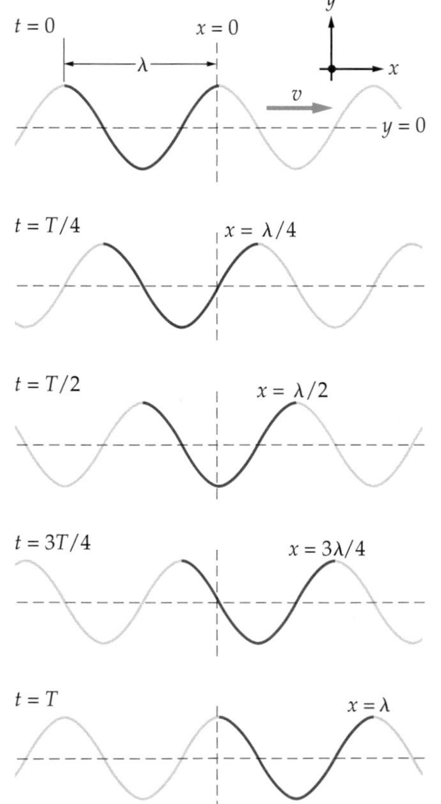
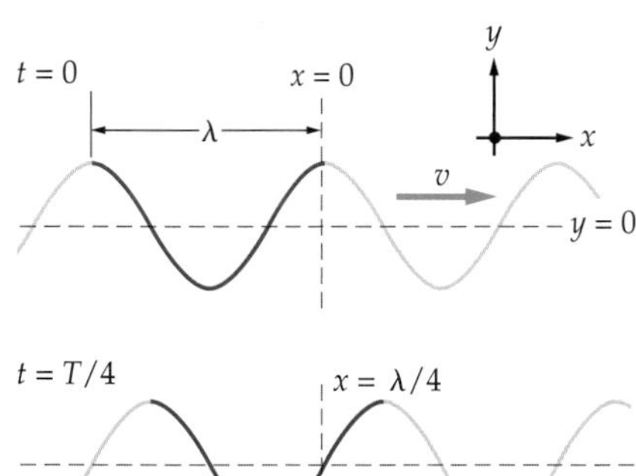
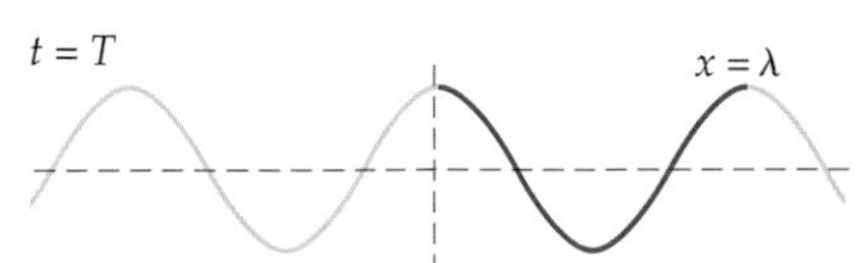

▲ **FIGURE 14–9 A harmonic wave moving to the right**

As a wave moves, the peak that was at $x = 0$ at time $t = 0$ moves to the position $x = \lambda t/T$ at the time t.

In general, if the position of a given point on a wave at $t = 0$ is $x(0)$, and its position at the time t is $x(t)$, the relation between these positions is $x(t) - \lambda t/T = x(0)$. Therefore, to take into account the time dependence of a wave, we replace $x = x(0)$ in Equation 14–3 with $x(0) = x - \lambda t/T$. This yields the harmonic wave function:

$$y(x, t) = A\cos\left[\frac{2\pi}{\lambda}\left(x - \lambda\frac{t}{T}\right)\right] = A\cos\left(\frac{2\pi}{\lambda}x - \frac{2\pi}{T}t\right) \qquad \text{14–4}$$

Note that the wave function, $y(x, t)$, depends on both time and position, and that the wave repeats whenever position increases by the wavelength, λ, or time increases by the period, T.

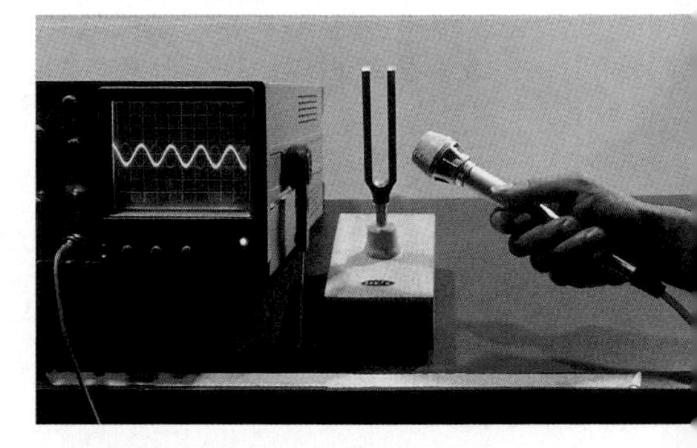

▲ An oscilloscope connected to a microphone can be used to display the wave form of a pure tone, created here by a tuning fork. The trace on the screen shows that the wave form is sinusoidal.

14–4 Sound Waves

The first thing we do when we come into this world is make a sound. It is many years later before we realize that sound is a wave propagating through the air at a speed of about 770 mi/h. More years are required to gain an understanding of the physics of a sound wave.

A useful mechanical model of a sound wave is provided by a slinky. If we oscillate one end of a slinky back and forth horizontally, as in **Figure 14–10**, we send out a longitudinal wave that also travels in the horizontal direction. The wave consists of regions where the coils of the slinky are compressed alternating with regions where the coils are more widely spaced.

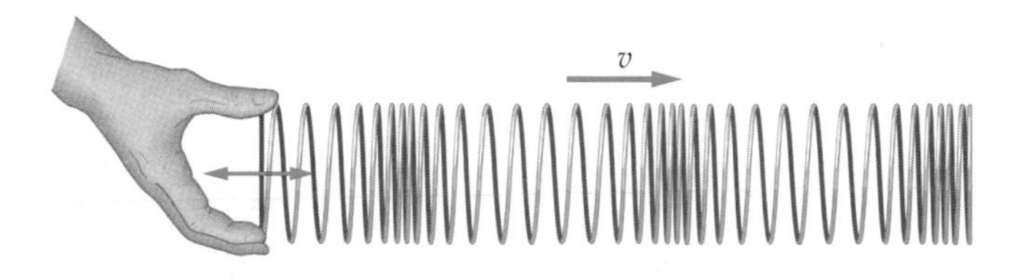

◀ **FIGURE 14–10 A wave on a slinky**
If one end of a slinky is oscillated back and forth, a series of longitudinal waves are produced. These slinky waves are analogous to sound waves.

In close analogy with the slinky model, a speaker produces sound waves by oscillating a diaphragm back and forth horizontally, as we saw in Figure 14–3. Just as with the slinky, a wave travels away from the source horizontally. The wave consists of compressed regions alternating with rarefied regions.

At first glance, the sound wave seems very different from the wave on a string. In particular, the sound wave doesn't seem to have the nice, sinusoidal shape of a wave. Certainly, Figure 14–3 gives no hint of such a wave-like shape.

If we plot the appropriate quantities, however, the classic wave shape emerges. For example, in **Figure 14–11 (a)** we plot the rarefactions and compressions of a typical sound wave, while in **Figure 14–11 (b)** we plot the fluctuations in the density of the air versus x. Clearly, the density oscillates in a wave-like fashion. Similarly, **Figure 14–11 (c)** shows a plot of the fluctuations in the pressure of the air as a function of x. In regions where the density is high the pressure is also high, and where the density is low, the pressure is low. Thus, pressure versus position again shows that a sound wave has the usual wave-like properties.

Just like the speed of a wave on a string, the speed of sound is determined by the properties of the medium through which it propagates. In air, under normal atmospheric pressure and temperature, the speed of sound is approximately the following:

Speed of Sound in Air

$v = 343\ \text{m/s} \approx 770\ \text{mi/h}$

SI unit: m/s

When we refer to the speed of sound in this text we will always assume the value is 343 m/s, unless stated specifically otherwise.

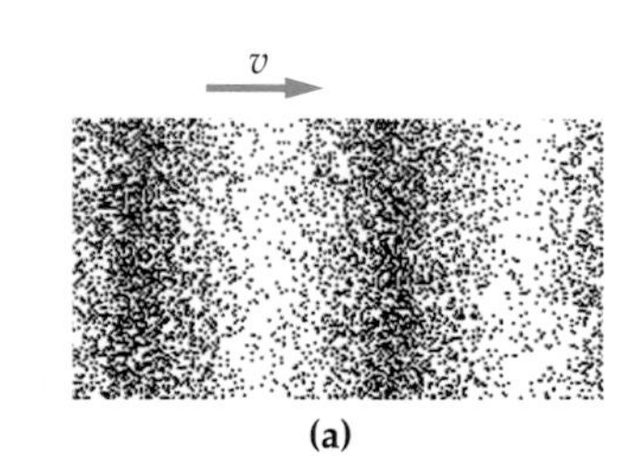

(a)

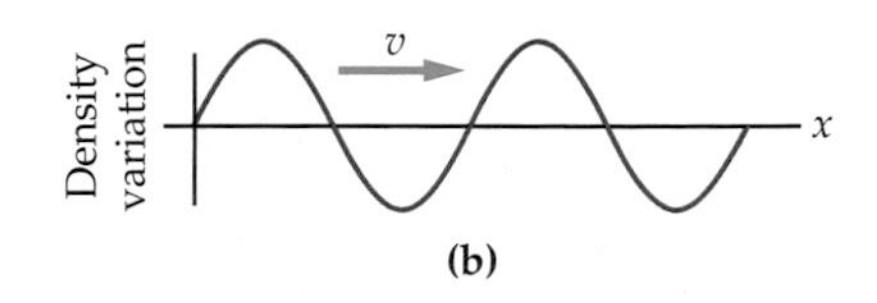

(b)

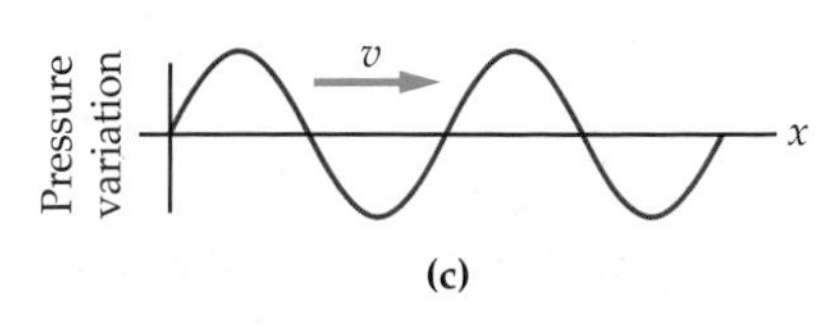

(c)

▲ **FIGURE 14–11 Wave properties of sound**
A sound wave moving through the air **(a)** produces a wave-like disturbance in the **(b)** density and **(c)** pressure of the air.

TABLE 14–1
Speed of Sound in Various Materials

Material	Speed (m/s)
Aluminum	6420
Granite	6000
Steel	5960
Pyrex glass	5640
Copper	5010
Plastic	2680
Fresh water (20 °C)	1482
Fresh water (0 °C)	1402
Hydrogen (0 °C)	1284
Helium (0 °C)	965
Air (20 °C)	343
Air (0 °C)	331

As we shall see in Chapter 17, where we study the kinetic theory of gases, the speed of sound in air is directly related to the speed of the molecules themselves. Did you know, for example, that the air molecules colliding with your body at this moment have speeds that are essentially the speed of sound? As the air is heated the molecules will move faster, and hence the speed of sound also increases with temperature.

In a solid, the speed of sound is determined in part by how stiff the material is. The stiffer the material, the faster the sound wave, just as having more tension in a string causes a faster wave. Thus the speed of sound in plastic is rather large (2680 m/s), and in steel it is greater still (5960 m/s). Both speeds are much higher than the speed in air, which is certainly a "squishy" material in comparison. Table 14–1 gives a sampling of sound speed in a range of different materials.

CONCEPTUAL CHECKPOINT 14–2

Five seconds after a brilliant flash of lightning, thunder shakes the house. Was the lightning **(a)** about a mile away, **(b)** much closer than a mile, or **(c)** much farther away than a mile?

Solution

As mentioned, the speed of sound is 343 m/s, which is just over 1,000 ft/s. Thus, in five seconds sound travels slightly more than one mile. This gives rise to the following popular rule of thumb: The distance to a lightning strike (in miles) is the time for the thunder to arrive (in seconds) divided by 5.

Notice that we have neglected the travel time of light in our discussion. This is because light propagates with such a large speed (approximately 186,000 mi/s) that its travel time is about a million times less than that of sound.

Answer:

(a) The lightning was about a mile away.

EXAMPLE 14–2 **Wishing Well**

You drop a stone into a well that is 7.35 m deep. How long does it take before you hear the splash?

Picture the Problem

Our sketch shows a well, with the depth indicated by d.

Strategy

The time until the splash is heard is the sum of (i) the time for the stone to drop a distance d, and (ii) the time for sound to travel a distance d.

For the time of drop we use kinematics with an acceleration g. Thus $d = \frac{1}{2}gt_1^2$, with $g = 9.81\ \text{m/s}^2$.

For the sound wave, we use $d = vt_2$, with $v = 343$ m/s.

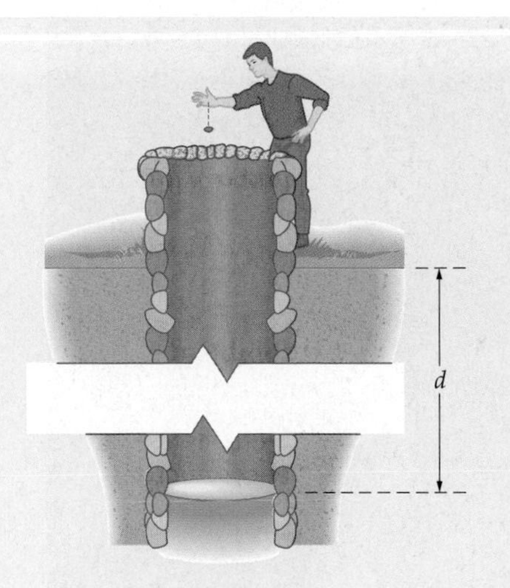

Solution

1. Calculate the time for the stone to drop:

$$d = \tfrac{1}{2}gt_1^2$$

$$t_1 = \sqrt{\frac{2d}{g}} = \sqrt{\frac{2(7.35\text{ m})}{9.81\text{ m/s}^2}} = 1.22\text{ s}$$

2. Calculate the time for sound to travel a distance d:

$$d = vt_2$$

$$t_2 = \frac{d}{v} = \frac{7.35\text{ m}}{343\text{ m/s}} = 0.0214\text{ s}$$

3. Sum the times found above:

$$t = t_1 + t_2 = 1.22\text{ s} + 0.0214\text{ s} = 1.24\text{ s}$$

Insight

Note that the time of travel for the sound is quite small, but still nonzero.

Practice Problem

You drop a stone into a well and hear the splash 1.47 s later. How deep is the well? [**Answer:** 10.2 m]

Some related homework problems: Problem 23, Problem 24

The Frequency of a Sound Wave

When we hear a sound, its frequency makes a great impression on us; in fact, the frequency determines the **pitch** of a sound. For example, the keys on a piano produce sound with frequencies ranging from 55 Hz for the key farthest to the left to 4187 Hz for the rightmost key. Similarly, as you hum a song you change the shape and size of your vocal chords slightly to change the frequency of the sound you produce.

The frequency range of human hearing extends well beyond the range of a piano, however. As a rule of thumb, humans can hear sounds between 20 Hz on the low frequency end and 20,000 Hz on the high frequency end. Sounds with frequencies above this range are referred to as **ultrasonic**, while those with frequencies lower than 20 Hz are classified as **infrasonic**. Though we are unable to hear ultrasound and infrasound, these frequencies occur commonly in nature, and are used in many technological applications as well.

For example, bats and dolphins produce ultrasound almost continuously as they go about their daily lives. By listening to the echoes of their calls—that is, by using *echolocation*—they are able to navigate about their environment and detect their prey. As a defense mechanism, some of the insects that are preyed upon by bats have the ability to hear the ultrasonic frequency of a hunting bat and take evasive action. For instance, the praying mantis has a specialized ultrasound receptor on its abdomen that allows it to take cover in response to an approaching bat. More dramatically, certain moths fold their wings in flight and drop into a precipitous dive toward the ground when they hear a bat on the prowl.

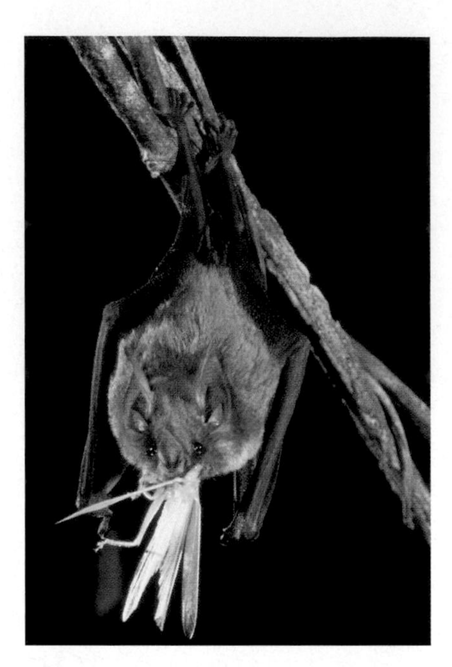

▲ Many animal species use sound waves that are too high (ultrasonic) or too low (infrasonic) for human ears to detect. Bats, for example, navigate in the dark and locate their prey by means of a system of biological sonar. They emit a continuous stream of ultrasonic sounds and detect the echoes from objects around them. Blue whales, by contrast, communicate over long distances by means of infrasonic sounds.

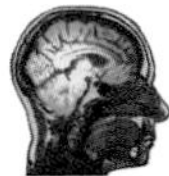

REAL WORLD PHYSICS: BIO

Medical applications of ultrasound: ultrasonic scans

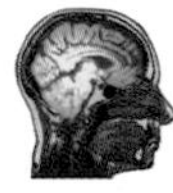

REAL WORLD PHYSICS: BIO

Medical applications of ultrasound: shock wave lithotripsy

Medical applications of ultrasound are also common. Perhaps the most familiar is the ultrasound scan that is used to image a fetus in the womb. By sending bursts of ultrasound into the body and measuring the time delay of the resulting echoes—the technological equivalent of echolocation— it is possible to map out the location of structures that lie hidden beneath the skin. In addition to imaging the interior of a body, ultrasound can also produce changes within the body that would otherwise require surgery. For example, in a technique called *shock wave lithotripsy* (SWL) an intense beam of ultrasound is concentrated onto a kidney stone that must be removed. After being hit with as many as 1000 to 3000 pulses of sound (at 23 joules per pulse), the stone is fractured into small pieces that the body can then eliminate on its own.

REAL WORLD PHYSICS

Infrasonic communication among animals

As for infrasound, it has been discovered in recent years that elephants can communicate with one another using sounds with frequencies as low as 15 Hz. In fact, it may be that *most* elephant communication is infrasonic. These sounds, which humans feel as vibration rather than hear as sound, can carry over an area of about thirty square kilometers on the dry African savanna. And elephants are not alone in this ability. Whales, such as the blue and the finback, produce powerful infrasonic calls as well. Since sound generally travels farther in water than in air, the whale calls can be heard by others of their species over distances of thousands of kilometers.

REAL WORLD PHYSICS

Infrasound produced by meteors

One final example of infrasound is related to a dramatic event that occurred not long ago in southern New Mexico. At 12:47 in the afternoon of October 10, 1997, a meteor shining as bright as the full Moon streaked across the sky for a few brief moments. The event was observed not just visually, however, but with infrasound as well. An array of special microphones at the Los Alamos National Laboratory—originally designed to listen for clandestine nuclear weapons tests—heard the infrasonic boom created by the meteor. By tracking the sonic signals of such meteors it may be possible to recover fragments that manage to reach the ground. The Los Alamos detector is in constant operation, and it detects about 10 rather large objects (2 m or more in diameter) entering the Earth's atmosphere each year.

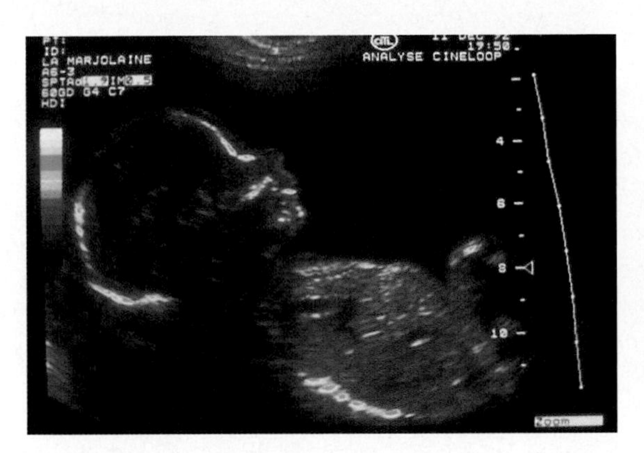

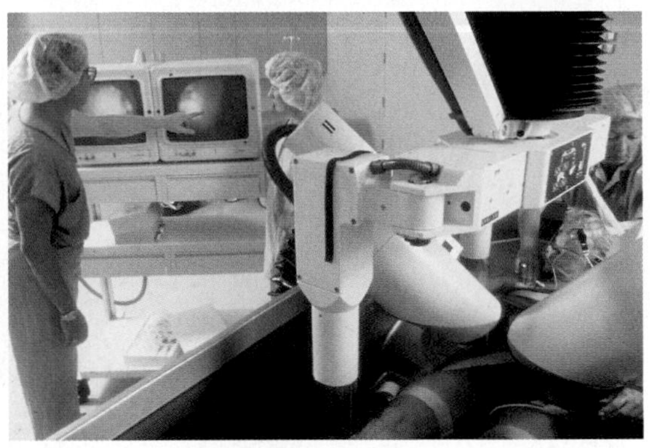

▲ Ultrasound is used in medicine both as an imaging medium and for therapeutic purposes. Ultrasound scans, or sonograms, are created by beaming ultrasonic pulses into the body and measuring the time required for the echoes to return. This technique is commonly used to evaluate heart function (echocardiograms) and to visualize the fetus in the uterus, as shown above (left). In shock wave lithotripsy (right), pulses of high-frequency sound waves are used to shatter kidney stones into fragments that can be passed in the urine.

It should be noted, in light of the wide range of frequencies observed in sound, that the speed of sound is the same for all frequencies. Thus, in the relation

$$v = \lambda f$$

the speed v remains fixed. For example, if the frequency of a wave is doubled, its wavelength is halved, so that the speed v stays the same. The fact that different frequencies travel with the same speed is evident when we listen to an orchestra in a large room. Different instruments are producing sounds of different frequencies, but we hear the sounds at the same time. Otherwise, listening to music from a distance would be quite a different and inharmonious experience.

14–5 Sound Intensity

The noise made by a jackhammer is much louder than the song of a sparrow. On this we can all agree. But how do we express such an observation physically? What physical quantity determines the loudness of a sound? We address these questions in this section, and we also present a quantitative scale by which loudness may be measured.

Intensity

The loudness of a sound is determined by its **intensity**; that is, by the amount of energy that passes through a given area in a given time. This is illustrated in **Figure 14–12**. If the energy E passes through the area A in the time t the intensity, I, of the wave carrying the energy is

$$I = \frac{E}{At}$$

Recalling that power is energy per time, $P = E/t$, we can express the intensity as follows:

Definition of Intensity, *I*

$$I = \frac{P}{A} \qquad \text{14–5}$$

SI unit: $\mathrm{W/m^2}$

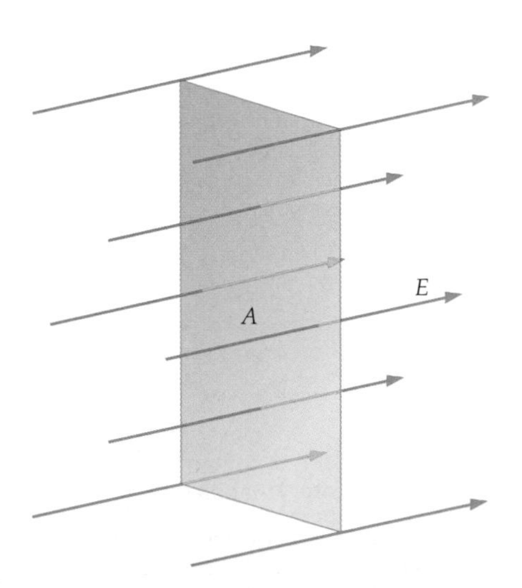

▲ **FIGURE 14–12 Intensity of a wave**

If a wave carries an energy E through an area A in the time t, the corresponding intensity is $I = E/At = P/A$, where $P = E/t$ is the power.

The units are those of power (watts, W) divided by area (meters squared, $\mathrm{m^2}$).

Though we have introduced the concept of intensity in terms of sound, it applies to all types of waves. For example, the intensity of light from the Sun as it

TABLE 14–2 Sound Intensities (W/m^2)

Loudest sound produced in laboratory	10^9
Saturn V rocket at 50 m	10^8
Rupture of the eardrum	10^4
Jet engine at 50 m	10
Threshold of pain	1
Rock concert	10^{-1}
Jackhammer at 1 m	10^{-3}
Heavy street traffic	10^{-5}
Conversation at 1 m	10^{-6}
Classroom	10^{-7}
Whisper at 1 m	10^{-10}
Normal breathing	10^{-11}
Threshold of hearing	10^{-12}

reaches the Earth's upper atmosphere is about 1380 W/m^2. If this intensity could be heard as sound it would be painfully loud—roughly the equivalent of four jet airplanes taking off simultaneously. By comparison, the intensity of microwaves in a microwave oven is even greater, about 6000 W/m^2, whereas the intensity of a whisper is an incredibly tiny 10^{-10} W/m^2. A selection of representative intensities is given in Table 14–2.

EXERCISE 14–3

A loudspeaker puts out 0.15 W of sound through a square area 2.0 m on each side. What is the intensity of this sound?

Solution

Applying Equation 14–5, with $A = (2.0 \text{ m})^2$, we find

$$I = \frac{P}{A} = \frac{0.15 \text{ W}}{(2.0 \text{ m})^2} = 0.038 \text{ W/m}^2$$

When we listen to a source of sound, such as a person speaking or a radio playing a song, we notice that the loudness of the sound decreases as we move away from the source. This means that the intensity of the sound is also decreasing. The reason for this reduction in intensity is simply that the energy emitted per time by the source spreads out over a larger area—just as spreading a certain amount of jam over a larger piece of bread reduces the intensity of the taste.

In **Figure 14–13** we show a source of sound (a bat) and two observers (moths) listening at the distances r_1 and r_2. Assuming no reflections of sound, and a power output by the bat equal to P, the intensity detected by the first moth is

$$I_1 = \frac{P}{4\pi r_1^2}$$

▲ **FIGURE 14–13 Echolocation**
Two moths, at distances r_1 and r_2, hear the sonar signals sent out by a bat. The intensity of the signal decreases with the square of the distance.

In writing this expression, we have used the fact that the area of a sphere of radius r is $4\pi r^2$. Similarly, the second moth hears the same sound with an intensity of

$$I_2 = \frac{P}{4\pi r_2^2}$$

The power P is the same in each case—it simply represents the amount of sound emitted by the bat. Solving for the intensity at moth 2 in terms of the intensity at moth 1 we find

$$I_2 = \left(\frac{r_1}{r_2}\right)^2 I_1 \qquad \textbf{14–6}$$

In words, the intensity falls off with the square of the distance; doubling the distance reduces the intensity by a factor of 4.

To summarize, the intensity a distance r from a point source of power P is

Intensity With Distance from a Point Source

$$I = \frac{P}{4\pi r^2} \qquad \textbf{14–7}$$

SI unit: W/m^2

This result assumes that no sound is reflected, which could increase the amount of energy passing through a given area, and that no sound is absorbed.

PROBLEM SOLVING NOTE
Intensity Variation with Distance

Suppose the intensity of a point source is I_1 at a distance r_1. This is enough information to find its intensity at any other distance. For example, to find the intensity I_2 at a distance r_2 we use the relation $I_2 = (r_1/r_2)^2 I_1$.

EXAMPLE 14–3 The Power of Song

Two people relaxing on a deck listen to a songbird sing. One person, only 1.00 m from the bird, hears sound with an intensity of 2.80×10^{-6} W/m^2. **(a)** What intensity is heard by the second person, who is 4.25 m from the bird? Assume that no reflected sound is heard by either person. **(b)** What is the power output of the bird's song?

Picture the Problem

Our sketch shows the two observers, one at a distance of $r_1 =$ 1.00 m, the other at a distance of $r_2 = 4.25$ m.

Strategy

(a) The two intensities are related by Equation 14–6, with $r_1 = 1.00$ m and $r_2 = 4.25$ m.

(b) The power output can be obtained from the definition of intensity, $I = P/A$. We can calculate P for either observer, noting that $A = 4\pi r^2$.

Solution

Part (a)

1. Substitute numerical values into Equation 14–6:

$$I_2 = \left(\frac{r_1}{r_2}\right)^2 I_1 = \left(\frac{1.00 \text{ m}}{4.25 \text{ m}}\right)^2 (2.80 \times 10^{-6} \text{ W/m}^2)$$
$$= 1.55 \times 10^{-7} \text{ W/m}^2$$

Part (b)

2. Solve $I = P/A$ for the power, P, using data for observer 1:

$$I_1 = P/A_1$$
$$P = I_1 A_1 = (2.80 \times 10^{-6} \text{ W/m}^2)[4\pi(1.00 \text{ m})^2]$$
$$= 3.52 \times 10^{-5} \text{ W}$$

3. As a check, repeat the calculation for observer 2:

$$I_2 = P/A_2$$
$$P = I_2 A_2 = (1.55 \times 10^{-7} \text{ W/m}^2)[4\pi(4.25 \text{ m})^2]$$
$$= 3.52 \times 10^{-5} \text{ W}$$

Insight

The intensity at observer 1 is $4.25^2 = 18.1$ times the intensity at observer 2. Even so, the bird only *seems* to be about 2.5 times louder to observer 1. The connection between intensity and perceived (subjective) loudness is discussed in detail later in this section.

Practice Problem

If the intensity at observer 2 were 7.40×10^{-7} W/m^2, how far would he be from the bird? [**Answer:** $r_2 = 1.95$ m]

Some related homework problems: Problem 28, Problem 29

ACTIVE EXAMPLE 14–1 The Big Hit

Ken Griffey, Jr. connects with a fast ball, and sends it out of the park. A fan in the outfield bleachers, 140 m away, hears the hit with an intensity of 3.80×10^{-7} W/m^2. Assuming no reflected sounds, what is the intensity heard by the first-base umpire, 90 ft (27.4 m) away from home plate?

Solution

1. Label the data given in the problem. Let the umpire be observer 1 and the fan be observer 2:

$$r_1 = 27.4 \text{ m}$$
$$I_2 = 3.80 \times 10^{-7} \text{ W/m}^2, r_2 = 140 \text{ m}$$

continued on the following page

continued from the previous page

2. Solve Equation 14–6 for I_1: $I_1 = (r_2/r_1)^2 I_2$

3. Substitute numerical values: $I_1 = 9.9 \times 10^{-6}\ \text{W/m}^2$

Insight
For the fan, the sound from the hit is somewhat less intense than normal conversation. For the umpire it is comparable to the sound of a busy street.

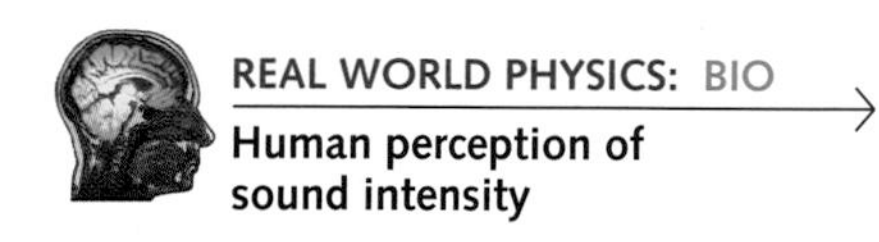
REAL WORLD PHYSICS: BIO
Human perception of sound intensity

Human Perception of Sound

Hearing, like most of our senses, is incredibly versatile and sensitive. We can detect sounds that are about a million times fainter than a typical conversation, and listen to sounds that are a million times louder before experiencing pain. In addition, we are able to hear sounds over a wide range of frequencies, from 20 Hz to 20,000 Hz.

When detecting the faintest of sounds our hearing is more sensitive than one would ever guess. For example, a faint sound, with an intensity of about $10^{-11}\ \text{W/m}^2$, causes a displacement of molecules in the air of about 10^{-10} m. This displacement is roughly the diameter of an atom!

Equally interesting is the way we perceive the loudness of a sound. As an example, suppose you hear a sound of intensity I_1. Next, you listen to a second sound of intensity I_2, and this sound seems to be "twice as loud" as the first. If the two intensities are measured, it turns out that I_2 is about 10 times I_1. Similarly, a third sound, twice as loud as I_2, has an intensity I_3 that is 10 times greater than I_2. Thus, $I_2 = 10\, I_1$ and $I_3 = 10\, I_2 = 100\, I_1$.

Our perception of sound, then, is such that uniform increases in loudness correspond to intensities that increase by multiplicative factors. For this reason, a convenient scale to measure loudness depends on the logarithm of intensity, as we discuss next.

Intensity Level and Decibels

In the study of sound, loudness is measured by the **intensity level** of a wave. Designated by the symbol β, the intensity level is defined as follows:

Definition of Intensity Level, β

$$\beta = 10 \log(I/I_0) \qquad \textbf{14–8}$$

SI unit: dimensionless

In this expression, log indicates the logarithm to the base 10, and I_0 is the intensity of the faintest sounds that can be heard. Experiments show this lowest detectable intensity to be

$$I_0 = 10^{-12}\ \text{W/m}^2$$

Note that β is dimensionless; the only dimensions that enter into the definition are those of intensity, and they cancel in the logarithm. Still, just as with radians, it is convenient to label the values of intensity level with a name. The name we use—the bel—honors the work of Alexander Graham Bell (1847–1922), the inventor of the telephone. Since the bel is a fairly large unit, it is more common to measure β in units that are 1/10 of a bel. This unit is referred to as the **decibel**, and its abbreviation is **dB**.

To get a feeling for the decibel scale, let's start with the faintest sounds. If a sound has an intensity $I = I_0$, the corresponding intensity level is

$$\beta = 10 \log(I_0/I_0) = 10 \log(1) = 0$$

Increasing the intensity by a factor of 10 makes the sound seem twice as loud. In terms of decibels, we have

$$\beta = 10\log(10I_0/I_0) = 10\log(10) = 10\text{ dB}$$

Going up in intensity by another factor of 10 doubles the loudness of the sound again, and yields

$$\beta = 10\log(100I_0/I_0) = 10\log(100) = 20\text{ dB}$$

Thus, *the loudness of a sound doubles with each increase in intensity level of 10 dB.* The *smallest* increase in intensity level that can be detected by the human ear is about 1 dB.

The intensity of a number of independent sound sources is simply the sum of the individual intensities. We use this fact in the following Example.

PROBLEM SOLVING NOTE

Calculating the Intesity Level

When determining the intensity level β be sure to use the base 10 logarithm (log), as opposed to the "natural," or base e, logarithm (ln).

EXAMPLE 14–4 Pass the Pacifier

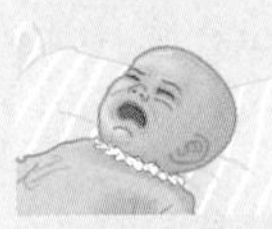

A crying child emits sound with an intensity of $8.0 \times 10^{-6}\text{ W/m}^2$. Find **(a)** the intensity level in decibels for the child's sounds, and **(b)** the intensity level for this child and its twin, both crying with identical intensities.

Picture the Problem

We consider the crying sounds of either one or two children. If two children are crying, each with an intensity I, the total intensity is $2I$.

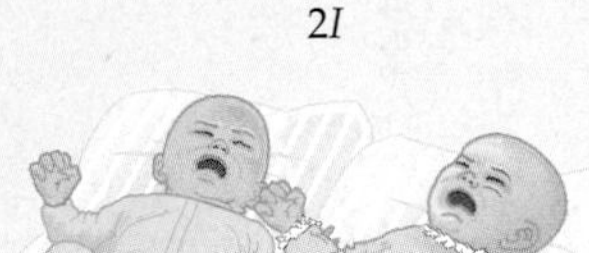

Strategy

The intensity level, β, is obtained by applying Equation 14–8.

Solution

Part (a)

1. Calculate β for $I = 8.0 \times 10^{-6}\text{ W/m}^2$:

$$\beta = 10\log(I/I_0)$$
$$= 10\log\left(\frac{8.0 \times 10^{-6}\text{ W/m}^2}{10^{-12}\text{ W/m}^2}\right) = 10\log(8.0 \times 10^6)$$
$$= 10\log(8.0) + 10\log(10^6) = 69\text{ dB}$$

Part (b)

2. Repeat the calculation with I replaced by $2I$:

$$\beta = 10\log(2I/I_0)$$
$$= 10\log(2) + 10\log(I/I_0)$$
$$= 3.0\text{ dB} + 69\text{ dB} = 72\text{ dB}$$

Insight

Note that the intensity level is increased by $10\log(2) = 3$ dB. This is a general rule: When the intensity is doubled, the intensity level, β, increases by 3 dB. Similarly, when the intensity is halved, β decreases by 3 dB.

Practice Problem

What is the intensity level of four identically crying quadruplets? [**Answer:** $\beta = 75$ dB]

Some related homework problems: Problem 30, Problem 31

ACTIVE EXAMPLE 14–2 Intensity Versus Intensity Level

Find the intensity that produces an intensity level of 60.0 dB.

Solution

1. Solve Equation 14–8 for I: $\quad I = 10^{\beta/10}I_0$

2. Substitute numerical values for β and I_0: $\quad I = 1.00 \times 10^{-6}\text{ W/m}^2$

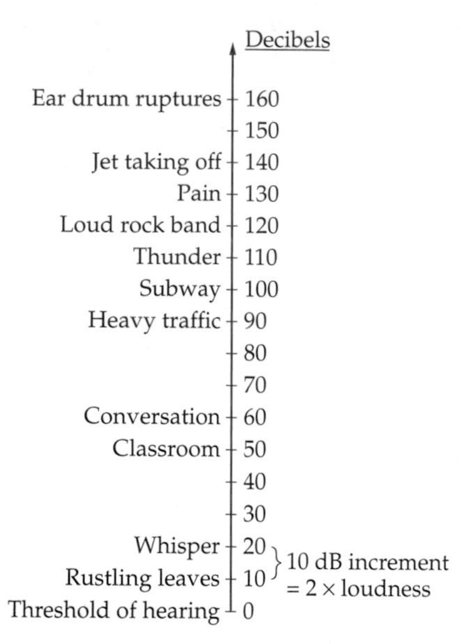

▲ **FIGURE 14–14 Representative intensity levels for common sounds**

Even though a change of 3 dB is relatively small—after all, a change of 10 dB is required to make a sound twice as loud—it still requires changing the intensity by a factor of two. For example, suppose a large nursery in a hospital has so many crying babies that the intensity level is 6 dB above the safe value, as determined by OSHA (Occupational Safety and Health Administration). To reduce the level by 6 dB it would be necessary to remove 3/4 of the children, leaving only 1/4 the original number. To our ears, however, the nursery will *sound* only 40 percent quieter!

Figure 14–14 shows the decibel scale with representative values indicated for a variety of common sounds.

14–6 The Doppler Effect

One of the most common physical phenomena involving sound is the change in pitch of a train whistle or a car horn as the vehicle moves past us. This change in pitch, due to the relative motion between a source of sound and the receiver, is called the **Doppler effect**, after the Austrian physicist Christian Doppler (1803–1853). If you listen carefully to the Doppler effect, you will notice that the pitch increases when the observer and the source are moving closer together, and decreases when the observer and source are separating.

One of the more fascinating aspects of the Doppler effect is the fact that it applies to all wave phenomena, not just to sound. In particular, the frequency of light is also Doppler shifted when there is relative motion between the source and receiver. For light, this change in frequency means a change in color. In fact, most distant galaxies are observed to be red-shifted in the color of their light, which means they are moving away from the Earth. Some galaxies, however, are moving toward us, and their light shows a blue shift.

In the remainder of this section, we focus on the Doppler effect in sound waves. We show that the effect is different depending on whether the observer or the source is moving. Finally, both observer and source may be in motion, and we present results for such cases as well.

Moving Observer

In **Figure 14–15** we see a stationary source of sound in still air. The radiated sound is represented by the circular patterns of compressions moving away from the source with a speed v. The distance between the compressions is the wavelength, λ, and the frequency of the sound is f. As for any wave, these quantities are related by

$$v = \lambda f$$

For an observer moving toward the source with a speed u, as in Figure 14–15, the sound *appears* to have a higher speed, $v + u$ (though, of course, the speed of sound relative to the air is always the same). As a result, more compressions move past the observer in a given time than if the observer had been at rest. To the observer, then, the sound has a frequency, f', that is higher than the frequency of the source, f.

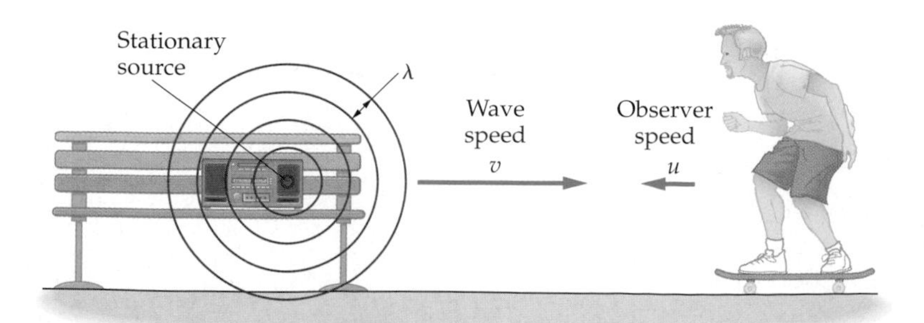

▶ **FIGURE 14–15 The Doppler effect: A moving observer**
Sound waves from a stationary source form concentric circles moving outward with a speed v. To the observer, who moves toward the source with a speed u, the waves are moving with a speed $v + u$.

We can find the frequency f' by first noting that the wavelength of the sound does not change—it is still λ. The speed, however, has increased to $v' = v + u$. Thus, we can solve $v' = \lambda f'$ for the frequency, which yields

$$f' = \frac{v'}{\lambda} = \frac{v + u}{\lambda}$$

Finally, we recall from Equation 14–1 that $\lambda = v/f$, hence

$$f' = \frac{v + u}{(v/f)} = \left(\frac{v + u}{v}\right)f = (1 + u/v)f$$

Note that f' is greater than f. This is the Doppler effect.

If the observer had been moving away from the source with a speed u, the sound would *appear* to the observer to have the reduced speed $v' = v - u$. Repeating the calculation just given, we find that

$$f' = \frac{v'}{\lambda} = \frac{v - u}{\lambda} = (1 - u/v)f$$

In this case the Doppler effect results in f' being less than f.

Combining these results we have

Doppler Effect for Moving Observer

$$f' = (1 \pm u/v)f \qquad \textbf{14–9}$$

SI unit: $1/\mathrm{s} = \mathrm{s}^{-1}$

In this expression u and v are speeds, and hence are always positive. The appropriate signs are obtained by using the *plus* sign when the observer moves *toward* the source, and the *minus* sign when the observer moves *away from* the source.

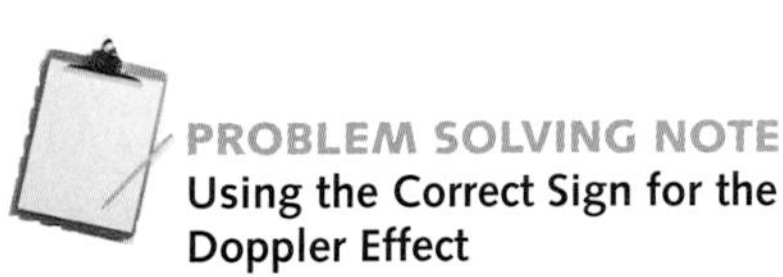

Using the Correct Sign for the Doppler Effect

When an observer approaches a source, the frequency heard by the observer is greater than the frequency of the source; that is, $f' > f$. This means that we must use the plus sign in $f' = (1 \pm u/v)f$. Similarly, we must use the minus sign when the observer moves away from the source.

EXAMPLE 14–5 A Moving Observer

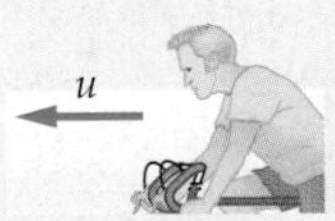

A street musician sounds the A string of his violin, producing a tone of 440 Hz. What frequency does a bicyclist hear as he **(a)** approaches and **(b)** recedes from the musician with a speed of 11.0 m/s?

Picture the Problem

The sketch shows a stationary source of sound and a moving observer. In part (a) the observer approaches the source with a speed u; in part (b) the observer has passed the source and is moving away.

Strategy

The frequency heard by the observer is given by Equation 14–9, with the plus sign for part (a) and the minus sign for part (b).

continued on the following page

continued from the previous page

Solution

Part (a)

1. Apply Equation 14–9 with the plus sign and $u = 11.0$ m/s: $f' = (1 + u/v)f = \left(1 + \frac{11.0\text{ m/s}}{343\text{ m/s}}\right)(440\text{ Hz}) = 454\text{ Hz}$

Part (b)

2. Now use the minus sign in Equation 14–9: $f' = (1 - u/v)f = \left(1 - \frac{11.0\text{ m/s}}{343\text{ m/s}}\right)(440\text{ Hz}) = 426\text{ Hz}$

Insight

As the bicyclist passes the musician the observed frequency of sound decreases, giving a "wow" effect. The difference in frequency is about 1 semitone.

Practice Problem

If the bicyclist hears a frequency of 451 Hz when approaching the musician, what is his speed? [**Answer:** $u = 8.58$ m/s]

Some related homework problems: Problem 36, Problem 41

Moving Source

With a stationary observer and a moving source, the Doppler effect is not due to the sound wave appearing to have a higher or lower speed, as when the observer moves. On the contrary, the speed of a wave is determined by the properties of the medium through which it propagates. Thus, once the source emits a sound wave, it travels through the medium with its characteristic speed v regardless of what the source is doing.

By way of analogy, consider a water wave. The speed of such waves is the same whether they are created by a rock dropped into the water or by a stick moved rapidly through the water. To take an extreme case, the waves coming to the beach from a slow-moving tugboat move with the same speed as the waves produced by a 100-mph speed boat. The same is true of sound waves.

Consider, then, a source moving toward an observer with a speed u, as shown in **Figure 14–16**. If the frequency of the source is f, it emits one compression every T seconds, where $T = 1/f$. Therefore, during one cycle of the wave a compression travels a distance vT while the source moves a distance uT. As a result, the next compression

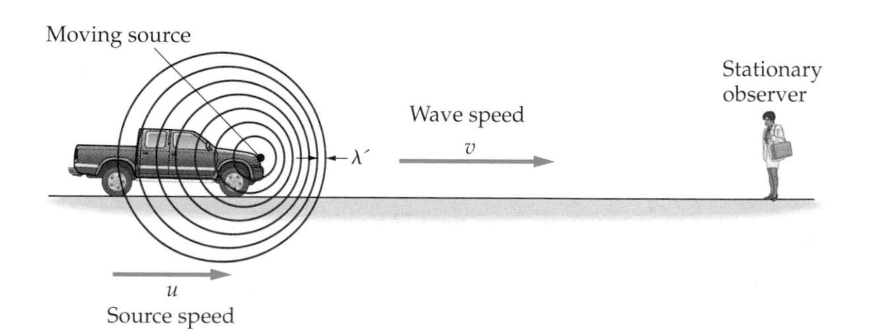

▶ **FIGURE 14–16 The Doppler effect: A moving source**

Sound waves from a moving source are bunched up in the forward direction, causing a shorter wavelength and a higher frequency.

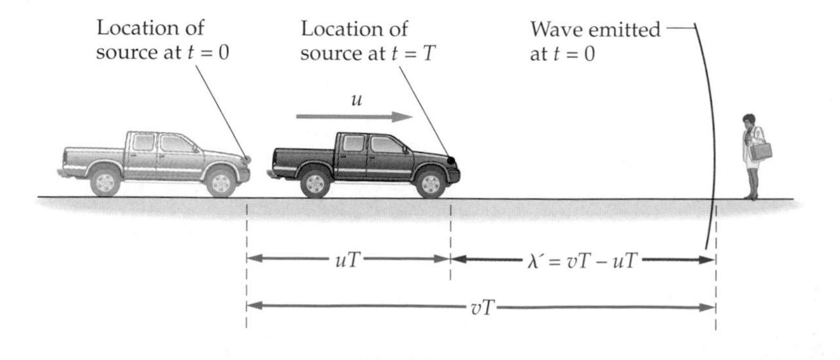

▶ **FIGURE 14–17 The Doppler-shifted wavelength**

During one period, T, the wave emitted at $t = 0$ moves through a distance vT. In the same time, the source moves toward the observer through the distance uT. At the time $t = T$ the next wave is emitted from the source, hence the distance between the waves (the wavelength) is $\lambda' = vT - uT$.

is emitted a distance $vT - uT$ behind the previous compression, as illustrated in **Figure 14–17**. This means that the wavelength in the forward direction is

$$\lambda' = vT - uT = (v - u)T$$

Now, the speed of the wave is still v, as mentioned, hence

$$v = \lambda' f'$$

Solving for the new frequency, f', we find

$$f' = \frac{v}{\lambda'} = \frac{v}{(v - u)T}$$

Finally, recalling that $T = 1/f$, we have

$$f' = \frac{v}{(v - u)(1/f)} = \frac{v}{v - u}f = \left(\frac{1}{1 - u/v}\right)f$$

Note that f' is greater than f, as expected.

In the reverse direction, the wavelength is increased by the amount uT. Thus,

$$\lambda' = vT + uT = (v + u)T$$

It follows that the Doppler-shifted frequency is

$$f' = \frac{v}{(v + u)\lambda} = \left(\frac{1}{1 + u/v}\right)f$$

This is less than the source frequency, f.

Finally, we can combine these results to yield

Doppler Effect for Moving Source

$$f' = \left(\frac{1}{1 \mp u/v}\right)f \qquad \text{14–10}$$

SI unit: $1/\text{s} = \text{s}^{-1}$

As before, u and v are positive quantities. The *minus* sign is used when the source moves *toward* the observer, and the *plus* sign when the source moves *away from* the observer.

PROBLEM SOLVING NOTE
Using Correct Signs

When a source approaches an observer, the frequency heard by the observer is greater than the frequency of the source; that is, $f' > f$. This means that we must use the minus sign in Equation 14–10, $f' = f/(1 - u/v)$, since this makes the denominator less than one. Similarly, use the plus sign when the source moves away from the observer.

EXAMPLE 14–6 Whistle Stop

A train sounds its whistle as it approaches a tunnel in a cliff. The whistle produces a tone of 650.0 Hz, and the train travels with a speed of 21.2 m/s. **(a)** Find the frequency heard by an observer standing near the tunnel entrance. **(b)** The sound from the whistle reflects from the cliff back to the engineer in the train. What frequency does the engineer hear?

Picture the Problem

The train emits sound at the frequency f. The observer near the tunnel hears the frequency f', and the reflected sound returns to the engineer with a frequency f''.

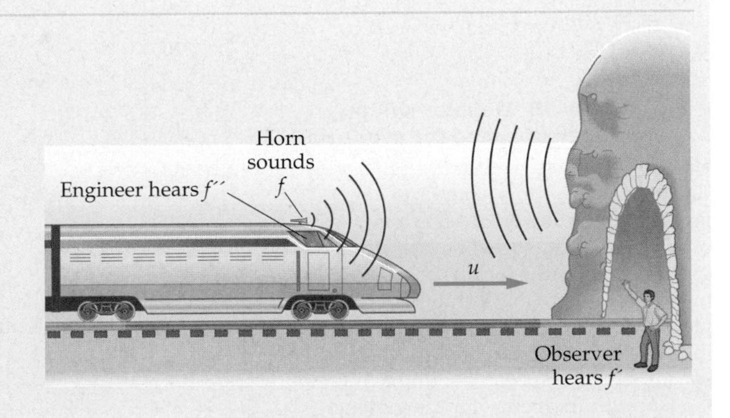

Strategy

Two Doppler shifts are involved in this problem. The first is due to the motion of the train toward the cliff. This shift causes the observer at the cliff to hear sound with a higher frequency f', given by Equation 14–10 with the minus sign. The reflected sound has the same frequency, f'.

The second shift is due to the moving engineer hearing the reflected sound. Thus, the engineer hears a frequency f'' that is greater than f'. We find f'' using Equation 14–9 with the plus sign.

continued on the following page

continued from the previous page

Solution

Part (a)

1. Use Equation 14–10, with the minus sign, to Doppler shift from f to f':

$$f' = \left(\frac{1}{1 - u/v}\right)f = \left(\frac{1}{1 - \dfrac{21.2 \text{ m/s}}{343 \text{ m/s}}}\right)(650.0 \text{ Hz})$$

$$= \left(\frac{1}{1 - 0.0618}\right)(650.0 \text{ Hz}) = 693 \text{ Hz}$$

Part (b)

2. Now use Equation 14–9, with the plus sign, to Doppler shift from f' to f'':

$$f'' = (1 + u/v)f' = \left(1 + \frac{21.2 \text{ m/s}}{343 \text{ m/s}}\right)(693 \text{ Hz})$$

$$= (1 + 0.0618)(693 \text{ Hz}) = 736 \text{ Hz}$$

Insight

Note that the reflected sound has the same frequency f' heard by the stationary observer, since all the cliff does is reverse the direction of motion of the sound heard at the cliff.

Practice Problem

If the stationary observer hears a frequency of 700.0 Hz, what is **(a)** the speed of the train and **(b)** the frequency heard by the engineer? [**Answer:** (a) $u = 24.5$ m/s, (b) $f'' = 750$ Hz]

Some related homework problems: Problem 35, Problem 80

Now that we have obtained the Doppler effect for both moving observers and moving sources, it is interesting to compare the results. **Figure 14–18** shows the Doppler-shifted frequency versus speed for a 400-Hz source of sound. The upper curve corresponds to a moving source, the lower curve to a moving observer. Notice that while the two cases give similar results for low speed, the high-speed behavior is quite different. In fact, the Doppler frequency for the moving source grows without limit for speeds near the speed of sound, while the Doppler frequency for the moving observer is relatively small.

These results can be understood both in terms of mathematics—by simply comparing Equations 14–9 and 14–10—and physically. In physical terms, recall that a moving observer encounters wave crests separated by the wavelength, as indicated in Figure 14–15. Doubling the speed of the observer simply reduces the time required to move from one crest to the next by a factor of 2, which doubles the frequency. Thus, in general, the frequency is proportional to the speed, as we see in

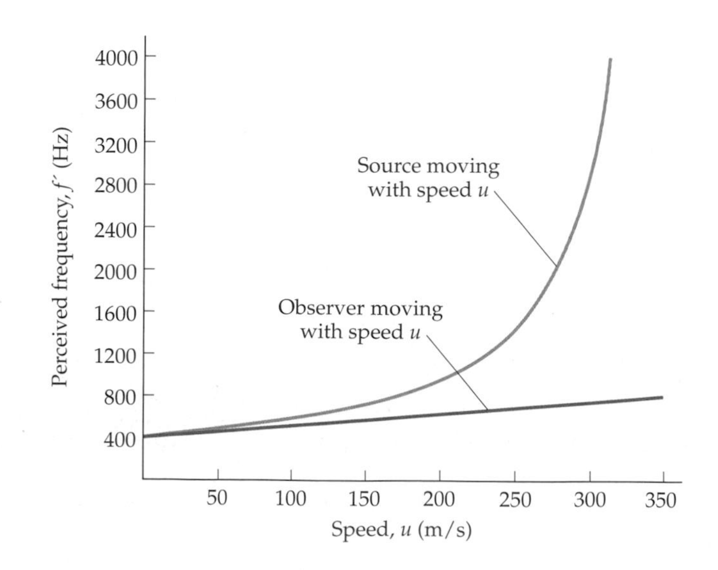

▶ **FIGURE 14–18 Doppler-shifted frequency versus speed for a 400-Hz sound source**

The upper curve corresponds to a moving source, the lower curve to a moving observer. Notice that while the two cases give similar results for low speed, the high-speed behavior is quite different. In fact, the Doppler frequency for the moving source grows without limit for speeds near the speed of sound, while the Doppler frequency for the moving observer is relatively small.

the lower curve in Figure 14–18. In contrast, when the source moves, as in Figure 14–16, the wave crests become "bunched up" in the forward direction, since the source is almost keeping up with the propagating waves. As the speed of the source approaches the speed of sound, the separation between crests approaches zero. Consequently, the frequency with which the crests pass a stationary observer approaches infinity, as indicated by the upper curve in Figure 14–18.

General Case

The results derived earlier in this section can be combined to give the Doppler effect for situations in which both observer and source move. Letting u_s be the speed of the source, and u_o be the speed of the observer, we have

Doppler Effect for Moving Source and Observer

$$f' = \left(\frac{1 \pm u_o/v}{1 \mp u_s/v}\right) f \qquad \text{14–11}$$

SI unit: $1/\text{s} = \text{s}^{-1}$

As in the previous cases, u_s, u_o, and v are positive quantities. In the numerator, the $+$ sign corresponds to the case in which the observer moves in the direction of the source, whereas the $-$ sign indicates motion in the opposite direction. In the denominator, the $-$ sign corresponds to the case in which the source moves in the direction of the observer, whereas the $+$ sign indicates motion in the opposite direction.

EXERCISE 14–4

A car moving at 18 m/s sounds its 550-Hz horn. A bicyclist on the sidewalk, moving with a speed of 7.2 m/s, approaches the car. What frequency is heard by the bicyclist?

Solution

We use Equation 14–11 with $u_s = 18$ m/s and $u_o = 7.2$ m/s. Since the source and observer are approaching, we use the plus sign in the numerator and the minus sign in the denominator:

$$f' = \left(\frac{1 + u_o/v}{1 - u_s/v}\right) f = \left(\frac{1 + 7.2/343}{1 - 18/343}\right)(550\text{ Hz}) = 590\text{ Hz}$$

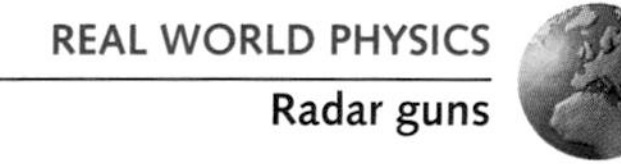

The Doppler effect is used in an amazing variety of technological applications. Perhaps one of the most familiar of these is the "radar gun," which is used to measure the speed of a pitched baseball or a car breaking the speed limit. Though the radar gun uses radio waves rather than sound waves, the basic physical principle is the same—by measuring the Doppler-shifted frequency of waves reflected from an object it is possible to determine its speed. Doppler radar, used in weather forecasting, applies this same technology to tracking the motion of precipitation caused by storm clouds.

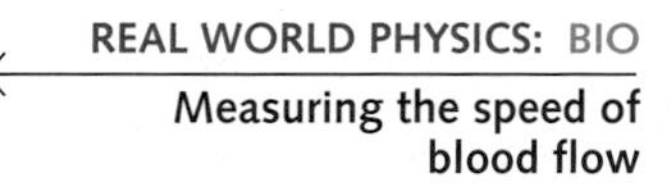

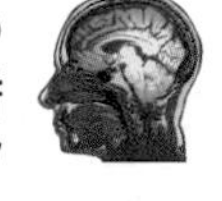

In medicine, the Doppler shift is used to measure the speed of blood flow in an artery, or in the heart itself. In this application, a beam of ultrasound is directed toward an artery in a patient. Some of the sound is reflected back by red blood cells moving through the artery. The reflected sound is detected, and its frequency is used to determine the speed of blood flow. If this information is color coded, with different colors indicating different speeds and directions of flow, an impressive image of blood flow can be constructed.

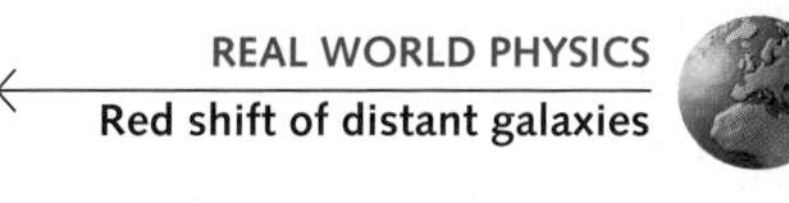

Finally, the Doppler effect applies to the light of distant galaxies as well. For example, if a galaxy moves away from us—as most do—the light we observe from that galaxy has a lower frequency than if the galaxy were at rest relative to our galaxy, the Milky Way. Since red light has the lowest frequency of visible light, we refer to this reduction in frequency as a "red shift." Thus, by measuring the red shift of a galaxy we can determine its speed relative to us.

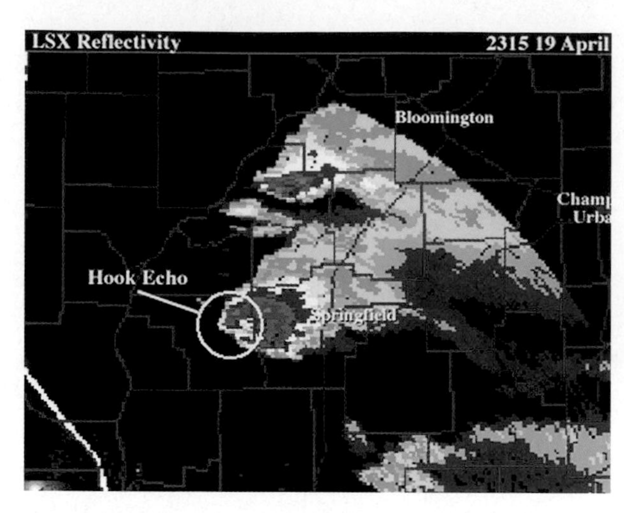

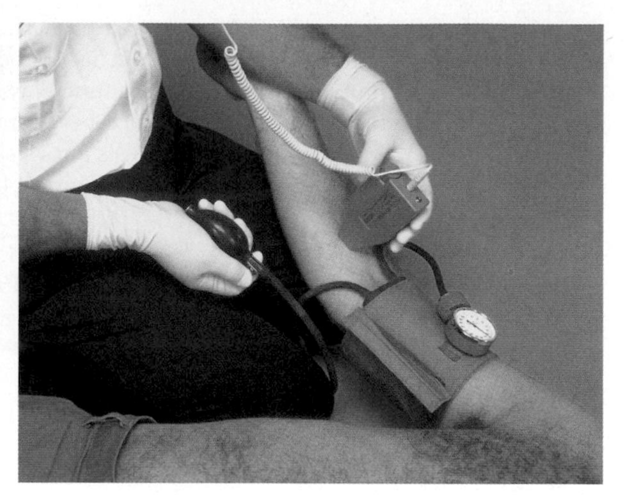

▲ Many familiar and not-so-familiar devices utilize the doppler effect. Doppler radar, now widely used at airports and for weather forecasting, makes it possible to determine the speed and direction of winds in a distant storm by measuring the radial velocity component of water droplets in clouds and rain. The image at left is the doppler radar scan of a severe thunderstorm that struck the town of Ogden, Illinois on April 19, 1996. The hook-shaped echo marked on the image is characteristic of tornadoes in the making. In the photo at right, a medical technician uses a doppler blood flow meter instead of a stethoscope while measuring the blood pressure of a patient.

14–7 Superposition and Interference

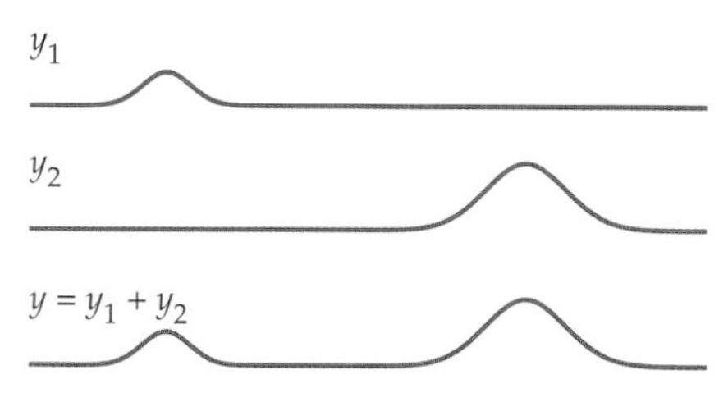

▲ **FIGURE 14–19 Wave superposition** Waves of small amplitude superpose (that is, combine) by simple addition.

So far we have considered only a single wave at a time. In this section we turn our attention to the way waves combine when more than one is present. As we shall see, the behavior of waves is quite simple in this respect.

Superposition

The combination of two or more waves to form a resultant wave is referred to as **superposition**. When waves are of small amplitude, they superpose in the simplest of ways—they just add. For example, consider two waves on a string, as in **Figure 14–19**, described by the wave functions y_1 and y_2. If these two waves are present on the same string at the same time, the result is a wave given by

$$y = y_1 + y_2$$

To see how superposition works as a function of time, consider a string with two wave pulses on it, one traveling in each direction as shown in **Figure 14–20 (a)**. When the pulses arrive in the same region, they add, as stated. This is illustrated in Figure 14–20 (a). The question is, "What do the pulses look like after they have passed through one another? Does their interaction change them in any way?"

The answer is that the waves are completely unaffected by their interaction. This is also shown in Figure 14–20 (a). After the wave pulses pass through one another they continue on as if nothing had happened. It is like listening to an orchestra, where many different instruments are playing simultaneously, and their sounds are combining throughout the concert hall. Even so, you can still hear individual instruments, each making its own sound as if the others were not present.

CONCEPTUAL CHECKPOINT 14–3

Since waves add, does the resultant wave y always have a greater amplitude than the individual waves y_1 and y_2?

Solution

The wave y is the sum of y_1 and y_2, but remember that y_1 and y_2 are sometimes positive and sometimes negative. Thus, if y_1 is positive at a given time, for example, and y_2 is negative, the sum $y_1 + y_2$ can be zero or even negative. For example, if y_1 and y_2 both have the amplitude A, the amplitude of y can take any value from 0 to $2A$.

Answer:

No. The amplitude of y can be greater than, less than, or equal to the amplitudes of y_1 and y_2.

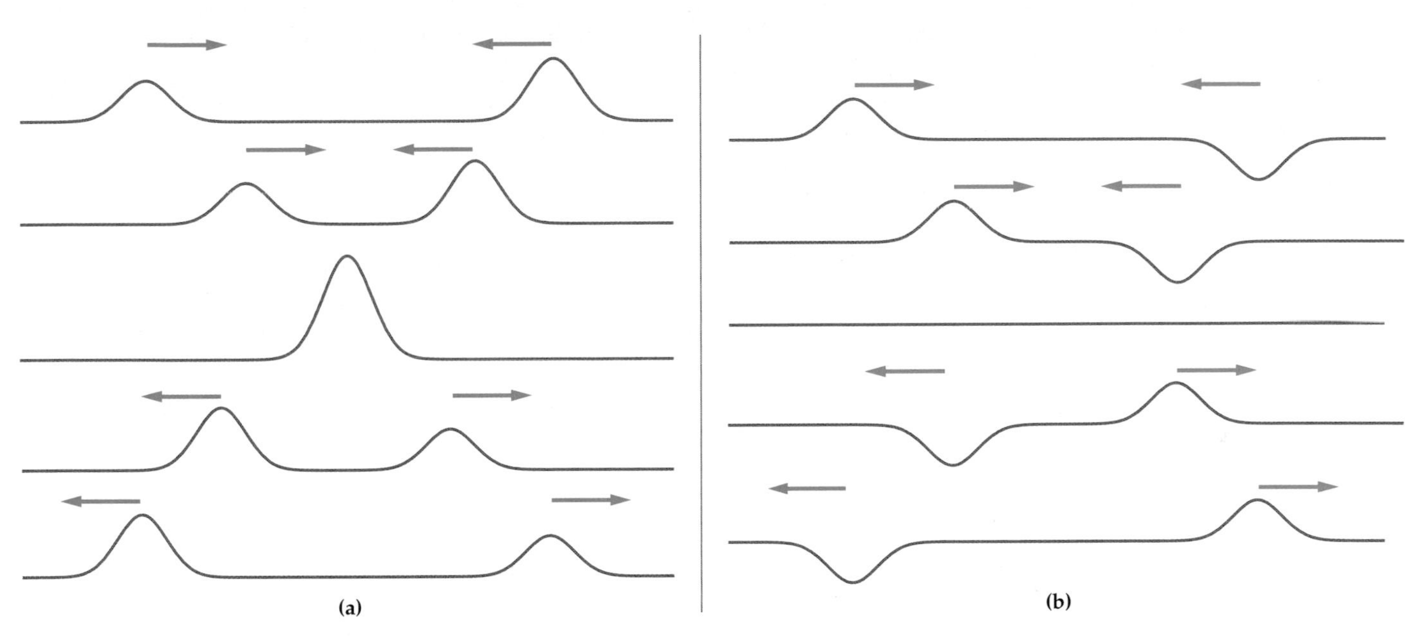

▲ FIGURE 14–20 Interference

Wave pulses superpose as they pass through one another. Afterwards, the pulses continue on unchanged. In **(a)**, the pulses combine to give a larger amplitude. This is an example of *constructive interference*. When a positive pulse superposes with a negative pulse **(b)**, the result is *destructive interference*. In this case, with symmetrical pulses, there is one moment of complete cancellation.

Interference

As simple as the principle of superposition is, it still leads to interesting consequences. For example, consider the wave pulses on a string shown in Figure 14–20 (a). When they combine, the resulting pulse has an amplitude equal to the sum of the amplitudes of the individual pulses. This is referred to as **constructive interference**.

On the other hand, two pulses like those in **Figure 14–20 (b)** may combine. When this happens the positive displacement of one wave adds to the negative displacement of the other to create a net displacement of zero. That is, the pulses momentarily cancel one another. This is **destructive interference**.

It is important to note that the waves don't simply disappear when they experience destructive interference. For example, in Figure 14–20 (b) the wave pulses continue on unchanged after they interact. This makes sense from an energy point of view—after all, each wave carries energy, hence the waves, along with their energy, cannot simply vanish. In fact, when the string is flat in Figure 14–20 (b) it has its greatest transverse speed—just like a swing has its highest speed when it is in its equilibrium position. Therefore, the energy of the wave is still present at this instant of time, it is just in the form of the kinetic energy of the string.

It should also be noted that interference is not limited to waves on a string; all waves exhibit interference effects. In fact, you could say that interference is one of the key characteristics that define waves. In general, when waves combine they form **interference patterns** that include regions of both constructive and destructive interference. An example is shown in **Figure 14–21**, where two water waves are interfering. Note the regions of constructive interference separated by regions of destructive interference.

▲ FIGURE 14–21 Interference pattern

Interference pattern formed by the superposition of two sets of circular water waves.

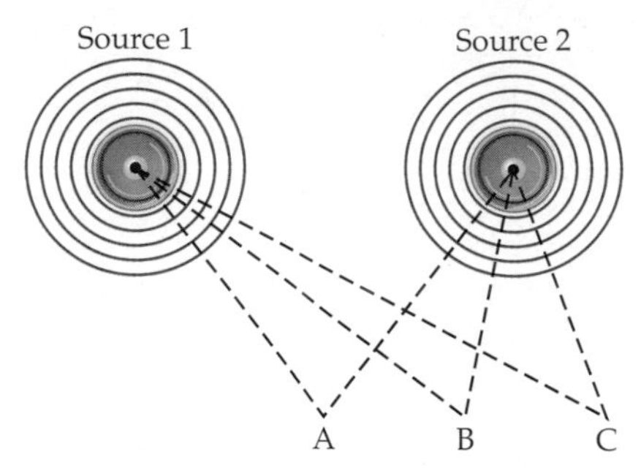

▲ **FIGURE 14–22 Interference with two sources**

Suppose the two sources emit waves in phase. At point A the distance to each source is the same, hence crest meets crest and constructive interference results. At B the distance from source 1 is greater than that from source 2 by half a wavelength. The result is crest meeting trough and destructive interference. Finally, at C the distance from source 1 is one wavelength greater than the distance from source 2. Hence, we find constructive interference at C. If the sources had been opposite in phase, then A and C would be points of destructive interference, and B would be a point of constructive interference.

To understand the formation of such patterns, consider a system of two identical sources, as in **Figure 14–22**. Each source sends out waves consisting of alternating crests and troughs. We set up the system so that when one source emits a crest, the other emits a crest as well. Sources that are synchronized like this are said to be **in phase**.

Now, at a point like A, the distance to each source is the same. Thus, if the wave from one source produces a crest at point A, so too does the wave from the other source. As a result, with crest combining with crest, the interference at A is constructive.

Next consider point B. At this location the wave from source 1 must travel a greater distance than the wave from source 2. If the extra distance is half a wavelength, it follows that when the wave from source 2 produces a crest at B the wave from source 1 produces a trough. As a result, the waves combine to give destructive interference at B. At point C, on the other hand, the distance from source 1 is one wavelength greater than the distance from source 2. Hence the waves are in phase again at C, with crest meeting crest for constructive interference.

In general, then, we can say that constructive and destructive interference occur under the following conditions for two sources that are in phase:

Constructive interference occurs when the path length from the two sources differs by $0, \lambda, 2\lambda, 3\lambda, \ldots$.

Destructive interference occurs when the path length from the two sources differs by $\lambda/2, 3\lambda/2, 5\lambda/2, \ldots$.

A specific example of interference patterns is provided by sound, using speakers that emit sound in phase with the same frequency. This situation is analogous to the two water-wave sources in Figure 14–21. As a result, constructive and destructive interference is to be expected, depending on the path length from each speaker. This is illustrated in the next Example.

EXAMPLE 14–7 Sound Off

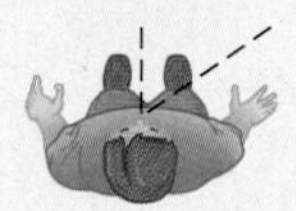

Two speakers separated by a distance of 4.30 m emit sound of frequency 221 Hz. The speakers are in phase with one another. A person listens from a location 2.80 m directly in front of one of the speakers. Does the person hear constructive or destructive interference?

Picture the Problem

In our sketch, we label the distance between the speakers with D, and the distance from each speaker with d_1 and d_2.

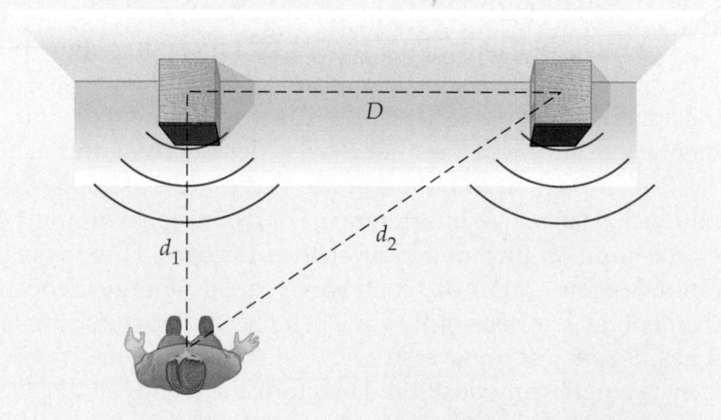

Strategy

The type of interference depends on whether the difference in path length, $d_2 - d_1$, is one or more wavelengths or an odd multiple of half a wavelength. Thus, we begin by calculating the wavelength, λ. Next, we find d_2, and compare the difference in path length to λ.

Solution

1. Calculate the wavelength of this sound, using $v = \lambda f$. As usual, let $v = 343$ m/s be the speed of sound:

$$\lambda = \frac{v}{f} = \frac{343 \text{ m/s}}{221 \text{ Hz}} = 1.55 \text{ m}$$

2. Find the path length d_2:

$$d_2 = \sqrt{D^2 + d_1^2} = \sqrt{(4.30 \text{ m})^2 + (2.80 \text{ m})^2} = 5.13 \text{ m}$$

3. Determine the difference in path length, $d_2 - d_1$:

$$d_2 - d_1 = 5.13 \text{ m} - 2.80 \text{ m} = 2.33 \text{ m}$$

4. Divide λ into $d_2 - d_1$ to find the number of wavelengths that fit into the path difference: $\dfrac{d_2 - d_1}{\lambda} = \dfrac{2.33 \text{ m}}{1.55 \text{ m}} = 1.50$

Insight

Since the path difference is $3\lambda/2$ we expect destructive interference. In the ideal case, the person would hear no sound. As a practical matter, some sound will be reflected from objects in the vicinity, resulting in a finite sound intensity.

Practice Problem

We know that 221 Hz gives destructive interference. What is the lowest frequency that gives constructive interference for the case described in this Example? [**Answer:** Set $\lambda = d_2 - d_1 = 2.33$ m. This gives $f = 147$ Hz.]

Some related homework problems: Problem 51, Problem 53

It is possible to connect a speaker with its wires reversed, which can result in a set of speakers that have **opposite phase**. In this case, as one speaker emits a compression the other sends out a rarefaction. When you set up a stereo system, it is important to be sure the wires are connected in a consistent fashion so that your speakers will be in phase.

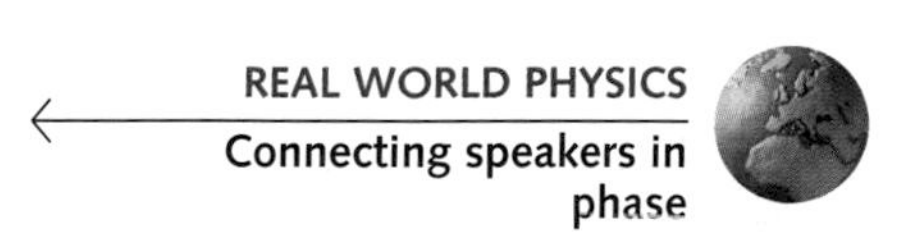

If the two speakers in Figure 14–22 have opposite phase, for example, the conditions for constructive and destructive interference are changed, as are the interference patterns. For example, at point A, where the distances from the two speakers are the same, the wave from one speaker is a compression when the wave from the other speaker is a rarefaction. Thus, point A is now a point of destructive interference rather than constructive interference. In general, then, the conditions for constructive and destructive interference are simply reversed—a path difference of 0, λ, 2λ, ... results in destructive interference, a path difference of $\lambda/2$, $3\lambda/2$, ... results in constructive interference.

ACTIVE EXAMPLE 14–3 Opposite Phase

The speakers shown below have opposite phase. They are separated by a distance of 5.20 m and emit sound with a frequency of 104 Hz. A person stands 3.00 m in front of the speakers and 1.30 m to one side of the center line between them. What type of interference occurs at the person's location?

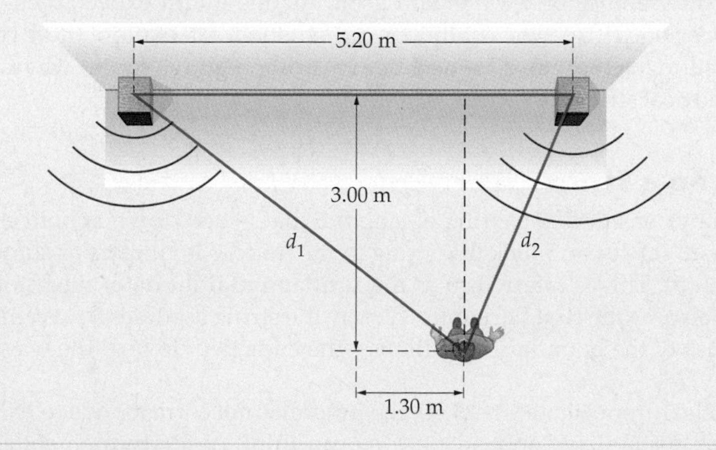

Solution

1. Calculate the wavelength: $\lambda = 3.30$ m

2. Find the path length d_1: $d_1 = 4.92$ m

3. Find the path length d_2: $d_2 = 3.27$ m

continued on the following page

continued from the previous page

4. Calculate the path length difference, $d_1 - d_2$: $d_1 - d_2 = 1.65\text{ m}$

5. Divide the path length difference by the wavelength: $(d_1 - d_2)/\lambda = 0.500$

Insight

The path difference is half a wavelength, and the speakers have opposite phase, therefore the result is constructive interference.

REAL WORLD PHYSICS

Active noise reduction

Destructive interference can be used to reduce the intensity of noise in a variety of situations, such as a factory, a busy office, or even the cabin of an airplane. The process, referred to as Active Noise Reduction (ANR), begins with a microphone that picks up the noise to be reduced. The signal from the microphone is then reversed in phase and sent to a speaker. As a result, the speaker emits sound that is opposite in phase to the incoming noise—in effect, the speaker produces "anti-noise." In this way, the noise is *actively* canceled by destructive interference, rather than simply reduced by absorption. The effect when wearing a pair of ANR headphones can be as much as a 30 dB reduction in the intensity level of noise.

14–8 Standing Waves

If you have ever plucked a guitar string, or blown across the mouth of a pop bottle to create a tone, you have generated **standing waves**. In general, a standing wave is one that oscillates with time, but remains fixed in its location. It is in this sense that the wave is said to be "standing."

In some respects, a standing wave can be considered as resulting from constructive interference of a wave with itself. As one might expect, then, standing waves occur only if specific conditions are satisfied. We explore these conditions in this section for two cases: (i) waves on a string, and (ii) sound waves in a hollow, cylindrical structure.

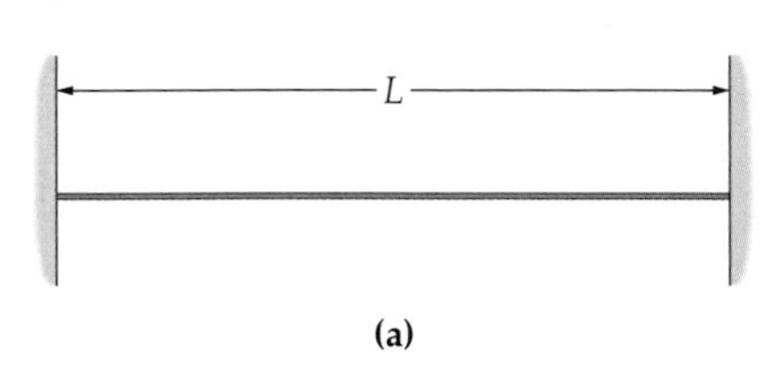

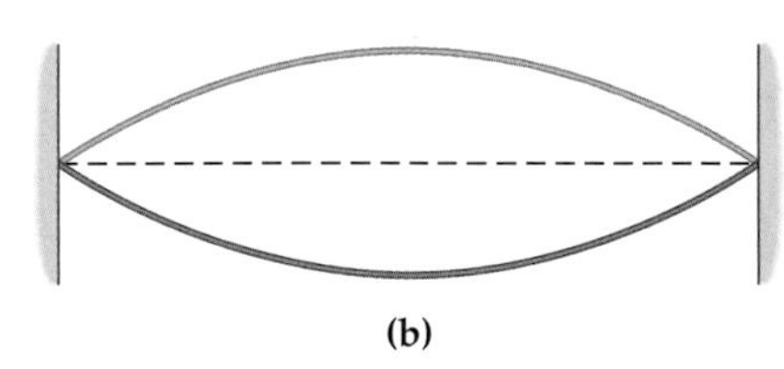

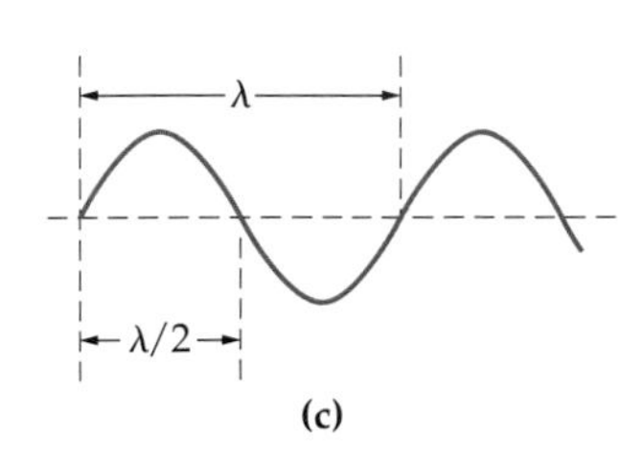

▲ **FIGURE 14–23 A standing wave**

(a) A string is tied down at both ends. **(b)** If the string is plucked in the middle a standing wave results. This is the fundamental mode of oscillation of the string. **(c)** The fundamental consists of one-half a wavelength between the two ends of the string. Hence, its wavelength is $2L$.

Waves on a String

We begin by considering a string of length L that is tied down at both ends, as in **Figure 14–23 (a)**. If you pluck this string in the middle it vibrates as shown in **Figure 14–23 (b)**. This is referred to as the **fundamental mode** of vibration for this string, or also, as the **first harmonic**. Clearly the string assumes a wave-like shape, but because of the boundary conditions—the ends tied down—the wave stays in place.

As is clear from **Figure 14–23 (c)**, the fundamental corresponds to half a wavelength of a usual wave on a string. One can think of the fundamental as being formed by this wave reflecting back and forth between the walls holding the string. If the frequency is just right the reflections combine to give constructive interference and the fundamental is formed; if the frequency differs from the fundamental frequency the reflections result in destructive interference and a standing wave does not result.

We can find the frequency of the fundamental as follows: First use the fact that the wavelength of the fundamental is twice the distance between the walls. Thus,

$$\lambda = 2L$$

If the speed of waves on the string is v, it follows that the frequency of the fundamental, f_1, is determined by $v = \lambda f_1 = (2L)f_1$. Therefore,

$$f_1 = \frac{v}{\lambda} = \frac{v}{2L}$$

Note that the fundamental frequency increases with the speed of the waves, and decreases as the string is lengthened.

The fundamental is not the only standing wave that can exist on a string, however. In fact, there are an infinite number of standing wave modes—or harmonics—for any given string. To see how to find higher harmonics, note that the two ends of the string must remain fixed. Points on a standing wave that stay fixed are referred to as **nodes**. Halfway between any two nodes is a point on the wave that has a maximum displacement, as indicated in **Figure 14–24**. Such a point is called an **anti-node**. Referring to Figure 14–24 (a), then, we see that the fundamental consists of two nodes (N) and one antinode (A); the sequence is N-A-N.

The second harmonic can be constructed by including one more half wavelength in the standing wave, as in Figure 14–24 (b). This mode has the sequence N-A-N-A-N, and has one complete wavelength between the walls. Therefore, its frequency, f_2, is

$$f_2 = \frac{v}{\lambda} = \frac{v}{L} = 2f_1$$

Similarly, the third harmonic again includes one more half wavelength, as in Figure 14–24 (c). Now there are one-and-a-half wavelengths in the length L; therefore $(3/2)\lambda = L$, or $\lambda = 2L/3$. The corresponding frequency, f_3, is

$$f_3 = \frac{v}{\lambda} = \frac{v}{\frac{2}{3}L} = 3\frac{v}{2L} = 3f_1$$

Note that the frequencies of the harmonics are increasing in integer steps. Clearly, then, the sequence of standing waves is characterized by the following:

Standing Waves on a String

Fundamental frequency and wavelength:

$$f_1 = \frac{v}{2L}$$
$$\lambda_1 = 2L \qquad \textbf{14–12}$$

Frequency and wavelength of the nth harmonic, with $n = 1, 2, 3, \ldots$:

$$f_n = nf_1$$
$$\lambda_n = 2L/n \qquad \textbf{14–13}$$

Note that the difference in frequency between any two successive harmonics is the fundamental frequency, f_1.

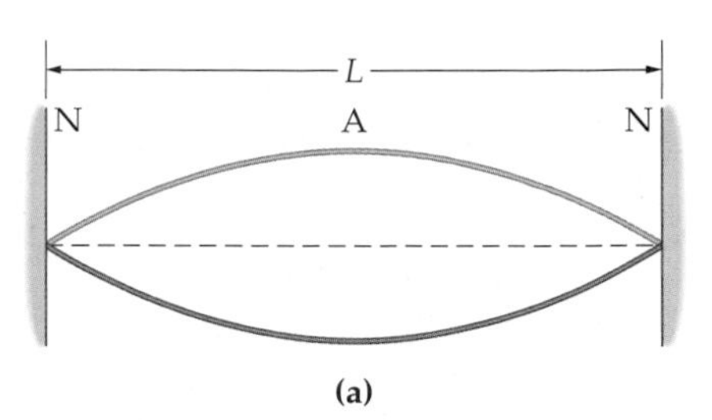

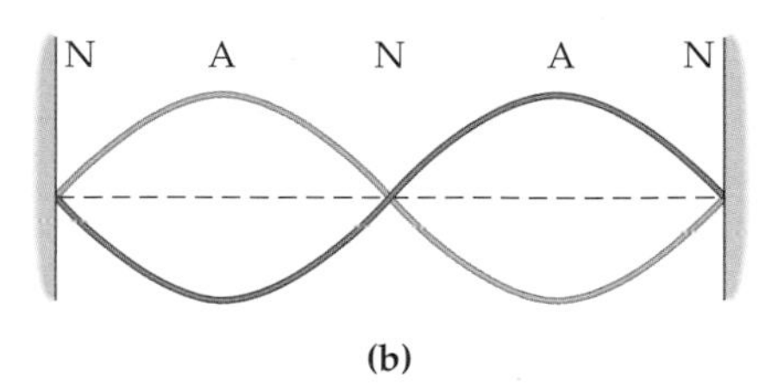

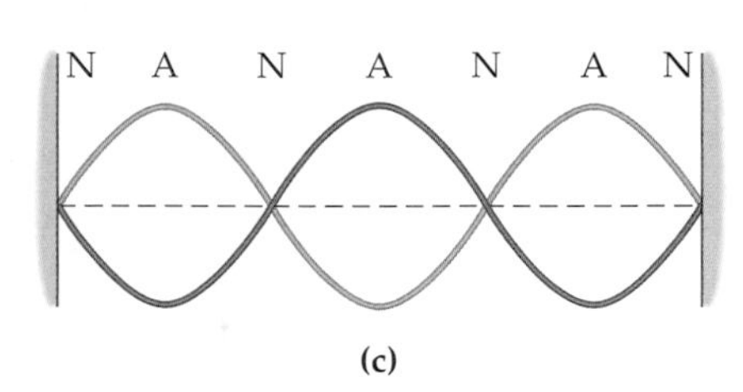

▲ **FIGURE 14–24 Harmonics**

The first three harmonics for a string tied down at both ends. Note that an extra half wavelength is added to go from one harmonic to the next. **(a)** $\lambda/2 = L$, $\lambda = 2L$; **(b)** $\lambda = L$; **(c)** $3\lambda/2 = L$, $\lambda = 2L/3$.

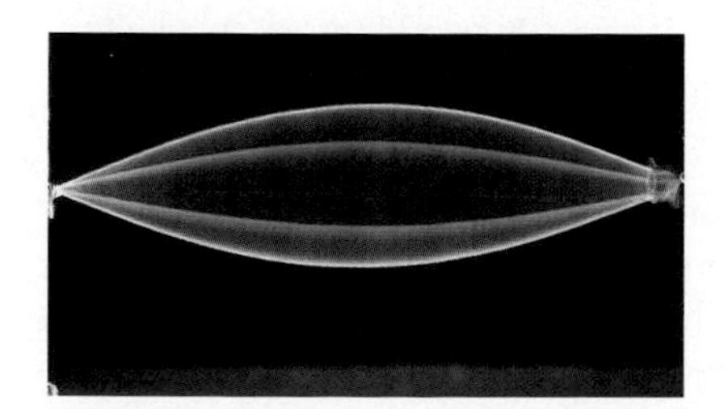

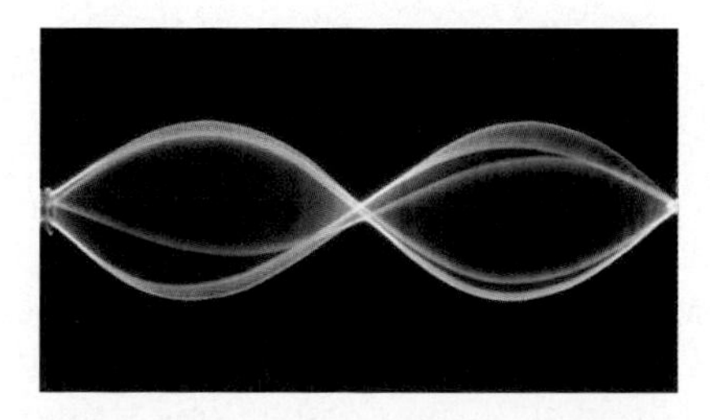

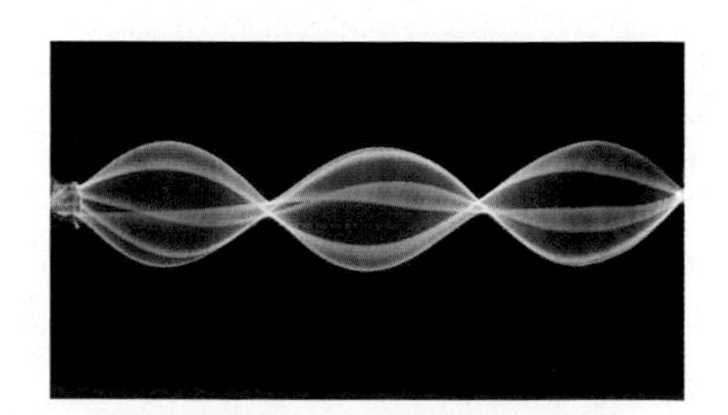

▲ The string in these multi-flash photographs vibrates in one of three different standing wave patterns, each with its own characteristic frequency. The lowest frequency standing wave—the fundamental, or first harmonic—is shown in the photograph at left. In this case, there are only two nodes, one at each end of the string where it is tied down. If the length of the string is L, we see that the wavelength of the fundamental is twice this length, or $\lambda_1 = 2L$. Thus, if waves have a speed v on this string, their frequency is $f_1 = v/2L$. Higher harmonics are produced by adding one node at a time to the standing wave pattern. The second harmonic, shown in the middle photograph, has a node at either end and one in the middle. In this case the wavelength is $\lambda_2 = L$ and the frequency is $f_2 = v/L = 2f_1$. The photograph at right shows the third harmonic, where $\lambda_3 = 2L/3$ and $f_3 = v/(2L/3) = 3v/2L = 3f_1$. In general, the nth harmonic on a string tied down at both ends is $f_n = nf_1$.

EXAMPLE 14–8 It's Fundamental

One of the harmonics on a string 1.30-m long has a frequency of 15.6 Hz. The next higher harmonic has a frequency of 23.4 Hz. Find **(a)** the fundamental frequency, and **(b)** the speed of waves on this string.

Picture the Problem

The problem statement does not tell us directly which two harmonics have the given frequencies. We do know, however, that they are successive harmonics of the string; thus, if the first harmonic has one node between the two ends of the string, the next harmonic has two nodes. Our sketch illustrates this case, which turns out to be appropriate for this problem.

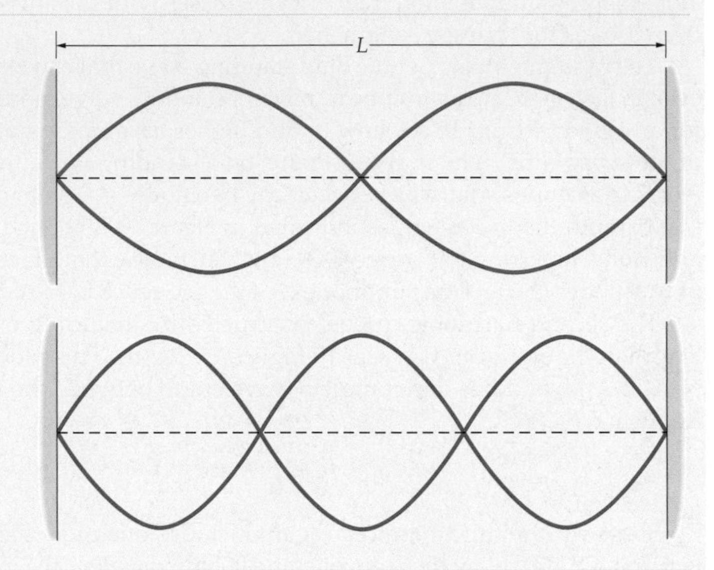

Strategy

(a) We know from Equation 14–13 that the frequencies of successive harmonics increase by f_1. That is, $f_2 = f_1 + f_1 = 2f_1$, $f_3 = f_2 + f_1 = 3f_1$, $f_4 = f_3 + f_1 = 4f_1, \ldots$. Therefore, we can find the fundamental by taking the difference between the given frequencies.

(b) Once the fundamental frequency is determined, we can find the speed of waves in the string from the relation $f_1 = v/2L$.

Solution

Part (a)

1. The fundamental frequency is the difference between the two given frequencies: $f_1 = 23.4\text{ Hz} - 15.6\text{ Hz} = 7.80\text{ Hz}$

Part (b)

2. Solve $f_1 = v/2L$ for the speed, v: $f_1 = v/2L$, $v = 2Lf_1$

3. Substitute numerical values: $v = 2Lf_1 = 2(1.30\text{ m})(7.80\text{ Hz}) = 20.0\text{ m/s}$

Insight

Now that we know the fundamental frequency, we can identify the harmonics given in the problem statement. First, 15.6 Hz = 2(7.80 Hz), so this is the second harmonic. The next mode, 23.4 Hz = 3(7.80 Hz), is the third harmonic, as expected.

Practice Problem

Suppose the tension in this string is increased until the speed of the waves is 22.0 m/s. What are the frequencies of the first three harmonics in this case? [**Answer:** $f_1 = 8.46$ Hz, $f_2 = 16.9$ Hz, $f_3 = 25.4$ Hz]

Some related homework problems: Problem 60, Problem 61

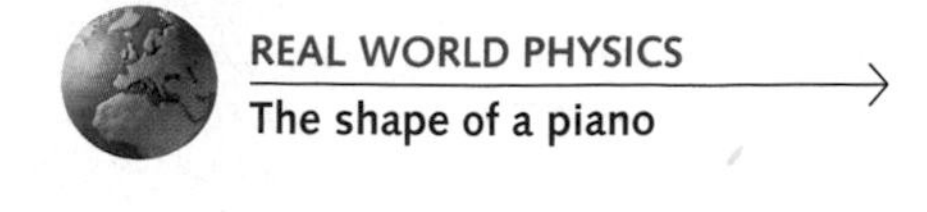

When a guitar string is plucked or a piano string is struck, it vibrates primarily in its fundamental mode, with smaller contributions coming from the higher harmonics. It follows that notes of different pitch can be produced by using strings of different length. Recalling that the fundamental frequency for a string of length L is $f_1 = v/2L$, we see that long strings produce low frequencies and short strings produce high frequencies—all other variables remaining the same.

This fact accounts for the general shape of a piano. Note that the strings shorten toward the right side of the piano, where the notes are of higher frequency. Similarly, a double bass is a larger instrument with longer strings than a violin, as one would expect by the different frequencies the instruments produce. To tune a stringed instrument, the tension in the strings is adjusted—since changing the length of the instrument is impractical. This in turn varies the speed v of waves on the string, and hence the fundamental frequency $f_1 = v/2L$ can be adjusted as desired.

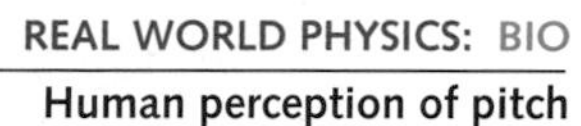

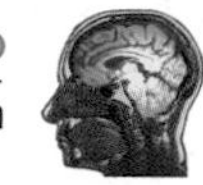

The human ear responds to frequency in a rather interesting and unexpected way. In particular, frequencies that seem to increase by the same amount are in fact increasing by the same multiplicative factor. For example, if three frequencies, f_1, f_2, and f_3, sound equally spaced to our ears you might think that f_2 is greater than f_1 by a certain amount, x, and that f_3 is greater than f_2 by the same amount. Mathematically, we would write this as $f_2 = f_1 + x$ and $f_3 = f_2 + x = f_1 + 2x$. In fact, when we measure the frequencies and compare, we find that f_2 is greater than f_1 by a multiplicative factor x, and that f_3 is greater than f_2 by the same factor; that is, $f_2 = xf_1$ and $f_3 = xf_2 = x^2f_1$.

For instance, middle C on the piano has a frequency of 261.7 Hz. If we move up one octave to the next C the frequency is 523.3 Hz, and going up one more octave the next C is 1047 Hz. Note that with each octave the frequency doubles; that is, it goes up by a multiplicative factor of 2. Since there are 12 semitones in one octave of the chromatic scale, the frequency increase from one semitone to the next is $(2)^{1/12}$. The frequencies for a full chromatic octave are given in Table 14–3.

TABLE 14–3 Chromatic Musical Scale

Note	Frequency (Hz)
Middle C	261.7
C#	277.2
D	293.7
D#	311.2
E	329.7
F	349.2
F#	370.0
G	392.0
G#	415.3
A	440.0
A#	466.2
B	493.9
C	523.3

On a guitar two full octaves and more can be produced on a single string by pressing the string down against frets to effectively change its length. Notice that the separation between frets is not uniform. In particular, suppose the unfretted string has a fundamental frequency of 250 Hz. Since one octave up on the scale would be twice the frequency, 500 Hz, the length of the string must be halved to produce that note. To go to the next octave, and double the frequency again to 1000 Hz, the string must be shortened by a factor of two again, to one quarter its original length. This is illustrated in **Figure 14–25**. Since the distance between successive octaves is decreasing—in this case from $L/2$ to $L/4$—it follows that the spacing between frets must decrease as one goes to higher notes. As a result, the frets on a guitar are always more closely spaced as one moves toward the base of the neck.

▲ Three factors determine the pitch of a vibrating string: mass per unit length, μ; tension, F; and length, L. In an instrument such as a guitar, the first of these factors is fixed once the strings are put on. (Note in the photos that the strings vary in thickness; other things being equal, the heavier the string, the lower the pitch.) The second factor, the tension, can be varied by means of pegs that the player uses to tune the instrument (left), adjusting the pitch of each "open" string to its correct value. The third factor, the length of the string, is the only one that the performer controls while playing. Pressing a string against one of the frets (right), changes its effective length—the length of string that is free to vibrate—and thus the note that is produced.

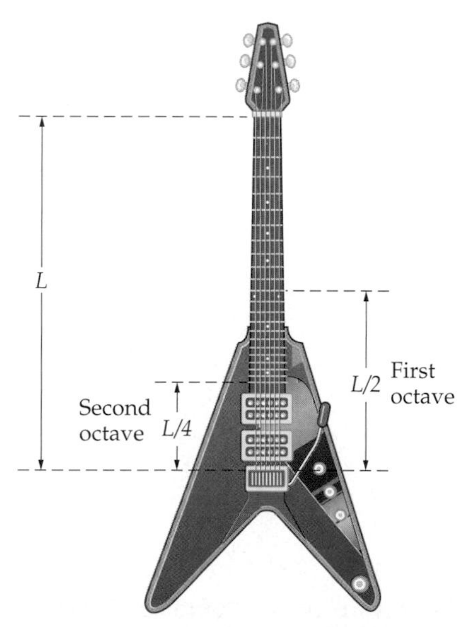

▲ FIGURE 14–25 Frets on a guitar
To go up one octave from the fundamental, the effective length of a guitar string must be halved. To increase one more octave it is necessary to halve the length of the string again. Thus, the distance between frets is not uniform; they are more closely spaced near the base of the neck.

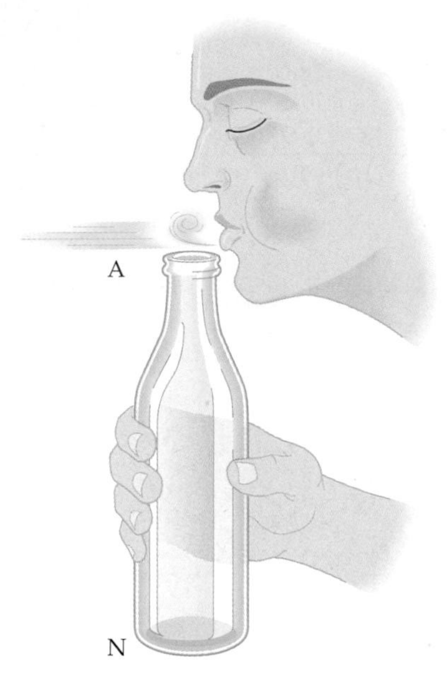

▲ FIGURE 14–26 Exciting a standing wave

When air is blown across the open top of a pop bottle the turbulent air flow can cause an audible standing wave. The standing wave will have an antinode, A, at the top (where the air is moving) and a node, N, at the bottom (where the air cannot move.)

Vibrating Columns of Air

If you blow across the open end of a pop bottle, as in **Figure 14–26**, you hear a tone of a certain frequency. If you pour some water into the bottle and repeat the experiment, the sound you hear has a higher frequency. In both cases you have excited the fundamental mode of the column of air within the bottle. When water was added to the bottle, however, the column of air was shortened, leading to a higher frequency—in the same way that a shortened string has a higher frequency.

Let's examine the situation more carefully. When you blow across the opening in the bottle the result is a swirling movement of air that excites rarefactions and compressions, as illustrated in the figure. For this reason, the opening is an antinode (A) for sound waves. On the other hand, the bottom of the bottle is closed, preventing movement of the air; hence it must be a node (N). Any standing wave in the bottle must have a node at the bottom and an antinode at the top.

The lowest frequency standing wave that is consistent with these conditions is shown in **Figure 14–27 (a)**. If we plot the density variation of the air for this wave, we see that one-quarter of a wavelength fits into the column of air in the bottle. Thus, if the length of the bottle is L, the fundamental has a wavelength satisfying the following:

$$\tfrac{1}{4}\lambda = L$$

$$\lambda = 4L$$

The fundamental frequency, f_1, is given by

$$v = \lambda f_1$$

Solving for f_1 we find

$$f_1 = \frac{v}{\lambda} = \frac{v}{4L}$$

This is half the corresponding fundamental frequency for a wave on a string.

The next harmonic is produced by adding half a wavelength, just as in the case of the string. Thus, if the fundamental is represented by N-A, the second harmonic can be written as N-A-N-A. Since the distance from a node to an antinode is a quarter of a wavelength, we see that $\frac{3}{4}$ of a wavelength fits into the bottle for this mode. This is shown in **Figure 14–27 (b)**. Therefore, $3\lambda/4 = L$, and hence

$$\lambda = \tfrac{4}{3}L$$

As a result, the frequency is

$$\frac{v}{\lambda} = \frac{v}{\frac{4}{3}L} = 3\frac{v}{4L} = 3f_1$$

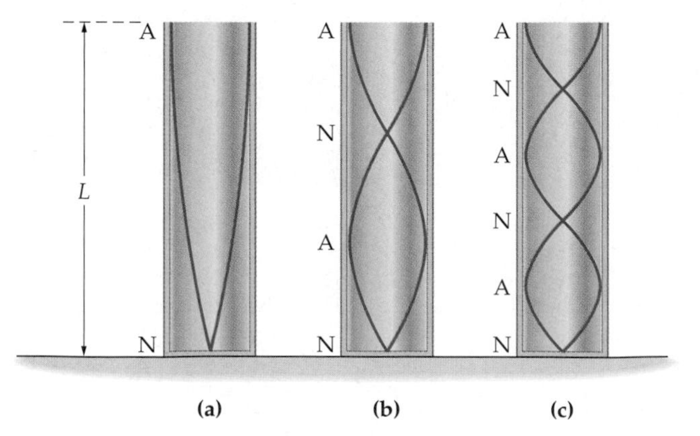

▶ FIGURE 14–27 Standing waves in a pipe that is open at one end

The first three harmonics for waves in a column of air of length L that is open at one end.
(a) $\lambda/4 = L, \lambda = 4L$;
(b) $3\lambda/4 = L, \lambda = 4L/3$;
(c) $5\lambda/4 = L, \lambda = 4L/5$.

Similarly, the next-higher harmonic is represented by N-A-N-A-N-A, as indicated in **Figure 14–27 (c)**. In this case, $5\lambda/4 = L$, and the frequency is

$$\frac{v}{\lambda} = \frac{v}{\frac{4}{5}L} = 5\frac{v}{4L} = 5f_1$$

Clearly, the progression of harmonics for a column of air that is closed at one end is described by the following frequencies and wavelengths:

Standing Waves in a Column of Air Closed at One End

$$f_1 = \frac{v}{4L}$$

$$f_n = nf_1 \qquad n = 1, 3, 5, \ldots \qquad \textbf{14–14}$$

$$\lambda_n = 4L/n$$

Note that only the odd harmonics are present in this case, as opposed to waves on a string, in which all integer harmonics occur.

EXAMPLE 14–9 Pop Music

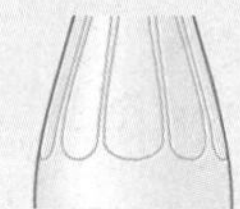

An empty pop bottle is to be used as a musical instrument in a band. In order to be tuned properly the fundamental frequency of the bottle must be 440.0 Hz. If the bottle is 26.0 cm tall, how high should it be filled with water to produce the desired frequency?

Picture the Problem

In our sketch, we label the height of the bottle with H, the height of water with h, and the length of the air column with L. Clearly, $L + h = H$.

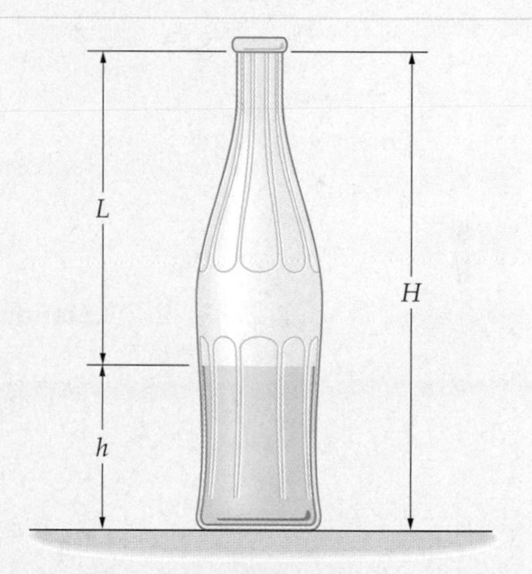

Strategy

Given the frequency of the fundamental ($f_1 = 440.0$ Hz) and the speed of sound in air ($v = 343$ m/s), we can use $f_1 = v/4L$ to solve for the length L of the air column. The height of water is then $h = H - L$.

Solution

1. Solve $f_1 = v/4L$ for the length L: $\quad f_1 = v/4L$, $\quad L = v/4f_1$

2. Substitute numerical values: $\quad L = \dfrac{v}{4f_1} = \dfrac{343 \text{ m/s}}{4(440.0 \text{ Hz})} = 0.195 \text{ m}$

3. Use $h = H - L$ to find the height of the water: $\quad h = H - L = 0.260 \text{ m} - 0.195 \text{ m} = 0.065 \text{ m} = 6.5 \text{ cm}$

Insight

If more water is added to the bottle the air column will shorten, and the fundamental frequency will become higher than 440.0 Hz.

Practice Problem

Calculate the fundamental frequency if the water level is increased to 7.00 cm. [**Answer:** $f_1 = 451$ Hz]

Some related homework problems: Problem 57, Problem 63

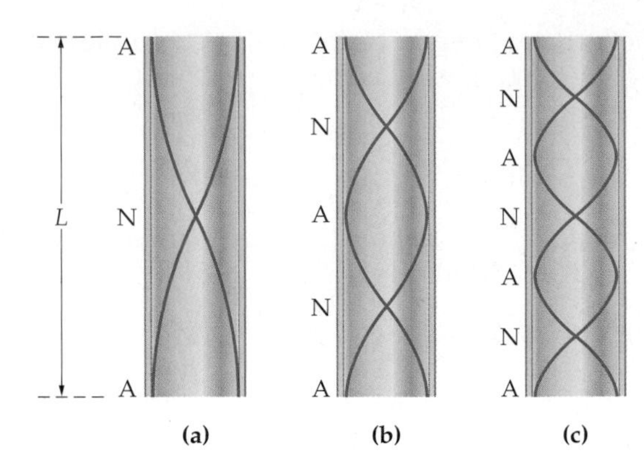

▶ **FIGURE 14–28 Standing waves in a pipe that is open at both ends**
The first three harmonics for waves in a column of air of length L that is open at both ends. **(a)** $\lambda/2 = L$, $\lambda = 2L$; **(b)** $\lambda = L$; **(c)** $3\lambda/2 = L$, $\lambda = 2L/3$.

It is also possible to excite standing waves in columns of air that are open at both ends, as illustrated in **Figure 14–28**. In this case there is an antinode at each end of the column. Hence, the fundamental is A-N-A, as shown in Figure 14–28 (a). Note that half a wavelength fits into the pipe, thus

$$f_1 = \frac{v}{2L}$$

This is the same as the corresponding result for a wave on a string.

The next harmonic is A-N-A-N-A, which fits one complete wavelength in the pipe. This harmonic is shown in Figure 14–28 (b). As a result, the second harmonic has the frequency

$$f_2 = \frac{v}{L} = 2f_1$$

The rest of the harmonics continue in exactly the same manner as for waves on a string. Thus, the frequencies and wavelengths in a column of air open at both ends are as follows:

Standing Waves in a Column of Air Open at Both Ends

$$f_1 = \frac{v}{2L}$$

$$f_n = nf_1 \qquad n = 1, 2, 3, \ldots \qquad \textbf{14–15}$$

$$\lambda_n = 2L/n$$

CONCEPTUAL CHECKPOINT 14–4

If you fill your lungs with helium and speak you sound something like Donald Duck. From this observation, we can conclude that the speed of sound in helium must be **(a)** less than, **(b)** the same as, or **(c)** greater than the speed of sound in air.

Solution

When we speak with helium our words are higher pitched. Looking at Equation 14–15, we see that for the frequency to increase, while the length of the vocal chords remains the same, the speed of sound must be higher.

Answer:

(c) The speed of sound is greater in helium than in air.

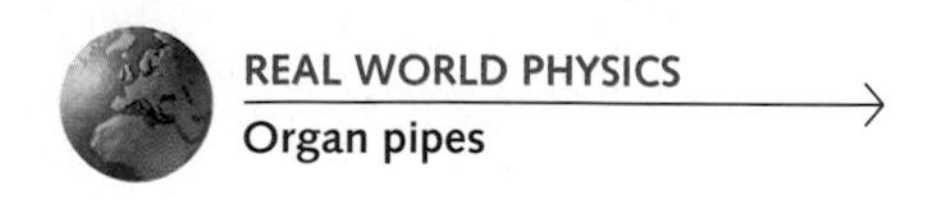

A pipe organ uses a variety of pipes of different length, with some being open at both ends, others open at one end only. When a key is pressed on the console of the organ, air is forced through a given pipe. By accurately adjusting the

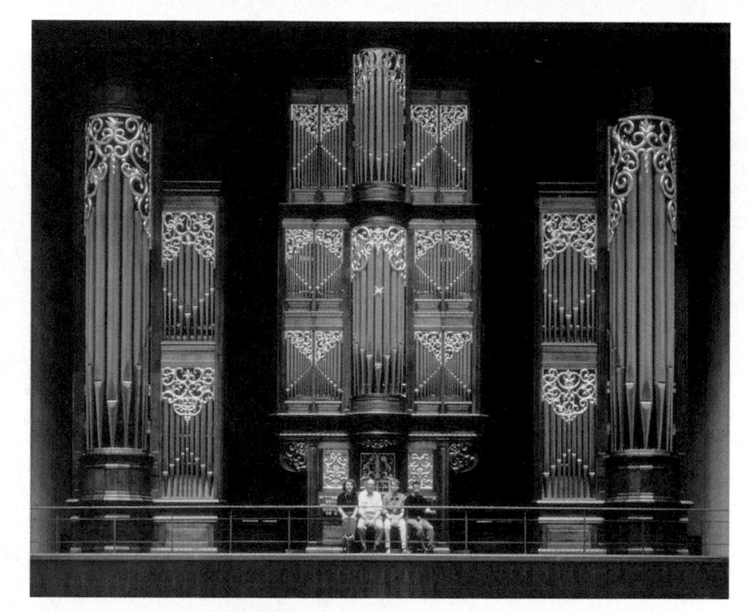

▲ Blowing across the mouth of a bottle (Figure 14–26) sets the air column within the bottle vibrating, producing a tone. This principle is put to use in the pipe organ. A large organ may have hundreds of pipes of different lengths, some open at both ends and some at only one, affording the performer great control over the tonal quality of the sound produced, as well as its pitch.

length of the pipe it can be given the desired tone. In addition, since open and closed pipes have different harmonic frequencies they sound distinctly different to the ear, even if they have the same fundamental frequency. Thus, by judiciously choosing both the length and the type of a pipe, an organ can be given a range of different sounds, allowing it to mimic a trumpet, a trombone, a clarinet, and so on.

14–9 Beats

An interference pattern, such as that shown in Figure 14–21, is a snapshot at a given time, showing locations where constructive and destructive interference occur. It is an interference pattern in space. **Beats**, on the other hand, can be thought of as an interference pattern in time.

To be specific, imagine plucking two guitar strings that have slightly different frequencies. If you listen carefully, you notice that the sound produced by the strings is not constant in time. In fact, the intensity increases and decreases with a definite period. These fluctuations in intensity are the beats, and the frequency of successive maximum intensities is the **beat frequency**.

As an example, suppose two waves, with frequencies $f_1 = 1/T_1$ and $f_2 = 1/T_2$, interfere at a given, fixed location. At this location, each wave moves up and down with simple harmonic motion, as described by Equation 13–2. Applying this result to the vertical position, y, of each wave yields the following:

$$y_1 = A\cos\left(\frac{2\pi}{T_1}t\right) = A\cos(2\pi f_1 t)$$

$$y_2 = A\cos\left(\frac{2\pi}{T_2}t\right) = A\cos(2\pi f_2 t) \qquad \text{14–16}$$

These equations are plotted in **Figure 14–29 (a)**, with $A = 1$, and their superposition, $y_{\text{total}} = y_1 + y_2$, is shown in **Figure 14–29 (b)**.

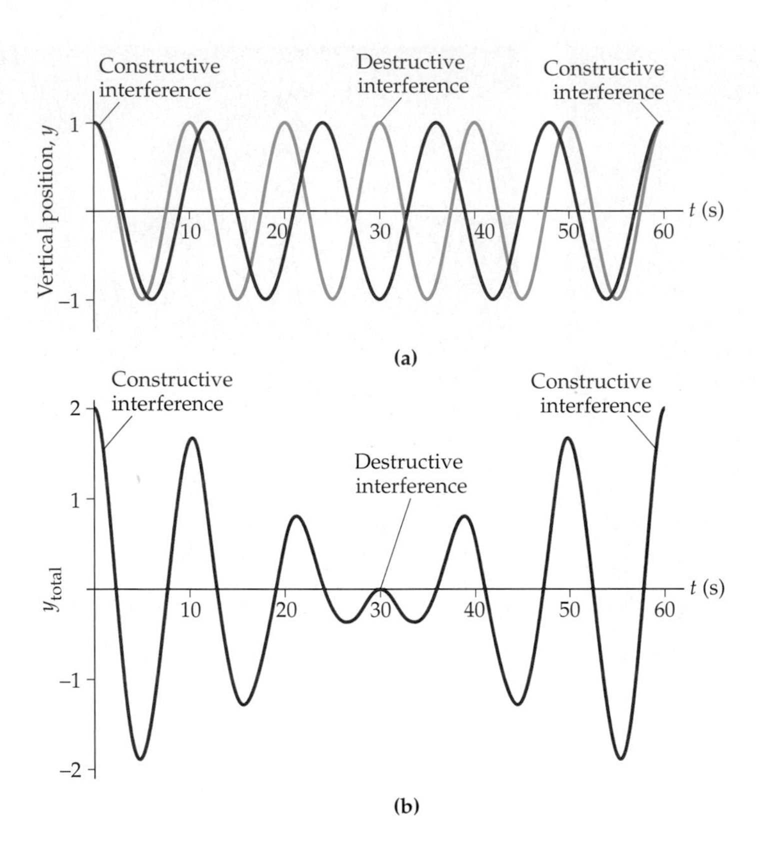

▶ **FIGURE 14–29 Interference of two waves with slightly different frequencies**

(a) A plot of the two waves, y_1(blue) and y_2 (red), given in Equation 14–16. **(b)** The resultant wave y_{total} for the two waves shown in part (a). Note the alternately constructive and destructive interference, leading to beats.

Note that at the time $t = 0$ both y_1 and y_2 are equal to A; thus their superposition gives $2A$. Since the waves have different frequencies, however, they do not stay in phase. At a later time, t_1, we find that $y_1 = A$ and $y_2 = -A$; their superposition gives zero at this time. At a still later time, $t_2 = 2\,t_1$, the waves are again in phase, and add to give $2A$. Thus, a person listening to these two waves hears a sound whose amplitude and loudness varies with time; that is, the person hears beats.

Superposing these waves mathematically, we find

$$\begin{aligned} y_{total} &= y_1 + y_2 \\ &= A\cos(2\pi f_1 t) + A\cos(2\pi f_2 t) \\ &= 2A\cos\left(2\pi\frac{f_1 - f_2}{2}t\right)\cos\left(2\pi\frac{f_1 + f_2}{2}t\right) \end{aligned} \qquad \textbf{14–17}$$

The final step in the expression follows from the trigonometric identities given in Appendix A. The first part of y_{total} is

$$2A\cos\left(2\pi\frac{f_1 - f_2}{2}t\right)$$

This gives the slowly-varying amplitude of the beats, as indicated in **Figure 14–30**. Since a loud sound is heard whenever this term is $2A$ or $-2A$, the beat frequency is

Definition of Beat Frequency

$$f_{beat} = |f_1 - f_2| \qquad \textbf{14–18}$$

SI unit: $1/\mathrm{s} = \mathrm{s}^{-1}$

Finally, the rapid oscillations within each beat are due to the second part of y_{total}:

$$\cos\left(2\pi\frac{f_1 + f_2}{2}t\right)$$

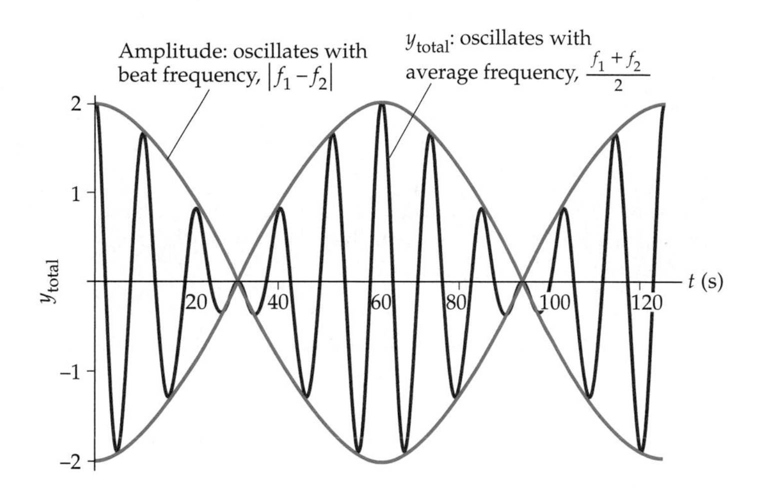

◀ **FIGURE 14–30 Beats**

Beats can be understood as oscillations at the average frequency, modulated by a slowly varying amplitude.

Note that these oscillations have a frequency that is an average of the two input frequencies.

These results apply to any type of wave. In particular, if two sound waves produce beats, your ear will hear the average frequency with a loudness that varies with the beat frequency. For example, suppose the two guitar strings mentioned at the beginning of this section have the frequencies 438 Hz and 442 Hz. If you sound them simultaneously you will hear the average frequency, 440 Hz, increasing and decreasing in loudness with a beat frequency of 4 Hz. This means that you will hear maximum loudness 4 times a second. If the frequencies are brought closer together, the beat frequency will be less and fewer maxima will be heard each second.

Clearly, beats can be used to tune a musical instrument to a desired frequency. To tune a guitar string to 440 Hz, for example, the string can be played simultaneously with a 440-Hz tuning fork. Listening to the beats, the tension in the string can be increased or decreased until the beat frequency becomes vanishingly small. This technique applies only to frequencies that are reasonably close to begin with, since the maximum beat frequencies the ear can detect are about 15 to 20 Hz.

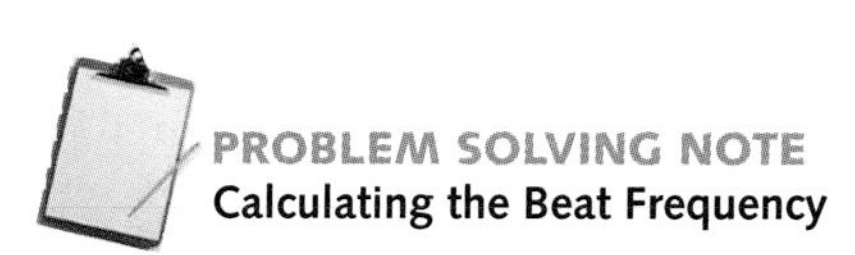

PROBLEM SOLVING NOTE
Calculating the Beat Frequency

The beat frequency of two waves is the *magnitude* of the difference in their frequencies. Thus, the beat frequency is always positive.

EXAMPLE 14–10 Getting a Tune-up

An experimental way to tune the pop bottle in Example 14–9 is to compare its frequency with that of a 440-Hz tuning fork. Initially, a beat frequency of 4 Hz is heard. As a small amount of water is added to that already present, the beat frequency increases steadily to 5 Hz. What were the initial and final frequencies of the bottle?

Picture the Problem

Our sketch shows the before and after situations for this problem. With the low water level the beat frequency is 4 Hz, with the higher level it is 5 Hz.

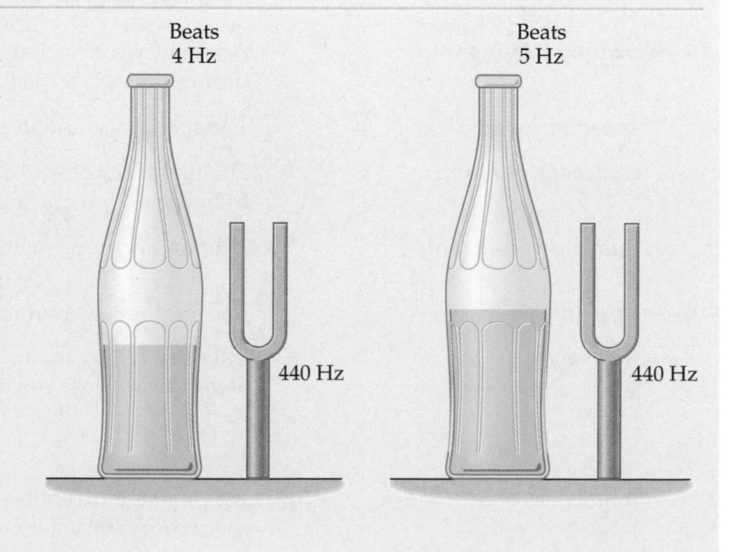

Strategy/Solution

The fact that the initial beat frequency is 4 Hz means the initial frequency of the bottle is either 336 Hz or 444 Hz.

As water is added we know from Example 14–9 that the frequency will increase. We also know that the new beat frequency is 5 Hz, hence the final frequency is either 335 Hz or 445 Hz. Only 445 Hz satisfies the condition that the frequency must have increased.

Hence, the initial frequency is 444 Hz, and the final frequency is 445 Hz.

continued on the following page

continued from the previous page

Insight

In this case, the initial frequency was too high. To tune the bottle properly it is necessary to lower the water level.

Practice Problem

Suppose the initial beat frequency was 4 Hz, and that adding a small amount of water caused the beat frequency to decrease steadily to 2 Hz. What were the initial and final frequencies in this case? [**Answer:** Initial frequency, 336 Hz; final frequency, 338 Hz]

Some related homework problems: Problem 68, Problem 70

Chapter Summary

Topic	Remarks and Relevant Equations	
14–1 Types of Waves	A wave is a propagating disturbance.	
transverse waves and longitudinal waves	In a transverse wave individual particles move at right angles to the direction of wave propagation. In a longitudinal wave individual particles move in the same direction as the wave propagation.	
wavelength, frequency, and speed	The wavelength, λ, frequency, f, and speed, v, of a wave are related by $$v = \lambda f$$	**14–1**
14–2 Waves on a String	Transverse waves can propagate on a string held taut with a tension force, F.	
mass per length	The mass per length of a string is $\mu = m/L$.	
speed of a wave of a string	The speed of a wave on a string with a tension force F and a mass per length μ is $$v = \sqrt{\frac{F}{\mu}}$$	**14–2**
reflections	If the end of a string is fixed, the reflection of a wave is inverted. If the end of a string is free to move transversely, waves are reflected with no inversion.	
***14–3 Harmonic Wave Functions**	A harmonic wave has the shape of a sine or a cosine.	
wave function	A harmonic wave of wavelength λ and period T is described by the following expression: $$y(x, t) = A\cos\left(\frac{2\pi}{\lambda}x - \frac{2\pi}{T}t\right)$$	**14–4**
14–4 Sound Waves	A sound wave is a longitudinal wave of compressions and rarefactions that can travel through the air, as well as other gases, liquids, and solids.	
speed of sound	The speed of sound in air, under typical conditions, is $v = 343$ m/s.	
frequency of sound	The frequency of sound determines its pitch. High-pitched sounds have high frequencies, low-pitched sounds have low frequencies.	
human hearing range	Human hearing extends from 20 Hz to 20,000 Hz.	
14–5 Sound Intensity	The loudness of a sound is determined by its intensity.	
intensity	Intensity, I, is a measure of the amount of energy per time that passes through a given area. Since energy per time is power, P, the intensity of a wave is $$I = \frac{P}{A}$$	**14–5**
point source	If a point source emits sound with a power P, and there are no reflections, the intensity a distance r from the source is $$I = \frac{P}{4\pi r^2}$$	**14–7**

human perception of loudness

The intensity of a sound must be increased by a factor of 10 in order for it to seem twice as loud to our ears.

intensity level and decibels

The intensity level, β, of a sound gives an indication of how loud it sounds to our ears. The intensity level is defined as follows:

$$\beta = 10 \log(I/I_0) \quad \text{14–8}$$

The value of β is given in decibels.

14–6 The Doppler Effect

The change in frequency due to relative motion between a source and a receiver is called the Doppler effect.

moving observer

Suppose an observer is moving with a speed u relative to a stationary source. If the frequency of the source is f, and the speed of the waves is v, the frequency f' detected by the observer is

$$f' = (1 \pm u/v)f \quad \text{14–9}$$

The plus sign applies to the observer approaching the source, and the minus sign to the observer receding from the source.

moving source

If the source is moving with a speed u and the observer is at rest, the observed frequency is

$$f' = \left(\frac{1}{1 \mp u/v}\right)f \quad \text{14–10}$$

The minus sign applies to the source approaching the observer, and the plus sign to the source receding from the observer.

general case

If the observer moves with a speed u_o and the source moves with a speed u_s, the Doppler effect gives

$$f' = \left(\frac{1 \pm u_o/v}{1 \mp u_s/v}\right)f \quad \text{14–11}$$

The meaning of the plus and minus signs is the same as for the moving-observer and moving-source cases given above.

14–7 Superposition and Interference

Waves can combine to give a variety of effects.

superposition

When two or more waves occupy the same location at the same time they simply add, $y_{total} = y_1 + y_2$.

constructive interference

Waves that add to give a larger amplitude exhibit constructive interference.

destructive interference

Waves that add to give a smaller amplitude exhibit destructive interference.

interference patterns

Waves that overlap can create patterns of constructive and destructive interference. These are referred to as interference patterns.

in phase/opposite phase

Two sources are in phase if they both emit crests at the same time. Sources have opposite phase if one emits a crest at the same time the other emits a trough.

14–8 Standing Waves

Standing waves oscillate in a fixed location.

waves on a string

The fundamental, or first harmonic, corresponds to half a wavelength fitting into the length of the string. The fundamental for waves of speed v on a string of length L is

$$f_1 = \frac{v}{2L} \quad \text{14–12}$$

$$\lambda_1 = 2L$$

The higher harmonics, with $n = 1, 2, 3, \ldots,$ are described by

$$f_n = nf_1 \quad \text{14–13}$$

$$\lambda_n = 2L/n$$

vibrating columns of air

The harmonics for a column of air closed at one end are

$$f_1 = \frac{v}{4L}$$

$$f_n = nf_1 \qquad n = 1, 3, 5, \ldots \qquad \textbf{14–14}$$

$$\lambda_n = 4L/n$$

The harmonics for a column of air open at both ends are

$$f_1 = \frac{v}{2L}$$

$$f_n = nf_1 \qquad n = 1, 2, 3, \ldots \qquad \textbf{14–15}$$

$$\lambda_n = 2L/n$$

In both of these expressions the speed of sound is v and the length of the column is L.

14–9 Beats

Beats occur when waves of slightly different frequencies interfere. They can be thought of as interference patterns in time. To the ear, beats are perceived as an alternating loudness and softness to the sound.

beat frequency

If waves of frequencies f_1 and f_2 interfere, the beat frequency is

$$f_{\text{beat}} = |f_1 - f_2| \qquad \textbf{14–18}$$

Problem-Solving Summary

Type of Problem	Relevant Physical Concepts	Related Examples
Find the speed of a wave on a string, or relate the speed of a wave to the mass of a string.	The speed of a wave on a string is related to the tension in the string, F, and the string's mass per length, $\mu = m/L$, by the expression $v = \sqrt{F/\mu}$.	Example 14–1
Relate the intensity of a sound wave to its intensity level.	The intensity level of a sound wave, β, depends on the logarithm of the wave's intensity, I. The relation between β and I is $\beta = 10\log(I/I_0)$, where $I_0 = 10^{-12}\ \text{W/m}^2$.	Active Example 14–2
Calculate the Doppler shift for a moving source or observer.	If an observer and a source of sound with frequency f approach one another, the frequency heard by the observer is greater than f. If the source and observer recede from one another, the frequency heard by the observer is less than f.	Examples 14–5, 14–6
Calculate the beat frequency.	The beat frequency produced when sounds of frequency f_1 and f_2 are heard simultaneously is the magnitude of the difference in frequencies; $f_{\text{beat}} = |f_1 - f_2|$.	Example 14–10

Conceptual Questions

1. A long nail has been driven halfway into the side of a barn. How should you hit the nail with a hammer to generate a longitudinal wave? How should you hit it to generate a transverse wave?
2. What type of wave is exhibited by "amber waves of grain?"
3. At a ball game, a "wave" circulating through the stands can be an exciting event. What type of wave (longitudinal or transverse) are we talking about? Is it possible to change the type of wave? Explain how people might move to accomplish this.
4. In a classic TV commercial a group of cats feed from bowls of cat food that are lined up side by side. Initially there is one cat for each bowl. When an additional cat is added to the scene, it runs to a bowl at the end of the line and begins to eat. The cat that was there originally moves to the next bowl, displacing that cat, which moves to the next bowl, and so on down the line. What type of wave have the cats created? Explain.
5. Consider a wave on a string with constant tension. How does the wavelength change if the frequency is doubled? How does the speed change?
6. To double the speed of a wave on a string, by what factor must you increase the tension?
7. A massive string and a light string have equal tensions. Compare the speed of waves on these strings.

8. A massive string and a light string have waves of equal speed. Compare the tensions in the two strings.
9. A harmonic wave travels along a string. Is the kinetic energy of the string a maximum where the displacement of the string is a maximum or a minimum? Explain. At what points along the string is the kinetic energy of the string a minimum?
10. A harmonic wave travels along a string. Is the potential energy of the string a maximum where the displacement of the string is a maximum or a minimum? Explain. At what points along the string is the potential energy of the string a minimum?
11. If the distance to a point source of sound is doubled, by what factor does the intensity decrease?
12. Describe how the sound of a symphony played by an orchestra might be altered if the speed of sound depended on the frequency of sound.
13. You are heading toward an island in your speedboat when you see a friend standing on shore at the base of a cliff. You sound the boat's horn to get your friend's attention. Compare the frequency of the horn with the frequency heard by your friend and the frequency of the echo when it returns to the boat. Which of these frequencies is highest? Which is lowest?
14. A "radar gun" is often used to measure the speed of a major league pitch by reflecting a beam of radio waves off a moving ball. Describe how the Doppler effect can give the speed of the ball from a measurement of the frequency of the reflected beam.
15. A moving source produces sound with a frequency f_0. A stationary observer hears a frequency $2f_0$. What is the speed of the source, in terms of the speed of sound?
16. A stationary source produces sound with a frequency f_0. An observer moving toward the source hears a frequency $2f_0$. What is the speed of the observer, in terms of the speed of sound?
17. When you drive a nail into a piece of wood you hear a tone with each blow of the hammer. In fact, the tone increases in pitch as the nail is driven further into the wood. Explain.
18. Explain the function of the sliding part of a trombone.
19. When you tune a violin string, what causes its frequency to change?
20. On a guitar, some strings are single wires, others are wrapped with another wire to increase the mass per length. Which type of string would you expect to be used for a low-frequency note? Explain.
21. As a string oscillates in its fundamental mode there are times when it is completely flat. Is the energy of oscillation zero at these times? Explain.
22. When you blow across the opening of a two-liter pop bottle that is partially filled you hear a tone. If you take a sip of the pop and blow across the opening again, is the tone you hear higher in frequency or lower in frequency? Explain.
23. When two guitar strings are plucked at the same time, a beat frequency of 2 Hz is heard. If string 1 is tightened, the beat frequency increases to 3 Hz. Which of the two strings had the lower frequency initially? Explain.
24. On a rainy day while driving your car, you notice that your windshield wipers are moving in synchrony with the wiper blades of the car in front of you. After several cycles, however, your wipers and the wipers of the other car are moving opposite to one another. A short time later the wipers are synchronous again. What wave phenomena do the wipers illustrate? Explain.
25. Compare the beat frequency produced by a 245 Hz tone and a 240 Hz tone with the beat frequency produced by a 140 Hz tone and a 145 Hz tone.
26. To play C major on the piano you hit the C, E, and G keys simultaneously. When you do so, you hear no beats. Why?

Problems

Note: **IP** *denotes an integrated conceptual/quantitative problem.* **BIO** *identifies problems of biological or medical interest. Blue bullets (•, ••, •••) are used to indicate the level of difficulty of each problem.*

Section 14–1 Types of Waves

1. • A wave travels along a stretched rope. The vertical distance from crest to trough for this wave is 15 cm and the horizontal distance from crest to trough is 25 cm. What is **(a)** the wavelength and **(b)** the amplitude of this wave?
2. • A surfer floating beyond the breakers notes 14 waves per minute passing his position. If the wavelength of these waves is 34 m, what is their speed?
3. • The speed of surface waves in water decreases as the water becomes shallower. Suppose waves travel across the surface of a lake with a speed of 2.0 m/s and a wavelength of 1.5 m. When these waves move into a shallower part of the lake their speed decreases to 1.6 m/s, though their frequency remains the same. Find the wavelength of the waves in the shallower water.
4. • A typical tidal wave (tsunami) can have a speed of 750 km/h and a wavelength of 310 km. What is the frequency of such a wave?
5. •• A 4.0-Hz wave with an amplitude of 12 cm and a wavelength of 30.0 cm travels along a stretched string. **(a)** How far does the wave travel in a time interval of 5.0 s? **(b)** How far does a knot on the string travel in the same time interval?
6. •• The speed of a deep water wave with a wavelength λ is approximately $v = \sqrt{g\lambda/2\pi}$. Find the speed and frequency of a deep water wave with a wavelength of 4.5 m.
7. •• In shallow water of depth d the speed of waves is approximately $v = \sqrt{gd}$. Find the speed and frequency of a wave with wavelength 0.75 cm in water that is 2.6 cm deep.

Section 14–2 Waves on a String

8. • Waves on a particular string travel with a speed of 16 m/s. By what factor should the tension in this string be changed to produce waves with a speed of 32 m/s?
9. •• A brother and sister try to communicate with a string tied between two tin cans (**Figure 14–31**, p. 456). If the string is 9.5 m long, has a mass of 32 g, and is pulled taut with a tension of 8.6 N, how long does it take a wave to travel from one end of the string to the other?
10. •• **IP (a)** Suppose the tension is increased in the previous problem. Does a wave take more, less, or the same time to travel from one end to the other? **(b)** Calculate the time of travel for tensions of 9.0 N and 10.0 N.

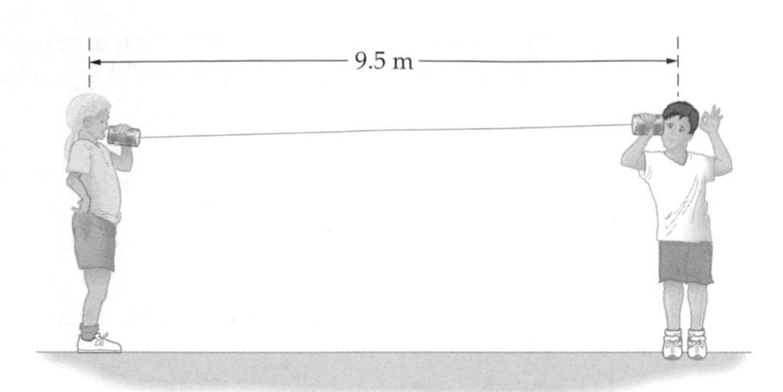

▲ **FIGURE 14–31** Problems 9 and 10

11. •• **IP** A 7.3-m wire with a mass of 85 g is attached to the mast of a sailboat. If the wire is given a "thunk" at one end, it takes 0.94 s for the resulting wave to reach the other end. **(a)** What is the tension in the wire? **(b)** Would the tension be larger or smaller if the mass of the wire were increased? **(c)** Calculate the tension for a 100.0-g wire.

12. •• Two steel guitar strings have the same length. String A has a diameter of 0.50 mm and is under 400.0 N of tension. String B has a diameter of 1.0 mm and is under a tension of 800.0 N. Find the ratio of the wave speeds, v_A/v_B, in these two strings.

13. ••• Use dimensional analysis to show how the speed v of a wave on a string of circular cross section depends on the tension in the string, T, the radius of the string, R, and its mass per volume, ρ.

*Section 14–3 Harmonic Wave Functions

14. • Write an expression for a harmonic wave with an amplitude of 0.21 m, a wavelength of 3.0 m, and a period of 2.0 s. The wave travels to the right.

15. • Write an expression for a harmonic wave that has a wavelength of 2.8 m and propagates to the left with a speed of 13.3 m/s. The amplitude of the wave is 0.12 m.

16. •• A wave on a string is described by the following equation:

$$y = (15\text{ cm})\cos\left(\frac{\pi}{5.0\text{ cm}}x - \frac{\pi}{12\text{ s}}t\right)$$

(a) What is the amplitude of this wave? **(b)** What is its wavelength? **(c)** What is its period? **(d)** What is its speed? **(e)** In which direction does the wave travel?

17. •• Consider the wave function given in the previous problem. Sketch this wave from $x = 0$ to $x = 10$ cm for the following times: **(a)** $t = 0$; **(b)** $t = 3.0$ s; **(c)** $t = 6.0$ s. **(d)** What is the least amount of time required for a given point on this wave to move from $y = 0$ to $y = 15$ cm? Verify your answer by referring to the sketches for parts (a), (b), and (c).

18. •• Four waves are described by the following equations, in which all distances are measured in centimeters and all times are measured in seconds:

$$y_A = 10\cos(3x - 4t)$$
$$y_B = 10\cos(5x + 4t)$$
$$y_C = 20\cos(-10x + 60t)$$
$$y_D = 20\cos(-4x - 20t)$$

(a) Which of these waves travel in the $+x$ direction? **(b)** Which of these waves travel in the $-x$ direction? **(c)** Which wave has the highest frequency? **(d)** Which wave has the greatest wavelength? **(e)** Which wave has the greatest speed?

Section 14–4 Sound Waves

19. • At Zion National Park a loud shout produces an echo 2.1 s later from a colorful sandstone cliff. How far away is the cliff?

20. • A dolphin sends out a series of high-pitched clicks that are reflected back from the bottom of the ocean 95.5 m below. How much time elapses before the dolphin hears the echoes of the clicks? (The speed of sound in seawater is approximately 1530 m/s.)

21. • The lowest note on a piano is A, four octaves below the A given in Table 14–3. The highest note on a piano is a C, four octaves above middle C. Find the frequencies and wavelengths of these notes.

22. •• **IP** A sound wave in air has a frequency of 400.0 Hz. **(a)** What is its wavelength? **(b)** If the frequency of the sound is increased, does its wavelength increase, decrease, or stay the same? Explain. **(c)** Calculate the wavelength for a sound wave with a frequency of 450 Hz.

23. •• **IP** When you drop a rock into a well you hear the splash 1.5 seconds later. **(a)** How deep is the well? **(b)** If the depth of the well were doubled, would the time required to hear the splash be greater than, less than, or equal to 3.0 seconds? Explain.

24. •• A rock is thrown downward into a well that is 8.8 m deep. If the splash is heard 1.2 seconds later, what was the initial speed of the rock?

Section 14–5 Sound Intensity

25. • The intensity level in a truck is 90.0 dB. What is the intensity of this sound?

26. • The distance to a point source is tripled. **(a)** By what factor does the intensity decrease? **(b)** By what amount does the intensity level decrease?

27. • Sound 1 has an intensity of 200.0 W/m^2. Sound 2 has an intensity level that is 2.5 dB greater than the intensity level of sound 1. What is the intensity of sound 2?

28. •• A bird watcher is hoping to add the song sparrow to the list of species she has seen. How far could she be from the sparrow in Example 14–3 and still hear it? Assume no reflections or absorption of the sparrow's sound.

29. •• Residents of Hawaii are warned of the approach of a tsunami (tidal wave) by sirens mounted on the top of towers. Suppose a siren produces a sound that has an intensity level of 120 dB at a distance of 2.0 m. Treating the siren as a point source of sound, and ignoring reflections and absorption, find the intensity level heard by an observer at a distance of **(a)** 12 m and **(b)** 21 m from the siren. **(c)** How far away can the siren be heard?

30. •• In a pig-calling contest, a caller produces a sound with an intensity level of 110 dB. How many such callers would be required to reach the pain level of 120 dB?

31. •• Twenty violins playing simultaneously with the same intensity combine to give an intensity level of 80.0 dB. What is the intensity level of each violin?

32. •• **BIO** The radius of a typical human eardrum is about 4.0 mm. Find the energy per second received by an eardrum when it listens to sound that is **(a)** at the threshold of hearing and **(b)** at the threshold of pain.

33. ••• A point source of sound that emits uniformly in all directions is located in the middle of a large, open field. The intensity at Brittany's location directly north of the source is twice that at Phillip's position due east of the source. What is the distance between Brittany and Phillip if Brittany is 10.0 m from the source?

Section 14–6 The Doppler Effect

34. • A person with perfect pitch sits on a bus bench listening to the 450-Hz horn of an approaching car. If the person detects a frequency of 470 Hz, how fast is the car moving?

35. • A train moving with a speed of 23.2 m/s sounds a 124-Hz horn. What frequency is heard by an observer standing near the tracks?

36. • In the previous problem, suppose the stationary observer sounds a horn that is identical to the one on the train. What frequency is heard from this horn by a passenger in the train?

37. • **BIO** A bat moving with a speed of 3.60 m/s and emitting sound of 35.0 kHz approaches a moth at rest on a tree trunk. **(a)** What frequency is heard by the moth? **(b)** If the speed of the bat is increased is the frequency heard by the moth higher or lower? **(c)** Calculate the frequency heard by the moth when the speed of the bat is 4.50 m/s.

38. • A motorcycle and a police car are moving toward one another. The police car emits sound with a frequency of 512 Hz and has a speed of 27.0 m/s. The motorcycle has a speed of 13.0 m/s. What frequency does the motorcyclist hear?

39. • In the previous problem, suppose that the motorcycle and the police car are moving in the same direction. What frequency does the motorcyclist hear in this case?

40. •• Hearing the siren of an approaching fire truck you pull over to the side of the road and stop. As the truck approaches, you hear a tone of 460 Hz; as the truck recedes, you hear a tone of 410 Hz. How long will it take for the truck to get to the fire 5.0 km away, assuming it maintains a constant speed?

41. •• With what speed must you approach a source of sound to observe a 10% change in frequency?

42. •• **IP** A particular jet engine produces a tone of 400.0 Hz. Suppose that one jet is at rest on the tarmac while a second identical jet flies overhead at 9/10 the speed of sound. The pilot of each jet listens to the sound produced by the engine of the other jet. **(a)** Which pilot hears a greater Doppler shift? Explain. **(b)** Calculate the frequency heard by the pilot in the moving jet. **(c)** Calculate the frequency heard by the pilot in the stationary jet.

43. •• **IP** Two bicycles approach one another, each traveling with a speed of 8.5 m/s. **(a)** If bicyclist A beeps a 300.0-Hz horn, what frequency is heard by bicyclist B? **(b)** Which of the following would cause the greater increase in the frequency heard by bicyclist B: (i) Bicyclist A speeds up by 1.5 m/s, or (ii) bicyclist B speeds up by 1.5 m/s? Explain.

44. •• A train on one track moves in the same direction as a second train on the adjacent track. The first train, which is ahead of the second train and moves with a speed of 32 m/s, blows a horn whose frequency is 125 Hz. If the frequency heard on the second train is 131 Hz, what is its speed?

45. •• Two cars traveling with the same speed move directly away from one another. One car sounds a horn whose frequency is 205 Hz and a person in the other car hears a frequency of 192 Hz. What is the speed of the cars?

46. ••• The Shinkansen, the Japanese "bullet" train, runs at high speed from Tokyo to Nagoya. Riding on the Shinkansen, you notice that the frequency of a crossing signal changes markedly as you pass the crossing. As you approach the crossing the frequency you hear is f; as you recede from the crossing the frequency you hear is $2f/3$. What is the speed of the train?

Section 14–7 Superposition and Interference

47. • Two wave pulses on a string approach one another at the time $t = 0$, as shown in **Figure 14–32**. Each pulse moves with a speed of 1.0 m/s. Make a careful sketch of the resultant wave at the times $t = 1.0$ s, 2.0 s, 2.5 s, 3.0 s, and 4.0 s, assuming that the superposition principle holds for these waves.

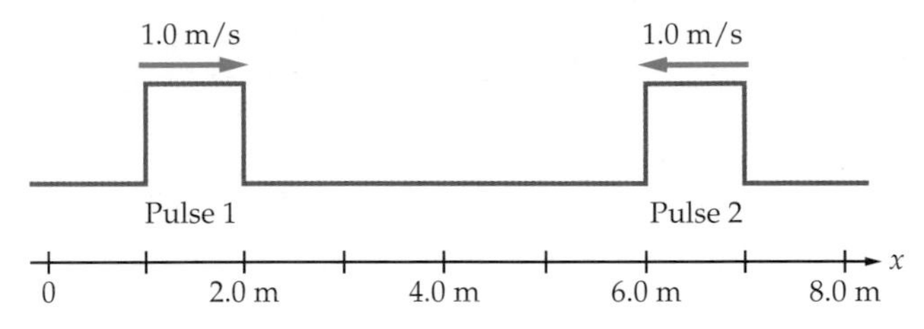

▲ **FIGURE 14–32** Problems 47 and 48

48. • Suppose pulse 2 in Problem 47 is inverted, so that it is a downward deflection of the string rather than an upward deflection. Repeat Problem 47 in this case.

49. • Two wave pulses on a string approach one another at the time $t = 0$, as shown in **Figure 14–33**. Each pulse moves with a speed of 1.0 m/s. Make a careful sketch of the resultant wave at the times $t = 1.0$ s, 2.0 s, 2.5 s, 3.0 s, and 4.0 s, assuming that the superposition principle holds for these waves.

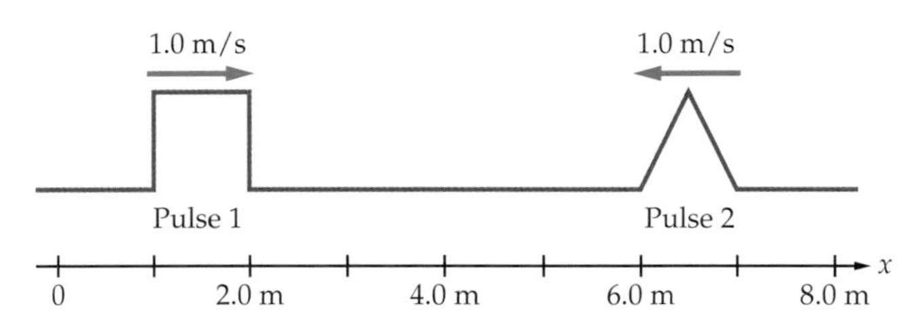

▲ **FIGURE 14–33** Problems 49 and 50

50. • Suppose pulse 2 in Problem 49 is inverted, so that it is a downward deflection of the string rather than an upward deflection. Repeat Problem 49 in this case.

51. •• A pair of in-phase stereo speakers are placed next to each other, 0.60 m apart. You stand directly in front of one of the speakers, 1.0 m from the speaker. What is the lowest frequency that will produce constructive interference at your location?

52. •• **IP** Two violinists, one directly behind the other, play for a listener directly in front of them. Both violinists sound concert A (440 Hz). **(a)** What is the smallest separation between the violinists that will produce destructive interference for the listener? **(b)** Does this smallest separation increase or decrease if the violinists produce a note with a higher frequency? **(c)** Repeat part (a) for violinists who produce sounds of 500.0 Hz.

53. •• Two loudspeakers are placed at either end of a gymnasium, both pointing toward the center of the gym and equidistant from it. The speakers emit 256-Hz sound that is in phase. An observer at the center of the gym experiences constructive interference. How far toward one speaker must the observer walk to first experience destructive interference?

54. •• **IP** **(a)** In the previous problem, does the required distance increase, decrease, or stay the same if the frequency of the speakers is lowered? **(b)** Calculate the distance to the first position of destructive interference if the frequency emitted by the speakers is lowered to 240 Hz.

55. •• Two speakers with opposite phase are positioned 3.5 m apart, both pointing toward a wall 5.0 m in front of them (Figure 14–34). An observer standing against the wall midway between the speakers hears destructive interference. If the observer hears constructive interference after moving 0.84 m to one side along the wall, what is the frequency of the sound emitted by the speakers?

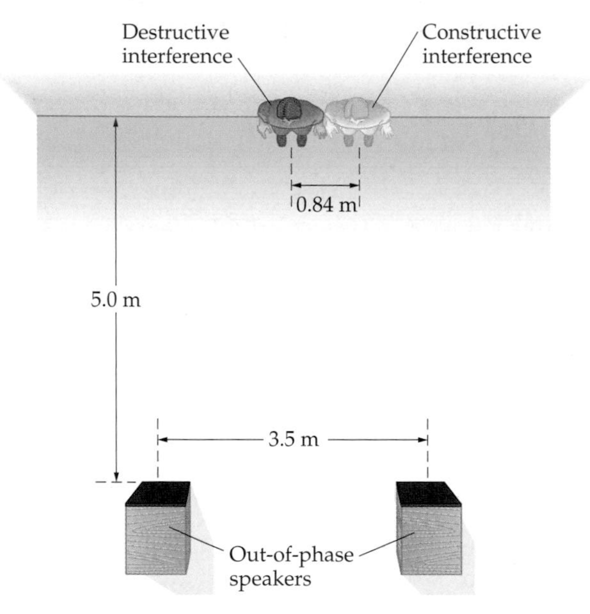

▲ **FIGURE 14–34** Problem 55

56. •• Suppose, in Example 14–7, that the speakers have opposite phase. What is the lowest frequency that gives destructive interference in this case?

Section 14–8 Standing Waves

57. • An open-ended organ pipe is 4.5 m long. What is its fundamental frequency?

58. • A string 1.5 m long with a mass of 2.1 g is stretched between two fixed points with a tension of 95 N. Find the frequency of the fundamental on this string.

59. •• **IP BIO** The human ear canal is much like an organ pipe that is closed at one end (at the tympanic membrane, or eardrum) and open at the other (see Figure 14–35). A typical ear canal has a length of about 2.8 cm. **(a)** What is the fundamental frequency of the ear canal? **(b)** Suppose a person has an ear canal that is longer than 2.8 cm. Is the fundamental frequency of that person's ear canal greater than or less than the value found in part (a)? Explain. **(c)** Find the length of an ear canal whose fundamental frequency is 2600 Hz. **(d)** Referring to the ear canal of part (c), what is the wavelength and frequency of its second harmonic?

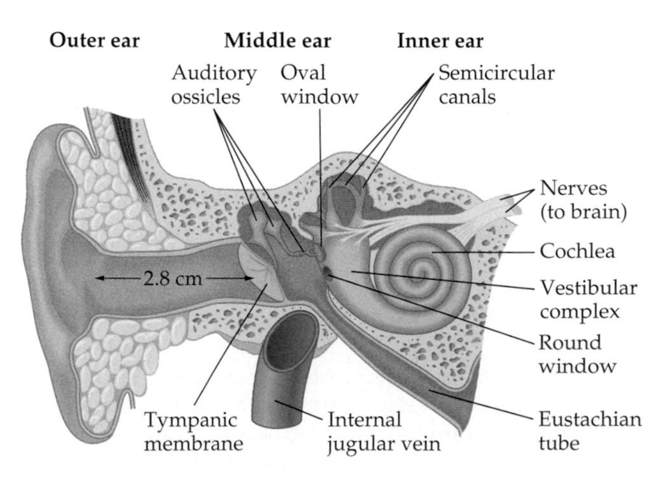

▲ **FIGURE 14–35** Problem 59

60. •• A guitar string 60 cm long vibrates with a standing wave that has three antinodes. **(a)** Which harmonic is this? **(b)** What is the wavelength of this wave?

61. •• A 12.5-g clothesline is stretched with a tension of 20.1 N between two poles 7.66 m apart. What is the frequency of **(a)** the fundamental and **(b)** the second harmonic?

62. •• **IP (a)** In the previous problem, will the frequencies increase, decrease, or stay the same if a more massive rope is used? **(b)** Repeat Problem 61 for a clothesline with a mass of 15.0 g.

63. •• The organ pipe in Figure 14–36 is 2.5 m long. **(a)** What is the frequency of the standing wave shown in the pipe? **(b)** What is the fundamental frequency of this pipe?

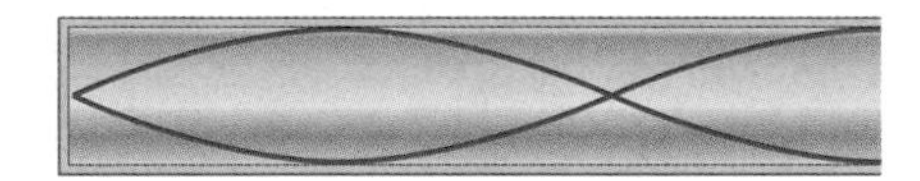

▲ **FIGURE 14–36** Problem 63

64. •• The frequency of the standing wave shown in Figure 14–37 is 232 Hz. **(a)** What is the fundamental frequency of this pipe? **(b)** What is the length of the pipe?

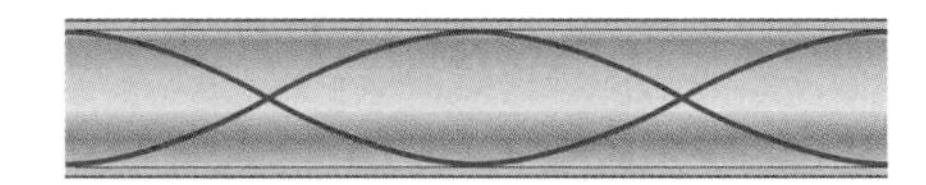

▲ **FIGURE 14–37** Problem 64

65. ••• An open organ pipe has a harmonic with a frequency of 440 Hz. The next higher harmonic in the pipe has a frequency of 522 Hz. Find **(a)** the frequency of the fundamental and **(b)** the length of the pipe.

Section 14–9 Beats

66. • Two tuning forks have frequencies of 275 Hz and 292 Hz. What is the beat frequency if both tuning forks are sounded simultaneously?

67. • To tune middle C on a piano, a tuner hits the key and at the same time sounds a 261-Hz tuning fork. If the tuner hears 3 beats per second, what are the possible frequencies of the piano key?

68. • Two musicians are comparing their clarinets. The first clarinet produces a tone that is known to be 441 Hz. When the two clarinets play together they produce eight beats every 2.00 seconds. If the second clarinet produces a higher pitched tone than the first clarinet, what is the second clarinet's frequency?

69. •• **IP** Two strings that are fixed at each end are identical, except that one is 0.560 cm longer than the other. Waves on these strings propagate with a speed of 34.2 m/s, and the fundamental frequency of the shorter string is 212 Hz. **(a)** What beat frequency is produced if each string is vibrating with its fundamental frequency? **(b)** Does the beat frequency in part (a) increase or decrease if the longer string is increased in length? **(c)** Repeat part (a) assuming that the longer string is 0.761 cm longer than the shorter string.

70. •• A tuning fork with a frequency of 310.0 Hz and a tuning fork of unknown frequency produce beats with a frequency of 4.5 Hz. If the frequency of the 310.0-Hz fork is lowered slightly by placing a bit of putty on one of its tines, the new beat frequency is 6.5 Hz. What is **(a)** the final frequency of the 310.0-Hz tuning fork and **(b)** the frequency of the other tuning fork?

71. •• Identical cellos are being tested. One is producing a fundamental frequency of 258 Hz on a string that is 1.15 m long and has a mass of 100.0 g. On the second cello the same string is fingered to reduce the length that can vibrate. If the beat frequency produced by these two strings is 4.0 Hz, what is the vibrating length of the second string?

72. ••• A friend in another city tells you that she has a pair of organ pipes, one open at both ends, the other open at one end only. In addition, she has determined that the beat frequency caused by the second-lowest frequency of each pipe is equal to the beat frequency caused by the third-lowest frequency of each pipe. Her challenge to you is to calculate the length of the organ pipe that is open at both ends, given that the length of the other pipe is 1.00 m.

General Problems

73. • Sitting peacefully in your living room one stormy day you see a flash of lightning through the windows. Five seconds later thunder shakes the house. Estimate the distance from your house to the bolt of lightning.

74. • The fundamental of a closed organ pipe is 261.6 Hz (middle C). The second harmonic of an open organ pipe has the same frequency. What are the lengths of these two pipes?

75. • A standing wave of 603 Hz is produced on a string that is 1.33 m long and fixed on both ends. If the speed of waves on this string is 402 m/s, how many antinodes are there in the standing wave?

76. •• A machine shop has 100 equally noisy machines that together produce an intensity level of 90 dB. If the intensity level must be reduced to 80 dB, how many machines must be turned off?

77. •• When you blow across the top of a pop bottle you hear a fundamental frequency of 182 Hz. If the bottle is now filled with helium, what is the new frequency of the fundamental? (Assume that the speed of sound in helium is three times that in air.)

78. •• Tsunamis (tidal waves) can have wavelengths between 100 and 400 km. Since this is much greater than the average depth of the oceans (about 4.3 km), the ocean can be considered as shallow water for these waves. Using the speed of waves in shallow water of depth d given in Problem 7, find the typical speed for a tsunami. (*Note:* In the open ocean, tsunamis generally have an amplitude of less than a meter, allowing them to pass ships unnoticed. As they approach shore, however, the water depth decreases and the waves slow down. This can result in an increase of amplitude to as much as 37 m or more.)

79. •• Two trains with 124-Hz horns approach one another. The slower of the two trains has a speed of 22 m/s. What is the speed of the fast train if an observer standing near the tracks between the trains hears a beat frequency of 4.4 Hz?

80. •• **IP** Jim is speeding toward James Island with a speed of 24 m/s when he sees Betsy standing on shore at the base of a cliff (**Figure 14–38**). Jim sounds his 330-Hz horn. **(a)** What frequency does Betsy hear? **(b)** Jim can hear the echo of his horn reflected back to him by the cliff. Is the frequency of this echo greater than or equal to the frequency heard by Betsy? Explain. **(c)** Calculate the frequency Jim hears in the echo from the cliff.

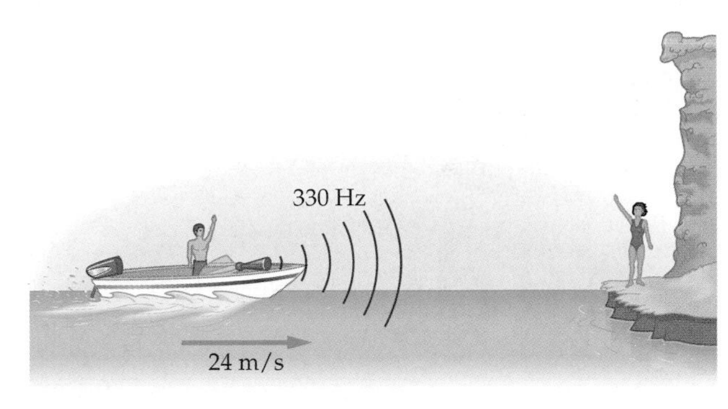

▲ **FIGURE 14–38** Problem 80

81. •• Two ships in a heavy fog are blowing their horns, both of which produce sound with a frequency of 165 Hz. One ship is at rest; the other moves on a straight line that passes through the one at rest. If people on the stationary ship hear a beat frequency of 3.0 Hz, what are the two possible speeds and directions of motion of the moving ship?

82. •• Give the factor by which the fundamental frequency of a guitar string changes under the following conditions: **(a)** The tension in the string is increased by a factor of 4. **(b)** The diameter of the string is increased by a factor of 3. **(c)** The length of the string is halved.

83. •• A slinky has a mass of 0.23 kg and negligible length. When it is stretched 1.5 m it is found that transverse waves travel the length of the slinky in 0.75 s. **(a)** What is the force constant, k, of the slinky? **(b)** If the slinky is stretched farther, will the time required for a wave to travel the length of the slinky increase, decrease, or stay the same? Explain. **(c)** If the slinky is stretched 3.0 m, how long does it take a wave to travel the length of the Slinky?

84. •• An organ pipe 1.5 m long is open at one end and closed at the other end. What is the linear distance between a node and the adjacent antinode for the third harmonic in this pipe?

85. •• Two identical strings with the same tension vibrate at 631 Hz. If the tension in one of the strings is increased by 2% what is the resulting beat frequency?

86. ••• A rope of length L and mass M hangs vertically from a ceiling. The tension in the rope is only that due to its own weight. Show that the speed of waves a height y above the bottom of the rope is $v = \sqrt{gy}$.

87. ••• Experiments on water waves show that the speed of waves in shallow water is independent of their wavelength. Using this observation and dimensional analysis, determine how the speed v of shallow-water waves depends on the depth of the water, d, the mass per volume of water, ρ, and the acceleration of gravity, g.

88. ••• A deep water wave of wavelength λ has a speed given approximately by $v = \sqrt{g\lambda/2\pi}$. Find an expression for the period of a deep water wave in terms of its wavelength. (Note the similarity of your result to the period of a pendulum.)

89. ••• In Problem 53, suppose the observer walks toward one speaker with a speed of 1.35 m/s. **(a)** What frequency does the observer hear from each speaker? **(b)** What beat frequency does the observer hear? **(c)** How far must the observer walk to go from one point of constructive interference to the next? (d) How many times per second does the observer hear maximum loudness from the speakers? Compare your result with the beat frequency from part (b).

15
Fluids

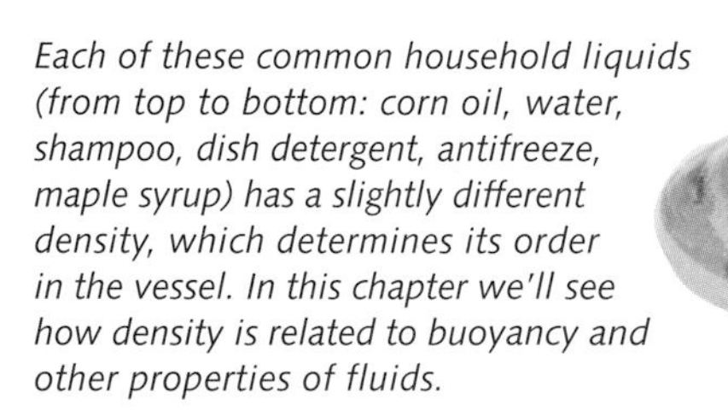

Each of these common household liquids (from top to bottom: corn oil, water, shampoo, dish detergent, antifreeze, maple syrup) has a slightly different density, which determines its order in the vessel. In this chapter we'll see how density is related to buoyancy and other properties of fluids.

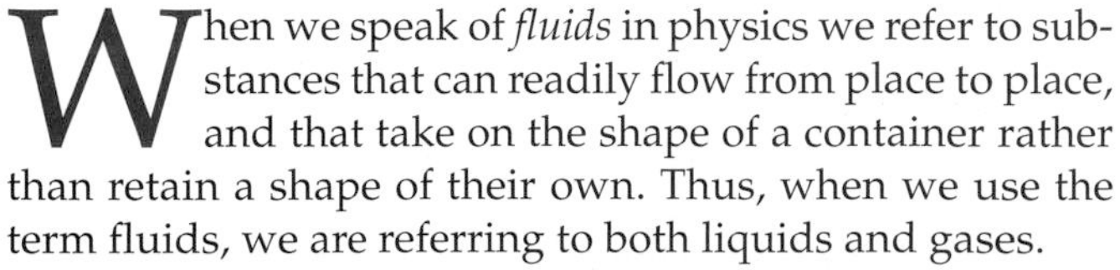

When we speak of *fluids* in physics we refer to substances that can readily flow from place to place, and that take on the shape of a container rather than retain a shape of their own. Thus, when we use the term fluids, we are referring to both liquids and gases.

It is hard to think of a subject more relevant to our everyday lives than fluids. After all, we begin life as a fluid-filled cell suspended in a fluid. We live our independent lives immersed in a fluid that we breathe. In fact, fluids coursing through our circulatory system are literally the lifeblood of our existence. If it were not for the gases in our atmosphere and the liquid water on the Earth's surface we could not exist.

In this chapter, we examine some of the fundamental physical principles that apply to fluids. All of these principles derive from the basic physics we have learned to this point. For example, straightforward considerations of force and weight lead to an understanding of buoyancy. Similarly, the work–energy theorem results in an understanding of how fluids behave when they flow. As such, fluids provide a wonderful opportunity for us to apply our knowledge of physics to a whole new array of interesting physical systems.

15–1 Density

The properties of a fluid can be hard to pin down, given that it can flow, change shape, and either split into smaller portions or combine into a larger system. Thus, one of the best ways to quantify a fluid is in terms of its **density**. Specifically, the density, ρ, of a material (fluid or not) is defined as the mass, M, per volume, V:

Definition of density, ρ

$$\rho = M/V \qquad 15\text{–}1$$

SI unit: $\mathrm{kg/m^3}$

The denser a material, the more mass it has in any given volume. Note, however, that the density of a substance is the same regardless of the total amount we have in a system.

To get a feel for densities in common substances, we start with water. For example, to fill a cubic container one meter on a side would take over 2000 pounds of water. More precisely, water has the following density:

$$\rho_{\mathrm{w}} = \text{density of water} = 1000\ \mathrm{kg/m^3}$$

A gallon (1 gallon $= 3.79\ \mathrm{L} = 3.79 \times 10^{-3}\ \mathrm{m^3}$) of water, then, has a mass of

$$M = \rho V = (1000\ \mathrm{kg/m^3})(3.79 \times 10^{-3}\ \mathrm{m^3}) = 3.79\ \mathrm{kg}$$

As a rule of thumb, a gallon of water weighs just over 8 pounds.

In comparison, the helium in a helium-filled balloon has a density of only about $0.179\ \mathrm{kg/m^3}$, and the density of the air in your room is roughly $1.29\ \mathrm{kg/m^3}$. On the higher end of the density scale, solid gold weighs in with a hefty $19{,}300\ \mathrm{kg/m^3}$. Further examples of densities for common materials are given in Table 15–1.

TABLE 15–1
Densities of Common Substances ($\mathrm{kg/m^3}$)

Substance	Density ($\mathrm{kg/m^3}$)
Air	1.29
Oxygen	1.43
Styrofoam	100
Balsa wood	120
Cherry (wood)	800
Ethyl alcohol	806
Olive oil	920
Ice	917
Fresh water	1000
Seawater	1025
Ebony (wood)	1220
Aluminum	2700
Iron	7860
Silver	10,500
Lead	11,300
Mercury	13,600
Gold	19,300

CONCEPTUAL CHECKPOINT 15–1

One day you look in your refrigerator and find nothing but a dozen eggs (44 g each). A quick measurement shows that the inside of the refrigerator is 1.0 m by 0.60 m by 0.75 m. Is the weight of the *air* in your refrigerator **(a)** much less than, **(b)** about the same as, or **(c)** much more than the weight of the *eggs*?

Reasoning and Discussion

At first it might seem that the "thin air" in the refrigerator weighs practically nothing compared with a carton full of eggs. A brief calculation shows this is not the case. For the eggs, we have

$$m_{\mathrm{eggs}} = 12(44\ \mathrm{g}) = 0.53\ \mathrm{kg}$$

For the air,

$$m_{\mathrm{air}} = \rho V = (1.29\ \mathrm{kg/m^3})(1.0\ \mathrm{m} \times 0.60\ \mathrm{m} \times 0.75\ \mathrm{m}) = 0.58\ \mathrm{kg}$$

Thus, the air, with a mass of 0.58 kg (1.28 lb), actually weighs slightly more than the eggs, which have a mass of 0.53 kg (1.17 lb)!

Answer:

(b) The air and the eggs weigh about the same.

15–2 Pressure

If you have ever pushed a button, or pressed a key on a keyboard, you have applied pressure. Now, you might object that you simply exerted a force on the button, or the key, which is correct. That force is spread out over an area, however. For example, when you press a button, the tip of your finger contacts the button over a small but finite area. **Pressure**, P, is a measure of the amount of force, F, per area A:

Definition of pressure, P

$$P = F/A \qquad \textbf{15–2}$$

SI unit: N/m^2

PROBLEM SOLVING NOTE
Pressure Is Force per Area

Remember that pressure is proportional to the applied force and *inversely* proportional to the area over which it acts.

Pressure is increased if the force applied to a given area is increased, or if a given force is applied to a smaller area. For example, if you press your finger against a balloon not much happens—your finger causes a small indentation. On the other hand, if you push a needle against the balloon with the same force you get an explosive pop. The difference is that the same force applied to the small area of a needle tip causes a large enough pressure to rupture the balloon.

EXAMPLE 15–1 Popping a Balloon

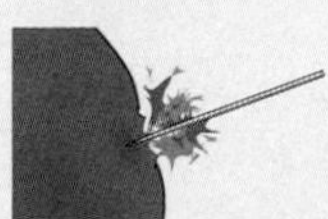

Find the pressure exerted on the skin of a balloon if you press with a force of 2.1 N using **(a)** your finger or **(b)** a needle. Assume the area of your fingertip is $1.0 \times 10^{-4}\ m^2$, and the area of the needle tip is $2.5 \times 10^{-7}\ m^2$. **(c)** Find the minimum force necessary to pop the balloon with the needle, given that the balloon pops with a pressure of $3.0 \times 10^5\ N/m^2$.

Picture the Problem

The same force is applied in either case. The difference is the area over which the force is spread.

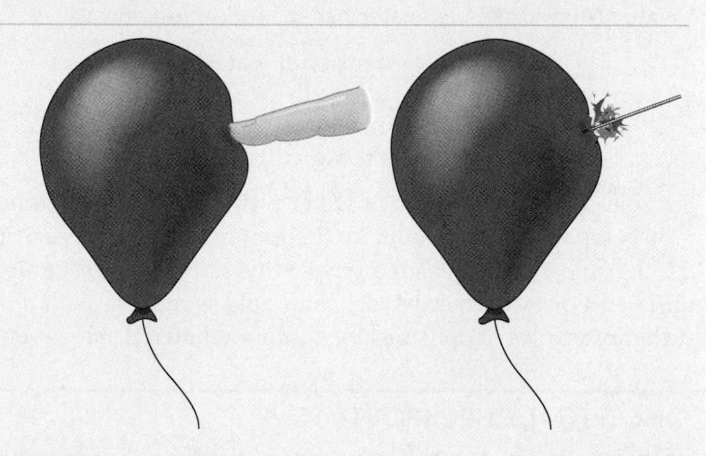

Strategy

(a), (b) Equation 15–2 can be used to find the pressure, given the force and area.

(c) Use Equation 15–2 to solve for the force corresponding to a given pressure and area.

Solution

Part (a)

1. Calculate the pressure exerted by the finger: $P = \frac{F}{A} = \frac{2.1\ N}{1.0 \times 10^{-4}\ m^2} = 2.1 \times 10^4\ N/m^2$

Part (b)

2. Calculate the pressure exerted by the needle: $P = \frac{F}{A} = \frac{2.1\ N}{2.5 \times 10^{-7}\ m^2} = 8.4 \times 10^6\ N/m^2$

Part (c)

3. Solve Equation 15–2 for the force: $F = PA$

4. Substitute numerical values: $F = (3.0 \times 10^5\ N/m^2)(2.5 \times 10^{-7}\ m^2) = 0.075\ N$

Insight

Note that the pressure exerted by the needle in part (b) is 400 times the pressure due to the finger in part (a).

Practice Problem

Find the area that a force of 2.1 N would have to act on to produce a pressure of $3.0 \times 10^5\ N/m^2$. [**Answer:** $A = 7.0 \times 10^{-6}\ m^2$]

Some related homework problems: Problem 8, Problem 9

An interesting example of force, area, and pressure in nature is provided by the family of small aquatic birds referred to as rails, and in particular by the gallinule. This bird has exceptionally long toes that are spread out over a large area. The result is that the weight of the bird causes only a relatively small pressure as it walks on the soft, muddy shorelines encountered in its habitat. In some species, the pres-

◀ This bird exerts only a small pressure on the lily pad on which it walks because its weight is spread out over a large area by its long toes. Since the pressure is not enough to sink a lily pad, the bird can seemingly "walk on water."

sure exerted while walking is so small that the birds can actually walk across lily pads without sinking into the water.

REAL WORLD PHYSICS: BIO
Walking on lily pads

Atmospheric Pressure and Gauge Pressure

We are all used to working under pressure—about 14.7 pounds per square inch to be precise. This is **atmospheric pressure**, P_{at}, a direct result of the weight of the air above us. In SI units, atmospheric pressure has the following value:

Atmospheric pressure, P_{at}

$$P_{at} = 1.01 \times 10^5\ \text{N/m}^2 \qquad \textbf{15–3}$$

SI unit: N/m^2

A shorthand unit for N/m^2 is the **pascal** (Pa):

$$1\ \text{Pa} = 1\ \text{N/m}^2 \qquad \textbf{15–4}$$

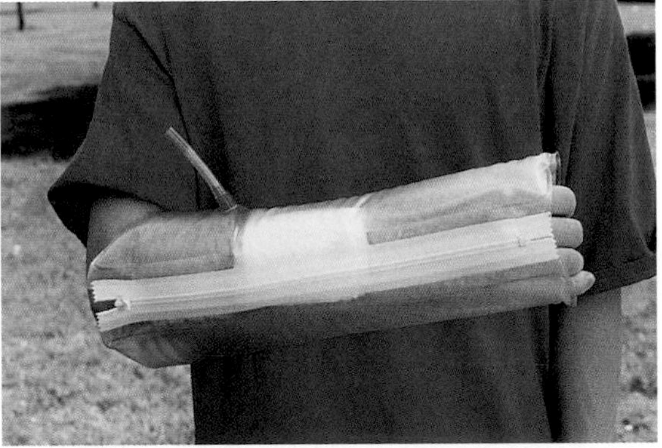

▲ The air around you exerts a force of about 14.7 pounds on every square inch of your body. Since this force is the same in all directions and is opposed by an equal pressure inside your body, you are generally unaware of it. However, when the air is pumped out of a sealed can (left), atmospheric pressure produces an inward force that is unopposed. The resulting collapse of the can vividly illustrates the pressure that is all around us. An air splint (right) utilizes the same principle of unequal pressure. A plastic sleeve is placed around an injured limb and inflated to a pressure greater than that of the atmosphere—and thus of the body's internal pressure. The increased external pressure retards bleeding from the injured area, and also tends to immobilize the limb in case it might be fractured. (Air splints are carried by hikers and others who might be in need of emergency treatment far from professional medical facilities. Before it is inflated, the air splint is about the size and weight of a credit card.)

▲ The pressure measured by a tire gauge is not atmospheric pressure but the 'gauge pressure': the difference between the pressure inside the tire and that of the atmosphere.

The pascal honors the pioneering studies of fluids by the French scientist Blaise Pascal (1623–1662). Thus, atmospheric pressure can be written as

$$P_{\text{at}} = 101\ \text{kPa}$$

In British units, pressure is measured in pounds per square inch, and

$$P_{\text{at}} = 14.7\ \text{lb/in}^2$$

Finally, a common unit for atmospheric pressure in weather forecasting is the **bar**, defined as follows:

$$1\ \text{bar} = 10^5\ \text{Pa} \approx 1\,P_{\text{at}}$$

EXERCISE 15–1

Find the force exerted on the palm of your hand by atmospheric pressure. Assume your palm measures 0.080 m by 0.10 m.

Solution

Applying Equations 15–3 and 15–2 we find

$$F = P_{\text{at}}A = (1.01 \times 10^5\ \text{Pa})(0.080\ \text{m})(0.10\ \text{m}) = 810\ \text{N}$$

Thus, the atmosphere pushes on the palm of your hand with a force of approximately 180 pounds! Of course, it also pushes on the back of your hand with essentially the same force, but in the opposite direction.

Figure 15–1 illustrates the forces exerted on your hand by atmospheric pressure. If your hand is vertical, atmospheric pressure pushes to the right and to the left equally, so your hand feels zero net force. If your hand is horizontal, atmospheric pressure exerts upward and downward forces on your hand that are essentially the same in magnitude, again giving zero net force. This cancellation of forces occurs no matter what the orientation of your hand; thus, we can conclude the following:

The pressure in a fluid acts equally in all directions, and acts at right angles to any surface.

In many cases we are interested in the difference between a given pressure and atmospheric pressure. For example, a flat tire does not have zero pressure in it; the pressure in the tire is atmospheric pressure. To inflate the tire to 241 kPa (35 lb/in^2), the pressure inside the tire must be greater than atmospheric pressure by this amount; that is, $P = 241\ \text{kPa} + P_{\text{at}} = 342\ \text{kPa}$.

To deal with such situations, we introduce the **gauge pressure**, P_g, defined as follows:

$$P_g = P - P_{\text{at}} \qquad \text{15–5}$$

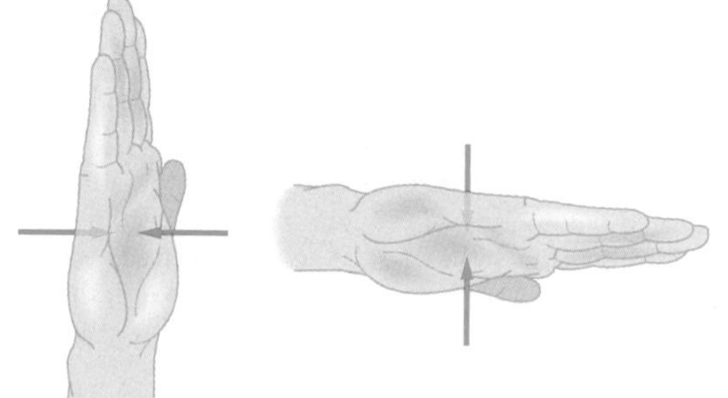

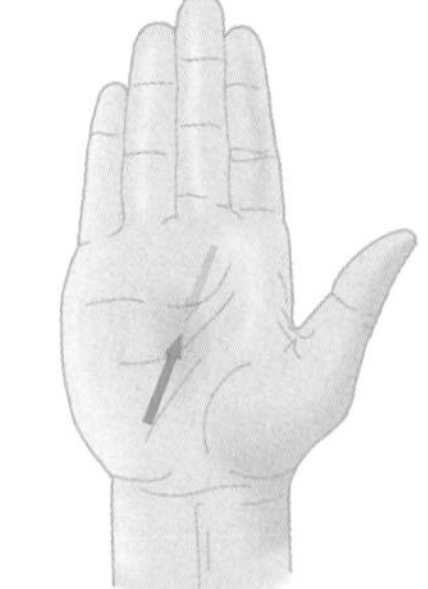

▶ **FIGURE 15–1 Pressure is the same in all directions**

The forces exerted on the two sides of a hand cancel, regardless of the hand's orientation. Hence, pressure acts equally in all directions.

It is the gauge pressure, then, that is determined by a tire gauge. Many problems in this chapter refer to the gauge pressure. Hence it must be remembered that the actual pressure in these cases is greater by the amount P_{at}.

If a problem gives you the gauge pressure, recall that the actual pressure is the gauge pressure *plus* atmospheric pressure.

EXAMPLE 15–2 Pressuring the Ball

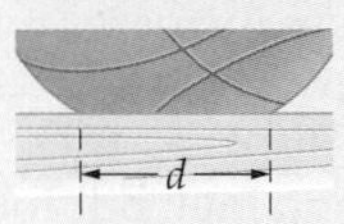

Estimate the gauge pressure in a basketball by pushing down on it and noting the area of contact it makes with the surface on which it rests.

Picture the Problem

Our sketch shows the basketball being pushed downward and flattening out on the bottom. The area of contact is a circle of diameter d.

Strategy

To solve this problem, we have to make reasonable estimates of the force applied to the ball and the area of contact.

Suppose, for example, that we push down with a moderate force of 22 N (about 5 lb). The circular area of contact will probably have a diameter of about 2.0 centimeters. This can be verified by carrying out the experiment. Thus, given $F = 22$ N and $A = \pi(d/2)^2$ we can find the gauge pressure.

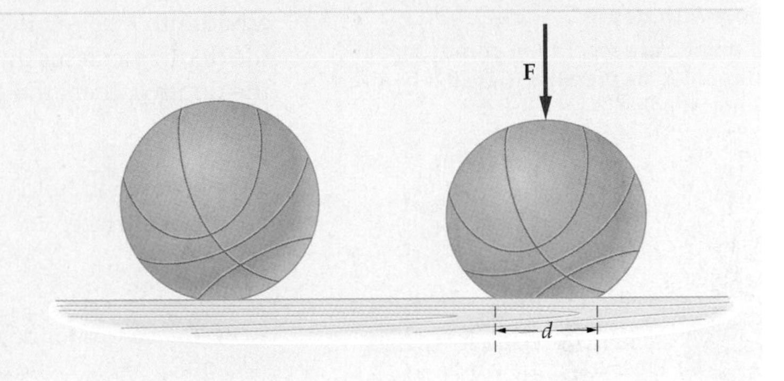

Solution

1. Using these estimates, calculate the gauge pressure, P_g:

$$P_g = \frac{F}{A} = \frac{22\text{ N}}{\pi\left(\frac{0.020\text{ m}}{2}\right)^2} = 7.0 \times 10^4\text{ Pa}$$

Insight

Given that 1.01×10^5 Pa $= 14.7$ lb/in^2, it follows that $P_g = 7.0 \times 10^4$ Pa ~ 10 lb/in^2. Thus, a basketball will typically have a gauge pressure in the neighborhood of 10 lb/in^2, and hence a total pressure inside the ball of about 25 lb/in^2.

Practice Problem

What is the diameter of the circular area of contact if a basketball with a 12 lb/in^2 gauge pressure is pushed down with a force of 44 N (about 10 lb)? [**Answer:** $d = 2.6$ cm]

Some related homework problems: Problem 11, Problem 12

15–3 Static Equilibrium in Fluids: Pressure and Depth

Countless war movies have educated us on the perils of taking a submarine too deep. The hull creaks and groans, rivets start to pop, water begins to spray into the ship, and the captain keeps a close eye on the depth gauge. But what causes the pressure to increase as a submarine dives, and how much does it go up for a given increase in depth?

The answer to the first question is that the increased pressure is due to the added weight of water pressing down on the submarine as it goes deeper. To see how this works, consider a cylindrical container filled to a height h with a fluid of density ρ, as in **Figure 15–2 (a)**. The top surface of the fluid is open to the atmosphere, with a pressure P_{at}. If the cross-sectional area of the container is A, the downward force exerted on the top surface by the atmosphere is

$$F_{top} = P_{at}A$$

Now, at the bottom of the container, the downward force is F_{top} *plus* the weight of the fluid. Recalling that $M = \rho V$, and that $V = hA$ for a cylinder of height h and area A, this weight is

$$W = Mg = \rho V g = \rho(hA)g$$

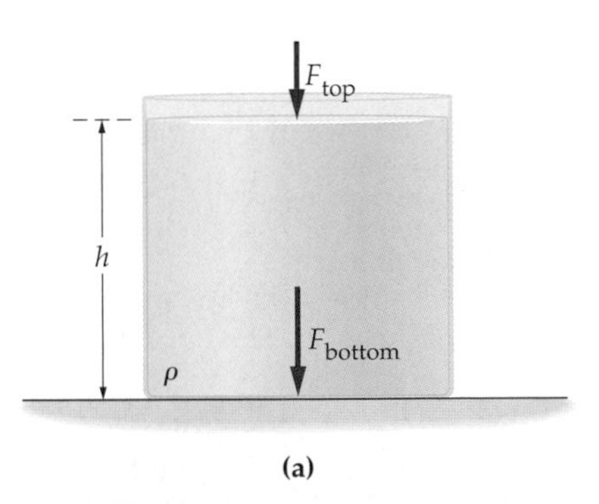

▲ **FIGURE 15–2 (a) Pressure and the weight of a fluid**
The force pushing down on the bottom of the flask is greater than the force pushing down on the surface of the fluid. The difference in force is the weight of fluid in the flask.

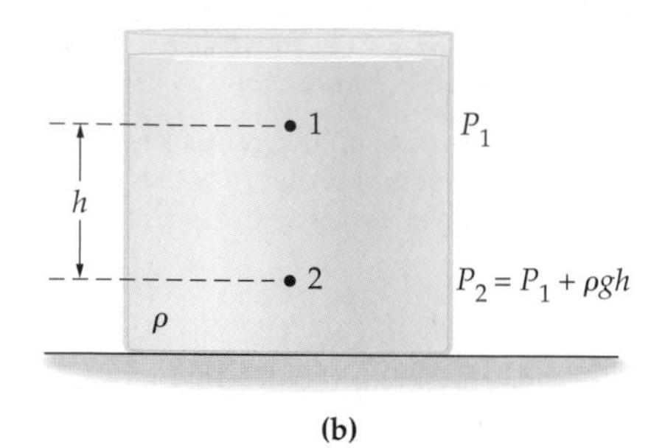

▲ **FIGURE 15–2 (b) Pressure variation with depth**

If point 2 is deeper than point 1 by the amount h, its pressure is greater by the amount ρgh.

Hence, we have

$$F_{\text{bottom}} = F_{\text{top}} + W = P_{\text{at}}A + \rho(hA)g$$

Finally, the pressure at the bottom is obtained by dividing F_{bottom} by the area A:

$$P_{\text{bottom}} = \frac{F_{\text{bottom}}}{A} = \frac{P_{\text{at}}A + \rho(hA)g}{A} = P_{\text{at}} + \rho gh$$

Thus, as we descend to a depth h in a fluid of density ρ, the pressure increases by

$$\rho gh$$

Of course, this relation holds not only for the bottom of the container, but for any depth h below the surface. Thus, the answer to the second question is that if the depth increases by the amount h, the pressure increases by the amount ρgh. At the depth h, then, the pressure P is given by

$$P = P_{\text{at}} + \rho gh \qquad \textbf{15–6}$$

This expression holds for any liquid with constant density ρ and a pressure P_{at} at its upper surface.

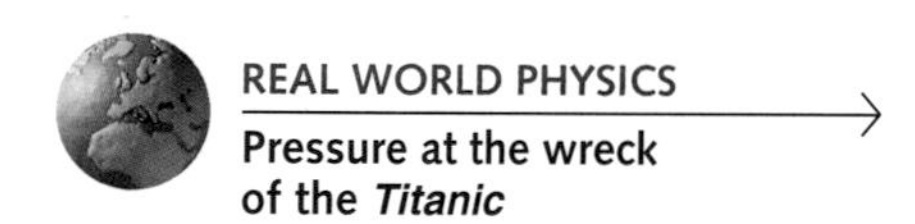

REAL WORLD PHYSICS

Pressure at the wreck of the *Titanic*

EXERCISE 15–2

The *Titanic* was found in 1985 lying on the bottom of the North Atlantic at a depth of 2.5 miles. What is the pressure at this depth?

Solution

Applying Equation 15–6 with $\rho = 1025\ \text{kg/m}^3$ we have

$$P = P_{\text{at}} + \rho gh = 1.01 \times 10^5\ \text{Pa} +$$

$$(1025\ \text{kg/m}^3)(9.81\ \text{m/s}^2)(2.5\ \text{mi})\left(\frac{1609\ \text{m}}{1\ \text{mi}}\right) = 4.1 \times 10^7\ \text{Pa}$$

This is about 400 atmospheres.

The relation $P = P_{\text{at}} + \rho gh$ can be applied to any two points in a fluid. For example, if the pressure at one point is P_1, the pressure P_2 at a depth h below that point is the following:

PROBLEM SOLVING NOTE

Pressure Depends Only on Depth

The pressure in a static fluid depends only on the depth of the fluid. It is independent of the shape of the container.

Dependence of pressure on depth

$$P_2 = P_1 + \rho gh \qquad \textbf{15–7}$$

This relation is illustrated in **Figure 15–2 (b)**, and utilized in the next Example.

EXAMPLE 15–3 Pressure and Depth

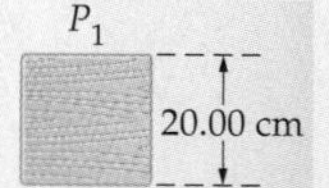

A cubical box 20.00 cm on a side is completely immersed in a fluid. At the top of the box the pressure is 105.0 kPa; at the bottom the pressure is 106.8 kPa. What is the density of the fluid?

Picture the Problem

Our sketch shows the box at an unknown depth d below the surface of the fluid. The important dimension for this problem is the height of the box, which is 20.00 cm.

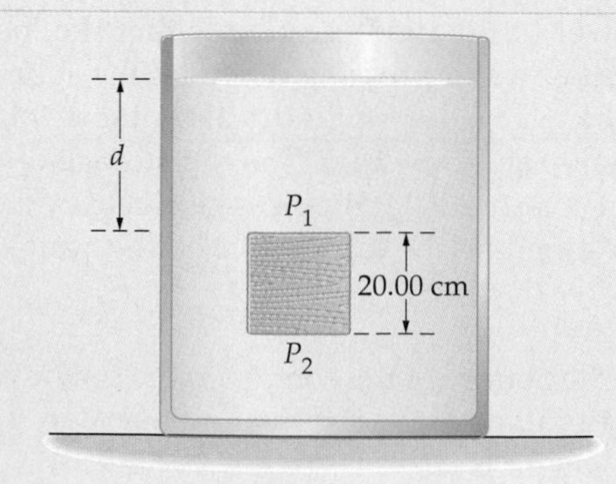

Strategy

The pressures at the top and bottom of the box are related by $P_2 = P_1 + \rho gh$. Since the pressures and the height of the box are given, this relation can be solved for the unknown density, ρ.

Solution

1. Solve $P_2 = P_1 + \rho g h$ for the density: $\rho = \dfrac{P_2 - P_1}{gh}$

2. Substitute numerical values: $\rho = \dfrac{1.068 \times 10^5\ \text{Pa} - 1.050 \times 10^5\ \text{Pa}}{(9.81\ \text{m/s}^2)(0.2000\ \text{m})} = 920\ \text{kg/m}^3$

Insight

Comparing with Table 15–1, it appears that the fluid in question is probably olive oil.

Practice Problem

Given the density obtained above, what is the depth d at the top of the box? [**Answer:** $d = 0.44$ m]

Some related homework problems: Problem 13, Problem 14

CONCEPTUAL CHECKPOINT 15–2

One day while swimming below the surface of the ocean you let out a small bubble of air from your mouth. As the bubble rises toward the surface, does its diameter **(a)** increase, **(b)** decrease, or **(c)** stay the same?

Reasoning and Discussion

As the bubble rises the pressure in the surrounding water decreases. This allows the air in the bubble to expand and occupy a larger volume.

Answer:

(a) The diameter of the bubble increases.

An interesting application of the variation of pressure with depth is the **barometer**, which can be used to measure atmospheric pressure. We consider here the simplest type of barometer, which was first proposed by Evangelista Torricelli (1608–1647) in 1643. First, fill a long glass tube—open at one end and closed at the other—with a fluid of density ρ. Next, invert the tube and place its open end below the surface of the same fluid in a bowl, as shown in **Figure 15–3**. Some of the fluid in the tube will flow into the bowl, leaving an empty space (vacuum) at the top. Enough will remain, however, to create a difference in level, h, between the fluid in the bowl and that in the tube.

The basic idea of the barometer is that this height difference is directly related to the atmospheric pressure that pushes down on the fluid in the bowl. To see how this works, first note that the pressure in the vacuum at the top of the tube is zero. Hence, the pressure in the tube at a depth h below the vacuum is $0 + \rho g h = \rho g h$. Now, at the level of the fluid in the bowl we know that the pressure is one atmosphere, P_{at}. Therefore, it follows that

$$P_{at} = \rho g h$$

If these pressures were not the same, there would be a pressure difference between the fluid in the tube and that in the bowl, resulting in a net force and a flow of fluid. Thus, a measurement of h immediately gives atmospheric pressure.

A fluid that is often used in such a barometer is mercury (Hg), with a density of $\rho = 1.3595 \times 10^4\ \text{kg/m}^3$. The corresponding height for a column of mercury is

$$h = \frac{P_{at}}{\rho g} = \frac{1.013 \times 10^5\ \text{Pa}}{(1.3595 \times 10^4\ \text{kg/m}^3)(9.81\ \text{m/s}^2)} = 760\ \text{mm}$$

In fact, atmospheric pressure is *defined* in terms of millimeters of mercury (mmHg):

$$1\ \text{atmoshpere} = P_{at} = 760\ \text{mmHg}$$

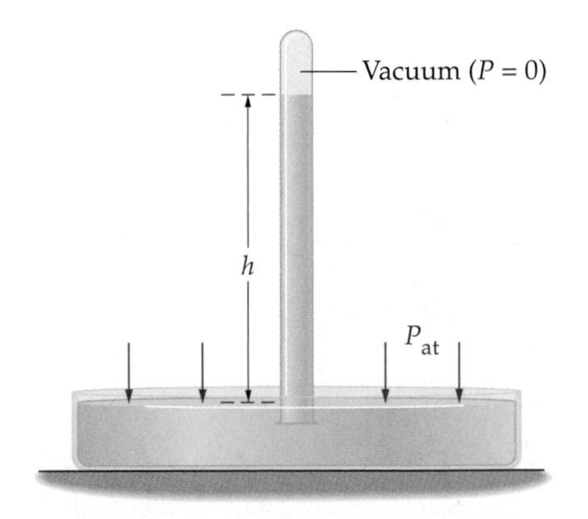

▲ **FIGURE 15–3 A simple barometer**
Atmospheric pressure, P_{at}, is related to the height of fluid in the tube by the relation $P_{at} = \rho g h$.

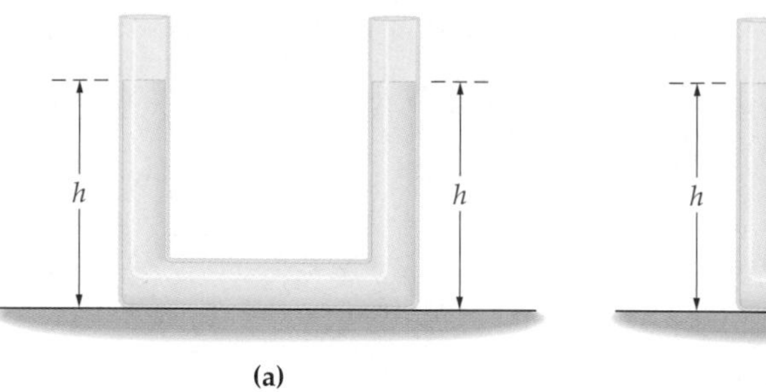

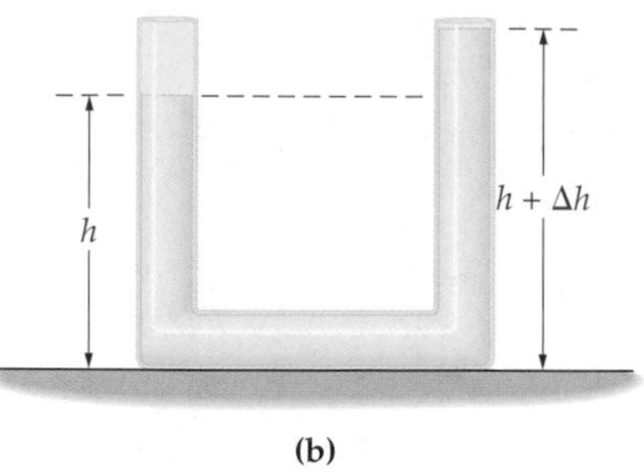

▶ **FIGURE 15–4 Fluids seek their own level**

(a) When the levels are equal, the pressure is the same at the base of each arm of the U tube. As a result, the fluid in the horizontal section of the U is in equilibrium. **(b)** With unequal heights, the pressures are different. In this case, the pressure is greater at the base of the right arm, hence fluid will flow toward the left and the levels will equalize.

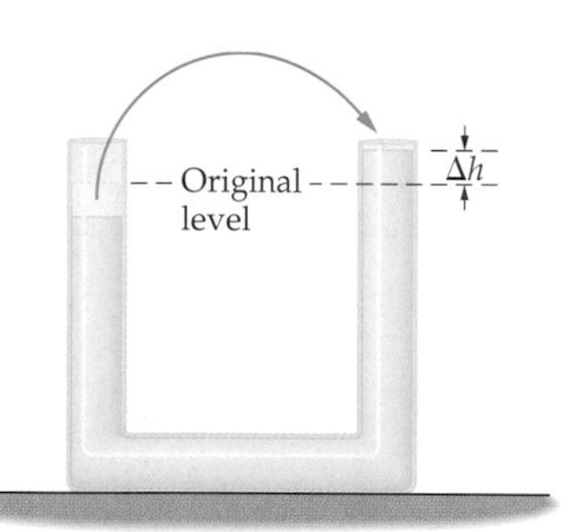

▲ **FIGURE 15–5 Gravitational potential energy of a fluid**

In order to create unequal levels in the two arms of the U tube, an element of fluid must be raised by the height Δh. This increases the gravitational potential energy of the system. The lowest potential energy corresponds to equal levels.

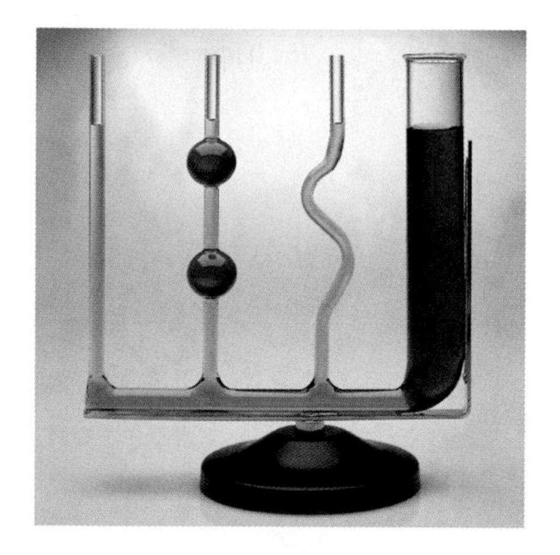

▲ The containers shown here are connected at the bottom by a hollow tube, which allows fluid to flow freely between them. As a result the fluid level is the same in each container regardless of its shape and size.

Water Seeks Its Own Level

We are all familiar with the aphorism that water seeks its own level. In order for this to hold true, however, it is necessary that the pressure at the surface of the water (or other fluid) be the same everywhere on the surface. This was not the case for the barometer just discussed, where the pressure was P_{at} on one portion of the surface, and zero on another. Let's take a moment, then, to consider the level assumed by a fluid with constant pressure on its surface. In doing so, we shall apply considerations involving force, pressure, and energy.

First, the force–pressure point of view. In **Figure 15–4 (a)** we show a U-shaped tube containing a quantity of fluid of density ρ. The fluid rises to the same level in each arm of the U, where it is open to the atmosphere. Therefore, the pressure at the base of each arm is the same; $P_{at} + \rho gh$. Thus, the fluid in the horizontal section of the U is pushed with equal force from each side, giving zero net force. As a result, the fluid remains at rest.

On the other hand, in **Figure 15–4 (b)** the two arms of the U are filled to different levels. Therefore, the pressure at the base of the two arms are different, with the greater pressure at the base of the right arm. The fluid in the horizontal section, then, will experience a net force to the left, causing it to move in that direction. This will tend to equalize the fluid levels of the two arms.

We can arrive at the same conclusion on the basis of energy minimization. Consider a U tube that is initially filled to the same level in both arms, as in Figure 15–4 (a). Now, consider moving a small element of fluid from one arm to the other, to create different levels, as in **Figure 15–5**. In moving this fluid element to the other arm, it is necessary to lift it upward. This, in turn, causes its potential energy to increase. Since nothing else in the system has changed its position, the only change in potential energy is the increase experienced by the element. Thus, we conclude that the system has a minimum energy when the fluid levels are the same and a higher energy when the levels are different. Just as a ball rolls to the bottom of a hill, where its energy is minimized, the fluid seeks its own level and a minimum energy.

If two different liquids, with different densities, are combined in the same U tube, the levels in the arms are not the same. Still, the pressures at the base of each arm must be equal, as before. This is discussed in the next Example.

EXAMPLE 15–4 **Oil and Water Don't Mix**

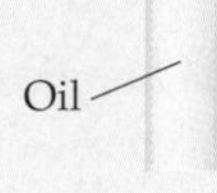

A U-shaped tube is filled mostly with water, but a small amount of vegetable oil has been added to one side, as shown in the sketch. The density of the water is 1000 kg/m^3, and the density of the vegetable oil is 920 kg/m^3. If the depth of the oil is 5.00 cm, what is the difference in level h between the top of the oil on one side of the U and the top of the water on the other side?

Picture the Problem

The U-shaped tube and the relevant dimensions are shown in the sketch. Note that the tube is open to the atmosphere on both sides of the U.

Strategy

For the system to be in equilibrium, it is necessary that the pressure be the same at the bottom of each side of the U; that is, at points C and D. If the pressure is the same at C and D, it will remain equal as one moves up through the water to the points A and B. Above this point the pressures will differ because of the presence of the oil.

Therefore, setting the pressure at point A equal to the pressure at point B determines the depth h_1 in terms of the known depth h_2. It follows that the difference in level between the two sides of the U is simply $h = h_2 - h_1$.

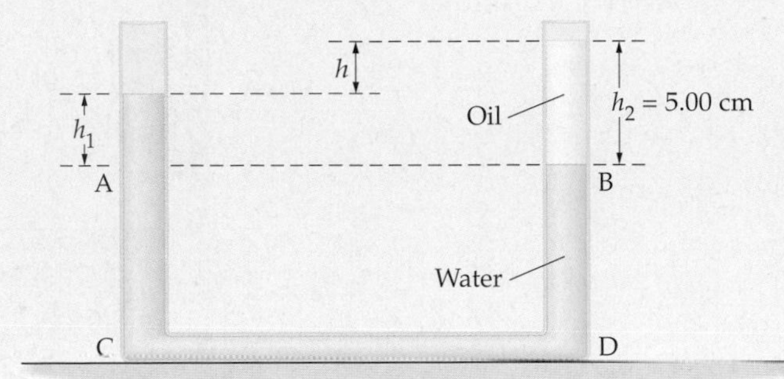

Solution

1. Find the pressure at point A, where the depth of the water is h_1: $P_A = P_{at} + \rho_{water} g h_1$
2. Find the pressure at point B, where the depth of the oil is $h_2 = 5.00$ cm: $P_B + P_{at} + \rho_{oil} g h_2$
3. Set P_A equal to P_B: $P_{at} + \rho_{water} g h_1 = P_{at} + \rho_{oil} g h_2$
4. Solve for the depth of the water, h_1, and substitute numerical values: $$h_1 = h_2\left(\frac{\rho_{oil}}{\rho_{water}}\right) = (5.00\text{ cm})\left(\frac{920\text{ kg/m}^3}{1000\text{ kg/m}^3}\right) = 4.60\text{ cm}$$
5. Calculate the difference in levels between the water and oil sides of the U: $h = h_2 - h_1 = 5.00\text{ cm} - 4.60\text{ cm} = 0.40\text{ cm}$

Insight

Note that the value of atmospheric pressure doesn't matter in this problem. What does matter is the increased pressure due to being submerged to a given depth in water or in oil.

Practice Problem

Find the pressure at points A and B. [**Answer:** $P_A = P_B = P_{at} + 45.1$ kPa]

Some related homework problems: Problem 16, Problem 22, Problem 23

Pascal's Principle

Recall from Equation 15–6 that if the surface of a fluid of density ρ is exposed to the atmosphere with a pressure P_{at}, the pressure at a depth h below the surface is

$$P = P_{at} + \rho g h$$

Suppose, now, that atmospheric pressure is increased from P_{at} to $P_{at} + \Delta P$. As a result, the pressure at the depth h is

$$P = P_{at} + \Delta P + \rho g h = (P_{at} + \rho g h) + \Delta P$$

Thus, by increasing the pressure at the top of the fluid by the amount ΔP, we have increased it by the same amount everywhere in the fluid. This is **Pascal's principle**:

> An external pressure applied to an enclosed fluid is transmitted unchanged to every point within the fluid.

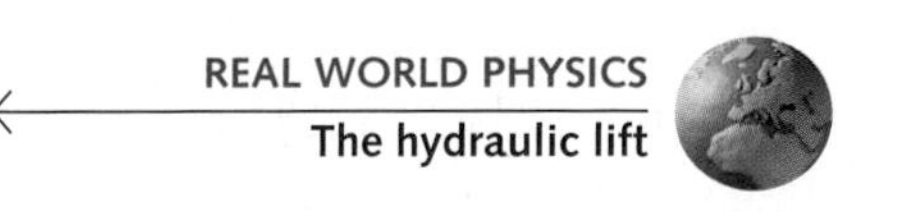

A classic example of Pascal's Principle at work is the **hydraulic lift**, which is shown schematically in **Figure 15–6**. Here we see two cylinders, one of cross-sectional area A_1, the other of cross-sectional area $A_2 > A_1$. The cylinders, each of which is fitted with a piston, are connected by a tube and filled with a fluid. Initially the pistons are at the same level and exposed to the atmosphere.

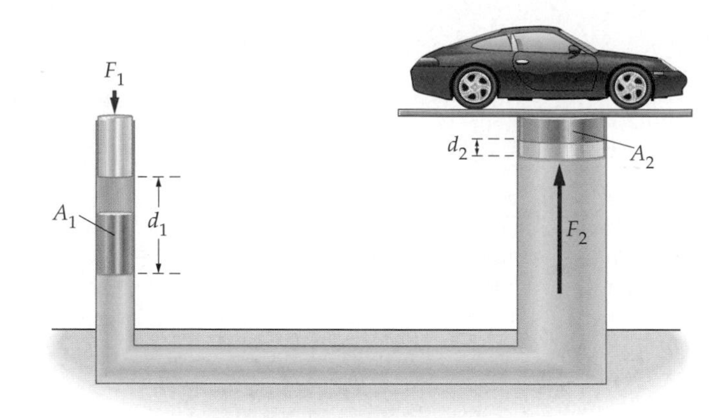

▶ FIGURE 15–6 A hydraulic lift
A force F_1 exerted on the small piston causes a much larger force, F_2, to act on the large piston.

▲ This flight simulator is supported and maneuvered by hydraulic pistons much like those that support cars in an auto repair shop. Similar devices are used in amusement parks to create exciting adventures in a simulated world. All such hydraulic lifts operate in accordance with Pascal's principle, which allows a force to be greatly magnified, but at a cost: The distance through which the force acts is reduced by the same factor.

Now, suppose we push down on piston 1 with the force F_1. This increases the pressure in that cylinder by the amount

$$\Delta P = F_1/A_1$$

By Pascal's Principle, the pressure in cylinder 2 increases by the *same* amount. Therefore, the increased upward force on piston 2 due to the fluid is

$$F_2 = (\Delta P)A_2$$

Substituting the increase in pressure, $\Delta P = F_1/A_1$, we find

$$F_2 = (F_1/A_1)A_2 = F_1\left(\frac{A_2}{A_1}\right) > F_1 \qquad \textbf{15–8}$$

To be specific, let's assume that A_2 is 100 times greater than A_1. Then, by pushing down on piston 1 with a force F_1 we push upward on piston 2 with a force of $100F_1$. Our force has been magnified 100 times!

If this sounds too good to be true, rest assured that we are not getting something for nothing. Just as with a lever, there is a tradeoff between the distance through which a force must be applied and the force magnification. This is illustrated in Figure 15–6, where we show piston 1 being pushed down through a distance d_1. This displaces a volume of fluid equal to A_1d_1. The same volume flows into cylinder 2, where it causes piston 2 to rise through a distance d_2. Equating the two volumes, we have

$$A_1d_1 = A_2d_2$$

or

$$d_2 = d_1\left(\frac{A_1}{A_2}\right)$$

Thus, in the example just given, if we move piston 1 down a distance d_1, piston 2 rises a distance $d_2 = d_1/100$. Our force at piston 2 has been magnified 100 times, but the distance it moves has been reduced 100 times.

EXERCISE 15–3

To inspect a 14,500-N car, it is raised with a hydraulic lift. If the radius of the small piston in Figure 15–6 is 4.0 cm, and the radius of the large piston is 17 cm, find the force that must be exerted on the small piston to lift the car.

Solution

Solving Equation 15–8 for F_1, and noting that the area is πr^2, we find

$$F_1 = F_2\left(\frac{A_1}{A_2}\right) = (14{,}500\text{ N})\left[\frac{\pi(0.040\text{ m})^2}{\pi(0.17\text{ m})^2}\right] = 800\text{ N}$$

15–4 Archimedes' Principle and Buoyancy

The fact that a fluid's pressure increases with depth leads to many interesting consequences. Among them is the fact that a fluid exerts a net upward force on any object it surrounds. This is referred to as a **buoyant force**.

To see the origin of buoyancy, consider a cubical block immersed in a fluid of density ρ, as in **Figure 15–7**. The surrounding fluid exerts normal forces on all of its faces. Clearly, the horizontal forces pushing to the right and to the left are equal, hence they cancel and have no effect on the block.

The situation is quite different for the vertical forces, however. Note, for example, that the downward force exerted on the top face is less than the upward force exerted on the lower face, since the pressure at the lower face is greater. This difference in forces gives rise to a net upward force—the buoyant force.

Let's calculate the buoyant force acting on the block. First, we assume that the cubical block is of length L on a side, and that the pressure on the top surface is P_1. The downward force on the block, then, is

$$F_1 = P_1A = P_1L^2$$

Note that we have used the fact that the area of a square face of side L is L^2. Next, we consider the bottom face. The pressure there is given by Equation 15–7, with a difference in depth of $h = L$:

$$P_2 = P_1 + \rho gL$$

Therefore, the upward force exerted on the bottom face of the cube is

$$F_2 = P_2A = (P_1 + \rho gL)L^2 = P_1L^2 + \rho gL^3 = F_1 + \rho gL^3$$

If we take upward as the positive direction, the net vertical force exerted by the fluid on the block—that is, the buoyant force, F_b—is

$$F_b = F_2 - F_1 = \rho gL^3$$

As expected, the block experiences a net upward force.

The precise value of the buoyant force is of some significance, as we now show. First, note that the volume of the cube is L^3. It follows that ρgL^3 is the weight of *fluid* that would occupy the same volume as the cube. Therefore, the buoyant force is equal to the weight of fluid that is displaced by the cube. This is a special case of **Archimedes' Principle**:

> An object completely immersed in a fluid experiences an upward buoyant force equal in magnitude to the weight of fluid displaced by the object.

More generally, if a volume V of an object is immersed in a fluid of density ρ, the buoyant force can be expressed thus:

Buoyant force when a volume V is submerged in a fluid of density ρ

$$F_b = \rho gV \qquad \textbf{15–9}$$

SI unit: N

The volume V may be the total volume of the object, or any fraction of the total volume.

To see that Archimedes' Principle is completely general, consider the submerged object shown in **Figure 15–8 (a)**. If we were to replace this object with an equivalent volume of fluid, as in **Figure 15–8 (b)**, the container would hold nothing but fluid and would be in static equilibrium. As a result, we conclude that the net force acting on this "fluid object" must be upward and equal in magnitude to its weight. Now here is the key idea: Since the original object and the "fluid object" occupy the same position, the forces acting on their surfaces are identical, and hence the net force is the same for both objects. Therefore, the original object experiences a buoyant force equal to the weight of fluid that it displaces—that is, equal to the weight of the "fluid object."

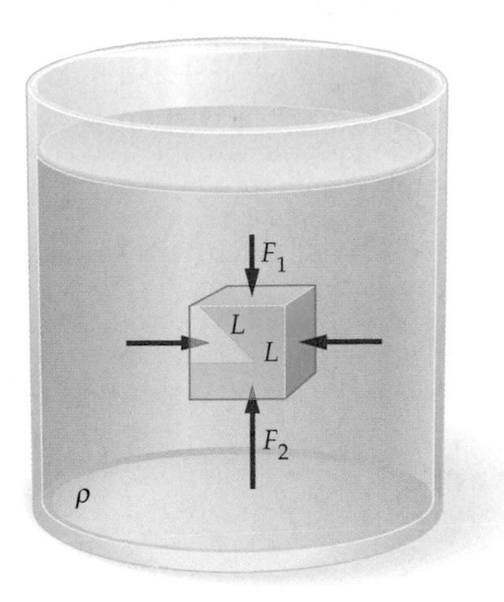

▲ **FIGURE 15–7 Buoyant force due to a fluid**

A fluid surrounding an object exerts a buoyant force in the upward direction. This is due to the fact that pressure increases with depth, hence the upward force on the object, F_2, is greater than the downward force, F_1. Forces acting to the left and to the right cancel.

PROBLEM SOLVING NOTE
The Buoyant Force

Note that the buoyant force is equal to the weight of displaced fluid. It does not depend on the weight of the object that displaces the fluid.

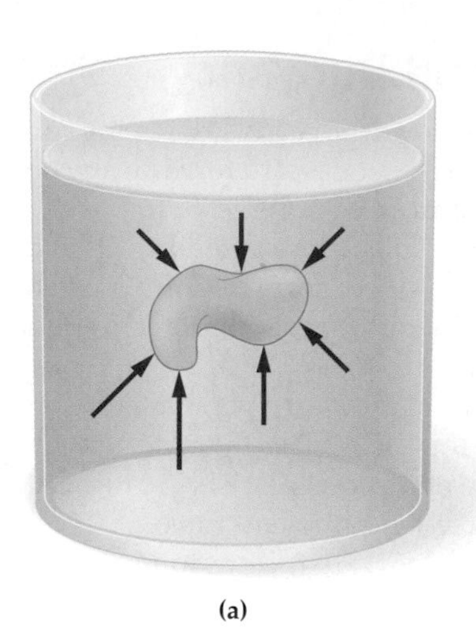

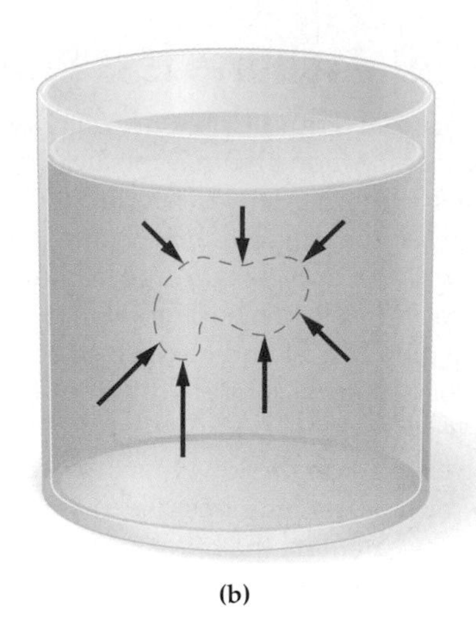

▶ **FIGURE 15–8 Buoyant force equals the weight of displaced fluid**
The buoyant force acting on the object in **(a)** is equal to the weight of the "fluid object" (with the same size and shape) in **(b)**. This follows from the fact that the fluid in (b) is at rest; hence the forces acting on it must cancel its weight. The forces acting on the original object, in (a), are exactly the same, so the net force it experiences is also the same as in (b).

CONCEPTUAL CHECKPOINT 15–3

A flask of water rests on a scale. If you dip your finger into the water, without touching the flask, does the reading on the scale **(a)** increase, **(b)** decrease, or **(c)** stay the same?

Reasoning and Discussion

Your finger experiences an upward buoyant force when it is dipped into the water. By Newton's third law, the water experiences an equal and opposite reaction force acting downward. This downward force is transmitted to the scale, which in turn gives a higher reading.

Another way to look at this result is to note that when you dip your finger into the water, its depth increases. This results in a greater pressure at the bottom of the flask, and hence a greater downward force on the flask. The scale reads this increased downward force.

Answer:

(a) The reading on the scale increases.

15–5 Applications of Archimedes' Principle

In this section we consider a variety of applications of Archimedes' principle. We begin with situations in which an object is fully immersed. Later we consider systems in which an object floats.

Complete Submersion

An interesting application of complete submersion can be found in an apparatus commonly used in determining a person's body fat percentage. We consider the basic physics of the apparatus and the measurement procedure in the next Example. Following the Example we derive the relation between overall body density and the body fat percentage.

EXAMPLE 15–5 Measuring the Body's Density

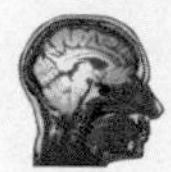

Real World Physics: Bio

A person who weighs 720.0 N in air is lowered into a tank of water to about chin level. He sits in a harness of negligible mass suspended from a scale that reads his apparent weight. He now exhales as much air as possible and dunks his head underwater, submerging his entire body. If his apparent weight while submerged is 34.3 N, find **(a)** his volume and **(b)** his density.

Picture the Problem

The scale, the tank of water, and the person are shown in the sketch. We also show the free-body diagram for the person. Note that the weight of the person in air is designated by $W = mg$, and the apparent weight in water, which is the upward force exerted by the scale on the person, is designated by W_a.

Strategy

To find the volume, V_p, and density, ρ_p, of the person, we must use two separate conditions. They are as follows:

(a) In water the person experiences an upward buoyant force given by Archimedes' principle: $F_b = \rho_{water}V_p g$.

(b) The weight of the person in air is $W = mg = \rho_p V_p g$.

Combining these two relations allows us to solve for both V_p and ρ_p.

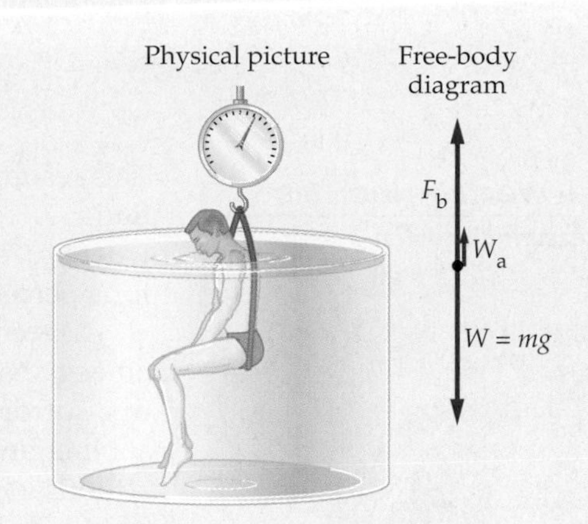

Solution

Part (a)

1. Apply Newton's second law to the person, and solve for the apparent weight, W_a:

$$W_a + F_b - W = 0$$
$$W_a = W - F_b$$

2. Substitute $F_b = \rho_{water}V_p g$ and solve for V_p:

$$W_a = W - \rho_{water}V_p g$$
$$V_p = \frac{W - W_a}{\rho_{water} g} = \frac{720.0\text{ N} - 34.3\text{ N}}{(1000\text{ kg/m}^3)(9.81\text{ m/s}^2)}$$
$$= 6.99 \times 10^{-2}\text{ m}^3$$

Part (b)

3. Use $W = \rho_p V_p g$ to solve for the density of the person, ρ_p:

$$W = \rho_p V_p g$$
$$\rho_p = \frac{W}{V_p g} = \frac{720.0\text{ N}}{(6.99 \times 10^{-2}\text{ m}^3)(9.81\text{ m/s}^2)}$$
$$= 1050\text{ kg/m}^3$$

Insight

As in Conceptual Checkpoint 15–3, the water exerts an upward buoyant force on the person, and an equal and opposite reaction force acts downward on the tank and water, making it press against the floor with a greater force.

Practice Problem

The person can float in water if his lungs are partially filled with air, increasing the volume of his body. What volume must his body have to just float? [**Answer:** $V_p = 7.34 \times 10^{-2}\text{ m}^3$]

Some related homework problems: Problem 33, Problem 34

Once the overall density of the body is determined, the percentage of body fat can be obtained by noting that body fat has a density of $\rho_f = 900\text{ kg/m}^3$, whereas the lean body mass (muscles and bone) has a density of $\rho_l = 1100\text{ kg/m}^3$. Suppose, for example, that a fraction x_f of the total body mass M is fat mass, and a fraction $(1 - x_f)$ is lean mass; that is, the fat mass is $m_f = x_f M$ and the lean mass is $m_l = (1 - x_f)M$. The total volume of the body is $V = V_f + V_l$. Using the fact that $V = m/\rho$, we can write the total volume as $V = m_f/\rho_f + m_l/\rho_l = x_f M/\rho_f + (1 - x_f)M/\rho_l$. Combining these results, the overall density of a person's body, ρ_p, is

$$\rho_p = \frac{M}{V} = \frac{1}{\dfrac{x_f}{\rho_f} + \dfrac{(1 - x_f)}{\rho_l}}$$

Solving for the body fat fraction, x_f, yields

$$x_f = \frac{1}{\rho_p}\left(\frac{\rho_l \rho_f}{\rho_l - \rho_f}\right) - \frac{\rho_f}{\rho_l - \rho_f}$$

Finally, substituting the values for ρ_f and ρ_l, we find

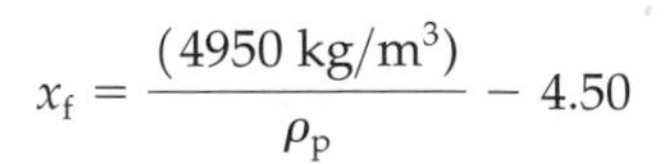

$$x_f = \frac{(4950\ \text{kg/m}^3)}{\rho_p} - 4.50$$

REAL WORLD PHYSICS: BIO
Measuring body fat

This result is known as *Siri's formula*. For example, if $\rho_p = 900\ \text{kg/m}^3$ (all fat), we find $x_f = 1$; if $\rho_p = 1100\ \text{kg/m}^3$ (no fat), we find $x_f = 0$. In the case of Example 15–5, where $\rho_p = 1050\ \text{kg/m}^3$, we find that this person's body fat fraction is $x_f = 0.214$, for a percentage of 21.4%. This is a reasonable value for a healthy adult male.

A recent refinement to the measurement of body fat percentage is the Bod Pod, an egg-shaped, air-tight chamber in which a person sits comfortably—high and dry, surrounded only by air. This device works on the same physical principle as submerging a person in water, only it uses air instead of water. Since air is about a thousand times less dense than water, the measurements of apparent weight must be roughly a thousand times more sensitive. Fortunately, this is possible with today's technology, allowing for a much more convenient means of measurement.

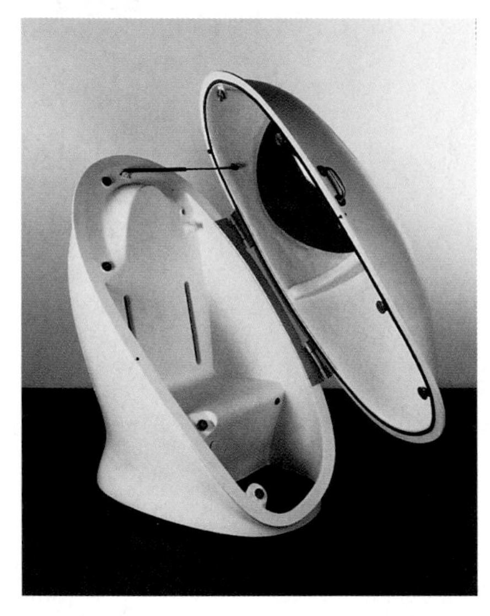

▲ This device, known as the Bod Pod, measures the body fat percentage of a person inside it by varying the air pressure in the chamber and measuring the corresponding changes in the person's apparent weight. Archimedes' principle is at work here, just as it is in Example 15–5.

Next we consider a low-density object, such as a piece of wood, held down below the surface of a denser fluid.

ACTIVE EXAMPLE 15–1 Tied Down

A piece of wood with a density of $706\ \text{kg/m}^3$ is tied with a string to the bottom of a water-filled flask. The wood is completely immersed, and has a volume of $8.00 \times 10^{-6}\ \text{m}^3$. What is the tension in the string?

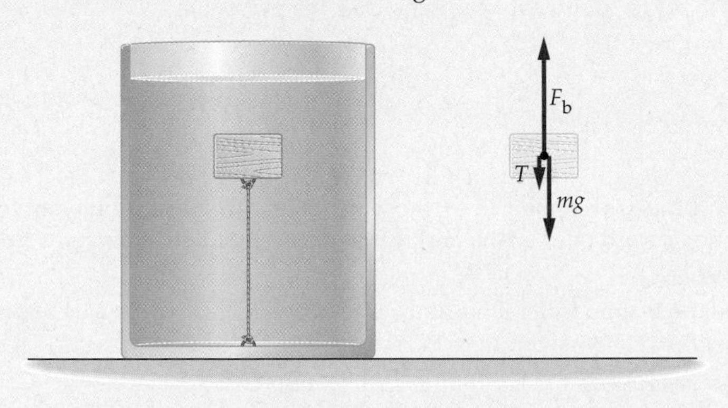

Solution

1. Apply Newton's second law to the wood: $F_b - T - mg = 0$
2. Solve for the tension, T: $T = F_b - mg$
3. Calculate the weight of the wood: $mg = 0.0554\ \text{N}$
4. Calculate the buoyant force: $F_b = 0.0785\ \text{N}$
5. Subtract to obtain the tension: $T = 0.0231\ \text{N}$

Insight

Since the wood floats in water, its buoyant force when completely immersed is greater than its weight.

Floatation

When an object floats, the buoyant force acting on it equals its weight. For example, suppose we slowly lower a block of wood into a flask of water. At first, as in **Figure 15–9 (a)**, only a small amount of water is displaced and the buoyant force

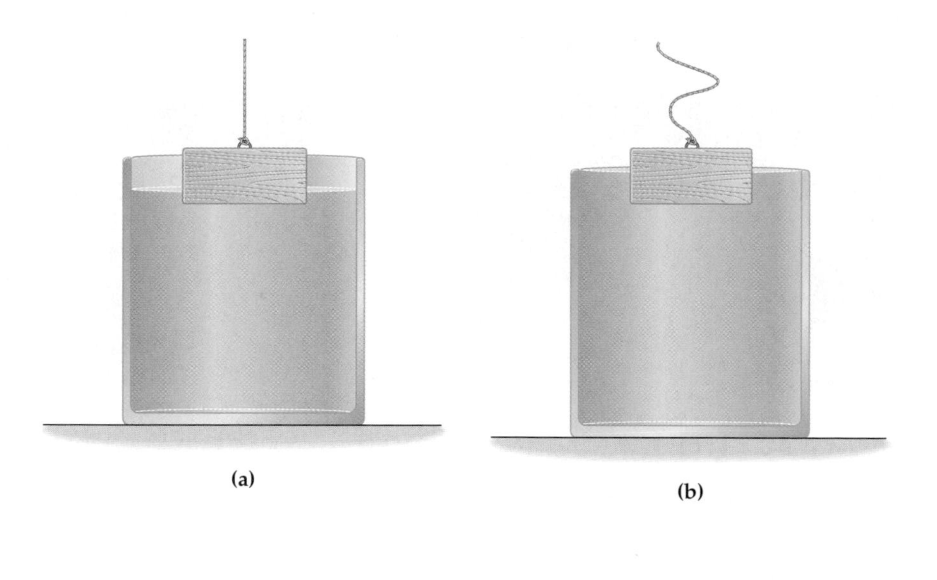

◀ **FIGURE 15–9 Floatation**
(a) The block of wood displaces some water, but not enough to equal its weight. Thus, the block would not float at this position. **(b)** The weight of displaced water equals the weight of the block in this case. The block floats now.

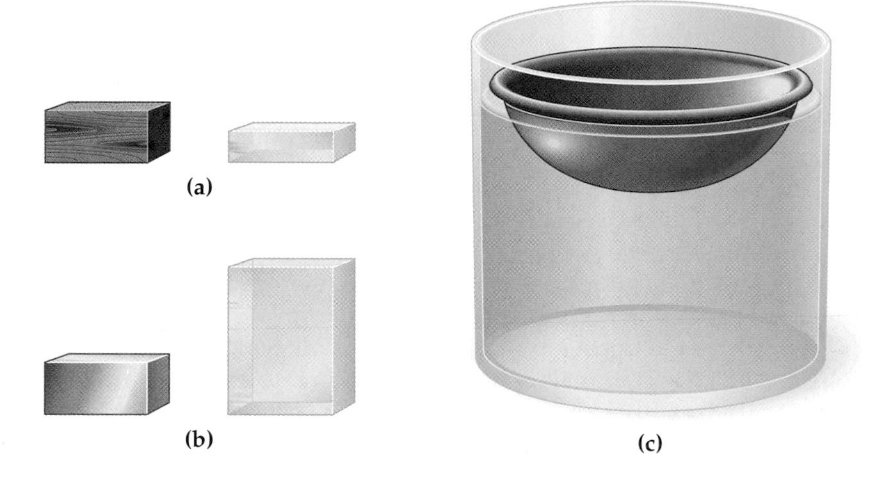

◀ **FIGURE 15–10 Floating an object that is more dense than water**
(a) A wood block and the volume of water that has the same weight. Since the wood displaces more water than this, it floats. **(b)** A metal block and the volume of water that has the same weight. Since the metal displaces less water than this, it sinks. **(c)** If the metal in (b) is shaped like a bowl, it can displace more water than the volume of the metal itself. In fact, it can displace enough water to float.

is a fraction of the block's weight. If we were to release the block now, it would drop further into the water. As we continue to lower the block, more water is displaced, increasing the buoyant force.

Eventually, we reach the situation pictured in **Figure 15–9 (b)**, where the block begins to float. In this case, the buoyant force equals the weight of the wood. This, in turn, means that the weight of the displaced water is equal to the weight of the wood. In general,

An object floats when it displaces an amount of fluid equal to its weight.

This is illustrated in **Figure 15–10 (a)**, where we show the volume of water equal to the weight of a block of wood. Similarly, in **Figure 15–10 (b)** we show the amount of water necessary to have the same weight as a block of metal. Clearly, if the metal is completely submerged the buoyant force is only a fraction of its weight, and so it sinks. On the other hand, if the metal is formed into the shape of a bowl, as in **Figure 15–10 (c)**, it can displace a volume of water equal to its weight and float.

Another way to change the buoyancy of an object is to alter its overall density. Consider, for example, the *Cartesian diver* shown in **Figure 15–11**. As illustrated, the diver is simply a small glass tube with an air bubble trapped inside. Initially, the overall density of the tube and the air bubble is less than the density of water, and the diver floats. When the bottle containing the diver is squeezed, however, the pressure in the water rises, and the air bubble is compressed to a smaller volume.

▲ **FIGURE 15–11 A Cartesian diver**
A Cartesian diver floats because of the bubble of air trapped within it. When the bottle is squeezed, increasing the pressure in the water, the bubble is reduced in size and the diver sinks.

▲ Although steel is denser than water, a ship's bowl-like hull (left) displaces enough water to allow it to float, so long as it is not too heavily loaded. (The boundary between the red and black areas of the hull is the Plimsoll line, which indicates where the ship should ride in the water when carrying its maximum safe load—see Conceptual Checkpoint 15–4.) For balloons (right), the key to buoyancy is the lower density and greater volume of hot air. As the air in the balloon is heated it expands the bag, increasing the volume of air that the balloon displaces. At the same time, heated air spills out of the balloon, decreasing its weight. Eventually, the average density of the balloon plus the hot air it contains becomes lower than that of the surrounding air, and it starts to float.

Now, the overall density of the tube and air bubble is greater than that of water, and the diver descends. By adjusting the pressure on the bottle, the diver can be made to float at any depth in the bottle.

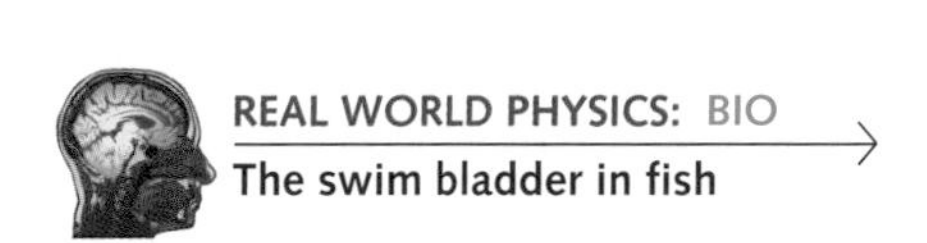

The same principle applies to the swim bladder of ray-finned bony fishes. The swim bladder is basically an air sac whose volume can be controlled by the fish. By adjusting the size of the swim bladder, the fish can give itself "neutral buoyancy"—that is, the fish can float without effort at a given depth, just like the Cartesian diver. Similar considerations apply to certain diving sea mammals, such as the bottlenose dolphin, Weddell seal, elephant seal, and blue whale. All of these animals are capable of diving to great depths—in fact some of the seals have been observed at depths of nearly 400 m. In order to conserve energy on their long dives they take advantage of the fact that the pressure of the surrounding water compresses their bodies and flattens the air sacs in their lungs. Just as with the Cartesian diver, this decreases their buoyancy to the point where they begin to sink. As a result, they can glide effortlessly to the desired depth, saving energy for the swim back to the surface.

▲ The water of the Dead Sea is unusually dense because of its great salt content. As a result, swimmers can float higher in the water than they are accustomed and engage in recreational activities that we don't ordinarily associate with a dip in the ocean.

ACTIVE EXAMPLE 15–2 **A Block of Wood**

How much water (density 1000 kg/m^3) must be displaced to float a cubical block of wood (density 655 kg/m^3) that is 15.0 cm on a side?

Solution

1. Calculate the volume of the wood: $V_{\text{wood}} = 3.38 \times 10^{-3}\ \text{m}^3$
2. Find the weight of the wood: $\rho_{\text{wood}} V_{\text{wood}} g = 21.7\ \text{N}$
3. Write an expression for the weight of a volume of water: $\rho_{\text{water}} V_{\text{water}} g$
4. Set the weight of water equal to the weight of the wood: $\rho_{\text{water}} V_{\text{water}} g = 21.7\ \text{N}$
5. Solve for the volume of water: $V_{\text{water}} = 2.21 \times 10^{-3}\ \text{m}^3$

Insight

As expected, only a fraction of the wood must be submerged in order for it to float.

CONCEPTUAL CHECKPOINT 15–4

On the side of a cargo ship you may see a horizontal line indicating "maximum load." (It is sometimes known as the "Plimsoll mark," after the nineteenth-century British legislator who caused it to be adopted.) When a ship is loaded to capacity, the maximum load line is at water level. The ship shown here has two maximum load lines, one for fresh water and one for salt water. Which line should be marked "maximum load for salt water": **(a)** the top line, or **(b)** the bottom line?

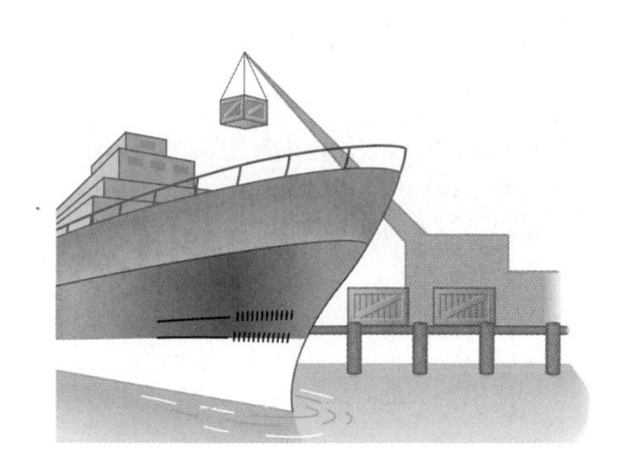

Reasoning and Discussion

If a ship sails from fresh water into salt water it floats higher, just as it is easier for you to float in an ocean than in a lake. The reason is that salt water is denser than fresh water, hence less of it needs to be displaced to provide a given buoyant force. Since the ship floats higher in salt water, the bottom line should be used to indicate maximum load.

Answer:

(b) The bottom line should be used in salt water.

Tip of the Iceberg

As we have seen, an object floats when its weight is equal to the weight of the fluid it displaces. Let's use this condition to determine just how much of a floating object is submerged. We will then apply our result to the classic case of an iceberg.

Consider, then, a solid of density ρ_s floating in a fluid of density ρ_f, as in **Figure 15–12**. If the solid has a volume V_s, its total weight is

$$W_s = \rho_s V_s g$$

Similarly, the weight of a volume V_f of displaced fluid is

$$W_f = \rho_f V_f g$$

Equating these weights, we find the following:

$$W_s = W_f$$

$$\rho_s V_s g = \rho_f V_f g$$

Canceling g, and solving for the volume of displaced fluid, we have

$$V_f = V_s(\rho_s/\rho_f)$$

Since, by definition, the volume of displaced fluid is the same as the volume of the solid that is submerged, V_{sub}, we find

Submerged volume V_{sub} for a solid of volume V_s and density ρ_s floating in a fluid of density ρ_f

$$V_{sub} = V_s(\rho_s/\rho_f) \qquad \textbf{15–10}$$

SI unit: m^3

Note that this relation agrees with the results of Active Example 15-2.

We now apply this result to ice floating in water.

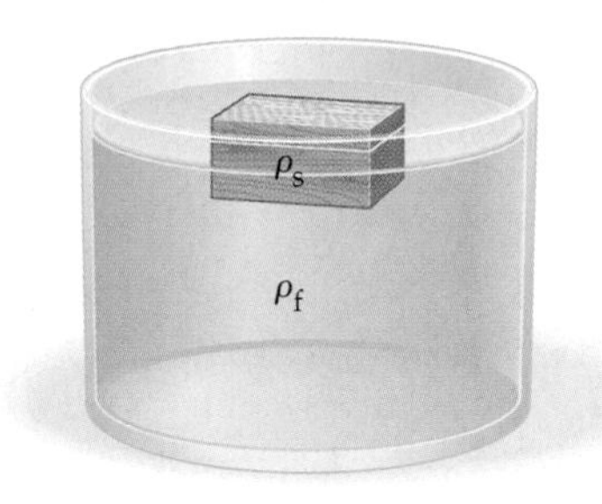

▲ **FIGURE 15–12 Submerged volume of a floating object**

A solid, of volume V_s and density ρ_s, floats in a fluid of density ρ_f. The volume of the solid that is submerged is $V_{sub} = V_s(\rho_s/\rho_f)$.

▲ Most people know that the bulk of an iceberg lies below the surface of the water. But as with ships and swimmers, the actual proportion that is submerged depends on whether the water is fresh or salt (see Example 15–6).

EXAMPLE 15–6 The Tip of the Iceberg

What percentage of a floating chunk of ice projects above the level of the water? Assume a density of 917 kg/m^3 for the ice, and 1000 kg/m^3 for the water.

continued on the following page

continued from the previous page

Picture the Problem

Our sketch shows the ice, with density ρ_s, floating in the water, with density ρ_f.

Strategy

We can apply Equation 15–10 to this system. First, the fraction of the total volume of the ice, V_s, that is submerged is $V_{sub}/V_s = \rho_s/\rho_f$. Hence the fraction that is above the water is $1 - V_{sub}/V_s = 1 - \rho_s/\rho_f$. Multiplying this fraction by 100 yields the percentage above water.

Solution

1. Calculate the fraction of the total volume of the ice that is submerged:

$$\frac{V_{sub}}{V_s} = \frac{\rho_s}{\rho_f} = \frac{917\ \text{kg/m}^3}{1000\ \text{kg/m}^3} = 0.917$$

2. Calculate the fraction of the ice that is above water:

$$1 - \frac{\rho_s}{\rho_f} = 1 - 0.917 = 0.0830$$

3. Multiply by 100 to obtain a percentage:

$$100\left(1 - \frac{\rho_s}{\rho_f}\right) = 100(0.0830) = 8.30\%$$

Insight

Since we seek a percentage, it is not necessary to know the total volume of the ice.

Practice Problem

What fraction of the ice is above water if it floats in sea water (density 1025 kg/m^3)? [**Answer:** 10.5%]

Some related homework problems: Problem 36, Problem 37

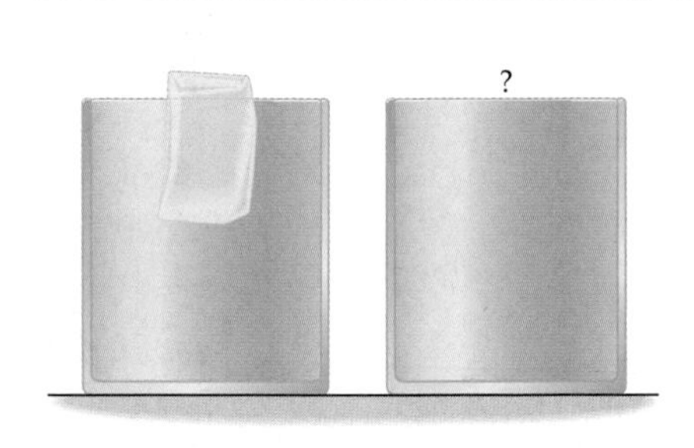

CONCEPTUAL CHECKPOINT 15–5

A cup is filled to the brim with water and a floating ice cube. When the ice melts, which of the following occurs? **(a)** Water overflows the cup, **(b)** the water level decreases, or **(c)** the water level remains the same.

Reasoning and Discussion

Since the ice cube floats, it displaces a volume of water equal to its weight. But when it melts, it becomes water, and its weight is the same. Hence, the melted water fills exactly the same volume that the ice cube displaced when floating. As a result, the water level is unchanged.

Answer:

(c) The water level remains the same.

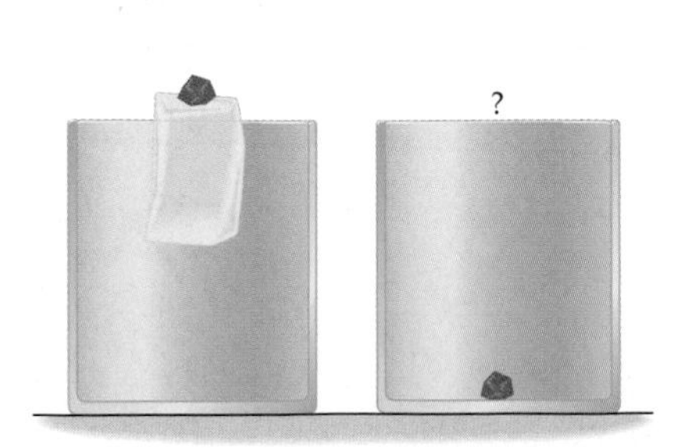

CONCEPTUAL CHECKPOINT 15–6

A cup is filled to the brim with water and a floating ice cube. Resting on top of the ice cube is a small pebble. When the ice melts, which of the following occurs? **(a)** Water overflows the cup, **(b)** the water level decreases, or **(c)** the water level remains the same.

Reasoning and Discussion

We know from the previous Conceptual Checkpoint that the ice itself makes no difference to the water level. As for the pebble, when it floats on the ice it displaces an amount of water equal to its *weight*. When the ice melts the pebble drops to the bottom of the cup, where it displaces a volume of water equal to its own *volume*. Since the volume of the pebble is less than the volume of water with the same weight, we conclude that less water is displaced after the ice melts. Hence, the water level decreases.

Answer:

(b) The water level decreases.

15–6 Fluid Flow and Continuity

Suppose you want to water the yard, but you don't have a spray nozzle for the end of the hose. Without a nozzle the water flows rather slowly from the hose and hits the ground within half a meter. But if you place your thumb over the end of the hose, narrowing the opening to a fraction of its original size, the water sprays out with a high speed and a large range. Why does decreasing the size of the opening have this effect?

To answer this question, we begin by considering a simple system that shows the same behavior. Imagine, then, that a fluid flows with a speed v_1 through a cylindrical pipe of cross-sectional area A_1, as in the left-hand portion of **Figure 15–13**. If the pipe narrows to a cross-sectional area A_2, as in the right-hand portion of Figure 15–13, the fluid will flow with a new speed, v_2.

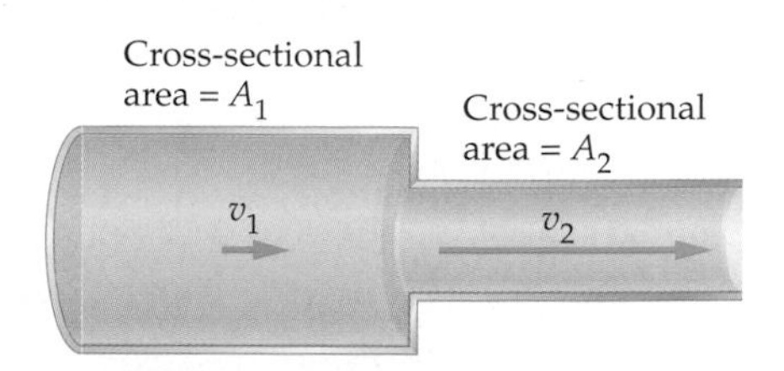

▲ FIGURE 15–13 Fluid flow through a pipe of varying diameter

As a fluid flows from a large pipe to a small pipe the same mass of fluid passes a given point in a given amount of time. Thus, the speed in the small pipe is greater than it is in the large pipe.

We can find the speed in the narrow section of pipe by assuming that any amount of fluid that passes point 1 in a given time, Δt, must also flow past the point 2 in the same time. If this were not the case, the system would be gaining or losing fluid. To find the mass of fluid passing point 1 in the time Δt, note that the fluid moves through a distance $v_1 \Delta t$ in this time. As a result, the volume of fluid going past point 1 is

$$\Delta V_1 = A_1 v_1 \Delta t$$

Hence, the mass of fluid passing point 1 is

$$\Delta m_1 = \rho_1 \Delta V_1 = \rho_1 A_1 v_1 \Delta t$$

Similarly, the mass passing point 2 in the time Δt is

$$\Delta m_2 = \rho_2 \Delta V_2 = \rho_2 A_2 v_2 \Delta t$$

Note that we have allowed for the possibility of the fluid having different densities at points 1 and 2.

Finally, equating these two masses yields the relation between v_1 and v_2:

$$\Delta m_1 = \Delta m_2$$

$$\rho_1 A_1 v_1 \Delta t = \rho_2 A_2 v_2 \Delta t$$

Canceling Δt we find

▲ Narrowing the opening in a hose with a nozzle (or thumb) increases the velocity of flow, as one would expect from the equation of continuity.

Equation of continuity

$$\rho_1 A_1 v_1 = \rho_2 A_2 v_2 \qquad \textbf{15–11}$$

This relation is referred to as the **equation of continuity**.

Most gases are readily compressed, which means that their densities can change. In contrast, most liquids are practically incompressible, so their densities are essentially constant. Unless stated otherwise, we will assume all liquids discussed in this text to be perfectly incompressible. Thus, for liquids, ρ_1 and ρ_2 are the same in Equation 15–11, and the equation of continuity reduces to the following:

Equation of continuity for an incompressible fluid

$$A_1 v_1 = A_2 v_2 \qquad \textbf{15–12}$$

We next apply this relation to the case of water flowing through the nozzle of a fire hose.

PROBLEM SOLVING NOTE
Continuity of Flow

The speed of an incompressible fluid is inversely proportional to the area through which it flows.

EXAMPLE 15–7 Spray I

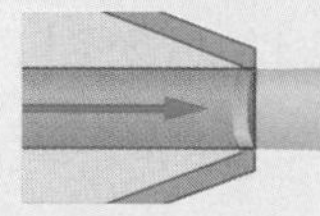

Water travels through a 9.6-cm diameter fire hose with a speed of 1.3 m/s. At the end of the hose the water flows out through a nozzle whose diameter is 2.5 cm. What is the speed of the water coming out of the nozzle?

continued on the following page

continued from the previous page

Picture the Problem

In the sketch, we label the speed of the water in the hose as v_1, and the speed of the water coming out the nozzle as v_2.

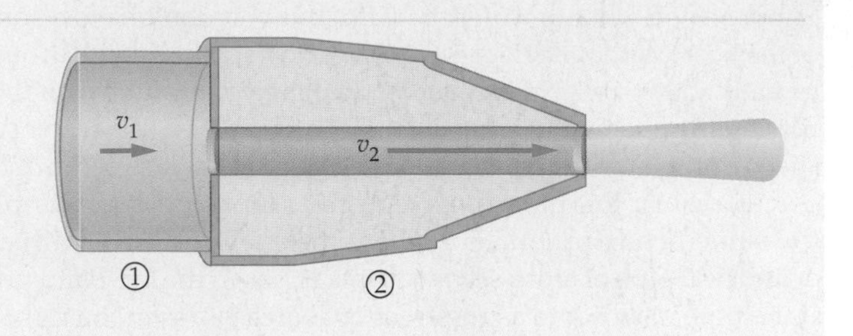

Strategy

We can find the water speed in the nozzle by applying Equation 15–12. In addition, we assume that the hose and nozzle are circular in cross section; hence their areas are given by $A = \pi d^2/4$, where d is the diameter.

Solution

1. Solve Equation 15–12 for v_2, the speed of the water in the nozzle:
$$v_2 = v_1(A_1/A_2)$$

2. Replace the areas with $A = \pi d^2/4$:
$$v_2 = v_1\left(\frac{\pi d_1^2/4}{\pi d_2^2/4}\right) = v_1\left(\frac{d_1^2}{d_2^2}\right)$$

3. Substitute numerical values:
$$v_2 = v_1\left(\frac{d_1^2}{d_2^2}\right) = (1.3\text{ m/s})\left(\frac{9.6\text{ cm}}{2.5\text{ cm}}\right)^2 = 19\text{ m/s}$$

Insight

Note that a small-diameter nozzle can give very high speeds. In fact, the speed depends on the diameter squared.

Practice Problem

What nozzle diameter would be required to give the water a speed of 21 m/s? [**Answer:** $d_2 = 2.4$ cm]

Some related homework problems: Problem 41, Problem 44, Problem 45

15–7 Bernoulli's Equation

In this section, we apply the work-energy theorem to fluids. The result is a relation between the pressure of a fluid, its speed, and its height. This relation is known as **Bernoulli's equation**.

Change in Speed

We begin by considering a system in which the speed of the fluid changes. To be specific, the system of interest is the same as that shown in Figure 15–13. We have already shown that the speed of the fluid increases as it flows from region 1 to region 2; we now investigate the corresponding change in pressure.

Our plan of attack is to first calculate the total work done on the fluid as it moves from one region to the next. This result will depend on the pressure in the fluid. Once the total work is obtained, the work-energy theorem allows us to equate it to the change in kinetic energy of the fluid. This will give the pressure-speed relationship we desire.

Consider an element of fluid of length Δx_1 (**Figure 15–14**). This element is pushed in the direction of motion by the pressure P_1. Thus, the pressure does positive work, ΔW_1, on the fluid element. Noting that the force exerted on the element is $F_1 = P_1 A_1$, and that work is force times distance, the work done on the element is

$$\Delta W_1 = F_1 \Delta x_1 = P_1 A_1 \Delta x_1$$

The volume of the fluid element is $\Delta V_1 = A_1 \Delta x_1$, so the work done by P_1 is

$$\Delta W_1 = P_1 \Delta V_1$$

Next, when the fluid element emerges into region 2 it experiences a force opposite to its direction of motion due to the pressure P_2. Thus, P_2 does negative work on the element. Following the same steps given previously we can write the work done by P_2 as

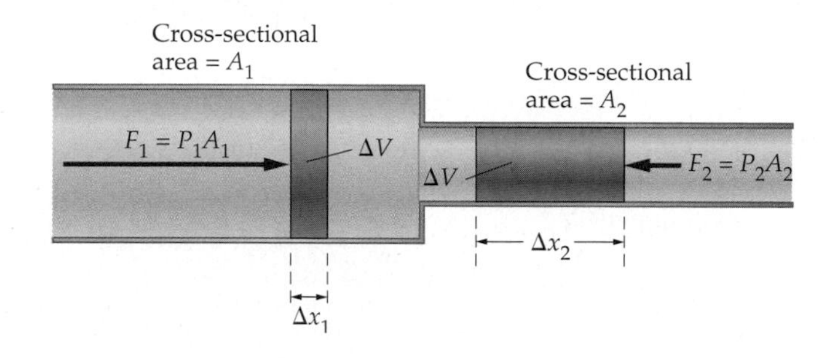

◀ FIGURE 15–14 Work done on a fluid element

As an incompressible fluid element of volume ΔV moves from pipe 1 to pipe 2 the pressure P_1 does a positive work $P_1\Delta V$ and the pressure P_2 does a negative work $P_2\Delta V$. Since P_1 is greater than P_2 the net result is that positive work is done, and the fluid element speeds up.

$$\Delta W_2 = -P_2\Delta V_2$$

Now, for an incompressible fluid, the volume of the element does not change as it goes from region 1 to region 2. Therefore,

$$\Delta V_1 = \Delta V_2 = \Delta V$$

Using this result, we can write the total work done on the fluid element as follows:

$$\Delta W_{\text{total}} = \Delta W_1 + \Delta W_2 = P_1\Delta V - P_2\Delta V = (P_1 - P_2)\Delta V$$

The final step is to equate the total work to the change in kinetic energy:

$$\Delta W_{\text{total}} = (P_1 - P_2)\Delta V = K_{\text{final}} - K_{\text{initial}} = K_2 - K_1 \qquad \textbf{15–13}$$

What is the kinetic energy of the fluid element? Well, the mass of the element is

$$\Delta m = \rho\Delta V$$

Thus, its kinetic energy is simply

$$K = \tfrac{1}{2}(\Delta m)v^2 = \tfrac{1}{2}(\rho\Delta V)v^2$$

Using this expression in Equation 15–13, we have

$$\Delta W_{\text{total}} = (P_1 - P_2)\Delta V = (\tfrac{1}{2}\rho v_2^2 - \tfrac{1}{2}\rho v_1^2)\Delta V$$

Canceling the common factor ΔV, and rearranging, we find

$$P_1 + \tfrac{1}{2}\rho v_1^2 = P_2 + \tfrac{1}{2}\rho v_2^2 \qquad \textbf{15–14}$$

Equation 15–14 is equivalent to saying that $P + \frac{1}{2}\rho v^2$ is constant. Thus, there is a tradeoff between the pressure in a fluid and its speed—as the fluid speeds up its pressure decreases. If this seems odd, recall that P_1 acts to increase the speed of the fluid element and P_2 acts to decrease its speed. The element will speed up, then, only if P_2 is less than P_1.

EXAMPLE 15–8 Spray II

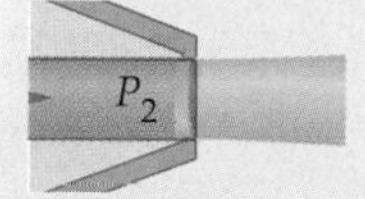

Referring to Example 15–7, suppose the pressure in the fire hose is 350 kPa. What is the pressure in the nozzle?

Picture the Problem

In our sketch, we use the same numbering system as in Example 15–7; that is, 1 refers to the hose, 2 to the nozzle.

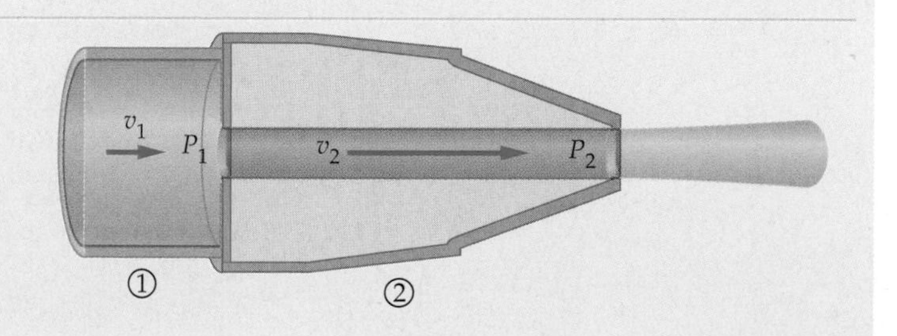

Strategy

We are given P_1, v_1, and the diameters of the hose and nozzle. From the equation of continuity, Equation 15–12, we find v_2 (as in Example 15–7). Now we use this result, plus Equation 15–14, to determine P_2.

continued on the following page

continued from the previous page

Solution

1. Solve Equation 15–14 for the pressure in the nozzle, P_2:

$$P_2 = P_1 + \tfrac{1}{2}\rho(v_1^2 - v_2^2)$$

2. Substitute numerical values, including v_2 from Example 15–7:

$$P_2 = 350\ \text{kPa} + \tfrac{1}{2}(1000\ \text{kg/m}^3)[(1.3\ \text{m/s})^2 - (19\ \text{m/s})^2]$$
$$= 170\ \text{kPa}$$

Insight

Note that the pressure in the nozzle is less than the pressure in the hose by roughly a factor of 2. What has happened is that part of the energy stored as pressure in the hose has been converted to kinetic energy as the water passes through the nozzle.

Practice Problem

What nozzle speed would be required to give a nozzle pressure of 110 kPa? [**Answer:** $v_2 = 22$ m/s]

Some related homework problems: Problem 48, Problem 49

Change in Height

If a fluid flows through the pipe shown in **Figure 15–15**, its height increases from y_1 to y_2 as it goes from one region to the next. Since the cross-sectional area of the pipe is constant, however, the speed of the fluid is unchanged, according to Equation 15–12. Thus, the change in kinetic energy of the fluid element shown in Figure 15–15 is zero.

The total work done on the fluid element is the sum of the works done by the pressure in each of the two regions, plus the work done by gravity. As before, the work done by pressure is

$$\Delta W_{\text{pressure}} = \Delta W_1 + \Delta W_2 = (P_1 - P_2)\Delta V$$

As the fluid element rises, gravity does negative work on it. Recalling that the mass of the element is

$$\Delta m = \rho \Delta V$$

the work done by gravity is

$$\Delta W_{\text{gravity}} = -\Delta m g(y_2 - y_1) = -\rho \Delta V g(y_2 - y_1)$$

Setting the total work equal to zero (since $\Delta K = 0$), yields

$$\Delta W_{\text{total}} = \Delta W_{\text{pressure}} + \Delta W_{\text{gravity}}$$
$$= (P_1 - P_2)\Delta V - \rho g(y_2 - y_1)\Delta V = 0$$

Canceling ΔV and rearranging gives

$$P_1 + \rho g y_1 = P_2 + \rho g y_2 \qquad \textbf{15–15}$$

In this case, it is $P + \rho g y$ that is constant—hence pressure decreases as the height within a fluid increases. Note, in fact, that Equation 15–15 is precisely the same as Equation 15–7, which was obtained using force considerations. Here we obtained the result using the work-energy theorem.

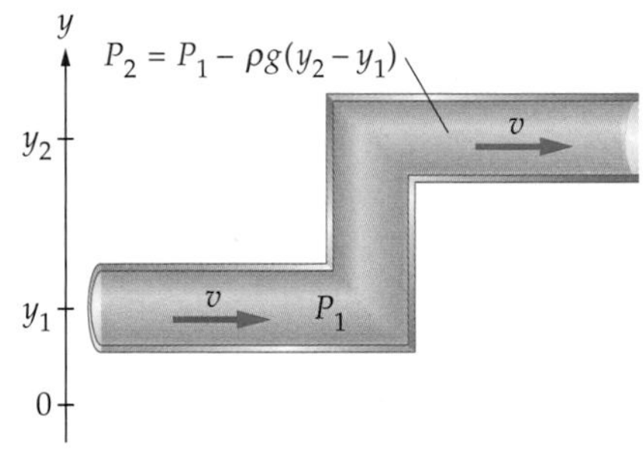

▲ **FIGURE 15–15 Fluid pressure in a pipe of varying elevation**
Fluid of density ρ flows in a pipe of uniform cross-sectional area from height y_1 to height y_2. As it does so, its pressure decreases by the amount $\rho g(y_2 - y_1)$.

EXERCISE 15–4

Water flows with constant speed through a garden hose that goes up a step 20.0-cm high. If the water pressure is 143 kPa at the bottom of the step, what is its pressure at the top of the step?

Solution

Apply Equation 15–15, letting subscript 1 refer to the bottom of the step and subscript 2 refer to the top of the step. Solve for P_2:

$$P_2 = P_1 + \rho g(y_1 - y_2) = 143\ \text{kPa} + (1000\ \text{kg/m}^3)(9.81\text{m/s}^2)(0 - 0.200\,\text{m}) = 141\ \text{kPa}$$

This precisely the pressure difference that would be observed if the water had been at rest.

General Case

In a more general case, both the height of a fluid and its speed may change. Combining the results obtained in Equations 15–14 and 15–15 yields the full form of Bernoulli's equation:

Bernoulli's equation

$$P_1 + \tfrac{1}{2}\rho v_1^2 + \rho g y_1 = P_2 + \tfrac{1}{2}\rho v_2^2 + \rho g y_2 \qquad \textbf{15–16}$$

Thus, in general, the quantity $P + \frac{1}{2}\rho v^2 + \rho g y$ is a constant within a fluid. This is basically a statement of energy conservation. For example, recalling the definition of density in Equation 15–1, we find that $\frac{1}{2}\rho v^2$ is $\frac{1}{2}(M/V)v^2 = (\frac{1}{2}Mv^2)/V$. Clearly, this term represents the kinetic energy per volume of the fluid. Similarly, the term $\rho g y$ can be written as $(M/V)gy = (Mgy)/V$, which is the gravitational potential energy per volume.

Finally, the first term in Bernoulli's equation—the pressure—can also be thought of as an energy per volume. Recall that $P = F/A$. If we multiply numerator and denominator by a distance, d, we have $P = Fd/Ad$. But, Fd is the work done by the force F as it acts through the distance d, and Ad is the volume swept out by an area A moved through a distance d. Therefore, the pressure can be thought of as work (energy) per volume: $P = W/V$.

As a result, Bernoulli's equation is simply a restatement of the work-energy theorem in terms of quantities per volume. Of course, this relation holds only as long as we can ignore frictional losses, which would lead to heating. We will consider the energy aspects of heat in the next chapter.

ACTIVE EXAMPLE 15–3 **A Step Up**

Repeat Exercise 15–4 with the following additional information: **(a)** the cross-sectional area of the hose on top of the step is half that at the bottom of the step, and **(b)** the speed of the water at the bottom of the step is 1.20 m/s.

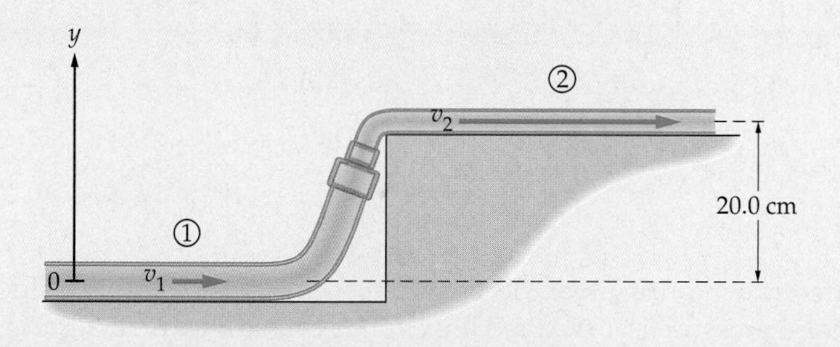

Solution

1. Use the continuity equation to find the water's speed on top of the step: $v_2 = 2v_1 = 2.40$ m/s
2. Solve Bernoulli's equation for P_2: $P_2 = P_1 - \rho g(y_2 - y_1) - \frac{1}{2}\rho(v_2^2 - v_1^2)$
3. Substitute numerical values: $P_2 = 139$ kPa

Insight

The pressure on top of the step is less than in Exercise 15–4. This is to be expected because in this case, the water speeds up as it rises over the step.

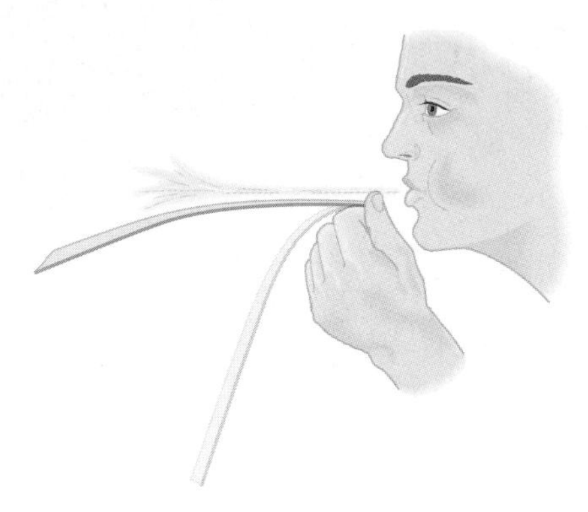

▲ **FIGURE 15–16 The Bernoulli effect on a sheet of paper**
If you hold a piece of paper by its end it will bend downward. Blowing across the top of the paper reduces the pressure there, resulting in a net upward force which lifts the paper to a nearly horizontal position.

15–8 Applications of Bernoulli's Equation

We now consider a variety of real-world examples that illustrate the application of Bernoulli's equation.

Pressure and Speed

As mentioned, it often seems counterintuitive that a fast-moving fluid should have less pressure than a slow-moving one. Remember, however, that pressure can be thought of as a form of energy. From this point of view there is an energy tradeoff between pressure and kinetic energy.

Perhaps the easiest way to demonstrate the dependence of pressure on speed is to blow across the top of a piece of paper. If you hold the paper as shown in **Figure 15–16**, then blow over the top surface, the paper will lift upward. The reason is that there is a difference in air speed between the top and the bottom of the paper, with the higher speed on top. As a result, the pressure above the paper is lower. This pressure difference, in turn, results in a net upward force, referred to as **lift**, and the paper rises.

A similar example of pressure and speed is provided by the airplane wing. A cross section of a typical wing is shown in **Figure 15–17**. The shape of the wing is designed so that air flows more rapidly over the top surface than the lower surface. As a result, the pressure is less on top. As with the piece of paper, the pressure difference results in a net upward force (lift) on the wing.

Note that lift is a dynamic effect; it requires a flow of air. The greater the speed difference, the greater the upward force.

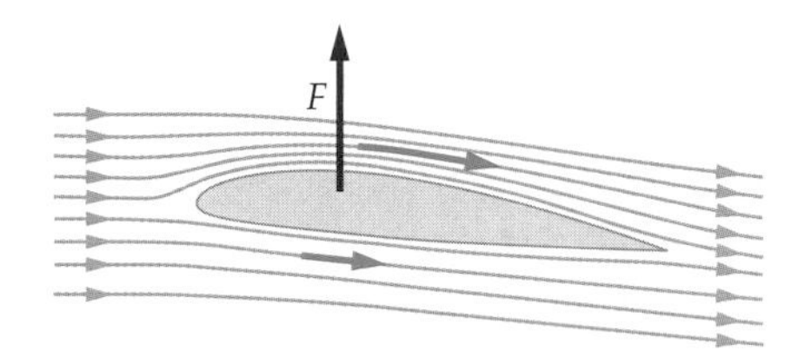

▲ **FIGURE 15–17 Airflow and lift in an airplane wing**
Cross section of an airplane wing with air flowing past it. The wing is shaped so that air flows more rapidly over the top of the wing than along the bottom. As a result, the pressure on top of the wing is reduced, and a net upward force (lift) is generated.

EXERCISE 15–5

During a windstorm, a 35.5 m/s wind blows across the flat roof of a small home, as in **Figure 15–18**. Find the difference in pressure between the air inside the home and the air just above the roof. (The density of air is 1.29 kg/m^3.)

Solution

Use Bernoulli's equation with point 1 just under the roof and point 2 just above the roof. Since there is little difference in elevation between these points, $y_1 = y_2 = y$. Thus,

$$P_1 + 0 + \rho g y = P_2 + \tfrac{1}{2}\rho v_2^2 + \rho g y$$

Solving for the pressure difference, $P_1 - P_2$, we find

$$P_1 - P_2 = \tfrac{1}{2}\rho v_2^2 = \tfrac{1}{2}(1.29 \text{ kg/m}^3)(35.5 \text{ m/s})^2 = 813 \text{ Pa}$$

▲ **FIGURE 15–18 Force on a roof due to wind speed**
Wind blows across the roof of a house, but the air inside is at rest. The pressure over the roof is therefore less than the pressure inside, resulting in a net upward force on the roof.

A difference in pressure of 813 Pa might seem rather small, considering that atmospheric pressure is 101 kPa. However, it can still cause a significant force on a relatively large area, such as a roof. If a typical roof has an area of about 150 m^2, for example, a pressure difference of 813 Pa results in an upward force of over 27,000 pounds! This is why roofs are often torn from houses during severe windstorms.

On a lighter note, prairie dogs seem to *benefit* from the effects of Bernoulli's equation. A schematic prairie dog burrow is pictured in **Figure 15–19**. Note that one of the entrance/exit mounds is higher than the other. This is significant because the speed of air flow due to the incessant prairie winds varies with height; the speed goes to zero right at ground level, and increases to its maximum value within a few feet above the surface. As a result, the speed of air over the higher mound is greater than that over the lower mound. This causes the pressure to be less over the higher mound. With a pressure difference between the two mounds, air is drawn through the burrow, giving a form of natural air conditioning.

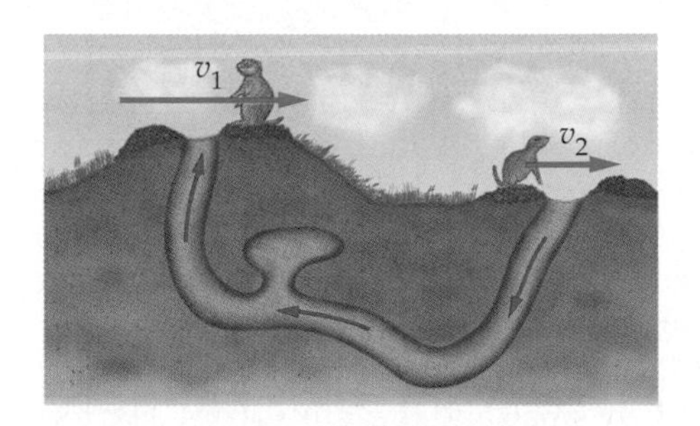

◀ **FIGURE 15–19 Air circulation in a prairie dog burrow**
A prairie dog burrow typically has a high mound on one end and a low mound on the other. Since the wind speed increases with height above the ground, the pressure is smaller at the high-mound end of the burrow. The result is a very convenient circulation of fresh air through the burrow.

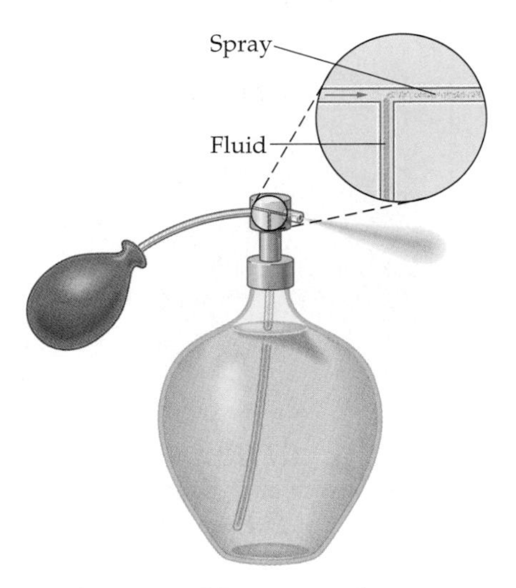

▲ **FIGURE 15–20 An atomizer**
The operation of an "atomizer" can be understood in terms of Bernoulli's equation. The high speed jet of air created by squeezing the bulb creates a low pressure at the top of the vertical tube. This causes fluid to be drawn up the tube and expelled with the air jet as a fine spray.

Similar effects are seem in an atomizer, which sprays perfume in a fine mist. As the bulb shoots a gust of air, as in **Figure 15–20**, it passes through a narrow orifice, which causes the air speed to increase. The pressure decreases as a result, and perfume is drawn up by the pressure difference into the stream of air.

CONCEPTUAL CHECKPOINT 15–7

A small ranger vehicle has a soft, ragtop roof. When the car is at rest the roof is flat. When the car is cruising at highway speeds with its windows rolled up, does the roof **(a)** bow upward, **(b)** remain flat, or **(c)** bow downward?

Reasoning and Discussion
When the car is in motion air flows over the top of the roof, while the air inside the car is at rest—since the windows are closed. Thus, there is less pressure over the roof than under it. As a result, the roof bows upward.

Answer:
(a) The roof bows upward.

▲ We often say that a hurricane or tornado "blew the roof off a house." However, the house at left lost its roof not because of the great pressure exerted by the wind, but rather the opposite. In accordance with the Bernoulli effect, the high speed of the wind passing over the roof created a region of reduced pressure. Normal atmospheric pressure *inside* the house then blew the roof off. The same phenomenon is exploited by prairie dogs to ventilate their burrows. One end of the burrow is always situated at a greater height than the other. Since the prairie wind blows much faster a few feet above ground level, the pressure at the elevated end of the burrow is reduced. The resulting pressure difference produces a flow of air through the burrow.

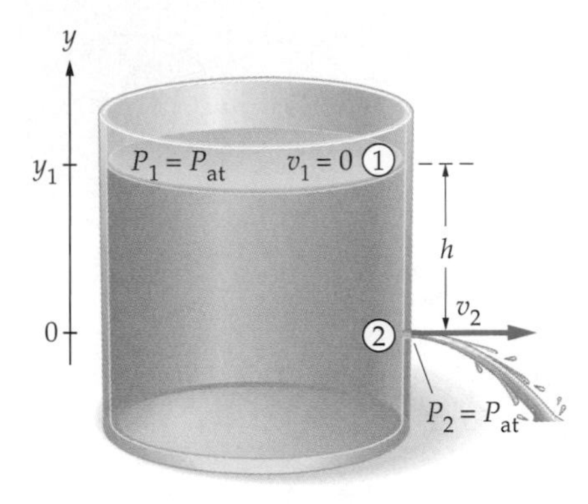

▲ FIGURE 15–21 Fluid emerging from a hole in a container

Since the fluid exiting the hole is in contact with the atmosphere, the pressure there is P_{at}, just as it is on the top surface of the fluid.

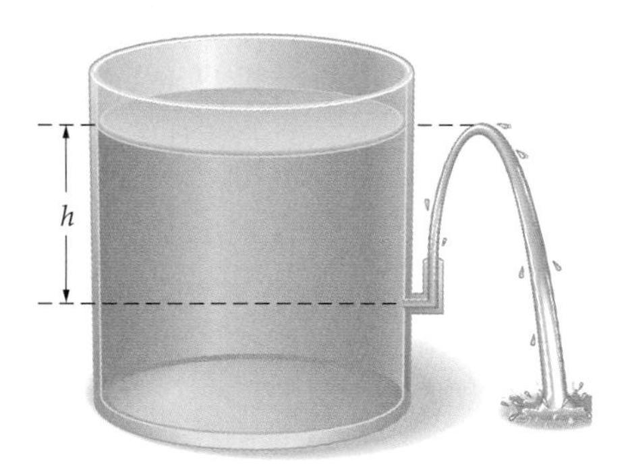

▲ FIGURE 15–22 Maximum height of a stream of water

If the fluid emerging from a hole in a container is directed upward, it has just enough speed to reach the surface level of the fluid. This is an example of energy conservation.

Torricelli's Law

Our final example of Bernoulli's equation deals with the speed of a fluid as it flows through a hole in a container. Consider, for example, the tank of water shown in **Figure 15–21**. If a hole is poked through the side of the tank at a depth h below the surface, what is the speed of the water as it emerges?

To answer this question, we apply Bernoulli's equation to the two points shown in the figure. First, at point 1 we note that the water is open to the atmosphere; thus $P_1 = P_{at}$. Next, with the origin at the level of the hole, the height of the water surface is $y_1 = h$. Finally, if the hole is relatively small and the tank is large, the top surface of the water will have essentially zero speed; thus we can set $v_1 = 0$. Collecting these results, we have the following for point 1:

$$P_1 + \tfrac{1}{2}\rho v_1^2 + \rho g y_1 = P_{at} + 0 + \rho g h$$

Now for point 2. At this point the height is $y_2 = 0$, by definition of the origin, and the speed of the escaping water is the unknown, v_2. Here is the key step: The pressure P_2 is *atmospheric pressure*, because the hole opens the water to the atmosphere. Thus, for point 2 we have the following:

$$P_2 + \tfrac{1}{2}\rho v_2^2 + \rho g y_2 = P_{at} + \tfrac{1}{2}\rho v_2^2 + 0$$

Equating these results yields

$$P_1 + \tfrac{1}{2}\rho v_1^2 + \rho g y_1 = P_2 + \tfrac{1}{2}\rho v_2^2 + \rho g y_2$$

$$P_{at} + \rho g h = P_{at} + \tfrac{1}{2}\rho v_2^2$$

Eliminating P_{at} and ρ we find

$$v_2 = \sqrt{2gh} \qquad \textbf{15–17}$$

This result is known as **Torricelli's law**.

This expression for v_2 should look familiar; it is the speed of an object that falls freely through a distance h. That is, the water emerges from the tank with the same speed as if it had fallen from the surface of the water to the hole. Similarly, if the emerging stream of water were to be directed upward, as in **Figure 15–22**, it would have just enough speed to rise through a height h—right back to the water's surface. This is precisely what one would expect on the basis of energy conservation.

EXAMPLE 15–9 A Water Fountain

In designing a backyard water fountain, a gardener wants a stream of water to exit from the bottom of one can and land in a second one, as shown in the sketch. The top of the second can is 0.500 m below the hole in the first can, which has water in it to a depth of 0.150 m. How far to the right of the first can must the second one be placed to catch the stream of water?

Picture the Problem

Our sketch labels the various pertinent quantities for this problem. We know that $h = 0.150$ m and $H = 0.500$ m. The distance D is to be found. We have also chosen an appropriate coordinate system.

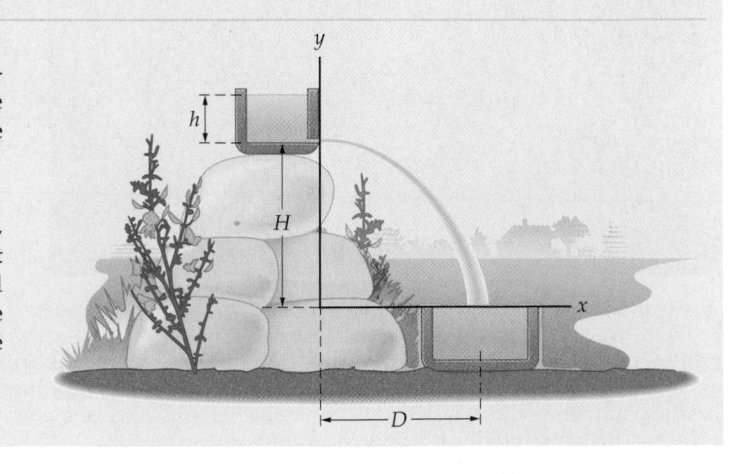

Strategy

This problem combines Torricelli's law and kinematics. First, we find the speed v of the stream of water as it leaves the first can, using Equation 15–17. Next, we find the time t required for the stream to fall freely through a distance H. Finally, since the stream moves with constant speed in the x direction, the distance D is given by $D = vt$.

Solution

1. Find the speed v of the stream when it leaves the first can: $v = \sqrt{2gh} = \sqrt{2(9.81\ \text{m/s}^2)(0.150\ \text{m})} = 1.72\ \text{m/s}$

2. Find the time t for free fall through a height H: $y = H - \frac{1}{2}gt^2 = 0$

$$t = \sqrt{\frac{2H}{g}} = \sqrt{\frac{2(0.500\ \text{m})}{9.81\ \text{m/s}^2}} = 0.319\ \text{s}$$

3. Multiply v times t to find the distance D: $x = vt = (1.72\ \text{m/s})(0.319\ \text{s}) = 0.548\ \text{m} = D$

Insight

Note that our solution for x can also be written as $x = vt = \sqrt{2gh}\,(\sqrt{2\,H/g}) = 2\sqrt{hH}$. Thus, if the values of h and H are interchanged, the distance D remains the same. Note also that x is independent of the acceleration of gravity.

Practice Problem

Find the distance D for $h = 0.500$ m and $H = 0.150$ m. [**Answer:** $D = 2\sqrt{hH} = 0.548$ m, as expected.]

Some related homework problems: Problem 52, Problem 83

*15–9 Viscosity and Surface Tension

To this point we have considered only "ideal" fluids. In particular, we have assumed that fluids flow without frictional losses and that the molecules in a fluid have no interaction with one another. In this section, we consider the consequences that follow from relaxing these assumptions.

Viscosity

When a block slides across a rough floor it experiences a frictional force opposing the motion. Similarly, a fluid flowing past a stationary surface experiences a force opposing the flow. This tendency to resist flow is referred to as the **viscosity** of a fluid. Fluids like air have low viscosities, thicker fluids like water are more viscous, and fluids like honey and molasses are characterized by their high viscosity.

To be specific, consider a situation of great practical importance—the flow of a fluid through a tube. Examples of this type of system include water flowing through a metal pipe in a house and blood flowing through an artery or a vein. If the fluid were ideal, with zero viscosity, it would flow through the tube with a speed that is the same throughout the fluid, as indicated in **Figure 15–23 (a)**. Real fluids with finite viscosity are found to have flow patterns like the one shown in **Figure 5–23 (b)**. In this case, the fluid is at rest next to the wall of the tube and flows with its greatest speed in the center of the tube. Because adjacent portions of the fluid flow past one another with different speeds, a force must be exerted on the fluid to maintain the flow, just as a force is required to keep a block sliding across a rough surface.

The force causing a viscous fluid to flow is provided by the pressure difference, $P_1 - P_2$, across a given length, L, of tube. Experiments show that the required pressure difference is proportional to the length of the tube and to the average speed, v, of the fluid. In addition, it is inversely proportional to the cross-sectional area, A, of the tube. Combining these observations, the pressure difference can be written in the following form:

$$P_1 - P_2 \propto \frac{vL}{A}$$

The constant of proportionality between the pressure difference and vL/A is related to the **coefficient of viscosity**, η, of a fluid. In fact, the viscosity is *defined* in such a way that the pressure difference is given by the following expression:

$$P_1 - P_2 = 8\pi\eta\frac{vL}{A} \qquad \textbf{15–18}$$

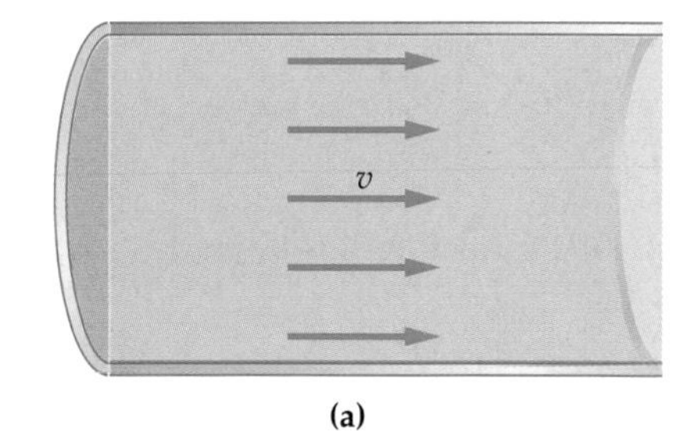

(a)

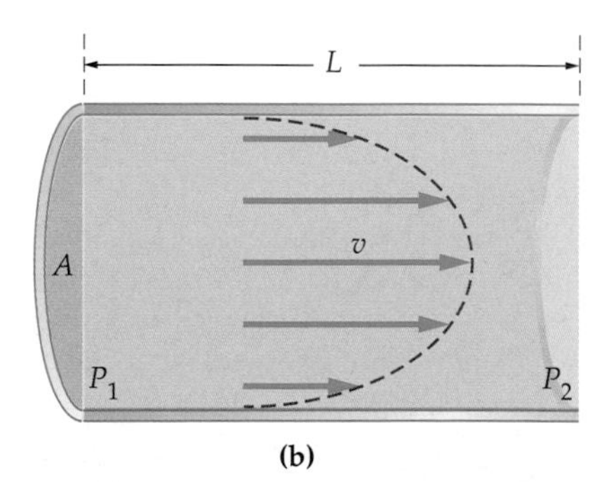

(b)

▲ **FIGURE 15–23 Fluid flow through a tube**

(a) An ideal fluid flows through a tube with a speed that is the same everywhere in the fluid. **(b)** In a fluid with finite viscosity, the speed of the fluid goes to zero on the walls of the tube and reaches its maximum value in the center of the tube. The average speed of the fluid depends on the pressure difference between the ends of the tube, $P_1 - P_2$, the length of the tube, L, the cross-sectional area of the tube, A, and the coefficient of viscosity of the liquid, η.

TABLE 15–2
Viscosities of Various Fluids (N·s/m²)

Glycerine (20 °C)	1.50
10-wt motor oil (30 °C)	0.250
Whole blood (37 °C)	0.0027
Water (0 °C)	0.00179
Water (100 °C)	0.00030

From this equation we can see that the dimensions of the coefficient of viscosity are $N \cdot s/m^2$. A common unit in the study of viscous fluids is the **poise**, named for the French physiologist Jean Louis Marie Poiseuille (1799–1869) and defined as:

$$1 \text{ poise} = 1 \text{ dyne} \cdot \text{s/cm}^2 = 0.1 \text{ N} \cdot \text{s/m}^2$$

For example, the viscosity of water at room temperature is $0.00101 \text{ N} \cdot \text{s/m}^2$ and the viscosity of blood at 37 °C is $0.0027 \text{ N} \cdot \text{s/m}^2$. A few additional viscosities are given in Table 15–2.

EXAMPLE 15–10 **Blood Speed in the Pulmonary Artery**

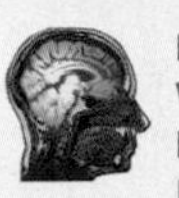
Real World Physics: Bio

The pulmonary artery, which connects the heart to the lungs, is 8.5 cm long and has a pressure difference over this length of 450 Pa. If the inside radius of the artery is 2.4 mm, what is the average speed of blood in the pulmonary artery?

Picture the Problem
Our sketch shows a schematic representation of the pulmonary artery, not drawn to scale. The relevant dimensions and the pressure difference are indicated.

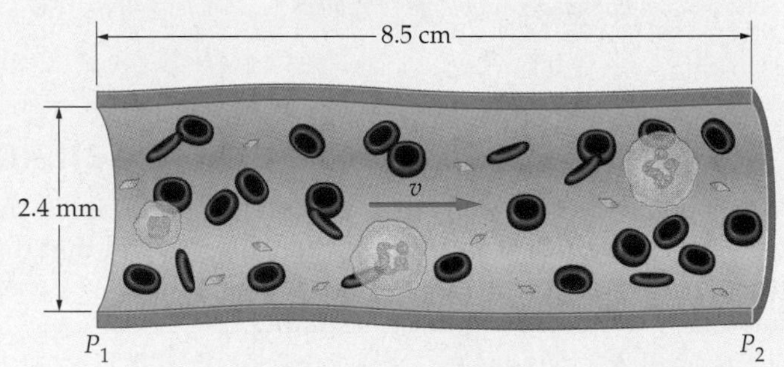

Strategy
The average speed can be found using Equation 15–18. Note that the pressure difference, $P_1 - P_2$, is given as 450 Pa = 450 N/m², and that the cross-sectional area of the blood vessel is $A = \pi r^2$.

Solution

1. Solve Equation 15–18 for the average speed, v: $$v = \frac{(P_1 - P_2)A}{8\pi\eta L}$$
2. Replace the cross-sectional area A with πr^2: $$v = \frac{(P_1 - P_2)r^2}{8\eta L}$$
3. Substitute numerical values: $$v = \frac{(450 \text{ Pa})(0.0024 \text{ m})^2}{8(0.0027 \text{ N} \cdot \text{s/m}^2)(0.085 \text{ m})} = 1.4 \text{ m/s}$$

Insight
The viscosity of blood increases rapidly with its hematocrit value; that is, with the concentration of red blood cells in the whole blood (see Chapter 10 for more information on the hematocrit value of blood). Thus, thick blood, with a high hematocrit value, requires a significantly larger pressure difference for a given rate of blood flow. This higher pressure must be provided by the heart, which consequently works harder with each beat.

Practice Problem
What pressure difference is required to give the blood in this pulmonary artery an average speed of 1.5 m/s? [**Answer:** 480 Pa]

Some related homework problems: Problem 74, Problem 75

A convenient way to characterize the flow of a fluid is in terms of its volume flow rate—the volume of fluid that passes a given point in a given amount of time. Referring to Section 15–6 we see that the volume flow rate of a fluid is simply vA, where v is the average speed of the fluid and A is the cross-sectional area of the tube through which it flows. Solving Equation 15–18 for the average speed gives $v = (P_1 - P_2)A/8\pi\eta L$. Multiplying this result by the cross-sectional area of the tube yields the volume flow rate:

$$\text{volume flow rate} = \frac{\Delta V}{\Delta t} = vA = \frac{(P_1 - P_2)A^2}{8\pi\eta L}$$

Using the fact that the cross-sectional area of the tube is $A = \pi r^2$, where r is its radius, we obtain the result known as **Poiseuille's equation**:

$$\frac{\Delta V}{\Delta t} = \frac{(P_1 - P_2)\pi r^4}{8\eta L} \qquad \textbf{15–19}$$

Note that the volume flow rate varies with the fourth power of the tube's radius, thus a small change in radius corresponds to a large change in volume flow rate.

To see the significance of the r^4 dependence, consider an artery that branches into an arteriole that has half the artery's radius. Letting r go to $r/2$ in Poiseuille's equation, and solving for the pressure difference, we find

$$P_1 - P_2 = \frac{8\eta L}{\pi (r/2)^4}\left(\frac{\Delta V}{\Delta t}\right) = 16\left[\frac{8\eta L}{\pi r^4}\left(\frac{\Delta V}{\Delta t}\right)\right]$$

Thus, the pressure difference across a given length of arteriole is 16 times what it is across the same length of artery. In fact, in the human body the pressure drop along an artery is small compared to the rather large pressure drop observed in the arterioles. This is a direct consequence of the increased viscous drag of the blood as it flows through narrower blood vessels. Similarly, a narrowing, or *stenosis*, of an artery can produce significant increases in blood pressure. For example, a reduction in radius of only 20%, from r to $0.8r$, causes an increase in pressure by a factor of $(1/0.8)^4 \sim 2.4$.

▲ Surface tension causes the surface of a liquid to behave like an elastic skin or membrane. When a small force is applied to the liquid surface it tends to stretch, resisting penetration. This phenomenon enables a fishing spider to launch itself vertically upward as much as 4 cm from the water surface, as if from a trampoline. Such an acrobatic maneuver can save the spider from the attack of a predatory fish.

Surface Tension

A small insect resting on the surface of a pond or a lake is a common sight in the summertime. If you look carefully, you can see that the insect creates tiny dimples in the water's surface, almost as if it were supported by a thin sheet of rubber. In fact, the surface of water and other fluids behaves in many respects as if it were an elastic membrane. This effect is known as **surface tension**.

To understand the origin of surface tension, we start by noting that the molecules in a fluid exert attractive forces on one another. Thus, a molecule deep within a fluid experiences forces in all directions, as indicated in **Figure 15–24**, due to the molecules that surround it on all sides. The net force on such a molecule is zero. As a molecule nears the surface, however, it experiences a net force away from the surface, since there are no fluid molecules on the other side of the surface to attract it in that direction. It follows that work must be done on a molecule to move it from within a fluid to the surface, and that the *energy* of a fluid is increased for every molecule on its surface.

In general, physical systems tend toward configurations of minimum energy. For example, a ball on a slope rolls downhill, lowering its gravitational potential energy. If the energy of a droplet of liquid is to be minimized it must have the smallest surface area possible for a given volume; that is, it must have the shape of a sphere. This is the reason small drops of dew are always spherical. Larger drops of water may be distorted by the downward pull of Earth's gravity, but in orbit drops of all sizes are spherical.

Since energy is required to increase the area of a liquid surface, the situation is similar to the energy required to stretch a spring or to stretch a sheet of rubber. Thus, the surface of a liquid behaves as if it were elastic, resisting tendencies to increase its area. For example, if a drop of dew is distorted into an ellipsoid it quickly returns to its original spherical shape. Similarly, when an insect alights on the surface of a pond it creates dimples that increase the surface area. The water resists this distortion with a force sufficient to support the weight of the insect—if the insect is not too large. In fact, even a needle or a razor blade can be supported on the surface of water if they are put into place gently, even though they have densities significantly greater than the density of water.

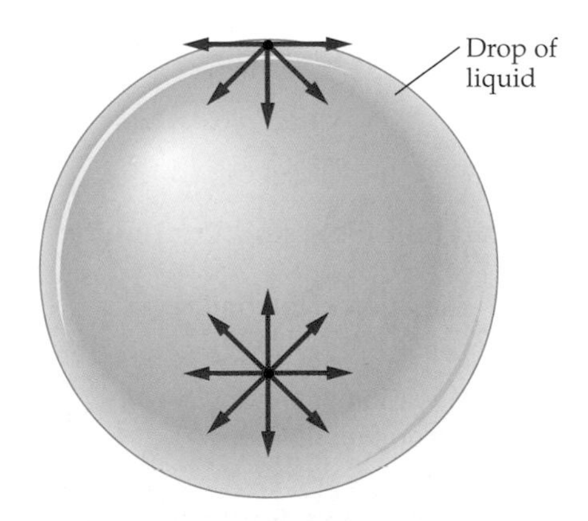

▲ **FIGURE 15–24 The origin of surface tension**

A molecule in the interior of a fluid experiences attractive forces of equal magnitude in all directions, giving a net force of zero. A molecule near the surface of the fluid experiences a net attractive force toward the interior of the fluid. This causes the surface to be pulled inward, resulting in a surface of minimum area.

Chapter Summary

Topic	Remarks and Relevant Equations	
15-1 Density	The density, ρ, of a material is its mass M per volume V: $\rho = M/V$	15-1
15-2 Pressure	Pressure, P, is force F per area A: $P = F/A$	15-2
atmospheric pressure	The pressure exerted by the atmosphere is $P_{at} = 1.01 \times 10^5\ \text{N/m}^2 \approx 14.7\ \text{lb/in}^2$.	
pascals	Pressure is often given in terms of the pascal (Pa), where $1\ \text{Pa} = 1\ \text{N/m}^2$.	
gauge pressure	Gauge pressure—as measured, for example, by a tire gauge—is the difference between the actual pressure and atmospheric pressure: $P_g = P - P_{at}$	15-5
15-3 Static Equilibrium in Fluids: Pressure and Depth	The pressure of a fluid in static equilibrium increases with depth.	
pressure with depth	If the pressure at one point in a fluid is P_1, the pressure at a depth h below that point is $P_2 = P_1 + \rho g h$	15-7
Pascal's principle	An external pressure applied to an enclosed fluid is transmitted unchanged to every point within the fluid.	
15-4 Archimedes' Principle and Buoyancy	The fact that the pressure in a fluid increases with depth leads to a net upward force on any object that is immersed in the fluid. This upward force is referred to as a buoyant force. The magnitude of the buoyant force is given by Archimedes' principle.	
Archimedes' principle	An object completely immersed in a fluid experiences an upward buoyant force equal in magnitude to the weight of fluid displaced by the object.	
15-5 Applications of Archimedes' Principle	Archimedes' principle applies equally well to objects that are completely immersed, partially immersed, or floating.	
floatation	An object floats when it displaces an amount of fluid equal to its weight.	
submerged volume	When a solid of volume V_s and density ρ_s floats in a fluid of density ρ_f, the volume of the solid that is submerged in the fluid, V_{sub}, is $V_{sub} = V_s(\rho_s/\rho_f)$	15-10
15-6 Fluid Flow and Continuity	The speed of a fluid changes as the cross-sectional area of the pipe through which it flows changes.	
equation of continuity, compressible flow	$\rho_1 A_1 v_1 = \rho_2 A_2 v_2$	15-11
equation of continuity, incompressible flow	$A_1 v_1 = A_2 v_2$	15-12
15-7 Bernoulli's Equation	Bernoulli's equation can be thought of as energy conservation per volume for a fluid.	
change in speed	If the speed of a fluid changes from v_1 to v_2, the corresponding pressures are related by: $P_1 + \frac{1}{2}\rho v_1^2 = P_2 + \frac{1}{2}\rho v_2^2$	15-14
change in height	If the height of a fluid changes from y_1 to y_2, the corresponding pressures are related by: $P_1 + \rho g y_1 = P_2 + \rho g y_2$	15-15
general case	If both the height and the speed of a fluid change, the pressures are related by: $P_1 + \frac{1}{2}\rho v_1^2 + \rho g y_1 = P_2 + \frac{1}{2}\rho v_2^2 + \rho g y_2$	15-16

15–8 Applications of Bernoulli's Equation

Bernoulli's equation applies to a wide range of everyday situations, including airplane wings and prairie dog burrows.

Torricelli's law

According to Torricelli's law, if a hole is poked in a container at a depth h below the surface, the fluid exits with the speed

$$v = \sqrt{2gh} \qquad \textbf{15–17}$$

Note that this is the same speed as an object in free fall for a distance h.

***15–9 Viscosity and Surface Tension**

Viscosity and surface tension are two features of real fluids that are not found in an "ideal" fluid.

viscosity

Viscosity in a fluid is similar to friction between two solid surfaces. A pressure difference, $P_1 - P_2$, is required to keep a viscous fluid flowing with a constant average speed, v. The relation between the volume flow rate of a fluid, $\Delta V/\Delta t$, and its coefficient of viscosity, η, is given by Poiseuille's equation:

$$\frac{\Delta V}{\Delta t} = \frac{(P_1 - P_2)\pi r^4}{8\eta L} \qquad \textbf{15–19}$$

This expression applies to a tube of radius r and length L.

surface tension

A fluid tends to pull inward on its surface, resulting in a surface of minimum area. The surface of the fluid behaves much like an elastic membrane enclosing the fluid.

Problem-Solving Summary

Type of Problem	Relevant Physical Concepts	Related Examples
Relate pressure to depth in a static fluid.	The pressure in a static fluid increases uniformly with depth by the amount ρgh. Thus, if point 2 is a depth h below point 1, the pressure there is $P_2 = P_1 + \rho gh$.	Examples 15–3, 15–4
Calculate the buoyant force.	According to Archimedes' principle, the buoyant force is equal to the weight of displaced fluid.	Example 15–5 Active Examples 15–1, 15–2
Calculate the speed of a fluid as the area through which it flows changes.	If a fluid is incompressible, its speed must vary inversely with the area through which it flows. Thus, if the area changes from A_1 to A_2 the speed of the fluid changes as follows: $A_1v_1 = A_2v_2$. This is the equation of continuity.	Example 15–7
Relate the pressure in a fluid to its height and speed.	Bernoulli's principle states that pressure, speed, and height in a fluid are related as follows: $P + \frac{1}{2}\rho v^2 + \rho g y = \text{constant}$. This is simply a statement of energy conservation per volume of the fluid.	Example 15–8 Active Example 15–3

Conceptual Questions

1. Suppose you drink a liquid through a straw. Explain why the liquid moves upward, against gravity, into your mouth.
2. Considering your answer to the previous question, is it possible to sip liquid through a straw on the Moon? Explain.
3. Two drinking glasses, with different cross sectional areas, are filled with water to the same depth. We know that the pressure is the same at the bottom of both glasses. How can this be, though, since the larger glass contains a greater mass of water?
4. What holds a suction cup in place?
5. Water towers on the roofs of buildings have metal bands wrapped around them for support. The spacing between bands is smaller near the base of a tower than near its top. Explain.
6. A helium-filled balloon for a birthday party is being brought home in a car. As the car accelerates away from a stop light, in which direction does the balloon move relative to the car?
7. Suppose a force of 400 N is required to blow the top off a wine barrel. In a famous experiment, Blaise Pascal attached a tall, thin tube to the top of a filled wine barrel, as shown in **Figure 15–25** (p. 492). Water was slowly added to the tube until the barrel burst. The puzzling result found by Pascal was that the barrel broke when the weight of water in the tube was much less than 400 N. Explain Pascal's observation.
8. Why is it more practical to use mercury in the barometer shown in Figure 15–3 than water?

▲ **FIGURE 15–25** Question 7

9. A weather glass, as shown in **Figure 15–26**, is used to give an indication of a change in the weather. Does the water level in the neck of the weather glass move up or down when a low pressure system approaches?

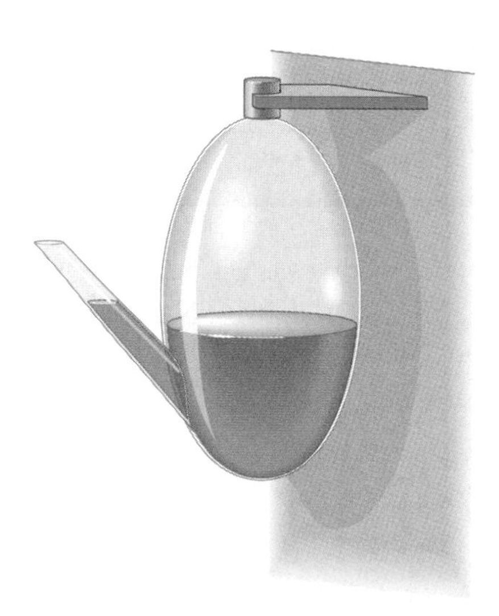

▲ **FIGURE 15–26** Question 9

10. How does a balloonist control the vertical motion of a hot air balloon?
11. A block of wood has a steel ball glued to one surface. The block can be floated with the ball "high and dry" on its top surface. When the block is inverted, and the ball immersed in water, does the volume of wood that is submerged increase, decrease, or stay the same?
12. In the preceding problem, suppose the block of wood with the ball "high and dry" is floating in a tank of water. When the block is inverted, does the water level in the tank increase, decrease, or stay the same?
13. A hydrometer, a device for measuring fluid density, is constructed as shown in **Figure 15–27**. If the hydrometer samples fluid A the small float inside the tube is submerged to the level A. When fluid B is sampled, the float is submerged to level B. Which of these fluids is more dense?
14. Referring to Active Example 15–1, suppose the flask with the wood tied to the bottom is placed on a scale. At some point the string breaks, and the wood rises to the surface where it floats. When the wood is floating, is the reading on the scale the same, more than, or less than it was before? Explain.
15. An object's density can be determined by first weighing it in air, then in water (provided the density of the object is greater than

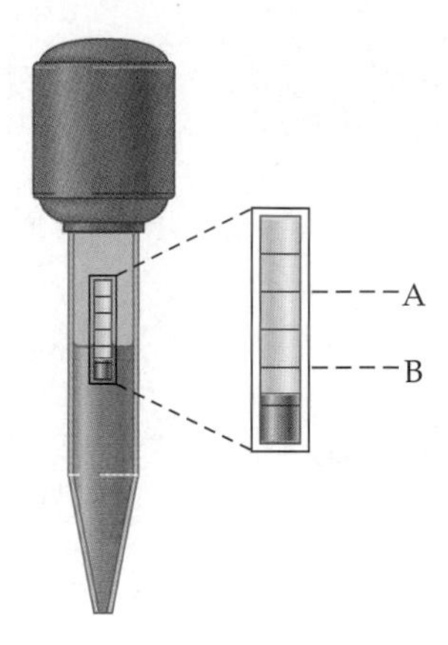

▲ **FIGURE 15–27** Question 13

the density of water, so that it is totally submerged when placed in water). Explain how these two measurements can give the desired result.

16. Does an ice cube float higher in salt water or fresh water? Explain.
17. A person floats in a boat in a small swimming pool. Inside the boat with the person are some bricks. If the person throws the bricks overboard into the pool, does the water level in the pool increase, decrease, or stay the same?
18. In the previous problem, how would the water level change if, instead of bricks, the person throws some blocks of wood overboard?
19. Why is it possible for people to float without effort in Utah's Great Salt Lake?
20. One day while snorkeling near the surface of a crystal-clear ocean, it occurs to you that you could go considerably deeper by simply lengthening the snorkel tube. Unfortunately, this doesn't work well at all. Why?
21. Lead is more dense than aluminum. Is the buoyant force on a solid lead sphere more than, less than, or the same as the buoyant force on a solid aluminum sphere of the same diameter? Does your answer depend on the fluid that is causing the buoyant force?
22. A fish carrying a pebble in its mouth swims with a small, constant velocity in a small bowl. When the fish drops the pebble to the bottom of the bowl, does the water level rise, fall, or stay the same? Explain.
23. On a planet in a different solar system the acceleration of gravity is greater than it is on Earth. If you float in the water on this planet, is it easier to float, harder, or the same as on Earth?
24. Since metal is more dense than water, how is it possible for a metal boat to float?
25. A sheet of water passing over a waterfall is thicker near the top than near the bottom. Similarly, a stream of water emerging from a water faucet becomes narrower as it falls. Explain.
26. It is a common observation that smoke rises more rapidly through a chimney when there is a wind blowing outside. Explain.
27. Is it best for an airplane to take off against the wind or with the wind? Explain.
28. If you have a hair dryer and a ping pong ball at home, try this demonstration. Direct the air from the dryer in a direction just above horizontal. Next, place the ping pong ball in the stream of air. If done just right, the ball will remain suspended in midair. Use the Bernoulli effect to explain this behavior.
29. Suppose a pitcher wants to throw a baseball so that it rises as it approaches the batter. How should the ball be spinning to accomplish this feat?

30. If you change altitude rapidly your eardrums may "pop." In which direction do your eardrums move if you have increased your altitude? In which direction do they move if you decrease your altitude?

31. When a person's blood pressure is taken, it is measured on the arm, at approximately the same level as the heart. How would the results differ if the measurement were to be made on the patient's leg?

32. A fluid accelerates as it flows from a large diameter hose to a small diameter hose. Explain the cause of the fluid's acceleration.

33. You jump into the ocean and swim downward as far as you can go. Assuming that your body and seawater are both incompressible, does the buoyant force exerted on you increase, decrease, or stay the same as you swim deeper?

34. Hold the top of a sheet of paper in each hand. Let the sheets hang freely downward, separated by an inch or two. Now blow between the sheets of paper. Do they move toward one another or do they move apart? Explain.

Problems

Note: **IP** *denotes an integrated conceptual/quantitative problem.* **BIO** *identifies problems of biological or medical interest. Blue bullets (•, ••, •••) are used to indicate the level of difficulty of each problem.*

Section 15–1 Density

1. • Estimate the weight of the air in your physics classroom.

2. • What weight of water is required to fill a 25-gallon aquarium?

3. • You buy a "gold" ring at a pawn shop. The ring has a mass of 0.016 g and a volume of 0.0020 cm^3. Is the ring solid gold?

4. • Estimate the weight of a treasure chest filled with gold doubloons.

5. • A cube of metal has a mass of 0.347 kg and measures 3.21 cm on a side. Calculate the density and identify the metal.

Section 15–2 Pressure

6. • What is the downward force exerted by the atmosphere on a football field?

7. • **BIO** Some species of dinoflagellate (a type of unicellular plankton) can produce light as the result of biochemical reactions within the cell. This light is an example of bioluminescence. It is found that bioluminescence in dinoflagellates can be triggered by deformation of the cell surface with a pressure as low as one dyne (10^{-5} N) per square centimeter. What is this pressure in **(a)** pascals and **(b)** atmospheres?

8. • A 75-kg person sits on a 3.5-kg chair. Each leg of the chair makes contact with the floor in a circle that is 1.5 cm in diameter. Find the pressure exerted on the floor by each leg of the chair, assuming the weight is evenly distributed.

9. • To prevent damage to floors (and to increase friction), a crutch will often have a rubber tip attached to its end. If the end of the crutch is a circle of radius 1.2 cm without the tip, and the tip is a circle of radius 2.5 cm, by what factor does the tip reduce the pressure exerted by the crutch?

10. • An inflated basketball has a gauge pressure of 8.0 lb/in^2. What is the actual pressure inside the ball?

11. •• Suppose that when you ride on your 7.70-kg bike the weight of you and the bike is supported equally by the two tires. If the gauge pressure in the tires is 70.5 lb/in^2 and the area of contact between each tire and the road is 7.13 cm^2, what is your weight?

12. •• **IP** The weight of your 1320-kg car is supported equally by its four tires, each inflated to a gauge pressure of 35.0 lb/in^2. **(a)** What is the area of contact each tire makes with the road? **(b)** If the gauge pressure is increased, does the area of contact increase, decrease, or stay the same? **(c)** What gauge pressure is required to give an area of contact of 116 cm^2 for each tire?

Section 15–3 Static Equilibrium in Fluids: Pressure and Depth

13. • Hoover Dam is 221 m deep. What is the water pressure at the base of the dam?

14. • As a storm front moves in you notice that the column of mercury in a barometer rises to only 740 mm. What is the air pressure?

15. • In the previous problem, to what height would water rise in a water barometer?

16. •• In the hydraulic system shown in Figure 15–28 the piston on the left has a diameter of 4.5 cm and a mass of 1.7 kg. The piston on the right has a diameter of 12 cm and a mass of 3.2 kg. If the density of the fluid is 750 kg/m^3, what is the height difference h between the two pistons?

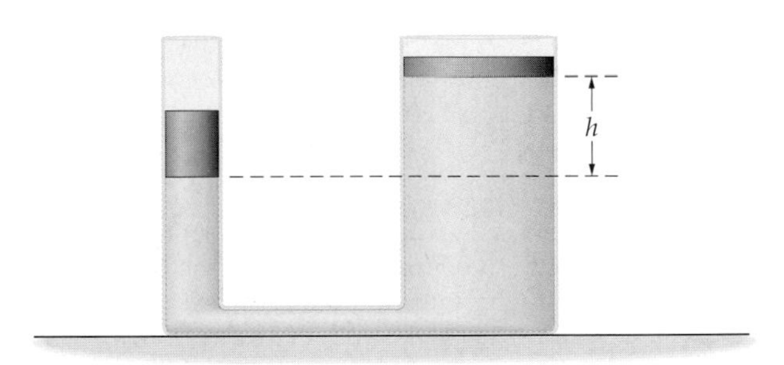

▲ **FIGURE 15–28** Problem 16

17. •• A circular wine barrel 75 cm in diameter will burst if the net upward force exerted on the top of the barrel is 6430 N. A tube 1.0 cm in diameter extends into the barrel through a hole in the top, as indicated in Figure 15–25. Initially, the barrel is filled to the top and the tube is empty above that level. What weight of water must be poured into the tube in order to burst the barrel?

18. •• A cylindrical container with a cross-sectional area of 65.2 cm^2 holds a fluid of density 806 kg/m^3. At the bottom of the container the pressure is 110 kPa. **(a)** What is the depth of the fluid? **(b)** Find the pressure at the bottom of the container after an additional 2.05×10^{-3} m^3 of this fluid is added to the container. Assume that no fluid spills out of the container.

19. •• In a classroom demonstration, the pressure inside a soft drink can is suddenly reduced to essentially zero. Assuming the can to be a cylinder of height 12 cm and diameter 6.5 cm, find the force exerted on the sides of the can due to atmospheric pressure.

▲ Roof-top water towers. (Problem 20)

20. •• **IP** A water tower, like those shown in the accompanying photo, is filled with fresh water to a depth of 5.5 m. What is the water pressure at a depth of **(a)** 4.0 m and **(b)** 5.0 m? **(c)** Why are the metal bands on such towers more closely spaced near the base of the tower?

21. •• Holding a glass of water filled to a depth of 6.5 cm you step into an elevator. The elevator moves upward with constant acceleration, increasing its speed from zero to 1.2 m/s in 2.7 s. Find the pressure exerted on the bottom of the glass during the time the elevator accelerates.

22. •• Suppose you pour water into a container until it reaches a depth of 12 cm. Next, you carefully pour in a 6.2-cm thickness of olive oil so that it floats on top of the water. What is the pressure at the bottom of the container?

23. •• Referring to Example 15–4, suppose that some vegetable oil has been added to both sides of the U-tube. On the right side of the tube the depth of oil is 5.00 cm, as before. On the left side of the tube the depth of the oil is 3.00 cm. Find the difference in fluid level between the two sides of the tube.

24. •• **IP** As a stunt, you want to sip some soda through a very long, vertical straw. **(a)** First, explain why the liquid moves upward, against gravity, into your mouth when you sip. **(b)** What is the tallest straw that you could, in principle, drink from?

25. •• **IP** BIO The patient in **Figure 15–29** is to receive an intravenous injection of medication. In order to work properly, the pressure of fluid containing the medication must be 109 kPa at the injection point. **(a)** If the fluid has a density of 1020 kg/m^3, find the height at which the bag of fluid must be suspended above the patient. Assume that the pressure inside the bag is one atmosphere. **(b)** If a less dense fluid is used instead, must the height of suspension be increased or decreased? Explain.

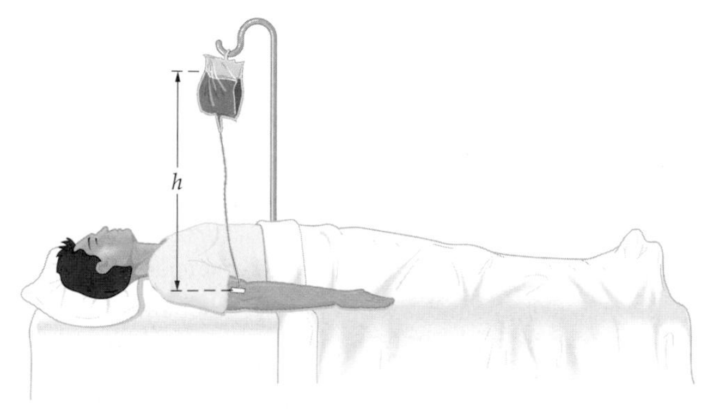

▲ **FIGURE 15–29** Problem 25

26. ••• A cylindrical container 1.0 m tall contains mercury to a certain depth, d. The rest of the cylinder is filled with water. If the pressure at the bottom of the cylinder is two atmospheres, what is the depth d?

Section 15–4 Archimedes' Principle and Buoyancy

27. • A raft is 4.2 m wide and 6.5 m long. When a horse is loaded onto the raft, it sinks 2.7 cm deeper into the water. What is the weight of the horse?

28. • To walk on water, all you need is a pair of water-walking boots shaped like boats. If each boot is 25-cm high and 35-cm wide, how long must they be to support a 75-kg person?

29. •• A 0.12-kg balloon is filled with helium (density = 0.179 kg/m^3). If the balloon is a sphere with a radius of 5.2 m, what is the maximum weight it can lift?

30. •• A hot air balloon plus cargo has a mass of 327 kg and a volume of 687 m^3. The balloon is floating at a constant height of 6.25 m above the ground. What is the density of the hot air in the balloon?

31. ••• In the lab you place a beaker that is half full of water (density ρ_w) on a scale. You now use a light string to suspend a piece of metal of volume V in the water. The metal is completely submerged, and none of the water spills out of the beaker. Give a symbolic expression for the change in reading of the scale.

Section 15–5 Applications of Archimedes' Principle

32. • An air mattress is 2.2 m long, 0.65 m wide, and 13 cm deep. If the air mattress itself has a mass of 0.22 kg, what is the maximum mass it can support in fresh water?

33. •• A solid block is attached to a spring scale. When the block is suspended in air the scale reads 20.0 N; when it is completely immersed in water the scale reads 17.7 N. What is the **(a)** volume and **(b)** density of the block?

34. •• As in the previous problem, a solid block is suspended from a spring scale. If the reading on the scale when the block is completely immersed in water is 25.0 N, and the reading when it is completely immersed in alcohol of density 806 kg/m^3 is 25.7 N, what is **(a)** the block's volume and **(b)** its density?

35. •• BIO A person weighs 768 N in air and has a body fat percentage of 18.4%. **(a)** What is the overall density of this person's body? **(b)** What is the volume of this person's body? **(c)** Find the apparent weight of this person when completely submerged in water.

36. •• **IP** A log floats in a river with one fourth of its volume above the water. **(a)** What is the density of the log? **(b)** If the river carries the log into the ocean, does the portion of the log above the water increase, decrease, or stay the same? Explain.

37. •• A person with a mass of 81 kg and a volume of 0.089 m^3 floats quietly in water. **(a)** What is the volume of the person that is above water? **(b)** If an upward force F is applied to the person, the volume above water increases by 0.0018 m^3. Find the force F.

38. •• **IP** A block of wood floats on water. A layer of oil is now poured on top of the water to a depth that more than covers the block, as shown in **Figure 15–30**. **(a)** Is the volume of wood submerged in water greater than, less than, or the same as before? **(b)** If 90% of the wood is submerged in water before the oil is added, find the fraction submerged when oil with a density of 875 kg/m^3 covers the block.

▲ **FIGURE 15–30** Problem 38

39. •• A piece of lead has the shape of a hockey puck, with a diameter of 7.5 cm and a height of 2.5 cm. If the puck is placed in a mercury bath it floats. How deep below the surface of the mercury is the bottom of the lead puck?

40. ••• **IP** A lead weight with a volume of 0.82×10^{-5} m^3 is lowered on a fishing line into a lake to a depth of 1.0 m. **(a)** What tension is required in the fishing line to give the weight an upward acceleration of 2.1 m/s^2? **(b)** If the depth of the weight is increased to 2.0 m, does the tension found in part (a) increase, decrease, or stay the same? Explain. **(c)** What acceleration will the weight have if the tension in the fishing line is 1.2 N? Give both direction and magnitude.

Section 15–6 Fluid Flow and Continuity

41. • To water the yard you use a hose with a diameter of 3.2 cm. Water flows from the hose with a speed of 1.3 m/s. If you partially block the end of the hose so the effective diameter is now 0.55 cm, with what speed does water spray from the hose?

42. • Water flows through a pipe with a speed of 1.6 m/s. Find the flow rate in kg/s if the diameter of the pipe is 4.6 cm.

43. • To fill a child's inflatable wading pool you use a garden hose with a diameter of 2.8 cm. Water flows from this hose with a speed of 1.1 m/s. How long will it take to fill the pool to a depth of 26 cm if it is circular and has a diameter of 2.0 m?

44. •• **IP** Water flows at the rate of 3.11 kg/s through a hose with a diameter of 3.10 cm. **(a)** What is the speed of water in this hose? **(b)** If the hose is attached to a nozzle with a diameter of 0.750 cm, what is the speed of water in the nozzle? **(c)** Is the number of kilograms per second flowing through the nozzle greater than, less than, or equal to 3.11 kg/s? Explain.

45. •• A river narrows at a rapids from a width of 12 m to a width of 4.0 m. The depth of the river before the rapids is 2.7 m; the depth in the rapids is 0.85 m. Find the speed of water flowing in the rapids, given that its speed before the rapids is 2.2 m/s. Assume the river has a rectangular cross section.

46. •• **BIO** The aorta has an inside diameter of approximately 0.50 cm, compared to that of a capillary, which is about 1.0×10^{-5} m (10 μm). In addition, the average speed of flow is approximately 1.0 m/s in the aorta and 1.0 cm/s in a capillary. Assuming that all the blood that flows through the aorta also flows through the capillaries, how many capillaries does the circulatory system have?

Section 15–7 Bernoulli's Equation

47. • **BIO** The buildup of plaque on the walls of an artery may decrease its diameter from 1.1 cm to 0.75 cm. If the speed of blood flow was 15 cm/s before reaching the region of plaque buildup, find **(a)** the speed of blood flow and **(b)** the pressure drop within the plaque region.

48. • A horizontal pipe contains water at a pressure of 110 kPa flowing with a speed of 1.4 m/s. When the pipe narrows to one-half its original diameter, what is **(a)** the speed and **(b)** the pressure of the water?

49. •• **BIO** Tests of lung capacity show that adults are able to exhale 1.5 liters of air through their mouths in as little as 1.0 seconds. **(a)** If a person blows air at this rate through a drinking straw with a diameter of 0.60 cm, what is the speed of air in the straw? **(b)** If the air from the straw in part (a) is directed horizontally across the upper end of a second straw that is vertical, as shown in **Figure 15–31**, to what height does water rise in the vertical straw?

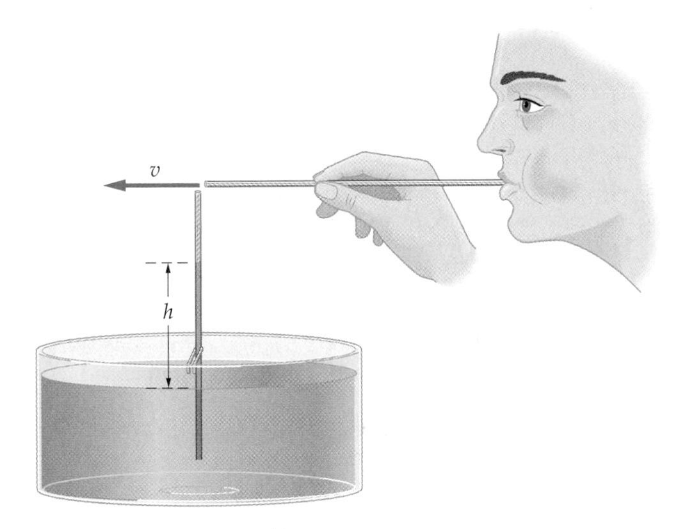

▲ **FIGURE 15–31** Problem 49

50. •• **IP** Water flows through a horizontal tube of diameter 2.8 cm that is joined to a second horizontal tube of diameter 1.6 cm. The pressure difference between the tubes is 7.5 kPa. **(a)** Which tube has the higher pressure? **(b)** Which tube has the higher speed of flow? **(c)** Find the speed of flow in the first tube.

51. •• A garden hose is attached to a water faucet on one end and a spray nozzle on the other end. The water faucet is turned on, but the nozzle is turned off so that no water flows through the hose. The hose lies horizontally on the ground, and a stream of water sprays vertically out of a small leak to a height of 0.75 m. What is the pressure inside the hose?

Section 15–8 Applications of Bernoulli's Equation

52. • A water tank springs a leak. Find the speed of water emerging from the hole if the leak is 3.2 m below the surface of the water, which is open to the atmosphere.

53. •• **(a)** Find the pressure difference on an airplane wing where air flows over the upper surface with a speed of 115 m/s, and along the bottom surface with a speed of 105 m/s. **(b)** If the area of the wing is 32 m^2, what is the net upward force exerted on the wing?

54. •• On a vacation flight, you look out the window of the jet and wonder about the forces exerted on the window. Suppose the air outside the window moves with a speed of approximately 150 m/s shortly after takeoff, and that the air inside the plane is at atmospheric pressure. **(a)** Find the pressure difference between the inside and outside of the window. **(b)** If the window is 25 cm by 42 cm, find the force exerted on the window by air pressure.

55. •• During a thunderstorm the wind speed reaches 45.2 m/s. Find the direction and magnitude of the force exerted on the roof of a flat-topped building as a result of this wind. The roof has an area of 578 m^2.

56. •• A garden hose with a diameter of 0.65 in has water flowing in it with a speed of 0.55 m/s and a pressure of 1.2 atmospheres. At the end of the hose is a nozzle with a diameter of 0.25 in. Find **(a)** the speed of water in the nozzle and **(b)** the pressure in the nozzle.

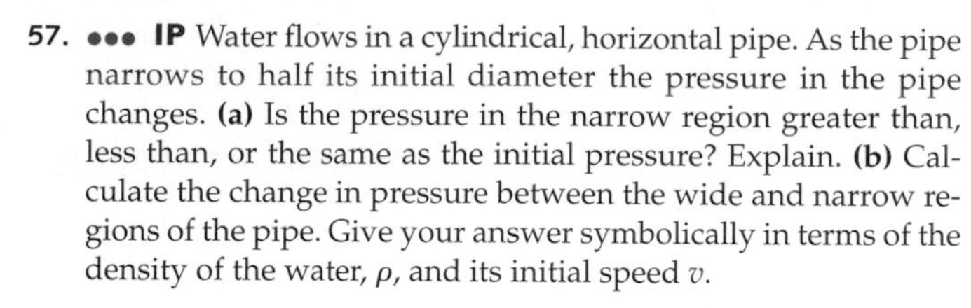

57. ••• **IP** Water flows in a cylindrical, horizontal pipe. As the pipe narrows to half its initial diameter the pressure in the pipe changes. **(a)** Is the pressure in the narrow region greater than, less than, or the same as the initial pressure? Explain. **(b)** Calculate the change in pressure between the wide and narrow regions of the pipe. Give your answer symbolically in terms of the density of the water, ρ, and its initial speed v.

General Problems

58. • At what depth in seawater is the pressure equal to two atmospheres?

59. •• An above-ground backyard swimming pool is shaped like a large hockey puck, with a circular bottom and a vertical wall forming its perimeter. The diameter of the pool is 5.2 m and its depth is 1.4 m. Find the total outward force exerted on the vertical wall of the pool.

60. •• A solid block is suspended from a spring scale. When the block is in air the scale reads 35.0 N, when immersed in water the scale reads 31.1 N, and when immersed in oil the scale reads 31.8 N. **(a)** What is the density of the block? **(b)** What is the density of the oil?

61. •• A wooden block with a density of 710 kg/m^3 and a volume of 0.012 m^3 is attached to the top of a vertical spring whose force constant is $k = 540$ N/m. Find the amount by which the spring is stretched or compressed if it and the wooden block are **(a)** in air or **(b)** immersed in water. [The density of air may be neglected in part (a).]

62. •• A 1.25-kg wooden block has an iron ball of radius 1.22 cm glued to one side. **(a)** If the block floats in water with the iron ball high and dry, what is the volume of wood that is submerged? **(b)** If the block is now inverted, so that the iron ball is completely immersed in the water, what volume of the wood block is submerged?

63. •• On a bet, you try to remove water from a glass by blowing across the top of a vertical straw immersed in the water. What is the minimum speed you must give the air at the top of the straw to draw water upward through a height of 1.5 cm?

64. •• Evangelista Torricelli (1608–1647) was the first to put forward the idea that we live at the bottom of an ocean of air. Given the value of atmospheric pressure at the surface of the Earth, and the fact that there is zero pressure in the vacuum of space, determine the depth of the atmosphere assuming that the density of air is a constant. Compare your result to the height of Mt. Everest. (In fact, the density of air decreases with altitude, so the result obtained here should be less than the actual depth of the atmosphere.)

65. •• Consider the glass containers shown in **Figure 15–32**. Both containers have bases of area $A_1 = 24$ cm^2 and depths of water equal to 18 cm. As a result, the downward force on the base of the containers is equal, even though the containers clearly have different weights. This is referred to as the hydrostatic paradox. **(a)** Given that the container in Figure 15–32 (b) has an annular (ring-shaped) region of area $A_2 = 12$ cm^2, determine the net downward force acting on the container. **(b)** Show that your result from part (a) is equal to the weight of water in the container.

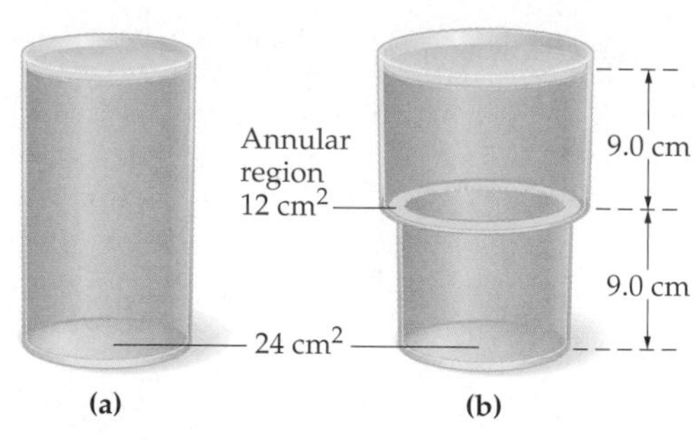

▲ **FIGURE 15–32** Problem 65

66. •• Consider the two containers shown in **Figure 15–33**. As in the previous problem, these containers have equal forces on their bases but contain different weights of water. **(a)** Determine the net downward force acting on the container in Figure 15–33 (b). Note that the base of the containers is of area $A_1 = 24$ cm^2, the annular region is of area $A_2 = 12$ cm^2, and the depth of the water is 18 cm. **(b)** Show that your result from part (a) is equal to the weight of water in the container. **(c)** If a hole is poked in the annular region of the container in Figure 15–33 (b), how fast will water exit the hole? How high will it rise?

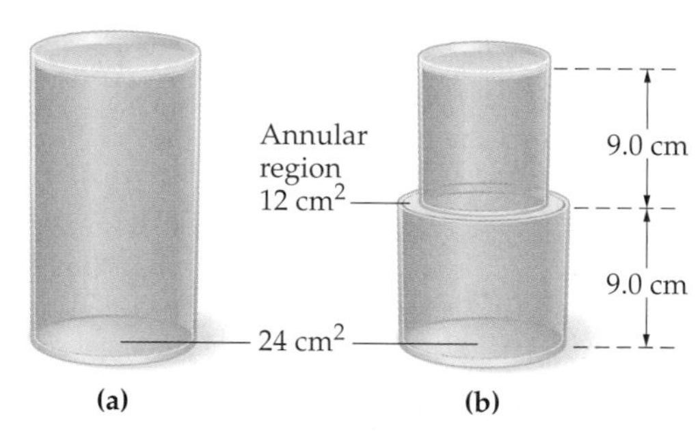

▲ **FIGURE 15–33** Problem 66

67. •• **IP** A backyard swimming pool is circular in shape and contains water to a uniform depth of 31 cm. It is 2.2 m in diameter and is not completely filled. **(a)** What is the pressure at the bottom of the pool? **(b)** If a person gets in the pool and floats peacefully, does the pressure at the bottom of the pool increase, decrease, or stay the same? **(c)** Calculate the pressure at the bottom of the pool if the floating person has a mass of 78 kg.

68. •• A prospector finds a solid rock composed of granite ($\rho = 2650$ kg/m^3) and gold. If the volume of the rock is 3.55×10^{-4} m^3 and its mass is 3.81 kg, how much gold is in the rock?

69. •• Consider the crustal rocks of the Earth to be a fluid of density 3.0×10^3 kg/m^3. Under this assumption, the pressure at a depth h within the crust is $P = P_{at} + \rho g h$. If the greatest pressure crustal rock can sustain before crumbling is 1.2×10^9 Pa, find the maximum depth of the Earth's crust. (Below this depth the crust changes from a solid to a plastic-like material.)

70. •• **IP** **(a)** If the tension in the string in Active Example 15–1 is 0.95 N, what is the volume of the wood? Assume that everything else remains the same. **(b)** If the string breaks and the wood floats on the surface, does the water level in the flask rise, drop, or stay the same? Explain. **(c)** Assuming the flask is cylindrical with a cross sectional area of 65 cm^2, find the change in water level after the string breaks.

71. •• **IP** A siphon is a device that allows water to flow from one level to another. The siphon shown in **Figure 15–34** delivers water from an irrigation canal to a field of crops. To operate the siphon, water is first drawn through the length of the tube. After the flow is started in this way it continues on its own. **(a)** Using points 1 and 3 in Figure 15-36, find the speed v of the water leaving the siphon at its lower end. Give a symbolic answer. **(b)** Is the speed of the water at point 2 greater than, less than, or the same as its speed at point 3? Explain.

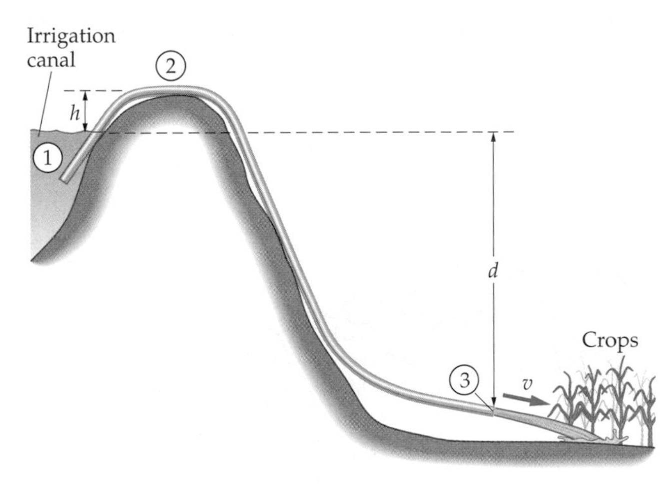

▲ **FIGURE 15–34** Problem 71

72. •• A tin can is filled with water to a depth of 22 cm. A hole in the bottom of the can produces a stream of water that is directed at an angle of 35° above the horizontal. Find **(a)** the range and **(b)** the maximum height of this stream of water.

73. •• **BIO** A person weighs 682 N in air but only 498 N when standing in water up to the hips. Find **(a)** the volume of each of the person's legs and **(b)** the mass of each leg, assuming they have a density that is 1.05 times the density of water.

* 74. •• A horizontal pipe carries oil whose coefficient of viscosity is $0.00012\ \mathrm{N \cdot s/m^2}$. The diameter of the pipe is 5.2 cm and its length is 55 m. **(a)** What pressure difference is required between the ends of this pipe if the oil is to flow with an average speed of 1.2 m/s? **(b)** What is the volume flow rate in this case?

* 75. •• **BIO** A patient is given an injection with a hypodermic needle 3.2 cm long and 0.28 mm in diameter. Assuming the solution being injected has the same density and viscosity as water at 20 °C, find the pressure difference needed to inject the solution at the rate of 1.5 g/s.

76. ••• A round wooden log with a diameter of 83 cm floats with one-half of its radius out of the water. What is the log's density?

77. ••• The hollow glass shell shown in **Figure 15–35** has an inner radius R and an outer radius $1.2\,R$. The density of the glass is ρ_g. What fraction of the shell is submerged when it floats in a liquid of density $\rho = 1.5\,\rho_g$?

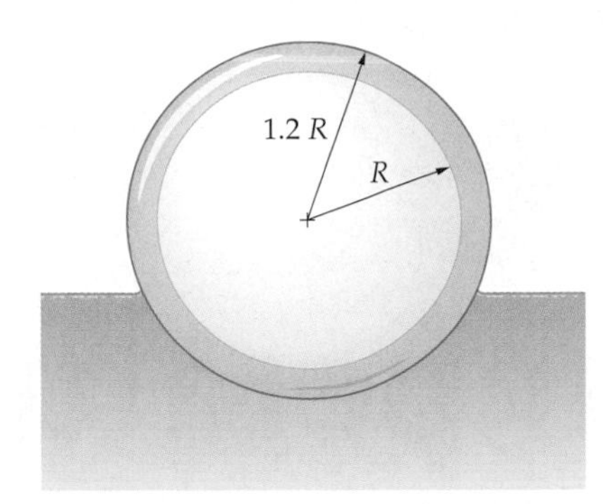

▲ **FIGURE 15–35** Problem 77

78. ••• **IP** A hollow sphere floats with 0.500 of its volume submerged in water. When the sphere floats in a particular oil it is found that only 0.400 of its volume is submerged. **(a)** Is the oil more or less dense than water? Explain. **(b)** Find the density of the oil and the sphere.

79. ••• A hollow cubical box, 0.23 m on a side, with walls of negligible mass and volume, floats with 0.45 of its volume submerged. How much water can be added to the box before it sinks?

80. ••• A geode is a hollow rock with a solid shell and an air-filled interior. Suppose a particular geode weighs twice as much in air as it does when completely submerged in water. If the density of the solid part of the geode is $2500\ \mathrm{kg/m^3}$, what fraction of the geode's volume is hollow?

81. ••• A tank of water filled to a depth d has a hole in its side a height h above the table on which it rests. Show that water emerging from the hole hits the table at a horizontal distance of $2\sqrt{(d-h)h}$ from the base of the tank.

82. ••• The water tank in **Figure 15–36** is open to the atmosphere and has two holes in it, one 0.80 m and one 3.6 m above the floor on which the tank rests. If the two streams of water strike the floor in the same place, what is the depth of water in the tank?

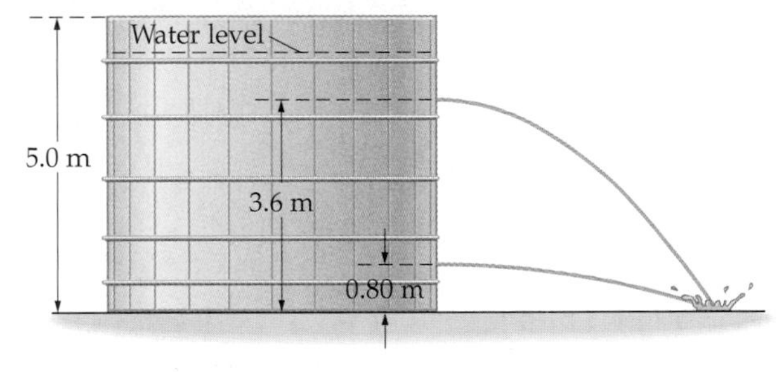

▲ **FIGURE 15–36** Problem 82

83. ••• A wooden block of cross-sectional area A and density ρ_1 floats in a fluid of density ρ_2. If the block is displaced downward and then released it will oscillate with simple harmonic motion. Find the period of its motion.

16 Temperature and Heat

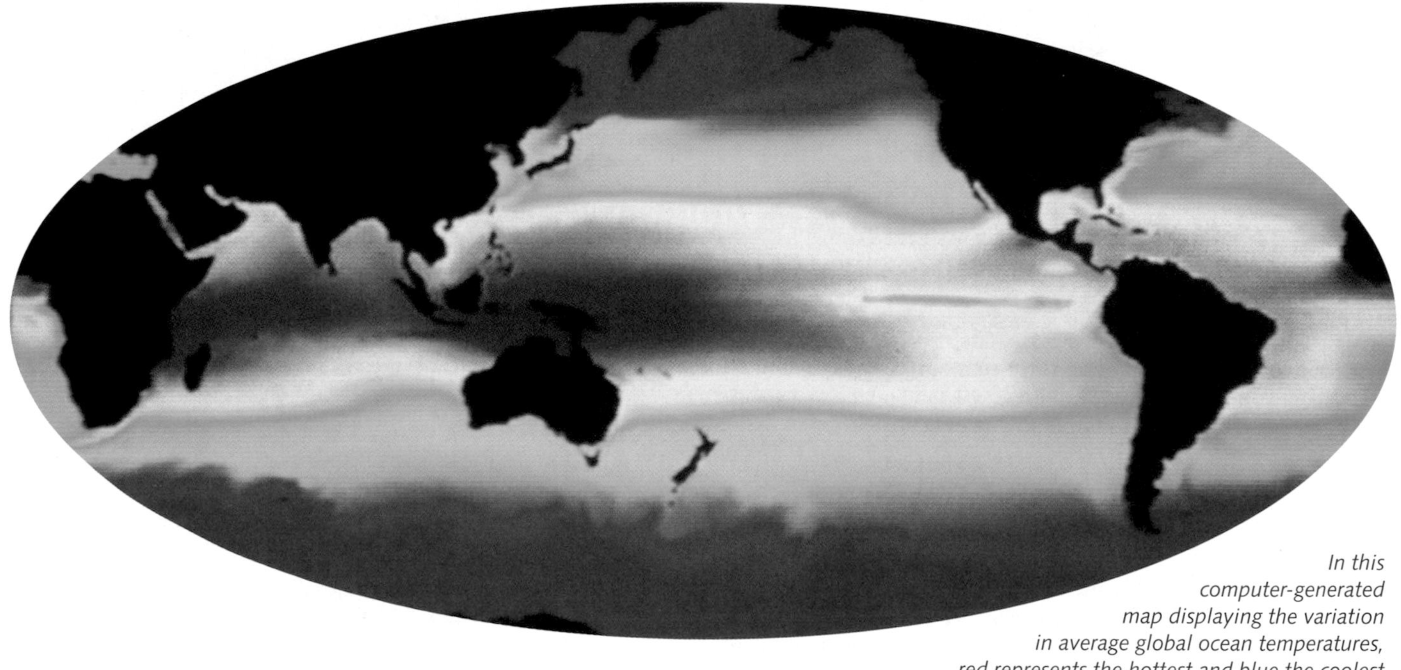

In this computer-generated map displaying the variation in average global ocean temperatures, red represents the hottest and blue the coolest areas. The relatively high-temperature water in equatorial regions warms the air above it, greatly influencing world-wide climate and weather patterns. But what exactly is temperature, and how does thermal energy pass from the warm water to the cooler air? This chapter explores such questions and others related to the phenomenon known as heat.

To this point our study of physics has involved just three physical quantities: mass, length, and time. Every measurement, every calculation, has been in terms of M, L, and T, or some combination of the three. We now add a fourth physical quantity—*temperature*. With the introduction of temperature we broaden the scope of physics, allowing the study of a wide variety of new physical situations that mechanics alone can not address.

In this chapter we introduce the concept of temperature, and discuss its effects on macroscopic systems. We begin by showing how differences in temperature relate to a particular type of energy transfer we call *heat*. We also discuss the connection between changes in temperature and changes in other physical quantities, such as length, pressure, and volume. Finally, we consider various mechanisms by which thermal energy is exchanged. Later in the text, when we consider the microscopic aspects of temperature, we shall see that temperature is ultimately related to the rapidity of molecular motion.

16–1 Temperature and the Zeroth Law of Thermodynamics

Even as small children we learn to avoid objects that are "hot." We also discover early on that if we forget to wear our coats outside we can become "cold." Later, we associate high values of something called "temperature" with hot objects, and low values with cold objects. These basic notions about temperature carry over into physics, though with a bit more precision.

Similarly, when we put a cool pan of water on a hot stove burner we say that "heat" flows from the hot burner to the cool water. To be specific, we will define *heat* as follows: Heat is the energy transferred between objects because of a temperature difference. Though a bit redundant, we will use common expressions such as "heat flow" and "heat transfer" to refer to the energy transfer associated with heat.

Next, objects are said to be in **thermal contact** if heat can flow between them. In general, when a hot object is brought into thermal contact with a cold object heat will be exchanged. The result is that the hot object cools off while the cold object warms up. After some time in thermal contact, the flow of heat ceases. At this point, we say that the objects are in **thermal equilibrium**.

Note that thermal contact and physical contact are not necessarily the same. For example, thermal contact can occur even when there is no physical contact at all—as when you warm your hands near a fire. Various types of thermal contact will be discussed in detail in Section 16–6.

We now introduce perhaps the most fundamental law obeyed by thermodynamic systems—referred to, appropriately enough, as the zeroth law of thermodynamics. The zeroth law spells out the basic properties of temperature. Later, in Chapter 18, we introduce the remaining three laws of thermodynamics. These laws enable us to analyze the behavior of engines and refrigerators, and to show, among other things, that perpetual motion machines are not possible.

Zeroth Law of Thermodynamics

The basic idea of the zeroth law of thermodynamics is that thermal equilibrium is determined by a single physical quantity—the **temperature**. For example, two objects in thermal contact are in equilibrium when they have the *same* temperature. If one or the other has a higher temperature, heat flows from that object to the other until their temperatures are equal.

This may seem almost too obvious to mention, at least until you give it more thought. Suppose, for example, that you have a piece of metal and a pool of water, and you want to know if heat will flow between them when you put the metal in the pool. You measure the temperature of each, and if they are the same you can conclude that no heat will flow. If the temperatures are different, however, it follows that there will be a flow of heat. Nothing else matters—not the type of metal, its mass, its shape, the amount of water, whether the water is fresh or salt, and so on—all that matters is one number, the temperature.

To summarize, the **zeroth law of thermodynamics** can be stated as follows:

> If object A is in thermal equilibrium with object B, and object C is also in thermal equilibrium with object B, then objects A and C will be in thermal equilibrium if brought into thermal contact.

This is illustrated in **Figure 16–1**. We begin with objects A and B in thermal contact and in equilibrium. Next, object B is separated from A, and placed in contact with C. Objects C and B are also found to be in equilibrium. Hence, by the zeroth law, we are assured that when A and C are placed in contact they also will be in equilibrium.

To apply this principle to our example, let object A be the piece of metal and object C be the pool of water. Object B, then, can be a thermometer, used to measure the temperature of the metal and the water. If A and C are each separately in equilibrium

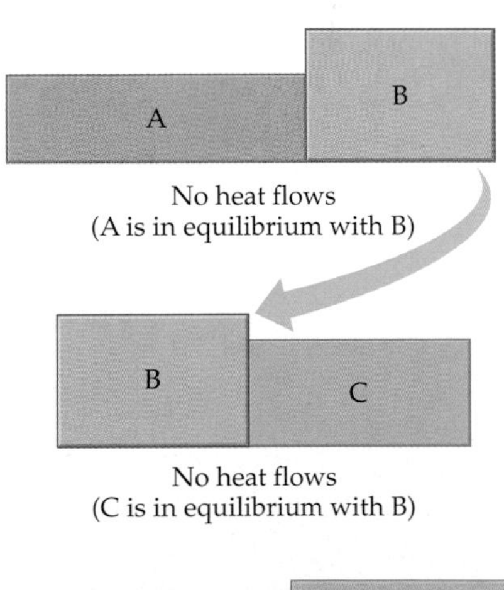

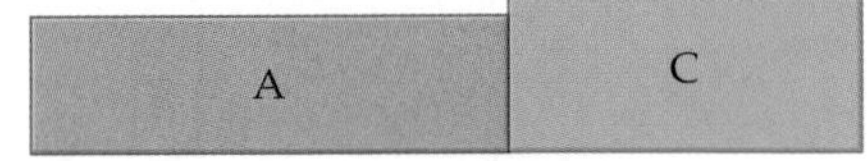

▲ **FIGURE 16–1 An illustration of the zeroth law of thermodynamics**
If A and C are each in thermal equilibrium with B, then A will be in thermal equilibrium with C when they are brought into thermal contact.

with B—which means that they have the same temperature—they will be in equilibrium with one another.

16–2 Temperature Scales

A variety of temperature scales are commonly used in both everyday situations and in physics. Some are related to familiar reference points, such as the temperature of boiling or freezing water. Others have more complex, historical rationale for their values. Here we consider three of the more frequently used temperature scales. We also examine the connections between them.

Later in this chapter we will discuss some of the physical phenomena—such as thermal expansion—that can be used to construct a thermometer. With a properly calibrated thermometer we can determine the temperature on any of these scales. In the next chapter, we will explore more fully the question of just what temperature means on a conceptual and microscopic level.

The Celsius Scale

Perhaps the easiest temperature scale to remember is the Celsius scale, named in honor of the Swedish astronomer Anders Celsius (1701–1744). Originally, Celsius assigned zero degrees to boiling water and 100 degrees to freezing water. These values were later reversed by the biologist Carolus Linnaeus (1707–1778). Thus, today we say that water freezes at zero degrees Celsius, which we abbreviate as 0 °C, and boils at 100 °C.

Note that the choice of zero level for a temperature scale is quite arbitrary, as is the number of degrees between any two reference points. In the Celsius scale, as in others, there is no upper limit to the value a temperature may have. There is a lower limit, however. For the Celsius scale, the lowest possible temperature is −273.15 °C, as we shall see later in this section.

One bit of notation should be pointed out before we continue. When we speak of a temperature, we give its value as shown with the degree symbol preceding the capital letter C. For example, 5 °C is the temperature five degrees Celsius. On the other hand, if the temperature of an object is *changed* by a given amount we use the notation C°. Thus, if we increase the temperature by five degrees on the Celsius scale, we say that the change in temperature is 5 C°; that is, five Celsius degrees. This is summarized below:

A temperature of five degrees is 5 °C
(five degrees Celsius)

A temperature change of five degrees is 5 C°
(five Celsius degrees)

The Fahrenheit Scale

The Fahrenheit scale was developed by Gabriel Fahrenheit (1686–1736), who chose zero to be the lowest temperature he was able to achieve in his laboratory. He also chose 96 degrees to be body temperature, though why he made this choice is not known. In the modern version of the Fahrenheit scale body temperature is 98.6 °F; in addition, water freezes at 32 °F and boils at 212 °F. Lastly, using the same convention as for °C and C°, we say that an increase of 180 F° is required to bring water from freezing to boiling.

Note that the Fahrenheit scale not only has a different zero than the Celsius scale, it also has a different "size" for its degree. As just noted, 180 Fahrenheit degrees are required for the same change in temperature as 100 Celsius degrees. Hence, the Fahrenheit degrees are smaller by a factor of 100/180 = 5/9.

To convert between a Fahrenheit temperature, T_F, and a Celsius temperature, T_C, we start by writing a linear relation between them. Thus, let

$$T_F = aT_C + b$$

We would like to determine the constants a and b. This requires two independent pieces of information, which we have in the freezing and boiling points of water. Using the freezing point, we find

$$32\,°\text{F} = a(0\,°\text{C}) + b = b$$

Thus, b is 32 °F. Next, the boiling point gives

$$212\,°\text{F} = a(100\,°\text{C}) + 32\,°\text{F}$$

Solving for the constant a we find

$$a = (212°\text{F} - 32°\text{F})/(100°\text{C}) = \frac{180\,°\text{F}}{100\,°\text{C}} = \tfrac{9}{5}\,°\text{F}/°\text{C}$$

Combining our results gives the following conversion relationship:

Conversion between degrees Celsius and degrees Fahrenheit

$$T_F = (\tfrac{9}{5}\,°\text{F}/°\text{C})T_C + 32\,°\text{F} \qquad \textbf{16–1}$$

Similarly, this relation can be rearranged to convert from Fahrenheit to Celsius:

Conversion between degrees Fahrenheit and degrees Celsius

$$T_C = (\tfrac{5}{9}\,°\text{C}/°\text{F})(T_F - 32\,°\text{F}) \qquad \textbf{16–2}$$

Since conversion factors like $\frac{9}{5}$°F/°C are a bit clumsy, and tend to clutter up an equation, we will generally drop the degree symbols until the final result. For example, to convert 10 °C to degrees Fahrenheit we write

$$T_F = \tfrac{9}{5}T_C + 32 = \tfrac{9}{5}(10) + 32 = 50\,°\text{F}$$

EXAMPLE 16–1 Temperature Conversions

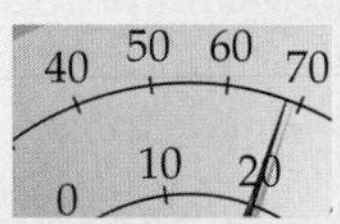

(a) On a fine spring day you notice that the temperature is 75 °F. What is the corresponding temperature on the Celsius scale? **(b)** If the temperature on a brisk winter morning is −2.0 °C, what is the corresponding Fahrenheit temperature?

Picture the Problem

The drawing shows the Celsius and Fahrenheit scales over the temperature range of interest.

Strategy

The conversions asked for in this problem are straightforward applications of the relations between T_F and T_C. In particular, for **(a)** we use Equation 16–2, and for **(b)** we use Equation 16–1.

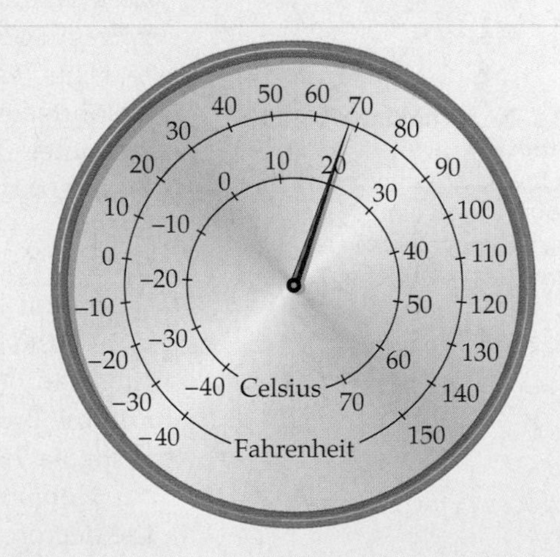

Solution

Part (a)

1. Substitute $T_F = 75$°F into Equation 16–2: $T_C = \tfrac{5}{9}(75 - 32) = 24$°C

Part (b)

2. Substitute $T_C = -2.0$°C in Equation 16–1: $T_F = \tfrac{9}{5}(-2.0) + 32 = 28$°F

Insight

Note that the results given here agree with the scales shown in the drawing.

Practice Problem

Find the Celsius temperature that corresponds to 110 °F. [**Answer:** $T_C = 43$°C]

Some related homework problems: Problem 1, Problem 3

ACTIVE EXAMPLE 16–1 Same Temperature

What temperature is the same on both the Celsius and Fahrenheit scales?

Solution

1. Set $T_F = T_C = t$ in Equation 16–1: $t = 9t/5 + 32$
2. Move all terms involving t to the left side of the equation: $-4t/5 = 32$
3. Solve for t: $t = -40$
4. As a check, substitute $T_F = -40\,°F$ in Equation 16–2: $T_C = (5/9)(-40 - 32) = -40\,°C$

Insight

Thus, $-40\,°F$ is the same as $-40\,°C$. This is consistent with the scale shown in Example 16–1.

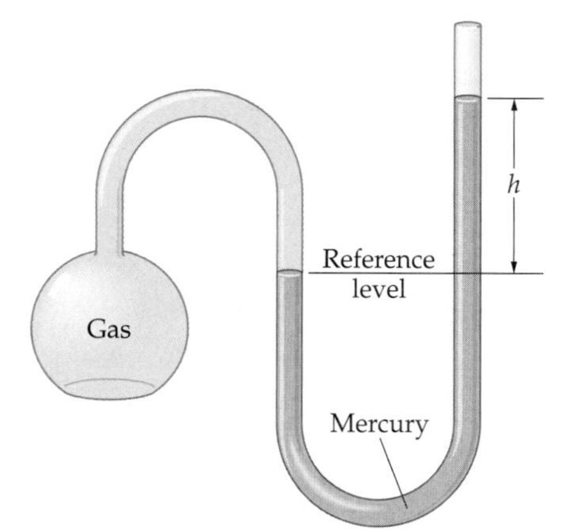

▲ **FIGURE 16–2 A constant-volume gas thermometer**
By adjusting the height of mercury in the right-hand tube, the level in the left-hand tube can be set at the reference level. This assures that the gas occupies a constant volume.

Absolute Zero

Experiments show conclusively that there is a lowest temperature below which it is impossible to cool an object. This is referred to as **absolute zero**. Though absolute zero may be approached arbitrarily closely, it may not be passed.

To give an idea of just where absolute zero is on the Celsius scale, we start with the following observation: If a given volume V of air—say the air in a balloon—is cooled from 100 °C to 0 °C its volume decreases by roughly $V/4$. Imagine this trend continuing uninterrupted. In this case, cooling from 0 °C to −100 °C would reduce the volume by another $V/4$, from −100 °C to −200 °C by another $V/4$, and finally, from −200 °C to −300 °C by another $V/4$, which brings the volume down to zero. Clearly, it doesn't make sense for the volume to be less than zero, hence absolute zero must be roughly −300 °C.

This result, though crude, is in the right ballpark. A precise determination of absolute zero can be made with a device known as a **constant-volume gas thermometer**. This instrument is shown in **Figure 16–2**. The basic idea is that by adjusting the level of mercury in the right-hand tube, the level of mercury in the left-hand tube can be set to a fixed reference level. With the mercury so adjusted, the gas occupies a constant volume and its pressure is simply $P_{gas} = P_{at} + \rho gh$ (Equation 15–7), where ρ is the density of mercury.

As the temperature of the gas is changed, the mercury level in the right-hand tube can be readjusted as described. The gas pressure can be determined again, and the process repeated. The results of a series of such measurements are shown in **Figure 16–3**.

Note that as a gas is cooled its pressure decreases, as one would expect. In fact, the decrease in pressure is approximately linear. At low enough temperatures the

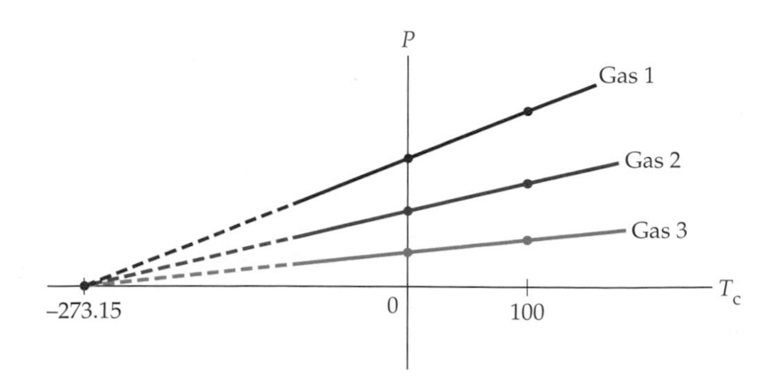

▶ **FIGURE 16–3 Determining absolute zero**
Different gases have different pressures at any given temperature. However, they all extend down to zero pressure at precisely the same temperature, −273.15 °C. This is the location of absolute zero.

gas eventually liquefies, and its behavior changes, but if we extrapolate the straight line obtained before liquefaction, we see that it reaches zero pressure (the lowest pressure possible) at −273.15 °C.

What is remarkable about this result is that it is independent of the gas we use in the thermometer. For example, gas 2 and gas 3 in Figure 16–3 have pressures that are different from one another, and from gas 1. Yet, all three gases extrapolate to zero pressure at precisely the same temperature. Thus, we conclude that there is indeed a *unique* value of absolute zero, below which further cooling is not possible.

EXAMPLE 16–2 Its a Gas

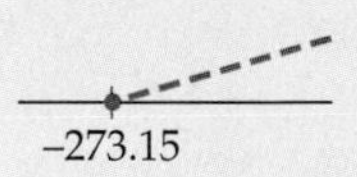

The gas in a constant-volume gas thermometer has a pressure of 80.0 kPa at 0.00 °C. Assuming ideal behavior, as in Figure 16–3, what is the pressure of this gas at 105 °C?

Picture the Problem

Our sketch shows the pressure as a function of temperature. Note that at $T_C = 0.00\,°C$ the pressure is 80.0 kPa, and that at $T_C = -273.15\,°C$ the pressure extrapolates to zero.

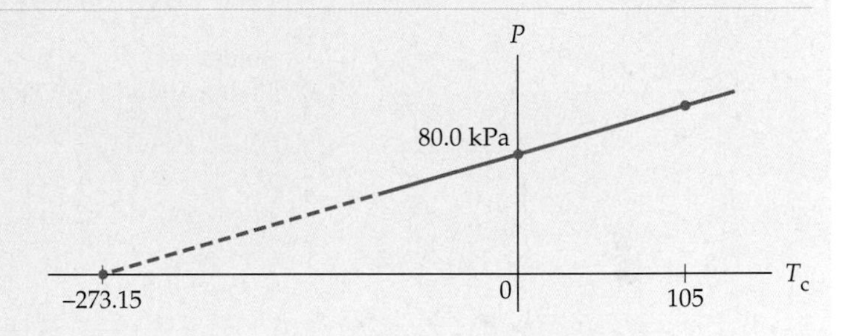

Strategy

We are to assume that the pressure lies on a straight line, as in Figure 16–3. To find the pressure at $T_C = 105\,°C$ we simply extend the straight line.

Thus, we start with the information that the pressure increases from 0 to 80.0 kPa when the temperature increases from −273.15 °C to 0.00 °C. This rate of increase in pressure must also apply to an increase in temperature from −273.15 °C to 105 °C. Using this rate we can find the desired pressure.

Solution

1. Calculate the rate at which pressure increases for this gas: $\text{rate} = \dfrac{80.0\text{ kPa}}{273.15\text{ C}°} = 293\text{ Pa/C}°$

2. Multiply this rate by the temperature change from −273.15 °C to 105 °C: $(293\text{ Pa/C}°)(378\text{ C}°) = 111\text{ kPa}$

Insight

The pressure of this gas increases from slightly less than one atmosphere at 0.00 °C to slightly more than one atmosphere at 105 °C.

Practice Problem

Find the temperature at which the pressure of the gas is 70.0 kPa. [**Answer:** $T_C = -34.2\,°C$]

Some related homework problems: Problem 5, Problem 6

The Kelvin Scale

The Kelvin temperature scale, named for the Scottish physicist William Kelvin (1824–1907), is based on the existence of absolute zero. In fact, the zero of the Kelvin scale, abbreviated 0 K, is set exactly at absolute zero. Thus, in this scale there are no negative equilibrium temperatures. The Kelvin scale is also chosen to have the same degree size as the Celsius scale.

As mentioned, absolute zero occurs at −273.15 °C, hence the conversion between a Kelvin-scale temperature, T, and a Celsius temperature, T_C, is as follows:

Conversion between a Celsius temperature and a Kelvin temperature

$$T = T_C + 273.15 \qquad \textbf{16–3}$$

Note that the difference between the Celsius and Kelvin scales is simply a difference in the zero level.

The notation for the Kelvin scale differs somewhat from that for the Celsius and Fahrenheit scales. In particular, by international agreement, the degree terminology and the degree symbol, °, are not used in the Kelvin scale. Instead, a temperature of 5 K is read simply as 5 kelvin. In addition, a change in temperature of 5 kelvin is written 5 K, the same as for a temperature of 5 kelvin.

Though the Celsius and Fahrenheit scales are the ones most commonly used in everyday situations, the Kelvin scale is used more than any other in physics. This stems from the fact that the Kelvin scale incorporates the significant concept of absolute zero. As a result, the thermal energy of a system depends in a very simple way on the Kelvin temperature. This will be discussed in detail in the next chapter.

EXERCISE 16–1

Convert 55 °F to the Kelvin temperature scale.

Solution

First, convert from °F to °C:

$$T_C = \tfrac{5}{9}(55 - 32) = 13\,°\text{C}$$

Next, convert °C to K:

$$T = 13 + 273.15 = 286\ \text{K}$$

The three temperature scales presented in this section are shown side by side in **Figure 16–4**. Temperatures of particular interest are indicated as well. This permits a useful visual comparison between the scales.

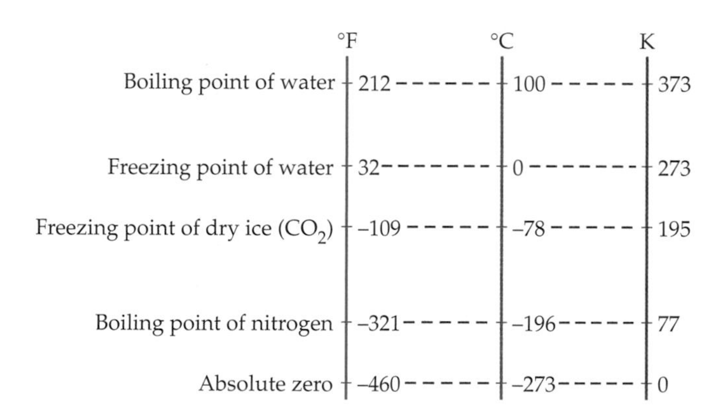

FIGURE 16–4 Temperature scales
A comparison of the Fahrenheit, Celsius, and Kelvin temperature scales. Physically significant temperatures, such as the freezing and boiling points of water, are indicated for each scale.

16–3 Thermal Expansion

Most substances expand when heated. For example, power lines on a hot summer day hang low compared to their appearance on a cold day in winter. In fact, thermal expansion is the basis for many thermometers, including the familiar thermometers used for measuring a fever. The expansion of a liquid, such as mercury or alcohol, results in a column of liquid of variable height within the glass neck of the thermometer. The height is read against markings on the glass, which gives the temperature.

You may wonder what bizarre substance could possibly be an exception to this common response to heating. The most important exception occurs in a substance you drink every day—water. This is just one of the many special properties that sets water apart from most other substances.

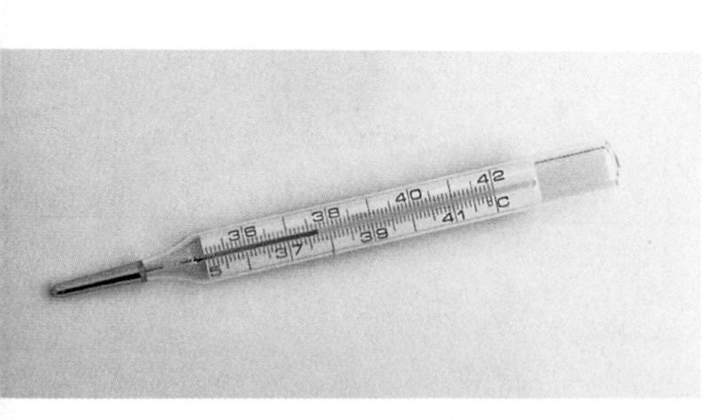

▲ Like nearly all substances, mercury expands with increasing temperature. This behavior is the basis for the familiar mercury clinical thermometer.

In this section we consider the physics of thermal expansion, including linear, area, and volume expansion. We also discuss briefly the unusual thermal behavior of water, and some of its more significant consequences.

Linear Expansion

Consider a rod whose length is L_0 at the temperature T_0. Experiments show that when the rod is heated or cooled, its length changes in direct proportion to the temperature change. Thus, if the change in temperature is ΔT, the change in length of the rod, ΔL, is

$$\Delta L = (\text{constant})\Delta T$$

The constant of proportionality depends, among other things, on the substance from which the rod is made.

CONCEPTUAL CHECKPOINT 16–1

When rod 1 is heated by an amount ΔT its length increases by ΔL. If rod 2, which is twice as long as rod 1, is heated by the same amount, does its length increase by **(a)** ΔL, **(b)** $2\Delta L$, or **(c)** $\Delta L/2$?

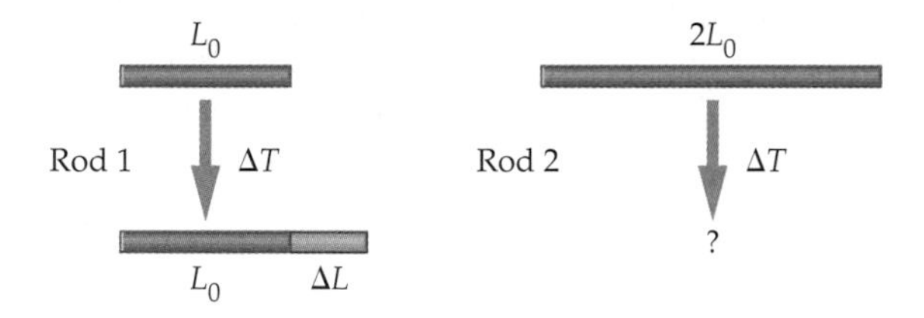

Reasoning and Discussion

We can imagine rod 2 to be composed of two copies of rod 1 placed end to end, as shown.

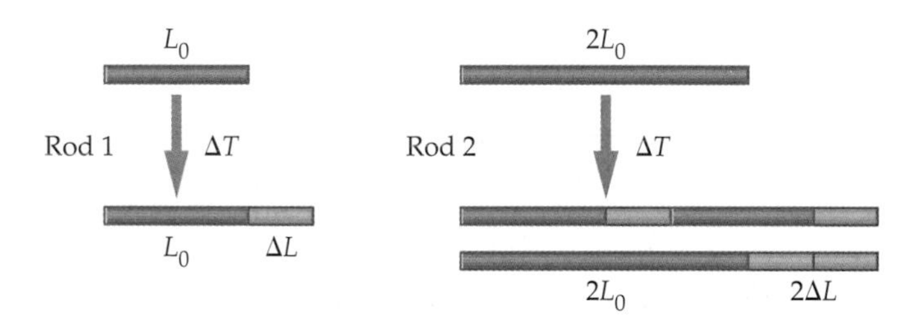

When the temperature is increased by ΔT, each copy of rod 1 expands by ΔL. Hence, the total expansion of the two copies is $2\Delta L$, as is the total expansion of rod 2.

Answer:

(b) The rod that is twice as long expands twice as much; $2\Delta L$.

We conclude, then, on the basis of the preceding Conceptual Checkpoint, that the change in length is proportional to *both* the initial length, L_0, and the temperature change, ΔT. The constant of proportionality is referred to as α, the **coefficient of linear expansion**, and is defined as follows:

Definition of coefficient of linear expansion, α

$$\Delta L = \alpha L_0 \Delta T \qquad 16\text{–}4$$

SI unit for α: $K^{-1} = (C°)^{-1}$

Table 16–1 gives values of α for a variety of substances.

TABLE 16–1
Coefficients of Thermal Expansion

Substance	Coefficient of linear expansion, α (K^{-1})
Lead	29×10^{-6}
Aluminum	24×10^{-6}
Brass	19×10^{-6}
Copper	17×10^{-6}
Iron (Steel)	12×10^{-6}
Concrete	12×10^{-6}
Window glass	11×10^{-6}
Pyrex glass	3.3×10^{-6}
Quartz	0.50×10^{-6}

Substance	Coefficient of volume expansion, β (K^{-1})
Ether	1.51×10^{-3}
Carbon tetrachloride	1.18×10^{-3}
Alcohol	1.01×10^{-3}
Gasoline	0.95×10^{-3}
Olive oil	0.68×10^{-3}
Water	0.21×10^{-3}
Mercury	0.18×10^{-3}

▲ The Eiffel Tower in Paris gains about a thirteenth of an inch in height for each Fahrenheit degree that the temperature rises.

EXERCISE 16–2

The Eiffel Tower, constructed in 1889 by Alexandre Eiffel, is an impressive latticework structure made of iron. If the tower is 301 m high on a 22 °C day, how much does its height decrease when the temperature cools to 0.0 °C?

Solution

We can calculate the change in height with Equation 16–4. Note that the coefficient of linear expansion for iron, given in Table 16–1, is $12 \times 10^{-6}\,\mathrm{K}^{-1}$ and the change in temperature is $\Delta T = -22\,\mathrm{C}° = -22\,\mathrm{K}$:

$$\Delta L = \alpha L_0 \Delta T = (12 \times 10^{-6}\,\mathrm{K}^{-1})(301\,\mathrm{m})(-22\,\mathrm{K}) = -7.9\,\mathrm{cm}$$

An interesting application of thermal expansion is in the behavior of a bimetallic strip. As the name suggests, a bimetallic strip consists of two metals bonded together to form a linear strip of metal. This is illustrated in **Figure 16–5**. Since two different metals will, in general, have different coefficients of linear expansion, α, the two sides of the strip will change lengths by different amounts when heated or cooled.

For example, suppose metal B in Figure 16–5 (a) has the larger coefficient of linear expansion. This means that its length will change by greater amounts than metal A for the same temperature change. Hence, if this strip is cooled, the B side will shrink more than the A side, resulting in the strip bending toward the B side, as in Figure 16–5 (b). On the other hand, if the strip is heated, the B side expands by a greater amount than the A side, and the strip curves toward the A side, as in Figure 16–5 (c). Thus, the shape of the bimetallic strip depends sensitively on temperature.

REAL-WORLD PHYSICS
Bimetallic strips

Because of this property, bimetallic strips are used in a variety of thermal applications. For example, a bimetallic strip can be used as a thermometer; as the strip changes its shape it can move a needle to indicate the temperature. Similarly, many thermostats have a bimetallic strip to turn on or shut off a heater. This is shown in Figure 16–5 (d). As the temperature of the room changes, the bimetallic strip deflects in one direction or the other, which either closes or breaks the electrical circuit connected to the heater.

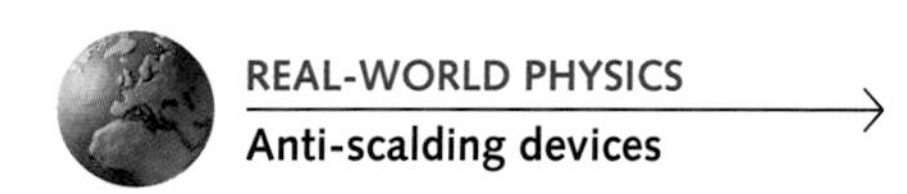

REAL-WORLD PHYSICS
Anti-scalding devices

Another common use of thermal expansion is the *anti-scalding device* for water faucets. An anti-scalding device is simply a valve inside a water faucet that is attached to a spring. When the water temperature is at a safe level, the valve permits water to flow freely through the faucet. If the temperature of the water reaches a dangerous level, however, the thermal expansion of the spring is enough to close the valve and stop the flow of water—thus preventing inadvertant scalding. When the water cools down again, the valve reopens and the flow of water resumes.

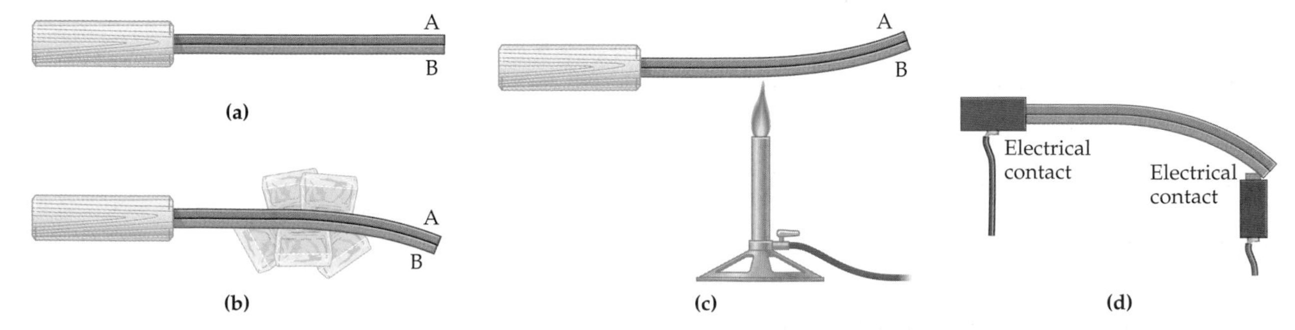

▲ **FIGURE 16–5 A bimetallic strip**
(a) A bimetallic strip composed of metals A and B. If metal B has a larger coefficient of linear expansion than metal A, it will shrink more when cooled **(b)** and expand more when heated **(c)**. A bimetallic strip can be used to construct a thermostat **(d)**. If the temperature falls, the strip bends downward and closes the electrical circuit, which then operates a heater. When the temperature rises, the strip deflects in the opposite direction, breaking the circuit and turning off the heater.

▲ Thermal expansion, though small, is far from negligible in many everyday situations. This is especially true when long objects such as railroad tracks, bridges, or pipelines are involved. Bridges and elevated highways (left), must include expansion joints to prevent the roadway from buckling when it expands in hot weather. Similarly, pipelines (right) typically include loops that allow for expansion and contraction when the temperature changes.

Finally, thermal expansion can have undesirable effects in some cases. For example, you may have noticed that bridges often have gaps between different sections of the structure. When the air temperature rises in the summer, the sections of the bridge can expand freely into these gaps. If the gaps were not present the expansion of the different sections could cause the bridge to buckle and warp. Thus, these gaps, referred to as *expansion joints*, are a way of avoiding this type of heat-related damage. Expansion joints can also be found in railroad tracks and oil pipelines, to name just two other examples.

REAL-WORLD PHYSICS
Thermal expansion joints

Area Expansion

Since the length of an object changes with temperature, it follows that its area changes as well. To see precisely how the area changes, consider a square piece of metal of length L on a side. The initial area of the square is $A = L^2$. If the temperature of the square is increased by ΔT the length of each side increases from L to $L + \Delta L = L + \alpha L \Delta T$. As a result, the square has an increased area, A', given by

$$A' = (L + \Delta L)^2 = (L + \alpha L \Delta T)^2$$
$$= L^2 + 2\alpha L^2 \Delta T + \alpha^2 L^2 \Delta T^2$$

Now, if $\alpha \Delta T$ is much less than one—which is certainly the case for typical changes in temperature, ΔT—then $\alpha^2 \Delta T^2$ is even smaller. Hence, if we ignore this small contribution, we find

$$A' \approx L^2 + 2\alpha L^2 \Delta T = A + 2\alpha A \Delta T$$

As a result, the change in area, ΔA, is

$$\Delta A = A' - A \approx 2\alpha A \Delta T \qquad \text{16–5}$$

Note the similarity between this relation and Equation 16–4; the length L has been replaced with the area A, and the expansion coefficient has been doubled to 2α.

Though this calculation was done for the simple case of a square, the result applies to an area of any shape. For example, a circular disk of radius r and area $A = \pi r^2$ will increase its area by the amount $2\alpha A \Delta T$ with an increase in temperature of ΔT.

What about a washer, however, which is a disk of metal with a circular hole cut out of its center? What happens to the area of the hole as the washer is heated? Does it expand along with everything else, or does the expanding washer "expand into the hole" to make it smaller? We consider this question in the next Conceptual Checkpoint.

CONCEPTUAL CHECKPOINT 16–2

A washer has a hole in the middle. As the washer is heated, does the hole **(a)** expand, **(b)** shrink, or **(c)** stay the same?

Reasoning and Discussion

To make a washer from a disk of metal, we can cut along a circular curve, as shown, and remove the inner disk. If we now heat the system, both the washer and the inner disk expand. On the other hand, if we had left the inner disk in place and heated the original disk it would also expand. Removing the *heated* inner disk would create an expanded washer, with an expanded hole in the middle. We obtain the same result whether we remove the inner disk and then heat, or heat first and then remove the inner disk.

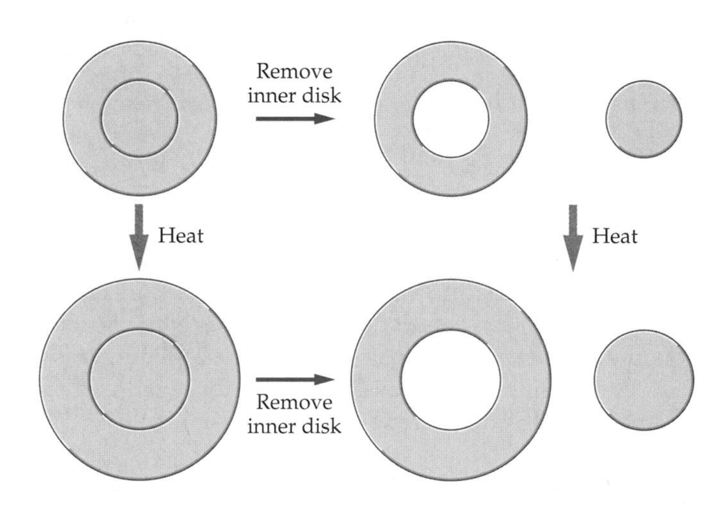

Thus, heating the washer causes both it and its hole to expand, and they both expand with the same coefficient of linear expansion. Basically, the system behaves the same as if we had produced a photographic enlargement—everything expands.

Answer:

(a) The hole expands along with everything else.

PROBLEM SOLVING NOTE
Expansion of a Hole

A hole in a material expands the same as if it were made of the material itself. Thus, to find the expansion of a hole in a steel plate we use the coefficient of linear expansion for steel.

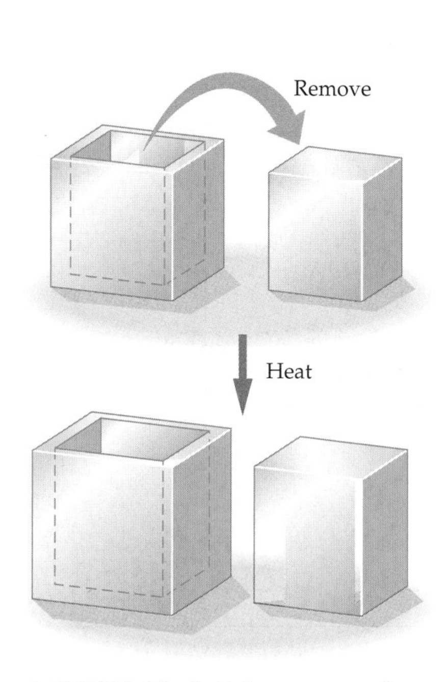

▲ **FIGURE 16–6 Volume expansion**
A portion of a cube is removed to create a container. When heated, the removed portion expands, just as the volume within the container expands.

Volume Expansion

Just as the hole in a washer increases in area with heating, so does the empty volume within a cup or other container. This is illustrated in **Figure 16–6**, where we show a block of material with a volume removed to convert it into a container. As the system is heated there will be an expansion of the container, the volume within it, and the volume that was removed. As with the area of a hole, the volume within a container expands with the same coefficient of expansion as the container itself.

To calculate the change in volume, consider a cube of length L on a side. The initial volume of the cube is $V = L^3$. Increasing the temperature results in an increased volume given by

$$V' = (L + \Delta L)^3 = (L + \alpha L \Delta T)^3$$
$$= L^3 + 3\alpha L^3 \Delta T + 3\alpha^2 L^3 \Delta T^2 + \alpha^3 L^3 \Delta T^3$$

Neglecting the smaller contributions, as we did with the area, we find

$$V' \approx L^3 + 3\alpha L^3 \Delta T = V + 3\alpha V \Delta T$$

PROBLEM SOLVING NOTE
Expansion of a Volume

The empty volume inside a container expands the same as if it were made of the same material as the container. For example, to find the increase in volume of a steel tank we use the coefficient of linear expansion for steel, just as if the tank were actually filled with steel.

Therefore, the change in volume, ΔV, is

$$\Delta V = V' - V \approx 3\alpha V \Delta T$$

This expression, though calculated for a cube, applies to any volume—even a volume of liquid with no shape of its own.

In general, volume expansion is described in the same way as linear expansion, but with a **coefficient of volume expansion**, β, defined as follows:

Definition of coefficient of volume expansion, β

$$\Delta V = \beta V \Delta T \qquad \text{16–6}$$

SI unit for β: $K^{-1} = (C°)^{-1}$

Typical values of β are given in Table 16–1. If Table 16–1 lists a value of α for a given substance, but not for β, its change in volume is calculated as follows:

$$\Delta V = \beta V \Delta T \approx 3\alpha V \Delta T \qquad \text{16–7}$$

That is, we simply make the identification $\beta = 3\alpha$. For substances that have a specific value for β listed in Table 16–1, we simply use Equation 16–6. This is illustrated in the next Example.

Temperature Change

Remember that a change in temperature of 1 C° is the same as a change in temperature of 1 K. Thus, when finding the thermal expansion of an object, the change in temperature ΔT can be expressed in terms of either the Celsius or the Kelvin temperature scale.

EXAMPLE 16–3 Oil Spill

Heat by 25 C°

A copper flask with a volume of 150 cm^3 is filled to the brim with olive oil. If the temperature of the system is increased from 6.0 °C to 31 °C, how much oil spills from the flask?

Picture the Problem

Our sketch shows the flask filled to the top initially, then spilling over when heated by 25 C°. Note: Since degrees have the same size on the Celsius and Kelvin scales, it follows that $\Delta T = 25\ C° = 25\ K$.

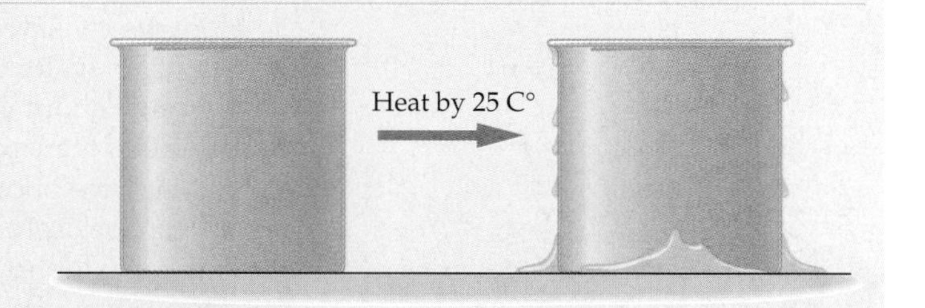

Strategy

As the system is heated, both the flask and the olive oil expand. Thus, we start by calculating the expansion of the oil and the flask separately. A quick glance at Table 16–1 shows that the olive oil will expand more, and the difference in expansion volumes is what spills out.

For the copper flask we find α in Table 16–1, then let $\beta = 3\alpha$.

Solution

1. Calculate the change in volume of the olive oil:

$$\Delta V_{oil} = \beta V \Delta T = (0.68 \times 10^{-3}\ K^{-1})(150\ cm^3)(25\ K) = 2.6\ cm^3$$

2. Calculate the change in volume of the flask:

$$\Delta V_{flask} = 3\alpha V \Delta T = 3(17 \times 10^{-6}\ K^{-1})(150\ cm^3)(25\ K) = 0.19\ cm^3$$

3. Find the difference in volume expansions. This is the volume of oil that spills out:

$$\Delta V_{oil} - \Delta V_{flask} = 2.6\ cm^3 - 0.19\ cm^3 = 2.4\ cm^3$$

Insight

If the system were cooled, the oil would lose volume more rapidly than the flask. This would result in a drop in oil level.

Practice Problem

What temperature change would result in 3.0 cm^3 of oil being spilled? [**Answer:** $\Delta T = 32\ K$]

Some related homework problems: Problem 15, Problem 19

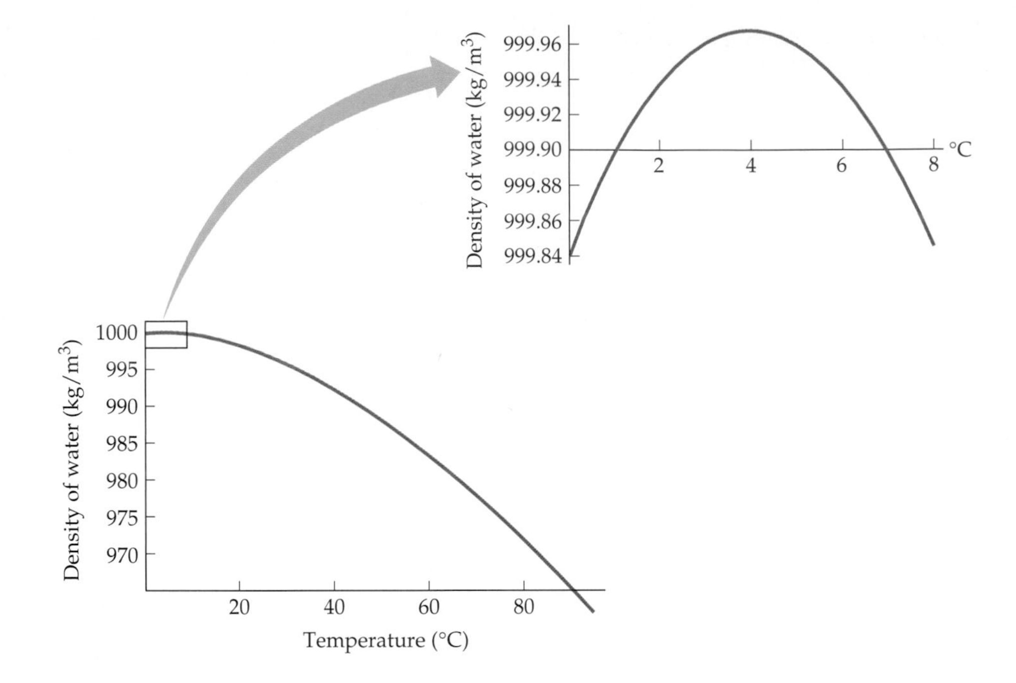

▶ **FIGURE 16–7 The unusual behavior of water near 4 °C**
The density of water actually *increases* as the water is heated between 0 °C and 4 °C. Maximum density occurs near 4 °C.

Special Properties of Water

REAL-WORLD PHYSICS
Floating icebergs

As we have mentioned, water is a substance rich with unusual behavior. For example, in the last chapter we discussed the fact that the solid form of water (ice) is less dense than the liquid form. Hence icebergs float. What is remarkable about icebergs floating is not that 90% is submerged, but that 10%, or any amount at all, is above the water. The solids of most substances are denser than their liquids; hence when they freeze their solids immediately sink.

Here we consider the unusual *thermal* behavior of water. **Figure 16–7** shows the density of water over a wide range of temperatures. Note that the density is a maximum at about 4 °C. Thus, when you *heat* water from 0 °C to 4 °C it actually *shrinks*, rather than expands, and becomes *more dense*. The reason is that water molecules that were once part of the rather open crystal structure of ice are now able to pack more closely together in the liquid.

REAL WORLD PHYSICS: BIO
The ecology of lakes

This behavior has significant consequences for the ecology of lakes in northern latitudes. When temperatures drop in the winter the surface waters of a lake cool first and sink, allowing warmer water to rise to the surface to be cooled in turn. Eventually, a lake can fill with water at 4 °C. Further drops in temperature result in cooler, less dense water near the surface, where it floats until it freezes. Thus, lakes freeze on the top surface first, with the bottom water staying relatively warm at about 4 °C. In addition, the ice and snow on top of a lake act as thermal insulation, slowing the continued growth of ice.

On the other hand, if water had more ordinary behavior—like shrinking when cooled, and a solid form that is more dense than the liquid—a lake would freeze from the bottom up. There would be no insulating layer of ice on top, and if the winter were long enough, and cold enough, the lake could freeze solid. This, of course, would be disastrous for fish and other creatures that live in the water.

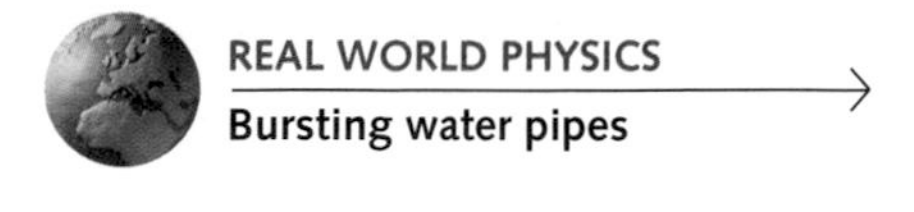

Finally, the same physics that is responsible for floating icebergs and ice-capped lakes is to blame for water pipes that burst in the winter. Even a water pipe made of steel is not strong enough to keep from rupturing when the ice forming within it expands outward. Later, when the temperature rises above freezing again, the burst pipe will make itself known by springing a leak.

16–4 Heat and Mechanical Work

In this section we consider the connection between heat and mechanical work. We also discuss the conservation of energy as it regards heat.

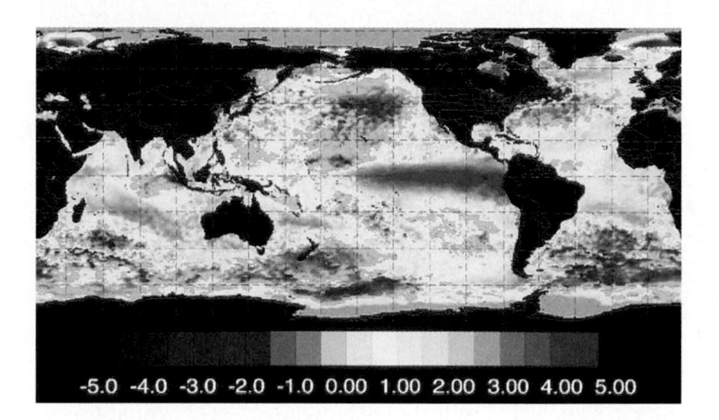

◀ Water exhibits unusual behavior around its freezing point, but above 4 °C, it expands with increasing temperature like any normal liquid. This photo shows the expansion of water on a grand scale. It depicts an El Niño event: the appearance of a mass of unusually warm water in the equatorial region of the eastern Pacific Ocean every few years. An El Niño causes worldwide changes in climate and weather patterns. In this satellite image, the warmest water temperatures are represented by red, the coolest by violet. The El Niño is the red spike west of the South American coast. In this region, sea levels are as much as 20 cm higher than their average values, due to thermal expansion of the water.

As mentioned previously, heat is the energy *transferred* from one object to another. At one time it was thought—erroneously—that an object contained a certain amount of "heat fluid," or caloric, that could flow from one place to another. This idea was overturned by the observations of Benjamin Thompson (1753–1814), also known as Count Rumford, the American-born physicist, spy, and social reformer who at one point in his eclectic career supervised the boring of cannon barrels by large drills. He observed that as long as mechanical work was done to turn the drill bits, they continued to produce heat in unlimited quantities. Clearly, the unlimited heat observed in boring the cannons was not present initially in the metal, but instead was produced by continually turning the drill bit.

With this observation, it became clear that heat was simply another form of energy that must be taken into account when applying conservation of energy. For example, if you rub sandpaper back and forth over a piece of wood you do work. The energy associated with that work is not lost; instead, it produces an increase in temperature. Taking into account the energy associated with this temperature change, we find that energy is indeed conserved. In fact, no observation has ever indicated a situation in which energy is not conserved.

The equivalence between work and heat was first explored quantitatively by James Prescott Joule (1818–1889), the British physicist. In one of his experiments, Joule observed the increase in temperature in a device similar to that shown in **Figure 16–8**. Here, a mass m falls through a certain distance h, during which gravity does the mechanical work mgh. As the mass falls it turns the paddles in the water, which results in a slight warming of the water. By measuring the mechanical work, mgh, and the increase in the water's temperature, ΔT, Joule was able to show that energy was indeed conserved. It had been converted from gravitational potential energy to heat, and an increased temperature.

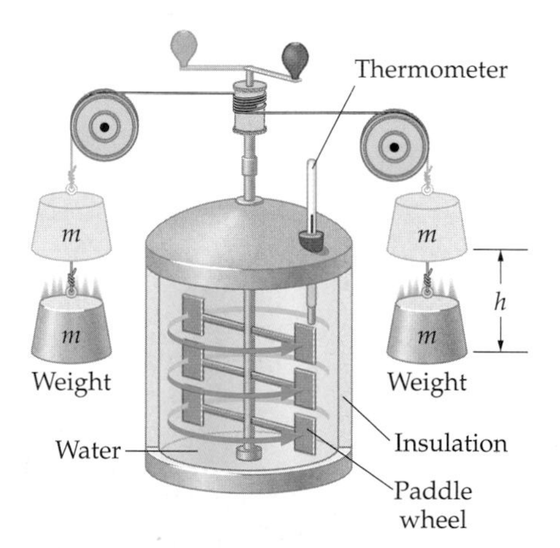

▲ FIGURE 16–8 The mechanical equivalent of heat
A device of this type was used by James Joule to measure the mechanical equivalent of heat.

Before Joule's work, heat was measured in a unit called the calorie (cal). In particular, one kilocalorie (kcal) was defined as the amount of heat needed to raise the temperature of 1 kg of water from 14.5 °C to 15.5 °C. With his experiments, Joule was able to show that 1 kcal = 4186 J, or equivalently, that one calorie is the equivalent of 4.186 J of mechanical work. This is referred to as the **mechanical equivalent of heat**:

The mechanical equivalent of heat

$$1 \text{ cal} = 4.186 \text{ J} \quad \textbf{16–8}$$

SI unit: J

In studies of nutrition a different calorie is used. It is the Calorie, with a capital C, and it is simply a kilocalorie; that is, 1 C = 1 kcal. Perhaps this helps people to feel a little better about their calorie intake. After all, a 250 C candy bar sounds a lot better than a 250,000 cal candy bar. They are equivalent, however.

Another common unit for measuring heat is the British thermal unit (Btu). By definition, a Btu is the energy required to heat 1 lb of water from 63 °F to 64 °F. In terms of calories and joules, a Btu is as follows:

$$1 \text{ Btu} = 0.252 \text{ kcal} = 1055 \text{ J} \quad \textbf{16–9}$$

Finally, we shall use the symbol Q to denote heat:

Heat, Q

Q = heat = energy transferred due to temperature differences **16–10**

SI unit: J

Using the mechanical equivalent of heat as the conversion factor, we will typically give heat in either calories or joules.

EXAMPLE 16–4 Stair Master

A 74.0-kg person drinks a thick, rich, 305-C milkshake. How many stairs must this person climb to work off the shake? Let the height of a stair be 20.0 cm.

Picture the Problem

We show the person and the stairs. The height of each stair is h, in contrast to the total height climbed, H. (Note that, as we saw in Conceptual Checkpoint 7–1, the horizontal distance is irrevelent here, as it does not affect the work done against gravity.)

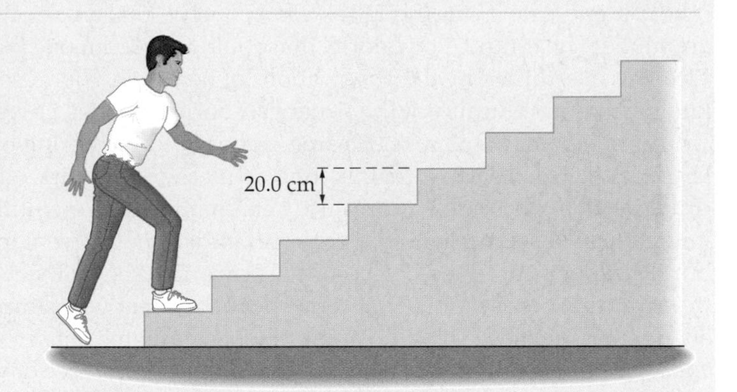

Strategy

We know that the energy intake by drinking the milkshake is $Q = 305{,}000$ cal. Using Equation 16–8, we can convert this to joules. Finally, we set the energy of the shake equal to the work done against gravity, mgH, in climbing to a height H.

Solution

1. Convert the energy of the milkshake to joules: $Q = 305{,}000 \text{ cal} = 305{,}000 \text{ cal}\left(\frac{4.186 \text{ J}}{1 \text{ cal}}\right) = 1.28 \times 10^6 \text{ J}$

2. Equate the energy of the milkshake with the work done against gravity: $Q = mgH$

   Solve for the height H: $H = Q/mg$

3. Substitute numerical values: $H = \frac{Q}{mg} = \frac{1.28 \times 10^6 \text{ J}}{(74.0 \text{ kg})(9.81 \text{ m/s}^2)} = 1760 \text{ m}$

4. Divide by the height of a stair to get the number of stairs: $\frac{1760 \text{ m}}{0.200 \text{ m/stair}} = 8800 \text{ stairs}$

Insight

This is clearly a lot of stairs, and a significant height. In fact, 1760 m is more than a mile. Even assuming a metabolic efficiency of 70 percent still leaves a height of about three-quarters of a mile that must be climbed to work off the shake.

Practice Problem

If the person in the problem climbs 100 stairs, how many Calories have been burned? Assume 100 percent efficiency. [**Answer:** 3.47 C]

Some related homework problems: Problem 21, Problem 22

16–5 Specific Heats

In the previous section, we mentioned that it takes 4186 J of heat to raise the temperature of 1 kg of water by 1 C°. We have to be specific about the fact that we are heating water, because the heat required varies considerably from one substance to another. For example, it takes only 128 J of heat to increase the temperature of

1 kg of lead by 1 C°. In general, the heat required for a given increase in temperature is given by the **heat capacity** of a substance.

Heat Capacity

Suppose we add the heat Q to a given object, and its temperature increases by the amount ΔT. The heat capacity, C, of this object is defined as follows:

Definition of heat capacity, *C*

$$C = \frac{Q}{\Delta T} \qquad \textbf{16–11}$$

SI unit: J/K = J/C°

Note that the units of heat capacity are joules per kelvin (J/K). Equivalently, since the degree size is the same for the Kelvin and Celsius scales, C can be expressed in units of joules per Celsius degree (J/C°).

The name "heat capacity" is perhaps a bit unfortunate. It derives from the mistaken idea of a "heat fluid," mentioned in the previous section. Objects were imagined to "contain" a certain amount of this nonexistent fluid. Today, we know that an object can readily gain or release heat when it is in thermal contact with other objects—objects cannot be thought of as holding a certain amount of heat.

Instead, the heat capacity should be viewed as the amount of heat necessary for a given temperature change. An object with a large heat capacity, like water, requires a large amount of heat for each increment in temperature. Just the opposite is true for an object with a small heat capacity, like a piece of lead.

To find the heat required for a given ΔT, we simply rearrange Equation 16–11 to solve for Q. This yields

$$Q = C\,\Delta T \qquad \textbf{16–12}$$

It should be noted that the *heat capacity is always positive*—just like a speed. Thus, Equation 16–12 shows that the heat Q and the temperature change ΔT have the same sign. This observation leads to the following sign conventions for Q:

Q is positive if ΔT is positive; that is, if heat is *added* to a system.

Q is negative if ΔT is negative; that is, if heat is *removed* from a system.

EXERCISE 16–3

The heat capacity of 1.00 kg of water is 4186 J/K. What is the temperature change of the water if **(a)** 505 J of heat is added to the system, or **(b)** 1010 J of heat is removed?

Solution

(a) Calculate ΔT for $Q = 505$ J:

$$\Delta T = \frac{Q}{C} = \frac{505\ \text{J}}{4186\ \text{J/K}} = 0.121\ \text{K}$$

(b) Since heat is removed in this case, $Q = -1010$ J:

$$\Delta T = \frac{Q}{C} = \frac{-1010\ \text{J}}{4186\ \text{J/K}} = -0.241\ \text{K}$$

Specific Heat

Since it takes 4186 J to increase the temperature of one kilogram of water by one degree Celsius, it takes twice that much to make the same temperature change in two kilograms of water, and so on. Thus, the heat capacity varies not only with the type of substance, but also with the amount of the substance.

We can therefore define a new quantity, the **specific heat**, c, that depends only on the substance, and not on the amount of the substance.

TABLE 16–2
Specific Heats at Atmospheric Temperature and Pressure

Substance	Specific heat, c [J/(kg·K)]
Water	4186
Ice	2090
Steam	2010
Beryllium	1820
Aluminum	900
Glass	837
Silicon	703
Iron (steel)	448
Copper	387
Silver	234
Gold	129
Lead	128

Definition of specific heat, c

$$c = \frac{Q}{m\,\Delta T} \qquad \textbf{16–13}$$

SI unit: J/(kg·K) = J/(kg·C°)

Thus, for example, the specific heat of water is

$$c_{\text{water}} = 4186\ \text{J/(kg}\cdot\text{K)}$$

Hence the heat required to raise 1.0 kg by 1.0 C° is

$$Q = mc\,\Delta T = (1.0\ \text{kg})[4186\ \text{J/(kg}\cdot\text{K)}](1.0\ \text{C}°) = 4200\ \text{J}$$

To produce the same temperature change in 2.0 kg of water requires twice as much heat:

$$Q = mc\,\Delta T = (2.0\ \text{kg})[4186\ \text{J/(kg}\cdot\text{K)}](1.0\ \text{C}°) = 8400\ \text{J}$$

Note that the specific heat is the same in both cases; the difference in heat is a reflection of the difference in mass m.

Specific heats for common substances are listed in Table 16–2. Note that the specific heat of water is by far the largest of any common material. This is just another of the many unusual properties of water. Having such a large specific heat means that water can give off or take in large quantities of heat with little change in temperature. It is for this reason that if you take a bite of a pie that is just out of the oven, you are much more likely to burn your tongue on the fruit filling (which has a high water content) than on the much drier crust.

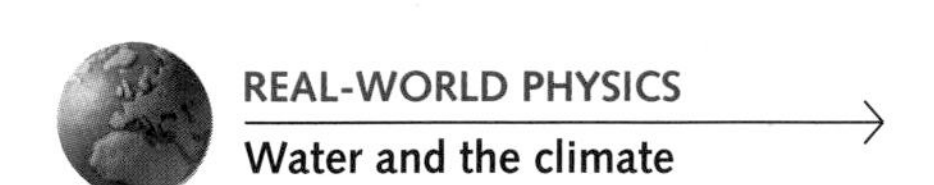

Water's unusually large specific heat also accounts for the moderate climates experienced in regions near great bodies of water. In particular, the enormous volume and large specific heat of an ocean serve to maintain a nearly constant temperature in the water, which in turn acts to even out the temperature of adjacent coastal areas. For example, the west coast of the United States benefits from the moderating effect of the Pacific Ocean, aided by the prevailing breezes that come from the ocean onto the coastal regions. In the midwest, on the other hand, temperature variations can be considerably greater as the land (with a relatively small specific heat) quickly heats up in the summer and cools off in the winter.

Calorimetry

Let's use the specific heat to solve a practical problem. Suppose a block of mass m_b, specific heat c_b, and initial temperature T_b is dropped into a **calorimeter** (basically, a lightweight, insulated flask) containing water. If the water has a mass m_w, a specific heat c_w, and an initial temperature T_w, find the final temperature of the block and the water. Assume that the calorimeter is light enough that it can be ignored, and that no heat is transferred from the calorimeter to its surroundings.

There are two basic ideas involved in solving this problem: (a) the final temperatures of the block and water are equal, since the system will be in thermal equilibrium; and (b) the total energy of the system is conserved. In particular, the second condition means that the amount of heat lost by the block is equal to the heat gained by the water—or vice versa if the water's initial temperature is higher.

Mathematically, we can write these conditions as follows:

$$Q_b + Q_w = 0 \qquad \textbf{16–14}$$

This means that the heat flow from the block is equal and opposite to the heat flow from the water; in other words, energy is conserved. If we write the heats Q in terms of the specific heats and temperatures, letting the final temperature be T, we have

$$m_b c_b (T - T_b) + m_w c_w (T - T_w) = 0$$

Note that for each heat the change in temperature is T_{final} minus $T_{initial}$, as it should be. Solving for the final temperature, T, we find

$$T = \frac{m_b c_b T_b + m_w c_w T_w}{m_b c_b + m_w c_w} \qquad \text{16–15}$$

This result need not be memorized, of course. Whenever solving a problem of this sort, one simply writes down energy conservation and solves for the desired unknown.

In some cases, we may wish to consider the influence of the container itself. This is illustrated in the following Active Example.

Heat Flow and Thermal Equilibriumn

To find the temperature of thermal equilibrium when two objects with different temperatures are brought into thermal contact, we simply use the idea that the heat that flows *out* of one of the objects flows *into* the other object.

ACTIVE EXAMPLE 16–2 **Canned Heat**

550 g of water at 32 °C is poured into a 210-g aluminum can with an initial temperature of 15 °C. Find the final temperature of the system, assuming no heat is exchanged with the surroundings.

Solution

1. Write an expression for the heat flow out of the water: $Q_w = m_w c_w (T - T_w)$
2. Write an expression for the heat flow into the aluminum: $Q_a = m_a c_a (T - T_a)$
3. Apply energy conservation: $Q_w + Q_a = 0$
4. Solve for the final temperature: $T = 31\ °C$

Insight

As one might expect from water's large specific heat, the final common temperature (T) is much closer to the initial temperature of the water (T_w) than that of the aluminum (T_a).

EXAMPLE 16–5 **Cooling Off**

A 0.500-kg block of metal with an initial temperature of 30.0 °C is dropped into a container holding 1.12 kg of water at 20.0 °C. If the final temperature of the block-water system is 20.4 °C, what is the specific heat of the metal? Assume the container can be ignored, and that no heat is exchanged with the surroundings.

Picture the Problem

Initially, when the block is first dropped into the water, the temperature of the block and water are T_b and T_w, respectively. When thermal equilibrium is established, both the block and the water have the same temperature T.

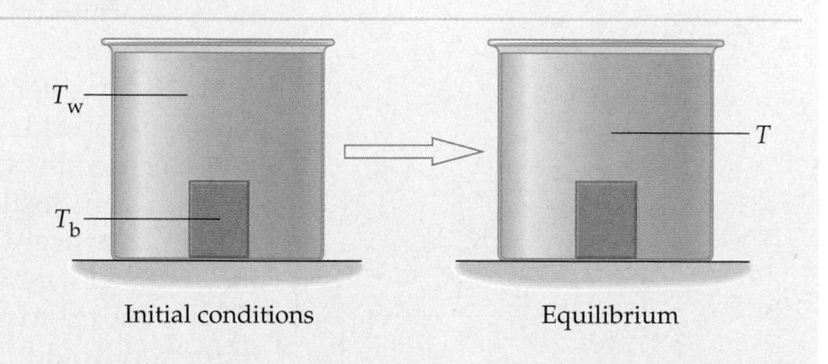

Strategy

Heat flows from the block to the water. Setting the heat flow *out of the block* plus the heat flow *into the water* equal to zero (conservation of energy) yields the block's specific heat.

Solution

1. Write an expression for the heat flow out of the block. Note that Q_{block} is negative, since T is less than T_b: $Q_{block} = m_b c_b (T - T_b)$
2. Write an expression for the heat flow into the water. Note that Q_{water} is positive, since T is greater than T_w: $Q_{water} = m_w c_w (T - T_w)$

continued on the following page

continued from the previous page

3. Set the sum of the heats equal to zero:

$$Q_{\text{block}} + Q_{\text{water}} = m_b c_b (T - T_b) + m_w c_w (T - T_w) = 0$$

4. Solve for the specific heat of the block, c_b:

$$c_b = \frac{m_w c_w (T - T_w)}{m_b (T_b - T)}$$

5. Substitute numerical values:

$$c_b = \frac{(1.12\ \text{kg})[4186\ \text{J}/(\text{kg}\cdot\text{K})](20.4\,°\text{C} - 20.0\,°\text{C})}{(0.500\ \text{kg})(30.0\,°\text{C} - 20.4\,°\text{C})}$$
$$= 391\ \text{J}/(\text{kg}\cdot\text{K})$$

Insight

We note from Table 16–2 that the block is probably made of copper.

In addition, note that the final temperature is much closer to the initial temperature of the water than to the initial temperature of the block. This is due in part to the fact that the mass of the water is about twice that of the block; more important, however, is the fact that the water's specific heat is more than 10 times greater than that of the block. In the following practice problem we set the mass of the water equal to the mass of the block, so that we can see clearly the effect of the different specific heats.

Practice Problem

If the mass of the water is also 0.500 kg, what is the equilibrium temperature? [**Answer:** $T = 20.9\,°\text{C}$. Still much closer to the water's initial temperature than to the block's.]

Some related homework problems: Problem 29, Problem 30

16–6 Conduction, Convection, and Radiation

Heat can be exchanged in a variety of ways. The Sun, for example, warms the Earth from across 93 million miles of empty space by a process known as radiation. As the sunlight strikes the ground and raises its temperature, the ground-level air gets warmer and begins to rise, producing a further exchange of heat by means of convection. Finally, if you walk across the ground in bare feet you will feel the warming effect of heat entering your body by conduction. In this section we consider each of these three mechanisms of heat exchange in detail.

Conduction

Perhaps the most familiar form of heat exchange is **conduction**, which is the flow of heat directly through a physical material. For example, if you hold one end of a metal rod and put the other end in a fire, it doesn't take long before you begin to feel warmth on your end. The heat you feel is transported along the rod by conduction.

Let's consider this observation from a microscopic point of view. To begin, when you placed one end of the rod into the fire the high temperature at that location caused the molecules to vibrate with an increased amplitude. These molecules in turn jostle their neighbors, and cause them to vibrate with greater amplitude as well. Eventually, the effect travels from molecule to molecule across the length of the rod, resulting in the macroscopic phenomenon of conduction.

If you were to repeat the experiment, this time with a wooden rod, the hot end of the rod would heat up so much that it might even catch on fire, but your end would still be comfortably cool. Thus, conduction depends on the type of material involved. Some materials conduct heat very well, whereas others are poor conductors. The latter are often referred to as **insulators**.

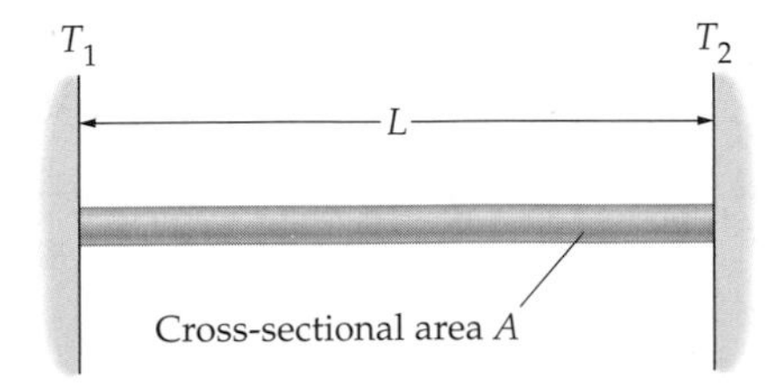

▲ FIGURE 16–9 Heat conduction through a rod
The amount of heat that flows through a rod of length L and cross-sectional area A per time is proportional to $A(T_2 - T_1)/L$.

Just how much heat flows as a result of conduction? To answer this question we consider the simple system shown in **Figure 16–9**. Here we show a rod of length L and cross-sectional area A, with one end at the temperature T_1 and the other at the temperature $T_2 > T_1$. Experiments show that the amount of heat Q that flows through this rod:

▲ Maintaining proper body temperature in an environment that is often too hot or too cold is a problem for many animals. When the sand is blazing hot, this lizard (left) keeps his contact with the ground to a minimum. By standing on two legs instead of four, he reduces conduction of heat from the ground to his body. Polar bears (right) have the opposite problem. The loss of precious body heat to their surroundings is retarded by their thick fur, which is actually made up of hollow fibers. Air trapped within these fibers provides enhanced insulation, just as it does in our thermal blankets and double-paned windows.

- increases in proportion to the rod's cross-sectional area, A;
- increases in proportion to the temperature difference, $\Delta T = T_2 - T_1$;
- increases steadily with time, t;
- decreases with the length of the rod, L.

Combining these observations in a mathematical expression gives:

Heat flow by conduction

$$Q = kA\left(\frac{\Delta T}{L}\right)t \qquad \text{16–16}$$

The constant k is referred to as the **thermal conductivity** of the rod. It varies from material to material, as indicated in Table 16–3.

PROBLEM SOLVING NOTE
Area and Length in Heat Conduction

When applying Equation 16–16, note that A is the area through which the heat flows. Thus, the plane of the area A is perpendicular to the direction of heat flow. The length L is the distance from the high-temperature side of an object to its low-temperature side.

CONCEPTUAL CHECKPOINT 16–3

You get up in the morning and walk barefoot from the bedroom to the bathroom. In the bedroom you walk on carpet, but in the bathroom the floor is tile. Does the tile feel **(a)** warmer, **(b)** cooler, or **(c)** the same temperature as the carpet?

Reasoning and Discussion

Everything in the house is at the same temperature, so it might seem that the carpet and the tile would feel the same. As you probably know from experience, however, the tile feels cooler. The reason is that tile has a much larger thermal conductivity than the carpet, which is actually a fairly good insulator. As a result, more heat flows from your skin to the tile than from your skin to the carpet. To your feet, then, it is as if the tile were much cooler than the carpet.

To get an idea of the thermal conductivities that would apply in this case, let's examine Table 16–3. For the tile, we might expect a thermal conductivity of roughly 0.84, the value appropriate for glass. For the carpet, the thermal conductivity might be as low as 0.04, the thermal conductivity of wool. Thus, the tile could have a thermal conductivity that is 20 times larger than that of the carpet.

Answer:

(b) The tile feels cooler.

Thermal conductivity is an important consideration when insulating a home. We consider some of these issues in the next Example.

TABLE 16–3
Thermal Conductivities

Substance	Thermal conductivity, k [W/(m·K)]
Silver	417
Copper	395
Gold	291
Aluminum	217
Steel, low carbon	66.9
Lead	34.3
Stainless steel—alloy 302	16.3
Ice	1.6
Concrete	1.3
Glass	0.84
Water	0.60
Asbestos	0.25
Wood	0.10
Wool	0.040
Air	0.0234

EXAMPLE 16–6 What a Pane

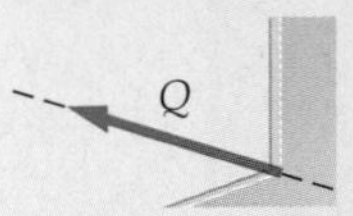

One of the windows in a house has the shape of a square 1.0 m on a side. The glass in the window is 0.50 cm thick. How much heat is lost through this window in one day if the temperature in the house is 21 °C and the temperature outside is 0.0 °C?

Picture the Problem

The glass from the window is shown in our sketch, along with its relevant dimensions. Heat flows from the 21 °C side of the window to the 0.0 °C side.

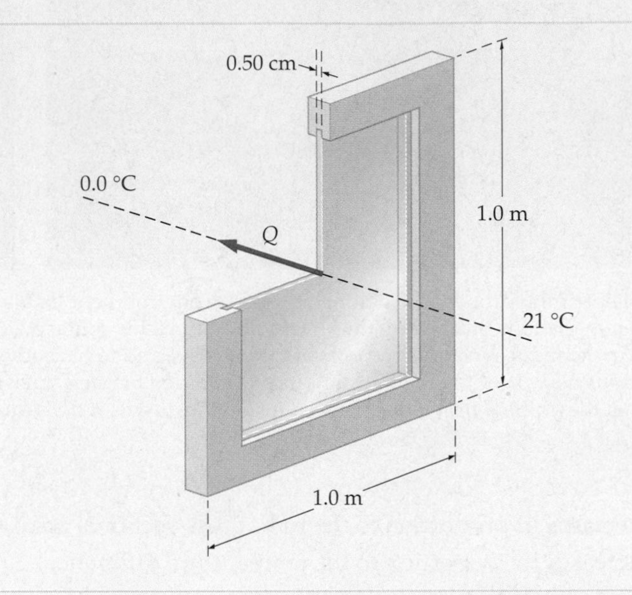

Strategy

The heat flow is given by Equation 16–16. Note that the area is $(1.0\text{ m})^2$ and that the length over which heat is conducted is, in this case, the thickness of the glass. Thus, $L = 0.0050$ m. The temperature difference is 21 C° = 21 K, and the thermal conductivity of glass (from Table 16–3) is 0.84 W/(m·K). (Recall from Section 7–4 that 1W = 1J/s.)

Solution

1. Calculate the heat flow for a given time, t:

$$Q = kA\left(\frac{\Delta T}{L}\right)t$$

$$= [0.84\text{ W/(m}\cdot\text{K)}](1.0\text{ m})^2\left(\frac{21\text{ K}}{0.0050\text{ m}}\right)t = (3500\text{ W})t$$

2. Substitute the number of seconds in a day, 86,400 s, for the time t in the expression for Q:

$$Q = (3500\text{ W})t = (3500\text{ W})(86{,}400\text{ s}) = 3.0 \times 10^8\text{ J}$$

Insight

This is a sizable amount of heat, roughly equivalent to the energy released in burning a gallon of gas.

Practice Problem

Suppose the window is replaced with a plate of solid silver. How thick must this plate be to have the same heat flow in a day as the glass? [**Answer:** The silver must have a thickness of $L = 2.5$ m.]

Some related homework problems: Problem 36, Problem 37

Now we consider the heat flow through a combination of two different materials with different thermal conductivities.

CONCEPTUAL CHECKPOINT 16–4

Two metal rods are to be used to conduct heat from a region at 100 °C to a region at 0 °C. The rods can be placed in parallel, as shown on the left, or in series, as on the right. Is the heat conducted in the parallel arrangement **(a)** greater than, **(b)** less than, or **(c)** the same as the heat conducted with the rods in series?

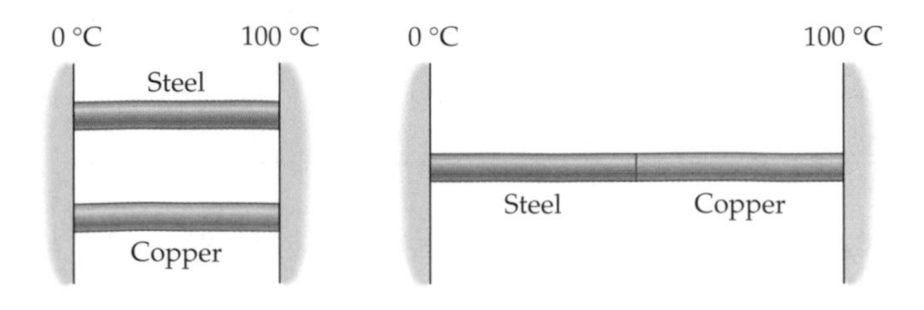

Reasoning and Discussion
The parallel arrangement conducts more heat for two reasons. First, the cross-sectional area available for heat flow is twice as large for the parallel rods. A greater cross-sectional area gives a greater heat flow—everything else being equal. Second, more heat flows through each rod in the parallel configuration because they both have the full temperature difference of 100 C° between their ends. In the series configuration each rod has a smaller temperature difference between its ends, so less heat flows.

Answer:
(a) More heat is conducted when the rods are in parallel.

In the next two Examples, we address numerical problems involving the same type of parallel and series arrangements considered in Conceptual Checkpoint 16–4.

EXAMPLE 16–7 Parallel Rods

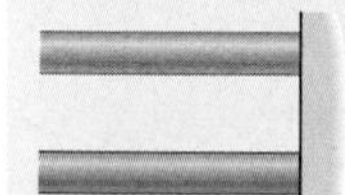

Two 0.525-m rods, one lead the other copper, are connected between metal plates held at 2.00 °C and 106 °C. The rods have a square cross section, 1.50 cm on a side. How much heat flows through the two rods in 1.00 s? Assume that no heat is exchanged between the rods and the surroundings, except at the ends.

Picture the Problem
The two rods, each 0.525 m long, are shown in the sketch. Note that both rods have a temperature difference of 104 C° = 104 K between their ends.

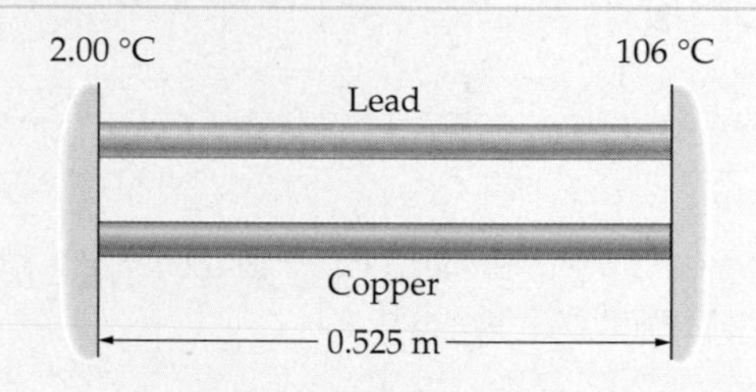

Strategy
The heat flowing through each rod can be calculated using Equation 16–16, and the value of k given in Table 16–3. The total heat flow is simply the sum of that calculated for each rod.

Solution

1. Calculate the heat flow in one second through the lead rod:

$$Q_l = k_l A\left(\frac{\Delta T}{L}\right)t$$
$$= [34.3\ \text{W}/(\text{m}\cdot\text{K})](0.0150\ \text{m})^2\left(\frac{104\ \text{K}}{0.525\ \text{m}}\right)(1.00\ \text{s})$$
$$= 1.53\ \text{J}$$

2. Calculate the heat flow in one second through the copper rod:

$$Q_c = k_c A\left(\frac{\Delta T}{L}\right)t$$
$$= [395\ \text{W}/(\text{m}\cdot\text{K})](0.0150\ \text{m})^2\left(\frac{104\ \text{K}}{0.525\ \text{m}}\right)(1.00\ \text{s})$$
$$= 17.6\ \text{J}$$

3. Sum the heats found in steps 1 and 2 to get the total heat:

$$Q_{\text{total}} = Q_l + Q_c$$
$$= 1.53\ \text{J} + 17.6\ \text{J} = 19.1\ \text{J}$$

Insight
The copper rod is by far the better conductor of heat. It is also a very good conductor of electricity, as we shall see in Chapter 21.

Practice Problem
What temperature difference would be required for the total heat flow in one second to be 15.0 J? [**Answer:** ΔT = 81.5 C° = 81.5 K]

Some related homework problems: Problem 39, Problem 40

As we shall see in the next Example, the results are different when the rods are connected in series.

EXAMPLE 16–8 Series Rods

The two rods in Example 16–7 are now placed in series. Find **(a)** the temperature at the lead-copper junction, and **(b)** the amount of heat that flows through the rods in 1.00 s. As before, assume that no heat is exchanged between the rods and the surroundings, except at the ends.

Picture the Problem

In this case, the rods are placed end-to-end. The temperature at the lead-copper junction is T.

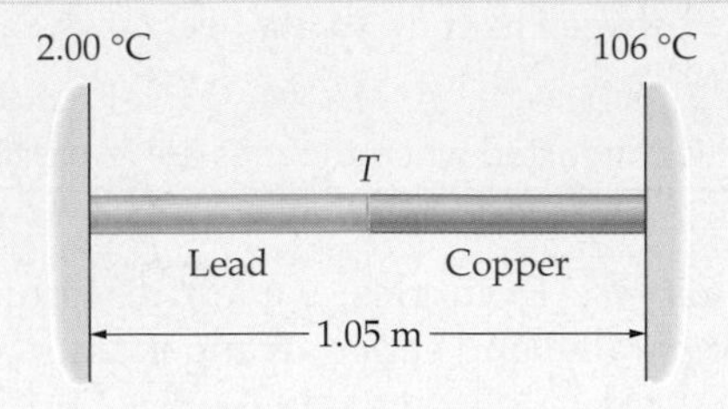

Strategy

(a) The basic idea in this problem is that the heat flow through the lead rod in one second must be the same as the heat flow through the copper rod in the same time. Setting these heat flows equal will determine T. Since lead has a smaller thermal conductivity, it must have a greater temperature difference to have the same heat flow. Thus, we expect T to be closer to 106 °C than to 2.00 °C.

(b) Once T is found, we use Equation 16–16 to find the heat flow.

Solution

Part (a)

1. Write an expression for the heat flow through the lead rod:

$$Q_l = k_l A\left(\frac{T - 2.00\,°\text{C}}{L}\right)t$$

2. Write an expression for the heat flow through the copper rod:

$$Q_c = k_c A\left(\frac{106\,°\text{C} - T}{L}\right)t$$

3. Set the heat flow in the lead equal to the heat flow in the copper. Cancel common terms:

$$k_l A\left(\frac{T - 2.00\,°\text{C}}{L}\right)t = k_c A\left(\frac{106\,°\text{C} - T}{L}\right)t$$

$$k_l(T - 2.00\,°\text{C}) = k_c(106\,°\text{C} - T)$$

4. Solve for T and substitute numerical values:

$$T = \frac{(106\,°\text{C})k_c + (2.00\,°\text{C})k_l}{k_c + k_l}$$

$$= \frac{(106\,°\text{C})[395\ \text{W}/(\text{m}\cdot\text{K})] + (2.00\,°\text{C})[34.3\ \text{W}/(\text{m}\cdot\text{K})]}{395\ \text{W}/(\text{m}\cdot\text{K}) + 34.3\ \text{W}/(\text{m}\cdot\text{K})}$$

$$= 97.7\,°\text{C}$$

Part (b)

5. Substitute T into the expression for heat flow through the lead rod:

$$Q_l = k_l A\left(\frac{T - 2.00\,°\text{C}}{L}\right)t$$

$$= [34.3\ \text{W}/(\text{m}\cdot\text{K})](0.0150\ \text{m})^2 \times \left(\frac{97.7\,°\text{C} - 2.00\,°\text{C}}{0.525\ \text{m}}\right)(1.00\ \text{s})$$

$$= 1.41\ \text{J}$$

6. As a check, substitute T into the expression for the heat flow through the copper rod. The heat flows should be equal:

$$Q_c = k_c A\left(\frac{106\,°\text{C} - T}{L}\right)t$$

$$= [395\ \text{W}/(\text{m}\cdot\text{K})](0.0150\ \text{m})^2 \times \left(\frac{106\,°\text{C} - 97.7\,°\text{C}}{0.525\ \text{m}}\right)(1.00\ \text{s})$$

$$= 1.41\ \text{J}$$

Insight

As expected, the heat flows are the same. Also, the temperature T is much closer to 106 °C than to 2.00 °C; thus, a temperature difference of 8.3 C° (106 °C − 97.7 °C) gives the same heat flow through copper as a temperature difference of 95.7 C° gives in lead.

Practice Problem

Repeat the problem for the case of an aluminum rod replacing the lead rod. [**Answer:** $T = 69.1\,°\text{C}$, $Q_{al} = Q_c = 6.24\ \text{J}$]

Some related homework problems: Problem 41, Problem 42

Note that the heat flow through the rods in series, 1.41 J, is much less than the heat flow through the rods in parallel, 19.1 J. This verifies the conclusion stated in Conceptual Checkpoint 16–4.

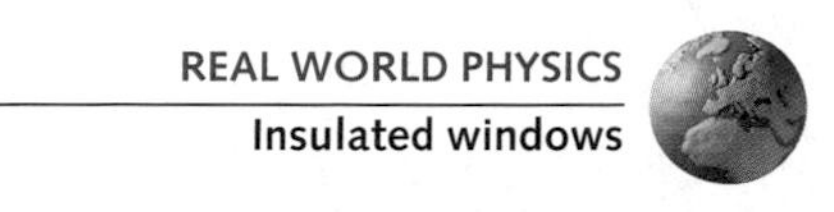

An application of thermal conductivity in series can be found in the *insulated window*. Most homes today have insulated windows as a means of increasing their energy efficiency. If you look closely at one of these windows you will see that it is actually constructed from two panes of glass separated by an air-filled gap. Thus, heat flows through three different materials in series as it passes into or out of a home. The fact that the thermal conductivity of air is about 40 times smaller than that of glass means that the insulated window results in significantly less heat flow than would be experienced with a single pane of glass.

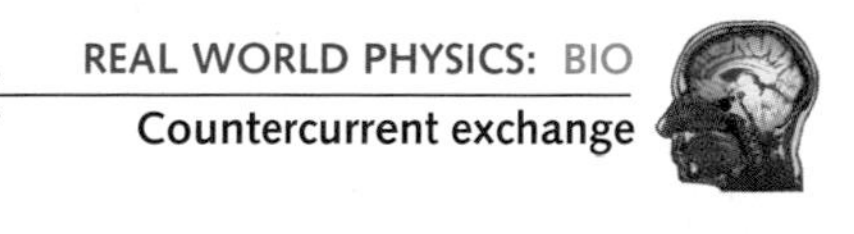

As a final example of conduction, we note that many biological systems transfer heat by a mechanism known as *countercurrent exchange*. Consider, for example, an egret or other wading bird standing in cool water all day. As warm blood flows through the arteries on its way to the legs and feet of the bird, it passes through constricted regions where the legs join to the body. Here, where the arteries and veins are packed closely together, the body-temperature arterial blood flowing into the legs transfers heat to the much cooler venous blood returning to the body. Thus, the counter-flowing streams of blood serve to maintain the core body temperature of the bird, while at the same time keeping the legs and feet at much cooler temperatures. The feet still receive the oxygen and nutrients carried by the blood, but they stay at a relatively low temperature to reduce the amount of heat lost to the water.

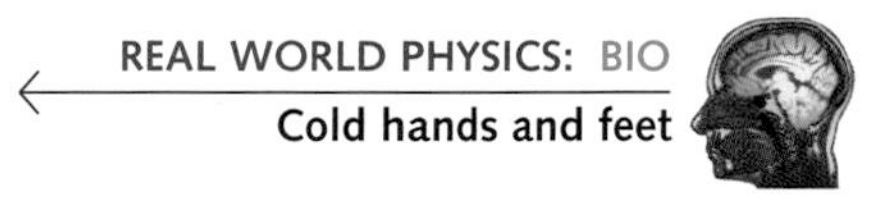

Similar effects occur in humans. It is common, for example, to hear complaints that a person's hands or feet are cold. There is good reason for this, since they are in fact much cooler than the core body temperature. Just as with the wading birds, the warm arterial blood flowing to the hands and feet exchanges heat with the cool venous blood flowing in the opposite direction (**Figure 16–10**). This helps to reduce the heat loss to our surroundings, and to maintain the desired temperature in the core of the body.

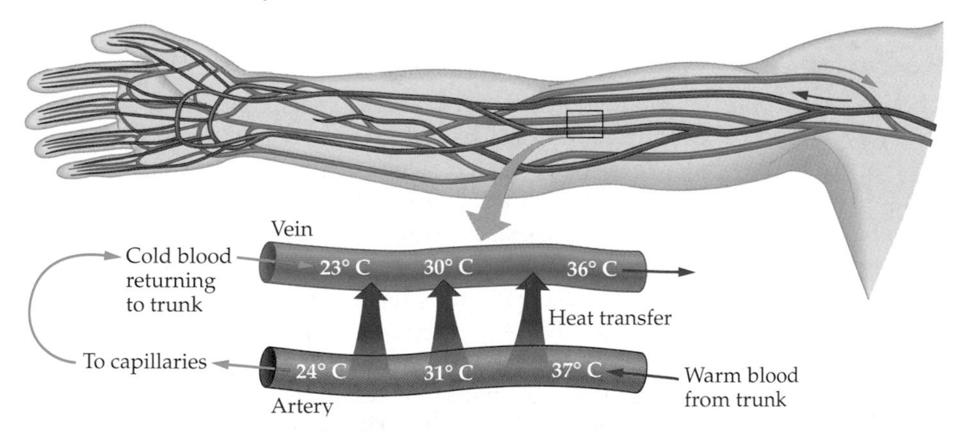

▲ **FIGURE 16–10 Countercurrent heat exchange in the human arm**
Arteries bringing warm blood to the limbs lie close to veins returning cooler blood to the body. This arrangement assures that a temperature gradient is maintained over the entire length that the vessels run parallel to one another, maximizing heat exchange between the warm arterial blood and the cooler venous blood.

▲ Many wading birds use countercurrent exchange in their circulatory systems. This mechanism allows them to keep the temperature of their legs well below that of their bodies. In this way they reduce the conductive loss of body heat to the water.

Convection

Suppose you want to heat a small room. To do so, you bring a portable electric heater into the room and turn it on. As the heating coils get red-hot they heat the air in their vicinity, and as this air warms, it expands, becoming less dense. Because of its lower density, the warm air rises, to be replaced by cold dense air descending from overhead. This sets up a circulating flow of air that transports heat from the heating coils to the air throughout the room. Heat exchange of this type is referred to as **convection**.

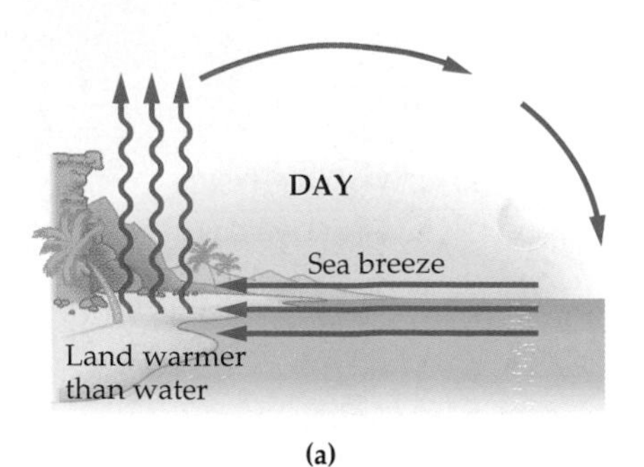

(a)

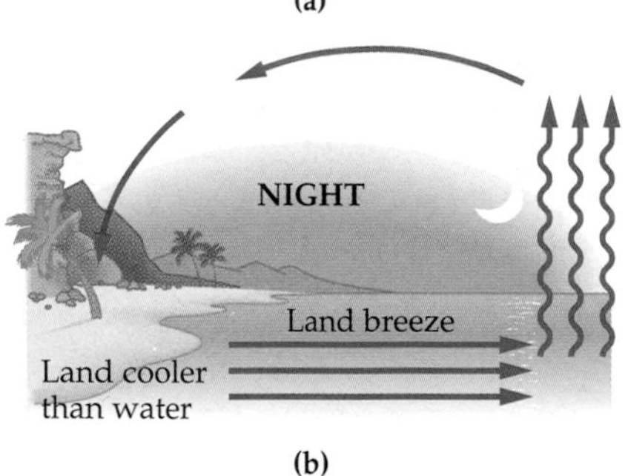

(b)

▲ **FIGURE 16–11 Alternating land and sea breezes**
(a) During the day, the sun warms the land more rapidly than the water. This is because the land, which is mostly rocks, has a lower specific heat than the water. The warm land heats the air above it, which becomes less dense and rises. Cooler air from over the water flows in to take its place, producing a "sea breeze." **(b)** At night, the land cools off more rapidly than the water—again because of its lower specific heat. Now it is the air above the relatively warm water that rises and is replaced by cooler air from over the land, producing a "land breeze."

◀ A dust devil is a column of whirling winds, like a much smaller and less dangerous version of a tornado. The winds are fed by convection—air warmed by the hot ground rises and is replaced by cooler air rushing in to take its place. As the swirling air is drawn closer to the center of the devil its speed increases, like an ice skater pulling in her arms during a spin (Section 11–7).

In general, convection occurs when a fluid is unevenly heated. As with the room heater, the warm portions of the fluid rise because of their lower density and the cool portions sink because of their higher density. Thus, in convection, temperature differences result in a flow of fluid. It is this physical flow of matter that carries heat throughout the system.

Convection occurs on an enormous range of length scales. For example, the same type of uneven heating produced by an electric heater in a room can occur in the atmosphere of the Earth as well. The common seashore occurrence of sea breezes during the day and land breezes in the evening is one such example. (See Figure 16–11 for an illustration of this effect.) On a larger scale, the Sun causes greater warming near the equator than near the poles; as a result warm equatorial air rises, cool polar air descends, and global convection patterns are established. Similar convection patterns occur in ocean waters; and plate tectonics is believed to be caused, at least in part, by convection currents in the Earth's mantle. The Sun also has convection currents, due to the intense heating that occurs in its interior, and disturbances in these currents are often visible as sunspots.

Radiation

Though convection and conduction occur primarily in specific situations, *all* objects give off energy as a result of **radiation**. The energy radiated by an object is in the form of electromagnetic waves (Chapter 25), which include visible light as well as infrared and ultraviolet radiation. Thus, unlike convection and conduction, radiation has no need for a physical material to mediate the energy transfer, since electromagnetic waves can propagate through empty space—that is, through a vacuum. Therefore, the heat you feel radiated from a hot furnace would reach you even if the air were suddenly removed—just as radiant energy from the Sun reaches the Earth across 150 million kilometers of vacuum.

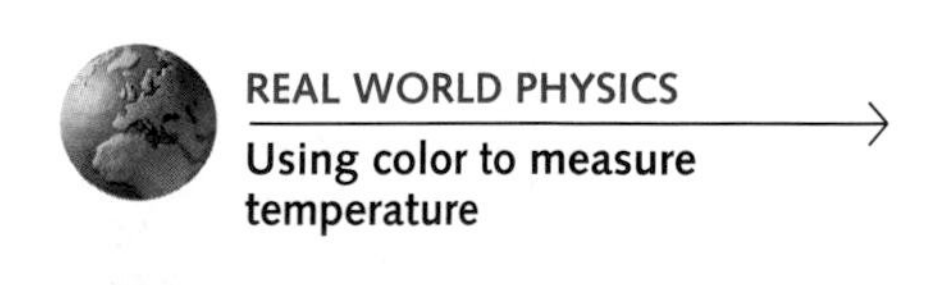

Since radiation can include visible light, it is often possible to "see" the temperature of an object. This is the physical basis of the **optical pyrometer**, invented by Josiah Wedgwood (1730–1795), the renowned English potter. When objects are about 800 °C they appear to be "red-hot." Examples include the heating coils in a range or oven. The filament in an incandescent lightbulb glows "white-hot" at about 3000 °C. For comparison, the surface of the Sun is about 6000 °C. Very hot stars, with surface temperatures in the vicinity of 20,000 to 30,000 °C, are "blue-hot" and actually appear bluish in the night sky. Rigel in the constellation Orion is an example of such a star.

The energy radiated per time by an object—that is, the radiated power, P—is proportional to the surface area, A, over which the radiation occurs. It also depends on the temperature of the object. In fact, the dependence is on the fourth power of the temperature, T^4, where T is the Kelvin-scale temperature. Thus, for instance, if T is doubled the radiated power increases by a factor of 16. All this behavior is contained in the **Stefan–Boltzmann law**:

▲ Red-hot volcanic lava is just hot enough (about 1000°C) to radiate in the visible range. Even when it cools enough to stop glowing it still emits energy, but most of it is in the form of invisible infrared radiation. Infrared radiation is also given off by the finned "radiators" attached to the supports of the Alaska pipeline. These fins are designed to prevent melting of the environmentally sensitive permafrost over which the pipeline runs. They function much like the radiator that cools your car engine, absorbing heat from the warm oil in the pipeline and dissipating it by radiation and convection into the atmosphere. (What design features can you see that facilitate this function?)

Stefan-Boltzmann Law for radiated power, P

$$P = e\sigma AT^4 \qquad \text{16–17}$$

SI unit: W

The constant σ in this expression is a fundamental physical constant, the **Stefan–Boltzmann constant**:

$$\sigma = 5.67 \times 10^{-8}\ \mathrm{W/(m^2 \cdot K^4)} \qquad \text{16–18}$$

The other constant in the Stefan–Boltzmann law is the **emissivity**, e. The emissivity is a dimensionless number between 0 and 1 that indicates how effective the object is in radiating energy. A value of 1 means that the object is a perfect radiator.

EXERCISE 16–4

Calculate the radiated power from a sphere with a radius of 5.00 cm at the temperature 355 K. Assume the emissivity is unity.

Solution

Using $A = 4\pi r^2$ for a sphere, we have

$$P = e\sigma AT^4 = (1)[5.67 \times 10^{-8}\ \mathrm{W/(m^2 \cdot K^4)}]4\pi(0.0500\ \mathrm{m})^2(355\ \mathrm{K})^4 = 28.3\ \mathrm{W}$$

Experiments show that objects absorb radiation from their surroundings according to the same law, the Stefan–Boltzmann law, by which they emit radiation. Thus, if the temperature of an object is T, and its surroundings are at the temperature T_s, the *net* power radiated by the object is

Net radiated power, P_{net}

$$P_{net} = e\sigma A(T^4 - T_s^4) \qquad \text{16–19}$$

SI unit: W

PROBLEM SOLVING NOTE
Radiated Power

To correctly calculate the radiated power, Equations 16–17 and 16–19, the temperatures must be expressed in the Kelvin scale.

If the object's temperature is greater than its surroundings it radiates more energy than it absorbs, and P_{net} is positive. On the other hand, if its temperature is less than the surroundings it absorbs more energy than it radiates, and P_{net} is negative. When the object has the same temperature as its surroundings it is in equilibrium, and the net power is zero.

EXAMPLE 16–9 Human Polar Bears

On New Year's Day, a group of human "polar bears" prepares for their annual dip into the icy waters of Narragansett Bay. One of these hardy souls has a surface area of 1.15 m^2 and a surface temperature of 303 K (~30 °C). Find the net radiated power from this person **(a)** in a dressing room where the temperature is 293 K (~20 °C), and **(b)** outside, where the temperature is 273 K (~0 °C). Assume an emissivity of 0.900 for the person's skin.

Picture the Problem

Our sketch shows the person radiating power in a room where the surroundings are at 293 K, and outside where the temperature is 273 K. The person also absorbs radiation from the surroundings; hence the net radiated power is greater when the surroundings are cooler.

Strategy

A straightforward application of Equation 16–19 applies to both parts (a) and (b).

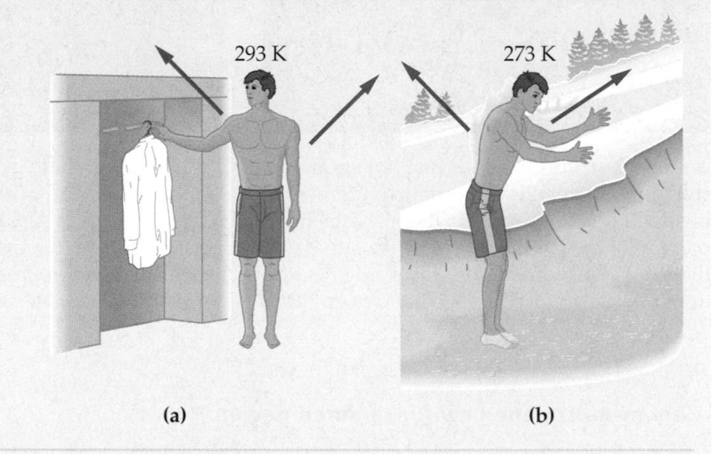

Solution

Part (a)

1. Calculate the net power using Equation 16–19 and $T_s = 293$ K:

$$P_{net} = e\sigma A(T^4 - T_s^4)$$
$$= (0.900)[5.67 \times 10^{-8}\ \text{W/(m}^2\cdot\text{K}^4)](1.15\ \text{m}^2) \times [(303\ \text{K})^4 - (293\ \text{K})^4]$$
$$= 62.1\ \text{W}$$

Part (b)

2. Calculate the net power using Equation 16–19 and $T_s = 273$ K:

$$P_{net} = e\sigma A(T^4 - T_s^4)$$
$$= (0.900)[5.67 \times 10^{-8}\ \text{W/(m}^2\cdot\text{K}^4)](1.15\ \text{m}^2) \times [(303\ \text{K})^4 - (273\ \text{K})^4]$$
$$= 169\ \text{W}$$

Insight

In the warm room the net radiated power is roughly that of a small lightbulb (about 60 W); outdoors the net radiated power has more than doubled, and is comparable to that of a 150-W lightbulb.

Practice Problem

What temperature must the surroundings have if the net radiated power is to be 155 W? [**Answer:** $T = 276\ \text{K} \approx 3\ °\text{C}$]

Some related homework problems: Problem 38, Problem 44

Note that the same emissivity e applies to both the emission and absorption of energy. Thus, a perfect emitter ($e = 1$) is also a perfect absorber. Such an object is referred to as a **blackbody**. As we shall see later, in Chapter 30, the study of blackbody radiation near the turn of the twentieth century ultimately lead to one of the most fundamental revolutions in science—the introduction of quantum physics.

The opposite of a blackbody is an ideal reflector, which absorbs *no* radiation ($e = 0$). It follows that an ideal reflector also radiates no energy. This is why the inside of a Thermos bottle is highly reflective. As an almost ideal reflector, the inside of the bottle radiates very little of the energy contained in the hot liquid that it holds. In addition to its shiny interior, a thermos bottle also has a vacuum between its inner and outer walls, as shown in **Figure 16–12**. This limits the flow of heat to radiation only, since convection and conduction cannot occur in a vacuum. This type of double-walled insulating container was invented by Sir James Dewar (1842–1923), a Scottish physicist and chemist.

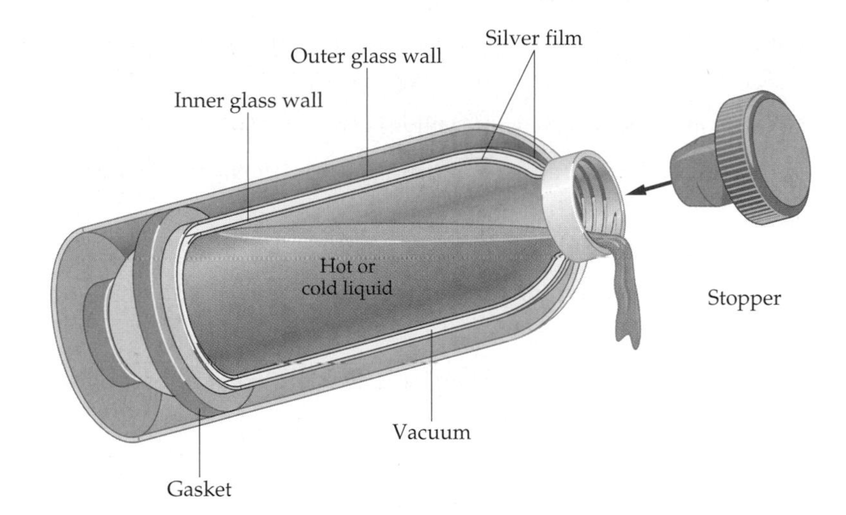

◀ **FIGURE 16–12 The Thermos bottle**
The hot or cold liquid stored in a Thermos bottle is separated from the outside world by a vacuum between the inner and outer walls. In addition, the inner wall has a reflective coating so that it is a good reflector and a poor radiator.

Chapter Summary

Topic	Remarks and Relevant Equations
16–1 Temperature and the Zeroth Law of Thermodynamics	This section defines several new terms dealing with heat and temperature.
heat	Heat is the energy transferred between objects because of a temperature difference.
thermal contact	Objects are in thermal contact if heat can flow between them.
thermal equilibrium	Objects that are in thermal contact, but have no heat exchange between them, are said to be in thermal equilibrium.
zeroth law of thermodynamics	If objects A and C are in thermal equilibrium with object B, they are in thermal equilibrium with each other.
temperature	Temperature is the quantity that determines whether or not two objects will be in thermal equilibrium.
16–2 Temperature Scales	Temperature is commonly measured in terms of several different scales.
Celsius scale	In the Celsius scale water freezes at 0 °C and boils at 100 °C.
Fahrenheit scale	In the Fahrenheit scale water freezes at 32 °F and boils at 212 °F.
absolute zero	The lowest temperature attainable is referred to as absolute zero. It is impossible to cool an object to a temperature lower than absolute zero, which is −273.15 °C.
Kelvin scale	In the Kelvin scale, absolute zero is 0 K. In addition, water freezes at 273.15 K and boils at 373.15 K. The degree size is the same for the Kelvin and Celsius scales.
conversion relations	The following relations convert between Celsius temperatures, T_C, Fahrenheit temperatures, T_F, and Kelvin temperatures, T:

$$T_F = \tfrac{9}{5}T_C + 32 \qquad \text{16–1}$$

$$T_C = \tfrac{5}{9}(T_F - 32) \qquad \text{16–2}$$

$$T = T_C + 273.15 \qquad \text{16–3}$$

16–3 Thermal Expansion

Most, though not all, substances expand when heated.

linear expansion

When an object of length L_0 is heated by the amount ΔT, its length increases by ΔL:

$$\Delta L = \alpha L_0 \Delta T \quad \text{16–4}$$

The constant α is the coefficient of linear expansion (Table 16–1).

volume expansion

When an object of volume V is heated by the amount ΔT, its volume increases by ΔV:

$$\Delta V = \beta V \Delta T \quad \text{16–6}$$

The constant β is the coefficient of volume expansion (Table 16–1).

relation between α and β

If β is not listed for a particular substance, but α is listed, the volume expansion can be calculated using

$$\beta = 3\alpha$$

special properties of water

Water is unusual in that it contracts when heated from 0 °C to 4 °C.

16–4 Heat and Mechanical Work

An important step forward in the understanding of heat was the recognition that it is a form of energy.

mechanical equivalent of heat

$$1 \text{ cal} = 4.186 \text{ J} \quad \text{16–8}$$

16–5 Specific Heats

heat capacity

The heat capacity of an object is the heat Q divided by the associated temperature change, ΔT:

$$C = \frac{Q}{\Delta T} \quad \text{16–11}$$

specific heat

The specific heat is the heat capacity per mass. Thus, the specific heat is independent of the quantity of a material in a given object:

$$c = \frac{Q}{m \Delta T} \quad \text{16–13}$$

conservation of energy

In a system in which no heat is exchanged with the surroundings, the heat that flows out of one object equals the heat that flows into a second object. This is energy conservation.

16–6 Conduction, Convection, and Radiation

This section considers three common mechanisms of heat exchange.

conduction

In conduction, heat flows through a material with no bulk motion. The heat flows as a result of the interactions of individual atoms with their neighbors.

If the thermal conductivity of a material is k, its cross-sectional area is A, and its length L, the heat exchanged in the time t is

$$Q = kA\left(\frac{\Delta T}{L}\right)t \quad \text{16–16}$$

convection

Convection is heat exchange due to the bulk motion of an unevenly heated fluid.

radiation

Radiation is the heat exchange due to electromagnetic radiation, such as infrared rays and light.

The energy per time, or power P, radiated by an object with a surface area A at the Kelvin temperature T is

$$P = e\sigma A T^4 \quad \text{16–17}$$

where e is the emissivity (a constant between 0 and 1) and σ is the Stefan–Boltzmann constant, $\sigma = 5.67 \times 10^{-8}\ \text{W}/(\text{m}^2 \cdot \text{K}^4)$.

Since an object will also absorb radiation from its surroundings at the temperature T_s, the net radiated power is

$$P_{net} = e\sigma A(T^4 - T_s^4) \quad \text{16–19}$$

Problem-Solving Summary

Type of Problem	Relevant Physical Concepts	Related Examples
Calculate the change in length of an object due to a change in temperature.	The change in an object's length is proportional to its initial length, L_0, and to the change in temperature, ΔT. In particular, $\Delta L = \alpha L_0 \Delta T$, where α is the coefficient of thermal expansion.	Exercise 16–2
Determine the change in volume of an object	If Table 16–1 gives a value of β for the substance in question, apply Equation 16–6. If a value of α is given instead, let $\beta = 3\alpha$ and use Equation 16–6.	Example 16–3
Relate the temperature change of an object to the amount of heat it absorbs or releases.	The heat capacity of an object determines the amount of heat, Q, it can absorb or give off for a given change in temperature, ΔT. The relationship is $Q = C\,\Delta T$.	Example 16–5, Active Example 16–2
Find the heat flow as a result of conduction.	Heat flow through a material is proportional to the temperature difference, the area through which the heat flows, and the time of flow; it is inversely proportional to the length L over which the heat flows. Specifically, $Q = kA\Delta T\, t/L$, where k is the thermal conductivity.	Examples 16–6, 16–7, 16–8
Calculate the power given off by radiation.	The power, P, radiated by an object is proportional to its area, A, and to the fourth power of its temperature, T. The complete expression is $P = e\sigma AT^4$, where e is the emissivity and σ is the Stefan–Boltzmann constant.	Example 16–9

Conceptual Questions

1. A cup of hot coffee is placed on the table. Is it in thermal equilibrium? What condition determines when the coffee is in equilibrium?
2. We know that $-40\,°F$ is the same as $-40\,°C$. Is there a temperature for which the Kelvin and Celsius scales agree? Explain.
3. To find the temperature at the core of the Sun you consult some Web sites on the Internet. One site says the temperature is about 15 million °C, another says it is 15 million kelvin. Is this a serious discrepancy? Explain.
4. Is it valid to say that a hot object contains more heat than a cold object?
5. Bimetallic strip 1 is made of copper and steel; bimetallic strip 2 is made of aluminum and steel. Which strip bends more for a given temperature change?
6. Suppose the glass in a glass thermometer expands more with temperature than the mercury it holds. What would happen to the mercury level as the temperature increased?
7. If the glass in a glass thermometer had the same coefficient of volume expansion as mercury, the thermometer would not be very useful. Explain.
8. When a mercury-in-glass thermometer is inserted into a hot liquid the mercury column first drops and then rises. Explain this behavior.
9. Which would be more accurate for all-season outdoor use, a tape measure made of steel or one made of aluminum? Explain.
10. Sometimes the metal lid on a glass jar has been screwed on so tightly that it is very difficult to open. Explain why holding the lid under hot running water often loosens it enough for easy opening.
11. A brass plate has a circular hole that is slightly smaller than the diameter of an aluminum ball. If the ball and the plate are always kept at the same temperature, should the temperature be increased or decreased in order for the ball to fit through the hole?
12. A steel tape measure is marked in such a way that it gives accurate length measurements at a normal room temperature of about 20 °C. If this tape measure is used outdoors on a cold day when the temperature is 0 °C, are its measurements **(a)** too long, **(b)** too short, or **(c)** accurate?
13. A pendulum is made from an aluminum rod with a mass attached to its free end. If the pendulum is cooled, does the pendulum's period increase, decrease, or stay the same?
14. Why do you hear creaking and groaning sounds in a house, particularly at night as the air temperature drops?
15. A copper ring stands on edge with a metal rod placed inside it, as shown in **Figure 16–13** (p. 528). As this system is heated, will the rod ever touch the top of the ring? Answer for the case of a rod that is made of **(a)** copper **(b)** aluminum, and **(c)** steel.
16. Referring to the copper ring in the previous problem, imagine that initially the ring is hotter than room temperature, and that an aluminum rod that is colder than room temperature fits snugly

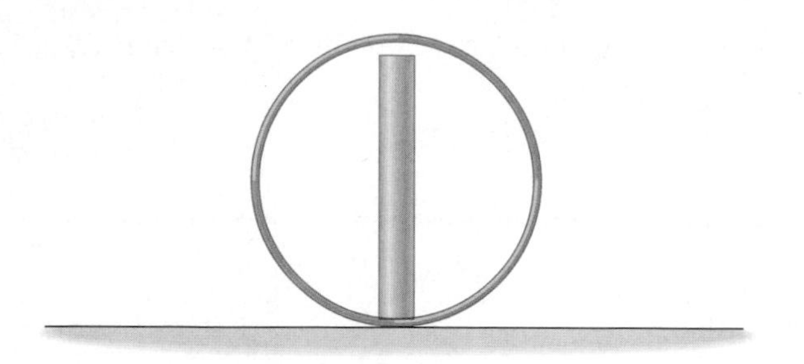

▲ FIGURE 16–13 Questions 15 and 16

inside the ring. When this system reaches thermal equilibrium at room temperature, is the rod firmly wedged in the ring, or can it be removed easily? Explain.

17. Two objects are made of the same material, but have different masses and temperatures. If the objects are brought into thermal contact, which object has the greater temperature change?
18. The specific heat of alcohol is about half that of water. Suppose you have 0.5 kg of alcohol at the temperature T_a in one container, and 0.5 kg of water at the temperature T_w in a second container. When these fluids are poured into the same container, and allowed to come to thermal equilibrium, is the final temperature closer to T_a or T_w? Explain.
19. Two different objects receive the same amount of heat. Give at least two reasons why their temperature changes may not be the same.
20. When you touch a piece of metal and a piece of wood that are both at room temperature the metal feels cooler. Why?
21. The specific heat of concrete is greater than that of soil. Given this fact, would you expect a major-league baseball field or the parking lot that surrounds it to cool off more in the evening following a sunny day?
22. Extending the result of the previous question to a larger scale, would you expect daytime winds to generally blow from a city to the surrounding suburbs or from the suburbs to the city? Explain.
23. After lighting a wooden match you can hold onto the end of it for some time, until the flame almost reaches your fingers. Why aren't you burned as soon as the match is lit?
24. In a popular lecture demonstration, a sheet of paper is wrapped around a rod that is made from wood on one half and metal on the other half. If held over a flame, the paper on one half of the rod is burned while the paper on the other half is unaffected. Which half of the rod has the burned paper? Explain.
25. If a lighted match is held beneath a balloon inflated with air, the balloon quickly bursts. If, instead, the lighted match is held beneath a balloon filled with water, the balloon remains intact, even if the flame comes in contact with the balloon. Explain.
26. The rate of heat flow through a slab does *not* depend on which of the following? **(a)** The temperature difference between opposite faces of the slab. **(b)** The thermal conductivity of the slab. **(c)** The thickness of the slab. **(d)** The cross-sectional area of the slab. **(e)** The specific heat of the slab.
27. Figure 16–14 shows a composite slab of three different materials with equal thickness but different thermal conductivities. The

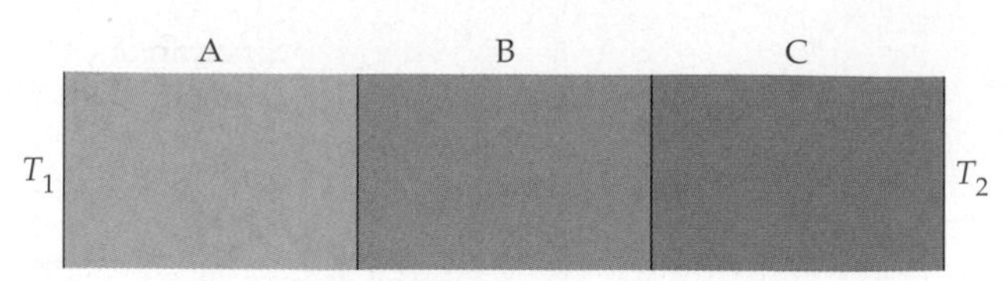

▲ FIGURE 16–14 Question 27

opposite sides of the composite slab are held at the fixed temperatures T_1 and T_2. Given that $k_B > k_A > k_C$, list the materials in order of the temperature difference across them, starting with the smallest.

28. Updrafts of air allow hawks and eagles to glide effortlessly, all the while gaining altitude. What causes the updrafts?
29. On a sunny day identical twins wear different shirts. One wears a dark shirt, the other a light colored shirt. Which twin has the warmer shirt?
30. Two bowls of soup with identical temperatures are placed on a table. One bowl has a metal spoon in it, the other does not. After a few minutes, how do the soup temperatures compare?
31. Hot tea is poured from the same pot into two identical mugs. Mug 1 is filled to the brim; mug 2 is filled only halfway. Which mug cools off more rapidly?
32. Explain why the following warning sign is often appropriate: "Caution: Bridge surface freezes before road surface."
33. The SR-71 reconnaissance aircraft is the fastest plane in the world. It is called "Blackbird" because of its distinctive black paint job (see Problem 10 on page 529). When in flight, air resistance makes the skin of the airplane too hot to touch. If the SR-71 were painted white instead, would its in-flight temperature be greater than, less than, or the same as with black paint? Explain.
34. The fur of polar bears consists of hollow fibers. (Sometimes algae will grow in the hollow regions, giving the fur a green cast.) Explain why hollow hairs can be beneficial to the polar bears.
35. Can the heat radiated by a group of people cause the temperature of a room to rise?
36. If people huddle together they will be warmer than if they are well separated. Explain.
37. Object 2 has twice the emissivity of object 1, though they have the same size and shape. If the two objects radiate the same power, what is the ratio of their Kelvin temperatures?
38. Though it may not be particularly appetizing, a potato will bake more rapidly if you poke a nail into it. Why?
39. An astronaut in the space shuttle conducts an experiment with two aluminum blocks of identical shape, size, and mass. One block is painted black, the other is painted white. Initially, inside the shuttle, both blocks are at a temperature of 300 K. The astronaut now ejects the blocks into space. According to thermometer probes installed in each of the blocks, which block cools off faster?

Problems

Note: **IP** *denotes an integrated conceptual/quantitative problem.* **BIO** *identifies problems of biological or medical interest. Blue bullets (•, ••, •••) are used to indicate the level of difficulty of each problem.*

Section 16–2 Temperature Scales

1. • Normal body temperature for humans is 98.6 °F. What is the corresponding temperature in **(a)** degrees Celsius and **(b)** kelvins?
2. • What is 1.0 K on the Fahrenheit scale?
3. • The temperature at the surface of the Sun is about 6000 K. Convert this temperature to the **(a)** Celsius and **(b)** Fahrenheit scales.

4. • One day you notice that the outside temperature increased by 25 F° between your early morning jog and your lunch at noon. What is the corresponding change in temperature in the **(a)** Celsius and **(b)** Kelvin scales?

5. • The gas in a constant-volume gas thermometer has a pressure of 95.0 kPa at 103 °C. **(a)** What is the pressure of the gas at 50.0 °C? **(b)** At what temperature does the gas have a pressure of 115 kPa?

6. • A constant-volume gas thermometer has a pressure of 80.3 kPa at −10.0 °C and a pressure of 86.4 kPa at 10.0 °C. **(a)** At what temperature is the pressure of this system equal to zero? **(b)** What are the pressures at the freezing and boiling points of water?

7. •• A world record for the greatest change in temperature was set in Spearfish, S. D., on January 22, 1943. At 7:30 a.m. the temperature was −4.0 °F; two minutes later the temperature was 45 °F. Find the rate of temperature change that day in kelvins per second.

8. •• We know that −40 °C corresponds to −40 K. What temperature has the same value in both the Fahrenheit and Kelvin scales?

9. •• When the bulb of a constant-volume gas thermometer is placed in a beaker of boiling water at 100 °C, the pressure of the gas is 227 mmHg. When the bulb is moved to an ice-salt mixture the pressure of the gas drops to 162 mmHg. Assuming ideal behavior, as in Figure 16–3, what is the Celsius temperature of the ice-salt mixture?

Section 16–3 Thermal Expansion

10. • The SR-71 Blackbird, which is 107 feet 5 inches long, is the world's fastest airplane. It can fly at three times the speed of sound (Mach 3) at altitudes of 80,000 ft. When it lands after a long flight it is too hot to be touched for about 30 minutes, and is 6.0 inches longer than at takeoff. How hot is the Blackbird when it lands, assuming its coefficient of linear expansion is $24 \times 10^{-6}\ \text{K}^{-1}$ and its temperature at takeoff if 23 °C?

▲ The SR-71 Blackbird reconnaissance aircraft can reach speeds of 2350 mph. (Problem 10)

11. • The world's longest suspension bridge is the Akashi Kaikyo Bridge in Japan. The bridge is 3910 m long and is constructed of steel. How much longer is the bridge on a warm summer day (30.0 °C) than on a cold winter day (−5.00 °C)?

12. •• A hole in an aluminum plate has a diameter of 1.178 cm at 23.00 °C. **(a)** What is the diameter of the hole at 199.0 °C? **(b)** At what temperature is the diameter of the hole equal to 1.176 cm?

13. •• **IP** It is desired to slip an aluminum ring over a steel bar (**Figure 16–15**). At 10.00 °C the inside diameter of the ring is 4.000 cm and the diameter of the rod is 4.040 cm. **(a)** In order for the ring to slip over the bar, should it be heated or cooled? Explain. **(b)** Find the temperature of the ring at which it fits over the bar.

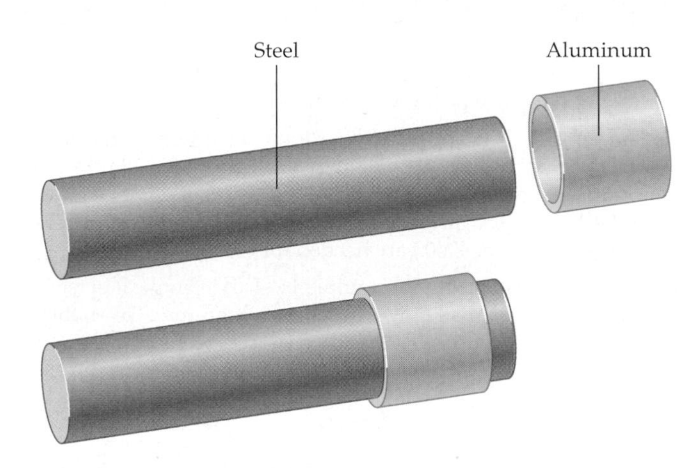

▲ **FIGURE 16–15** Problem 13

14. •• At 12.25 °C a brass sleeve has an inside diameter of 2.196 cm and a steel shaft has a diameter of 2.199 cm. It is desired to shrink-fit the sleeve over the steel shaft. **(a)** To what temperature must the sleeve be heated in order for it to slip over the shaft? **(b)** Alternatively, to what temperature must the shaft be cooled before it is able to slip through the sleeve?

15. •• Early in the morning, when the temperature is 5.0 °C, gasoline is pumped into a car's 51-L steel gas tank until it is filled to the top. Later in the day the temperature rises to 25 °C. Since the volume of gasoline increases more for a given temperature increase than the volume of the steel tank, gasoline will spill out of the tank. How much gasoline spills out in this case?

16. •• Some cookware has a stainless steel interior ($\alpha = 17.3 \times 10^{-6}\ \text{K}^{-1}$) and a copper bottom ($\alpha = 17.0 \times 10^{-6}\ \text{K}^{-1}$) for better heat distribution. Suppose an 8.0-inch pot of this construction is heated to 610 °C on the stove. If the initial temperature of the pot is 22 °C, what is the difference in diameter change for the copper and the steel?

17. •• **IP** You construct two wire-frame cubes, one using copper wire, the other using aluminum wire. At 23 °C the cubes enclose equal volumes. **(a)** If the temperature of the cubes is increased, which cube encloses the greater volume? **(b)** Find the difference in volume between the cubes when their temperature is 95 °C.

18. •• A copper ball with a radius of 1.3 cm is heated until its diameter has increased by 0.20 mm. Assuming a room temperature of 22 °C, find the final temperature of the ball.

19. ••• **IP** An aluminum saucepan with a diameter of 23 cm and a height of 6.0 cm is filled to the brim with water. The initial temperature of the pan and water is 19 °C. The pan is now placed on a stove burner and heated to 88 °C. **(a)** Will water overflow from the pan, or will the water level in the pan decrease? Explain. **(b)** Calculate the volume of water that overflows, or the drop in water level in the pan, whichever is appropriate.

Section 16–4 Heat and Mechanical Work

20. • **BIO** When people sleep, their metabolic rate is about $2.6 \times 10^{-4}\ \text{C/(s·kg)}$. How many Calories does a 75-kg person metabolize while getting a good night's sleep of 8.0 hr?

21. •• **BIO** An exercise machine indicates that you have worked off 2.5 Calories in a minute-and-a-half of running in place. What was your power output during this time? Give your answer in both watts and horsepower.

22. •• **BIO** During a workout, a person repeatedly lifts a 15-lb barbell through a distance of 1.5 ft. How many "reps" of this lift are required to burn off 120 C?

23. •• Consider the apparatus that Joule used in his experiments on the mechanical equivalent of heat, shown in Figure 16–9. Suppose the block has a mass of 0.75 kg and that it falls through a distance of 0.50 m. Find the expected rise in temperature of the water given that 6200 J are needed for every 1.0 C° increase.

24. •• **BIO** It was shown in Example 16–9 that a typical person radiates about 62 W of power at room temperature. Given this result, how long does it take for a person to radiate away the energy acquired by consuming a 230-Calorie donut?

Section 16–5 Specific Heats

25. • 53.0 J of heat is added to a 108-g piece of aluminum at 25.0 °C. What is the final temperature of the aluminum?

26. • How much heat is required to raise the temperature of a 55-g glass ball by 15 C°?

27. • Estimate the heat required to heat a 0.13-kg apple from 15 °C to 35 °C. (Assume the apple is mostly water.)

28. • A 5.0-g lead bullet is fired into a fence post. The initial speed of the bullet is 250 m/s, and when it comes to rest half its kinetic energy goes into heating the bullet. How much does the bullet's temperature increase?

29. •• 1.0-g lead pellets at 75 °C are to be added to 180 g of water at 22 °C. How many pellets are needed to increase the equilibrium temperature to 25 °C?

30. •• A 223-g lead ball at a temperature of 83.2 °C is placed in a light calorimeter containing 178 g of water at 24.5 °C. Find the equilibrium temperature of the system.

31. •• If 2200 J of heat are added to a 190-g object its temperature increases by 12 C°. **(a)** What is the heat capacity of this object? **(b)** What is the object's specific heat?

32. •• An 87.6-g lead ball is dropped from rest from a height of 5.43 m. The collision between the ball and the ground is totally inelastic. Assuming all the ball's kinetic energy goes into heating the ball, find its change in temperature.

33. •• To determine the specific heat of an object, a student heats it to 100 °C in boiling water. She then places the 38.0-g object in a 155-g aluminum calorimeter containing 103 g of water. The aluminum and water are initially at a temperature of 20.0 °C, and are thermally insulated from their surroundings. If the final temperature is 22.0 °C, what is the specific heat of the object. Referring to Table 16–2, identify the material in the object.

34. •• A blacksmith drops a 0.50-kg horseshoe into a bucket containing 25 kg of water. If the initial temperature of the horseshoe is 450 °C, and the initial temperature of the water is 23 °C, what is the equilibrium temperature of the system? Assume no heat is exchanged with the surroundings.

35. •• The ceramic coffee cup in **Figure 16–16**, with $m = 116$ g and $c = 1090$ J/(kg·K), is initially at room temperature (24.0 °C). If 225 g of 80.3 °C coffee and 12.2 g of 5.00 °C cream are added to the cup, what is the equilibrium temperature of the system? Assume that no heat is exchanged with the surroundings, and that the specific heat of coffee and cream are the same as the specific heat of water.

Section 16–6 Convection, Conduction, and Radiation

36. • A glass window 0.60 cm thick measures 75 cm by 42 cm. How much heat flows through this window per minute if the inside and outside temperatures differ by 15 C°?

▲ **FIGURE 16–16** Problem 35

37. • Find the heat that flows in 1 s through a lead brick 15 cm long if the temperature difference between the ends of the brick is 8.5 C°. The cross-sectional area of the brick is 12 cm^2.

38. • **BIO** Assuming your skin temperature is 37.2 °C and the temperature of your surroundings is 21.8 °C, determine the length of time required for you to radiate away the energy gained by eating a 306-Calorie ice cream cone. Let the emmissivity of your skin be 0.915 and its area be 1.22 m^2.

39. •• Two metal rods of equal length, one aluminum the other stainless steel, are connected in parallel with a temperature of 10.0 °C at one end and 108 °C at the other end. Both rods are 2.50 cm in diameter. Determine the length the rods must have if the combined rate of heat flow through them is to be 27.5 J/s.

40. •• Two metal rods, one copper, the other lead, are connected in parallel with a temperature of 21.0 °C at one end and 112 °C at the other end. Both rods are 0.750 m in length, and the lead rod is 2.60 cm in diameter. If the combined rate of heat flow through the two rods is 33.2 J/s, what is the diameter of the copper rod?

41. •• **IP** Two metal rods, one aluminum the other copper, are connected in series with a temperature of 0.0 °C at the aluminum end and 110 °C at the copper end. **(a)** If the rods are of equal length, is the temperature at the aluminum-copper junction greater than, less than, or equal to 55 °C? Explain. **(b)** Calculate the temperature at the aluminum-copper interface.

42. •• Two metal rods, one lead the other copper, are connected in series with a temperature of 20.0 °C at the lead end and 80.0 °C at the copper end. The copper rod is 0.750 m long. Determine the length the lead rod must have if the temperature at the lead-copper interface is to be 50.0 °C.

43. •• A copper rod 95 cm long is used to poke a fire. The hot end of the rod is maintained at 87 °C and the cool end has a constant temperature of 24 °C. What is the temperature of the rod 23 cm from the cool end?

44. •• Two identical objects are placed in a room at 23 °C. Object 1 has a temperature of 85 °C and object 2 has a temperature of 35 °C. What is the ratio of the net power emitted by object 1 to that radiated by object 2?

45. ••• A block has the dimensions L, $2L$, and $3L$. When one of the $L \times 2L$ faces is maintained at the temperature T_1 and the other $L \times 2L$ face is held at the temperature T_2 the rate of heat conduction through the block is P. Answer the following questions in terms of P. **(a)** What is the rate of heat conduction in this block if one of the $L \times 3L$ faces is held at the temperature T_1 and the other $L \times 3L$ face is held at the temperature T_2? **(b)** What is the rate of heat conduction in this block if one of the $2L \times 3L$ faces

is held at the temperature T_1 and the other $2L \times 3L$ face is held at the temperature T_2?

General Problems

46. • In the continuous-caster process, steel sheets 25.4 cm thick, 2.03 m wide, and 10.0 m long are produced at a temperature of 872 °C. What are the dimensions of a steel sheet once it has cooled to 20.0 °C?

47. • Two objects at the same initial temperature absorb equal amounts of heat. If the final temperature of the objects is different, it may be because they differ in which of the following properties: **(a)** mass, **(b)** coefficient of expansion, **(c)** thermal conductivity, or **(d)** specific heats. Explain.

48. •• If heat is transferred to 150 g of water at a constant rate for 2.5 min its temperature increases by 13 C°. When heat is transferred at the same rate for the same amount of time to a 150-g object of unknown material its temperature increases by 61 C°. **(a)** From what material is the object made? **(b)** What is the heating rate?

49. •• **IP** A pendulum consists of a large weight suspended by a steel wire that is 0.9500 m long. **(a)** If the temperature increases, does the period of the pendulum increase, decrease, or stay the same? Explain. **(b)** Calculate the change in length of the pendulum if the temperature increase is 150.0 C°. **(c)** Calculate the period of the pendulum before and after the temperature increase. (Assume that the coefficient of linear expansion for the wire is $12.00 \times 10^{-6}\ \text{K}^{-1}$.)

50. •• **IP** Once the aluminum ring in Problem 13 is slipped over the bar, the ring and bar are allowed to equilibrate at a temperature of 22 °C. The ring is now stuck on the bar. **(a)** If the temperature of both the ring and the bar are changed together, should the system be heated or cooled to remove the ring? **(b)** Find the temperature at which the ring can be removed.

51. •• A steel plate has a circular hole with a diameter of 1.000 cm. In order to drop a glass marble 1.003 cm in diameter through the hole in the plate, how much must the temperature of the system be raised? (Assume the plate and the marble are always at the same temperature.)

52. •• A 206-kg rock sits in full sunlight on the edge of a cliff 5.00 m high. The temperature of the rock is 30.2 °C. If the rock falls from the cliff into a pool containing 6.00 m^3 of water at 15.5 °C, what is the final temperature of the rock-water system? Assume that the specific heat of the rock is 1010 J/(kg·K).

53. •• Water going over Iguacu Falls on the border of Argentina and Brazil drops through a height of about 72 m. Suppose that all the gravitational potential energy of the water goes into raising its temperature. Find the increase in water temperature at the bottom of the falls as compared with the top.

54. •• **IP** A 0.20-kg steel pot on a stove contains 1.2 L of water at 21 °C. When the burner is turned on, the water begins to boil after 7.5 minutes. **(a)** At what rate is heat being transferred from the burner to the pot of water? **(b)** At this rate of heating, would it take more time or less time for the water to start to boil if the pot were made of gold rather than steel?

55. •• **BIO** Suppose you could convert the 525 Calories in the cheeseburger you ate for lunch into mechanical energy with 100% efficiency. **(a)** How high could you throw a 0.145-kg baseball with the energy contained in the cheeseburger? **(b)** How fast would the ball be moving at the moment of release?

56. •• You turn a crank on a device similar to that shown in Figure 16–8 and produce a power of 0.12 hp. If the paddles are immersed in 0.65 kg of water, how long must you turn the crank to increase the temperature of the water by 5.0 C°?

57. ••• Bars of two different metals are bolted together, as shown in **Figure 16–17**. Show that the distance D does not change with temperature if the length of the two bars have the following ratio: $L_A/L_B = \alpha_B/\alpha_A$.

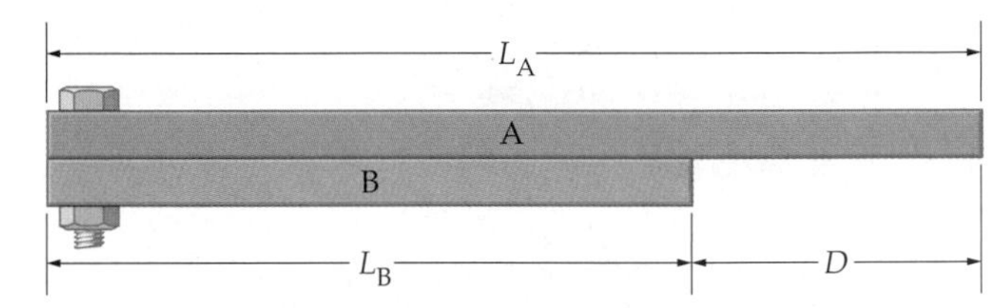

▲ **FIGURE 16–17** Problem 57

58. ••• A grandfather clock has a simple brass pendulum of length L. One night, the temperature in the house is 23.0 °C and the period of the pendulum is 1.00 s. The clock keeps correct time at this temperature. If the temperature in the house quickly drops to 19.1 °C just after 10 p.m., and stays at that value, what is the actual time when the clock indicates that it is 10 a.m. the next morning?

59. ••• **IP** A sheet of aluminum has a circular hole with a diameter of 10.0 cm. A 9.99 cm long steel rod is placed inside the hole, along a diameter of the circle, as shown in **Figure 16–18**. It is desired to change the temperature of this system until the steel rod just touches both sides of the circle. **(a)** Should the temperature of the system be increased or decreased? Explain. **(b)** By how much should the temperature be changed?

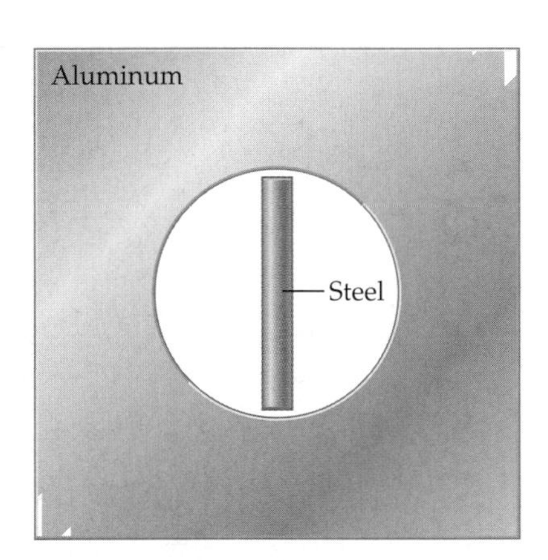

▲ **FIGURE 16–18** Problem 59

60. ••• A layer of ice has formed on a small pond. The air just above the ice is at −5.4 °C, the water-ice interface is at 0 °C, and the water at the bottom of the pond is at 4.0 °C. If the total depth from the top of the ice to the bottom of the pond is 1.4 m, how thick is the layer of ice? *Note*: The thermal conductivity of ice is 1.6 W/(m·C°) and that of water is 0.60 W/(m·C°).

61. ••• An energy-efficient double-paned window consists of two panes of glass, each with thickness L_1 and thermal conductivity k_1, separated by a layer of air of thickness L_2 and thermal conductivity k_2. Show that the equilibrium rate of heat flow through this window per unit area, A, is

$$\frac{Q}{At} = \frac{(T_2 - T_1)}{2L_1/k_1 + L_2/k_2}$$

In this expression, T_1 and T_2 are the temperatures on either side of the window.

17 Phases and Phase Changes

Though many people are surprised to learn that a liquid can boil and freeze at the same time, it is in fact true. At a unique combination of temperature and pressure called the triple point, *all three phases of matter—solid, liquid, and gas—coexist in equilibrium. In this chapter we'll learn more about the three phases of matter, their differences and similarities, and what happens when a substance changes from one phase to another.*

The matter we come into contact with every day is in one of three forms: solid, liquid, or gas. The air we breathe is a gas, the water we drink is a liquid, and the salt on our popcorn is a crystalline solid. In this chapter we consider these three *phases* of matter in detail.

We begin by considering the behavior of a gas, which is the easiest phase to describe physically. In particular, we show that an "ideal gas"—one that has no interactions between its molecules—is a good approximation to real gases. In addition, we show how the kinetic theory of gases allows us to relate the temperature of a substance to the kinetic energy of its molecules.

Next, we discuss some of the mechanical properties associated with the solid phase of matter. In particular, we explore the relationship between a force applied to a solid and the resulting deformation. For example, we shall determine the amount of stretch in your arm bones when you carry a heavy suitcase through an airport.

Finally, we turn our attention to the behavior of a substance when it changes from one phase to another—as when solid ice melts to form liquid water or water boils to produce gaseous clouds of steam. Remarkably, when a material is changing phase the addition of heat does not result in a higher temperature. A significant amount of heat must first be absorbed by the material so that it can complete its phase change. Only after the change is accomplished can its temperature begin to rise again.

17–1 Ideal Gases

In Chapter 16 we discussed the thermal behavior of gases. In particular, we saw that the pressure of a constant volume of gas decreases linearly with decreasing temperature over a wide range of temperatures. Eventually, however, when the temperature is low enough, real gases change to liquid and then solid form, and their behavior changes. These changes are due to interactions between the molecules in a gas. Hence, the weaker these interactions the wider the range of temperatures over which the simple, linear gas behavior persists. We now consider a simplified model of a gas, the **ideal gas**, in which intermolecular interactions are vanishingly small.

Though ideal gases do not actually exist in nature, the behavior of real gases is generally quite well approximated by that of an ideal gas. By studying the simple "ideal" case, we can gain considerable insight into the workings of a real gas. This is similar to the kind of idealizations we made in mechanics. For example, we often considered a surface to be perfectly frictionless, when real surfaces have at least some friction, and we imagined springs to obey Hooke's law exactly, though real springs show some deviations. The ideal gas plays an analogous role in our study of thermodynamics.

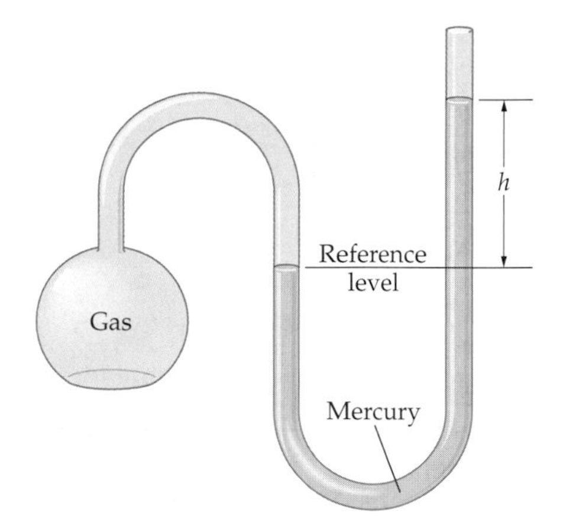

▲ **FIGURE 17–1 A constant-volume gas thermometer**
A device like this can be used as a thermometer. Note that both the volume of the gas, and the number of molecules in the gas, remain constant.

Equation of State

We can describe the way the pressure, P, of an ideal gas depends on temperature, T, number of molecules, N, and volume, V, from just a few simple observations. First, imagine holding the number of molecules and the volume of a gas constant, as in the constant-volume gas thermometer shown in **Figure 17–1**. As we have already noted, the pressure of a gas under these conditions varies linearly with temperature. Therefore, we conclude the following:

$$P = (\text{constant})T$$

(fixed volume, V; fixed number of molecules, N)

In this expression the constant multiplying the temperature depends on the number of molecules in the gas and its volume; the temperature T is measured on the Kelvin scale.

Second, imagine you have a basketball that is just slightly underinflated, as in **Figure 17–2**. The ball has the size and shape of a basketball, but is a bit too squishy. To increase the pressure you "pump" more molecules from the atmosphere into the ball. Thus, while the temperature and volume of the gas in the ball remain constant, its pressure increases as the number of molecules increases:

$$P = (\text{constant})N$$

(fixed volume, V; fixed temperature, T)

▲ **FIGURE 17–2 Inflating a basketball**
A hand pump can be used to increase the number of molecules inside a basketball. This raises the pressure of the gas in the ball, causing it to inflate.

Our third and final observation concerns the volume dependence of pressure. Returning to the basketball, suppose that instead of pumping it up you sit

▲ **FIGURE 17–3 Increasing pressure by decreasing volume**
Sitting on a basketball reduces its volume and increases the pressure of the gas it contains.

PROBLEM SOLVING NOTE
Ideal Gas Law

When using the equation of state for an ideal gas, remember that the temperature must always be given in terms of the Kelvin scale.

on it, as pictured in **Figure 17–3**. This deforms it a bit, *reducing* its volume. At the same time, the pressure of the gas in the ball *increases*. Thus, when the number of molecules and temperature remain constant, the pressure varies *inversely* with volume:

$$P = \frac{\text{(constant)}}{V}$$

(fixed number of molecules, N; fixed temperature, T)

Combining these observations we arrive at the following mathematical expression for the pressure of a gas:

$$P = k\frac{NT}{V}$$

The constant k in this expression is a fundamental constant of nature, the **Boltzmann constant**, named for the Austrian physicist Ludwig Boltzmann (1844–1906):

Boltzmann constant, *k*

$$k = 1.38 \times 10^{-23}\ \text{J/K} \qquad \textbf{17–1}$$

SI unit: J/K

In general, a relationship between the thermal properties of a substance is referred to as an **equation of state**. Rearranging slightly, the equation of state for an ideal gas can be written as

Equation of state for an ideal gas

$$PV = NkT \qquad \textbf{17–2}$$

We apply this result to the gas contained in a person's lungs in the next Example.

EXAMPLE 17–1 Take a Deep Breath

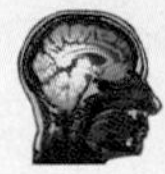

Real World Physics: Bio

A person's lungs might hold 6.0 L ($1\ \text{L} = 10^{-3}\ \text{m}^3$) of air at body temperature (310 K) and atmospheric pressure (101 kPa). Given that the air is 21% oxygen, find the number of oxygen molecules in the lungs.

Picture the Problem
Our sketch shows the relevant physical quantities for this problem.

Strategy
We will treat the air in the lungs as an ideal gas. Given the volume, temperature, and pressure of the gas, we can use the equation of state, $PV = NkT$, to solve for the number, N.

Finally, since only 21% of the molecules are oxygen, we multiply N by 0.21.

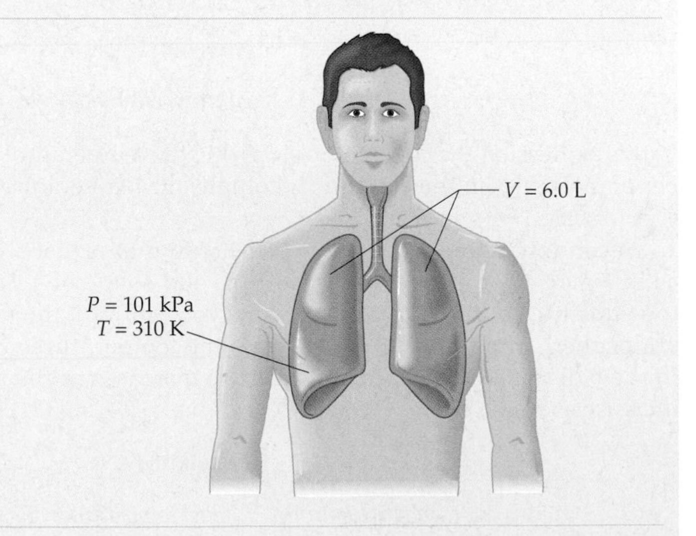

Solution

1. Solve the equation of state for the number of molecules: $N = PV/kT$

2. Substitute numerical values to find the number of molecules in the lungs:

$$N = \frac{PV}{kT}$$

$$= \frac{(1.01 \times 10^5\ \text{Pa})(6.0 \times 10^{-3}\ \text{m}^3)}{(1.38 \times 10^{-23}\ \text{J/K})(310\ \text{K})} = 1.4 \times 10^{23}$$

3. Multiply N by 0.21 to find the number of molecules that are oxygen, O_2:

$$0.21N = 0.21(1.4 \times 10^{23}) = 2.9 \times 10^{22}$$

Insight

As one might expect, the number of molecules in an "everyday-size" volume is enormous.

Practice Problem

If the person takes a particularly deep breath, so that the lungs hold a total of 1.5×10^{23} molecules, what is the new volume of the lungs? [**Answer:** $V = 6.4 \times 10^{-3}\ \text{m}^3 = 6.4\ \text{L}$]

Some related homework problems: Problem 1, Problem 2, Problem 58

Another common way to write the ideal gas equation of state is in terms of the number of **moles** in a gas—as opposed to using the number of molecules, N, as in Equation 17–2. In the SI system of units, the mole (mol) is defined in terms of the most abundant isotope of carbon, which is referred to as carbon-12. The definition is as follows:

> A mole is the amount of a substance that contains as many elementary entities as there are atoms in 12 g of carbon-12.

The phrase "elementary entities" refers to molecules, which may contain more than one atom, as in water (H_2O), or just a single atom, as in carbon (C) and helium (He). Experiments show that the number of atoms in a mole of carbon-12 is 6.022×10^{23}. This number is known as **Avogadro's number**, N_A, named for the Italian physicist and chemist Amedeo Avogadro (1776–1856).

Avogadro's number, N_A

$$N_A = 6.022 \times 10^{23}\ \text{molecules/mol} \qquad 17\text{–}3$$

SI unit: mol^{-1}; the number of molecules is dimensionless

▲ **FIGURE 17–4 Moles of various substances**

Counting atoms or molecules is obviously hard to do, but the mole concept provides a useful way of dealing with the difficulty. A mole of any substance contains the same number of elementary entities (atoms or molecules). This number, referred to as Avogadro's number, has the value $N_A = 6.022 \times 10^{23}$. To count out N_A molecules of a gas at standard temperature and pressure, for example, all you need do is measure out a volume of gas equal to 22.4 liters. Similarly, you can effectively count one mole of any substance by measuring out a weight in grams that is equal to the atomic weight of that substance. Thus, the mole provides a convenient bridge between the realm of atoms and molecules and the macroscopic world of observable weights and volumes. This photo shows molar amounts of four different substances: hydrogen, copper, mercury, and sulfur.

Examples of one mole of various substances are shown in Figure 17–4.

Now, if we let n denote the number of moles in a gas, the number of molecules it contains is

$$N = nN_A$$

Substituting this into the ideal-gas equation of state yields

$$PV = nN_AkT$$

The constants N_A and k are combined to form the **universal gas constant**, R, defined as follows:

Universal gas constant, R

$$R = N_Ak = (6.022 \times 10^{23}\ \text{molecules/mol})(1.38 \times 10^{-23}\ \text{J/K})$$

$$= 8.31\ \text{J/(mol}\cdot\text{K)} \qquad 17\text{–}4$$

SI unit: $\text{J/(mol}\cdot\text{K)}$

Thus, an alternative form of the equation of state is

Equation of state for an ideal gas

$$PV = nRT \qquad 17\text{–}5$$

This relation is used in the following Active Example.

ACTIVE EXAMPLE 17–1 The Amount of Air In a Basketball

How many moles of air are in an inflated basketball? Assume that the pressure in the ball is 171 kPa, the temperature is 293 K, and the diameter of the ball is 30.0 cm.

Solution

1. Solve $PV = nRT$ for the number of moles, n: $n = PV/RT$
2. Calculate the volume of the ball: $V = 4\pi r^3/3 = 0.0141\ \text{m}^3$
3. Substitute numerical values: $n = 0.990\ \text{mol}$

Insight

Thus, an inflated basketball contains approximately one mole of air.

As mentioned before, a mole of anything has precisely the same number (N_A) of particles. What differs from substance to substance is the mass of one mole. For example, one mole of helium atoms has a mass of 4.00260 g, and one mole of copper atoms has a mass of 63.546 g. In general, we define the **atomic or molecular mass, *M*,** of a substance to be the mass in grams of one mole of that substance. Thus, the molecular mass for helium is $M = 4.00260$ g/mol and for copper it is $M = 63.546$ g/mol. The periodic table in Appendix D gives the atomic masses for all elements.

Note that the molecular mass provides a convenient bridge between the macroscopic world, where we measure the mass of a substance in grams, and the microscopic world, where the number of molecules is typically 10^{23} or greater. As we have seen, if you measure out a mass of copper equal to 63.546 g you have, in effect, counted out $N_A = 6.022 \times 10^{23}$ atoms of copper. It follows that the mass of an individual copper atom, m, is the molecular mass of copper divided by Avogadro's number; that is

$$m = \frac{M}{N_A} \qquad \text{17–6}$$

We use this relation to find the mass of a copper atom and an oxygen molecule in the following Exercise.

EXERCISE 17–1

Find the mass of **(a)** a copper atom and **(b)** a molecule of oxygen, O_2. Molecular masses are listed in Appendix D.

Solution

(a) The atomic mass of copper is 63.546 g/mol. Thus, a copper atom has the mass

$$m_{\text{Cu}} = \frac{M}{N_A} = \frac{63.546\ \text{g/mol}}{6.022 \times 10^{23}\ \text{molecules/mol}} = 1.055 \times 10^{-22}\ \text{g/molecule}$$

(b) Since the atomic mass of oxygen is 16.00 g/mol (Appendix D), the molecular mass of O_2 is 32.00 g/mol. Therefore, the mass of O_2 is

$$m_{O_2} = \frac{M}{N_A} = \frac{32.00\ \text{g/mol}}{6.022 \times 10^{23}\ \text{molecules/mol}} = 5.314 \times 10^{-23}\ \text{g/molecule}$$

An interesting medical application of the ideal-gas equation of state is the air splint (p. 463). This device is a lightweight, compact plastic sleeve that can be packed in the first aid kit of a hiker or kayaker. If a person suffers a broken leg or arm, the air splint can be pulled over the affected limb like the sleeve of a coat. The air splint is then inflated, as one would inflate an air mattress, until the pressure is sufficient to immobilize the broken bone. As can be seen from Equation 17–5, the amount of air that must be forced into the splint to produce a given pressure is proportional to the volume of the splint and inversely proportional to the temperature.

Finally, we consider a common situation to further illustrate the general features of the ideal-gas equation of state.

CONCEPTUAL CHECKPOINT 17–1

Feeling a bit cool, you turn up the thermostat in your house or apartment. A short time later the air is warmer. Assuming the room is well sealed, is the pressure of the air **(a)** greater than, **(b)** less than, or **(c)** the same as before you turned up the heat?

Reasoning and Discussion

We assume that the number of molecules and the volume occupied by the molecules is approximately constant. Thus, increasing T, while holding N and V fixed, leads to an increased pressure, P.

Answer:

(a) The air pressure increases.

Isotherms

Historically, the ideal-gas equation of state was arrived at piece by piece, as a result of the combined efforts of a number of researchers. For example, the English scientist Robert Boyle (1627–1691) established the fact that the pressure of a gas varies inversely with volume—as long as temperature and the number of molecules are held constant. This is known as **Boyle's law**:

$$P_iV_i = P_fV_f$$

(fixed number of molecules, N; fixed temperature, T) **17–7**

To see that Boyle's law is consistent with $PV = NkT$, note that if N and T are constant, then so too is PV. Another way of saying that PV is a constant is to say that its initial value must be equal to its final value. This is Boyle's law.

When N and T are constant, the equation of state $PV = NkT$ implies that

$$P = \frac{NkT}{V} = \frac{\text{constant}}{V}$$

This result is plotted in **Figure 17–5**, where we show P as a function of V. Note that the larger the temperature, T, the larger the constant in the numerator. Therefore, the curves farther from the origin correspond to higher temperatures, as indicated.

Each of the curves in Figure 17–5 corresponds to a different, fixed temperature. As a result, these curves are known as **isotherms**, which means, literally, "constant temperature." We shall return to the topic of isotherms in the next chapter, when we consider various thermal processes.

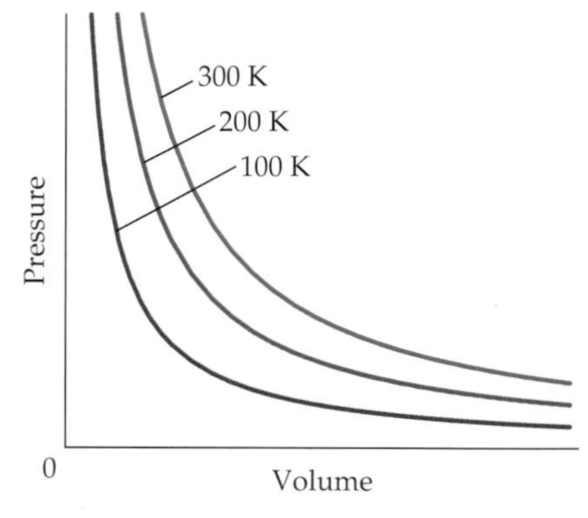

▲ **FIGURE 17–5 Ideal-gas isotherms** Pressure versus volume isotherms for an ideal gas. Each isotherm is of the form $P = NkT/V = \text{constant}/V$. The three isotherms shown are for the temperatures 100 K, 200 K, and 300 K.

PROBLEM SOLVING NOTE
Pressure and Volume in an Isotherm

If the temperature of an ideal gas is constant, the pressure and volume change in such a way that their product has a constant value. This is true no matter what the value of the temperature.

EXAMPLE 17–2 Under Pressure I

A cylindrical flask of cross-sectional area A is fitted with an airtight piston that is free to slide up and down. Contained within the flask is an ideal gas. Initially the pressure applied by the piston is 130 kPa and the height of the piston above the base of the flask is 25 cm. When additional mass is added to the piston, the pressure increases to 170 kPa. Assuming the system is always at the temperature 290 K, find the new height of the piston.

continued on the following page

continued from the previous page

Picture the Problem
The physical situation is shown in the sketch. The initial and final pressures and the initial and final heights are indicated.

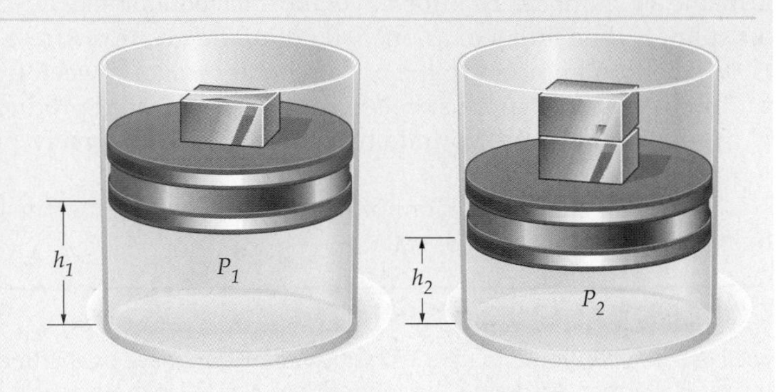

Strategy
Since the temperature is held constant in this system, it follows that $PV = NkT$ is also constant. Thus, as in Boyle's law, $P_1V_1 = P_2V_2$. This relation can be used to find V_2 (since we are given P_1, P_2, and V_1).

Next, the cylindrical volume is related to the height by $V = Ah$. Note that the area A is the same in both cases, hence it will cancel, allowing us to solve for the height, h_2.

Solution

1. Set the initial and final values of PV equal to one another, then solve for V_2:

$$P_1V_1 = P_2V_2$$
$$V_2 = V_1(P_1/P_2)$$

2. Substitute $V_1 = Ah_1$ and $V_2 = Ah_2$, and solve for h_2:

$$Ah_2 = Ah_1(P_1/P_2)$$
$$h_2 = h_1(P_1/P_2)$$

3. Substitute numerical values:

$$h_2 = h_1\frac{P_1}{P_2} = (25\text{ cm})\frac{130\text{ kPa}}{170\text{ kPa}} = 19\text{ cm}$$

Insight
Note that the pressure exerted on the gas is the sum of the atmospheric pressure, P_{at}, and the pressure caused by the weight it supports.

Practice Problem
What pressure would be required to change the height of the piston to 29 cm? [**Answer:** $P = 110$ kPa]

Some related homework problems: Problem 13, Problem 14, Problem 70

Finally, recall that in Conceptual Checkpoint 15–2 we considered the behavior of an air bubble rising from a swimmer in shallow water. At the time, we said that the diameter of the bubble increases as it rises because the pressure of the surrounding water is decreasing. This is certainly the case, but now we can be more precise. If we assume that the water temperature is constant, and that no gas molecules are added to or removed from the bubble, then the volume of the bubble has the following dependence:

$$V = \frac{NkT}{P} = \frac{\text{constant}}{P}$$

This is just our isotherm, again, and we see that the volume indeed increases as pressure decreases.

▲ Charles's law states that the volume of a gas at constant pressure is proportional to its Kelvin temperature. For example, this balloon is fully inflated at room temperature (293 K) and atmospheric pressure. When cooled by vapors from liquid nitrogen (77 K), however, the volume of the air inside the balloon decreases markedly, causing it to shrivel.

Constant Pressure

Another aspect of ideal-gas behavior was discovered by the French scientist Jacques Charles (1746–1823), and later studied in greater detail by fellow Frenchman Joseph Gay-Lussac (1778–1850). Known today as **Charles's law**, their result is that the volume of a gas divided by its temperature is constant, as long as the pressure and number of molecules is constant:

$$\frac{V_i}{T_i} = \frac{V_f}{T_f} \qquad \text{17–8}$$

(fixed number of molecules, N; fixed pressure, P)

As with Boyle's law, this result follows immediately from the ideal-gas equation of state. Solving Equation 17–2 for V/T, we have

$$\frac{V}{T} = \frac{Nk}{P}$$

If N and P are constant, then so is the quantity V/T.

Charles's law can be rewritten as a linear relation between volume and temperature:

$$V = (\text{constant})T$$

The constant in this expression is Nk/P. This result is illustrated in **Figure 17–6**, where we see the volume of an ideal gas vanishing as the temperature approaches absolute zero.

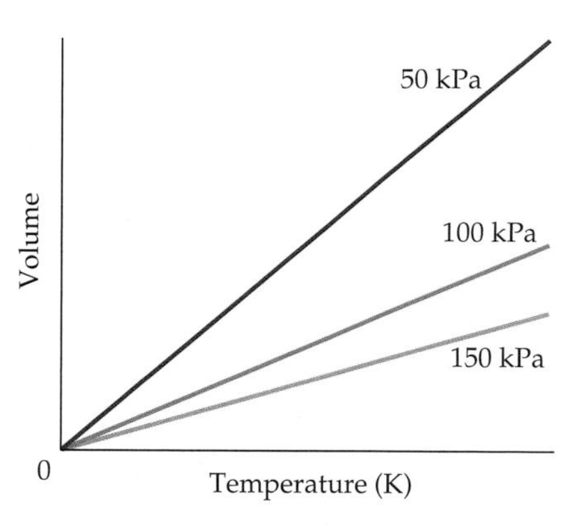

▲ FIGURE 17–6 Volume versus temperature for an ideal gas at constant pressure
The temperature in this plot is the Kelvin temperature, T, hence $T = 0$ corresponds to absolute zero. At absolute zero the volume attains its lowest possible value—zero.

ACTIVE EXAMPLE 17–2 Under Pressure II

Consider again the system described in Example 17–2. In this case the temperature is changed from an initial value of 290 K to a final value of 330 K. The pressure exerted on the gas remains constant at 130 kPa, and the initial height of the piston is 25 cm. Find the final height of the piston.

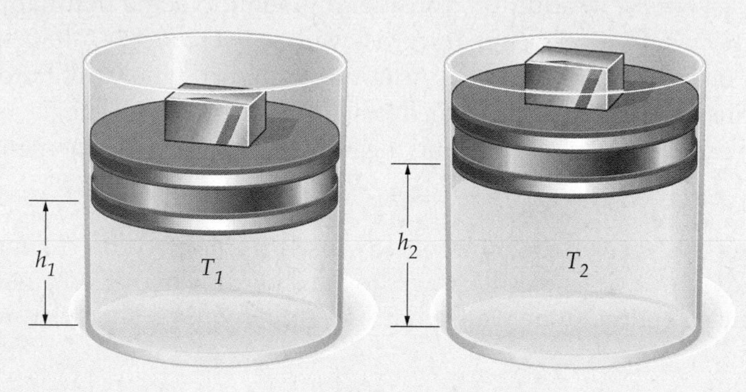

Solution

1. Set the initial value of V/T equal to the final value of V/T: $V_1/T_1 = V_2/T_2$
2. Solve for V_2: $V_2 = V_1(T_2/T_1)$
3. Use $V = Ah$ to solve for the height, h_2: $h_2 = h_1(T_2/T_1)$
4. Substitute numerical values: $h_2 = 28$ cm

Insight

The volume of the gas, and hence the height of the piston, increased in direct proportion to the increase in temperature. This is to be expected when the pressure is constant.

PROBLEM SOLVING NOTE
Volume and Temperature in an Ideal Gas

If the pressure of an ideal gas is constant, the ratio of the volume and temperature remains constant. This is true no matter what the value of the pressure.

17–2 Kinetic Theory

We can readily measure the pressure and temperature of a gas using a pressure gauge and a thermometer. These are *macroscopic* quantities that apply to the gas as a whole. It is not so easy, however, to measure *microscopic* quantities, such as the position or velocity of an individual molecule. Still, there must be some connection between what happens on the microscopic level and what we observe on the macroscopic level. This connection is described by the **kinetic theory of gases**.

In the kinetic theory, we imagine a gas to be made up of a collection of molecules moving about inside a container of volume V. In particular, we assume the following:

- The container holds a very large number N of identical molecules. Each molecule has a mass m, and behaves as a point particle.
- The molecules move about the container in a random manner. They obey Newton's laws of motion at all times.
- When molecules hit the walls of the container or collide with one another, they bounce elastically. Other than these collisions, the molecules have no interactions.

With these basic assumptions we can relate the pressure of a gas to the behavior of the molecules themselves.

The Origin of Pressure

As we shall see, the pressure exerted by a gas is due to the innumerable collisions between gas molecules and the walls of their container. Each collision results in a change of momentum for a given molecule, just like throwing a ball at a wall and having it bounce back. The total change in momentum of the molecules in a given time, divided by the time, is simply the force a wall must exert on the gas to contain it (see Section 9–2). The average of this force over time, and over the area of a wall, is the pressure of the gas.

To be specific, imagine a container that is a cube of length L on a side. Its volume, then, is $V = L^3$. In addition, consider a given molecule that happens to be moving in the negative x direction toward a wall, as in **Figure 17–7**. If its speed is v_x, its initial momentum is $p_{i,x} = -mv_x$. After bouncing from the wall it moves in the positive x direction with the same speed (since the collision is elastic), hence its final momentum is $p_{f,x} = +mv_x$. As a result, the molecule's change in momentum is

$$\Delta p_x = p_{f,x} - p_{i,x} = mv_x - (-mv_x) = 2mv_x$$

The wall exerts a force on the molecule to cause this momentum change.

After the bounce, the molecule travels to the other side of the container and back before bouncing off the same wall again. The time required for this round trip of length $2L$ is

$$\Delta t = 2L/v_x$$

Thus, by Newton's second law, the average force exerted by the wall on the molecule is

$$F = \frac{\Delta p}{\Delta t} = \frac{2mv_x}{2L/v_x} = \frac{mv_x^2}{L}$$

The average pressure exerted by this wall, then, is simply the force divided by the area. Since the area of the wall is $A = L^2$, we have

$$P = \frac{|F|}{A} = \frac{(mv_x^2/L)}{L^2} = \frac{mv_x^2}{L^3} = \frac{mv_x^2}{V} \quad \text{17–9}$$

Note that we have used the fact that the volume of the container is L^3.

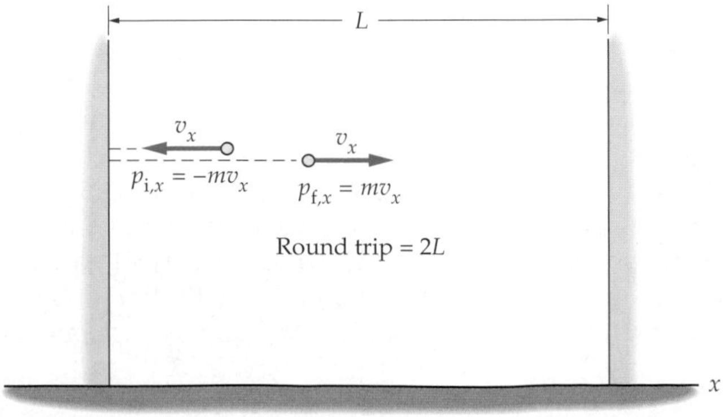

L

v_x v_x

$p_{i,x} = -mv_x$ $p_{f,x} = mv_x$

Round trip = 2L

x

▶ **FIGURE 17–7 Force exerted by a molecule on the wall of a container**

A molecule bounces off a wall of a container, changing its momentum from $-mv_x$ to $+mv_x$; the change in momentum is $2mv_x$. A round trip will be completed in the time $\Delta t = 2L/v_x$, so the average force exerted on the molecule by the wall is $F = \Delta p/\Delta t = 2mv_x/(2L/v_x) = mv_x^2/L$.

In this calculation we assumed that the molecule moves in the x direction. This was merely to simplify the derivation. If, instead, the molecule moves at some angle to the x-axis, the calculation applies to its x component of motion. The final conclusions are unchanged.

Speed Distribution of Molecules

In deriving Equation 17–9, we considered a single molecule with a particular speed. Other molecules, of course, will have different speeds. In addition, the speed of any given molecule changes with time as it collides with other molecules in the gas. What remains constant, however, is the overall **distribution of speeds**.

This is illustrated in Figure 17–8, where we present the results obtained by the Scottish physicist James Clerk Maxwell (1831–1879). This plot shows the relative probability that an O_2 molecule will have a given speed. For example, on the curve labeled 300 K, the most probable speed is about 390 m/s. When the temperature is increased to 1100 K the most probable speed increases to roughly 750 m/s. Other speeds occur as well, from speeds near zero to those that are very large, but these have much lower probabilities.

Thus, in Equation 17–9, the term v_x^2 should be replaced with the average of v_x^2 over all the molecules in the gas. Writing this average as $(v_x^2)_{av}$, we have

$$P = \frac{m(v_x^2)_{av}}{V}$$

Since all N molecules in the gas follow the same distribution, the pressure exerted by the gas as a whole is N times this result:

$$P = N\frac{m(v_x^2)_{av}}{V} = \left(\frac{N}{V}\right)m(v_x^2)_{av} \qquad \textbf{17–10}$$

Of course, there is nothing special about the x direction—Equation 17–10 applies equally well with $(v_y^2)_{av}$ or $(v_z^2)_{av}$ in place of $(v_x^2)_{av}$. Thus it would be preferable to express the pressure of the gas in terms of the overall speed of the molecules, rather than in terms of a single component. The speed squared of a molecule is

$$v^2 = v_x^2 + v_y^2 + v_z^2$$

Hence, the average of v^2 is

$$(v^2)_{av} = (v_x^2)_{av} + (v_y^2)_{av} + (v_z^2)_{av}$$

Since the x, y, and z directions are equivalent, it follows that

$$(v_x^2)_{av} = (v_y^2)_{av} = (v_z^2)_{av}$$

As a result,

$$(v^2)_{av} = (v_x^2)_{av} + (v_y^2)_{av} + (v_z^2)_{av} = 3(v_x^2)_{av}$$

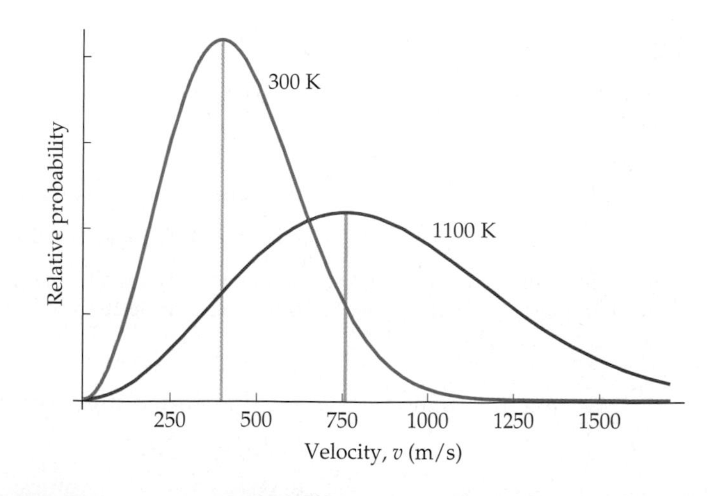

◀ **FIGURE 17–8 The Maxwell speed distribution**

The Maxwell distribution of molecular speeds for O_2 at the temperatures $T = 300$ K, and T = 1100 K. Note that the most probable speed increases with increasing temperature.

or, equivalently

$$(v_x^2)_{av} = \tfrac{1}{3}(v^2)_{av}$$

Using this replacement in Equation 17–10 yields

$$P = \tfrac{1}{3}\left(\frac{N}{V}\right)m(v^2)_{av}$$

The last part of this expression, $m(v^2)_{av}$, is simply twice the average kinetic energy of a molecule. Thus, using K for the kinetic energy, as in Chapters 7 and 8, we can write the pressure as follows:

Pressure in the kinetic theory of gases

$$P = \tfrac{1}{3}\left(\frac{N}{V}\right)2K_{av} = \tfrac{2}{3}\left(\frac{N}{V}\right)(\tfrac{1}{2}mv^2)_{av} \qquad \textbf{17–11}$$

To summarize, using the kinetic theory we have shown that the pressure of a gas is proportional to the number of molecules and inversely proportional to the volume. We discussed both of these dependences before when considering the ideal gas. In addition, we see that

> The pressure of a gas is directly proportional to the average kinetic energy of its molecules.

This is the key connection between microscopic behavior and macroscopic observables.

Kinetic Energy and Temperature

If we compare the ideal-gas equation of state, $PV = NkT$, with the result from kinetic theory, Equation 17–11, we find

$$PV = NkT = \tfrac{2}{3}N(\tfrac{1}{2}mv^2)_{av}$$

As a result,

$$\tfrac{2}{3}(\tfrac{1}{2}mv^2)_{av} = kT$$

Equivalently,

Kinetic energy and temperature

$$(\tfrac{1}{2}mv^2)_{av} = K_{av} = \tfrac{3}{2}kT \qquad \textbf{17–12}$$

This is one of the most important results of kinetic theory. It says that the average kinetic energy of a gas molecule is directly proportional to the Kelvin temperature, T. Thus, when we heat a gas, what happens on the microscopic level is that the molecules move with speeds that are, on average, greater. Similarly, cooling a gas causes the molecules to slow.

EXERCISE 17–2

Find the average kinetic energy of oxygen molecules in the air. Assume the air is at a temperature of 20.0 °C.

Solution

Using Equation 17–12, and the fact that 20.0 °C = 293 K, we find

$$K_{av} = \tfrac{3}{2}kT = \tfrac{3}{2}(1.38 \times 10^{-23}\ \text{J/K})(293\ \text{K}) = 6.07 \times 10^{-21}\ \text{J}$$

Note that this is also the average kinetic energy of nitrogen molecules in the air. In fact, the type of molecule is not important; all that matters is the temperature of the gas.

Returning to Equation 17–12, we find with a slight rearrangement that

$$(v^2)_{av} = 3kT/m$$

Now, the square root of $(v^2)_{av}$ is given a special name—it is called the **root mean square** (rms) speed. For gas molecules, then, the rms speed, v_{rms}, is the following:

RMS speed of a gas molecule

$$v_{rms} = \sqrt{(v^2)_{av}} = \sqrt{\frac{3kT}{m}} \qquad \textbf{17–13}$$

SI unit: m/s

Rewriting this in terms of the molecular mass, M, we have

$$v_{rms} = \sqrt{\frac{3kT}{m}} = \sqrt{\frac{3kT}{(M/N_A)}} = \sqrt{\frac{3N_AkT}{M}}$$

Finally, using $N_Ak = R$ yields

$$v_{rms} = \sqrt{\frac{3RT}{M}} \qquad \textbf{17–14}$$

The rms speed is one of the characteristic speeds of the Maxwell speed distribution. As shown in **Figure 17–9**, v_{rms} is slightly greater than the most probable speed, v_{mp}, and the average speed, v_{av}.

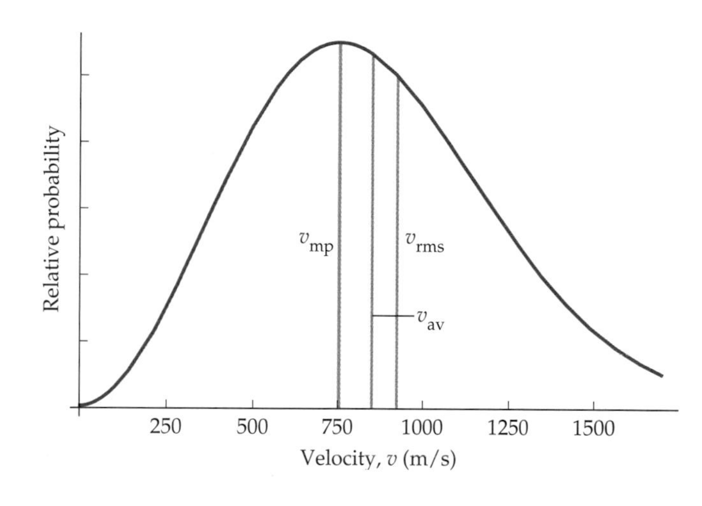

◀ **FIGURE 17–9 Most probable, average, and rms speeds**
Characteristic speeds for O_2 at the temperature $T = 1100$ K. From left to right, the indicated speeds are the most probable speed, v_{mp}, the average speed, v_{av}, and the rms speed, v_{rms}, respectively.

EXAMPLE 17–3 Fresh Air

The atmosphere is composed primarily of nitrogen N_2 (78%) and oxygen O_2 (21%). **(a)** Is the rms speed of N_2 (28.0 g/mol) greater than, less than, or the same as the rms speed of O_2 (32.0 g/mol)? **(b)** Find the rms speed of N_2 and O_2 at 293 K.

Picture the Problem
Our sketch shows molecules in the air bouncing off walls and people. The molecules move in all directions. In general, they all have different speeds.

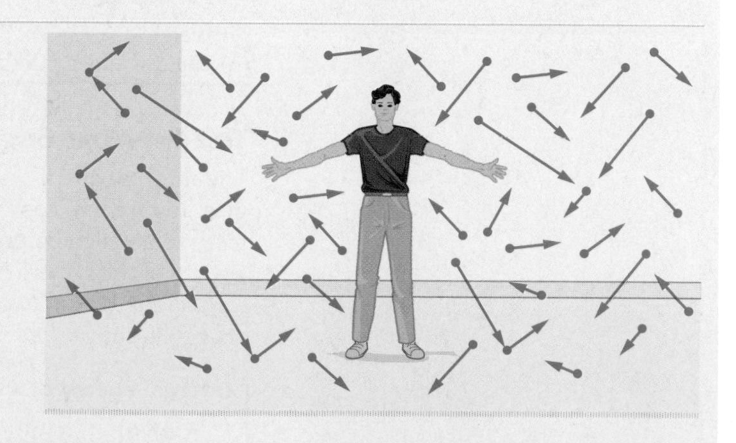

Strategy
The rms speeds are calculated by straightforward substitution in Equation 17–14. The only point to be careful about is the molecular mass of each molecule. In particular, note that for nitrogen the molecular mass is 28.0 g/mol = 0.0280 kg/mol, and similarly for oxygen, the molecular mass is 32.0 g/mol = 0.0320 kg/mol.

continued on the following page

continued from the previous page

Solution

Part (a)

Since both molecules are at the same temperature, they have the same kinetic energy. The nitrogen molecule has less mass, however; thus if it is to have the same kinetic energy it must have a higher speed. This is also what one would expect on the basis of Equation 17–14.

Part (b)

1. To find the rms speed of N_2, substitute $M = 0.0280\ \text{kg/mol}$ in Equation 17–14:

$$v_{rms} = \sqrt{\frac{3[8.31\ \text{J/(mol}\cdot\text{K)}](293\ \text{K})}{0.0280\ \text{kg/mol}}} = 511\ \text{m/s}$$

2. To find the rms speed of O_2, substitute $M = 0.0320\ \text{kg/mol}$ in Equation 17–14:

$$v_{rms} = \sqrt{\frac{3[8.31\ \text{J/(mol}\cdot\text{K)}](293\ \text{K})}{0.0320\ \text{kg/mol}}} = 478\ \text{m/s}$$

Insight

As expected, N_2 has the higher rms speed. For comparison, the speed of sound at 293 K is 343 m/s, which is just over 750 mi/h. Thus, the molecules in the air are bouncing off your skin with the speed of a supersonic jet.

Practice Problem

One of the molecules found in air is monatomic, and has an rms speed of 428 m/s at 293 K. What is this molecule? [**Answer:** $M = 0.0399\ \text{kg/mol}$; thus the molecule is argon, Ar. Argon comprises about 0.94% of the atmosphere.]

Some related homework problems: Problem 18, Problem 19

CONCEPTUAL CHECKPOINT 17–2

Two containers of equal volume each hold an ideal gas. Container A has twice as many molecules as container B. If the gas pressure is the same in the two containers, is the rms speed of the molecules in container A **(a)** greater than, **(b)** less than, or **(c)** the same as the rms speed of the molecules in container B?

Reasoning and Discussion

Since P and V are the same, you might think the rms speeds are the same as well. Recall, however, that *the pressure of a gas is caused by the collisions of gas molecules with the walls of the container*. If more molecules occupy a given volume, and they bounce off the walls with the same speed as in a container with fewer molecules, the pressure will be greater. Therefore, in order for the pressure in the two containers to be the same, it is necessary that the rms speed of the molecules in container A be less than the rms speed in container B.

Mathematically, we note that $P_A = P_B$ and $V_A = V_B$; hence $P_AV_A = P_BV_B$. Using the ideal-gas relation, $PV = NkT$, we can write this condition as $N_AkT_A = N_BkT_B$. Thus, if $N_A = 2N_B$ it follows that $T_A = T_B/2$. Since the temperature is lower in container A, its molecules have the smaller rms speed.

Answer:

(b) The rms speed is less in container A.

The Internal Energy of an Ideal Gas

The internal energy of a substance is the sum of all its potential and kinetic energies. In an ideal gas there are no interactions between molecules, other than perfectly elastic collisions; hence there is no potential energy. As a result, the total energy of the system is the sum of the kinetic energy of each of its molecules. Thus, for an ideal gas of N point-like molecules—that is, a monatomic gas—the internal energy is simply

Internal energy of a monatomic ideal gas

$$U = \tfrac{3}{2}NkT \qquad 17\text{–}15$$

SI unit: J

In terms of moles, this result is

$$U = \frac{3}{2}nRT \qquad \text{17–16}$$

We shall return to this result in the next chapter.

EXERCISE 17–3

A basketball at 290 K holds 0.95 mol of air molecules. What is the internal energy of the air in the ball?

Solution

Applying Equation 17–16 we find

$$U = \frac{3}{2}nRT = \frac{3}{2}(0.95\text{ mol})[8.31\text{ J}/(\text{mol}\cdot\text{K})](290\text{ K}) = 3400\text{ J}$$

This is roughly the kinetic energy a basketball would have if you dropped it from a height of 700 m.

If an ideal gas is diatomic, there are additional contributions to the internal energy. For example, a diatomic molecule, which is shaped somewhat like a dumbbell, can have rotational kinetic energy. Diatomic molecules can also vibrate along the line joining the two atoms, which is yet another contribution to the total energy. Thus, the result in Equation 17–16 applies only to the simplest case, the ideal monatomic gas.

17–3 Solids and Elastic Deformation

The defining characteristic of a solid object is that it has a particular shape. For example, when Michelangelo carved the statue of David from a solid block of marble he was confident it would retain its shape long after his work was done. Liquids and gases do not behave in this way. In contrast, they assume the shape of the container into which they are placed. On a molecular level, these differences arise from the fact that the intermolecular forces in a solid are strong enough to practically immobilize its molecules, whereas the intermolecular forces in liquids and gases are so weak the molecules can move about relatively freely.

Even so, the shape of a solid *can* be changed—though usually only slightly—if it is acted on by a force. In this section we consider various types of deformations that can occur in solids and the way these deformations are related to the forces that cause them.

Changing the Length of a Solid

A useful physical model for a solid is a lattice of small balls representing molecules connected to one another by springs representing the intermolecular forces. Pulling on a solid rod with a force F, for example, causes each "intermolecular spring" in the direction of the force to expand by an amount proportional to F. The net result is that the entire solid increases its length by an amount $\Delta L \propto F$, as indicated in **Figure 17–10**.

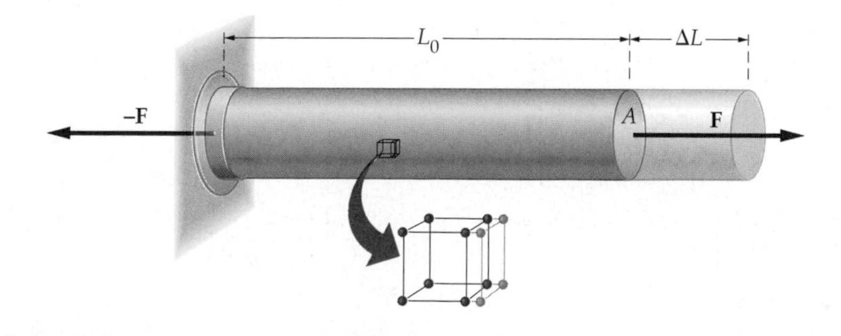

◀ **FIGURE 17–10 Stretching a rod**
Equal and opposite forces applied to the ends of a rod cause it to stretch. On the atomic level, the forces stretch the "intermolecular springs" in the solid, resulting in an overall increase in length. The stretch is proportional to the force F, the initial length L_0, and inversely proportional to the cross-sectional area A.

TABLE 17–1
Young's Modulus for Various Materials

Material	Young's modulus, Y (N/m^2)
Tungsten	36×10^{10}
Steel	20×10^{10}
Copper	11×10^{10}
Brass	9.0×10^{10}
Aluminum	6.9×10^{10}
Pyrex glass	6.2×10^{10}
Lead	1.6×10^{10}
Bone	
Tension	1.6×10^{10}
Compression	0.94×10^{10}
Nylon	0.37×10^{10}

PROBLEM SOLVING NOTE
Area and Young's Modulus

When using Young's modulus, remember that the area A is the area that is at right angles to the applied force.

The stretch ΔL is also proportional to the initial length of the rod, L_0. To see why, we first note that each intermolecular spring expands by the same amount, giving a total stretch that is proportional to the number of such springs. But the number of intermolecular springs is proportional to the total initial length of the rod. Thus, it follows that $\Delta L \propto FL_0$.

Finally, the amount of stretch for a given force F is inversely proportional to the cross-sectional area A of the rod. For example, a rod with a cross-sectional area $2A$ is like two rods of cross-sectional area A placed side by side. Thus, applying a force F to a rod of area $2A$ is equivalent to applying a force $F/2$ to two rods of area A. The result is half the stretch when the area is doubled; that is, $\Delta L \propto FL_0/A$. Solving this relation for the amount of force F required to produce a given stretch ΔL we find $F \propto (\Delta L/L_0)A$, or as an equality

$$F = Y\left(\frac{\Delta L}{L_0}\right)A \qquad \textbf{17–17}$$

The proportionality constant Y in this expression is **Young's modulus**, named for the English physicist Thomas Young (1773–1829). Comparing the two sides of Equation 17–17, we see that Young's modulus has the units of force per area (N/m^2).

Typical values for Young's modulus are given in Table 17–1. Notice that the values vary from material to material, but are all rather large. This means that a large force is required to cause even a small stretch in a solid. Of course, Equation 17–17 applies equally well to a compression or a stretch. However, some materials have a slightly different Young's modulus for compression and stretching. For example, human bones under tension (stretching) have a Young's modulus of 1.6×10^{10} N/m^2, while bones under compression have a slightly smaller Young's modulus of 9.4×10^{9} N/m^2.

EXAMPLE 17–4 **Stretching a Bone**

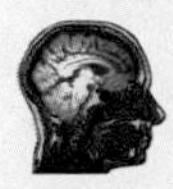

Real World Physics: Bio

At the local airport, a person carries a 21-kg suitcase in one hand. Assuming the humerus (the upper arm bone) supports the entire weight of the suitcase, determine the amount by which it stretches. (The humerus may be assumed to be 33 cm in length and to have an effective cross-sectional area of 5.2×10^{-4} m^2.)

Picture the Problem

The humerus is oriented vertically, hence the weight of the suitcase applies tension to the bone. The Young's modulus in this case is 1.6×10^{10} N/m^2. Other relevant physical quantities are indicated in the sketch.

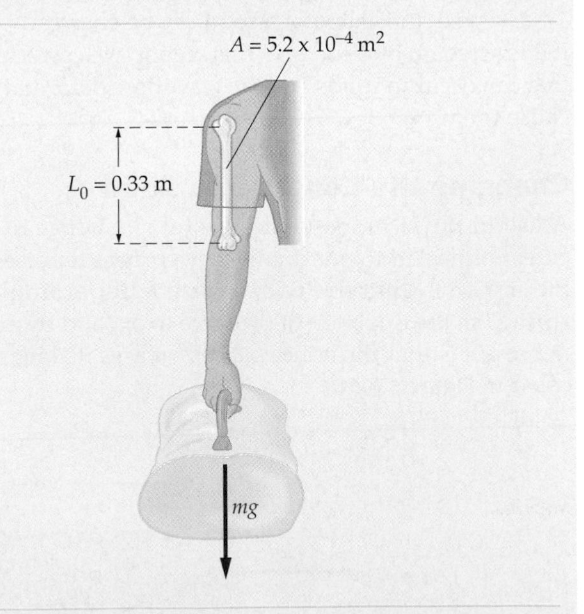

Strategy

We can find the amount of stretch by solving Equation 17–17 for the quantity ΔL. For the force, F, we use the weight of the suitcase, mg.

Solution

1. Solve Equation 17–17 for the amount of stretch, ΔL:

$$\Delta L = \frac{FL_0}{YA}$$

2. Calculate the force applied to the humerus: $F = mg = (21\ \text{kg})(9.81\ \text{m/s}^2) = 210\ \text{N}$

3. Substitute numerical values into the expression for ΔL:

$$\Delta L = \frac{FL_0}{YA} = \frac{(210\ \text{N})(0.33\ \text{m})}{(1.6 \times 10^{10}\ \text{N/m}^2)(5.2 \times 10^{-4}\ \text{m}^2)} = 8.3 \times 10^{-6}\ \text{m}$$

Insight

Since Young's modulus has such a large value, the amount of stretch is imperceptibly small.

Practice Problem

Suppose the humerus is reduced uniformly in size by a factor of two. This means that both the length and diameter are halved. What is the amount of stretch in this case? [**Answer:** Since $L_0 \rightarrow L_0/2$ and $A \rightarrow A/4$, the stretch is doubled. Thus, $\Delta L \rightarrow 2\Delta L = 1.7 \times 10^{-5}\ \text{m}$.]

Some related homework problems: Problem 27, Problem 28

There is a straightforward connection between Equation 17–17 and Hooke's law for a spring (Equation 6–4). To see the connection, we rewrite Equation 17–17 as follows:

$$F = \left(\frac{YA}{L_0}\right)\Delta L$$

Notice that the force required to cause a certain stretch is proportional to the stretch—just as in Hooke's law. In fact, if we identify ΔL with the displacement x of a spring from equilibrium and YA/L_0 with the force constant k we have Hooke's law:

$$F = \left(\frac{YA}{L_0}\right)\Delta L = kx$$

Thus, we see that the force constant of a spring depends on the Young's modulus, Y, of the material from which it is made, the cross-sectional area A of the wire, and the length of the wire, L_0.

CONCEPTUAL CHECKPOINT 17–3

Two identical springs are connected end to end. Is the force constant of the resulting compound spring **(a)** greater than, **(b)** less than, or **(c)** equal to that of a single spring?

Reasoning and Discussion

It might seem that the force constant would be the same, since we simply have twice the length of the same spring. On the other hand, it might seem that the force constant is greater, since two springs are exerting a force rather than just one. In fact, the force constant decreases.

The reason for the reduced force constant is that if we apply a force F to the compound spring, each individual spring stretches by a certain amount. The compound spring, then, stretches by twice this amount. By Hooke's law, $F = kx$, a spring that stretches twice as far for the same applied force has half the force constant.

We can also obtain this result by recalling that the force constant is $k = YA/L_0$. Thus, if the length of a spring is doubled—as in the compound spring—the force constant is halved.

Answer:

(b) The force constant of the compound spring is less, by a factor of 2.

Changing the Shape of a Solid

Another type of deformation, referred to as a **shear deformation**, changes the shape of a solid. Consider a book of thickness L_0 resting on a table, as shown in **Figure 17–11**. A force F is applied to the right on the top cover of the book, and static friction applies

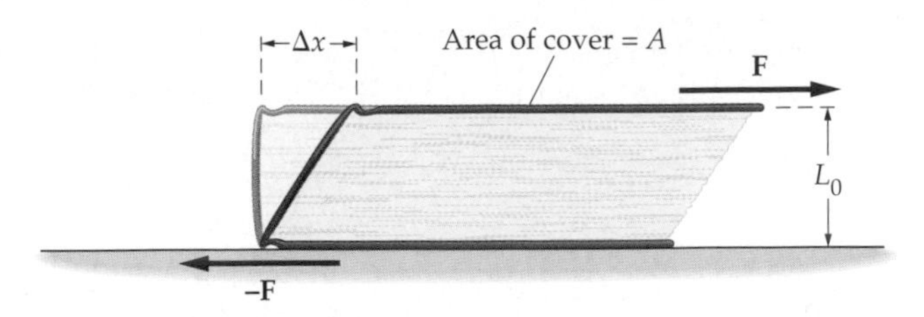

▶ **FIGURE 17–11 Shear deformation**
Equal and opposite forces applied to the top and bottom of a book result in a shear deformation. The amount of deformation is proportional to the force F, the thickness of the book L_0, and inversely proportional to the area A.

TABLE 17–2
Shear Modulus for Various Materials

Material	Shear modulus, S (N/m^2)
Tungsten	15×10^{10}
Steel	8.1×10^{10}
Bone	8.0×10^{10}
Copper	4.2×10^{10}
Brass	3.5×10^{10}
Aluminum	2.4×10^{10}
Lead	0.54×10^{10}

PROBLEM SOLVING NOTE
Area and the Shear Modulus

When using the shear modulus, remember that the area A is the area that is parallel to the applied force.

a force F to the left on the bottom cover of the book. The result is that the book remains at rest, but becomes slanted by the amount Δx. The force required to cause a given amount of slant is proportional to Δx, inversely proportional to the thickness of the book L_0, and proportional to the surface area A of the book's cover; that is, $F \propto A\Delta x/L_0$. Writing this as an equality, we have

$$F = S\left(\frac{\Delta x}{L_0}\right)A \qquad \textbf{17–18}$$

The constant of proportionality in this case is the **shear modulus**, S. Like Young's modulus, the shear modulus has the units N/m^2. Typical values of the shear modulus are collected in Table 17–2. As with Young's modulus, the shear modulus is large in magnitude, meaning that most solids require a large force to cause even a small amount of shear.

Equations 17–17 and 17–18 are similar in structure, but it is important to be aware of their differences as well. For example, the term L_0 in the Young's modulus equation refers to the length of a solid measured in the direction of the applied force. In contrast, L_0 in the shear modulus equation refers to the thickness of the solid as measured in a direction perpendicular to the applied force. Similarly, the area A in Equation 17–17 is the cross-sectional area of the solid, and this area is perpendicular to the applied force. On the other hand, the area A in Equation 17–18 is the area of the solid in the plane of the applied force.

ACTIVE EXAMPLE 17–3 A Stack of Pancakes

A horizontal force of 1.2 N is applied to the top of a stack of pancakes 13 cm in diameter and 9.0 cm high. The result is a shear deformation of 2.5 cm. What is the shear modulus for these pancakes?

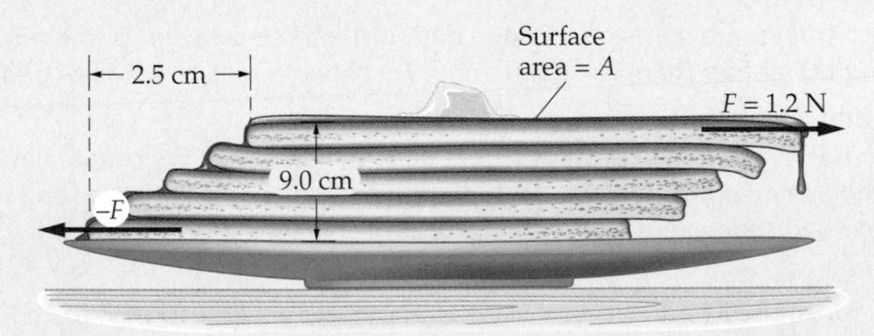

Solution

1. Solve Equation 17–18 for the shear modulus: $S = FL_0/A\Delta x$
2. Calculate the area of the pancakes: $A = \pi d^2/4 = 0.013\ \text{m}^2$
3. Substitute numerical values: $S = 330\ \text{N/m}^2$

Insight

Notice the small value of the pancake's shear modulus, especially when compared to the shear modulus of a typical metal. This is a reflection of the fact that the pancake stack is easily deformed.

Changing the Volume of a Solid

If a piece of Styrofoam is taken deep into the ocean, the tremendous pressure of the water causes it to shrink to a fraction of its original volume. This is an extreme example of the volume change that occurs in all solids when the pressure of their surroundings is changed. The general situation is illustrated in **Figure 17–12**, where we show a spherical solid whose volume decreases by the amount ΔV when the pressure acting on it increases by the amount ΔP. Experiments show that the pressure difference required to cause a given change in volume, ΔV, is proportional to ΔV and inversely proportional to the initial volume of the object, V_0. Therefore, we can write ΔP as follows:

$$\Delta P = -B\left(\frac{\Delta V}{V_0}\right) \qquad \text{17–19}$$

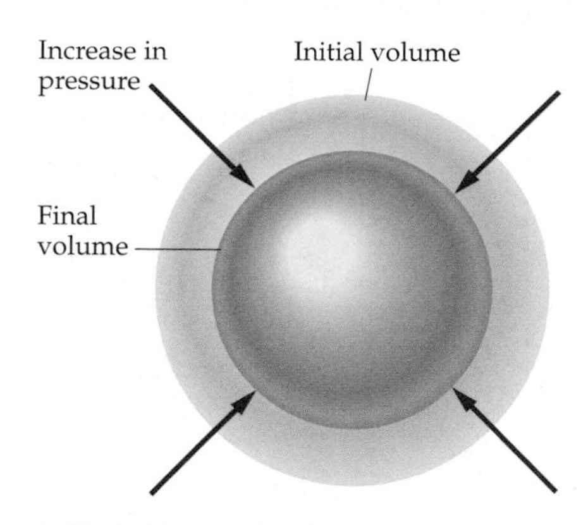

▲ FIGURE 17–12 Changing the volume of a solid
As the pressure surrounding an object increases, its volume decreases. The amount of volume change is proportional to the initial volume and to the change in pressure.

The constant of proportionality in this case, B, is called the **bulk modulus**. As with Young's modulus and the shear modulus, the bulk modulus is defined to be a positive quantity; hence the minus sign in Equation 17–19. For example, if the pressure increases ($\Delta P > 0$) the volume will decrease ($\Delta V < 0$) and the quantity $-\Delta V$ will be positive. Since ΔP is equal to $B(-\Delta V/V_0)$, and V_0 is always positive, it follows that B must be positive as well.

Table 17–3 gives a list of representative values of the bulk modulus. Note the large magnitudes of B, indicating that even small volume changes require large changes in pressure. Finally, the units of B are N/m^2, as is clear from Equation 17–19.

TABLE 17–3
Bulk Modulus for Various Materials

Material	Bulk modulus, B (N/m^2)
Gold	22×10^{10}
Tungsten	20×10^{10}
Steel	16×10^{10}
Copper	14×10^{10}
Aluminum	7.0×10^{10}
Brass	6.1×10^{10}
Ice	0.80×10^{10}
Water	0.22×10^{10}
Oil	0.17×10^{10}

ACTIVE EXAMPLE 17–4 A Gold Doubloon

A gold doubloon 6.1 cm in diameter and 2.0 mm thick is dropped over the side of a pirate ship. When it comes to rest on the ocean floor at a depth of 770 m, how much has its volume changed?

Solution

1. Solve Equation 17–19 for the change in volume: $\Delta V = -V_0\Delta P/B$
2. Calculate the initial volume of the doubloon (a cylinder): $V_0 = 5.8 \times 10^{-6}\ m^3$
3. Find the change in pressure due to the depth of sea water. Use Equation 15-7: $\Delta P = \rho g h = 7.7 \times 10^6\ N/m^2$
4. Substitute numerical values: $\Delta V = -2.0 \times 10^{-10}\ m^3$

Insight

The change in volume is imperceptibly small. Clearly, enormous pressures must be applied to metals to cause a significant change in volume.

Stress and Strain

Equation 17–19 appears to differ from Equations 17–17 and 17–18 because of the absence of the area A on the right-hand side of the equation. When one recalls that pressure is a force per area, however, we can see that the area in Equation 17–19 is contained in the denominator of the left-hand side. Similarly, we can rewrite Equation 17–17 as $F/A = Y(\Delta L/L_0)$ and Equation 17–18 as $F/A = S(\Delta x/L_0)$. Written in this way, each of these equations states that a deformation of a particular type is proportional to a corresponding applied force per area.

In general, we refer to an applied force per area as a **stress** and the resulting deformation as a **strain**. If the stress applied to an object is not too large the proportional relationship between the strain and stress is found to hold, and a plot of

▲ Styrofoam has a very small bulk modulus, which means that even a relatively small increase in pressure can cause a large decrease in volume. The Styrofoam cup at right was immersed to a depth of 1955 m, where the water pressure is nearly 200 atm.

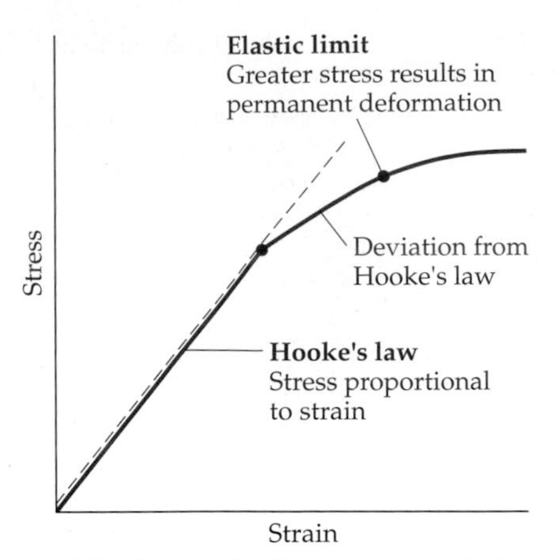

▲ **FIGURE 17–13 Stress versus strain**
When stress and strain are small they are proportional, as given by Hooke's law. Larger stresses can result in deviations from Hooke's law. Stresses greater than the elastic limit cause permanent deformation of the material.

strain versus stress gives a straight line, as indicated in **Figure 17–13**. This straight-line relationship is simply a generalization of Hooke's law—extending the result for a spring to any solid object. As the stress becomes larger the strain eventually begins to increase at a rate that is greater than the straight line.

A change in behavior occurs when the stress reaches the elastic limit. For stresses less than the elastic limit, the deformation of an object is reversible; that is, the deformation vanishes as the stress that caused it is reduced to zero. This is just like a spring returning to its equilibrium position when the applied force vanishes. When a deformation is reversible, we say that it is an **elastic deformation**. For stresses greater than the elastic limit, on the other hand, the object becomes permanently deformed, and it will not return to its original size and shape when the stress is removed. This is like a spring that has been stretched too far, or a car fender that has been dented. If the stress on an object is increased even further, the object eventually tears apart or fractures.

17–4 Phase Equilibrium and Evaporation

To see what is meant by phases of matter in equilibrium, consider a closed container that is partially filled with a liquid, as in **Figure 17–14**. The container is kept at the constant temperature T_0, and initially the volume above the liquid is empty of molecules—it is a vacuum. Soon, however, some of the faster molecules in the liquid begin to escape the relatively strong intermolecular forces of their neighbors in the liquid and start to form a low-density gas. Occasionally a gas molecule collides with the liquid and reenters it, but initially, more molecules are entering the gas than returning to the liquid.

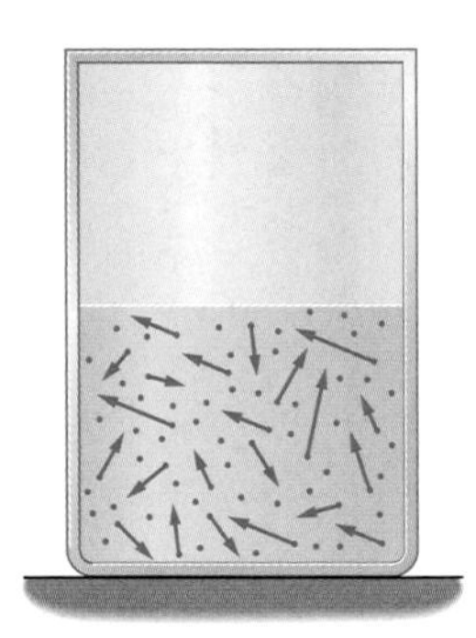
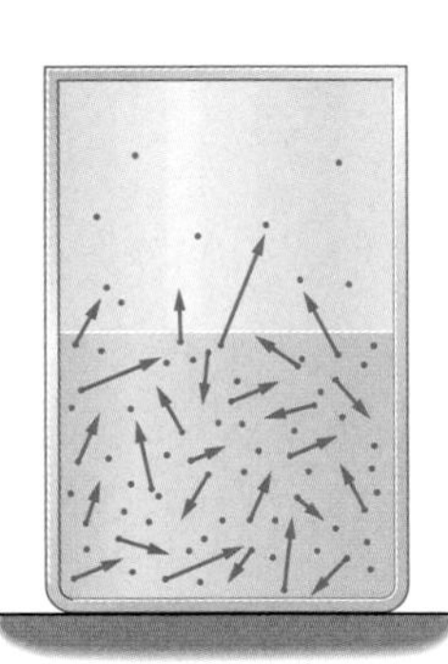
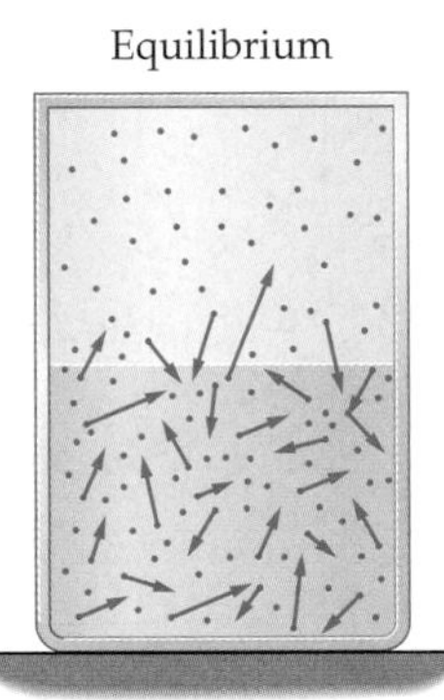

▶ **FIGURE 17–14 A liquid in equilibrium with its vapor**
Initially, a liquid is placed in a container, and the volume above it is a vacuum. High-speed molecules in the liquid are able to escape into the upper region of the container, forming a low-density gas. As the gas becomes more dense, the number of molecules leaving the liquid is balanced by the number returning to the liquid. At this point the system is in equilibrium.

This process continues until the gas is dense enough that the number of molecules returning to the liquid equals the number entering the gas. There is a constant "flow" of molecules in both directions, but when **phase equilibrium** is reached these flows cancel, and the number of molecules in each phase remains constant. The pressure of the gas when equilibrium is established is referred to as the **equilibrium vapor pressure**. In **Figure 17–15** we plot the equilibrium vapor pressure of water.

What happens when we change the temperature? Well, if we increase the temperature there will be more high-speed molecules in the liquid that can escape into the gas. Thus, to have an equal number of gas molecules returning to the liquid, it will be necessary for the pressure of the vapor to be greater. Thus, the equilibrium vapor pressure increases with temperature. This is also illustrated in Figure 17–15.

For each temperature there is just one equilibrium vapor pressure—that is, just one pressure where the precise balance between the phases is established. Thus, when we plot the equilibrium vapor pressure versus temperature, as in Figure 17–15,

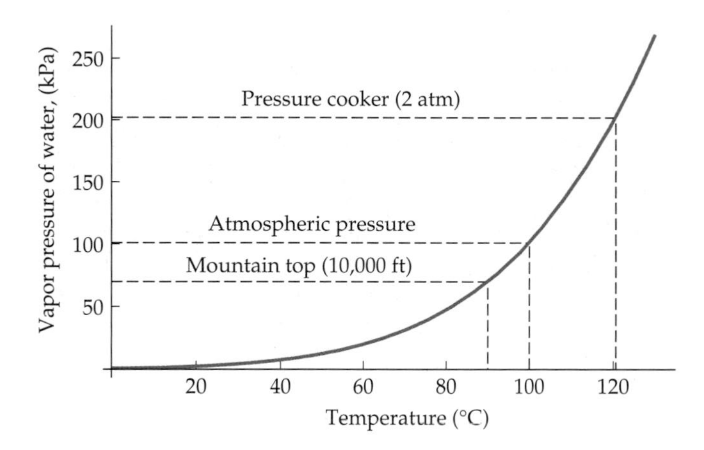

◀ **FIGURE 17–15 The vapor-pressure curve for water**

The vapor pressure of water increases with increasing temperature. In particular, at $T = 100\,^{\circ}\mathrm{C}$ the vapor pressure is one atmosphere, 101 kPa.

the result is a curve—the **vapor pressure curve**. The significance of this curve is that it determines the boiling point of a liquid. In particular:

> A liquid boils at the temperature at which its vapor pressure equals the external pressure.

Note in Figure 17–15 that a vapor pressure equal to atmospheric pressure, $P_{\mathrm{at}} = 101$ kPa, occurs when the temperature of the water is 100 °C, as expected.

CONCEPTUAL CHECKPOINT 17–4

When water boils at the top of a mountain, is its temperature **(a)** more than, **(b)** less than, or **(c)** equal to 100 °C?

Reasoning and Discussion

At the top of a mountain, air pressure is less than it is at sea level. Therefore, according to Figure 17–15, the boiling temperature will be less as well.

For example, the atmospheric pressure at the top of a 10,000-ft mountain is roughly three-quarters what it is at sea level. The corresponding boiling point, as shown in Figure 17–15, is about 90 °C. Thus, cooking food in boiling water may be very difficult at such altitudes. Similarly, by measuring the boiling temperature of a pot of water on the summit of an uncharted mountain, it is possible to gain a rough estimate of its altitude.

Answer:

(b) The temperature of the water is less than 100 °C.

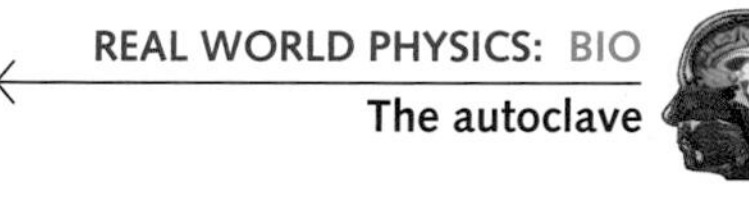

An autoclave (which is basically an elaborate pressure cooker) sterilizes surgical tools by using the same principle discussed in Conceptual Checkpoint 17–4, only in the opposite direction. If surgical tools were heated in boiling water open to the atmosphere, they would experience a temperature of 100 °C. In the autoclave, which is a sealed vessel, the pressure rises to values significantly greater than atmospheric pressure. As a result, the water has a much higher boiling temperature, and the sterilization is more effective. In the pressure cooker, this elevated temperature results is a reduced cooking time.

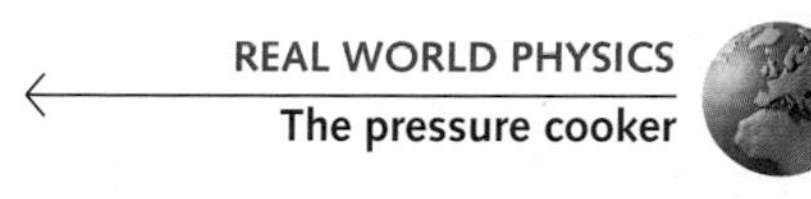

Another way to increase the boiling temperature of water is to add a pinch of salt. The salt dissolves into sodium and chloride ions in the water. These ions have a strong interaction with the water molecules, making it harder for them to break free of the liquid and enter the vapor phase. As a result, the boiling temperature rises. This is why salt is often added to water when boiling eggs; the result is a higher temperature of boiling, which helps to solidify any material that might leak out of small cracks in the eggs.

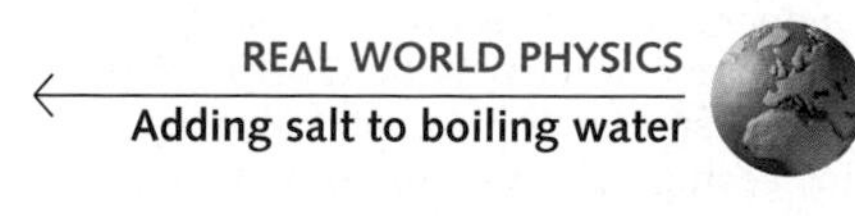

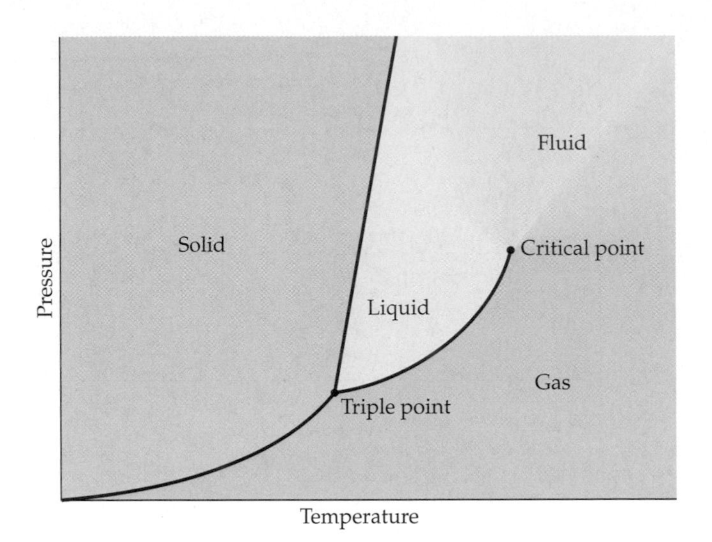

▶ FIGURE 17–16 A typical phase diagram

This phase diagram shows the regions corresponding to each of the three common phases of matter. Note that all three phases are in equilibrium at the triple point. In addition, the liquid and gas phases become indistinguishable beyond the critical point, where they are referred to as a fluid.

In **Figure 17–16**, note that the vapor-pressure curve comes to an end at a finite temperature and pressure. This end point is called the **critical point**. Beyond the critical point there is no longer a distinction between liquid and gas. They are one and the same, and are referred to, simply, as a fluid. Thus, as mentioned before, the liquid and gas phases are very similar—the liquid is just more dense than the gas and its molecules are somewhat less mobile.

A curve similar to the vapor-pressure curve indicates where the solid and liquid phases are in equilibrium. It is referred to as the **fusion curve**. Similarly, equilibrium between the solid and gas phases occurs along the **sublimation curve**. All three equilibrium curves are shown in Figure 17–16, on a plot that is referred to as a **phase diagram**.

Note that the phase diagram also indicates that there is one particular temperature and pressure where all three phases are in equilibrium. This point is called the **triple point**. In water, the triple point occurs at the temperature $T = 273.16$ K and the pressure $P = 611.2$ Pa. At this temperature and pressure, ice, water, and steam are all in equilibrium with one another.

Finally, there is one feature of the fusion line that is of particular interest. In most substances, this line has a positive slope, as in Figure 17–16. This means that as the pressure is increased, the melting temperature of the substance also increases. This is sensible, because a solid is generally more dense than the corresponding liquid. Hence, if you apply pressure to a liquid—with the temperature held constant—the system will tend to become more dense and eventually solidify. This is indicated in **Figure 17–17 (a)**, where we see that an increase in pressure at constant temperature results in crossing the fusion curve.

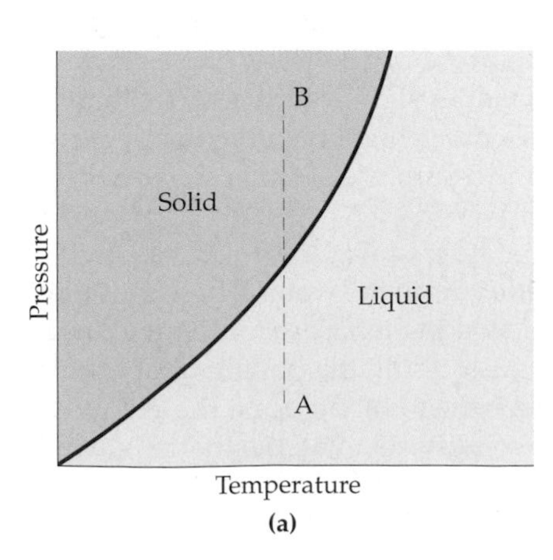

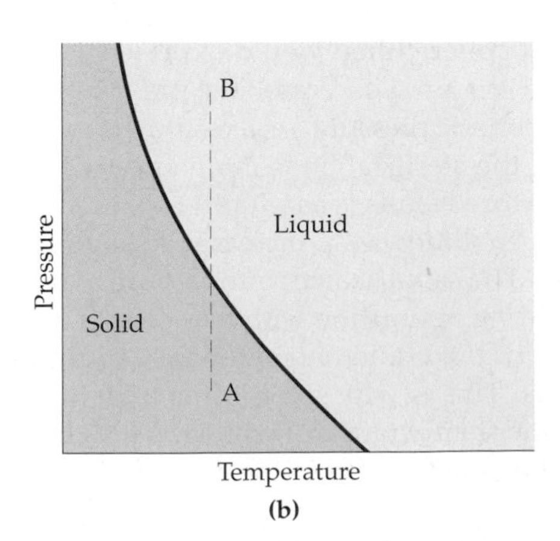

▶ FIGURE 17–17 The solid-liquid phase boundary

Fusion curves for a typical substance **(a)** and for water **(b)**. In (a), we note that an increase in pressure at constant temperature results in a liquid being converted to a solid. For the case of water, increasing the pressure exerted on ice at constant temperature can result in the ice melting.

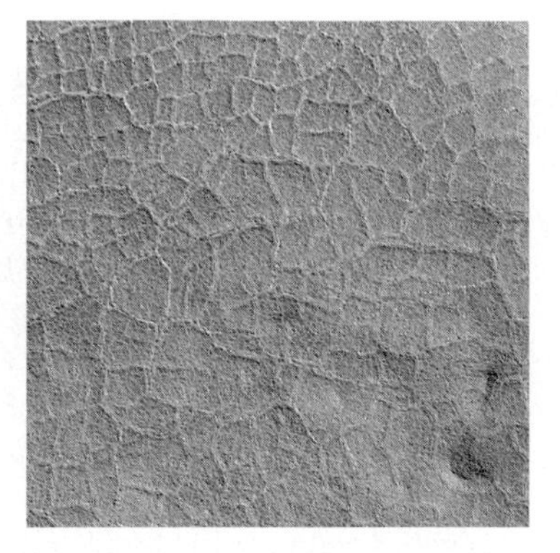

◀ The repeated expansion and contraction of water as it freezes and thaws over long periods of time can cause the ground to crack and buckle. Eventually, the resulting cracks may combine to form polygonal networks referred to as patterned ground (left), a common feature in arctic regions of the Earth. The same type of patterned ground has recently been observed on the surface of Mars (right), giving a strong indication that water may exist just below the surface in certain parts of the planet.

There are exceptions to this rule, however, and it will probably come as no surprise that water is one such exception. This is related to the fact that ice is less dense than water. As a result, the fusion curve for water has a negative slope coming out of the triple point. This is illustrated in **Figure 17–17 (b)**. What this means is that if the temperature is held constant, ice will melt when the pressure applied to it is increased. Thus, the pressure due to an ice skating blade, for example, can cause the ice beneath it to melt.

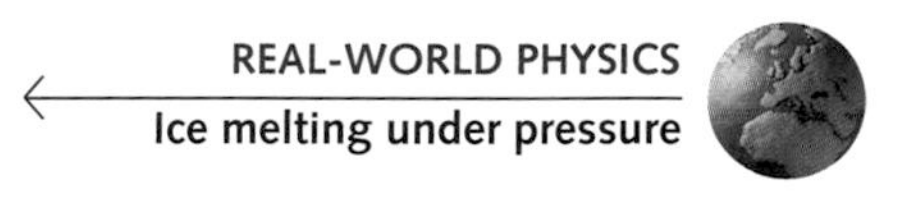

The expansion of water as it freezes has important implications in geology as well. Suppose, for example, that water finds its way into cracks and fissures on a rocky cliff. If the temperature dips below freezing, the water in these cracks will begin to freeze and expand. This tends to split the rock farther apart in a process known as *frost wedging*. Over time, the repeated freezing and melting of water can break a "solid" rock cliff into a pile of debris, forming a talus slope at the base of the cliff. Similar effects occur in *frost heaving*, where expanding ice lifts rocks that were buried in the soil to the surface—a phenomenon well known in areas with long, cold winters.

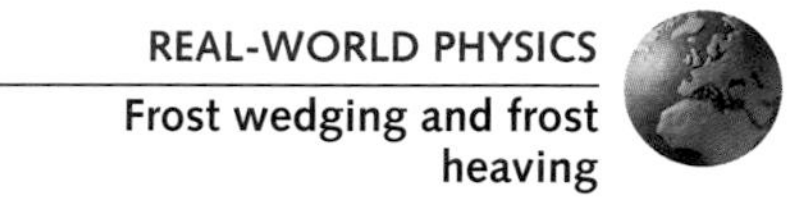

Finally, the expansion of ice can have devastating effects in living systems. If the blood cells of an organism were to freeze, for example, the resulting expansion could rupture the cells. The icefish, a transparent fish found in southern polar seas, has a high concentration of glycoproteins in its bloodstream that serve as a biological antifreeze, inhibiting the formation and growth of ice crystals.

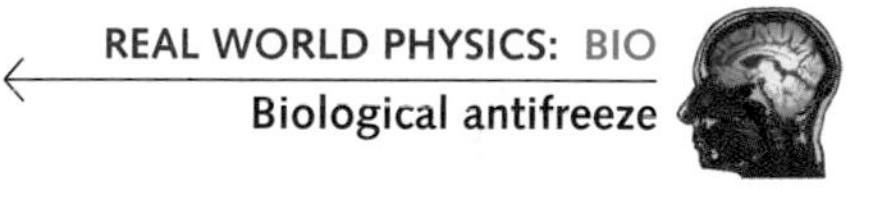

Evaporation

If you think about it a moment, you realize that evaporation is a bit odd. After all, when you are hot and sweaty from physical exertion, steam rises from your skin. But how can this be, since water boils at 100 °C? How can a skin temperature of only 30 or 35 °C result in steam?

Recall that if a liquid is placed in a closed container with a vacuum above it, some of its fastest molecules will break loose and form a gas. When the gas becomes dense enough its pressure rises to the equilibrium vapor pressure of the liquid, and the system attains equilibrium. But what if you open the container and let a breeze blow across it, removing much of the gas? In that case, the release of molecules from the liquid into the gas—that is, **evaporation**—continues without reaching equilibrium. As the molecules are continually removed, the liquid progressively evaporates until none is left. This is the basic mechanism of evaporation.

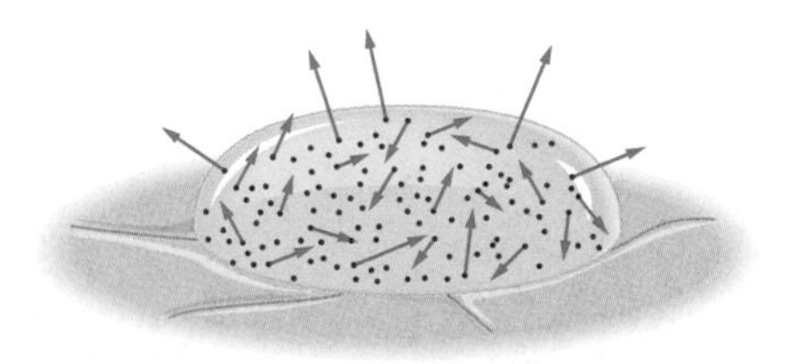

▲ **FIGURE 17–18 A droplet of sweat resting on the skin**
High-speed molecules in a droplet of sweat are able to escape the droplet and become part of the atmosphere. The average speed of the molecules that remain in the droplet is reduced, so the temperature of the droplet is reduced as well.

Let's investigate how evaporation helps to cool us when we exercise or work up a sweat. First, consider a droplet of sweat on the skin, as illustrated in **Figure 17–18**. As mentioned before, the high-speed molecules in the drop are the ones that will escape from the liquid into the surrounding air. The breeze takes these molecules away as soon as they escape, hence the chance of their reentering the drop of sweat is very small.

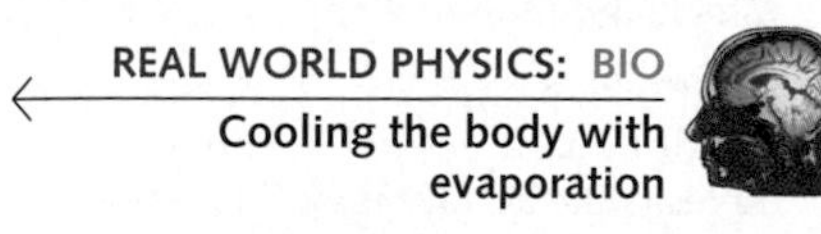

▲ Although the body temperature of these bison is well below the boiling point of water, water still evaporates rapidly from their skin. (Since water vapor is invisible, the "steam" we see in such photos, though it indicates the presence of evaporation, is not the water vapor itself. Rather, it is a cloud of tiny droplets that form when the water vapor loses heat to the cold air around it and condenses back to the liquid state.)

Now what does this mean for the molecules that are left behind in the sweat droplet? Well, since the droplet is preferentially losing high-speed molecules, the average kinetic energy of the remaining molecules must decrease. As we know from kinetic theory, this means that the temperature of the droplet must also decrease. Since the droplet is now cooler than its surroundings, including the skin on which it rests, it draws in heat from the body. This warms the droplet, increasing the speed of its molecules, and continuing the evaporation process at more or less the same rate. Thus, sweat droplets are an effective means of drawing heat from the body, and releasing it into the surrounding air in the form of high-speed water molecules.

To look at this a bit more quantitatively, consider the Maxwell speed distribution for water molecules in a sweat droplet at 30 °C. This is shown in **Figure 17–19**. The rms speed at this temperature is 648 m/s. Also shown in Figure 17–19 is the speed distribution for water molecules at 100 °C. In this case the rms speed is 719 m/s, only slightly greater than that at 30 °C. Thus, it is clear that if water molecules at 100 °C have enough speed to escape into the gas phase, many molecules at 30 °C will be able to escape as well.

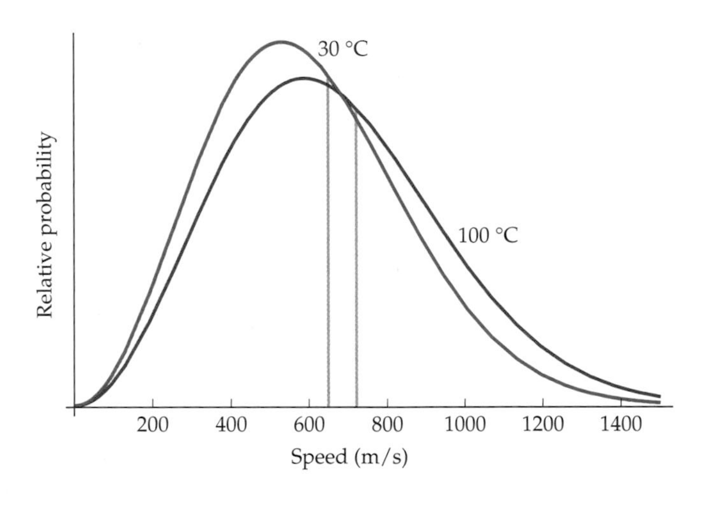

▶ **FIGURE 17–19 Speed distribution for water**

The red curve shows the speed distribution for water at 30 °C, the blue curve corresponds to 100 °C. Note that the rms speed for 100 °C is only slightly greater than that for 30 °C.

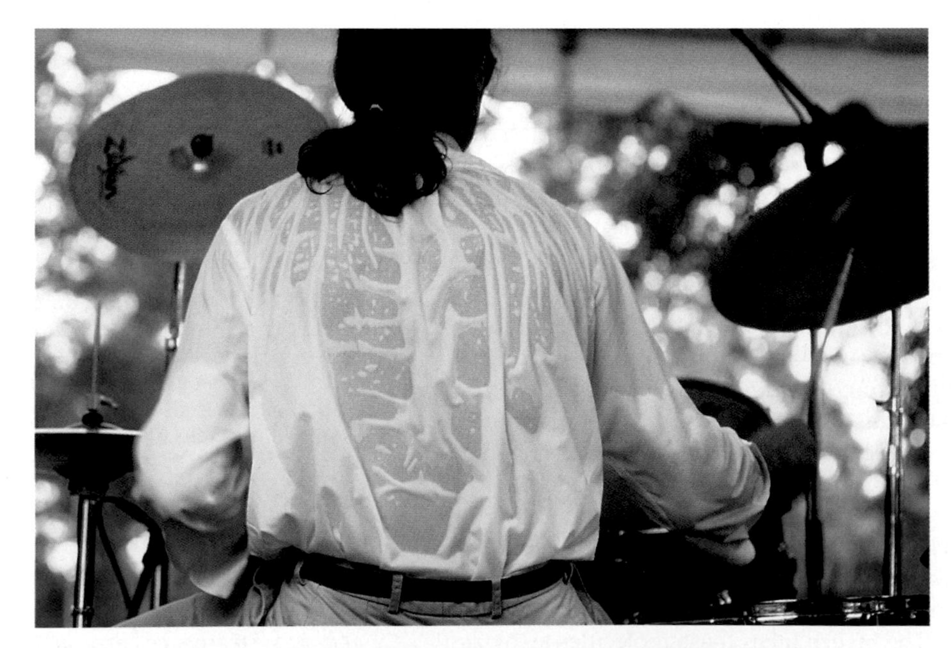

▲ Human beings keep cool by sweating. Because the most energetic molecules are the ones most likely to escape by evaporation, significant quantities of heat are removed from the body when perspiration evaporates from the skin. Dogs, lacking sweat glands, nevertheless take advantage of the same mechanism to help regulate body temperature. In hot weather they pant to promote evaporation from the tongue. Of course, sitting in a cool pond can also help. (What mechanism is involved here?)

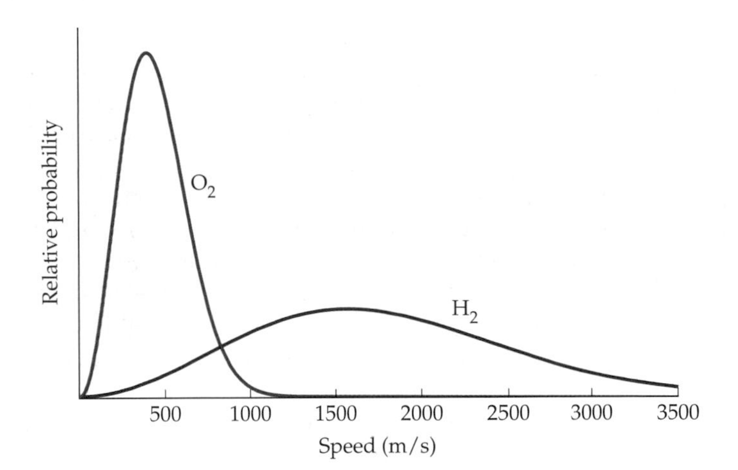

FIGURE 17–20 Speed distribution for O_2 and H_2 at 20 °C
The typical speeds of an H_2 molecule (red curve) are much greater than those for an O_2 molecule (blue curve). In fact, some H_2 molecules move fast enough to escape from the Earth's atmosphere.

The Evaporating Atmosphere

The atmosphere of a planet or a moon can evaporate in much the same way a drop of sweat evaporates from your forehead. In the case of an atmosphere, however, it is the force of gravity that must be overcome. If an astronomical body has a weak gravitational field, some molecules in its atmosphere may be moving rapidly enough to escape; that is, some molecules may have speeds in excess of the escape speed for that body.

Consider the Earth, for example. If a rocket is to escape the Earth it must have a speed of at least 11,200 m/s. Recall from Chapter 12, however, that the escape speed is independent of the mass of rocket. Thus, it applies equally well to molecules and rockets. As a result, molecules moving faster than 11,200 m/s may escape the Earth.

Now, let's compare the escape speed to the speeds of some of the molecules in the Earth's atmosphere. We have already seen in this chapter that the speed of nitrogen and oxygen are on the order of the speed of sound; that is, several hundred meters per second. This is much below the Earth's escape speed. Thus, the odds against a nitrogen or oxygen molecule having enough speed to escape the Earth is truly astronomical. Good thing, too, since this is why these molecules have persisted in our atmosphere for billions of years.

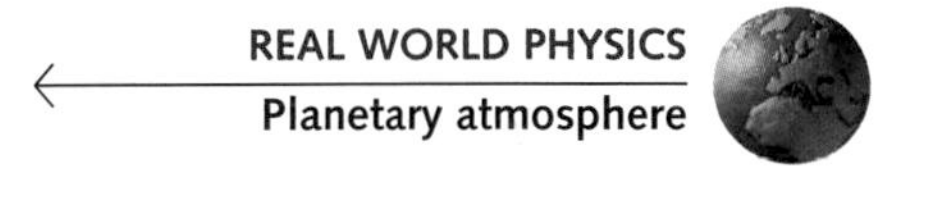

On the other hand, consider a lightweight molecule like hydrogen, H_2. The fact that its average kinetic energy is the same as that of all the other molecules in the air (see Equation 17–12) means that its speed will be much greater than the speed of, say, oxygen or nitrogen. This is illustrated in **Figure 17–20**, where we show the speed distribution for both O_2 and H_2. Clearly, it is very likely to find an H_2 molecule with a speed on the order of a couple thousand meters per second. Because of the higher speeds for H_2, the probability that an H_2 molecule will have enough speed to escape the Earth is about 300 orders of magnitude greater than the corresponding probability for O_2. It is no surprise, then, that Earth's atmosphere contains essentially no hydrogen.

On Jupiter, however, gravity is more intense than on Earth and the temperature is less. As a result, not even hydrogen moves quickly enough to escape. In fact, Jupiter's atmosphere is composed mostly of H_2 and He.

At the other extreme, the Moon has a rather weak gravitational field. In fact, it is unable to maintain any atmosphere at all. Whatever atmosphere it may have had early in its history has long since evaporated. You might say that the Moon's atmosphere is "lost in space."

17–5 Latent Heats

When two phases coexist something surprising happens—the temperature remains the same even when you add a small amount of heat. How can that be?

To understand this behavior, let's start by considering an ice cube initially at the temperature $-10\,°C$. Adding an amount of heat Q results in a temperature increase ΔT given by the relation $Q = mc_{ice}\,\Delta T$, as discussed in Chapter 16. When the ice cube's temperature reaches $0\,°C$, however, adding more heat does not cause an additional increase in temperature. Instead, the heat goes into converting some of the ice into water. On a microscopic level, the added heat causes some of the molecules in the solid ice to vibrate with enough energy to break loose from neighboring molecules and become part of the liquid.

Thus, as long as any ice remains in a cup of water, *and* the water and ice are in equilibrium, you can be sure that both the ice and the water are at $0\,°C$. If heat is added to the system the amount of ice decreases; if heat is removed the amount of ice increases. The amount of heat required to completely convert 1 kg of ice to water is referred to as the **latent heat**, L. In general, the latent heat is defined as follows:

> The latent heat, L, is the heat that must be added to or removed from one kilogram of a substance to convert it from one phase to another. During the conversion process, the temperature of the system remains constant.

Just as with the specific heat, the latent heat is always a positive quantity.

In mathematical form, we can say that the heat Q required to convert a mass m from one phase to another is mL. This gives the following relation:

Definition of Latent Heat, *L*

$$Q = mL \qquad \text{17–20}$$

SI unit: J/kg

The value of the latent heat depends on which phases are involved. For example, the latent heat to melt (or fuse) a substance is referred to as the **latent heat of fusion**, L_f. Similarly, the latent heat required to convert a liquid to a gas is the **latent heat of vaporization**, L_v, and the latent heat needed to convert a solid directly to a gas is the **latent heat of sublimation**, L_s. Typical latent heats are given in Table 17–4.

The relationship between the temperature of a substance and the heat added to it is illustrated in **Figure 17–21**. Initially 1 kg of H_2O is in the form of ice at $-20\,°C$. As heat is added to the ice its temperature rises until it begins to melt at $0\,°C$. The temperature then remains constant until the latent heat of fusion is supplied to the system. When all the ice has melted to water at $0\,°C$, continued heating results in a renewed increase in temperature. When the temperature of the water rises to $100\,°C$, boiling begins and the temperature again remains constant—this time until an amount of heat equal to the latent heat of vaporization is added to the system. Finally, with the entire system converted to steam, continued heating again produces an increasing temperature.

TABLE 17–4 Latent Heats for Various Materials

Material	Latent heat of fusion (J/kg)	Latent heat of vaporization (J/kg)
Water	33.5×10^4	22.6×10^5
Ammonia	33.2×10^4	13.7×10^5
Copper	20.7×10^4	47.3×10^5
Benzene	12.6×10^4	3.94×10^5
Ethyl alcohol	10.8×10^4	8.55×10^5
Gold	6.28×10^4	17.2×10^5
Nitrogen	2.57×10^4	2.00×10^5
Lead	2.32×10^4	8.59×10^5
Oxygen	1.39×10^4	2.13×10^5

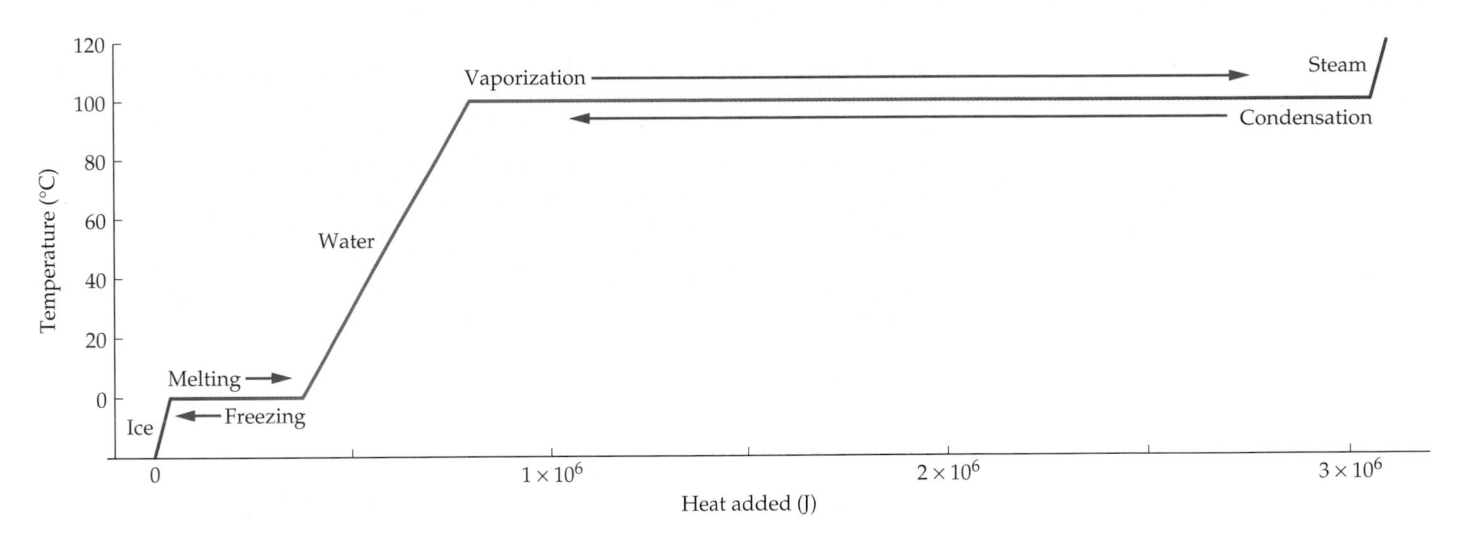

▲ **FIGURE 17–21 Temperature versus heat added or removed**
The temperature of 1.0 kg of water as heat is added to or removed from the system. Note that the temperature stays the same—even as heat is added—when the system is changing from one phase to another.

CONCEPTUAL CHECKPOINT 17–5

Both water at 100 °C and steam at 100 °C can cause serious burns. Is a burn produced by steam likely to be **(a)** more severe, **(b)** less severe, or **(c)** the same as a burn produced by water?

Reasoning and Discussion

As the water or steam comes into contact with the skin it cools from 100 °C to a skin temperature of something like 35 °C. For the case of water, this means that a certain amount of heat is transferred to the skin, which can cause a burn. The steam, on the other hand, must first give off the heat required for it to condense to water at 100 °C. After that, the condensed water cools to body temperature, as before. Thus, the heat transferred to the skin will be larger in the case of steam, resulting in a more serious burn.

Answer:

(a) The steam burn is worse.

As a numerical example of using latent heat, let's calculate the heat energy required to raise the temperature of 0.550 kg of ice from −20.0 °C to water at 20.0 °C; that is, from point A to point B in **Figure 17–22**. The way to approach a problem like this is to take it one step at a time—that is, each phase, and each conversion from one phase to another, should be treated separately.

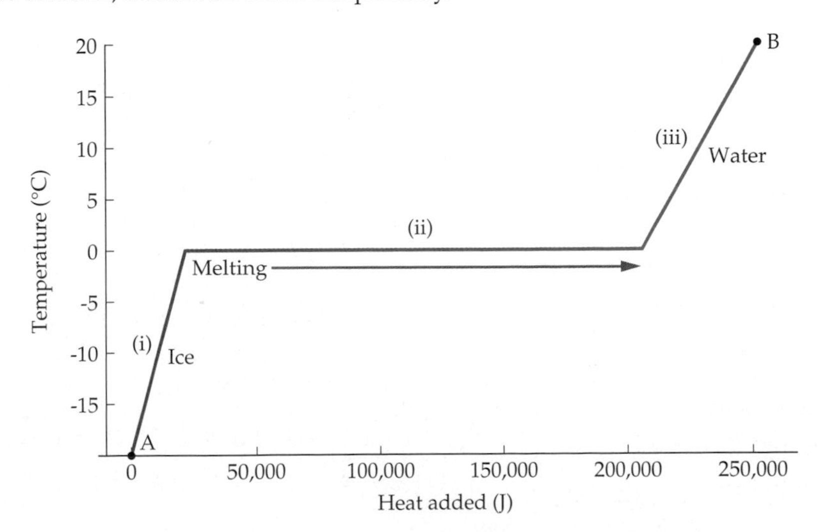

◀ **FIGURE 17–22 Heat required for a given change in temperature**
The amount of heat required to raise 0.550 kg of H_2O from ice at −20.0 °C to water at 20.0 °C is the heat difference between points A and B. To calculate this heat we sum the following three heats: (i) the heat to warm the ice from −20.0 °C to 0 °C; (ii) the heat to melt all the ice; (iii) the heat to warm the water from 0 °C to 20.0 °C.

Thus, the first step is to find the heat necessary to warm the ice from −20.0 °C to 0 °C. Using the specific heat of ice, $c_{ice} = 2090\ \text{J}/(\text{kg}\cdot\text{C}°)$, we find

$$Q_1 = mc_{ice}\Delta T = (0.550\ \text{kg})[2090\ \text{J}/(\text{kg}\cdot\text{C}°)](20.0\ \text{C}°) = 23{,}000\ \text{J}$$

The second step in the process is to melt the ice at 0 °C. The heat required for this is found using the latent heat of fusion for water ($L_f = 33.5 \times 10^4\ \text{J/kg}$):

$$Q_2 = mL_f = (0.550\ \text{kg})(33.5 \times 10^4\ \text{J/kg}) = 184{,}000\ \text{J}$$

PROBLEM SOLVING NOTE

Specific Heats versus Latent Heats

In solving problems involving specific heats and latent heats, recall that specific heats give the heat related to a *change in temperature* in a given phase, and latent heats give the heat related to a *change in phase* at a given temperature.

Finally, the third step is to heat the water at 0 °C to 20.0 °C. This time we use the specific heat for water, $c_{water} = 4186\ \text{J}/(\text{kg}\cdot\text{C}°)$:

$$Q_3 = mc_{water}\Delta T = (0.550\ \text{kg})[4186\ \text{J}/(\text{kg}\cdot\text{C}°)](20.0\ \text{C}°) = 46{,}000\ \text{J}$$

The total heat required for this process, then, is

$$\begin{aligned} Q_{total} &= Q_1 + Q_2 + Q_3 \\ &= 23{,}000\ \text{J} + 184{,}000\ \text{J} + 46{,}000\ \text{J} = 253{,}000\ \text{J} \end{aligned}$$

We consider a similar problem in the following Example.

EXAMPLE 17–5 Steam Heat

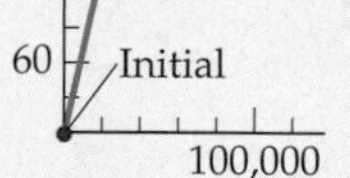

To make steam, you add 5.60×10^5 J of heat to 0.220 kg of water at an initial temperature of 50.0 °C. Find the final temperature of the steam.

Picture the Problem

Our sketch shows the temperature-versus-heat-added curve for water. The initial point for this system, placed at the origin, is at 50.0 °C. As we shall see, adding the given amount of heat, $Q = 5.60 \times 10^5$ J, raises the temperature to the point labeled "final" in the plot.

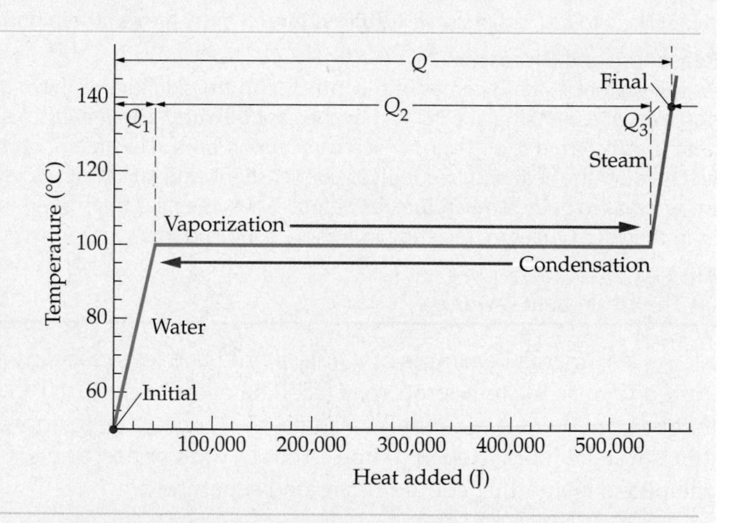

Strategy

To find the final temperature, we first calculate the amount of heat that must be added to heat the water to 100 °C. If this is less than the total heat added, we continue by calculating the amount of heat needed to vaporize all the water. If the sum of these two heats is still less than the total heat added to the water, we calculate the increase in temperature when the remaining heat is added to the steam.

Solution

1. Calculate the heat that must be added to the water to heat it to 100 °C. Call the result Q_1:

$$\begin{aligned} Q_1 &= mc_{water}\Delta T \\ &= (0.220\ \text{kg})[4186\ \text{J}/(\text{kg}\cdot\text{C}°)](50.0\text{C}°) \\ &= 4.60 \times 10^4\ \text{J} \end{aligned}$$

2. Next, calculate the heat that must be added to the water to convert it to steam. Let this result be Q_2:

$$\begin{aligned} Q_2 &= mL_f = (0.220\ \text{kg})(22.6 \times 10^5\ \text{J/kg}) \\ &= 4.97 \times 10^5\ \text{J} \end{aligned}$$

3. Determine the heat that is still to be added to the system. Let this remaining heat be Q_3:

$$\begin{aligned} Q_3 &= 5.60 \times 10^5\ \text{J} - Q_1 - Q_2 \\ &= 5.60 \times 10^5\ \text{J} - 4.60 \times 10^4\ \text{J} - 4.97 \times 10^5\ \text{J} \\ &= 17{,}000\ \text{J} \end{aligned}$$

4. Use Q_3 to find the increase in temperature of the steam:

$$Q_3 = mc_{steam}\Delta T$$

$$\Delta T = \frac{Q_3}{mc_{steam}} = \frac{17{,}000\ \text{J}}{(0.220\ \text{kg})(2010\ \text{J}/(\text{kg}\cdot\text{C}°))} = 38\text{C}°$$

Insight
Thus, the system ends up completely converted to steam at a temperature of 138 °C.

Practice Problem
Find the final temperature if the amount of heat added to the system is 3.40×10^5 J. [**Answer:** In this case, only 0.130 kg of water vaporizes into steam. Hence the final temperature is 100 °C, with water and steam in equilibrium.]

Some related homework problems: Problem 45, Problem 48

A pleasant application of latent heat is found in the homemade ice cream maker. As you may know, it is necessary to add salt to the ice-water mixture surrounding the ice cream. The dissolved salt molecules interact with water molecules in the liquid, impairing their ability to interact with one another and freeze. This means that ice and water are no longer in equilibrium at 0 °C; a lower temperature is required. The result is that ice begins to melt in the ice-water mixture, and in the process of melting it draws the required latent heat from its surroundings—which include the ice cream. Thus, the salt together with the ice produce a temperature lower than the melting temperature of ice alone.

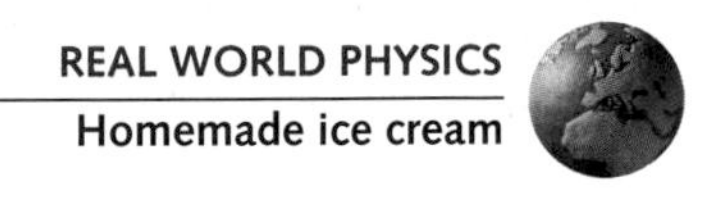

17–6 Phase Changes and Energy Conservation

In the last section we considered problems in which a given amount of heat is simply added to or removed from a system. We now turn to a more interesting type of problem involving energy conservation. In these problems, heat is exchanged within a system—that is, between its parts—but not with the external world. As a result, the total energy of the system is constant. Still, the heat flow within the system can cause changes in temperatures and phases.

The basic idea in solving energy conservation problems is the following:

> Set the magnitude of the heat *lost* by one part of the system equal to the magnitude of the heat *gained* by another.

For example, consider the following problem: A 0.0420-kg ice cube at −10.0 °C is placed in a Styrofoam cup containing 0.350 kg of water at 0 °C. Assuming the cup can be ignored, and that no heat is exchanged with the surroundings, find the mass of ice in the system when the equilibrium temperature of 0 °C is reached.

The initial setup for this problem is illustrated in **Figure 17–23 (a)**. Since the water is warmer than the ice, it follows that heat flows into the ice from the water, as indicated in **Figure 17–23 (b)**. The amount of heat will be just enough to raise the temperature of the ice from −10.0 °C to 0 °C. Thus,

$$Q = \text{heat gained by the ice}$$
$$= mc_{\text{ice}}\Delta T = (0.0420\text{ kg})[2090\text{ J}/(\text{kg}\cdot\text{C}°)](10.0\text{ C}°) = 878\text{ J}$$

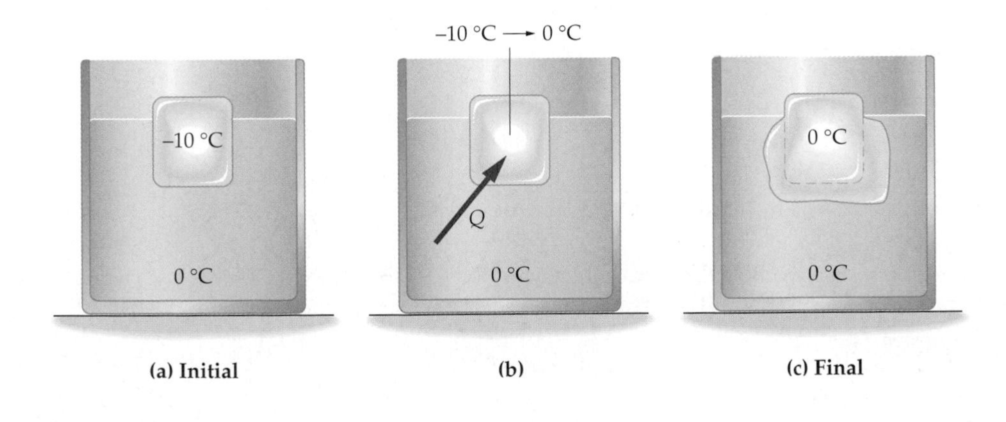

◀ **FIGURE 17–23 Water freezing to form ice**
An ice cube at −10.0 °C is placed in water at 0 °C. Since the ice is colder than the water, heat flows *from* the water *into* the ice. Heat that leaves the water results in the formation of additional ice in the system.

PROBLEM SOLVING NOTE
Determining Equilibrium

In problems involving two different phases—like solid and liquid—it may not be clear in advance whether both phases or only one is present in the equilibrium state. It may be necessary to assume one case or the other and proceed with the calculation based on that assumption. If the result obtained in this way is not physical, the assumption must be changed.

By energy conservation we can say that the heat lost by the water has the same magnitude as the heat gained by the ice. (In general, in these problems it is simplest to calculate the magnitude of each heat, so that all the heats are positive, and then set the "lost" and "gained" magnitudes equal.) In this case, then, we have

$$\text{heat lost by the water} = \text{heat gained by the ice} = 878\text{ J}$$

Thus, 878 J of heat are removed from the water.

Now, since the water is already at 0 °C, removing heat from it does not lower its temperature—instead it merely converts some of the water to ice. How much is converted? The amount is determined by the latent heat of fusion. In particular, we have

$$Q = \text{heat lost by the water}$$
$$= mL_f = 878\text{ J}$$

In this expression, m is the mass of water that has been converted to ice. Solving for the mass yields

$$m = \frac{Q}{L_f} = \frac{878\text{ J}}{33.5 \times 10^4\text{ J/kg}} = 0.00262\text{ kg}$$

Thus, the final amount of ice in the system is 0.0420 kg + 0.00262 kg = 0.0446 kg. This is illustrated in **Figure 17–23 (c)**.

A similar problem is considered in the next Example.

EXAMPLE 17–6 A COOL DRINK

A 0.072-kg ice cube at 0 °C is dropped into a Styrofoam cup holding 0.35 kg of water at 15 °C. Find the final temperature of the system, and the amount of ice (if any) remaining. Assume that the cup and the surroundings can be ignored.

Picture the Problem

Initially, a 0 °C ice cube floats in water at 15 °C. Since the water is warmer than the ice, it loses heat to the ice. During this process, the water cools and some of the ice is melted.

Strategy

First, we calculate the heat lost by the water as it cools to 0 °C. By energy conservation, the heat *lost* by the water is *gained* by the ice. Thus, we can calculate the amount of ice that this heat gain converts to water. If some ice remains, the final temperature of the system is 0 °C.

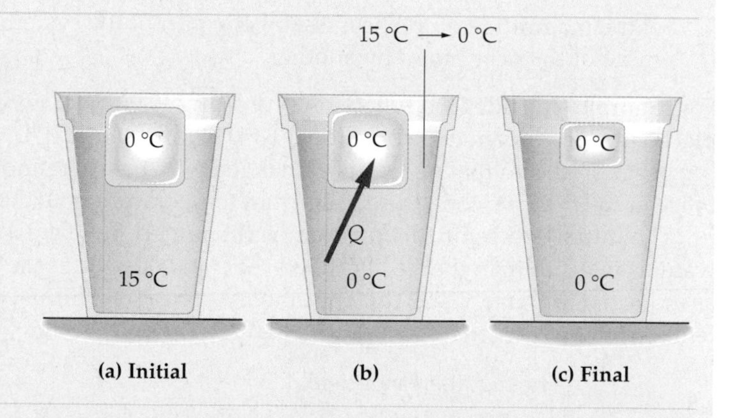

Solution

1. Calculate the heat lost by the water as it cools to 0 °C. Call this heat Q:

$$Q = m_{\text{water}} c_{\text{water}} \Delta T$$
$$= (0.35\text{ kg})[4186\text{ J/(kg}\cdot\text{C}^\circ)](15\text{ C}^\circ) = 2.2 \times 10^4\text{ J}$$

2. Calculate the amount of ice that the heat Q will melt:

$$Q = mL_f$$
$$m = \frac{Q}{L_f} = \frac{2.2 \times 10^4\text{ J}}{33.5 \times 10^4\text{ J/kg}} = 0.066\text{ kg}$$

Insight

Thus, a small amount of ice still remains in the system. As a result, the final temperature is 0 °C.

Practice Problem

Find the initial temperature of the water that would *just barely* melt all the ice. [**Answer:** $T_i = 16$°C. If the water starts with a temperature above this amount, the final temperature of the system will be greater than 0 °C.]

Some related homework problems: Problem 51, Problem 54

Finally, we consider a system in which all of the ice melts, and the final temperature is greater than 0 °C. Still, the basic idea is to apply energy conservation.

EXAMPLE 17–7 Warm Punch

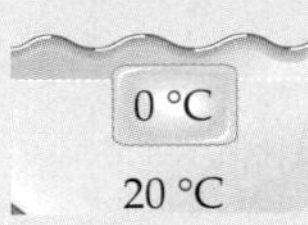

A large punch bowl holds 3.95 kg of lemonade (which is essentially just water) at 20.0 °C. A 0.0450-kg ice cube at 0 °C is placed in the lemonade. What is the final temperature of the system, and the amount of ice (if any) remaining? Ignore any heat exchange with the bowl or the surroundings.

Picture the Problem

Our sketch shows the various heat exchanges imagined for this problem. As one might guess, from the large amount of lemonade and the small amount of ice, all the ice melts.

Strategy

We approach this problem much as we did the problem in the previous Example. We first calculate the heat that would be lost by the water, Q_w, if we cooled it to 0 °C. We then imagine using part of this heat, Q_{ice}, to melt the ice.

As we shall see, a great deal of heat, $Q = Q_w - Q_{ice}$, is left after the ice is melted. We imagine adding this heat back into the system, which now contains 3.95 kg + 0.0450 kg of water at 0 °C. Calculating the increase in temperature caused by the heat Q gives us the final temperature of the system.

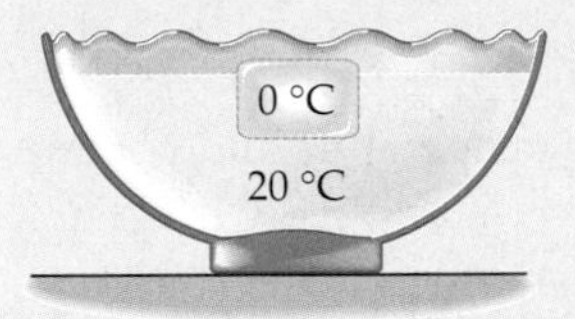

Initial condition

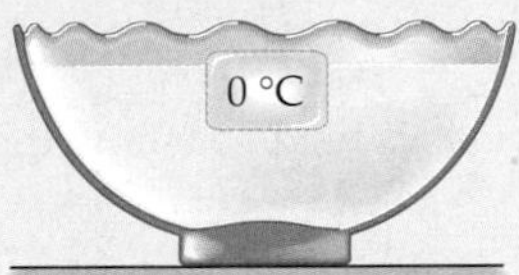

Calculate heat Q_w lost by water, 20 °C ⟶ 0 °C

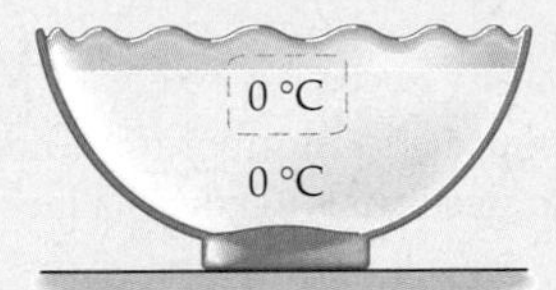

Calculate heat Q_{ice} needed to melt ice cube

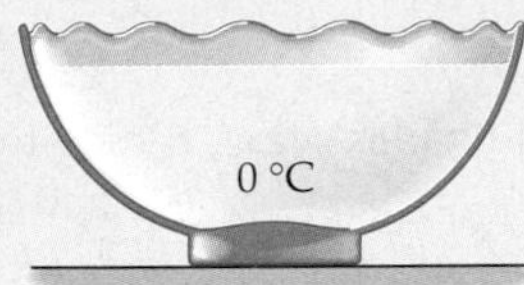

Calculate remaining heat $Q = Q_w - Q_{ice}$

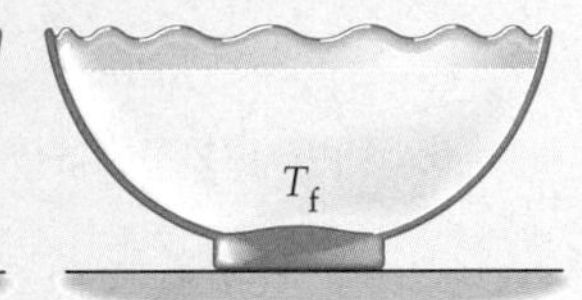

Calculate final temperature when heat Q is added to water and melted ice

Solution

1. Find the heat lost by the water, Q_w, if it is cooled to 0 °C:

$$Q_w = m_{water}c_{water}\Delta T = (3.95\ \text{kg})[4186\ \text{J/(kg}\cdot\text{C°)}](20.0\ \text{C°}) = 3.31 \times 10^5\ \text{J}$$

2. Calculate the amount of heat, Q_{ice}, needed to melt all the ice:

$$Q_{ice} = m_{ice}L_f = (0.0450\ \text{kg})(33.5 \times 10^4\ \text{J/kg}) = 1.51 \times 10^4\ \text{J}$$

3. Determine the amount of heat, Q, that is left:

$$Q = Q_w - Q_{ice} = 3.31 \times 10^5\ \text{J} - 1.51 \times 10^4\ \text{J} = 3.16 \times 10^5\ \text{J}$$

4. Use the heat Q to warm the 3.95 kg + 0.0450 kg of water at 0 °C to its final temperature:

$$Q = (m_{water} + m_{ice})c_{water}\Delta T$$

$$\Delta T = \frac{Q}{(m_{water} + m_{ice})c_{water}} = \frac{3.16 \times 10^5\ \text{J}}{(3.95\ \text{kg} + 0.0450\ \text{kg})[4186\ \text{J/(kg}\cdot\text{C°)}]} = 18.9\ \text{C°}$$

Insight

Therefore, the final temperature of the system is 18.9 °C. As expected, the relatively small ice cube did not lower the temperature of the system very much.

Practice Problem

What would the final temperature of the system be if the ice cube's mass were 0.0750 kg? **[Answer:** $T_f = 18.2$ °C]

Some related homework problems: Problem 51, Problem 54

Chapter Summary

Topic	Remarks and Relevant Equations	
17–1 Ideal Gases	An ideal gas is a simplified model of a real gas in which interactions between molecules are ignored.	
equation of state	The equation of state for an ideal gas is $$PV = NkT$$	17–2
	In this expression, N is the number of molecules, T is the Kelvin temperature, and $k = 1.38 \times 10^{-23}$ J/K is Boltzmann's constant.	
	In terms of the universal gas constant, $R = 8.31$ J/(mol·K), and the number of moles in the gas, n, the ideal-gas equation of state is $$PV = nRT$$	17–5
Avogadro's number and moles	The number of molecules in a mole (mol) is Avogadro's number, $N_A = 6.022 \times 10^{23}$.	
molecular mass	If the mass of an individual molecule is m, its molecular mass, M, is $$M = N_A m$$	17–6
isotherms and Boyle's law	If the temperature and number of molecules is held constant, the pressure and volume of an ideal gas satisfy Boyle's law: $$PV = \text{constant}$$	17–7
constant pressure and Charles's law	If the pressure and number of molecules is held constant, the temperature and volume of an ideal gas satisfy Charles's law: $$\frac{V}{T} = \text{constant}$$	17–8
17–2 Kinetic Theory	In kinetic theory, a gas is imagined to be comprised of a large number of point-like molecules bouncing off the walls of a container.	
the origin of pressure	The pressure exerted by a gas is a result of the momentum transfers that occur every time a molecule bounces off a wall of a container.	
speed distribution of molecules	The molecules in a gas have a range of speeds. The Maxwell distribution indicates which speeds are most likely to occur in a given gas.	
kinetic energy and temperature	Kinetic theory relates the average kinetic energy of the molecules in a gas to the Kelvin temperature of the gas, T: $$\left(\tfrac{1}{2}mv^2\right)_{av} = K_{av} = \tfrac{3}{2}kT$$	17–12
rms speed	The rms (root mean square) speed of the molecules in a gas at the Kelvin temperature T is $$v_{rms} = \sqrt{\frac{3kT}{m}} = \sqrt{\frac{3RT}{M}}$$	17–13
internal energy of an ideal gas	The internal energy of a monatomic ideal gas is $$U = \tfrac{3}{2}NkT = \tfrac{3}{2}nRT$$	17–15, 17–16
17–3 Solids and Elastic Deformation	When a force is applied to a solid, its size and shape may change.	
changing the length of a solid	The force required to change the length of a solid by the amount ΔL is $$F = Y\left(\frac{\Delta L}{L_0}\right)A$$	17–17
	In this expression, Y is Young's modulus, L_0 is the initial length parallel to the applied force, and A is the cross-sectional area perpendicular to the applied force.	
shear deformation	The force required to shear, or deform, a solid by the amount Δx is	

$$F = S\left(\frac{\Delta x}{L_0}\right)A \qquad \textbf{17–18}$$

In this expression, S is the shear modulus, L_0 is the initial length perpendicular to the applied force, and A is the cross-sectional area parallel to the applied force.

changing the volume of a solid

The pressure required to change the volume of a solid by the amount ΔV is

$$\Delta P = -B\left(\frac{\Delta V}{V_0}\right) \qquad \textbf{17–19}$$

In this expression, B is the bulk modulus and V_0 is the initial volume.

stress and strain

The applied force per area is the stress; the resulting deformation is the strain.

elastic deformation

An elastic deformation is one in which a solid returns to its original size and shape when the stress is removed—for example; stretching or compressing an ideal spring.

17–4 Phase Equilibrium and Evaporation

The three most common phases of matter are the solid, liquid, and gas. The solid has a well-defined long-range order due to its crystalline structure. Liquids and gases are more random. They lack the definite structure exhibited by solids, and thus are able to flow and assume the shape of their container.

equilibrium between phases

When phases are in equilibrium the number of molecules in each phase remains constant.

evaporation

Evaporation occurs when some molecules in a liquid have speeds great enough to allow them to escape into the gas phase.

17–5 Latent Heats

The latent heat, L, is the amount of heat that must be added to or removed from a substance to convert it from one phase to another.

latent heat of fusion

The heat required for melting or freezing is called the latent heat of fusion, L_f.

latent heat of vaporization

The heat required for vaporizing or condensing is the latent heat of vaporization, L_v.

latent heat of sublimation

The heat required to sublime a solid directly to a gas, or to condense a gas to a solid, is the latent heat of sublimation, L_s.

17–6 Phase Changes and Energy Conservation

When heat is exchanged within a system, with no exchanges with the surroundings, the energy of the system is conserved.

Problem-Solving Summary

Type of Problem	Relevant Physical Concepts	Related Examples
Find pressure, volume, or temperature in an ideal gas.	These basic quantities are related by the ideal gas equation of state, $PV = NkT = nRT$.	Examples 17–1, 17–2 Active Examples 17–1, 17–2
Relate the rms speed of a gas to the absolute temperature T.	Absolute temperature is directly related to the average kinetic energy, $3kT/2 = K_{av}$. From this connection we obtain the result $v_{rms} = \sqrt{3kT/m}$.	Example 17–3
Find the strain produced by a given stress.	Strain is proportional to stress, at least for small stress. The basic relations are $F/A = Y(\Delta L/L_0)$, $F/A = S(\Delta x/L_0)$, and $\Delta P = -B(\Delta V/V_0)$, where Y, S, and B are the Young's modulus, the shear modulus, and the bulk modulus.	Example 17–4 Active Examples 17–3, 17–4
Calculate the heat associated with a change in phase.	To change from one phase to another a certain amount of heat, called the latent heat L, must be added to or taken from a system. The process occurs at constant temperature. The amount of heat involved in a change of phase is given by $Q = mL$, where m is the mass that changes phase.	Examples 17–5, 17–6, 17–7

Conceptual Questions

1. Is the number of molecules in one mole of N_2 greater than, less than, or equal to the number of molecules in one mole of O_2? Which of these has the greater mass? Explain.
2. Is the number of atoms in one mole of carbon greater than, less than, or the same as the number of atoms in one mole of oxygen gas (O_2)?
3. If you put a helium-filled balloon in the refrigerator, will its volume increase, decrease, or stay the same?
4. Plastic bubble wrap is used as packing material. Why is the bubble wrap less effective on a cold day than on a warm day?
5. Two containers hold ideal gases at the same temperature. Container B has twice the volume and half the number of molecules as container A. How does the pressure of container B compare with that of container A?
6. How is the air pressure in a tightly sealed house affected by operating the furnace?
7. At the beginning of a typical airline flight you are instructed about the proper use of oxygen masks that will fall from the ceiling if the cabin pressure suddenly drops. You are advised that the oxygen masks are working properly, even if the bags do not fully inflate. In fact, the bags expand to their fullest if cabin pressure is lost at high altitude, but expand only partially if the plane is at low altitude. Explain.
8. Is it possible to change both the pressure and the volume of an ideal gas and still not change the average kinetic energy of its molecules? Explain.
9. The average speed of air molecules in your room is on the order of the speed of sound. What is their average velocity?
10. The air in your room is composed mostly of oxygen (O_2) and nitrogen (N_2) molecules. The oxygen molecules are more massive than the nitrogen molecules. How do their rms speeds compare?
11. Two adjacent rooms in a hotel are equal in size and connected by an open door. One of the rooms is warmer than the other. Which room contains more air?
12. As you go up in altitude, do you expect the ratio of oxygen to nitrogen in the atmosphere to increase, decrease, or stay the same? Explain.
13. If the translational speed of molecules in an ideal gas were to double, by what factor has the Kelvin temperature increased? Explain.
14. One of the highest airports in the world is located in La Paz, Bolivia. Pilots prefer to take off from this airport in the morning or the evening, when the air is quite cold. Explain.
15. If the temperature of an ideal gas is doubled from 100 °C to 200 °C does the average kinetic energy of the molecules double?
16. Repeat the above question, this time assuming the temperature is doubled from 100 K to 200 K.
17. A brick has faces that are 1 cm by 2 cm, 2 cm by 3 cm, and 1 cm by 3 cm. On which face should the brick be placed if it is to have the smallest change in dimensions due to its own weight? Explain.
18. A hollow cylindrical rod and a solid cylindrical rod are made of the same material. The two rods have the same length and outer radius. If the same compressional force is applied to each rod, which has the greater change in length? Explain.
19. An autoclave is a device used to sterilize medical instruments. It is essentially a pressure cooker that heats the instruments in water under high pressure. This ensures that the sterilization process occurs at temperatures greater than the normal boiling point of water. Explain why the autoclave produces such high temperatures.
20. A camping stove just barely boils water on a mountain top. When the stove is used at sea level, will it be able to boil water? Explain your answer.
21. Isopropyl alcohol is sometimes rubbed onto a patient's arms and legs to lower their body temperature. Why is this effective?
22. As the temperature of ice is increased it changes first into a liquid and then into a vapor. On the other hand, dry ice, which is solid carbon dioxide, changes directly from a solid to a vapor as its temperature is increased. How might one produce liquid carbon dioxide?
23. A drop of water on a kitchen counter evaporates in a matter of minutes. However, only a relatively small fraction of the molecules in the drop move rapidly enough to escape through the drop's surface. Why, then, does the entire drop evaporate rather than just a small fraction of it?
24. Why do you feel so much cooler in a wet shirt than in a dry one?
25. As we have seen in this chapter, the molecules in a substance at a given temperature move with a range of different speeds. Describe a physical observation that supports this fact.
26. If you toss an ice cube into a swimming pool, is the water in the pool now at 0 °C? Explain.
27. If the latent heat for the vaporization of water into steam were doubled, would the time required to boil away a pot of water increase, decrease, or stay the same? Explain.
28. If the latent heat for the vaporization of water into steam were halved, would sweating be more or less effective as a means of cooling? Explain.

Problems

Note: **IP** *denotes an integrated conceptual/quantitative problem.* **BIO** *identifies problems of biological or medical interest. Blue bullets (•, ••, •••) are used to indicate the level of difficulty of each problem.*

Section 17–1 Ideal Gases

1. • Standard temperature and pressure (STP) is defined as a temperature of 0 °C and a pressure of 101.3 kPa. What is the volume occupied by one mole of an ideal gas at STP?
2. • **BIO** After emptying her lungs, a person inhales 4.2 L of air at 0.0 °C and holds her breath. How much does the volume of the air increase as it warms to her body temperature of 37 °C?
3. • In the morning, when the temperature is 288 K, a bicyclist finds that the absolute pressure in his tires is 505 kPa. That afternoon he finds that the pressure in the tire has increased to 552 kPa. Ignoring expansion of the tires, find the afternoon temperature.
4. • An automobile tire has a volume of 0.0185 m^3. At a temperature of 294 K the absolute pressure in the tire is 212 kPa. How many moles of air must be pumped into the tire to increase its pressure to 252 kPa, given that the temperature and volume of the tire remain constant?
5. • The Goodyear blimp *Spirit of Akron* is 62.6 m long and contains 7023 m^3 of helium. When the temperature of the helium is

285 K its absolute pressure is 112 kPa. Find the mass of the helium in the blimp.

6. • A compressed-air tank holds 0.500 m^3 of air at a temperature of 285 K and a pressure of 850 kPa. What volume would the air occupy if it were released into the atmosphere, where the pressure is 101 kPa and the temperature is 303 K?

7. • A typical region of interstellar space may contain 10^6 atoms per cubic meter (primarily hydrogen) at a temperature of 100 K. What is the pressure exerted by this gas?

8. •• A balloon contains 2.0 liters of nitrogen gas at a temperature of 77 K and a pressure of 101 kPa. If the temperature of the gas is allowed to increase to 23 °C and the pressure remains constant, what volume will the gas occupy?

9. •• A balloon is filled with helium at a pressure of 1.2×10^5 Pa. The balloon is at a temperature of 23 °C and has a radius of 0.15 m. How many helium atoms are contained in the balloon?

10. •• **IP** A gas has a temperature of 310 K and a pressure of 101 kPa. **(a)** Find the volume occupied by 1.25 mol of this gas, assuming it is ideal. **(b)** Assuming the gas molecules can be approximated as small spheres of diameter 2.5×10^{-10} m, determine the fraction of the volume found in part (a) that is occupied by the molecules. **(c)** In determining the properties of an ideal gas, we assume that molecules are points of zero volume. Discuss the validity of this assumption for the case considered here.

11. •• A 515-cm^3 flask contains 0.460 g of a gas at a pressure of 153 kPa and a temperature of 322 K. What is the molecular mass of this gas?

12. •• **IP** On Mars, the average temperature is −64 °F and the average atmospheric pressure is 0.92 kPa. **(a)** What is the number of molecules per volume in the Martian atmosphere? **(b)** Is the number of molecules per volume on the Earth greater than, less than, or equal to the number per volume on Mars? Explain your reasoning. **(c)** Estimate the number of molecules per volume in Earth's atmosphere.

13. •• A cylindrical flask is fitted with an airtight piston that is free to slide up and down, as shown in **Figure 17–24**. A mass rests on top of the piston. Contained within the flask is an ideal gas at a constant temperature of 313 K. Initially the pressure applied by the piston and the mass is 137 kPa and the height of the piston above the base of the flask is 23.4 cm. When additional mass is added to the piston, the height of the piston decreases to 20.0 cm. Find the new pressure applied by the piston.

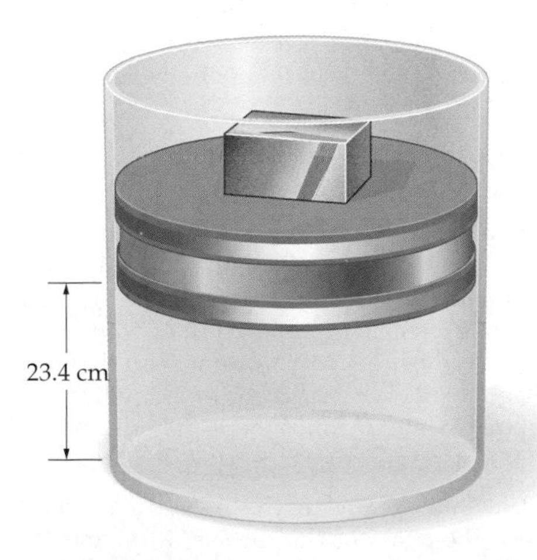

▲ **FIGURE 17–24** Problems 13, 14 and 70

14. •• Consider the system described in Problem 13. The initial temperature of the system is 313 K and the pressure of the gas is held constant at 137 kPa. The temperature is now increased until the height of the piston rises from 23.4 cm to 26.0 cm. What is the final temperature of the gas?

15. •• The air inside a hot air balloon has an average temperature of 80.0 °C. The outside air has a temperature of 20.0 °C. What is the ratio of the density of air in the balloon to the density of air in the surrounding atmosphere?

16. ••• One mole of a monatomic ideal gas has an initial pressure of 210 kPa, an initial volume of 1.2×10^{-3} m^3, and an initial temperature of 350 K. The gas now undergoes three separate processes: (i) a constant-temperature expansion that triples its volume; (ii) a constant-pressure compression to its initial volume; (iii) a constant-volume increase in pressure to its initial pressure. At the end of these three processes the gas is back at its initial pressure, volume, and temperature. Plot these processes on a pressure versus volume graph, showing the values of P and V at the end points of each process.

Section 17–2 Kinetic Theory

17. • If the molecules in a tank of hydrogen have the same rms speed as the molecules in a tank of oxygen, we may be sure that: **(a)** the pressures are the same; **(b)** the hydrogen is at the higher temperature; **(d)** the hydrogen is at the higher pressure; **(c)** the temperatures are the same; **(e)** the oxygen is at the higher temperature. Justify your answer.

18. • At what temperature is the rms speed of H_2 equal to the rms speed that O_2 has at 273 K?

19. • Suppose a planet has an atmosphere of pure ammonia at 0.0 °C. What is the rms speed of the ammonia molecules?

20. •• Two moles of oxygen gas (O_2) are placed in a portable container with a volume of 0.0025 m^3. If the temperature of the gas is 305 °C, find **(a)** the pressure of the gas and **(b)** the average kinetic energy of an oxygen molecule.

21. •• **IP** The rms speed of O_2 is 1550 m/s at a given temperature. **(a)** Is the rms speed of H_2O at this temperature greater than, less than, or equal to 1550 m/s? Explain. **(b)** Find the rms speed of H_2O at this temperature.

22. •• **IP** An ideal gas is kept in a container of constant volume. The pressure of the gas is also kept constant. **(a)** If the number of molecules in the gas is doubled, does the rms speed increase, decrease, or stay the same? Explain. **(b)** If the initial rms speed is 1200 m/s, what is the final rms speed?

23. •• What is the temperature of a gas of CO_2 molecules whose rms speed is 449 m/s?

24. •• The rms speed of a sample of gas is increased by one percent. **(a)** What is the percent change in the temperature of the gas? **(b)** What is the percent change in the pressure of the gas, assuming its volume is held constant?

25. •• In naturally occurring uranium atoms, 99.3% are ^{238}U (atomic mass = 238 u, where $u = 1.6605 \times 10^{-27}$ kg) and only 0.7% are ^{235}U (atomic mass = 235 u). Uranium-fueled reactors require an enhanced proportion of ^{235}U. Since both isotopes of uranium have identical chemical properties, they can be separated only by methods that depend on their differing masses. One such method is gaseous diffusion, in which uranium hexaflouride (UF_6) gas diffuses through a series of porous barriers. The lighter $^{235}UF_6$ molecules have a slightly higher rms speed at a given temperature than the heavier $^{238}UF_6$ molecules, and this allows the two isotopes to be separated. Find the ratio of the rms speeds of the two isotopes at 23.0 °C.

26. ••• A 450-mL spherical flask contains 0.050 mol of an ideal gas at a temperature of 293 K. What is the average force exerted on the walls of the flask by a single molecule?

Section 17–3 Solids and Elastic Deformation

27. • A rock climber hangs freely from a nylon rope that is 12 m long and has a diameter of 5.5 mm. If the rope stretches 4.7 cm, what is the mass of the climber?

28. • **BIO** To stretch a relaxed biceps muscle 2.5 cm requires a force of 25 N. Find the Young's modulus for the muscle tissue, assuming it to be a uniform cylinder of length 0.24 m and cross-sectional area 47 cm^2.

29. • A 22-kg monkey hangs from the end of a horizontal, broken branch 1.1 m long, as shown in **Figure 17–25**. The branch is a uniform cylinder 4.6 cm in diameter, and the end of the branch supporting the monkey sags downward through a vertical distance of 13 cm. What is the shear modulus for this branch?

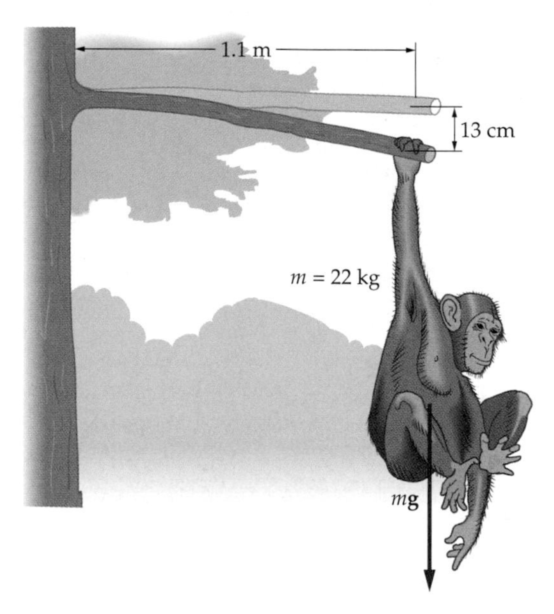

▲ **FIGURE 17–25** Problem 29

30. • The deepest place in all the oceans is the Marianas Trench, where the depth is 10.9 km and the pressure is 1.10×10^8 Pa. If a copper ball 10.0 cm in diameter is taken to the bottom of the trench, by how much does its volume decrease?

31. •• **IP** A steel wire 4.5 m long stretches 0.15 cm when it is given a tension of 370 N. **(a)** What is the diameter of the wire? **(b)** If it is desired that the stretch be less than 0.15 cm, should its diameter be increased or decreased? Explain.

32. •• **BIO** An orb weaver spider with a mass of 0.26 g hangs vertically by one of its threads. The thread has a Young's modulus of 4.7×10^9 N/m^2 and a radius of 9.8×10^{-6} m. **(a)** What is the fractional increase in the thread's length caused by the spider? **(b)** Suppose a 76-kg person hangs vertically from a nylon rope. What radius must the rope have if its fractional increase in length is to be the same as that of the spider's thread?

33. •• Two rods of equal length (0.25 m) and diameter (0.75 cm) are placed end to end. One rod is aluminum, the other is brass. If a compressive force of 7600 N is applied to the rods, how much does their combined length decrease?

34. •• A piano wire 0.84 m long and 0.95 mm in diameter is fixed on one end. The other end is wrapped around a tuning peg 3.5 mm in diameter. Initially the wire, whose Young's modulus is 2.4×10^{10} N/m^2, has a tension of 12 N. Find the tension in the wire after the tuning peg has been turned through one complete revolution.

35. •• The American naturalist Charles William Beebe (1877–1962) set a world record in 1934 when he made a dive to a depth of 923 m below the surface of the ocean. The dive was made in a device known as the bathysphere, which was basically a steel sphere 4.75 ft in diameter. How much did the volume of the sphere change as it was lowered to its record depth?

36. ••• Referring to the previous problem, suppose the bathysphere and its occupants had a combined mass of 12,700 kg and that they were lowered into the ocean on a steel cable with a radius of 1.85 cm. How much did the cable stretch when the bathysphere was at the depth of 923 m? (Neglect the weight of the cable itself, but include the effects of buoyancy.)

Section 17–4 Phase Equilibrium and Evaporation

37. • The formation of ice from water is accompanied by: **(a)** an absorption of heat by the water; **(b)** an increase in temperature; **(c)** a decrease in volume; **(d)** a removal of heat from the water; **(e)** a decrease in temperature. Justify your answer.

38. • **Figure 17–26** shows a portion of the vapor pressure curve for water. Referring to the figure, estimate the pressure that would be required for water to boil at 30 °C.

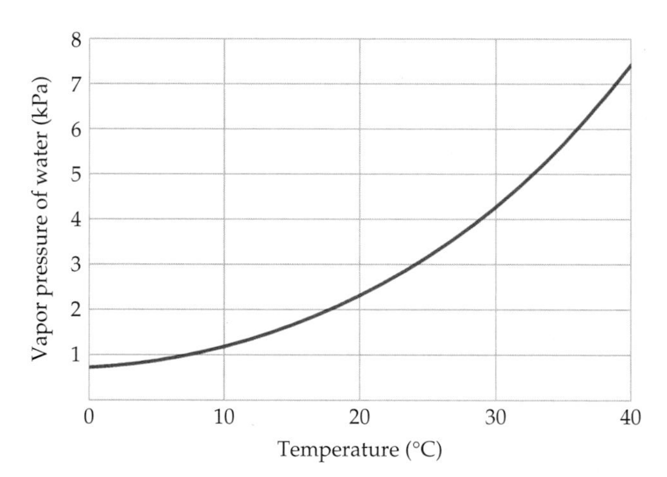

FIGURE 17–26 Problems 38 and 39

39. • Using the vapor pressure curve given in Problem 38, find the temperature at which water boils when the pressure is 1.5 kPa.

40. • A portion of the vapor pressure curve for carbon dioxide is given in **Figure 17–27**. What pressure is necessary for CO_2 to boil at 0 °C?

41. • Referring to Problem 40, find the temperature at which carbon dioxide boils when the pressure is 1.5×10^6 Pa.

42. •• The phase diagram for water is shown in **Figure 17–28**. **(a)** What is the temperature T_1 on the phase diagram? **(b)** What is the temperature T_2 on the phase diagram? **(c)** What happens to the melting/freezing temperature of water if atmospheric pressure is *decreased*? Justify your answer by referring to the phase diagram. **(d)** What happens to the boiling/condensation temperature of water if atmospheric pressure is *increased*? Justify your answer by referring to the phase diagram.

43. •• The phase diagram for CO_2 is shown in **Figure 17–29**. **(a)** What is the phase of CO_2 at $T = 20$ °C and $P = 500$ kPa? **(b)** What is the phase of CO_2 at $T = -80$ °C and $P = 120$ kPa?

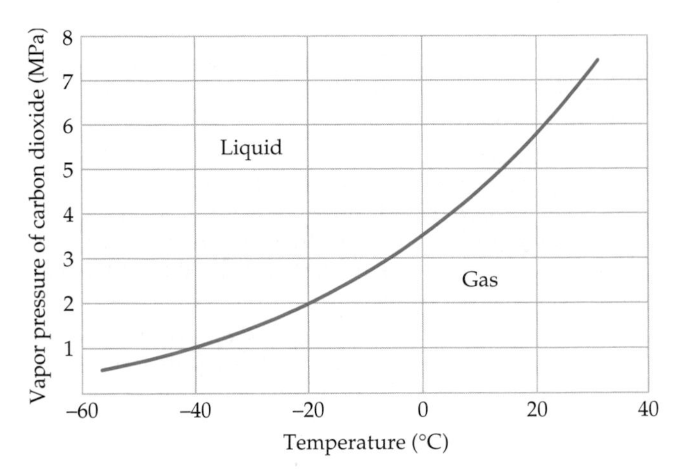

▲ **FIGURE 17–27** Problems 40 and 41

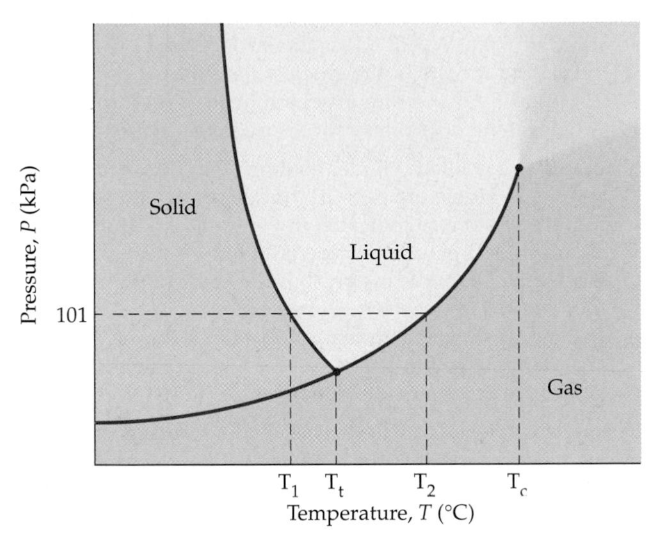

▲ **FIGURE 17–28** Problem 42

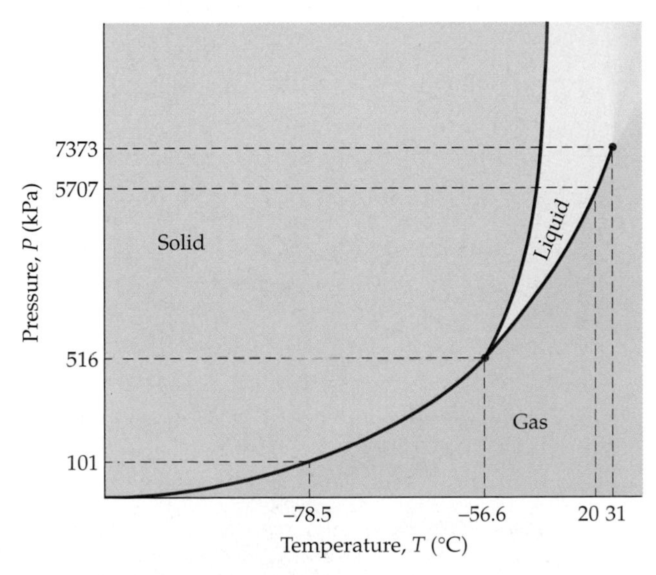

▲ **FIGURE 17–29** Problem 43

(c) For reasons of economy and convenience, bulk CO_2 is often transported in liquid form in pressurized tanks. Using the phase diagram, determine the minimum pressure required to keep CO_2 in the liquid phase at 20 °C.

44. •• A sample of pure water is initially at atmospheric pressure and has a temperature that is below the freezing point. **(a)** If the temperature of the sample is increased while the pressure is held constant, what phase changes occur? **(b)** Suppose, instead, that the pressure of the sample is decreased while the temperature is held constant. What phase changes occur now?

Section 17–5 Latent Heats

45. • How much heat must be removed from 0.64 kg of water at 0 °C to make ice cubes at 0 °C?

46. • A heat transfer of 8.5×10^5 J is required to convert a block of ice at −12 °C to water at 12 °C. What was the mass of the block of ice?

47. • How much heat must be added to 1.2 kg of copper to change it from a solid at 1356 K to a liquid at 1356 K?

48. •• A 1.1-kg block of ice is initially at a temperature of −5.0 °C. If 5.3×10^5 J of heat are added to the ice, what is the final temperature of the resulting liquid water?

49. •• In the previous problem, find the final temperature of the system if 2.6×10^5 J of heat are added to the ice. Find the amount of ice, if any, that remains.

50. •• When you go out to your car one cold winter morning you discover a 0.50-cm thick layer of ice on the windshield, which has an area of 1.6 m^2. If the temperature of the ice is −2.0 °C, and its density is 917 kg/m^3, find the heat required to melt all the ice.

Section 17–6 Phase Changes and Energy Conservation

51. •• A large punch bowl holds 3.95 kg of lemonade (which is essentially water) at 20.0 °C. A 0.0450-kg ice cube at −10.2 °C is placed in the lemonade. What is the final temperature of the system, and the amount of ice (if any) remaining? Ignore any heat exchange with the bowl or the surroundings.

52. •• A 155-g aluminum cylinder is removed from a liquid nitrogen bath, where it has been cooled to −196 °C. The cylinder is immediately placed in an insulated cup containing 80.0 g of water at 15.0 °C. What is the equilibrium temperature of this system? If your answer is 0 °C, determine the amount of water that has frozen. The average specific heat of aluminum over this temperature range is 653 J/(kg·K).

53. •• An 825-g iron block is heated to 352 °C and placed in an insulated container (of negligible heat capacity) containing 40.0 g of water at 20.0 °C. What is the equilibrium temperature of this system? If your answer is 100 °C, determine the amount of water that has vaporized. The average specific heat of iron over this temperature range is 560 J/(kg·K).

54. •• A 35-g ice cube at 0.0 °C is added to 110 g of water in a 62-g aluminum cup. The cup and the water have an initial temperature of 23 °C. Find the equilibrium temperature of the cup and its contents.

55. •• To help keep her barn warm on cold days, a farmer stores 845 kg of warm water in the barn. How many hours would a 2.00-kilowatt electric heater have to operate to provide the same amount of heat as is given off by the water as it cools from 20.0 °C to 0 °C and then freezes at 0 °C?

General Problems

56. • A drop of water has a radius of 9.2×10^{-4} m. How many molecules are in the drop?

57. •• A reaction vessel contains 8.06 g of H_2 and 64.0 g of O_2 at a temperature of 125 °C and a pressure of 101 kPa. (a) What is the volume of the vessel? (b) The hydrogen and oxygen are now ignited by a spark, initiating the reaction $2H_2 + O_2 \rightarrow 2H_2O$. This reaction consumes all the hydrogen and oxygen in the vessel. What is the pressure of the resulting water vapor when it returns to its initial temperature of 125 °C?

58. •• A bicycle tire with a radius of 0.66 m has a gauge pressure of 62 lb/in^2. Treating the tire as a hollow cylinder with a cross-sectional area of 0.0028 m^2, find the number of air molecules in the tire when its temperature is 34 °C.

59. •• Peter catches a 4.8-kg striped bass on a fishing line 0.54 mm in diameter and begins to reel it in. He fishes from a pier well above the water, and his fish hangs vertically from the line out of the water. The fishing line has a Young's modulus of 5.1×10^9 N/m^2. **(a)** What is the fractional increase in length of the fishing line if the fish is at rest? **(b)** What is the fractional increase in the fishing line's length when the fish is pulled upward with a constant speed of 1.5 m/s? **(c)** What is the fractional increase in the fishing line's length when the fish is pulled upward with a constant acceleration of 1.5 m/s^2?

60. •• You use a steel socket wrench 28 cm long to loosen a rusty bolt, applying a force F at the end of the handle. The handle undergoes a shear deformation of 0.11 mm. If the cross-sectional area of the handle is 2.3 cm^2, what is the magnitude of the applied force F?

61. •• A steel ball (density = 7860 kg/m^3) with a diameter of 6.4 cm is tied to an aluminum wire 82 cm long and 2.5 mm in diameter. The ball is whirled about in a vertical circle with a tangential speed of 7.8 m/s at the top of the circle and 9.3 m/s at the bottom of the circle. Find the amount of stretch in the wire **(a)** at the top and **(b)** at the bottom of the circle.

62. •• A lead brick with the dimensions shown in **Figure 17–30** rests on a rough, solid surface. A force of 2400 N is applied as indicated. Find **(a)** the change in height of the brick and **(b)** the amount of shear deformation.

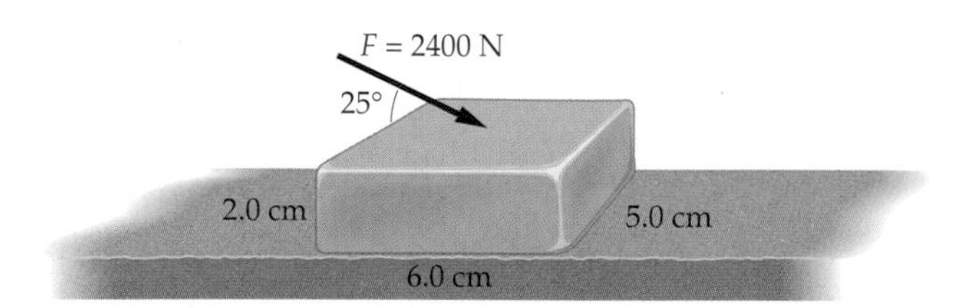

▲ **FIGURE 17–30** Problem 62

63. •• **IP** Five molecules have the following speeds: 221 m/s, 301 m/s, 412 m/s, 44.0 m/s, and 182 m/s. **(a)** Find v_{av} for these molecules. **(b)** Do you expect $(v^2)_{av}$ to be greater than, less than, or equal to $(v_{av})^2$? Explain. **(c)** Calculate $(v^2)_{av}$ and comment on your results.

64. •• **(a)** Find the amount of heat that must be extracted from 1.0 kg of steam at 120 °C to convert it to ice at 0.0 °C. **(b)** What speed would this 1.0-kg block of ice have if its translational kinetic energy were equal to the thermal energy calculated in part (a)?

65. •• **(a)** Calculate the internal energy of 21.3 g of water vapor at 100.0 °C. **(b)** Compare the internal energy found in part (a) to the energy required to change 21.3 g of liquid water at 100 °C to water vapor at 100 °C.

66. •• A 55-g ice cube is dropped from rest at a height of 1.0 m above the floor. If 10.0 percent of its initial gravitational potential energy is converted into thermal energy when it lands, how much of the ice cube melts?

67. •• When water freezes into ice it expands in volume by 9.05 percent. Suppose a volume of water is in a household water pipe or a cavity in a rock. If the water freezes, what pressure must be exerted on it to keep its volume from expanding? (If the pipe or rock cannot supply this pressure the pipe will burst and the rock will split.)

68. ••• A 5.0-kg block of ice at −1.5 °C slides on a horizontal surface with a coefficient of kinetic friction equal to 0.062. The initial speed of the block is 6.9 m/s and its final speed is 5.5 m/s. Assuming that all the energy dissipated by kinetic friction goes into melting part of the ice, determine the mass of ice that melts.

69. ••• Students on a spring break picnic bring a cooler that contains 5.5 kg of ice at 0.0 °C. The cooler has walls that are 4.2 cm thick and are made of Styrofoam, which has a thermal conductivity of 0.030 W/(m·C°). The surface area of the cooler is 1.5 m^2 and it rests in the shade where the air temperature is 21 °C. **(a)** Find the rate at which heat flows into the cooler. **(b)** How long does it take for the ice in the cooler to melt?

70. ••• **IP** A gas-filled cylinder holding 1.7 mol of an ideal gas with an initial volume of 3.1×10^{-3} m^3 is fitted with a frictionless, movable piston, as indicated in Figure 17–24. The diameter of the piston is 12 cm and its mass is 0.14 kg. As heat is added to the gas, the height of the piston above the base of the cylinder, h, is found to increase at the rate of 6.4 cm per minute. It is desired to find the rate at which the temperature of the gas increases. **(a)** Describe a strategy you can use to solve this problem. **(b)** Use your strategy to determine the rate of temperature increase.

71. ••• A cylindrical copper rod 37 cm long and 7.5 cm in diameter is placed upright on a hot plate held at a constant temperature of 120 °C, as indicated in **Figure 17–31**. A small depression on top of the rod holds a 25-g ice cube at an initial temperature of 0.0 °C. How long does it take for the ice cube to melt? Assume there is no heat loss through the vertical surface of the rod, and that the thermal conductivity of copper is 390 W/(m·C°).

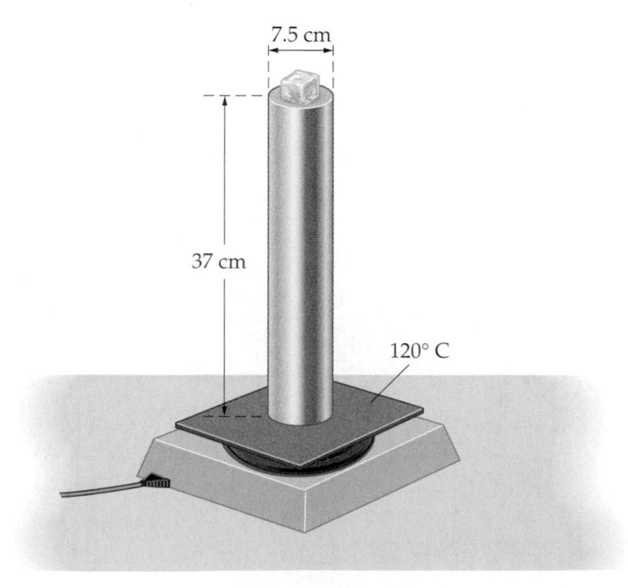

▲ **FIGURE 17–31** Problem 71

18 The Laws of Thermodynamics

Every day, green plants take small, simple molecules and use them to create large, complex molecules such as proteins and DNA. These in turn are assembled into even more highly ordered structures: cells and their components. Yet one of the most profound and far-reaching of physical laws holds that all processes decrease the amount of order in the universe. Are living things an exception to this principle? This chapter explores the laws of thermodynamics, and considers the question of whether they apply to biological organisms.

In this chapter we discuss the fundamental laws of nature that govern thermodynamics. One of these laws—the one dealing with temperature—was first introduced in Chapter 17. The others are presented here for the first time.

Of particular interest are the first and second laws of thermodynamics. The first law extends the basic principle of energy conservation to include the type of energy transfer we call heat. The real heart of thermodynamics, however, is embodied in the second law. This law introduces a fundamentally new concept to physics—the idea that there is a directionality to the behavior of nature. Melting ice, cooling lava, and the crumbling ruins of the Parthenon all illustrate the second law in action.

Finally, the third law of thermodynamics states that absolute zero is, in fact, the lowest temperature possible. Great efforts have been made over the years to reach ever lower temperatures, and with great success, but the third law sets the absolute limit that cannot be exceeded.

18–1 The Zeroth Law of Thermodynamics

Though the zeroth law of thermodynamics has already been presented in Chapter 17, we repeat it here so that all the laws of thermodynamics can be collected together in one chapter. As you recall, the zeroth law states the conditions under which objects will be in thermal equilibrium with one another. To be precise:

Zeroth law of thermodynamics

If object A is in thermal equilibrium with object C, and object B is separately in thermal equilibrium with object C, then objects A and B will be in thermal equilibrium if they are placed in thermal contact.

The physical quantity that is equal when two objects are in thermal equilibrium is the temperature. In particular, if two objects have the same temperature, we can be assured that *no heat will flow* when they are placed in thermal contact. On the other hand, if heat does flow between two objects, it follows that they are not in thermal equilibrium, and they do not have the same temperature.

Any temperature scale can be used to determine whether objects will be in thermal equilibrium. As we saw in the previous chapter, however, the Kelvin scale is particularly significant in physics. For example, the average kinetic energy of a gas molecule is directly proportional to the Kelvin temperature, as is the volume of an ideal gas. In this chapter we present additional illustrations of the special significance of the Kelvin scale.

18–2 The First Law of Thermodynamics

The first law of thermodynamics is essentially a statement of energy conservation that specifically includes heat. For example, consider the system shown in **Figure 18–1**. The internal energy of this system—that is, the sum of all its potential and kinetic energies—has the initial value U_i. If an amount of heat Q flows into the system, its internal energy increases to the final value $U_f = U_i + Q$. Thus,

$$\Delta U = U_f - U_i = Q \qquad \text{18–1}$$

Of course, if heat is removed from the system its internal energy decreases. We can take this into account by giving Q a *positive* value when the system *gains* heat, and a *negative* value when it *loses* heat.

Similarly, suppose the system under consideration does a work W on the external world, as in **Figure 18–2**. If the system is insulated so that no heat can flow in or out, the energy to do the work must come from the internal energy of the system. Thus, if the initial internal energy is U_i, the final internal energy is $U_f = U_i - W$. Therefore,

$$\Delta U = U_f - U_i = -W \qquad \text{18–2}$$

On the other hand, if work is done *on* the system its internal energy increases. Thus, we use the following sign convention for the work: W has a *positive* value when the system *does work on the external world*, and it has a *negative* value when *work is done on the system*. These sign conventions are summarized in Table 18–1.

TABLE 18–1 Signs of Q and W

Q positive	System *gains* heat
Q negative	System *loses* heat
W positive	Work done *by* system
W negative	Work done *on* system

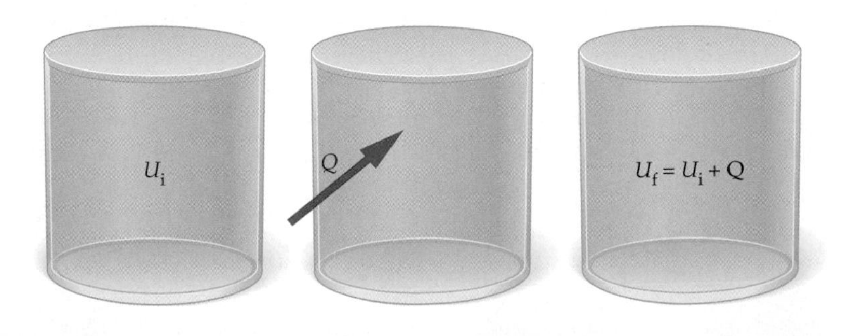

▶ FIGURE 18–1 The internal energy of a system

A system initially has the internal energy U_i (left). After the heat Q is added, the system's new internal energy is $U_f = U_i + Q$ (right). The system has rigid walls, hence it can do no work on the external world.

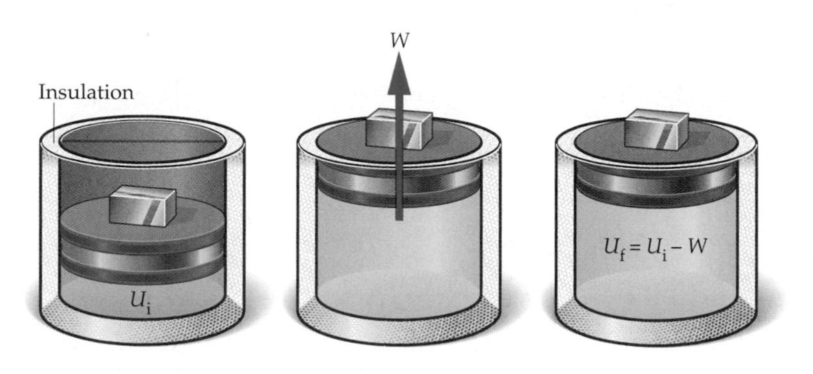

◀ FIGURE 18–2 Work and internal energy
A system initially has the internal energy U_i (left). After the system does the work W on the external world, its remaining internal energy is $U_f = U_i - W$ (right). Note that the insulation guarantees that no heat is gained or lost by the system.

Combining the results in Equations 18–1 and 18–2 yields the first law of thermodynamics:

First law of thermodynamics

The change in a system's internal energy, ΔU, is related to the heat Q and the work W as follows:

$$\Delta U = Q - W \qquad \textbf{18–3}$$

Applying the sign conventions given here, it is straightforward to verify that adding heat to a system, and/or doing work on it, increases the internal energy. On the other hand, if the system does work, and/or heat is removed, its internal energy decreases. Example 18–1 gives a specific numerical application of the first law.

PROBLEM SOLVING NOTE
Proper Signs for *Q* and *W*

When applying the first law of thermodynamics it is important to determine the proper signs for Q, W, and ΔU.

EXAMPLE 18–1 Heat, Work, and Internal Energy

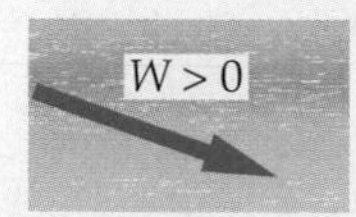

(a) Jogging along the beach one day you do 4.3×10^5 J of work and give off 3.8×10^5 J of heat. What is the change in your internal energy? **(b)** Switching over to walking, you give off 1.2×10^5 J of heat and your internal energy decreases by 2.6×10^5 J. How much work have you done while walking?

Picture the Problem

Our sketch shows a person jogging along the beach. The fact that the person does work on the external world means that W is positive. As for the heat, the fact that heat is given off by the person means that Q is negative.

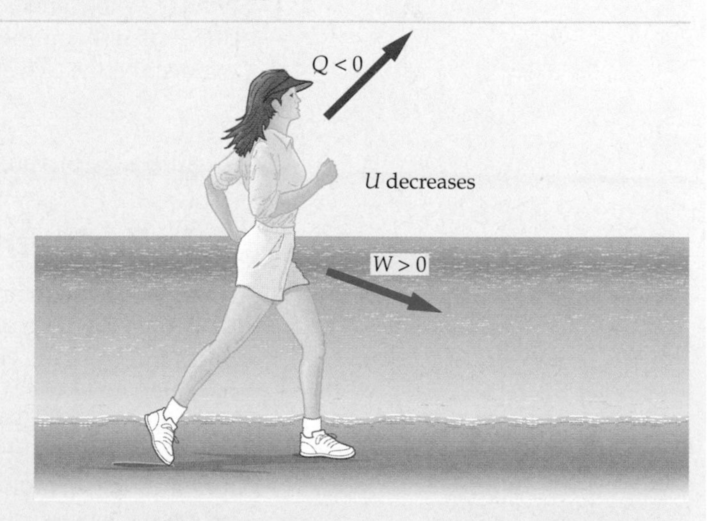

Strategy

The signs of W and Q have been determined in our sketch, and the magnitudes are given in the problem statement. To find ΔU for part (a) we simply use the first law of thermodynamics, $\Delta U = Q - W$. To find the work W for part (b) we solve the first law for W, which yields $W = Q - \Delta U$.

Solution

Part (a)

1. Calculate ΔU, using $Q = -3.8 \times 10^5$ J and $W = 4.3 \times 10^5$ J:

$$\Delta U = Q - W$$
$$= (-3.8 \times 10^5 \text{ J}) - 4.3 \times 10^5 \text{ J} = -8.1 \times 10^5 \text{ J}$$

Part (b)

2. Solve $\Delta U = Q - W$ for W:

$$W = Q - \Delta U$$

Substitute $Q = -1.2 \times 10^5$ J and $\Delta U = -2.6 \times 10^5$ J:

$$W = -1.2 \times 10^5 \text{ J} - (-2.6 \times 10^5 \text{ J}) = 1.4 \times 10^5 \text{ J}$$

continued on the following page

continued from the previous page

Insight

Note the importance of using the correct sign for Q, W, and ΔU.

Practice Problem

After walking for a few minutes you begin to run, doing 5.1×10^5 J of work and decreasing your internal energy by 8.8×10^5 J. How much heat have you given off? [**Answer:** $Q = -3.7 \times 10^5$ J. The negative sign means you have given off heat, as expected.]

Some related homework problems: Problem 1, Problem 3

Just looking at the first law, $\Delta U = Q - W$, it is easy to get the false impression that U, Q, and W are basically the same type of physical quantity. Certainly, they are all measured in the same units (J). In other respects, however, they are quite different. For example, the heat Q represents energy that flows through thermal contact. In contrast, the work W indicates a transfer of energy by the action of a force through a distance.

The most important distinction between these quantities, however, is the way they depend on the **state of a system**, which is determined by its temperature, pressure, and volume. The internal energy, for example, depends only on the state of a system, and not on how the system is brought to that state. A simple example is the ideal gas, where U depends only on the temperature T, and not on any previous values T may have had. Since U depends only on the state of a system—whether the system is an ideal gas or something more complicated—it is referred to as a **state function**.

On the other hand, Q and W are not state functions; they depend on the precise way—that is, the process—by which a system is changed from one state to another. For example, one process connecting two states may result in a heat Q_1 and a work W_1. On a different process connecting the *same two states* we may find a heat, $Q_2 \neq Q_1$, and a work, $W_2 \neq W_1$. Still, the difference in internal energy—which depends only on the initial and final states—must be the same for all processes connecting the two states. It follows that

$$\Delta U = Q_1 - W_1 = Q_2 - W_2$$

We turn now to a study of thermal processes.

18–3 Thermal Processes

In this section we consider a variety of thermodynamic processes that can be used to change the state of a system. Among these are ones that take place at constant pressure, constant volume, or constant temperature. We also consider processes in which no heat is allowed to flow into or out of the system.

All of the processes discussed in this section are assumed to be **quasi-static**, which is a fancy way of saying they occur so slowly that at any given time the system and its surroundings are essentially in equilibrium. Thus, in a quasi-static process, the pressure and temperature are always uniform throughout the system. Furthermore, we assume that the system in question is free from friction or other dissipative forces.

These assumptions can be summarized by saying that the processes we consider are **reversible**. To be precise, a reversible process is defined by the following condition:

> For a process to be reversible, it must be possible to return both the system and its surroundings to exactly the same states they were in before the process began.

The fact that both the system and its surroundings must be returned to their initial states is the key element of this definition. For example, if there is friction between a piston and the cylinder in which it slides, a reversible process is not

▲ **FIGURE 18–3 An idealized reversible process**
(a) A piston is slowly moved downward, compressing a gas. In order for its temperature to remain constant, a heat Q goes from the gas into the constant-temperature heat bath, which may be nothing more than a large volume of water. **(b)** As the piston is allowed to slowly rise back to its initial position, it draws the same heat Q from the heat bath that it gave to the bath in (a). Hence, both the system (gas) and the surroundings (heat bath) return to their initial states.

possible. Even if the piston is returned to its original location, the heat generated by friction will warm the cylinder, and eventually flow into the surrounding air. Thus, it is not possible to "undo" the effects of friction, and such a process is said to be **irreversible**. Even without friction, a process can be irreversible if it occurs rapidly enough to cause effects such as turbulence, or if the system is far from equilibrium.

In practice, then, all real processes are irreversible to some extent. It is still possible, however, to have a process that closely approximates a perfectly reversible process, just as we can have systems that are practically free of friction. For example, in **Figure 18–3 (a)**, we consider a "frictionless" piston that is slowly forced downward, while the gas in the cylinder is kept at constant temperature. An amount of heat, Q, goes from the gas to its surroundings in order to keep the temperature from rising. In **Figure 18–3 (b)** the piston is slowly moved back upward, drawing in the same heat Q from its surroundings to keep the temperature from dropping. In an "ideal" case like this, the process is reversible, since the system and its surroundings are left unchanged. We shall assume that all the processes described in this section are reversible.

Constant Pressure/Constant Volume

We begin by considering a process that occurs at **constant pressure**. To be specific, suppose that a gas with the pressure P_0 is held in a cylinder of cross-sectional area A, as in **Figure 18–4**. If the piston moves outward, so that the volume of the gas increases from an initial value V_i to a final value V_f, the process can be represented

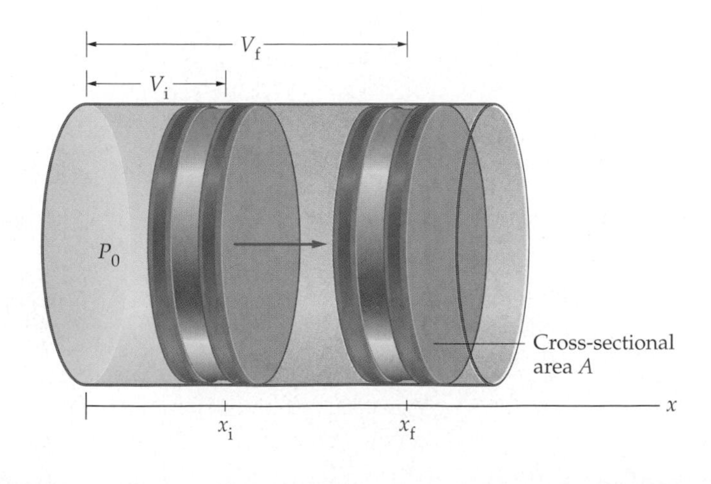

◀ **FIGURE 18–4 Work done by an expanding gas**
A gas in a cylinder of cross-sectional area A expands with a constant pressure of P_0 from an initial volume $V_i = Ax_i$ to a final volume $V_f = Ax_f$. As it expands, it does the work $W = P_0(V_f - V_i)$.

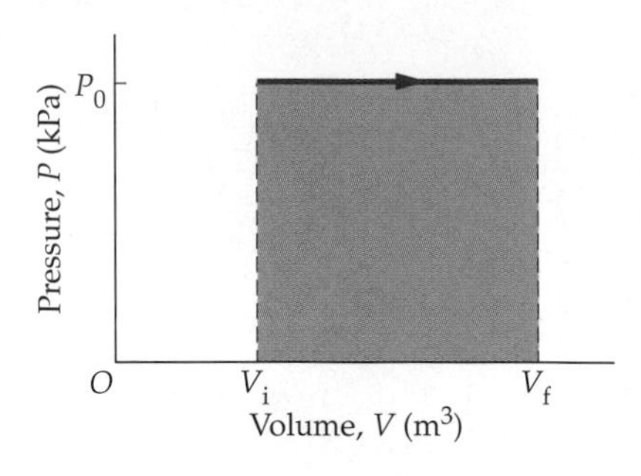

▲ FIGURE 18–5 A constant-pressure process
A PV plot representing the constant-pressure process shown in Figure 18–4. The area of the shaded region, $P_0(V_f - V_i)$, is equal to the work done by the expanding gas in Figure 18–4.

graphically as shown in **Figure 18–5**. Here we plot pressure P versus volume V in a PV plot; the process just described is the horizontal line segment.

As the gas expands, it does work on the piston. First, the gas exerts a force on the piston equal to the pressure times the area:

$$F = P_0 A$$

Second, the gas moves the piston from the position x_i to the position x_f, where $V_i = Ax_i$ and $V_f = Ax_f$, as indicated in Figure 18–4. Thus, the work done by the gas is the force times the distance through which the piston moves:

$$W = F(x_f - x_i) = P_0 A(x_f - x_i) = P_0(Ax_f - Ax_i) = P_0(V_f - V_i)$$

In general, if the pressure, P, is constant, and the volume changes by the amount ΔV, the work done by the gas is

$$W = P\Delta V \qquad (\textit{constant pressure}) \qquad \textbf{18–4}$$

EXERCISE 18–1

A gas with a constant pressure of 150 kPa expands from a volume of 0.76 m^3 to a volume of 0.92 m^3. How much work does the gas do?

Solution

Applying Equation 18–4 we find

$$W = P\Delta V = (150 \text{ kPa})(0.92 \text{ m}^3 - 0.76 \text{ m}^3) = 24{,}000 \text{ J}$$

This is the energy required to raise the temperature of 1.0 kg of water by 5.7 C°.

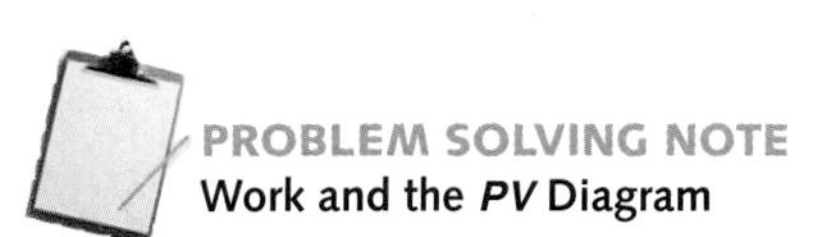

PROBLEM SOLVING NOTE
Work and the *PV* Diagram

When finding the work done by calculating the area on a PV diagram, recall that a pressure of 1 Pa times a volume of 1 m^3 gives an energy equal to 1 J.

Looking closely at the PV plot in Figure 18–5, we see that the work done by the gas is equal to the area under the horizontal line representing the constant-pressure process. In particular, the shaded region is a rectangle of height P_0 and width $V_f - V_i$, hence its area is

$$\text{area} = P_0(V_f - V_i) = W$$

Though this result was obtained for the special case of constant pressure, it applies to any process; that is, *the work done by an expanding gas is equal to the area under the curve representing the process in a PV plot*. This result is applied in the next Example.

EXAMPLE 18–2 Work Area

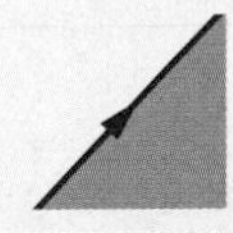

A gas expands from an initial volume of 0.40 m^3 to a final volume of 0.62 m^3 as the pressure increases linearly from 110 kPa to 230 kPa. Find the work done by the gas.

Picture the Problem

The sketch shows the process for this problem. As the volume increases from $V_i = 0.40 \text{ m}^3$ to $V_f = 0.62 \text{ m}^3$ the pressure increases linearly from $P_i = 110$ kPa to $P_f = 230$ kPa.

Strategy

The work done by this gas is equal to the shaded area in the sketch. We can calculate this area as the sum of the area of a rectangle plus the area of a triangle.

In particular, the rectangle has a height P_i and a width $(V_f - V_i)$. Similarly, the triangle has a height $(P_f - P_i)$ and a base of $(V_f - V_i)$.

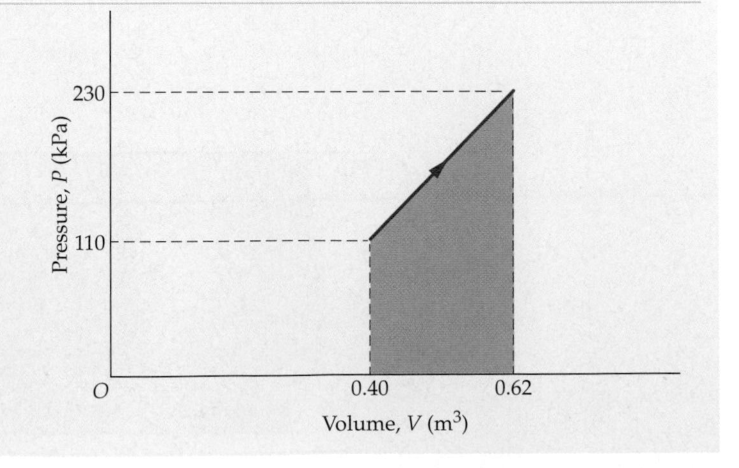

Solution

1. Calculate the area of the rectangular portion of the total area:

$$A_{\text{rectangle}} = P_i(V_f - V_i) = (110\text{ kPa})(0.62\text{ m}^3 - 0.40\text{ m}^3) = 2.4 \times 10^4\text{ J}$$

2. Next, calculate the area of the triangular portion of the total area:

$$A_{\text{triangle}} = \tfrac{1}{2}(P_f - P_i)(V_f - V_i) = \tfrac{1}{2}(230\text{ kPa} - 110\text{ kPa})(0.62\text{ m}^3 - 0.40\text{ m}^3) = 1.3 \times 10^4\text{ J}$$

3. Sum these areas to find the work done by the gas:

$$W = A_{\text{rectangle}} + A_{\text{triangle}} = 2.4 \times 10^4\text{ J} + 1.3 \times 10^4\text{ J} = 3.7 \times 10^4\text{ J}$$

Insight

We could also have solved this problem by noting that since the pressure varies linearly, its average value is simply $P_{av} = \frac{1}{2}(P_f + P_i) = 170$ kPa. The work done by the gas, then, is $W = P_{av}\Delta V = (170\text{ kPa})(0.22\text{ m}^3) = 3.7 \times 10^4$ J, as before.

Practice Problem

Suppose the pressure varies linearly from 110 kPa to 260 kPa. How much work does the gas do in this case?
[**Answer:** $W = 4.1 \times 10^4$ J]

Some related homework problems: Problem 14, Problem 15

◀ **FIGURE 18–6 Adding heat to a system of constant volume**
Heat is added to a system of constant volume, increasing its pressure from P_i to P_f. Since there is no displacement of the walls, there is no work done in this process.

Next, we consider a **constant-volume** process. Suppose, for example, that heat is added to a gas in a container of fixed volume, as in **Figure 18–6**, causing the pressure to increase. Since there is no displacement of any of the walls, it follows that the force exerted by the gas does no work. Thus, for *any* constant-volume process, we have

$$W = 0 \qquad (\textit{constant volume})$$

Note that zero work in a constant-volume process is consistent with our earlier statement that the work is equal to the area under the curve representing the process. For example, in **Figure 18–7** we show a constant volume process in which the pressure is increased from P_i to P_f. Since this line is vertical, the area under it is zero; that is, $W = 0$ as expected.

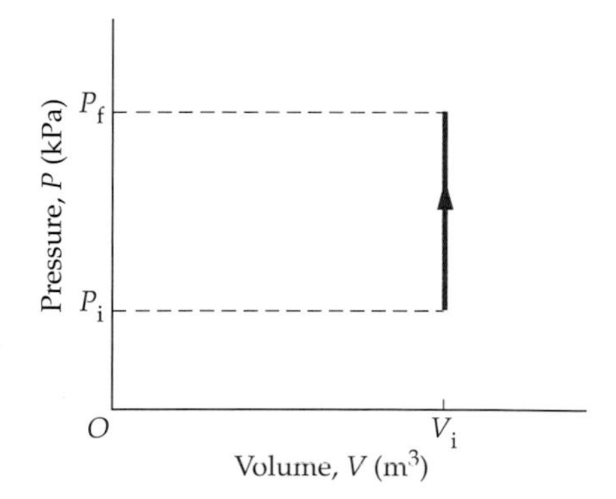

▲ **FIGURE 18–7 A constant-volume process**
The pressure increases from P_i to P_f, just as in Figure 18–6, while the volume remains constant at its initial value, V_i. The area under this process is zero, as is the work.

ACTIVE EXAMPLE 18–1 Three-Part Process

A gas undergoes the three-part process shown on p. 576, connecting points A and B. Find the total work done by the gas during this process.

continued on the following page

continued from the previous page

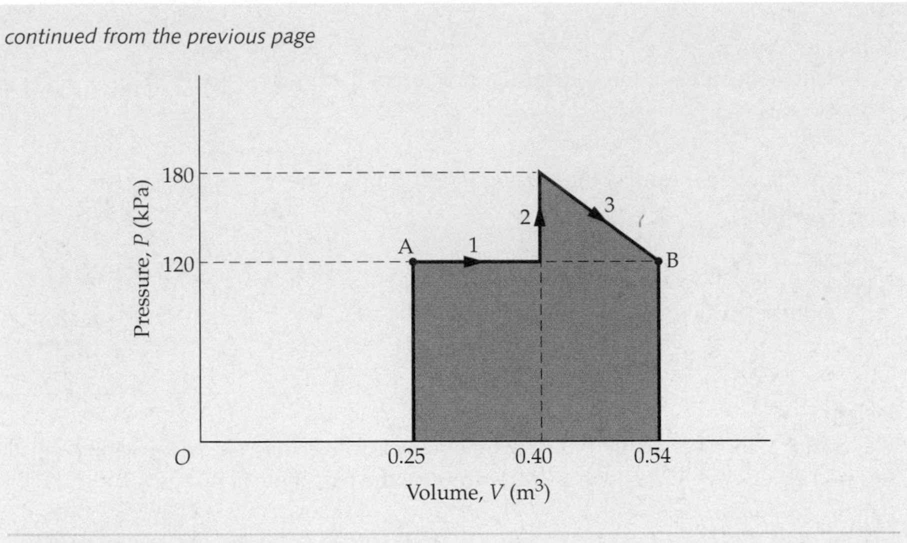

Solution

1. Calculate the work done during part 1 of the total process: 18,000 J

2. Calculate the work done during part 2 of the total process: 0

3. Calculate the work done during part 3 of the total process: 21,000 J

4. Sum the results to find the total work: 39,000 J

Insight

Note that the work done by the gas would be different if the process connecting A and B were a constant-pressure expansion with a pressure of 120 kPa. In that case, the total work done would be 35,000 J. Thus, as expected, the work W depends on the process.

Isothermal Processes

Another common process is one that takes place at constant temperature; that is, an **isothermal process**. For an ideal gas the isotherm has a relatively simple form. In particular, if T is constant, it follows that PV is a constant as well:

$$PV = NkT = \text{constant}$$

Thus, ideal-gas isotherms have the following pressure-volume relationship:

$$P = \frac{NkT}{V} = \frac{\text{constant}}{V}$$

This is illustrated in **Figure 18–8,** where we show several isotherms corresponding to different temperatures.

For any given isotherm, such as the one shown in **Figure 18–9,** the work done by an expanding gas is equal to the area under the curve, as usual. In particular, the work in expanding from V_i to V_f in Figure 18–9 is equal to the shaded area. This area may be derived by using the methods of calculus. The result is found to be

$$W = NkT \ln\left(\frac{V_f}{V_i}\right) = nRT \ln\left(\frac{V_f}{V_i}\right) \quad (\textit{constant temperature}) \qquad \textbf{18–5}$$

Note that "ln" stands for the natural logarithm; that is, log to the base e. This result is utilized in the next Example.

▲ FIGURE 18–8 Isotherms on a *PV* plot
These isotherms are for one mole of an ideal gas at the temperatures 300 K, 500 K, 700 K, and 900 K. Notice that each isotherm has the shape of a hyperbola. As the temperature is increased, however, the isotherms move farther from the origin. Thus, the pressure corresponding to a given volume increases with temperature, as one would expect.

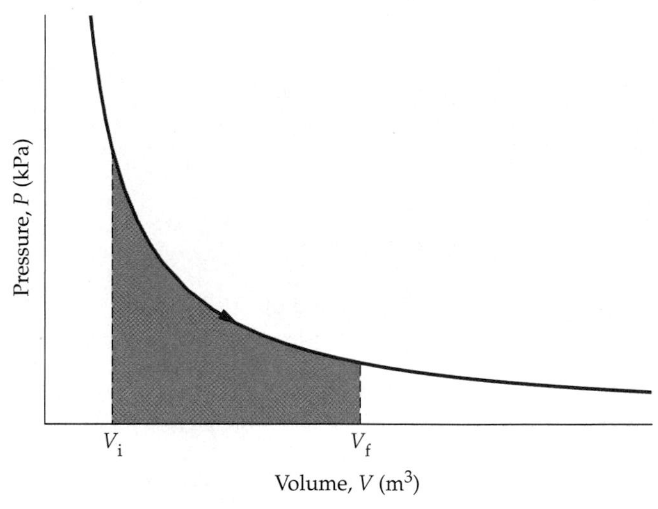

▲ FIGURE 18–9 An isothermal expansion
In an isothermal expansion from the volume V_i to the volume V_f, the work done is equal to the shaded area. For n moles of an ideal gas at the temperature T, the work by the gas is $W = nRT \ln(V_f/V_i)$.

EXAMPLE 18–3 Heat Flow

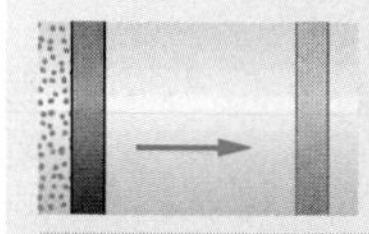

A cylinder holds 0.50 mol of an ideal gas at a temperature of 310 K. As the gas expands isothermally from an initial volume of 0.31 m^3 to a final volume of 0.45 m^3, determine the amount of heat that must be added to the gas to maintain a constant temperature.

Picture the Problem

The physical system is illustrated in the drawing at left. Note that heat flows into the gas as it expands in order to keep its temperature from dropping. The graph at right is a *PV* plot representation of the process. The work done by the expanding gas is equal to the shaded area.

Strategy

We can use the first law, $\Delta U = Q - W$, to find the heat Q in terms of W and ΔU. We can find W and ΔU as follows:
First, the work W is found using $W = nRT \ln(V_f/V_i)$.
Next, recall that the internal energy of an ideal gas depends only on the temperature. For example, the internal energy of a monatomic ideal gas is $U = \frac{3}{2}nRT$. Since the temperature is constant in this process, there is no change in internal energy; that is, $\Delta U = 0$.

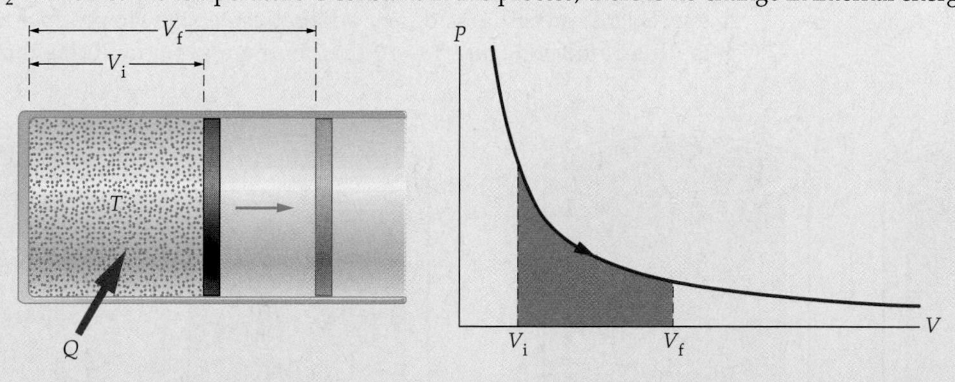

Solution

1. Solve the first law of thermodynamics for the heat, Q:

$$\Delta U = Q - W$$
$$Q = \Delta U + W$$

continued on the following page

continued from the previous page

2. Calculate the work done by the expanding gas:

$$W = nRT \ln\left(\frac{V_f}{V_i}\right)$$
$$= (0.50 \text{ mol})[8.31 \text{ J}/(\text{mol}\cdot\text{K})](310 \text{ K}) \ln\left(\frac{0.45 \text{ m}^3}{0.31 \text{ m}^3}\right)$$
$$= 480 \text{ J}$$

3. Calculate the change in the gas's internal energy:

$$\Delta U = nR(T_f - T_i) = 0$$

4. Substitute numerical values to find Q:

$$Q = \Delta U + W = 0 + 480 \text{ J} = 480 \text{ J}$$

Insight

We find, then, that when an ideal gas undergoes an isothermal expansion, the work done by the gas is equal to the heat it gains. This is a direct result of the fact that U is unchanged in this case.

Practice Problem

Find the final volume of this gas when it has expanded enough to do 590 J of work. [**Answer:** $V_f = 0.49 \text{ m}^3$]

Some related homework problems: Problem 18, Problem 19

Note that if a gas is compressed at constant temperature, work is done *on* the gas, rather than *by* the gas. As a result, we expect the work to be negative. This is consistent with Equation 18–9, since in a compression the final volume is less than the initial volume. Thus, $V_f/V_i < 1$, and hence $W = nRT \ln(V_f/V_i)$ is negative.

Finally, we consider one last point regarding isothermal processes.

CONCEPTUAL CHECKPOINT 18–1

The internal energy of a certain gas increases when it is compressed isothermally. Is the gas ideal?

Reasoning and Discussion

In an isothermal process, the internal energy of an ideal gas is unchanged, since the temperature is constant. Hence, if the internal energy of this gas changes during an isothermal compression, it must not be ideal.

Answer:

No, the gas is not ideal.

Adiabatic Processes

The final process we consider is one in which no heat flows into or out of the system. Such a process is said to be **adiabatic**. One way to produce an adiabatic process is illustrated in **Figure 18–10 (a)**. Here we see a cylinder that is insulated well

▲ **FIGURE 18–10 An adiabatic process**
In adiabatic processes no heat flows into or out of the system. In the cases shown in this figure, heat flow is prevented by insulation. **(a)** An adiabatic compression increases both the pressure and the temperature. **(b)** An adiabatic expansion results in a decrease in pressure and temperature.

enough that no heat can pass through the insulation (adiabatic means, literally, "not passable"). When the piston is pushed downward in the cylinder—decreasing the volume—the gas heats up and its pressure increases. Similarly, in **Figure 18–10 (b)**, an adiabatic expansion causes the temperature of the gas to decrease, as does the pressure.

What does an adiabatic process look like on a *PV* plot? Certainly, its general shape must be similar to that of an isotherm; in particular, as the volume is decreased the pressure increases. However, it can't be identical to an isotherm because, as we have pointed out, the temperature changes during an adiabatic process. The comparison between an adiabatic curve and an isothermal curve is the subject of the next Conceptual Checkpoint.

CONCEPTUAL CHECKPOINT 18–2

A certain gas has an initial volume and pressure given by point A in the *PV* plot. If the gas is compressed isothermally, its pressure rises as indicated by the curve labeled "isotherm." If, instead, the gas is compressed adiabatically from point A, does its pressure follow **(a)** curve i, **(b)** curve ii, or **(c)** curve iii?

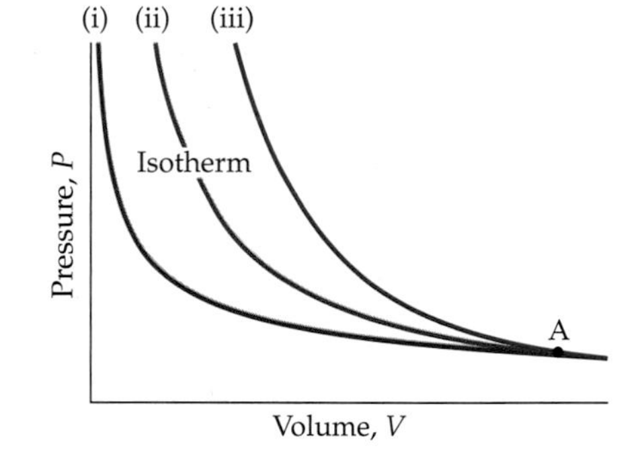

Reasoning and Discussion

In an isothermal compression, some heat flows out of the system in order for its temperature to remain constant. No heat flows out of the system in the adiabatic process, however, thus the temperature rises. As a result, the pressure is greater for any given volume when the compression is adiabatic, as compared to isothermal. Therefore, the adiabat is curve iii.

Answer:

(c) As volume is decreased, the pressure on an adiabat rises more rapidly than on an isotherm.

As we have seen, then, an adiabatic curve is similar to an isotherm, only steeper. The precise mathematical relationship describing an adiabat will be presented in the next section.

EXAMPLE 18–4 Work into Energy

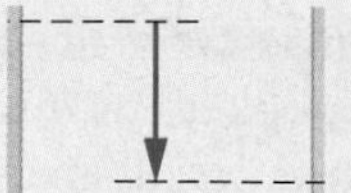

When a certain gas is compressed adiabatically, the amount of work done on it is 640 J. Find the change in internal energy of the gas.

Picture the Problem

Our sketch shows a piston being pushed downward, compressing a gas in an insulated cylinder.

Strategy

We know the work done on the gas, and we know that no heat is exchanged—since the process is adiabatic. Thus, we can find ΔU by substituting Q and W into the first law of thermodynamics. One note of caution: Be careful to use the correct sign for the work. In particular, recall that work done *on* a system is negative.

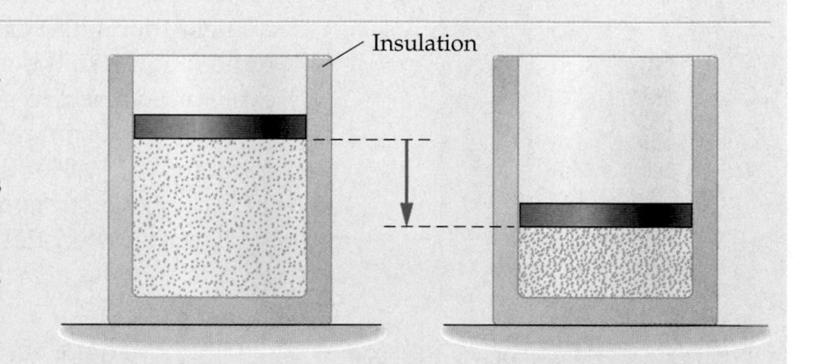

Solution

1. Identify the work and heat for this process: $W = -640\text{ J}$, $Q = 0$

2. Substitute Q and W into the first law of thermodynamics to find the change in internal energy, ΔU: $\Delta U = Q - W = 0 - (-640\text{ J}) = 640\text{ J}$

continued on the following page

continued from the previous page

Insight

Since no energy can enter or leave the system in the form of heat, all the work done on the system goes into increasing its internal energy. As a result, the temperature of the gas increases.

A familiar example of this type of effect is the heating that occurs when you pump air into a tire or a ball—the work done on the pump appears as an increased temperature. The effect occurs in the reverse direction as well. When air is let out of a tire, for example, it does work on the atmosphere as it expands, producing a cooling effect that can be quite noticeable. In extreme cases, the cooling can be great enough to create frost on the valve stem of the tire.

Practice Problem

If a system's internal energy decreases by 470 J in an adiabatic process, how much work was done by the system?
[**Answer:** $W = +470$ J]

Some related homework problems: Problem 20, Problem 21

An adiabatic process can occur when the system is thermally insulated, as in Figure 18–10, or in a system where the change in volume occurs rapidly. For example, if an expansion or compression happens quickly enough, there is no time for heat flow to occur. As a result, the process is adiabatic, even if there is no insulation.

An example of a rapid process is shown in **Figure 18–11**. Here, a piston is fitted into a cylinder that contains a certain volume of gas and a small piece of tissue paper. If the piston is driven downward rapidly, by a sharp impulsive blow, for example, the gas is compressed before heat has a chance to flow. As a result, the temperature of the gas rises rapidly. In fact, the rise in temperature can be enough for the paper to burst into flames.

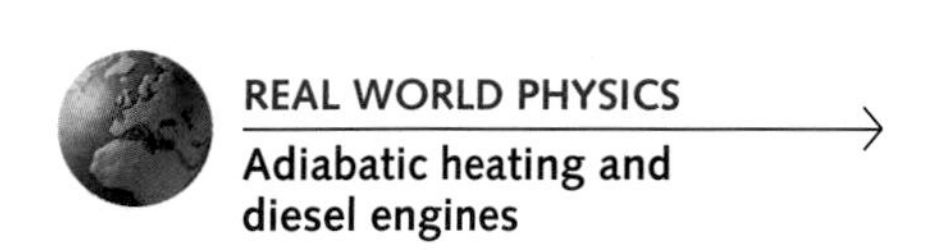

The same principle applies to the operation of a diesel engine. As you may know, a diesel differs from a standard internal combustion engine in that it has no spark plugs. It doesn't need them. Instead of using a spark to ignite the fuel in a cylinder, it uses adiabatic heating. Fuel and air is admitted into the cylinder, then the piston rapidly compresses the air-fuel mixture. Just as with the piece of paper in Figure 18–11, the rising temperature is sufficient to ignite the fuel and run the engine.

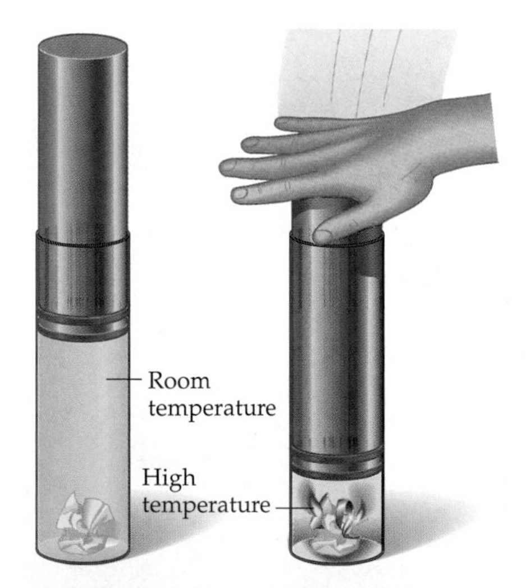

▲ **FIGURE 18–11 Adiabatic heating** When a rod that fits snugly into a cylinder is pushed downward rapidly, the temperature of the gas increases before there is time for heat to flow out of the system. Thus, the process is essentially adiabatic. As a result, the temperature of the gas can increase enough to ignite bits of paper in the cylinder. In a diesel engine the same principle is used to ignite an air-gasoline mixture.

Adiabatic heating is one of the mechanisms being considered to explain the fascinating and enigmatic phenomenon known as *sonoluminescence*. Sonoluminescence occurs when an intense, high frequency sound wave causes a small gas bubble in water to pulsate. When the sound wave collapses the bubble to its minimum size, which is about a thousandth of a millimeter, the bubble gives off an extremely short burst of light. The light is mostly in the ultraviolet (see Chapter 25), but enough is in the visible range of light to make the bubble appear blue to the eye. For an object to give off light in the ultraviolet it must be extremely hot (see Chapter 30). Estimates for the temperature in a collapsing bubble range from a minimum of 10,000 °F, somewhat hotter than the surface of the Sun, to as high as 10^9 K, hot enough to initiate thermonuclear fusion.

The characteristics of constant-pressure, constant-volume, isothermal, and adiabatic processes are summarized in Table 18–2.

TABLE 18–2
Thermodynamic Processes and Their Characteristics

Process		
Constant pressure	$W = P\Delta V$	$Q = \Delta U + P\Delta V$
Constant volume	$W = 0$	$Q = \Delta U$
Isothermal (constant temperature)	$W = Q$	$\Delta U = 0$
Adiabatic (no heat flow)	$W = \Delta U$	$Q = 0$

18–4 Specific Heats for an Ideal Gas: Constant Pressure, Constant Volume

Recall that the specific heat of a substance is a measure of the amount of heat required to raise its temperature by a given number of degrees. As we know, however, the amount of heat depends on the type of process used to raise the temperature. Thus, we should specify, for example, whether a specific heat applies to a process at constant pressure or constant volume.

If a substance is heated or cooled while open to the atmosphere, the process occurs at constant (atmospheric) pressure. This has been the case in all the specific heat discussions to this point; thus it was not necessary to make a distinction between different types of specific heats. We now wish to consider constant-volume processes as well, and the relationship between constant-volume and constant-pressure specific heats.

A constant-volume process is illustrated in **Figure 18–12.** Here we see an ideal gas of mass m in a container of fixed volume V. A heat Q flows into the container. As a result of this added heat, the temperature of the gas rises by the amount ΔT, and its pressure increases as well. Now, the specific heat at constant volume, c_{v}, is defined by the following relation:

$$Q_{\mathrm{v}} = mc_{\mathrm{v}}\,\Delta T$$

In what follows, it will be more convenient to use the **molar specific heat**, denoted by a capital letter C, which is defined in terms of the number of moles rather than the mass of the substance. Thus, if a gas contains n moles, its molar specific heat at constant volume is given by

$$Q_{\mathrm{v}} = nC_{\mathrm{v}}\,\Delta T$$

Similarly, a constant-pressure process is illustrated in **Figure 18–13**. In this case, the gas is held in a container with a moveable piston that applies a constant

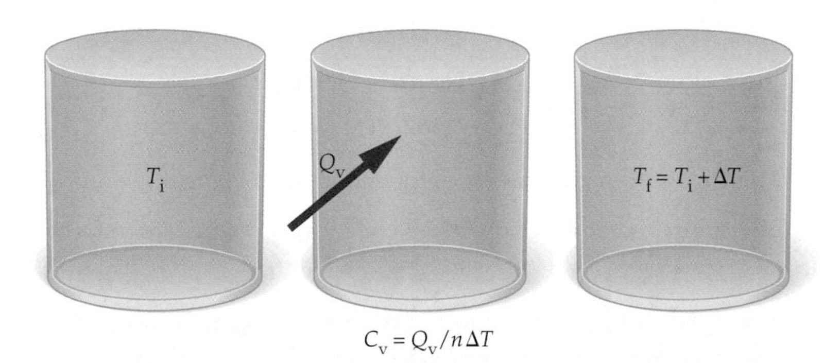

◀ **FIGURE 18–12 Heating at constant volume**
If the heat Q_{v} is added to n moles of gas, and the temperature rises by ΔT, the molar specific heat at constant volume is $C_{\mathrm{v}} = Q_{\mathrm{v}}/n\,\Delta T$. No mechanical work is done at constant volume.

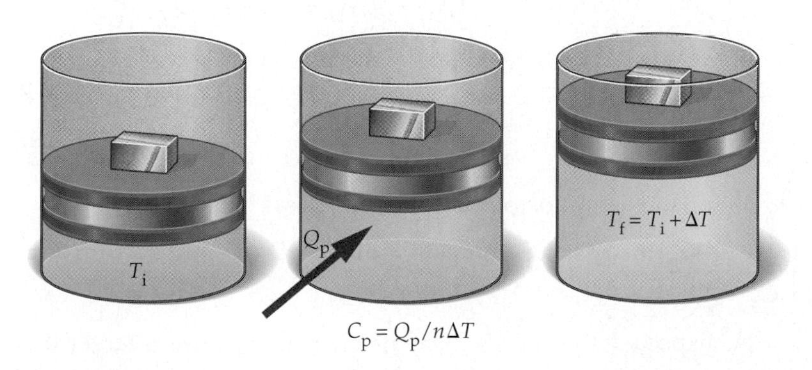

◀ **FIGURE 18–13 Heating at constant pressure**
If the heat Q_{p} is added to n moles of gas, and the temperature rises by ΔT, the molar specific heat at constant pressure is $C_{\mathrm{p}} = Q_{\mathrm{p}}/n\,\Delta T$. Note that the heat Q_{p} must increase the temperature *and* do mechanical work by lifting the piston.

pressure P. As a heat Q is added to the gas its temperature increases, which causes the piston to rise—after all, if the piston didn't rise, the pressure of the gas would increase. If the temperature of the gas increases by the amount ΔT, the molar specific heat at constant pressure is given by

$$Q_{\text{p}} = nC_{\text{p}}\Delta T$$

We would now like to obtain a relation between C_{p} and C_{v}.

Before we carry out the mathematics, let's consider the qualitative relationship between these specific heats. This is addressed in the following Conceptual Checkpoint.

CONCEPTUAL CHECKPOINT 18–3

How does the molar specific heat at constant pressure, C_{p}, compare with the molar specific heat at constant volume, C_{v}? **(a)** $C_{\text{p}} > C_{\text{v}}$; **(b)** $C_{\text{p}} = C_{\text{v}}$; **(c)** $C_{\text{p}} < C_{\text{v}}$.

Reasoning and Discussion

In a constant-volume process, as in Figure 18–12, the heat that is added to a system goes entirely into increasing the temperature, since no work is done. On the other hand, at constant pressure the heat added to a system increases the temperature *and* does mechanical work. This is illustrated in Figure 18–13 where we see that the heat must not only raise the temperature, but also supply enough energy to lift the piston. Thus, more heat is required in the constant-pressure process, and hence that specific heat is greater.

Answer:

(a) The specific heat at constant pressure is greater than the specific heat at constant volume.

We turn now to a detailed calculation of C_{v} for a monatomic, ideal gas. To begin, rearrange the first law of thermodynamics, $\Delta U = Q - W$, to solve for the heat, Q:

$$Q = \Delta U + W$$

Recall from the previous section, however, that the work is zero, $W = 0$, for any constant volume process. Hence, for constant volume we have

$$Q_{\text{v}} = \Delta U$$

Finally, noting that $U = \frac{3}{2}NkT = \frac{3}{2}nRT$, yields

$$Q_{\text{v}} = \Delta U = \tfrac{3}{2}nR\,\Delta T$$

Comparing with the definition of the molar specific heat, we find

Molar specific heat for a monatomic ideal gas at constant volume

$$C_{\text{v}} = \tfrac{3}{2}R \qquad \text{18–6}$$

Now, we perform a similar calculation for constant pressure. In this case, referring again to the previous section, we find that $W = P\,\Delta V$. Since we consider an ideal gas, in which $PV = nRT$, it follows that

$$W = P\,\Delta V = nR\,\Delta T$$

Combining this with the first law of thermodynamics yields

$$Q_{\text{p}} = \Delta U + W$$
$$= \tfrac{3}{2}nR\,\Delta T + nR\,\Delta T = \tfrac{5}{2}nR\,\Delta T$$

PROBLEM SOLVING NOTE
Constant Volume versus Constant Pressure

The heat required to increase the temperature of an ideal gas depends on whether the process is at constant pressure or constant volume. More heat is required when the process occurs at constant pressure.

Applying the definition of molar specific heats, yields

Molar specific heat for a monatomic ideal gas at constant pressure

$$C_{\text{p}} = \tfrac{5}{2}R \qquad \text{18–7}$$

As expected, the specific heat at constant pressure is larger than the specific heat at constant volume, and the difference is precisely the extra contribution due to the work done in lifting the piston in the constant-pressure case. In particular, we see that

$$C_p - C_v = R \qquad \textbf{18–8}$$

Though this relation was derived for a monatomic ideal gas, it holds for all ideal gases, regardless of the structure of their molecules. It is also a good approximation for most real gases, as can be seen in Table 18–3.

TABLE 18–3 $C_p - C_v$ for Various Gases

Helium	0.995 R
Nitrogen	1.00 R
Oxygen	1.00 R
Argon	1.01 R
Carbon dioxide	1.01 R
Methane	1.01 R

EXERCISE 18–2

Find the heat required to raise the temperature of 0.200 mol of a monatomic ideal gas by 5.00 C° at **(a)** constant volume and **(b)** constant pressure.

Solution

Applying Equations 18–6 and 18–7, we find

(a) $Q_v = \frac{3}{2}nR\Delta T = \frac{3}{2}(0.200 \text{ mol})[8.31 \text{ J}/(\text{mol}\cdot\text{K})](5.00 \text{ K}) = 12.5 \text{ J}$

(b) $Q_p = \frac{5}{2}nR\Delta T = \frac{5}{2}(0.200 \text{ mol})[8.31 \text{ J}/(\text{mol}\cdot\text{K})](5.00 \text{ K}) = 20.8 \text{ J}$

Adiabatic Processes

We return now briefly to a consideration of adiabatic processes. As we shall see, the relationship between C_p and C_v is important in determining the behavior of a system undergoing an adiabatic process.

Figure 18–14 shows an adiabatic curve and two isotherms. As mentioned before, the adiabatic curve is steeper, and it cuts across the isotherms. For the isotherms, we recall that the curves are described by the equation

$$PV = \text{constant} \qquad (\textit{isothermal})$$

A similar equation applies to adiabats. In this case, using calculus, it can be shown that the appropriate equation is

$$PV^{\gamma} = \text{constant} \qquad (\textit{adiabatic}) \qquad \textbf{18–9}$$

In this expression, the constant γ is the ratio C_p/C_v:

$$\gamma = \frac{C_p}{C_v} = \frac{\frac{5}{2}R}{\frac{3}{2}R} = \frac{5}{3}$$

This value of γ applies to monatomic ideal gases—and is a good approximation for monatomic real gases as well. The value of γ is different, however, for gases that are diatomic, triatomic, and so on.

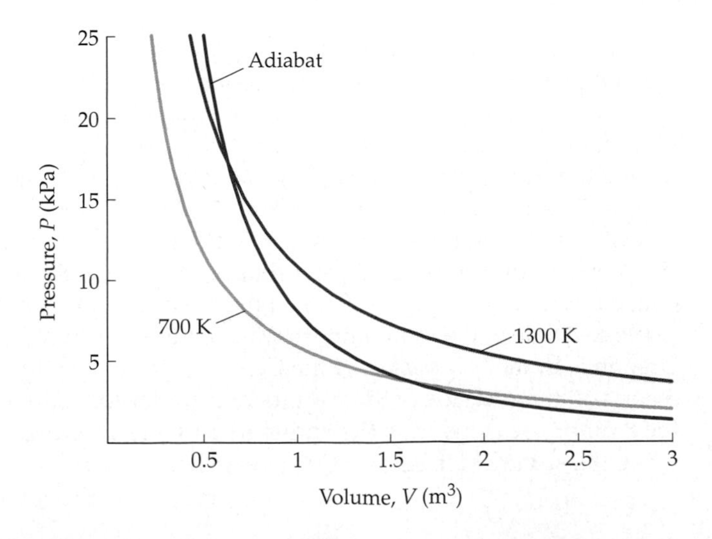

◀ **FIGURE 18–14 A comparison between isotherms and adiabats**
Two isotherms are shown, one for 700 K and one for 1300 K. An adiabat is also shown. Note that the adiabat is a steeper curve than the isotherms.

EXAMPLE 18–5 Hot Air

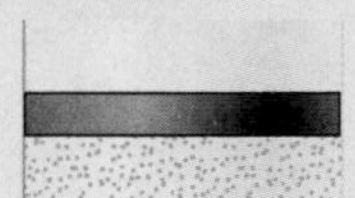

A container with an initial volume of 0.0625 m^3 holds 2.50 moles of a monatomic ideal gas at a temperature of 315 K. The gas is now compressed adiabatically to a volume of 0.0350 m^3. Find **(a)** the final pressure and **(b)** the final temperature of the gas.

Picture the Problem

The physical system is shown in the sketch. As the gas is compressed both its pressure and temperature increase.

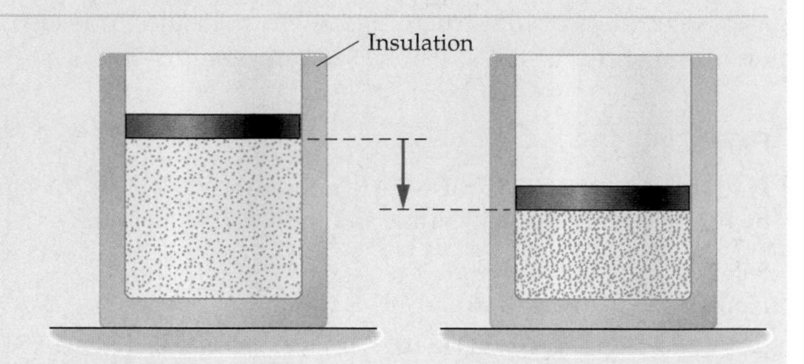

Strategy

(a) We can find the final pressure as follows:
First, find the initial pressure using the ideal-gas equation of state, $P_iV_i = nRT_i$. Next, let $P_iV_i^\gamma = P_fV_f^\gamma$, since this is an adiabatic process. Solve this relation for the final pressure.

(b) Use the final pressure and volume to find the final temperature, using the ideal-gas relation, $P_fV_f = nRT_f$.

Solution

Part (a)

1. Find the initial pressure, using $PV = nRT$:

$$P_i = \frac{nRT_i}{V_i} = \frac{(2.50\text{ mol})[8.31\text{ J/(mol}\cdot\text{K)}](315\text{ K})}{0.0625\text{ m}^3} = 105\text{ kPa}$$

2. Use PV^γ = constant to find a relation for P_f:

$$P_iV_i^\gamma = P_fV_f^\gamma$$
$$P_f = P_i(V_i/V_f)^\gamma$$

3. Substitute numerical values:

$$P_f = (105\text{ kPa})(0.0625\text{ m}^3/0.0350\text{ m}^3)^{5/3} = 276\text{ kPa}$$

Part (b)

4. Use $PV = nRT$ to solve for the final temperature:

$$T_f = \frac{P_fV_f}{nR} = \frac{(276\text{ kPa})(0.0350\text{ m}^3)}{(2.50\text{ mol})[8.31\text{ J/(mol}\cdot\text{K)}]} = 465\text{ K}$$

Insight

Thus, decreasing the volume of the gas by a factor of roughly two has increased its pressure from 105 kPa to 276 kPa, and increased its temperature from 315 K to 465 K. This is a specific example of adiabatic heating.

Adiabatic cooling is the reverse effect, where the temperature of a gas decreases as its volume increases. For example, if the gas in this system is expanded back to its initial volume of 0.0625 m^3, its temperature will drop from 465 K to 315 K.

Practice Problem

To what volume must the gas be compressed to yield a final pressure of 425 kPa? [**Answer:** $V_f = 0.0270\text{ m}^3$]

Some related homework problems: Problem 36, Problem 37

Adiabatic heating and cooling can have important effects on the climate of a given region. For example, moisture-laden winds blowing from the Pacific Ocean into western Oregon are deflected upward when they encounter the Cascade Mountains. As the air rises the atmospheric pressure decreases (see Chapter 15), allowing the air to expand and undergo adiabatic cooling. The result is that the moisture in the air condenses to form clouds and precipitation on the west side of the mountains. (In some cases, where the air holds relatively little moisture, this mechanism may result in isolated, lens-shaped clouds just above the peak of a mountain, as shown in the photograph on the next page.) When the winds continue on to the east side of the mountains they have little moisture remaining; thus, eastern Oregon is in the *rain shadow* of the Cascade Mountains. In addition, as the air descends on the east side of the mountains it undergoes adiabatic heating. These are the primary reasons why the summers in western Oregon are moist and mild, while the summers in eastern Oregon are hot and dry.

18–5 The Second Law of Thermodynamics

Have you ever warmed your hands by pressing them against a block of ice? Probably not. But if you think about it, you might wonder why it doesn't work. After all, the first law of thermodynamics would be satisfied if energy simply flowed from the ice to your hands. The ice would get colder while your hands got warmer, and the energy of the universe would remain the same.

As we know, however, this sort of thing just doesn't happen—the spontaneous flow of heat is *always* from warmer objects to cooler objects, and never in the reverse direction. This simple observation, in fact, is one of many ways of expressing the **second law of thermodynamics**:

Second law of thermodynamics: heat flow

When objects of different temperatures are brought into thermal contact, the spontaneous flow of heat that results is always from the high temperature object to the low temperature object. Spontaneous heat flow never proceeds in the reverse direction.

Thus, the second law of thermodynamics is more restrictive than the first law; it says that of all processes that conserve energy, only those that proceed in a certain direction actually occur. In a sense, the second law enforces a certain "directionality" on the behavior of nature. For this reason, the second law is sometimes referred to as the "arrow of time."

For example, suppose you saw a movie that showed a snowflake landing on a person's hand and melting to a small drop of water. Nothing would seem particularly noteworthy about the scene, from a physics point of view. But if the movie showed a drop of water on a person's hand suddenly freeze into the shape of a snowflake, then lift off the person's hand into the air, it wouldn't take long to realize the film was running backwards. It is clear in which direction time should "flow."

We shall study further consequences of the second law in the next few sections. As we do so, we shall find other more precise, but equivalent, ways of stating the second law.

▲ A spectacular lenticular (lens-shaped) cloud floats above a mountain in Tierra del Fuego, at the southern tip of Chile. Lenticular clouds are often seen "parked" above and just downwind of high mountain peaks, even when there are no other clouds in the sky. The reason is that as moisture-laden winds are deflected upward by the mountain, the moisture they contain cools due to adiabatic expansion and condenses to form a cloud.

18–6 Heat Engines and the Carnot Cycle

A **heat engine**, simply put, is a device that converts heat into work. The classic example of this type of device is the steam engine, whose basic elements are illustrated in **Figure 18–15**. First, some form of fuel (oil, wood, coal, etc.) is used to vaporize water in the boiler. The resulting steam is then allowed to enter the engine

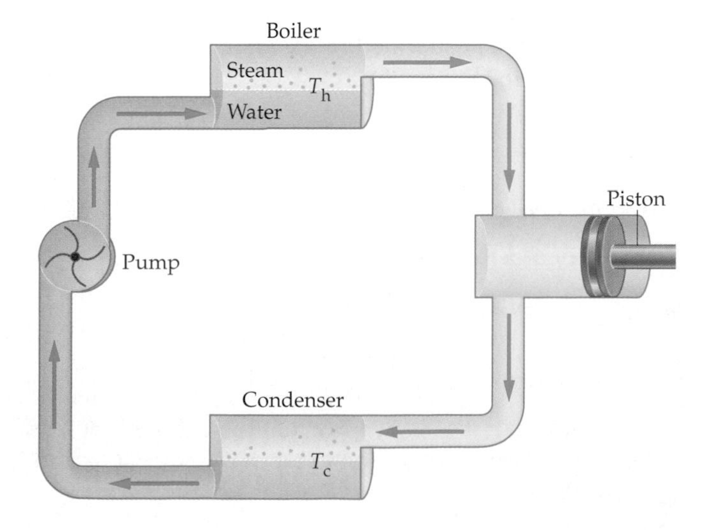

◀ **FIGURE 18–15 A schematic steam engine**

The basic elements of a steam engine are a boiler, where heat converts water to steam, and a piston that can be displaced by the expanding steam. In some engines, the steam is simply exhausted into the atmosphere after it has expanded against the piston. More sophisticated engines send the exhaust steam to a condenser, where it is cooled and condensed back to liquid water, then recycled to the boiler.

▲ At left, a modern version of Hero's engine, invented by the Greek mathematician and engineer Hero of Alexandria. In this simple heat engine, the steam that escapes from a heated vessel of water is directed radially, causing the vessel to rotate. This converts the thermal energy supplied to the water into mechanical energy, in the form of rotational motion. At right, a steam engine of slightly more recent design hauls passengers up and down Mt. Washington in New Hampshire. Note in the photo that the locomotive is belching two clouds, one black and one white. Can you explain their origin?

itself, where it expands against a piston, doing mechanical work. As the piston moves it causes gears or wheels to rotate, which delivers the mechanical work to the external world. After leaving the engine, the steam proceeds to the condenser where it gives off heat to the cool air in the atmosphere and condenses to liquid form.

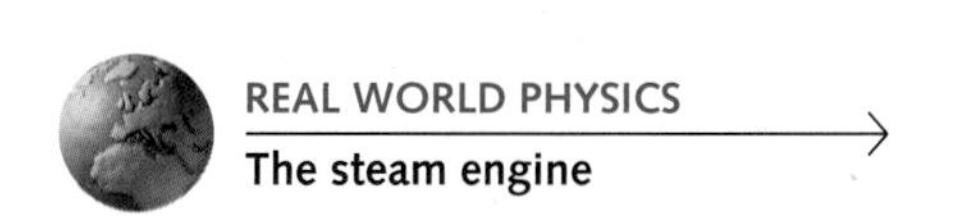

REAL WORLD PHYSICS
The steam engine

What all heat engines have in common are: (i) A high temperature region, or reservoir, that supplies heat to the engine (the boiler in the steam engine); (ii) a low temperature reservoir where "waste" heat is released (the condenser in the steam engine); and (iii) an engine that operates in a cyclic fashion. These features are illustrated schematically in **Figure 18–16**. In addition, though not shown in the figure, heat engines have a working substance (steam in the steam engine) that causes the engine to operate.

Notice that a certain amount of heat, Q_h, is supplied to the engine from the high temperature or "hot" reservoir. Of this heat, a fraction appears as work, W, and the rest is given off as waste heat, Q_c, at a relatively low temperature to the "cold" reservoir. Letting Q_h and Q_c denote magnitudes, so that both quantities are positive, energy conservation can be written as follows:

$$W = Q_h - Q_c \quad \text{18–10}$$

As we shall see, the second law of thermodynamics requires that heat engines must *always* exhaust a finite amount of heat to a cold reservoir.

Of particular interest for any engine is its **efficiency**, e, which is simply the fraction of the heat supplied to the engine that appears as work. Thus, we define the efficiency to be

$$e = \frac{W}{Q_h} \quad \text{18–11}$$

Using the energy-conservation result just derived in Equation 18–10, we find

Efficiency of a heat engine, e

$$e = \frac{W}{Q_h} = \frac{Q_h - Q_c}{Q_h} = 1 - \frac{Q_c}{Q_h} \quad \text{18–12}$$

SI unit: dimensionless

For example, if $e = 0.80$, we say that the engine is 80% efficient. In this case, 80% of the input heat is converted to work, $W = 0.80Q_h$, and 20% goes to waste heat, $Q_c = 0.20Q_h$.

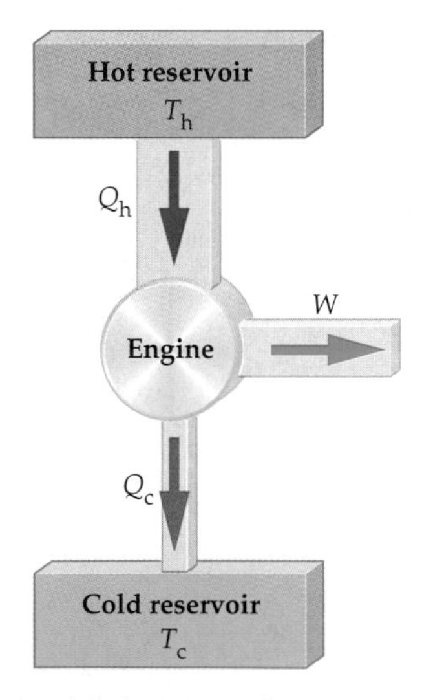

▲ **FIGURE 18–16 A schematic heat engine**
The engine absorbs a heat Q_h from the hot reservoir, performs the work W, and gives off the heat Q_c to the cold reservoir. Energy conservation gives $Q_h = W + Q_c$, where Q_h and Q_c are the magnitudes of the hot and cold heats.

EXAMPLE 18–6 Heat into Work

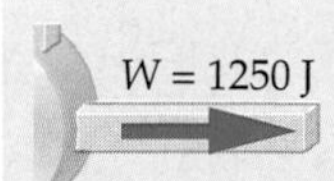

A heat engine with an efficiency of 24.0% performs 1250 J of work. Find **(a)** the heat absorbed from the hot reservoir, and **(b)** the heat given off to the cold reservoir.

Picture the Problem

Our sketch shows a schematic of the heat engine. We know the amount of work that is done, and the efficiency of the engine. We seek the heats Q_h and Q_c.

Note that an efficiency of 24.0% means that $e = 0.240$.

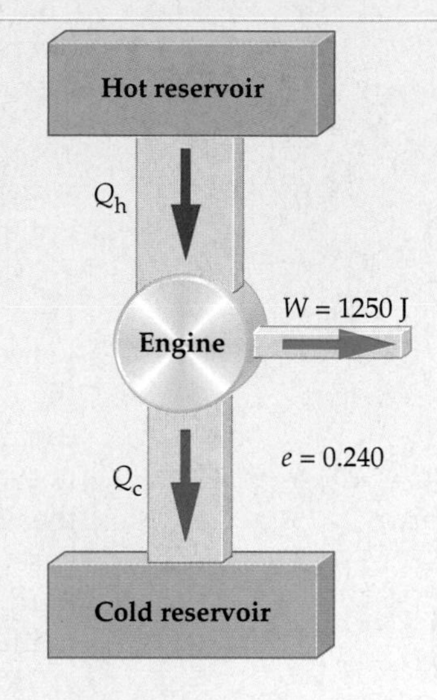

Strategy

(a) We can find the heat absorbed from the hot reservoir directly from the definition of efficiency, $e = W/Q_h$.

(b) We can find Q_c by using energy conservation, $W = Q_h - Q_c$, or by using the expression for efficiency in terms of the heats, $e = 1 - Q_c/Q_h$.

Solution

Part (a)

1. Use $e = W/Q_h$ to solve for the heat Q_h:

$$e = W/Q_h$$
$$Q_h = \frac{W}{e} = \frac{1250\text{ J}}{0.240} = 5210\text{ J}$$

Part (b)

2. Use energy conservation to solve for Q_c:

$$W = Q_h - Q_c$$
$$Q_c = Q_h - W = 5210\text{ J} - 1250\text{ J} = 3960\text{ J}$$

3. Use the efficiency, expressed in terms of Q_h and Q_c, to find Q_c:

$$e = 1 - Q_c/Q_h$$
$$Q_c = (1 - e)Q_h = (1 - 0.240)(5210\text{ J}) = 3960\text{ J}$$

Insight

Note that when the efficiency of a heat engine is less than 1/2, as in this case, the amount of heat given off as waste to the cold reservoir is more than the amount of heat converted to work.

Practice Problem

What is the efficiency of a heat engine that does 1250 J of work and gives off 5250 J of heat to the cold reservoir? [**Answer:** $e = 0.192$]

Some related homework problems: Problem 41, Problem 42

A temperature difference is essential to the operation of a heat engine. As heat flows from the hot to the cold reservoir in Figure 18–16, for example, the heat engine is able to tap into that flow and convert part of it to work—the greater the efficiency of the engine, the more heat converted to work. The second law of thermodynamics imposes limits, however, on the maximum efficiency a heat engine can have. We explore these limits next.

The Carnot Cycle and Maximum Efficiency

In 1824, Sadi Carnot (1796–1832) published a book entitled *Reflections on the Motive Power of Fire* in which he considered the following question: Under what conditions will a heat engine have maximum efficiency? To address this question, let's consider a heat engine that operates between a single hot reservoir at the fixed temperature T_h and a single cold reservoir at the fixed temperature T_c. Carnot's result, known today as **Carnot's Theorem**, can be expressed as follows:

Carnot's theorem

If an engine operating between two constant-temperature reservoirs is to have maximum efficiency, it must be an engine in which all processes are reversible. In addition, all reversible engines operating between the same two temperatures, T_c and T_h, have the same efficiency.

We should point out that no real engine can ever be perfectly reversible, just as no surface can be perfectly frictionless. Nonetheless, the concept of a reversible engine is a useful idealization.

Carnot's Theorem is remarkable for a number of reasons. First, consider what the theorem says: No engine, no matter how sophisticated or technologically advanced, can exceed the efficiency of a reversible engine. We can strive to improve the technology of heat engines, but there is an upper limit to the efficiency that can never be exceeded. Second, the theorem is just as remarkable for what it does not say. It says nothing, for example, about the working substance that is used in the engine—it is as valid for a liquid or solid working substance as for one that is gaseous. Furthermore, it says nothing about the type of reversible engine that is used, what the engine is made of, or how it is constructed. None of these things matter. In fact all that *does* matter are the two temperatures, T_c and T_h.

Recall that the efficiency of a heat engine can be written as follows:

$$e = 1 - \frac{Q_c}{Q_h}$$

Since the efficiency e depends only on the temperatures T_c and T_h, according to Carnot's theorem, it follows that Q_c/Q_h must also depend only on T_c and T_h. In fact, Lord Kelvin used this observation to propose that, instead of using a thermometer to measure temperature, we measure the efficiency of a heat engine and from this determine the temperature. Thus, he suggested that we *define* the ratio of the temperatures of two reservoirs, T_c/T_h, to be equal to the ratio of the heats Q_c/Q_h:

$$\frac{Q_c}{Q_h} = \frac{T_c}{T_h}$$

If we choose the size of a degree in this temperature scale to be equal to 1 C°, then we have, in fact, defined the Kelvin temperature scale discussed in Chapter 17. Thus, if T_h and T_c are given in kelvins, the maximum efficiency of a heat engine is:

Maximum efficiency of a heat engine

$$e_{max} = 1 - \frac{T_c}{T_h} \qquad \text{18–13}$$

Suppose for a moment that we *could* construct an ideal engine, perfectly reversible and free from all forms of friction. Would this ideal engine have 100% efficiency? No, it would not. From Equation 18–13, we can see that the only way the efficiency of a heat engine could be 100% (that is, $e_{max} = 1$) would be for T_c to be 0 K. As we shall see in the last section of this chapter, this is ruled out by the third law of thermodynamics. Hence, the maximum efficiency will always be less than 100%. No matter how perfect the engine, some of the input heat will always be wasted—given off as Q_c—rather than converted to work.

Since the efficiency is defined to be $e = W/Q_h$, it follows that the maximum work a heat engine can do with the input heat Q_h is

$$W_{max} = e_{max}Q_h = \left(1 - \frac{T_c}{T_h}\right)Q_h \qquad \text{18–14}$$

If the hot and cold reservoirs have the same temperature, so that $T_c = T_h$, it follows that the maximum efficiency is zero. As a result, the amount of work that such an engine can do is also zero. As mentioned before, a heat engine requires different temperatures in order to operate. For example, for a fixed T_c, the higher the temperature of T_h the greater the efficiency.

Finally, even though Carnot's theorem may seem quite different from the second law of thermodynamics, they are, in fact, equivalent. It can be shown, for example, that if Carnot's theorem were violated it would be possible for heat to flow spontaneously from a cold object to a hot object.

PROBLEM SOLVING NOTE
Maximum Efficiency

The maximum efficiency a heat engine can have is determined solely by the temperature of the hot (T_h) and cold (T_c) reservoirs. The numerical value of this efficiency is $e = 1 - T_c/T_h$. Remember, however, that the temperatures must be expressed in the kelvin scale for this expression to be valid.

CONCEPTUAL CHECKPOINT 18–4

Suppose you have a heat engine that can operate in one of two different modes. In mode 1, the temperatures of the two reservoirs are $T_c = 200$ K and $T_h = 400$ K; in mode 2, the temperatures are $T_c = 400$ K and $T_h = 600$ K. Is the efficiency of mode 1 **(a)** greater than, **(b)** less than, or **(c)** equal to the efficiency of mode 2?

Reasoning and Discussion

At first, you might think that since the temperature difference is the same in the two modes the efficiency is the same as well. This is not the case, however, since efficiency depends on the *ratio* of the two temperatures ($e = 1 - T_c/T_h$) rather than on their difference. In this case, the efficiency of mode 1 is $e_1 = 1 - 1/2 = 1/2$ and the efficiency of mode 2 is $e_2 = 1 - 2/3 = 1/3$. Thus, mode 1, even though it operates at the lower temperatures, is more efficient.

Answer:

(a) The efficiency of mode 1 is greater than the efficiency of mode 2.

ACTIVE EXAMPLE 18–2 Maximum Efficiency

If the heat engine in Example 18–6 is operating at maximum efficiency, and its cold reservoir is at a temperature of 295 K, what is the temperature of the hot reservoir?

Solution

1. Write the efficiency, e, in terms of the hot and cold temperatures:	$e = 1 - T_c/T_h$
2. Solve for T_h:	$T_h = T_c/(1 - e)$
3. Substitute numerical values for T_c and e to find T_h:	$T_h = 388$ K

Insight

Though an efficiency of 24% may seem low, it is characteristic of many real engines.

We can summarize the conclusions of this section as follows: The first law of thermodynamics states that you cannot get something for nothing. To be specific, you cannot get more work out of a heat engine than the amount of heat you put in. The best you can do is break even. The second law of thermodynamics is more restrictive than the first law; it says that you can't even break even— some of the input heat must be wasted. It's a law of nature.

18–7 Refrigerators, Air Conditioners, and Heat Pumps

When we stated the second law of thermodynamics in Section 18–5, we said that the spontaneous flow of heat is always from high temperature to low temperature. The key word here is "spontaneous." It *is* possible for heat to flow "uphill," from a cold object to a hot one, but it doesn't happen spontaneously—work must be done on the system to make it happen, just as work must be done to pump water from a well. Refrigerators, air conditioners, and heat pumps are devices that use work to make heat flow against its natural tendency.

Let us first compare the operation of a heat engine, **Figure 18–17 (a)**, and a **refrigerator**, **Figure 18–17 (b)**. Note that all the arrows are reversed in the refrigerator; in effect, a refrigerator is a heat engine running backward. In particular, the refrigerator uses a work W to remove a certain amount of heat, Q_c, from the cold reservoir (the interior of the refrigerator). It then exhausts an even larger amount of heat, Q_h, to the hot reservoir (the air in the kitchen). By energy conservation, it follows that

$$Q_h = Q_c + W$$

Thus, as a refrigerator operates, it warms the kitchen at the same time that it cools the food stored within it.

To design an effective refrigerator, you would like it to remove as much heat from its interior as possible for the smallest amount of work. After all, the work is supplied by electrical energy that must be paid for each month. Thus, we define the **coefficient of performance**, COP, for a refrigerator as an indicator of its effectiveness:

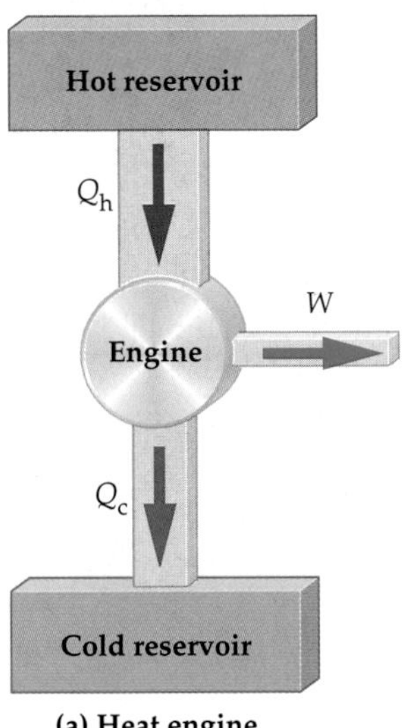

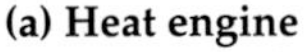
(a) Heat engine

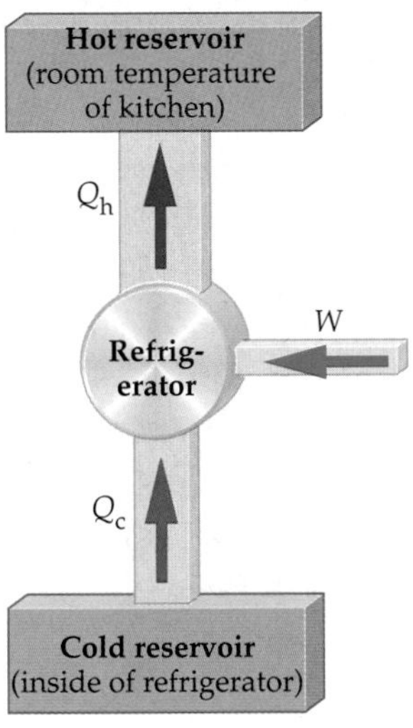

(b) Refrigerator

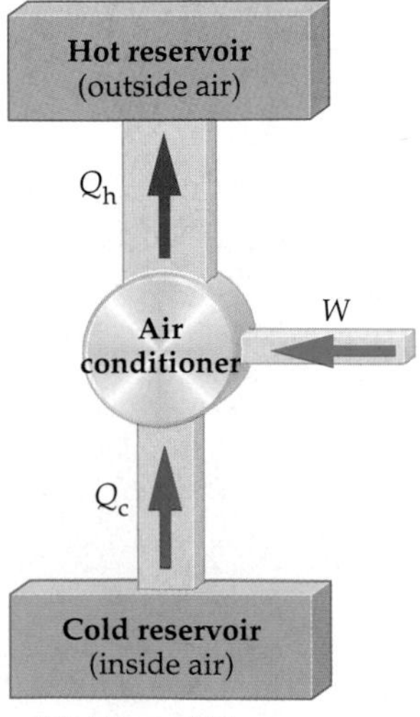

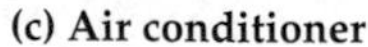
(c) Air conditioner

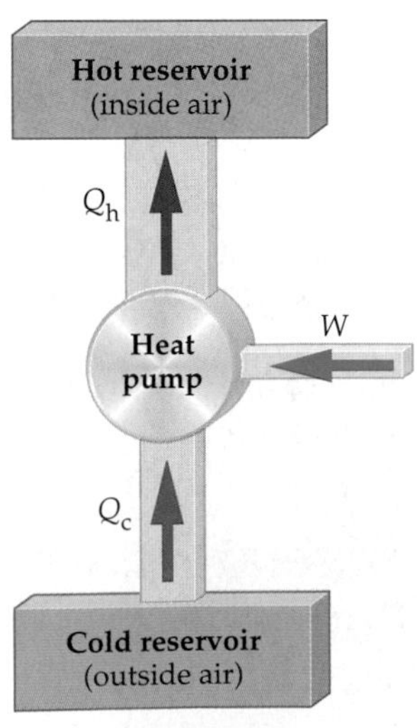

(d) Heat pump

▶ **FIGURE 18–17 Cooling and heating devices**

Schematic comparison of a generalized heat engine **(a)** compared with a refrigerator, an air conditioner, and a heat pump. In the refrigerator, **(b)** a work W is done to remove a heat Q_c from the cold reservoir (the inside of the refrigerator). By energy conservation, the heat given off to the hot reservoir (the kitchen) is $Q_h = W + Q_c$. An air conditioner **(c)** removes heat from the cool air inside a house and exhausts it into the hot air of the atmosphere. A heat pump **(d)** moves heat from a cold reservoir to a hot reservoir, just like an air conditioner. The difference is that the reservoirs are switched, so that heat is pumped into the house rather than out to the atmosphere.

Coefficient of performance for a refrigerator, COP

$$\text{COP} = \frac{Q_c}{W} \quad \textbf{18–15}$$

SI unit: dimensionless

Typical values for the coefficient of performance are in the range 2 to 6.

EXERCISE 18–3

A refrigerator has a coefficient of performance of 2.50. How much work must be supplied to this refrigerator in order to remove 225 J of heat from its interior?

Solution

Solving Equation 18–15 for the work yields

$$W = \frac{Q_c}{\text{COP}} = \frac{225\text{ J}}{2.50} = 90.0\text{ J}$$

Thus, 90.0 J of work removes 225 J of heat from the refrigerator, and exhausts 90.0 J + 225 J = 315 J of heat into the kitchen.

An **air conditioner**, **Figure 18–17 (c)**, is basically a refrigerator in which the cold reservoir is the room that is being cooled. To be specific, the air conditioner uses electrical energy to "pump" heat from the cool room to the warmer air outside. As with the refrigerator, more heat is exhausted to the hot reservoir than is removed from the cold reservoir; that is, $Q_h = Q_c + W$, as before.

REAL WORLD PHYSICS
Air conditioners

CONCEPTUAL CHECKPOINT 18–5

You haven't had time to install your new air conditioner in the window yet, so as a short-term measure you decide to place it on the dining room table and turn it on to cool things off a bit. As a result, does the air in the dining room **(a)** get warmer, **(b)** get cooler, or **(c)** stay at the same temperature?

Reasoning and Discussion

You might think the room stays at the same temperature since the air conditioner draws heat from the room as usual, but then exhausts heat back into the room which would normally be sent outside. However, the motor of the air conditioner is doing work in order to draw heat from the room, and the heat that would normally be exhausted outdoors is equal to the heat drawn from the room *plus* the work done by the motor: $Q_h = Q_c + W$. Thus, the net effect is that the motor of the air conditioner is continually adding heat to the room, causing it to get warmer.

Answer:

(a) The air in the dining room gets warmer.

▲ An air conditioner will cool your room—but only if part of the unit is kept outside. Why is this necessary?

Finally, a **heat pump** can be thought of as an air conditioner with the reservoirs switched. As we see in **Figure 18–17 (d)**, a heat pump does a work W to remove an amount of heat Q_c from the cold reservoir of outdoor air, then exhausts a heat Q_h into the hot reservoir of air in the room. Just as with the refrigerator and the air conditioner, the heat going to the hot reservoir is $Q_h = Q_c + W$.

REAL WORLD PHYSICS
Heat pumps

In an **ideal**, reversible heat pump with only two operating temperatures, T_c and T_h, the Carnot relationship $Q_c/Q_h = T_c/T_h$ holds, just as it does for a heat engine. Thus, if you want to add a heat Q_h to a room, the work that must be done to accomplish this is

$$W = Q_h - Q_c = Q_h\left(1 - \frac{Q_c}{Q_h}\right) = Q_h\left(1 - \frac{T_c}{T_h}\right) \quad \textbf{18–16}$$

We use this result in the next Example.

EXAMPLE 18–7 Pumping Heat

Heat pump

An ideal heat pump, one that satisfies the Carnot relation given in Equation 18–16, is used to heat a room that is at 293 K. If the pump does 275 J of work, how much heat does it supply to the room if the outdoor temperature is **(a)** 273 K or **(b)** 263 K?

Picture the Problem

Our sketch shows the operation of the heat pump, along with the relevant quantities for this problem. The quantity to be determined is Q_h.

Strategy

Since we know that Equation 18–16 applies to this system, it is straightforward to use it to calculate Q_h.

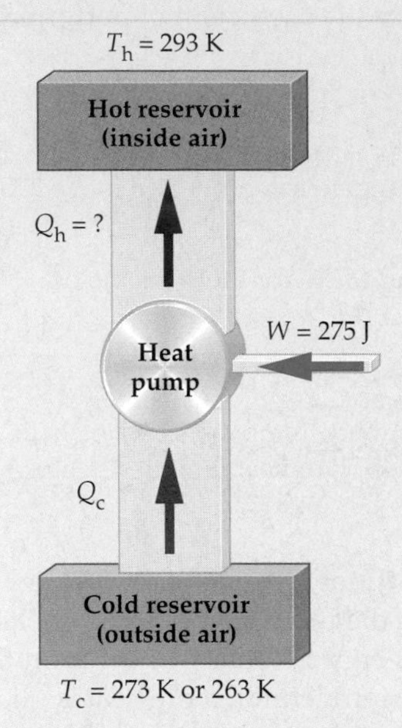

Solution

1. Solve Equation 18–16 for the heat Q_h: $Q_h = W/(1 - T_c/T_h)$

Part (a)

2. Substitute $W = 275$ J, $T_c = 273$ K, and $T_h = 293$ K into the expression for Q_h:

$$Q_h = \frac{W}{1 - \dfrac{T_c}{T_h}} = \frac{275\text{ J}}{1 - \dfrac{273\text{ K}}{293\text{ K}}} = 4030\text{ J}$$

Part (b)

3. Substitute $W = 275$ J, $T_c = 263$ K, and $T_h = 293$ K into the expression for Q_h:

$$Q_h = \frac{W}{1 - \dfrac{T_c}{T_h}} = \frac{275\text{ J}}{1 - \dfrac{263\text{ K}}{293\text{ K}}} = 2690\text{ J}$$

Insight

As one might expect, the same amount of work provides less heat when the outside temperature is lower. That is, more work must be done on a colder day to provide the same heat to the inside air.

In addition, note that if 275 J of heat is supplied to an electric heater, then 275 J of heat is given to the air in the room. When that same energy is used to run a heat pump, a good deal more than 275 J of heat is added to the room.

Practice Problem

How much work must be done by this heat pump to supply 2550 J of heat on a day when the outside temperature is 253 K? [**Answer:** $W = 348$ J]

Some related homework problems: Problem 55, Problem 56

Since the purpose of a heat pump is to add heat to a room, and we want to add as much heat as possible for the least work, the **coefficient of performance**, COP, for a heat pump is defined as follows:

Coefficient of performance for a heat pump, COP

$$\text{COP} = \frac{Q_h}{W} \qquad \textbf{18–17}$$

SI unit: dimensionless

The COP for a heat pump, which is usually in the range of 3 to 4, depends on the inside and outside temperatures. We use the COP in the next Exercise.

EXERCISE 18–4

A heat pump with a coefficient of performance equal to 3.5 supplies 2500 J of heat to a room. How much work is required?

Solution

Solving Equation 18–17 for the work, W, we find

$$W = \frac{Q_h}{COP} = \frac{2500 \text{ J}}{3.5} = 710 \text{ J}$$

18–8 Entropy

In this section we introduce a new quantity that is as fundamental to physics as energy or temperature. This quantity is referred to as the entropy, and it is related to the amount of disorder in a system. For example, a messy room has more entropy than a neat one, a pile of bricks has more entropy than a building constructed from the bricks, a freshly laid egg has more entropy than one that is just about to hatch, and a puddle of water has more entropy than the block of ice from which it melted. We begin by considering the connection between entropy and heat, and later develop more fully the connection between disorder and entropy.

When discussing heat engines, we saw that if an engine is reversible it satisfies the following relation:

$$\frac{Q_c}{Q_h} = \frac{T_c}{T_h}$$

Rearranging slightly, we can rewrite this as

$$\frac{Q_c}{T_c} = \frac{Q_h}{T_h}$$

Notice that the quantity Q/T is the same for both the hot and the cold reservoirs. This relationship prompted the German physicist Rudolf Clausius (1822–1888) to propose the following definition: The **entropy**, S, is a quantity whose change is given by the heat Q divided by the absolute temperature T:

Definition of entropy change, ΔS

$$\Delta S = \frac{Q}{T} \qquad \textbf{18–18}$$

SI unit: J/K

For this definition to be valid, the heat Q must be transferred reversibly at the fixed Kelvin temperature T. Note that if heat is added to a system ($Q > 0$) the entropy of the system increases; if heat is removed from a system ($Q < 0$) its entropy decreases.

The entropy is a state function, just like the internal energy, U. This means that the value of S depends only on the state of a system, and not on how the system gets to that state. It follows, then, that the *change* in entropy, ΔS, depends only on the initial and final states of a system. Thus, if a process is irreversible—so that Equation 18–18 *does not* hold—we can still calculate ΔS by using a reversible process to connect the same initial and final states.

EXAMPLE 18–8 Melts in Your Hand

Q

Calculate the change in entropy when a 0.125-kg chunk of ice melts at 0 °C. Assume the melting occurs reversibly.

continued on the following page

continued from the previous page

Picture the Problem
In order for the ice to melt, it must absorb a heat Q from its surroundings. Since it absorbs heat, its entropy increases.

Strategy
The entropy change is Q/T, where $T = 0\,°\text{C} = 273\text{ K}$. To find the heat Q, we note that to melt the ice we must add to it the latent heat of fusion. Thus, the heat is $Q = mL_f$, where $L_f = 33.5 \times 10^4\text{ J/kg}$.

Solution

1. Find the heat that must be absorbed by the ice for it to melt: $Q = mL_f = (0.125\text{ kg})(33.5 \times 10^4\text{ J/kg}) = 4.19 \times 10^4\text{ J}$
2. Calculate the change in entropy: $\Delta S = \dfrac{Q}{T} = \dfrac{4.19 \times 10^4\text{ J}}{273\text{ K}} = 153\text{ J/K}$

Insight
This example emphasizes the fact that the temperature in $\Delta S = Q/T$ must be given in kelvins.

Practice Problem
Find the mass of ice that would be required to give an entropy change of 275 J/K. [**Answer:** $m = 0.224$ kg]

Some related homework problems: Problem 61, Problem 62

Let's apply the definition of entropy change to the case of a reversible heat engine. First, a heat Q_h leaves the hot reservoir at the temperature T_h. Thus, the entropy of this reservoir decreases by the amount Q_h/T_h:

$$\Delta S_h = -\frac{Q_h}{T_h}$$

Recall that Q_h is the magnitude of the heat leaving the hot reservoir, hence the minus sign is used to indicate a decrease in entropy. Similarly, heat is added to the cold reservoir, hence its entropy increases by the amount Q_c/T_c:

$$\Delta S_c = \frac{Q_c}{T_c}$$

The total entropy change for this system is

$$\Delta S_{\text{total}} = \Delta S_h + \Delta S_c = -\frac{Q_h}{T_h} + \frac{Q_c}{T_c}$$

Since we know that $Q_h/T_h = Q_c/T_c$ it follows that the total entropy change vanishes:

$$\Delta S_{\text{total}} = -\frac{Q_h}{T_h} + \frac{Q_c}{T_c} = 0$$

What is special about a reversible engine, then, is the fact that its entropy does not change.

On the other hand, a real engine will always have a lower efficiency than a reversible engine operating between the same temperatures. This means that in a real engine less of the heat from the hot reservoir is converted to work; hence more heat is given off as waste heat to the cold reservoir. Thus, for a given value of Q_h, the heat Q_c is greater in an irreversible engine than in a reversible one. As a result, instead of $Q_c/T_c = Q_h/T_h$, we have

$$\frac{Q_c}{T_c} > \frac{Q_h}{T_h}$$

Therefore, if an engine is irreversible the total entropy change is positive:

$$\Delta S_{\text{total}} = -\frac{Q_{\text{h}}}{T_{\text{h}}} + \frac{Q_{\text{c}}}{T_{\text{c}}} > 0$$

In general, *any irreversible process results in an increase of entropy.*

These results can be summarized in the following general statement:

Entropy in the universe

The total entropy of the universe *increases* whenever an *irreversible* process occurs.

The total entropy of the universe is *unchanged* whenever a *reversible* process occurs.

Since all *real* processes are irreversible (with reversible processes being a useful idealization), the total entropy of the universe continually increases. Thus, in terms of the entropy, the universe moves in only one direction—toward an ever-increasing entropy. This is quite different from the behavior with regard to energy, which remains constant no matter what type of process occurs.

In fact, this statement about entropy in the universe is yet another way of expressing the second law of thermodynamics. Recall, for example, that our original statement of the second law said that heat flows spontaneously from a hot object to a cold object. During this flow of heat the entropy of the universe increases, as we show in the next Example. Hence, the direction in which heat flows is seen to be the result of the general principle of entropy increase in the universe. Again, we see a directionality in nature; the ever-present "arrow of time."

Though it is tempting to treat entropy like energy, setting the final value equal to the initial value, this is not the case in general. Only in a reversible process is the entropy unchanged—otherwise it increases. Still, the entropy of part of a system can decrease, as long as the entropy of other parts increases by the same amount or more.

EXAMPLE 18–9 Entropy Is Not Conserved!

A hot reservoir at the temperature 576 K transfers 1050 J of heat irreversibly to a cold reservoir at the temperature 305 K. Find the change in entropy of the universe.

Picture the Problem

The relevant physical situation is shown in the sketch, and the known quantities are all identified. Since the same amount of heat leaves the hot reservoir as enters the cold reservoir, we use the single label Q to denote its magnitude.

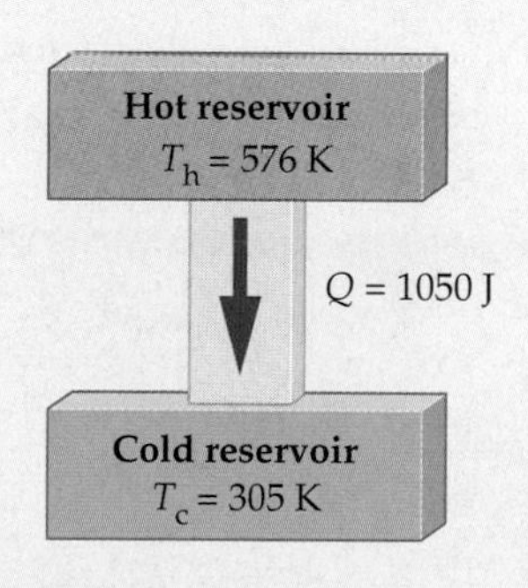

Strategy

As the heat Q leaves the hot reservoir its entropy *decreases* by Q/T_{h}. When the same heat Q enters the cold reservoir its entropy *increases* by the amount Q/T_{c}. Summing these two contributions gives the entropy change of the universe.

Solution

1. Calculate the entropy change of the hot reservoir: $\Delta S_{\text{h}} = -\dfrac{Q}{T_{\text{h}}} = -\dfrac{1050\text{ J}}{576\text{ K}} = -1.82\text{ J/K}$

2. Calculate the entropy change of the cold reservoir: $\Delta S_{\text{c}} = \dfrac{Q}{T_{\text{c}}} = \dfrac{1050\text{ J}}{305\text{ K}} = 3.44\text{ J/K}$

3. Sum these contributions to obtain the entropy change of the universe: $\Delta S_{\text{universe}} = \Delta S_{\text{h}} + \Delta S_{\text{c}} = -\dfrac{Q}{T_{\text{h}}} + \dfrac{Q}{T_{\text{c}}}$
$= -1.82\text{ J/K} + 3.44\text{ J/K} = 1.62\text{ J/K}$

Insight

Note that the decrease in entropy of the hot reservoir is more than made up for by the increase in entropy of the cold reservoir. This is a general result.

continued on the following page

continued from the previous page

Practice Problem

What amount of heat must be transferred between these reservoirs for the entropy of the universe to increase by 1.50 J/K? [**Answer:** $Q = 972$ J]

Some related homework problems: Problem 63, Problem 66

When certain processes occur, it sometimes appears as if the entropy of the universe has decreased. On closer examination, however, it can always be shown that there is a larger increase in entropy elsewhere that results in an overall increase. This issue is addressed in the next Conceptual Checkpoint.

CONCEPTUAL CHECKPOINT 18–6

You put a tray of water in the kitchen freezer, and some time later it has turned to ice. Has the entropy of the universe **(a)** increased, **(b)** decreased, or **(c)** stayed the same?

Reasoning and Discussion

It might seem that the entropy of the universe has decreased. After all, heat is removed from the water to freeze it, and, as we know, removing heat from an object lowers its entropy. On the other hand, we also know that the freezer does work to draw heat from the water, and hence it exhausts more heat into the kitchen than it absorbs from the water. Detailed calculations always show that the entropy of the heated air in the kitchen increases by more than the entropy of the water decreases, thus the entropy of the universe increases—as it must for *any* real process.

Answer:

(a) The entropy of the universe has increased.

As the entropy in the universe increases, the amount of work that can be done is diminished. For example, the heat flow in Example 18–9 resulted in an increase in the entropy of the universe by the amount 1.62 J/K. If this same heat had been used in a reversible engine, however, it could have done work, and since the engine was reversible, the entropy of the universe would have stayed the same. In the next Active Example we calculate the work that could be done with a reversible engine.

ACTIVE EXAMPLE 18–3 Doing Work

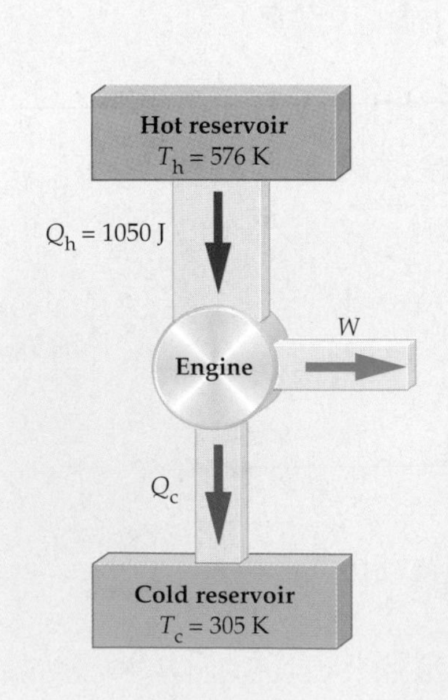

Suppose a reversible heat engine operates between the two heat reservoirs described in Example 18–9. Find the amount of work done by such an engine when 1050 J of heat is drawn from the hot reservoir.

Solution

1. Calculate the efficiency of this engine: $e = 1 - T_c/T_h = 0.470$
2. Multiply the efficiency by Q_h to find the work done: $W = eQ_h = 494$ J

Insight

Since this engine is reversible, its total entropy change must be zero. The decrease in entropy of the hot reservoir is $-(1050\text{ J})/576\text{ K} = -1.82$ J/K. It follows that the increase in entropy of the cold reservoir must have the same magnitude. The amount of heat that flows into the cold reservoir, Q_c, is $Q_h - W = 1050\text{ J} - 494\text{ J} = 556$ J. This heat causes an entropy increase equal to $556\text{ J}/305\text{ K} = +1.82$ J/K, as expected. Thus, the reason the engine exhausts the heat Q_c is to produce zero net change in entropy. If the engine were irreversible it would exhaust a heat greater than 556 J, and this would create a net increase in entropy and a reduction in the amount of work done. Clearly, then, a reversible engine produces the maximum amount of work.

Note that when 1050 J of heat is simply transferred from the hot reservoir to the cold reservoir, as in Example 18–9, the entropy of the universe increases by 1.62 J/K. When this same heat is transferred reversibly, with an ideal engine, the entropy of the universe stays the same, but 494 J of work is done. The connection between the entropy increase and the work done is very simple:

$$W = T_c \Delta S_{\text{universe}} = (305\text{ K})(1.62\text{ J/K}) = 494\text{ J}$$

To see that this expression is valid in general, recall that in Example 18–9 the total change in entropy is $\Delta S_{\text{universe}} = Q/T_c - Q/T_h$. That is, the heat $Q_h = Q$ is withdrawn from the hot reservoir (lowering the entropy by the amount Q/T_h), and the same heat $Q_c = Q$ is added to the cold reservoir (increasing the entropy by the larger amount, Q/T_c). If we multiply this increase in entropy by the temperature of the cold reservoir, T_c, we have $T_c \Delta S_{\text{universe}} = Q - QT_c/T_h = Q(1 - T_c/T_h)$. Recalling that the efficiency of an ideal engine is $e = 1 - T_c/T_h$, we see that $T_c \Delta S_{\text{universe}} = Qe$. Finally, the work done by an ideal engine is $W = eQ_h$, or in this case, $W = eQ$, since $Q_h = Q$. Therefore, we see that $W = T_c \Delta S_{\text{universe}}$, as expected.

In general, a process in which the entropy of the universe increases is one in which less work is done than if the process had been reversible. Thus, we lose forever the ability for that work to be done, because to restore the universe to its former state would mean lowering its entropy, which cannot be done. Thus, with every increase in entropy, there is that much less work that can be done by the universe.

For this reason, entropy is sometimes referred to as a measure of the "quality" of energy. When an irreversible process occurs, and the entropy of the universe increases, we say that the energy of the universe has been "degraded" because less of it is available to do work. This process of increasing degradation of energy and increasing entropy in the universe is a continuing aspect of nature.

18–9 Order, Disorder, and Entropy

In the previous section we considered entropy from the point of view of thermodynamics. We saw that as heat flows from a hot object to a cold object the entropy of the universe increases. In this section we show that entropy can also be thought of as a measure of the amount of **disorder** in the universe.

We begin with the situation of heat flow from a hot to a cold object. In **Figure 18–18 (a)** we show two bricks, one hot and the other cold. As we know from kinetic theory, the molecules in the hot brick have more kinetic energy than the molecules in the cold brick. This means that the system is rather orderly, in that all the high kinetic energy molecules are grouped together in the hot brick, and all the low kinetic energy molecules are grouped together in the cold brick. There is a definite regularity, or order, to the distribution of the molecular speeds.

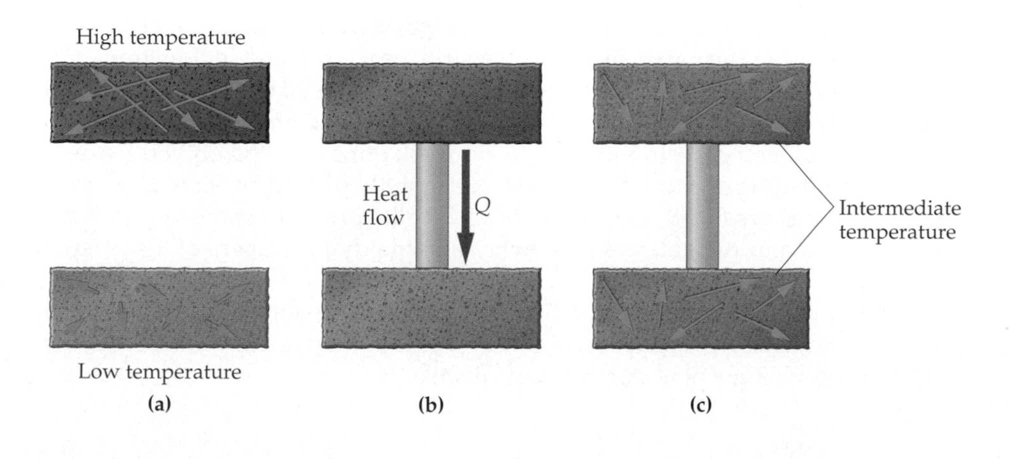

◀ **FIGURE 18–18 Heat flow and disorder**
(a) Initially, two bricks have different temperatures, and hence different average kinetic energies. **(b)** Heat flows from the hot brick to the cold brick. **(c)** The final result is that both bricks have the same intermediate temperature, and all the molecules have the same average kinetic energy. Thus, the initial orderly segregation of molecules by kinetic energy has been lost.

▲ All processes that occur spontaneously increase the entropy of the universe. In this case, the movements of the water molecules become more random and chaotic when they reach the tumultuous, swirling pool at the bottom of the falls. In addition, some of their kinetic energy is converted into heat—the most disordered and degraded form of energy.

Now, bring the bricks into thermal contact, as in **Figure 18–18 (b)**. The result is a flow of heat from the hot brick to the cold brick until the temperatures become equal. The final result is indicated in **Figure 18–18 (c)**. During the heat transfer the entropy of the universe increases, as we know, and the system loses the nice orderly distribution it had in the beginning. Now, all the molecules have the same average kinetic energy; hence, the system is randomized, or disordered. We are lead to the following conclusion:

> As the entropy of a system increases, its disorder increases as well; that is, an *increase* in entropy is the same as a *decrease* in order.

Note that if heat had flowed in the opposite direction—from the cold brick to the hot brick—the ordered distribution of molecules would have been reinforced, rather than lost.

To take another example, consider the 0.12-kg cube of ice discussed in Example 18–8. As we saw there, the entropy of the universe increases as the ice melts. Now, let's consider what happens on the molecular level. To begin, the molecules are well ordered in their crystalline positions. As heat is absorbed, however, the molecules begin to free themselves from the ice and move about randomly in the growing puddle of water. Thus, the regular order of the solid is lost. Again, we see that as entropy increases, so too does the disorder of the molecules.

Similarly, suppose you open a bottle of perfume. Initially, the perfume molecules are contained within the bottle, but as they begin to collide randomly with the molecules in the air they start to move about and diffuse throughout the room. The entropy of the system increases because the orderly confinement of the perfume molecules in the beginning has given way to a random distribution of perfume molecules about the room. Note also that the random motion of the perfume molecules means that they will never again concentrate within the bottle—irreversible physical processes occur in a direction that increases the entropy, and disorder, of the universe.

Thus, the second law of thermodynamics can be stated as the principle that the disorder of the universe is continually increasing. Everything that happens in the universe is simply making it a more disorderly place. And there is nothing you can do to prevent it—nothing you can do will result in the universe being more ordered. Just as freezing a tray of water to make ice actually results in more entropy—and more disorder—in the universe, so does any action you take.

Heat Death

If one carries the previous discussion to its logical conclusion, it seems that the universe is "running down." That is, the disorder of the universe constantly increases, and as it does, the amount of energy available to do work decreases. If this process continues, might there come a day when no more work can be done? And if that day does come, what then?

This is one possible scenario for the fate of the universe, and it is referred to as the "heat death" of the universe. In this scenario, heat continues to flow from hotter regions in space (like stars) to cooler regions (like planets) until, after many billions of years, all objects in the universe have the same temperature. With no temperature differences left, there can be no work done, and the universe would cease to do anything of particular interest. Not a pretty picture, but certainly a possibility. The universe may simply continue with its present expansion until the stars burn out and the galaxies fade away like the dying embers of a scattered campfire.

Another possibility, however, is that the universe will stop expanding at some point, after which it will collapse in a "big crunch," perhaps setting off another big bang, and providing a new start for the universe.

Living Systems

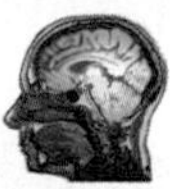

So far we have focused on the rather gloomy prospect of the universe constantly evolving toward greater disorder. Is it possible, however, that life is an exception to this rule? After all, we know that an embryo utilizes simple raw materials to produce a complex, highly ordered living organism. Similarly, the well-known biological aphorism "ontogeny recapitulates phylogeny," while not strictly correct, expresses the fact that the development of an individual organism from embryo to adult often reflects certain aspects of the evolutionary development of the species as a whole. Thus, over time species often evolve toward more complex forms. Finally, living systems are able to use disordered raw materials in the environment to produce orderly structures in which to live. It seems, then, that there are many ways in which living systems produce increasing order.

This conclusion is flawed, however, since it fails to take into account the entropy of the environment in which the organism lives. It is similar to the conclusion that a freezer violates the second law of thermodynamics because it reduces the entropy of water as it freezes it into ice. This analysis neglects the fact that the freezer exhausts heat into the room, increasing the entropy of the air by an amount that is greater than the entropy decrease of the water. In the same way, living organisms constantly give off heat to the atmosphere as a by-product of their metabolism, increasing its entropy. Thus, if we build a house from a pile of bricks—decreasing the entropy of the bricks—the heat we give off during our exertions increases the entropy of the atmosphere more than enough to give a net increase in entropy.

Finally, all living organisms can be thought of as heat engines, tapping into the flow of energy from one place to another to produce mechanical work. Plants, for example, tap into the flow of energy from the high temperature of the Sun to the cold of deep space and use a small fraction of this energy to sustain themselves and reproduce. Animals consume plants and generate heat within their bodies as they metabolize their food. A fraction of the energy released by the metabolism is in turn converted to mechanical work. Living systems, then, obey the same laws of physics as steam engines and refrigerators—they simply produce different results as they move the universe toward greater disorder.

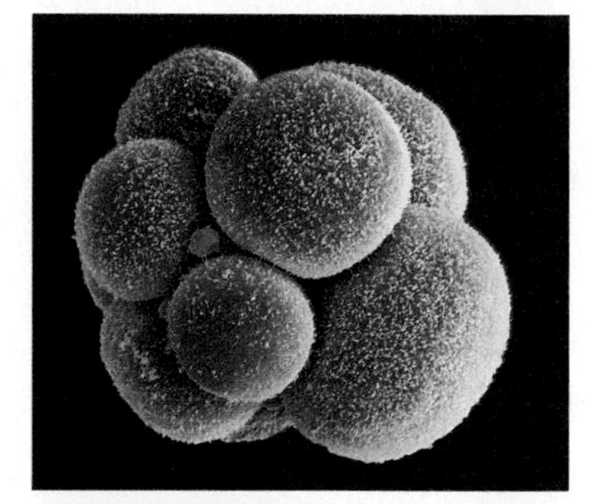

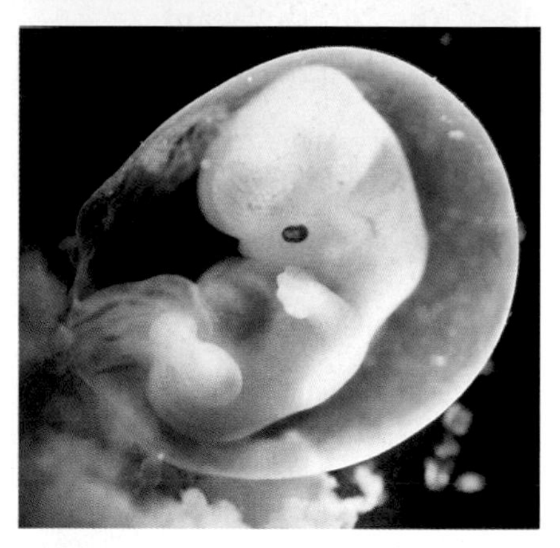

▲ Many species, including humans, develop from a single fertilized egg into a complex multicellular organism, In the process they create large, intricately ordered molecules such as proteins and DNA from smaller, simpler precursors. In the metabolic processes of living things, however, heat is produced, increasing the entropy of the universe as a whole. Thus, the second law is not violated.

18–10 The Third Law of Thermodynamics

Finally, we consider the **third law of thermodynamics**, which states that there is no temperature lower than absolute zero, and that absolute zero is unattainable. It is possible to cool an object to temperatures arbitrarily close to absolute zero—experiments have reached temperatures as low as 2.0×10^{-8} K—but no object can ever be cooled to precisely 0 K.

As an analogy to cooling toward absolute zero, imagine walking toward a wall, with each step half the distance between you and the wall. Even if you take an infinite number of steps, you will still not reach the wall. You can get arbitrarily close, of course, but you never get all the way there.

The same sort of thing happens when cooling. To cool an object, you can place it in thermal contact with an object that is colder. Heat transfer will occur, with your object ending up cooler and the other object ending up warmer. In particular, suppose you had a collection of objects at 0 K to use for cooling. You put your object in contact with one of the 0-K objects and your object cools, while the 0-K object warms slightly. You continue this process, each time throwing away the "warmed up" 0-K object and using a new one. Each time you cool your object it gets closer to 0 K, without ever actually getting there.

In light of this discussion, we can express the third law of thermodynamics as follows:

The third law of thermodynamics

It is impossible to lower the temperature of an object to absolute zero in a finite number of steps.

As with the second law of thermodynamics, this law can be expressed in a number of different but equivalent ways. The essential idea, however, is always the same: Absolute zero is the limiting temperature and, though it can be approached arbitrarily closely, it can never be attained.

Chapter Summary

Topic	Remarks and Relevant Equations
18–1 The Zeroth Law of Thermodynamics	When two objects have the same temperature they are in thermal equilibrium.
18–2 The First Law of Thermodynamics	The first law of thermodynamics is a statement of energy conservation that includes heat. If U is the internal energy of an object, Q is the heat added to it, and W is the work done by the object, the first law of thermodynamics can be written as follows: $\Delta U = Q - W$ **18–3**
state function	The internal energy U depends only on the state of a system; that is, on its temperature, pressure, and volume. The value of U is independent of how a system is brought to a certain state.
18–3 Thermal Processes	
quasi-static	A quasi-static process is one in which a system is moved slowly from one state to another. The change of state is so slow that the system may be considered to be in equilibrium at any given time during the process.
reversible	In a reversible process it is possible to return the system and its surroundings to their initial states.
irreversible	Irreversible processes cannot be "undone." When the system is returned to its initial state the surroundings have been altered.
work	In general, the work done during a process is equal to the area under the process in a PV plot.
constant pressure	In a PV plot a constant-pressure process is represented by a horizontal line. The work done at constant pressure is $W = P\,\Delta V$.
constant volume	In a PV plot a constant-volume process is represented by a vertical line. The work done at constant volume is zero; $W = 0$.
isothermal process	In a PV plot an adiabtic process is represented by PV = constant. The work done in an isothermal expansion from V_i to V_f is $W = nRT\ln(V_f/V_i)$.
adiabatic process	An adiabatic process occurs with no heat transfer; that is, $Q = 0$.
18–4 Specific Heats for an Ideal Gas: Constant Pressure, Constant Volume	Specific heats have different values depending on whether they apply to a process at constant pressure or a process at constant volume.
molar specific heat	The molar specific heat, C, is defined by $Q = nC\,\Delta T$, where n is the number of moles.
constant volume	The molar specific heat for an ideal monatomic gas at constant volume is

$$C_v = \tfrac{3}{2}R \qquad \text{18–6}$$

constant pressure

The molar specific heat for an ideal monatomic gas at constant pressure is

$$C_p = \tfrac{5}{2}R \qquad \text{18–7}$$

adiabatic process

In a PV plot an adiabatic process is represented by PV^{γ} = constant, where γ is the ratio C_p/C_v. For a monatomic, ideal gas, $\gamma = 5/3$.

18–5 The Second Law of Thermodynamics

When objects of different temperatures are brought into thermal contact, the spontaneous flow of heat that results is always from the high temperature object to the low temperature object. Spontaneous heat flow never proceeds in the reverse direction.

18–6 Heat Engines and the Carnot Cycle

A heat engine is a device that converts heat into work; for example, a steam engine.

efficiency

The efficiency e of a heat engine that takes in the heat Q_h from a hot reservoir, exhausts a heat Q_c to a cold reservoir, and does the work W is

$$e = \frac{W}{Q_h} = \frac{Q_h - Q_c}{Q_h} = 1 - \frac{Q_c}{Q_h} \qquad \text{18–12}$$

Carnot's theorem

If an engine operating between two constant-temperature reservoirs is to have maximum efficiency, it must be an engine in which all processes are reversible. In addition, all reversible engines operating between the same two temperatures have the same efficiency.

maximum efficiency

The maximum efficiency of a heat engine operating between the Kelvin temperatures T_h and T_c is

$$e_{max} = 1 - \frac{T_c}{T_h} \qquad \text{18–13}$$

maximum work

If a heat engine takes in the heat Q_h from a hot reservoir, the maximum work it can do is

$$W_{max} - e_{max}Q_h = \left(1 - \frac{T_c}{T_h}\right)Q_h \qquad \text{18–14}$$

18–7 Refrigerators, Air Conditioners, and Heat Pumps

Refrigerators, air conditioners, and heat pumps are devices that use work to make heat flow from a cold region to a hot region.

coefficient of performance

The coefficient of performance of a refrigerator or air conditioner doing the work W to remove a heat Q_c from a cold reservoir is

$$\text{COP} = \frac{Q_c}{W} \qquad \text{18–15}$$

heat pump

In an ideal heat pump, the work W that must be done to deliver a heat Q_h to a hot reservoir at the temperature T_h by extracting a heat Q_c from a cold reservoir at the temperature T_c is

$$W = Q_h - Q_c = Q_h\left(1 - \frac{Q_c}{Q_h}\right) = Q_h\left(1 - \frac{T_c}{T_h}\right) \qquad \text{18–16}$$

The coefficient of performance for a heat pump is

$$\text{COP} = \frac{Q_h}{W} \qquad \text{18–17}$$

18–8 Entropy

Like the internal energy U, the entropy S is a state function.

change in entropy

The change in entropy during a reversible exchange of the heat Q at the Kelvin temperature T is

$$\Delta S = \frac{Q}{T} \qquad \text{18–18}$$

entropy in the universe	The total entropy of the universe increases whenever an irreversible process occurs. *Note:* Entropy is not conserved. In an idealized reversible process the entropy of the universe is unchanged.
18–9 Order, Disorder, and Entropy	Entropy is a measure of the disorder of a system. As entropy increases, a system becomes more disordered.
heat death	A possible fate of the universe is heat death, in which everything is at the same temperature and no more work can be done.
18–10 The Third Law of Thermodynamics	It is impossible to lower the temperature of an object to absolute zero in a finite number of steps.

Problem-Solving Summary

Type of Problem	Relevant Physical Concepts	Related Examples
Relate the heat and work exchanged by a system to its change in internal energy.	The heat, work, and internal energy of a system are related by the first law of thermodynamics; $\Delta U = Q - W$.	Example 18–1
Calculate the work done by a system during a given process in terms of the corresponding PV plot.	The work done by an expanding system is equal to the area under the curve that represents the process on a PV plot. If the system is compressed the work is equal to the negative of the area, meaning that work is done on the system rather than by it.	Example 18–2 Active Example 18–1
Find the efficiency of a heat engine.	If a heat engine takes in a heat Q_h and does the work W its efficiency, e, is given by $e = W/Q_h$. Since W is the difference between the heat going into the engine, Q_h, and the heat leaving the engine, Q_c, the efficiency can also be written as $e = (Q_h - Q_c)/Q_h = 1 - Q_c/Q_h$. Finally, the efficiency can also be related to the absolute temperature of the hot (T_h) and cold (T_c) reservoirs as follows: $e = 1 - T_c/T_h$.	Example 18–6 Active Example 18–2
Determine the change in entropy of a system.	If a system exchanges a heat Q at the absolute temperature T its entropy S changes by the following amount: $\Delta S = Q/T$. If heat is added to the system its entropy increases; if heat is removed from the system its entropy decreases.	Examples 18–8, 18–9

Conceptual Questions

1. If an engine has a reverse gear, does this make it reversible?
2. A gas expands, doing 100 J of work. How much heat must be added to this system for its internal energy to increase by 200 J?
3. A substance is thermally insulated, so that no heat can flow between it and its surroundings. Is it possible for the temperature of this substance to rise? Explain.
4. Heat is added to a substance. Is it safe to conclude that the temperature of the substance rises? Explain.
5. Is it possible to convert a given amount of mechanical work completely into heat? Explain.
6. Are there thermodynamic processes in which all the heat absorbed by an ideal gas goes completely into mechanical work? If so, give an example.
7. Give the change in internal energy if **(a)** $W = 50$ J, $Q = 50$ J, **(b)** $W = -50$ J, $Q = -50$ J, or **(c)** $W = 50$ J, $Q = -50$ J.
8. An ideal gas is held in an insulated container at the temperature T. All the gas is initially in one-half of the container, with a partition separating the gas from the other half of the container, which is a vacuum. If the partition ruptures, and the gas expands to fill the entire container, what is its final temperature?
9. Which of the following processes are approximately reversible? **(a)** Lighting a match. **(b)** Pushing a block up a frictionless inclined plane. **(c)** Frying an egg. **(d)** Swimming from one end of a pool to the other. **(e)** Stretching a spring by a small amount. **(f)** Writing a report for class.
10. Which law of thermodynamics would be violated if heat were to spontaneously flow between two objects of equal temperature?
11. You plan to add a certain amount of heat to a gas in order to raise its temperature. If you add the heat at constant volume, is the increase in temperature more than, less than, or the same as if you add the heat at constant pressure?

12. Consider the three-process cycle shown in **Figure 18–19**. For each process in the cycle, state whether the work done by the system is positive, negative, or zero.

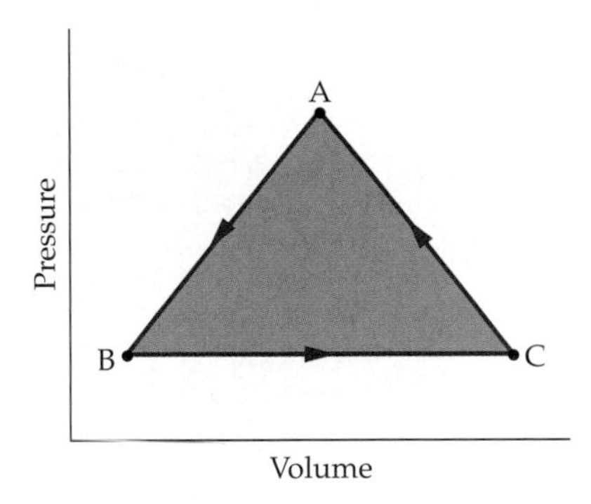

▲ **FIGURE 18–19** Question 12

13. Rank the three ideal-gas expansions shown in **Figure 18–20** according to the amount of work done by the gas, starting with the least.

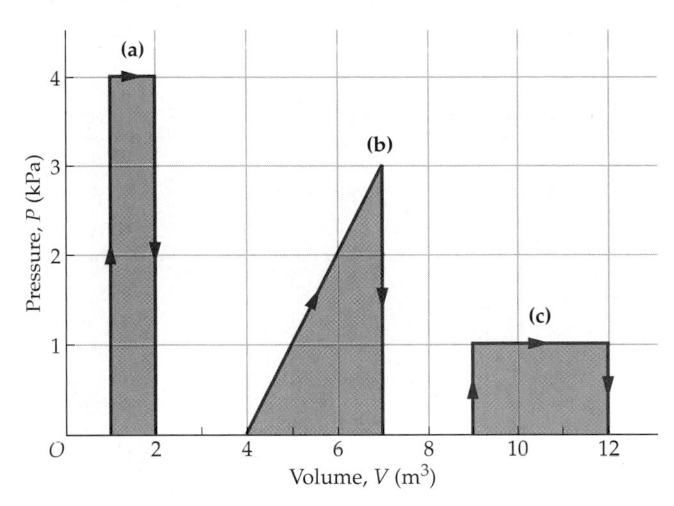

▲ **FIGURE 18–20** Question 13

14. Heat engines always give off a certain amount of heat to a low-temperature reservoir. Would it be possible to use this "waste" heat as the heat input to a second heat engine, and then use the "waste" heat of the second engine to run a third engine, and so on?
15. A Carnot engine operates between a hot reservoir at the Kelvin temperature T_h and a cold reservoir at the Kelvin temperature T_c. If both temperatures are doubled, does the efficiency increase, decrease, or stay the same?
16. If the temperature in the kitchen is decreased, does it cost more, less, or the same amount of money to freeze a dozen ice cubes? Explain.
17. A heat pump uses 100 J of energy as it operates for a given time. Is it possible for the heat pump to deliver more than 100 J of heat to the inside of the house in this same time?
18. If you rub your hands together, does the entropy of the universe increase, decrease, or stay the same? Explain.
19. A gas is expanded isothermally. Does its entropy increase, decrease, or stay the same?
20. A gas is expanded adiabatically. Does its entropy increase, decrease, or stay the same?
21. If you clean up a messy room, putting things back where they belong, you decrease the room's entropy. Does this violate the second law of thermodynamics?
22. Which law of thermodynamics is most pertinent to the statement that "all the king's horses and all the king's men couldn't put Humpty Dumpty back together again?"
23. Which has more entropy: **(a)** popcorn kernels, or the resulting popcorn; **(b)** two eggs in a carton, or an omelet made from the eggs; **(c)** a pile of bricks, or the resulting house; **(d)** a piece of paper, or the piece of paper after it has been burned?

Problems

Note: **IP** *denotes an integrated conceptual/quantitative problem.* **BIO** *identifies problems of biological or medical interest. Blue bullets (•, ••, •••) are used to indicate the level of difficulty of each problem.*

Section 18–2 The First Law of Thermodynamics

1. • A swimmer does 5.7×10^5 J of work and gives off 3.2×10^5 J of heat during a workout. Determine ΔU, W, and Q for the swimmer.
2. • When 1210 J of heat are added to one mole of an ideal monatomic gas, its temperature increases from 272 K to 276 K. Find the work done by the gas during this process.
3. • Three different processes act on a system. **(a)** In process A, 42 J of work are done on the system and 77 J of heat are added to the system. Find the change in the system's internal energy. **(b)** In process B, the system does 42 J of work and 77 J of heat are added to the system. What is the change in the system's internal energy? **(c)** In process C, the system's internal energy decreases by 120 J while the system performs 120 J of work on its surroundings. How much heat was added to the system?
4. • An ideal gas is taken through the four processes shown in **Figure 18–21**. The change in internal energy for three of these processes are as follows: $\Delta U_{AB} = +82$ J; $\Delta U_{BC} = +15$ J; $\Delta U_{DA} = -56$ J. Find the change in internal energy for the process from C to D.

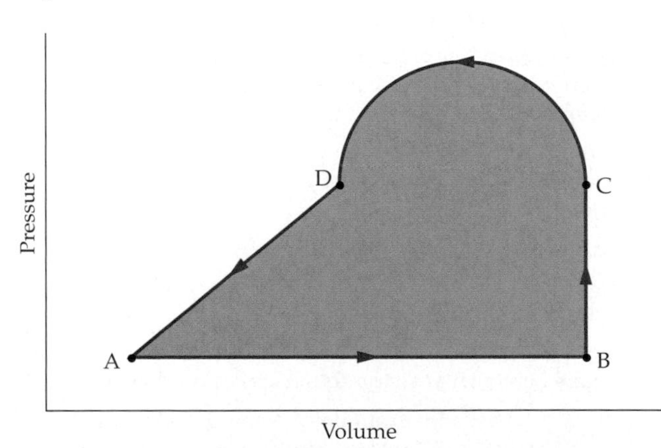

▲ **FIGURE 18–21** Problem 4

5. •• An ideal gas is taken through the three processes shown in **Figure 18–22**. Fill in the missing entries in the following table:

	Q	W	ΔU
A → B	−53 J	(a)	(b)
B → C	−280 J	−130 J	(c)
C → A	(e)	150 J	(d)

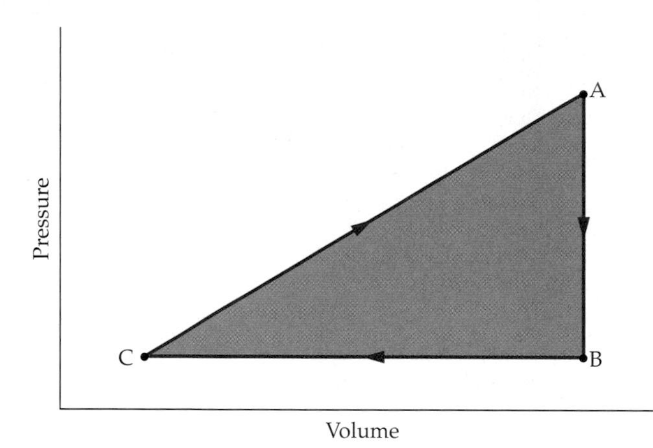

▲ **FIGURE 18–22** Problem 5

6. •• One mole of an ideal monatomic gas is initially at a temperature of 323 K. Find the final temperature of the gas if 2250 J of heat are added to it and it does 834 J of work.

7. •• Burning a gallon of gasoline releases 1.19×10^8 J of internal energy. If a certain car requires 5.20×10^5 J of work to drive one mile, how much heat is given off to the atmosphere each mile when the car gets 25.0 miles to the gallon?

8. •• A cylinder contains 4.0 moles of a monatomic gas at an initial temperature of 27 °C. The gas is compressed by doing 560 J of work on it, and its temperature increases by 130 C°. How much heat flows into or out of the gas?

9. •• A basketball player does 2.13×10^5 J of work during her time in the game, and evaporates 0.110 kg of water. Assuming a latent heat of 2.26×10^6 J/kg for the perspiration (the same as for water) determine **(a)** the change in the player's internal energy and **(b)** the number of nutritional calories the player has converted to work and heat.

Section 18–3 Thermal Processes

10. • A system consisting of an ideal gas at the constant pressure of 110 kPa gains 820 J of heat. Find the change in volume of the system if the internal energy of the gas increases by **(a)** 820 J or **(b)** 360 J.

11. • An ideal gas is compressed at constant pressure to one-half its initial volume. If the pressure of the gas is 120 kPa, and 760 J of work is done on it, find the initial volume of the gas.

12. • As an ideal gas expands at constant pressure from a volume of 0.64 m^3 to a volume of 1.3 m^3 it does 83 J of work. What is the gas pressure during this process?

13. • The volume of a monatomic ideal gas doubles in an adiabatic expansion. By what factor does its pressure change?

14. •• **IP** A fluid expands from point A to point B along the path shown in **Figure 18–23**. **(a)** How much work is done by the fluid during this expansion? **(b)** Does your answer to part (a) depend on whether the fluid is an ideal gas? Explain.

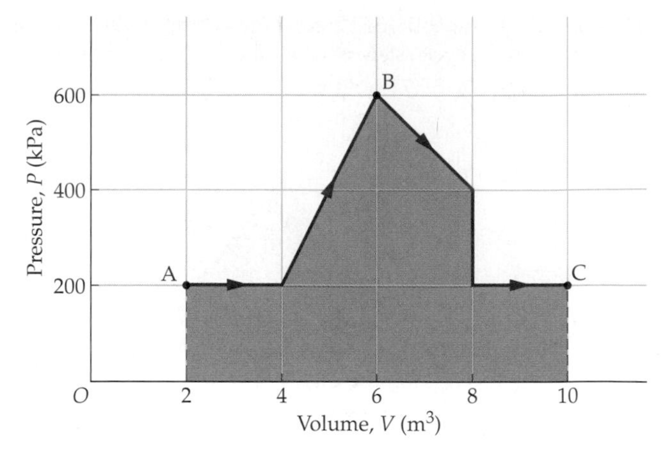

▲ **FIGURE 18–23** Problems 14 and 15

15. •• **(a)** Find the work done by a monatomic ideal gas as it expands from point A to point C along the path shown in Figure 18–23. **(b)** If the temperature of the gas is 220 K at point A, what is its temperature at point C? **(c)** How much heat has been added to or removed from the gas during this process?

16. •• **IP** Consider a system in which 2.00 mol of an ideal monatomic gas is expanded at a constant pressure of 101 kPa from an initial volume of 2.15 L to a final volume of 3.30 L. **(a)** Is heat added to or removed from the system during this process? Explain. **(b)** Find the change in temperature for this process. **(c)** Determine the amount of heat added to or removed from the system during this process.

17. •• 5.00 moles of a monatomic ideal gas at a temperature of 345 K are expanded isothermally from a volume of 1.32 L to a volume of 3.66 L. Calculate **(a)** the work done and **(b)** the heat flow into or out of the gas.

18. •• One mole of a monatomic ideal gas undergoes an isothermal expansion from 1.00 m^3 to 4.00 m^3, as shown in **Figure 18–24**. Find the work done by the gas during this expansion.

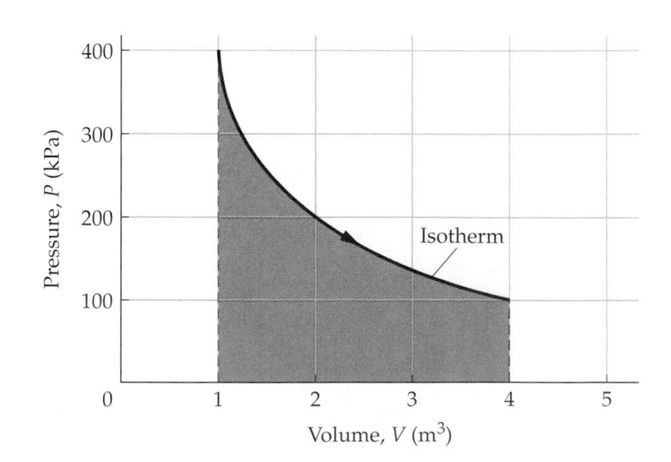

▲ **FIGURE 18–24** Problem 18

19. •• Three moles of an ideal monatomic gas expand isothermally at a temperature of 34 °C. If the volume of the gas quadruples during this process, what is **(a)** the work done by the gas and **(b)** the heat flow into the gas?

20. •• **IP** **(a)** If the internal energy of a system increases as the result of an adiabatic process, is work done on the system or by the

system? **(b)** Calculate the work done on or by the system in part (a) if its internal energy increases by 670 J.

21. •• During an adiabatic process, the temperature of 6.00 moles of a monatomic ideal gas drops from 505 °C to 255 °C. Find **(a)** the work done and **(b)** the change in volume if the expansion is at a constant pressure of one atmosphere.

22. •• **IP** **(a)** A monatomic ideal gas expands at constant pressure. Is heat added to the system or taken from the system during this process? **(b)** Find the heat added to or taken from the gas in part (a) if it expands at a pressure of 160 kPa from a volume of 0.76 m^3 to a volume of 0.93 m^3.

23. •• With the pressure held constant at 210 kPa, 49 mol of a monatomic ideal gas expands from an initial volume of 0.75 m^3 to a final volume of 1.9 m^3. **(a)** How much work was done by the gas during the expansion? **(b)** What were the initial and final temperatures of the gas? **(c)** What was the change in the internal energy of the gas? **(d)** How much heat was added to the gas?

24. •• An ideal gas follows the three-part process shown in **Figure 18–25**. At the completion of one full cycle, find **(a)** the net work done by the system, **(b)** the net change in internal energy of the system, and **(c)** the net heat absorbed by the system.

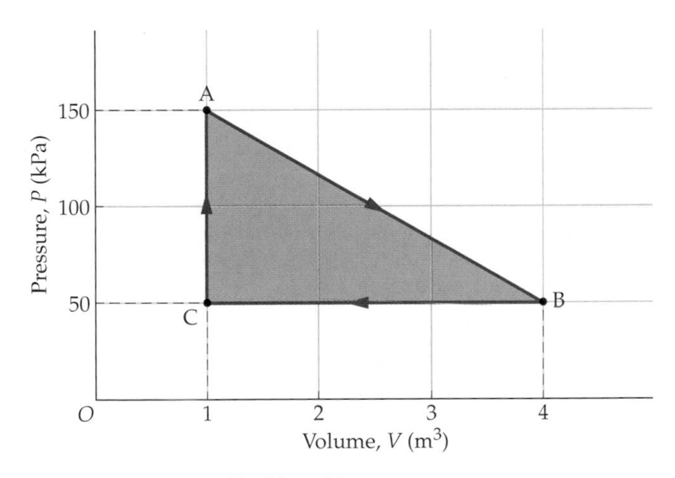

▲ **FIGURE 18–25** Problem 24

25. •• A gas is contained in a cylinder with a pressure of 160 kPa and an initial volume of 0.83 m^3. How much work is done by the gas as it **(a)** expands at constant pressure to twice its initial volume, or **(b)** is compressed to one-third its initial volume?

26. •• A system expands by 0.75 m^3 at a constant pressure of 105 kPa. Find the heat that flows into or out of the system if its internal energy **(a)** increases by 65 J or **(b)** decreases by 1850 J. In each case, give the direction of heat flow.

27. •• An ideal monatomic gas is held in a perfectly insulated cylinder fitted with a moveable piston. The initial pressure of the gas is 130 kPa, and its initial temperature is 310 K. By pushing down on the piston you are able to increase the pressure to 180 kPa. Find the temperature of the gas at this higher pressure.

28. ••• A certain amount of a monatomic ideal gas undergoes the process shown in **Figure 18–26**, in which its pressure doubles and its volume triples. In terms of the number of moles, n, the initial pressure, P_i, and the initial volume, V_i, determine **(a)** the work done by the gas, W, **(b)** the change in internal energy of the gas, U, and **(c)** the heat added to the gas, Q.

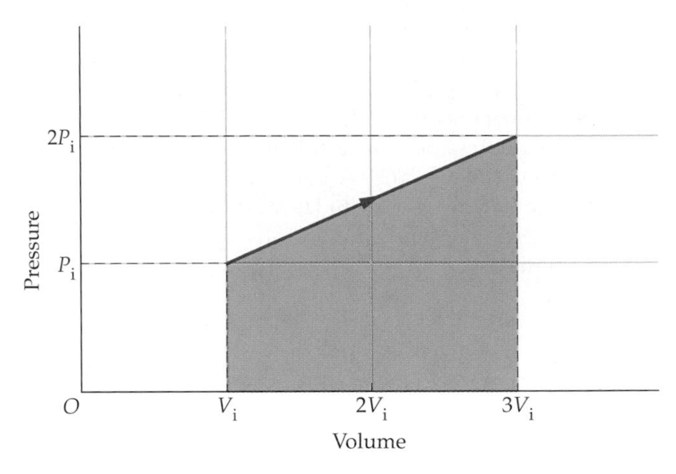

▲ **FIGURE 18–26** Problem 28

29. ••• An ideal gas doubles its volume in one of three different ways: **(i)** at constant pressure; **(ii)** at constant temperature; **(iii)** adiabatically. Explain your answers to each of the following questions: **(a)** In which expansion does the gas do the most work? **(b)** In which expansion does the gas do the least work? **(c)** Which expansion results in the highest final temperature? **(d)** Which expansion results in the lowest final temperature?

Section 18–4 Specific Heats for an Ideal Gas: Constant Pressure, Constant Volume

30. • Find the amount of heat needed to increase the temperature of 2.5 mol of an ideal monatomic gas by 15 K if **(a)** the pressure or **(b)** the volume is held constant.

31. • **(a)** If 570 J of heat are added to 45 moles of a monatomic gas at constant volume, how much does the temperature of the gas increase? **(b)** Repeat part (a), this time for a constant-pressure process.

32. • A system consists of 1.5 mol of an ideal monatomic gas at 375 K. How much heat must be added to the system to double its internal energy at **(a)** constant pressure or **(b)** constant volume.

33. • Find the change in temperature if 130 J of heat are added to 2.8 mol of an ideal monatomic gas at **(a)** constant pressure or **(b)** constant volume.

34. •• At constant pressure, it takes 210 J of heat to raise the temperature of an ideal monatomic gas 17 K. How much heat is required for the same temperature change at constant volume?

35. •• **IP** A cylinder contains 15 moles of a monatomic ideal gas at a constant pressure of 140 kPa. **(a)** How much work does the gas do as it expands 3200 cm^3, from 5400 cm^3 to 8600 cm^3? **(b)** If the gas expands by 3200 cm^3 again, this time from 2200 cm^3 to 5400 cm^3, is the work it does greater than, less than, or equal to the work found in part (a)? Explain. **(c)** Calculate the work done as the gas expands from 2200 cm^3 to 5400 cm^3.

36. •• The volume of a monatomic ideal gas doubles in an adiabatic expansion. By what factor does the **(a)** pressure and **(b)** temperature change?

37. •• A monatomic ideal gas is held in a thermally insulated container with a volume of 0.0750 m^3. The pressure of the gas is 115 kPa and its temperature is 325 K. **(a)** To what volume must the gas be compressed to increase its pressure to 145 kPa? **(b)** At what volume will the gas have a temperature of 295 K?

38. ••• Consider the expansion of 60.0 moles of a monatomic ideal gas along processes 1 and 2 in **Figure 18–27**. On process 1 the gas is heated at constant volume from an initial pressure of 106 kPa to a final pressure of 212 kPa. On process 2 the gas expands at constant pressure from an initial volume of 1.00 m^3 to a final volume of 3.00 m^3. **(a)** How much heat is added to the gas during these two processes? **(b)** How much work does the gas do during this expansion? **(c)** What is the change in the internal energy of the gas?

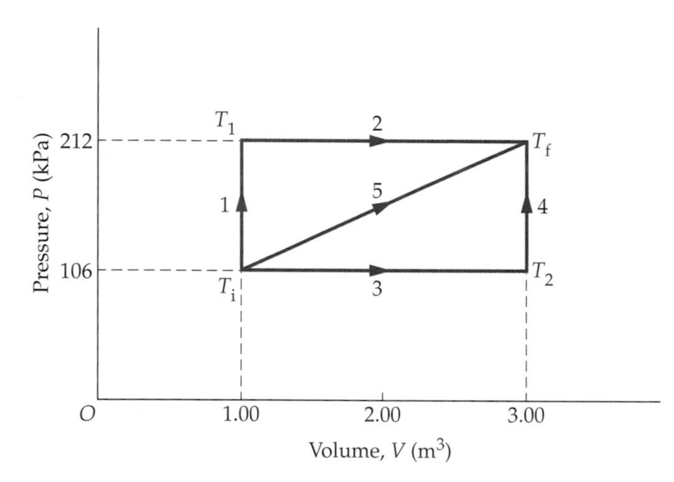

▲ **FIGURE 18–27** Problems 38, 39, and 67

39. ••• Referring to Problem 38, suppose the gas is expanded along processes 3 and 4 in Figure 18–27. On process 3 the gas expands at constant pressure from an initial volume of 1.00 m^3 to a final volume of 3.00 m^3. On process 4 the gas is heated at constant volume from an initial pressure of 106 kPa to a final pressure of 212 kPa. **(a)** How much heat is added to the gas during these two processes? **(b)** How much work does the gas do during this expansion? **(c)** What is the change in the internal energy of the gas?

Section 18–6 Heat Engines and the Carnot Cycle

40. • A heat engine takes in 1300 J of heat from the hot reservoir and exhausts 760 J of heat to the cold reservoir. How much work is done by the engine?

41. • What is the efficiency of an engine that exhausts 870 J of heat in the process of doing 340 J of work?

42. • An engine receives 690 J of heat from a hot reservoir, and gives off 430 J of heat to a cold reservoir. What is **(a)** the work done and **(b)** the efficiency of this engine?

43. • A Carnot engine operates between the temperatures 440 K and 310 K. **(a)** How much heat must be given to the engine to produce 1500 J of work? **(b)** How much heat is discarded to the cold reservoir as this work is done?

44. • A nuclear power plant has a reactor that produces heat at the rate of 820 MW. This heat is used to produce 250 MW of mechanical power to drive an electrical generator. **(a)** At what rate is heat discarded to the environment by this power plant? **(b)** What is the thermal efficiency of the plant?

45. • At a coal-burning power plant a steam turbine is operated with a power output of 548 MW. The thermal efficiency of the power plant is 32.0%. **(a)** At what rate is heat discarded to the environment by this power plant? **(b)** At what rate must heat be supplied to the power plant by burning coal?

46. • A research group plans to construct an engine that extracts energy from the ocean, using the fact that the water temperature near the surface is greater than the temperature in deep water. If the surface water temperature is 25.0 °C and the deep-water temperature is 12.0 °C, what is the maximum efficiency such an engine could have?

47. •• If an engine does 3200 J of work with an efficiency of 0.14, find **(a)** the heat taken in from the hot reservoir, and **(b)** the heat given off to the cold reservoir.

48. •• The efficiency of a Carnot engine with a cold reservoir at a temperature of 295 K is 21.0%. Assuming the temperature of the hot reservoir remains the same, find the temperature the cold reservoir must have in order for the engine's efficiency to be 25.0%.

49. •• **IP** The efficiency of a particular Carnot engine is 0.300. **(a)** If the high-temperature reservoir is at a temperature of 545 K, what is the temperature of the low-temperature reservoir? **(b)** To increase the efficiency of this engine to 40.0%, must the temperature of the low-temperature reservoir be increased or decreased? Explain. **(c)** Find the temperature of the low-temperature reservoir that gives an efficiency of 0.400.

50. •• During each cycle a reversible engine absorbs 1500 J of heat from a high-temperature reservoir and performs 1200 J of work. **(a)** What is the efficiency of this engine? **(b)** How much heat is exhausted to the low-temperature reservoir during each cycle? **(c)** What is the ratio, T_h/T_c, of the two reservoir temperatures?

51. ••• The operating temperatures for a Carnot engine are T_c and $T_h = T_c + 55$ K. The efficiency of the engine is 11%. Find T_c and T_h.

52. ••• A certain Carnot engine takes in the heat Q_h and exhausts the heat $Q_c = 2Q_h/3$, as indicated in **Figure 18–28**. **(a)** What is the efficiency of this engine? **(b)** Using the Kelvin temperature scale, find the ratio T_c/T_h.

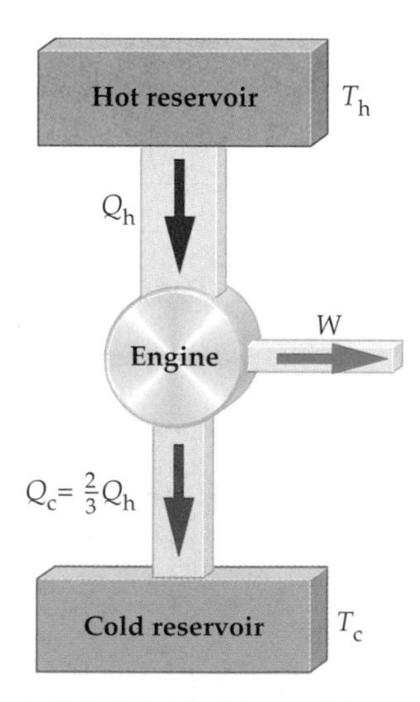

▲ **FIGURE 18–28** Problem 52

Section 18–7 Refrigerators, Air Conditioners, and Heat Pumps

53. • The refrigerator in your kitchen does 420 J of work to remove 130 J of heat from its interior. **(a)** How much heat does the refrigerator exhaust into the kitchen? **(b)** What is the refrigerator's coefficient of performance?

54. • A refrigerator with a coefficient of performance of 1.75 absorbs 3.45×10^4 J of heat from the low temperature reservoir during each cycle. **(a)** How much mechanical work is required to operate the refrigerator for a cycle? **(b)** How much heat does the refrigerator discard to the high-temperature reservoir during each cycle?

55. •• A Carnot air conditioner operates between an indoor temperature of 21.0 °C and an outdoor temperature of 32.0 °C. **(a)** How much work must the air conditioner do to remove 1550 J of heat from inside the house? **(b)** How much heat is exhausted to the outside?

56. •• To keep a room at a comfortable 21.0 °C a Carnot heat pump does 345 J of work and supplies it with 3240 J of heat. **(a)** How much heat is removed from the outside air by the heat pump? **(b)** What is the temperature of the outside air?

57. •• An air conditioner is used to keep the interior of a house at a temperature of 21 °C while the outside temperature is 32 °C. If heat leaks into the house at the rate of 11 kW, and the air conditioner has the efficiency of a Carnot engine, what is the mechanical power required to keep the house cool?

58. •• A reversible refrigerator has a coefficient of performance equal to 10.0. What is its efficiency?

59. •• A freezer has a coefficient of performance equal to 4.0. How much electrical energy must this freezer use to produce 1.5 kg of ice at −5.0 °C from water at 15 °C?

60. •• If a Carnot engine has an efficiency of 0.23, what is its coefficient of performance if it is run backward as a heat pump?

Section 18–8 Entropy

61. • Find the change in entropy when 1.25 kg of water at 100 °C is boiled away to steam at 100 °C.

62. • Determine the change in entropy that occurs when 1.1 kg of water freezes at 0 °C.

63. •• On a cold winter's day heat leaks slowly out of a house at the rate of 20.0 kW. If the inside temperature is 22 °C, and the outside temperature is −14.5 °C, find the rate of entropy increase.

64. •• A 72-kg parachutist descends through a vertical height of 460 m with constant speed. Find the increase in entropy produced by the parachutist, assuming the air temperature is 21 °C.

65. •• **IP** Consider the air conditioning system described in Problem 57. **(a)** Is the entropy of the universe increasing or decreasing as the air conditioner operates? Explain. **(b)** At what rate is entropy changing during this process?

66. •• A heat engine operates between a high-temperature reservoir at 610 K and a low-temperature reservoir at 320 K. In one cycle, the engine absorbs 6400 J of heat from the high-temperature reservoir and does 2200 J of work. What is the net change in entropy as a result of this cycle?

General Problems

67. •• Referring to Figure 18–27, suppose 60.0 moles of a monatomic ideal gas are expanded along process 5. **(a)** How much work does the gas do during this expansion? **(b)** What is the change in the internal energy of the gas? **(c)** How much heat is added to the gas during this process?

68. •• **IP** Engine A has an efficiency of 66%. Engine B absorbs the same amount of heat from the hot reservoir and exhausts twice as much heat to the cold reservoir. **(a)** Which engine has the greater efficiency? Explain. **(b)** What is the efficiency of engine B?

69. •• A freezer with a coefficient of performance of 3.88 is used to convert 1.75 kg of water to ice in one hour. The water starts at a temperature of 20.0 °C and the ice that is produced is cooled to a temperature of −5.00 °C. **(a)** How much heat must be removed from the water for this process to occur? **(b)** How much electrical energy does the freezer use during this hour of operation? **(c)** How much heat is discarded into the room that houses the freezer?

70. •• 1400 J of heat are added to 3.5 mol of argon gas at a constant pressure of 120 kPa. Find the change in **(a)** internal energy and **(b)** temperature for this gas. **(c)** Calculate the change in volume of the gas. (Assume that the argon can be treated as an ideal monatomic gas.)

71. •• A system consists of 1.5 mol of the monatomic gas neon (which may be treated as an ideal gas). Initially the system has a temperature of 320 K and a volume of 1.8 m^3. Heat is added to this system at constant pressure until the volume triples, then more heat is added at constant volume until the pressure doubles. Find the total heat added to the system.

72. •• A cylinder with a moveable piston holds 2.50 mol of argon at a constant temperature of 295 K. As the gas is compressed isothermally, its pressure increases from 101 kPa to 121 kPa. Find **(a)** the final volume of the gas, **(b)** the work done by the gas, and **(c)** the heat added to the gas.

73. •• An inventor claims a new cyclic engine that uses organic grape juice as its working material. According to the claims, the engine absorbs 1250 J of heat from a 1010 K reservoir and performs 1120 J of work each cycle. The waste heat is exhausted to the atmosphere at a temperature of 302 K. **(a)** What is the efficiency that is implied by these claims? **(b)** What is the efficiency of a reversible engine operating between the same high and low temperatures used by this engine? (Should you invest in this invention?)

74. •• A nonreversible heat engine operates between a high-temperature reservoir at $T_h = 810$ K and a low-temperature reservoir at $T_c = 320$ K. During each cycle the engine absorbs 660 J of heat from the high-temperature reservoir and performs 250 J of work. **(a)** Calculate the total entropy change ΔS_{tot} for one cycle. **(b)** How much work would a reversible heat engine perform in one cycle if it operated between the same two temperatures and absorbed the same amount of heat? **(c)** Show that the difference in work between the non-reversible engine and the reversible engine is equal to $T_c \Delta S_{tot}$.

75. •• **IP** A small dish containing 510 g of water is placed outside for the birds. During the night the outside temperature drops to −5.0 °C and stays at that value for several hours. **(a)** When the water in the dish freezes, does its entropy increase, decrease, or stay the same? Explain. **(b)** Calculate the change in entropy that occurs as the water freezes. **(c)** When the water freezes, is there an entropy change anywhere else in the universe? If so, specify where the increase occurs.

76. ••• One mole of an ideal gas follows the three-part cycle shown in **Figure 18–29**. **(a)** Provide the missing entries in the following table:

	Q	W	ΔU
A → B			
B → C			
C → A			

(b) What is the efficiency of this cycle?

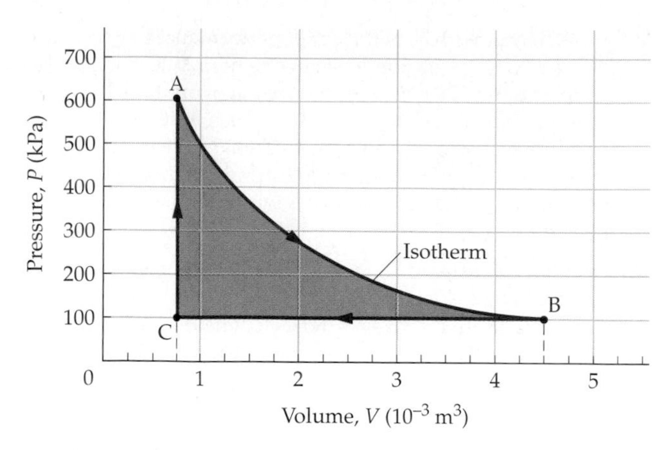

▲ **FIGURE 18–29** Problem 76

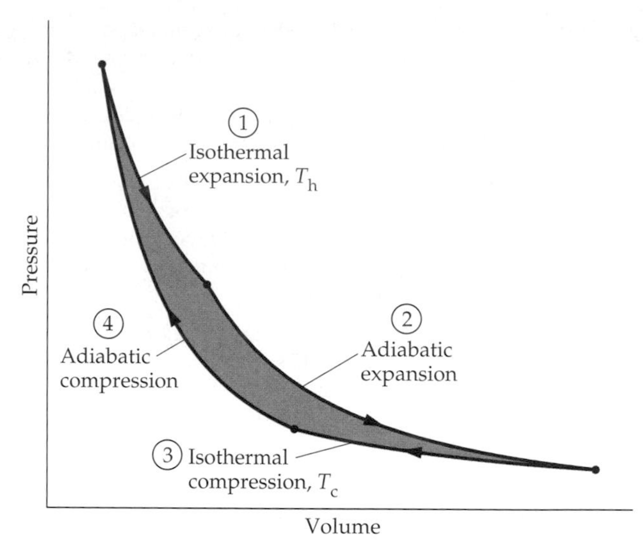

▲ **FIGURE 18-30** Problem 79

77. ••• When a heat Q is added to a monatomic ideal gas at constant pressure, the gas does a work W. Find the ratio, W/Q.

78. ••• A diesel engine uses adiabatic compression to heat the air in its cylinders to a temperature sufficiently high to cause combustion of the injected fuel. If the volume of air in a cylinder is reduced by a factor of 25, by what factor does its absolute temperature increase?

79. ••• A Carnot cycle is shown in **Figure 18–30**. The cycle consists of the following four processes: (1) an isothermal expansion from V_1 to V_2 at the temperature T_h; (2) an adiabatic expansion from V_2 to V_3 during which the temperature drops from T_h to T_c. (3) An isothermal compression from V_3 to V_4 at the temperature T_c; and (4) An adiabatic compression from V_4 to V_1 during which the temperature increases from T_c to T_h. Show that the efficiency of this cycle is $e = 1 - T_c/T_h$, as expected.

80. ••• A Carnot engine and a Carnot refrigerator operate between the same two temperatures. Show that the coefficient of performance, COP, for the refrigerator is related to the efficiency, e, of the engine by the following expression; COP $= (1 - e)/e$.

81. ••• Which would make the greater change in the efficiency of a Carnot heat engine; **(i)** raising the temperature of the high temperature reservoir by ΔT, or **(ii)** lowering the temperature of the low temperature reservoir by ΔT? Justify your answer.

Appendix A Basic mathematical tools

This text is designed for students with a working knowledge of basic algebra and trigonometry. Even so, it is useful to review some of the mathematical tools that are of particular importance in the study of physics. In this Appendix we cover a number of topics related to mathematical notation, trigonometry, algebra, and mathematical expansions.

Mathematical Notation

Common mathematical symbols

In Table A–1 we present some of the more common mathematical symbols, along with a translation into English. Though these symbols are probably completely familiar, it is worthwhile to be sure we all interpret them in the same way.

TABLE A–1
Mathematical Symbols

Symbol	Meaning
$=$	is equal to
$\neq$	is not equal to
$\approx$	is approximately equal to
$\propto$	is proportional to
$>$	is greater than
$\geq$	is greater than or equal to
$\gg$	is much greater than
$<$	is less than
$\leq$	is less than or equal to
$\ll$	is much less than
$\pm$	plus or minus
$\mp$	minus or plus
x_{av} or $\overline{x}$	average value of x
Δx	change in x $(x_f - x_i)$
$\|x\|$	absolute value of x
Σ	sum of
$\rightarrow 0$	approaches 0
∞	infinity

A couple of the symbols in Table A–1 warrant further discussion. First, Δx, which means "change in x," is used frequently, and in many different contexts. Pronounced "delta x," it is defined as the final value of x, x_f, minus the initial value of x, x_i:

$$\Delta x = x_f - x_i \qquad \text{A–1}$$

Thus, Δx is not Δ times x; it is a shorthand way of writing $x_f - x_i$.

The same delta notation can be applied to any quantity—it doesn't have to be x. In general, we can say that

$$\Delta(\textit{anything}) = (\textit{anything})_f - (\textit{anything})_i$$

For example, $\Delta t = t_f - t_i$ is the change in time, $\Delta \mathbf{v} = \mathbf{v}_f - \mathbf{v}_i$ is the change in velocity, and so on. Throughout this text, we use the delta notation whenever we want to indicate the change in a given quantity.

Second, the Greek letter Σ (capital sigma) is also encountered frequently. In general, Σ is shorthand for "sum." For example, suppose we have a system comprised of nine masses, m_1 through m_9. The total mass of the system, M, is simply

$$M = m_1 + m_2 + m_3 + m_4 + m_5 + m_6 + m_7 + m_8 + m_9$$

This is a rather tedious way to write M, however, and would be even more so if the number of masses were larger. To simplify our equation, we use the Σ notation:

$$M = \sum_{i=1}^{9} m_i \qquad \text{A–2}$$

With this notation we could sum over any number of masses, simply by changing the upper limit of the sum.

In addition, Σ is often used to designate a general summation, where the number of terms in the sum may not be known, or may vary from one system to another. In a case like this we would simply write Σ without specific upper and lower limits. Thus, a general way of writing the total mass of a system is as follows:

$$M = \sum m \qquad \text{A–3}$$

Vector notation

When we draw a vector to represent a physical quantity, we typically use an arrow whose length is proportional to the magnitude of the quantity, and whose direction is the direction of the quantity. (This and other aspects of vector notation are discussed in Chapter 3.) A slight problem arises, however, when a physical quantity points into or out of the page. In such a case, we use the conventions illustrated in **Figure A–1.**

Figure A–1 (a) shows a vector pointing out of the page. Note that we see only the tip. Below, we show the corresponding convention, which is a dot set off by a circle. The dot represents the point of the vector's arrow coming out of the page toward you.

A similar convention is employed in Figure A–1 (b) for a vector pointing into the page. In this case, the arrow moves directly away from you, giving a view of its "tail feathers." The feathers are placed in an X-shaped pattern, so we represent the vector as an X set off by a circle.

These conventions are used in Chapter 22 to represent the magnetic field vector, **B**, and in other locations in the text as well.

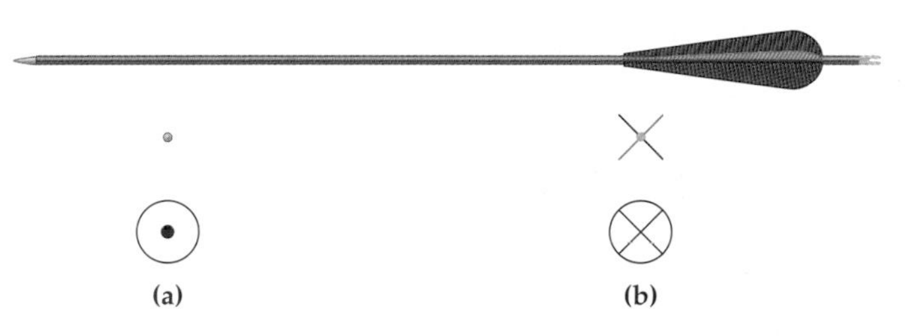

▲ **FIGURE A–1 Vectors pointing out of and into the page**
(a) A vector pointing out of the page is represented by a dot in a circle. The dot indicates the tip of the vector's arrow. **(b)** A vector pointing into the page is represented by an X in a circle. The X indicates the "tail feathers" of the vector's arrow.

Scientific notation

In physics, the numerical value of a physical quantity can cover an enormous range, from the astronomically large to the microscopically small. For example, the mass of the Earth is roughly

$$M_{\mathrm{E}} = 5970000000000000000000000 \text{ kg}$$

In contrast, the mass of a hydrogen atom is approximately

$$M_{\text{hydrogen}} = 0.00000000000000000000000000167 \text{ kg}$$

Clearly, representing such large and small numbers with a long string of zeroes is clumsy and prone to error.

The preferred method for handling such numbers is to replace the zeros with the appropriate power of ten. For example, the mass of the Earth can be written as follows:

$$M_{\mathrm{E}} = 5.97 \times 10^{24} \text{ kg}$$

The factor of 10^{24} simply means that the decimal point for the mass of the Earth is 24 places to the right of its location in 5.97. Similarly, the mass of a hydrogen atom is

$$M_{\text{hydrogen}} = 1.67 \times 10^{-27} \text{ kg}$$

In this case, the correct location of the decimal point is 27 places to the left of its location in 1.67. This type of representation, using powers of ten, is referred to as **scientific notation**.

Scientific notation also simplifies various mathematical operations, such as multiplication and division. For example, the product of the mass of the Earth and the mass of a hydrogen atom is

$$\begin{aligned} M_{\mathrm{E}} M_{\text{hydrogen}} &= (5.97 \times 10^{24} \text{ kg})(1.67 \times 10^{-27} \text{ kg}) \\ &= (5.97 \times 1.67)(10^{24} \times 10^{-27}) \text{kg}^2 \\ &= 9.99 \times 10^{24-27} \text{ kg}^2 \\ &= 9.99 \times 10^{-3} \text{ kg}^2 \end{aligned}$$

Similarly, the mass of a hydrogen atom divided by the mass of the Earth is

$$\begin{aligned} \frac{M_{\text{hydrogen}}}{M_{\mathrm{E}}} &= \frac{1.67 \times 10^{-27} \text{ kg}}{5.97 \times 10^{24} \text{ kg}} = \frac{1.67}{5.97} \times \frac{10^{-27}}{10^{24}} \\ &= 0.280 \times 10^{-27-24} \\ &= 0.280 \times 10^{-51} = 2.80 \times 10^{-52} \end{aligned}$$

Note the change in location of the decimal point in the last two expressions, and the corresponding change in the power of ten.

Exponents and their manipulation are discussed in greater detail later in this Appendix.

Trigonometry

Degrees and radians

We all know the definition of a degree; there are 360 degrees in a circle. The definition of a radian is somewhat less well known; there are 2π radians in a circle. An equivalent definition of the radian is the following:

> A radian is the angle for which the corresponding arc length is equal to the radius.

To visualize this definition, consider a pie with a piece cut out, as shown in **Figure A–2 (a)**. Note that a piece of pie has three sides—two radial lines from the center, and an arc of crust. If a piece of pie is cut with an angle of one radian, all three sides are equal in length, as shown in **Figure A–2 (b)**. Since a radian is about 57.3°, this amounts to a fairly good-sized piece of pie. Thus, if you want a healthy helping of pie, just tell the server, "One radian, please."

Now, radians are particularly convenient when we are interested in the length of an arc. In **Figure A–3** we show a circular arc corresponding to the radius r and the angle θ. *If the angle θ is measured in radians*, the length of the arc, s, is given by

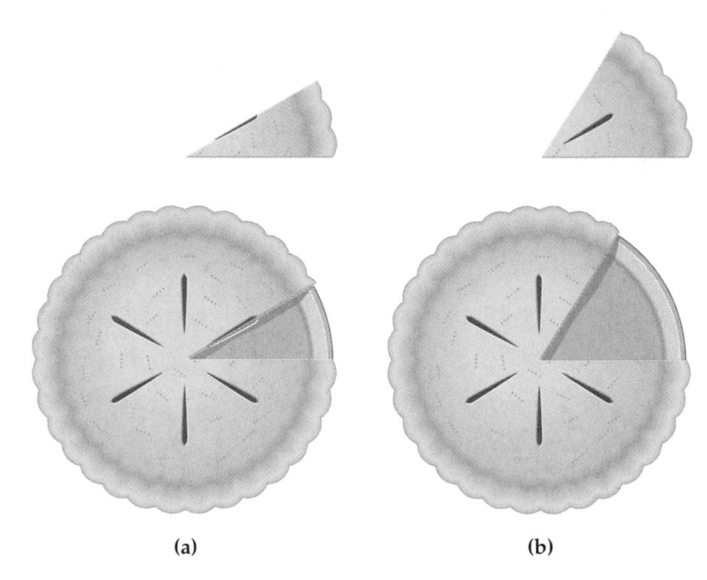

▲ **FIGURE A–2 The definition of a radian**
(a) This piece of pie is cut with an angle less than a radian. Thus, the two radial sides (coming out from the center) are longer than the arc of crust. **(b)** The angle for this piece of pie is equal to one radian (about 57.3°). Thus, all three sides of the piece are of equal length.

$$s = r\theta \qquad \text{A–4}$$

Note that this simple relation is *not valid* when θ is measured in degrees. For a full circle, in which case $\theta = 2\pi$, the length of the arc (which is the circumference of the circle) is $2\pi r$, as expected.

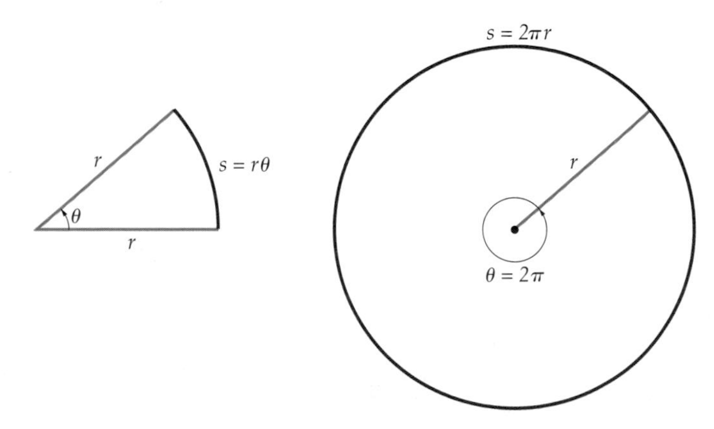

▲ **FIGURE A–3 The Length of an arc**
The arc created by a radius r rotated through an angle θ has a length $s = r\theta$. If the radius is rotated through a full circle, the angle is $\theta = 2\pi$, and the arc length is the circumference of a circle, $s = 2\pi r$.

Trigonometric functions and the Pythagorean theorem

Next, we consider some of the more important and frequently used results from trigonometry. We start with the right triangle, shown in **Figure A–4**, and the basic **trigonometric functions**, $\sin\theta$ (sine theta), $\cos\theta$ (cosine theta), and $\tan\theta$ (tangent theta). The cosine of an angle θ is defined to be the side adjacent to the angle divided by the hypotenuse; $\cos\theta = x/r$. Similarly, the sine is defined to be the opposite side divided by the hypotenuse, $\sin\theta = y/r$, and the tangent is the opposite side divided by the adjacent side, $\tan\theta = y/x$. These relations are summarized in the following equations:

$$\cos\theta = \frac{x}{r}$$

$$\sin\theta = \frac{y}{r} \qquad \text{A–5}$$

$$\tan\theta = \frac{y}{x} = \frac{\sin\theta}{\cos\theta}$$

Note that each of the trigonometric functions is the ratio of two lengths, and hence is dimensionless.

According to the **Pythagorean theorem**, the sides of the right triangle in Figure A–4 are related as follows:

$$x^2 + y^2 = r^2 \qquad \text{A–6}$$

Dividing by r^2 yields

$$\frac{x^2}{r^2} + \frac{y^2}{r^2} = 1$$

This can be re-written in terms of sine and cosine to give

$$\sin^2\theta + \cos^2\theta = 1$$

Figure A–4 also shows how sine and cosine are used in a typical calculation. In many cases, the hypotenuse of a triangle, r, and one of its angles, θ, are given. To find the short sides of the triangle we rearrange the relations given in Equation A–5. For example, in Figure A–4 we see that $x = r\cos\theta$ is the length of the short side adjacent to the angle, θ, and $y = r\sin\theta$ is the length of the short side opposite the angle. The following Example applies this type of calculation to the case of an inclined roadway.

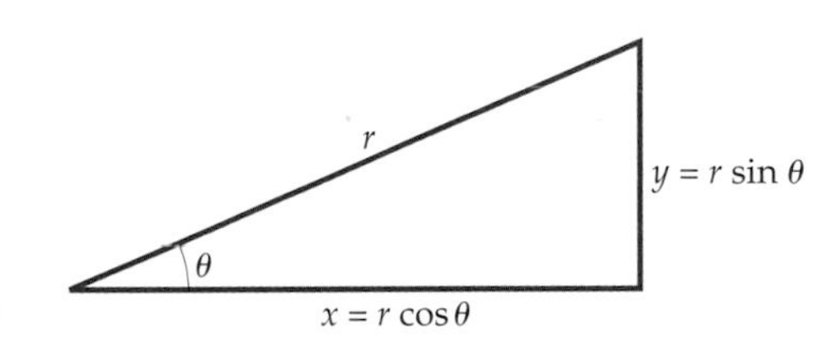

▲ **FIGURE A–4 Relating the short sides of a right triangle to its hypotenuse**

The trigonometric functions $\sin\theta$ and $\cos\theta$ and the Pythagorean theorem are useful in relating the lengths of the short sides of a right triangle to the length of its hypotenuse.

EXAMPLE A–1 Highway to Heaven

You are driving on a long straight road that slopes uphill at an angle of 6.4° above the horizontal. At one point you notice a sign that reads, "Elevation 1500 feet." What is your elevation after you have driven another 1.0 mi?

Picture the Problem

From our sketch, we see that the car is moving along the hypotenuse of a right triangle. The length of the hypotenuse is one mile.

Strategy

The elevation gain is the vertical side of the triangle, y. We find y by multiplying the hypotenuse, r, by the sine of theta. That is, since $\sin\theta = y/r$ it follows that $y = r\sin\theta$.

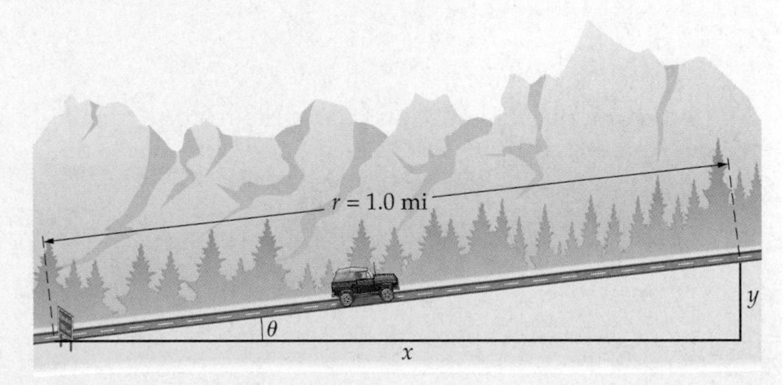

Solution

1. Calculate the elevation gain, y: $y = r\sin\theta$
$= (1.0\text{ mi})\sin 6.4° = (1.0\text{ mi})(0.11) = 0.11\text{ mi}$

2. Convert y from miles to feet: $y = (0.11\text{ mi})\left(\frac{5280\text{ ft}}{1\text{ mi}}\right) = 580\text{ ft}$

3. Add the elevation gain to the original elevation to obtain the new elevation: elevation $= 1500\text{ ft} + 580\text{ ft} = 2100\text{ ft}$

Insight

As surprising as it may seem, the horizontal distance covered by the car is $r\cos\theta = (5280\text{ ft})\cos 6.4° = 5200\text{ ft}$, only about 80 ft less than the total distance driven by the car. At the same time, the car rises a distance of 580 ft.

Practice Problem

How far up the road from the first sign should the road crew put another sign reading "Elevation 3500 ft"?
[**Answer:** 18,000 ft = 3.4 mi]

In some problems, the sides of a triangle (x and y) are given and it is desired to find the corresponding hypotenuse, r, and angle, θ. For example, suppose that $x = 5.0$ m and $y = 2.0$ m. Using the Pythagorean theorem, we find $r = \sqrt{x^2 + y^2} = \sqrt{(5.0 \text{ m})^2 + (2.0 \text{ m})^2} = 5.4$ m. Similarly, to find the angle we use the definition of tangent: $\tan\theta = y/x$. The inverse of this relation is $\theta = \tan^{-1}(y/x) = \tan^{-1}(2.0 \text{ m}/5.0 \text{ m}) = \tan^{-1}(0.40)$. Note that the expression $\tan^{-1}$ is the *inverse tangent function*—it does not mean 1 divided by tangent, but rather "the angle whose tangent is ——." Your calculator should have a button on it labeled $\tan^{-1}$. If you enter 0.40 and then press $\tan^{-1}$, you should get 22° (to two significant figures), which means that $\tan 22° = 0.40$. Inverse sine and cosine functions work in the same way.

Trigonometric identities

In addition to the basic definitions of sine, cosine, and tangent just given, there are a number of useful relationships involving these functions referred to as **trigonometric identities**. First, consider changing the sign of an angle. This corresponds to flipping the triangle in Figure A–4 upside-down, which changes the sign of y but leaves x unaffected. The result is that sine changes its sign, but cosine does not. Specifically, for a general angle A we find the following:

$$\sin(-A) = -\sin A \qquad \cos(-A) = \cos A \qquad \text{A–7}$$

Next, we consider trigonometric identities relating to the sum or difference of two angles. For example, consider two general angles A and B. The sine and cosine of the sum of these angles, $A + B$, are given below:

$$\sin(A + B) = \sin A \cos B + \sin B \cos A$$
$$\cos(A + B) = \cos A \cos B - \sin A \sin B \qquad \text{A–8}$$

By changing the sign of B, and using the results given in Equation A–7, we obtain the corresponding results for the difference between two angles:

$$\sin(A - B) = \sin A \cos B - \sin B \cos A$$
$$\cos(A - B) = \cos A \cos B + \sin A \sin B \qquad \text{A–9}$$

Applications of these relations can be found in Chapters 4, 14, 23, and 24.

To see how one might use a relation like $\sin(A + B) = \sin A \cos B + \sin B \cos A$, consider the case where $A = B = \theta$. With this substitution we find

$$\sin(\theta + \theta) = \sin\theta\cos\theta + \sin\theta\cos\theta$$

Simplifying somewhat yields the commonly used double-angle formula

$$\sin 2\theta = 2\sin\theta\cos\theta \qquad \text{A–10}$$

This expression is used in deriving Equation 4–16.

As a final example of using trigonometric identities, let $A = 90°$ and $B = \theta$. Making these substitutions in Equations A–9 yields

$$\sin(90° - \theta) = \sin 90° \cos\theta - \sin\theta\cos 90° = \cos\theta$$
$$\cos(90° - \theta) = \cos 90° \cos\theta + \sin 90° \sin\theta = \sin\theta \qquad \text{A–11}$$

Algebra

The quadratic equation

A well-known result that finds many uses in physics is the solution to the **quadratic equation**

$$ax^2 + bx + c = 0 \qquad \text{A–12}$$

In this equation, a, b, and c are constants and x is a variable. When we refer to the solution of the quadratic equation, we mean the values of x that satisfy Equation A–12. These values are given by the following expression:

Solutions to the Quadratic Equation

$$x = \frac{-b \pm \sqrt{b^2 - 4ac}}{2a} \qquad \text{A–13}$$

Note that there are two solutions to the quadratic equation, in general, corresponding to the plus and minus sign in front of the square root. In the special case that the quantity under the square root vanishes, there will be only a single solution. If the quantity under the square root is negative the result for x is not physical, which means a mistake has probably been made in the calculation.

To illustrate the use of the quadratic equation and its solution, we consider a standard one-dimensional kinematics problem, such as one might encounter in Chapter 2:

> A ball is thrown straight upward with an initial speed of 11 m/s. How long does it take for the ball to first reach a height of 4.5 m above its launch point?

The first step in solving this problem is to write the equation giving the height of the ball, y, as a function of time. Referring to Equation 2–11, we have

$$y = y_0 + v_0 t - \tfrac{1}{2}gt^2$$

To make this look more like a quadratic equation, we move all the terms onto the left-hand side, which yields

$$\tfrac{1}{2}gt^2 - v_0 t + y - y_0 = 0$$

This is the same as Equation A–12 if we make the following identifications: $x = t$; $a = \tfrac{1}{2}g$; $b = -v_0$; $c = y - y_0$. The desired solution, then, is given by making these substitutions in Equation A–13:

$$t = \frac{v_0 \pm \sqrt{v_0^2 - 2g(y - y_0)}}{g}$$

The final step is to use the appropriate numerical values; $g = 9.81 \text{ m/s}^2$, $v_0 = 11$ m/s, $y - y_0 = 4.5$ m. Straightforward calculation gives $t = 0.54$ s and $t = 1.7$ s. Therefore, the time it takes to first reach a height of 4.5 m is 0.54 s; the second solution is the time when the ball is again at a height of 4.5 m, this time on its way down.

Two equations in two unknowns

In some problems, two unknown quantities are determined by two interlinked equations. In such cases it often seems at first that you have not been given enough information to obtain a solution. By patiently writing out what is known, however, you can generally use straightforward algebra to solve the problem.

As an example, consider the following problem: A father and daughter share the same birthday. On one birthday the father announces to his daughter, "Today I am four times older than you, but in 5 years I will be only three times older." How old are the father and daughter now?

You might be able to solve this problem by guessing, but here's how to approach it systematically. First, write what is given in the form of equations. Letting F be the father's age in years, and D the daughter's age in years, we know that on this birthday

$$F = 4D \qquad \text{A–14}$$

In 5 years, the father's age will be $F + 5$, the daughter's age will be $D + 5$, and the following will be true:

$$F + 5 = 3(D + 5)$$

Multiplying through the parenthesis gives

$$F + 5 = 3D + 15 \qquad \text{A–15}$$

Now if we subtract Equation A–15 from Equation A–14 we can eliminate one of the unknowns, F:

$$\begin{array}{rrl} & F &= 4D \\ - & F + 5 &= 3D + 15 \\ \hline & -5 &= D - 15 \end{array}$$

The solution to this new equation is clearly $D = 10$, and thus the father's age is $F = 4D = 40$.

The following Example investigates a similar problem. In this case, we use the fact that if you drive with a speed v for a time t the distance covered is $d = vt$.

EXAMPLE A–2 Hit the Road

It takes 1.50 h to drive with a speed v from home to a nearby town, a distance d away. Later, on the way back, the traffic is lighter, and you are able to increase your speed by 15 mi/h. With this higher speed, you get home in just 1.00 h. Find your initial speed v, and the distance to the town, d.

Picture the Problem

Our sketch shows home and the town, separated by a distance d. Going to town the speed is v, returning home the speed is v + 15 mi/h.

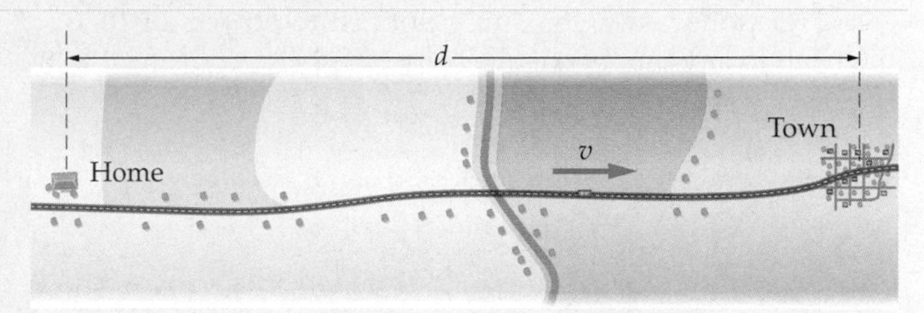

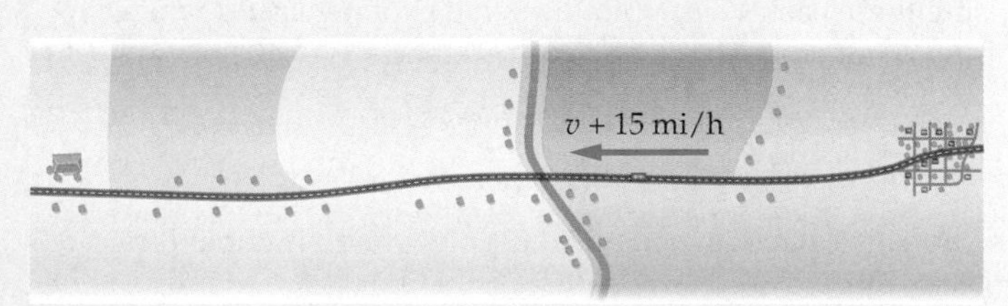

Strategy

To determine the two unknowns, v and d, we need two separate equations. One equation corresponds to what we know about the trip to the town, the second equation corresponds to what we know about the return trip.

Solution

1. Write an equation for the trip to the town. Recall that this trip takes one and a half hours:

$$d = vt = v(1.50\ \text{h})$$

2. Write an equation for the trip home. This trip takes one hour, and covers the same distance d:

$$d = (v + 15\ \text{mi/h})t = (v + 15\ \text{mi/h})(1.00\ \text{h})$$

3. Subtract these two equations to eliminate d:

$$\begin{array}{rl} & d = v(1.50\ \text{h}) \\ - & d = (v + 15\ \text{mi/h})(1.00\ \text{h}) \\ \hline & 0 = v(1.50\ \text{h}) - v(1.00\ \text{h}) - (15\ \text{mi/h})(1.00\ \text{h}) \end{array}$$

4. Solve this new equation for v:

$$0 = v(1.50\ \text{h}) - v(1.00\ \text{h}) - (15\ \text{mi/h})(1.00\ \text{h})$$
$$0 = v(0.50\ \text{h}) - (15\ \text{mi/h})(1.00\ \text{h})$$
$$v = \frac{(15\ \text{mi/h})(1.00\ \text{h})}{(0.50\ \text{h})} = 30\ \text{mi/h}$$

5. Use the first equation to solve for d:

$$d = vt = (30\ \text{mi/h})(1.50\ \text{h}) = 45\ \text{mi}$$

continued on the following page

continued from the previous page

Insight

We could also use the second equation to solve for d. The algebra is a bit messier, but it provides a good double check on our results: $d = (30\ \text{mi/h} + 15\ \text{mi/h})(1.00\ \text{h}) = (45\ \text{mi/h})(1.00\ \text{h}) = 45\ \text{mi}$.

Practice Problem

Suppose the speed going home is 12 mi/h faster than the speed on the way to town, but the times are the same as above. What are v and d in this case? [**Answer:** $v = 24$ mi/h, $d = 36$ mi]

Exponents and logarithms

An **exponent** is the power to which a number is raised. For example, in the expression 10^3, we say that the exponent of 10 is 3. To evaluate 10^3 we simply multiply 10 by itself three times:

$$10^3 = 10 \times 10 \times 10 = 1000$$

Similarly, a negative exponent implies an inverse, as in the relation $10^{-1} = 1/10$. Thus, to evaluate a number like 10^{-4}, for example, we multiply 1/10 by itself four times:

$$10^{-4} = \frac{1}{10} \times \frac{1}{10} \times \frac{1}{10} \times \frac{1}{10} = \frac{1}{10{,}000} = 0.0001$$

The relations just given apply not just to powers of 10, of course, but to any number at all. Thus, x^4 is

$$x^4 = x \times x \times x \times x$$

and x^{-3} is

$$x^{-3} = \frac{1}{x} \times \frac{1}{x} \times \frac{1}{x} = \frac{1}{x^3}$$

Using these basic rules, it follows that exponents add when two or more numbers are multiplied together:

$$x^2x^3 = (x \times x)(x \times x \times x) = x \times x \times x \times x \times x = x^5 = x^{2+3}$$

On the other hand, exponents multiply when a number is raised to a power:

$$(x^2)^3 = (x \times x) \times (x \times x) \times (x \times x) = x \times x \times x \times x \times x \times x = x^6 = x^{2\times 3}$$

In general, the rules obeyed by exponents can be summarized as follows:

$$x^n x^m = x^{n+m}$$
$$x^{-n} = \frac{1}{x^n}$$
$$\frac{x^n}{x^m} = x^{n-m} \qquad \textbf{A–16}$$
$$(xy)^n = x^n y^n$$
$$(x^n)^m = x^{nm}$$

Fractional exponents, such as $1/n$, indicate the nth root of a number. Specifically, the square root of x is written as

$$\sqrt{x} = x^{1/2}$$

For n greater than 2 we write the nth root in the following form:

$$\sqrt[n]{x} = x^{1/n} \qquad \textbf{A–17}$$

Thus, the nth root of a number, x, is the value that gives x when multiplied by itself n times: $(x^{1/n})^n = x^{n/n} = x^1 = x$.

A general method for calculating the exponent of a number is provided by the **logarithm**. For example, suppose x is equal to 10 raised to the power n:

$$x = 10^n$$

In this expression, 10 is referred to as the *base*. The exponent, n, is equal to the logarithm (log) of x:

$$n = \log x$$

The notation "log" is known as the *common logarithm*, and it refers specifically to base 10.

As an example, suppose that $x = 1000 = 10^n$. Clearly, we can write x as 10^3, which means that the exponent of x is 3:

$$\log x = \log 1000 = \log 10^3 = 3$$

When dealing with a number this simple, the exponent can be determined without a calculator. Suppose, however, that $x = 1205 = 10^n$. To find the exponent for this value of x we use the "log" button on a calculator. The result is

$$n = \log 1205 = 3.081$$

Thus, 10 raised to the 3.081 power gives 1205.

Another base that is frequently used for calculating exponents is $e = 2.718\ldots$. To represent $x = 1205$ in this base we write

$$x = 1205 = e^m$$

The logarithm to base e is known as the *natural logarithm*, and it is represented by the notation "ln." Using the "ln" button on a calculator, we find

$$m = \ln 1205 = 7.094$$

Thus, e raised to the 7.094 power gives 1205. The connection between the common and natural logarithms is as follows:

$$\ln x = 2.3026 \log x \qquad \textbf{A–18}$$

In the example just given, we have $\ln 1205 = 7.094 = 2.3026 \log 1205 = 2.3026\,(3.081)$.

The basic rules obeyed by logarithms follow directly from the rules given for exponents in Equation A–16. In particular,

$$\ln(xy) = \ln x + \ln y$$
$$\ln\left(\frac{x}{y}\right) = \ln x - \ln y \qquad \textbf{A–19}$$
$$\ln x^n = n \ln x$$

Though these rules are stated in terms of natural logarithms, they are satisfied by logarithms with any base.

Mathematical Expansions

We conclude with a brief consideration of small quantities in mathematics. Consider the following equation:

$$(1 + x)^3 = 1 + 3x + 3x^2 + x^3$$

This expression is valid for all values of x. However, if x is much smaller than one, $x \ll 1$, we can say to a good approximation that

$$(1 + x)^3 \approx 1 + 3x$$

Now, just how good is this approximation? After all, it ignores two terms that would need to be included to produce an equality. In the case $x = 0.001$, for example, the two terms that are neglected, $3x^2$ and x^3, have a combined contribution of only about 3 ten-thousandths of a percent! Clearly, then, little error is made in the approximation $(1 + 0.001)^3 \sim 1 + 3(0.001) = 1.003$. This can be seen visually in **Figure A–5 (a)**, where we plot $(1 + x)^3$ and $1 + 3x$ for x ranging from 0 to 1. Note that there is little difference in the two expressions for x less than about 0.1.

This is just one example of a general result in mathematics that can be derived from the **binomial expansion**. In general, we can say that the following approximation is valid for $x \ll 1$:

$$(1 + x)^n \approx 1 + nx \qquad \textbf{A–20}$$

This result holds for arbitrary n, not just for the case of $n = 3$. For example, if $n = -1$ we have

$$(1 + x)^{-1} = \frac{1}{1 + x} \approx 1 - x$$

We plot $(1 + x)^{-1}$ and $1 - x$ in **Figure A–5 (b)**, and again we see that the results are in good agreement for x less than about 0.1

An example of an expansion that arises in the study of relativity concerns the following quotient:

$$\frac{1}{\sqrt{1 - \dfrac{v^2}{c^2}}}$$

In this expression v is the speed of an object and c is the speed of light. Since objects we encounter generally have speeds much less than the speed of light, the ratio v/c is much less than one, and v^2/c^2 is even smaller than v/c. Thus, if we let $x = v^2/c^2$ we have

$$\frac{1}{\sqrt{1 - x}}$$

We can apply the binomial expansion to this result if we replace n with $-1/2$ and x with $-x$ in Equation A–20. This yields

$$\frac{1}{\sqrt{1 - \dfrac{v^2}{c^2}}} \approx 1 + \tfrac{1}{2}\frac{v^2}{c^2}$$

The two sides of this approximate equality are plotted in **Figure A– 5 (c)**, showing the accuracy of the approximation for small v/c.

Another type of mathematical expansion leads to the following useful results:

$$\sin\theta \approx \theta$$
$$\cos\theta \approx 1 - \tfrac{1}{2}\theta^2 \qquad \textbf{A–21}$$

These expansions are valid for small angles θ measured in radians. Note that the result $\sin\theta \approx \theta$ is used to derive Equations 6–13 and 13–19. (See Table 6–2, p. 158, and Figure 13–15, p. 400, for more details on this expansion.)

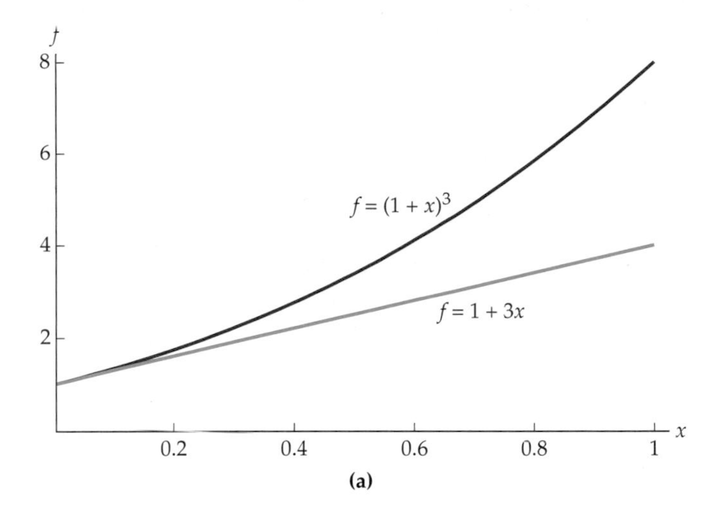

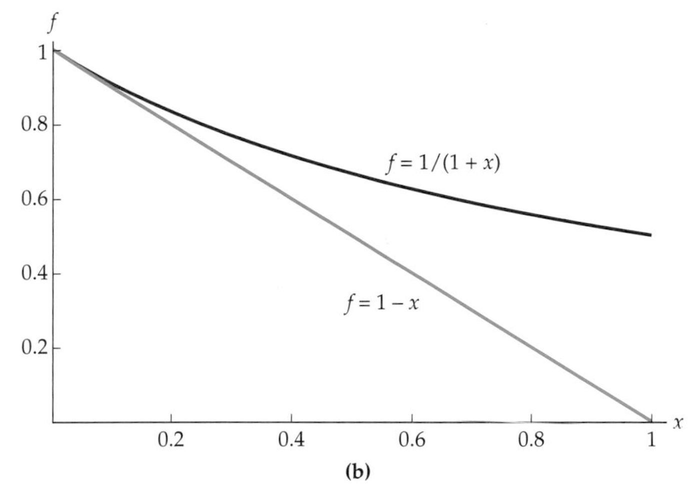

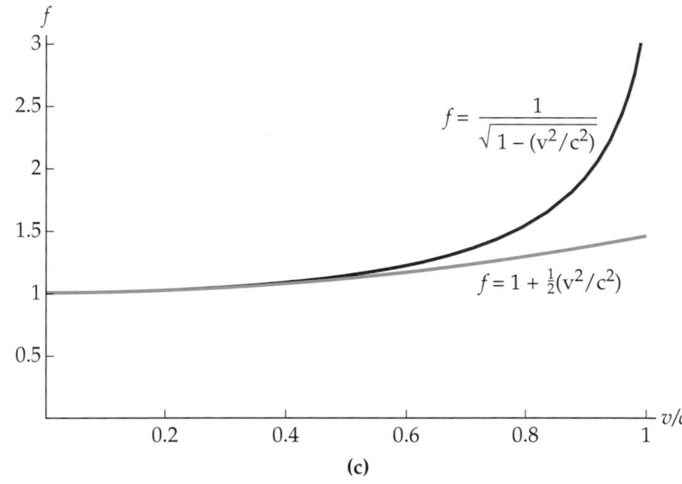

▲ **FIGURE A–5 Examples of mathematical expansions**
(a) A comparison between $(1 + x)^3$ and the result obtained from the binomial expansion, $1 + 3x$. **(b)** A comparison between $1/(1 + x)$ and the result obtained from the binomial expansion, $1 - x$. **(c)** A comparison between $1/\sqrt{1 - (v^2/c^2)}$ and the result obtained from the binomial expansion, $1 + \frac{1}{2}(v^2/c^2)$.

Appendix B Typical values

Mass

Sun	2.00×10^{30} kg
Earth	5.97×10^{24} kg
Moon	7.35×10^{22} kg
747 airliner (maximum takeoff weight)	3.5×10^{5} kg
blue whale	178,000 kg = 197 tons
elephant	5400 kg
mountain gorilla	180 kg
human	70 kg
bowling ball	7 kg
half gallon of milk	1.81 kg = 4 lbs
baseball	0.141–0.148 kg
golf ball	0.045 kg
female calliope hummingbird (smallest bird in North America)	3.5×10^{-3} kg = $\frac{1}{8}$ oz
raindrop	3×10^{-5} kg
antibody molecule (IgG)	2.5×10^{-22} kg
hydrogen atom	1.67×10^{-27} kg

Length

orbital radius of Earth (around Sun)	1.5×10^{8} km
orbital radius of Moon (around Earth)	3.8×10^{5} km
altitude of geosynchronous satellite	35,800 km = 22,300 mi
radius of Earth	6370 km
altitude of Earth's ozone layer	50 km
height of Mt. Everest	8848 m
height of Washington Monument	169 m = 555 ft
pitcher's mound to home plate	18.44 m
baseball bat	1.067 m
CD (diameter)	120 mm
aorta (diameter)	18 mm
period in sentence (diameter)	0.5 mm
red blood cell	$7.8\ \mu m = \frac{1}{3300}$ in.
typical bacterium (*E. coli*)	$2\ \mu m$
wavelength of green light	550 nm
virus	20–300 nm
large protein molecule	25 nm
diameter of DNA molecule	2.0 nm
radius of hydrogen atom	5.29×10^{-11} m

Time

estimated age of Earth	approx. 4.6 billion y $\approx 10^{17}$ s
estimated age of human species	approx. 150,000 y $\approx 5 \times 10^{12}$ s
half life of carbon-14	5730 y = 1.81×10^{11} s
period of Halley's comet	76 y = 2.40×10^{9} s
half life of technetium-99	6 h = 2.16×10^{4} s
time for driver of car to apply brakes	0.46 s
human reaction time	60–180 ms
air bag deployment time	10 ms
period of middle C sound wave	3.9 ms
collision time for batted ball	2 ms
decay of excited atomic state	10^{-8} s
period of green light wave	1.8×10^{-15} s

Speed

light	3×10^8 m/s
meteor	35–95 km/s
space shuttle (orbital velocity)	8.5 km/s = 19,000 mi/h
rifle bullet	700–750 m/s
sound in air (STP)	340 m/s
fastest human nerve impulses	140 m/s
747 at takeoff	80.5 m/s
kangaroo	18.1 m/s = 40.5 mi/h
200-m dash (Olympic record)	10.1 m/s
butterfly	1 m/s
blood speed in aorta	0.35 m/s
giant tortoise	0.076 m/s = 0.170 mi/h
Mer de Glace glacier (French Alps)	4×10^{-6} m/s

Acceleration

protons in particle accelerator	9×10^{13} m/s^2
ultracentrifuge	3×10^6 m/s^2
meteor impact	10^5 m/s^2
baseball struck by bat	3×10^4 m/s^2
loss of consciousness	7.14 g = 70 m/s^2
acceleration of gravity on Earth (g)	9.81 m/s^2
braking auto	8 m/s^2
acceleration of gravity on the moon	1.62 m/s^2
rotation of Earth at equator	3.4×10^{-2} m/s^2

Appendix C Planetary data

Name	Equatorial Radius (km)	Mass (Relative to Earth's)*	Mean Density (kg/m^3)	Surface Gravity (Relative to Earth's)	Orbital Semimajor Axis $\times 10^6$ km	Orbital Semimajor Axis A. U.	Escape Speed (km/s)	Orbital Period (Years)	Orbital Eccentricity
Mercury	2440	0.0553	5430	0.38	57.9	0.387	4.2	0.240	0.206
Venus	6052	0.816	5240	0.91	108.2	0.723	10.4	0.615	0.007
Earth	6370	1	5510	1	149.6	1	11.2	1.000	0.017
Mars	3394	0.108	3930	0.38	227.9	1.523	5.0	1.881	0.093
Jupiter	71,492	318	1360	2.53	778.4	5.203	60	11.86	0.048
Saturn	60,268	95.1	690	1.07	1427.0	9.539	36	29.42	0.054
Uranus	25,559	14.5	1270	0.91	2871.0	19.19	21	83.75	0.047
Neptune	24,776	17.1	1640	1.14	4497.1	30.06	24	163.7	0.009
Pluto	1137	0.0021	2060	0.07	5906	39.84	1.2	248.0	0.249

*Mass of Earth = 5.97×10^{24} kg

Appendix D Periodic table of the elements

Key: 26 **Fe** (Atomic number — Symbol) / 58.85 (Atomic mass) / $3d^64s^2$ (Outer electron configuration)

PERIODS	GROUP I	GROUP II	Transition elements										GROUP III	GROUP IV	GROUP V	GROUP VI	GROUP VII	GROUP VIII
	s		d										p					
1	1 **H** 1.01 $1s^1$																	2 **He** 4.00 $1s^2$
2	3 **Li** 6.94 $2s^1$	4 **Be** 9.01 $2s^2$											5 **B** 10.81 $2p^1$	6 **C** 12.01 $2p^2$	7 **N** 14.01 $2p^3$	8 **O** 16.00 $2p^4$	9 **F** 19.00 $2p^5$	10 **Ne** 20.18 $2p^6$
3	11 **Na** 22.99 $3s^1$	12 **Mg** 24.31 $3s^2$											13 **Al** 26.98 $3p^1$	14 **Si** 28.09 $3p^2$	15 **P** 30.97 $3p^3$	16 **S** 32.07 $3p^4$	17 **Cl** 35.45 $3p^5$	18 **Ar** 39.95 $3p^6$
4	19 **K** 39.10 $4s^1$	20 **Ca** 40.08 $4s^2$	21 **Sc** 44.96 $3d^14s^2$	22 **Ti** 47.88 $3d^24s^2$	23 **V** 50.94 $3d^34s^2$	24 **Cr** 52.00 $3d^54s^1$	25 **Mn** 54.94 $3d^54s^2$	26 **Fe** 55.85 $3d^64s^2$	27 **Co** 58.93 $3d^74s^2$	28 **Ni** 58.69 $3d^84s^2$	29 **Cu** 63.55 $3d^{10}4s^1$	30 **Zn** 65.39 $3d^{10}4s^2$	31 **Ga** 69.72 $4p^1$	32 **Ge** 72.61 $4p^2$	33 **As** 74.92 $4p^3$	34 **Se** 78.96 $4p^4$	35 **Br** 79.90 $4p^5$	36 **Kr** 83.80 $4p^6$
5	37 **Rb** 85.47 $5s^1$	38 **Sr** 87.62 $5s^2$	39 **Y** 88.96 $4d^15s^2$	40 **Zr** 91.22 $4d^25s^2$	41 **Nb** 92.91 $4d^45s^1$	42 **Mo** 95.94 $4d^55s^1$	43 **Tc** (98) $4d^55s^2$	44 **Ru** 101.07 $4d^75s^1$	45 **Rh** 102.91 $4d^85s^1$	46 **Pd** 106.42 $4d^{10}5s^6$	47 **Ag** 107.87 $4d^{10}5s^1$	48 **Cd** 112.41 $4d^{10}5s^2$	49 **In** 114.82 $5p^1$	50 **Sn** 118.71 $5p^2$	51 **Sb** 121.76 $5p^3$	52 **Te** 127.60 $5p^4$	53 **I** 126.90 $5p^5$	54 **Xe** 131.29 $5p^6$
6	55 **Cs** 132.91 $6s^1$	56 **Ba** 137.33 $6s^2$	57 **La** 138.91 * $5d^16s^2$	72 **Hf** 178.49 $5d^26s^2$	73 **Ta** 180.95 $5d^36s^2$	74 **W** 183.85 $5d^46s^2$	75 **Re** 186.21 $5d^56s^2$	76 **Os** 190.2 $5d^66s^2$	77 **Ir** 192.22 $5d^76s^2$	78 **Pt** 195.08 $5d^96s^1$	79 **Au** 196.97 $5d^{10}6s^1$	80 **Hg** 200.59 $5d^{10}6s^2$	81 **Tl** 204.36 $6p^1$	82 **Pb** 207.2 $6p^2$	83 **Bi** 208.98 $6p^3$	84 **Po** (209) $6p^4$	85 **At** (210) $6p^5$	86 **Rn** (222) $6p^6$
7	87 **Fr** (223) $7s^1$	88 **Ra** 226.03 $7s^2$	89 **Ac** 227.03 † $6d^17s^2$	104 **Rf** (261) $6d^27s^2$	105 **Db** (262) $6d^37s^2$	106 **Sg** (266) $6d^47s^2$	107 **Bh** (264) $6d^57s^2$	108 **Hs** (269) $6d^67s^2$	109 **Mt** (268) $6d^77s^2$	110 (271)	111 (272)	112 (277)		114 (289)		116 (289)		118 (293)

	f														
*	58 **Ce** 140.12 $5d^14f^16s^2$	59 **Pr** 140.91 $4f^36s^2$	60 **Nd** 144.24 $4f^46s^2$	61 **Pm** (145) $4f^56s^2$	62 **Sm** 150.36 $4f^66s^2$	63 **Eu** 151.96 $4f^76s^2$	64 **Gd** 157.25 $5d^14f^76s^2$	65 **Tb** 158.93 $5d^14f^86s$	66 **Dy** 162.50 $4f^{10}6s^2$	67 **Ho** 164.93 $4f^{11}6s^2$	68 **Er** 167.26 $4f^{12}6s^2$	69 **Tm** 168.93 $4f^{13}6s^2$	70 **Yb** 173.04 $4f^{14}6s^2$	71 **Lu** 174.97 $5d^14f^{14}6s^2$	Lanthanides
†	90 **Th** 232.04 $6d^27s^2$	91 **Pa** 231.04 $5f^26d^17s^2$	92 **U** 238.03 $5f^36d^17s^2$	93 **Np** 237.05 $5f^46d^17s^2$	94 **Pu** (244) $5f^66d^07s^2$	95 **Am** (243) $5f^76d^07s^2$	96 **Cm** (247) $5f^76d^17s^2$	97 **Bk** (247) $5f^86d^17s^2$	98 **Cf** (251) $5f^{10}6d^07s^2$	99 **Es** (252) $5s^{11}6d^07s^2$	100 **Fm** (257) $5f^{12}6d^07s^2$	101 **Md** (258) $5f^{13}6d^07s^2$	102 **No** (259) $5f^{14}6d^07s^2$	103 **Lr** (262) $5f^{14}6d^17s^2$	Actinides

Appendix E Properties of selected isotopes

Atomic Number (Z)	Element	Symbol	Mass Number (A)	Atomic Mass*	Abundance (%) or Decay Mode† (if radioactive)	Half-Life (if radioactive)
0	(Neutron)	n	1	1.008 665	β^-	10.6 min
1	Hydrogen	H	1	1.007 825	99.985	
	Deuterium	D	2	2.014 102	0.015	
	Tritium	T	3	3.016 049	β^-	12.33 y
2	Helium	He	3	3.016 029	0.000 14	
			4	4.002 603	$\approx$100	
3	Lithium	Li	6	6.015 123	7.5	
			7	7.016 005	92.5	
4	Beryllium	Be	7	7.016 930	EC, γ	53.3 d
			8	8.005 305	2α	6.7×10^{-17} s
			9	9.012 183	100	
5	Boron	B	10	10.012 938	19.8	
			11	11.009 305	80.2	
			12	12.014 353	β^-	20.4 ms
6	Carbon	C	11	11.011 433	β^+, EC	20.4 ms
			12	12.000 000	98.89	
			13	13.003 355	1.11	
			14	14.003 242	β^-	5730 y
7	Nitrogen	N	13	13.005 739	β^-	9.96 min
			14	14.003 074	99.63	
			15	15.000 109	0.37	
8	Oxygen	O	15	15.003 065	β^+, EC	122 s
			16	15.994 915	99.76	
			18	17.999 159	0.204	
9	Fluorine	F	19	18.998 403	100	
10	Neon	Ne	20	19.992 439	90.51	
			22	21.991 384	9.22	
11	Sodium	Na	22	21.994 435	β^+, EC, γ	2.602 y
			23	22.989 770	100	
			24	23.990 964	β^-, γ	15.0 h
12	Magnesium	Mg	24	23.985 045	78.99	
13	Aluminum	Al	27	26.981 541	100	
14	Silicon	Si	28	27.976 928	92.23	
			31	30.975 364	β^-, γ	2.62 h
15	Phosphorus	P	31	30.973 763	100	
			32	31.973 908	β^-	14.28 d
16	Sulfur	S	32	31.972 072	95.0	
			35	34.969 033	β^-	87.4 d
17	Chlorine	Cl	35	34.968 853	75.77	
			37	36.965 903	24.23	

Atomic Number (Z)	Element	Symbol	Mass Number (A)	Atomic Mass*	Abundance (%) or Decay Mode† (if radioactive)	Half-Life (if radioactive)
18	Argon	Ar	40	39.962 383	99.60	
19	Potassium	K	39	38.963 708	93.26	
			40	39.964 000	β^-, EC, γ, β^+	1.28×10^9 y
20	Calcium	Ca	30	39.962 591	96.94	
24	Chromium	Cr	52	51.940 510	83.79	
25	Manganese	Mn	55	54.938 046	100	
26	Iron	Fe	56	55.934 939	91.8	
27	Cobalt	Co	59	58.933 198	100	
			60	59.933 820	β^-, γ	5.271 y
28	Nickel	Ni	58	57.935 347	68.3	
			60	59.930 789	26.1	
			64	63.927 968	0.91	
29	Copper	Cu	63	62.929 599	69.2	
			64	63.929 766	β^-, β^+	12.7 h
			65	64.927 792	30.8	
30	Zinc	Zn	64	63.929 145	48.6	
			66	65.926 035	27.9	
33	Arsenic	As	75	74.921 596	100	
35	Bromine	Br	79	78.918 336	50.69	
36	Krypton	Kr	84	83.911 506	57.0	
			89	88.917 563	β^-	3.2 min
38	Strontium	Sr	86	85.909 273	9.8	
			88	87.905 625	82.6	
			90	89.907 746	β^-	28.8 y
39	Yttrium	Y	89	89.905 856	100	
43	Technetium	Tc	98	97.907 210	β^-, γ	4.2×10^6 y
47	Silver	Ag	107	106.905 095	51.83	
			109	108.904 754	48.17	
48	Cadmium	Cd	114	113.903 361	28.7	
49	Indium	In	115	114.903 88	95.7; β^-	5.1×10^{14} y
50	Tin	Sn	120	119.902 199	32.4	
53	Iodine	I	127	126.904 477	100	
			131	130.906 118	β^-, γ	8.04 d
54	Xenon	Xe	132	131.904 15	26.9	
			136	135.907 22	8.9	
55	Cesium	Cs	133	132.905 43	100	
56	Barium	Ba	137	136.905 82	11.2	
			138	137.905 24	71.7	
			144	143.922 73	β^-	11.9 s
61	Promethium	Pm	145	144.912 75	EC, α, γ	17.7 y
74	Tungsten (Wolfram)	W	184	183.950 95	30.7	
76	Osmium	Os	191	190.960 94	β^-, γ	15.4 d
			192	191.961 49	41.0	
78	Platinum	Pt	195	194.964 79	33.8	
79	Gold	Au	197	196.966 56	100	

Atomic Number (Z)	Element	Symbol	Mass Number (A)	Atomic Mass*	Abundance (%) or Decay Mode† (if radioactive)	Half-Life (if radioactive)
80	Mercury	Hg	202	201.970 63	29.8	
81	Thallium	Tl	205	204.974 41	70.5	
			210	209.990 069	β^-	1.3 min
82	Lead	Pb	204	203.973 044	β^-, 1.48	1.4×10^{17} y
			206	205.974 46	24.1	
			207	206.975 89	22.1	
			208	207.976 64	52.3	
			210	209.984 18	α, β^-, γ	22.3 y
			211	210.988 74	β^-, γ	36.1 min
			212	211.991 88	β^-, γ	10.64 h
			214	213.999 80	β^-, γ	26.8 min
83	Bismuth	Bi	209	208.980 39	100	
			211	210.987 26	α, β^-, γ	2.15 min
84	Polonium	Po	210	209.982 86	α, γ	138.38 d
			214	213.995 19	α, γ	164 μs
86	Radon	Rn	222	222.017 574	α, β	3.8235 d
87	Francium	Fr	223	223.019 734	α, β^-, γ	21.8 min
88	Radium	Ra	226	226.025 406	α, γ	1.60×10^3 y
			228	228.031 069	β^-	5.76 y
89	Actinium	Ac	227	227.027 751	α, β^-, γ	21.773 y
90	Thorium	Th	228	228.028 73	α, γ	1.9131 y
			232	232.038 054	100; α, γ	1.41×10^{10} y
92	Uranium	U	232	232.037 14	α, γ	72 y
			233	233.039 629	α, γ	1.592×10^5 y
			235	235.043 925	0.72; α, γ	7.038×10^8 y
			236	236.045 563	α, γ	2.342×10^7 y
			238	238.050 786	99.275; α, γ	4.468×10^9 y
			239	239.054 291	β^-, γ	23.5 min
93	Neptunium	Np	239	239.052 932	β^-, γ	2.35 d
94	Plutonium	Pu	239	239.052 158	α, γ	2.41×10^4 y
95	Americium	Am	243	243.061 374	α, γ	7.37×10^3 y
96	Curium	Cm	245	245.065 487	α, γ	8.5×10^3 y
97	Berkelium	Bk	247	247.070 03	α, γ	1.4×10^3 y
98	Californium	Cf	249	249.074 849	α, γ	351 y
99	Einsteinium	Es	254	254.088 02	α, γ, β^-	276 d
100	Fermium	Fm	253	253.085 18	EC, α, γ	3.0 d
101	Mendelevium	Md	255	255.0911	EC, α	27 min
102	Nobelium	No	255	255.0933	EC, α	3.1 min
103	Lawrencium	Lr	257	257.0998	α	≈35 s
104	Rutherfordium	Rf	261	261.1087	α	1.1 min
105	Hahnium	Ha	262	262.1138	α	0.7 min

*The masses given throughout this table are those for the neutral atom, including the Z electrons.

†EC stands for electron capture.

Answers to Odd-Numbered Conceptual Questions

Chapter 1

1. No. The factor of 2 is dimensionless.

3. Each of the equations is dimensionally consistent.

5. (a) Not possible, since units have dimensions. For example, seconds can only have the dimension of time. **(b)** Possible, since different units can be used to measure the same dimensions. For example, time can be measured in seconds, minutes, or hours.

7. To the nearest power of ten: **(a)** 1 m; **(b)** 10^{-2} m; **(c)** 10 m; **(d)** 100 m; **(e)** 10^7 m.

Chapter 2

1. The displacement is the same for you and your dog; the distance covered by the dog is greater.

3. Yes, if you have driven in a straight line.

5. Their velocities are different because they travel in different directions.

7. Since the car circles the track its direction of motion must be changing. Therefore, its velocity changes as well. Its speed, however, can be constant.

9. Your average speed is 20 m/s since you spent equal amounts of time at 15 m/s and 25 m/s.

11. Constant-velocity motion; that is, straight-line motion with constant speed.

13. (a) The time required to stop is doubled. **(b)** The distance required to stop increases by a factor of four.

15. Yes, if it moves with constant velocity.

17. (a) If air resistance can be ignored, the acceleration of the ball is the same at each point on its flight. **(b)** Same answer as part (a).

19. Ignoring air resistance, the two gloves have the same acceleration.

21. Ball A has the greater increase in speed because it takes longer to land. As a result, it is accelerated for the greater length of time.

23. The initial speed of the first ball (at the height h) is zero; the initial speed of the second ball is v_0. Therefore, the second ball covers more distance in a given amount of time, and the balls meet at a height greater than $h/2$. In fact, they meet at the height $3h/4$.

Chapter 3

1. (a) scalar; **(b)** vector; **(c)** vector; **(d)** scalar.

3. (a) A and **B** have the same magnitude. **(b) A** and **B** have opposite directions.

5. (a) The magnitude of the vector doubles. **(b)** The direction of the vector is unchanged.

7. Yes, if they have the same magnitude and point in opposite directions.

9. Note that the magnitudes A, B, and C satisfy the Pythagorean theorem. It follows that **A**, **B**, and **C** form a right triangle, with **C** as the hypotenuse. Thus, **A** and **B** are perpendicular to one another.

11. A and **B** must be collinear and point in the same direction.

13. The direction angle for **A** is between 270° and 360°.

15. Two vectors of unequal magnitude cannot add to zero, even if they point in opposite directions. Three vectors of unequal magnitude can add to zero if they can form a triangle.

17. Tilt the umbrella forward so that it points in the opposite direction of the rain's velocity relative to you.

Chapter 4

1. Ignoring air resistance, the acceleration of a projectile is vertically downward at all times.

3. The projectile was launched at an angle of 30° above the positive x axis; when it landed its direction of motion was 30° below the positive x axis. Hence, its change in direction was 60° clockwise.

5. The skateboarder and the skateboard have the same horizontal component of velocity. Therefore, the skateboarder will come down on the skateboard.

7. At its highest point, the projectile has a speed equal to its horizontal component of velocity, $v_0 \cos \theta$, where θ is the launch angle. The angle for which $\cos \theta = 1/2$ is $\theta = 60°$.

9. The divers are in free fall for the same amount of time. It follows that diver 2 covers twice the distance.

11. (a) Projectile c, projectile b, projectile a. **(b)** Projectile a, projectile b, projectile c.

13. Since the penguin lands at an elevation that is above the water, its speed will be less.

15. (a) From the child's point of view the scoop of ice cream falls straight downward. **(b)** From the point of view of the parents the scoop of ice cream follows a parabolic trajectory.

17. Ignoring air resistance, the two snow balls have the same acceleration as soon as they are released, regardless of the direction they are thrown.

Chapter 5

1. The force exerted on the car by the brakes causes it to slow down, but your body continues to move forward with the same velocity (due to inertia) until the seat belt exerts a force on it to decrease its speed.

3. The thrust from the engines accelerates the plane. For your body to have the same acceleration a force must act on it; this is the force you feel as the seat pushes on your back.

5. When the magnitude of the force exerted on the girl by the rope equals the magnitude of her weight, the net force acting on her is zero. As a result, she moves with constant velocity.

7. (a) The upper string breaks, because the tension in it is equal to the applied force on the lower string plus the weight of the block. When the tension in the upper string reaches the breaking point, the tension in the lower string is below this value. **(b)** The lower string breaks in this case, because of the inertia of the block. As you move your hand downward rapidly the lower string stretches and breaks before the block can move a significant distance and stretch the upper string.

9. Each time the astronauts throw or catch the ball they exert a force on it, and it exerts an equal and opposite force on them. This causes the astronauts to move farther apart from one another with increasing speed as the game progresses.

11. No. You are at rest relative to your immediate surroundings, but you are in motion relative to other objects in the universe.

13. Mr. Ed's reasoning is incorrect because he is adding two action-reaction forces that act on *different* objects. Wilbur should point out that the net force exerted on the cart is simply the force exerted on it by Mr. Ed. Thus the cart will accelerate. The equal and opposite reaction force acts on Mr. Ed, and does not cancel the force acting on the cart.

15. Yes. An object with zero net force acting on it has a constant velocity. This velocity may or may not have zero magnitude.

17. The whole brick has twice the force acting on it, but it also has twice the mass. Since the acceleration of an object is proportional to the force exerted on it and inversely proportional to its mass, the whole brick has the same acceleration as the half brick.

19. The acceleration of an object is inversely proportional to its mass. The diver and the Earth experience the same force, but the Earth—with its much larger mass—has a much smaller acceleration.

21. Since the car is at rest, we can conclude that it has zero *net* force acting on it. This is not the same as having no forces at all acting on the car. In fact, gravity exerts a downward force, and the road exerts an equal and opposite net upward force (summed over the four tires.)

23. No adjustment is necessary because you are moving with constant velocity. If you throw an object in the elevator it falls with the same downward acceleration relative to you as it would if you were to throw it while standing on the surface of the Earth.

25. Yes, if by heavy or light we mean whether an object has a large or a small mass. If the astronaut pushes on the object its acceleration will be inversely proportional to its mass. Thus a massive object can be recognized by its small acceleration.

27. On solid ground you come to rest in a much smaller distance than when you plunge into water. The smaller the distance over which you come to rest the greater the acceleration, and the greater the acceleration the greater the force exerted on you. Thus, when you land on solid ground the large force it exerts on you may be enough to cause injury.

29. Yes, in fact it happens all the time. Whenever you throw a ball upward it moves in the opposite direction to the net force acting on it, until it reaches the top of its trajectory.

31. **(a)** northward; **(b)** northward; **(c)** zero; **(d)** zero; **(e)** downward.

33. The net force acting on the plane is zero, since its acceleration is zero.

Chapter 6

1. The clothesline has a finite mass, and so the tension in the line must have an upward component to oppose the downward force of gravity. Thus, the line sags much the same as if a weight were hanging from it.

3. The shape of the object's path is circular. The object's speed remains constant since there is no component of acceleration in the direction of motion.

5. The force that ultimately is responsible for stopping a train is the frictional force between its metal wheels and the metal track. These are fairly smooth surfaces. In contrast, the frictional force that stops a car is between the rubber tires and the concrete roadway. These are rougher surfaces with a greater coefficient of friction.

7. As you brake harder your car has a greater acceleration. The greater the acceleration of the car the greater the force required to give the flat of strawberries the same acceleration. When the required force exceeds the maximum force of static friction the strawberries begin to slide.

9. The braking distance of a skidding car depends on its initial speed and the coefficient of kinetic friction. Thus, if the coefficient of friction is known reasonably well, the initial speed can be determined from the length of the skid marks.

11. For a drop of water to stay on a rotating wheel an inward force is required to give the drop the necessary centripetal acceleration. Since the force between a drop of water and the wheel is small, the drop will separate from the wheel rather than follow its circular path.

13. The net force acting on the masses remains the same. Their acceleration will decrease, however, since the total mass of the system has increased.

15. A centripetal force is required to make the motorcycle follow a circular path, and this force increases rapidly with the speed of the cycle. If the necessary centripetal force exceeds the weight of the motorcycle, because its speed is high enough, then the track must exert a downward force on the cycle at the top of the circle. This keeps the cycle in firm contact with the track.

17. Since the passengers are moving in a circular path a centripetal force must be exerted on them. This force, which is radially inward, is supplied by the wall of the cylinder.

19. One should allow the wheels to continue turning. The reason is that in this case the force stopping the car is static friction which, in general, is greater than the force of kinetic (skidding) friction.

21. This helps because the students sitting on the trunk increase the normal force between your tires and the road. Since the force of friction is proportional to the normal force, this increases the frictional force enough (one hopes) to allow your car to move.

23. Yes. The steering wheel can accelerate a car—even if its speed remains the same—by changing its direction of motion.

25. **(a)** The car's velocity is not constant since its direction of motion changes. **(b)** The magnitude of the car's acceleration is v^2/r. Since the speed and radius are constant, so is the magnitude of the acceleration. **(c)** The direction of the car's acceleration is toward the center of the circular path. Thus the direction of the acceleration changes as the car moves around the circle.

Chapter 7

1. No. Work requires that a force acts through a distance.

3. True. To do work on an object a force must have a nonzero component along its direction of motion.

5. Yes, you must do work against the force of gravity to raise your body upward out of bed.

7. The catcher does negative work on the ball. This is because the force exerted by the catcher is opposite in direction to the motion of the ball. Since the work done on the ball is negative, its speed decreases.

9. The frictional force between your shoes and the ground does positive work on you whenever you begin to walk.

11. The work required is $8W_0$

13. Gravity exerts an equal and opposite force on the package, hence the net work done on it is zero. The result is no change in kinetic energy.

15. Since the kinetic energy of the youngster increased, the work done was positive.

17. The speed of car 1 is the smaller of the two. In fact, the speed of car 1 is equal to the speed of car 2 divided by the square root of two.

19. **(a)** If v is doubled the kinetic energy of the block increases by a factor of four. For the spring to do four times the work it must be compressed by the distance $2\Delta x$. **(b)** If m is doubled the kinetic energy of the block doubles. For the spring to do twice the work it must be compressed by the distance $\sqrt{2}\,\Delta x$.

21. Since power is the rate at which work is done, F_2 produces more power (0.6 J/s) than F_1 (0.5 J/s).

23. No. Engine 1 may do the same amount of work as engine 2 in half the time.

Chapter 8

1. The kinetic energy cannot be negative, since m and v^2 are always positive or zero. The gravitational potential energy can be negative since any level can be chosen to be zero.

3. Your answers will disagree on **(a)**, the initial and final potential energy, but will agree on **(b)**, the change in potential energy.

5. No. The change in gravitational potential energy depends only on the change in elevation, not on how the change in elevation comes about.

7. The potential energy of the spring increases, the gravitational potential energy of the mass-Earth system decreases.

9. Your answers will disagree on **(a)**, but agree on **(b)** and **(c)**.

11. Since there is no friction, we conclude that energy is conserved. It follows that the block will leave the spring with the same speed, v.

13. The two balls have the same change in gravitational potential energy. Thus, ignoring air resistance, they will also have the same change in kinetic energy.

15. The jumper's initial kinetic energy is largely converted to a compressional, spring-like potential energy as the pole bends. The pole straightens out, converting its potential energy into gravitational potential energy. As the jumper falls, the gravitational energy is converted into kinetic energy, and finally, the kinetic energy is converted to compressional potential energy as the cushioning pad on the ground is compressed.

17. Ignoring any type of frictional force, the speed of the object is the same at points A and G. Similarly, the speed of the object at point B is the same as its speed at points D and F.

19. The total mechanical energy decreases with time if air resistance is present.

21. As the leaf falls, air resistance exerts a force on it opposite to its direction of motion. Hence, a negative work is done on it and its final mechanical energy is less than its initial mechanical energy.

23. With a smaller acceleration of gravity, it would be necessary for the skier to rise to a greater height in order to gain the same amount of gravitational potential energy. Thus, a higher hill would be needed for part **(a)**. The final speed in part **(b)** would be the same because the values of the initial kinetic energy and the final gravitational potential energy are the same, assuming we are using the higher hill discussed for part (a).

25. Mechanical energy is conserved for the bicycle that is coasting. The bicycle that is pedaled has a positive nonconservative work done on it. This causes its mechanical energy to increase.

27. A variety of conservative and nonconservative forces are involved in the situation shown in the photo. First, the engine of the earth mover does positive nonconservative work as it digs out and lifts a load of rocks. At the same time, gravity does negative conservative work on the rocks as the gravitational potential energy of the system increases. Next, the earth mover does positive nonconservative work to transport the rocks to the dump truck. When the rocks are released, gravity does positive conservative work as the gravitational potential energy of the system is converted to kinetic energy. Nonconservative frictional forces do negative work to convert the kinetic energy of the rocks into sound and heat when they land in the truck. Finally, the increased load in the truck does conservative work as it compresses the springs, storing part of the system's energy in the form of spring potential energy.

Chapter 9

1. The momentum of the keys increases as they fall because a net force acts on them. The momentum of the universe is unchanged because an equal and opposite force acts on the Earth.

3. No, their kinetic energies are not necessarily the same. Suppose one object has a mass m and a speed v; a second object has a mass $m/2$ and a speed $2v$. The momenta are the same, $mv = (m/2)(2v)$, but the second object has twice as much kinetic energy, $\frac{1}{2}(m/2)(2v)^2 = 2(\frac{1}{2}mv^2)$.

5. If the kinetic energy is zero the speeds must be zero as well. This means that the momentum is zero.

7. The rubber bullet is more likely to knock the block over. The reason is that the change in momentum is twice as great when an object rebounds as it is when the object is simply brought to rest.

9. No, the impulse is the product of the force and the time interval over which the force acts. Thus a smaller force can deliver a greater impulse than a larger force if it acts for a longer time.

11. The boulder and the pebble have the same rate of momentum change, since the same force acts on both objects. Force, in fact, is the rate of change of momentum.

13. Yes, in much the same way that a propeller in water can power a speedboat.

15. When a heavy object and a light object collide they exert equal and opposite forces on one another. Since the light object has less mass, its acceleration is greater. This can result in more severe injuries for the light vehicle.

17. No. The fact that the initial momentum of the system is nonzero means that the final momentum must also be nonzero. Thus, it is not possible for both objects to be at rest after the collision.

19. Yes. Suppose two objects have momenta of equal magnitude. If these objects collide in a head-on, completely inelastic collision, the two objects will come to rest. In this case, all of the initial kinetic energy is lost to other forms of energy.

21. The two balls have the same momentum. However, the first ball has a kinetic energy equal to $\frac{1}{2}mv^2$ and the second ball has half that much kinetic energy; $\frac{1}{2}(2m)(v/2)^2 = \frac{1}{2}(\frac{1}{2}mv^2)$. Thus, less energy is dissipated in stopping the second ball, and so it has less "sting."

23. The kinetic energy of the bullet is much greater than that of the gun. Thus, less energy is dissipated in stopping the gun.

25. The answer to both questions is no. In an elastic collision we know that the total kinetic energy of the system is conserved, as is the total momentum of the system. There is no reason the individual kinetic energies or individual momenta should be conserved—and in general they are not. Conservation applies only to the system as a whole.

27. Your center of mass is somewhere directly above the area of contact between your foot and the ground.

29. The center of mass of a baseball bat is nearer to the thick end of the bat, since most of the mass is at that end. Only if the bat were uniform would its center of mass be midway between the two ends.

31. (a) If we take the direction of fall to be positive, the horizontal component of the contact force is in the positive direction. This is because as the pencil falls its center of mass accelerates both downward and in the positive direction. **(b)** Since the center of mass of the pencil has a nonzero, downward acceleration, it follows that the net vertical force acting on it is downward. Thus, the upward force exerted by the table is less than the weight of the pencil.

33. The plane weighs the same whether the fly lands on the dashboard or flies about the cockpit. The reason is that the fly must exert a downward force on the air in the cockpit equal in magnitude to its own weight in order to stay aloft. This force ultimately acts downward on the plane, just as if the fly had landed.

Chapter 10

1. In the time $t/2$ the object rotates through the angle $\theta/4$. This follows because of the t^2 time dependence associated with constant angular acceleration.

3. All points on the rigid object have the same angular speed. Not all points have the same linear speed, however. The farther a given point is from the axis of rotation the greater its linear speed.

5. (a) Betsy and Jason have the same angular speed, ω. **(b)** Jason's linear speed ($v_t = r\omega$) is one half of Betsy's linear speed. **(c)** Jason's centripetal acceleration ($a_{cp} = v^2/r = r\omega^2$) is one half of Betsy's centripetal acceleration.

7. No. As long as you are driving in a circular path you will have a nonzero centripetal acceleration.

9. Your linear velocity changes with time due to your changing direction of motion. Similarly, your linear acceleration changes direction with time. On the other hand, your angular velocity and angular acceleration are constant.

11. Your angular speed due to the Earth's rotation is the same regardless of your elevation.

13. Yes. Imagine picking up a bicycle and spinning one of its tires. If you increase the angular speed of the tire you increase its rotational kinetic energy. The translational kinetic energy of the tire remains constant at zero, however, regardless of its angular speed.

15. In principle, the Earth's moment of inertia increases because both its mass and radius increase.

17. The reading on the speedometer gives the speed of the axles of your car. This is the same as the speed of the occupants inside the car. If the top of the tires have a speed v, the axles have a speed $v/2$. (See Figure 10–11.)

19. The hard-boiled egg spins longer, because it is a rigid body. When you give a hard-boiled egg a spin the entire egg rotates with the same angular speed. When you spin a raw egg, however, the outside rotates more rapidly than the liquid interior. The drag between the interior and the exterior results in the raw egg slowing down more rapidly than the solid hard-boiled egg.

21. The center of the outer quarter moves in a circle that has twice the radius of a quarter. As a result, the linear distance covered by the center of the outer quarter is twice the circumference of a quarter. Therefore, if the outer quarter rolls without slipping, it must complete two revolutions.

23. The moment of inertia of the object is greatest when it is rotated about the z axis. It is least when rotated about the x axis.

Chapter 11

1. No. Torque depends both on the magnitude of the force and on the distance from the axis of rotation, or moment arm, at which it is applied. A small force can produce the same torque as a large force if it is applied farther from the axis of rotation.

3. The long pole has a large moment of inertia, which means that for a given applied torque the walker and pole have a small angular acceleration. This allows more time for the walker to "correct" his balance.

5. Since the mass accelerates downward, we conclude that the upward acting tension in the string is less than the weight of the mass.

7. A force applied radially to a wheel produces zero torque, though the net force is nonzero.

9. No. In most cars the massive engine is located in the front, thus the car's center of mass is not in the middle of the car, but is closer to the front end. This means that the force exerted on the front tires is greater than the force exerted on the rear tires. (See Active Example 11–1.)

11. You are in static equilibrium as you sit in your chair; so is the building where you have your physics class.

13. Yes. When an airplane's engine starts up from rest the propeller has a nonzero rotational acceleration, though its translational acceleration is zero.

15. The angular speed of the dust cloud increases, just like a skater pulling in her arms, due to conservation of angular momentum.

17. Since the string is always at right angles to the motion of the puck, the tension does no work on it. Therefore, the linear speed of the puck, v, remains constant. As the string becomes shorter the angular speed, $\omega = v/r$, increases. The angular momentum of the puck is $L = I\omega = (mr^2)(v/r) = rmv$. As r decreases, the angular momentum of the puck decreases as well.

19. His angular speed is unaffected. The angular momentum of the student-weights system decreases when the floor exerts a torque on the weights to bring them to rest.

21. The disk rotates with its bottom moving to the right and its top moving to the left. The center of the disk moves to the right, since that is the direction of the net force exerted on the disk.

23. Since there are no external torques acting on the system, its angular momentum must remain constant. Thus, as the beetle walks toward the axis of rotation, which reduces the moment of inertia of the system, the angular speed of the turntable increases.

25. If the Earth were to expand, keeping the same mass and the same mass distribution, its moment of inertia would increase. Since the angular momentum of the Earth would stay the same (no external torques), its angular speed would decrease, like a skater extending her arms. As a result, the length of the day would increase.

27. When the 100 cm end of the stick is on the ground the moment of inertia of the system and the torque applied to it both increase. Since the moment of inertia ($\propto r^2$) increases more than the torque ($\propto r$), the angular acceleration of the system is reduced and the time of fall increases.

29. The block moves to the left. The reason is that the hoop, with the larger moment of inertia, has the smaller angular acceleration. Since less string unwinds from the hoop, the block moves in that direction.

31. The angular acceleration is inversely proportional to the moment of inertia. Hence, the greatest angular acceleration occurs when the object is rotated about the x axis, and the smallest angular acceleration occurs when the object is rotated about the z axis.

Chapter 12

1. No. The force of Earth's gravity is practically as strong in orbit as it is on the surface of the Earth. The astronauts experience weightlessness because they are in constant free fall.

3. No. The amount of area swept out per time varies from planet to planet.

5. Once the period of Charon is determined, the mass of the body it orbits (Pluto) can be calculated using Equation 12–7.

7. In one complete orbit, whether circular or elliptical, a satellite returns to its starting point. Since gravity is a conservative force, it does zero net work when there is zero net displacement.

9. On the Moon, where there is no atmosphere, a rock can orbit at any altitude where it clears the mountains—as long as it has sufficient speed. Thus, if you could give the rock enough speed, it would orbit the Moon and come up to you from behind.

11. No. In the weightless environment of the Shuttle there would be no convection, which is needed to bring fresh oxygen to the flame. Without convection a flame usually goes out very quickly. In carefully controlled experiments on the Shuttle, however, small flames have been maintained for considerable times. These "weightless" flames are spherical in shape, as opposed to the tear-shaped flames here on Earth.

13. Your weight is slightly less than on the surface of the Earth, due to the decrease in Earth's gravitational force with distance.

15. Yes, there is a slight difference. If you fly to the east, which is the direction of the Earth's rotation, you have a greater speed relative to the center of the Earth than if you fly to the west. As a result, the centripetal force required to maintain your circular motion is greater, and your apparent weight is less.

17. **(a)** The satellite's orbit becomes elliptical. **(b)** The apogee distance increases. **(c)** The perigee distance stays the same. (See Figure 12–13.)

19. No. There is a direct relationship between a satellite's speed and its orbit, just as there is a direct relationship between a satellite's period and its orbit. For example, the speed of the shuttle in its orbit is determined by the laws of physics, not by NASA.

21. The Earth moves faster the closer it comes to the Sun. Thus the Earth is closest to the Sun around January 4 and farthest from the Sun around July 4.

Chapter 13

1. The motion is periodic. It is not simple harmonic, however, because the position and velocity of the ball do not vary sinusoidally with time.

3. This motion is not simple harmonic because the position and velocity of the cart do not vary sinusoidally with time.

5. The mass travels a distance $4A$ in the time T and a distance $9A$ in the time $5T/2$.

7. At $x = 0$ the speed of the mass is v_{max} and its acceleration is zero. At $x = A$ the speed of the mass is zero and its acceleration is $-a_{max}$.

9. **(a)** No change in angular frequency; **(b)** no change in frequency; **(c)** no change in period; **(d)** maximum speed doubles; **(e)** maximum acceleration doubles; **(f)** total mechanical energy quadruples.

11. The period remains the same because, even though the distance traveled by the object is doubled, its speed at any given time is also doubled.

13. The longer spring has a smaller force constant, hence the period of oscillation is greater.

15. Since gravity is generally weaker on top of a mountain, the period of the pendulum would increase. As a result, the clock would run slow.

17. The length of the pendulum must be increased to $4L$.

19. The tie should be moved downward. This decreases the moment of inertia of the oscillator which, in turn, decreases the period of oscillation.

21. **(a)** When the elevator moves upward with constant acceleration, the pendulum experiences an effective acceleration of gravity greater than g. Hence, the period decreases. **(b)** The situation is just the opposite of that described in part (a), hence the period increases.

23. The period of a mass on a spring is independent of the acceleration of gravity. Hence, the period is the same whether the elevator accelerates upward, downward, or not at all.

25. The period stays the same since it depends only on the mass and the force constant, not on the acceleration of gravity.

27. If soldiers march in synchrony, the bridge will oscillate with the frequency of their step. If this frequency is near a resonance frequency of the bridge, the amplitude of oscillation could increase to dangerous levels.

Chapter 14

1. To generate a longitudinal wave, hit the nail on the head in a direction parallel to its length. To generate a transverse wave, hit the nail in a direction that is perpendicular to its length.

3. Typical waves at stadiums are transverse, since people move vertically up and down while the wave moves horizontally. To produce a longitudinal wave people could move to their left or right.

5. Since the tension in the string is constant the wave speed does not change. Thus, doubling the frequency results in the wavelength decreasing by a factor of two.

7. Waves have lower speeds on the more massive string.

9. The kinetic energy of the string is a maximum where the displacement of the string is zero, because the string moves with its greatest speed at this point. The kinetic energy is a minimum (zero) where the displacement is a maximum, because the string is instantaneously at rest at this point.

11. The intensity decreases with the square of the distance. Hence, doubling the distance reduces the intensity by a factor of $2^2 = 4$.

13. The lowest frequency is the frequency of the horn itself. The person on the island hears a higher frequency due to the motion of the boat toward the island. This same frequency goes out as the echo. The person in the boat hears the echo, but Doppler shifted up to yet a higher frequency.

15. The speed of the source is one-half the speed of sound.

17. The part of the nail that vibrates most freely is the portion not yet in the wood. As you drive the nail farther into the wood, therefore, the part that vibrates becomes shorter and shorter. The vibrating portion of the nail is similar to the vibrating air column in an organ pipe, hence the frequency goes up as the length decreases.

19. When you tune a violin you change the tension in the string. This causes the speed of waves in the string to change. The wavelength of a given mode of oscillation is unchanged, however, due to the fixed length of string. Thus, changing the speed while keeping the wavelength constant results in a change in frequency, according to the relation $v = f\lambda$.

21. No, the energy of oscillation is the same at all times. When the string is flat, the energy of oscillation is purely kinetic.

23. Tightening string 1 increases its frequency, and at the same time increases the beat frequency. Hence, string 2 has the lower frequency.

25. The beat frequencies are the same. All that matters as far as the beat frequency is concerned is the absolute value of the difference between the two frequencies.

Chapter 15

1. To draw a liquid up a straw you expand your lungs, which reduces the air pressure inside your mouth to less than atmospheric pressure. The resulting difference in pressure produces a net upward force on the liquid in the straw.

3. Pressure is force per area, and this is the same in both cases. Since the larger glass has a greater area on its base, the net force acting downward on it is larger than the force acting downward on the smaller glass, as one would expect.

5. The pressure in a tank of water increases with depth, hence the pressure is greatest near the bottom. To provide sufficient support there, the metal bands must be spaced more closely together.

7. This experiment shows that a certain pressure is needed at the bottom of the water column and not just a certain weight of water. To blow the top off the barrel it is necessary to increase the pressure in the barrel enough so that the increase in pressure times the surface area of the top exceeds 400 N. Thus, the required height of water is the height that gives the necessary increase in pressure. But the increase in pressure, ρgh, depends only on the height of the water in the tube, not on its weight.

9. Different water levels in the tube and the main body of the weather glass indicate different pressures in the atmosphere and the interior of the weather glass. As a low-pressure system approaches, atmospheric pressure drops. This allows the water level in the tube to rise.

11. The volume of wood that is submerged decreases when the steel ball is immersed in the water. In the original situation, the block had to be submerged far enough to provide enough buoyant force to support both the block and the ball. In the inverted situation, the ball experiences a buoyant force that partially supports its weight. Thus, the buoyant force that must be provided by the block is less, and so is its submerged volume.

13. In both cases, the weight of displaced fluid is equal to the weight of the small float. Since less of fluid B is displaced, it must be more dense.

15. Two quantities are unknown; the object's density and its volume. The two weight measurements provide two independent conditions that can be solved for the two unknowns.

17. When a brick is in the boat it displaces a volume of water equal to its weight. When the brick is at the bottom of the pool it displaces a volume of water equal to its own volume. Thus, the brick displaces a greater volume of water when it is in the boat. As a result, the water level decreases when the brick is thrown overboard.

19. The Great Salt Lake has water with a higher salinity, and hence a higher density, than ocean water. In fact, the density of its water is somewhat greater than the density of a typical human body. This means that a person can float in the Great Salt Lake much like a block of wood floats in fresh water.

21. The buoyant force, which is equal to the weight of displaced fluid, is the same for both the lead and aluminum balls since they have equal diameters. This result is true for any fluid (as long as both objects sink in the fluid), though the magnitude of the buoyant force varies from fluid to fluid.

23. As far as floating goes, this planet and the Earth are the same. The increased acceleration of gravity increases your weight and the weight of water by the same factor.

25. As the water falls it speeds up. Still, the amount of water that passes a given point in a given time is the same at any height. If the thickness of the water stayed the same, and its speed increased, the amount of water per time would increase. Hence, the thickness of the water must decrease to offset the increase in speed.

27. If you takeoff into the wind the air speed over the wings is greater than if you takeoff with the wind. This means that more lift is produced when taking off into the wind, which is clearly the preferable situation.

29. The ball should spin so that the side facing the batter is moving upward and the side facing the pitcher is moving downward.

31. If blood pressure is measured at a lower elevation than the heart, the result will be a higher value.

33. The buoyant force stays the same because the weight of water you displace is the same at any depth.

Chapter 16

1. The coffee is not in equilibrium because its temperature is different from that of its surroundings. Over time the temperature of the coffee will decrease, until finally it is the same as room temperature. At this point it will be in equilibrium—as long as the room stays at the same temperature.

3. The discrepancy is not serious. After all, a temperature of 15,000,000 °C is equivalent to a temperature of 15,000,273.15 K. A difference of 273.15 out of 15 million is generally insignificant.

5. The amount of bend in a bimetallic strip depends on the difference in the coefficients of thermal expansion for the two metals—the greater the difference in thermal expansion, the greater the bend. For this reason, strip 2 (aluminum-steel) bends more than strip 1 (copper-steel).

7. If the glass and the mercury had the same coefficient of volume expansion, the level of mercury in the glass would not change with temperature. This is because the volume of the cavity in the glass would expand by the same amount as the volume of mercury.

9. The steel tape measure would be better because its coefficient of thermal expansion is smaller. This means that its length would change less with temperature.

11. Aluminum has the greater coefficient of thermal expansion, hence its diameter changes more with temperature than does the diameter of the hole in the brass plate. It follows that the temperature should be lowered, thus shrinking the aluminum ball enough to fit through the hole.

13. The period of the pendulum decreases, since its length is reduced as the temperature is lowered.

15. **(a)** A copper rod will never touch the top of the ring, because both the ring and the rod expand with the same coefficient of thermal expansion. **(b)** Aluminum expands more with temperature than copper, hence if the temperature is increased enough the rod can touch the top of the ring. **(c)** Steel expands less with temperature than copper, thus heating will not cause the rod to reach the top of the ring.

17. The less massive object, which has the smaller heat capacity, will have the greatest change in temperature.

19. If the objects have different masses, the less massive object will have a greater temperature change, since it has the smaller heat capacity. On the other hand, the objects may have the same mass but differ in the material from which they are made. In this case, the object with the smaller specific heat will have the greater temperature change.

21. The soil in the field cools off faster than the concrete parking lot, since its temperature changes more for a given amount of heat loss.

23. Even though the flame at the far end of the match is very hot, the wood from which it is made is a poor conductor of heat. The air between the flame and your finger is an even poorer conductor of heat.

25. Two important factors work in favor of the water-filled balloon. First, the water has a large heat capacity, hence it can take on a large amount of heat with little change in temperature. Second, water is a better conductor of heat than air, hence the heat from the flame is conducted into a large volume of water—which gives it a larger effective heat capacity.

27. The smallest temperature difference occurs in the material with the greatest thermal conductivity. Hence, in order of increasing temperature difference, we have the following: material B, material A, material C.

29. The darker shirt is a more effective absorber of radiation, hence it will be warmer.

31. The tea in the mug that is filled only halfway cools off more rapidly. The reason is that its surface area is greater in proportion to its mass than the tea in the full mug, hence it loses heat more rapidly. (In general, small objects have more surface area per mass than larger objects of the same basic shape.)

33. With white paint, the airplane would be a less effective radiator of energy than with black paint; that is, its emmissivity, e, would be less. As a result, a white SR-71 would be significantly hotter in flight.

35. Most definitely! This is one reason an igloo can be quite comfy.

37. Object 1 must have the higher temperature, to compensate for object 2's greater emissivity. Since radiated power depends on temperature raised to the fourth power, the temperature of object 1 must be greater by a factor of 2 to the one fourth power.

39. The black block, which is the more efficient radiator of energy, cools off faster. Note that the black block also absorbs energy more efficiently, but since it is in the cold of space there is little energy to be gained from absorption.

Chapter 17

1. By the definition of a mole, the number of molecules in one mole of N_2 is the same as the number of molecules in one mole of O_2. The mole of O_2 has the greater mass, however, since each of its molecules is more massive than an N_2 molecule.

3. Treating the helium as an ideal gas, it satisfies the relation $PV = nRT$. When we put the balloon in the refrigerator, the pressure and the number of moles remain constant. Under these conditions, a decrease in temperature implies a decrease in volume.

5. The pressure of an ideal gas is given by $P = nRT/V$. If $P_A = n_A RT/V_A$, it follows that $P_B = (n_A/2)RT/(2V_A) = P_A/4$.

7. The volume of an ideal gas is given by $V = nRT/P$. Thus, we expect the volume of the oxygen bags to vary inversely with pressure—the lower the pressure in the cabin the larger the bags.

9. Recall that velocity takes into account both the speed of an object and its direction of motion. Therefore, the average *velocity* of air molecules in a room is zero, since they move randomly in all directions.

11. Assuming the air in the rooms is an ideal gas, we have that the number of moles is given by $n = PV/RT$. The two rooms have the same pressure (atmospheric) and the same volume. Hence, the room with the lower temperature is the one with the greater number of molecules.

13. The Kelvin temperature is proportional to the average kinetic energy of the molecules, which in turn is proportional to the speed of the molecules squared. Hence, doubling the speed implies an increase in temperature by a factor of $2^2 = 4$.

15. No. The average kinetic energy is proportional to the Kelvin temperature, not the Celsius temperature.

17. For a given force, the change in length is proportional to the initial length, L_0, and inversely proportional to the cross-sectional area A. Hence, the change in dimensions of the brick will be least when it is placed on the side that is 2 cm by 3 cm. This gives the smallest height ($L_0 = 1$ cm) and the largest cross-sectional area ($A = 2\text{ cm} \times 3\text{ cm} = 6\text{ cm}^2$).

19. The boiling temperature of water depends on the pressure at which the boiling occurs—the higher the pressure the higher the boiling temperature. Thus, in the autoclave, where the pressure is greater than atmospheric pressure, the temperature of boiling water is greater than 100 °C.

21. The alcohol, besides having antiseptic qualities, evaporates readily. As it evaporates, it draws heat from the body.

23. As high-speed molecules leave the drop, it draws heat from its surroundings to keep its temperature constant. As long as the temperature of the drop is constant, it will continue to have the same fraction of molecules moving quickly enough to escape.

25. The slow evaporation of a drop of water is evidence that only some of its molecules are moving fast enough to escape at any given time, and that the other molecules are moving more slowly.

27. Twice as much time would be required because twice as much heat would be needed for the vaporization to occur.

Chapter 18

1. No. If the engine has friction it is not reversible.

3. Yes. As an example, consider a gas in a thermally insulated cylinder. If this gas is compressed, its temperature will rise.

5. Yes. You can convert mechanical work completely into heat by rubbing your hands together.

7. Recall that $\Delta U = Q - W$. Therefore, we have the following results: **(a)** 0; **(b)** 0; **(c)** -100 J.

9. **(a)** No; **(b)** yes; **(c)** no; **(d)** no; **(e)** yes; **(f)** no.

11. You get a greater increase in temperature when you add the heat at constant volume. At constant pressure, the gas expands and does work as the heat is added. Hence, only part of the heat goes into increasing the internal energy. When heat is added at constant volume no work is done. In this case, all the heat goes into increasing the internal energy, and hence the temperature.

13. The work done by the gas is equal to the area under the process. Hence, the process with the least work is **(c)**, with 3 J, followed by **(a)**, with 4 J, and finally **(b)**, with 4.5 J.

15. The efficiency ($e = 1 - T_c/T_h$) depends on the ratio of T_c to T_h. If both temperatures are doubled there is no change in the ratio, and no change in the efficiency.

17. Yes. In fact, the heat delivered to a room is typically 3 to 4 times the work done by the heat pump.

19. The entropy of the gas increases because heat must be added to it to keep its temperature constant.

21. No. As you do work to put things in order you give off heat to the atmosphere. This increases the entropy of the air by more than the decrease in entropy of the room, for a net increase in entropy.

23. **(a)** Popped popcorn; **(b)** an omelet; **(c)** a pile of bricks; **(d)** a burned piece of paper.

Answers to Odd-Numbered Problems

Note: In cases where an ambiguity might arise, numbers in this text are assumed to have the smallest possible number of significant figures. For example, a number like 150 is assumed to have just two significant figures–the zero simply indicates the location of the decimal point. To represent this number with three significant figures, we would use the form 1.50×10^2.

Chapter 1

1. **(a)** 1.27 gigadollars
   (b) 1.27×10^{-3} teradollar
3. 3×10^8 m/s
5. $p = 1$
9. $\frac{[M]}{[T^2]}$
11. 3.00×10^8 m/s
13. 30.0 lb
15. **(a)** 89.65 m^2 **(b)** 23 m^2
17. **(a)** 2.70×10^4 ft^3 **(b)** 764 m^3
19. 1.087828×10^{-4} s
21. 75,600
23. 0.23 lb
25. 6.71×10^8 mi/h
27. 2.0×10^6 m^2
29. **(a)** 65.6 ft/s **(b)** 44.7 mi/h
31. 10^4 seats
33. **(a)** 1000 mi/h **(b)** 24,000 mi **(c)** 4000 mi
35. **(a)** 39 ft/s^2 **(b)** 1.6×10^5 km/h^2
37. $p = 1; q = -1$
39. 38 mi/h

Chapter 2

1. **(a)** 1.95 mi **(b)** 0.75 mi
3. **(a)** 15 m **(b)** 10 m
5. **(a)** 130 m; 100 m **(b)** 260 m; 0
7. 10.1 m/s; 22.7 mi/h
9. 2.2 km
11. 1.28 s
13. 10 ms
15. 6.0 m/s
17. 12 m
19. **(a)**

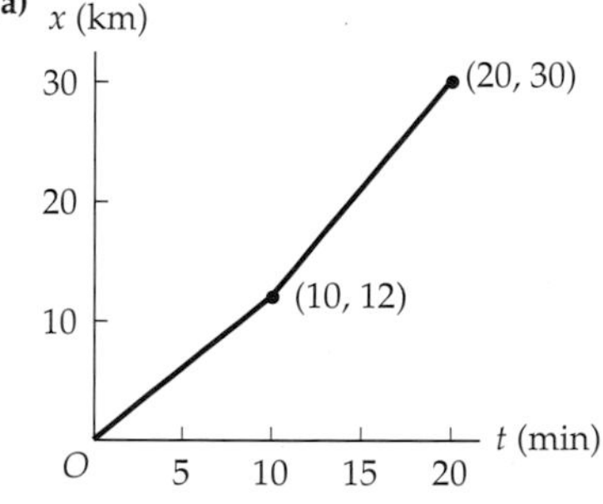

   (b) 23.3 m/s
21. **(a)**

A	B	C	D
+	0	+	−

   (b) 2 m/s, +; 0; 1 m/s, +; −1.5 m/s, −
23. **(a)**

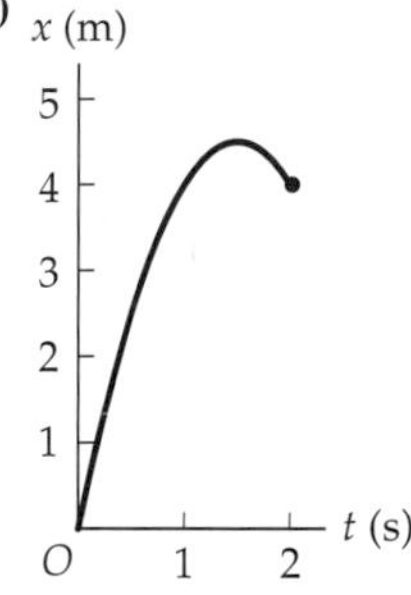

   (b) 4 m/s **(c)** 4 m/s
25. 35 mi/h
27. **(a)**

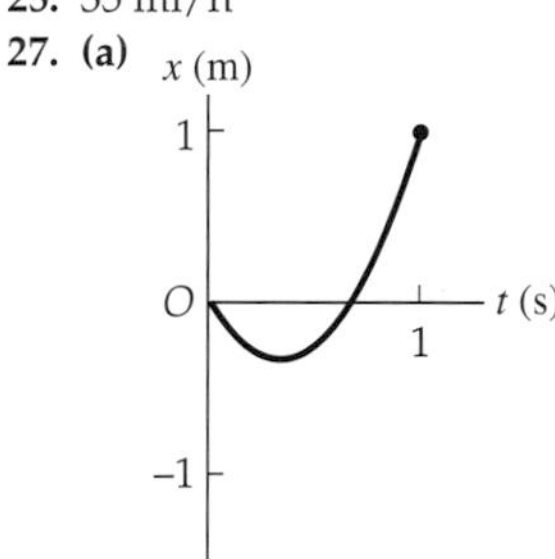

   (b) −0.80 m/s **(c)** −0.8 m/s
29. **(a)** 3.8 m/s **(b)** 4.2 m/s
31. **(a)** 29.4 m/s north **(b)** 8.90 m/s north
33. 10 m; 20 m; 40 m
35. **(a)** a factor of two **(b)** 3.8 s; 7.6 s
37. 11.4 m/s
39. 9.45 m/s^2, due north
41. 5.8 s
43. **(a)** 78 m **(b)** 19 m
45. 20.1 m/s^2
47. $x_1 = (20.0\text{ m/s})t + (1.25\text{ m/s}^2)t^2$;
    $x_2 = 1000\text{ m} + (-30.0\text{ m/s})t + (1.60\text{ m/s}^2)t^2$
49. **(a)** 18 m/s^2 **(b)** 53 m/s
51. **(a)** 3.4 s **(b)** 6.0 m/s
53. 1.5 m/s^2, due east
55. 5.3 s
57. **(a)** 5.0 m/s^2, toward third base **(b)** 3.6 m
59. **(a)**

angle	10.0°	20.0°	30.0°
acceleration, $\frac{m}{s^2}$	1.71	3.37	4.88

61. The statement is accurate.
63. 4.9 m/s
65. **(a)** 3.4 m/s, 3.4 m/s
    (b) 6.4 m/s, −6.4 m/s
67. 0.10 s
69. $x_B = (1/2)gt^2$
    $x_T = 2.0\text{ m} + (-4.2\text{ m/s})t + (1/2)gt^2$
71. more than half as fast; 12 m/s
73. **(a)** 58 m/s **(b)** 5.9 s
75. **(a)** 1.2 s **(b)** 0.61 s
77. 24.8 m/s
79. **(a)** 2.8 s **(b)** −30 m/s
81. 11 m/s
83. 1.8 m/s
85. **(a)** 2 **(b)** 4
    (c) 0.41 s, 0.82 s; 0.20 m, 0.82 m
87. **(a)**

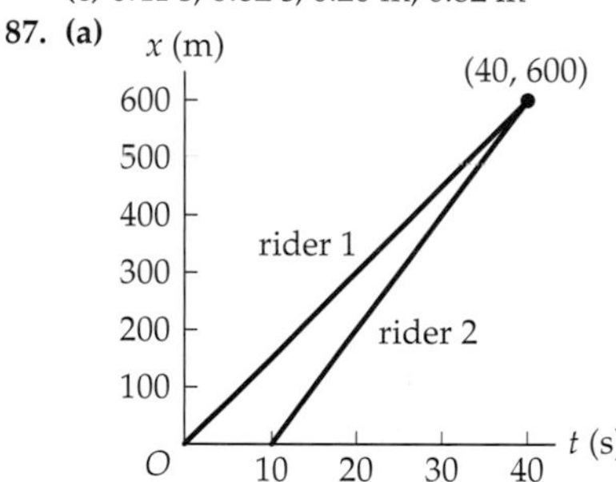

   (b) 40 s **(c)** 600 m
89. 2.73 s
91. **(a)** 9.81 m/s^2, downward **(b)** 15 m
    (c) 1.1 s **(d)** 5.4 m/s
93. **(a)** 13.4 m **(b)** 16.2 m/s
95. 3.3 m
97. 6.67 cm
99. **(a)** 1.0 m; 4.4 m/s **(b)** 130 drops/min
101. 5.5 m; 11 m/s
103. **(a)** 10 ms **(b)** 4.5 m/s **(c)** 0.10 m

Chapter 3

1. 142 ft
3. 3.43°
5. **(a)** $(90\text{ ft})\hat{\mathbf{x}} + (90\text{ ft})\hat{\mathbf{y}}$ **(b)** $(90\text{ ft})\hat{\mathbf{y}}$
   (c) $(0\text{ ft})\hat{\mathbf{x}} + (0\text{ ft})\hat{\mathbf{y}}$
7. 1.5 Å
9. 1.1 m
11. **(a)** **A** **(b)** **A**
13. **(a)** 51 m deep **(b)** 140 m
15. **(a)**

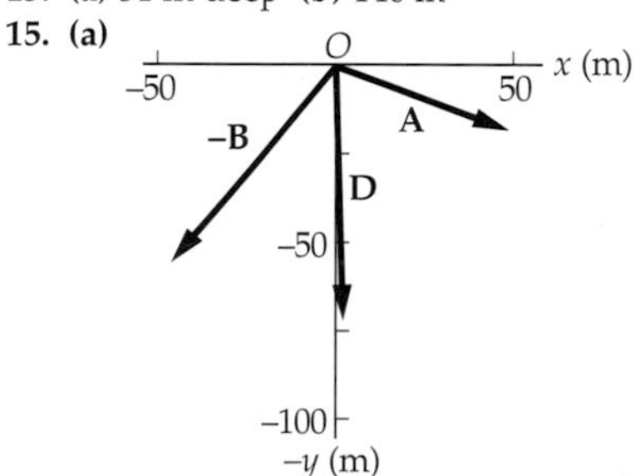

   (b) 71 m; −88°

17. (a)

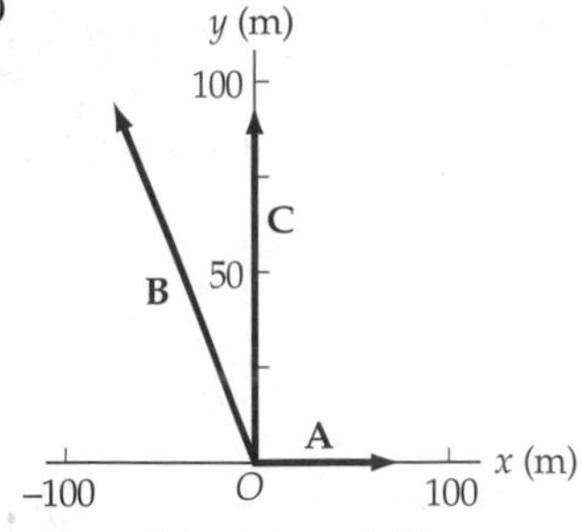

(b) about 120 m; about 130°
(c) 120 m; 130°

19. (a) $-26.6°, 5\sqrt{5}$ **(b)** $-153°, 5\sqrt{5}$
(c) $26.6°, 5\sqrt{5}$

21. $(48\text{ m})\hat{\mathbf{x}} - (28\text{ m})\hat{\mathbf{y}}$

23. (a) $-22°$, 5.4 m **(b)** 110°, 5.4 m
(c) 45°, 4.2 m

25. (a) $(23\text{ m})\hat{\mathbf{x}} - (27\text{ m})\hat{\mathbf{y}}$
(b) $-(23\text{ m})\hat{\mathbf{x}} + (27\text{ m})\hat{\mathbf{y}}$

27. $(2.1\text{ m})\hat{\mathbf{x}} + (0.74\text{ m})\hat{\mathbf{y}}$

29. 140 m, 31° east of south

31. 21° south of west, 2.6 km

33. (a) 3.3 m/s, 1.2 m/s
(b) The components will be halved.

37. 14.3 m/s

39. 25 s

41. (a) 14° west of north
(b)

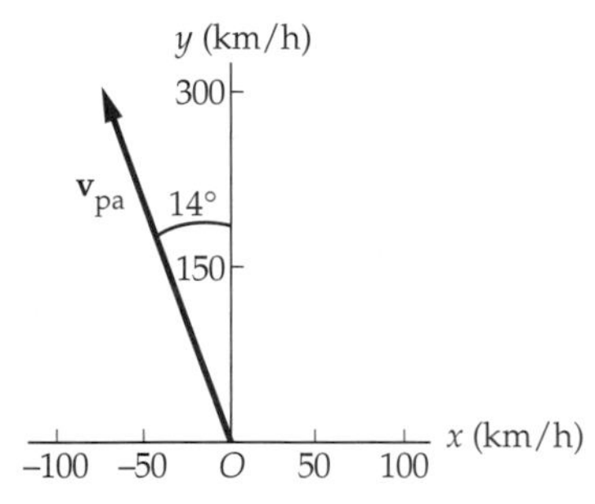

(c) increased

43. 11 m/s

45. (a) Jet ski A **(b)** $\frac{t_A}{t_B} = 0.82$

47. 50 m, 35 m, 43 m

49. (a) $\approx$38 ft **(b)** 39 ft

51. (a) 3.7 **(b)** 67 m/s

53. 3.5 m/s, 45° left of shoreward;
3.5 m/s, 45° left of shoreward

55. 28 m; 19 m

Chapter 4

1. (a) 5.2 km **(b)** 3.6 km

3. (a) 100 m **(b)** 9.6 m

5. (a) 2.95×10^{-9} s **(b)** 2.31 cm

7. 46 m/s

9. 7.87 m/s

11. (a) 1.6 m **(b)** decreases **(c)** decreases

13. (a) 2.70 m/s **(b)** -20.6 m/s
(c) The speed of the clam in the x-direction would increase with the speed of the crow, but the speed in the y-direction would stay the same. The speed of the crow determines v_x and gravity determines v_y.

15. (a) 0.378 m **(b)** decreases

17. (a) $-66°$, 8.1 m/s
(b) $-76°$, 14 m/s

19. (a) 3.01 m **(b)** the same time

21. 1.1 m

23. (a) 13.9 m/s **(b)** 1.99 s

25. 1.57 m/s

27. 1.3 m

29. (a) the same as **(b)** 18 m/s, 18 m/s

31. (a) 92 m **(b)** 21 m/s

33. 24.8 m/s

35. (a) 12 m/s, 1.3°
(b) No; the y-component of the velocity is still positive.

37. 10.5 m

39. 62.8 m/s

41. 50.7 m

43. (a) 29.8 m/s **(b)** 4.29 s

45. 46.2°

47. (a) 849 m/s **(b)** less than

49. 4.6 m/s

51. (a) 5.4 m/s at 22° above the horizontal
(b) 6.2 m **(c)** 1.3 s

53. (a) Use the vector magnitude formula and $v_y = v_{0y} - gt$ to determine v_{0y}. Then use the relationship of v_0 and v_{0y} to determine θ. **(b)** 22.3°

55. (a) 0.75 s **(b)** 1.0 m

57. (a) 2.3 m **(b)** 2.3 m **(c)** 2.2 m

59. 14.5 m/s

61. (a) To find the initial speed of the puck, eliminate t from the equations
$x = (v_0 \cos\theta)t$ and
$y = (v_0 \sin\theta)t - (1/2)gt^2$,
then solve for v_0.
(b) 25.1 m/s

63. (a) 2.7 cm **(b)** 0.021 s

65. 76°

Chapter 5

1. $(1.24\text{ m/s}^2)\hat{\mathbf{y}}$

3. 37 kg

5. 260 N

7. (a) 81 N **(b)** increased

9. (a) 5.1 kN opposite to the direction of motion **(b)** 15.3 m

11. (a) -22 kN
(b) The strategy is to determine the acceleration from the speeds and displacement and then determine the force from the acceleration and mass.

13. (a) two forces **(b)** The forces acting on the brick are due to gravity and your hand.
(c) No **(d)** No

15. (a) the same as **(b)** more than
(c) 0.87 m/s²

17. (a) 1.04 N **(b)** 3.59 N

19. 15 kN

21. 28 N

23. 1.6 m/s²

25. (a) 240 N, downhill, parallel to the slope
(b) increases

27. 6.5 s

29. (a) 11 kg **(b)** 110 N

31. 3.6 kN

33. (a) 0.0119 N **(b)** the same

35. (a) upward **(b)** 1.9 m/s²

37. 110 N

39. (a) ; 88 N

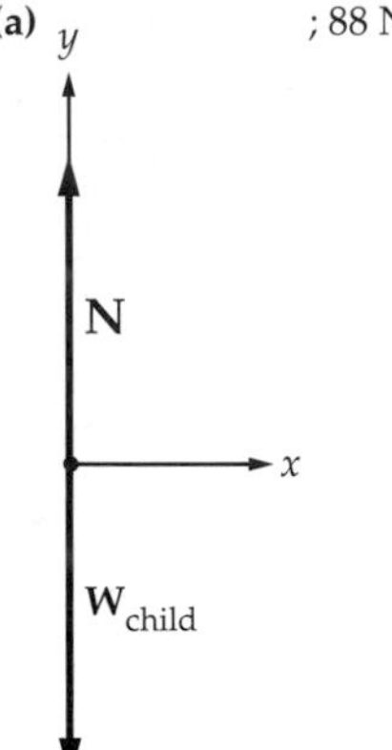

(b) ; 110 N

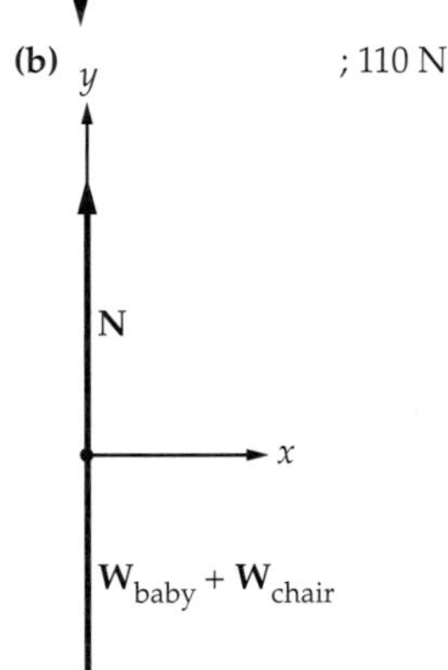

41. ; The free-body diagram does not change.

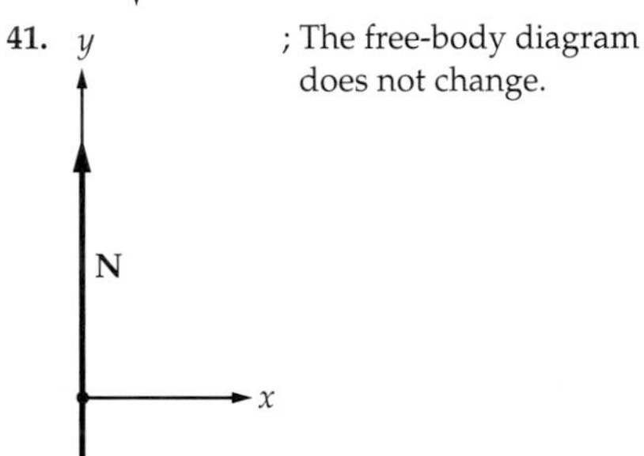
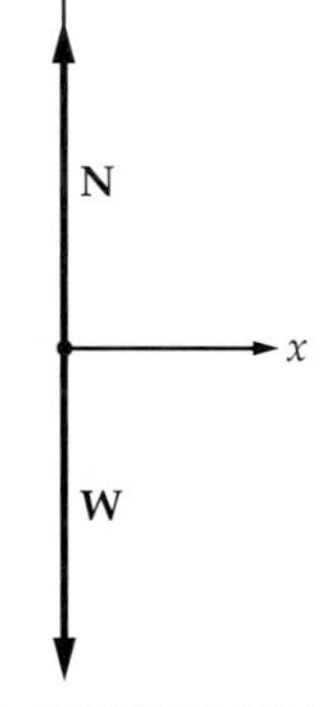

43. (a) 340 N **(b)** increases

45. 54 N

47. 9.1 kg

49. (a) -180 N
(b) The penetration depth doubles.

51. 5.1 kN

53. (a) 280 kN
(b) Determine the acceleration during takeoff using the given data (x, t). Then calculate the force using $F = ma$.

55. 74 kg

57. $F_1 = (m/2)(a_1 - a_2)$;
$F_2 = (m/2)(a_1 + a_2)$

Chapter 6

1. 7.8 m

3. 1.2

5. 0.75 N opposite the direction of the push
7. 250 N
9. **(a)** 0.42 **(b)** $\theta = 23°$; $\mu_s = f(\theta)$ only
11. **(a)** 2.4 m/s^2 **(b)** 6.4 s
13. 53.6 N
15. 9.8 kN/m
17. **(a)** 3.0 N **(b)** No
19. **(a)** 260 N **(b)** 510 N
21. **(a)** −4.8 cm
(b) Yes; the spring displacement is proportional to the block's mass.
23. 0.0985 N
25. **(a)** 50 N **(b)** No
27. 41.6 lb
29. 42.5 N
31. 0.85 N
33. **(a)** 60° **(b)** 106 N
35. −0.076 m/s^2 upward
37. 4.9 N; 15 N
39. **(a)** 2.6 m/s^2 **(b)** 2.4 N **(c)** decrease
41. 4.5 kN
43. 36 m/s
45. 650 N
47. **(b)** 520 N; 560 N
49. 19 m/s
51. 1.9 kg
53. 0.14
55. 6.9 kg
57. **(a)** 14.9 N **(b)** 57.6 N
59. **(a)** 0.025 N down the incline
(b) 0.025 N up the incline
61. $F_1 = mg \sin 20°$; $F_2 = mg \cos 20°$
63. **(a)** 22.1 N **(b)** stays the same
65. 19°; 0.78 N
67. **(a)** 32 N **(b)** increase
69. **(a)** greater than **(b)** 1.88 kN
71. **(a)** the coefficient of static friction
(b) 0.88
73. 29 m/s^2
75. 7.7°; The mass of the dice drops out of the equations.
77. 28°
79. $\mu g(2m + M)$
81. **(a)** $\dfrac{\mu_s}{1 + \mu_s}$
83. **(a)** $\dfrac{\mu_s mg}{\cos\theta - \mu_s \sin\theta}$
85. **(a)** 0.160 s **(b)** 0.104 m
87. $T = mg \sin\theta$

Chapter 7

1. 240 J
3. **(a)** 38 J **(b)** 0
5. **(a)** 74 J **(b)** 0 **(c)** −74 J
7. **(a)** negative **(b)** −5.8 kJ
9. 480 J
11. 57.4°
13. 3.3 kJ
15. **(a)** 7.20 kJ **(b)** 1.80 kJ **(c)** 28.8 kJ
17. **(a)** −3.3 J **(b)** 0.28 N upward
19. **(a)** −530 J **(b)** 0.30
21. **(a)** less than **(b)** −1.4 m/s
23. **(a)** 31 J **(b)** 31 J
25. 65 N/m
27. **(a)** 0.85 m **(b)** 0.10 m
29. **(a)** 26 kN/m **(b)** more than 130 J; 390 J
31. **(a)** 820 J **(b)** 1.1 kJ
33. 53 kW
35. 3.6×10^6 J
37. 5.1 min
39. 0.119 hp
41. **(a)** 2.3 MJ **(b)** 2 Snickers
43. 0.9 hp
45. $\sqrt{\dfrac{P}{b}}$
47. 6.3 km/s
49. 240 W **(b)** 8.4 kJ
51. **(a)** 0.18 **(b)** less than
53. $\dfrac{K}{2}$
55. **(a)** $mg(h - h_{max})$ **(b)** mgh
(c) $\dfrac{1}{2}mv_0^2 + mgh$
57. 0.23 J
59. 0.25 mW
61. **(a)** 6.0 W **(b)** 540 J
63. **(a)** 2.9×10^4 kg **(b)** 38 MJ/s
65. $\left(\dfrac{P}{b}\right)^{1/3}$
67. $\dfrac{1}{2}(k_1 + k_2)x^2$
69. **(a)** 42.2° **(b)** 14.8 kg

Chapter 8

1. −51 J; −51 J; −51 J
3. **(a)** −0.11 J; −0.11 J **(b)** The results have no dependence on the mass of the block.
5. 6.94 MJ
7. 0.92 cm
9. 36.7 J
11. 7.16 m/s
13. **(a)** 0.942 m **(b)** Doubling the ball's mass would cause no change to (a).
15.

y (m)	4.0	3.0	2.0	1.0	0
U (J)	8.2	6.2	4.1	2.1	0
K (J)	0	2.1	4.1	6.2	8.2
E (J)	8.2	8.2	8.2	8.2	8.2

17. **(a)** 15 m/s **(b)** 43 m
19. **(a)** 0; 98.1 J; 98.1 J **(b)** 98.1 J; 196 J; 98.1 J
21. 4.0 cm
23. 1.3 m/s
25. 57 MJ
27. 1.8 m/s
29. 0, 17.2 J, 0, 17.2 J; −2.05 J, 8.58 J, 6.54 J, 15.1 J; −4.10 J, 0, 13.1 J, 13.1 J
31. **(a)** uphill **(b)** 3.70 m
33. **(a)** −29.4 MJ **(b)** 5.51 MJ **(c)** No
35. At point A, the object is at rest. As the object travels from point A to point B, some of its potential energy is converted into kinetic energy and the object's speed increases. As the object travels from point B to point C, some of its kinetic energy is converted back into potential energy and its speed decreases. From point C to point D, the speed increases again, and from Point D to point E, the speed decreases.
37. Just to the right of point A and just to the left of point E.
39. ±12°

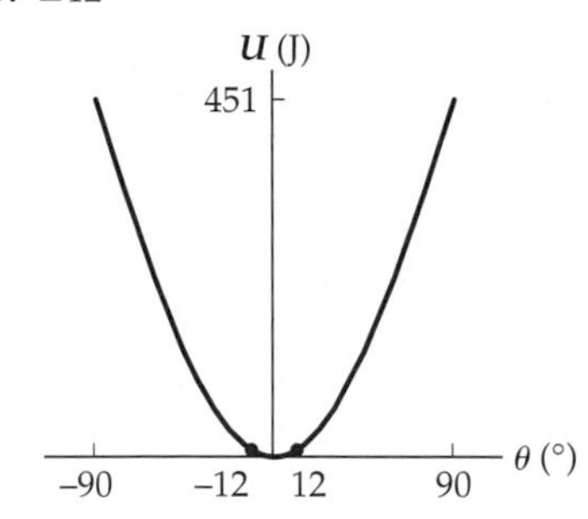

41. $\pm v_{max}\sqrt{\dfrac{m}{k}}$
43. 3.68 m
45. 190 N/m
47. **(a)** 24.0 MJ **(b)** 218 m/s
49. 1.04 m
51. **(a)** 634 N **(b)** decrease
53. 7.20 m/s
55. 0.14 m
57. 975 N
59. **(a)** ±0.081 m **(b)** $1/\sqrt{2}$
61. 48 cm
65. 48.2°
67. 0.46 m/s
69. 68.8 cm

Chapter 9

1. 2.49×10^5 mi/h
3. 73.3°; 0.745 m/s
5. 1.05 m
7. 2.9 kN
9. 0.3 kg
11. 2.7 kg · m/s
13. 150°; 5.5 kg · m/s
15. 31 kg
17. The piece with the smaller kinetic energy has the larger mass.
19. **(a)** 2.2 m/s **(b)** greater than **(c)** 2.3 m/s
21. $\dfrac{1}{4}mv^2$
23. 1.38 m/s $\hat{\mathbf{x}}$ + 2.38 m/s $\hat{\mathbf{y}}$
25. **(a)** 100 m/s **(b)** less than
(c) 850 J; 47 J
27. 3.71 cm
29. $v_{truck} = 5.56$ m/s; $v_{car} = 20.1$ m/s
31. 1.31 m/s; 74.8°
33. **(a)** −17 m/s
(b) Kinetic energy has been transferred from the elephant to the ball.
35. The half gallon of milk should be placed 0.23 m from the center of the basket, opposite the cartons of cereal.
37. $L/10$ units below the center of the box
39. $(0, 3.6 \times 10^{-11}$ m$)$

41. $Y_{cm} = (0.207\ m/s^2)t^2$

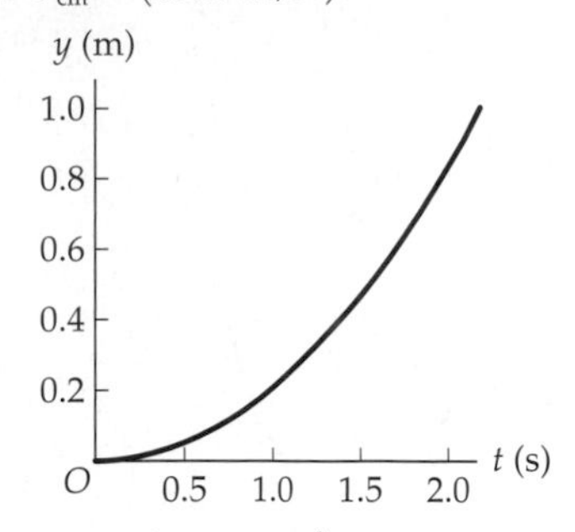

$V_{cm} = (0.414\ m/s^2)t$

V_{cm} (m/s)

1.0 0.8 0.6 0.4 0.2 O 0.5 1.0 1.5 2.0 t (s)

43. **(a)** 13.2 N **(b)** 13.2 N
45. **(a)** 0.333 m/s **(b)** 0.333 m/s **(c)** 0.0776 J; 0.0568 J
47. 900 kg/s
49. 0.503 m/s
51. **(a)** 12.8 N **(b)** 12.7 N
53. 6.51×10^{-21} ft
55. 2.16 m
57. 0.24 N
59. 0.354 mm
61. **(a)** greater than **(b)** 3.6 m
63. **(a)** 13.2 N **(b)** 12.9 N
65. $(6.5 \times 10^{-12}\ m, 0)$
67. $2m$
73. 141°

Chapter 10

1. $\frac{\pi}{6}; \frac{\pi}{4}; \frac{\pi}{2}; \pi$
3. **(a)** 1 rev/min **(b)** 1 rev/h
5. tire, propeller, drill
7. 190 rad/s
9. 1.9×10^{-6} rev/min
11. 110 min
13. 5.1 rad
15. **(a)** 3.0 rev **(b)** 2.2 rev
17. just past 3:08
19. $-4.01 \times 10^{-22}\ rad/s^2$
21. **(a)** $-49\ rev/s^2$ **(b)** 510 ft **(c)** 17 in.
23. 0.012 mm/s
25. **(a)** 1.18 rad/s **(b)** $10.0\ m/s^2$
27. **(a)** $0.50\ m/s^2$ **(b)** $0.50\ m/s^2$
29. **(a)** 69.7 cm/s **(b)** It would also double.
31. $2.12\ m/s^2$; 170°
33. **(a)** 69 m/s **(b)** 7.2×10^{-5} s
35. **(a)** 4.12×10^{16} rad/s **(b)** 6.56×10^{15} orbits **(c)** $8.98 \times 10^{22}\ m/s^2$
37. 81 rad/s
39. 29 rad/s
41. **(a)** $2.0\ rad/s^2$ **(b)** greater than
43. 4.81 J
45. **(a)** 1.19 J **(b)** 1.19 J **(c)** 2.39 J
47. $K_t = 120$ J; $K_r = 0.069$ J
49. -2.34×10^{12} W, -3.14×10^9 hp
51. $0.17\ kg \cdot m^2$
53. **(a)** 3.8 m/s **(b)** 3.3 m/s
55. 0.54 m
57. **(a)** 3.0 m/s **(b)** decrease
59. **(a)** 15 J **(b)** 4.9 J **(c)** 9.8 J
61. 1.2 rev/s
63. 910 m
65. **(a)** 22 m/s **(b)** 200 m/s **(c)** 63 cm **(d)** 6.9 N
67. about 2:44 P.M.
69. 0.30 rev
71. **(a)** 72 s **(b)** 150° ccw from North
73. **(a)** $7.9\ rad/s^2$ **(b)** It doubles.
75. **(a)** 0.334 m **(b)** 1.0 kN
77. **(a)** $15\ rad/s^2$ **(b)** 4.0×10^3 rad
79. **(a)** 0.50 m **(b)** 2.4 rev
81. **(a)** 1.5 m **(b)** 2.7 rev

Chapter 11

1. 60 N
3. **(a)** $8.30\ N \cdot m$ **(b)** $7.80\ N \cdot m$
5. **(a)** $-2.14\ N \cdot m$ **(b)** downward **(c)** increase
7. $11\ rad/s^2$
9. $0.0072\ N \cdot m$
11. **(a)** $0.25\ N \cdot m$ **(b)** decrease
13. **(a)** $7.4\ rad/s^2$ **(b)** $5.1\ rad/s^2$ **(c)** $3.0\ rad/s^2$
15. **(a)** $54\ rad/s^2$ **(b)** 0.092 m
17. 61 N
19. **(a)** 90.3 N **(b)** 13.2 N; The biceps' force is much greater.
21. 1.1 m
23. 7.85 cm
25. 3.1 kN; -2.1 kN
27. **(a)** 32 N **(b)** 32 N; 30 N
29. 900 N; 170 N; 170 N
31. **(a)** 2.26 N perpendicular to the meter stick and away from the bowling ball **(b)** 1.14 N, 7.38° from the horizontal
33. **(a)** 1.10 m **(b)** 76.8 N
35. 0.10 m
37. $(11/12)L$
39. **(a)** $8.68\ m/s^2$ **(b)** $71.7\ rad/s^2$ **(c)** 9.77 m
41. 213 kg
43. $14\ m/s^2$
45. $\left(\frac{m_2 - m_1}{m_1 + m_2 + \frac{1}{2}M}\right)g$
47. $0.0067\ kg \cdot m^2/s$
49. **(a)** $235\ kg \cdot m/s$ **(b)** $1.17 \times 10^3\ kg \cdot m^2/s$
51. **(a)** 0 **(b)** $2.55 \times 10^3\ kg \cdot m^2/s$ **(c)** $2.55 \times 10^3\ kg \cdot m^2/s$
53. $210\ N \cdot m$
55. 0.573
57. 2
59. 2.84 rad/s
61. 0.37 rad/s
63. **(a)** faster **(b)** 3.5 rad/s
65. **(a)** $\frac{Iv}{I + mR^2}$ **(b)** As $I \to 0$, $v_g \to 0$; as $I \to \infty$, $v_g \to v$.
67. $0.076\ N \cdot m$
69. 21 J
71. **(a)** 34 J **(b)** 38 J **(c)** 72 J
73. **(a)** $11\ N \cdot m$ **(b)** 270 rad/s
75. 3.0 m
77. **(a)** The force from the index finger. **(b)** 0.49 N; 0.25 N
79. **(a)** 229 N **(b)** 216 N; 78.5 N
81. $F_1 = -(0.59\ kN/m)x + 0.33$ kN; $F_2 = (0.59\ kN/m)x + 1.4$ kN
83. **(a)** increase **(b)** 261 s
85. **(a)** 7.5 N
89. $3.87Mg$
91. $(2/3)L$

Chapter 12

1. **(a)** 1.1×10^{-11} N **(b)** 2.7×10^{-12} N
3. **(a)** 2.5 kN **(b)** 81 N
5. **(a)** 2.8×10^{-12} N **(b)** 2.8×10^{-12} N
7. **(a)** 3.56×10^{22} N toward the Sun **(b)** 2.40×10^{20} N toward the Sun **(c)** 3.58×10^{22} N toward the Earth-Moon system
9. 3.58×10^{22} N, 0.00178° toward the Moon off the ray from the Sun to the Earth
11. **(a)** 4.7×10^{-8} N, 74° below horizontal, down and to the left **(b)** All forces are reduced by a factor of $2^2 = 4$; the directions of the forces are unchanged.
13. **(a)** $3.70\ m/s^2$ **(b)** $8.88\ m/s^2$
15. 5.5×10^{-8} N
17. $1.36\ m/s^2$
19. $(1/96)M_E$
21. 1.98 h
23. 1.4×10^{11} m
25. 1.90×10^{27} kg
27. **(a)** 12 h **(b)** 3.9 km/s
29. **(a)** 3.98 h **(b)** 7.31 h
31. 1.91×10^{30} kg
33. -4.94×10^8 J
35. **(a)** -3.12558×10^8 J **(b)** -3.12509×10^8 J **(c)** 4.9×10^4 J
37. **(a)** 8.2×10^{10} J **(b)** 1.8×10^{12} J
39. 5.04 km/s
41. 7.91 km/s
43. **(a)** 4.25 km/s **(b)** 10.4 km/s
45. 1.73 km/s
47. It is 10 times that of the Earth.
49. 2.96 km
51. $\sqrt{\frac{3gh}{4\pi G\rho}}$
53. **(a)** 1.98×10^{20} N **(b)** 4.36×10^{20} N **(c)** It makes more sense to think of the moon as orbiting the sun, with a small effect due to the earth.
55. **(a)** less than **(b)** $4.91\ m/s^2$
59. $\sqrt{\frac{192\pi^2 r_1^3}{Gm_1}}$
61. 8.69×10^7 m
63. 232 m/s
67. **(a)** No; m drops out of the expression for v. **(b)** 7.8 km/s **(c)** 1.5 h
69. $2\sqrt[4]{3}\pi\sqrt{\frac{R^3}{GM}}$

71. $-\dfrac{(1.17 \times 10^{10}\ \text{N}\cdot\text{m}^2)x}{[x^2 + (1500\ \text{m})^2]^{3/2}}$

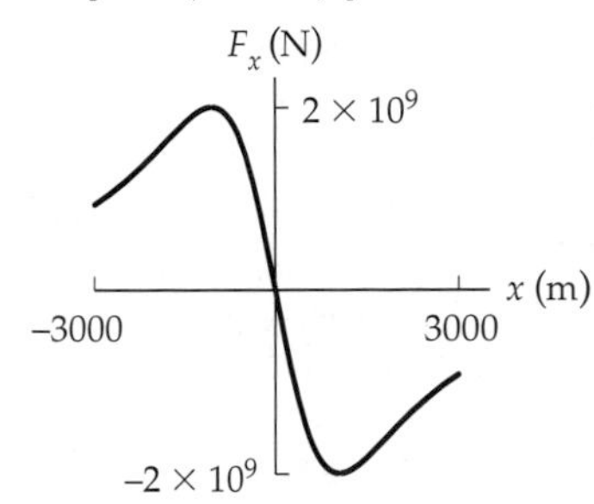

73. **(a)** $G\dfrac{m^2}{4a^2}$ **(b)** $\left(\dfrac{16a^3 M_E}{m}\right)^{1/3}$
(c) 2.8×10^7 m; 1.3×10^8 m

Chapter 13

1. 13 s; 0.075 Hz
3. 0.36 s
5. 0.81 s; 1.2 Hz
7. **(a)** 0.024 s; 42 Hz **(b)** 1800 rpm
9. **(a)** 1.3 Hz **(b)** 0.39 s
11. $x = (3.00\ \text{nm})\cos[(6.00\pi \times 10^{14}\ \text{s}^{-1})t]$
13. **(a)** 0.88 s **(b)** −1.4 cm
15. $T/3$
17. **(a)** $v_{max}{}^2/a_{max}$ **(b)** $2\pi v_{max}/a_{max}$
19. **(a)** 17 m **(b)** 32 s
21. $0.0484g$
23. **(a)** 1100 m/s^2 **(b)** 6.2 m/s
25. **(a)** The rider must begin hanging on when a_{max} (at the top of the cycle) equals g, the downward acceleration due to gravity. **(b)** 0.14 m
27. 27 N/m
29. 0.47 kg
31. **(a)** 1060 kg **(b)** 932 kg
33. 0.0799 m
35. **(a)** more than **(b)** $\sqrt{2}T$
37. 1.37 m/s
39. 0.97 J
41. **(a)** 0.26 m/s **(b)** 2.8 cm
43. 0.41 m/s
45. **(a)** 6.43 m/s **(b)** 0.187 s
47. 9.6 m/s^2
49. 24.8 cm
51. $2\sqrt{2}\pi\sqrt{\dfrac{R}{g}}$
53. 2/3 m
55. 1.2 s; 0.86 m/s
57. 0.58 m/s
59. $(4 \times 10^{13})g$
61. $T/6$
63. **(a)** increase **(b)** 0.96 s
65. 3
67. 1100 kg/s
69. **(a)** 28 N/m **(b)** 0.19 kg
71. **(a)** less than **(b)** $\pi\sqrt{\ell/g} + \pi\sqrt{L/g}$ **(c)** 1.5 s
73. $\dfrac{1}{2\pi}\sqrt{\dfrac{g}{A}}$

Chapter 14

1. **(a)** 50 cm **(b)** 7.5 cm
3. 1.2 m
5. **(a)** 6.0 m **(b)** 9.6 m
7. 0.51 m/s; 67 Hz
9. 0.19 s
11. **(a)** 0.70 N **(b)** larger **(c)** 0.83 N
13. $v = \sqrt{\dfrac{T}{\pi R^2 \rho}}$
15. $y = (0.12\ \text{m})\cos\left(\dfrac{\pi}{1.4\ \text{m}}x + \dfrac{\pi}{0.11\ \text{s}}t\right)$
17. **(a)**

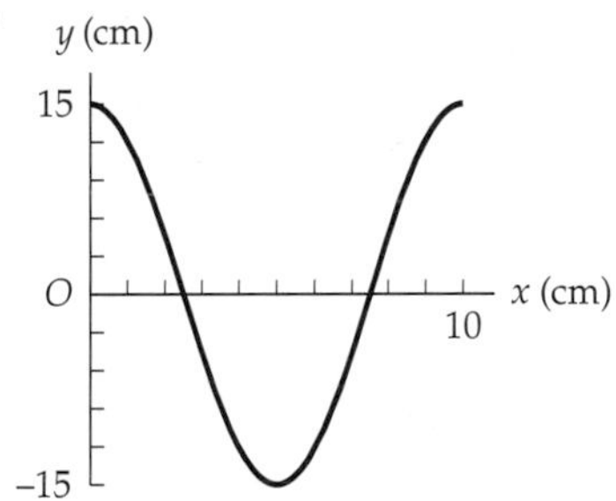

(b)

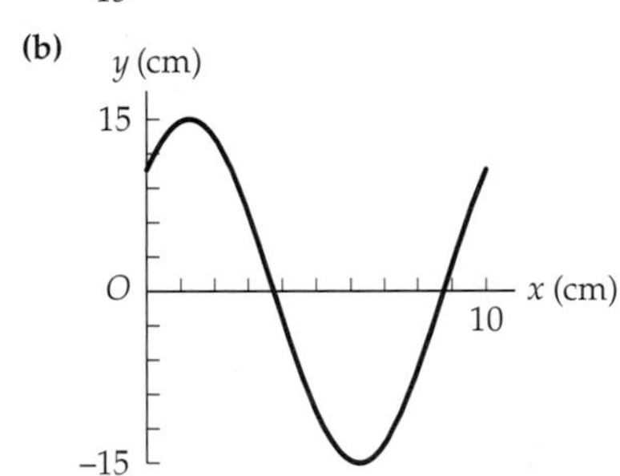

(c)

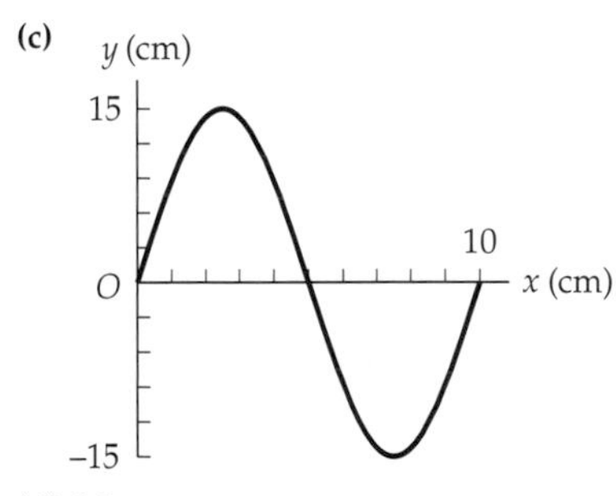

(d) 6.0 s
19. 360 m
21. 27.5 Hz, 12.5 m; 4.19 kHz, 8.18 cm
23. **(a)** 11 m **(b)** less than
25. 1.00 mW/m^2
27. 360 W/m^2
29. **(a)** 104 dB **(b)** 99.6 dB **(c)** 2.0×10^6 m
31. 67 dB
33. $10\sqrt{3}$ m
35. 133 Hz
37. **(a)** 35.4 kHz **(b)** higher **(c)** 35.5 kHz
39. 535 Hz
41. 34.3 m/s
43. **(a)** 320 Hz **(b)** Bicyclist A speeding up.
45. 11.2 m/s
47.

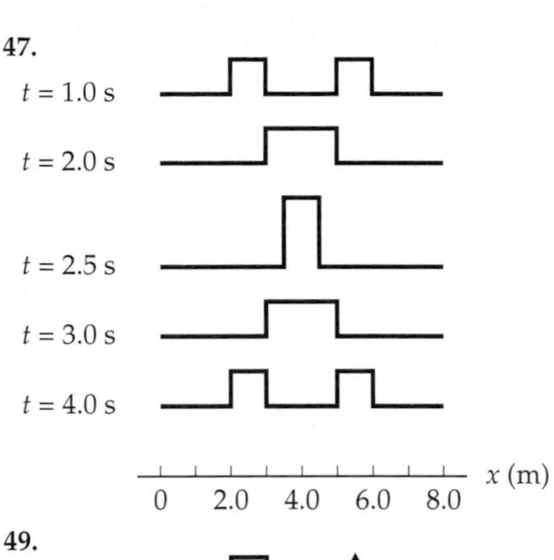

49.

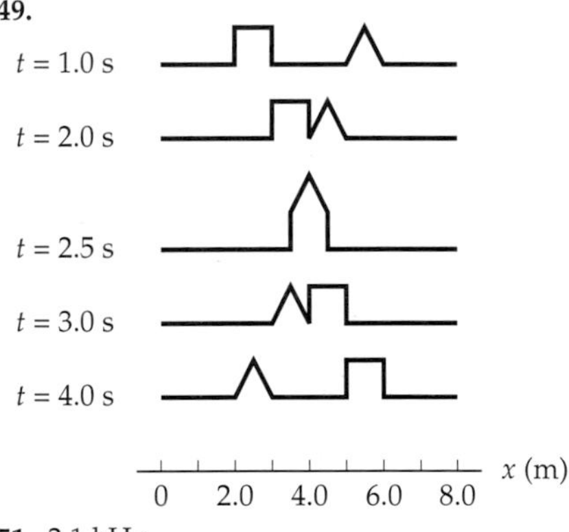

51. 2.1 kHz
53. 0.335 m
55. 0.31 kHz
57. 38 Hz
59. 3.1 kHz
61. **(a)** 7.24 Hz **(b)** 14.5 Hz
63. **(a)** 0.10 kHz **(b)** 34 Hz
65. **(a)** 82 Hz **(b)** 2.1 m
67. 264 Hz or 258 Hz
69. **(a)** 13.8 Hz **(b)** increase **(c)** 18.3 Hz
71. 1.13 m
73. 1.7 km
75. 4
77. 546 Hz
79. 32 m/s
81. 6.13 m/s; 6.35 m/s
83. **(a)** 0.41 N/m **(b)** stay the same **(c)** 0.75 s
85. 6.28 Hz
87. $v = \sqrt{gd}$
89. **(a)** 257 Hz, 255 Hz **(b)** 2 Hz **(c)** 0.670 m **(d)** 2 times per second

Chapter 15

1. 10^3 N
3. The ring is not solid gold.
5. 1.05×10^4 kg/m^3; silver
7. **(a)** 10^{-1} Pa **(b)** 10^{-6} atm
9. 0.23
11. 615 N
13. 2.27×10^6 Pa
15. 10 m
17. 1.1 N
19. 3.1 kN
21. 1.02×10^5 Pa
23. 0.16 cm
25. **(a)** 0.770 m **(b)** Increased; the gauge pressure in a fluid is proportional to its density.

27. 7.2 kN
29. 5.7 kN
31. $\rho_w V_{metal} g$
33. **(a)** 2.34×10^{-4} m^3 **(b)** 8.70×10^3 kg/m^3
35. **(a)** 1060 kg/m^3 **(b)** 0.0741 m^3 **(c)** 41 N
37. **(a)** 0.0080 m^3 **(b)** 18 N
39. 2.1 cm
41. 44 m/s
43. 1.2×10^3 s
45. 21 m/s
47. **(a)** 32 cm/s **(b)** 40 Pa
49. **(a)** 53 m/s **(b)** 0.31 cm
51. 7.4 kPa
53. **(a)** 1.42 kPa **(b)** 45 kN
55. 762 kN upward
57. **(a)** less than **(b)** $\Delta P = -\frac{15}{2}\rho v^2$
59. 160 kN
61. **(a)** -0.15 m **(b)** 0.063 m
63. 15 m/s
65. **(a)** 5.3 N
67. **(a)** 1.04×10^5 Pa **(b)** increases **(c)** 1.05×10^5 Pa
69. 41 km
71. **(a)** $v = \sqrt{2gd}$ **(b)** the same
73. **(a)** 9.38×10^{-3} m^3 **(b)** 9.85 kg
75. 320 kPa
77. 0.67
79. 6.7 kg
83. $2\pi\sqrt{\frac{\rho_1 H}{\rho_2 g}}$

Chapter 16

1. **(a)** 37.0°C **(b)** 310.2 K
3. **(a)** 5727°C **(b)** 10,340°F
5. **(a)** 81.6 kPa **(b)** 182°C
7. 0.23 K/s
9. -6.85°C
11. 1.6 m
13. **(a)** heated **(b)** 430°C
15. 0.93 L
17. **(a)** aluminum **(b)** $0.0015V_0$
19. **(a)** The water will overflow. **(b)** 24 cm^3
21. 120 W or 0.16 hp
23. 5.9×10^{-4}°C
25. 25.5°C
27. 11 kJ
29. 350
31. **(a)** 0.18 kJ/°C **(b)** 0.96 kJ/kg · °C
33. 385 J/kg · °C, copper
35. 70.5°C
37. 2.3 J
39. 0.408 m
41. **(a)** greater than **(b)** 71.0°C
43. 39°C
45. **(a)** $(9/4)P$ **(b)** $9P$
47. mass and/or specific heat
49. **(a)** increase **(b)** 0.1710 cm **(c)** 1.955 s, 1.957 s
51. 350°C
53. 0.17°C
55. **(a)** 1.54×10^6 m **(b)** 5.51 km/s
59. **(a)** decreased **(b)** -83 K

Chapter 17

1. 0.0224 m^3
3. 315 K
5. 1.33×10^3 kg
7. 10^{-15} Pa
9. 4.2×10^{23} atoms
11. 15.6 g/mol
13. 160 kPa
15. 0.830
17. (e)
19. 630 m/s
21. **(a)** The rms speed of H_2O is more than the speed of O_2, since it is less massive but has the same average kinetic energy. **(b)** 2.07 km/s
23. 356 K
25. $v_{238}/v_{235} = 0.996$
27. 35 kg
29. 1.1×10^6 N/m^2
31. **(a)** 0.27 cm **(b)** The diameter should be increased because a wire's cross-sectional area and its elongation are inversely related.
33. 1.1 mm
35. -9.2×10^{-5} m^3
37. (d)
39. about 14°C
41. about -28°C
43. **(a)** gas **(b)** solid **(c)** 5707 kPa
45. 210 kJ
47. 250 kJ
49. 0°C; 0.36 kg
51. 18.8°C
53. 123°C; All of the water has vaporized.
55. 49.1 h
57. **(a)** 0.196 m^3 **(b)** 67.5 kPa
59. **(a)** 0.040 **(b)** 0.040 **(c)** 0.046
61. **(a)** 0.16 mm **(b)** 0.29 mm (including the radius of the ball)
63. **(a)** 232 m/s **(b)** $(v^2)_{av}$ will be greater than $(v_{av})^2$ because, in general, $v_{rms} > v_{av}$. **(c)** 6.88×10^4 m^2/s^2, 6.88×10^4 m^2/s^2 > (232 m/s)^2
65. **(a)** 5.50 kJ **(b)** 48.1 kJ, $Q_v > U$
67. 6.6×10^8 Pa
69. **(a)** 23 W **(b)** 23 h
71. 15 s

Chapter 18

1. -8.9×10^5 J, 5.7×10^5 J, -3.2×10^5 J
3. **(a)** 119 J **(b)** 35 J **(c)** 0
5. **(a)** 0 **(b)** -53 J **(c)** -150 J **(d)** 200 J **(e)** 350 J
7. 4.24 MJ/mi
9. **(a)** -462 kJ **(b)** 110 kcal
11. 1.3×10^{-2} m^3
13. 0.315
15. **(a)** 2.6 MJ **(b)** 1100 K **(c)** 5.0 MJ
17. **(a)** 14.6 kJ **(b)** 14.6 kJ
19. **(a)** 11 kJ **(b)** 11 kJ
21. **(a)** 18.7 kJ **(b)** 0.185 m^3
23. **(a)** 240 kJ **(b)** 390 K, 980 K **(c)** 360 kJ **(d)** 600 kJ
25. **(a)** 130 kJ **(b)** -89 kJ
27. 350 K
29. **(a)** The area under the constant-pressure curve is the greatest, so that process does the most work. **(b)** The area under the adiabatic curve is the smallest, so that process does the least work. **(c)** Constant-pressure expansion, because for a given final volume, temperature varies with pressure. **(d)** Adiabatic expansion, since at final volume, temperature varies with pressure.
31. **(a)** 1.0 K **(b)** 0.61 K
33. **(a)** 2.2 K **(b)** 3.7 K
35. **(a)** 0.45 kJ **(b)** equal to **(c)** 0.45 kJ
37. **(a)** 0.0653 m^3 **(b)** 0.0867 m^3
39. **(a)** 530 kJ, 477 kJ **(b)** 212 kJ **(c)** 795 kJ
41. 0.28
43. **(a)** 5.1 kJ **(b)** 3.6 kJ
45. **(a)** 1.16 GW **(b)** 1.71 GW
47. **(a)** 23 kJ **(b)** 20 kJ
49. **(a)** 382 K **(b)** The efficiency of a heat engine increases as the difference in temperature of the hot and cold reservoirs increases. Therefore, the temperature of the low temperature reservoir must be decreased. **(c)** 327 K
51. 450 K, 500 K
53. **(a)** 550 J **(b)** 0.31
55. **(a)** 58.0 J **(b)** 1610 J
57. 0.41 kW
59. 150 kJ
61. 7.57 kJ/K
63. 9.6 W/K
65. **(a)** The entropy of the universe stays the same because the air conditioner has Carnot efficiency. **(b)** 0
67. **(a)** 318 kJ **(b)** 795 kJ **(c)** 1113 kJ
69. **(a)** 751 kJ **(b)** 194 kJ **(c)** 945 kJ
71. 38 kJ
73. **(a)** 0.896 **(b)** 0.701, Don't invest.
75. **(a)** The entropy decreases because heat flows out of the water. **(b)** -0.63 kJ/K **(c)** Wherever the heat flows, there will be an increase in entropy.
77. 2/5
81. Lowering the temperature of the low temperature reservoir will result in a greater change in efficiency.

Photo Credits

Chapter 1—p. 1 Frans Lanting/Minden Pictures **p. 3 (left)** Oliver Meckes/E.O.S./MPI-Tubingen/Photo Researchers, Inc. **p. 3 (right)** National Optical Astronomy Observatories/Association of Universities for Research in Astronomy **p. 4 (margin)** International Bureau of Weights and Measures **p. 4 (left)** U.S. Department of Commerce **p. 4 (right)** © 2000 by Sidney Harris **p. 5** PhotoDisc, Inc., **p. 7** Don Smetzer/Stone **p. 8** AP/Wide World Photos **p. 9** David Frazier Photolibrary/Photo Researchers, Inc. **p. 10** CNRI/Science Photo Library/Photo Researchers, Inc. **p. 11** Amaldi Archives, Dipartimento di Fisica, Universita "La Sapienza," Rome, courtesy AIP Emilo Segre Visual Archives **p. 14** Kevin Schafer/Stone **p. 15 (left)** Michael Mancuso/Omni-Photo Communications, Inc. **p. 15 (right)** University of Texas at Austin

Chapter 2—p. 16 Michael Probst/AP/Wide World Photos **p. 22** Paul Silverman/Fundamental Photographs **p. 25** NASA **p. 28** Chris Urso/AP/Wide World Photos **p. 38 (top)** James Sugar/Black Star **p. 38 (bottom)** Chris Arend/Stone **p. 42** Douglas Peebles Photography **p. 49** Worlds of Fun Photo by Dan Feicht **p. 50 (left)** Stephen Dalton/Photo Researchers, Inc. **p. 50 (right)** Sidney Harris **p. 51** Frank Labua/Pearson Education **p. 52** Nancy Wolff/Omni-Photo Communications, Inc.

Chapter 3—p. 53 Mark Downey/PhotoDisc, Inc. **p. 54** Jeff Greenberg/Photo Researchers, Inc. **p. 61** Art Wolfe/Stone **p. 65** George Whiteley/Photo Researchers, Inc. **p. 68** Lauren Rebours/AP/Wide World Photos **p. 71** Thomas Kitchin/Tom Stack & Associates **p. 73** U.S. Air Force photo by Master Sergeant Joe Cupido.

Chapter 4—p. 71 Stephen Dalton/Photo Researchers, Inc. **p. 79** Tom & Pat Leeson/Photo Researchers, Inc. **p. 82 (left)** Loren M. Winters, North Carolina School of Science and Mathematics **p. 82 (right)** NASA **p. 85 (top)** Elaine Mayes/Bruce Coleman Inc. **p. 85 (bottom)** John Wang/PhotoDisc, Inc. **p. 88 (left)** Michael Quinton/Minden Pictures **p. 88 (center)** Martha Swope/Time Life Syndication **p. 88 (right)** Galen Rowell/Mountain Light/Explorer/Photo Researchers, Inc. **p. 92** Courtesy of the Archives, California Institute of Technology **p. 95** Stephen Dalton/Photo Researchers, Inc. **p. 98** Gregory G. Dimijian/Photo Researchers, Inc. **p. 102** NASA **p. 103** Keith Gunnar/FPG International LLC

Chapter 5—p. 104 Klaus Lahnstein/Stone **p. 106** James S. Walker **p. 108 (top)** Dave Lawrence/The Stock Market **p. 108 (bottom)** Bruce Bennett Studios, Inc. **p. 114** William J. Hughes Technical Center, Federal Aviation Administration **p. 116** Ken Fisher/Stone **p. 122** David Madison/Stone **p. 126** NASA **p. 131** Norbert Wu/DRK Photo **p. 136** Michel Euler/AP/Wide World Photos

Chapter 6—p. 137 Stuart McClymont/Stone **p. 138 (top)** Kelly-Mooney Photography/Corbis **p. 138 (bottom)** Ernest Braun/Stone **p. 143** Jack Dykinga/Stone **p. 145 (left)** Index Stock Imagery, Inc. **p. 145 (right)** W. Kenneth Hamblin **p. 146** Ph. Bourseiller/Hoaqui/Photo Researchers, Inc. **p. 153** Galen Rowell/Corbis **p. 158** Doug Wilson/Corbis **p. 160 (left)** AP/Wide World Photos **p. 160 (center)** Official U.S. Navy photo by Mark Meyer. **p. 160 (right)** Adtranz Sweden **p. 162** Chris Priest/Science Photo Library/Photo Researchers, Inc. **p. 165** Photofest **p. 169** Color Box/FPG International LLC

Chapter 7—p. 173 Fotopic/Omni-Photo Communications Inc. **p. 174** Don Tremain/PhotoDisc, Inc. **p. 176 (left)** Robert E. Daemmrich/Stone **p. 176 (right)** Mimmo Jodice/Corbis **p. 186** DaimlerChrysler Rail Systems Adtranz **p. 190** Philippe Achache/Liaison Agency, Inc. **p. 197** John E. Bortle

Chapter 8—p. 198 Lawrence M. Sawyer/PhotoDisc, Inc. **p. 203 (left)** Grant Heilman/Grant Heilman Photography, Inc. **p. 203 (right)** Jay Brousseau/The Image Bank **p. 204** Philip H. Coblentz/Stone **p. 206** Hulton-Deutsch Collection/Corbis **p. 208 (top)** Bill Nation/Corbis/Sygma **p. 208 (bottom)** Salt River Project, Phoenix, Arizona **p. 214** Jeff Greenberg/Omni-Photo Communications, Inc. **p. 217** Alex L. Fradkin/PhotoDisc, Inc. **p. 211** Federal Motor Carrier Safety Administration, Federal Highway Administration, Colorado Division **p. 226 (left)** Stephen Derr/The Image Bank **p. 226 (right)** Rob Badger/Rob Badger Photography

Chapter 9—p. 232 Stephen Dunn/Allsport Photography (USA), Inc. **p. 238 (top)** © Harold & Esther Edgerton Foundation, 1999, courtesy of Palm Press, Inc. **p. 238 (bottom)** Wally McNamee/Corbis **p. 239** J. Scott Altenbach, University of New Mexico **p. 242** NASA **p. 244** NASA **p. 246 (left)** J. McIsaac/Bruce Bennett Studios, Inc. **p. 246 (right)** Franck Seguin/Corbis **p. 253** Richard Megna/Fundamental Photographs **p. 256** Alexander Calder, *Antennae with Red and Blue Dots*, mobile, 1960. © ARS, N.Y. Tate Gallery, London/Art Resource, N.Y **p. 260** Loren M. Winters, North Carolina School of Science and Mathematics. **p. 262 (bottom)** Fred Bavendam/Peter Arnold, Inc. **p. 262 (top)** NASA **p. 265** © Jose Azel/Aurora/PNI **p. 269** Michael Rinaldi/U.S. Navy News Photo

Chapter 10—p. 271 Hisham F. Ibrahim/PhotoDisc, Inc. **p. 272 (left)** NASA **p. 272 (center)** Novastock/PhotoEdit **p. 272 (right)** Barry Dowsett/Photo Researchers, Inc. **p. 274** Fred Espenak/Photo Researchers, Inc. **p. 281 (left)** Loren M. Winters, North Carolina School of Science and Mathematics **p. 281 (right)** Karl Weatherly/PhotoDisc, Inc. **p. 282 (left)** SovFoto/EastFoto **p. 282 (right)** Bill Gallery/Stock Boston **p. 285** Richard Megna/Fundamental Photographs **p. 289** Mehau Kulyk/Science Photo Library/Photo Researchers, Inc. **p. 298 (left)** Jeff Hester/Arizona State University/NASA **p. 298 (right)** A. Ramey/Stock Boston **p. 301** Loren M. Winters, North Carolina School of Science and Mathematics **p. 302** Yomega

Chapter 11—p. 303 Kurt Stier/Corbis **p. 304 (left)** Michael Newman/PhotoEdit **p. 306** Index Stock Imagery, Inc. **p. 310** Richard Megna/Fundamental Photographs **p. 315** PhotoDisc, Inc., **p. 319** Murray Close/Paramount Pictures/Photofest **p. 320 (left)** Dana White/PhotoEdit **p. 320 (right)** Allan Price/Omni-Photo Communications, Inc. **p. 327** Glyn Kirk/Stone **p. 328 (left)** Photo by Arthur Harvey, courtesy of the Miami Herald. **p. 328 (right)** Hasler, Pierce, Palaniappan, Manyin/NASA Goddard Laboratory for Atmospheres **p. 329** Chandra X-Ray Center/Smithsonian Astrophysical Observatory/NASA/Marshall Space Flight Center **p. 334 (left)** Pearson Education **p. 334 (right)** AIP Emilio Segre Visual Archives **p. 335** Gravity Probe B at Stanford University

Chapter 12—p. 345 Courtesy of NASA/JPL/Caltech **p. 346** The Burndy Library, Dibner Institute for the History of Science and Technology, and the Grace K. Babson Collection of the Works of Sir Isaac Newton. **p. 350** Graphics by Jim Frawley, Raytheon/STX; model developed by NASA Goddard Space Flight Center, Space Geodesy Branch, National Imagery and Mapping Agency, and Ohio State University. **p. 353 (top)** NASA **p. 353 (bottom)** Phil Degginger/NASA/JPL **p. 358 (left)** NASA **p. 358 (right)** JPL/NASA **p. 360** NASA **p. 365 (left)** E.R. Degginger/Color-Pic, Inc. **p. 365 (right)** © Mark Pilkington/Geological Survey of Canada/Science Photo Library/Photo Researchers, Inc. **p. 368** David Nunuk/Science Photo Library/Photo Researchers, Inc. **p. 373** W. Kenneth Hamblin **p. 377** NASA

Chapter 13—p. 380 Loren M. Winters, North Carolina School of Science and Mathematics **p. 381** Index Stock Imagery, Inc. **p. 391** NASA **p. 394 (left)** Loren M. Winters, North Carolina School of Science and Mathematics **p. 394 (right)** Terje Rakke/The Image Bank **p. 402** GeoSpectra Corporation **p. 404 (left)** Globe Photos, Inc. **p. 404 (right)** Thomas D. Mangelsen/Peter Arnold, Inc. **p. 405** Taylor Devices, Inc. **p. 407 (left)** © Phil Schermeister/Corbis **p. 407 (right)** Farqhuarson/University of Washington Libraries **p. 410** Harmony On-Line

Chapter 14—p. 415 Andy Sacks/Stone **p. 416 (top)** Corbis Digital Stock **p. 415 (bottom)** Index Stock Imagery, Inc. **p. 423** Leonard Lessin/Peter Arnold, Inc. **p. 426 (left)** Merlin D. Tuttle/Bat Conservation International/Photo Researchers, Inc. **p. 426 (right)** Katherine L. Welsher **p. 427 (left)** Eurelios/Phototake NYC **p. 427 (right)** Visuals Unlimited **p. 438 (left)** Department of Atmospheric Sciences, University of Illinois at Urbana-Champaign **p. 438 (right)** Michal Heron/Pearson Education **p. 443** Richard Megna/Fundamental Photographs **p. 445 (left)** Neal Preston/Corbis **p. 445 (right)** Index Stock Imagery, Inc. **p. 449 (left)** PhotoEdit **p. 449 (right)** Wada Mitsumiro/ Liaison Agency, Inc.

Chapter 15—p. 460 Richard Megna/Fundamental Photographs **p. 463 (top)** Hamman/Heldring/Animals Animals/Earth Scenes **p. 463 (left)** Richard Megna/Fundamental Photographs **p. 463 (right)** Michal Heron/Pearson Education **p. 464** Michael Newman/PhotoEdit **p. 468** Richard Megna/Fundamental Photographs **p. 470** James King-Holmes/Photo Researchers, Inc. **p. 474** Life Measurement Instruments **p. 476 (left)** E.R. Degginger/Color-Pic, Inc. **p. 476 (right)** Richard Rowan/Photo Researchers, Inc. **p. 476 (bottom)** Carl Purcell/Photo Researchers, Inc. **p. 477** Brock May/Photo Researchers, Inc. **p. 479** Kent Knudson/Index Stock Imagery, Inc. **p. 485 (left)** Tom Myers/Photo Researchers, Inc. **p. 485 (right)** Tom McHugh/Photo Researchers, Inc. **p. 489** Robert B. Suter, Vassar College **p. 494** Leonard Lessin/Peter Arnold, Inc.

Chapter 16—p. 498 Ray Nelson/Phototake NYC **p. 504** Poehlmann/ Mauritius, GmbH/Phototake NYC **p. 506** SuperStock, Inc. **p. 507 (left)** Bob Daemmrich/Stock Boston **p. 507 (right)** Joe Sohm/The Image Works **p. 511** Phil Degginger/Color-Pic, Inc. **p. 517 (left)** Michael Fogden/Animals Animals/Earth Scenes **p. 517 (right)** E.R. Degginger/Color-Pic, Inc. **p. 521** Robert Maier/Animals Animals/Earth Scenes **p. 522** Buff Corsi/Visuals Unlimited **p. 523 (left)** Ron Dahlquist/Stone **p. 523 (right)** Thomas Kitchin/Tom Stack & Associates **p. 529** Jim Ross/NASA/Dryden Flight Research Center

Chapter 17—p. 532 Richard Megnal/Fundamental Photographs **p. 535** Carey Van Loon **p. 538** Tom Bochsler/Pearson Education **p. 549** Woods Hole Oceanographic Institute **p. 553 (left)** Steve McCutcheon/Visuals Unlimited **p. 553 (right)** Malin Space Science Systems/Jet Propulsion Laboratory/NASA **p. 554 (top)** Gerald & Buff Corsi/Visuals Unlimited **p. 554 (left)** Philip Gould/Corbis **p. 554 (right)** Frank Oberle/Stone

Chapter 18—p. 569 Ken McGraw/Index Stock Imagery, Inc. **p. 585** Olaf Soot/Stone **p. 586 (left)** /Sargent-Welch/VWR Scientific Products **p. 586 (right)** Porterfield/Chickering/Photo Researchers, Inc. **p. 591** Stephen Ferry/Liaison Agency, Inc. **p. 598** Jeff Hunter/The Image Bank **p. 599 (top)** Yorgos Nikas/Biol. Reprod./Stone **p. 599 (middle)** Photo Lennart Nilsson/Bonnier Alba AB, A CHILD IS BORN, Dell Publishing Company. **p. 599 (bottom)** Tim Brown/Stone

Index

MULTIPLES AND PREFIXES FOR METRIC UNITS

Multiple	Prefix (Abbreviation)	Pronunciation
10^{24}	yotta- (Y)	yot′ta (*a* as in *a*bout)
10^{21}	zetta- (Z)	zet′ta (*a* as in *a*bout)
10^{18}	exa- (E)	ex′a (*a* as in *a*bout)
10^{15}	peta- (P)	pet′a (as in *peta*l)
10^{12}	tera- (T)	ter′a (as in *terra*ce)
10^{9}	giga- (G)	ji′ga (*ji* as in *ji*ggle, *a* as in *a*bout)
10^{6}	mega- (M)	meg′a (as in *mega*phone)
10^{3}	kilo- (k)	kil′o (as in *kilo*watt)
10^{2}	hecto- (h)	hek′to (*heck-toe*)
10	deka- (da)	dek′a (*deck* plus *a* as in *a*bout)
10^{-1}	deci- (d)	des′i (as in *deci*mal)
10^{-2}	centi- (c)	sen′ti (as in *senti*mental)
10^{-3}	milli- (m)	mil′li (as in *mili*tary)
10^{-6}	micro- (μ)	mi′kro (as in *micro*phone)
10^{-9}	nano- (n)	nan′oh (*an* as in *an*nual)
10^{-12}	pico- (p)	pe′ko (*peek-oh*)
10^{-15}	femto- (f)	fem′toe (*fem* as in *fem*inine)
10^{-18}	atto- (a)	at′toe (as in an*atom*y)
10^{-21}	zepto- (z)	zep′toe (as in *zep*pelin)
10^{-24}	yocto- (y)	yock′toe (as in *sock*)

THE GREEK ALPHABET

Alpha	A	α
Beta	B	β
Gamma	Γ	γ
Delta	Δ	δ
Epsilon	E	ε
Zeta	Z	ζ
Eta	H	η
Theta	Θ	θ
Iota	I	ι
Kappa	K	κ
Lambda	Λ	λ
Mu	M	μ
Nu	N	ν
Xi	Ξ	ξ
Omicron	O	o
Pi	Π	π
Rho	P	ρ
Sigma	Σ	σ
Tau	T	τ
Upsilon	Υ	υ
Phi	Φ	ϕ
Chi	X	χ
Psi	Ψ	ψ
Omega	Ω	ω

SI BASE UNITS

Physical Quantity	Name of Unit	Symbol
Length	meter	m
Mass	kilogram	kg
Time	second	s
Electric current	ampere	A
Temperature	kelvin	K
Amount of substance	mole	mol
Luminous intensity	candela	cd

SOME SI DERIVED UNITS

Physical Quantity	Name of Unit	Symbol	SI Unit
Frequency	hertz	Hz	s^{-1}
Energy	joule	J	$kg \cdot m^2/s^2$
Force	newton	N	$kg \cdot m/s^2$
Pressure	pascal	Pa	$kg/(m \cdot s^2)$
Power	watt	W	$kg \cdot m^2/s^3$
Electric charge	coulomb	C	$A \cdot s$
Electric potential	volt	V	$kg \cdot m^2/(A \cdot s^3)$
Electric resistance	ohm	Ω	$kg \cdot m^2/(A^2 \cdot s^3)$
Capacitance	farad	F	$A^2 \cdot s^4/(kg \cdot m^2)$
Inductance	henry	H	$kg \cdot m^2/(A^2 \cdot s^2)$
Magnetic field	tesla	T	$kg/(A \cdot s^2)$
Magnetic flux	weber	Wb	$kg \cdot m^2/(A \cdot s^2)$

SI UNITS OF SOME OTHER PHYSICAL QUANTITIES

Physical Quantity	SI Unit
Density (ρ)	kg/m^3
Speed (v)	m/s
Acceleration (a)	m/s^2
Momentum, impulse (p)	$kg \cdot m/s$
Angular speed (ω)	rad/s
Angular acceleration (α)	rad/s^2
Torque (τ)	$kg \cdot m^2/s^2$ *or* $N \cdot m$
Specific heat (c)	$J/(kg \cdot K)$
Thermal conductivity (k)	$W/(m \cdot K)$ *or* $J/(s \cdot m \cdot K)$
Entropy (S)	J/K *or* $kg \cdot m^2/(K \cdot s^2)$ *or* $N \cdot m/K$
Electric field (E)	N/C *or* V/m

APPENDICES IN THE TEXT

FUNDAMENTAL CONSTANTS

Quantity	Symbol	Approximate Value
Speed of light	c	3.00×10^{8} m/s $= 3.00 \times 10^{10}$ cm/s $= 186{,}000$ mi/s
Universal gravitational constant	G	$6.67 \times 10^{-11}\ \mathrm{N \cdot m^2/kg^2}$
Stefan-Botzmann constant	σ	$5.67 \times 10^{-8}\ \mathrm{W/(m^2 \cdot K^4)}$
Boltzmann's constant	k	1.38×10^{-23} J/K
Avogadro's number	N_A	$6.022 \times 10^{23}\ \mathrm{mol^{-1}}$
Gas constant	$R = N_A k$	$8.31\ \mathrm{J/(mol \cdot K)} = 1.99\ \mathrm{cal/(mol \cdot K)}$
Coulomb's law constant	$k = 1/4\pi\varepsilon_o$	$8.99 \times 10^{9}\ \mathrm{N \cdot m^2/C^2}$
Electron charge	e	1.60×10^{-19} C
Permittivity of free space	ε_o	$8.85 \times 10^{-12}\ \mathrm{C^2/(N \cdot m^2)}$
Permeability of free space	μ_o	$4\pi \times 10^{-7}\ \mathrm{T \cdot m/A} = 1.26 \times 10^{-6}\ \mathrm{T \cdot m/A}$
Planck's constant	h	$6.63 \times 10^{-34}\ \mathrm{J \cdot s}$
	$\hbar = h/2\pi$	$1.05 \times 10^{-34}\ \mathrm{J \cdot s}$
Atomic mass unit	u	1.66×10^{-27} kg $\leftrightarrow$ 931 MeV
Electron mass	m_e	9.10939×10^{-31} kg $= 5.49 \times 10^{-4}$ u $\leftrightarrow$ 0.511 MeV
Neutron mass	m_n	$1.675\,00 \times 10^{-27}$ kg $= 1.008\,665$ u $\leftrightarrow$ 939.57 MeV
Proton mass	m_p	$1.672\,65 \times 10^{-27}$ kg $= 1.007\,267$ u $\leftrightarrow$ 938.28 MeV

USEFUL PHYSICAL DATA

Acceleration due to gravity (surface of Earth)	$9.81\ \mathrm{m/s^2} = 32.2\ \mathrm{ft/s^2}$
Absolute zero	$0\ \mathrm{K} = -273.15\,°\mathrm{C} = -459.67\,°\mathrm{F}$
Standard temperature & pressure (STP)	$0\,°\mathrm{C} = 273.15$ K 1 atm = 101.325 kPa
Density of air (STP)	$1.29\ \mathrm{kg/m^3}$
Speed of sound in air (20 °C)	343 m/s
Density of water (4 °C)	$1.000 \times 10^{3}\ \mathrm{kg/m^3}$
Latent heat of fusion of water	3.35×10^{5} J/kg
Latent heat of vaporization of water	2.26×10^{6} J/kg
Specific heat of water	$4186\ \mathrm{J/(kg \cdot K)}$

SOLAR SYSTEM DATA*

Equatorial radius of Earth	6.37×10^{3} km = 3950 mi
Mass of Earth	5.97×10^{24} kg
Radius of Moon	1740 km = 1080 mi
Mass of Moon	7.35×10^{22} kg $\approx \frac{1}{81}$ mass of Earth
Average distance of Moon from Earth	3.84×10^{5} km $= 2.39 \times 10^{5}$ mi
Radius of Sun	6.95×10^{5} km = 432,000 mi
Mass of Sun	2.00×10^{30} kg
Average distance of Earth from Sun	1.50×10^{8} km $= 93.0 \times 10^{6}$ mi

*See Appendix C for additional planetary data.